AN ATLAS OF FUNCTIONS

AN ATLAS OF FUNCTIONS

Jerome Spanier

The Claremont Graduate School
Claremont, California, U.S.A.

Keith B. Oldham

Trent University
Peterborough, Ontario, Canada

HEMISPHERE PUBLISHING CORPORATION
A member of the Taylor & Francis Group

New York Washington Philadelphia London

AN ATLAS OF FUNCTIONS

2 3 4 5 6 7 8 9 0 EBEB 8 9 8

This book was set in Times Roman by Edwards Brothers, Inc.
The editor was Sandra Tamburrino. Edwards Brothers, Inc., was printer and binder.

Library of Congress Cataloging in Publication Data

Spanier, Jerome, date
An atlas of functions.

Bibliography: p.
Includes indexes.
1. Oldham, Keith B. II. Title.
QA331.S685 1987 515 86-18294
ISBN 0-89116-573-8 Hemisphere Publishing Corporation

CONTENTS

PREFACE

The majority of engineers and physical scientists must consult reference books containing information on a variety of mathematical functions. This reflects the fact that all but the most trivial quantitative work involves relationships that are best described by functions of various complexities. Of course, the need will depend on the user, but all will require information about the general behavior of the function in question and its values at a number of arguments.

Historically, this latter need has been met primarily by tables of function values. However, the ubiquity of computers and programmable calculators presents an opportunity to provide reliable, fast and accurate function values without the need to interpolate. Computer technology also enables graphical presentations of information to be made with digital accuracy. *An Atlas of Functions* exploits these opportunities by presenting algorithms for the calculation of most functions to more than seven-digit precision and computer-generated maps that may be read to two or three figures. Of course, the need continues for ready access to the many formulas and properties that characterize a specific function. This need is met in the *Atlas* through the display of the most important definitions, relationships, expansions and other properties of the 400 functions covered in this book.

The *Atlas* is organized into 64 chapters, each of which is devoted to one function or to a number of closely related functions; these appear roughly in order of increasing complexity. A standardized format has been adopted for each chapter to minimize the time required to locate a sought item of information. A description of how the chapters are sectioned is included in Chapter 0. Two appendixes, a references/bibliography section and two indexes complete the volume.

It is a pleasure to acknowledge our gratitude to Diane Kaiser and Charlotte Oldham who prepared the typescript and to Jan Myland who created the figures. Their skill and unfailing good humor in these exacting tasks have exceeded all expectations. We would also like to record our indebtedness to our families for their support and patience during the many years that the preparation of the *Atlas* took time that was rightfully theirs. To them, and especially to Bunny and Charlotte, we dedicate this book.

Jerome Spanier
Keith B. Oldham

CHAPTER 0

GENERAL CONSIDERATIONS

In this chapter are collected some considerations that relate to all, or most, functions. The general organization of the *Atlas* is also explained here. Thus, this could be a good starting point for the reader. However, the intent of the authors is that the information in the *Atlas* be immediately available to an unprepared reader. There are no special codes that must be mastered in order to use the book, and the only conventions that we adopt are those that are customary in scientific writing.

Each chapter in the *Atlas* is devoted to a single function or to a small number of intimately related functions. The preamble to the chapter exposes any such relationships and introduces special features of the subject function.

0:1 NOTATION

The *nomenclature* and *symbolism* of mathematical functions are bedevilled by ambiguities and inconsistencies. Several names may be attached to a single function, and one symbol may be used to denote several functions. In the first section of each chapter the reader is alerted to such sources of confusion.

For the sake of standardization, we have imposed certain conventions relating to symbols. Thus, we have eschewed boldface and similar typographical niceties on the grounds that they are difficult to reproduce by pencil on paper, or with office machines. We reserve the use of italics to represent numbers (such as constants, c; coefficients, a_1, a_2, a_3, ... a_n; and function arguments, x) and avoid their use in symbolizing functions. Another instance of a convention intended to add clarity is our distinction in using commas and semicolons as separators, as in $\mathrm{F}(a,b;c;x)$. Elements in a string may be interchanged when they are separated by commas but not where a semicolon serves as separator.

The symbol $z(=x + iy)$ is reserved to denote a *complex variable*. All other variables are implicitly real.

0:2 BEHAVIOR

Some functions are defined for all values of their variable(s). For other functions there are restrictions, such as $-1 < x \leq 1$ or $n = 1, 2, 3, \ldots$, on these values that specify the *range* of each variable and thereby the *domain* of the function. Likewise, the function itself may be restricted in range and may be real valued, complex valued or each of these in different domains. Such considerations are discussed in the second section of each chapter.

This section also reveals how the function changes in value as its variables change throughout their ranges, thereby exposing the general "shape" of the function. This information is conveyed by a verbal description, supplemented by a graphical "map." All diagrams have been computer generated and are precisely drawn. Each graph has been scaled so that the graticule spacing is either ten or fifteen millimetres (2.00 or 3.00 mm between adjacent dots), which permits interpolation to an accuracy of two or three significant digits. Much greater precision is available from the algorithms of Section 8.

A *univariate function* (i.e., a function depending only on a single variable, its *argument*) is generally illustrated by a simple graph of that function versus its argument. Bivariate functions of two continuous variables are often diagrammed by *contour maps* whose curves correspond to specified values of the function and whose axes are the argument and the second variable.

In discussing the behavior of some functions it is convenient to adopt the terminology *quadrant*. There are four quadrants defined as illustrated in Figure 0-1.

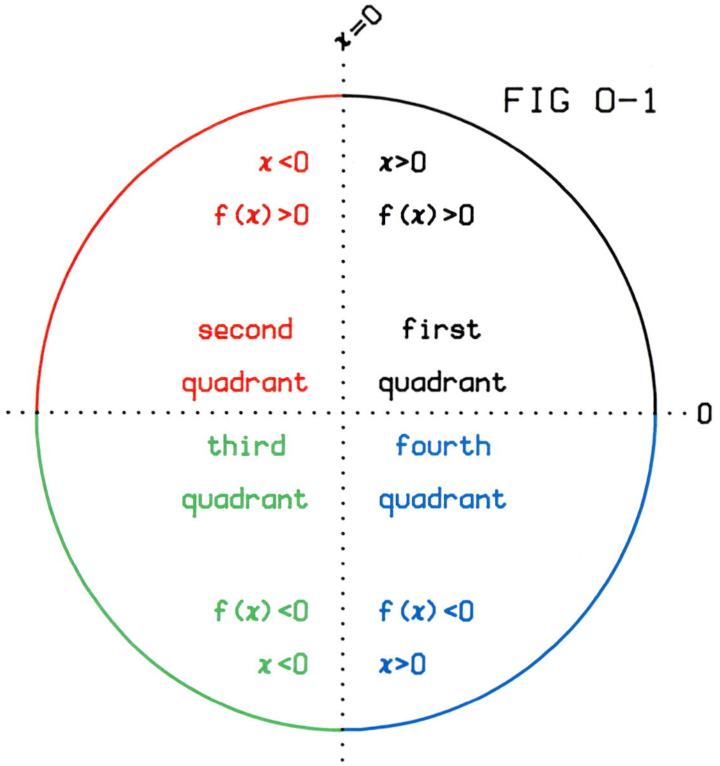

FIG 0-1

0:3 DEFINITIONS

Often there are several formulas relating a function to its variable(s), although they may not all apply over the entire domain of the function. These various interrelationships are listed in the third section of each chapter under the heading Definitions even though, from a strictly logical viewpoint, some might prefer to select one as the unique definition and cite the others as "equivalences."

Several types of definition are encountered in Section 3. For example, a function may be defined:

(a) by an equation that explicitly defines the function in terms of "simpler" functions and algebraic operations;
(b) by a formula relating the function to its variable(s) through a finite or an infinite number of arithmetic or algebraic operations;
(c) as the derivative or indefinite integral of a simpler function;
(d) as an *integral transform* of the form

0:3:1
$$f(x) = \int_{t_0}^{t_1} g(x;t)\,dt$$

where g is a function having one more variable than f, t_0 and t_1 being specified limits of integration;

(e) through a *generating function*, $G(x;t)$, that defines a family of functions $f_j(x)$ via the expansion

0:3:2
$$G(x;t) = \sum_j f_j(x)g_j(t)$$

where $g_j(t)$ is a simpler set of functions such as t^j;

(f) as the *inverse* of another function $F(x)$ so that the implicit equation

0:3:3
$$F(f(x)) = x$$

is used to define $f(x)$;

(g) as a special case or a limiting case of a more general function;

(h) parametrically through a pair of equations that separately relate the function $f(x)$ and its argument x to a third variable;

(i) implicitly via a differential equation, the solution (or one of the solutions) of which is the subject function; and

(j) through concepts borrowed from geometry or trigonometry.

Some definitions generate two or more values of a function from a single value of the argument. The *Atlas* minimizes this complication by concentrating, wherever possible, on single-valued functions.

0:4 SPECIAL CASES

If the function reduces to a simpler function under special circumstances, this is noted in the fourth section of each chapter.

0:5 INTRARELATIONSHIPS

An equation linking the two functions $f(x)$ and $g(x)$ is an *interrelationship* between them. In contrast, one speaks of an *intrarelationship* in case there exists a formula that provides a link between values of a single function at two or more values of one of its variables, for example, between $f(x_1)$ and $f(x_2)$. In this *Atlas* intrarelationships will be found in Section 5 of each chapter, interrelationships mainly in Sections 3 and 13.

An equation expressing the relationship between $f(-x)$ and $f(x)$ is called a *reflection formula*. Less commonly there exist reflection formulas relating $f(a - x)$ to $f(a + x)$ for nonzero values of a.

A second class of intrarelationships are *translation formulas;* these relate $f(x + a)$ to $f(x)$. The most general translation formula, in which a is free to vary continuously, becomes an *argument-addition formula* that relates $f(x + y)$ to $f(x)$ and $f(y)$. However, many translation formulas are restricted to special values of a such as $a = 1$ or $a = n\pi$; the relationships are then known as *recurrence relations* or *recursion formulas*. Such relationships are common in bivariate functions; a recursion formula then normally relates $f(\nu;x)$ to $f(\nu - 1;x)$ or to both $f(\nu - 1;x)$ and $f(\nu - 2;x)$.

A very general argument-addition formula is provided by the *Taylor expansion:*

0:5:1
$$f(y \pm x) = f(y) \pm x\frac{df}{dx}(y) + \frac{x^2}{2!}\frac{d^2f}{dx^2}(y) \pm \frac{x^3}{3!}\frac{d^3f}{dx^3}(y) + \cdots$$

Expressions for the remainder after the series is truncated to a finite number of terms are provided by Abramowitz and Stegun [Section 3.6].

A third class of intrarelationships are *argument-multiplication formulas* that relate $f(nx)$ to $f(x)$. More rarely there exist *function-multiplication formulas* or *function-addition formulas* that provide expressions for $f(x)f(y)$ and $f(x) + f(y)$, respectively.

Yet other intrarelationships are those provided by finite and infinite series. With bivariate and multivariate functions there may be a great number of such formulas, and functions other than f may be involved.

0:6 EXPANSIONS

The sixth section of each chapter is devoted to ways in which the function(s) may be expressed as a finite or infinite array of terms. Such arrays are normally series, products or continued fractions.

Notation such as

0:6:1 $$f(x) = \sum_{j=0}^{\infty} g(j;x)$$

is used to represent a *convergent infinite series*, where g is a function of j and x. Unless otherwise qualified, 0:6:1 implies that, for values of x in a specified range, the numerical value of the finite sum

0:6:2 $$g(0;x) + g(1;x) + g(2;x) + \cdots + g(j;x) + \cdots + g(J;x)$$

can be brought indefinitely close to f(x) by choosing J to be a large enough integer.

Frequently encountered are convergent series whose successive terms, for sufficiently large j, decrease in magnitude and alternate in sign. We shall loosely call such series *alternating series*. A valuable property of such alternating series that is often used in this *Atlas* enables the remainder after a finite number of terms are summed to be estimated in terms of the first omitted term. When $\Sigma(-1)^j g(j;x)$ is used to represent an alternating series, this result:

0:6:3 $$\left| \sum_{j=0}^{\infty} (-1)^j g(j;x) - \sum_{j=0}^{J} (-1)^j g(j;x) \right| < |g(J+1;x)|$$

plays a particularly important role in the design of many algorithms of Sections 8.

In contrast to 0:6:1, the symbolism

0:6:4 $$f(x) \sim \sum_j g(j;x) \qquad j = 0, 1, 2, \ldots \qquad x \to \infty$$

which is reserved for *asymptotic series*, implies that for every J the numerical value of 0:6:2 can be brought indefinitely close to f(x) by making x, *not* J, sufficiently large. It is this restriction on the magnitude of x that makes an asymptotic expansion, though of great utility in many applications, rather treacherous for the incautious user.

If the function $g(j;x)$ in 0:6:1 or 0:6:4 can be written as the product $c_j x^{\alpha+\beta j}$, where c_j is independent of x while α and β are constants, then the expansions 0:6:1 and 0:6:4 are called *power series*. The evaluation of the finite analog of such series is aided by rewriting 0:6:2 in the form

0:6:5 $$(((\cdots((c_J x^\beta + c_{J-1})x^\beta + c_{J-2})x^\beta + \cdots + c_2)x^\beta + c_1)x^\beta + c_0)x^\alpha$$

which is termed a *concatenation* or *nested sum*. Such nested sums will also frequently be found in algorithms appearing in Section 8.

The *infinite product* notation

0:6:6 $$f(x) = \prod_{j=0}^{\infty} g(j;x)$$

implies that the numerical value of the finite product

0:6:7 $$g(0;x)g(1;x)g(2;x) \cdots g(J;x)$$

approaches f(x) indefinitely closely as J takes larger and larger integer values.

The notation

0:6:8 $$\alpha_0 + \frac{\alpha_1}{\beta_1+}\,\frac{\alpha_2}{\beta_2+}\,\frac{\alpha_3}{\beta_3+}\cdots$$

is a standard abbreviation for the *continued fraction*

0:6:9
$$\alpha_0 + \cfrac{\alpha_1}{\beta_1 + \cfrac{\alpha_2}{\beta_2 + \cfrac{\alpha_3}{\beta_3 + \cfrac{\alpha_4}{\beta_4 + \cdots}}}}$$

in which each α_j and β_j may denote constants or variables. Continued fractions may be infinite, as denoted in 0:6:8, or finite (or "terminated"):

0:6:10
$$\alpha_0 + \frac{\alpha_1}{\beta_1+}\frac{\alpha_2}{\beta_2+}\frac{\alpha_3}{\beta_3+}\cdots\frac{\alpha_{J-1}}{\beta_{J-1}+}\frac{\alpha_J}{\beta_J}$$

Of great utility in working with continued fractions is the equivalence

0:6:11
$$\alpha_0 + \frac{\alpha_1}{\beta_1+}\frac{\alpha_2}{\beta_2+}\frac{\alpha_3}{\beta_3+}\cdots\frac{\alpha_n}{\beta_n} = \alpha_0 + \frac{\gamma_1\alpha_1}{\gamma_1\beta_1+}\frac{\gamma_1\gamma_2\alpha_2}{\gamma_2\beta_2+}\frac{\gamma_2\gamma_3\alpha_3}{\gamma_3\beta_3+}\cdots\frac{\gamma_{n-1}\gamma_n\alpha_n}{\gamma_n\beta_n}$$

Demonstrating the interchangeability of continued fractions and power series is the identity

0:6:12
$$\frac{1}{a_0} + \frac{x}{a_0a_1} + \frac{x^2}{a_0a_1a_2} + \cdots + \frac{x^n}{a_0a_1a_2\cdots a_n} = \frac{1}{a_0-}\frac{a_0x}{a_1+x-}\frac{a_1x}{a_2+x-}\frac{a_2x}{a_3+x}\cdots\frac{a_{n-1}x}{a_n+x}$$

0:7 PARTICULAR VALUES

If certain values of the variable(s) of a function generate noteworthy function values, these are cited in the seventh section of each chapter, often as a table.

In Section 7 of many chapters we include information about those arguments that lead to inflections, minima, maxima and zeros of the subject function $f(x)$. The term *extremum* is used to mean either a maximum or a minimum.

An *inflection* of a function occurs at a value of its argument at which the second derivative of the function is zero; that is:

0:7:1
$$\frac{d^2f}{dx^2}(x_i) = 0 \qquad f(x_i) = \text{inflection of } f(x)$$

A *minimum* and a *maximum* of a function are characterized respectively by

0:7:2
$$\frac{df}{dx}(x_m) = 0,\ \frac{d^2f}{dx^2}(x_m) > 0 \qquad f(x_m) = \text{minimum of } f(x)$$

and

0:7:3
$$\frac{df}{dx}(x_M) = 0,\ \frac{d^2f}{dx^2}(x_M) < 0 \qquad f(x_M) = \text{maximum of } f(x)$$

A *zero* of a function is a value of its argument at which the function vanishes; that is, if

0:7:4
$$f(r) = 0 \text{ then } r = \text{a zero of } f(x)$$

Equivalent to the phrase "a zero of $f(x)$" is "a *root* of the equation $f(x) = 0$." A *double zero* or a *repeated root* occurs at a value r of the argument such that

0:7:5
$$f(r) = \frac{df}{dx}(r) = 0 \qquad r = \text{a double zero of } f(x)$$

The concept extends to multiple zeros; thus, if

0:7:6
$$f(r) = \frac{df}{dx}(r) = \cdots = \frac{d^nf}{dx^n}(r) = 0 \qquad r = \text{a zero of } f(x) \text{ of multiplicity } n+1$$

A value r of x that satisfies 0:7:4 but not 0:7:5 corresponds to a *simple zero* or a simple root.

0:8 NUMERICAL VALUES

The prime goal of this section is to enable the reader to calculate precise values of the function in question for most argument values. Usually this is accomplished with the aid of an algorithm, it being assumed that the reader has at his disposal a programmable calculator or computer of modest power.

Because such devices invariably incorporate means of evaluating the functions $\sqrt{x}$, ln(x), exp(x), sin(x), cos(x), tan(x), and arctan(x), no algorithms are presented for these seven functions. By the same token, free use is made of these seven ubiquitous functions in the algorithms of other functions. Several other functions (x^ν, Int(x), frac(x) and $|x|$) that are usually available are also assumed in the algorithms, but short subalgorithms for these are included to cover cases in which they may be absent from a particular device.

No programs, as such, are presented because this would restrict the applicability of the algorithms to a particular programming language. Instead we report the algorithms as a list of straightforward commands that can easily be converted to statements in any programming language or into a sequence of calculator keystrokes in either reversed Polish or algebraic logic.

Traditionally, the goal in algorithm design has been to minimize the computation time required to generate a function value to some specified precision. The algorithms of this *Atlas* are designed to meet quite a different criterion: to minimize the programmer's time.

The algorithms of Section 8 are mostly based on the mathematical properties of the function in question. Usually, a series is summed, the number of summands being determined by an empirical formula or by making some test(s) after each addition. There are two advantages, other than sacrificing computer time in favor of "people time," to this style of algorithm. First, most of our algorithms may be easily modified to provide more, or less, precision. Second, since there are few multidigit numbers to be keyed, there is less chance of a mistake in programming.

Many algorithms in this *Atlas* claim a *precision* of "24-bits." By this is meant that the algorithm gives an approximation $\hat{f}(x)$ that is related to the true value f(x) of the function by

0:8:1 $$-2^{-24} \le \frac{f(x) - \hat{f}(x)}{f(x)} \le 2^{-24}$$

That is, the *relative error* in the algorithm never exceeds about 6×10^{-8}. It should be emphasized that the precision claimed is that of the *algorithm* and may not necessarily be reflected in a calculation based on the algorithm. Degraded precision could result, for example, from inadequate accuracy in the built-in functions of the computing device being used, or from variations in *rounding error* from computer to computer. A related way in which the precision of implementation of an algorithm may be degraded is through the subtraction of two nearly equal numbers. Where possible we have designed algorithms to minimize this loss of *significance*.

A typical *algorithm* has the following features:

(a) provision for loading the argument and other input parameters, if any;
(b) indication of the restrictions, if any, on the input values;
(c) a list of the constants and variables used by the algorithm;
(d) a list of easily understood sequential commands, such as "Set $f = 0$" or "Replace g by $\sqrt{g}/(1 + x)$";
(e) labels in the form of parenthesized numbers, such as "(2)," preceding certain commands;
(f) commands in the form "If . . . go to (#)" where (#) is a label and . . . is some condition, it being understood that if the condition is *not* met, one proceeds to the next command in sequence;
(g) provision for output; and
(h) test values to be used in confirming that a program derived from the algorithm is functioning correctly.

The *test values* supplied with each algorithm have been carefully chosen to exercise the algorithm maximally. Test outputs are usually reported with two or more terminal digits italicized. A correctly programmed algorithm should always generate a test output that reproduces at least those digits that are not italicized. On the other hand, italicized digits may fail to be reproduced by a correctly programmed algorithm. This is because changes in these digits correspond to errors that lie within our permitted range of $\pm 6 \times 10^{-8}$. For example, our algorithm from Section 43:8 lists the test value $\Gamma(0.6) = 1.489192$*25*; however, an output of 1.48919228, having a relative error of only 2×10^{-8}, is no cause for concern.

The techniques employed in the design of the algorithms for this *Atlas* are so diverse that it is inappropriate to attempt to summarize them here. However, each algorithm is accompanied by at least a brief synopsis of its mode of operation.

Throughout the *Atlas* there are also a number of general-purpose algorithms. These are recapitulated in Appendix A, which also contains several additional utility algorithms of wide applicability.

0:9 APPROXIMATIONS

Entries in this section enable function values to be calculated for stated ranges of the argument. Often, the precision is specified to be "8-bits," which implies

0:9:1 $$-2^{-8} \leq \frac{\mathrm{f}(x) - \hat{\mathrm{f}}(x)}{\mathrm{f}(x)} \leq 2^{-8}$$

(i.e., a relative error of no more than about $\pm 4 \times 10^{-3}$), where $\hat{\mathrm{f}}(x)$ is the value given by the approximation and $\mathrm{f}(x)$ is the value of the function.

0:10 OPERATIONS OF THE CALCULUS

Some of the most important properties of functions are associated with their behavior when subjected to the various operations of the calculus. Accordingly, the tenth is often one of the largest sections of a chapter.

For most functions defined for a continuous range of argument x, the *derivative*

0:10:1 $$\frac{\mathrm{df}}{\mathrm{d}x}$$

exists and is reported in Section 10. The derivatives of function combinations such as $\mathrm{f}(x)\mathrm{g}(x)$, $\mathrm{f}(x)/\mathrm{g}(x)$ or $\mathrm{f}(\mathrm{g}(x))$ are seldom reported because they may be readily evaluated via the *product rule*:

0:10:2 $$\frac{\mathrm{d}}{\mathrm{d}x}\mathrm{f}(x)\mathrm{g}(x) = \mathrm{f}(x)\frac{\mathrm{dg}}{\mathrm{d}x} + \mathrm{g}(x)\frac{\mathrm{df}}{\mathrm{d}x}$$

the *quotient rule*:

0:10:3 $$\frac{\mathrm{d}}{\mathrm{d}x}\frac{\mathrm{f}(x)}{\mathrm{g}(x)} = \frac{1}{\mathrm{g}(x)}\frac{\mathrm{df}}{\mathrm{d}x} - \frac{\mathrm{f}(x)}{\mathrm{g}^2(x)}\frac{\mathrm{dg}}{\mathrm{d}x}$$

or the *chain rule*:

0:10:4 $$\frac{\mathrm{d}}{\mathrm{d}x}\mathrm{f}(\mathrm{g}(x)) = \frac{\mathrm{df}}{\mathrm{dg}}\frac{\mathrm{dg}}{\mathrm{d}x} \qquad \mathrm{g} = \mathrm{g}(x)$$

If $\mathrm{f}(x)$ is differentiated n times, the nth derivative

0:10:5 $$\frac{\mathrm{d}^n\mathrm{f}}{\mathrm{d}x^n}$$

is generated, and expressions for this result are sometimes reported. The nth derivative of a product is given by the *Leibniz theorem*:

0:10:6 $$\frac{\mathrm{d}^n}{\mathrm{d}x^n}\mathrm{f}(x)\mathrm{g}(x) = \mathrm{f}(x)\frac{\mathrm{d}^n\mathrm{g}}{\mathrm{d}x^n} + \frac{n}{1!}\frac{\mathrm{df}}{\mathrm{d}x}\frac{\mathrm{d}^{n-1}\mathrm{g}}{\mathrm{d}x^{n-1}} + \frac{n(n-1)}{2!}\frac{\mathrm{d}^2\mathrm{f}}{\mathrm{d}x^2}\frac{\mathrm{d}^{n-2}\mathrm{g}}{\mathrm{d}x^{n-2}} + \cdots + \frac{n}{1!}\frac{\mathrm{d}^{n-1}\mathrm{f}}{\mathrm{d}x^{n-1}}\frac{\mathrm{dg}}{\mathrm{d}x} + \frac{\mathrm{d}^n\mathrm{f}}{\mathrm{d}x^n}\mathrm{g}(x)$$
$$= \sum_{j=0}^{n}\binom{n}{j}\frac{\mathrm{d}^{n-j}\mathrm{f}}{\mathrm{d}x^{n-j}}\frac{\mathrm{d}^j\mathrm{g}}{\mathrm{d}x^j}$$

where $\binom{n}{j}$ denotes a binomial coefficient [Chapter 6] and $n!$ the factorial function [Chapter 2]. Another useful operation bearing the name of Leibniz gives a rule for *differentiating an integral*:

0:10:7
$$\frac{d}{dx}\int_{t_0(x)}^{t_1(x)} f(x;t)dt = \int_{t_0(x)}^{t_1(x)} \frac{\partial}{\partial x} f(x;t)dt + \frac{dt_1}{dx} f(x;t_1(x)) - \frac{dt_0}{dx} f(x;t_0(x))$$

Reference books often express the results of *indefinite integration* in a form such as

0:10:8
$$\int f(x)dx = c + \phi(x)$$

where $\phi(x)$ is a function that gives $f(x)$ on differentiation (i.e., $d\phi/dx = f(x)$), and c is an arbitrary constant. To achieve closer unity with the representation

0:10:9
$$\int_{x_0}^{x_1} f(t)dt$$

of a *definite integral*, the *Atlas* adopts the formulation

0:10:10
$$\int_{x_0}^{x} f(t)dt = \phi(x) - \phi(x_0)$$

for indefinite integration. In 0:10:9 and 0:10:10, x_0 and x_1 are specific lower and upper limits, x_0 in 0:10:10 often being chosen to make $\phi(x_0)$ vanish. Of course, the information contained in 0:10:8 is also present in 0:10:10. An invaluable tool for integration is the formula for *integration by parts*:

0:10:11
$$\int_{x_0}^{x_1} f(t)g(t)dt = \int_{t=x_0}^{t=x_1} f(t)dG(t) = f(x_1)G(x_1) - f(x_0)G(x_0) - \int_{x_0}^{x_1} \frac{df}{dt}(t)G(t)dt$$

where g is the derivative of G.

Frequently, integrals cannot be expressed in terms of known functions. In such cases it is convenient to evaluate the integral numerically, for example by utilizing the algorithm presented in Appendix A [see Section A:7].

Perhaps the most general operation of the calculus is that named *differintegration* and denoted by the symbolism

0:10:12
$$\frac{d^\nu f}{[d(x-x_0)]^\nu}$$

for the νth differintegral of $f(x)$ with respect to x, using x_0 as the lower limit. The value of ν is unrestricted: it may be positive or negative, integer or noninteger. Differintegration, as its name suggests, generalizes the operations of differentiation and integration so that 0:10:1, 0:10:5 and 0:10:10 are the $\nu = 1$, $\nu = n$ and $\nu = -1$ instances of 0:10:12. Note that the differintegral 0:10:12 is independent of the lower limit x_0 only when $\nu = 0, 1, 2, \ldots$. For further information on differintegration, the reader is referred to a monograph on the subject [Oldham and Spanier]. The symbol 0:10:12 collapses to

0:10:13
$$\frac{d^\nu f}{dx^\nu}$$

when x_0 is chosen to be zero.

The most important noninteger order instances of differintegration are generated when $\nu = \frac{1}{2}$ or $\nu = -\frac{1}{2}$ and are termed *semidifferentiation* and *semiintegration*, respectively. Some examples of semiderivatives and semiintegrals are reported in Sections 10, and most often with a lower limit of zero or minus infinity.

An algorithm that generates approximate values of the differintegral 0:10:12 may be found in Section A:7.

0:11 COMPLEX ARGUMENT

Other than in the eleventh section of each chapter, the argument of a function is treated as real. Moreover, we usually restrict the domain of the function to ensure function values that are real.

In Sections 11, however, the effect of replacing the real argument x by the imaginary iy or the complex $x + iy$ is often reported. The real and imaginary components of $f(x)$ or $f(x + iy)$ are sometimes exposed as well.

0:12 GENERALIZATIONS

Functions can be arranged in a hierarchy in which lower members are subsumed by more general higher members. In the chapter dealing with the $f(x)$ function, special cases of $f(x)$—that is, functions lower in the hierarchy—are reported in Section 4. Conversely, functions that occur higher than $f(x)$ in some hierarchy are reported in Section 12; that is, the functions reported here are ones of which $f(x)$ is a special case.

0:13 COGNATE FUNCTIONS

The thirteenth section of each chapter is devoted to citing functions—other than those reported in Sections 4 and 12—that are related to the subject function, and to exploring the properties of some of them.

0:14 SPECIAL TOPICS

An application, or some other special feature of the function, is elaborated in the final section of many chapters. Such applications often amount to brief discourses on important topics in science, engineering and applied mathematics. For example, certain topics in approximation and curve-fitting, Laplace and Fourier transforms, facts about permutations, combinations and distributions, and geometric properties of various sorts are to be found among the many exposed in Section 14.

CHAPTER 1

THE CONSTANT FUNCTION *c*

The constant function is the simplest, and an almost trivial, function.

1:1 NOTATION

Constants are also known as *invariants* and are represented by a wide variety of symbols, mostly letters drawn from the early members of the English and Greek alphabets.

1:2 BEHAVIOR

The graphical representation (see Figure 1-1) of the constant function $f(x) = c$ is a horizontal line, extending indefinitely in the $x \to \pm\infty$ directions.

FIG 1-1

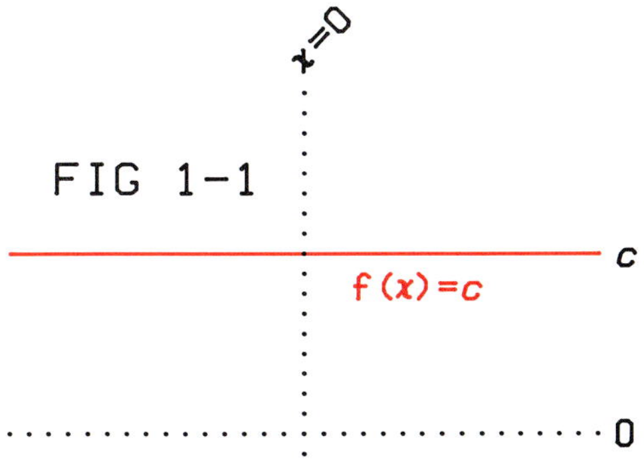

1:3 DEFINITIONS

The constant function is defined for all values of its argument x and has the same value, c, irrespective of x.

1:4 SPECIAL CASES

When c is zero, the constant function is sometimes termed the *zero function*. Likewise, the function $f(x) = c = 1$ is known as the *unity function*.

1:5 INTRARELATIONSHIPS

Being relations between function values at different values of the argument, intrarelationships are of no consequence for the constant function.

1:6 EXPANSIONS

A constant may be represented as a finite sum by utilizing the formulas for an *arithmetic series*:

1:6:1 $$c = \alpha + (\alpha + \delta) + (\alpha + 2\delta) + \cdots + (\alpha + J\delta) = \sum_{j=0}^{J} (\alpha + j\delta) \qquad \alpha = \frac{c}{J+1} - \frac{J\delta}{2}$$

a *geometric series*:

1:6:2 $$c = \alpha + \alpha\beta + \alpha\beta^2 + \cdots + \alpha\beta^J = \sum_{j=0}^{J} \alpha\beta^j \qquad \alpha = \frac{c(\beta - 1)}{\beta^{J+1} - 1}$$

or an *arithmetic-geometric series*:

1:6:3 $$c = \alpha + \beta(\alpha + \delta) + \beta^2(\alpha + 2\delta) + \cdots + \beta^J(\alpha + J\delta) = \sum_{j=0}^{J} \beta^j(\alpha + j\delta)$$

$$\alpha = \frac{c(\beta - 1) - J\delta\beta^{J+1} + \delta\beta(\beta^J - 1)/(\beta - 1)}{\beta^{J+1} - 1}$$

In these formulas β and δ may take any values and J may be any positive integer.

Any constant greater than $\frac{1}{2}$ may be expanded as the infinite geometric sum

1:6:4 $$c = 1 + \left(\frac{c-1}{c}\right) + \left(\frac{c-1}{c}\right)^2 + \left(\frac{c-1}{c}\right)^3 + \cdots = \sum_{j=0}^{\infty} \left(\frac{c-1}{c}\right)^j \qquad c > \frac{1}{2}$$

or as the infinite product

1:6:5 $$c = \left[1 + \left(\frac{c-1}{c}\right)\right]\left[1 + \left(\frac{c-1}{c}\right)^2\right]\left[1 + \left(\frac{c-1}{c}\right)^4\right] \cdots = \prod_{j=0}^{\infty} \left[1 + \left(\frac{c-1}{c}\right)^{2^j}\right] \qquad c > \frac{1}{2}$$

Similarly, a constant is expansible as the infinite continued fraction

1:6:6 $$c = \frac{\alpha}{\beta+}\,\frac{\alpha}{\beta+}\,\frac{\alpha}{\beta+}\cdots$$

[see Section 0:6] in a variety of ways, some of which are indicated in Table 1.6.1.

1:7 PARTICULAR VALUES

Four irrational constants that are frequently encountered are *Archimedes' number*, *Catalan's constant*, the *base of natural logarithms* and *Euler's constant*. Archimedes' number π, also known as *Rudolf's number* or simply as *pi*, may be defined by the infinite sum

1:7:1 $$\pi = 4 - \frac{4}{3} + \frac{4}{5} - \frac{4}{7} + \cdots = 4\sum_{j=0}^{\infty} \frac{(-1)^j}{2j+1} = 3.141592654$$

Table 1.6.1

α	β	Constraint
c	$1 - c$	$-1 \le c < 1$
1	$\dfrac{1 - c^2}{c}$	$0 < c^2 < 1$
$c^2 + c$	1	$c \ge -\frac{1}{2}$
$c^2 - c$	$c - 1$	$c \le \frac{1}{2}$

or the infinite product

1:7:2 $$\pi = 2 \times \frac{4}{3} \times \frac{16}{15} \times \frac{36}{35} \times \frac{64}{63} \times \cdots = 2 \prod_{j=1}^{\infty} \frac{j^2}{j^2 - {}^1/_4} = 3.141592654$$

The definition of Catalan's constant G is similar to 1:7:1:

1:7:3 $$G = 1 - \frac{1}{9} + \frac{1}{25} - \frac{1}{49} + \cdots = \sum_{j=0}^{\infty} \frac{(-1)^j}{(2j+1)^2} = 0.9159655942$$

The base of natural logarithms e and Euler's constant γ are defined by the limit operations

1:7:4 $$e = \lim_{n\to\infty} \left(1 + \frac{1}{n}\right)^n = 2.718281828$$

and

1:7:5 $$\gamma = \lim_{n\to\infty} \left(1 + \frac{1}{2} + \frac{1}{3} + \cdots + \frac{1}{n} - \ln(n)\right) = \lim_{n\to\infty} \left(\sum_{j=1}^{n} \frac{1}{j} - \ln(n)\right) = 0.5772156649$$

The latter is also known as *Mascheroni's constant* and is often denoted by the symbol C. There are many alternative formulations of these four constants [see Gradshteyn and Ryzhik, Chapter 0, for some of these]. Also of widespread occurrence throughout this *Atlas* is the *ubiquitous constant*

1:7:6 $$U = \text{common mean [Section 61:8] of 1 and } 1/\sqrt{2} = 0.8472130848$$

The most important family of rational constants is the family of natural numbers, $n = 1, 2, 3, \ldots$ [see Section 1:14]. Other families that occur principally in coefficients of series expansions are the factorials $n!$ [Chapter 2], Bernoulli numbers B_n [Chapter 4] and Euler numbers E_n [Chapter 5].

1:8 NUMERICAL VALUES

This simplest of all functions has the same numerical value at all arguments.

1:9 APPROXIMATIONS

None are needed for the constant function.

1:10 OPERATIONS OF THE CALCULUS

Differentiation gives

1:10:1 $$\frac{d}{dx} c = 0$$

while indefinite and definite integration produce

1:10:2
$$\int_0^x c\mathrm{d}t = cx$$

and

1:10:3
$$\int_{x_0}^{x_1} c\mathrm{d}t = c(x_1 - x_0)$$

respectively.

The results of semidifferentiation and semiintegration [Section 0:10] with a lower limit of zero are

1:10:4
$$\frac{\mathrm{d}^{1/2}}{\mathrm{d}x^{1/2}}c = \frac{c}{\sqrt{\pi x}}$$

and

1:10:5
$$\frac{\mathrm{d}^{-1/2}}{\mathrm{d}x^{-1/2}}c = 2c\sqrt{\frac{x}{\pi}}$$

Differintegration [see Section 0:10] with a lower limit of zero yields

1:10:6
$$\frac{\mathrm{d}^{\nu}}{\mathrm{d}x^{\nu}}c = \frac{cx^{-\nu}}{\Gamma(1-\nu)}$$

where Γ is the gamma function [see Chapter 43]. In fact, equations 1:10:1, 1:10:2, 1:10:4 and 1:10:5 are the $\nu = 1, -1, \frac{1}{2}$ and $-\frac{1}{2}$ instances of 1:10:6.

1:11 COMPLEX ARGUMENT

A complex constant can be expressed as

1:11:1
$$c = \alpha + i\beta$$

where α and β are real constants and $i = \sqrt{-1}$. The rules for performing arithmetic operations upon the pair of complex constants $\alpha + i\beta$ and $A + iB$ are

1:11:2
$$(\alpha + i\beta) \pm (A + iB) = (\alpha \pm A) + i(\beta \pm B)$$

1:11:3
$$(\alpha + i\beta)(A + iB) = (\alpha A - \beta B) + i(\alpha B + A\beta)$$

1:11:4
$$\left(\frac{\alpha + i\beta}{A + iB}\right) = \left(\frac{\alpha A + \beta B}{A^2 + B^2}\right) + i\left(\frac{A\beta - \alpha B}{A^2 + B^2}\right)$$

Powers may be handled making use of the identity

1:11:5
$$(\alpha + i\beta)^{A+iB} = \exp[(A + iB)\mathrm{Ln}(\alpha + i\beta)]$$

where exp and Ln are the exponential and logarithmic functions treated in Chapters 26 and 25, respectively. Applying equations 25:11:1, 25:11:2 and 25:11:3 produces the multiple-valued function

1:11:6
$$(\alpha + i\beta)^{A+iB} = (\alpha^2 + \beta^2)^{A/2}[\cos(\phi) + i\sin(\phi)]\exp[-B(\theta + 2k\pi)]$$

where k is any integer and

1:11:7
$$\theta = \begin{cases} \mathrm{sgn}(\beta)\mathrm{arccot}(\alpha/|\beta|) & \beta \neq 0 \\ \dfrac{\pi}{2}[1 - \mathrm{sgn}(\alpha)] & \beta = 0 \end{cases}$$

1:11:8
$$\phi = \frac{B}{2}\ln(\alpha^2 + \beta^2) + A(\theta + 2k\pi)$$

Setting $k = 0$ in equations 1:11:6 and 1:11:8 defines a single-valued function. When this is done and when $A + iB = \nu$, a real number, the result is

1:11:9 $$(\alpha + i\beta)^{\nu} = (\alpha^2 + \beta^2)^{\nu/2}[\cos(\nu\theta) + i\sin(\nu\theta)]$$

an expression of de Moivre's theorem [see Section 32:11].

1:12 GENERALIZATIONS

The constant function is the special $b = 0$ case of the linear function discussed in Chapter 7.

1:13 COGNATE FUNCTIONS

Whereas the constant function has the same value for all x, the related *pulse function* is zero at values of the argument outside a "window" of width h, and is a nonzero constant, c, within this window (see Figure 1-2):

FIG 1-2

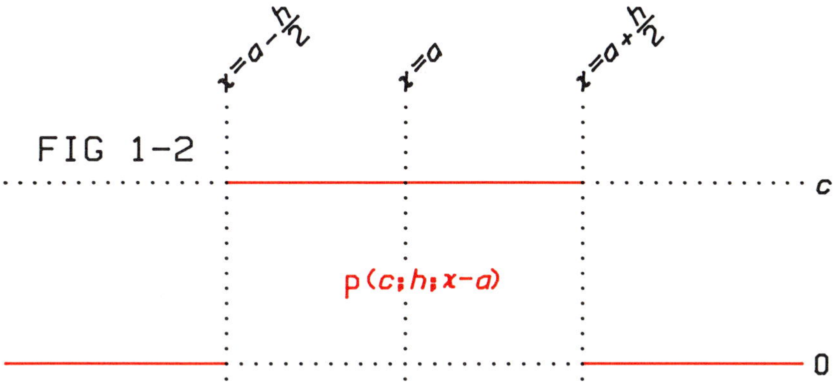

1:13:1 $$\mathrm{p}(c;h;x-a) = \begin{cases} 0 & x < a - \dfrac{h}{2} \\ c & a - \dfrac{h}{2} < x < a + \dfrac{h}{2} \\ 0 & x > a + \dfrac{h}{2} \end{cases}$$

The value of a establishes the "location" of the pulse, while c and h are termed the *pulse height* and *pulse width*, respectively. The pulse function can be expressed as

1:13:2 $$\mathrm{p}(c;h;x-a) = c\left[\mathrm{u}\left(x - a + \frac{h}{2}\right) - \mathrm{u}\left(x - a - \frac{h}{2}\right)\right]$$

in terms of the unit-step function of Chapter 8.

The concept of a general "window function" is discussed in Section 8:12.

The addition of a number of pulse functions, having various locations, heights and widths, produces a function whose map consists of horizontal straight line segments. Such a function, known as a *piecewise-constant function* and illustrated in Figure 1-3, may be used to approximate a more complicated or incompletely known function. It is the approximation usually adopted, for example, whenever a varying quantity is measured by a digital instrument.

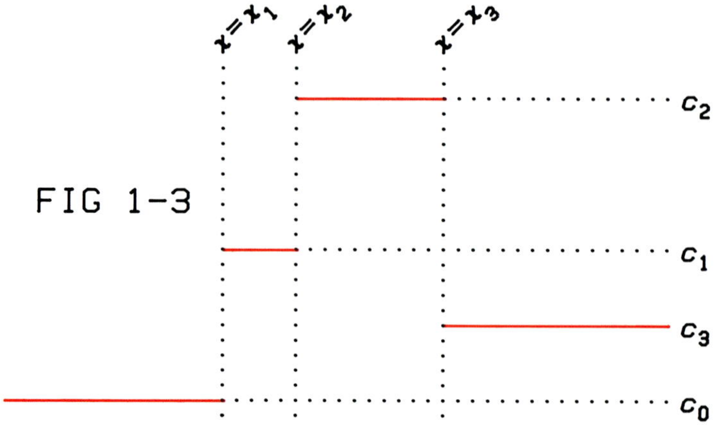

FIG 1-3

1:14 RELATED TOPICS

The *natural numbers* 1, 2, 3, ... are ubiquitous in mathematics and science. We record here several results for finite sums of their powers.

1:14:1 $$1+2+3+\cdots+n=\sum_{j=1}^{n} j=\frac{n(n+1)}{2} \qquad n=1,2,3,\ldots$$

1:14:2 $$1^2+2^2+3^2+\cdots+n^2=\sum_{j=1}^{n} j^2=\frac{n(n+1)(2n+1)}{6} \qquad n=1,2,3,\ldots$$

1:14:3 $$1^3+2^3+3^3+\cdots+n^3=\sum_{j=1}^{n} j^3=\frac{n^2(n+1)^2}{4} \qquad n=1,2,3,\ldots$$

Similarly, the sums of fourth and fifth powers of the first n natural numbers are $n(n+1)(2n+1)(3n^2+3n-1)/4$ and $n^2(n+1)^2(2n^2+2n-1)/12$, respectively. The general case is

1:14:4 $$1^m+2^m+\cdots+n^m=\sum_{j=1}^{n} j^m=\frac{\mathrm{B}_{m+1}(n+1)-\mathrm{B}_{m+1}}{m+1} \qquad n,m=1,2,3,\ldots$$

where B_m denotes a Bernoulli number [see Chapter 4] and $\mathrm{B}_m(x)$ denotes a Bernoulli polynomial [Chapter 19]. The sum of the reciprocals of the first n natural numbers is

1:14:5 $$\frac{1}{1}+\frac{1}{2}+\frac{1}{3}+\cdots+\frac{1}{n}=\sum_{j=1}^{n}\frac{1}{j}=\gamma+\psi(n+1) \qquad n=1,2,3,\ldots$$

where γ is Euler's constant [Section 1:7] and $\psi(x)$ denotes the digamma function [Chapter 44]. When continued indefinitely, the sum 1:14:5 defines the divergent *harmonic series*. If m is not an integer, summation 1:14:4 may be evaluated, for large n, by use of equation 13:5:5.

The corresponding expressions when the signs alternate are

1:14:6 $$1-2+3-4+\cdots\pm n=-\sum_{j=1}^{n}(-1)^j j=\begin{cases}(n+1)/2 & n=1,3,5,\ldots\\ -n/2 & n=2,4,6,\ldots\end{cases}$$

1:14:7 $$1^2-2^2+3^2-4^2+\cdots\pm n^2=-\sum_{j=1}^{n}(-1)^j j^2=\begin{cases}n(n+1)/2 & n=1,3,5,\ldots\\ -n(n+1)/2 & n=2,4,6,\ldots\end{cases}$$

1:14:8 $$1^3-2^3+3^3-4^3+\cdots\pm n^3=-\sum_{j=1}^{n}(-1)^j j^3=\begin{cases}(2n^3+3n^2-1)/4 & n=1,3,5,\ldots\\ -n^2(2n+3)/4 & n=2,4,6,\ldots\end{cases}$$

1:14:9 $$1^m-2^m+3^m-4^m+\cdots\pm n^m=-\sum_{j=1}^{n}(-1)^j j^m=-\frac{\mathrm{E}_m(0)}{2}-\frac{(-1)^n\mathrm{E}_m(n+1)}{2} \qquad n,m=1,2,3,\ldots$$

where $E_m(x)$ denotes an Euler polynomial [Chapter 20], and

1:14:10 $$\frac{1}{1}-\frac{1}{2}+\frac{1}{3}-\frac{1}{4}+\cdots\pm\frac{1}{n}=-\sum_{j=1}^{n}\frac{(-1)^j}{j}=\begin{cases}\psi(n+1)-\psi\left(\dfrac{n+1}{2}\right) & n=1,3,5,\ldots\\ \psi(n+1)-\psi\left(\dfrac{n}{2}+1\right) & n=2,4,6,\ldots\end{cases}$$

The numbers 2, 4, 6, ... are called the *even numbers*. Sums of their powers are easily found by using the identity

1:14:11 $$2^m+4^m+6^m+\cdots+n^m=2^m\left[1^m+2^m+3^m+\cdots+\left(\frac{n}{2}\right)^m\right] \qquad n=2,4,6,\ldots$$

in conjunction with equations 1:14:1–1:14:5. Likewise, use of these equations together with the identity

1:14:12 $$1^m+3^m+5^m+\cdots+n^m=[1^m+2^m+3^m+\cdots+n^m]$$
$$-2^m\left[1^m+2^m+3^m+\cdots+\left(\frac{n-1}{2}\right)^m\right] \qquad n=1,3,5,\ldots$$

permits sums of powers of the *odd numbers*, 1, 3, 5, ..., to be evaluated.

For the *infinite* sums $\Sigma j^{-\nu}$ where j runs from 1 to ∞, see Chapter 3. The same chapter also addresses the related infinite sums $\Sigma(-1)^j j^{-\nu}$, $\Sigma(2j-1)^{-\nu}$ and $\Sigma(-1)^j(2j-1)^{-\nu}$.

CHAPTER 2

THE FACTORIAL FUNCTION $n!$ AND ITS RECIPROCAL

Factorials occur widely in function theory: for example, in the power series expansions of many algebraic and transcendental functions. They also play an important role in combinatorics [see Section 2:14].

2:1 NOTATION

The factorial function of n is symbolized $\underline{n|}$ in some older literature. The symbol $\Pi(n)$ is occasionally encountered.

2:2 BEHAVIOR

The factorial function is defined only for nonnegative integer arguments and is itself a positive integer. Figure 2-1 (with a logarithmic vertical scale) and the table of rounded values (Table 2.2.1) demonstrate the explosive increase of $n!$ as n increases beyond $n = 2$.

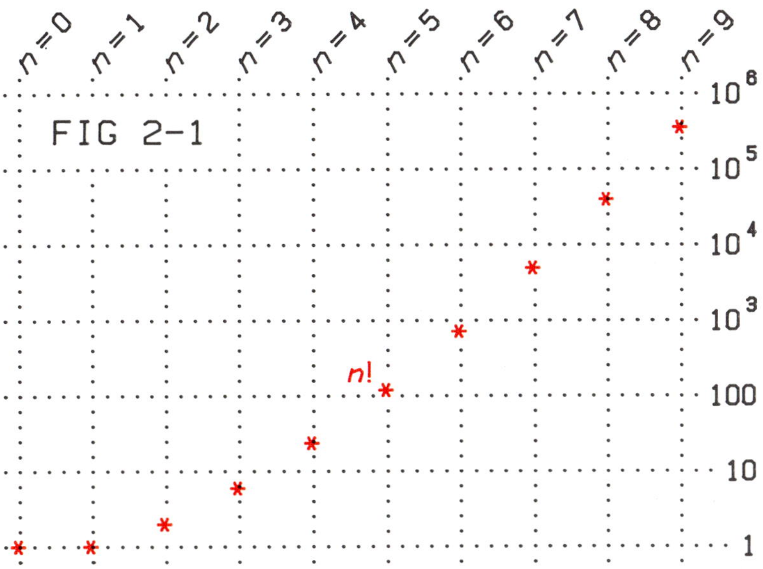

Table 2.2.1

n	$n!$
0	1
10	4×10^{6}
20	2×10^{18}
30	3×10^{32}
40	8×10^{47}
50	3×10^{64}
60	8×10^{81}
70	1×10^{100}
80	7×10^{118}
90	1×10^{138}
100	9×10^{157}

2:3 DEFINITIONS

The factorial of the positive integer n equals the product of all positive integers up to and including n:

2:3:1 $$n! = 1 \times 2 \times 3 \times \cdots \times (n-1) \times n = \prod_{j=1}^{n} j \qquad n = 1, 2, 3, \ldots$$

This definition is supplemented by the value

2:3:2 $$0! = 1$$

conventionally accorded to factorial zero.

The exponential function [Chapter 26] is a generating function [see Section 0:3] for the reciprocal of the factorial

2:3:3 $$\exp(t) = \sum_{n=0}^{\infty} \frac{t^n}{n!}$$

2:4 SPECIAL CASES

There are none.

2:5 INTRARELATIONSHIPS

The most important property of the factorial function is its recurrence

2:5:1 $$(n+1)! = n!(n+1) \qquad n = 0, 1, 2, \ldots$$

which may be iterated to produce the argument-addition formula

2:5:2 $$(n+m)! = n!(n+1)(n+2)\cdots(n+m) = n!(n+1)_m \qquad n, m = 0, 1, 2, \ldots$$

where $(n+1)_m$ is a Pochhammer polynomial [see Chapter 18]. Formula 2:5:2 leads to an expression for the ratio of two factorials. An alternative expression is

2:5:3 $$\frac{n!}{(n-m)!} = n(n-1)(n-2)\cdots(n-m+1) = \sum_{j=0}^{m} \mathrm{S}_m^{(j)} n^j$$

where $\mathrm{S}_m^{(j)}$ is a Stirling number of the first kind, of which a short table is included in Chapter 18.

Because reciprocals of factorials occur frequently as coefficients of power series, many expressions exist for infinite sums involving these reciprocals. For example:

2:5:4 $$\frac{1}{0!} + \frac{1}{1!} + \frac{1}{2!} + \cdots = \sum_{j=0}^{\infty} \frac{1}{j!} = \exp(1) = 2.718281828$$

2:5:5 $$\left(\frac{1}{0!}\right)^2 + \left(\frac{1}{1!}\right)^2 + \left(\frac{1}{2!}\right)^2 + \cdots = \sum_{j=0}^{\infty} \frac{1}{(j!)^2} = \mathrm{I}_0(2) = 2.279585302$$

2:5:6 $$\frac{1}{0!} + \frac{1}{2!} + \frac{1}{4!} + \cdots = \sum_{j=0}^{\infty} \frac{1}{(2j)!} = \cosh(1) = 1.543080635$$

2:5:7 $$\frac{1}{1!} + \frac{1}{3!} + \frac{1}{5!} + \cdots = \sum_{j=0}^{\infty} \frac{1}{(2j+1)!} = \sinh(1) = 1.175201194$$

where exp is the exponential function of Chapter 26, I_0 is a hyperbolic Bessel function [see Chapter 49] and cosh and sinh are the hyperbolic cosine and sine functions of Chapter 28.

The corresponding summations with alternating plus and minus signs give $\exp(-1)$, $\mathrm{J}_0(2)$, $\cos(1)$ and $\sin(1)$, where the Bessel coefficient J_0 is discussed in Chapter 52, and the cosine and sine functions, cos and sin, appear in Chapter 32.

2:6 EXPANSIONS

Stirling's formula provides an asymptotic expansion [see Section 0:6]

2:6:1 $$n! \sim \sqrt{2\pi n}\exp(-n)n^n\left[1+\frac{1}{12n}+\frac{1}{288n^2}-\frac{139}{51840n^3}-\cdots\right] \qquad n\to\infty$$

for the factorial $n!$ as n tends to infinity. This formula and a similar one for the logarithm of $n!$:

2:6:2 $$\ln(n!) \sim n\ln(n) - n + \frac{1}{2}\ln(2\pi n) + \frac{1}{12n} - \frac{1}{360n^3} + \frac{1}{1260n^5} - \cdots \qquad n\to\infty$$

produce remarkable accuracy, even for small n values.

2:7 PARTICULAR VALUES

2:7:1 $$0! = 1! = 1$$

2:8 NUMERICAL VALUES

Many calculators compute $n!$ from n by a single keystroke, and some computer languages have similar facilities. The simple algorithm

```
                Set f = 1
Input n >>>>>>>
                If n = 0 go to (2)
            (1) Replace f by fn
                Replace n by n - 1
                If n ≠ 0 go to (1)
            (2) Output f
f = n! <<<<<<<
```

Storage needed: n, f

Input restrictions: n must be a nonnegative integer.

Test values:
$0! = 1$
$(12)! = 479001600$
$(50)! = 3.04140932 \times 10^{64}$

based on definition 2:3:1, is precise, but may be tediously slow. Alternatively, one may use a second algorithm

```
Input n >>>>>>>
                Set f = [1 - (1/30n²)]/12n
                Replace f by exp(f + n[ln(n) - 1])
                Replace f by 1 + Int(f√(2πn))
                Output f
f ≃ n! <<<<<<<
```

Storage needed: n, f

Input restrictions: n must be a positive integer.

Test value:
$12! = 479001600$

which is based on exponentiating 2:6:2, truncation after the n^{-3} term and rounding to the next higher integer. This algorithm has a precision [see Section 0:8] better than 6×10^{-8}. Because of identities 2:12:1 and 2:12:2, the algorithms of Sections 18:8 and 43:8 may also be used to calculate factorials.

2:9 APPROXIMATIONS

Based on the asymptotic expansion 2:6:1 is the approximation

2:9:1 $$n! \simeq \sqrt{2\pi n}\exp(-n)n^n\left(1+\frac{1}{12n}\right) \qquad \text{8-bit precision} \qquad n\geq 2$$

2:10 OPERATIONS OF THE CALCULUS

No operations of the calculus are possible on a function such as $n!$ that is defined only for discrete arguments.

2:11 COMPLEX ARGUMENT

In view of relation 2:12:2, the gamma function formulas given in Section 43:11 may be used to generate expressions for $(im)!$ and $(n + im)!$.

2:12 GENERALIZATIONS

The factorial is a special case of Pochhammer's polynomial $(x)_n$ [Chapter 18]:

2:12:1 $$n! = (1)_n \qquad n = 0, 1, 2, \ldots$$

and of the gamma function [Chapter 43]:

2:12:2 $$n! = \Gamma(n + 1) \qquad n = 0, 1, 2, \ldots$$

2:13 COGNATE FUNCTIONS

The factorial function and the binomial coefficient [Chapter 6] are closely related.

The *double factorial* or *semifactorial function* is defined by

2:13:1 $$n!! = \begin{cases} 1 & n = -1, 0 \\ n(n-2) \times (n-4) \times \cdots \times 5 \times 3 \times 1 & n = 1, 3, 5, \ldots \\ n(n-2) \times (n-4) \times \cdots \times 6 \times 4 \times 2 & n = 2, 4, 6, \ldots \end{cases}$$

For even argument it reduces to

2:13:2 $$n!! = 2^{n/2}\left(\frac{n}{2}\right)! \qquad n = 0, 2, 4, \ldots$$

while for odd n it may be expressed in terms of factorials, or as a gamma function [Chapter 43] or as a Pochhammer polynomial [Chapter 18]

2:13:3 $$n!! = \frac{n!}{2^{(n-1)/2}\left(\dfrac{n-1}{2}\right)!} = 2^{n/2}\sqrt{\frac{2}{\pi}}\,\Gamma\left(1 + \frac{n}{2}\right) = 2^{(n+1)/2}\left(\frac{1}{2}\right)_{(n+1)/2} \qquad n = 1, 3, 5, \ldots$$

Of frequent occurrence [for example in Sections 6:4, 32:5, 61:6 and 62:12] is the ratio $(n - 1)!!/n!!$ of the double factorials of consecutive integers. For odd n the ratio is expressible by the integral

2:13:4 $$\frac{(n-1)!!}{n!!} = \frac{2^{n-1}}{n!}\left[\left(\frac{n-1}{2}\right)!\right]^2 = \int_0^{\pi/2} \sin^n(t)\,dt \qquad n = 1, 3, 5, \ldots$$

while for even n it is given by *Wallis' formula*

2:13:5 $$\frac{(n-1)!!}{n!!} = \frac{n!}{2^n[(n/2)!]^2} = \frac{2}{\pi}\int_0^{\pi/2} \sin^n(t)\,dt \qquad n = 0, 2, 4, \ldots$$

This important ratio has the asymptotic expansion

2:13:6 $$\frac{(n-1)!!}{n!!} \sim \begin{cases} \sqrt{\dfrac{2}{\pi n}}\left[1 - \dfrac{1}{4n} + \dfrac{1}{32n^2} - \cdots\right] & \text{even } n \to \infty \\ \sqrt{\dfrac{\pi}{2n}}\left[1 - \dfrac{1}{4n} + \dfrac{1}{32n^2} - \cdots\right] & \text{odd } n \to \infty \end{cases}$$

Finite sums of such ratios obey the simple rule

2:13:7 $$1 + \frac{1}{2} + \frac{3}{8} + \frac{15}{48} + \cdots + \frac{(2n-1)!!}{(2n)!!} = \sum_{j=0}^{n} \frac{(2j-1)!!}{(2j)!!} = \frac{(2n+1)!!}{(2n)!!} \qquad n = 0, 1, 2, \ldots$$

and there is the related infinite summation due to Ross:

2:13:8 $$\frac{1}{2} + \frac{3}{16} + \frac{15}{144} + \frac{105}{1536} + \cdots = \sum_{j=1}^{\infty} \frac{(2j-1)!!}{j(2j)!!} = \ln(4)$$

The *triple factorial* is defined analogously

2:13:9 $$n!!! = \begin{cases} 1 & n = -2, -1, 0 \\ n(n-3) \times (n-6) \times \cdots \times 7 \times 4 \times 1 & n = 1, 4, 7, \ldots \\ n(n-3) \times (n-6) \times \cdots \times 8 \times 5 \times 2 & n = 2, 5, 8, \ldots \\ n(n-3) \times (n-6) \times \cdots \times 9 \times 6 \times 3 & n = 3, 6, 9, \ldots \end{cases}$$

and finds application, for example, in connection with Airy functions [Chapter 56]. The extension to a *quadruple factorial* $n!!!!$ is obvious; it is useful in Sections 43:4 and 59:7.

2:14 RELATED TOPICS

The factorial function appears very often in applications involving combinatorics. For example, the number of *permutations* (arrangements) of n objects, all different, is $n!$. If not all of the n objects are different, the number of permutations is reduced to

2:14:1 $$\frac{n!}{(n_1)!(n_2)! \cdots (n_J)!} \qquad n = n_1 + n_2 + \cdots + n_J$$

where there are n_1 samples of object 1, n_2 samples of object 2, ..., n_J samples of object J.

If from a group of n objects, all different, one withdraws m objects, one at a time, the number of *variations* (possible withdrawal sequences) is

2:14:2 $$\frac{n!}{(n-m)!} \qquad m \le n$$

If one ignores the order of withdrawal, 2:14:2 is reduced to

2:14:3 $$\frac{n!}{(n-m)!m!} \qquad m \le n$$

which then represents the number of ways in which m objects can be chosen from among n, all different, and is known as the number of *combinations*.

The number of *partitions* (different ways in which n distinct objects may be placed in m identical boxes so that each box contains at least one object) is given by a *Stirling number of the second kind*:

2:14:4 $$\sigma_n^{(m)} = \sum_{j=0}^{m} \frac{(-1)^{m-j} j^n}{(m-j)!j!}$$

Table 2.14.1

	$\sigma_0^{(m)}$	$\sigma_1^{(m)}$	$\sigma_2^{(m)}$	$\sigma_3^{(m)}$	$\sigma_4^{(m)}$	$\sigma_5^{(m)}$	$\sigma_6^{(m)}$	$\sigma_7^{(m)}$	$\sigma_8^{(m)}$	$\sigma_9^{(m)}$	$\sigma_{10}^{(m)}$	$\sigma_{11}^{(m)}$	$\sigma_{12}^{(m)}$	$\sigma_{13}^{(m)}$	$\sigma_{14}^{(m)}$
$m = 0$	1	0	0	0	0	0	0	0	0	0	0	0	0	0	0
$m = 1$	0	1	1	1	1	1	1	1	1	1	1	1	1	1	1
$m = 2$	0	0	1	3	7	15	31	63	127	255	511	1023	2047	4095	8191
$m = 3$	0	0	0	1	6	25	90	301	966	3025	9330	28501	86526	261625	788970
$m = 4$	0	0	0	0	1	10	65	350	1701	7770	34105	145750	611501	2532530	10391745
$m = 5$	0	0	0	0	0	1	15	140	1050	6951	42525	246730	1379400	7508501	40075035
$m = 6$	0	0	0	0	0	0	1	21	266	2624	22827	179487	1323652	9321312	63436373
$m = 7$	0	0	0	0	0	0	0	1	28	462	5880	63987	627396	5715424	49329280

Some examples of Stirling numbers of the second kind are given in Table 2.14.1. Others may be calculated by the exact (but rather slow) algorithm

Set $a_0 = 1$
Input m >>>>>>>
Set $a_1 = a_2 = a_3 = \cdots = a_m = 0$
Input n >>>>>>>
If $n = 0$ go to (3)
Set $a_0 = 0$
Set $a_1 = k = 1$
If $n = 1$ go to (3)
(1) Set j = the lesser of m and $k + 1$
(2) Replace a_j by ja_j [or by $-ka_j$]
Replace a_j by $a_j + a_{j-1}$
Replace j by $j - 1$
If $j(m + k - n - j) \neq 0$ go to (2)
Replace k by $k + 1$
If $k \neq n$ go to (1)
(3) Output a_m
$a_m = \sigma_n^{(m)}$ <<<<<<<
(or $S_n^{(m)}$)

Storage needed: $m + 5$ registers are required for a_0, a_1, a_2, ..., a_m, m, k, n and j.

Test values:
$\sigma_1^{(0)} = \sigma_0^{(1)} = 0$
$\sigma_0^{(0)} = \sigma_1^{(1)} = 1$
$\sigma_7^{(3)} = 301$
$\sigma_{17}^{(9)} = 9528822303$
$S_8^{(3)} = -13132$
$S_{10}^{(6)} = 63273$

which utilizes the recursion formula

2:14:5 $$\sigma_n^{(m)} = \sigma_{n-1}^{(m-1)} + m\sigma_{n-1}^{(m)} \qquad m = 1, 2, 3, \ldots \qquad n = 1, 2, 3, \ldots$$

When the alternative command shown in green is selected, the algorithm will generate $S_n^{(m)}$, a Stirling number of the first kind [see Section 18:6].

CHAPTER 3

THE ZETA NUMBERS AND RELATED FUNCTIONS

As detailed in Section 3:14, the four number families $\zeta(n)$, $\lambda(n)$, $\eta(n)$ and $\beta(n)$ occur as coefficients in many power series expansions. In this context it is only positive integer orders, $n = 1, 2, 3, \ldots$, that are encountered, and this chapter therefore emphasizes these cases. However, one is able to extend the definitions of all four functions to accept noninteger and negative orders, and these possibilities are also addressed here.

The first three functions are interrelated by the simple proportionalities

3:0:1
$$\frac{\zeta(\nu)}{2^\nu} = \frac{\lambda(\nu)}{2^\nu - 1} = \frac{\eta(\nu)}{2^\nu - 2}$$

and by the consequential identity

3:0:2
$$\zeta(\nu) + \eta(\nu) = 2\lambda(\nu)$$

but there are no corresponding relationships involving $\beta(\nu)$. Because they are so easily related to $\zeta(\nu)$ via 3:0:1, few formulas for $\lambda(\nu)$ and $\eta(\nu)$ are exhibited in this chapter.

3:1 NOTATION

The numbers $\zeta(n)$, $\lambda(n)$, $\eta(n)$ and $\beta(n)$ do not appear to have acquired definitive names; we shall call them *zeta numbers*, *lambda numbers*, *eta numbers* and *beta numbers*. When the order is unrestricted, we adopt the symbolism $\zeta(\nu)$, $\lambda(\nu)$, $\eta(\nu)$ and $\beta(\nu)$ and the names *zeta function*, *lambda function*, *eta function* and *beta function*. The last should not be confused with the complete beta function [Section 43:13] or the incomplete beta function [Chapter 62] to which it is totally unrelated.

The symbols σ and L have been used, respectively, in place of η and β. Riemann's name is associated with the zeta function, which is often known as "Riemann's function" or "Riemann's zeta function." We avoid these names to prevent confusion with the bivariate function of Chapter 64, with which Riemann's name is also commonly associated.

3:2 BEHAVIOR

The zeta and lambda functions are infinite only at $\nu = 1$, whereas $\eta(\nu)$ and $\beta(\nu)$ are always finite. Figure 3-1 maps the behaviors of the four functions. All approach unity rapidly as the order ν increases. For $\nu \leq 0$, the four functions

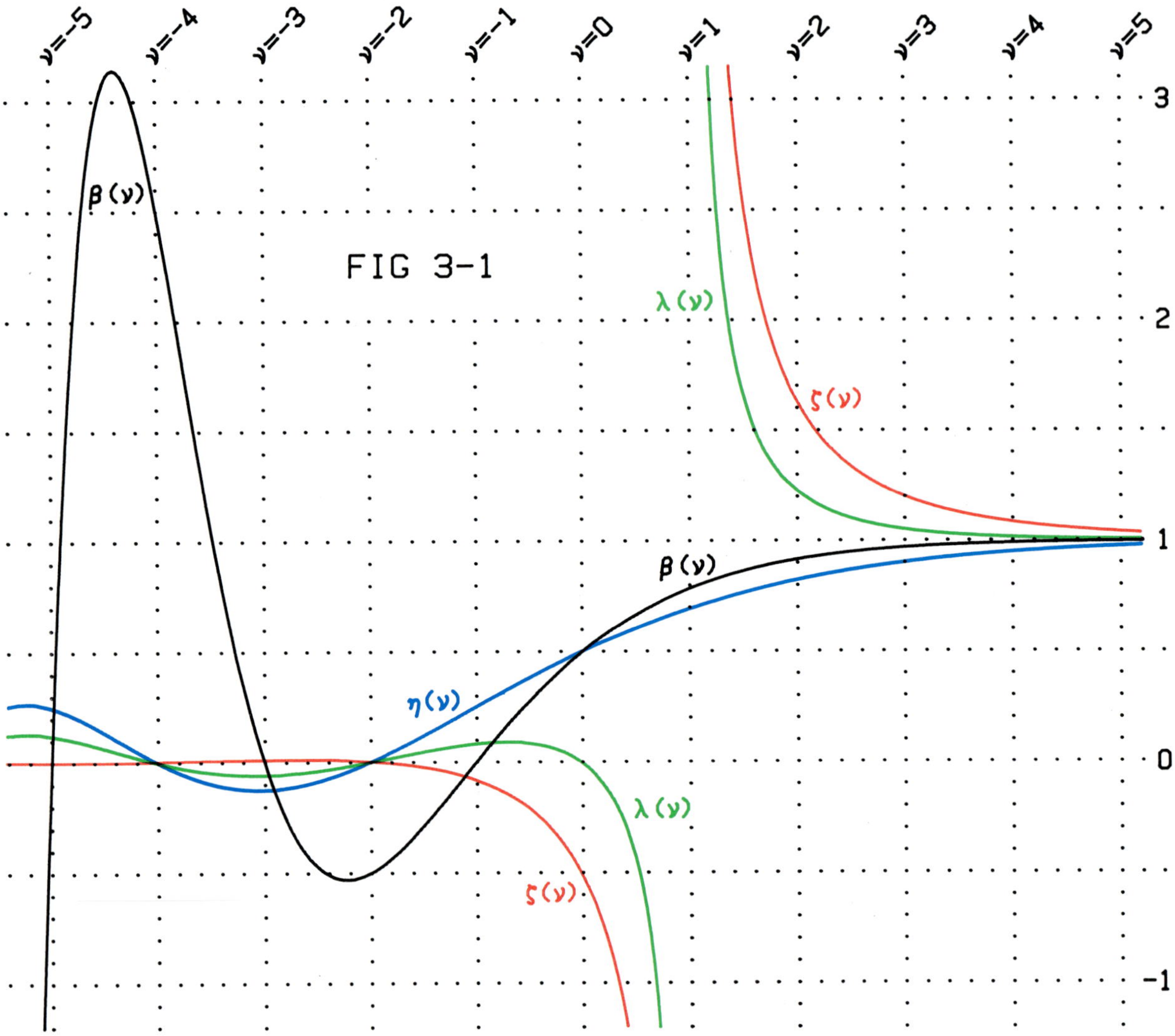

are oscillatory, with ever-increasing amplitudes as $\nu \to -\infty$. The zeros of $\zeta(\nu)$, $\lambda(\nu)$ and $\eta(\nu)$ occur when ν is an even negative integer, whereas $\beta(\nu) = 0$ at $\nu = -1, -3, -5, \ldots$.

3:3 DEFINITIONS

The four functions may be defined by the definite integrals

3:3:1 $$\zeta(\nu) = \frac{1}{\Gamma(\nu)} \int_0^\infty \frac{t^{\nu-1}\mathrm{d}t}{\exp(t) - 1} \qquad \nu > 1$$

3:3:2 $$\lambda(\nu) = \frac{1}{2\Gamma(\nu)} \int_0^\infty t^{\nu-1}\,\mathrm{csch}(t)\,\mathrm{d}t \qquad \nu > 1$$

3:3:3 $$\eta(\nu) = \frac{1}{\Gamma(\nu)} \int_0^\infty \frac{t^{\nu-1}\mathrm{d}t}{\exp(t) + 1} \qquad \nu > 0$$

and

3:3:4 $$\beta(\nu) = \frac{1}{2\Gamma(\nu)} \int_0^\infty t^{\nu-1} \operatorname{sech}(t)\, dt \qquad \nu > 0$$

involving functions discussed in Chapters 13, 26, 29 and 43.

The most commonly encountered definitions of zeta, lambda, eta and beta numbers are via the infinite series 3:6:1–3:6:4. These series may be reformulated as limits, for example:

3:3:5 $$\zeta(\nu) = \lim_{J\to\infty} \left\{ 1 + \frac{1}{2^\nu} + \frac{1}{3^\nu} + \cdots + \frac{1}{(J-1)^\nu} + \frac{1}{J^\nu} \right\} = \lim_{J\to\infty} \sum_{j=1}^{J} j^{-\nu}$$

and apply generally for $\nu > 1$ in the cases of $\zeta(\nu)$ and $\lambda(\nu)$, or for $\nu > 0$ in the $\eta(\nu)$ and $\beta(\nu)$ cases. However, when sufficient additional terms are included in series 3:3:5, to give

3:3:6 $$\zeta(\nu) = \lim_{J\to\infty} \left\{ 1 + \frac{1}{2^\nu} + \frac{1}{3^\nu} + \cdots + \frac{1}{(J-1)^\nu} + \frac{1}{(\nu-1)J^{\nu-1}} + \frac{1}{2J^\nu} + \frac{\nu}{12J^{\nu+1}} - \frac{\nu(\nu+1)(\nu+2)}{720J^{\nu+3}} + \cdots \right\}$$

the limit provides a definition of the zeta function for any order ν: positive, negative or zero. The general expression for the kth appended term in 3:3:6 involves functions from Chapters 2, 4 and 18 and is $(\nu)_{k-1}\mathrm{B}_k/(k!J^{\nu+k-1})$. It is necessary to include only those appended terms for which $\nu + k - 1$ is negative or zero, but a few extra will speed the approach to the limit. The corresponding definition of the beta function is

3:3:7 $$\beta(\nu) = \lim_{J\to\infty} \left\{ 1 - \frac{1}{3^\nu} + \frac{1}{5^\nu} - \cdots - \frac{1}{(J-2)^\nu} + \frac{1}{2J^\nu} + \frac{\nu}{2J^{\nu+1}} - \frac{\nu(\nu+1)(\nu+2)}{6J^{\nu+3}} + \cdots \right\}$$

the kth appended term being $(\nu)_{k-1}2^k(2^k - 1)\mathrm{B}_k/(2k!J^{k+\nu-1})$ in this case, J being a number equal to $4n + 1$ where n is a natural number.

3:4 SPECIAL CASES

The zeta, lambda, eta and beta numbers may be regarded as special cases of the corresponding functions. No further specialization is fruitful.

3:5 INTRARELATIONSHIPS

The zeta and beta functions satisfy the reflection formulas

3:5:1 $$\zeta(1-\nu) = \frac{2\Gamma(\nu)\zeta(\nu)}{(2\pi)^\nu} \cos\left(\frac{\nu\pi}{2}\right)$$

3:5:2 $$\beta(1-\nu) = \left(\frac{2}{\pi}\right)^\nu \Gamma(\nu)\beta(\nu) \sin\left(\frac{\nu\pi}{2}\right)$$

involving the gamma function [Chapter 43] and the functions of Chapter 32.

With $\mathrm{f}(n)$ representing any one of the four numbers $\zeta(n)$, $\lambda(n)$, $\eta(n)$ or $\beta(n)$, one may sum the infinite series $\Sigma(-1)^n\mathrm{f}(n)/n$, as well as the series of complements $\Sigma[1 - \mathrm{f}(n)]$, $\Sigma(-1)^n[1 - \mathrm{f}(n)]$, $\Sigma[1 - \mathrm{f}(n)]/n$ and $\Sigma(-1)^n[1 - \mathrm{f}(n)]/n$. With the lower summation limit taken as $n = 2$, these sums are as shown in Table 3.5.1. These sums involve the logarithmic function [Chapter 25] of various constants, Archimedes number π [Section 1:7], Euler's constant γ [Section 1:7] and the ubiquitous constant $U = \Gamma^2(\frac{3}{4})/\sqrt{\pi} = 0.8472130848$ [see Section 1:7]. Also listed in Table 3.5.1 are the sums $\Sigma(-1)^n\mathrm{f}(n)$. Strictly, these particular series do not converge, the tabulated entries being the limits

3:5:3 $$\lim_{J\to\infty} \left\{ \mathrm{f}(2) - \mathrm{f}(3) + \mathrm{f}(4) - \cdots \mp \mathrm{f}(J-1) \pm \frac{1}{2}\mathrm{f}(J) \right\}$$

which *do* converge and whose values may be associated with $\Sigma(-1)^n\mathrm{f}(n)$.

Table 3.5.1

	$\sum_{n=2}^{\infty}(-1)^n f(n)$	$\sum_{n=2}^{\infty}(-1)^n \frac{f(n)}{n}$	$\sum_{n=2}^{\infty}[1-f(n)]$	$\sum_{n=2}^{\infty}(-1)^n[1-f(n)]$	$\sum_{n=2}^{\infty}\frac{1-f(n)}{n}$	$\sum_{n=2}^{\infty}(-1)^n\frac{1-f(n)}{n}$
$f=\zeta$	1	γ	-1	$\frac{-1}{2}$	$\gamma-1$	$1-\ln(2)-\gamma$
$f=\lambda$	$\ln(2)$	$\frac{\gamma}{2}+\ln\left(\frac{2}{\sqrt{\pi}}\right)$	$\ln(2)-1$	$\frac{1}{2}-\ln(2)$	$\frac{\gamma}{2}+\ln(\sqrt{\pi})-1$	$1-\ln\left(\frac{4}{\sqrt{\pi}}\right)-\frac{\gamma}{2}$
$f=\eta$	$\ln(4)-1$	$\ln\left(\frac{4}{\pi}\right)$	$\ln(4)-1$	$\frac{3}{2}-\ln(4)$	$\ln(\pi)-1$	$1-\ln\left(\frac{8}{\pi}\right)$
$f=\beta$	$\frac{\pi}{4}-\ln(\sqrt{2})$	$\frac{\pi}{4}-\ln\left(\frac{\sqrt{2}}{U}\right)$	$\frac{\pi}{4}+\ln(\sqrt{2})-1$	$\frac{1}{2}+\ln(\sqrt{2})-\frac{\pi}{4}$	$\frac{\pi}{4}+\ln\left(\frac{\pi}{\sqrt{8}U}\right)-1$	$1-\ln(\sqrt{2}U)-\frac{\pi}{4}$

3:6 EXPANSIONS

The series

3:6:1 $$\zeta(\nu)=1+\frac{1}{2^\nu}+\frac{1}{3^\nu}+\cdots=\sum_{j=1}^{\infty}j^{-\nu}\qquad \nu>1$$

3:6:2 $$\lambda(\nu)=1+\frac{1}{3^\nu}+\frac{1}{5^\nu}+\cdots=\sum_{j=1}^{\infty}(2j-1)^{-\nu}\qquad \nu>1$$

3:6:3 $$\eta(\nu)=1-\frac{1}{2^\nu}+\frac{1}{3^\nu}-\cdots=\sum_{j=1}^{\infty}(-1)^{j+1}j^{-\nu}\qquad \nu>0$$

3:6:4 $$\beta(\nu)=1-\frac{1}{3^\nu}+\frac{1}{5^\nu}-\cdots=\sum_{j=1}^{\infty}(-1)^{j+1}(2j-1)^{-\nu}\qquad \nu>0$$

are the most useful representations of the four functions and serve as definitions of the zeta and lambda numbers, $\zeta(n)$ and $\lambda(n)$, for $n = 2, 3, 4, \ldots$ as well as for the eta and beta numbers, $\eta(n)$ and $\beta(n)$, for $n = 1, 2, 3, \ldots$.

Zeta and lambda functions are expansible as the infinite products

3:6:5 $$\zeta(\nu)=\frac{1}{1-2^{-\nu}}\,\frac{1}{1-3^{-\nu}}\,\frac{1}{1-5^{-\nu}}\,\frac{1}{1-7^{-\nu}}\cdots=\prod_{j=1}^{\infty}\frac{1}{1-\pi_j^{-\nu}}\qquad \nu>1$$

3:6:6 $$\lambda(\nu)=\frac{1}{1-3^{-\nu}}\,\frac{1}{1-5^{-\nu}}\,\frac{1}{1-7^{-\nu}}\,\frac{1}{1-11^{-\nu}}\cdots=\prod_{j=2}^{\infty}\frac{1}{1-\pi_j^{-\nu}}\qquad \nu>1$$

where π_j is the jth prime number.

3:7 PARTICULAR VALUES

In Table 3.7.1 Z has the value

3:7:1 $$Z=\zeta(3)=1.202056903$$

and G is Catalan's constant [see Section 1:7].

For $n = 2, 4, 6, \ldots$ all zeta, lambda and eta numbers equal π^n multiplied by a proper fraction [the fraction is related by equation 3:13:1 to the Bernoulli number B_n of Chapter 4]. Similarly, for $n = 1, 3, 5, \ldots$, $\beta(n)$ is proportional to π^n, the proportionality constant being a proper fraction related to the Euler number E_{n-1} [see Chapter

Table 3.7.1

	$\nu=-5$	$\nu=-4$	$\nu=-3$	$\nu=-2$	$\nu=-1$	$\nu=0$	$\nu=1$	$\nu=2$	$\nu=3$	$\nu=4$	$\nu=\infty$
$\zeta(\nu)$	$\frac{-1}{252}$	0	$\frac{1}{120}$	0	$\frac{-1}{12}$	$\frac{-1}{2}$	$\pm\infty$	$\frac{\pi^2}{6}$	Z	$\frac{\pi^4}{90}$	1
$\lambda(\nu)$	$\frac{31}{252}$	0	$\frac{-7}{120}$	0	$\frac{1}{12}$	0	$\pm\infty$	$\frac{\pi^2}{8}$	$\frac{7Z}{8}$	$\frac{\pi^4}{96}$	1
$\eta(\nu)$	$\frac{1}{4}$	0	$\frac{-1}{8}$	0	$\frac{1}{4}$	$\frac{1}{2}$	$\ln(2)$	$\frac{\pi^2}{12}$	$\frac{3Z}{4}$	$\frac{7\pi^4}{720}$	1
$\beta(\nu)$	0	$\frac{5}{2}$	0	$\frac{-1}{2}$	0	$\frac{1}{2}$	$\frac{\pi}{4}$	G	$\frac{\pi^3}{32}$	$\beta(4)$	1

5] by equation 3:13:2. For negative integer orders, the zeta and beta functions are related much more simply to the Bernoulli and Euler numbers; thus:

3:7:2
$$\zeta(-n) = \frac{-\mathrm{B}_{n+1}}{n+1} \qquad n = 1, 2, 3, \ldots$$

and

3:7:3
$$\beta(-n) = \frac{-\mathrm{E}_n}{2} \qquad n = 0, 1, 2, \ldots$$

3:8 NUMERICAL VALUES

We present two algorithms. The first is designed to calculate zeta, lambda and eta numbers for $n \geq 5$, so that, in conjunction with the table in Section 3:7, one may evaluate $\zeta(n)$, $\lambda(n)$ or $\eta(n)$ for all positive integer orders. This algorithm is based on expansion 3:6:6, using primes as large as 43, and on relation 3:0:1. The precision exceeds 24 bits.

Input n >>>>>>>
    Set $k = 7 + 2\,\mathrm{Int}(9/n)^5$
Input c >>>>>>>
    Set $f = (c - 2^n)/(1 - 2^n)$
(1) If $\mathrm{frac}\left(\frac{k}{3}\right)\mathrm{frac}\left(\frac{k}{5}\right) = 0$ go to (3)
(2) Replace f by $f/(1 - k^{-n})$
(3) Replace k by $k - 2$
    If $k > 8$ go to (1)
    If $k > 2$ go to (2)
    Output f
$f_0 \simeq \zeta(n)$, $f_1 \simeq \lambda(n)$, $f_2 \simeq \eta(n)$ <<<<<

Storage needed: n, k and f

Input restrictions: $n \geq 5$; code $c = 0$, 1 or 2

code, c	function
0	$\zeta(n)$
1	$\lambda(n)$
2	$\eta(n)$

Test values:
$\zeta(6) = 1.01734306$
$\lambda(6) = 1.00144708$
$\eta(6) = 0.985551091$

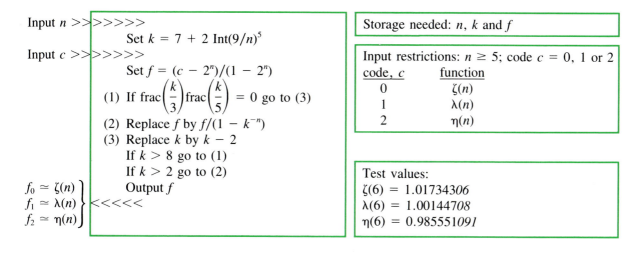

The second algorithm is much more sophisticated, being designed to calculate $\beta(\nu)$, $\zeta(\nu)$, $\lambda(\nu)$ or $\eta(\nu)$ according to the value given to a code c that is input prior to ν. Because $\zeta(1) = \lambda(1) = \pm\infty$, the algorithm returns the value 10^{99} when $c = 0$ or 1 and $\nu = 1$. Otherwise, the absolute algorithmic accuracy is better than $\pm 6 \times 10^{-8}$ for all four functions and for all orders. For $\nu \geq \frac{1}{2}$, the algorithm utilizes equation 3:3:7 (or the λ-analog of 3:3:6, together with 3:0:1) with J set equal to 13 and with appended terms up to $k = 9$. For $\nu < \frac{1}{2}$, $\beta(1-\nu)$ or $\lambda(1-\nu)$ is computed and this value is transformed into $\beta(\nu)$ or $\lambda(\nu)$ [and hence to $\zeta(\nu)$ or $\eta(\nu)$ via 3:0:1] by use of equation 3:5:2 or the λ-analog of 3:5:1. The gamma function $\Gamma(1-\nu)$ needed for this transformation is computed by a routine described in Section 43:8. To avoid "divide by zero" errors, special measures are adopted if $\nu = 0$ or 1.

Storage needed: c, f, ν, w, g, k and J

Use degree mode or change 90 to $\pi/2$.

Input restriction: c is a code that must equal -1, 0, 1 or 2, as follows:

code, c	function
-1	$\beta(\nu)$
0	$\zeta(\nu)$
1	$\lambda(\nu)$
2	$\eta(\nu)$

Input c >>>>>>>

Set $f = \left[\dfrac{1+c}{3}\ln(2)\right] + \left[1 - \dfrac{c}{2}\right]\left[\dfrac{\pi}{6} + (1+c)10^{99}\right]$

Input ν >>>>>>>

If $\nu = 1$ go to (7)
Set $f = (3|c| - 2 - c)/4$
If $\nu = 0$ go to (7)
Set $f = (|c| - 2^{\nu})/(1 - 2^{\nu})$
If $\nu \geq \dfrac{1}{2}$ go to (3)
Replace ν by $1 - \nu$
Set $w = \nu$
If $c < 0$ go to (1)
Replace f by $f[(2 - 2^{\nu})/(2^{\nu} - 1)]\cos(90\nu)$
Go to (2)
(1) Set $f = 2\sin(90\nu)$
(2) Replace f by f/w
Replace w by $w + 1$
If $w < 3$ go to (2)
Set $g = \left[1 + \dfrac{2}{7w^2}\left(\dfrac{2}{3w^2} - 1\right)\right] \Big/ 30w^2$
Replace g by $\dfrac{1-g}{12w} + w[\ln(w) - 1]$
Replace f by $[f\exp(g)\sqrt{2\pi/w}]/\pi^{\nu}$
(3) If $c < 0$ go to (4)
Set $c = 1$
(4) Set $J = 13$
Set $k = 11$
Set $g = 0$
(5) Replace g by $8g/(297 - 92.6k + 13k^2 - 0.6k^3)$
Replace k by $k - 2$
Replace g by $[\nu + k - 1]\left[2^k(1-c) + c - \dfrac{g(\nu+k)}{J}\right] \Big/ J$
If $k \neq 1$ go to (5)
Replace g by $\dfrac{g}{3} + 1 + \dfrac{(1+c)^J}{2(\nu - 1)} \Big/ 2J^{\nu}$
(6) Replace J by $J - 4$
Replace g by $g + [c/(J+2)^{\nu}] + [1/J^{\nu}]$
If $J \neq 1$ go to (6)
Replace f by fg
(7) Output f

$f_{-1} \simeq \beta(\nu)$
$f_0 \simeq \zeta(\nu)$
$f_1 \simeq \lambda(\nu)$
$f_2 \simeq \eta(\nu)$

<<<<

Test values:
$\beta(1) = 0.785398164$
$\eta(0) = 0.500000000$
$\beta(-3) = 0$
$\zeta(-3) = 0.00833333333$
$\lambda(-3) = -0.0583333333$
$\eta(-3) = -0.125000000$
$\beta(5) = 0.996157828$
$\zeta(5) = 1.03692776$
$\lambda(5) = 1.00452376$
$\eta(5) = 0.972119770$
$\beta(\frac{1}{2}) = 0.667691457$
$\zeta(\frac{1}{2}) = -1.46035451$
$\lambda(\frac{1}{2}) = -0.427727933$
$\eta(\frac{1}{2}) = 0.604898644$

3:9 APPROXIMATIONS

The formulas

3:9:1 $$\zeta(\nu) \simeq 1 + 2^{-\nu} + 3^{-\nu} \qquad \text{8-bit precision} \qquad \nu \geq 5$$

3:9:2 $$\lambda(\nu) \simeq 1 + 3^{-\nu} \qquad \text{8-bit precision} \qquad \nu \geq 4$$

3:9:3 $$\eta(\nu) \simeq 1 - 2^{-\nu} \qquad \text{8-bit precision} \qquad \nu \geq 5$$

3:9:4 $$\beta(\nu) \simeq 1 - 3^{-\nu} \qquad \text{8-bit precision} \qquad \nu \geq 4$$

are based upon the truncation of expansions 3:6:1 through 3:6:4 and apply for large positive orders. For large *negative* orders

3:9:5 $$\zeta(-\nu) \simeq -2\left(\frac{\nu}{2\pi}\right)^{\nu+1/2} \exp(-\nu)\sin\left(\frac{\nu\pi}{2}\right) \qquad \nu \to \infty$$

3:9:6 $$\lambda(-\nu) \simeq \frac{\eta(-\nu)}{2} \simeq \sqrt{2}\left(\frac{\nu}{\pi}\right)^{\nu+1/2} \exp(-\nu)\sin\left(\frac{\nu\pi}{2}\right) \qquad \nu \to \infty$$

3:9:7 $$\beta(-\nu) \simeq 2\left(\frac{2\nu}{\pi}\right)^{\nu+1/2} \exp(-\nu)\cos\left(\frac{\nu\pi}{2}\right) \qquad \nu \to \infty$$

In the vicinity of $\nu = 1$, the zeta function is well approximated by

3:9:8 $$\zeta(\nu) \simeq \gamma + \frac{1}{\nu - 1} \simeq 0.577 + \frac{1}{\nu - 1} \qquad \text{8-bit precision} \qquad 0.79 \leq \nu \leq 1.24$$

where γ is Euler's constant [Section 1:7].

3:10 OPERATIONS OF THE CALCULUS

The derivatives and indefinite integrals of the four functions may be expressed as infinite series, for example:

3:10:1 $$\frac{d}{d\nu}\zeta(\nu) = -\sum_{j=2}^{\infty} \frac{\ln(j)}{j^{\nu}} \qquad \nu > 1$$

and

3:10:2 $$\int_0^{\nu} \beta(t)dt = \nu + \sum_{j=1}^{\infty} \frac{(-1)^j}{\ln(2j + 1)}[1 - (2j + 1)^{-\nu}] \qquad \nu \geq 0$$

but not as established functions. At $\nu = 0$ the derivative of the zeta function equals $-\ln\sqrt{2\pi}$.

3:11 COMPLEX ARGUMENT

The representation of $\zeta(\nu + i\mu)$ in terms of its real and imaginary parts is given by

3:11:1 $$\zeta(\nu + i\mu) = \sum_{k=1}^{\infty} \frac{\cos\{\mu \ln(k)\}}{k^{\nu}} - i\sum_{k=2}^{\infty} \frac{\sin\{\mu \ln(k)\}}{k^{\nu}} \qquad \nu > 1$$

3:12 GENERALIZATIONS

The four functions of this chapter are special cases of the Hurwitz function of Chapter 64. Thus:

3:12:1 $$\zeta(\nu) = \zeta(\nu;1)$$

3:12:2 $$\lambda(\nu) = 2^{-\nu}\zeta\left(\nu;\frac{1}{2}\right)$$

3:12:3 $$\eta(\nu) = 2^{1-\nu}\zeta\left(\nu;\frac{1}{2}\right) - \zeta(\nu;1) = \eta(\nu;1)$$

3:12:4 $$\beta(\nu) = 4^{-\nu}\left[\zeta\left(\nu;\frac{1}{4}\right) - \zeta\left(\nu;\frac{3}{4}\right)\right] = 2^{-\nu}\eta\left(\nu;\frac{1}{2}\right)$$

The last pair of equations shows that the bivariate eta function [see Section 64:13] is also a generalization of the eta and beta functions.

3:13 COGNATE FUNCTIONS

When n is even $\zeta(n)$ is related to the Bernoulli number B_n [Chapter 4] by

3:13:1 $$\zeta(n) = \frac{(2\pi)^n|B_n|}{2n!} \qquad n = 2, 4, 6, \ldots$$

For odd n the zeta number is related to the Bernoulli polynomial [Chapter 19] via the integral

3:13:2 $$\zeta(n) = \frac{(2\pi)^n}{2n!}\left|\int_0^1 B_n(t)\cot(\pi t)dt\right| \qquad n = 1, 3, 5, \ldots$$

The beta number of odd argument is related to the Euler number E_{n-1} [Chapter 5] by

3:13:3 $$\beta(n) = \left(\frac{\pi}{2}\right)^n \frac{|E_{n-1}|}{2(n-1)!} \qquad n = 1, 3, 5, \ldots$$

while for even n the relationship is to the integral of an Euler polynomial [Chapter 20]

3:13:4 $$\beta(n) = \frac{\pi^n}{4(n-1)!}\left|\int_0^1 E_{n-1}(t)\sec(\pi t)dt\right| \qquad n = 2, 4, 6, \ldots$$

3:14 RELATED TOPICS

The four number families occur as coefficients in power series expansions of trigonometric and hyperbolic functions of argument πx or $\pi x/2$:

3:14:1 $$\cot(\pi x) = \frac{1}{\pi x} - \frac{2}{\pi x}\sum_{n=1}^{\infty}\zeta(2n)x^{2n} \qquad -1 < x < 1$$

3:14:2 $$\csc(\pi x) = \frac{1}{\pi x} + \frac{2}{\pi x}\sum_{n=1}^{\infty}\eta(2n)x^{2n} \qquad -1 < x < 1$$

3:14:3 $$\tan\left(\frac{\pi x}{2}\right) = \frac{4}{\pi x}\sum_{n=1}^{\infty}\lambda(2n)x^{2n} \qquad -1 < x < 1$$

3:14:4 $$\sec\left(\frac{\pi x}{2}\right) = \frac{4}{\pi x}\sum_{n=1}^{\infty}\beta(2n-1)x^{2n-1} \qquad -1 < x < 1$$

3:14:5 $$\coth(\pi x) = \frac{1}{\pi x} - \frac{2}{\pi x}\sum_{n=1}^{\infty}(-1)^n\zeta(2n)x^{2n} \qquad -1 < x < 1$$

3:14:6 $$\text{csch}(\pi x) = \frac{1}{\pi x} + \frac{2}{\pi x}\sum_{n=1}^{\infty}(-1)^n\eta(2n)x^{2n} \qquad -1 < x < 1$$

3:14:7 $$\tanh\left(\frac{\pi x}{2}\right) = \frac{-4}{\pi x}\sum_{n=1}^{\infty}(-1)^n\lambda(2n)x^{2n} \qquad -1 < x < 1$$

and

3:14:8 $$\mathrm{sech}\left(\frac{\pi x}{2}\right) = \frac{4}{\pi x}\sum_{n=1}^{\infty}(-1)^n\beta(2n-1)x^{2n-1} \qquad -1 < x < 1$$

Similar series represent the logarithms of trigonometric functions

3:14:9 $$\ln\{\csc(\pi x)\} = -\ln\{\sin(\pi x)\} = -\ln(\pi x) + \sum_{n=1}^{\infty}\frac{\zeta(2n)}{n}x^{2n} \qquad -1 < x < 1$$

3:14:10 $$\ln\left\{\tan\left(\frac{\pi x}{2}\right)\right\} = -\ln\left\{\cot\left(\frac{\pi x}{2}\right)\right\} = \ln\left(\frac{\pi x}{2}\right) + \sum_{n=1}^{\infty}\frac{\eta(2n)}{n}x^{2n} \qquad -1 < x < 1$$

3:14:11 $$\ln\left\{\sec\left(\frac{\pi x}{2}\right)\right\} = -\ln\left\{\cos\left(\frac{\pi x}{2}\right)\right\} = \sum_{n=1}^{\infty}\frac{\lambda(2n)}{n}x^{2n} \qquad -1 < x < 1$$

Notice that it is invariably the zeta, eta and lambda numbers of even argument, and the beta numbers of odd argument, that appear in such expansions.

CHAPTER 4

THE BERNOULLI NUMBERS, B_n

Bernoulli numbers constitute a family of rational numbers that occur principally as coefficients in power series.

4:1 NOTATION

There are two systems for indexing and assigning signs to Bernoulli numbers. Unfortunately, both systems are in widespread use, although the one we adopt is more generally used than its rival. Table 4.1.1 compares the two systems. Some authors employ both systems, using B_n for one set of Bernoulli numbers and some modified symbolism such as B_n^* or $\bar{B}_n$ for the other. These latter are sometimes called *auxiliary Bernoulli numbers*.

4:2 BEHAVIOR

We define B_n for all nonnegative integers n. The accompanying map, Figure 4-1, plots the first sixteen Bernoulli numbers.

In our notation, all Bernoulli numbers with odd index, except B_1, are zero. All B_n for which the index n is a multiple of 4, except B_0, are negative rational numbers. All other even-indexed Bernoulli numbers are positive rational numbers. The absolute values of the Bernoulli numbers of even index, $|B_{2n}|$, acquire a minimum value of $\frac{1}{42}$ when $2n = 6$. Table 4.2.1 shows that the magnitude of even-indexed Bernoulli numbers increases rapidly with increasing even n.

B_0 is the only Bernoulli number that is a nonzero integer but

4:2:1 $$\frac{2(n+1)!B_n}{(n/2)!2^{n/2}} = \text{odd integer} \qquad n = 2, 4, 6, \ldots$$

4:3 DEFINITIONS

The Bernoulli numbers are defined through the generating function

4:3:1 $$\frac{x}{\exp(x) - 1} = \sum_{n=0}^{\infty} B_n \frac{x^n}{n!}$$

Table 4.1.1

Value	Our system	Rival system
1	B_0	
$\frac{-1}{2}$	B_1	
$\frac{1}{6}$	B_2	B_1
0	B_3	
$\frac{-1}{30}$	B_4	$-B_2$
0	B_5	
$\frac{1}{42}$	B_6	B_3
0	B_7	
$\frac{-1}{30}$	B_8	$-B_4$

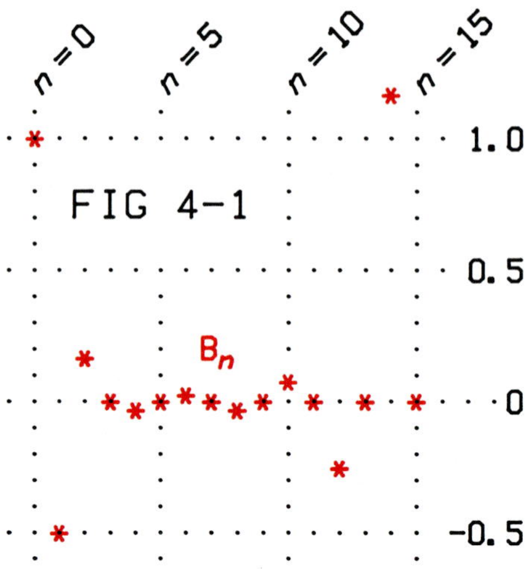

Table 4.2.1

Bernoulli number	Rounded value
B_{10}	8×10^{-2}
B_{20}	-5×10^{2}
B_{30}	6×10^{8}
B_{40}	-2×10^{16}
B_{50}	8×10^{24}
B_{60}	-2×10^{34}
B_{70}	3×10^{44}
B_{80}	-2×10^{55}
B_{90}	4×10^{66}
B_{100}	-3×10^{78}

The integral representations of the even-indexed Bernoulli numbers include

4:3:2 $$B_n = (-1)^{(n+2)/2}\pi^{-n}\int_0^\infty t^n \operatorname{csch}^2(t)\,dt \qquad n = 2, 4, 6, \ldots$$

where csch denotes the hyperbolic cosecant function [Chapter 29]. For others, the reader is directed to Erdelyi, Magnus, Oberhettinger and Tricomi [*Higher Transcendental Functions*, Volume 1, pages 38–39].

4:4 SPECIAL CASES

When n is an even integer, the Bernoulli number B_n is expressible as a zeta number [Chapter 3]

4:4:1 $$B_n = (-1)^{(n+2)/2}\frac{2n!}{(2\pi)^n}\zeta(n) \qquad n = 2, 4, 6, \ldots$$

4:5 INTRARELATIONSHIPS

Bernoulli numbers satisfy the recursion formula

4:5:1 $$B_n = -n!\sum_{j=0}^{n-1}\frac{B_j}{j!(n-1-j)!} \qquad n = 2, 3, 4, \ldots$$

with $B_0 = 1$. Recursion 4:5:1 is easily derived from the more compact expression

4:5:2 $$\sum_{j=0}^{n}\binom{n}{j}B_j = 0 \qquad n = 2, 3, 4, \ldots$$

where $\binom{n}{j}$ is the binomial coefficient [Chapter 6].

4:6 EXPANSIONS

Even-indexed Bernoulli numbers are expansible as

4:6:1 $$B_n = (-1)^{(n+2)/2} \frac{2n!}{(2\pi)^n}\left[1 + \frac{1}{2^n} + \frac{1}{3^n} + \cdots\right] = (-1)^{(n+2)/2}\, 2n! \sum_{j=1}^{\infty} (2j\pi)^{-n} \qquad n = 2, 4, 6, \ldots$$

If the multiplier $(-1)^{(n+2)/2}$ is replaced by $-\cos(n\pi/2)$, the expansion is valid for all values of n except 0 and 1.

4:7 PARTICULAR VALUES

B_0	B_1	B_2	B_3	B_4	B_5	B_6	B_7	B_8	B_9	B_{10}	B_{11}	B_{12}	B_{13}	B_{14}	B_{15}	B_{16}	B_{17}	B_{18}	B_{19}	B_{20}	B_{21}
1	$\frac{-1}{2}$	$\frac{1}{6}$	0	$\frac{-1}{30}$	0	$\frac{1}{42}$	0	$\frac{-1}{30}$	0	$\frac{5}{66}$	0	$\frac{-691}{2730}$	0	$\frac{7}{6}$	0	$\frac{-3617}{510}$	0	$\frac{43867}{798}$	0	$\frac{-174611}{330}$	0

4:8 NUMERICAL VALUES

The algorithm below produces Bernoulli numbers with a precision [see Section 0:8] better than 6×10^{-8}. For $n = 2, 4, 6, \ldots$ the algorithm uses relationship 4:4:1 with the very crude approximation $\zeta(n) \simeq (1 + 3^{-n})/(1 - 2^{-n})$. The crude B_n value that results is then corrected by a rounding procedure based on 4:2:1.

Input n >>>>>>>

If frac$(n/2) \neq 0$ go to (1)

Set $f = 1$

If $n = 0$ go to (2)

Set $g = (n + 1)!/[(n/2)!2^{n/2}]$

Replace f by $(f + 3^{-n})n!/[\pi^n(2^n - 1)]$

Replace f by $\frac{1}{2}$ + Int$(2fg)$

Replace f by $f[4\text{frac}(n/4) - 1]/g$

Go to (2)

(1) Replace n by $n - 2$

Set $f = [(n/|n|) - 1]/4$

(2) Output f

$f \simeq B_n$ <<<<<<<

Storage needed: n, f and g

Input restriction: n must be a nonnegative integer.

Test values:
$B_0 = 1$
$B_1 = -0.5$
$B_9 = 0$
$B_{14} = 1.16666667$
$B_{44} = -4.03380719 \times 10^{19}$

4:9 APPROXIMATIONS

Based on expansion 4:6:1 the approximation

4:9:1 $$B_n \simeq \frac{(-1)^{(n+2)/2} 2n!}{(2\pi)^n}\left[1 + \frac{1}{2^n}\right] \qquad \text{8-bit precision} \qquad n = 6, 8, 10, \ldots$$

is useful for all even n exceeding 4. It may be supplemented by values drawn from the table in Section 4:7. The approximate recursion formula

4:9:2 $$B_{2k} \simeq k\left(k - \frac{1}{2}\right) B_{2k-2}/\pi^2 \qquad \text{8-bit precision} \qquad k = 5, 6, 7, \ldots$$

becomes increasingly accurate as $k \to \infty$.

4:10 OPERATIONS OF THE CALCULUS

Neither differentiation nor integration may be applied to discretely defined functions such as B_n.

4:11 COMPLEX ARGUMENT

The Bernoulli number B_n has been defined only for real integer n.

4:12 GENERALIZATIONS

Bernoulli numbers are the special $x = 0$ cases of the Bernoulli polynomials $B_n(x)$ [Chapter 19]

4:12:1 $$B_n = B_n(0) \qquad n = 0, 1, 2, \ldots$$

Apart possibly from sign, the Bernoulli numbers are also the $x = 1$ values of the Bernoulli polynomials

4:12:2 $$B_n = (-1)^n B_n(1) \qquad n = 0, 1, 2, \ldots$$

4:13 COGNATE FUNCTIONS

Bernoulli numbers are closely related to Euler numbers [Chapter 5] and to the number families discussed in Chapter 3.

4:14 RELATED TOPICS

In addition to their roles as coefficients in power series expansions of trigonometric and hyperbolic functions [see Chapters 28–34] Bernoulli numbers occur as coefficients in the *Euler-Maclaurin formula,* which provides a valuable link between a definite integral and a sum. To exhibit this formula, let $x_0, x_1, x_2, \ldots, x_J$ be evenly spaced arguments with $x_{j+1} - x_j = h$ and $j = 0, 1, 2, \ldots, J - 1$. Then

4:14:1 $$\begin{aligned} h\sum_{j=0}^{J-1} f(x_j) - \int_{x_0}^{x_J} f(t)\,dt &= -\frac{h}{2}[f(x_J) - f(x_0)] + \frac{h^2}{12}\left[\frac{df}{dx}(x_J) - \frac{df}{dx}(x_0)\right] \\ &\quad - \frac{h^4}{720}\left[\frac{d^3f}{dx^3}(x_J) - \frac{d^3f}{dx^3}(x_0)\right] + \frac{h^6}{30240}\left[\frac{d^5f}{dx^5}(x_J) - \frac{d^5f}{dx^5}(x_0)\right] + \cdots \\ &= \sum_{n=1}^{\infty} \frac{h^n B_n}{n!}\left[\frac{d^{n-1}f}{dx^{n-1}}(x_J) - \frac{d^{n-1}f}{dx^{n-1}}(x_0)\right] \end{aligned}$$

provided that the function $f(x)$ is sufficiently differentiable.

CHAPTER 5

THE EULER NUMBERS, E_n

Euler numbers occur as coefficients in the power series expansions of the secant [Chapter 33] and hyperbolic secant [Chapter 29] functions.

5:1 NOTATION

As with Bernoulli numbers [Section 4:1], there are (at least) two notational systems in use for Euler numbers. Table 5.1.1 explains the differences between the two systems. Some authors use both systems and introduce supplementary notation such as E_n^* or $\bar{E}_n$ to distinguish from E_n. The former symbols are sometimes called *auxiliary Euler numbers*.

5:2 BEHAVIOR

The Euler number E_n is defined for all nonnegative integer index n. All Euler numbers are themselves integers.

Euler numbers of odd index are invariably zero. Even-indexed Euler numbers are positive integers, or negative integers, according as n is, or is not, a multiple of 4. In the accompanying map, Figure 5-1, either E_n (red points)

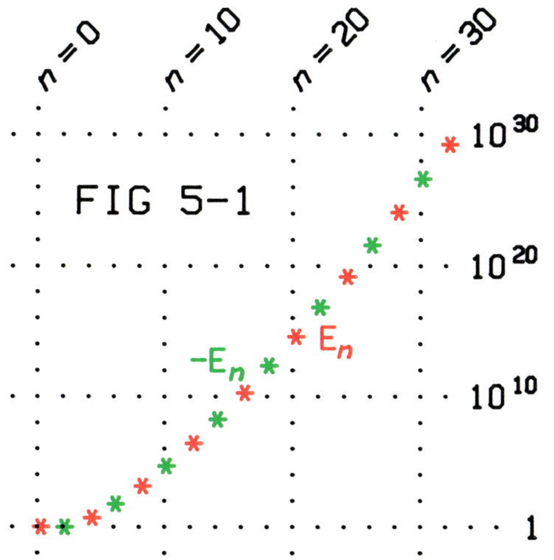

Table 5.1.1

Value	Our system	Rival system
1	E_0	E_0
0	E_1	
−1	E_2	$-E_1$
0	E_3	
5	E_4	E_2
0	E_5	
−61	E_6	$-E_3$
0	E_7	
1385	E_8	E_4

or $-E_n$ (green points) is plotted logarithmically versus n for even n values. Note the very rapid increase in the absolute value $|E_n|$ with increasing even n.

The least significant digit of each negative Euler number is a "1"; that is:

5:2:1 $$\text{frac}(-E_n/10) = \frac{1}{10} \qquad n = 2, 6, 10, \ldots$$

where frac denotes the fractional value function [Chapter 9]. The least significant digit of each positive Euler number, except E_0, is a "5"; that is:

5:2:2 $$\text{frac}(E_n/10) = \frac{1}{2} \qquad n = 4, 8, 12, \ldots$$

5:3 DEFINITIONS

The generating function

5:3:1 $$\text{sech}(t) = \sum_{n=0}^{\infty} E_n \frac{t^n}{n!}$$

may be used to define the Euler numbers. Here sech is the hyperbolic secant discussed in Chapter 29.

An integral definition of even-indexed Euler numbers is

5:3:2 $$E_n = (-1)^{n/2}\left(\frac{2}{\pi}\right)^{n+1}\int_0^{\infty} t^n\,\text{sech}(t)\,dt \qquad n = 0, 2, 4, \ldots$$

5:4 SPECIAL CASES

The Euler numbers of even index are related by the formula

5:4:1 $$E_n = (-1)^{n/2}\,2\left(\frac{2}{\pi}\right)^{n+1} n!\,\beta(n+1) \qquad n = 0, 2, 4, \ldots$$

to the beta numbers discussed in Chapter 3.

5:5 INTRARELATIONSHIPS

The compact expression

5:5:1 $$\sum_{j=0}^{n}\binom{2n}{2j} E_{2j} = 0 \qquad n = 1, 2, 3, \ldots$$

where $\binom{2n}{2j}$ is the binomial coefficient treated in Chapter 6, gives rise to the recursion formula for even-indexed Euler numbers

5:5:2 $$E_n = -n!\sum_{j=0}^{J}\frac{E_{2j}}{(n-2j)!(2j)!} \qquad n = 2, 4, 6, \ldots \qquad J = \frac{n}{2} - 1$$

with $E_0 = 1$.

5:6 EXPANSIONS

It follows from equations 5:4:1 and 3:6:4 that Euler numbers of even index may be expanded as

5:6:1 $$E_n = (-1)^{n/2}\, 2n! \left(\frac{2}{\pi}\right)^{n+1} \left[1 - \frac{1}{3^{n+1}} + \frac{1}{5^{n+1}} - \cdots\right] = (-1)^{n/2} \frac{2n!}{\pi^{n+1}} \sum_{j=0}^{\infty} \frac{(-1)^j}{(j + {}^1/_2)^{n+1}} \qquad n = 0, 2, 4, \ldots$$

5:7 PARTICULAR VALUES

E_0	E_1	E_2	E_3	E_4	E_5	E_6	E_7	E_8	E_9	E_{10}	E_{11}	E_{12}	E_{13}	E_{14}	E_{15}	E_{16}
1	0	−1	0	5	0	−61	0	1385	0	−50521	0	2702765	0	−199360981	0	19391512145

5:8 NUMERICAL VALUES

For $n = 2, 4, 6, \ldots$ the accompanying algorithm begins by using equation 5:4:1 with $\beta(n + 1)$ replaced by unity. This produces a very crude estimate of E_n, which is then refined by making use of rules 5:2:1 and 5:2:2. The final output has a precision better than 6×10^{-8}.

Set $f = 0$
Input n >>>>>>>
If frac($n/2$) $\neq 0$ go to (1)
Set $f = (2/\pi)^{n+1}\, n!/5$
Replace f by $-1 - 10\,\mathrm{Int}(f)$
If frac($n/4$) $\neq 0$ go to (1)
Replace f by $4 - f$
If $n \neq 0$ go to (1)
Set $f = 1$
(1) Output f
$f \simeq E_n$ <<<<<<<

Storage needed: n, f

Input restriction: n must be a nonnegative integer

Test values:
$E_0 = 1$
$E_5 = 0$
$E_{10} = -50521$
$E_{20} = 3.70371188 \times 10^{14}$

5:9 APROXIMATIONS

Based on 5:6:1, one is led to

5:9:1 $$E_n \simeq (-1)^{n/2}\, 2n!\, (2/\pi)^{n+1} \qquad \text{8-bit precision} \qquad n = 4, 6, 8, \ldots$$

5:10 OPERATIONS OF THE CALCULUS

Neither differentiation nor integration may be applied to discretely defined functions such as E_n.

5:11 COMPLEX ARGUMENT

Euler numbers have been defined only for real integer argument.

5:12 GENERALIZATIONS

Euler polynomials $E_n(x)$ [Chapter 20] represent a generalization of Euler numbers, to which they are related by

5:12:1 $$E_n = 2^n E_n(1/2) \qquad n = 0, 1, 2, \ldots$$

5:13 COGNATE FUNCTIONS

Euler numbers have much in common with beta numbers [Chapter 3] and Bernoulli numbers [Chapter 4].

CHAPTER 6

THE BINOMIAL COEFFICIENTS $\binom{\nu}{m}$

Binomial coefficients occur widely throughout mathematics: for example, in the expansions discussed in Section 6:14 and in the Leibniz theorem, equation 0:10:7.

6:1 NOTATION

When ν is the positive integer n, the binomial coefficient $\binom{n}{m}$ is sometimes denoted ${}_nC_m$ or C_n^m. These symbols have their origins in the role played by the binomial coefficient $\binom{n}{m}$ in expressing the number of combinations of m objects selected from a group of n different objects [see Section 2:14].

6:2 BEHAVIOR

The lower index m of the binomial coefficient $\binom{\nu}{m}$ is invariably a nonnegative integer, whereas the upper index ν may take any value. Positive integer values of the upper index are common, however, and we shall write $\binom{n}{m}$ as the general expression for these cases.

The binomial coefficient $\binom{n}{m}$ is always a positive integer when $m = 0, 1, 2, \ldots, n$ and zero when $m = n + 1, n + 2, n + 3, \ldots$. Values of $\binom{n}{m}$ for n values up to 16 and m up to 8 are tabulated in Section 6:8 (Table 6.8.1). For a given n, $\binom{n}{m}$ assumes its maximum value when $m = n/2$ if n is even, or when $m = (n \pm 1)/2$ if n is odd; this is illustrated in Figure 6-1.

When ν is not a positive integer or zero, the binomial coefficient $\binom{\nu}{m}$ may be positive or negative, integer or noninteger. Its values are zero when $\nu = 0, 1, 2, \ldots, m - 1$ and, as illustrated in Figure 6-2, are close to zero for all other ν values in the range $-\frac{1}{2} < \nu < m - \frac{1}{2}$. The binomial coefficient $\binom{\nu}{m}$ increases without limit (except when $m = 0$) as $\nu \to \infty$.

6:3 DEFINITIONS

The binomial coefficient is defined by the finite product

6:3:1 $$\binom{\nu}{m} = \left(\frac{\nu - m + 1}{1}\right)\left(\frac{\nu - m + 2}{2}\right)\left(\frac{\nu - m + 3}{3}\right)\cdots\left(\frac{\nu}{m}\right) = \prod_{j=1}^{m}\left(1 + \frac{\nu - m}{j}\right) = \prod_{j=0}^{m-1}\frac{\nu - j}{m - j}$$

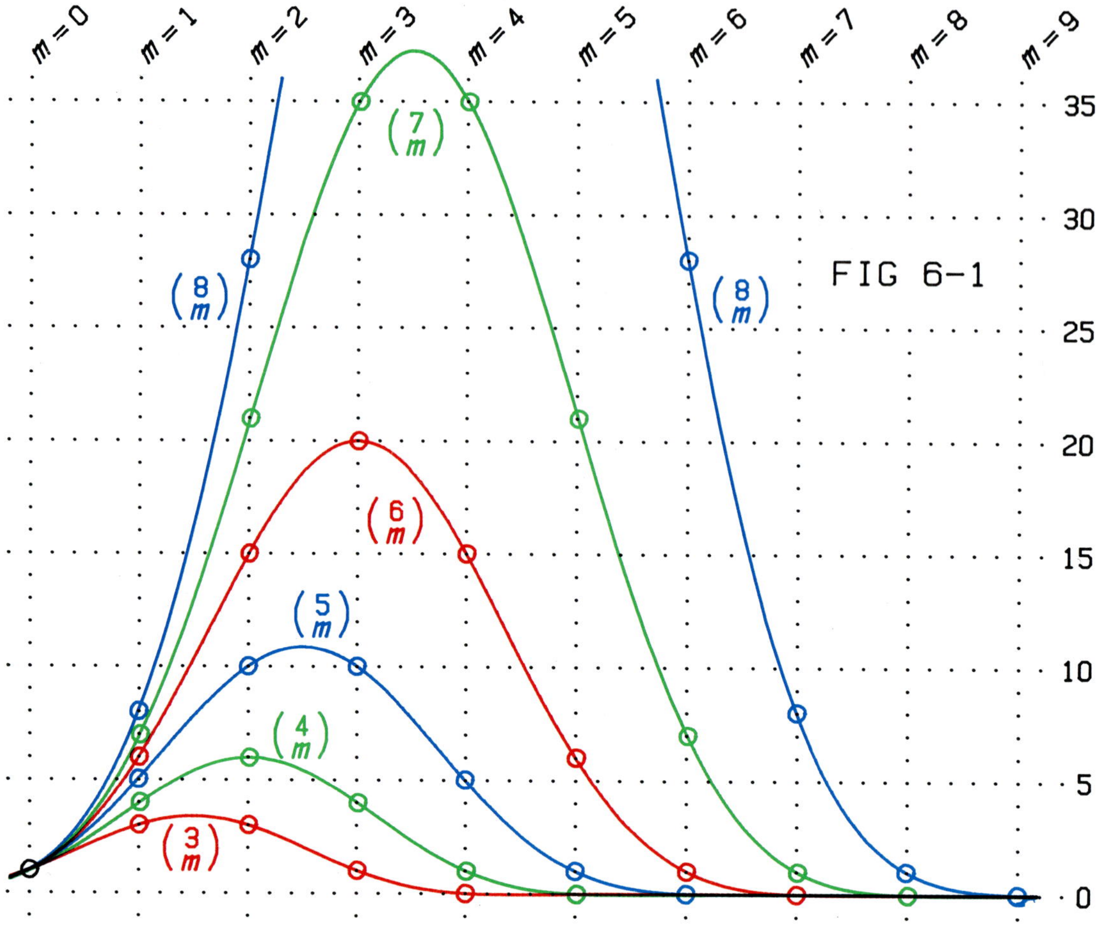

supplemented by the definition

6:3:2 $$\binom{\nu}{0} = 1$$

of a binomial coefficient with zero lower index. Binomial coefficients may also be defined by the generating function [Section 0:3]

6:3:3 $$(1+t)^\nu = \sum_{m=0}^{\infty} \binom{\nu}{m} t^m \qquad -1 < t < 1$$

For positive integer upper index the expression

6:3:4 $$\binom{n}{m} = \frac{n!}{m!(n-m)!} \qquad n \geq m$$

in terms of factorials [see Chapter 2], also serves as a definition.

6:4 SPECIAL CASES

Reduction to an expression involving the double factorial [Section 2:13] occurs when the upper index is equal to twice the lower index

6:4:1 $$\binom{2m}{m} = \frac{2^m(2m-1)!!}{m!} = \frac{4^m(2m-1)!!}{(2m)!!}$$

or differs by unity from twice the lower index

6:4:2 $$\binom{2m-1}{m-1} = \binom{2m-1}{m} = \frac{2^{m-1}(2m-1)!!}{m!} = \frac{2^{2m-1}(2m-1)!!}{(2m)!!} \qquad m = 1, 2, 3, \ldots$$

6:5 INTRARELATIONSHIPS

There exist reflection formulas for the upper

6:5:1 $$\binom{-\nu}{m} = (-1)^m \binom{\nu+m-1}{m}$$

and lower

6:5:2 $$\binom{n}{n-m} = \binom{n}{m}$$

indices, as well as recursion formulas

6:5:3 $$\binom{\nu+1}{m} = \binom{\nu}{m} + \binom{\nu}{m-1}$$

6:5:4 $$\binom{\nu}{m+1} = \frac{(\nu-m)}{(m+1)}\binom{\nu}{m}$$

for each index. The addition formula

6:5:5 $$\binom{\nu+\omega}{m} = \sum_{j=0}^{m} \binom{\nu}{j}\binom{\omega}{m-j}$$

known as *Vandermonde's convolution*, applies to the upper index.

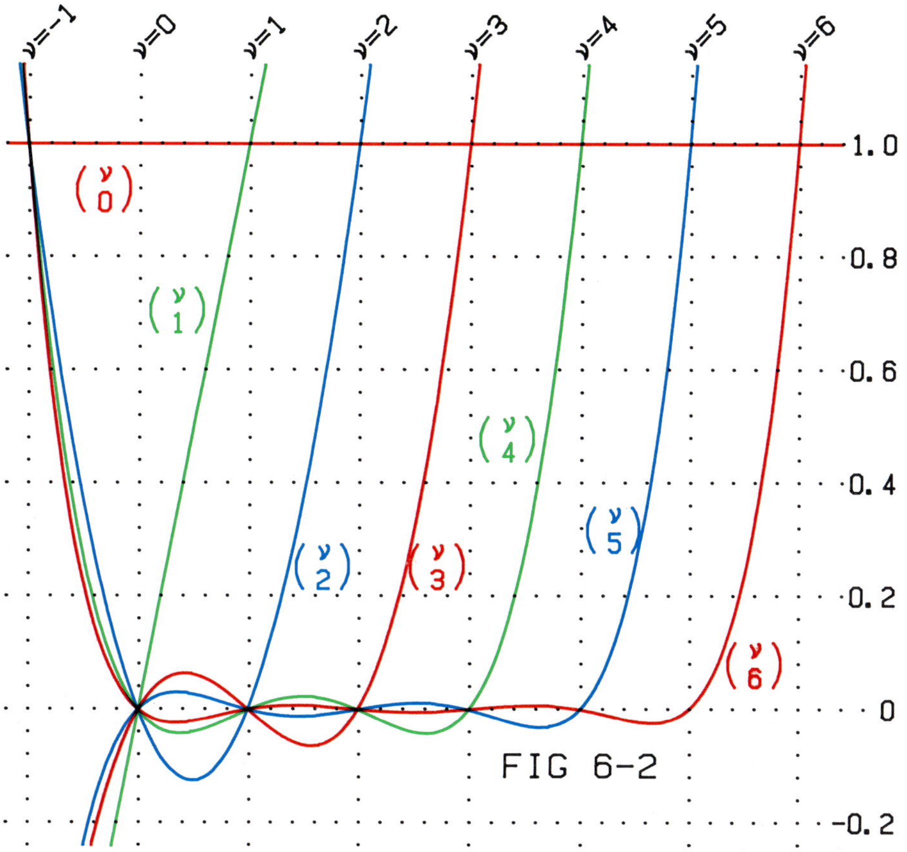

FIG 6-2

There are a number of intrarelationships involving sums of binomial coefficients and including

6:5:6 $$\sum_{j=0}^{n-m}\binom{j+m}{m}=\binom{n+1}{m+1} \qquad n>m$$

6:5:7 $$\sum_{j=0}^{m}(-1)^j\binom{\nu}{j}=(-1)^m\binom{\nu-1}{m}$$

and

6:5:8 $$\sum_{j=0}^{J}\binom{j+\nu}{j+m}=\binom{J+1+\nu}{J+m}$$

Formula 6:5:5 provides an expression for a sum of products of binomial coefficients; a second such formula is

6:5:9 $$\sum_{j=0}^{J}\binom{j}{m}\binom{j+\nu}{j}=\binom{m+\nu}{m}\binom{J+1+\nu}{J-m} \qquad J\geq m$$

When the upper index is an integer, finite or infinite series of binomial coefficients frequently have simple expressions; examples include

6:5:10 $$\sum_{j=0}^{J}\binom{n}{j}=2^n \qquad J\geq n$$

6:5:11 $$\sum_{j=0}^{J}(-1)^j\binom{n}{j}=0 \qquad J\geq n$$

6:5:12 $$\binom{n}{0}+\binom{n}{2}+\binom{n}{4}+\cdots+\binom{n}{J}=2^{n-1} \qquad n\leq J=2,4,6,\ldots$$

6:5:13 $$\binom{n}{1}+\binom{n}{3}+\binom{n}{5}+\cdots+\binom{n}{J}=2^{n-1} \qquad n\leq J=1,3,5,\ldots$$

and

6:5:14 $$\sum_{j=0}^{J}j\binom{n}{j}=2^{n-1}n \qquad J\geq n$$

Similarly, the sum of squares of binomial coefficients gives

6:5:15 $$\sum_{j=0}^{J}\binom{n}{j}^2=\binom{2n}{n} \qquad J\geq n$$

6:6 EXPANSIONS

A binomial coefficient may be expanded as a power series in its upper index by the formula

6:6:1 $$\binom{\nu}{m}=\frac{1}{m!}\sum_{j=0}^{m}S_m^{(j)}\nu^j$$

in which $S_m^{(j)}$ is a Stirling number of the first kind [see Section 18:6]. As well, definition 6:3:1 constitutes the expansion of $\binom{\nu}{m}$ as a finite product.

When ν, but not m, is large in magnitude, the asymptotic expansion

6:6:2 $$\binom{\nu}{m}\sim\frac{\nu^m}{m!}\left[1-\frac{m(m-1)}{2\nu}+\frac{m(m-1)(m-2)(3m-1)}{24\nu^2}+\cdots\right] \qquad \nu\to\pm\infty$$

applies. Conversely, when m, but not ν, is large, we have

6:6:3 $$\binom{\nu}{m} \sim \frac{(-1)^m}{m^{\nu+1}\Gamma(-\nu)}\left[1 + \frac{\nu(\nu+1)}{2m} + \frac{\nu(\nu+1)(\nu+2)(3\nu+1)}{24m^2} + \cdots\right] \qquad m \to \infty$$

where Γ is the gamma function [Chapter 43].

As noted in Section 6:2, the maximum value of $\binom{n}{m}$, for a given even value of n, occurs when $m = n/2$. An asymptotic series expansion for this maximum is

6:6:4 $$\binom{n}{n/2} \sim 2^n\sqrt{\frac{2}{\pi n}}\left[1 - \frac{1}{4n} + \frac{1}{32n^2} + \cdots\right] \qquad \text{even } n \to \infty$$

Similarly, the asymptotic expansion for the maximum value of $\binom{n}{m}$ when n is odd is

6:6:5 $$\binom{n}{n/2 \pm 1/2} \sim 2^n\sqrt{\frac{2}{\pi n}}\left[1 - \frac{3}{4n} + \frac{25}{32n^2} + \cdots\right] \qquad \text{odd } n \to \infty$$

6:7 PARTICULAR VALUES

The rules

6:7:1 $$\binom{\nu}{0} = 1$$

and

6:7:2 $$\binom{\nu}{1} = \nu$$

are valid for all values of the upper index. For integer values of this index, as well as for $\nu = \pm\frac{1}{2}$, there are additional particular values of the binomial coefficient as given in Table 6.7.1.

6:8 NUMERICAL VALUES

For integer upper index, values of the binomial coefficient $\binom{n}{m}$ are given in Table 6.8.1 for $0 \le n \le 16$ and $0 \le m \le 8$.

Table 6.7.1

	$\binom{-1}{m}$	$\binom{-1/2}{m}$	$\binom{0}{m}$	$\binom{1/2}{m}$	$\binom{1}{m}$	$\binom{n}{m}$	$\binom{\infty}{m}$
$m = 0$	1	1	1	1	1	1	1
$m = 1$	-1	$\frac{-1}{2}$	0	$\frac{1}{2}$	1	n	∞
$m = 2, 3, \ldots, n-2$	$(-1)^m$	$\frac{(-1)^m(2m-1)!!}{(2m)!!}$	0	$\frac{-(-1)^m(2m-3)!!}{(2m)!!}$	0	$\frac{n!}{m!(n-m)!}$	∞
$m = n-1$	$-(-1)^n$	$\frac{-(-1)^n(2n-3)!!}{(2n-2)!!}$	0	$\frac{(-1)^n(2n-5)!!}{(2n-2)!!}$	0	n	∞
$m = n$	$(-1)^n$	$\frac{(-1)^n(2n-1)!!}{(2n)!!}$	0	$\frac{-(-1)^n(2n-3)!!}{(2n)!!}$	0	1	∞
$m = n+1, n+2, \ldots$	$(-1)^m$	$\frac{(-1)^m(2m-1)!!}{(2m)!!}$	0	$\frac{-(-1)^m(2m-3)!!}{(2m)!!}$	0	0	∞

Table 6.8.1

	$\binom{0}{m}$	$\binom{1}{m}$	$\binom{2}{m}$	$\binom{3}{m}$	$\binom{4}{m}$	$\binom{5}{m}$	$\binom{6}{m}$	$\binom{7}{m}$	$\binom{8}{m}$	$\binom{9}{m}$	$\binom{10}{m}$	$\binom{11}{m}$	$\binom{12}{m}$	$\binom{13}{m}$	$\binom{14}{m}$	$\binom{15}{m}$	$\binom{16}{m}$	
$m = 0$	1	1	1	1	1	1	1	1	1	1	1	1	1	1	1	1	1	$m = 0$
$m = 1$	0	1	2	3	4	5	6	7	8	9	10	11	12	13	14	15	16	$m = 1$
$m = 2$	0	0	1	3	6	10	15	21	28	36	45	55	66	78	91	105	120	$m = 2$
$m = 3$	0	0	0	1	4	10	20	35	56	84	120	165	220	286	364	455	560	$m = 3$
$m = 4$	0	0	0	0	1	5	15	35	70	126	210	330	495	715	1001	1365	1820	$m = 4$
$m = 5$	0	0	0	0	0	1	6	21	56	126	252	462	792	1287	2002	3003	4368	$m = 5$
$m = 6$	0	0	0	0	0	0	1	7	28	84	210	462	924	1716	3003	5005	8008	$m = 6$
$m = 7$	0	0	0	0	0	0	0	1	8	36	120	330	792	1716	3432	6435	11440	$m = 7$
$m = 8$	0	0	0	0	0	0	0	0	1	9	45	165	495	1287	3003	6435	12870	$m = 8$

Recursion 6:5:3 provides a useful way of evaluating binomial coefficients. In a construction known as *Pascal's triangle*, binomial coefficients are listed in such a way that each entry is the sum of the two above.

$$
\begin{array}{c}
0 \quad 1 \quad 0 \\
0 \quad 1 \quad 1 \quad 0 \\
0 \quad 1 \quad 2 \quad 1 \quad 0 \\
0 \quad 1 \quad 3 \quad 3 \quad 1 \quad 0 \\
0 \quad 1 \quad 4 \quad 6 \quad 4 \quad 1 \quad 0 \\
0 \quad 1 \quad 5 \quad 10 \quad 10 \quad 5 \quad 1 \quad 0 \\
0 \quad 1 \quad 6 \quad 15 \quad 20 \quad 15 \quad 6 \quad 1 \quad 0
\end{array}
$$

The simple algorithm,

Input ν >>>>>>>

Set $f = 1$

Input m >>>>>>>

If $m = 0$ go to (2)

Set $j = m$

Replace m by $m - \nu$

(1) Replace f by $f\left(1 - \dfrac{m}{j}\right)$

Replace j by $j - 1$

If $j \neq 0$ go to (1)

(2) Output f

$f = \binom{\nu}{m}$ <<<<<<<

Storage needed: ν, f, m and j

Input restrictions: m must be a nonnegative integer; ν is unrestricted

Test values:
$\binom{4}{0} = 1$
$\binom{-5}{9} = -715$
$\binom{3/2}{4} = 0.0234375$

which is based on definitions 6:3:1 and 6:3:2, is exact.

6:9 APPROXIMATIONS

When both indices are small, the binomial coefficient is so easily calculated that no approximations are needed. When only one of ν and m is large in magnitude, the asymptotic expansion 6:6:2 or 6:6:3 may be truncated to provide an approximation to $\binom{\nu}{m}$.

When ν is large, $\binom{\nu}{m}$ is very large for values in the vicinity of $\nu/2$. Values of such binomial coefficients are approximated by the *Laplace-de Moivre formula*

6:9:1 $$\binom{\nu}{m} \simeq 2^{\nu}\sqrt{\frac{2}{\nu\pi}}\exp\left\{\frac{-2}{\nu}\left(m - \frac{\nu}{2}\right)^2\right\} \qquad \nu \text{ large}$$

which finds statistical applications [see Sokolnikoff and Redheffer, pages 623–626]. Even for ν as small as 10, the approximation leads to small absolute errors, as illustrated in Figure 6-3.

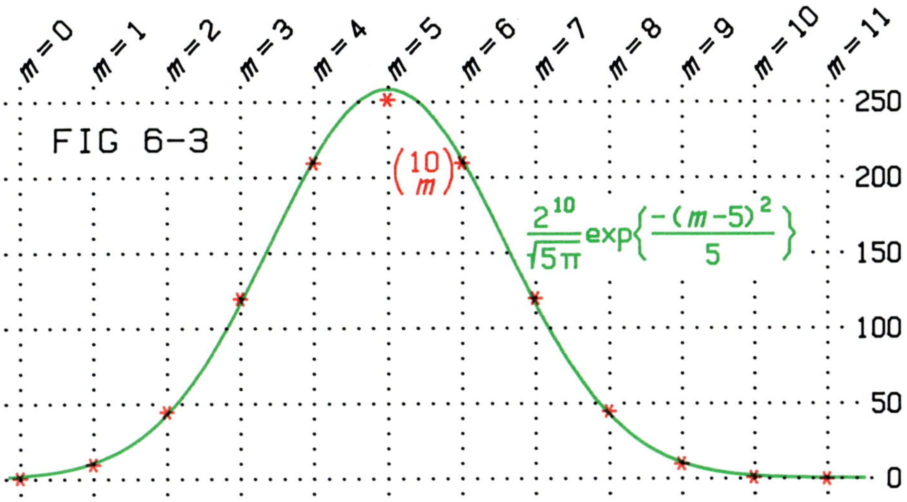

6:10 OPERATIONS OF THE CALCULUS

Differentiation with respect to the upper index gives a derivative involving a difference of two digamma functions [Chapter 44]:

6:10:1 $$\frac{\partial}{\partial\nu}\binom{\nu}{m} = \binom{\nu}{m}\left[\frac{1}{\nu} + \frac{1}{\nu-1} + \frac{1}{\nu-2} + \cdots + \frac{1}{\nu-m+1}\right] = \binom{\nu}{m}[\psi(-\nu) - \psi(m-\nu)]$$

6:11 COMPLEX ARGUMENT

The equivalence presented in equation 6:13:2 permits the formulas of Section 43:11 to be used to evaluate a binomial coefficient when one or both of the indices are imaginary or complex.

6:12 GENERALIZATIONS

The reciprocal of the complete beta function B [Chapter 43] generalizes the binomial coefficient to noninteger lower index because of the relationship

6:12:1 $$\binom{\nu}{m} = \frac{1}{m\mathrm{B}(m,\nu-m+1)} \qquad m \neq 0$$

between the two functions. If the upper index is an integer, this generalization simplifies to

6:12:2 $$\frac{1}{\mu\mathrm{B}(\mu,n-\mu+1)} = \frac{-n!\sin(\pi\mu)}{\pi(-\mu)_{n+1}}$$

where $(-\mu)_{n+1}$ denotes a Pochhammer polynomial [Chapter 18]. This formulation was used to draw the curves linking the binomial coefficient points in Figure 6-1.

A generalization in a different direction is provided by *multinomial coefficients* which are the integer coefficients that arise in such expansions as

6:12:3 $$\begin{aligned}(x+y+z)^5 = {} & [x^5+y^5+z^5] + 5[x^4y + x^4z + y^4x + z^4y + y^4z + z^4y] \\ & + 10[x^3y^2 + x^3z^2 + y^3x^2 + z^3x^2 + y^3z^2 + z^3y^2] \\ & + 20[x^3yz + y^3xz + z^3xy] + 30[x^2y^2z + x^2z^2y + y^2z^2x]\end{aligned}$$

The general expression may be written

6:12:4 $$(x_1 + x_2 + x_3 + \cdots + x_n)^N = \Sigma\, \mathrm{M}(N;m_1,m_2,m_3,\ldots,m_n)[x_1^{m_1}x_2^{m_2} \cdots x_n^{m_n} + x_2^{m_1}x_1^{m_2} \cdots x_n^{m_n} + \cdots]$$

where the summation embraces all combinations of nonnegative integer m's that satisfy

6:12:5 $$m_1 + m_2 + m_3 + \cdots + m_n = N$$

The multinomial coefficient is given by

6:12:6 $$\mathrm{M}(N;m_1,m_2,m_3,\ldots,m_n) = \frac{N!}{(m_1)!(m_2)!(m_3)! \cdots (m_n)!}$$

and the number of terms multiplied by that coefficient is

6:12:7 $$\#\text{ of terms} = \frac{n!}{(\mu(m_1))!(\mu(m_2))!(\mu(m_3))! \cdots (\mu(m_n))!}$$

where $\mu(m_j)$ is the number of occurrences of a specific value of m_j. (Thus, the denominator in 6:12:7 is unity if all the m values are distinct.) Tables 6.12.1 and 6.12.2 give a small number of *trinomial coefficients* (as exemplified in 6:12:3) and *quadrinomial coefficients*; see Abramowitz and Stegun [pages 831–832] for a more comprehensive tabulation.

6:13 COGNATE FUNCTIONS

The formulas

6:13:1 $$\binom{\nu}{m} = \frac{(\nu - m + 1)_m}{m!}$$

and

6:13:2 $$\binom{\nu}{m} = \frac{\Gamma(\nu + 1)}{m!\Gamma(\nu - m + 1)} = \frac{\Gamma(\nu + 1)}{\Gamma(m + 1)\Gamma(\nu - m + 1)}$$

relate the binomial coefficient to the Pochhammer polynomial [Chapter 18] and to the gamma function [Chapter 43].

6:14 RELATED TOPICS

The *binomial theorem*, also known as *Newton's formula*, permits a sum or difference to be raised to any power:

6:14:1 $$(x \pm y)^\nu = x^\nu\left[1 \pm \binom{\nu}{1}\frac{y}{x} + \binom{\nu}{2}\frac{y^2}{x^2} \pm \binom{\nu}{3}\frac{y^3}{x^3} + \cdots\right] = \sum_{m=0}^{\infty}\binom{\nu}{m}x^{\nu-m}(\pm y)^m \qquad -x < y < x$$

The series terminates if ν is a positive integer but is infinite otherwise. Some commonly encountered binomial expansions are

6:14:2 $$(1 \pm x)^4 = 1 \pm 4x + 6x^2 \pm 4x^3 + x^4$$

6:14:3 $$(1 \pm x)^3 = 1 \pm 3x + 3x^2 \pm x^3$$

6:14:4 $$(1 \pm x)^2 = 1 \pm 2x + x^2$$

6:14:5 $$(1 \pm x)^{3/2} = 1 \pm \frac{3}{2}x + \frac{3}{8}x^2 \mp \frac{1}{16}x^3 + \frac{3}{128}x^4 \mp \frac{3}{256}x^5 + \cdots \qquad -1 \le x \le 1$$

Table 6.12.1

N	m_1, m_2, m_3	$M(N;m_1,m_2,m_3)$	No. of terms
0	0, 0, 0	1	1
1	1, 0, 0	1	3
2	2, 0, 0	1	3
2	1, 1, 0	2	3
3	3, 0, 0	1	3
3	2, 1, 0	3	6
3	1, 1, 1	6	1
4	4, 0, 0	1	3
4	3, 1, 0	4	6
4	2, 2, 0	6	3
4	2, 1, 1	12	3
5	5, 0, 0	1	3
5	4, 1, 0	5	6
5	3, 2, 0	10	6
5	3, 1, 1	20	3
5	2, 2, 1	30	3
6	6, 0, 0	1	3
6	5, 1, 0	6	6
6	4, 2, 0	15	6
6	4, 1, 1	30	3
6	3, 3, 0	20	3
6	3, 2, 1	60	6
6	2, 2, 2	90	1
7	7, 0, 0	1	3
7	6, 1, 0	7	6
7	5, 2, 0	21	6
7	5, 1, 1	42	3
7	4, 3, 0	35	6
7	4, 2, 1	105	6
7	3, 3, 1	140	3
7	3, 2, 2	210	3

Table 6.12.2

N	m_1, m_2, m_3, m_4	$M(N;m_1,m_2,m_3,m_4)$	No. of terms
0	0, 0, 0, 0	1	1
1	1, 0, 0, 0	1	4
2	2, 0, 0, 0	1	4
2	1, 1, 0, 0	2	4
3	3, 0, 0, 0	1	4
3	2, 1, 0, 0	3	12
3	1, 1, 1, 0	6	4
4	4, 0, 0, 0	1	4
4	3, 1, 0, 0	4	12
4	2, 2, 0, 0	6	6
4	2, 1, 1, 0	12	12
4	1, 1, 1, 1	24	1
5	5, 0, 0, 0	1	4
5	4, 1, 0, 0	5	12
5	3, 2, 0, 0	10	12
5	3, 1, 1, 0	20	12
5	2, 2, 1, 0	30	12
5	2, 1, 1, 1	60	4
6	6, 0, 0, 0	1	4
6	5, 1, 0, 0	6	12
6	4, 2, 0, 0	15	12
6	4, 1, 1, 0	30	12
6	3, 3, 0, 0	20	6
6	3, 2, 1, 0	60	24
6	3, 1, 1, 1	120	4
6	2, 2, 2, 0	90	4
6	2, 2, 1, 1	180	6
7	7, 0, 0, 0	1	4
7	6, 1, 0, 0	7	12
7	5, 2, 0, 0	21	12
7	5, 1, 1, 0	42	12
7	4, 3, 0, 0	35	12
7	4, 2, 1, 0	105	24
7	4, 1, 1, 1	210	4
7	3, 3, 1, 0	140	12
7	3, 2, 2, 0	210	12
7	3, 2, 1, 1	420	12
7	2, 2, 2, 1	630	4

6:14:6 $$(1 \pm x)^{1/2} = 1 \pm \frac{1}{2}x - \frac{1}{8}x^2 \pm \frac{1}{16}x^3 - \frac{5}{128}x^4 \pm \frac{7}{256}x^5 + \cdots \qquad -1 \le x \le 1$$

6:14:7 $$(1 \pm x)^{-1/2} = 1 \mp \frac{1}{2}x + \frac{3}{8}x^2 \mp \frac{5}{16}x^3 + \frac{35}{128}x^4 \mp \frac{63}{256}x^5 + \cdots \qquad -1 < \pm x \le 1$$

6:14:8 $$(1 \pm x)^{-1} = 1 \mp x + x^2 \mp x^3 + x^4 \mp x^5 + \cdots \qquad -1 < x < 1$$

6:14:9 $$(1 \pm x)^{-3/2} = 1 \mp \frac{3}{2}x + \frac{15}{8}x^2 \mp \frac{35}{16}x^3 + \frac{315}{128}x^4 \mp \frac{693}{256}x^5 + \cdots \qquad -1 < x < 1$$

6:14:10 $$(1 \pm x)^{-2} = 1 \mp 2x + 3x^2 \mp 4x^3 + 5x^4 \mp 6x^5 + \cdots \qquad -1 < x < 1$$

6:14:11 $$(1 \pm x)^{-3} = 1 \mp 3x + 6x^2 \mp 10x^3 + 15x^4 \mp 21x^5 + \cdots \qquad -1 < x < 1$$

6:14:12 $$(1 \pm x)^{-4} = 1 \mp 4x + 10x^2 \mp 20x^3 + 35x^4 \mp 56x^5 + \cdots \qquad -1 < x < 1$$

Equation 6:14:1 may be regarded as a special case of the Taylor expansion, equation 0:5:1. The function $(1 \pm x)^\nu$, or sometimes $(y + x)^\nu$, is known as the *binomial function*.

CHAPTER 7

THE LINEAR FUNCTION $bx + c$ AND ITS RECIPROCAL

Many relationships in science and engineering take the form $f(x) = bx + c$, and many others can be manipulated into this form by a redefinition of the variables. Data analysis problems can therefore often be reduced to the determination of the coefficients b and c from pairs $(x,f(x))$ of experimental values. The *least squares* method for performing this analysis is presented in Section 7:14.

7:1 NOTATION

The constant c of the linear function $bx + c$ is termed the *intercept*, whereas b is known as the *slope* or *gradient*. These terms derive, of course, from the observation that, if the linear function $bx + c$ is plotted, its graph is a straight line that cuts the vertical axis at altitude c and whose inclination from the horizontal is characterized by the number b [see Figure 7-1]. Occasionally the word slope is used to mean $\arctan(b)$ [see Chapter 35]. The symbol m frequently replaces b.

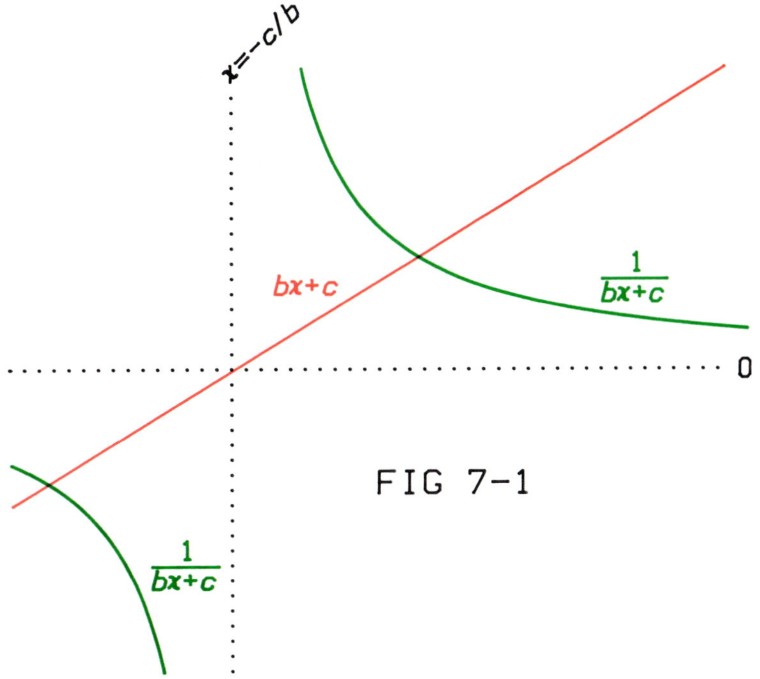

FIG 7-1

The name *inverse linear function* is sometimes given to the function $1/(bx + c)$. Throughout this *Atlas*, however, we reserve the phrase *inverse function* for the relationship described in Section 0:3. The name *reciprocal linear function* unambiguously denotes $1/f(x)$ where $f(x) = bx + c$.

7:2 BEHAVIOR

The linear function $bx + c$ is defined for all values of x and (unless b is zero) itself assumes all values. The same is true of the reciprocal linear function $1/(bx + c)$ which, unlike the linear function, is discontinuous at $x = -c/b$. The accompanying Figure 7-1 is a map showing typical behavior of the linear function and its reciprocal when both b and c are positive. Figure 7-2 illustrates how the linear function is positioned with respect to the axes when b and/or c changes sign.

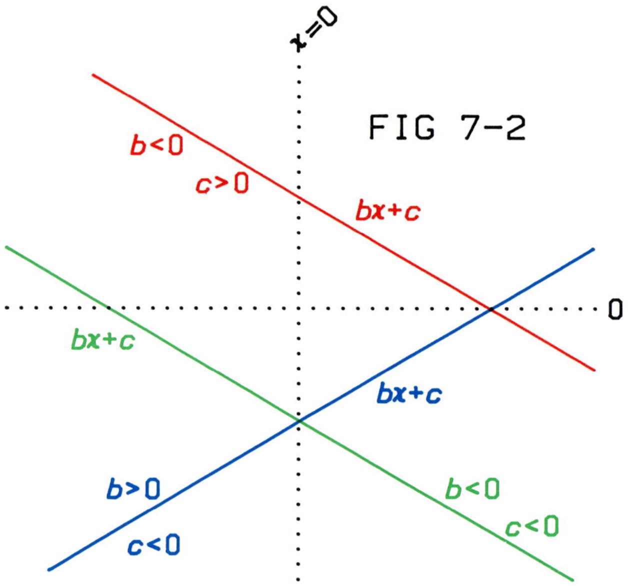

7:3 DEFINITIONS

The arithmetic operations of multiplication by b and addition of c define $f(x) = bx + c$ for all values of x. The function $f(x) = 1/(bx + c)$ is defined by the same operations followed by division into unity. The linear function is completely characterized when its values, f_1 and f_2, are known at two (distinct) arguments, x_1 and x_2. The slope and intercept may be inferred from the formula

7:3:1 $$bx + c = \left(\frac{f_2 - f_1}{x_2 - x_1}\right)x + \frac{x_2 f_1 - x_1 f_2}{x_2 - x_1}$$

7:4 SPECIAL CASES

When $b = 0$, both the linear function and its reciprocal reduce to a constant function [Chapter 1].

7:5 INTRARELATIONSHIPS

Both the linear function and its reciprocal obey the reflection formula

7:5:1 $$f(-x) = -f\left(x - \frac{2c}{b}\right)$$

The sum or difference of two linear functions is a third linear function

7:5:2 $$(b_1x + c_1) \pm (b_2x + c_2) = (b_1 \pm b_2)x + (c_1 \pm c_2)$$

The product of two linear functions is a quadratic function [Chapter 16]

7:5:3 $$(b_1x + c_1)(b_2x + c_2) = b_1b_2x^2 + (b_1c_2 + b_2c_1)x + c_1c_2$$

and (unless $b_2c_1 = b_1c_2$) the quotient of two linear functions is an infinite series

7:5:4 $$\frac{b_1x + c_1}{b_2x + c_2} = \begin{cases} \dfrac{c_1}{c_2} + \dfrac{b_2c_1 - b_1c_2}{b_2c_2} \displaystyle\sum_{j=1}^{\infty} \left(\frac{-b_2x}{c_2}\right)^j & |x| < \left|\dfrac{c_2}{b_2}\right| \\ \dfrac{b_1}{b_2} - \dfrac{b_2c_1 - b_1c_2}{b_2c_2} \displaystyle\sum_{j=1}^{\infty} \left(\frac{-c_2}{b_2x}\right)^j & |x| > \left|\dfrac{c_2}{b_2}\right| \end{cases}$$

The inverse function [Section 0:3] of the linear function $f(x) = bx + c$ is another linear function

7:5:5 $$F(x) = \frac{x}{b} - \frac{c}{b} \qquad b \neq 0$$

while the inverse of the $f(x) = 1/(bx + c)$ function is

7:5:6 $$F(x) = \frac{1}{bx} - \frac{c}{b} \qquad b \neq 0$$

As explained in Section 16:3, the difference of two reciprocal linear functions that share a common b parameter produces a reciprocal quadratic function. Finite and infinite sums of certain reciprocal linear functions may be evaluated in terms of the digamma or Bateman's G functions [Chapter 44] as follows:

7:5:7 $$\frac{1}{c} + \frac{1}{x+c} + \frac{1}{2x+c} + \cdots + \frac{1}{Jx+c} = \sum_{j=0}^{J} \frac{1}{jx+c} = \frac{1}{x}\left[\psi\left(J + 1 + \frac{c}{x}\right) - \psi\left(\frac{c}{x}\right)\right]$$

7:5:8 $$\frac{1}{c} - \frac{1}{x+c} + \frac{1}{2x+c} - \cdots \pm \frac{1}{Jx+c} = \sum_{j=0}^{J} \frac{(-1)^j}{jx+c} = \frac{1}{2x}\left[\psi\left(\frac{1}{2} + \frac{c}{2x}\right) - \psi\left(\frac{c}{2x}\right) \pm \psi\left(\frac{J}{2} + 1 + \frac{c}{2x}\right) \mp \psi\left(\frac{J}{2} + \frac{1}{2} + \frac{c}{2x}\right)\right] = \frac{1}{2x}\left[G\left(\frac{c}{x}\right) \pm G\left(J + 1 + \frac{c}{x}\right)\right]$$

where the upper signs are taken if J is even, the lower if J is odd, and

7:5:9 $$\frac{1}{c} - \frac{1}{x+c} + \frac{1}{2x+c} - \cdots = \sum_{j=0}^{\infty} \frac{(-1)^j}{jx+c} = \frac{1}{2x}\left[\psi\left(\frac{1}{2} + \frac{c}{2x}\right) - \psi\left(\frac{c}{2x}\right)\right] = \frac{1}{2x} G\left(\frac{c}{x}\right)$$

7:6 EXPANSIONS

The reciprocal linear function is expansible as a so-called *geometric series* for either small

7:6:1 $$\frac{1}{bx+c} = \frac{1}{c} - \frac{bx}{c^2} + \frac{b^2x^2}{c^3} - \frac{b^3x^3}{c^4} + \cdots = \frac{1}{c}\sum_{j=0}^{\infty}\left(\frac{-bx}{c}\right)^j \qquad |x| < \left|\frac{c}{b}\right|$$

or large

7:6:2 $$\frac{1}{bx+c} = \frac{1}{bx} - \frac{c}{b^2x^2} + \frac{c^2}{b^3x^3} - \frac{c^3}{b^4x^4} + \cdots = \frac{1}{bx}\sum_{j=0}^{\infty}\left(\frac{-c}{bx}\right)^j \qquad |x| > \left|\frac{c}{b}\right|$$

values of the argument. The linear function may be expanded

7:6:3 $$bx + c = c + 2J_1(bx) + 6J_3(bx) + 10J_5(bx) + \cdots = c + 2\sum_{j=1}^{\infty}(2j-1)J_{2j-1}(bx)$$

as an infinite series of Bessel coefficients [Chapter 52] of odd order, although this representation is rarely employed. The reciprocal linear function may also be expanded as an infinite product

7:6:4 $$\frac{1}{bx+c} = \begin{cases} \dfrac{c-bx}{c^2}\displaystyle\prod_{j=1}^{\infty}\left[1+\left(\frac{bx}{c}\right)^{2^j}\right] & -1 < \dfrac{bx}{c} < 1 \\ \dfrac{bx-c}{b^2x^2}\displaystyle\prod_{j=1}^{\infty}\left[1+\left(\frac{c}{bx}\right)^{2^j}\right] & \left|\dfrac{bx}{c}\right| > 1 \end{cases}$$

For example, if $|x| < 1$, $(1 \pm x)^{-1} = (1 \mp x)(1 + x^2)(1 + x^4)(1 + x^8)(1 + x^{16}) \ldots$.

7:7 PARTICULAR VALUES

7:7:1 $$bx + c = 0 \qquad x = -c/b \qquad b \neq 0$$

7:7:2 $$bx + c = c \qquad x = 0$$

7:8 NUMERICAL VALUES

These are easily calculated by direct substitution.

7:9 APPROXIMATIONS

For small absolute values of x, the reciprocal linear function may be approximated by a linear function

7:9:1 $$\frac{1}{bx+c} \simeq \frac{1}{c} - \frac{bx}{c^2} \qquad \text{8-bit precision} \qquad |x| < \frac{|c|}{16|b|}$$

Similarly, the approximation

7:9:2 $$\frac{1}{bx+c} \simeq \frac{bx-c}{b^2x^2} \qquad \text{8-bit precision} \qquad |x| > \frac{16|c|}{|b|}$$

is useful for large arguments of either sign. These two results follow directly from expansions 7:6:1 and 7:6:2, respectively.

7:10 OPERATIONS OF THE CALCULUS

Differentiation of the linear function produces

7:10:1 $$\frac{d}{dx}(bx + c) = b$$

while the derivative of the reciprocal linear function is

7:10:2 $$\frac{d}{dx}\left(\frac{1}{bx+c}\right) = \frac{-b}{(bx+c)^2}$$

Integration of the same pair results in

7:10:3
$$\int_0^x (bt+c)\mathrm{d}t = \frac{bx^2}{2} + cx$$

and

7:10:4
$$\int_0^x \frac{1}{bt+c}\,\mathrm{d}t = \frac{1}{b}\ln\left(\left|1+\frac{bx}{c}\right|\right)$$

respectively. If $0 < (-c/b) < x$, the integrand in 7:10:4 (and in some of the integrals below) encounters an infinity; in this case the integral is to be interpreted as a Cauchy limit, as explained later in this section. Other important integrals include

7:10:5
$$\int_0^x \frac{\mathrm{d}t}{(Bt+C)(bt+c)} = \frac{1}{bC-Bc}\ln\left(\left|1+\frac{bx}{c}\right| \Big/ \left|1+\frac{Bx}{C}\right|\right) \qquad Bc \neq bC$$

7:10:6
$$\int_0^x \frac{Bt+C}{bt+c}\,\mathrm{d}t = \frac{Bx}{b} + \frac{bC-Bc}{b^2}\ln\left(\left|1+\frac{bx}{c}\right|\right)$$

and

7:10:7
$$\int_0^x \frac{t^n\mathrm{d}t}{bt+c} = \frac{(-c)^n}{b^{n+1}}\left[\ln\left(\left|1+\frac{bx}{c}\right|\right) + \sum_{j=1}^{n}\frac{1}{j}\left(\frac{-bx}{c}\right)^j\right] \qquad n = 0, 1, 2, \ldots$$

Semidifferentiation and semiintegration [see Section 0:10] with lower limit zero give

7:10:8
$$\frac{\mathrm{d}^{1/2}}{\mathrm{d}x^{1/2}}(bx+c) = \frac{2bx+c}{\sqrt{\pi x}}$$

and

7:10:9
$$\frac{\mathrm{d}^{-1/2}}{\mathrm{d}x^{-1/2}}(bx+c) = \sqrt{\frac{x}{\pi}}\left(\frac{4bx}{3} + 2c\right)$$

when applied to the linear function, and

7:10:10
$$\frac{\mathrm{d}^{1/2}}{\mathrm{d}x^{1/2}}\frac{1}{bx+c} = \frac{1-\sqrt{x}\,\phi}{\sqrt{\pi x}(bx+c)}$$

7:10:11
$$\frac{\mathrm{d}^{-1/2}}{\mathrm{d}x^{-1/2}}\frac{1}{bx+c} = \frac{2\phi}{\sqrt{\pi b}}$$

when applied to the reciprocal linear function, where the form of the function ϕ depends on the magnitude of c/b, as follows:

7:10:12
$$\phi = \begin{cases} \dfrac{\operatorname{arcsin}(\sqrt{-bx/c})}{\sqrt{-x-(c/b)}} & -\infty < \dfrac{c}{b} < -x \\[2ex] \dfrac{\operatorname{arcosh}(\sqrt{-bx/c})}{\sqrt{x+(c/b)}} & -x < \dfrac{c}{b} < 0 \\[2ex] \dfrac{\operatorname{arsinh}(\sqrt{bx/c})}{\sqrt{x+(c/b)}} & 0 < \dfrac{c}{b} < \infty \end{cases}$$

In formulas 7:10:8–7:10:12 it is understood that $x > 0$.

The semiderivative, with lower limit $-\infty$, of $1/(bx + c)$ is

7:10:13
$$\frac{d^{1/2}}{[d(x+\infty)]^{1/2}} \frac{1}{bx+c} = \begin{cases} \dfrac{1}{2}\sqrt{\dfrac{-\pi b}{(bx+c)^3}} & x < \dfrac{-c}{b} \\ 0 & x > \dfrac{-c}{b} \end{cases}$$

while the semiintegral is

7:10:14
$$\frac{d^{-1/2}}{[d(x+\infty)]^{-1/2}} \frac{1}{bx+c} = \begin{cases} \dfrac{-\sqrt{\pi}}{\sqrt{-b(bx+c)}} & x < \dfrac{-c}{b} \\ 0 & x > \dfrac{-c}{b} \end{cases}$$

Definite integrals of the form

7:10:15
$$\int_{-\infty}^{\infty} \frac{f(t)}{bt+c} dt$$

can be evaluated by a *Hilbert transform* defined as

7:10:16
$$\bar{f}_H(s) = \frac{1}{\pi} \int_{-\infty}^{\infty} f(t) \frac{dt}{t-s}$$

Therefore, integral 7:10:15 can be evaluated by means of the identity

7:10:17
$$\int_{-\infty}^{\infty} \frac{f(t)}{bt+c} dt = \frac{\pi}{b} \bar{f}_H(-c/b)$$

For most functions f the integral

7:10:18
$$\int_{-\infty}^{x} f(t) \frac{dt}{t-s}$$

encounters an infinity at $x = s$ and therefore the definition of the Hilbert transform requires a so-called *Cauchy limit* interpretation of the integral in 7:10:16; that is:

7:10:19
$$\int_{-\infty}^{\infty} f(t) \frac{dt}{t-s} = \lim_{\varepsilon \to 0} \left\{ \int_{-\infty}^{s-\varepsilon} f(t) \frac{dt}{t-s} + \int_{s+\varepsilon}^{\infty} f(t) \frac{dt}{t-s} \right\} \qquad \varepsilon > 0$$

Occasionally in this *Atlas* [e.g., in Section 22:10] a Hilbert transform will be encountered with integration limits other than $\pm\infty$. The shorthand notation

7:10:20
$$\bar{f}_H(s) = \frac{1}{\pi} \int_{t_0}^{t_1} f(t) \frac{dt}{t-s} = \frac{1}{\pi} \int_{-\infty}^{\infty} F(t) \frac{dt}{t-s} \qquad \text{where} \qquad F(t) = \begin{cases} 0 & t < t_0 \\ f(t) & t_0 \le t \le t_1 \\ 0 & t > t_1 \end{cases}$$

is then intended. The Hilbert transforms of many functions are tabulated in Erdelyi, Magnus, Oberhettinger and Tricomi [*Tables of Integral Transforms*, Volume 2, Chapter 15]. For example, the Hilbert transform of a pulse function [Section 1:13] is a logarithm [Chapter 25]

7:10:21
$$\frac{1}{\pi} \int_{-\infty}^{\infty} p(1;2;t) \frac{dt}{t-s} = \frac{1}{\pi} \ln\left(\left| \frac{s-1}{s+1} \right| \right)$$

7:11 COMPLEX ARGUMENT

The formulas of Section 1:11 can easily be applied to $bx + c$ or $1/(bx + c)$ if x (or, indeed, if b or c) is complex.

7:12 GENERALIZATIONS

The linear function is the special $a = 0$ case of the quadratic function discussed in Chapter 16.

A linear function is a member of most of the families of functions discussed in Chapters 17 through 24; these polynomials may therefore be regarded as generalizations of the linear function.

7:13 COGNATE FUNCTIONS

A frequent need in science and engineering is to approximate a function whose values $f(x_0)$, $f(x_1)$, $f(x_2)$, ..., $f(x_n)$ are known only at a limited number of arguments $x_0, x_1, x_2, \ldots, x_n$. One of the simplest ways of making such an approximation, but one that is nevertheless adequate in many applications, is by using the *piecewise-linear function*. For arguments between two adjacent points, x_j and x_{j+1}, at which the function has known values, the interpolation formula

7:13:1 $$f(x) = \frac{(x_{j+1} - x)f(x_j) + (x - x_j)f(x_{j+1})}{x_{j+1} - x_j} \qquad x_j \le x \le x_{j+1}$$

defines the piecewise-linear approximation. In graphical terms, this piecewise-linear function is constructed by "connecting the dots," as in Figure 7-3.

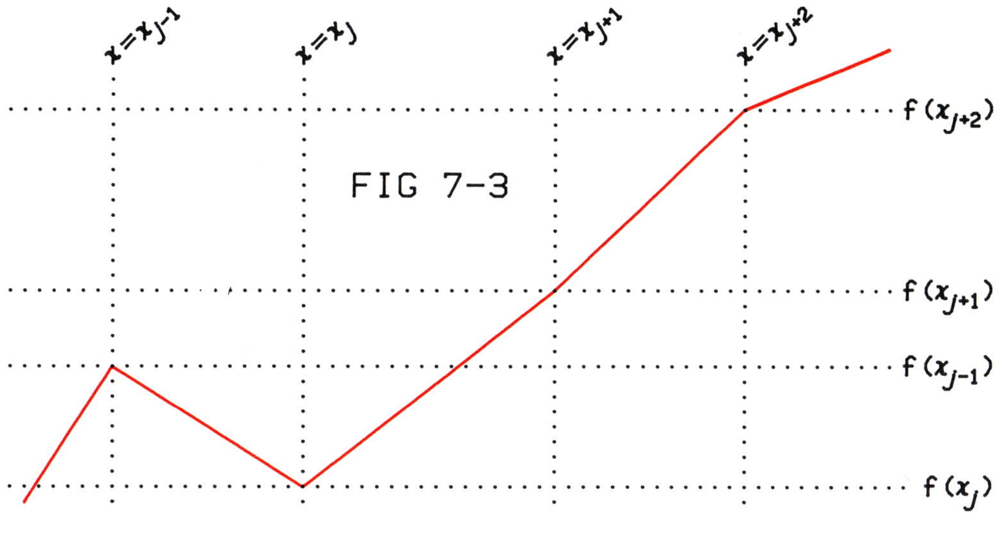

Usually a piecewise-linear function has discontinuous derivatives at all the interior values $x_1, x_2, x_3, \ldots, x_{n-1}$. The alternative interpolation formulas discussed in Section 17:14 do not suffer from this defect.

The reciprocal linear function is related to the function discussed in Section 15:4 because the shape of each is a rectangular hyperbola. Thus, counterclockwise rotation through an angle $\pi/4$ of the curve $1/(bx + c)$ about the point $x = -c/b$ on the x axis produces a new function

7:13:2 $$\pm\sqrt{\left(x + \frac{c}{b}\right)^2 + \frac{2}{b}}$$

that is a rectangular hyperbola of the class discussed in Section 15:4.

7:14 RELATED TOPICS

Frequently, experimenters collect data that are known, or believed, to obey the equation $f = f(x) = bx + c$ but that incorporate errors. From the data, which consist of n pairs (x_1,f_1) (x_2,f_2), (x_3,f_3), ..., (x_n,f_n) of numbers, the scientist needs to determine the "best" values of b and c. If the errors obey a Gaussian distribution [Section 27:14] and are entirely associated with the measurement of f (that is, the x values are exact), then the formulas

7:14:1
$$b = \frac{n\,\Sigma\,xf - \Sigma\,x\,\Sigma\,f}{n\,\Sigma\,x^2 - (\Sigma x)^2}$$

and

7:14:2
$$c = \frac{\Sigma x^2 \Sigma f - \Sigma x \Sigma xf}{n\Sigma x^2 - (\Sigma x)^2} = \frac{\Sigma f - b\Sigma x}{n}$$

give these best values. An abbreviated notation, exemplified by

7:14:3
$$\Sigma xf = \sum_{j=1}^{n} x_j f_j$$

is used in the formulas of this section. The term "best" is employed here in the sense of producing the minimum sum of squared deviations $\Sigma(bx + c - f)^2$, and the procedure is known as *least squares* fitting or *linear regression* [see also Appendix A, Section A:4]. Figure 7-4 shows an example of five data pairs and the "best straight line" through them.

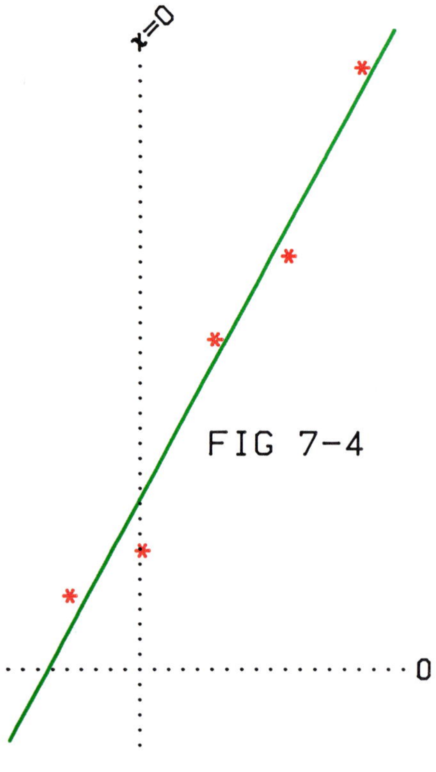

FIG 7-4

A measure of how well the data obey the linear relationship is provided by the *correlation coefficient*, given by

7:14:4
$$r = \frac{n\Sigma xf - \Sigma x \Sigma f}{\sqrt{[n\Sigma x^2 - (\Sigma x)^2][n\Sigma f^2 - (\Sigma f)^2]}} = b\sqrt{\frac{n\,\Sigma\,x^2 - (\Sigma x)^2}{n\,\Sigma\,f^2 - (\Sigma f)^2}}$$

Values of r close to ± 1 imply a good fit of data to the linear function, whereas r will be close to zero if there is little or no correlation between f and x.

Frequently, there is a need to know not only what the best values are of the slope b and intercept c but also what uncertainties are associated with these best values. Such uncertainties are often expressed in terms of the *standard error* [see Section 40:14] in b and c, these being reported as

7:14:5 $$\text{slope} = b \pm d \qquad \text{where} \qquad d = \text{standard error in } b$$

and

7:14:6 $$\text{intercept} = c \pm e \qquad \text{where} \qquad e = \text{standard error in } c$$

The significance to be attached to these statements is that the probability is approximately 68% that the true slope lies between $b - d$ and $b + d$. Similarly, there is approximately a 68% probability that the true intercept lies between $c - e$ and $c + e$. The formulas

7:14:7 $$d = \sqrt{\frac{n\Sigma f^2 - (\Sigma f)^2}{n\Sigma x^2 - (\Sigma x)^2} - \left[\frac{n\Sigma xf - \Sigma x\Sigma f}{n\Sigma x^2 - (\Sigma x)^2}\right]^2} = b\sqrt{\frac{1}{r^2} - 1}$$

and

7:14:8 $$e = \sqrt{\frac{\Sigma x^2\Sigma f^2}{n\Sigma x^2 - (\Sigma x)^2} - \frac{(\Sigma x^2\Sigma f)^2 - 2\Sigma x^2\Sigma x\Sigma xf\Sigma f + \Sigma x^2(\Sigma xf)^2}{[n\Sigma x^2 - (\Sigma x)^2]^2}} = d\sqrt{\frac{\Sigma x^2}{n}}$$

permit these standard errors to be estimated.

Commonly, data are gathered with equal spacing, that is, $x_2 - x_1 = x_3 - x_2 = x_4 - x_3 = \cdots = x_n - x_{n-1}$. The formulas of this section simplify considerably under these circumstances, and these simplified formulas are the basis of the algorithm presented here for the calculation of the best values of b and c. If the portions printed in green are included, the algorithm also generates the correlation coefficient, as well as values of the standard errors of b and c.

Input n >>>>>>>
Set $b = c = r = j = 0$
Input x_1 >>>>>>>
Set $d = x_1$
Input x_n >>>>>>>
Set $e = (x_n + d)/2$
Replace d by $(e - d)/3$
(1) Replace j by $j + 1$
Output j
j (as cue) <<<<<<
Input f_j >>>>>>>
Replace c by $c + f_j$
Replace r by $r + f_j^2$
Replace b by $b + jf_j$
If $j < n$ go to (1)
Replace b by $\left(\frac{2b}{n+1} - c\right)/nd$
Set $j = d\sqrt{3(n+1)/(n-1)}$
Replace r by $nbj/\sqrt{nr - c^2}$
Output r
r = correlation coefficient <<<
Output b
b = slope <<<<<<
Set $d = b\sqrt{(1/r^2) - 1}$
Output d
d = slope error <<<<<<
Replace c by $\frac{c}{n} - be$

Storage needed: n, b, c, r, j, d and e.
Several of these registers are also used for temporary storage.

Test values:

j	x_j	f_j
1	−0.1	1.0
2		1.6
3		4.4
4		5.5
$n = 5$	0.3	8.0

Output c

c = intercept <<<<

Replace e by $d\sqrt{e^2 + j^2}$

Output e

$e = \begin{matrix}\text{intercept}\\ \text{error}\end{matrix}$ <<<<

Output:
$r = 0.984$
$b \pm d = 17.9 \pm 3.3$
$c \pm e = 2.31 \pm 0.57$

A related, but simpler, problem that occurs frequently is the construction of the best straight line through the points (x_1,f_1), (x_2,f_2), (x_3,f_3), . . ., (x_n,f_n) with the added constraint that the line must pass through the point (x_0,f_0). In practical problems this obligatory point is often the coordinate origin. The slope and intercept in this case are given by

7:14:9 $$b = \frac{\Sigma(x - x_0)(f - f_0)}{\Sigma(x - x_0)^2}$$

and

7:14:10 $$c = f_0 - bx_0$$

Equations 7:14:1–7:14:8 are based on the assumption that all data points are known with equal reliability, a circumstance that is not always valid. Variable reliability can be treated by assigning different *weights* to the points; thus, if the value of f_1 is more reliable than f_2, one assigns a larger weight w_1 to the data pair (x_1,f_1) than the weight w_2 assigned to pair (x_2,f_2). The weights then appear as multipliers in all summations, so that, for example, equations 7:14:1 and 7:14:2 become replaced by

7:14:11 $$b = \frac{\Sigma w \Sigma wxf - \Sigma wx \Sigma wf}{\Sigma w \Sigma wx^2 - (\Sigma wx)^2}$$

and

7:14:12 $$c = \frac{\Sigma wx^2 \Sigma wf - \Sigma wx \Sigma wxf}{\Sigma w \Sigma wx^2 - (\Sigma wx)^2} = \frac{\Sigma wf - b\Sigma wx}{\Sigma w}$$

Only the *relative* weights are of importance; absolute magnitudes of w_1, w_2, w_3, . . ., w_n have no significance beyond this. In practice, one attempts to attach a weight w_j to point (x_j,f_j) such that w_j is inversely proportional to the square of the error or uncertainty associated with f_j. Notice that 7:14:1 and 7:14:2 are the special cases of equations 7:14:11 and 7:14:12 in which all weights equal unity, and that 7:14:9 and 7:14:10 are similarly special cases in which the weight of one point, (x_0,f_0), is overwhelming in comparison with the other n weights, which are uniform.

CHAPTER 8

THE UNIT–STEP $u(x - a)$ AND RELATED FUNCTIONS

This chapter concerns the unit-step functions $u(x - a)$ and $u(x)$, the signum function $\mathrm{sgn}(x)$ and the absolute value function $|x|$. These functions are closely interrelated through the simple formulas

8:0:1 $$\mathrm{sgn}(x) = 2u(x) - 1$$

and

8:0:2 $$|x| = x\,\mathrm{sgn}(x) = 2x\,u(x) - x$$

and it is therefore appropriate to treat them together.

The main utility of the unit-step and related functions is in modifying other functions, and they are usually encountered as in the following expressions:

8:0:3 $$u(x - a)f(x)$$

8:0:4 $$\mathrm{sgn}(f(x))$$

8:0:5 $$f(|x|)$$

and

8:0:6 $$|f(x)|$$

8:1. NOTATION

The unit-step function is also known as the *Heaviside function,* or *Heaviside's step function,* and is sometimes denoted by $H(x - a)$ or $S_a(x)$. It should not be confused with the unity function [Section 1:4]. The notation $u(x)$ replaces $u(x - a)$ when a is zero.

The signum function is also called the *sign function* and symbolized $\mathrm{sign}(x)$ or $\mathrm{sg}(x)$.

Alternative names for the absolute value function are the *magnitude* of $f(x)$ or, especially when $f(x)$ is complex, its *modulus*. The term "modulus" should not be confused with "modulo" [see Section 9:13]. The symbolism $\mathrm{ABS}(x)$ sometimes replaces $|x|$, especially in computer applications.

8:2 BEHAVIOR

Figures 8-1 through 8-4 illustrate the effect of modifying f(x) to each of the expressions 8:0:3–8:0:6. The modification is shown in red and the original (arbitrary) function f(x) in green. In Figures 8-3 and 8-4, r is a zero of the function f. Figure 8-4 relates to a function f that is real because, as explained in Section 8:11, a special significance attaches to $|f(x)|$ when f is complex.

FIG 8-1

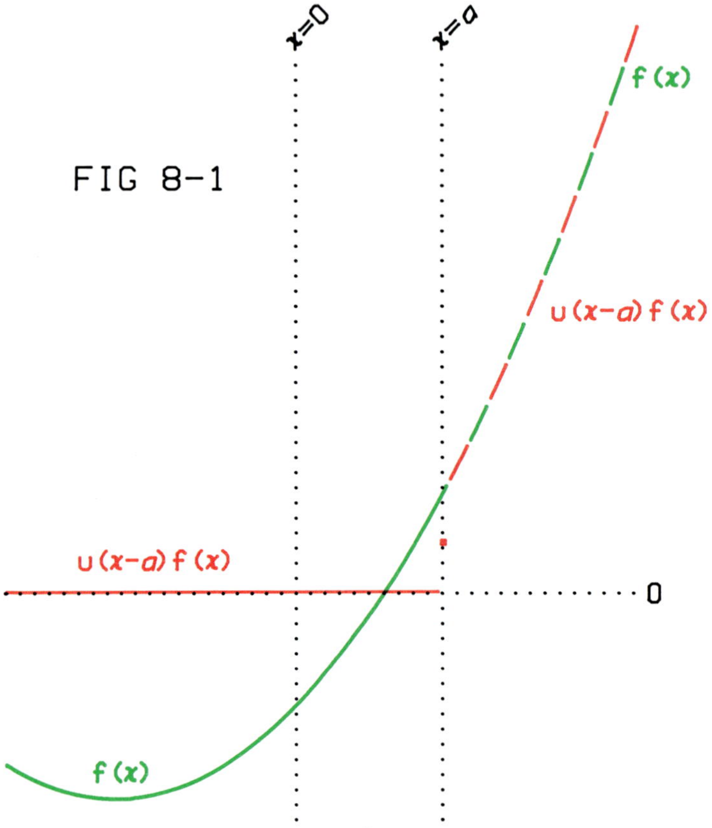

8:3 DEFINITIONS

The unit-step function is defined by

8:3:1 $$u(x-a) = \begin{cases} 0 & x < a \\ 1/2 & x = a \\ 1 & x > a \end{cases}$$

and therefore the effect of multiplying a function by u($x - a$) is to nullify the function for arguments less than a and leave it unchanged for arguments greater than a:

8:3:2 $$u(x-a)f(x) = \begin{cases} 0 & x < a \\ f(x)/2 & x = a \\ f(x) & x > a \end{cases}$$

The signum function extracts the sign of its argument so that

8:3:3 $$\mathrm{sgn}(f(x)) = \begin{cases} -1 & f(x) < 0 \\ 0 & f(x) = 0 \\ 1 & f(x) > 0 \end{cases}$$

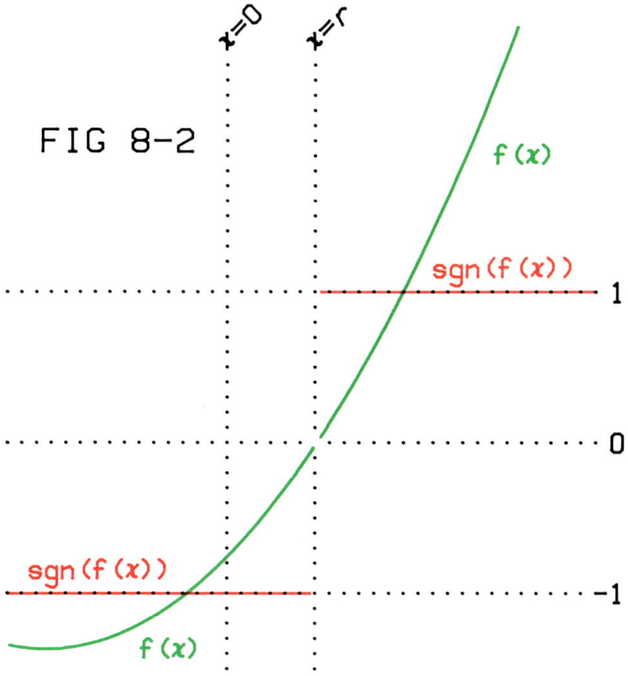

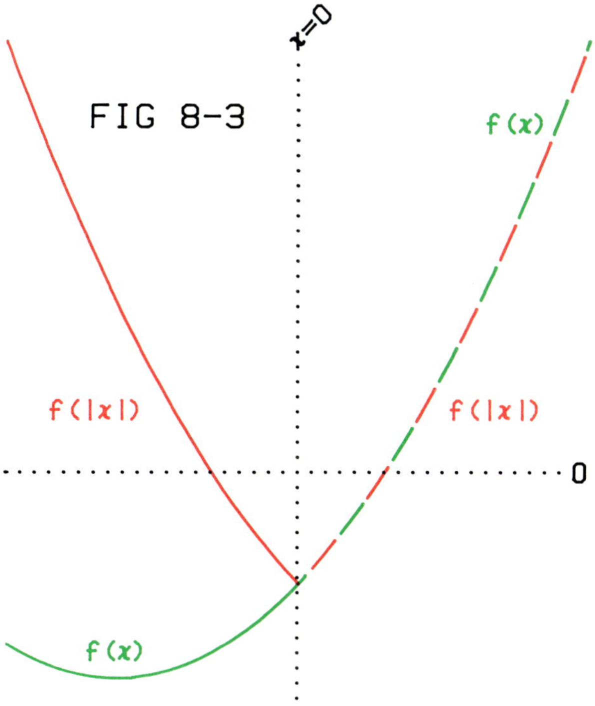

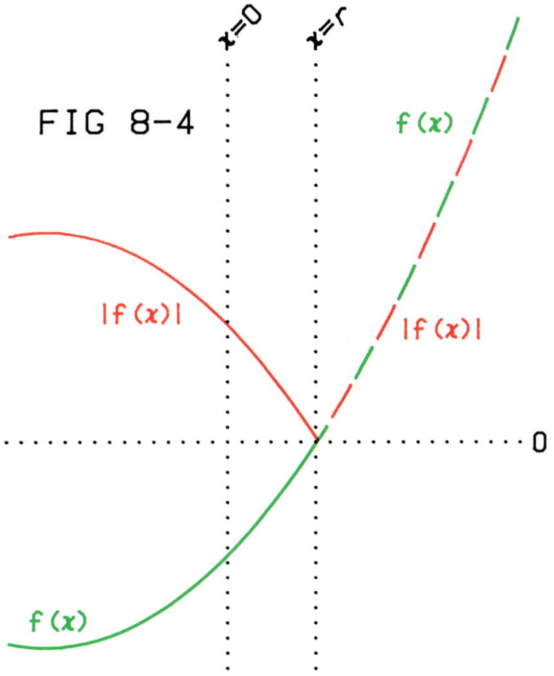

An integral may also be used to define the signum function

8:3:4 $$\mathrm{sgn}(x) = \frac{2}{\pi}\int_0^\infty \frac{\sin(xt)}{t}\,\mathrm{d}t$$

The absolute value function equals its argument when the latter is positive and equals the negative of its argument when the latter is negative. Hence

8:3:5 $$f(|x|) = \begin{cases} f(-x) & x \le 0 \\ f(x) & x \ge 0 \end{cases}$$

and

8:3:6 $$|f(x)| = \begin{cases} -f(x) & f(x) \le 0 \\ f(x) & f(x) \ge 0 \end{cases}$$

with the requirement that, for 8:3:5 to be valid, $f(x)$ must be real.

8:4 SPECIAL CASES

The unit-step function may be used to construct many special discontinuous functions. For example, the pulse function [Section 1:13] can be represented by a difference of two unit-step functions:

8:4:1 $$p(c;h;x-a) = c\left[u\left(x-a+\frac{h}{2}\right) - u\left(x-a-\frac{h}{2}\right)\right]$$

Similarly, the functions

8:4:2 $$\sum_{j=1}^{\infty} u(x-j) \quad \text{and} \quad \sum_{j=1}^{\infty} (x-j)[u(x-j) + u(j+1-x)]$$

are examples of the staircase and sawtooth functions, respectively [see Section 9:13].

Acting on a sinusoidal function [Chapter 32], the signum function generates a "square wave" such as $\mathrm{sgn}(\sin(x))$, while the absolute value function effects "full rectification" as in $|\cos(x)|$. See Section 36:14 for an explanation of both of these concepts.

8:5 INTRARELATIONSHIPS

The definitions lead to the following reflection formulas:

8:5:1 $$u(-x) = 1 - u(x)$$

8:5:2 $$\mathrm{sgn}(-x) = -\mathrm{sgn}(x)$$

and

8:5:3 $$|-x| = |x|$$

From 8:5:1 it follows that $u(x-a)f(x)$ and $u(a-x)f(x)$ are complementary functions in the sense that

8:5:4 $$u(x-a)f(x) + u(a-x)f(x) = f(x)$$

8:6 EXPANSIONS

If the function $f(x)$ is expansible as a power series [Section 11:14], the unit-step function may be distributed through the series

8:6:1 $$u(x-a)f(x) = u(x-a)\sum_{j=0}^{\infty} a_j x^j = \sum_{j=0}^{\infty} a_j\, u(x-a)x^j$$

but the same is not true of the signum or absolute-value function.

8:7 PARTICULAR VALUES

	$x < 0$	$x = 0$	$x > 0$
$\mathrm{u}(x)$	0	$\frac{1}{2}$	1
$\mathrm{sgn}(x)$	−1	0	1
$\lvert x\rvert$	$-x$	0	x

8:8 NUMERICAL VALUES

Many programmable calculators have a key that directly extracts the absolute value of the register's contents. Similarly, computer languages usually permit the finding of an absolute value via an operation such as ABS(x). When this is not so, the operations of squaring followed by taking the square root will give the absolute value, because

8:8:1 $$\sqrt{x^2} = |x|$$

8:9 APPROXIMATIONS

The unit-step and signum functions may be approximated by a variety of continuous functions. For example, as b takes ever larger positive values, the approximations

8:9:1 $$\mathrm{u}(x-a) \simeq \frac{1}{2}[1 + \tanh(bx - ba)]$$

and

8:9:2 $$\mathrm{sgn}(x) \simeq \mathrm{erf}(bx)$$

become increasingly valid. The functions in 8:9:1 and 8:9:2 are the hyperbolic tangent [Chapter 30] and the error function [Chapter 40], respectively.

8:10 OPERATIONS OF THE CALCULUS

Differentiation produces

8:10:1 $$\frac{\mathrm{d}}{\mathrm{d}x}\mathrm{u}(x-a) = \delta(x-a)$$

8:10:2 $$\frac{\mathrm{d}}{\mathrm{d}x}[\mathrm{u}(x-a)\mathrm{f}(x)] = \begin{cases} 0 & x < a \\ \dfrac{\mathrm{df}}{\mathrm{d}x}(x) & x > a \end{cases}$$

8:10:3 $$\frac{\mathrm{d}}{\mathrm{d}x}\mathrm{sgn}(\mathrm{f}(x)) = 0 \qquad x \neq r_n \qquad n = 1, 2, 3, \ldots$$

8:10:4 $$\frac{\mathrm{d}}{\mathrm{d}x}\mathrm{f}(|x|) = \begin{cases} \dfrac{\mathrm{df}}{\mathrm{d}x}(x) & x > 0 \\ -\dfrac{\mathrm{df}}{\mathrm{d}x}(-x) & x < 0 \end{cases}$$

and

8:10:5 $$\frac{d}{dx}|f(x)| = \mathrm{sgn}(f(x))\frac{df}{dx}(x) \qquad x \neq r_n$$

where $r_1, r_2, r_3, \ldots$ are the zeros of f(x), that is, f(r_n) = 0, n = 1, 2, 3, ... The delta function, δ, occurring in formula 8:10:1 is the Dirac delta function [see Chapter 10].

Integration is also easily accomplished:

8:10:6 $$\int_{x_0}^{x} u(t-a)f(t)\,dt = \int_{a}^{x} f(t)\,dt \qquad x_0 \le a \le x$$

8:10:7 $$\int_{x_0}^{x} \mathrm{sgn}(f(t))\,dt = 2\,\mathrm{sgn}(f(x_0))\left[\frac{-x_0}{2} + r_1 - r_2 + r_3 - \cdots - (-1)^n\left(r_n - \frac{x}{2}\right)\right] \qquad f(x_0) \neq 0$$

8:10:8 $$\int_{x_0}^{x} f(|t|)\,dt = \int_{0}^{x} f(t)\,dt - \mathrm{sgn}(x_0)\int_{0}^{|x_0|} f(t)\,dt$$

8:10:9 $$\int_{x_0}^{x} |f(t)|\,dt = \left|\int_{x_0}^{r_1} f(t)\,dt\right| + \left|\int_{r_1}^{r_2} f(t)\,dt\right| + \left|\int_{r_2}^{r_3} f(t)\,dt\right| + \cdots + \left|\int_{r_n}^{x} f(t)\,dt\right|$$

where $r_1, r_2, r_3, \ldots, r_n$ are those zeros of f(t) that lie between x_0 and x.

Differintegration with a lower limit x_0 gives

8:10:10 $$\frac{d^\nu}{[d(x-x_0)]^\nu}u(x-a) = \frac{[x-a]^{-\nu}}{\Gamma(1-\nu)} \qquad x_0 < a < x$$

and

8:10:11 $$\frac{d^\nu}{[d(x-x_0)]^\nu}[u(x-a)\,f(x)] = u(x-a)\frac{d^\nu}{[d(x-a)]^\nu}f(x) \qquad x_0 < a < x$$

This last result is a generalization of 8:10:1 and 8:10:5.

8:11 COMPLEX ARGUMENT

A special significance is accorded to the absolute value function of complex argument; thus

8:11:1 $$|x + iy| = \sqrt{x^2 + y^2}$$

8:12 GENERALIZATIONS

The function defined by

8:12:1 $$[u(x-x_0) - u(x-x_1)]\,f(x) \qquad x_0 < x_1$$

is equal to f(x) inside the "window" $x_0 < x < x_1$ and to zero outside. Such a *window function* is exhibited in Figure 8-5.

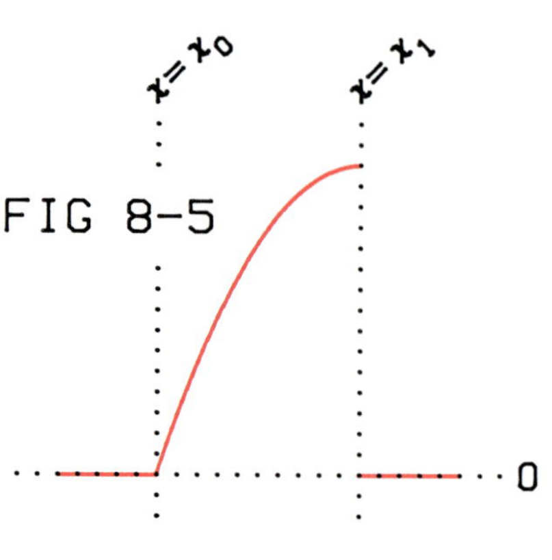

FIG 8-5

The pulse function [Section 1:13] is the simplest example of such a window function:

8:12:2 $$\mathrm{p}\left(c; x_1 - x_0; x - \frac{x_0 + x_1}{2}\right) = [\mathrm{u}(x - x_0) - \mathrm{u}(x - x_1)]c$$

The concept may be generalized to

8:12:3 $$\sum_{j=0}^{J-1} [\mathrm{u}(x - x_j) - \mathrm{u}(x - x_{j+1})]\, \mathrm{f}_j(x) \qquad x_0 < x_1 < x_2 < \cdots < x_J$$

which constitutes a *piecewise-defined function,* as illustrated in Figure 8-6. The properties of such a function may be deduced in a straightforward fashion from those of u($x - a$)f(x) as discussed throughout this chapter.

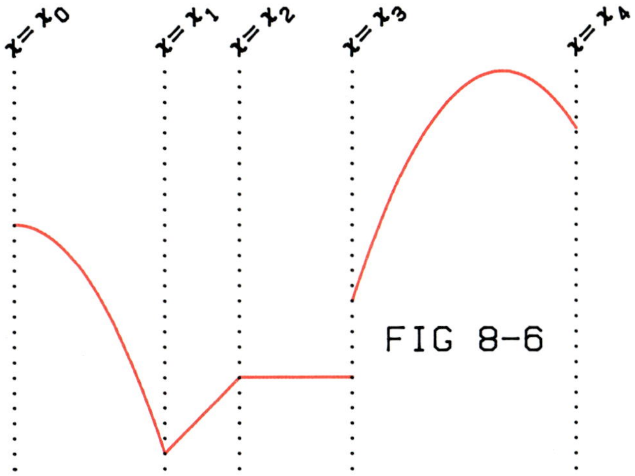

FIG 8-6

8:13 COGNATE FUNCTIONS

The value of the unit-step function at the point where the step rises has little effect on the properties of the function. Hence, the function defined by

8:13:1 $$\upsilon(x - a) = \begin{cases} 0 & x < a \\ 1 & x \geq a \end{cases}$$

is very similar to u($x - a$): it may be called the *alternative unit-step function* and occurs in the expansions of the functions discussed in Chapter 9.

CHAPTER 9

THE INTEGER–VALUE Int(x) AND FRACTIONAL–VALUE frac(x) FUNCTIONS

The two discontinuous functions of this chapter satisfy the relationship

9:0:1 $$\mathrm{Int}(x) + \mathrm{frac}(x) = x$$

and have their main utility as modifiers of other functions. Therefore, in addition to the simple integer-value and fractional-value functions themselves, this chapter will also discuss f(Int(x)), f(frac(x)), Int(f(x)) and frac(f(x)), f being some unspecified function.

9:1 NOTATION

The integer-value function is also symbolized [x], E(x) or int(x). The fractional-value function is sometimes symbolized (x). See Section 9:13 for the very similar *integer-part* and *fractional-part* functions; these are sometimes also denoted Int(x) and frac(x).

9:2 BEHAVIOR

Figures 9-1, 9-2, 9-3 and 9-4 depict maps of f(Int(x)), f(frac(x)), Int(f(x)) and frac(f(x)). They are sketched for a function f that is monotonic and continuous in the region graphed: a *monotonic function* has no maximum or minimum. Several of the formulas in this chapter are restricted to values of x for which f is monotonic.

The functions f(Int(x)) and Int(f(x)) are piecewise constant [Section 1:13]. Because its segments are of equal width, the f(Int(x)) function has a graph that resembles a *histogram*.

The f(frac(x)) and frac(f(x)) functions assume values in restricted ranges. Thus, provided f is monotonic in the $0 \le x \le 1$ range, f(frac(x)) takes values between f(0) and f(1). Similarly, $0 \le \mathrm{frac}(\mathrm{f}(x)) < 1$.

9:3 DEFINITIONS

The integer-value function extracts from its argument the largest integer that does not exceed the argument; that is:

9:3:1 $$\mathrm{Int}(x) = n \qquad n \le x < n + 1 \qquad n = 0, \pm 1, \pm 2, \ldots$$

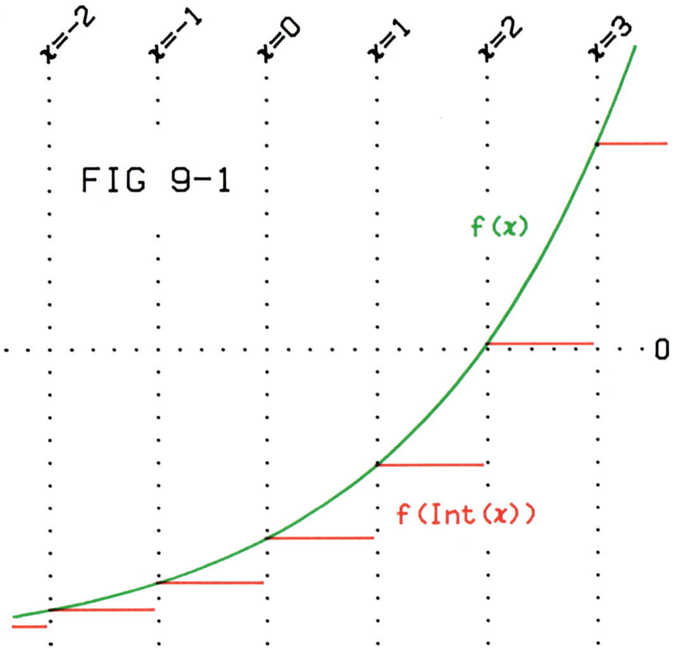

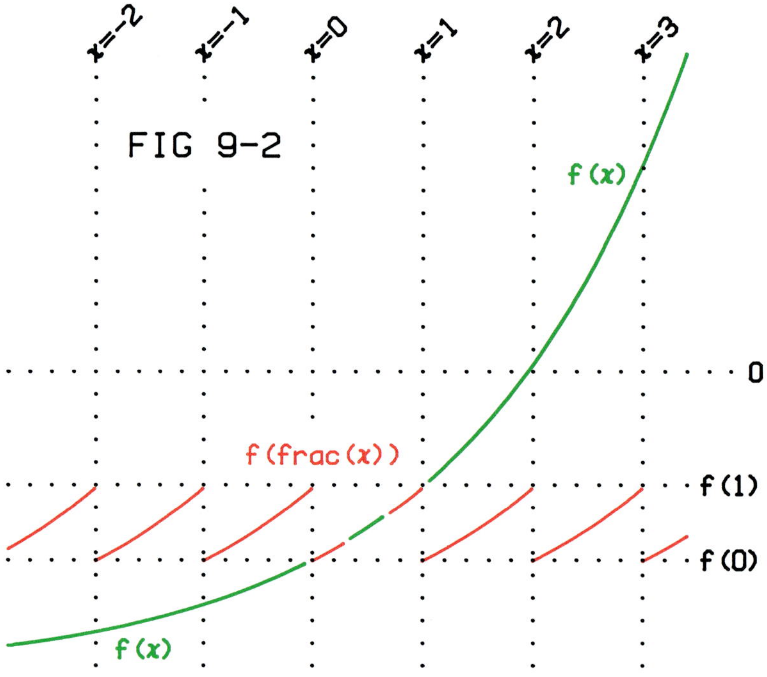

so that

9:3:2 $$f(\mathrm{Int}(x)) = f(n) \qquad n \le x < n + 1 \qquad n = 0, \pm 1, \pm 2, \ldots$$

and

9:3:3 $$\mathrm{Int}(f(x)) = n \qquad \text{f monotonic} \qquad n \le f(x) < n + 1 \qquad n = 0, \pm 1, \pm 2, \ldots$$

The fractional-value function is defined by

9:3:4 $$\mathrm{frac}(x) = x - n \qquad n \le x < n + 1 \qquad n = 0, \pm 1, \pm 2, \ldots$$

and therefore

9:3:5 $$f(\mathrm{frac}(x)) = f(x - n) \qquad 0 \le x - n < 1 \qquad n = 0, \pm 1, \pm 2, \ldots$$

and

9:3:6 $\quad \text{frac}(\text{f}(x)) = \text{f}(x) - n \qquad \text{f}(x) - n \qquad \text{f monotonic} \qquad 0 \le \text{f}(x) - n < 1 \qquad n = 0, \pm1, \pm2, \ldots$

9:4 SPECIAL CASES

When f(x) is the linear function, $bx + c$ [Chapter 7], f(Int(x)) and Int(f(x)) are similar in shape and each is known by the descriptive name *staircase function*. Equally descriptive is the name *sawtooth function* given to $b$frac(x) + c or frac($bx + c$). Figure 9-5 depicts this pair of functions.

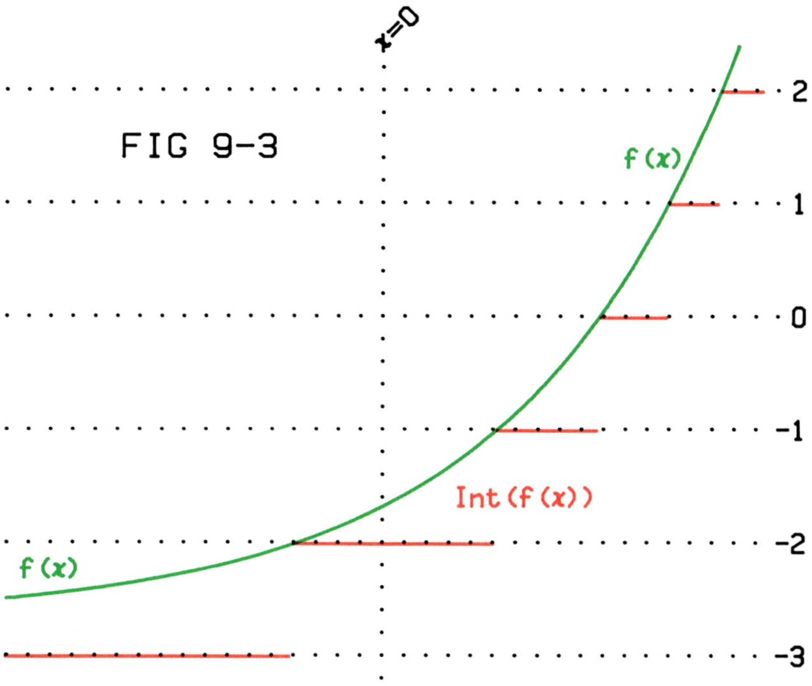

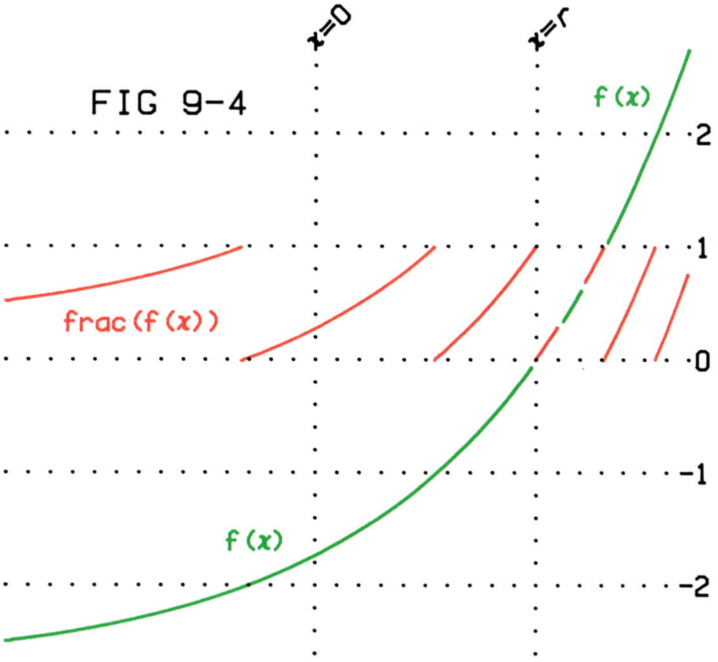

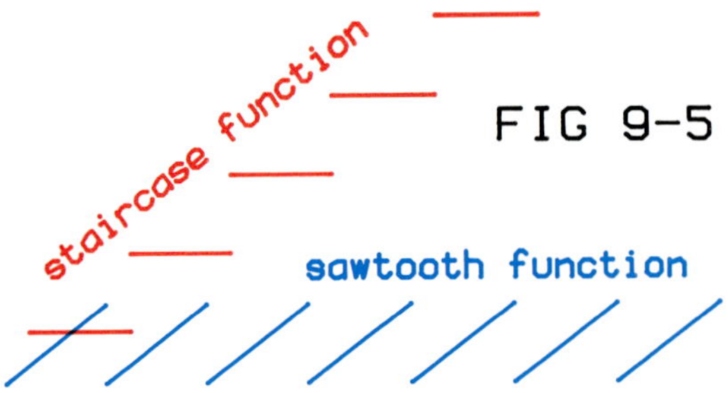

FIG 9-5

9:5 INTRARELATIONSHIPS

It follows from equation 9:0:1 that Int and frac are complementary functions in the same sense that

9:5:1 $$f(\mathrm{Int}(x) + \mathrm{frac}(x)) = f(x)$$

and

9:5:2 $$\mathrm{Int}(f(x)) + \mathrm{frac}(f(x)) = f(x)$$

The following recurrence formulas apply:

9:5:3 $$f(\mathrm{Int}(x+n)) = f(\mathrm{Int}(x) + n) \qquad n = 0, \pm1, \pm2, \ldots$$

9:5:4 $$f(\mathrm{frac}(x+n)) = f(\mathrm{frac}(x)) \qquad n = 0, \pm1, \pm2, \ldots$$

9:6 EXPANSIONS

When f is a monotonic function, Int(f(x)) may be expanded in terms of the alternative unit-step function [Section 8:13]

9:6:1 $$\mathrm{Int}(f(x)) = \sum_{j=1}^{\infty} \upsilon(x - F(j)) + \sum_{j=0}^{\infty} [\upsilon(x + F(-j)) - 1] \qquad \text{f monotonic} \qquad \frac{df}{dx} > 0$$

9:6:2 $$\mathrm{Int}(f(x)) = \sum_{j=0}^{\infty} [1 - \upsilon(x - F(j))] - \sum_{j=1}^{\infty} \upsilon(x + F(-j)) \qquad \text{f monotonic} \qquad \frac{df}{dx} < 0$$

where F is the inverse function of f, that is, F(f(x)) = x.

Because f(frac(x)) is a periodic function of x, it may be expanded by the methods described in Chapter 36.

9:7 PARTICULAR VALUES

For integer arguments one has the simplification

9:7:1 $$f(\mathrm{Int}(n)) = f(n) \qquad n = 0, \pm1, \pm2, \ldots$$

and

9:7:2 $$f(\mathrm{frac}(n)) = f(0) \qquad n = 0, \pm1, \pm2, \ldots$$

If, as before, F is the inverse function of f, then

9:7:3 $$\mathrm{Int}(f(x)) = n = 0, \pm1, \pm2, \ldots \qquad x = F(n) \qquad \text{f monotonic}$$

and

9:7:4 $$\mathrm{frac}(f(x)) = 0 \qquad x = F(n) \qquad \text{f monotonic} \qquad n = 0, \pm1, \pm2, \ldots$$

9:8 NUMERICAL VALUES

Many calculators and computers are able to implement the Int(x) and frac(x) functions directly. Beware, however, that such operations are not actually implementing the "integer-part" or "fractional-part" functions discussed in Section 9:13, rather than those defined in Section 9:3. In view of relationship 9:13:1, the rounding facilities of other calculators can be adapted to give Int(x). In yet other devices "shifting" to the right or left may be used to generate Int(x) or frac(x), respectively. If all else fails the short algorithm

```
                    Set f = 0.5
Input x >>>>>>>
                    If cos(180x) = 0 go to (1)
                    Set f = arctan(tan(180x))/180
                    If f ≥ 0 go to (1)
                    Replace f by 1 + f
                (1) Output f
f = frac(x) <<<<<
```

Storage needed: f, x

Use degree mode or change 180 to π.

Test values: frac(3.5) = 0.5 frac(1.7) = 0.7 frac(−1.7) = 0.3

may be used to find frac(x). The difference x − frac(x) then gives Int(x). This algorithm uses functions from Chapters 32, 34 and 35 with the identity

9:8:1 $$\arctan(\tan(\pi[n+x])) = \begin{cases} \pi x & 0 \le x < \frac{1}{2} \\ \pi(x-1) & \frac{1}{2} < x \le 1 \end{cases} \qquad n = 0, \pm 1, \pm 2, \ldots$$

The commands shown in green guard against overflow; they may be omitted with many computing devices.

9:9 APPROXIMATIONS

There are none that are sufficiently useful to be included here.

9:10 OPERATIONS OF THE CALCULUS

With F designating the inverse function to f, as before, the following formulas indicate how the functions Int and frac affect differentiation:

9:10:1 $$\frac{d}{dx}\mathrm{f}(\mathrm{Int}(x)) = 0 \qquad x \neq 0, \pm 1, \pm 2, \ldots$$

9:10:2 $$\frac{d}{dx}\mathrm{f}(\mathrm{frac}(x)) = \frac{df}{dx}(x-n) \qquad n \le x < n+1 \qquad n = 0, \pm 1, \pm 2, \ldots$$

9:10:3 $$\frac{d}{dx}\mathrm{Int}(\mathrm{f}(x)) = 0 \qquad x \neq \mathrm{F}(0), \mathrm{F}(\pm 1), \mathrm{F}(\pm 2), \ldots$$

9:10:4 $$\frac{d}{dx}\mathrm{frac}(\mathrm{f}(x)) = \frac{df}{dx}(x) \qquad x \neq \mathrm{F}(0), \mathrm{F}(\pm 1), \mathrm{F}(\pm 2), \ldots$$

Integration leads to formulas that are more complicated:

9:10:5 $$\int_a^x \mathrm{f}(\mathrm{Int}(t))dt = (n+1-a)\mathrm{f}(n) + (x-N)\mathrm{f}(N) + \sum_{j=n+1}^{N-1} \mathrm{f}(j) \qquad n = \mathrm{Int}(a) \qquad N = \mathrm{Int}(x)$$

9:10:6 $$\int_a^x \mathrm{f}(\mathrm{frac}(t))\mathrm{d}t = \int_{a-n}^{1} \mathrm{f}(t)\mathrm{d}t + (N-n-2)\int_0^1 \mathrm{f}(t)\mathrm{d}t + \int_0^{x-N} \mathrm{f}(t)\mathrm{d}t \qquad n = \mathrm{Int}(a) \qquad N = \mathrm{Int}(x)$$

9:10:7 $$\int_a^x \mathrm{Int}(\mathrm{f}(t))\mathrm{d}t = \begin{cases} Xx - Aa - \sum_{j=1}^{X-A} \mathrm{F}(A+j) & \mathrm{f}\text{ monotonic} \quad \frac{\mathrm{df}}{\mathrm{d}x} > 0 \\ Xx - Aa + \sum_{j=1}^{A-X} \mathrm{F}(X+j) & \mathrm{f}\text{ monotonic} \quad \frac{\mathrm{df}}{\mathrm{d}x} < 0 \end{cases}$$

where $A = \mathrm{Int}(\mathrm{f}(a))$, $X = \mathrm{Int}(\mathrm{f}(x))$ and F is the inverse function of f. Finally, by virtue of 9:5:2

9:10:8 $$\int_a^x \mathrm{frac}(\mathrm{f}(t))\mathrm{d}t = \int_a^x \mathrm{f}(t)\mathrm{d}t - \int_a^x \mathrm{Int}(\mathrm{f}(t))\mathrm{d}t$$

9:11 COMPLEX ARGUMENT

The integer-value and fractional-value functions are not defined for imaginary or complex arguments.

9:12 GENERALIZATIONS

The function f(frac(x)) is an example of a periodic function [see Chapter 36]. Its period is unity. Likewise f(Int(x)) and Int(f(x)) are examples of piecewise-constant functions [see Section 1:13].

9:13 COGNATE FUNCTIONS

The *rounding function* (or *nearest-integer function*) $\langle x\rangle$ is defined as the integer nearest to x, with rounding upward to $n+1$ in case $x = n + \frac{1}{2}$. It is related by

9:13:1 $$\langle x\rangle = \mathrm{Int}(x + {}^1/_2)$$

to the integer-value function.

The *integer-part function* Ip and *fractional-part function* Fp are very similar to the integer-value and fractional-value functions. In fact, when x is positive, $\mathrm{Ip}(x) = \mathrm{Int}(x)$ and $\mathrm{Fp}(x) = \mathrm{frac}(x)$. The distinction for negative argument is evident from the following example:

9:13:2 $$\mathrm{Int}(-2.34) = -3$$

9:13:3 $$\mathrm{Ip}(-2.34) = -2 = 1 + \mathrm{Int}(-2.34)$$

9:13:4 $$\mathrm{frac}(-2.34) = 0.66$$

9:13:5 $$\mathrm{Fp}(-2.34) = -0.34 = -1 + \mathrm{frac}(-2.34)$$

This *Atlas* makes no use of the integer-part or fractional-part functions.

The *modulo function* (or *remainder function*) is a bivariate function of two integer variables. This function, denoted $n(\mathrm{mod}\ m)$, is the remainder upon subtraction of m from n as many times as possible without leaving a negative difference. The modulo function is related to the fractional-value function by

9:13:6 $$n(\mathrm{mod}\ m) = m\ \mathrm{frac}\left(\frac{n}{m}\right)$$

Occasionally the symbol $n(m)$ is encountered and the name *modulus* is used to describe m.

9:14 RELATED TOPICS

The integer-value and fractional-value functions occur widely in number theory. In addition they lie at the heart of such devices as analog-to-digital convertors that digitize measured signals of various kinds. They also play roles in the conversion of a number x from one number system to another; from the decimal system to the binary, for example, or from the hexadecimal to the decimal.

Any positive number can be represented as

9:14:1
$$x = N_0 \cdot N_{-1}N_{-2}N_{-3} \cdots \times \beta^n$$

where n is an integer and each N_j takes one of the integer values 0, 1, 2, ..., $(\beta - 1)$, where β is the *base* of the number system and where $N_0 \neq 0$. Such a representation is termed *floating point* or, if $\beta = 10$, *scientific notation*. An alternative is the *fixed point* representation

9:14:2
$$x = N_nN_{n-1} \cdots N_0 \cdot N_{-1}N_{-2} \cdots$$

where n is a sufficiently large integer and, as before, each N_j is an integer drawn from the set 0, 1, 2, ..., $(\beta - 1)$.

The algorithm below digitizes any positive number x by generating a sequence of integers. The first integer is n, a decimal integer of either sign. The other integers are digits of the base β_2 number system. These may be interpreted as the N_0, N_{-1}, N_{-2}, ... digits of the floating point representation 9:14:1, or as the N_n, N_{n-1}, N_{n-2}, ... digits of the fixed point representation 9:14:2.

```
Input β2 >>>>>>>>
              Set n = 0
Input x >>>>>>>>
          (3) If Int(x) ≠ 0 go to (4)
              Replace x by β2x
              Replace n by n − 1
              Go to (3)
          (4) Replace x by x/β2
              If Int(x) = 0 go to (5)
              Replace n by n + 1
              Go to (4)
          (5) Output n
n <<<<<<<<<<
          (6) Replace x by β2x
              Output Int(x)
digit <<<<<<<<
              Replace x by frac(x)
              If x = 0 stop
              Go to (6)
```

Storage needed: β_2, x and n

Intervention to halt output is needed if x is recurrent in base β_2.

Test values:
$x = \pi$, $\beta_2 = 2$
Output: $n = 1$
followed by the digits
1,1,0,0,1,0,
0,1,0,0,0,0,
1,1,1,1,1,1,
0,1,1,0,1,0,
1,0,1,0,0,0

The second algorithm accepts a number in base β_1 representation and converts it to a decimal number x. The input is in the form of the signed decimal number n followed by a sequence of digits in base β_1. All these digits must, of course, be positive. Following the entry of the last true digit a negative digit is input as a cue that the data entry is complete. This negative digit is ignored in the computation of x.

The two algorithms are designed so that they can be combined, with the second preceding the first. The conjoined routine will then convert any positive number from base β_1 to base β_2.

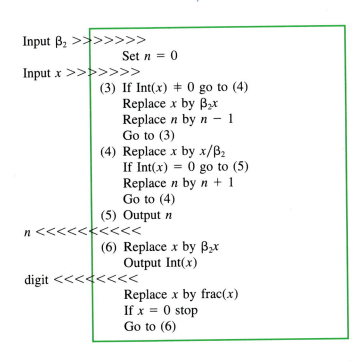

Input β_1 >>>>>>>>

Set $x = 0$

Input n >>>>>>>>

Set $g = \beta_1^n$

(1) Wait for input

Input digit N >>>>

If $N < 0$ go to (2)

Replace x by $x + Ng$

Replace g by g/β_1

Go to (1)

(2) Output x

x <<<<<<<<<<

Storage needed: β_1, x and g

Test values:
$\beta_1 = 16$, $n = 2$
followed by the digits 7,12,0,8,−1
Output: $x = 1984.5$

CHAPTER 10

THE DIRAC DELTA FUNCTION $\delta(x - a)$

This function, strictly speaking, is not a function at all because it violates rules that would be valid if it were a function. For example, while it takes the value zero at all arguments but one, its integral has the value unity. Nevertheless, the Dirac delta "function" is a useful concept that has found widespread application, especially in classical and quantum mechanics, where it occurs mainly as a multiplier of an integrand, as in

10:0:1 $$\int_{-\infty}^{\infty} \delta(x - a)\mathrm{f}(x)\mathrm{d}x$$

10:1 NOTATION

The *impulse function* or *unit-impulse function* is an alternative name for $\delta(x - a)$.

Dirac's name is associated with this function to avoid confusion with the Kronecker delta function [Section 10:13].

10:2 BEHAVIOR

The Dirac delta function cannot be graphed: $\delta(x - a)$ is zero for all values of x except $x = a$, where it is infinite.

10:3 DEFINITIONS

The Dirac delta function may be defined in terms of a limit in many ways, for example:

10:3:1 $$\delta(x - a) = \lim_{h\to 0} \mathrm{p}\left(\frac{1}{h}; h; x - a\right)$$

where p is the pulse function [Section 1:13], or

10:3:2 $$\delta(x - a) = \lim_{b\to\infty} \left[\sqrt{\frac{b}{\pi}} \exp(-b(x - a)^2)\right]$$

[see Chapter 27], or

10:3:3 $$\delta(x-a) = \lim_{c\to 0}\left[\frac{1}{2c}\,\text{sech}^2\!\left(\frac{x-a}{c}\right)\right]$$

[see Chapter 29]. All of these definitions—and many others—describe a function peaked at $x = a$ that, as the limit is approached, becomes infinitely high and infinitesimally wide but whose area remains constant and equal to unity. Figure 10-1 illustrates the progress toward the limit in the case of definition 10:3:3. It follows that

10:3:4 $$\int_{x_0}^{x_1} \delta(t-a)\mathrm{d}t = 1 \qquad x_0 < a < x_1$$

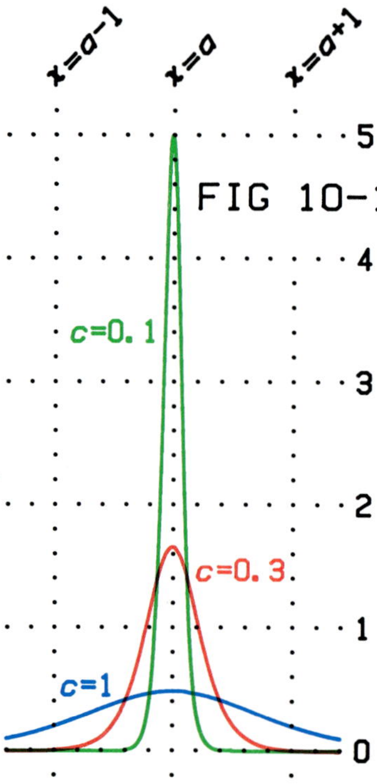

FIG 10-1

Another interpretation of the Dirac delta function is as the derivative

10:3:5 $$\delta(x-a) = \frac{\mathrm{d}}{\mathrm{d}x}\,\mathrm{u}(x-a)$$

of the unit-step function [Chapter 8]. Yet another representation is as the definite integral

10:3:6 $$\delta(x-a) = \int_{-\infty}^{\infty} \cos[2\pi(x-a)t]\mathrm{d}t$$

which is important in the evaluation of certain Fourier transforms [see Section 32:14].

10:4 SPECIAL CASES

When $a = 0$, the $\delta(x - a)$ symbol is replaced by $\delta(x)$.

10:5 INTRARELATIONSHIPS

One may derive the reflection formula

10:5:1 $$\delta(a-x) = \delta(x-a)$$

and the multiplication formula

10:5:2 $$\delta(\nu x) = \frac{\delta(x)}{\nu} \qquad \nu > 0$$

from the definition of the Dirac delta function. Another result is

10:5:3 $$\delta(x^2 - a^2) = \frac{1}{2a}[\delta(x-a) + \delta(x+a)] \qquad a > 0$$

10:6 EXPANSIONS

There are none.

10:7 PARTICULAR VALUES

If the Dirac delta function is interpreted conventionally, then

10:7:1 $$\delta(x-a) = 0 \qquad x \neq a$$

10:7:2 $$\delta(x-a) = \infty \qquad x = a$$

10:8 NUMERICAL VALUES

No finite nonzero value attaches to $\delta(x-a)$ for any value of x.

10:9 APPROXIMATIONS

No approximation of $\delta(x-a)$ itself is of any use. However, expressions 10:3:1–10:3:3 can sometimes be of value in approximating integrals in which $\delta(x-a)$ appears as a factor of the integrand.

10:10 OPERATIONS OF THE CALCULUS

Differentiation of the Dirac delta function produces the unit-moment function discussed in Section 10:12. Integration gives the unit-step function [Chapter 8]

10:10:1 $$\int_{x_0}^{x} \delta(t-a)dt = u(x-a) \qquad -\infty \le x_0 < a$$

and, with f any function,

10:10:2 $$\int_{x_0}^{x} \delta(t-a)f(t)dt = u(x-a)f(a) \qquad -\infty \le x_0 \le a$$

of which a special case is

10:10:3 $$\int_{-\infty}^{\infty} \delta(t-a)f(t)dt = f(a)$$

It is this last relationship, known as its *sifting property*, that renders the Dirac delta function so useful.

The notation $f(x) * g(x)$ and the definition

10:10:4 $$f(x) * g(x) = \int_{-\infty}^{\infty} f(t)g(x-t)dt = \int_{-\infty}^{\infty} f(x-t)g(t)dt$$

relate to the so-called *convolution* of the functions f and g. The convolution of two Dirac delta functions obeys the rule

10:10:5 $$\delta(x-a) * \delta(x-b) = \delta(x-a-b)$$

10:11 COMPLEX ARGUMENT

The Dirac delta function of complex argument $\delta(x + yi - a - bi)$ is nonzero only when $x = a$ and $y = b$. It obeys

10:11:1 $$\int_{-\infty}^{\infty}\int_{-\infty}^{\infty} \delta(x + yi - a - bi)dx\, dy = 1$$

as well as other relations that parallel its behavior as a function of a real argument.

10:12 GENERALIZATIONS

By utilizing the concept of differintegration [Section 0:10] it is possible to define a continuum of functions of which the unit-step function and the Dirac delta function are respectively the $\nu = 0$ and $\nu = 1$ instances. The general definition is

10:12:1 $$\frac{d^\nu}{[d(x+\infty)]^\nu} u(x-a)$$

which evaluates to

10:12:2 $$u(x-a)\frac{(x-a)^{-\nu}}{\Gamma(1-\nu)} \qquad \nu \neq 1, 2, 3, \ldots$$

except when ν is a positive integer. The $\nu = 2$ case, symbolized $\delta'(x-a)$, may be regarded as the limit

10:12:3 $$\delta'(x-a) = \lim_{h\to 0} \frac{\delta\left(x - a - \frac{h}{2}\right) - \delta\left(x - a + \frac{h}{2}\right)}{h}$$

and is named the *unit-moment function*. It satisfies the integral identity

10:12:4 $$\int_{-\infty}^{\infty} \delta'(t-a)f(t)dt = -f'(a) = -\frac{df}{dx}(a)$$

10:13 COGNATE FUNCTIONS

The Dirac delta function may be regarded as a bivariate function of the variables x and a, each of which can adopt any real value: it is zero except when these two variables are equal. The *Kronecker delta function* $\delta(n,m)$ or δ_{nm} is an analogous bivariate function but its two variables are restricted to integer values. It is defined by

10:13:1 $$\delta(n,m) = \begin{cases} 0 & n \neq m \\ 1 & n = m \end{cases}$$

CHAPTER 11

THE INTEGER POWERS $(bx + c)^n$ AND x^n

With $n = 0, \pm 1, \pm 2, \ldots$ this chapter concerns the function $(bx + c)^n$ and its special $b = 1$, $c = 0$ case. The powers $1, x, x^2, \ldots$ and the reciprocal powers $1, x^{-1}, x^{-2}, \ldots$ are the units from which most expansions are built; such expansions are the subjects of Section 11:14.

11:1 NOTATION

The two symbolisms $(bx + c)^{-n}$ and $1/(bx + c)^n$ are equivalent in all respects. The powers x^2 and x^3 are termed the *square* and the *cube* of x, respectively.

In the general notation β^α, β is known as the *base* and α as the *power* or *exponent*. In this chapter [and in Chapter 13] the family of functions in which the base is the primary variable is treated. In contrast, Chapter 19 is concerned with functions in which the exponent varies and the base is held constant.

11:2 BEHAVIOR

The power function is defined for all values of x and for all integer n except that x^n is undefined when both x and n are zero. Figures 11-1 and 11-2 illustrate the behavior of x^n for $n = 0, \pm 1, \pm 2, \pm 3, \pm 4, \pm 7$ and ± 12. Notice the contrasting behavior of the positive and negative powers. Note also how the reflection properties depend on whether n is even or odd.

11:3 DEFINITIONS

The function $(bx + c)^n$ for $n = 1, 2, 3, \ldots$ is the product of n factors, each equal to $bx + c$:

11:3:1 $$(bx + c)^n = \prod_{j=1}^{n} (bx + c) \qquad n = 1, 2, 3, \ldots$$

When n is a negative integer, the definition is the quotient

11:3:2 $$(bx + c)^n = \prod_{j=1}^{-n} \frac{1}{bx + c} \qquad n = -1, -2, -3, \ldots$$

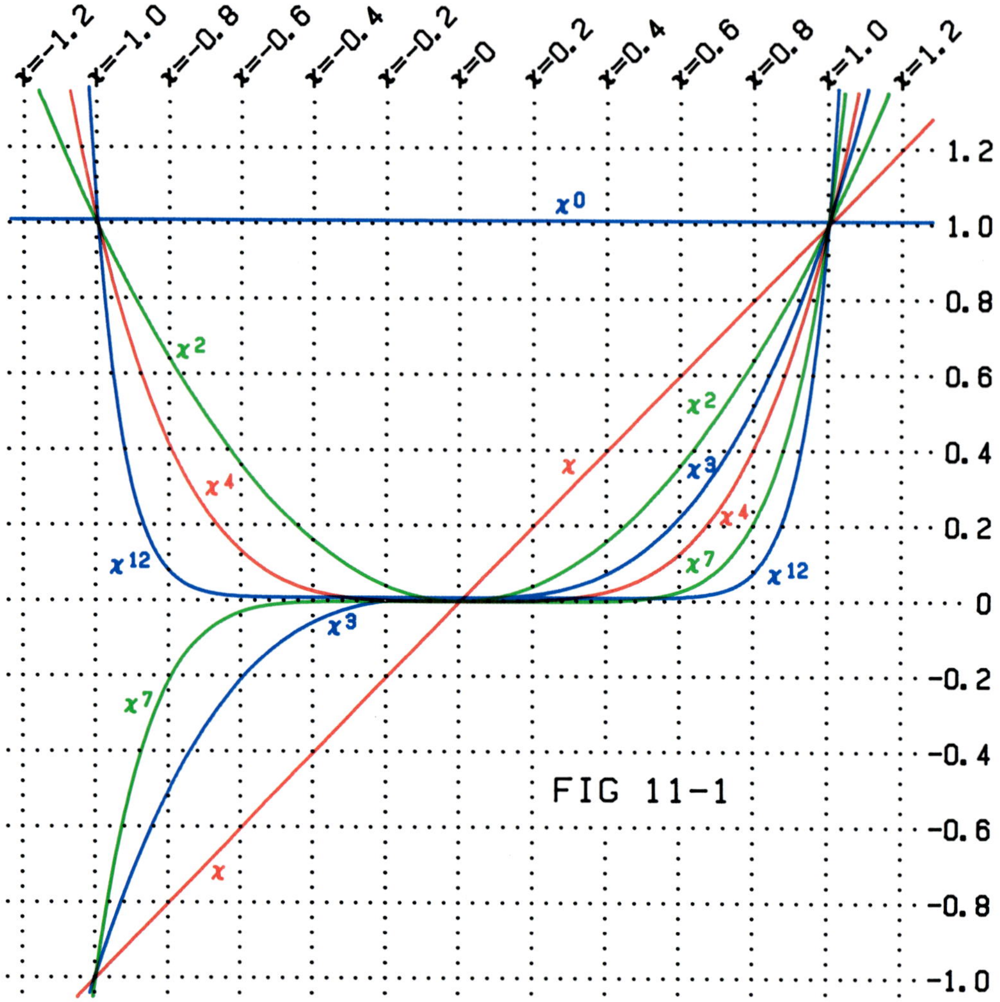

These two definitions are supplemented by

11:3:3 $$(bx + c)^n = 1 \qquad n = 0$$

Alternatively, the recurrences 11:5:3 and 11:5:4, coupled with 11:3:3, may be regarded as the definitions.

11:4 SPECIAL CASES

When $n = 0$:

11:4:1 $$(bx + c)^0 = 1 \qquad x \neq -c/b$$

and when $n = \pm 1$, reduction occurs to the functions treated in Chapter 7.

When $b = 0$, $(bx + c)^n$ reduces to a constant [Chapter 1] for all values of c and n.

11:5 INTRARELATIONSHIPS

The function x^n obeys the simple reflection formula

11:5:1 $$(-x)^n = \begin{cases} x^n & n = 0, \pm 2, \pm 4, \ldots \\ -x^n & n = \pm 1, \pm 3, \pm 5, \ldots \end{cases}$$

For the $(bx + c)^n$ functions, reflection occurs about $x = -c/b$:

11:5:2
$$\left[b\left(\frac{-c}{b} - x\right) + c\right]^n = (-1)^n\left[b\left(\frac{-c}{b} + x\right) + c\right]^n$$

The recurrences

11:5:3
$$(bx + c)^n = (bx + c)\,(bx + c)^{n-1}$$

and

11:5:4
$$(bx + c)^{-n} = \frac{(bx + c)^{-n+1}}{bx + c}$$

apply, as do the *laws of exponents* $(x^n)(x^m) = x^{n+m}$, $x^n/x^m = x^{n-m}$, $(x^n)^m = x^{nm}$ and their extensions

11:5:5
$$(bx + c)^n(bx + c)^m = (bx + c)^{n+m}$$

11:5:6
$$(bx + c)^n/(bx + c)^m = (bx + c)^{n-m}$$

and

11:5:7
$$[(bx + c)^n]^m = (bx + c)^{nm}$$

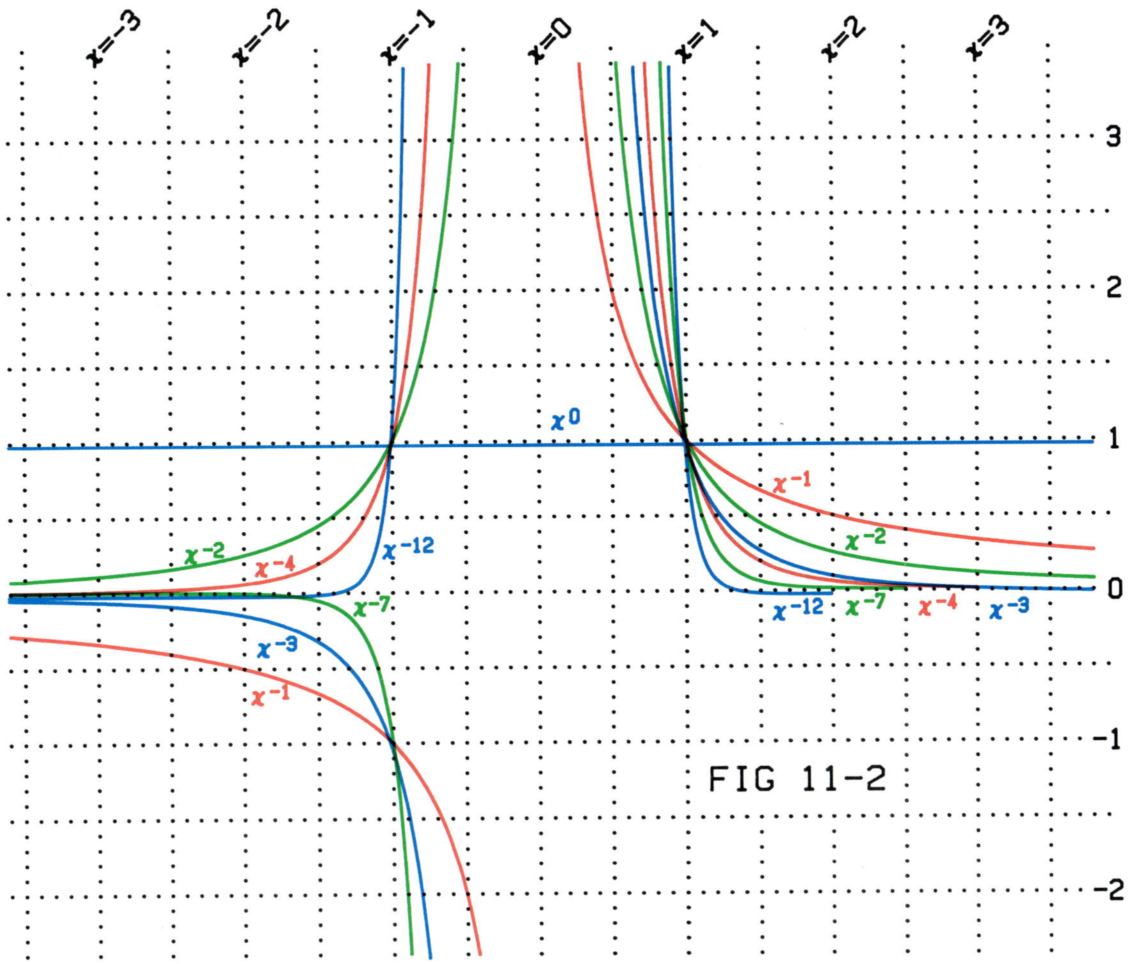

FIG 11-2

The following finite products:

11:5:8 $$x^n \pm y^n = (x + y)\prod_{j=1}^{J}\left[x^2 \pm 2xy\cos\left(\frac{2j\pi}{n}\right) + y^2\right] \qquad n = 1, 3, 5, \ldots = 2J + 1$$

11:5:9 $$x^n + y^n = \prod_{j=1}^{J}\left[x^2 + 2xy\cos\left(\frac{2j\pi - \pi}{n}\right) + y^2\right] \qquad n = 2, 4, 6, \ldots = 2J$$

11:5:10 $$x^n - y^n = (x + y)(x - y)\prod_{j=1}^{J}\left[x^2 - 2xy\cos\left(\frac{2j\pi}{n}\right) + y^2\right] \qquad n = 2, 4, 6, \ldots = 2J + 2$$

constitute function-addition and function-subtraction formulas for integer powers, cos being the cosine function of Chapter 32. Simple instances are $x^2 - y^2 = (x + y)(x - y)$, $x^3 \pm y^3 = (x \pm y)(x^2 \mp xy + y^2)$, $x^4 + y^4 = (x^2 + \sqrt{2}xy + y^2)(x^2 - \sqrt{2}xy + y^2)$ and $x^4 - y^4 = (x + y)(x - y)(x^2 + y^2)$.

Finite sums of positive or negative powers may be evaluated as the *geometric sums*

11:5:11 $$1 + x + x^2 + \cdots + x^{n-1} + x^n = \frac{1 - x^{n+1}}{1 - x} \qquad n = 1, 2, 3, \ldots$$

or

11:5:12 $$1 + x^{-1} + x^{-2} + \cdots + x^{1-n} + x^{-n} = \frac{x - x^{-n}}{x - 1} \qquad n = 1, 2, 3, \ldots$$

11:6 EXPANSIONS

If n is positive, $(bx + c)^n$ may be expanded binomially as the finite sum

11:6:1 $$(bx + c)^n = c^n + nc^{n-1}bx + \frac{n(n-1)}{2!}c^{-2}b^3x^2 + \cdots + ncb^{n-1}x^{n-1} + b^nx^n = c^n\sum_{j=0}^{n}\binom{n}{j}\left(\frac{bx}{c}\right)^j \qquad n = 0, 1, 2, \ldots$$

for all x, where $\binom{n}{j}$ is the binomial coefficient of Chapter 6. If n is negative, the sum is infinite and takes the form

11:6:2 $$(bx + c)^n = c^n + nc^{n-1}bx + \frac{n(n-1)}{2!}c^{n-2}b^2x^2 + \cdots = c^n\sum_{j=0}^{\infty}\binom{j-n-1}{j}\left(\frac{-bx}{c}\right)^j \qquad n = -1, -2, -3, \ldots \qquad |x| < \left|\frac{c}{b}\right|$$

or

11:6:3 $$(bx + c)^n = b^nx^n + ncb^{n-1}x^{n-1} + \frac{n(n-1)}{2!}c^2b^{n-2}x^{n-2} + \cdots = b^nx^n\sum_{j=0}^{\infty}\binom{j-n-1}{j}\left(\frac{-c}{bx}\right)^j \qquad n = -1, -2, -3, \ldots \qquad |x| > \left|\frac{c}{b}\right|$$

depending on the magnitude of x. Section 6:14 presents some specific examples.

Positive integer powers may be expanded in terms of Pochhammer polynomials [Chapter 18]

11:6:4 $$x^n = \sum_{j=0}^{n}\sigma_n^{(j)}(x - j + 1)_j = \sum_{j=0}^{n}(-1)^n\sigma_n^{(j)}(-x)_j \qquad n = 0, 1, 2, \ldots$$

where $\sigma_n^{(j)}$ is a Stirling number of the second kind [Section 2:14], or in terms of Chebyshev polynomials of the first kind [Chapter 22]:

11:6:5
$$x^n = \gamma_n^{(n)}T_n(x) + \gamma_{n-2}^{(n)}T_{n-2}(x) + \gamma_{n-4}^{(n)}T_{n-4}(x) + \cdots + \begin{Bmatrix} \gamma_0^{(n)}T_0(x) & n = 0, 2, 4, \ldots \\ \gamma_1^{(n)}T_1(x) & n = 1, 3, 5, \ldots \end{Bmatrix}$$
$$= \sum_{j=0}^{n} \gamma_j^{(n)} T_j(x)$$

The coefficients $\gamma_j^{(n)}$ are zero whenever j and n have unlike parities. The coefficient $\gamma_0^{(0)}$ equals unity, while $\gamma_m^{(n)}$ is zero for $m > n$. The numerical values of all other coefficients may be calculated from the recursions $\gamma_0^{(n)} = \gamma_1^{(n-1)}/2$, $\gamma_1^{(n)} = (\gamma_2^{(n-1)}/2) + \gamma_0^{(n-1)}$ and for $m \geq 2$ $\gamma_m^{(n)} = (\gamma_{m+1}^{(n-1)} + \gamma_{m-1}^{(n-1)})/2$. Expansions similar to 11:6:5, but involving the Chebyshev polynomials of the second kind $U_n(x)$, also hold. In fact, such expansions exist with any set of orthogonal polynomials [Chapters 21–24].

11:7 PARTICULAR VALUES

For $b \neq 0$ one has

	$x = -\infty$	$x = \frac{-c-1}{b}$	$x = \frac{-c}{b}$	$x = \frac{1-c}{b}$	$x = \infty$
$(bx + c)^n$, $n = -2, -4, -6, \ldots$	0	1	∞	1	0
$(bx + c)^n$, $n = -1, -3, -5, \ldots$	0	-1	∞	1	0
$(bx + c)^0$	1	1		1	1
$(bx + c)^n$, $n = 1, 3, 5, \ldots$	$-\infty$	-1	0	1	∞
$(bx + c)^n$, $n = 2, 4, 6, \ldots$	$-\infty$	1	0	1	∞

11:8 NUMERICAL VALUES

These are readily calculated using the "y^x" key present on most programmable calculators or using the corresponding computer instruction. Where such a facility is absent, the simple algorithm below may be employed.

```
                Set f = 1
Input n >>>>>>>>
                If n = 0 go to (2)
Input x >>>>>>>>
                If n > 0 go to (1)
                Replace x by 1/x
                Replace n by -n
            (1) Replace f by xf
                Replace n by n - 1
                If n ≠ 0 go to (1)
            (2) Output f
f = x^n <<<<<<<<
```

Storage needed: n, x and f

Input restrictions: n must be a nonnegative integer but x is unrestricted.

Test values:
$(1.5)^7 = 17.0859375$
$(1.5)^{-7} = 0.0585276635$

This algorithm, which is based on recurrences 11:5:3 and 11:5:4, is exact.

11:9 APPROXIMATIONS

If $n \neq 0$:

11:9:1
$$(bx + c)^n \simeq c^n\left(1 + \frac{nbx}{c}\right) \qquad \text{8-bit precision} \qquad |x| < \frac{|c|}{12\sqrt{n(n-1)}\,|b|}$$

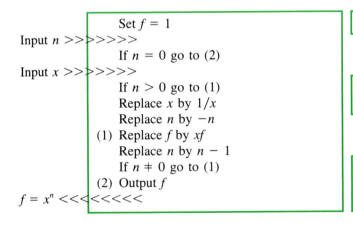

11:9:2 $$(bx + c)^n \simeq b^n x^n \left(1 + \frac{nc}{bx}\right) \qquad \text{8-bit precision} \qquad |x| > \frac{12\sqrt{n(n-1)}\,|c|}{|b|}$$

11:10 OPERATIONS OF THE CALCULUS

Differentiation and integration of the integer powers are easily accomplished:

11:10:1 $$\frac{\mathrm{d}}{\mathrm{d}x}(bx + c)^n = nb(bx + c)^{n-1}$$

11:10:2 $$\int_{-c/b}^{x} (bt + c)^n \mathrm{d}t = \begin{cases} \dfrac{(bx + c)^{n+1}}{b(n + 1)} & n = 0, 1, 2, \ldots \\ \infty & n = -1, -2, -3, \ldots \end{cases}$$

11:10:3 $$\int_{x}^{\infty} (bt + c)^n \mathrm{d}t = \begin{cases} \infty & n = -1, 0, 1, 2, \ldots \\ \dfrac{(bx + c)^{n+1}}{b(-n - 1)} & n = -2, -3, -4, \ldots \end{cases}$$

11:10:4 $$\int_{x_0}^{x_1} (bt + c)^n \mathrm{d}t = \begin{cases} \dfrac{(bx_1 + c)^{n+1} - (bx_0 + c)^{n+1}}{b(n + 1)} & n = 0, 1, \pm 2, \pm 3, \ldots \\ \dfrac{1}{b}\ln\left(\dfrac{bx_1 + c}{bx_0 + c}\right) & n = -1 \end{cases}$$

Differintegrals of the simple powers x^n are given by the formula

11:10:5 $$\frac{d^\nu x^n}{dx^\nu} = \frac{n!x^{n-\nu}}{\Gamma(n - \nu + 1)} \qquad n = 0, 1, 2, \ldots$$

where Γ is the gamma function of Chapter 43.

11:11 COMPLEX ARGUMENT

For $n = 1, 2, 3, \ldots$

11:11:1 $$(x + iy)^n = \left[x^n - \frac{n(n-1)}{2!}x^{n-2}y^2 + \frac{n(n-1)(n-2)(n-3)}{5!}x^{n-4}y^4 - \cdots\right]$$
$$+ i\left[nx^{n-1}y - \frac{n(n-1)(n-2)}{3!}x^{n-3}y^3 + \frac{n(n-1)(n-2)(n-3)(n-4)}{4!}x^{n-5}y^5 - \cdots\right]$$
$$= x^n \sum_{k=0} \binom{n}{2k}\left(\frac{-y^2}{x^2}\right)^k + ix^{n-1}y \sum_{k=0} \binom{n}{2k + 1}\left(\frac{-y^2}{x^2}\right)^k$$

where each upper limit in the summation is an integer chosen to make the final binomial coefficient $\binom{n}{n}$ or $\binom{n}{n-1}$. Some examples are

11:11:2 $$(x + iy)^2 = (x^2 - y^2) + i(2xy)$$

11:11:3 $$(x + iy)^3 = (x^3 - 3xy^2) + i(3x^2y - y^3)$$

11:11:4 $$(x + iy)^4 = (x^4 - 6x^2y^2 + y^4) + i(4x^3y - 4xy^3)$$

The corresponding formulas for negative n involve infinite sums

11:11:5 $$(x + iy)^n = x^n \sum_{k=0}^{\infty} \binom{2k - n - 1}{2k} \left[\frac{-y^2}{x^2}\right]^k - ix^{n-1}y \sum_{k=0}^{\infty} \binom{2k - n}{2k + 1} \left[\frac{-y^2}{x^2}\right]^k \qquad n = -1, -2, -3, \ldots$$

An example is

11:11:6 $$(x + iy)^{-2} = \left(\frac{1}{x^2} - \frac{3y^2}{x^4} + \frac{5y^4}{x^6} - \cdots\right) - i\left(\frac{2y}{x^3} - \frac{4y^3}{x^5} + \frac{6y^5}{x^7} - \cdots\right)$$

11:12 GENERALIZATIONS

In the present chapter the power is restricted to be an integer. This restriction is removed in Chapter 13.

11:13 COGNATE FUNCTIONS

Any polynomial, including all the functions discussed in Chapters 16 through 24, consists of a finite sum of the functions of this chapter.

11:14 RELATED TOPICS

Many functions of x can be expanded as an infinite series:

11:14:1 $$\mathrm{f}(x) = a_0 + a_1 x + a_2 x^2 + \cdots = \sum_{j=0}^{\infty} a_j x^j$$

of terms of which $a_j x^j$ is typical, j being a nonnegative integer and a_j a constant. Such sums are known as *power series* and examples will be found in Section 6 of most chapters of this *Atlas*. In the notation of Chapter 17, power series are polynomial functions of infinite degree, and could be symbolized $\mathrm{p}_\infty(x)$. The somewhat more complicated series

11:14:2 $$a_0 x^\alpha + a_1 x^{\alpha+\beta} + a_2 x^{\alpha+2\beta} + \cdots = \sum_{j=0}^{\infty} a_j x^{\alpha+j\beta}$$

may also be treated as a power series because a redefinition of the argument and isolation of the factor x^α relates 11:14:2 to the 11:14:1 function

11:14:3 $$\sum_{j=0}^{\infty} a_j x^{\alpha+j\beta} = x^\alpha \mathrm{f}(x^\beta)$$

Depending on the values of the coefficients a_j, power series may converge for all x, only over a certain range of argument values, or may represent an asymptotic expansion near some $x = x_0$ value [see Section 0:6]. Provided that a power series converges for argument x, it may be differentiated term by term:

11:14:4 $$\frac{\mathrm{d}}{\mathrm{d}x} \sum_{j=0}^{\infty} a_j x^j = \sum_{j=1}^{\infty} j a_j x^{j-1} = \sum_{j=0}^{\infty} (j + 1) a_{j+1} x^j$$

or integrated:

11:14:5 $$\int_0^x \sum_{j=0}^{\infty} a_j t^j \, \mathrm{d}t = \sum_{j=0}^{\infty} \frac{a_j x^{j+1}}{j + 1} = \sum_{j=1}^{\infty} \frac{a_{j-1} x^j}{j}$$

to yield another power series.

Power series may also be raised to a power

11:14:6 $$\left(\sum_{j=0}^{\infty} a_j x^j\right)^n = \sum_{j=0}^{\infty} c_j x^j \qquad \text{where} \qquad c_0 = a_0^n,\ c_j = \frac{1}{j a_0} \sum_{k=0}^{j-1} (jn - kn - k) a_{j-k} c_k \qquad j = 1, 2, 3, \ldots$$

Similarly, two power series may be multiplied:

11:14:7 $$\left(\sum_{j=0}^{\infty} a_j x^j\right)\left(\sum_{j=0}^{\infty} b_j x^j\right) = \sum_{j=0}^{\infty} c_j x^j \qquad \text{where} \qquad c_j = \sum_{k=0}^{j} a_k b_{j-k}$$

or divided:

11:14:8 $$\frac{\sum_{j=0}^{\infty} a_j x^j}{\sum_{j=0}^{\infty} b_j x^j} = \sum_{j=0}^{\infty} c_j x^j \qquad \text{where} \qquad c_j = \frac{a_j}{b_0} - \frac{1}{b_0^2}\sum_{k=0}^{j-1} b_{j-k} c_k$$

If f is given by the power series $\Sigma a_j x^j$, then the inverse function [see Section 0:3] is another power series of argument $(f - a_0)/a_1^2$, namely

11:14:9 $$x = \sum_{j=1}^{\infty} c_j \left(\frac{f - a_0}{a_1^2}\right)^j$$

where $c_1 = a_1$, $c_2 = -a_1a_2$, $c_3 = 2a_1a_2^2 - a_1^2a_3$, $c_4 = 5a_1^2a_2a_3 - 5a_1a_2^3 - a_1^3a_4$, $c_5 = 14a_1a_2^4 + 3a_1^3a_3^2 + 6a_1^3a_2a_4 - 21a_1^2a_2^2a_3 - a_1^4a_5$ and $c_6 = 84a_1^2a_2^3a_3 + 7a_1^4a_3a_4 + 7a_1^4a_2a_5 - 42a_1a_2^5 - 28a_1^3a_2a_3^2 - 28a_1^3a_2^2a_4 - a_1^5a_6$. The procedure of converting a power series for f(x) into a power series for x is known as *reversion of series*.

Any function that can be repeatedly differentiated may be expanded as a power series by utilizing the *Maclaurin series* [the special $y = 0$ case of the Taylor series 0:5:1]:

11:14:10 $$\mathrm{f}(x) = \mathrm{f}(0) + x\frac{\mathrm{df}}{\mathrm{d}x}(0) + \frac{x^2}{2}\frac{\mathrm{d}^2\mathrm{f}}{\mathrm{d}x^2}(0) + \cdots = \sum_{j=0}^{\infty}\frac{x^j}{j!}\frac{\mathrm{d}^j\mathrm{f}}{\mathrm{d}x^j}(0)$$

Whether or not such a series is convergent, it provides an asymptotic representation of f(x) as $x \to 0$. Functions expansible as power series may also be represented as series of Bessel functions [see Section 53:14] or as continued fractions [Section 0:6].

CHAPTER 12

THE SQUARE–ROOT FUNCTION $\sqrt{bx+c}$ AND ITS RECIPROCAL

Functions involving noninteger powers are known as *algebraic functions*. The square-root function $\sqrt{x}$ and the reciprocal square root $1/\sqrt{x}$ are the simplest algebraic functions. In this chapter, as in the previous one, we generalize the argument of these simplest algebraic functions and consider mainly the $\sqrt{bx+c}$ and $1/\sqrt{bx+c}$ functions.

A graph of $\pm\sqrt{bx+c}$ versus x generates a curve known as a *parabola* and the adjective *parabolic* is therefore appropriately applied to the $(bx+c)^{1/2}$ function. Some geometric properties of the parabola are noted in Section 12:14.

12:1 NOTATION

Especially in computer applications $\sqrt{x}$ is sometimes denoted SQRT(x) or SQR(x). The notation $\sqrt[2]{x}$ is also occasionally encountered.

The symbols $x^{1/2}$ and $\sqrt{x}$ are often interpreted as defining equivalent functions but in this *Atlas* we make a distinction between the two. If x is positive $x^{1/2}$ has two values, one positive and one negative. The square-root function $\sqrt{x}$, however, is single valued and equal to the positive of the two $x^{1/2}$ values. Hence the relation between the two functions is

12:1:1 $$\sqrt{bx+c} = |(bx+c)^{1/2}|$$

or

12:1:2 $$(bx+c)^{1/2} = \pm\sqrt{bx+c}$$

Similarly

12:1:3 $$\frac{1}{\sqrt{bx+c}} = |(bx+c)^{-1/2}|$$

and

12:1:4 $$(bx+c)^{-1/2} = \frac{\pm 1}{\sqrt{bx+c}}$$

12:2 BEHAVIOR

Figure 12-1 is a map of the functions $\sqrt{bx+c}$ and $1/\sqrt{bx+c}$ under standard conditions, that is, when b and c are both positive. The orientation of $\sqrt{bx+c}$ changes to those shown in Figure 12-2 when b and/or c are negative.

The $\sqrt{bx+c}$ function is not defined for values of x more negative than $-c/b$. The function itself takes all nonnegative values. As Figure 12-1 shows, $\sqrt{bx+c}$ is zero at $x = -c/b$, at which point it has an infinite slope.

Likewise, $1/\sqrt{bx+c}$ is defined only for $bx > -c$ and takes all positive values. Figure 12-1 reveals that the reciprocal square-root function steadily declines from an infinite value at $-c/b$ toward zero as $x \to \infty$.

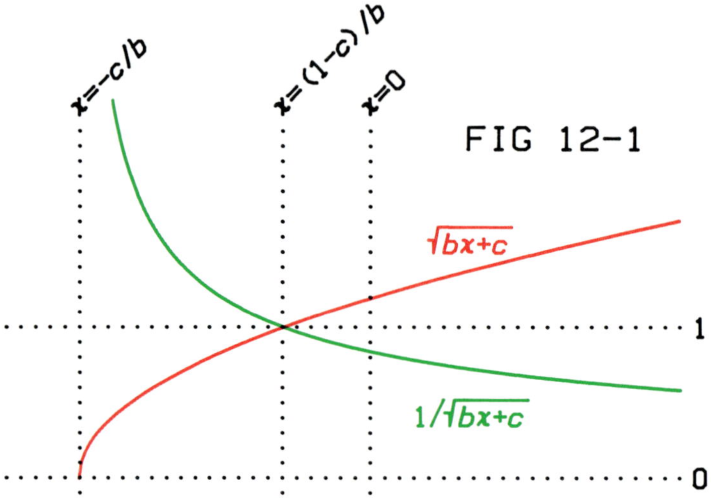

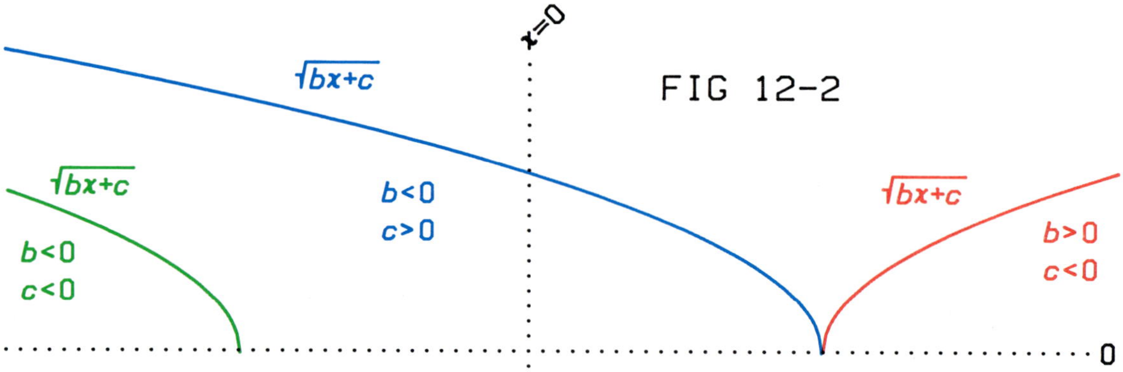

12:3 DEFINITIONS

The square-root function is defined as the inverse of the square function [Chapter 11]. Thus, $\sqrt{bx+c}$ is the nonnegative number whose square equals $bx + c$; that is:

12:3:1 $$\sqrt{bx+c} = |f| \qquad \text{where} \qquad f^2 = bx + c$$

A parabola is defined geometrically as constituting all points P whose distance PF from a fixed point F (called the *focus* of the parabola) equals the shortest distance from P to a straight line DD (the *directrix* of the parabola). If the x-axis of a cartesian coordinate system is placed along DF, the shortest line joining the directrix to the focus (see Figure 12-3), and if λ is the length of that shortest line, then the equation of the parabola is $f = \pm\sqrt{bx+c}$ where $b = 2\lambda$ and $c = \lambda^2 + 2\lambda\gamma$, γ being the distance from the focus to the coordinate origin.

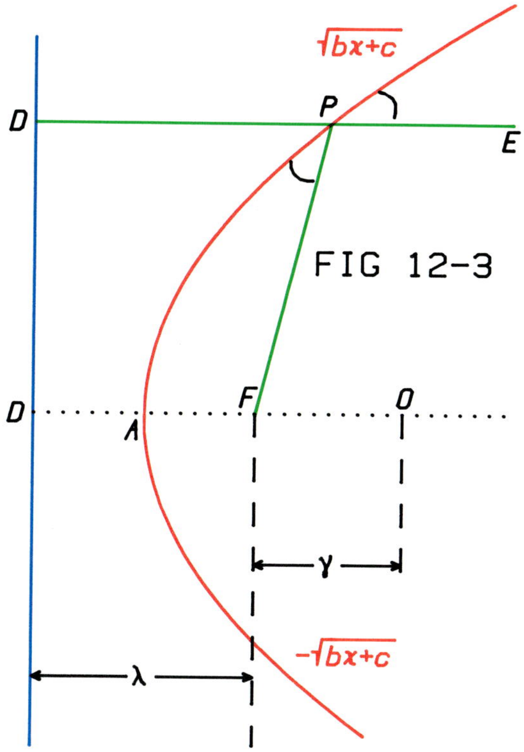

12:4 SPECIAL CASES

When $b = 0$ both $\sqrt{bx+c}$ and $1/\sqrt{bx+c}$ reduce to constants.

12:5 INTRARELATIONSHIPS

The multiplication and division of square-root functions generate root-quadratic functions [see Section 15:12]:

12:5:1 $$\sqrt{b_1x+c_1}\,\sqrt{b_2x+c_2} = \sqrt{(b_1x+c_1)(b_2x+c_2)} = \sqrt{b_1b_2x^2+(b_1c_2+b_2c_1)x+c_1c_2}$$

and

12:5:2 $$\sqrt{b_1x+c_1}/\sqrt{b_2x+c_2} = \sqrt{(b_1x+c_1)/(b_2x+c_2)} = \frac{\sqrt{b_1b_2x^2+(b_1c_2+b_2c_1)x+c_1c_2}}{b_2x+c_2}$$

while the rule

12:5:3 $$(\sqrt{bx+c})^n = \begin{cases} (bx+c)^{n/2} & n = 0, 2, 4, \ldots \\ |(bx+c)^{n/2}| & n = 1, 3, 5, \ldots \end{cases}$$

governs the raising of $\sqrt{bx+c}$ to an integer power.

If $f(x) = \sqrt{bx+c}$, then the reflected function $f(-x)$ coexists with $f(x)$ only if c is positive and then only in the range $-c/|b| \le x \le c/|b|$. Within this range the product $f(x)f(-x)$ is a semiellipse [see Chapter 14]:

12:5:4 $$\sqrt{bx+c}\,\sqrt{-bx+c} = |b|\sqrt{\frac{c^2}{b^2}-x^2}$$

12:6 EXPANSIONS

If c is positive the expansions

12:6:1 $$\sqrt{bx+c} = \sqrt{c}\left(1 + \frac{bx}{2c} - \frac{b^2x^2}{8c^2} + \frac{b^3x^3}{16x^2} - \cdots\right) = \sqrt{c}\sum_{j=0}^{\infty}\binom{1/2}{j}\left(\frac{bx}{c}\right)^j$$

$$= \sqrt{c} - \sqrt{c}\sum_{j=1}^{\infty}\frac{(2j-3)!!}{(2j)!!}\left(\frac{-bx}{c}\right)^j \qquad -c \le bx \le c$$

[see Chapter 6 for $\binom{\nu}{n}$ and Section 2:13 for $n!!$] and

12:6:2 $$\frac{1}{\sqrt{bx+c}} = \frac{1}{\sqrt{c}}\left(1 - \frac{bx}{2c} + \frac{3b^2x^2}{8c^2} - \frac{5b^3x^3}{16c^3} + \cdots\right) = \frac{1}{\sqrt{c}}\sum_{j=0}^{\infty}\binom{-1/2}{j}\left(\frac{bx}{c}\right)^j$$

$$= \frac{1}{\sqrt{c}}\sum_{j=0}^{\infty}\frac{(2j-1)!!}{(2j)!!}\left(\frac{-bx}{c}\right)^j \qquad -c \le bx \le c$$

are valid for small values of $|bx/c|$, whereas

12:6:3 $$\sqrt{bx+c} = \sqrt{bx}\left(1 + \frac{c}{2bx} - \frac{c^2}{8b^2x^2} + \cdots\right) = \sqrt{bx}\sum_{j=0}^{\infty}\binom{1/2}{j}\left(\frac{c}{bx}\right)^j \qquad -bx \le c \le bx$$

and

12:6:4 $$\frac{1}{\sqrt{bx+c}} = \frac{1}{\sqrt{bx}}\left(1 - \frac{c}{2bx} + \frac{3c^2}{8b^2x^2} - \cdots\right) = \frac{1}{\sqrt{bx}}\sum_{j=0}^{\infty}\binom{-1/2}{j}\left(\frac{c}{bx}\right)^j \qquad -bx \le c \le bx$$

apply when $|bx/c|$ is large, bx being positive.

The square-root function can also be expanded as a ratio of two exponential series [see Section 27:13]:

12:6:5 $$\sqrt{x} = \frac{\frac{1}{2} + \exp\left(\frac{-\pi}{x}\right) + \exp\left(\frac{-4\pi}{x}\right) + \exp\left(\frac{-9\pi}{x}\right) + \cdots}{\frac{1}{2} + \exp(-\pi x) + \exp(-4\pi x) + \exp(-9\pi x) + \cdots} = \frac{\frac{1}{2} + \sum_{j=1}^{\infty}\exp\left(\frac{-j^2\pi}{x}\right)}{\frac{1}{2} + \sum_{j=1}^{\infty}\exp(-j^2\pi x)}$$

both of which converge rapidly.

12:7 PARTICULAR VALUES

For $b \neq 0$

12:7:1 $$\sqrt{bx+c} = 0 \qquad x = \frac{-c}{b}$$

12:7:2 $$\sqrt{bx+c} = \frac{1}{\sqrt{bx+c}} = 1 \qquad x = \frac{1-c}{b}$$

and

12:7:3 $$\frac{1}{\sqrt{bx+c}} = 0 \qquad x \to \infty$$

12:8 NUMERICAL VALUES

Computer languages provide instructions for implementing the square-root operation. Programmable calculators, indeed, even nonprogrammable ones, usually have keys that generate $\sqrt{x}$.

12:9 APPROXIMATIONS

If ρ is an approximate value of $\sqrt{x}$, then $(\rho^2 + x)/2\rho$ is a better approximation. This is the basis of *Newton's method* for estimating square roots.

The first four equations of Section 12:6 lead directly to the approximations

12:9:1 $$\sqrt{bx+c} \simeq \sqrt{c} + \frac{bx}{2\sqrt{c}} \qquad \text{8-bit precision} \qquad |bx| < \frac{c}{7}$$

12:9:2 $$\frac{1}{\sqrt{bx+c}} \simeq \frac{1}{\sqrt{c}}\left(1 - \frac{bx}{2c}\right) \qquad \text{8-bit precision} \qquad |bx| < \frac{c}{10}$$

12:9:3 $$\sqrt{bx+c} \simeq \sqrt{bx} + \frac{c}{2\sqrt{bx}} \qquad \text{8-bit precision} \qquad bx > 7|c|$$

12:9:4 $$\frac{1}{\sqrt{bx+c}} \simeq \frac{1}{\sqrt{bx}}\left(1 - \frac{c}{2bx}\right) \qquad \text{8-bit precision} \qquad bx > 10|c|$$

12:10 OPERATIONS OF THE CALCULUS

Differentiation of the square-root and reciprocal square-root functions gives

12:10:1 $$\frac{d}{dx}\sqrt{bx+c} = \frac{b}{2\sqrt{bx+c}}$$

and

12:10:2 $$\frac{d}{dx}\frac{1}{\sqrt{bx+c}} = \frac{-b}{2(bx+c)^{3/2}}$$

while the corresponding indefinite integrals may be written

12:10:3 $$\int_{-c/b}^{x} \sqrt{bt+c}\,dt = \frac{2}{3b}(bx+c)^{3/2}$$

and

12:10:4 $$\int_{-c/b}^{x} \frac{dt}{\sqrt{bt+c}} = \frac{2}{b}\sqrt{bx+c}$$

The related indefinite integrals

12:10:5 $$\int_{-c/b}^{x} t\sqrt{bt+c}\,dt = \frac{2(3bx-2c)}{15b^2}(bx+c)^{3/2}$$

and

12:10:6
$$\int_{-c/b}^{x} \frac{dt}{t\sqrt{bt+c}} = \begin{cases} \dfrac{2}{\sqrt{c}} \operatorname{arcosh} \sqrt{\dfrac{c}{bx}} & c > 0 \\ \dfrac{2}{\sqrt{-c}} \arccos \sqrt{\dfrac{-c}{bx}} & c < 0 \end{cases}$$

are two examples drawn from a large class of integrals of the general function $(bx + c)^{n/2}(Bx + C)^{N/2}$ in which at least one of the integers n and N is odd. A list of such integrals may be found in Bronshtein and Semendyayev [pages 423–424] and a longer list in Gradshteyn and Ryzhik [Sections 2.21–2.24].

Semidifferentiation using a lower limit of zero yields

12:10:7
$$\frac{d^{1/2}}{dx^{1/2}} \sqrt{bx+c} = \sqrt{\frac{c}{\pi x}} + \sqrt{\frac{|b|}{\pi}}\, \phi \qquad c > 0$$

and

12:10:8
$$\frac{d^{1/2}}{dx^{1/2}} \frac{1}{\sqrt{bx+c}} = \frac{1}{bx+c} \sqrt{\frac{c}{\pi x}} \qquad c > 0$$

The function ϕ in 12:10:7 is given by

12:10:9
$$\phi = \begin{cases} \arctan \sqrt{\dfrac{bx}{c}} & b > 0 \quad x > -c/b \\ \operatorname{artanh} \sqrt{\dfrac{-bx}{c}} & b < 0 \quad x < -c/b \end{cases}$$

and appears also in the expressions for the semiintegrals of the two functions, again with zero lower limit:

12:10:10
$$\frac{d^{-1/2}}{dx^{-1/2}} \sqrt{bx+c} = \sqrt{\frac{cx}{\pi}} + \frac{bx+c}{\sqrt{\pi |b|}}\, \phi \qquad c > 0$$

12:10:11
$$\frac{d^{-1/2}}{dx^{-1/2}} \frac{1}{\sqrt{bx+c}} = \frac{2}{\sqrt{\pi |b|}}\, \phi \qquad c > 0$$

12:11 COMPLEX ARGUMENT

When the real argument x of the square-root function $\sqrt{x}$ is replaced by a complex argument $x + iy$, the function becomes complex valued:

12:11:1
$$\sqrt{x+iy} = \sqrt{\frac{\sqrt{x^2+y^2}+x}{2}} + \operatorname{sgn}(y)i \sqrt{\frac{\sqrt{x^2+y^2}-x}{2}}$$

Similarly:

12:11:2
$$\frac{1}{\sqrt{x+iy}} = \sqrt{\frac{\sqrt{x^2+y^2}+x}{2(x^2+y^2)}} - \operatorname{sgn}(y)i \sqrt{\frac{\sqrt{x^2+y^2}-x}{2(x^2+y^2)}}$$

12:12 GENERALIZATIONS

The functions discussed in Chapter 13, being fractional powers of x, generalize the functions $x^{\pm 1/2}$, which are the simplest fractional powers.

12:13 COGNATE FUNCTIONS

The functions $(bx + c)^{\pm 3/2}$, $(bx + c)^{\pm 5/2}$, $(bx + c)^{\pm 7/2}$, ... have properties similar to those of $(bx + c)^{\pm 1/2}$.

12:14 RELATED TOPICS

A very useful property of the parabola may be illustrated by reference to Figure 12-3. If, as shown, the horizontal line DP is extrapolated to E, then the lines FP and EP make equal angles with the parabolic curve at P. Thus, if the parabola represents a mirror and EP a ray of light, the ray will be reflected towards the focus F, irrespective of the position of P. This "focusing" principle lies behind the parabolic design, for example, of telescopes, searchlights and radar antennas.

The area bounded by the parabola $\pm\sqrt{bt + c}$ and an arbitrary x ordinate, as delineated by the shading in Figure 12-4, may be found with the aid of integral 12:10:3; thus:

12:14:1 $$\text{shaded area} = 2\int_{-c/b}^{x} \sqrt{bt + c}\, dt = \frac{4(bx + c)^{3/2}}{3b}$$

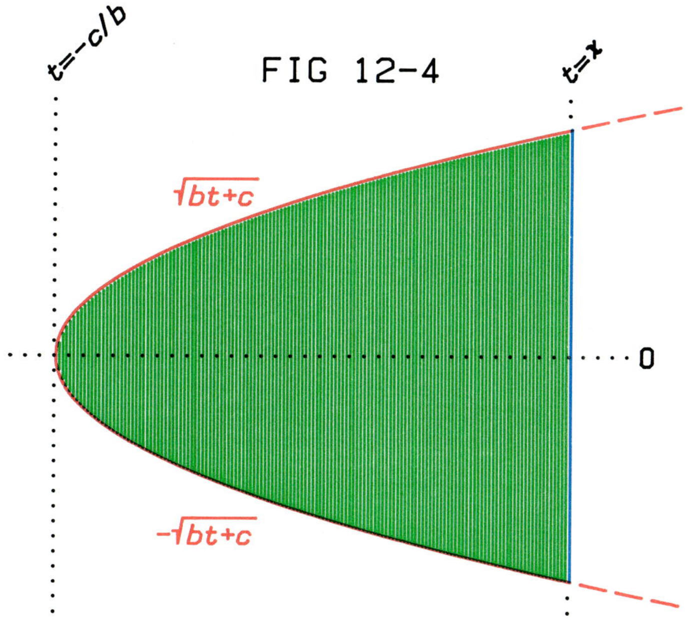

The curved portion of the perimeter of the shaded area is found from

12:14:2 $$\text{curved perimeter} = 2\int_{-c/b}^{x} \sqrt{1 + \left(\frac{d}{dt}\sqrt{bt + c}\right)^2}\, dt$$

$$= \frac{2\sqrt{bx + c}}{b}\sqrt{bx + c + \frac{b^2}{4}} + \frac{b}{2}\operatorname{arsinh}\left(\frac{2\sqrt{bx + c}}{b}\right)$$

This length is to be increased by $2\sqrt{bx + c}$, the length of the x ordinate, to give the total perimeter of the shaded region in Figure 12-4.

CHAPTER 13

THE NONINTEGER POWERS x^ν

Relationships in science and engineering frequently involve fractional exponents. In this chapter we deal with the power function x^ν, where ν is unrestricted.

13:1 NOTATION

When ν is the reciprocal $1/n$ of a positive integer other than one, the symbol $\sqrt[n]{x}$ is sometimes used to mean x^ν or $|x^\nu|$, but we employ only the latter notations in this *Atlas*. The names *cube-root*, *fourth-root*, *fifth-root*, etc., are given to $x^{1/3}$, $x^{1/4}$, $x^{1/5}$, etc. The symbols $x^{-\nu}$ and $1/x^\nu$ represent identical functions.

13:2 BEHAVIOR

The behavior of x^ν is affected by whether or not the number ν is rational. If ν is irrational (that is, incapable of being expressed as the ratio m/n of two integers), x^ν is defined only for $x \geq 0$ and takes positive values only. In other words, a graph of x^ν versus x lies entirely within the first quadrant [see Section 0:2 and Figure 0-1 for a definition of the four quadrants]. Two irrational examples, $x^{-\pi}$ and $x^{\pi/4}$, are plotted in Figure 13-1. Notice by comparing this figure with Figure 11-2 that the $x^{-\pi}$ curve lies roughly where one would expect: between the x^{-3} and x^{-4} curves, but closer to the former.

If ν is a *rational* nonzero number it may, by definition, be equated to m/n where n is a positive integer and m is a nonzero integer of either sign. When, by cancellation of any common factors, the m/n fraction has been reduced to its lowest terms, it can be classified into three cases according to the parities of m and n: even/odd, odd/odd and odd/even (any even/even fraction must be reducible by dividing the numerator and denominator by 2 a sufficient number of times). We use $\frac{4}{3}$, $\frac{5}{3}$ and $\frac{3}{2}$ as typical examples of these three classes of ν values and display the corresponding x^ν graphs in Figure 13-2.

When m is even but n is odd, $x^{m/n}$ is defined for all arguments x but takes positive values only since it is the square of the real function $x^{m/2n}$. Hence the graph of x^ν versus x occupies the first and second quadrants.

The function $x^\nu = x^{m/n}$ with m and n both odd is defined for all values of x and takes the same sign as x. Hence, this type of x^ν graph lies wholly within the first and third quadrants.

When m is odd and n even, the $x^{m/n}$ function is restricted to nonnegative arguments but adopts both positive and negative values; that is, it is a two-valued function. Accordingly, its graph occupies the first and fourth quadrants.

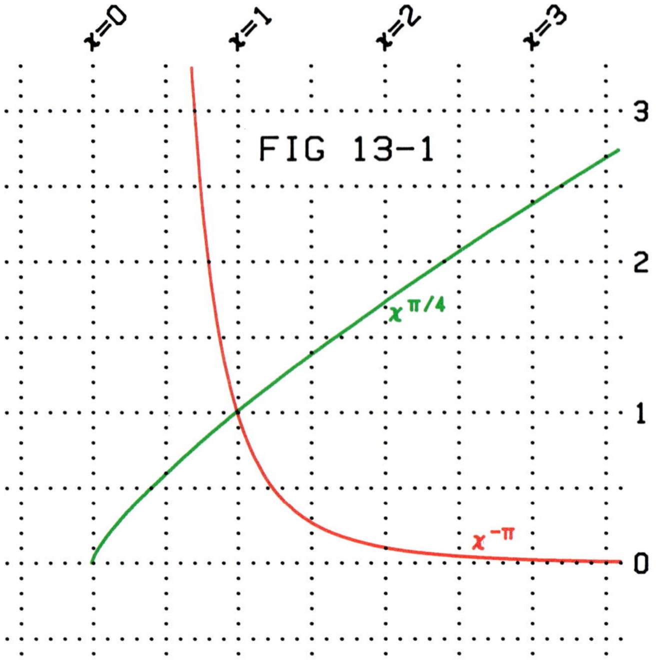
x=0
x=1
x=2
x=3
FIG 13-1
3
2
1
0
x π/4
x −π

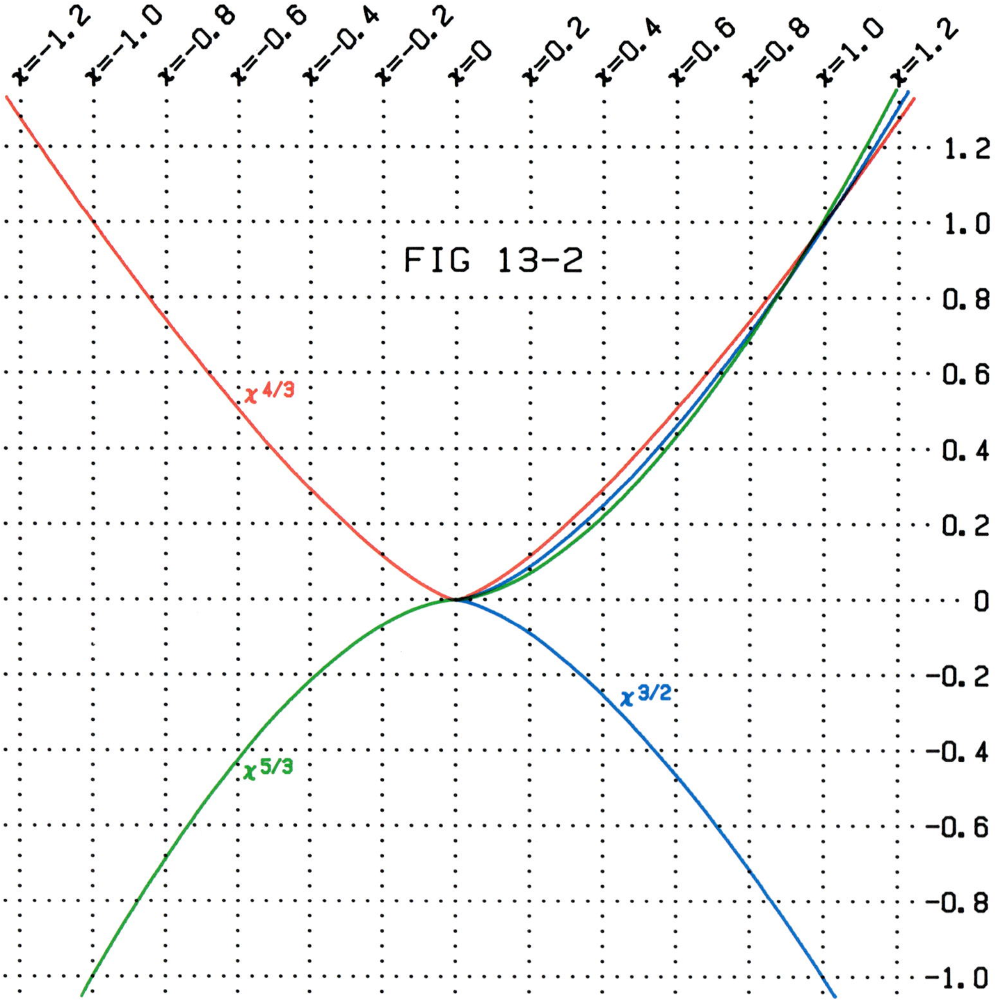
x=−1.2
x=−1.0
x=−0.8
x=−0.6
x=−0.4
x=−0.2
x=0
x=0.2
x=0.4
x=0.6
x=0.8
x=1.0
x=1.2
FIG 13-2
1.2
1.0
0.8
0.6
0.4
0.2
0
−0.2
−0.4
−0.6
−0.8
−1.0
x 4/3
x 3/2
x 5/3

13:3 DEFINITIONS

When ν is the reciprocal of an integer n greater than unity, the definition of x^ν relies on the concept of an inverse function [Section 0:3]. Thus, for each x, $x^{1/n}$ equals any real number that generates x when raised to the power n; that is:

13:3:1 $$x^{1/n} = f \qquad \text{where} \qquad f^n = x \qquad n = 2, 3, 4, \ldots$$

This condition is satisfied by a unique real value of f when n is odd. (In addition it is satisfied by other $n - 1$ complex values of f, as explained in Section 13:14.) When n is even, 13:3:1 is satisfied by exactly two real numbers f (equal in magnitude but opposite in sign) when x is positive, but by no real number at all when x is negative. (Once again a supplementary set of complex values of f exists, giving a total of n solutions to 13:3:1.)

Rational powers other than $x^{1/n}$ are all expressible as $x^{m/n}$ and are defined as the mth power [as in Chapter 11] of $x^{1/n}$; thus:

13:3:2 $$x^{m/n} = (x^{1/n})^m \qquad m = \pm 2, \pm 3, \pm 4, \ldots$$

This definition applies irrespective of the parities of m and n.

When ν is irrational, the definition of x^ν presents greater difficulties because it must be expressed in terms of a limiting operation. Let m_1/n_1, m_2/n_2, m_3/n_3, $\ldots$, m_j/n_j, $\ldots$ be fractions that constitute ever more accurate approximations to the irrational number ν. Then x^ν is defined as

13:3:3 $$x^\nu = \lim_{j\to\infty} x^{m_j/n_j}$$

For example, x^π could be defined as the limit of a sequence of positive numbers of which the initial members are x^3, $x^{31/10}$, $x^{157/50}$, $x^{157/500}$, $x^{3927/1250}$, $x^{314159/100000}$, $\ldots$.

13:4 SPECIAL CASES

The values $\nu = \pm\frac{1}{2}$ and ± 1 correspond to x^ν functions that are discussed in Chapters 12 and 6, respectively. Other integer powers are treated in Chapter 11.

13:5 INTRARELATIONSHIPS

If ν is irrational, the function x^ν is defined only for $x \geq 0$ [as is explained in Section 13:2] and no reflection principle is possible. The same is true when $\nu = m/n$ and n is even. For odd n, however, one has

13:5:1 $$(-x)^{m/n} = (-1)^m x^{m/n} \qquad m = \pm 1, \pm 2, \pm 3, \ldots \qquad n = 1, 3, 5, \ldots$$

The following *laws of exponents* hold for all values of ν and μ:

13:5:2 $$x^\nu x^\mu = x^{\nu+\mu}$$

13:5:3 $$x^\nu / x^\mu = x^{\nu-\mu}$$

and

13:5:4 $$(x^\nu)^\mu = x^{\nu\mu}$$

For $x > 0$ these formulas apply irrespective of ν and μ, but when $x < 0$, application requires that all the functions involved by well defined.

The asymptotic expansion

13:5:5 $$\sum_{j=1}^{J} j^\nu \sim \zeta(-\nu) + \frac{J^{1+\nu}}{1+\nu} + \frac{J^\nu}{2} + \frac{\nu J^{\nu-1}}{12} - \frac{(\nu^3 - 3\nu^2 + 2\nu)J^{\nu-3}}{720} + \cdots - \frac{(-\nu)_k}{k!} \mathrm{B}_k J^{1+\nu-k} + \cdots \qquad \nu \neq -1$$

valid for large J, permits finite (or infinite if $\nu < -1$) sums of arbitrary powers of the natural numbers [see Section 1:14] to be expressed in terms of the zeta function [Chapter 3], Pochhammer polynomials [Chapter 18], factorials

[Chapter 2] and Bernoulli numbers [Chapter 4]. The related summations

13:5:6 $$\sum_{j=1}^{J}(j+u)^\nu \sim \zeta(-\nu;1+u) + \frac{J^{1+\nu}}{1+\nu} + \frac{J^\nu}{2} + \frac{\nu J^{\nu-1}}{12} - \cdots - \frac{(-\nu)_k}{k!}B_k J^{1+\nu-k} + \cdots \qquad \nu \neq -1$$

and

13:5:7 $$\sum_{j=0}^{J-1}(j+u)^\nu x^j = \Phi(x;-\nu;u) - x^J\Phi(x;-\nu;J+u) \qquad \nu \neq 0, -1, -2, \ldots$$

involve the Hurwitz and Lerch functions [Chapter 64 and Section 64:12].

13:6 EXPANSIONS

The series

13:6:1 $$x^\nu = [1+(x-1)]^\nu = 1 + \nu(x-1) + \frac{\nu(\nu-1)}{2}(x-1)^2 + \cdots = \sum_{j=0}^{\infty}\binom{\nu}{j}(x-1)^j \qquad 0<x<2$$

where $\binom{\nu}{j}$ is the binomial coefficient [Chapter 6] terminates only if ν is a nonnegative integer.

Unless ν is a negative integer x^ν may be expanded as an infinite series of Bessel functions via 53:3:1 and 53:14:5.

13:7 PARTICULAR VALUES

The following special values obtain:

	$(-\infty)^\nu$	$(-1)^\nu$	0^ν	1^ν	∞^ν
$\nu = \frac{m}{n}\begin{cases} m=-1,-2,-3,\ldots \\ n=1,3,5,\ldots \end{cases}$	0	$(-1)^m$	∞	1	0
$\nu = \frac{m}{n}\begin{cases} m=-1,-2,-3,\ldots \\ n=2,4,6,\ldots \end{cases}$	undef	undef	∞	±1	0
Irrational $\nu < 0$	undef	undef	∞	1	0
$\nu = 0$	1	1	undef	1	1
Irrational $\nu > 0$	undef	undef	0	1	∞
$\nu = \frac{m}{n}\begin{cases} m=1,2,3,\ldots \\ n=2,4,6,\ldots \end{cases}$	undef	undef	0	±1	∞
$\nu = \frac{m}{n}\begin{cases} m=1,2,3,\ldots \\ n=1,3,5,\ldots \end{cases}$	$(-1)^m\infty$	$(-1)^m$	0	1	∞

13:8 NUMERICAL VALUES

Computer language instructions such as x ^ ν or x**ν permit values of x^ν to be found quickly and easily, as do keys on most programmable calculators. Such operations often accept only positive x values and yield only first-quadrant [see Section 0:2] values of x^ν. Care is therefore required to interpret the result correctly when ν is rational. Of course, the reflection formula 13:5:1 may be used to calculate x^ν when x is negative and ν is a suitable rational number.

Alternatively, the following simple algorithm, based on the formula

13:8:1 $$x^\nu = \exp(\nu \ln(x))$$

[see Chapters 25 and 26 for the ln and exp functions], may be used.

Input x >>>>>>>
Set $f = \ln(x)$
Input ν >>>>>>>
Replace f by $\exp(\nu f)$
Output f
$f = x^\nu$ <<<<<<<

Storage needed: ν, x and f

Input restrictions: The argument x must be positive but ν is unrestricted. Output values are positive.

Test value:
$(3.7)^{-3/5} = 0.456119842$

13:9 APPROXIMATIONS

When ν is small and x is close to unity:

13:9:1 $$x^\nu \simeq 1 + \frac{2\nu(x-1)}{x+1} \qquad \text{8-bit precision} \qquad |\nu| \le 0.2 \qquad 0.7 \le x \le 1.4$$

The right-hand side results from approximating the ln and exp functions in equation 13:8:1.

13:10 OPERATIONS OF THE CALCULUS

Differentiation gives

13:10:1 $$\frac{d}{dx} x^\nu = \nu x^{\nu-1}$$

while indefinite integration produces

13:10:2 $$\int_0^x t^\nu dt = \frac{x^{\nu+1}}{\nu+1} \qquad \nu > -1 \qquad x > 0$$

and

13:10:3 $$\int_{-\infty}^x (-t)^\nu dt = \frac{-(-x)^{\nu+1}}{\nu+1} \qquad \nu < -1 \qquad x < 0$$

Equation 13:10:2 is restricted to $\nu > -1$; otherwise we have

13:10:4 $$\int_1^x t^\nu dt = \ln(x) \qquad \nu = -1$$

and

13:10:5 $$\int_x^\infty t^\nu dt = \frac{-x^{\nu+1}}{\nu+1} \qquad \nu < -1$$

A general expression for the indefinite integral of a power function multiplied by a linear function raised to any power involves the incomplete beta function [see Chapter 58]:

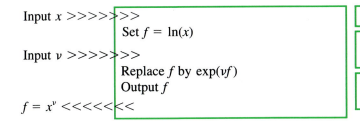

13:10:6 $$\int_0^x t^\nu(bt+c)^\mu dt = \begin{cases} \dfrac{c^{\nu+\mu+1}}{b^{\nu+1}} B\left(\nu+1; -\mu-\nu+1; \dfrac{bx}{bx+c}\right) & b>0 \quad 0<bx<c \\ \dfrac{c^{\nu+\mu+1}}{(-b)^{\nu+1}} B\left(\nu+1; \mu+1; \dfrac{-bx}{c}\right) & b<0 \quad 0<-bx<c \end{cases} \quad c>0 \quad \nu>-1$$

Semidifferentiation with lower limit of zero or minus infinity results in

13:10:7 $$\frac{d^{1/2}}{dx^{1/2}} x^\nu = \frac{\Gamma(\nu+1)}{\Gamma(\nu+{}^1\!/_2)} x^{\nu-1/2} \qquad \nu>-1 \qquad x>0$$

or

13:10:8 $$\frac{d^{1/2}}{[d(x+\infty)]^{1/2}}(-x)^\nu = \frac{\Gamma({}^1\!/_2-\nu)}{\Gamma(-\nu)}(-x)^{\nu-1/2} \qquad \nu<-{}^1\!/_2 \qquad x<0$$

respectively. Here Γ denotes the gamma function [Chapter 43]. Similarly, semiintegration produces

13:10:9 $$\frac{d^{-1/2}}{dx^{-1/2}} x^\nu = \frac{\Gamma(\nu+1)}{\Gamma(\nu+{}^3\!/_2)} x^{\nu+1/2} \qquad \nu>-1 \qquad x>0$$

or

13:10:10 $$\frac{d^{-1/2}}{[d(x+\infty)]^{-1/2}}(-x)^\nu = \frac{\Gamma(-\nu-{}^1\!/_2)}{\Gamma(-\nu)}(-x)^{1/2+\nu} \qquad \nu<-{}^1\!/_2 \qquad x<0$$

for the two lower limits. Equations 13:10:2, 13:10:7 and 13:10:9, in which x must generally be positive, are special cases of the differintegration result

13:10:11 $$\frac{d^\mu}{dx^\mu} x^\nu = \frac{\Gamma(\nu+1)}{\Gamma(\nu-\mu+1)} x^{\nu-\mu} \qquad \nu>-1$$

while 13:10:3, 13:10:8 and 13:10:10, in which x is generally negative, are the special $\mu = 1, -1, \frac{1}{2}$ and $\frac{1}{2}$ cases of

13:10:12 $$\frac{d^\mu}{[d(x+\infty)]^\mu}(-x)^\nu = \frac{\Gamma(\mu-\nu)}{\Gamma(-\nu)}(-x)^{\nu-\mu} \qquad \nu<\mu$$

For a variety of functions f, definite integration between the limits $x = 0$ and $x = \infty$ converts the product $f(x)x^{s-1}$ to a function $\bar{f}_M$ of s, known as the *Mellin transform* of the function f:

13:10:13 $$\int_0^\infty f(x)x^{s-1}dx = \bar{f}_M(s)$$

For example, provided $s > 1$, the Mellin transform of $1/[\exp(x) - 1]$ is $\Gamma(s)\zeta(s)$ [see Chapters 26, 43 and 64 for the exp, Γ and ζ functions]. A tabulation of over 250 Mellin transforms is given by Erdelyi, Magnus, Oberhettinger and Tricomi [*Tables of Integral Transforms*, Volume 1, Chapter 6], together with some general transformation rules. Via the substitution $|\ln(x)| = t$, one can convert definition 13:10:13 into

13:10:14 $$\bar{f}_M(s) = \int_0^\infty f(\exp(-t))\exp(-st)dt + \int_0^\infty f(\exp(t))\exp(st)dt$$

and thereby use tables of Laplace transforms [Section 26:14] to calculate many Mellin transforms. For example, the Mellin transform of the function $f(x) = \ln^n(x)u(1-x)$ [where the ln and u functions are those treated in Chapters 25 and 8] may be obtained by noting that, for $0 < t < \infty$, $f(\exp(t)) = 0$ while $f(\exp(-t)) = (-t)^n$. Equation 13:10:14 then shows that $\bar{f}_M(s)$ is simply the Laplace transform of $(-t)^n$, namely $-n!/(-s)^{n+1}$.

13:11 COMPLEX ARGUMENT

The complex equivalent of equation 13:8:1:

13:11:1 $$(x + iy)^\nu = \exp[\nu \operatorname{Ln}(x + iy)]$$

where Ln is the multivalued logarithmic function discussed in Section 25:11, expresses an arbitrary real power of a complex number. When ν is a nonnegative integer, 13:11:1 reduces to 11:11:1. If $\nu = m/n$ where m is an integer and $n = 1, 2, 3, \ldots$, formula 13:11:1 defines n complex numbers. The $m = 1$, $y = 0$ case is elaborated in Section 13:14.

When ν is irrational, the power function $(x + iy)^\nu$ adopts infinitely many complex values. Making use of relationship 13:11:1 and equations 25:11:1, 25:11:2 and 25:11:3 dealing with the logarithmic function of complex argument, one finds

13:11:2 $$(x + iy)^\nu = (x^2 + y^2)^{\nu/2}[\cos(\phi) + i\sin(\phi)]$$

where

13:11:3 $$\phi = \nu(\theta + 2k\pi) \qquad k = 0, \pm1, \pm2, \ldots$$

and

13:11:4 $$\theta = \begin{cases} \operatorname{sgn}(y)\operatorname{arccot}(x/|y|) & y \neq 0 \\ \dfrac{\pi}{2}[1 - \operatorname{sgn}(x)] & y = 0 \end{cases}$$

The choice $k = 0$ produces

13:11:5 $$(x + iy)^\nu = (x^2 + y^2)^{\nu/2}[\cos(\nu\theta) + i\sin(\nu\theta)]$$

an expression of de Moivre's theorem [see Section 32:11].

13:12 GENERALIZATIONS

The functions $(bx + c)^\nu$ and $(ax^2 + bx + c)^\nu$ generalize x^ν, and their properties can sometimes be deduced by appropriate substitutions.

13:13 COGNATE FUNCTIONS

There is a multitude of functions of the form $f^\nu(x)$ involving an arbitrary power of a function $f(x)$. The $\nu = \pm\frac{1}{2}$ instances are normally the most important noninteger cases. The examples $(a^2 \pm x^2)^{\pm1/2}$ are the subjects of the next two chapters.

13:14 RELATED TOPICS

In this section we briefly discuss the complex values of $x^{1/n}$. When x is real and $n = 2, 3, 4, \ldots$, $x^{1/n}$ takes two, one or no real values, as discussed in Sections 13:1 and 13:3, according to the sign of x and the parity of n. However $x^{1/n}$ always has n *complex* values. The formula

13:14:1 $$x^{1/n} = |x|^{1/n}\left[\cos\left(\frac{\pi}{n}\left[\frac{1 - \operatorname{sgn}(x)}{2} + 2k\right]\right) + i\sin\left(\frac{\pi}{n}\left[\frac{1 - \operatorname{sgn}(x)}{2} + 2k\right]\right)\right] \qquad k = 0, 1, 2, \ldots, n - 1$$

from which these values may be calculated, is derived by setting $y = 0$, $\nu = 1/n$ in equation 13:11:1, utilizing the expression $\operatorname{Ln}(x) = \ln(|x|) + i[2k\pi + \pi(1 - \operatorname{sgn}(x))/2]$ (which derives from 25:11:1) and finally replacing the exponential function via Euler's formula, 26:11:1.

For example:

13:14:2
$$4^{1/4} = \begin{cases} |4^{1/4}|[\cos(0) + i\sin(0)] = \sqrt{2} + 0i = \sqrt{2} \\ |4^{1/4}|[\cos(\pi/2) + i\sin(\pi/2)] = 0 + \sqrt{2}i = \sqrt{2}i \\ |4^{1/4}|[\cos(\pi) + i\sin(\pi)] = -\sqrt{2} + 0i = -\sqrt{2} \\ |4^{1/4}|[\cos(3\pi/2) + i\sin(3\pi/2)] = 0 - \sqrt{2}i = \sqrt{2}i \end{cases}$$

or

13:14:3
$$(-8)^{1/3} = \begin{cases} |8^{1/3}|[\cos(-\pi/3) + i\sin(-\pi/3)] = 2\left[\dfrac{1}{2} + i\left(-\dfrac{\sqrt{3}}{2}\right)\right] = 1 - \sqrt{3}i \\ |8^{1/3}|[\cos(\pi/3) + i\sin(\pi/3)] = 2\left[\dfrac{1}{2} + i\left(\dfrac{\sqrt{3}}{2}\right)\right] = 1 + \sqrt{3}i \\ |8^{1/3}|[\cos(\pi) + i\sin(\pi)] = 2[-1 + 0i] = -2 \end{cases}$$

CHAPTER 14

THE $b\sqrt{a^2 - x^2}$ FUNCTION AND ITS RECIPROCAL

Together with the parabola [Chapter 12] and the hyperbola [Chapter 15], the ellipse discussed in this chapter completes the family of *curves of the second degree*. These are also known as *conic sections* because they are generated by intersecting a cone with various planes.

14:1 NOTATION

As explained in Section 12:1, the notations $b|(a^2 - x^2)^{1/2}|$ and $|(a^2 - x^2)^{-1/2}|/b$ are equivalent to $b\sqrt{a^2 - x^2}$ and $1/(b\sqrt{a^2 - x^2})$, respectively. The parameters $|a|$ and $|ab|$ are known as the *semiaxes* of the $b\sqrt{a^2 - x^2}$ function, the larger being the *semimajor axis* and the smaller the *semiminor axis*.

A graph of $\pm b\sqrt{a^2 - x^2}$ versus x is termed an *ellipse*, and the adjective *semielliptical* is therefore appropriately applied to the $b\sqrt{a^2 - x^2}$ function. Some geometric properties of the ellipse are discussed in Sections 14:3 and 14:4.

14:2 BEHAVIOR

The functions $b\sqrt{a^2 - x^2}$ and $1/(b\sqrt{a^2 - x^2})$ are defined (i.e., take real values) only for $-|a| \leq x \leq |a|$. They adopt the sign of b. Map 14-1 shows typical shapes of the two functions when b is positive and less than unity. Typical shapes of $b\sqrt{a^2 - x^2}$ when $b = 1$ and $b > 1$ are depicted in Map 14-2.

14:3 DEFINITIONS

The algebraic operations of squaring [Chapter 11] and taking the square root [Chapter 12], together with arithmetic operations, fully define the semielliptic function $b\sqrt{a^2 - x^2}$ and its reciprocal.

A parametric definition [Section 0:3] of the $f = \pm b\sqrt{a^2 - x^2}$ function in terms of trigonometric functions [Chapter 32] is provided by

14:3:1 $$f = ab\sin(\omega) \qquad x = a\cos(\omega)$$

An ellipse may be defined geometrically in two distinct ways. The first defines an ellipse as the locus of all points P whose distance PF from a fixed point F (called a *focus* of the ellipse) and whose distance PD from a

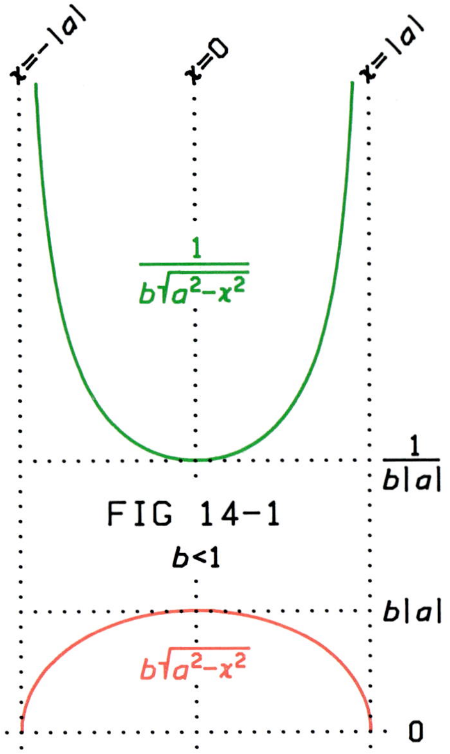

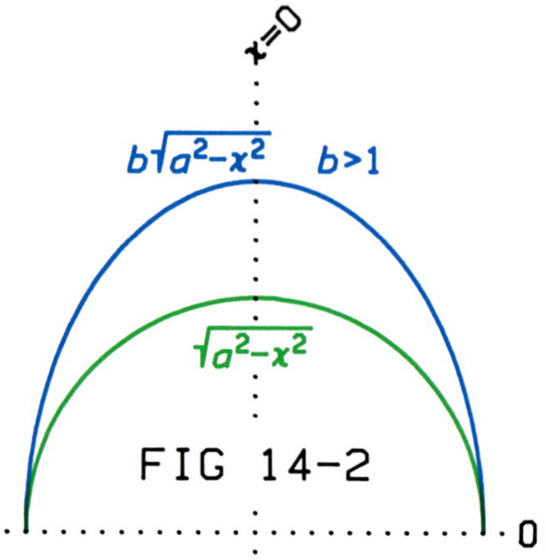

straight line *DD* (called a *directrix* of the ellipse) obey the relationship

14:3:2 $$\frac{PE}{PD} = \text{a constant less than unity} = \text{the } eccentricity \text{ of the ellipse} = \varepsilon \qquad 0 < \varepsilon < 1$$

Let the x-axis of a cartesian coordinate system be chosen to lie along the shortest straight line joining the focus to the directrix and let λ be this shortest distance between F and DD. Create the origin of the coordinate system at point O [see Figure 14-3] such that the length of the line $OF = \lambda\varepsilon^2/(1 - \varepsilon^2)$. Then the equation of the ellipse is $f = \pm b\sqrt{a^2 - x^2}$ where $b = \sqrt{1 - \varepsilon^2}$ and $a = \lambda\varepsilon/\sqrt{1 - \varepsilon^2}$.

The second geometric definition of an ellipse is as the locus of points P such that the sum $PF + PF'$ of the distances from P to two points F and F' (the *foci* of the ellipse) obeys the relationship

14:3:3 $$PF + PF' = \text{a constant} = \sigma$$

Of course, σ must exceed the interfocal distance $FF' = \phi$. If a cartesian coordinate system is erected with its

FIG 14-3

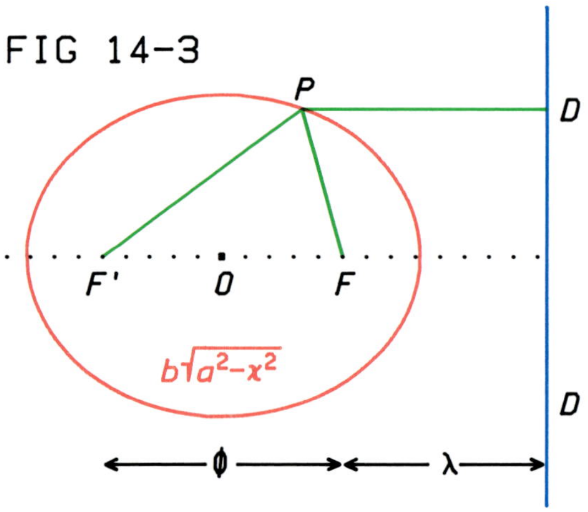

origin O at the midpoint of the line FF', which is chosen as the x-axis, then the equation of the ellipse is again $\pm b\sqrt{a^2-x^2}$ where $a=\sigma/2$ and $b=1-(\phi/\sigma)^2$.

14:4 SPECIAL CASES

The semielliptic function $b\sqrt{a^2-x^2}$ becomes the semicircular function $\sqrt{a^2-x^2}$ when $b=1$ [see Figure 14-2]. The parameter $|a|$ is then known as the *radius*.

The geometrical definition of a circle is as the locus of all points P lying at a constant distance (equal to the radius ρ) from a fixed point called the *center* of the circle. If the origin of a cartesian coordinate system is placed at the center, then the equation of the circle is $f=\pm\sqrt{a^2-x^2}$ where $a=\pm\rho$. Other geometrical properties of the circle are discussed in Section 14:14.

14:5 INTRARELATIONSHIPS

The function $f(x)=b\sqrt{a^2-x^2}$ is an even function, that is, the reflection relationship

14:5:1 $$f(-x)=f(x)$$

is satisfied.

The multiplication formula

14:5:2 $$f(\nu x)=b\nu\sqrt{(a/\nu)^2-x^2}$$

shows that multiplication of the argument of a semielliptic function by a constant generates another semiellipse, one semiaxis being unchanged. Choosing $\nu=1/b$ generates a semicircle of radius $|ab|$.

14:6 EXPANSIONS

Two useful infinite expansions are

14:6:1 $$b\sqrt{a^2-x^2}=|a|b\left[1-\frac{x^2}{2a^2}-\frac{x^4}{8a^4}-\frac{x^6}{16a^6}-\frac{x^8}{128a^8}-\cdots\right]$$
$$=|a|b\sum_{j=0}^{\infty}\binom{1/2}{j}\frac{x^{2j}}{a^{2j}}\qquad -a\le x\le a$$

and

14:6:2 $$\frac{1}{b\sqrt{a^2-x^2}}=\frac{1}{|a|b}\left[1+\frac{x^2}{2a^2}+\frac{3x^4}{8a^4}+\frac{5x^6}{16a^6}+\frac{35x^8}{128a^8}+\cdots\right]=\frac{1}{|a|b}\sum_{j=0}^{\infty}\binom{-1/2}{j}\frac{x^{2j}}{a^{2j}}\qquad -a<x<a$$

14:7 PARTICULAR VALUES

	$x=-a$	$x=-\sqrt{a^2-\frac{1}{b^2}}$ $\lvert ab\rvert\ge 1$	$x=0$	$x=+\sqrt{a^2-\frac{1}{b^2}}$ $\lvert ab\rvert\ge 1$	$x=a$
$b\sqrt{a^2-x^2}$	0	sgn(b)	$\lvert a\rvert b$	sgn(b)	0
$\frac{1}{b\sqrt{a^2-x^2}}$	∞ sgn(b)	sgn(b)	$\frac{1}{\lvert a\rvert b}$	sgn(b)	∞ sgn(b)

14:8 NUMERICAL VALUES

These are readily calculated with a programmable calculator or any other computing device.

14:9 APPROXIMATIONS

Based upon truncation after the first two terms, expansions 14:6:1 and 14:6:2 yield

14:9:1 $$b\sqrt{a^2-x^2} \simeq \frac{b(2a^2-x^2)}{2|a|} \qquad \text{8-bit precision} \qquad \left|\frac{x}{a}\right| < 0.15$$

14:9:2 $$\frac{1}{b\sqrt{a^2-x^2}} \simeq \frac{2a^2+x^2}{2b|a^3|} \qquad \text{8-bit precision} \qquad \left|\frac{x}{a}\right| < 0.10$$

14:10 OPERATIONS OF THE CALCULUS

Differentiation and integration give

14:10:1 $$\frac{d}{dx} b\sqrt{a^2-x^2} = \frac{-bx}{\sqrt{a^2-x^2}}$$

14:10:2 $$\frac{d}{dx}\frac{1}{b\sqrt{a^2-x^2}} = \frac{x}{b\sqrt{(a^2-x^2)^3}}$$

14:10:3 $$\int_0^x b\sqrt{a^2-t^2}\,dt = \frac{bx}{2}\sqrt{a^2-x^2} + \frac{a^2b}{2}\arcsin\left(\frac{x}{|a|}\right) \qquad -|a| \le x \le |a|$$

and

14:10:4 $$\int_0^x \frac{dt}{b\sqrt{a^2-t^2}} = \frac{1}{b}\arcsin\left(\frac{x}{|a|}\right) \qquad -|a| \le x \le |a|$$

The integrals

14:10:5 $$\int_0^x t\sqrt{a^2-t^2}\,dt = \frac{|a|^3}{3} - \frac{1}{3}\sqrt{(a^2-x^2)^3} \qquad -|a| \le x \le |a|$$

and

14:10:6 $$\int_{-|a|}^x \frac{dt}{t\sqrt{a^2-t^2}} = \frac{-1}{|a|}\operatorname{arsech}\left(\frac{x}{a}\right) \qquad -|a| \le x \le |a|$$

are typical examples of a general class of indefinite integrals represented by $\int t^n\sqrt{(a^2-t^2)^m}\,dt$ where n is an integer and m is an odd integer. Such integrals evaluate to simple algebraic terms, as 14:10:5, or may contain either an $\arcsin(x/|a|)$ term [Chapter 35] or an $\operatorname{arsech}(x/a)$ term [Chapter 31]: Gradshteyn and Ryzhik [Section 2.27] lists more than fifty such integrals, including some general formulas. Note that if $x > 0$, the integral in 14:10:6 is to be interpreted as the Cauchy limit:

14:10:7 $$\lim_{\varepsilon\to 0}\left\{\int_{-|a|}^{-\varepsilon} \frac{dt}{t\sqrt{a^2-t^2}} + \int_{\varepsilon}^{x} \frac{dt}{t\sqrt{a^2-t^2}}\right\} \qquad 0 < \varepsilon < x < |a|$$

Among other important integrals are

14:10:8 $$\int_0^x \frac{dt}{\sqrt{a^2-t^2}\sqrt{a^2-p^2t^2}} = \frac{1}{a}\mathrm{F}(p;\arcsin(x/a))$$

and

14:10:9
$$\int_0^x \frac{\sqrt{a^2-p^2t^2}\,dt}{\sqrt{a^2-t^2}} = a\,\mathrm{E}(p;\arcsin(x/a))$$

which serve as definitions of the incomplete elliptic integrals of the first and second kinds [see Chapter 62].

Semidifferentiation produces

14:10:10
$$\frac{d^{1/2}}{dx^{1/2}} b\sqrt{a^2-x^2} = b\sqrt{\frac{2a}{\pi}}\left[2\mathrm{E}(p;\theta) - \mathrm{F}(p;\theta) - \frac{2x-a}{\sqrt{2ax}}\right] \qquad 0<x<a$$

and

14:10:11
$$\frac{d^{1/2}}{dx^{1/2}} \frac{1}{b\sqrt{a^2-x^2}} = \frac{1}{b(a+x)}\sqrt{\frac{a}{2\pi}}\left[\frac{2\mathrm{E}(p;\theta)}{a-x} - \frac{\mathrm{F}(p;\theta)}{a} + \sqrt{\frac{2}{ax}}\right] \qquad 0<x<a$$

The functions $\mathrm{F}(p;\theta)$ and $\mathrm{E}(p;\theta)$ that occur in the above semiderivatives are incomplete elliptic integrals [Chapter 62] of argument $\theta = \arcsin(\sqrt{2x/(a+x)})$ and parameter $p = \sqrt{(a+x)/2a}$. The same functions occur in the expressions

14:10:12
$$\frac{d^{-1/2}}{dx^{-1/2}} b\sqrt{a^2-x^2} = \frac{b}{3}\sqrt{\frac{8a}{\pi}}\left[(a-x)\mathrm{F}(p;\theta) + 2x\mathrm{E}(p;\theta) - (2x-a)\sqrt{\frac{x}{2a}}\right] \qquad 0<x<a$$

and

14:10:13
$$\frac{d^{-1/2}}{dx^{-1/2}} \frac{1}{b\sqrt{a^2-x^2}} = \frac{1}{b}\sqrt{\frac{2}{\pi a}}\,\mathrm{F}(p;\theta) \qquad 0<x<a$$

for the semiintegrals with zero lower limit.

14:11 COMPLEX ARGUMENT

Replacement of the real argument x, in the semielliptic function or its reciprocal, by the complex argument $x + iy$ yields

14:11:1
$$b\sqrt{a^2-(x+iy)^2} = \frac{b}{\sqrt{2}}\sqrt{\sqrt{(a+x)^2+y^2}\sqrt{(a-x)^2+y^2}+a^2-x^2+y^2}$$
$$-\frac{ib\,\mathrm{sgn}(xy)}{\sqrt{2}}\sqrt{\sqrt{(a+x)^2+y^2}\sqrt{(a-x)^2+y^2}-a^2+x^2-y^2}$$

or

14:11:2
$$\frac{1}{b\sqrt{a^2-(x+iy)^2}} = \frac{1}{b\sqrt{2}}\sqrt{\frac{\sqrt{(a+x)^2+y^2}\sqrt{(a-x)^2+y^2}+a^2-x^2+y^2}{[(a+x)^2+y^2][(a-x)^2+y^2]}}$$
$$+\frac{i\,\mathrm{sgn}(xy)}{b\sqrt{2}}\sqrt{\frac{\sqrt{(a+x)^2+y^2}\sqrt{(a-x)^2+y^2}-a^2+x^2-y^2}{[(a+x)^2+y^2][(a-x)^2+y^2]}}$$

where the $\mathrm{sgn}(xy)$ function [Chapter 8] equals the sign of the xy product, or is zero if either x or y is zero.

When the argument is purely imaginary, equation 14:11:1 reduces to

14:11:3
$$b\sqrt{a^2-(iy)^2} = b\sqrt{a^2+y^2}$$

which corresponds to a semihyperbola [see Chapter 15].

14:12 GENERALIZATIONS

The ellipse, $b(a^2 - x^2)^{1/2}$, is the special $n = 2$ case of the more general function $b(a^n - x^n)^{1/n}$. When n is even and greater than 2, the curves obtained by plotting $b(a^n - x^n)^{1/n}$ versus x have been called *superellipses*: the cases $n = 2$, 4 and 8 are depicted in Figure 14-4. As n approaches infinity the superellipse shape becomes more and more rectangular.

FIG 14-4

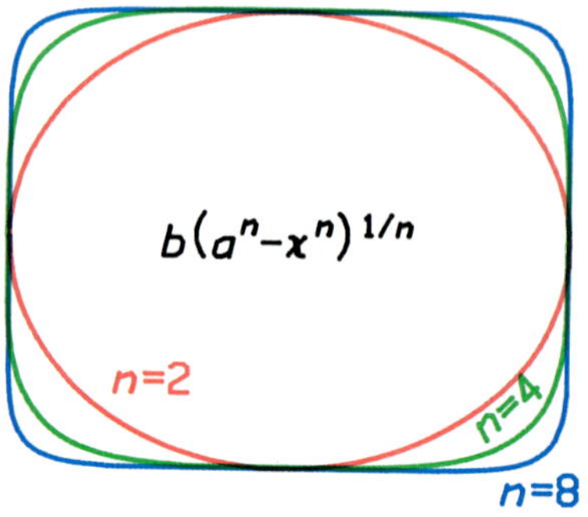

The reader is referred to Section 15:12 for a discussion of the circumstances under which the function $\sqrt{ax^2 + bx + c}$ is a generalization of the semiellipse $b\sqrt{a^2 - x^2}$.

14:13 COGNATE FUNCTIONS

The other *conic sections*, discussed in Chapters 12 and 15, resemble the semielliptic functions in many respects.

14:14 RELATED TOPICS

In this section we discuss geometric properties of the semicircle and the ellipse.

The area of the shaded segment of a semicircle of radius a [see Figure 14-5] is easily found from 14:10:3 to be

14:14:1 $$\begin{matrix}\text{area of}\\ \text{shaded segment}\end{matrix} = \frac{a^2}{2}\left[\frac{\pi}{2} + \arcsin\left(\frac{x}{a}\right)\right] + \frac{x}{2}\sqrt{a^2 - x^2}$$

FIG 14-5

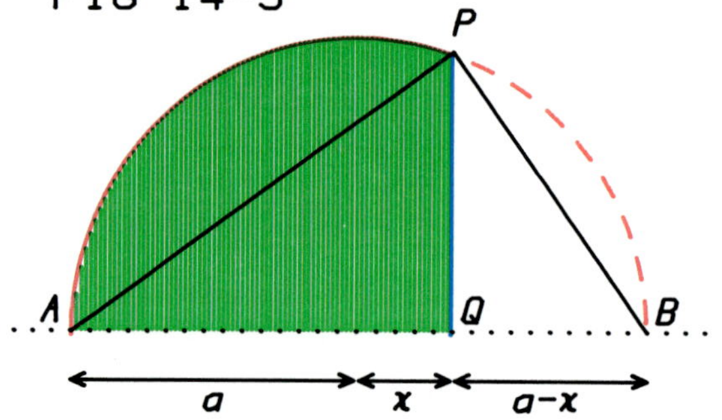

the total area of the semicircle being $\pi a^2/2$.

The curved perimeter of the shaded region has a length

14:14:2 $$\begin{matrix}\text{curved}\\ \text{perimeter}\end{matrix} = a\left[\frac{\pi}{2} + \arcsin\left(\frac{x}{a}\right)\right]$$

while the entire semicircumference has a length πa. An important property of a semicircle is that the angle APB in Figure 14-5 is a right angle for any point P on the semicircle. Thus the triangles APB, AQP and PQB are similar.

As explained in Section 14:3, the ellipse $\pm b\sqrt{a^2 - x^2}$ for $b < 1$ has foci F and F' located on the x-axis at $x = \pm a\sqrt{1 - b^2}$. The sum of the distances to each focus from any point P on the ellipse is a constant, $2a$. Moreover, the lines PF and PF' make equal angles with the ellipse itself at point P, as is illustrated in Figure 14-6. This means that any wave motion that originates at one focus F' and is reflected at the surface of the ellipse will be "focused" at the other focus F and all the reflected waves will arrive at F at the same time, though from a variety of directions [Figure 14-6]. This property of an ellipse is exploited in furnace design and in several acoustic and optical devices.

FIG 14-6

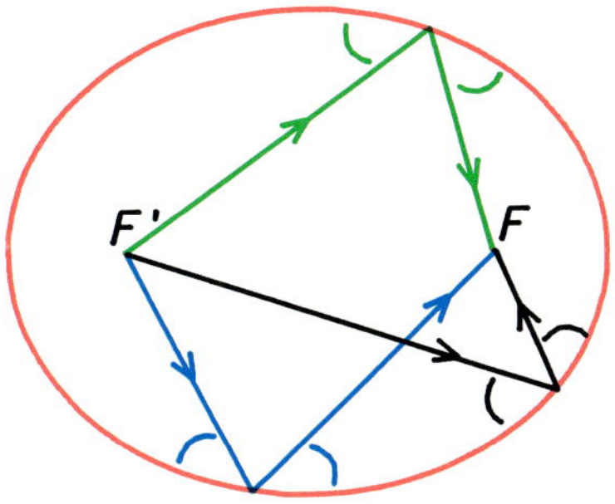

A segment of an ellipse, such as that shaded in Figure 14-7, has an area that can be determined with the help of integral 14:10:3 as

14:14:3 $$\begin{matrix}\text{area of}\\ \text{shaded segment}\end{matrix} = 2\int_{-a}^{x} f\,dt = a^2|b|\left[\frac{\pi}{2} + \arcsin\left(\frac{x}{a}\right) + \frac{x\sqrt{a^2 - x^2}}{a}\right] \qquad f = b\sqrt{a^2 - t^2}$$

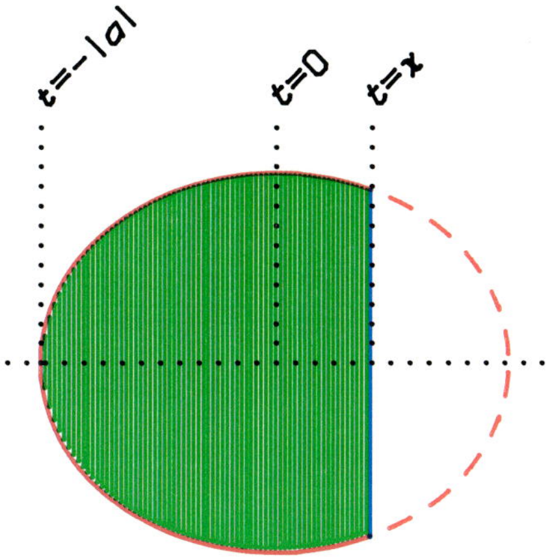

FIG 14-7

The entire ellipse thus has an area $\pi a^2|b|$. The curved perimeter of the shaded area in Figure 14-7 may be evaluated in terms of an incomplete elliptic integral by utilizing equations 14:10:1 and 14:10:13:

14:14:4 $$\begin{matrix}\text{curved}\\ \text{perimeter}\end{matrix} = 2\int_{-a}^{x} \sqrt{1 + \left(\frac{df}{dt}\right)^2}\,dt = 2a\left[E\left(\sqrt{1 - b^2};\arcsin\left(\frac{x}{a}\right)\right) - E\left(\sqrt{1 - b^2};\frac{-\pi}{2}\right)\right]$$

Because [see Chapter 62] $E(p;\pi/2) = -E(p;-\pi/2) = E(p)$ where $E(p)$ denotes the complete elliptic integral [Chapter 61] of p, the entire perimeter of an ellipse with semiaxes $|a|$ and $|ab|$ equals $4aE(\sqrt{1 - b^2})$.

CHAPTER 15

THE $b\sqrt{x^2+a}$ FUNCTION AND ITS RECIPROCAL

Although the properties of the $b\sqrt{x^2+a}$ function depend on the sign of the constant a, a graph of $\pm b\sqrt{x^2+a}$ is a curve called a *hyperbola* whether a is positive or negative, the sign of a determining only the orientation of the hyperbola.

15:1 NOTATION

The notations $b|(x^2+a)^{1/2}|$ and $|(x^2+a)^{-1/2}|/b$ are equivalent to $b\sqrt{x^2+a}$ and $1/(b\sqrt{x^2+a})$, respectively. Because the curve $\pm b\sqrt{x^2+a}$ is a hyperbola, the adjective *semihyperbolic* is appropriately applied to the $b\sqrt{x^2+a}$ function.

15:2 BEHAVIOR

When a is positive, $b\sqrt{x^2+a}$ and its reciprocal are defined for all values of the argument x and both functions share the sign of b. Figure 15-1 shows typical behaviors of these functions when b is positive.

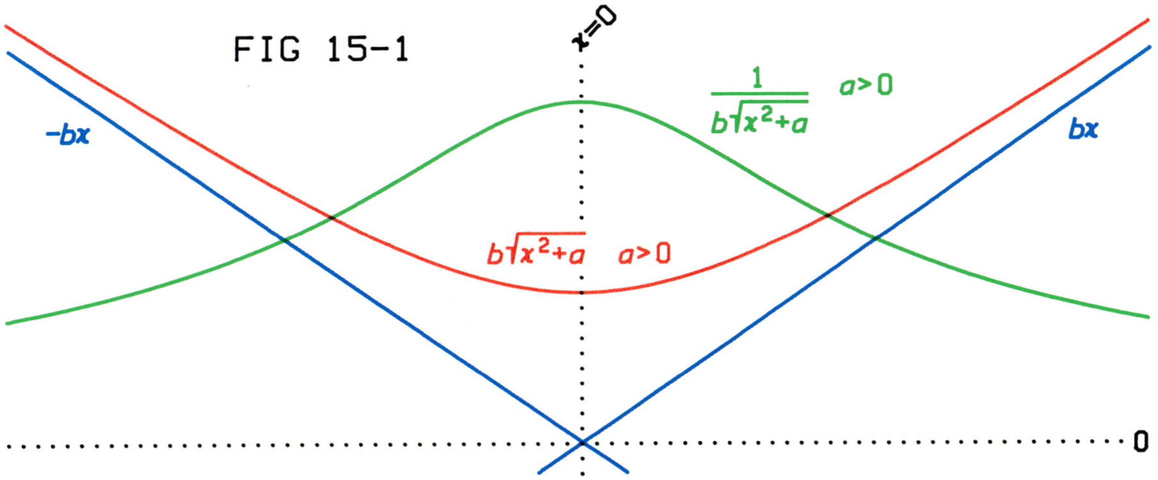

When a is negative, neither $b\sqrt{x^2+a}$ nor its reciprocal is defined in the interval $-\sqrt{-a} < x < \sqrt{-a}$ and therefore each of these functions has two branches, as illustrated in Figure 15-2 for positive b.

As $x \to \infty$, the function $b\sqrt{x^2+a}$ acquires values ever closer to bx. The linear function bx is said to be an *asymptote* of the $b\sqrt{x^2+a}$ function. Similarly, $-bx$ is an asymptote of $b\sqrt{x^2+a}$ as $x \to -\infty$. These asymptotes are shown in Figures 15-1 and 15-2.

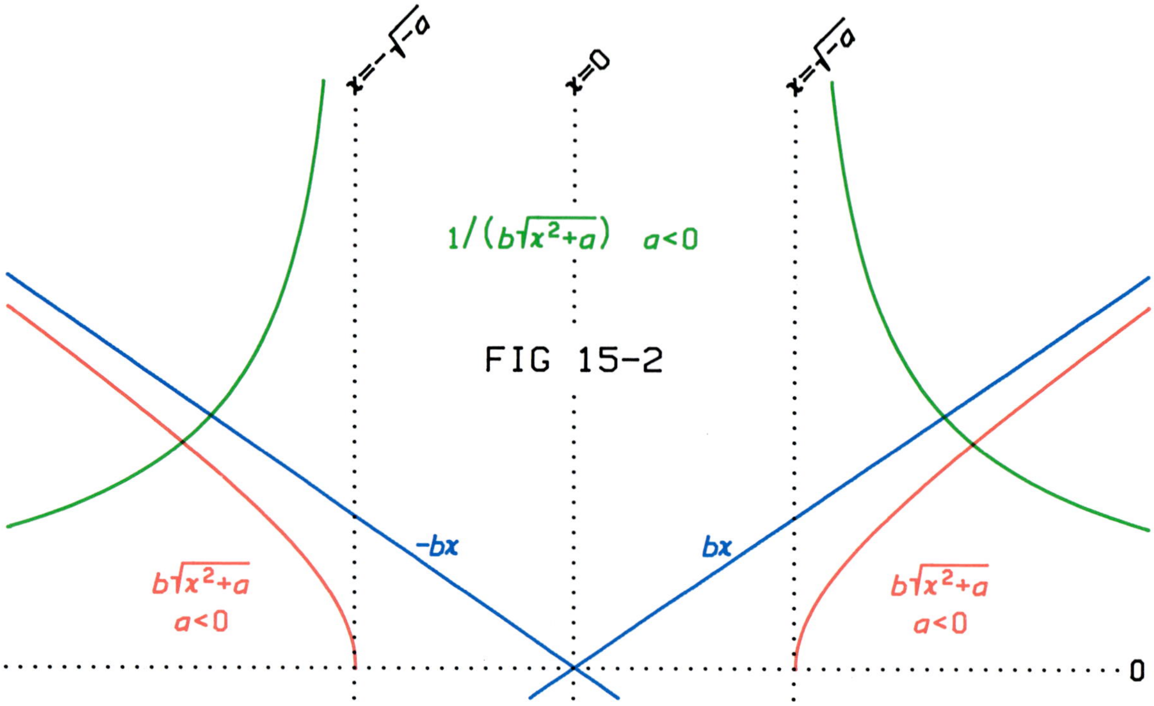

15:3 DEFINITIONS

Algebraic and arithmetic operations define $b\sqrt{x^2+a}$ and its reciprocal in a straightforward fashion.

A hyperbola may be defined geometrically as the locus of all points P such that the length of the line PF connecting P to a fixed point F (a *focus* of the hyperbola) obeys the relationship

15:3:1 $$\frac{PF}{PD} = \frac{\text{a constant}}{\text{greater than unity}} = \frac{\text{the } eccentricity}{\text{of the hyperbola}} = \varepsilon \qquad 1 < \varepsilon < \infty$$

with respect to the length PD of the shortest line from P to the straight line DD (called a *directrix* of the hyperbola). The eccentricity of the hyperbola $\pm b\sqrt{x^2+a}$ is $\sqrt{1+b^2}$. An illustration of relationship 15:3:1 is shown in Figure 15-3. Note that the hyperbola of Figure 15-3 has two branches and that there exists a second directrix $D'D'$ and a second focus F' symmetrically positioned with respect to a line parallel to both directrices and midway between them.

An alternative geometric definition of a hyperbola is as the locus of all points P such that

15:3:2 $$|PF - PF'| = \text{constant}$$

where F and F' are the foci.

15:4 SPECIAL CASES

When $a = 0$, the $b\sqrt{x^2+a}$ and $1/(b\sqrt{x^2+a})$ functions reduce to the linear bx and reciprocal linear $1/(bx)$ functions, respectively. When $b = 1$, the function $\pm b\sqrt{x^2+a}$ becomes $\pm\sqrt{x^2+a}$ and is termed a *rectangular hyperbola* or an *equilateral hyperbola* and has the interesting rotational properties now to be described.

FIG 15-3

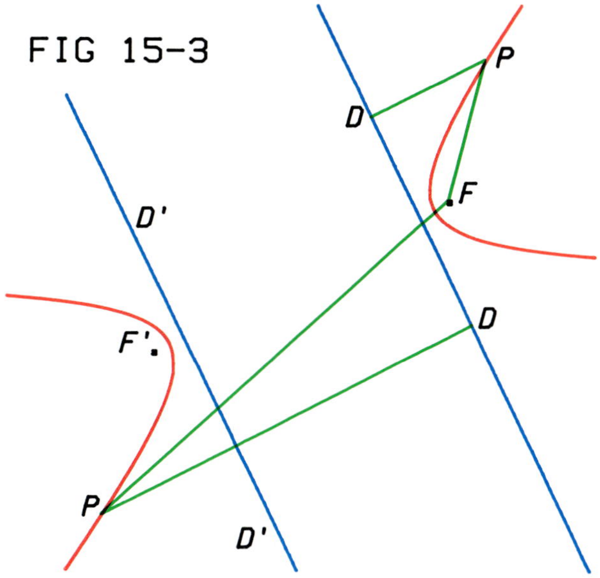

If a function g(x), interpreted as a cartesian graph of g(x) versus x, is rotated through a positive angle θ in a counterclockwise direction about the point (X,G) then a new function f = f(x) is created that is related to g(x) by the general *rotation formula*

15:4:1 $$f\cos(\theta) = [x - X]\sin(\theta) - [1 - \cos(\theta)]G + g(y) \qquad y = X + [x - X]\cos(\theta) + [f - G]\sin(\theta)$$

When rotation occurs about the origin, so that $X = G = 0$, the simpler relationship

15:4:2 $$f\cos(\theta) = x\sin(\theta) + g(x\cos(\theta) + f\sin(\theta))$$

holds. This formula will be used to rotate the function g(x) = $\pm\sqrt{x^2+a}$ about the origin through angles of $-\pi/4$ and $-\pi/2$.

The equation $f/\sqrt{2} = -x/\sqrt{2} \pm\sqrt{[(x/\sqrt{2} - f/\sqrt{2})]^2 + a}$, obtained by setting g($x$) = $\pm\sqrt{x^2+a}$ and θ = $-\pi/4$ in 15:4:2, simplifies to

15:4:3 $$f = \frac{a}{2x}$$

Similarly, the equation $0 = -x \pm \sqrt{(-f)^2 + a}$, which arises from setting g(x) to $\pm\sqrt{x^2+a}$ and θ to $-\pi/2$ in equation 15:4:2, reduces to

15:4:4 $$f = \pm\sqrt{x^2 - a}$$

Thus, the three functions $\pm\sqrt{x^2+a}$, $a/2x$ and $\pm\sqrt{x^2-a}$ all have exactly the same shape and differ only in their orientations, as illustrated in Figure 15-4. The function $-a/2x$ is also a member of the quartet of identically shaped rectangular hyperbolas.

It may be apparent from Figure 15-4 that the curve f(x) = $\pm\sqrt{x^2-a}$ can be obtained from g(x) = $\pm\sqrt{x^2+a}$ not only by rotation but also by reflection in their common asymptote. A very general result states that the reflection of the function g(x) in the straight line $bx + c$ generates the function f = f(x) where

15:4:5 $$[1 - b^2]f = 2bx + 2c - [1 + b^2]g(y) \qquad y = \frac{2bf + x[1 - b^2] - 2bc}{1 + b^2}$$

When $c = 0$ and $b = 1$, this general *reflection formula* collapses to

15:4:6 $$x = g(f(x))$$

showing that the reflection of a function g(x) in the straight line x generates the inverse function [see Section 0:3] of g(x). Hence, the rectangular hyperbola $\pm\sqrt{x^2-a}$ is the inverse function of the rectangular hyperbola $\pm\sqrt{x^2+a}$, and conversely.

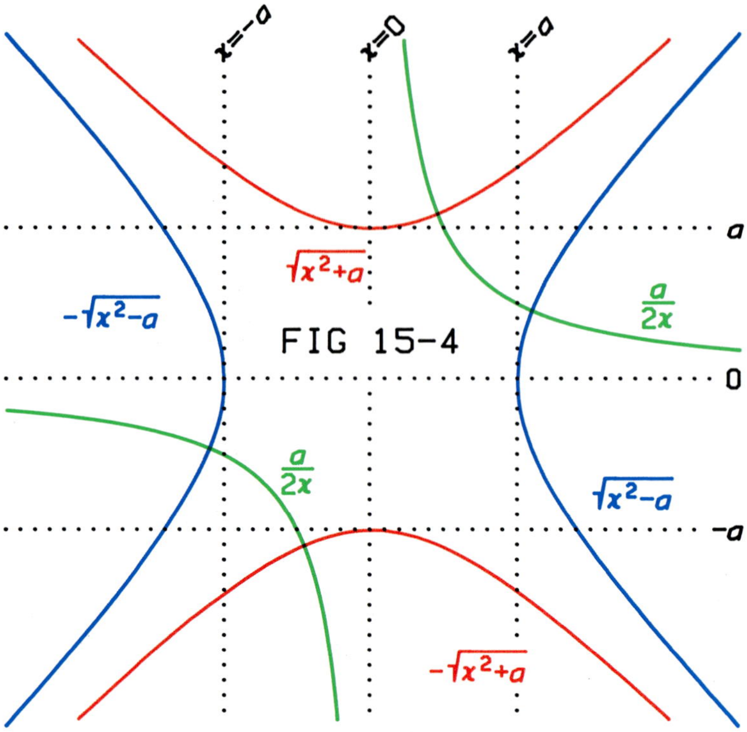

15:5 INTRARELATIONSHIPS

The f(x) = $b\sqrt{x^2+a}$ function is even and therefore satisfies the reflection formula

15:5:1 $$f(-x) = f(x)$$

The multiplication formula

15:5:2 $$f(\nu x) = b|\nu|\sqrt{x^2 + a/(\nu^2)}$$

shows that multiplication of the argument x of the semihyperbolic function by a constant generates another semihyperbolic function. The inverse function of $\pm b\sqrt{x^2+a}$ is also another hyperbola

15:5:3 $$f = \frac{\pm 1}{b}\sqrt{x^2 - ab^2} \qquad \text{where} \qquad x = \pm b\sqrt{f^2+a}$$

The two hyperbolas $\pm b\sqrt{x^2+a}$ and $\pm b\sqrt{x^2-a}$ are said to be *conjugate* to each other. As shown in Section 15:4, they share the same asymptotes.

15:6 EXPANSIONS

The semihyperbolic function may be expanded binomially as

15:6:1 $$b\sqrt{x^2+a} = b\sqrt{a}\left(1 + \frac{x^2}{2a} - \frac{x^4}{8a^2} + \frac{x^6}{16a^3} - \frac{5x^8}{128a^4} + \cdots\right) = b\sqrt{x}\sum_{j=0}^{\infty}\binom{1/2}{j}\left(\frac{x^2}{a}\right)^j \qquad 0 \le x^2 \le a$$

when x is small, and as

15:6:2 $$b\sqrt{x^2+a} = b\left(x + \frac{a}{2x} - \frac{a^2}{8x^3} + \frac{a^3}{16x^5} - \frac{5a^4}{128x^7} + \cdots\right) = bx\sum_{j=0}^{\infty}\binom{1/2}{j}\left(\frac{a}{x^2}\right)^j \qquad |x| \ge \sqrt{|a|}$$

when x is large. It is this latter equation that describes the approach of the semihyperbola to its linear asymptote.

Similarly, the reciprocal function is expansible in the two following ways:

15:6:3 $$\frac{1}{b\sqrt{x^2+a}} = \frac{1}{b\sqrt{a}}\left(1 - \frac{x^2}{2a} + \frac{3x^4}{8a^2} - \frac{5x^6}{16a^3} + \frac{35x^8}{128a^4} - \cdots\right) = \frac{1}{b\sqrt{a}}\sum_{j=0}^{\infty}\binom{-1/2}{j}\left(\frac{x^2}{a}\right)^j \qquad 0 \le x^2 < a$$

or

15:6:4 $$\frac{1}{b\sqrt{x^2+a}} = \frac{1}{b|x|}\left(1 - \frac{a}{2x^2} + \frac{3a^2}{8x^4} - \frac{5a^3}{16x^6} + \frac{35a^4}{128x^8} - \cdots\right) = \frac{1}{b|x|}\sum_{j=0}^{\infty}\binom{-1/2}{j}\left(\frac{a}{x^2}\right)^j \qquad |x| > \sqrt{|a|}$$

15:7 PARTICULAR VALUES

	$x = -\infty$	$x = -\sqrt{\frac{1}{b^2} - a}$ $ab^2 \le 1$	$x = -\sqrt{\lvert a\rvert}$	$x = 0$	$x = \sqrt{\lvert a\rvert}$	$x = \sqrt{\frac{1}{b^2} - a}$ $ab^2 \le 1$	$x = \infty$
$b\sqrt{x^2+a}$ $a < 0$	$\infty\,\mathrm{sgn}(b)$	$\mathrm{sgn}(b)$	0	undef	0	$\mathrm{sgn}(b)$	$\infty\,\mathrm{sgn}(b)$
$b\sqrt{x^2+a}$ $a > 0$	$\infty\,\mathrm{sgn}(b)$	$\mathrm{sgn}(b)$	$b\sqrt{2a}$	$b\sqrt{a}$	$b\sqrt{2a}$	$\mathrm{sgn}(b)$	$\infty\,\mathrm{sgn}(b)$
$\frac{1}{b\sqrt{x^2+a}}$ $a < 0$	0	$\mathrm{sgn}(b)$	$\infty\,\mathrm{sgn}(b)$	undef	$\infty\,\mathrm{sgn}(b)$	$\mathrm{sgn}(b)$	0
$\frac{1}{b\sqrt{x^2+a}}$ $a > 0$	0	$\mathrm{sgn}(b)$	$\frac{1}{b\sqrt{2a}}$	$\frac{1}{b\sqrt{a}}$	$\frac{1}{b\sqrt{2a}}$	$\mathrm{sgn}(b)$	0

15:8 NUMERICAL VALUES

These are readily calculated with a programmable calculator or other computing device.

15:9 APPROXIMATE VALUES

The following approximations are obtained by truncating expansions 15:6:1, 15:6:3, 15:6:2 and 15:6:4, respectively.

15:9:1 $$b\sqrt{x^2+a} \simeq \frac{b(2a+x^2)}{2\sqrt{a}} \qquad \text{8-bit precision} \qquad |x| \le 0.4\sqrt{a} \qquad a > 0$$

15:9:2 $$\frac{1}{b\sqrt{x^2+a}} \simeq \frac{2a - x^2}{2b\sqrt{a^3}} \qquad \text{8-bit precision} \qquad |x| \le 0.3\sqrt{a} \qquad a > 0$$

15:9:3 $$b\sqrt{x^2+a} \simeq bx + \frac{ab}{2x} \qquad \text{8-bit precision} \qquad |x| \ge 2.5\sqrt{|a|}$$

15:9:4 $$\frac{1}{b\sqrt{x^2+a}} \simeq \frac{2x^2 - a}{2b|x^3|} \qquad \text{8-bit precision} \qquad |x| \ge 3.3\sqrt{|a|}$$

15:10 OPERATIONS OF THE CALCULUS

Differentiation gives

15:10:1
$$\frac{d}{dx}b\sqrt{x^2+a} = \frac{bx}{\sqrt{x^2+a}}$$

and

15:10:2
$$\frac{d}{dx}\frac{1}{b\sqrt{x^2+a}} = \frac{-x}{b\sqrt{(x^2+a)^3}}$$

The simplest formulas for indefinite integration must employ differing limits according to the sign of a. Thus:

15:10:3
$$\int_0^x b\sqrt{t^2+a}\,dt = \frac{bx}{2}\sqrt{x^2+a} + \frac{ab}{2}\operatorname{arsinh}\left(\frac{x}{\sqrt{a}}\right) \qquad a>0$$

15:10:4
$$\int_{\sqrt{-a}}^x b\sqrt{t^2+a}\,dt = \frac{bx}{2}\sqrt{x^2+a} + \frac{ab}{2}\operatorname{arcosh}\left(\frac{x}{\sqrt{-a}}\right) \qquad a<0$$

15:10:5
$$\int_0^x \frac{dt}{b\sqrt{t^2+a}} = \frac{1}{b}\operatorname{arsinh}\left(\frac{x}{\sqrt{a}}\right) \qquad a>0$$

15:10:6
$$\int_{\sqrt{-a}}^x \frac{dt}{b\sqrt{t^2+a}} = \frac{1}{b}\operatorname{arcosh}\left(\frac{x}{\sqrt{-a}}\right) \qquad a<0$$

15:10:7
$$\int_x^\infty \frac{dt}{t\sqrt{t^2+a}} = \frac{1}{\sqrt{a}}\operatorname{arcsch}\left(\frac{x}{\sqrt{a}}\right) \qquad a>0 \qquad x>0$$

15:10:8
$$\int_{\sqrt{-a}}^x \frac{dt}{t\sqrt{t^2+a}} = \frac{1}{\sqrt{-a}}\operatorname{arcsec}\left(\frac{x}{\sqrt{-a}}\right) \qquad a<0$$

where the functions arsinh, arcosh, arcsch and arcsec are discussed in Chapters 31 and 35. These six indefinite integrals are simple members of the class $\int t^n(\sqrt{t^2+a})^m\,dt$ where $n = 0, \pm1, \pm2, \ldots$ and $m = \pm1, \pm3, \pm5, \ldots$: a long list of such integrals will be found in Gradshteyn and Ryzhik [Section 2.27].

15:11 COMPLEX ARGUMENT

Replacement of the real argument x in $b\sqrt{x^2+a}$ by the complex argument $x+iy$ yields

15:11:1
$$b\sqrt{(x+iy)^2+a} = \frac{b}{\sqrt{2}}\sqrt{x^2-y^2+a+\sqrt{y^4+2(x^2-a)y^2+(x^2+a)^2}} + \frac{ib\,\operatorname{sgn}(xy)}{\sqrt{2}}\sqrt{y^2-x^2-a+\sqrt{y^4+2(x^2-a)y^2+(x^2+a)^2}}$$

where sgn(xy) equals the sign of the xy product, or is zero if either x or y is zero. Similarly:

15:11:2
$$\frac{1}{b\sqrt{(x+iy)^2+a}} = \frac{1}{b\sqrt{2}}\sqrt{\frac{x^2-y^2+a+\sqrt{y^4+2(x^2-a)y^2+(x^2+a)^2}}{y^4+2(x^2-a)y^2+(x^2+a)^2}} - \frac{i\,\operatorname{sgn}(xy)}{b\sqrt{2}}\sqrt{\frac{y^2-x^2+a+\sqrt{y^4+2(x^2-a)y^2+(x^2+a)^2}}{y^4+2(x^2-a)y^2+(x^2+a)^2}}$$

If the argument in 15:11:1 is purely imaginary, the function reduces to a semiellipse [Chapter 14]

15:11:3 $$b\sqrt{(iy)^2+a} = b\sqrt{(\sqrt{a})^2 - y^2}$$

provided that a is positive.

15:12 GENERALIZATIONS

The function $b\sqrt{(x-\alpha)^2+a}$ differs from $b\sqrt{x^2+a}$ only by being translated a distance α along the x-axis; it is, therefore, a semihyperbola centered at $x = \alpha$ rather than at $x = 0$. The function $\sqrt{ax^2+bx+c}$ may similarly represent a translated semihyperbola; however, this so-called *root-quadratic function* may represent any *conic section* (or degenerate instances of the conic sections) according to the magnitudes of the coefficients a, b and c. The situation is summarized in Table 15.12.1.

Table 15.12.1

	$b^2 > 4ac$	$b^2 = 4ac$	$b^2 < 4ac$
$a > 0$ $a \neq 1$	$\sqrt{a}\sqrt{\left(x+\frac{b}{2a}\right)^2 - \left(\frac{\sqrt{b^2-4ac}}{2a}\right)^2}$ Semihyperbola of two branches centered at $x = -b/2a$ [see this chapter]	$\pm(\sqrt{a}x + \sqrt{c})$ Two straight lines [see Chapter 7]	$\sqrt{a}\sqrt{\left(x+\frac{b}{2a}\right)^2 + \frac{4ac-b^2}{4a^2}}$ Semihyperbola of one branch centered at $x = -b/2a$ [see this chapter]
$a = 1$	$\sqrt{\left(x+\frac{b}{2}\right)^2 - \left(\frac{b^2}{4} - c\right)}$ Rectangular semihyperbola [see Section 15:4] of two branches centered at $x = -b/2$	$\pm(x + \sqrt{c})$ Two straight lines of slope unity	$\sqrt{\left(x+\frac{b}{2}\right)^2 + \left(c - \frac{b^2}{4}\right)}$ Rectangular semihyperbola of one branch [see Section 15:4] centered at $x = -b/2$
$a = 0$	$\sqrt{bx+c}$ Semiparabola [see Chapter 12]	$\sqrt{c}$ Constant [see Chapter 1]	Not possible
$a = -1$	$\sqrt{\frac{b^2}{4} + c - \left(x - \frac{b}{2}\right)^2}$ Semicircle of radius $\sqrt{(b/2)^2 + c}$ centered at $x = b/2$ [see Section 14:4]	Not possible	Not possible
$a < 0$ $a \neq -1$	$\sqrt{-a}\sqrt{\left(\frac{\sqrt{b^2-4ac}}{-2a}\right)^2 - \left(x+\frac{b}{2a}\right)^2}$ Semiellipse with semiaxes of $\sqrt{b^2-4ac}/(-2a)$ and $\sqrt{c-(b^2/4a)}$ centered at $x = -b/2a$ [see Chapter 14]	Not possible	Not possible

15:13 COGNATE FUNCTIONS

The functions $b(x^n + a)^{1/n}$ for $n = 3, 4, 5, \ldots$ are hyperbola-like, especially if n is even. The straight line bx is an asymptote in all cases, as is $-bx$ when n is even.

15:14 SPECIAL TOPICS

In this section we treat geometrical properties of the hyperbola.

The area enclosed by the rightmost branch of the function $\pm b\sqrt{t^2+a}$ and an x ordinate, for $0 < \sqrt{-a} < x$, is shown shaded in Figure 15-5 and may be evaluated directly from the integral 15:10:4 as

15:14:1 $$\begin{matrix}\text{shaded}\\ \text{area}\end{matrix} = 2\int_{\sqrt{-a}}^{x} |b\sqrt{t^2+a}|dt = |b|x\sqrt{x^2+a} + a|b|\text{arcosh}\left(\frac{x}{\sqrt{-a}}\right) \qquad a<0$$

The curved perimeter of the shaded area has a length given by

15:14:2 $$\begin{matrix}\text{curved}\\ \text{perimeter}\end{matrix} = 2\int_{\sqrt{-a}}^{x}\sqrt{1+\left(\frac{d}{dt}b\sqrt{t^2+a}\right)^2}\,dt = 2\sqrt{-a-ab^2}\left[\frac{b^2}{1+b^2}F\left(\frac{1}{\sqrt{1+b^2}};\phi\right) - E\left(\frac{1}{\sqrt{1+b^2}};\phi\right) + \frac{x}{\sqrt{-a}}\sin(\phi)\right] \qquad \phi = \arctan\left(\frac{\sqrt{x^2+a}\sqrt{1+b^2}}{b\sqrt{-a}}\right) \qquad a<0$$

where F and E denote incomplete elliptic integrals [Chapter 62].

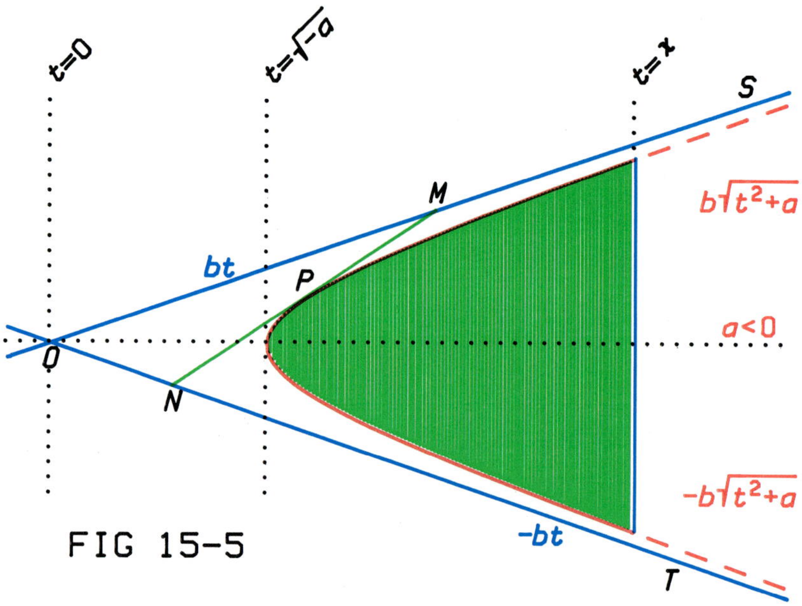

FIG 15-5

Also shown in Figure 15-5 are the asymptotes OS and OT of the hyperbola and the tangent MN to the hyperbolic branch at an arbitrary point P (M and N being the points at which the tangent meets the asymptotes, as depicted in Figure 15-5). Two remarkable properties of the hyperbola are that P bisects the line MN, so that $MP = PN$, and that the area of the triangle MNO equals $|ab|$ and is independent of the position of P along the hyperbola.

CHAPTER 16

THE QUADRATIC FUNCTION $ax^2 + bx + c$ AND ITS RECIPROCAL

Next to the constant function [Chapter 1] and the linear function [Chapter 7], the quadratic function is the simplest of the large class of polynomial functions; others are discussed in the next eight chapters.

16:1 NOTATION

The parameters a, b and c are called the *coefficients* of the quadratic function $ax^2 + bx + c$. The value of the quantity

16:1:1 $$\Delta = 4ac - b^2$$

called the *discriminant* of the quadratic function, strongly influences the properties of the function and its reciprocal. Some authors define the discriminant to have the opposite sign to that in 16:1:1.

16:2 BEHAVIOR

Graphs of $ax^2 + bx + c$ are shown in Figure 16-1 for $a > 0$ and $a < 0$. The function is defined for all x and, for positive a, assumes values that are not less than $c - (b^2/4a)$. Whether the value zero is part of the range of the quadratic function is determined by the sign of its discriminant.

The behavior of the reciprocal quadratic function is profoundly affected by the sign of Δ. Figure 16-2 shows, for $a > 0$, typical graphs of $1/(ax^2 + bx + c)$ when the discriminant is positive, zero or negative.

16:3 DEFINITIONS

The quadratic function is defined by the arithmetic and algebraic operations signified in the expression $ax^2 + bx + c$. The reciprocal quadratic function is defined as the quotient of unity by the quadratic function.

A reciprocal quadratic function may also be defined as the difference of two reciprocal linear functions [see

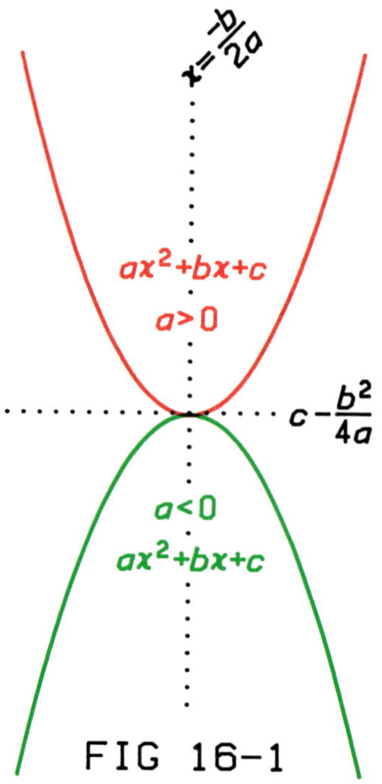

FIG 16-1

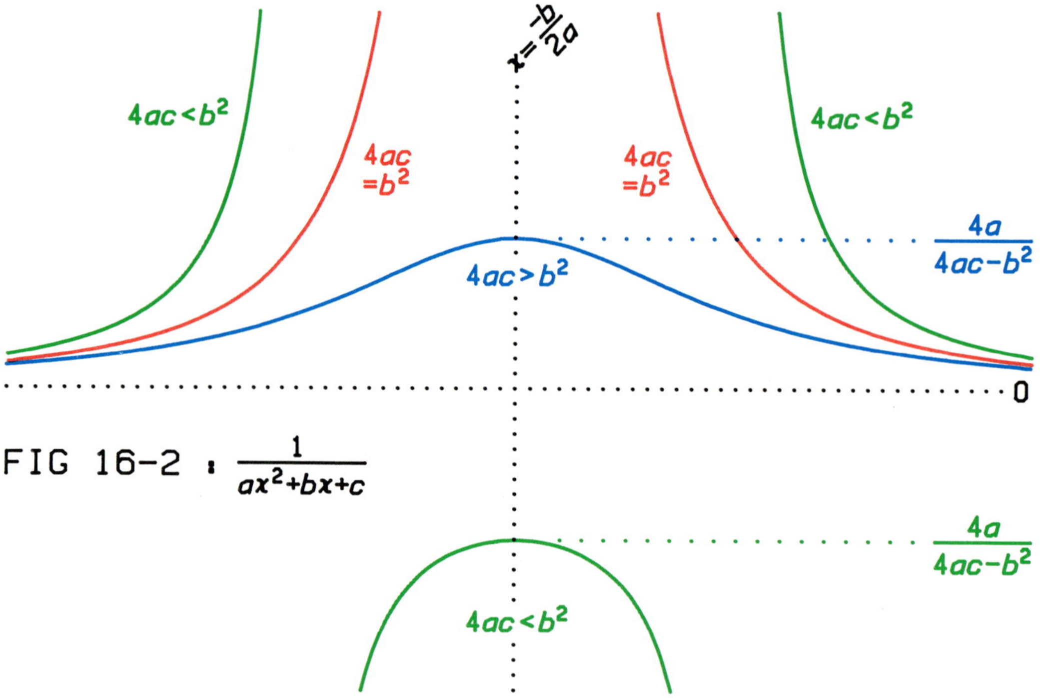

FIG 16-2 : $\dfrac{1}{ax^2+bx+c}$

Chapter 7] that share the same b coefficient:

16:3:1
$$\frac{1}{bx + C} - \frac{1}{bx + c} = \frac{1}{\left(\dfrac{b^2}{c - C}\right)x^2 + b\left(\dfrac{c + C}{c - C}\right)x + \left(\dfrac{Cc}{c - C}\right)}$$

Not every reciprocal quadratic function may be defined in this way, however, since the decomposition

16:3:2
$$\frac{1}{ax^2 + bx + c} = \frac{1}{\sqrt{b^2 - 4ac}\,x + 2c + \dfrac{b}{2a}(\sqrt{b^2 - 4ac} - b)} - \frac{1}{\sqrt{b^2 - 4ac}\,x - 2c + \dfrac{b}{2a}(\sqrt{b^2 - 4ac} + b)}$$

is possible (using real arithmetic) only if the discriminant $4ac - b^2$ is negative.

16:4 SPECIAL CASES

The linear function [Chapter 7] is the special $a = 0$ case of the quadratic function; further specialization to the constant function [Chapter 1] occurs when $a = b = 0$. When $c = b^2/4a$, the quadratic function reduces to $\{\sqrt{a}x + (b/2\sqrt{a})\}^2$, a square function [Chapter 11].

Similarly, the reciprocal quadratic function reduces to a reciprocal linear function when $a = 0$, to a constant when $a = b = 0$ and to a power function of exponent -2 when $c = b^2/4a$. The decomposition

16:4:1
$$\frac{1}{ax^2 + bx} = \frac{1}{bx} - \frac{a}{abx + b^2}$$

is always possible when $c = 0$. Equation 16:4:1 is a special case of a more general *partial fraction* decomposition that provides a powerful tool for integration [see Section 17:13].

16:5 INTRARELATIONSHIPS

Unless $A = \mp a$, the sum or difference of two quadratic functions

16:5:1
$$(ax^2 + bx + c) \pm (Ax^2 + Bx + C) = (a \pm A)x^2 + (b \pm B)x + c \pm C$$

is another quadratic function, while their product

16:5:2
$$(ax^2 + bx + c)(Ax^2 + Bx + C) = aAx^4 + (aB + Ab)x^3 + (aC + bB + Ac)x^2 + (bC + Bc)x + cC$$

is a quartic function [Chapter 17]. The quotient of two quadratic functions can be simplified, provided the discriminant of the denominator is negative, via the following formula into a constant and two reciprocal linear functions:

16:5:3
$$\frac{Ax^2 + Bx + C}{ax^2 + bx + c} = \frac{A}{a} + \frac{Ar_1^2 - Br_1 + C}{\sqrt{b^2 - 4ac}\,(x - r_1)} - \frac{Ar_2^2 + Br_2 + C}{\sqrt{b^2 - 4ac}\,(x - r_2)} \qquad b^2 > 4ac$$

where $r_1 = (-b + \sqrt{b^2 - 4ac})/2a$ and $r_2 = (-b - \sqrt{b^2 - 4ac})/2a$.

The inverse [Section 0:3] of the quadratic function is a constant combined with the square-root function [Chapter 12]

16:5:4
$$f = \frac{-b}{2a} \pm \frac{1}{\sqrt{a}}\sqrt{x + \frac{b^2}{4a} - c} \qquad x = af^2 + bf + c$$

16:6 EXPANSIONS

For small x we have

16:6:1
$$\frac{1}{ax^2+bx+c} = \frac{1}{c} - \frac{bx}{c^2} + \frac{(b^2-ac)x^2}{c^3} - \frac{(b^3-2abc)x^3}{c^4} + \frac{(b^4-3ab^2c+a^2c^2)x^4}{c^5} - \cdots$$
$$= \frac{1}{c}\sum_{j=0}^{\infty}\left[\frac{-x(ax+b)}{c}\right]^j \qquad \left|\frac{ax^2+bx}{c}\right| < 1$$

while for large x

16:6:2
$$\frac{1}{ax^2+bx+c} = \frac{1}{ax^2} - \frac{b}{a^2x^3} + \frac{b^2-ac}{a^3x^4} - \frac{b^3-2abc}{a^4x^5} + \frac{b^4-3ab^2c+a^2c^2}{a^5x^6} - \cdots$$
$$= \frac{1}{ax^2}\sum_{j=0}^{\infty}\left(\frac{-bx-c}{ax^2}\right)^j \qquad \left|\frac{bx+c}{ax^2}\right| < 1$$

holds.

16:7 PARTICULAR VALUES

	$x = -\infty$	$x = r_2 = \dfrac{-b-\sqrt{b^2-4ac}}{2a}$ when $b^2 > 4ac$	$x = \dfrac{-b}{2a}$	$x = 0$	$x = r_1 = \dfrac{-b+\sqrt{b^2-4ac}}{2a}$ when $b^2 > 4ac$	$x = \infty$
$ax^2 + bx + c$	$\infty\,\mathrm{sgn}(a)$	0	$(4ac - b^2)/4a = \begin{cases}\text{min if } a > 0\\ \text{max if } a < 0\end{cases}$	c	0	$\infty\,\mathrm{sgn}(a)$
$\dfrac{1}{ax^2+bx+c}$	0	$\pm\infty$	$4a/(4ac - b^2) = \begin{cases}\text{max if } a > 0\\ \text{min if } a < 0\end{cases}$	$\dfrac{1}{c}$	$\pm\infty$	0

The values of x that make the quadratic function equal to zero are known as the *zeros* of the quadratic function or as the *roots* of the quadratic equation $ax^2 + bx + c = 0$. Note that "root" is used here in a sense different from that of Chapter 13. There are two zeros

16:7:1
$$r_1 = \frac{-b+\sqrt{b^2-4ac}}{2a} \quad \text{and} \quad r_2 = \frac{-b-\sqrt{b^2-4ac}}{2a}$$

if $4ac < b^2$, one zero

16:7:2
$$r = \frac{-b}{2a}$$

if $4ac = b^2$ and no (real) zeros if $4ac > b^2$. The reciprocal quadratic function becomes infinite at the zeros of the quadratic function, the so-called *poles* of the reciprocal quadratic function.

16:8 NUMERICAL VALUES

These are easily calculated from the definitions.

16:9 APPROXIMATE VALUES

16:9:1
$$\frac{1}{ax^2+bx+c} \simeq \frac{c-bx}{c^2} \qquad \text{8-bit precision} \qquad |x| < \frac{|c|}{16\sqrt{|b^2-ac+abx|}}$$

16:9:2
$$\frac{1}{ax^2+bx+c} \simeq \frac{ax-b}{a^2x^3} \qquad \text{8-bit precision} \qquad |x| > \frac{16\sqrt{\left|b^2-ac+\dfrac{bc}{x}\right|}}{|a|}$$

16:10 OPERATIONS OF THE CALCULUS

Differentiation produces

16:10:1
$$\frac{\mathrm{d}}{\mathrm{d}x}(ax^2+bx+c) = 2ax+b$$

and

16:10:2
$$\frac{\mathrm{d}}{\mathrm{d}x}\left(\frac{1}{ax^2+bx+c}\right) = \frac{-2ax-b}{(ax^2+bx+c)^2}$$

while integration results in

16:10:3
$$\int_0^x (at^2+bt+c)\mathrm{d}t = \frac{2ax^3+3bx^2+6cx}{6}$$

and

16:10:4
$$\int_{-b/2a}^x \frac{\mathrm{d}t}{at^2+bt+c} = \begin{cases} \dfrac{2}{\sqrt{\Delta}}\arctan\left(\dfrac{2ax+b}{\sqrt{\Delta}}\right) & \Delta = 4ac-b^2>0 \\ \dfrac{-2}{\sqrt{-\Delta}}\operatorname{artanh}\left(\dfrac{2ax+b}{\sqrt{-\Delta}}\right) & \Delta<0 \quad x < \dfrac{\sqrt{-\Delta}-b}{2a} \end{cases}$$

The 16:10:4 integral is infinite if $\Delta = 0$ but

16:10:5
$$\int_x^\infty \frac{\mathrm{d}t}{at^2+bt+c} = \frac{2}{2ax+b} \qquad b^2 = 4ac$$

An important integral is

16:10:6
$$\int_0^x \frac{t\mathrm{d}t}{at^2+bt+c} = \begin{cases} \dfrac{1}{2a}\ln\left(\dfrac{ax^2+bx+c}{c}\right) - \dfrac{b}{a\sqrt{\Delta}}\arctan\left(\dfrac{x\sqrt{\Delta}}{bx+2c}\right) & \Delta = 4ac-b^2>0 \\ \dfrac{1}{2a}\ln\left(\dfrac{ax^2+bx+c}{c}\right) - \dfrac{b}{a\sqrt{-\Delta}}\operatorname{artanh}\left(\dfrac{x\sqrt{-\Delta}}{bx+2c}\right) & \Delta<0 \end{cases}$$

and others of the general form $\int t^n(at^2+bt+c)^m\mathrm{d}t$, where n and m are integers, will be found in Gradshteyn and Ryzhik [Section 2.17].

16:11 COMPLEX ARGUMENT

Replacement of the real argument x of the quadratic and reciprocal quadratic functions by the complex argument $x + iy$ leads to

16:11:1
$$a(x + iy)^2 + b(x + iy) + c = (ax^2 - ay^2 + bx + c) + i(2axy + by)$$

16:11:2
$$\frac{1}{a(x + iy)^2 + b(x + iy) + c} = \frac{(ax^2 - ay^2 + bx + c) - i(2axy + by)}{(ax^2 + ay^2 + bx + c)^2 + y^2(b^2 - 4ac)}$$

16:12 GENERALIZATIONS

The quadratic function is a special case of the functions treated in Chapter 17. A quadratic function is a member of all the function families addressed in Chapters 18–24.

16:13 COGNATE FUNCTIONS

The *root-quadratic function* $\sqrt{ax^2 + bx + c}$ and its reciprocal are functions of some importance. It is demonstrated in Section 15:12 that the root-quadratic function is equivalent to various conic sections depending on the magnitudes of the a, b and c coefficients. Some integrals involving the reciprocal root-quadratic function are

16:13:1
$$\int_{-b/2a}^{x} \frac{dt}{\sqrt{at^2 + bt + c}} = \begin{cases} \dfrac{1}{\sqrt{a}} \operatorname{arsinh}\left(\dfrac{2ax + b}{\sqrt{\Delta}}\right) & a > 0 \quad \Delta = 4ac - b^2 > 0 \\ \dfrac{-1}{\sqrt{-a}} \arcsin\left(\dfrac{2ax + b}{\sqrt{-\Delta}}\right) & a < 0 \quad \Delta < 0 \quad \left|\dfrac{2ax + b}{\sqrt{-\Delta}}\right| \le 1 \end{cases}$$

16:13:2
$$\int_{x_0}^{x} \frac{dt}{\sqrt{at^2 + bt + c}} = \frac{1}{\sqrt{a}} \operatorname{arcosh}\left(\frac{2ax + b}{\sqrt{-\Delta}}\right) \qquad a > 0 \quad \Delta < 0 \quad x_0 = \frac{\sqrt{-\Delta} - b}{2a}$$

16:13:3
$$\int_{-2c/b}^{x} \frac{dt}{t\sqrt{at^2 + bt + c}} = \begin{cases} \dfrac{-1}{\sqrt{c}} \operatorname{arsinh}\left(\dfrac{bx + 2c}{x\sqrt{\Delta}}\right) & c > 0 \quad \Delta = 4ac - b^2 > 0 \\ \dfrac{1}{\sqrt{-c}} \arcsin\left(\dfrac{bx + 2c}{x\sqrt{-\Delta}}\right) & c < 0 \quad \left|\dfrac{bx + 2c}{x\sqrt{-\Delta}}\right| \le 1 \quad \Delta < 0 \end{cases}$$

and

16:13:4
$$\int_{x_0}^{x} \frac{dt}{t\sqrt{at^2 + bt + c}} = \frac{-1}{\sqrt{c}} \operatorname{arcosh}\left(\frac{bx + 2c}{x\sqrt{-\Delta}}\right) \qquad c > 0 \quad \Delta < 0 \quad x_0 = \frac{2c}{\sqrt{-\Delta} - b}$$

Others are given in Gradshteyn and Ryzhik [Section 2.26].

16:14 RELATED TOPICS

A quadratic function describes the trajectory of a body that travels at constant speed in one direction, while experiencing a constant acceleration (or force) at right angles to that direction. Thus, ignoring air resistance, the path of a projectile launched at $x = x_0$ with initial velocity v at an angle θ to the horizontal travels along the arc illustrated

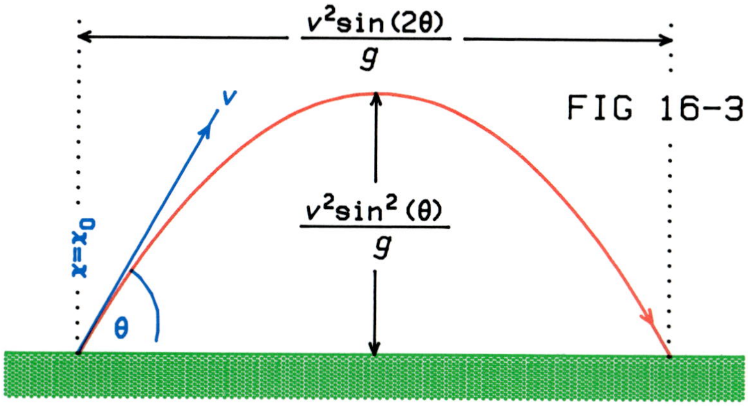

in Figure 16-3 and is airborne for a time $2v\,\sin(\theta)/g$ where g is the gravitational acceleration [see Appendix B:6]. The equation of the trajectory is

16:14:1 $$\text{height} = -\frac{g\sec^2(\theta)}{2v^2}x^2 + \left(\frac{gx_0\sec^2(\theta)}{v^2} + \tan(\theta)\right)x - \left(\frac{gx_0^2\sec^2(\theta)}{2v^2} + x_0\tan(\theta)\right)$$

CHAPTER 17

THE CUBIC FUNCTION $x^3 + ax^2 + bx + c$ AND HIGHER POLYNOMIALS

The function $a_nx^n + a_{n-1}x^{n-1} + \cdots + a_1x + a_0$, in which the a multipliers are specified constants, is known as a *polynomial function of degree n*. This chapter treats such functions generally and pays particular attention to the $n = 3$ case with, for added simplicity, a_3 taken as unity.

17:1 NOTATION

A polynomial function is often simply called a *polynomial;* sometimes the term *integral function* is used to describe the same family.

We shall use the general notation

17:1:1 $$p_n(x) = a_nx^n + a_{n-1}x^{n-1} + \cdots + a_1x + a_0 = \sum_{j=0}^{n} a_jx^j \qquad a_n \neq 0$$

to denote a polynomial of argument x and of degree n. The constants $a_n, a_{n-1}, \ldots, a_1, a_0$ are called the *coefficients* of the polynomial function; there are $n + 1$ such coefficients, some of which may equal zero.

Special names are given to polynomials of degrees 5, 4, 3, 2, 1 or 0; they are known as *quintic functions, quartic functions, cubic functions, quadratic functions, linear functions* or *constant functions*.

The quantity

17:1:2 $$D = Q^2 - P^3 \qquad \text{where} \qquad P = \frac{a^2}{9} - \frac{b}{3} \qquad \text{and} \qquad Q = \frac{ab}{6} - \frac{c}{2} - \frac{a^3}{27}$$

is known as the *discriminant* of the cubic function $x^3 + ax^2 + bx + c$. Its value affects the number of (real) zeros of the function as explained in Section 17:7.

17:2 BEHAVIOR

Polynomial functions are defined for all values $-\infty < x < \infty$ of their arguments. Those of odd degree acquire all real values, whereas polynomials of even degree have a restricted range.

Figure 17-1 shows the three possible shapes that the cubic function $x^3 + ax^2 + bx + c$ may exhibit according to the sign of $a^2 - 3b^2$. Notice that an inflection [see Section 0:7] occurs at $x = -a/3$ in all three cases. A local maximum and minimum occur on either side of this inflection if, but only if, $a^2 > 3b$. (i.e, if P, given in 17:1:2,

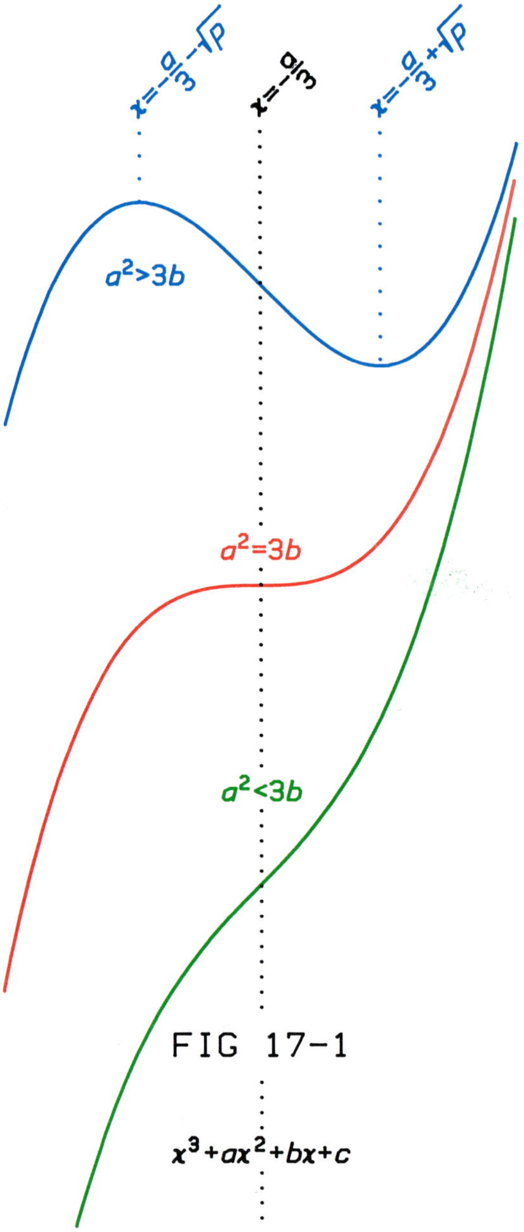

FIG 17-1

x^3+ax^2+bx+c

is positive). The cubic function always acquires the value zero at least once; that is, there is at least one real root to the equation $p_3(x) = 0$. Section 17:7 discusses the circumstances under which a second or third real root exists.

The behavior of a polynomial of higher degree is determined in a complicated fashion by its coefficients and by its degree. Generally, a graph of a polynomial $p_n(x)$ exhibits a number of inflections (at the zeros of the polynomial of degree $n - 2$ that is the second derivative of $p_n(x)$—see Section 0:7), and local maxima and minima (at the zeros of the polynomial $dp_n(x)/dx$ of degree $n - 1$). Some general rules concerning the number of zeros of $p_n(x)$ (i.e., the number of real roots of the equation $p_n(x) = 0$) are given in Table 17.2.1.

17:3 DEFINITIONS

The operations of addition, multiplication and raising to a power fully define a polynomial. Written as a concatenation

Table 17.2.1

	$n = 3, 5, 7, \ldots$	$n = 2, 4, 6, \ldots$ $a_n > 0$	$n = 2, 4, 6, \ldots$ $a_n < 0$
Number N_i of inflections	$1 \le N_i \le n - 2$	$0 \le N_i \le n - 2$	$0 \le N_i \le n - 2$
Number N_m of minima	$0 \le N_m \le \frac{n-1}{2}$	$1 \le N_m \le \frac{n}{2}$	$N_m = N_M - 1$
Number N_M of maxima	$N_M = N_m$	$N_M = N_m - 1$	$1 \le N_M \le \frac{n}{2}$
Number N_r of zeros	$1 \le N_r \le n$	$0 \le N_r \le n$	$0 \le N_r \le n$

17:3:1 $$p_n(x) = ((\cdots((a_nx + a_{n-1})x + a_{n-2})x + \cdots + a_2)x + a_1)x + a_0$$

the operations of addition and multiplication suffice.

The product of n linear functions [Chapter 7] with real coefficients

17:3:2 $$\prod_{j=1}^{n}(b_jx + c_j)$$

defines a polynomial of degree n, but not every polynomial function of degree n may be defined in this way. However, every polynomial of degree n can be defined as a product of l linear functions and $(n - l)/2$ quadratic functions [Chapter 16]:

17:3:3 $$p_n(x) = \sum_{j=0}^{n} a_jx^j = \prod_{j=1}^{l}(b_jx + c_j)\prod_{k=1}^{K}(A_kx^2 + B_kx + C_k) \qquad K = \frac{n}{2} - \frac{l}{2}$$

where $l = N_r$, the number of (real) zeros of $p_n(x)$. The identification of a suitable set of linear and quadratic coefficients $\{b_j, c_j, A_k, B_k, C_k\}$ from a given set of polynomial coefficients a_j will generally require iteration (or inspiration!).

17:4 SPECIAL CASES

The functions $p_n(x)$ for $n = 0$, 1 and 2 are treated in Chapters 1, 7 and 16. If the polynomial coefficients obey the relationship

17:4:1 $$\frac{a_{n-j}}{a_n} = \binom{n}{j}\left(\frac{a_0}{a_n}\right)^{j/n} \qquad \text{for} \qquad j = 1, 2, 3, \ldots, n - 1$$

then the polynomial Σa_jx^j reduces to the power function $(bx + c)^n$ of Chapter 11 with $b = (a_n)^{1/n}$ and $c = (a_0)^{1/n}$

Each of Chapters 18–24 deals with a family of polynomials with special coefficients.

17:5 INTRARELATIONSHIPS

The cubic function $p_3(x) = x^3 + ax^2 + bx + c$ obeys the reflection formula

17:5:1 $$p_3\left(\frac{-a}{3} - x\right) = \frac{4a^3}{27} - \frac{2ab}{3} + 2c - p_3\left(\frac{-a}{3} + x\right)$$

but no comparable formula exists for higher polynomials.

The polynomial function obeys the simple argument-multiplication formula

17:5:2 $$p_n(vx) = \sum_{j=0}^{n}(v^ja_j)x^j \qquad \text{where} \qquad p_n(x) = \sum_{j=0}^{n} a_jx^j$$

showing that multiplying (or, of course, dividing) the argument of any polynomial by a constant generates another polynomial function of the same degree.

The addition or subtraction of two polynomials of degrees n and m yields a polynomial of degree equal to the larger of n and m (unless $m = n$ and the highest degree coefficients cancel); thus:

17:5:3
$$\sum_{j=0}^{n} a_j x^j \pm \sum_{j=0}^{m} A_j x^j = \sum_{j=0}^{J} (a_j \pm A_j) x^j \begin{cases} J = n & n \geq m \\ J = m & n \leq m \end{cases}$$

while their multiplication

17:5:4
$$\left(\sum_{j=0}^{n} a_j x^j\right)\left(\sum_{j=0}^{m} A_j x^j\right) = \sum_{j=0}^{n+m} \alpha_j x^j \qquad \text{where} \qquad \alpha_j = \sum_{k=0}^{j} a_k A_{j-k}$$

yields a polynomial function of degree $n + m$. The quotient of two polynomials is termed a *rational function* and is discussed in Section 17:13.

The converse of 17:5:4—the factoring of a polynomial $p_n(x)$ into two polynomials of strictly lower degrees—is always possible for $n \geq 3$ because of 17:3:3. Thus, a cubic function may be factored into the product of a linear function and a quadratic function

17:5:5
$$x^3 + ax^2 + bx + c = (x - r)[x^2 + (a + r)x - c/r]$$

where r is a (real) zero of the cubic function. (Further factorization of the quadratic function into a pair of linear functions is possible if its discriminant [see Chapter 16] is nonpositive, i.e., if it has two real zeros). This factorization of higher polynomials into ones of lower degree is the key to the decomposition of reciprocal polynomials [see Section 17:13].

17:6 EXPANSIONS

A polynomial function may be expanded as a continued fraction

17:6:1
$$\sum_{j=0}^{n} a_j x^j = \frac{a_0}{1-}\;\frac{a_1 x}{a_0 + a_1 x-}\;\frac{a_0 a_2 x}{a_1 + a_2 x-}\;\frac{a_1 a_3 x}{a_2 + a_3 x-}\cdots\frac{a_{n-2} a_n x}{a_{n-1} + a_n x}$$

A polynomial function may be written as the product of n linear functions with complex coefficients

17:6:2
$$p_n(x) = a_n \prod_{j=1}^{n} (x - \rho_j)$$

Each ρ_j is known as a *complex zero* of $p_n(x)$; those that are not real occur as *complex conjugate pairs* [see Section 17:11]. If ρ_j and ρ_k are such a conjugate pair then the product $(x - \rho_j)(x - \rho_k)$ is the quadratic function $x^2 - (\rho_j + \rho_k)x + \rho_j\rho_k$ with real coefficients inasmuch as $\rho_j + \rho_k = (\lambda + i\mu) + (\lambda - i\mu) = 2\lambda$ and $\rho_j\rho_k = (\lambda + i\mu)(\lambda - i\mu) = \lambda^2 + \mu^2$. Real and/or complex zeros may be "repeated" in expansion 17:6:2, that is, ρ_j and ρ_k may be identical for $j \neq k$. The zeros ρ_j are related to the a_j coefficients by the formulas

17:6:3
$$\sum_{j=1}^{n} \rho_j = \rho_1 + \rho_2 + \cdots + \rho_{n-1} + \rho_n = \frac{-a_{n-1}}{a_n}$$

17:6:4
$$\sum_{j=1}^{n} \sum_{k=j+1}^{n} \rho_j \rho_k = \rho_1\rho_2 + \rho_1\rho_3 + \rho_1\rho_4 + \cdots + \rho_{n-2}\rho_{n-1} + \rho_{n-1}\rho_n = \frac{a_{n-2}}{a_n}$$

17:6:5
$$\sum_{j=1}^{n} \sum_{k=j+1}^{n} \sum_{l=k+1}^{n} \rho_j \rho_k \rho_l = \rho_1\rho_2\rho_3 + \rho_1\rho_2\rho_4 + \rho_1\rho_2\rho_5 + \cdots + \rho_{n-3}\rho_{n-2}\rho_{n-1} + \rho_{n-2}\rho_{n-1}\rho_n = \frac{-a_{n-3}}{a_n}$$

$$\vdots$$

17:6:6
$$\rho_1\rho_2\cdots\rho_{n-1} + \rho_1\rho_2\cdots\rho_{n-2}\rho_n + \rho_1\rho_2\cdots\rho_{n-3}\rho_{n-1}\rho_n + \cdots + \rho_1\rho_3\rho_4\cdots\rho_n + \rho_2\rho_3\cdots\rho_n = \prod_{j=1}^{n}\rho_j \sum_{k=1}^{n}\frac{1}{\rho_k} = \frac{-(-1)^n a_1}{a_n}$$

17:6:7
$$\rho_1\rho_2\rho_3\cdots\rho_n = \prod_{j=1}^{n}\rho_j = \frac{(-1)^n a_0}{a_n}$$

The functions occurring on the left-hand sides of equations 17:6:3–17:6:7 are the so-called *elementary symmetric functions* of $\rho_1, \rho_2, \rho_3, \ldots, \rho_n$ [see Korn and Korn, page 11]. An advantage in expressing the polynomial as in 17:6:2 is that its reciprocal can be decomposed thereby into a sum of n terms, known as *partial fractions:*

17:6:8
$$\frac{1}{\mathrm{p}_n(x)} = \frac{1}{a_n}\sum_{j=1}^{n}\frac{C_j}{x-\rho_j} \qquad \text{where} \qquad C_j = \prod_{\substack{k=1\\k\neq j}}^{n}\frac{1}{\rho_j-\rho_k}$$

Alternatively, C_j may be equated to $a_n/\mathrm{p}_n'(\rho_j)$ where $\mathrm{p}_n'(x)$ denotes the derivative of $\mathrm{p}_n(x)$ [see 17:10:1 or 17:10:4]. This decomposition must be modified if any of the zeros of $\mathrm{p}_n(x)$ are repeated.

17:7 PARTICULAR VALUES

For the general polynomial function of degree n, as defined in equation 17:1:1, we have

	$x = -\infty$	$x = -1$	$x = 0$	$x = 1$	$x = \infty$
$\mathrm{p}_n(x)$	$(-1)^n\,\infty\,\mathrm{sgn}(a_n)$	$\sum_{j=0}^{n}(-1)^j a_j$	a_0	$\sum_{j=0}^{n}a_j$	$\infty\,\mathrm{sgn}(a_n)$

The polynomial displays an inflection at any argument x_i that satisfies the equation

17:7:1
$$\sum_{j=0}^{n-2}(j+1)(j+2)a_{j+2}x_i^j = 0 \qquad \mathrm{p}_n(x_i) = \text{inflection}$$

Minima and maxima occur at arguments x_m and x_M that satisfy the conditions

17:7:2
$$\sum_{j=0}^{n-1}(j+1)a_{j+1}x_m^j = 0 \quad \text{and} \quad \sum_{j=0}^{n-2}(j+1)(j+2)a_{j+2}x_m^j > 0 \qquad \mathrm{p}_n(x_m) = \text{minimum}$$

17:7:3
$$\sum_{j=0}^{n-1}(j+1)a_{j+1}x_M^j = 0 \quad \text{and} \quad \sum_{j=0}^{n-2}(j+1)(j+2)a_{j+2}x_M^j < 0 \qquad \mathrm{p}_n(x_M) = \text{maximum}$$

Equation 17:7:1 is equivalent to the condition $\mathrm{d}^2\mathrm{p}_n(x_i)/\mathrm{d}x^2 = 0$, while equations 17:7:2 and 17:7:3 state that $\mathrm{dp}_n(x_m)/\mathrm{d}x = 0$, $\mathrm{d}^2\mathrm{p}_n(x_m)/\mathrm{d}x^2 > 0$ and $\mathrm{dp}_n(x_M)/\mathrm{d}x = 0$, $\mathrm{d}^2\mathrm{p}_n(x_M)/\mathrm{d}x^2 < 0$, respectively [see Section 0:7]. Zeros of $\mathrm{p}_n(x)$ correspond to the roots of the equation $\mathrm{p}_n(x) = 0$, that is, r is a zero if

17:7:4
$$\sum_{j=0}^{n}a_j r^j = 0 \qquad \mathrm{p}_n(r) = 0$$

The numbers of inflections, minima, maxima and zeros depend not only on the degree of the polynomial but on the values of the coefficients; see Table 17.2.1.

For the cubic function $x^3 + ax^2 + bx + c$, an inflection occurs at $x = -a/3$ [see Figure 17-1]. Provided a^2 exceeds $3b$, a maximum and a minimum occur at arguments $(-\sqrt{a^2-3b}-a)/3$ and $(\sqrt{a^2-3b}-a)/3$, respectively. This cubic function has the single zero

17:7:5
$$r = (Q+\sqrt{D})^{1/3} + (Q-\sqrt{D})^{1/3} - \frac{a}{3} \qquad D > 0$$

if the discriminant D [defined in equation 17:1:2] is positive, and three distinct zeros

17:7:6 $$\left\{\begin{aligned} r_1 &= 2\sqrt{|P|}\cos(\phi) - \frac{a}{3} \\ r_2 &= -2\sqrt{|P|}\cos\left(\phi + \frac{\pi}{3}\right) - \frac{a}{3} \\ r_3 &= 2\sqrt{|P|}\cos\left(\phi - \frac{\pi}{3}\right) - \frac{a}{3} \end{aligned}\right\} \quad \text{where} \quad \phi = 3\arccos\left(\frac{Q}{\sqrt{|P|^3}}\right) \qquad D < 0$$

if the discriminant is negative. The constants P and Q are defined in equation 17:1:2. When the discriminant equals zero, r_2 and r_3 of 17:7:6 become identical and there are then two zeros:

17:7:7 $$\left\{\begin{aligned} r_1 &= 2Q - \frac{a}{3} \\ r_2 &= -Q - \frac{a}{3} \end{aligned}\right\} \qquad D = 0$$

but these coalesce to a single value if $a^2 = 3b$.

Formulas exist [see Abramowitz and Stegun, pages 17–18, for example] that permit the zeros of a quartic function to be determined algebraically, but no such formulas are available for quintics or higher polynomials. Numerical methods must therefore be used in these cases and, in practice, iterative numerical methods are applied for these higher polynomials and, in fact, for quartics and even cubics as well.

A popular method for determining the zeros of a polynomial function $p_n(x) = \Sigma a_j x^j$ requires that each zero first be located approximately (often from a crudely drawn graph of the polynomial versus x). The approximation is then improved iteratively. The accompanying algorithm may be used to carry out the iteration; it requires values of the $n + 1$ coefficients of the polynomial as well as a value of the degree n and r_0, the approximate zero. Improvement ceases when two successive approximants to the sought zero differ by less than one part in 10^8. The algorithm may fail to locate a multiple zero [see Section 0:2] and will fail to locate a zero at $x = 0$. It is based on *Newton's formula* that gives

17:7:8 $$r_0 - f(r_0) \Big/ \frac{df}{dx}(r_0)$$

as an improved approximation to a zero of the $f(x)$ function (which actually need not be restricted to a polynomial). For more information on *Newton's method* (as this iterative technique is called) and on alternative techniques for finding zeros, see Bronshtein and Semendyayev [pages 164–171]. Section A:5 presents a program for finding zeros of an arbitrary function f that also avoids the need to evaluate the derivative of f. If the input parameters n, a_n, a_{n-1}, ..., a_0 of the algorithm are replaced by $n - 1$, na_n, $(n - 1)a_{n-1}$, ..., a_1, then this algorithm will yield a zero of $dp_n(x)/dx$, that is, a minimum or a maximum of $p_n(x)$.

```
Input n >>>>>>>>
        Set j = n + 1
    (1) Replace j by j − 1
        Output j
j (as cue) <<<<<<<
Input a_j >>>>>>>>
        If j ≠ 0 go to (1)
Input r_0 >>>>>>>>
        Set r = r_0
    (2) Set j = n
        Set f = a_j
```

Storage needed: $n + 6$ registers are required for n, r, j, f, g and the coefficients a_n, a_{n-1}, ..., a_1, a_0.

```
            (3) Replace j by j − 1
                Replace f by fr + a_j
                If j ≠ 0 go to (3)
                Set j = n
                Set g = ja_j
            (4) Replace j by j − 1
                Replace g by gr + ja_j
                If j ≠ 0 go to (4)
                Replace r by r[1 − (f/g)]
                If |f| ≥ 10^-8 go to (1)
                Output r
r <<<<<<<<<<
```

Test values: Input $n = 4$, $a_4 = 2$, $a_3 = 5$, $a_2 = 2$, $a_1 = -9$ and $a_0 = -18$. There are two zeros: $r_1 = -2$ and $r_2 = \frac{3}{2}$. Any input $r_0 < 0.5$ will give r_1; any input $r_0 > 0.6$ will give r_2.

17:8 NUMERICAL VALUES

These are easily calculated for the polynomial function, especially with the aid of the concatenation formula 17:3:1. The simple algorithm

```
Input n >>>>>>>>
                Set j = n + 1
            (1) Replace j by j − 1
                Output j
j (as cue) <<<<<<
Input a_j >>>>>>>>
                If j ≠ 0 go to (1)
Input x >>>>>>>>
                Set j = n
                Set f = a_j
            (2) Replace j by j − 1
                Replace f by fx + a_j
                If j ≠ 0 go to (2)
                Output f
f = p_n(x) <<<<<<
```

Storage needed: $n + 5$ registers are required for n, j, x, f and a_n, a_{n-1}, $\ldots$, a_1, a_0.

Test values: Input $n = 4$, $a_4 = 2$, $a_3 = 5$, $a_2 = 2$, $a_1 = -9$, $a_0 = -18$, $x = 2$ or -1.
Output: $p_4(2) = 44$, $p_4(-1) = -10$

uses this formula.

17:9 APPROXIMATIONS

One may construct a polynomial $p_m(x)$ to approximate a polynomial $p_n(x)$ of higher degree over a specified interval $x_0 \le x \le x_1$ by a process known as *economization*. The procedure is fully described in Section 22:14 and a versatile algorithm is presented there for calculating the economized coefficients. The degree m of the economized polynomial is determined by the acceptable error ε of the approximation

17:9:1 $$|p_n(x) - p_m(x)| \le \varepsilon \qquad x_0 \le x \le x_1$$

The economization procedure has five steps, each of which would usually be carried out numerically.

First, one replaces the argument of $p_n(x)$ by a new argument $y = (2x - x_1 - x_0)/(x_1 - x_0)$

17:9:2 $$p_n(x) = \sum_{j=0}^{n} a_j x^j = \sum_{j=0}^{n} b_j y^j \qquad \text{where} \qquad b_j = \left(\frac{x_1 - x_0}{x_1 + x_0}\right)^j \sum_{k=j}^{n} \binom{k}{j} \left(\frac{x_1 + x_0}{2}\right)^k a_k$$

so that the region of interest comes to lie in the standardized range $-1 \le y \le 1$.

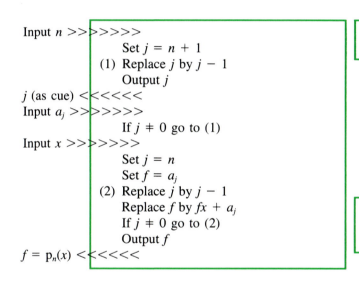

Second, the given polynomial of degree n is replaced by a sum of Chebyshev polynomials $T_j(y)$, where j takes values up to n. The replacement uses equation 11:6:5 and leads to

17:9:3 $$p_n(x) = \sum_{j=0}^{n} b_j y^j = \sum_{j=0}^{n} c_j T_j(y) \qquad \text{where} \qquad c_j = \sum_{k=0}^{n} b_k \gamma_j^{(k)}$$

and where $\gamma_j^{(k)}$ is the coefficient of $T_j(y)$ in the expansion of y^k. About half of these coefficients equal zero.

Now, because for $-1 \le y \le 1$ the absolute value of $T_j(y)$ never exceeds unity, it follows that

17:9:4 $$\Sigma c_j T_j(y) \le \Sigma |c_j||T_j(y)| \le \Sigma |c_j| \qquad -1 \le y \le 1$$

for any set of j values. Therefore, if m is the smallest integer that satisfies

17:9:5 $$\sum_{j=m+1}^{n} |c_j| < \varepsilon$$

it follows that, to within the acceptable error specified by 17:9:1:

17:9:6 $$p_n(x) \simeq \sum_{j=0}^{m} c_j T_j(y) = p_m(x)$$

The third step in the economization procedure consists of finding the smallest integer m that satisfies 17:9:5.

The fourth and fifth steps achieve the reverse of the second and first steps, using the economized polynomial. Thus, in the fourth step, one enumerates the coefficients B_j defined by

17:9:7 $$p_m(x) = \sum_{j=0}^{m} c_j T_j(y) = \sum_{j=0}^{m} B_j y^j \qquad \text{where} \qquad B_j = \sum_{k=0}^{m} c_j \beta_j^{(k)}$$

where $\beta_j^{(k)}$ denotes the coefficient of y^j in the polynomial expansion of $T_k(y)$ [see Section 22:5]. Finally, in the fifth step, one determines

17:9:8 $$p_m(x) = \sum_{j=0}^{m} B_j y^j = \sum_{j=0}^{m} A_j x^j \qquad \text{where} \qquad A_j = \left(\frac{-2}{x_1 + x_0}\right)^j \sum_{k=j}^{m} (-1)^k \left(\frac{x_1 + x_0}{x_1 - x_0}\right)^k B_k$$

and evaluates the coefficients A_j of the economized polynomial.

17:10 OPERATIONS OF THE CALCULUS

The derivative, indefinite integral and differintegral of the polynomial function:

17:10:1 $$\frac{d}{dx}\sum_{j=0}^{n} a_j x^j = \sum_{j=1}^{n} j a_j x^{j-1}$$

17:10:2 $$\int_0^x \sum_{j=0}^{n} a_j t^j dt = \sum_{j=0}^{n} \frac{a_j}{j+1} x^{j+1}$$

and

17:10:3 $$\frac{d^\nu}{dx^\nu}\sum_{j=0}^{n} a_j x^j = \sum_{j=0}^{n} \frac{j! a_j}{\Gamma(j - \nu + 1)} x^{j-\nu}$$

follow directly from the results of Section 11:10. The cubic function provides an exemplary subject for the three operations, yielding $3x^2 + 2ax + b$ on differentiation, $(3x^4 + 4ax^3 + 6bx^2 + 12cx)/12$ on integration and $(((3x/(3 - \nu) + a)2x/(2 - \nu) + b)x/(1 - \nu) + c)/x^\nu\Gamma(1 - \nu)$ on differintegration.

Operations of the calculus may also be carried out on the polynomial function and on its reciprocal via the expansion 17:6:2. Thus:

17:10:4 $$\frac{d}{dx} p_n(x) = p_n(x) \sum_{j=1}^{n} \frac{1}{x - \rho_j} = p_n'(x)$$

17:10:5
$$\frac{d}{dx}\frac{1}{p_n(x)} = \frac{-1}{p_n(x)}\sum_{j=1}^{n}\frac{1}{x-\rho_j} = \frac{-p_n'(x)}{p_n^2(x)}$$

where $p_n'(x)$ is thereby defined, and

17:10:6
$$\int_0^x \frac{dt}{p_n(t)} = \frac{1}{a_n}\sum_{j=1}^{n} C_j \ln\left(\left|\frac{\rho_j - x}{\rho_j}\right|\right) = \sum_{j=1}^{n}\frac{\ln\left(\left|\frac{\rho_j - x}{\rho_j}\right|\right)}{p_n'(x_j)} \qquad \text{where} \qquad C_j = \sum_{\substack{k=1\\k\neq j}}^{n}\frac{1}{\rho_j - \rho_k}$$

as in 17:6:8. Similarly, we have the Laplace inversion formula [see Section 26:14]:

17:10:7
$$\frac{1}{p_n(s)} = \int_0^\infty \sum_{j=1}^{n}\frac{\exp(\rho_j t)}{p_n'(\rho_j)}\exp(-st)dt$$

sometimes known as the *Heaviside expansion theorem*. Recall that, in general, the ρ_j need not be real. The four equations above are useful only if each ρ_j is a simple zero. This limitation does not affect the alternative methods of handling reciprocal polynomial functions that are discussed in Section 17:13.

Historically, it was the need to integrate the *root-cubic* $\sqrt{p_3(x)}$ and *root-quartic* functions $\sqrt{p_4(x)}$, as well as their reciprocals, that led to the discovery of elliptic integrals [Chapters 61 and 62]. Section 62:14 may be consulted for further details and for explicit formulas for $\int dt/\sqrt{\pm p_3(t)}$.

17:11 COMPLEX ARGUMENT

If the argument of the polynomial function is complex, the coefficients remaining real, then

17:11:1
$$p_n(x+iy) = \sum_{j=0}^{n} a_j x^j - \sum_{j=2}^{n}\binom{j}{2}a_j x^{j-2}y^2 + \sum_{j=4}^{n}\binom{j}{4}a_j x^{j-4}y^4 - \cdots$$
$$+ i\left[\sum_{j=1}^{n} j a_j x^{j-1}y - \sum_{j=3}^{n}\binom{j}{3}a_j x^{j-3}y^3 + \sum_{j=5}^{n}\binom{j}{5}a_j x^{j-5}y^5 - \cdots\right]$$

In the case of the cubic:

17:11:2 $\quad (x+iy)^3 + a(x+iy)^2 + b(x+iy) + c = x^3 + ax^2 + bx + c - y^2(3x+a) + iy(3x^2 - y^2 + 2ax + b)$

A zero ρ_j of the polynomial $p_n(x) = \Sigma a_j x^j$ may have real and imaginary components even when the argument x and the coefficients a_j are real. If $\rho_j = \lambda + i\mu$ where λ and μ are real, then, invariably, there will also be a zero $\rho_k = \lambda - i\mu$. The complex numbers $(\lambda + i\mu)$ and $(\lambda - i\mu)$ are said to *conjugate* to each other.

17:12 GENERALIZATIONS

The power series discussed in Section 11:14 may in many respects be regarded as polynomials of infinite degree.

Sums of negative or fractional powers can often be expressed as polynomial functions of altered argument, possibly with a power multiplier. For example:

17:12:1
$$a_n x^{-n} + a_{n-1}x^{-n+1} + a_{n-2}x^{-n+2} + \cdots + a_0 = p_n\left(\frac{1}{x}\right)$$

and

17:12:2
$$a_n x^{(2n-1)/4} + a_{n-1}x^{(2n-3)/4} + a_{n-2}x^{(2n-5)/4} + \cdots + a_0 x^{-1/4} = x^{-1/4}p_n(\sqrt{x})$$

17:13 COGNATE FUNCTIONS

The ratio of two polynomial functions is termed a *rational function*. We use the special notation

17:13:1
$$\mathrm{R}_n^m(x) = \mathrm{p}_m(x)/\mathrm{p}_n(x) = \sum_{j=0}^{m} A_j x^j \Big/ \sum_{j=0}^{n} a_j x^j$$

If $m \geq n$ the rational function is said to be *improper* and may be resolved into a polynomial and a *proper rational function* (i.e., one whose numeratorial degree is less than its denominatorial degree) by the procedure of algebraic division [see Bronshtein and Semendyayev, pages 150–151].

To simplify a proper rational function, one first factorizes its denominatorial polynomial into linear and quadratic functions, thus:

17:13:2
$$\mathrm{p}_n(x) = \sum_{j=0}^{n} a_j x^j = a_n \prod_{j=1}^{l} (x - r_j) \prod_{j=1}^{J} (x^2 - p_j x + q_j) \qquad J = \frac{n}{2} - \frac{l}{2}$$

where each r_j is a zero of $\mathrm{p}_n(x)$ and the discriminant $4q_j - p_j^2$ of each quadratic factor is positive. Next, the r_j values are inspected to see if there are duplicates. Similarly, the p_j, q_j values are searched to see if there are any repeated complex zeros. If all zeros of $\mathrm{p}_n(x)$ are *simple* (i.e., of multiplicity one) then the proper rational function may be decomposed into $(n + l)/2$ so-called *partial fractions* as follows:

17:13:3
$$\frac{\sum_{j=0}^{m} A_j x^j}{\sum_{j=0}^{n} a_j x^j} = \sum_{j=1}^{l} \frac{\alpha_j}{x - r_j} + \sum_{j=1}^{J} \frac{\beta_j x + \gamma_j}{x^2 - p_j x + q_j} \qquad J = \frac{n}{2} - \frac{l}{2} \qquad m < n$$

The new constants in this relationship (the α_j, β_j, γ_j, of which there are n all told) are then determined, completing the simplification. Most often the *method of undetermined coefficients* [see Bronshtein and Semendyayev, pages 151–153 for details and examples] is used to identify the constants, although other methods can be employed.

Equation 17:13:3 is appropriate only if there are no repeated zeros of $\mathrm{p}_n(x)$. If two r values, say r_1 and r_2, are identical, then in 17:13:3 the terms

17:13:4
$$\frac{\alpha_1}{x - r_1} + \frac{\alpha_2}{x - r_2} \quad \text{should be replaced by} \quad \frac{\alpha_1}{x - r_1} + \frac{\alpha_2}{(x - r_1)^2}$$

If the three r values r_1, r_2 and r_3 are identical, that is, if $\mathrm{p}_n(x)$ has a triple zero at r_1 [corresponding to $\mathrm{d}^2\mathrm{p}_n(r_1)/\mathrm{d}x^2 = \mathrm{dp}_n(r_1)/\mathrm{d}x = \mathrm{p}_n(r_1) = 0$, see Section 0:2], then the terms

17:13:5
$$\frac{\alpha_1}{x - r_1} + \frac{\alpha_2}{x - r_2} + \frac{\alpha_3}{x - r_3} \quad \text{should be replaced by} \quad \frac{\alpha_1}{x - r_1} + \frac{\alpha_2}{(x - r_1)^2} + \frac{\alpha_3}{(x - r_1)^3}$$

Similarly, if $p_1 = p_2$ and $q_1 = q_2$, 17:13:3 is to be modified so that

17:13:6
$$\frac{\beta_1 x + \gamma_1}{x^2 - p_1 x + q_1} + \frac{\beta_2 x + \gamma_2}{x^2 - p_2 x + q_2} \quad \text{should be replaced by} \quad \frac{\beta_1 x + \gamma_1}{x^2 - p_1 x + q_1} + \frac{\beta_2 x + \gamma_2}{(x^2 - p_1 x + q_1)^2}$$

Otherwise the partial fraction decomposition is unchanged.

With $\mathrm{R}_n^m(x)$ defined as in 17:13:1, the infinite series $\mathrm{R}_n^m(0) + \mathrm{R}_n^m(1) + \mathrm{R}_n^m(2) + \cdots$ can be summed provided $n \geq m + 2$. The procedure is described in Section 44:14.

A rational function frequently serves as a convenient approximation to a transcendental function. For specified numeratorial and denominatorial degrees, m and n, the *Padé approximant* is the best such approximation for small values of the argument x. If the transcendental function $\mathrm{f}(x)$ can be represented by the power series $\Sigma A_j x^j$, then the R_0^m Padé approximant is simply the sum of the first $m + 1$ terms in this series

17:13:7
$$\mathrm{R}_0^m(x) = \mathrm{f}_m(x) = t_0 + t_1 + t_2 + \cdots + t_j + \cdots + t_m \qquad t_j = A_j x^j$$

Below is shown an assemblage of Padé approximants to the exponential function [$\exp(x)$, see Chapter 26].

$R_0^0(x) = 1$	$R_0^1(x) = 1 + x$	$R_0^2(x) = 1 + x + \dfrac{x^2}{2}$	$R_0^3(x) = 1 + x + \dfrac{x^2}{2} + \dfrac{x^3}{6}$
$R_1^0(x) = \dfrac{1}{1 - x}$	$R_1^1(x) = \dfrac{2 + x}{2 - x}$	$R_1^2(x) = \dfrac{6 + 4x + x^2}{6 - 2x}$	$R_1^3(x) = \dfrac{24 + 18x + 6x^2 + x^3}{24 - 6x}$
$R_2^0(x) = \dfrac{2}{2 - 2x + x^2}$	$R_2^1(x) = \dfrac{6 + 2x}{6 - 4x + x^2}$	$R_2^2(x) = \dfrac{12 + 6x + x^2}{12 - 6x + x^2}$	$R_2^3(x) = \dfrac{60 + 36x + 9x^2 + x^3}{60 - 24x + 3x^2}$
$R_3^0(x) = \dfrac{6}{6 - 6x + 3x^2 - x^3}$	$R_3^1(x) = \dfrac{24 + 6x}{24 - 18x + 6x^2 - x^3}$	$R_3^2(x) = \dfrac{60 + 24x + 3x^2}{60 - 36x + 9x^2 - x^3}$	$R_3^3(x) = \dfrac{120 + 60x + 12x^2 + x^3}{120 - 60x + 12x^2 - x^3}$

Such an assemblage, extended indefinitely, is known as a *Padé table* [see Wall, Chapter XX]. Though they may not always be optimal, the *diagonal Padé approximants* $R_0^0(x)$, $R_1^1(x)$, $R_2^2(x)$, ..., $R_k^k(x)$, ... will be our especial concern [see Section A:6]; we will report a general procedure for determining successive members of this family. For the exponential function example, the diagonal Padé approximants are given by the formula

17:13:8
$$R_k^k(x) = \frac{p_k(x)}{p_k(-x)} \qquad \text{where} \qquad p_k(x) = \sum_{j=1}^{k} \frac{(2k - j)!x^j}{(k - j)!j!}$$

For the present purpose it is necessary to augment the Padé table by including entries corresponding to $R_n^m(x)$, where m and n are odd multiples of $\frac{1}{2}$. These new entries are *not* approximations to $f(x)$, though they *are* rational functions. Continuing with our $f(x) = \exp(x)$ example, we list below part of the *augmented Padé table:*

$$R_{-1/2}^{1/2}(x) = 0 \qquad R_{-1/2}^{3/2}(x) = 0 \qquad R_{-1/2}^{5/2}(x) = 0 \qquad R_{-1/2}^{7/2}(x) = 0$$

$$R_0^0(x) = 1 \qquad R_0^1(x) = 1 + x \qquad R_0^2(x) = 1 + x + \frac{x^2}{2} \qquad R_0^3(x) = 1 + x + \frac{x^2}{2} + \frac{x^3}{6}$$

$$R_{1/2}^{1/2}(x) = \frac{1}{x} \qquad R_{1/2}^{3/2}(x) = \frac{2}{x^2} \qquad R_{1/2}^{5/2}(x) = \frac{6}{x^3} \qquad R_{1/2}^{7/2}(x) = \frac{24}{x^4}$$

$$R_1^1(x) = \frac{2 + x}{2 - x} \qquad R_1^2(x) = \frac{6 + 4x + x^2}{6 - 2x} \qquad R_1^3(x) = \frac{24 + 18x + 6x^2 + x^3}{24 - 6x}$$

$$R_{3/2}^{3/2}(x) = \frac{-12 + 12x - 2x^2}{x^3} \qquad R_{3/2}^{5/2}(x) = \frac{-72 + 48x - 6x^2}{x^4} \qquad R_{3/2}^{7/2}(x) = \frac{-480 + 240x - 24x^2}{x^5}$$

$$R_2^2(x) = \frac{12 + 6x + x^2}{12 - 6x + x^2} \qquad R_2^3(x) = \frac{60 + 36x + 9x^2 + x^3}{60 - 24x + 3x^2}$$

Note that the identity

17:13:9
$$R_{-1/2}^m(x) = 0 \qquad m = \frac{1}{2}, \frac{3}{2}, \frac{5}{2}, \ldots$$

is true for all functions. Use of this rule and rule 17:13:7 enables the top two rows of the augmented Padé table to be completed for any function $f(x)$ whose power series is known. Every other entry in the augmented table may then be found from those immediately above it by use of the rule [see Macdonald for details]

17:13:10
$$R_n^m(x) = R_{n-1}^m(x) + \frac{1}{R_{n-1/2}^{m+1/2}(x) - R_{n-1/2}^{m-1/2}(x)} \qquad m = \frac{1}{2}, 1, \frac{3}{2}, \ldots \qquad n = \frac{1}{2}, 1, \frac{3}{2}, \ldots, m$$

We will use the phrase "Padé operation on $R_n^m(x)$" to describe the procedure, indicated by equation 17:13:10, by which R_n^m is constructed from R_{n-1}^m, $R_{n-1/2}^{m+1/2}$ and $R_{n-1/2}^{m-1/2}$.

By carefully sequencing Padé operations, it is possible to generate an unlimited number of diagonal Padé approximants corresponding to any power series. If we assign the verb "papoo" to mean "perform a Padé operation on" then our goal may be achieved economically by the following sequence: load R_0^0, load R_0^1, null $R_{-1/2}^{1/2}$, papoo $R_{1/2}^{1/2}$, load R_0^2, null $R_{-1/2}^{3/2}$, papoo $R_{1/2}^{3/2}$, papoo R_1^1, load R_0^3, null $R_{-1/2}^{5/2}$, papoo $R_{1/2}^{5/2}$, papoo R_1^2, papoo $R_{3/2}^{3/2}$, load R_0^4, null $R_{-1/2}^{7/2}$, papoo $R_{1/2}^{7/2}$, papoo R_1^3, papoo $R_{3/2}^{5/2}$, papoo R_2^2, load R_0^5, null $R_{-1/2}^{9/2}$, papoo $R_{1/2}^{9/2}$, papoo R_1^4, papoo $R_{3/2}^{7/2}$, papoo R_2^3, papoo $R_{5/2}^{5/2}$, load R_0^6, null $R_{-1/2}^{11/2}$, papoo $R_{1/2}^{11/2}$, papoo R_1^5, papoo $R_{3/2}^{9/2}$, papoo R_2^4, papoo $R_{5/2}^{7/2}$, papoo R_3^3, etc.

Because the number of papoo steps increases (0,3,7,11,15,...) for each successive diagonal Padé approximant, the algebra soon becomes daunting, imposing a limit to the usefulness of this particular procedure for the evaluation of expressions for $R_k^k(x)$. On the other hand, the procedure is extremely useful for calculating *numerical* values of the diagonal Padé approximants, for some argument x of interest, because, unlike the algebra, the arithmetic may be performed efficiently by a computing device. The algorithm below calculates and outputs numerical values of R_0^0, R_1^1, R_2^2, ..., R_k^k, ..., R_J^J from input values of t_0, t_1, t_2, ..., t_{2j}, ... t_{2J}.

```
                    Set k = s = 0
                (1) Set j = 2k
                    Output j
j (as cue) <<<<<<<
Input t_j >>>>>>>>
                    If j = 0 go to (2)
                    Set s = r_{j-1}
                (2) Set r_j = s + t
                    If j = 0 go to (6)
                    Set t = 0
                (3) Replace j by j - 1
                    Set s = r_j
                    If s ≠ r_{j+1} go to (4)
                    Set r_j = 10^99
                    If s ≠ 10^99 go to (5)
                    Set r_j = t
                    Go to (5)
                (4) Set r_j = t + [1/(r_{j+1} - s)]
                (5) Set t = s
                    If j ≠ 0 go to (3)
                    If frac(k) ≠ 0 go to (7)
                (6) Output r_0
r_0 = R_k^k <<<<<<<<
                (7) Replace k by k + 1/2
                    Go to (1)
```

Storage needed:
k, j, t (stores all t_j and serves as a temporary store), s and r_0, r_1, ..., r_{2J-1}, r_{2J}

Input restriction:
$j \le J$ where $2J + 1$ is the size of the allocated storage array.

Test values (correspond to exp(x) with $x = 1$):
input $t_0 = 1$
output $R_0^0 = 1.00000000$
inputs $t_1 = 1$, $t_2 = 1/2$
output $R_1^1 = 3.00000000$
inputs $t_3 = 1/6$, $t_4 = 1/24$
output $R_2^2 = 2.71428571$
inputs $t_5 = 1/120$, $t_6 = 1/720$
output $R_3^3 = 2.71830986$
inputs $t_7 = 1/5040$, $t_8 = 1/40320$
output $R_4^4 = 2.71828172$
inputs $t_9 = 1/362880$, $t_{10} = 1/3628800$
output $R_5^5 = 2.71828183$

The algorithm makes use of two running indices, k and j, that have simple interpretations in terms of the augmented Padé table. It may be noticed that the entries in the augmented table define two sets of diagonals. Each diagonal in one set is composed of R_n^m members such that $m + n$ is a constant; this constant equals $2k$. The other set comprises diagonals that are infinitely long, each diagonal being characterized by a constant difference $m - n$ between the degrees of the table members; this difference equals the index j. Thus, the identities

17:13:11 $$m = k + \frac{j}{2} \quad \text{and} \quad n = k - \frac{j}{2} \qquad k = 0, \frac{1}{2}, 1, \ldots \qquad j = 0, 1, 2, \ldots$$

relate the indices to the degrees.

Though the algorithm is slow, it is so economical in storage requirements that it may be implemented by most programmable calculators. The commands shown in green guard against a "divide by zero" error, should the terms in the denominator of the Padé operation

17:13:12 $$\text{Set } r_j = t + \frac{1}{r_{j-1} - s}$$

be equal (either coincidentally or by loss of arithmetic significance); often these five commands may be omitted.

The diagonal Padé approximants $R_0^0(x)$, $R_1^1(x)$, $R_2^2(x)$, ... converge toward $f(x)$ much more rapidly than do $f_0(x)$, $f_1(x)$, $f_2(x)$, ..., where $f_j(x)$ is the partial sum defined by 17:13:7. This is illustrated by the test values that accompany

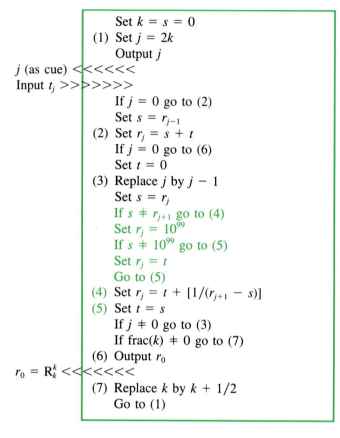

Table 17.13.1

j	t_j	$f_j(x)$	$R_k^k(x)$	k
0	$3.602598514 \times 10^{-2}$	0.0360259851	0.0360259851	0
1	$-1.501082714 \times 10^{-3}$	0.0345249024		
2	$2.814530089 \times 10^{-4}$	0.0348063554	0.0347619155	1
3	$-9.772673920 \times 10^{-5}$	0.0347086287		
4	$4.988135647 \times 10^{-5}$	0.0347585101	0.0347405098	2
5	$-3.366991560 \times 10^{-5}$	0.0347248402		
6	$2.829208187 \times 10^{-5}$	0.0347531322	0.0347395777	3
7	$-2.846048712 \times 10^{-5}$	0.0347246717		
8	$3.335213335 \times 10^{-5}$	0.0347580239	0.0347395117	4
9	$-4.462391917 \times 10^{-5}$	0.0347133999		
10	$6.712181177 \times 10^{-5}$	0.0347805218	0.0347395053	5
11	$-1.121239355 \times 10^{-4}$	0.0346683978		
12	$2.059498676 \times 10^{-4}$	0.0348743477	0.0347395045	6
13	$-4.125598309 \times 10^{-4}$	0.0344617878		
14	$8.951074902 \times 10^{-4}$	0.0353568953	0.0347395044	7
15	$-2.091070553 \times 10^{-3}$	0.0332658248		
16	$5.233121879 \times 10^{-3}$	0.0384989467	0.0347395044	8

the algorithm. Here, $t_j = 1/j!$ and so R_k^k is an approximant to exp(1), equal to 2.718281828. The output data show that R_k^k has converged to nine significant figures already by $k = 5$. It requires a partial sum of twelve terms, $t_0 + t_1 + t_2 + \ldots + t_{11}$ to achieve comparable accuracy. This advantage of the Padé approach is largely illusory, however, because terms up to t_{10} were needed to calculate the R_5^5 approximant! In fact, for the summation of convergent series [Section 0:6] such as this, the Padé procedure will seldom be worth the effort.

The great virtue of the Padé procedure comes from its ability to handle asymptotic series [see Section 0:6], and the *Atlas* takes advantage of this in a few of the algorithms in Sections 8 [for instance in Section 51:8]. A demonstration of the efficacy of this procedure is shown in Table 17.13.1, which lists approximations to the zero-order Basset function [Chapter 51] of argument 3:

17:13:13 $$K_0(3) = 0.03473950439$$

calculated as partial sums of the asymptotic series 51:6:8 and as the corresponding diagonal Padé approximants. Evidently, whereas the former procedure can never give a precision better than about three significant figures, the latter has achieved nine significant figures of convergence by $k = 7$.

Even more remarkable is the effect of the Padé procedure on series that are not even asymptotic. Consider the Gauss function [Chapter 60] $F(\frac{1}{2},\frac{1}{2};1;-2)$ expanded via 60:6:1:

17:13:14 $$F\left(\frac{1}{2},\frac{1}{2};1;-2\right) = \sum_{j=0}^{\infty}\left[\frac{(2j-1)!!}{(2j)!!}\right]^2(-2)^j = 1 - \frac{1}{2} + \frac{9}{16} - \frac{225}{288} + \frac{11025}{9216} - \cdots$$

This series is so wildly divergent that 17:13:14 would not usually be considered a valid representation of $F(\frac{1}{2},\frac{1}{2};1;-2)$. Nevertheless, the diagonal Padé approximants converge to 0.745749189, a value that can be identified with $(1/\sqrt{3})F(\frac{1}{2},\frac{1}{2};1;\frac{2}{3})$ [or with $(2/\pi\sqrt{3})K(\sqrt{2/3})$, where K denotes the complete elliptic integral of Chapter 61], to which $F(\frac{1}{2},\frac{1}{2};1;-2)$ is equivalent via transformation 60:5:3.

17:14 RELATED TOPICS

Technologists and engineers commonly collect extensive lists of values of a function f(x) without knowing the form of the relationship between f(x) and its argument x. Table 17.14.1 shows such a list, with f(x_j) abbreviated to f_j. A frequent need is to use these data so that a value of the function may be estimated at an argument where no measurement was made. Two situations arise in this setting. In the first, the tabular data are regarded as exact and the problem is one of *interpolation*. This means the selection of a relationship between f(x) and x that is satisfied by all (or by a subset of) the data. The relationship is then taken to apply equally well between the given data

Table 17.14.1

x	f (x)
x_0	f_0
x_1	f_1
.	.
.	.
.	.
x_{j-1}	f_{j-1}
x_j	f_j
x_{j+1}	f_{j+1}
x_{j+2}	f_{j+2}
.	.
.	.
.	.
x_J	f_J

points. In the second situation, error is assumed to contaminate the $f(x_j)$ data and a (usually rather simple) relationship $f(x)$ is sought that does not exactly reproduce $f(x)$ at the $x = x_0, x_1, x_2, \ldots, x_J$ points but is close. Such a procedure is known as *regression*. Polynomials are commonly employed for both interpolation and regression.

There exist many interpolative schemes, but one that provides a smooth interpolation without undue complexity is the *sliding cubic,* or *Lagrange four-point interpolate*. In this technique a cubic expression is fitted to the data at x_{j-1}, x_j, x_{j+1} and x_{j+2} but is used to describe the relationship only between the middle two points, x_j and x_{j+1}. The cubic is

17:14:1
$$f(x) \simeq p_3(x) = \sum_k \frac{(x - x_l)(x - x_m)(x - x_n)f_k}{(x_k - x_l)(x_k - x_m)(x_k - x_n)} \qquad x_j \le x \le x_{j+1}$$

where, in turn, k takes the values $j - 1$, j, $j + 1$ and $j + 2$, while l, m and n are the three of those four integers other than k. Observe that $p_3(x_k) = f_k$ for $x_k = x_{j-1}, x_j, x_{j+1}, x_{j+2}$.

Figure 17-2 illustrates this interpolation scheme in the frequently encountered case in which the data are evenly spaced so that $x_{j+2} - x_{j+1} = x_{j+1} - x_j = x_j - x_{j-1} = h$. The sliding cubic then takes the form

17:14:2
$$f(x) \simeq f_j + \left(\frac{x - x_j}{h}\right)\left[\frac{-f_{j+2}}{6} + f_{j+1} - \frac{f_j}{2} - \frac{f_{j-1}}{3}\right] + \left(\frac{x - x_j}{h}\right)^2\left[\frac{f_{j+1}}{2} - f_j + \frac{f_{j-1}}{2}\right]$$
$$+ \left(\frac{x - x_j}{h}\right)^3\left[\frac{f_{j+2}}{6} - \frac{f_{j+1}}{2} + \frac{f_j}{2} - \frac{f_{j-1}}{6}\right] \qquad x_j \le x \le x_{j+1}$$

which is convenient for computation. When the argument passes from the $x_j \le x \le x_{j+1}$ range into the $x_{j+1} \le x \le x_{j+2}$ range, the cubic "slides" with it and the interpolation now uses the f_j, f_{j+1}, f_{j+2}, f_{j+3} data.

An alternative, and popular, interpolation scheme is provided by the *cubic spline*. This is also a piecewise cubic polynomial that is, in general, even "smoother" than the sliding cubic. For this reason it is preferred for treating data that display discontinuities. For further information on the cubic spline, reference may be made to Hamming [Section 20.9].

The remainder of this section will deal with the technique of *polynomial fitting* in which a K-degree polynomial

17:14:3
$$\hat{f}(x) = a_K x^K + a_{K-1}x^{K-1} + \cdots + a_1 x + a_0$$

is fitted to the evenly spaced data $(x_0,f_0),(x_1,f_1), \ldots, (x_{J-1},f_{J-1}), (x_J,f_J)$. If K is chosen to equal J, the fit is exact and the procedure is one of interpolation. If K is chosen to be less than J, the procedure minimizes the sum of squares

17:14:4
$$\sum_{j=0}^{J} [f_j - \hat{f}(x_j)]^2$$

and is said to produce a *least-squares regression*. The integer K is restricted to values not exceeding J since, with

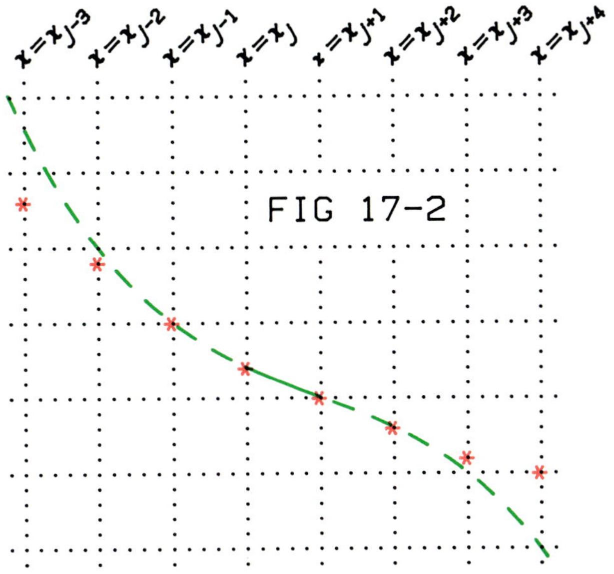

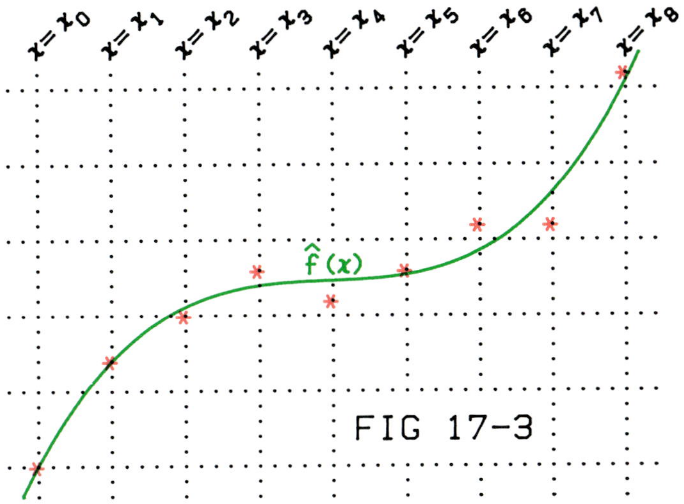

$K = J$, the fit is already exact. Figure 17-3 shows an example of the least-squares cubic function fitted to a set of nine data pairs.

An algorithm follows that permits the coefficients a_K, a_{K-1}, ..., a_1, a_0 to be found. There is no limit to $J + 1$, the number of data that may be input. The degree K of the fitted polynomial is limited only by the registers available and by the need for K not to exceed J. There is no lower limit to K: $K = 1$ will return the coefficients $a_1 = b$, $a_0 = c$ of the best straight line fit [see Section 7:14]; $K = 0$ will return the constant that is the average of all the f_j values.

The algorithm is lengthy because it has five phases.

The first phase simply allows the input of J, x_0, x_J and K. The x_0 and x_J parameters are not stored as such but as $(x_0 + x_J)/2 = u$ and $(x_J - x_0)/2 = h$. The independent variable x is subsequently regarded as having been replaced by $y = (u - x)/h$ so that all the data lie in the range $1 \geq y \geq -1$ and are characterized by $y_j = 1 - (2j/J)$.

In the second phase x_j is output to prompt the operator to load f_j. These input values are each multiplied by values that the algorithm calculates for $t_k(y_j)$ for $k = 0, 1, 2, \ldots, K$, and are accumulated as the sums

17:14:5 $$e_k = \sum_{j=0}^{J} f_j t_k(y_j) \qquad y_j = 1 - \frac{2j}{J} \qquad k = 0, 1, 2, \ldots, K$$

in registers $b_0, b_1, b_2, \ldots, b_K$. Here $\mathrm{t}_k(y)$ denotes the discrete Chebyshev polynomial of degree k and argument y; its definition and properties are given in Section 22:13. The algorithm uses recursion formula 22:13:8 to calculate $\mathrm{t}_k(y_j)$.

As explained in Section 22:13, the "best" K-degree polynomial through the points $(f_0,y_0), (f_1,y_1), \ldots, (f_{J-1},y_{J-1}), (f_J,y_J)$ is given by

17:14:6 $$\mathrm{p}_K(y) = \sum_{k=0}^{K} d_k\mathrm{t}_k(y) \qquad \text{where} \qquad d_k = \frac{(2k+1)(J-k+1)_k}{(J+1)_{k+1}} e_k$$

and where $(J - k + 1)_k$ and $(J + 1)_{k+1}$ are Pochhammer polynomials [see Chapter 18]. The third phase of the algorithm is devoted to replacing each e_k value by the corresponding d_k.

The fourth phase effects the replacement of the stored d_k values by b_k values where

17:14:7 $$\sum_{k=0}^{K} b_k y^k = \sum_{k=0}^{K} d_k\mathrm{t}_k(y)$$

A second set of registers $a_0, a_1, a_2, \ldots, a_K$ is employed for this purpose. For $j = 0, 1, 2, \ldots, K$ the register is set equal to $c_k^{(j)}$, the coefficient of y^j in the polynomial expansion of $\mathrm{t}_k(y)$. These coefficients are zero for $j < k$ and if j and k have unlike parities. They are constructed using the recursion formula

17:14:8 $$c_k^{(j)} = \frac{(2k-1)J}{k(J-k+1)} c_{k-1}^{(j-1)} - \frac{(k-1)(J+k)}{k(J-k+1)} c_{k-2}^{(j)}$$

with $c_0^{(0)} = c_1^{(1)} = 0$.

In the fifth phase the coefficients a_k that satisfy

17:14:9 $$\sum_{k=0}^{K} a_k x^k = \sum_{k=0}^{K} b_k y^k = \sum_{k=0}^{K} \frac{b_k}{(-h)^k}(x-u)^k$$

Input J >>>>>>>
Input x_0 >>>>>>>
Set $h = x_0$
Input x_J >>>>>>>
Set $u = (x_J + h)/2$
Replace h by $u - h$
Input K >>>>>>>
Set $k = K$
(1) Set $b_k = 0$
Replace k by $k - 1$
If $k \geq 0$ go to (1)

first phase

Set $j = -1$
(2) Replace j by $j + 1$
If $j > J$ go to (4)
Set $k = r = 0$
Set $s = 1$
Output $[(2j/J) - 1]h + u$
x_j(as cue) <<<<<<
Input f_j >>>>>>>
Set $f = f_j$
Replace b_k by $b_k + f$
(3) Replace k by $k + 1$
If $k > K$ go to (2)
Set $t = r + (sJ - rJ - 2sj)(2k - 1)/[k(J - k + 1)]$
Replace b_k by $b_k + tf$
Set $r = s$
Set $s = t$
Go to (3)

second phase

Storage needed: As written, the algorithm requires $2K + 12$ registers: for the parameters $J, h, u, K, k, b_0, b_1, b_2, \ldots, b_K, j, r, s, f, t, a_0, a_1, a_2, \ldots, a_K$. However, the minimum number of registers required for $K \geq 3$ is $2K + 8$ because r and a_0 may share the same register, as may s and a_1, f and a_2 and t and a_3.

third phase

(4) Set $k = 0$
Replace b_k by $b_k/(J + 1)$
(5) Replace k by $k + 1$
If $k > K$ go to (7)
Replace b_k by $b_k(2k + 1)$
Set $j = 0$
(6) Replace j by $j + 1$
Replace b_k by $b_k/(J + j)$
If $k < j$ go to (5)
Replace b_k by $b_k(J - k + j)$
Go to (6)

fourth phase

(7) Set $a_0 = k = 1$
(8) Set $a_k = 0$
Replace k by $k + 1$
If $k \leq K$ go to (8)
Set $j = k = 0$
Go to (10)
(9) Replace j by $j + 1$
If $j > K$ go to (11)
Set $k = j$
Set $a_k = a_{k-1}(2k - 1)(J/k)/(J - k + 1)$
Replace b_k by $b_k a_k$
(10) Replace k by $k + 2$
If $k > K$ go to (9)
Set $a_k = a_{k-2} - (a_{k-2} - a_{k-1})(2k - 1)(J/k)/(J - k + 1)$
Replace b_j by $b_j + b_k a_k$
Go to (10)

fifth phase

(11) Set $k = K + 1$
(12) Replace k by $k - 1$
Set $a_k = k!b_k$
Set $j = k$
(13) Replace j by $j + 1$
If $j > K$ go to (14)
Replace a_k by $a_k + b_j j!(u/h)^{j-k}/(j - k)!$
Go to (13)
(14) Replace a_k by $a_k/[k!(-h)^k]$
Output a_k — $a_K, a_{K-1}, \ldots, a_0$ <<<
If $k \neq 0$ go to (12)

Test values:
$J = 8$, $x_0 = 0$, $x_J = 80$, $K = 3$, and

x	f
0	0
10	14
20	20
30	26
40	22
50	26
60	32
70	32
80	52

Output: $a_3 = 0.0003754$, $a_2 = -0.04464$, $a_1 = 1.803$ and $a_0 = -0.1010$

are calculated via a binomial expansion of the $(x - u)^k$ term. The expression used is

17:14:10 $$a_k = \frac{1}{k!(-h)^k}\left[k!b_k + \sum_{j=k+1}^{K} \frac{b_j j!}{(j-k)!}\left(\frac{u}{h}\right)^{j-k}\right]$$

The a_k coefficients are output in the sequence a_K, a_{K-1}, …, a_1, a_0 and also remain in the a_k registers. They may be used in conjunction with the Section 17:8 algorithm to compute values of $\hat{f}(x)$.

CHAPTER 18

THE POCHHAMMER POLYNOMIALS $(x)_n$

This family of polynomials has very simple recursion properties [Section 18:5] and plays an important role as coefficients in the series that define hypergeometric functions, discussed in Section 18:14.

18:1 NOTATION

These polynomial functions were studied in 1730 by Stirling and later by Appell, who used the symbolism (x,n). The name *Pochhammer polynomial* recognizes the inventor of the now conventional $(x)_n$ notation. An alternative name is *shifted factorial function*.

18:2 BEHAVIOR

The Pochhammer polynomial $(x)_n$ is defined for all arguments x and for all nonnegative integer values of n [but see Section 18:12 for a generalization]. In common with other polynomial functions of degree n, the Pochhammer polynomial adopts all real values when n is odd, but has a restricted range for even n.

The Pochhammer polynomial $(x)_n$ has exactly $\mathrm{Int}[(n-1)/2]$ maxima, $\mathrm{Int}(n/2)$ minima and n zeros, these latter occurring at $x = 0, -1, -2, \ldots, -n+1$ as illustrated in Figure 18-1. Outside the range $-n < x < 1$, $(x)_n$ generally increases rapidly in magnitude.

18:3 DEFINITIONS

The Pochhammer polynomial is defined as the n-fold product

18:3:1 $$(x)_n = x(x+1)(x+2)\cdots(x+n-1) = \prod_{j=0}^{n-1}(x+j) \qquad n = 1, 2, 3, \ldots$$

supplemented by the definition

18:3:2 $$(x)_0 = 1$$

of the Pochhammer polynomial of zero degree.

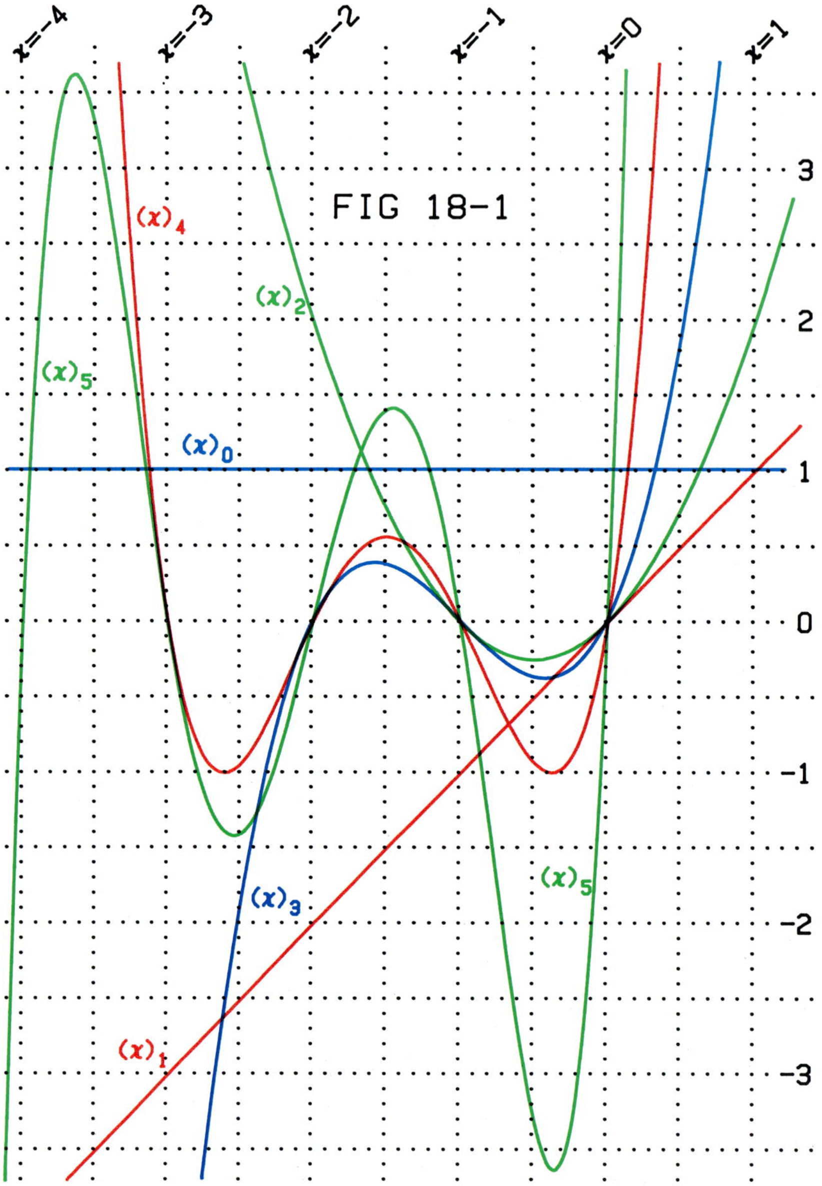

An alternative definition

18:3:3 $$(x)_n = n!\binom{x+n-1}{n} \qquad n = 0, 1, 2, \ldots$$

relates Pochhammer's polynomial to a factorial function [Chapter 2] and a binomial coefficient [Chapter 6].

The exponential function [Chapter 26] is used to define a generating function [see Section 0:3] for $(x)_n$:

18:3:4 $$(1-t)^{-x} = \exp[-x\ln(1-t)] = \sum_{n=0}^{\infty} (x)_n \frac{t^n}{n!}$$

18:4 SPECIAL CASES

The first eight Pochhammer polynomials are

18:4:1 $$(x)_0 = 1$$

18:4:2 $$(x)_1 = x$$

18:4:3 $$(x)_2 = x^2 + x$$

18:4:4 $$(x)_3 = x^3 + 3x^2 + 2x$$

18:4:5 $$(x)_4 = x^4 + 6x^3 + 11x^2 + 6x$$

18:4:6 $$(x)_5 = x^5 + 10x^4 + 35x^3 + 50x^2 + 24x$$

18:4:7 $$(x)_6 = x^6 + 15x^5 + 85x^4 + 225x^3 + 274x^2 + 120x$$

18:4:8 $$(x)_7 = x^7 + 21x^6 + 175x^5 + 735x^4 + 1624x^3 + 1764x^2 + 720x$$

18:5 INTRARELATIONSHIPS

Pochhammer polynomials obey the reflection formula

18:5:1 $$(-x)_n = (-1)^n(x - n + 1)_n$$

and the duplication formula

18:5:2 $$(2x)_n = \begin{cases} 2^n(x)_{n/2}\left(x + \frac{1}{2}\right)_{n/2} & n = 0, 2, 4, \ldots \\ 2^n(x)_{(n+1)/2}\left(x + \frac{1}{2}\right)_{(n-1)/2} & n = 1, 3, 5, \ldots \end{cases}$$

Formulas similar to 18:5:2 may be derived for $(3x)_n$, $(4x)_n$ and generally for $(mx)_n$ where m is an integer.

Equation 18:5:2 may be reformulated as a duplication formula for the degree of the Pochhammer polynomial

18:5:3 $$(x)_{2n} = 4^n\left(\frac{x}{2}\right)_n\left(\frac{1 + x}{2}\right)_n$$

and similarly

18:5:4 $$(x)_{2n+1} = 4^n x\left(\frac{1 + x}{2}\right)_n\left(1 + \frac{x}{2}\right)_n = 2^{2n+1}\left(\frac{x}{2}\right)_{1+n}\left(\frac{1 + x}{2}\right)_n$$

Likewise, formulas for $(x)_{3n}$, $(x)_{3n+1}$, $(x)_{3n+2}$, $(x)_{4n}$, etc., may be derived.

Simple recursion formulas exist for both the argument

18:5:5 $$(x + 1)_n = \left(1 + \frac{n}{x}\right)(x)_n$$

and the degree

18:5:6 $$(x)_{n+1} = (n + x)(x)_n = x(x + 1)_n$$

of a Pochhammer polynomial. The last formula generalizes to the expression

18:5:7 $$\frac{(x)_n}{(x)_m} = \begin{cases} (x + m)_{n-m} & n \geq m \\ \dfrac{1}{(x + n)_{m-n}} & n \leq m \end{cases}$$

for the quotient of two Pochhammer polynomials of common argument. Likewise, 18:5:5 may be generalized to the formulas

18:5:8 $$\frac{(x + m)_n}{(x)_n} = \frac{(x + n)_m}{(x)_m} \qquad m = 0, 1, 2, \ldots$$

and

18:5:9 $$\frac{(x-m)_n}{(x)_n} = \frac{(x-m)_m}{(x-m+n)_m} = \frac{(1-x)_m}{(1-n-x)_m} \qquad m = 0, 1, 2, \ldots$$

for the quotient of two Pochhammer polynomials of common degree whose arguments differ by an integer. These last formulas are useful when m is small.

Addition formulas exist for both the argument and degree of a Pochhammer polynomial. The expression

18:5:10 $$(x+y)_n = \sum_{j=0}^{n} \binom{n}{j} (x)_j (y)_{n-j}$$

is known as *Vandermonde's theorem,* while the rule

18:5:11 $$(x)_{n+m} = (x)_n (x+n)_m$$

is a simple consequence of the definition 18:3:1 (or of the recursion formula 18:5:6).

18:6 EXPANSIONS

The polynomial expansion of a Pochhammer function

18:6:1 $$(x)_n = (-1)^n \sum_{m=1}^{n} \mathrm{S}_n^{(m)} (-x)^m = \sum_{m=0}^{n} |\mathrm{S}_n^{(m)}| x^m$$

involves the numbers $\mathrm{S}_n^{(m)}$, known as the *Stirling numbers of the first kind.* A small display of these integers appears in Table 18.6.1. It may be extended by use of the recursion formula for these Stirling numbers

18:6:2 $$\mathrm{S}_{n+1}^{(m)} = \mathrm{S}_n^{(m-1)} - n\mathrm{S}_n^{(m)} \qquad n = 0, 1, 2, \ldots \qquad m = 1, 2, 3, \ldots$$

The same formula is utilized in our algorithm for $\mathrm{S}_n^{(m)}$, which may be found in Section 2:14 where it is incorporated into the algorithm for Stirling numbers of the *second* kind.

Stirling numbers of the first kind satisfy the summations

18:6:3 $$\sum_{m=1}^{n} \mathrm{S}_n^{(m)} = 0 \qquad n = 2, 3, 4, \ldots$$

18:6:4 $$\sum_{m=1}^{n} |\mathrm{S}_n^{(m)}| = n! \qquad n = 1, 2, 3, \ldots$$

For further information and an extended table see Abramowitz and Stegun [Chapter 24].

Table 18.6.1

	$\mathrm{S}_0^{(m)}$	$\mathrm{S}_1^{(m)}$	$\mathrm{S}_2^{(m)}$	$\mathrm{S}_3^{(m)}$	$\mathrm{S}_4^{(m)}$	$\mathrm{S}_5^{(m)}$	$\mathrm{S}_6^{(m)}$	$\mathrm{S}_7^{(m)}$	$\mathrm{S}_8^{(m)}$	$\mathrm{S}_9^{(m)}$	$\mathrm{S}_{10}^{(m)}$
$m = 0$	1	0	0	0	0	0	0	0	0	0	0
$m = 1$	0	1	−1	2	−6	24	−120	720	−5040	40320	−362880
$m = 2$	0	0	1	−3	11	−50	274	−1764	13068	−109584	1026576
$m = 3$	0	0	0	1	−6	35	−225	1624	−13132	118124	−1172700
$m = 4$	0	0	0	0	1	−10	85	−735	6769	−67284	723680
$m = 5$	0	0	0	0	0	1	−15	175	−1960	22449	−269325
$m = 6$	0	0	0	0	0	0	1	−21	322	−4536	63273
$m = 7$	0	0	0	0	0	0	0	1	−28	546	−9450

18:7 PARTICULAR VALUES

	$(-\infty)_n$	$(-n)_n$	$\left(\frac{1-n}{2}\right)_n$	$(-m)_n$, $m = 1, 2, 3, \ldots, n-1$	$\left(\frac{-1}{2}\right)_n$	$(0)_n$	$\left(\frac{1}{2}\right)_n$	$(1)_n$	$(2)_n$	$(n)_n$	$(\infty)_n$
$n = 0$	1	1	1	1	1	1	1	1	1	1	1
$n = 1, 3, 5, \ldots$	$-\infty$	$-n!$	0	0	$\frac{-(2n-3)!!}{2^n}$	0	$\frac{(2n)!}{4^n n!}$	$n!$	$(n+1)!$	$\frac{(2n)!}{2n!}$	∞
$n = 2, 4, 6, \ldots$	∞	$n!$	$\frac{(-1)^{n/2}(n!)^2}{4^n\left(\frac{n}{2}!\right)^2}$	0	$\frac{-(2n-3)!!}{2^n}$	0	$\frac{(2n)!}{4^n n!}$	$n!$	$(n+1)!$	$\frac{(2n)!}{2n!}$	∞

18:8 NUMERICAL VALUES

The following algorithm, based on definition 18:3:1, is exact:

```
Input x >>>>>>>
        Set f = 1
Input n >>>>>>>
        If n = 0 go to (2)
    (1) Replace f by fx
        Replace x by x + 1
        Replace n by n − 1
        If n ≠ 0 go to (1)
    (2) Output f
f = (x)_n <<<<<<<
```

Storage needed: x, f and n

Input restriction: The degree n must be a nonnegative integer.

Test values:
$(17)_0 = 1$
$(-0.1)_7 = -55.9122291$

18:9 APPROXIMATIONS

When x or n or both are of large magnitude, relationship 18:12:1, together with results from Section 40:9, may be used to derive asymptotic expressions for Pochhammer polynomials. Thus, for large n and modest x, the asymptotic approximation

18:9:1 $$(x)_n \sim \frac{n^{x-1}n!}{\Gamma(x)}\left[1 + \frac{x(x-1)}{2n} + \frac{x(x-1)(x-2)(3x-1)}{24n^2} + \cdots\right] \qquad n \to \infty$$

is useful and shows, for example, that

18:9:2 $$({}^1\!/_2)_n \sim \frac{n!}{\sqrt{\pi n}} \qquad n \to \infty$$

A second asymptotic approximation for $(\frac{1}{2})_n$

18:9:3 $$({}^1\!/_2)_n \sim \sqrt{2}\left(\frac{n}{e}\right)^n \qquad n \to \infty$$

follows from 18:7:7, providing a link between what are probably the three most important irrational numbers: π, $\sqrt{2}$ and e.

For large n and x close to $-n/2$, the Pochhammer polynomial approximates a sine function [see Chapter 32]:

18:9:4 $$(x)_n \simeq \frac{n^n \exp(-n)}{2^{n-1}} \sin(\pi x) \qquad \left|x + \frac{n}{2}\right| << \sqrt{\frac{n}{2}} \to \infty$$

The development of this sinusoidal behavior is evident in Figure 18-1, even for n as small as 4.

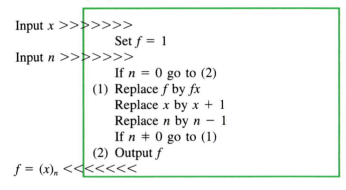

18:10 OPERATIONS OF THE CALCULUS

The rules given in Section 17:10 apply since $(x)_n$ is, after all, a polynomial function. In particular, the effects of differentiation and indefinite integration are

18:10:1
$$\frac{d}{dx}(x)_n = (x)_n \sum_{j=0}^{n-1} \frac{1}{x+j} = (x)_n[\psi(n+x) - \psi(x)]$$

where ψ is the digamma function [Chapter 44] and

18:10:2
$$\int_0^x (t)_n dt = (-1)^{n+1} \sum_{j=1}^{n} S_n^{(j)} \frac{(-x)^{j+1}}{j+1} = \sum_{j=2}^{n+1} \left| S_n^{(j-1)} \right| \frac{x^j}{j}$$

where $S_n^{(j)}$ is a Stirling number of the first kind, as discussed in Section 18:6.

18:11 COMPLEX ARGUMENT

Because of 18:6:1, an expression for $(x + iy)_n$ may be derived from equation 17:11:1 on replacement of a_j by the Stirling number $(-1)^j S_n^{(j)}$.

18:12 GENERALIZATIONS

Pochhammer polynomials can be expressed as the quotient of two gamma functions [Chapter 43]

18:12:1
$$(x)_n = \frac{\Gamma(n+x)}{\Gamma(x)}$$

whose arguments differ by the nonnegative integer n. Hence, the quotient $\Gamma(y + x)/\Gamma(x)$, with y unrestricted, is a generalization of the Pochhammer polynomial. A more restricted generalization, permitting n to be a negative integer, may be developed from recursion formula 18:5:4.

18:13 COGNATE FUNCTIONS

Factorial functions [Chapter 2], binomial coefficients [Chapter 6], gamma functions [Chapter 43], complete beta functions [Chapter 43] and Pochhammer polynomials are all closely interrelated.

A *factorial polynomial,* as defined by Tuma [Section 1.03]:

18:13:1
$$x(x-h)(x-2h)\cdots(x-nh+h) = X_h^{(n)}(x) \qquad n = 1, 2, 3, \ldots$$

is a Pochhammer polynomial of changed argument. One has $(x)_n = X_{-1}^{(n)}(x)$ and $X_h^{(n)}(x) = (-h)^n(-x/h)_n$. The symbol $x^{[n]}$ is also used for the factorial polynomial defined as

18:13:2
$$x^{[n]} = x(x-1)(x-2)\cdots(x-n+1) = X_1^{(n)}(x) = (-1)^n(-x)_n$$

Yet another confusing symbolism, due to Kramp, is

18:13:3
$$x^{n/c} = x(x+c)(x+2c)\cdots(x+nc-c) = c^n(x/c)_n$$

None of these notations is employed in this *Atlas*.

18:14 RELATED TOPICS

Pochhammer polynomials occur in the coefficients of so-called *hypergeometric functions*. The most general representation of such a function is as the sum

18:14:1
$$\sum_{j=0}^{\infty} \frac{(a_1)_j(a_2)_j(a_3)_j \cdots (a_K)_j}{(c_1)_j(c_2)_j(c_3)_j \cdots (c_L)_j} x^j$$

Table 18.14.1

K, number of numeratorial parameters	L, number of denominatorial parameters	Values of numeratorial parameters	Values of denominatorial parameters	Name of function and usual notation	Chapter or section no.
0	0			infinite geometric sum $1/(1-x)$	1:6
0	1		1	exponential function $\exp(x)$	26
0	1		ν	incomplete gamma function $\Gamma(\nu)\exp(x)\gamma^*(\nu-1;x)$	45
1	1	ν	1	binomial function $(1-x)^{-\nu}$	6:14
1	1	a	c	incomplete beta function $(c-1)x^{1-c}(1-x)^{c-a-1}\mathrm{B}(c-1;1+a-c;x)$	58
1	2	a	$1, c$	Kummer function $\mathrm{M}(a;c;x)$	47
2	1	a, b	1	Tricomi function $(-x)^a\mathrm{U}(a;1+a-b;-1/x)$	48
2	2	a, b	$1, c$	Gauss function $\mathrm{F}(a,b;c;x)$	60
3	3	a_1, a_2, a_3	$1, c_2, c_3$	Clausen functions	
K	L	$a_1, a_2, a_3, \ldots, a_K$	$1, c_2, c_3, \ldots, c_L$	generalized hypergeometric function ${}_KF_{L-1}(a_1,\ldots,a_K;c_2,\ldots,c_L;x)$	60:13

where x is the argument, $a_1, a_2, a_3, \ldots, a_K$ are prescribed *numeratorial parameters* and $c_1, c_2, c_3, \ldots, c_L$ are prescribed *denominatorial parameters*. If one of the numeratorial parameters is a negative integer, the 18:14:1 series may terminate, but generally hypergeometric functions are represented by infinite series.

If the number of denominatorial parameters exceeds the number of numeratorial parameters, that is, if $L > K$, the hypergeometric series necessarily converges for all finite values of x. If $L = K$, convergence is generally limited to the argument range $|x| < 1$. If $L < K$ the series diverges (unless it terminates) for all nonzero arguments, but it may nevertheless usefully represent a function asymptotically [see equation 37:6:5 for an example].

The name "hypergeometric" arises because the 18:14:1 series can be regarded as an extension of the geometric series [see equation 1:6:5] to which it reduces when $K = L = 0$. Other small values of K and L give rise to certain well-known families of functions, discussed elsewhere in this *Atlas,* as well as a number of generic classes of functions. Table 18.14.1 summarizes these types of functions. Notice that the so-called *generalized hypergeometric function* is not, in fact, fully general because one of its denominatorial parameters is constrained to be unity.

Hypergeometric functions are important because most of the so-called *special functions* of mathematical physics (i.e., functions that arise as solutions to differential equations of practical importance) are instances of this class. The cylinder functions of Chapters 49–56 provide examples, all being $K = 0$, $L = 2$ hypergeometric functions. Cases with $L = 4$ and $K = 0$, 2 or 4 occur in the theory of elasticity and are encountered in Sections 47:11, 53:6 and 55:6 of this *Atlas*.

Functions that are expressible as hypergeometric series may be evaluated by exploiting convenient features of expansion 18:14:1. If the abbreviation

18:14:2 $$G_j = \frac{(a_1+j)(a_2+j)(a_3+j)\cdots(a_K+j)}{(c_1+j)(c_2+j)(c_3+j)\cdots(c_L+j)}$$

is adopted, then a hypergeometric function of argument X may be represented as

18:14:3 $$1 + G_0X + G_0G_1X^2 + G_0G_1G_2X^3 + \cdots + G_0G_1G_2\cdots G_{J-2}X^{J-1} + R_J$$

where R_J is the remainder when the hypergeometric series 18:14:1 is truncated after the $j = J - 1$ term. If J is chosen to be larger in magnitude than any of the a or c parameters, and also to exceed $|X|^{1/(L-K)}$, then the dominant term in G_J is J^{K-L}. Unless $L < K$, the R_J remainder will become steadily smaller in magnitude as J increases. For large enough J, the R_J term may be well approximated by a geometric series

18:14:4 $$R_J = G_0G_1G_2\cdots G_{J-1}X^J[1 + G_JX + G_JG_{J+1}X^2 + G_JG_{J+1}G_{J+2}X^3 + \cdots$$

$$\simeq G_0G_1G_2\cdots G_{J-1}X^J[1 + J^{K-L}X + J^{2K-2L}X^2 + J^{3K-3L}X^3 + \cdots] = \frac{G_0G_1G_2\cdots G_{J-1}X^J}{1 - J^{K-L}X}$$

With approximation 18:14:4 incorporated when appropriate, expression 18:14:3 forms the basis of a *universal hypergeometric algorithm* that may be used to calculate numerical values of a wide variety of functions. The algorithm, which is detailed below, requires the following inputs:

(a) K, the number of numeratorial parameters;
(b) L, the number of denominatorial parameters;
(c) $a_1, a_2, a_3, \ldots, a_K$; $c_1, c_2, c_3, \ldots, c_L$, the values of the parameters (these are stored serially in the algorithm as $b_1, b_2, b_3, \ldots, b_{K+L}$); and
(d) X, either the value of the argument x or some simple variant of it, such as $1 - x$ or $x^2/4$.

Input K >>>>>>>
Set $i = 0$
Input L >>>>>>>
(1) Replace i by $i + 1$
If $i > K + L$ go to (2)
Input b_i parameter >>>>>>
Go to (1)
(2) Set $j = 0$
Input X >>>>>>>
Set $t = f = 1$
Go to (4)
(3) Replace t by $t/(b_i + j)$
(4) Replace i by $i - 1$
If $i > K$ go to (3)
If $i = 0$ go to (5)
Replace t by $t(b_i + j)$
Go to (4)
(5) Replace t by tX
Replace f by $f + t$
Replace j by $j + 1$
If frac$(j/5) \ne 0$ go to (7)
If $1 \ge j^{L-K}/X$ go to (6)
Set $i = t/[1 - (j^{L-K}/X)]$
(6) Output $f - i$
Approximate f(x) value <<<<
(7) Set $i = K + L + 1$
Go to (4)

Storage needed: $7 + K + L$ registers are required to store the following quantities: $K, L, i, b_1, b_2, b_3, \ldots, b_{K+L}, X, t, f$ and j.

Intervention to halt output is necessary.

Test values:
input: $K = 1$, $L = 1$;
parameters: $1, \frac{3}{2}$; $X = \frac{1}{2}$
output: 1.57229437, 1.57081694, 1.57079671, 1.57079633, 1.57079633($=\pi/2$), ...

input: $K = 0$, $L = 2$;
parameters: $\frac{1}{2}$, 1; $X = 1$
output: 3.76219871, 3.76219569, 3.76219569(=cosh(2)), ...

input: $K = 1$, $L = 0$;
parameter: $\frac{1}{2}$; $X = 0.04$
output: 1.021340580, 1.021340743 1.021340745, 1.021340744 1.021340744(=10 daw(5)), ...

Because the performance of the algorithm depends on many factors, it is not possible generally to predict the value of J required to achieve a sought precision in f(X). Accordingly, the universal hypergeometric algorithm sets $J = 5, 10, 15, \ldots$ and generates an output for each of these J values, leaving it to the operator to observe when convergence has been achieved, or to judge when the output value has sufficient accuracy for his purpose. In some cases, convergence may occur but be tediously slow. Note that when $L < K$, the algorithm may fail to produce an acceptable value unless X is sufficiently small; in this case the output values may initially tend toward the correct answer but later diverge.

Table 18.14.2 lists some of the functions that may be evaluated using the universal hypergeometric algorithm. For many of these functions alternative algorithms will be found in the appropriate chapter of this *Atlas*. Such "custom" algorithms are generally more satisfactory than the universal algorithm, but the latter has the supreme advantage of versatility.

Table 18.14.2

No. of parameters K	L	Parameter values $b_1, b_2, b_3, \ldots, b_{K+L}$	X	Sought function f(x)	Restrictions	Chapter or section no.
1	1	$1; \frac{3}{2}$	$\frac{1}{2}$	$\frac{\pi}{2}$		1:7
0	1	1	1	e		1:7
2	2	$\frac{1}{2}, \frac{1}{2}; \frac{3}{2}, \frac{3}{2}$	-1	G		1:7
n	n	$1, \ldots, 1; 2, \ldots, 2$	-1	$\eta(n)$	$n = 1, 2, 3, \ldots$	3
n	n	$\frac{1}{2}, \ldots, \frac{1}{2}; \frac{3}{2}, \ldots, \frac{3}{2}$	-1	$\beta(n)$	$n = 1, 2, 3, \ldots$	3
1	1	$\nu; 1$	$\mp x$	$(1 \pm x)^{-\nu}$	$-1 < x < 1$	6:14
0	0		$\frac{-bx}{c}$	$\frac{c}{bx + c}$	$\|x\| < \|c/b\|$	7
0	0		$\frac{-c}{bx}$	$\frac{bx}{bx + c}$	$\|x\| > \|c/b\|$	7
1	1	$n; 1$	$\frac{-bx}{c}$	$(bx + c)^n/c^n$	$\|x\| < \|c/b\|$	11
1	1	$n; 1$	$\frac{-c}{bx}$	$(bx + c)^n/b^n x^n$	$\|x\| > \|c/b\|$	11
1	1	$\frac{1}{2}; 1$	$\frac{-bx}{c}$	$\sqrt{bx + c}/\sqrt{c}$	$\|x\| < \|c/b\|$	12
1	1	$\frac{1}{2}; 1$	$\frac{-c}{bx}$	$\sqrt{bx + c}/\sqrt{bx}$	$\|x\| > \|c/b\|$	12
1	1	$\frac{-1}{2}; 1$	$\frac{-bx}{c}$	$\frac{\sqrt{c}}{\sqrt{bx + c}}$	$\|x\| < \|c/b\|$	12
1	1	$\frac{-1}{2}; 1$	$\frac{-c}{bx}$	$\frac{\sqrt{c}}{\sqrt{bx + c}}$	$\|x\| > \|c/b\|$	12
1	1	$\nu; 1$	$\frac{x - 1}{x}$	x^ν	$\frac{1}{2} \le x < \infty$	13
1	1	$\frac{-1}{2}; 1$	$\pm \frac{x^2}{a^2}$	$\sqrt{a^2 \mp x^2}/a$	$-a < x < a$	14, 15
2	2	$-n, n + 1; 1, 1$	$\frac{1 - x}{2}$	$P_n(x)$	$-1 < x < 3$	21
2	2	$-n, -n; -2n, 1$	$\frac{2}{1 - x}$	$n!P_n(x)/(2n - 1)!!(x - 1)^n$	$x < -1$ or $x > 3$	21
2	2	$\frac{-n}{2}, \frac{1 + n}{2}; \frac{1}{2}, 1$	x^2	$(-1)^{n/2} n!!P_n(x)/(n - 1)!!$	$\|x\| < 1;\ n = 0, 2, 4, \ldots$	21
2	2	$\frac{1 - n}{2}, \frac{2 + n}{2}; 1, \frac{3}{2}$	x^2	$(-1)^{(n-1)/2}\left(\frac{n - 1}{2}\right)!!P_n(x)/n!!$	$\|x\| < 1;\ n = 1, 3, 5, \ldots$	21
2	2	$\frac{-n}{2}, \frac{1 - n}{2}; \frac{1 - 2n}{2}, 1$	$\frac{1}{x^2}$	$n!P_n(x)/(2n - 1)!!x^n$	$\|x\| > 1$	21
2	2	$-n, n; \frac{1}{2}, 1$	$\frac{1 - x}{2}$	$T_n(x)$	$-1 < x < 3$	22
2	2	$-n, \frac{1}{2} - n; \frac{1}{2}, 1$	$\frac{x - 1}{x + 1}$	$\left(\frac{2}{1 + x}\right)^n T_n(x)$	$x > 0$	22
2	2	$-n, n + 2; 1, \frac{3}{2}$	$\frac{1 - x}{2}$	$U_n(x)/n + 1$	$-1 < x < 3$	22

Table 18.14.2 (Continued)

No. of parameters K	L	Parameter values $b_1, b_2, b_3, \ldots, b_{K+L}$	X	Sought function f(x)	Restrictions	Chapter or section no.
2	2	$-n-\frac{1}{2}, -n; 1, \frac{3}{2}$	$\frac{x-1}{x+1}$	$\left(\frac{2}{1+x}\right)^n U_n(x)/n+1$	$x>0$	22
2	2	$-n, n+\nu+\mu+1; 1, 1+\nu$	$\frac{1-x}{2}$	$P_n^{(\nu;\mu)} \Big/ \binom{n+\nu}{n}$	$-1<x<3$	22:12
2	2	$-n-\nu, -n; 1, 1+\nu$	$\frac{x-1}{x+1}$	$\left(\frac{2}{1+x}\right)^n P_n^{(\nu;\mu)}(x) \Big/ \binom{n+\nu}{n}$	$x>0$	22:12
2	2	$-n, n+2\lambda; 1, \frac{1}{2}+\lambda$	$\frac{1-x}{2}$	$C_n^{(\lambda)}(x) \Big/ \binom{n+2\lambda-1}{n}$	$-1<x<3$	22:12
1	2	$-n; 1, 1$	x	$L_n(x)$		23
1	2	$1+n; 1, 1$	$-x$	$\exp(-x)L_n(x)$		23
1	2	$-n; 1, 1+\mu$	x	$L_n^{(\mu)}(x) \Big/ \binom{n+\mu}{n}$	$\mu>-1$	23:12
1	2	$-\nu; 1, 1$	x	$L_\nu(x)$		23:14
2	1	$\frac{1-n}{2}, \frac{-n}{2}; 1$	$\frac{-1}{x^2}$	$H_n(x)/(2x)^n$		24
1	2	$\frac{-n}{2}; \frac{1}{2}, 1$	x^2	$H_n(x)/(-2)^{n/2}(n-1)!!$	$n=0, 2, 4, \ldots$	24
1	2	$\frac{1-n}{2}; 1, \frac{3}{2}$	x^2	$-H_n(x)/(-2)^{(n+1)/2}n!!$	$n=1, 3, 5, \ldots$	24
2	1	$\frac{1-n}{2}, \frac{-n}{2}; 1$	$\frac{-2}{x^2}$	$He_n(x)/x^n$		24:13
1	1	1; 2	$1-x$	$\frac{\ln(x)}{x-1}$	$0<x\le 2$	25
1	1	1; 2	$\frac{x-1}{x}$	$\frac{x}{x-1}\ln(x)$	$x\ge\frac{1}{2}$	25
1	2	1; 2, 2	$\ln(x)$	$\frac{\mathrm{li}(x)-\gamma-\ln(\lvert\ln(x)\rvert)}{\ln(x)}$	$0<x\ne 1$	25, 37
1	0	1	$\frac{1}{\ln(x)}$	$\frac{\ln(x)\mathrm{li}(x)}{x}$	positive x, large or small	25, 37
1	1	$1; 1+\nu$	$\frac{x-1}{x}$	$\frac{\nu}{x}\left(\frac{x}{x-1}\right)^\nu \ln_\nu(x)$	$x\ge\frac{1}{2}$	25:12
2	2	1, 1; 2, 2	$1-x$	$\frac{\mathrm{diln}(x)}{x-1}$	$0<x\le 2$	25:13
3	3	1, 1, 1; 2, 2, 2	$1-x$	$\frac{\mathrm{triln}(x)}{x-1}$	$0<x\le 2$	25:13
0	1	1	x	$\exp(x)$		26
1	0	$-n$	$\frac{-1}{x}$	$n!e_n(x)/x^n$		26:13
2	1	$-n, n+1; 1$	$\frac{-1}{2x}$	$\sqrt{2x/\pi}\exp(x)\,K_{n+1/2}(x)$	$n=0, 1, 2, \ldots$	26:13
0	2	$1, \frac{3}{2}$	$\frac{x^2}{4}$	$\frac{\sinh(x)}{x}$		28
0	2	$\frac{1}{2}, 1$	$\frac{x^2}{4}$	$\cosh(x)$		28
0	2	$1, \frac{1}{2}+n$	$\frac{x^2}{4}$	$\sqrt{\pi/2}\,(2n-1)!!I_{n-1/2}(x)/x^{n-1/2}$	$x>0$; $n=0, 1, 2, \ldots$	28:13

Table 18.14.2 (Continued)

No. of parameters K	L	Parameter values $b_1, b_2, b_3, \ldots, b_{K+L}$	X	Sought function f(x)	Restrictions	Chapter or section no.
2	2	$\frac{1}{2}, \frac{1}{2}; 1, \frac{3}{2}$	$-x^2$	$\frac{\text{arsinh}(x)}{x}$	$-1 < x \le 1$	31
2	2	$1, \frac{3}{2}; 2, 2$	$\frac{-1}{x^2}$	$4x^2[\text{arsinh}(x) - \ln(2x)]$	$x \ge 1$	31
2	2	$1, \frac{3}{2}; 2, 2$	$\frac{1}{x^2}$	$4x^2[\ln(2x) - \text{arcosh}(x)]$	$x > 1$	31
2	2	$1, \frac{3}{2}; 2, 2$	x^2	$4[\ln(2/x) - \text{arsech}(x)]/x^2$	$0 < x < 1$	31
2	2	$\frac{1}{2}, \frac{1}{2}; 1, \frac{3}{2}$	$\frac{-1}{x^2}$	$x\ \text{arcsch}(x)$	$\lvert x\rvert > 1$	31
1	1	$\frac{1}{2}; \frac{3}{2}$	x^2	$\frac{\text{artanh}(x)}{x}$	$-1 < x < 1$	31
1	1	$\frac{1}{2}; \frac{3}{2}$	$\frac{1}{x^2}$	$x\ \text{arcoth}(x)$	$\lvert x\rvert > 1$	31
0	2	$1, \frac{3}{2}$	$\frac{-x^2}{4}$	$\frac{\sin(x)}{x} = \frac{1}{x \csc(x)}$		32, 33
0	2	$\frac{1}{2}, 1$	$\frac{-x^2}{4}$	$\cos(x) = \frac{1}{\sec(x)}$		32, 33
0	2	$1, \frac{3}{2}$	$\frac{-\pi^2x^2}{4}$	$\text{sinc}(x)$		32:13
0	2	$1, \frac{1}{2} + n$	$\frac{-x^2}{4}$	$\sqrt{\pi/2}\,(2n-1)!!J_{n-1/2}(x)/x^{n-1/2}$	$x > 0;\ n = 0, 1, 2, \ldots$	32:13
0	2	$1, \frac{1}{2} - n$	$\frac{-x^2}{4}$	$-\sqrt{\pi/2}\,x^{n+1/2}Y_{n+1/2}(x)/(2n-1)!!$	$x > 0;\ n = 0, 1, 2, \ldots$	32:13
1	1	$\frac{1}{2}; \frac{3}{2}$	$\tan^2\left(\frac{x}{2}\right)$	$\text{invgd}(x)/2\tan\left(\frac{x}{2}\right)$	$0 < x < \frac{\pi}{2}$	33:14
1	1	$\frac{1}{2}; \frac{3}{2}$	$-x^2$	$\frac{\arctan(x)}{x}$	$-1 \le x \le 1$	35
2	2	$\frac{1}{2}, \frac{1}{2}; 1, \frac{3}{2}$	x^2	$\frac{\arcsin(x)}{x}$	$-1 < x < 1$	35
2	2	$\frac{1}{2}, \frac{1}{2}; 1, \frac{3}{2}$	$\frac{1}{x^2}$	$x\ \text{arccsc}(x)$	$\lvert x\rvert > 1$	35
1	1	$\frac{1}{2}; \frac{3}{2}$	$\frac{-1}{x^2}$	$x\ \text{arccot}(x)$	$\lvert x\rvert \ge 1$	35
2	2	$\frac{1}{2}, \frac{1}{2}; 1, \frac{3}{2}$	$\frac{1-x}{2}$	$\arccos(x)/\sqrt{2-2x}$	$-1 < x \le 1$	35
2	2	$\frac{1}{2}, \frac{1}{2}; 1, \frac{3}{2}$	$\frac{2x-1}{2x}$	$\sqrt{\frac{x}{2x-2}}\ \text{arcsec}(x)$	$\lvert x\rvert \ge 1$	35
1	2	$1; 2, 2$	$-x$	$\frac{\text{Ein}(x)}{x}$		37
1	0	1	$\frac{1}{x}$	$x \exp(-x)\ \text{Ei}(x)$	large $\lvert x\rvert$	37
1	3	$\frac{1}{2}; 1, \frac{3}{2}, \frac{3}{2}$	$\frac{-x^2}{4}$	$\frac{\text{Si}(x)}{x}$		38
1	3	$\frac{1}{2}; 1, \frac{3}{2}, \frac{3}{2}$	$\frac{x^2}{4}$	$\frac{\text{Shi}(x)}{x}$		38

Table 18.14.2 (Continued)

No. of parameters		Parameter values				Chapter or section
K	L	$b_1, b_2, b_3, \ldots, b_{K+L}$	X	Sought function f(x)	Restrictions	no.
1	3	$1; \frac{3}{2}, 2, 2$	$\frac{-x^2}{4}$	$4\,\mathrm{Cin}(x)/x^2$		38
1	3	$1; \frac{3}{2}, 2, 2$	$\frac{x^2}{4}$	$4\,\mathrm{Chin}(x)/x^2$		38
2	0	$\frac{1}{2}, 1$	$\frac{-4}{x^2}$	$x\,\mathrm{fi}(x)$	large x	38:13
2	0	$1, \frac{3}{2}$	$\frac{-4}{x^2}$	$x^2\,\mathrm{gi}(x)$	large x	38:13
1	3	$\frac{3}{4}; 1, \frac{3}{2}, \frac{7}{4}$	$\frac{-x^4}{4}$	$\sqrt{9\pi/2}\,\mathrm{S}(x)/x^3$		39
1	3	$\frac{1}{4}; \frac{1}{2}, 1, \frac{5}{4}$	$\frac{-x^4}{4}$	$\sqrt{\pi/2}\,\mathrm{C}(x)/x$		39
1	3	$\frac{1+\nu}{2}; 1, \frac{3}{2}, \frac{3+\nu}{2}$	$\frac{-x^2}{4}$	$\frac{1+\nu}{x^{1+\nu}}[\mathrm{S}(0;\nu) - \mathrm{S}(x;\nu)]$	$\nu < 1$, $x > 0$	39:12
1	3	$\frac{\nu}{2}; \frac{1}{2}, 1, \frac{2+\nu}{2}$	$\frac{-x^2}{4}$	$\frac{\nu}{x^\nu}[\mathrm{C}(0;\nu) - \mathrm{C}(x;\nu)]$	$\nu < 1$, $x > 0$	39:12
2	0	$\frac{1-\nu}{2}, \frac{2-\nu}{2}$	$\frac{-4}{x^2}$	$x^{1-\nu}[\cos(x)\mathrm{S}(x;\nu) - \sin(x)\mathrm{C}(x;\nu)]$	$\nu < 1$, large x	39:12
2	0	$\frac{2-\nu}{2}, \frac{3-\nu}{2}$	$\frac{-4}{x^2}$	$\frac{x^{2-\nu}}{1-\nu}[\sin(x)\mathrm{S}(x;\nu) + \cos(x)\mathrm{C}(x;\nu)]$	$\nu < 1$, large x	39:12
0	2	$\frac{3}{4}, \frac{5}{4}$	$\frac{-x^4}{4}$	$\sqrt{\pi}\left[\sin\left(x^2 + \frac{\pi}{4}\right) - \sqrt{2}\,\mathrm{Gres}(x)\right] \Big/ 2x$		39:13
0	2	$\frac{5}{4}, \frac{7}{4}$	$\frac{-x^4}{4}$	$3\sqrt{\pi}\left[\sqrt{2}\,\mathrm{Fres}(x) - \cos\left(x^2 + \frac{\pi}{4}\right)\right] \Big/ 4x^3$		39:13
2	0	$\frac{1}{4}, \frac{3}{4}$	$\frac{-4}{x^4}$	$\sqrt{2\pi}\,x\,\mathrm{Fres}(x)$	large x	39:13
2	0	$\frac{3}{4}, \frac{5}{4}$	$\frac{-4}{x^4}$	$\sqrt{8\pi}\,x^3\,\mathrm{Gres}(x)$	large x	39:13
1	2	$\frac{1}{2}; 1, \frac{3}{2}$	$-x^2$	$\sqrt{\pi}\,\mathrm{erf}(x)/2x$		40
0	1	$\frac{3}{2}$	x^2	$\sqrt{\pi}\,\exp(x^2)\mathrm{erf}(x)/2x$		40
1	0	$\frac{1}{2}$	$\frac{-1}{x^2}$	$\sqrt{\pi}\,x\,\exp(x^2)\mathrm{erfc}(x)$	large x	40
1	2	$\frac{-1}{2}; \frac{1}{2}, 1$	$-x^2$	$\sqrt{\pi}\,[x + \mathrm{ierfc}(x)]$		40:13
1	2	$\frac{-1}{2}; \frac{1}{2}, \frac{3}{2}$	$-x^2$	$\frac{\sqrt{\pi}}{x}\left[\frac{1}{4} + \frac{x^2}{2} - \mathrm{i}^2\mathrm{erfc}(x)\right]$		40:13
1	2	$\frac{-3}{2}; \frac{1}{2}, 1$	$-x^2$	$6\sqrt{\pi}\left[\frac{x}{4} + \frac{x^3}{3} + \mathrm{i}^3\mathrm{erfc}(x)\right]$		40:13
1	0	$\frac{3}{2}$	$\frac{-1}{x^2}$	$2\sqrt{\pi}\,x^2\exp(x^2)\mathrm{ierfc}(x)$	large x	40:13
1	0	$\frac{1}{2}$	$\frac{-1}{x}$	$\sqrt{\pi x}\,\exp(x)\mathrm{erfc}(\sqrt{x})$	large x	41
0	1	$\frac{1}{2}$	x	$1 + \sqrt{\pi x}\,\exp(x)\mathrm{erf}(\sqrt{x})$	$x > 0$	41

Table 18.14.2 (Continued)

No. of parameters K	L	Parameter values $b_1, b_2, b_3, \ldots, b_{K+L}$	X	Sought function f(x)	Restrictions	Chapter or section no.
0	1	$\frac{3}{2}$	$-x^2$	$\frac{\mathrm{daw}(x)}{x}$		42
1	0	$\frac{1}{2}$	$\frac{1}{x^2}$	$2x\,\mathrm{daw}(x)$	large x	42
1	2	$\frac{1}{2}; 1, \frac{3}{2}$	x	$\frac{\exp(x)}{\sqrt{x}}\mathrm{daw}(\sqrt{x})$		42
2	2	$1-y, x; 1, 1+x$	1	$x\,\mathrm{B}(x,y)$	$y > 0$	43:13
2	2	$1, \nu; 1+\nu, 2$	1	$\frac{\nu}{\nu-1}[\psi(\nu)+\gamma]$		44
2	2	$1, 1+x-y; 1+x, 2$	1	$\frac{x}{x-y}[\psi(x)-\psi(y)]$	$y > 0$	44
2	2	$x, x; 1+x, 1+x$	1	$x^2\psi'(x)$		44:12
n	n	$x, \ldots, x; 1+x, \ldots, 1+x$	1	$(-x)^n\psi^{(n-1)}(x)/(n-1)!$	$n = 2, 3, 4, \ldots$	44:12
1	1	$\nu; \nu+1$	-1	$\nu G(\nu)/2$		44:13
1	1	$\nu; 1+\nu$	$\frac{1}{2}$	$\nu^3 G(\nu)/2^{2\nu+1}$		44:13
0	1	$1+\nu$	x	$\nu x^{-\nu}\exp(x)\gamma(\nu;x)$		45
1	2	$\nu; 1, 1+\nu$	$-x$	$\nu x^{-\nu}\gamma(\nu;x)$		45
0	1	$1+\nu$	x	$\Gamma(1+\nu)\exp(x)\gamma^*(\nu;x)$	$\nu \neq -1, -2, -3, \ldots$	45
1	2	$\nu; 1, 1+\nu$	$-x$	$\Gamma(1+\nu)\gamma^*(\nu;x)$	$\nu \neq -1, -2, -3, \ldots$	45
1	0	$1-\nu$	$\frac{-1}{x}$	$x^{\nu-1}\exp(x)\Gamma(\nu;x)$	large x	45
1	2	$\nu; 1, 1+\nu$	$-x$	$\nu x^{-\nu}\gamma(\nu;x)$	large ν	45
1	2	$\frac{-\nu}{2}; \frac{1}{2}, 1$	$\frac{x^2}{2}$	$\Gamma\left(\frac{1-\nu}{2}\right)\exp\left(\frac{x^2}{4}\right)[D_\nu(-x)+D_\nu(x)]/2\sqrt{2^\nu\pi}$	$\nu \neq 1, 3, 5, \ldots$	46
1	2	$\frac{1-\nu}{2}; 1, \frac{3}{2}$	$\frac{x^2}{2}$	$\Gamma\left(\frac{-\nu}{2}\right)\exp\left(\frac{x^2}{4}\right)[D_\nu(-x)-D_\nu(x)]/x\sqrt{2^{\nu+3}\pi}$	$\nu \neq 0, 2, 4, \ldots$	46
1	2	$\frac{1+\nu}{2}; \frac{1}{2}, 1$	$\frac{-x^2}{2}$	$\Gamma\left(\frac{1-\nu}{2}\right)\exp\left(\frac{-x^2}{4}\right)[D_\nu(-x)+D_\nu(x)]/2\sqrt{2^\nu\pi}$	$\nu \neq 1, 3, 5, \ldots$	46
1	2	$1+\frac{\nu}{2}; 1, \frac{3}{2}$	$\frac{-x^2}{2}$	$\Gamma\left(\frac{-\nu}{2}\right)\exp\left(\frac{-x^2}{4}\right)[D_\nu(-x)-D_\nu(x)]/x\sqrt{2^{\nu+3}\pi}$	$\nu \neq 0, 2, 4, \ldots$	46
2	1	$\frac{-\nu}{2}, \frac{1-\nu}{2}; 1$	$\frac{-1}{8x^2}$	$x^{-\nu}\exp\left(\frac{x^2}{4}\right)D_\nu(x)$	large x	46
1	2	$a; c, 1$	x	$M(a;c;x)$		47
1	2	$c-a; c, 1$	$-x$	$\exp(-x)M(a;c;x)$		47
2	1	$a, 1+a-c; 1$	$\frac{-1}{x}$	$x^a U(a;c;x)$	large x	48
1	2	$\frac{1}{2}+\mu-\nu; 1+2\mu, 1$	x	$x^{-1/2-\mu}\exp\left(\frac{x}{2}\right)M_{\nu;\mu}(x)$		48:13
2	1	$\frac{1}{2}+\mu-\nu, \frac{1}{2}-\mu-\nu; 1$	$\frac{-1}{x}$	$x^{-\nu}\exp\left(\frac{x}{2}\right)W_{\nu;\mu}(x)$	large x	48:13
0	2	$1, 1$	$\frac{x^2}{4}$	$I_0(x)$		49
0	2	$1, 2$	$\frac{x^2}{4}$	$\frac{2}{x}I_1(x)$		49

Table 18.14.2 (Continued)

No. of parameters K	L	Parameter values $b_1, b_2, b_3, \ldots, b_{K+L}$	X	Sought function f(x)	Restrictions	Chapter or section no.
0	2	$1, 1+n$	$\frac{x^2}{4}$	$n!(2/x)^n I_n(x)$	$n = 0, 1, 2, \ldots$	49
2	1	$\frac{1}{2}, \frac{1}{2}; 1$	$\frac{1}{2x}$	$\sqrt{2\pi x}\exp(-x)I_0(x)$	large x	49
2	1	$\frac{-1}{2}, \frac{3}{2}; 1$	$\frac{1}{2x}$	$\sqrt{2\pi x}\exp(-x)I_1(x)$	large x	49
0	2	$1, 1+\nu$	$\frac{x^2}{4}$	$\Gamma(1+\nu)(2/x)^\nu I_\nu(x)$	$\nu \neq -1, -2, -3, \ldots$	50
1	2	$\frac{1}{2}+\nu; 1, 1+2\nu$	$2x$	$\Gamma(1+\nu)(2/x)^\nu \exp(x)I_\nu(x)$	$\nu \neq -1, -2, -3, \ldots$	50
2	1	$\frac{1}{2}-\nu, \frac{1}{2}+\nu; 1$	$\frac{1}{2x}$	$\sqrt{2\pi x}\exp(-x)I_\nu(x)$	large	50
0	2	$1, \frac{3}{2}+n$	$\frac{x^2}{4}$	$(2n+1)!!x^{-n}i_n(x)$	$n = -1, 0, 1, 2, \ldots$	50:4
2	1	$\frac{1}{2}-\nu, \frac{1}{2}+\nu; 1$	$\frac{-1}{2x}$	$\sqrt{\frac{2x}{\pi}}\exp(x)K_\nu(x)$	large x	51
2	1	$-n, 1+n; 1$	$\frac{-1}{2x}$	$\frac{2x}{\pi}\exp(x)k_n(x)$	$n = 0, 1, 2, \ldots$	51:4
0	2	$1, 1$	$\frac{-x^2}{4}$	$J_0(x)$		52
0	2	$1, 2$	$\frac{-x^2}{4}$	$\frac{2}{x}J_1(x)$		52
0	2	$1, 1+n$	$\frac{-x^2}{4}$	$n!(2/x)^n J_n(x)$	$n = 0, 1, 2, \ldots$	52
0	2	$1, 1+\nu$	$\frac{-x^2}{4}$	$\Gamma(1+\nu)(2/x)^\nu J_\nu(x)$	$\nu \neq -1, -2, -3, \ldots$	53
1	3	$\frac{1}{2}+\nu; 1, 1+\nu, 1+2\nu$	$-x^2$	$\Gamma^2(1+\nu)(4/x^2)^\nu J_\nu^2(x)$	$\nu \neq -1, -2, -3, \ldots$	53
1	3	$\frac{1}{2}; 1-\nu, 1, 1+\nu$	$-x^2$	$\pi\nu\csc(\pi\nu)J_\nu(x)J_{-\nu}(x)$	$\nu \neq \pm1, \pm2, \pm3, \ldots$	53
0	2	$1, \frac{3}{2}+n$	$\frac{-x^2}{4}$	$(2n+1)!!x^{-n}j_n(x)$	$n = -1, 0, 1, 2, \ldots$	53:4
0	2	$1, 1+\nu$	$-x$	$\Gamma(1+\nu)C_\nu(x)$	$\nu \neq -1, -2, -3, \ldots$	53:6
4	2	$\alpha-\frac{1}{2}, \alpha, 1-\alpha, \frac{3}{2}-\alpha; \frac{1}{2}, 1$	$\frac{-1}{x^2}$	$P(\nu;x)$ with $\alpha = \frac{3}{4}-\frac{\nu}{2}$	large x	53:6
4	2	$\alpha, \alpha+\frac{1}{2}, \frac{3}{2}-\alpha, 2-\alpha; 1, \frac{3}{2}$	$\frac{-1}{x^2}$	$\frac{2xQ(\nu;x)}{\nu^2-1/4}$ with $\alpha = \frac{3}{4}-\frac{\nu}{2}$	$\nu \neq \pm\frac{1}{2}$, large x	53:6
0	4	$\frac{1}{2}, \frac{1}{2}, 1, 1$	$\frac{-x^4}{256}$	$\mathrm{ber}(x)$		55
0	4	$1, 1, \frac{3}{2}, \frac{3}{2}$	$\frac{-x^4}{256}$	$\frac{4}{x^2}\mathrm{bei}(x)$		55
0	4	$\frac{1}{2}, 1, 1, 1$	$\frac{x^4}{16}$	$\sqrt{\pi}\,[\mathrm{ber}^2(x)+\mathrm{bei}^2(x)]$		55
0	4	$\frac{1}{2}, 1, \frac{1+\nu}{2}, 1+\frac{\nu}{2}$	$\frac{-x^4}{256}$	$\Gamma(1+\nu)\left(\frac{2}{x}\right)^\nu \mathrm{Fe}_\nu(x)$		55:6

Table 18.14.2 (Continued)

No. of parameters K	L	Parameter values $b_1, b_2, b_3, \ldots, b_{K+L}$	X	Sought function f(x)	Restrictions	Chapter or section no.
0	4	$1, \frac{3}{2}, 1+\frac{\nu}{2}, \frac{3+\nu}{2}$	$\frac{-x^4}{256}$	$\Gamma(2+\nu)\left(\frac{2}{x}\right)^{2+\nu}\mathrm{Ge}_\nu(x)$		55:6
0	2	$\frac{2}{3}, 1$	$\frac{x^3}{9}$	$\mathrm{fai}(x) = \frac{3^{2/3}\Gamma(\frac{2}{3})}{2}\left[\frac{\mathrm{Bi}(x)}{\sqrt{3}} + \mathrm{Ai}(x)\right]$		56
0	2	$1, \frac{4}{3}$	$\frac{x^3}{9}$	$\frac{\mathrm{gai}(x)}{x} = \frac{3^{1/3}\Gamma(\frac{1}{3})}{2x}\left[\frac{\mathrm{Bi}(x)}{\sqrt{3}} - \mathrm{Ai}(x)\right]$		56
0	2	$\frac{4}{3}, \frac{5}{3}$	$\frac{x^3}{9}$	$\frac{2\mathrm{hai}(x)}{x^2} = \frac{2}{x^2}[\mathrm{Hi}(x) - \mathrm{fai}(x) - \mathrm{gai}(x)]$		56:13
2	1	$\frac{1}{6}, \frac{5}{6}; 1$	$\frac{-3}{4x^{3/2}}$	$2\sqrt{\pi\sqrt{x}}\exp\left(\frac{2}{3}x^{3/2}\right)\mathrm{Ai}(x)$	large positive x	56
2	1	$\frac{1}{6}, \frac{5}{6}; 1$	$\frac{3}{4x^{3/2}}$	$\sqrt{\pi\sqrt{x}}\exp\left(-\frac{2}{3}x^{3/2}\right)\mathrm{Bi}(x)$	large positive x	56
4	2	$\frac{1}{12}, \frac{5}{12}, \frac{7}{12}, \frac{11}{12}; \frac{1}{2}, 1$	$\frac{-9}{4x^3}$	$\sqrt{\pi\sqrt{x}}\,[s\mathrm{Ai}(-x) + c\mathrm{Bi}(-x)],\ s = \sin\left(\frac{2}{3}x^{3/2} + \frac{\pi}{4}\right)$	large positive x	56
4	2	$\frac{7}{12}, \frac{11}{12}, \frac{13}{12}, \frac{17}{12}; 1, \frac{3}{2}$	$\frac{-9}{4x^3}$	$\sqrt{\pi\sqrt{x}}\,[s\mathrm{Bi}(-x) - c\mathrm{Ai}(-x)],\ c = \cos\left(\frac{2}{3}x^{3/2} + \frac{\pi}{4}\right)$	large positive x	56
3	1	$\frac{1}{6}, \frac{1}{2}, \frac{5}{6}; 1$	$\frac{-9}{4x^3}$	$\pi\sqrt{x}\,[\mathrm{Ai}^2(-x) + \mathrm{Bi}^2(-x)]$	large positive x	56
0	2	$\frac{1}{2}, \frac{3}{2}$	$\frac{-x^2}{4}$	$\frac{\pi}{2}h_{-1}(x)$		57
0	2	$\frac{3}{2}, \frac{3}{2}$	$\frac{-x^2}{4}$	$\frac{\pi}{2x}h_0(x)$		57
0	2	$\frac{3}{2}, \frac{5}{2}$	$\frac{-x^2}{4}$	$\frac{3\pi}{2x^2}h_1(x)$		57
0	2	$\frac{3}{2}, \frac{3}{2}+\nu$	$\frac{-x^2}{4}$	$\frac{\sqrt{\pi}}{2}\Gamma\left(\frac{3}{2}+\nu\right)\left(\frac{2}{x}\right)^{1+\nu}h_\nu(x)$	$x > 0$	57
2	0	$\frac{1}{2}, \frac{3}{2}$	$\frac{-4}{x^2}$	$\frac{\pi x^2}{2}[\mathrm{Y}_{-1}(x) - h_{-1}(x)]$	large x	57
2	0	$\frac{1}{2}, \frac{1}{2}$	$\frac{-4}{x^2}$	$\frac{\pi x}{2}[h_0(x) - \mathrm{Y}_0(x)]$	large x	57
2	0	$\frac{-1}{2}, \frac{1}{2}$	$\frac{-4}{x^2}$	$\frac{\pi}{2}[h_1(x) - \mathrm{Y}_1(x)]$	large x	57
2	0	$\frac{1}{2}-\nu, \frac{1}{2}$	$\frac{-4}{x^2}$	$\sqrt{\pi}\,\Gamma\left(\frac{1}{2}+\nu\right)\left(\frac{2}{x}\right)^{\nu-1}[h_\nu(x) - \mathrm{Y}_\nu(x)]$	large x	57
0	2	$\frac{3}{2}, \frac{3}{2}+\nu$	$\frac{x^2}{4}$	$\frac{\sqrt{\pi}}{4}\Gamma\left(\frac{3}{2}+\nu\right)\left(\frac{2}{x}\right)^{1+\nu}\mathrm{l}_\nu(x)$	$x > 0$	57:13
2	0	$\frac{1}{2}-\nu, \frac{1}{2}$	$\frac{4}{x^2}$	$\sqrt{\pi}\,\Gamma\left(\frac{1}{2}+\nu\right)\left(\frac{2}{x}\right)^{\nu-1}[\mathrm{I}_{-\nu}(x) - \mathrm{l}_\nu(x)]$	large x	57:13
1	1	$\nu; 1+\nu$	x	$\nu x^{-\nu}\mathrm{B}(\nu;0;x)$	$0 \le x < 1$	58:4
1	1	$\nu+\mu; 1+\nu$	x	$\nu x^{-\nu}(1-x)^{-\mu}\mathrm{B}(\nu;\mu;x)$	$0 \le x < 1$	58
1	1	$\nu+\mu; 1+\nu$	$1-x$	$\mu x^{-\nu}[\mathrm{B}(\nu,\mu) - \mathrm{B}(\nu;\mu;x)]$	$0 < x \le 1$	58
2	2	$-\nu, 1+\nu; 1, 1$	$\frac{1-x}{2}$	$\mathrm{P}_\nu(x)$	$-1 < x < 3$	59
2	2	$\frac{1+\nu}{2}, 1+\frac{\nu}{2}; 1, \frac{3}{2}+\nu$	$\frac{1}{x^2}$	$\frac{\Gamma(\frac{3}{2}+\nu)}{\sqrt{\pi}\,\Gamma(1+\nu)}(2x)^{1+\nu}\mathrm{Q}_\nu(x)$	$x > 1$	59

Table 18.14.2 (Continued)

No. of parameters K	L	Parameter values $b_1, b_2, b_3, \ldots, b_{K+L}$	X	Sought function f(x)	Restrictions	Chapter or section no.
2	2	$\frac{-\nu}{2}, \frac{1+\nu}{2}; \frac{1}{2}, 1$	x^2	$\mathrm{lef}_\nu(x)$	$-1 < x < 1$	59:6
2	2	$\frac{1-\nu}{2}, 1+\frac{\nu}{2}; 1, \frac{3}{2}$	x^2	$\frac{1}{2x}\,\mathrm{leg}_\nu(x)$	$-1 < x < 1$	59:6
2	2	$\frac{-\nu-\mu}{2}, \frac{1+\nu-\mu}{2}; \frac{1}{2}, 1$	x^2	$\mathrm{lef}_\nu^{(\mu)}(x)$	$-1 < x < 1$	59:13
2	2	$\frac{1-\nu-\mu}{2}, 1+\frac{\nu-\mu}{2}; 1, \frac{3}{2}$	x^2	$\frac{1}{2x}\,\mathrm{leg}_\nu^{(\mu)}(x)$	$-1 < x < 1$	59:13
1	1	$a; c$	x	$\mathrm{F}(1,a;c;x) = \frac{(c-1)\mathrm{B}(c-1;1+a-c;x)}{x^{c-1}(1-x)^{1+a-c}}$	$-\infty < x < 1$	60:4
2	2	$a, b; 1, c$	x	$\mathrm{F}(a,b;c;x)$	$-\infty < x < 1$	60
2	2	$a, b; 1+a-b, 1$	-1	$\sqrt{\pi}\,\Gamma(1+a-b) \Big/ \left[2^a\Gamma\left(1+\frac{a}{2}-b\right)\Gamma\left(\frac{1+a}{2}\right)\right]$	$a - b \neq -1, -2, -3, \ldots$	60:7
2	2	$a, b; \frac{1+a+b}{2}, 1$	$\frac{1}{2}$	$\sqrt{\pi}\,\Gamma\left(\frac{1+a+b}{2}\right) \Big/ \left[\Gamma\left(\frac{1+a}{2}\right)\Gamma\left(\frac{1+b}{2}\right)\right]$	$a + b \neq -1, -3, -5, \ldots$	60:7
2	2	$a, 1-a; c, 1$	$\frac{1}{2}$	$\sqrt{\pi}\,\Gamma(c) \Big/ \left[2^{c-1}\Gamma\left(\frac{a+c}{2}\right)\Gamma\left(\frac{1+c-a}{2}\right)\right]$	$c \neq 0, -1, -2, \ldots$	60:7
n	$d+1$	$a_1, a_2, \ldots, a_n; 1, c_1, \ldots c_d$	x	${}_n\mathrm{F}_d(a_1,a_2,\ldots,a_n;c_1,c_2,\ldots,c_d;x)$		60:13
2	2	$\frac{1}{2}, \frac{1}{2}; 1, 1$	x^2	$\frac{2}{\pi}\,\mathrm{K}(x)$	$-1 < x < 1$	61
2	2	$\frac{-1}{2}, \frac{1}{2}; 1, 1$	x^2	$\frac{2}{\pi}\,\mathrm{E}(x)$	$-1 < x < 1$	61
2	2	$\frac{1}{2}, \frac{1}{2}; 1, 1$	$\frac{x^2}{x^2-1}$	$\frac{2\sqrt{1-x^2}}{\pi}\,\mathrm{K}(x)$	$\frac{1}{\sqrt{2}} < x < 1$	61
2	2	$\frac{1}{2}, \frac{3}{2}; 1, 1$	x^2	$\frac{2}{\pi(1-x^2)}\,\mathrm{E}(x)$	$-1 < x < 1$	61
3	3	$\frac{1}{2}, \frac{1}{2}, \frac{1}{2}; 1, 1, \frac{3}{2}$	x^2	$\frac{2}{\pi x}\int_0^x \mathrm{K}(t)\mathrm{d}t$	$0 \le x \le 1$	61:10
3	3	$\frac{-1}{2}, \frac{1}{2}, \frac{1}{2}; 1, 1, \frac{3}{2}$	x^2	$\frac{2}{\pi x}\int_0^x \mathrm{E}(t)\mathrm{d}t$	$0 \le x \le 1$	61:10
n	n	$u, \ldots, u; 1+u, \ldots, 1+u$	x	$u^n\Phi(x;n;u)$	$-1 < x < 1;\ n = 1, 2, 3, \ldots$	64:12
n	n	$u, \ldots, u; 1+u, \ldots, 1+u$	-1	$\eta(n;u)$	$n = 1, 2, 3, \ldots$	64:13

Operations of the calculus are easily applied to functions that are expressible as hypergeometric functions. For example, values of the derivative of any function f(x) that appears in Table 18.14.2 may be calculated by taking advantage of the rule

18:14:5
$$\frac{\mathrm{d}}{\mathrm{d}x}\sum_{j=0}^{\infty}\frac{(a_1)_j(a_2)_j\cdots(a_K)_j}{(c_1)_j(c_2)_j\cdots(c_L)_j}X^j = \frac{a_1a_2\cdots a_K}{c_1c_2\cdots c_L}\frac{\mathrm{d}X}{\mathrm{d}x}\sum_{j=0}^{\infty}\frac{(2)_j(a_1+1)_j(a_2+1)_j\cdots(a_K+1)_j}{(c_1+1)_j(c_2+1)_j\cdots(c_L+1)_j(1)_j}X^j$$

which shows the derivative to be another hypergeometric function with an altered and augmented list of numeratorial and denominatorial parameters. The first two rows in Table 18.14.3 show explicitly how the universal hypergeometric algorithm may be adapted to calculate values of $\mathrm{d}f(x)/\mathrm{d}x$. To economize on space, the table makes use of abbreviation 18:14:2.

Similarly, but more usefully, the universal hypergeometric algorithm may often be adapted to generate values of the indefinite integral of f(x). The situation is more complicated than for differentiation because the form adopted by the hypergeometric series for the integral depends on the relationship between x, the argument of the function, and X, the argument of the algorithm. The most commonly encountered relationships are $X = \pm x$, $X = bx^n$ (b is a constant such as $\frac{1}{4}$ or -2; n is a positive integer greater than unity), $X = b/x^n$, $X = b/x$ and $X = b(1-x)$. These five possibilities are all addressed in Table 18.14.3, which itemizes the changes needed to enable the universal hypergeometric algorithm to generate values of an indefinite integral of f(x). For example, using the $J_0^2(x)$ entry from Table 18.14.2 and the fourth row of Table 18.14.3, one sees that

18:14:6 $$\sum_{j=0}^{\infty} \frac{(\frac{1}{2})_j(\frac{1}{2})_j}{(1)_j(1)_j(1)_j(\frac{3}{2})_j}(-x^2)^j = \frac{1}{x}\int_0^x J_0^2(t)\,dt$$

Other operations of the calculus may also be applied to functions expressible as hypergeometric functions. Table 18.14.3 reveals the effects of semidifferentiation and semiintegration; see Oldham and Spanier for other cases of generalized differintegration. Moreover, the Laplace transform [Section 26:14] of the hypergeometric function 18:14:1 is yet another hypergeometric function

18:14:7 $$\frac{1}{s}\sum_{j=0}^{\infty} \frac{(1)_j(a_1)_j(a_2)_j(a_3)_j \cdots (a_K)_j}{(c_1)_j(c_2)_j(c_3)_j \cdots (c_L)_j}\left(\frac{1}{s}\right)^j$$

that has an extra numeratorial parameter.

The table suggests that operations of the calculus invariably increase the number of parameters. This is not necessarily so because, for example, a "new" numeratorial parameter may equal an "old" denominatorial parameter, permitting the two to be "cancelled."

Table 18.14.3

No. of parameters		Parameter values	X	Algorithm evaluates
K	L	$a_1, a_2, \ldots, a_K; c_1, c_2, \ldots, c_L$	X	$f(x)$
$K+1$	$L+1$	$2, a_1+1, a_2+1, \ldots, a_K+1; c_1+1, c_2+1, \ldots, c_L+1, 1$	X	$\frac{df}{dx}(x) \Big/ G_0\frac{dX}{dx}$
$K+1$	$L+1$	$1, a_1, a_2, \ldots, a_K; c_1, c_2, \ldots, c_L, 2$	$\pm x$	$\frac{1}{x}\int_0^x f(t)dt$
$K+1$	$L+1$	$\frac{1}{n}, a_1, a_2, \ldots, a_K; c_1, c_2, \ldots, c_L, \frac{n+1}{n}$	bx^n	$\frac{1}{x}\int_0^x f(t)dt$
$K+1$	$L+1$	$\frac{n-1}{n}, a_1+1, a_2+1, \ldots, a_K+1; c_1+1, c_2+1, \ldots, c_L+1, \frac{2n-1}{n}$	$\frac{b}{x^n}$	$\frac{(n-1)x^{n-1}}{bG_0}\int_x^\infty [f(t)-1]dt$
$K+1$	$L+1$	$1, a_1+2, a_2+2, \ldots, a_K+2; c_1+2, c_2+2, \ldots, c_L+2, 2$	$\frac{b}{x}$	$\frac{x}{b^2G_0G_1}\int_x^\infty \left[f(t)-1-\frac{bG_0}{t}\right]dt$
$K+1$	$L+1$	$1, a_1, a_2, \ldots, a_K; c_1, c_2, \ldots, c_L, 2$	$b(1-x)$	$\frac{1}{1-x}\int_x^1 f(t)dt$
$K+1$	$L+1$	$1, a_1, a_2, \ldots, a_K; c_1, c_2, \ldots, c_L, \frac{1}{2}$	bx	$\sqrt{\pi x}\frac{d^{1/2}}{dx^{1/2}}f(x)$
$K+1$	$L+1$	$1, a_1, a_2, \ldots, a_K; c_1, c_2, \ldots, c_L, \frac{3}{2}$	bx	$\frac{1}{2}\sqrt{\frac{\pi}{x}}\frac{d^{-1/2}}{dx^{-1/2}}f(x)$

CHAPTER 19

THE BERNOULLI POLYNOMIALS $B_n(x)$

Included among the useful properties of the Bernoulli polynomials is their utility for expressing sums of powers of the integers [see Section 19:14].

19:1 NOTATION

$B_n(x)$ is the usual symbol for a Bernoulli polynomial of degree n and argument x, but $\bar{B}_n(x)$ is sometimes used by authors who adopt the "rival notation" [see Section 4:1] for Bernoulli numbers. The name "Bernoulli polynomial" and the symbol $\Phi_n(x)$ have been used for the quantity that we represent by $B_n(x) - B_n$, where B_n is the nth Bernoulli number [Chapter 4].

19:2 BEHAVIOR

Bernoulli polynomials are defined for all nonnegative degrees n and all real values of the argument x. However, $0 \le x \le 1$ is the most important range of the argument for $B_n(x)$.

For even n, $B_n(x)$ never exceeds the Bernoulli number B_n in magnitude in the range $0 \le x \le 1$:

19:2:1 $$|B_n(x)| \le |B_n| \qquad 0 \le x \le 1 \qquad n = 0, 2, 4, \ldots$$

and equals B_n at both ends of this range. Apart from $B_0(x)$, all Bernoulli polynomials of even degree display a local extremum [see Section 0:7] at $x = \frac{1}{2}$, as exemplified in Figure 19-1. Bernoulli polynomials of odd degree (except $B_1(x)$) vanish at $x = 0$, $x = \frac{1}{2}$ and $x = 1$.

The magnitude of $B_n(x)$ for $2 \le n \le 11$ and $0 \le x \le 1$ is small, never exceeding $\frac{1}{6}$. Figure 19-2 shows the typical behavior of Bernoulli polynomials outside these ranges.

19:3 DEFINITIONS

The Bateman manuscript [Erdélyi, Magnus, Oberhettinger and Tricomi, *Higher Transcendental Functions*, Volume 1, pages 38 and 39] gives several integral representations of the Bernoulli polynomials, but their usual definition is via the generating function

19:3:1 $$\frac{t\exp(xt)}{\exp(t) - 1} = \sum_{n=0}^{\infty} B_n(x)\frac{t^n}{n!}$$

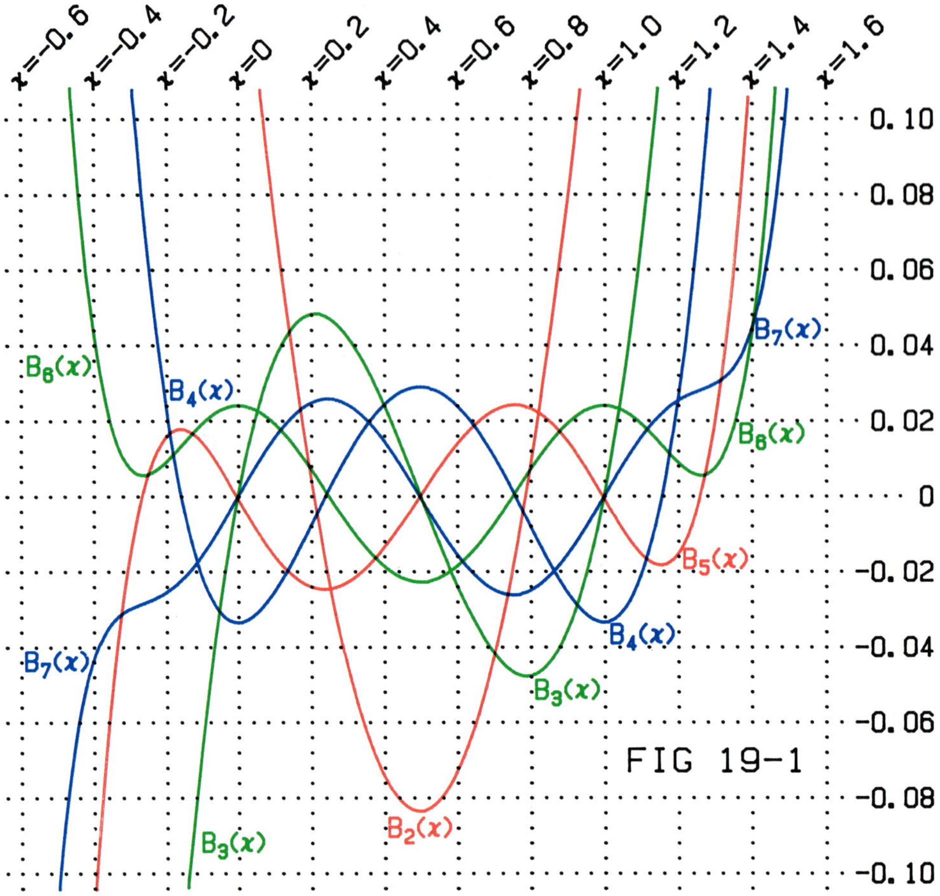

FIG 19-1

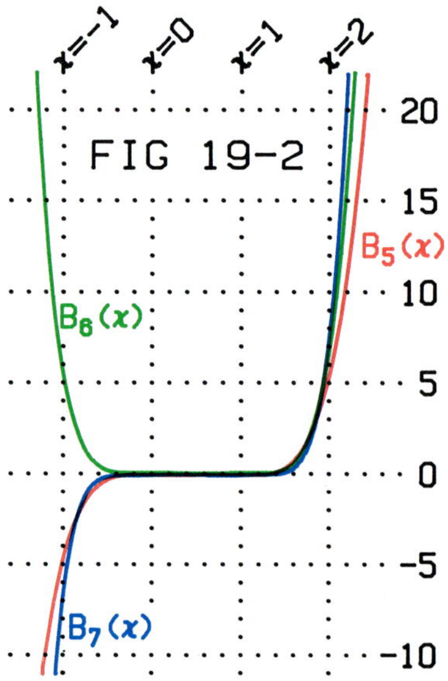

FIG 19-2

Another definition

19:3:2 $$B_n(x) = \sum_{j=0}^{n} \binom{n}{j} B_j x^{n-j} \qquad n = 0, 1, 2, \ldots$$

relates the polynomial coefficients to binomial coefficients [Chapter 6] and Bernoulli numbers [Chapter 4].

19:4 SPECIAL CASES

The first eight Bernoulli polynomials are

19:4:1 $$B_0(x) = 1$$

19:4:2 $$B_1(x) = x - \frac{1}{2}$$

19:4:3 $$B_2(x) = x^2 - x + \frac{1}{6}$$

19:4:4 $$B_3(x) = x^3 - \frac{3x^2}{2} + \frac{x}{2}$$

19:4:5 $$B_4(x) = x^4 - 2x^3 + x^2 - \frac{1}{30}$$

19:4:6 $$B_5(x) = x^5 - \frac{5x^4}{2} + \frac{5x^3}{3} - \frac{x}{6}$$

19:4:7 $$B_6(x) = x^6 - 3x^5 + \frac{5x^4}{2} - \frac{x^2}{2} + \frac{1}{42}$$

and

19:4:8 $$B_7(x) = x^7 - \frac{7x^6}{2} + \frac{7x^5}{2} - \frac{7x^3}{6} + \frac{x}{6}$$

Coefficients of all terms in $B_n(x)$ for $n \le 15$ are listed in Abramowitz and Stegun [Table 23.1].

19:5 INTRARELATIONSHIPS

The Bernoulli polynomials have even or odd symmetry about $x = 1/2$

19:5:1 $$B_n\left(\frac{1}{2} - x\right) = (-1)^n B_n\left(\frac{1}{2} + x\right) \qquad n = 0, 1, 2, \ldots$$

and obey the reflection formula

19:5:2 $$B_n(-x) = (-1)^n[B_n(x) + nx^{n-1}] \qquad n = 0, 1, 2, \ldots$$

The general argument-addition and argument-multiplication formulas

19:5:3 $$B_n(x + y) = \sum_{j=0}^{n} \binom{n}{j} B_j(x) y^{n-j}$$

and

19:5:4 $$B_n(mx) = m^{n-1} \sum_{j=0}^{m-1} B_n\left(x + \frac{j}{m}\right) \qquad n = 0, 1, 2, \ldots \qquad m = 1, 2, 3, \ldots$$

have the important special cases

19:5:5 $$B_n(x + 1) = \sum_{j=0}^{n} \binom{n}{j} B_j(x) = B_n(x) + nx^{n-1}$$

and

19:5:6 $$B_n(2x) = 2^{n-1}\left[B_n(x) + B_n\left(x + \frac{1}{2}\right)\right]$$

Equation 19:5:5 leads to the summation formula

19:5:7 $$\sum_{j=0}^{n} \binom{n+1}{j} B_j(x) = (n+1)x^n$$

when n is replaced by $n+1$.

19:6 EXPANSIONS

Bernoulli polynomials may be expanded as finite power series in x:

19:6:1 $$B_n(x) = x^n - \frac{n}{2}x^{n-1} + \frac{n(n-1)}{12}x^{n-2} - \frac{1}{30}\binom{n}{4}x^{n-4} + \frac{1}{42}\binom{n}{6}x^{n-6} - \cdots + \frac{n(n-1)}{2!}B_{n-2}x^2 + nB_{n-1}x + B_n$$

in $(x - \frac{1}{2})$:

19:6:2 $$B_n(x) = \left(x-\frac{1}{2}\right)^n - \frac{n(n-1)}{24}\left(x-\frac{1}{2}\right)^{n-2} + \frac{7}{240}\binom{n}{4}\left(x-\frac{1}{2}\right)^{n-4} - \cdots - \left[\left(1-\frac{1}{2^{n-3}}\right)\frac{n(n-1)}{2!}B_{n-2}\right]\left(x-\frac{1}{2}\right)^2 - \left[\left(1-\frac{1}{2^{n-2}}\right)nB_{n-1}\right]\left(x-\frac{1}{2}\right) - \left(1-\frac{1}{2^{n-1}}\right)B_n$$

or in $(x-1)$:

19:6:3 $$B_n(x) = (x-1)^n + \frac{n}{2}(x-1)^{n-1} + \frac{n(n-1)}{12}(x-1)^{n-2} - \frac{1}{30}\binom{n}{4}(x-1)^{n-4} + \cdots + \left[\frac{n(n-1)}{2!}B_{n-2}\right](x-1)^2 + [nB_{n-1}](x-1) + B_n$$

In these series many terms vanish because the Bernoulli number B_n [see Chapter 4] is zero when $n = 3, 5, 7, \ldots$. Expansion 19:6:2 can be written concisely in terms of the eta numbers [Chapter 3] if $\eta(0)$ is interpreted as $\frac{1}{2}$:

19:6:4 $$B_n(x) = 2n! \sum_j (-1)^{j/2} \frac{\eta(j)\left(x-\frac{1}{2}\right)^{n-j}}{(n-j)!(2\pi)^j} \qquad j = 0, 2, 4, \ldots, n - \frac{1}{2} \pm \frac{1}{2}$$

A Fourier expansion [see Section 36:6] of a Bernoulli polynomial leads to the general formula

19:6:5 $$B_n(x) = -2n! \sum_{j=1}^{\infty} (2j\pi)^{-n} \cos\left(2j\pi x - \frac{n\pi}{2}\right) \qquad n = 2, 3, 4, \ldots \qquad 0 \le x \le 1$$

which simplifies to

19:6:6 $$B_n(x) = (-1)^{(n/2)+1} 2n! \sum_{j=1}^{\infty} (2j\pi)^{-n} \cos(2j\pi x) \qquad n = 2, 4, 6, \ldots \qquad 0 \le x \le 1$$

or to

19:6:7 $$B_n(x) = (-1)^{(n+1)/2} 2n! \sum_{j=1}^{\infty} (2j\pi)^{-n} \sin(2j\pi x) \qquad n = 3, 5, 7, \ldots \qquad 0 \le x \le 1$$

according as n is even or odd. The last formula may be extended to include $n = 1$ if the points $x = 0$ and $x = 1$ are excluded.

Table 19.7.1

	$B_n(-\infty)$	$B_n(-1)$	$B_n(\frac{-1}{2})$	$B_n(0)$	$B_n(\frac{1}{4})$	$B_n(\frac{1}{2})$	$B_n(\frac{3}{4})$	$B_n(1)$	$B_n(\frac{3}{2})$	$B_n(2)$	$B_n(\infty)$
$n = 0$	1	1	1	1	1	1	1	1	1	1	1
$n = 1, 3, 5, \ldots$	$-\infty$	$-n$	$\frac{-n}{2^{n-1}}$	0	$\frac{-nE_{n-1}}{4^n}$	0	$\frac{nE_{n-1}}{4^n}$	0	$\frac{n}{2^{n-1}}$	n	∞
$n = 2, 4, 6, \ldots$	∞	$n + B_n$	$\frac{n+B_n}{2^{n-1}} - B_n$	B_n	$\left(\frac{1}{2^n} - \frac{2}{4^n}\right)B_n$	$\left(\frac{2}{2^n} - 1\right)B_n$	$\left(\frac{1}{2^n} - \frac{2}{4^n}\right)B_n$	B_n	$\frac{n+B_n}{2^{n-1}} - B_n$	$n + B_n$	∞

19:7 PARTICULAR VALUES

The particular values of $B_n(x)$ depend on the parity of n and are mostly related to the corresponding Bernoulli number B_n [see Chapter 4] or the Euler number E_{n-1} [see Chapter 5], as in Table 19.7.1.

19:8 NUMERICAL VALUES

For small values of n, it is easiest to calculate $B_n(x)$ using the exact formulas of Section 19:4. For $n \geq 5$, the following algorithm may be used. The green segments may be omitted if the argument lies in the $0 \leq x \leq 1$ range.

Input n >>>>>>>
  Set $f = g = 0$
Input x >>>>>>>
  (1) If $x \leq 1$ go to (3)
  Replace x by $x - 1$
  Replace g by $g + nx^{n-1}$
  Go to (1)
  (2) Replace g by $g - nx^{n-1}$
  Replace x by $x + 1$
  (3) If $x < 0$ go to (2)
  Set $j = 3 + \text{Int}[30/(3n - 13)]$
  (4) Replace f by $f + \cos[90(4jx - n)]/j^n$
  Replace j by $j + 1$
  If $j \neq 0$ go to (4)
  Replace f by $[-2n!/(2\pi)^n]f$
  Replace f by $f + g$
  Output f
$f \simeq B_n(x)$ <<<<<<<

Storage needed: n, x, f, g and j

Input restrictions: The degree n must be an integer not less than 5. The argument x is unrestricted if the green commands are included.

Use degree mode or change 90 to $\pi/2$.

Test values:
$B_{11}(0.25) = 0.13249660$
$B_{20}(1/6) = -264.561616$
$B_5\left(\frac{1}{2} \pm \sqrt{\frac{7}{12}}\right) = 0$
$B_7(\pi) = 690.849557$

The algorithm uses expansion 19:6:5 with the infinite summation limit replaced by an empirical function of n that ensures an *absolute* algorithmic accuracy of 6×10^{-8} for $n \geq 5$. The relative error may exceed 6×10^{-8}, especially near the zeros of the Bernoulli polynomials. The colored portions of the algorithm make use of the recursion 19:5:5, or the equivalent

19:8:1 $$B_n(x - 1) = B_n(x) - n(x - 1)^{n-1}$$

to bring an arbitrary argument into the range of applicability of the expansion.

19:9 APPROXIMATIONS

Approximate values for many Bernoulli polynomials in the range $-0.1 \leq B_n(x) \leq 0.1$ can be read directly from Figure 19-1. For large n and arguments between zero and unity, the leading term of expansion 19:6:5 provides the approximation

19:9:1 $$B_n(x) \simeq \frac{-2n!}{(2\pi)^n}\cos\left(2\pi x - \frac{n\pi}{2}\right) \qquad n \to \infty \qquad 0 \le x \le 1$$

For x values close to 0, $\frac{1}{2}$ or 1, approximation formulas are readily derived from expansions 19:6:1–19:6:4. For x values of magnitude large in comparison with n, the approximation

19:9:2 $$B_n(x) \simeq \left(x - \frac{1}{2}\right)^n \qquad \text{8-bit precision} \qquad \left|x - \frac{1}{2}\right| \ge 2.3(n-1)$$

follows from 19:6:2.

19:10 OPERATIONS OF THE CALCULUS

Additional to the general formulas of Section 17:10 are the following special results for the operations of differentiation and integration of Bernoulli polynomials

19:10:1 $$\frac{d^n}{dx^n} B_n(x) = n!$$

19:10:2 $$\frac{d}{dx} B_n(x) = nB_{n-1}(x)$$

19:10:3 $$\int_0^x B_n(t)dt = \frac{B_{n+1}(x) - B_{n+1}}{n+1}$$

19:10:4 $$\int_x^{x+1} B_n(t)dt = x^n \qquad nx \neq 0$$

19:10:5 $$\int_0^1 B_n(t)dt = \begin{cases} 1 & n = 0 \\ 0 & n = 1, 2, 3, \ldots \end{cases}$$

19:10:6 $$\int_0^1 B_n(t)B_m(t)dt = (-1)^{(2+n+m)/2}\frac{m!n!}{(m+n)!}B_{m+n} \qquad m, n = 1, 2, 3, \ldots$$

19:11 COMPLEX ARGUMENT

Bernoulli polynomials are defined for real arguments only.

19:12 GENERALIZATIONS

A class of functions defined by the generating function

19:12:1 $$\frac{t^m \exp(xt)}{[\exp(t) - 1]^m} = \sum_{n=0}^{\infty} B_n^{(m)}(x)\frac{t^n}{n!} \qquad m = 0, 1, 2, \ldots$$

is known as *Bernoulli polynomials of order m* [see Korn and Korn, page 824]. They represent a generalization of the Bernoulli polynomials discussed in this chapter, which are of first order. A still more general set of *higher order Bernoulli polynomials* is discussed by Erdélyi, Magnus, Oberhettinger and Tricomi [*Higher Transcendental Functions*, Volume 1, pages 39 and 40].

19:13 COGNATE FUNCTIONS

Bernoulli functions are closely related to the Euler polynomials that are the subject of Chapter 20. Equations 20:3:3 and 20:3:4 make this connection explicit.

19:14 RELATED TOPICS

From integrals 19:10:4 and 19:10:3 it follows that

19:14:1 $$x^n + (1+x)^n + (2+x)^n + \cdots + (m-1+x)^n = \frac{B_{n+1}(m+x) - B_{n+1}(x)}{n+1}$$

The most important applications of this general formula arise by setting x equal to unity;

19:14:2 $$1^n + 2^n + 3^n + \cdots + m^n = \frac{B_{n+1}(m+1) + (-1)^n B_{n+1}}{n+1} \qquad n = 0, 1, 2, \ldots$$

or a moiety:

19:14:3 $$1^n + 3^n + 5^n + \cdots + (2m-1)^n = \frac{B_{n+1}(2m-1) - 2^n B_{n+1}(m) - (2^n - 1)B_{n+1}}{n+1} \qquad n = 0, 1, 2, \ldots$$

Further simplification occurs if $n = 2, 4, 6, \ldots$ because B_{n+1} is then zero. Some examples of these finite summations are presented in Section 1:14. See Section 20:14 for cases of alternating signs.

CHAPTER 20

THE EULER POLYNOMIALS $E_n(x)$

Certain series involving the natural numbers can be conveniently expressed using Euler polynomials, as explained in Section 20:14.

20:1 NOTATION

Euler polynomials of degree n and argument x are generally denoted $E_n(x)$, although the symbol $\bar{E}_n(x)$ is occasionally encountered. The $E_n(x)$ symbolism is also used to denote Schlömilch functions, which are quite unrelated to Euler polynomials. The Schlömilch functions are briefly discussed in Section 37:14, but they are not used elsewhere in this *Atlas*.

20:2 BEHAVIOR

The Euler polynomials $E_n(x)$ are defined for all nonnegative integer n and for all real values of x. Figure 20-1 shows the behavior of the first few Euler polynomials in the range $0 \le x \le 1$, which is the most important range for these functions. Over this range the inequality

20:2:1 $$0 \le \frac{E_n(x)}{E_n} \le 2^{-n} \qquad 0 \le x \le 1 \qquad n = 0, 2, 4, \ldots$$

is valid for even n, E_n being the nth Euler number [Chapter 5].

As is evident from the diagram, all Euler polynomials of even degree (except $E_0(x)$) are zero at $x = 0$ and $x = 1$ and exhibit a local extremum [see Section 0:7] at $x = \frac{1}{2}$. In complementary fashion, all Euler polynomials of odd degree are zero at $x = \frac{1}{2}$ and (except for $E_1(x)$) display a local maximum or minimum at $x = 0$ and $x = 1$.

Figure 20-2 demonstrates the behavior of typical Euler polynomials outside the $0 \le x \le 1$ range.

20:3 DEFINITIONS

Euler polynomials are defined by the generating function [Section 0:3]

20:3:1 $$\frac{2\exp(xt)}{1 + \exp(t)} = \sum_{n=0}^{\infty} E_n(x) \frac{t^n}{n!}$$

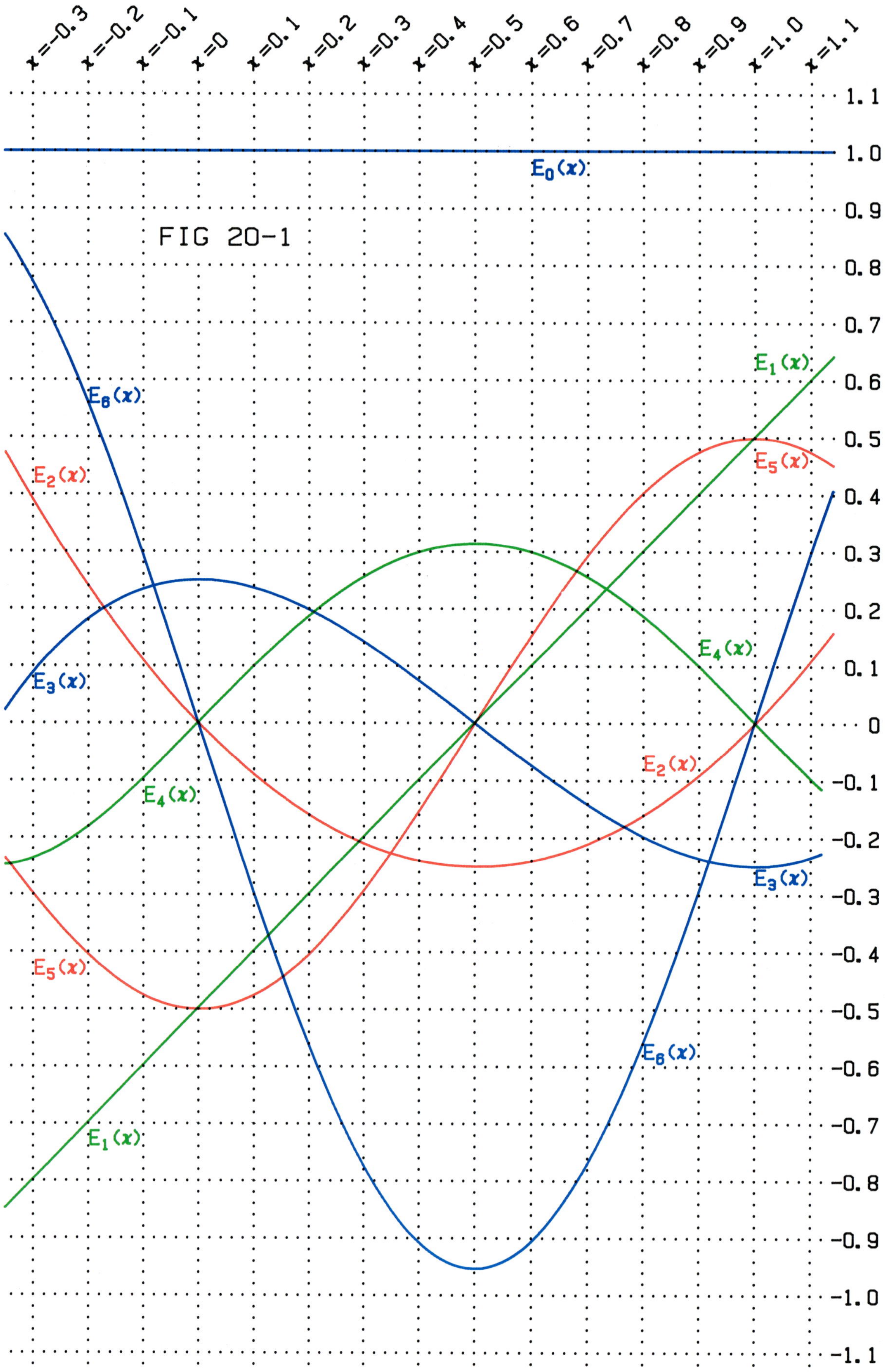
x=-0.3
x=-0.2
x=-0.1
x=0
x=0.1
x=0.2
x=0.3
x=0.4
x=0.5
x=0.6
x=0.7
x=0.8
x=0.9
x=1.0
x=1.1
1.1
1.0
0.9
0.8
0.7
0.6
0.5
0.4
0.3
0.2
0.1
0
-0.1
-0.2
-0.3
-0.4
-0.5
-0.6
-0.7
-0.8
-0.9
-1.0
-1.1
E0(x)
E1(x)
E2(x)
E3(x)
E4(x)
E5(x)
E6(x)

FIG 20-1

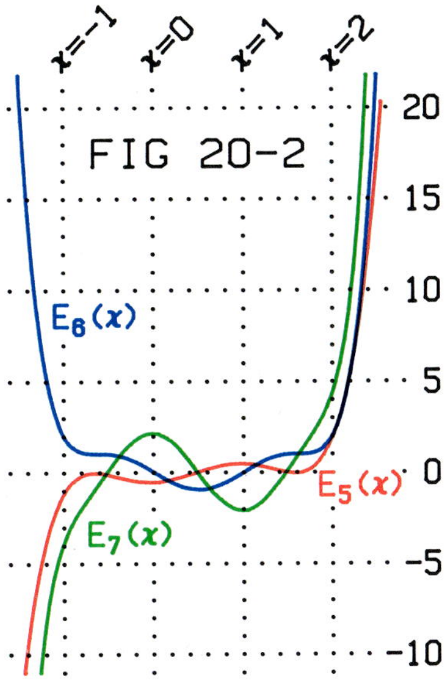

An alternative definition in terms of Euler numbers E_j [Chapter 5] is

20:3:2
$$E_n(x) = 2^{-n} \sum_{j=0}^{n} \binom{n}{j} (2x-1)^{n-j} E_j$$

The formulas

20:3:3
$$E_n(x) = \frac{2^{n+1}}{n+1}\left[B_{n+1}\left(\frac{x+1}{2}\right) - B_{n+1}\left(\frac{x}{2}\right)\right] \qquad n = 1, 2, 3, \ldots$$

and

20:3:4
$$E_n(x) = \frac{2}{n+1}\left[B_{n+1}(x) - 2^{n+1}B_{n+1}\left(\frac{x}{2}\right)\right] \qquad n = 1, 2, 3, \ldots$$

relating Euler polynomials to Bernoulli polynomials [Chapter 19] also serve as definitions.

20:4 SPECIAL CASES

The first eight Euler polynomials are

20:4:1
$$E_0(x) = 1$$

20:4:2
$$E_1(x) = x - \frac{1}{2}$$

20:4:3
$$E_2(x) = x^2 - x$$

20:4:4
$$E_3(x) = x^3 - \frac{3x^2}{2} + \frac{1}{4}$$

20:4:5
$$E_4(x) = x^4 - 2x^3 + x$$

20:4:6
$$E_5(x) = x^5 - \frac{5x^4}{2} + \frac{5x^2}{2} - \frac{1}{2}$$

20:4:7
$$E_6(x) = x^6 - 3x^5 + 5x^3 - 3x$$

and

20:4:8 $$E_7(x) = x^7 - \frac{7x^6}{2} + \frac{35x^4}{4} - \frac{21x^2}{2} + \frac{17}{8}$$

Coefficients of all terms in $E_n(x)$ for $n \le 15$ are listed by Abramowitz and Stegun [Table 23.1].

20:5 INTRARELATIONSHIPS

The Euler polynomials possess even or odd symmetry about $x = \frac{1}{2}$:

20:5:1 $$E_n\left(\frac{1}{2} - x\right) = (-1)^n E_n\left(\frac{1}{2} + x\right)$$

and obey the reflection formula

20:5:2 $$E_n(-x) = (-1)^n[2x^n - E_n(x)]$$

The argument-addition formula

20:5:3 $$E_n(x + y) = \sum_{j=0}^{n} \binom{n}{j} E_j(x) y^{n-j}$$

has the special cases

20:5:4 $$E_n(1 + x) = \sum_{j=0}^{n} \binom{n}{j} E_j(x) = 2x^n - E_n(x) = (-1)^n E_n(-x)$$

and

20:5:5 $$E_n\left(\frac{1}{2} + x\right) = x^n \sum_{j=0}^{n} \binom{n}{j} \frac{E_j}{(2x)^j}$$

Because all Euler numbers of odd index are zero, about half of the terms in the last summation [which is equivalent to 20:3:2] vanish.

The argument-multiplication formula takes different forms:

20:5:6 $$E_n(mx) = \frac{-2m^n}{n+1} \sum_{j=0}^{m-1} (-1)^j B_{n+1}\left(x + \frac{j}{m}\right) \qquad m = 2, 4, 6, \ldots$$

20:5:7 $$E_n(mx) = m^n \sum_{j=0}^{m-1} (-1)^j E_n\left(x + \frac{j}{m}\right) \qquad m = 1, 3, 5, \ldots$$

according to whether the multiplier is even or odd. Definition 20:3:3 is a special case of formula 20:5:6.

20:6 EXPANSIONS

Definition 20:3:2 leads immediately to the polynomial

20:6:1 $$E_n(x) = \left(x - \frac{1}{2}\right)^n - \frac{n(n-1)}{8}\left(x - \frac{1}{2}\right)^{n-2} + \frac{5n(n-1)(n-2)(n-3)}{384}\left(x - \frac{1}{2}\right)^{n-4} + \cdots$$
$$+ \frac{n(n-1)}{2}\frac{E_{n-2}}{2^{n-2}}\left(x - \frac{1}{2}\right)^2 + \frac{nE_{n-1}}{2^{n-1}}\left(x - \frac{1}{2}\right) + \frac{E_n}{2^n}$$

in $x - \frac{1}{2}$; all terms containing odd-indexed Euler numbers are zero. Polynomials in x:

20:6:2 $$E_n(x) = x^n - \frac{n}{2}x^{n-1} + \frac{1}{4}\binom{n}{3}x^{n-3} - \frac{1}{2}\binom{n}{5}x^{n-5} + \frac{17}{8}\binom{n}{7}x^{n-7} + \cdots$$

or $x - 1$:

20:6:3 $$E_n(x) = (x-1)^n + \frac{n}{2}(x-1)^{n-1} - \frac{1}{4}\binom{n}{3}(x-1)^{n-3} + \frac{1}{2}\binom{n}{5}(x-1)^{n-5} - \frac{17}{8}\binom{n}{7}(x-1)^{n-7} + \cdots$$

may also be written; these contain $(n + 2)/2$ terms if n is even, or $(n + 3)/2$ terms if n is odd; see Section 20:4 for the coefficients in expansion 20:6:2 when n adopts specific small values.

The Fourier expansion [see Section 36:6]

20:6:4 $$E_n(x) = \frac{4n!}{\pi^{n+1}} \sum_{j=0}^{\infty} (2j+1)^{-n-1} \sin\left[(2j+1)\pi x - \frac{n\pi}{2}\right] \qquad n = 1, 2, 3, \ldots \qquad 0 \le x \le 1$$

reduces to

20:6:5 $$E_n(x) = \frac{(-1)^{n/2} 4n!}{\pi^{n+1}} \sum_{j=0}^{\infty} \frac{\sin[(2j+1)\pi x]}{(2j+1)^{n+1}} \qquad n = 2, 4, 6, \ldots \qquad 0 \le x \le 1$$

when the degree is even, and to

20:6:6 $$E_n(x) = \frac{(-1)^{(n+1)/2} 4n!}{\pi^{n+1}} \sum_{j=0}^{\infty} \frac{\cos[(2j+1)\pi x]}{(2j+1)^{n+1}} \qquad n = 1, 3, 5, \ldots \qquad 0 \le x \le 1$$

when n is odd.

20:7 PARTICULAR VALUES

As is the case with the Bernoulli polynomials [Chapter 19], particular values of the Euler polynomials depend on the parity of n and are most easily related to the Bernoulli numbers B_n [see Chapter 4] and the Euler numbers [see Chapter 5]. Some values of $E_n(x)$ may be found in Table 20.7.1.

Table 20.7.1

	$E_n(-\infty)$	$E_n(-1)$	$E_n(\frac{-1}{2})$	$E_n(0)$	$E_n(\frac{1}{2})$	$E_n(1)$	$E_n(\frac{3}{2})$	$E_n(2)$	$E_n(\infty)$
$n = 0$	1	1	1	1	1	1	1	1	1
$n = 1, 3, 5, \ldots$	$-\infty$	$\frac{2^{n+2}-2}{n+1}B_{n+1} - 2$	$\frac{-1}{2^{n-1}}$	$\frac{2-2^{n+2}}{n+1}B_{n+1}$	0	$\frac{2^{n+2}-2}{n+1}B_{n+1}$	$\frac{1}{2^{n-1}}$	$2 - \frac{2^{n+2}-2}{n+1}B_{n+1}$	∞
$n = 2, 4, 6, \ldots$	∞	2	$\frac{2-E_n}{2^n}$	0	$\frac{E_n}{2^n}$	0	$\frac{2-E_n}{2^n}$	2	∞

20:8 NUMERICAL VALUES

When n is small, exact values of $E_n(x)$ are best found by using the polynomial coefficients [Section 20:4] in conjunction with the general algorithm of Section 17:8. The algorithm presented in this section may be more convenient if n is large.

Input n >>>>>>>
Input x >>>>>>>
Set $s = f = 1$
If $n = 0$ go to (5)
Set $f = g = 0$
(1) If $x \le 1$ go to (3)
Replace x by $x - 1$
Replace g by $g + 2sx^n$
Replace s by $-s$
Go to (1)

Storage needed: n, x, s, f, g and j

Input restrictions: n must be an integer not less than 5. x is unrestricted if the green commands are included.

Use degree mode or change 90 to $\pi/2$.

(2) Replace g by $g + 2sx^n$
Replace s by $-s$
Replace x by $x + 1$
(3) If $x < 0$ go to (2)
Set $j = 1.5 + \text{Int}(30/n)$
(4) Replace f by $f + \sin[90(4jx - n)]/j^{n+1}$
Replace j by $j - 1$
If $j > 0$ go to (4)
Replace f by $[4sn!/(2\pi)^{n+1}]f$
Replace f by $f + g$
(5) Output f
$f \simeq E_n(x)$ <<<<<<

Test values:
$E_6(0.4) = -0.906624$
$E_{13}(1) = 2730.5$
$E_8(-1.5) = 56.66015625$
$E_9(\pi) = 1902.63483$

For arguments in the range $0 \le x \le 1$, the portions of the algorithm shown in green may be omitted. The algorithm is based on the Fourier expansion 20:6:4, truncated at $j = 1 + \text{Int}(30/n)$. This choice ensures that the *absolute* error in $E_n(x)$ does not exceed 6×10^{-8} for $n \ge 5$, although the *relative* error may be greater than this, particularly near a zero of $E_n(x)$. When the green portions of the algorithm are included, arguments of any magnitude can be accommodated. Recursion 20:5:4, or its equivalent:

20:8:1 $$E_n(x - 1) = 2(x - 1)^n - E_n(x)$$

is employed to shift the argument into the range covered by the Fourier expansion.

20:9 APPROXIMATIONS

For small n and arguments between zero and unity, Figure 20-1 can provide approximate values of $E_n(x)$. For large n and arguments between zero and unity, the leading term of expansion 20:6:4 provides the approximation

20:9:1 $$E_n(x) \simeq \frac{4n!}{\pi^{n+1}} \sin\left(\pi x - \frac{n\pi}{2}\right) \qquad n \to \infty \qquad 0 \le x \le 1$$

Approximation formulas may be derived from 20:6:2, 20:6:1 or 20:6:3 for x values close to 0, $\frac{1}{2}$ or 1, respectively. For x values of large magnitude, the approximation

20:9:2 $$E_n(x) \simeq \left(x - \frac{1}{2}\right)^n \qquad \text{8-bit precision} \qquad \left|x - \frac{1}{2}\right| > 4(n - 1)$$

follows from 20:6:1.

20:10 OPERATIONS OF THE CALCULUS

In addition to results that follow from the general formulas of Section 17:10, there are several special results from applying the differentiation and integration operators to the Euler polynomial. These include

20:10:1 $$\frac{d^n}{dx^n} E_n(x) = n!$$

20:10:2 $$\frac{d}{dx} E_n(x) = nE_{n-1}(x)$$

20:10:3 $$\int_{x_0}^{x} E_n(t)dt = \frac{E_{n+1}(x) - E_{n+1}(x_0)}{n + 1} = \frac{E_{n+1}(x)}{n + 1} \qquad \begin{cases} n = 1, 3, 5, \ldots & x_0 = 0 \\ n = 0, 2, 4, \ldots & x_0 = \dfrac{1}{2} \end{cases}$$

and

20.10:4 $$\int_0^1 E_n(t)E_m(t)dt = (-1)^{(m+n)/2}(2^{m+n+2} - 1)\frac{4m!n!}{(m+n+2)!}B_{m+n+2}$$

20:11 COMPLEX ARGUMENT

Euler polynomials are defined for real argument only.

20:12 GENERALIZATIONS

Euler polynomials may be generalized in the manner described in Erdelyi, Magnus, Oberhettinger and Tricomi [*Higher Transcendental Functions*, Volume 1, page 43].

20:13 COGNATE FUNCTIONS

Bernoulli polynomials [Chapter 19] are closely related to Euler polynomials. Equations 20:3:3 and 20:3:4 establish a direct link.

20:14 RELATED TOPICS

Iteration of recurrence formula 20:5:4 leads to

20:14:1 $$x^n - (1+x)^n + (2+x)^n - \cdots \mp (m-2+x)^n \pm (m-1+x)^n = \frac{E_n(x) \pm E_n(m+x)}{2}$$

where the upper/lower signs correspond to odd/even m. Setting $x = 1$ in this relationship yields the useful result

20:14:2 $$1^n - 2^n + 3^n - \cdots \mp (m-1)^n \pm m^n = \frac{2^{n+1}-1}{n+1}B_{n+1} \pm \frac{E_n(m+1)}{2}$$

while setting $x = \frac{1}{2}$ generates

20:14:3 $$1^n - 3^n + 5^n - \cdots \mp (2m-3)^n \pm (2m-1)^n = \frac{E_n \pm 2^n E_n(m+{}^1/_2)}{2}$$

The first of these results is presented in Chapter 1 as equation 1:14:9, and equations 1:14:6–1:14:8 are special cases.

CHAPTER

21

THE LEGENDRE POLYNOMIALS $P_n(x)$

The Legendre polynomials arise in several branches of applied mathematics (as, for example, in describing electrostatic or other fields), especially in problems involving spherical symmetry. The Legendre polynomials are among the simplest of the families of *orthogonal polynomials* [see Section 21:14].

21:1 NOTATION

The symbol $P_n(x)$ is standard for the Legendre polynomial of degree n and argument x, although the P is often italicized. The name *spherical polynomial* is also encountered. Sometimes *normalized* Legendre polynomials are specified; the polynomial $P_n(x)$ of this chapter is normalized through multiplication by $\sqrt{(2n+1)/2}$ [see Section 21:14]. Yet another name is zonal surface harmonic [see Section 59:14].

The symbol $P_\nu(x)$, where ν is not necessarily an integer, denotes a member of the function family known as *Legendre functions of the first kind*, which are a subject of Chapter 59. When ν takes any integer value, positive, negative or zero, $P_\nu(x)$ reduces to a Legendre polynomial

21:1:1 $$P_\nu(x) = P_n(x) \qquad \nu = n = 0, 1, 2, \ldots$$

21:1:2 $$P_\nu(x) = P_{n-1}(x) \qquad \nu = -n = -1, -2, -3, \ldots$$

The polynomials of the present chapter are thus special instances of Legendre functions of the first kind and sometimes are so identified. The reader is referred to Sections 21:12 and 21:13 for explanations of the related notations $P_n^*(x)$, $P_\nu^\mu(x)$, $P_n^{(\alpha,\beta)}(x)$ and $P''(x)$.

For $-1 \le x \le 1$ the argument x is often replaced by the cosine of a subsidiary variable θ:

21:1:3 $$P_n(\cos(\theta)) = P_n(x) \qquad \theta = \arccos(x)$$

as a reflection of the fact that Legendre polynomials often arise naturally in scientific and engineering applications with arguments that are cosines of salient angles.

21:2 BEHAVIOR

In this chapter we restrict the degree n to be a nonnegative integer; $n = 0, 1, 2, \ldots$. Although the argument x of the Legendre polynomial $P_n(x)$ may take any real value, the most important range is $-1 \le x \le 1$; this is the only range accessible when x is replaced by $\cos(\theta)$. Within this range the polynomials never exceed unity in magnitude

21:2:1 $$|P_n(x)| \leq 1 \qquad -1 \leq x \leq 1$$

as illustrated in Figure 21-1. The Legendre polynomial $P_n(x)$ has exactly n real zeros, all of which lie in the interval $-1 < x < 1$.

Outside the $-1 \leq x \leq 1$ range, the Legendre polynomial $P_n(x)$ increases in magnitude without limit (except

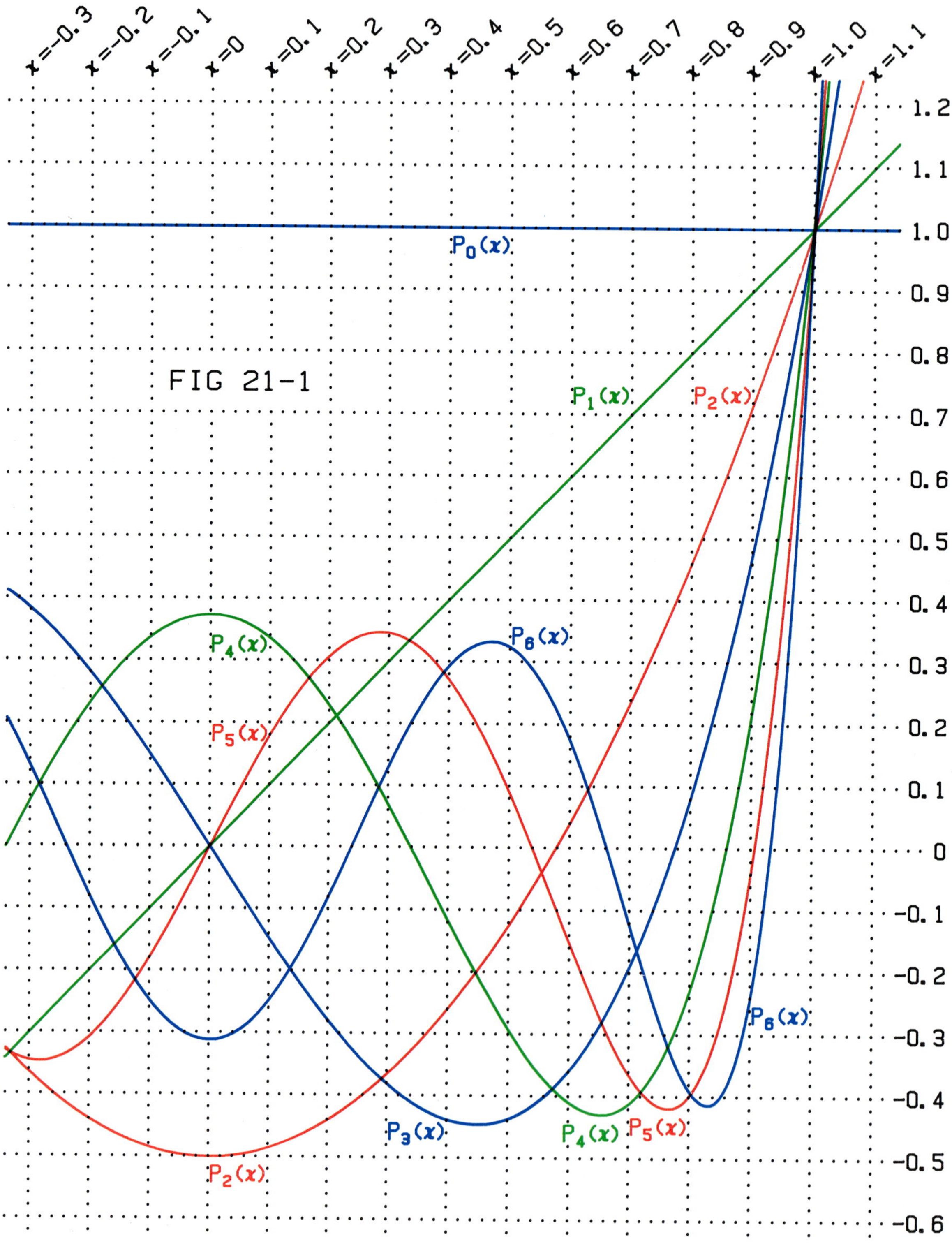

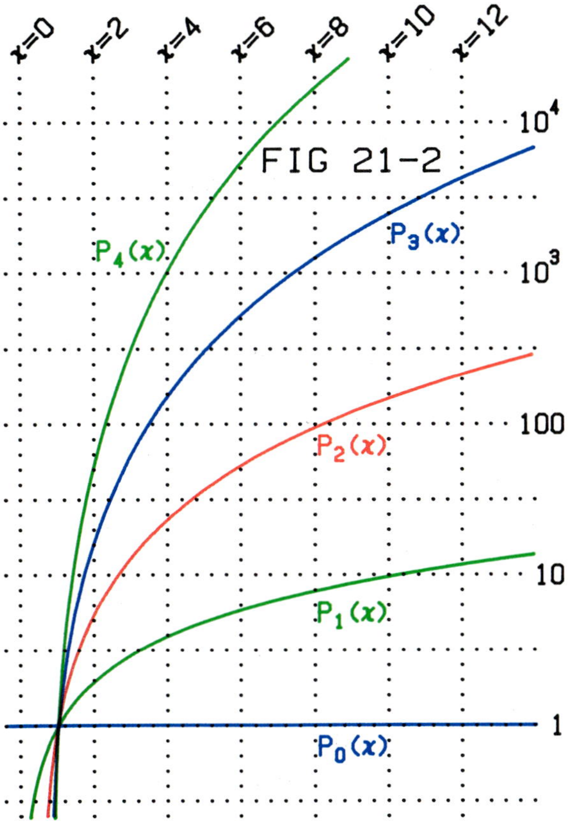

for $P_0(x)$) as $x \to \pm\infty$. This behavior is exemplified in Figure 21-2 and conforms to the rules

21:2:2 $$1 = P_0(x) < P_1(x) < P_2(x) < \cdots \qquad x > 1$$

21:2:3 $$1 = P_0(x) < -P_1(x) < P_2(x) < -P_3(x) < \cdots \qquad x < -1$$

21:3 DEFINITIONS

Legendre polynomials may be defined by means of the generating function [see Section 0:3]

21:3:1 $$\frac{1}{\sqrt{1 - 2xt + t^2}} = \begin{cases} \sum_{n=0}^{\infty} P_n(x)t^n & -1 < t < 1 \qquad -1 \le x \le 1 \\ \sum_{n=0}^{\infty} P_n(x)t^{-n-1} & |t| > 1 \qquad -1 \le x \le 1 \end{cases}$$

The *Rodrigues' formula* for Legendre polynomials

21:3:2 $$P_n(x) = \frac{1}{2^n n!}\frac{d^n}{dx^n}(x^2 - 1)^n$$

also serves as a definition.

The explicit representation as a polynomial

21:3:3 $$P_n(x) = \sum_j (-1)^{j/2}\frac{(2n - j - 1)!!}{j!!(n - j)!}x^{n-j} \qquad j = 0, 2, 4, \ldots, n - \frac{1}{2} \pm \frac{1}{2}$$

shows that Legendre polynomials contain $(n + 2)/2$ terms if n is even and $(n + 1)/2$ terms if n is odd. An equivalent expression for $|x| < 1$ is

21:3:4 $$P_n(\cos(\theta)) = \frac{1}{2^n}\sum_{j=0}^{n}\frac{(2j-1)!!(2n-2j-1)!!}{j!(n-j)!}\cos[(n-2j)\theta]$$

Two representations of Legendre polynomials by integrals are

21:3:5 $$P_n(\cos(\theta)) = \frac{\sqrt{2}}{\pi}\int_0^{\pi}\frac{\sin\left[\left(n+\frac{1}{2}\right)t\right]dt}{\cos(\theta)-\cos(t)}$$

and

21:3:6 $$P_n(x) = \frac{1}{\pi}\int_0^{\pi}(x+\sqrt{x^2-1}\cos(t))^n dt \qquad x > 1$$

the latter being known as *Laplace's integral representation*.

The Legendre polynomial function $f = P_n(x)$ satisfies *Legendre's differential equation*

21:3:7 $$(1-x^2)\frac{d^2f}{dx^2} - 2x\frac{df}{dx} + n(n+1)f = 0$$

The most general solution of this equation is $c_1P_n(x) + c_2Q_n(x)$, where c_1 and c_2 are arbitrary constants and $Q_n(x)$ is the function discussed in Section 21:13.

Figure 21-3 illustrates a geometric definition that explains how Legendre polynomials arise in certain applications. The triangle shown has two of its sides, one of unity length and one of length r, enclosing the angle θ. By the *cosine law* [see Section 34:14], the length z of the third side equals $\sqrt{1-2r\cos(\theta)+r^2}$. In some physical problems it is necessary to expand z^{-1} as a power series in r^{-1}. Formula 21:3:1 shows this series to be

21:3:8 $$\frac{1}{z} = \frac{P_0(\cos(\theta))}{r} + \frac{P_1(\cos(\theta))}{r^2} + \frac{P_2(\cos(\theta))}{r^3} + \cdots$$

if $r > 1$, or a similar series if $r < 1$.

FIG 21-3

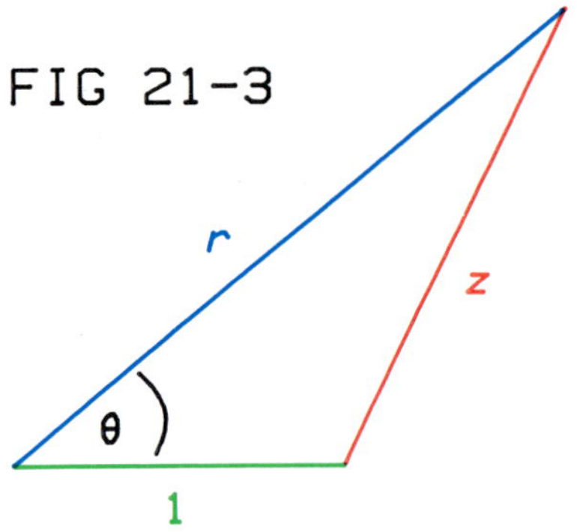

21:4 SPECIAL CASES

The first eight polynomials are

21:4:1 $$P_0(x) = 1 = P_0(\cos(\theta))$$

21:4:2 $$P_1(x) = x = \cos(\theta)$$

21:4:3 $$P_2(x) = \frac{3x^2-1}{2} = \frac{1+3\cos(2\theta)}{4}$$

21:4:4 $$P_3(x) = \frac{5x^3 - 3x}{2} = \frac{3\cos(\theta) + 5\cos(3\theta)}{8}$$

21:4:5 $$P_4(x) = \frac{35x^4 - 30x^2 + 3}{8} = \frac{9 + 20\cos(2\theta) + 35\cos(4\theta)}{64}$$

21:4:6 $$P_5(x) = \frac{63x^5 - 70x^3 + 15x}{8} = \frac{30\cos(\theta) + 35\cos(3\theta) + 63\cos(5\theta)}{128}$$

21:4:7 $$P_6(x) = \frac{231x^6 - 315x^4 + 105x^2 - 5}{16} = \frac{50 + 105\cos(2\theta) + 126\cos(4\theta) + 231\cos(6\theta)}{512}$$

21:4:8 $$P_7(x) = \frac{429x^7 - 693x^5 + 315x^3 - 35x}{16} = \frac{175\cos(\theta) + 189\cos(3\theta) + 231\cos(5\theta) + 429\cos(7\theta)}{1024}$$

The formulas given above for $P_n(x)$ apply for all values of x, whereas those for $P_n(\cos(\theta))$ are restricted to $-1 \le \cos(\theta) \le 1$.

21:5 INTRARELATIONSHIPS

Legendre polynomials are even or odd

21:5:1 $$P_n(-x) = (-1)^n P_n(x)$$

according to the parity of n. The recurrence formula

21:5:2 $$P_n(x) = \frac{2n-1}{n} x P_{n-1}(x) - \frac{n-1}{n} P_{n-2}(x) \qquad n = 2, 3, 4, \ldots$$

relates three polynomials of consecutive degrees.

Positive integer powers can be written as finite series of Legendre polynomials. For example:

21:5:3 $$x^2 = \frac{1}{3} P_0(x) + \frac{2}{3} P_2(x)$$

21:5:4 $$x^3 = \frac{3}{5} P_1(x) + \frac{2}{5} P_3(x)$$

21:5:5 $$x^4 = \frac{1}{5} P_0(x) + \frac{4}{7} P_2(x) + \frac{8}{35} P_4(x)$$

Other well-behaved functions of x can be expressed as infinite series of Legendre polynomials $P_n(x)$, and a collection of such formulas is presented by Gradshteyn and Ryzhik [Section 8.92]. Many such series expansions may be obtained by the procedure explained in Section 21:14.

Among summation formulas for Legendre polynomials are

21:5:6 $$(2n-1)P_{n-1}(x) + (2n-3)P_{n-2}(x) + (2n-5)P_{n-3}(x) + \cdots + 3P_1(x) + P_0(x) = \frac{n[P_{n-1}(x) - P_n(x)]}{1-x} \qquad x \neq 1$$

and

21:5:7 $$(2n-1)P_{n-1}(x) - (2n-3)P_{n-2}(x) + (2n-5)P_{n-3}(x) - \cdots \pm 3P_1(x) \mp P_0(x) = \frac{n[P_{n-1}(x) + P_n(x)]}{1+x} \qquad x \neq -1$$

21:6 EXPANSIONS

Definition 21:3:3 leads to the finite series

21:6:1
$$P_n(x) = \begin{cases} (-1)^{n/2}\dfrac{(n-1)!!}{(n/2)!}\left[1 - \dfrac{n(n+1)}{2!}x^2 + \dfrac{(n-2)n(n+1)(n+3)}{4!}x^4 - \cdots\right] & n = 0, 2, 4, \ldots \\ (-1)^{(n-1)/2}\dfrac{n!!}{\left(\dfrac{n-1}{2}\right)!}\left[x - \dfrac{(n-1)(n+2)}{3!}x^3 + \dfrac{(n-3)(n-1)(n+2)(n+4)}{5!}x^5 - \cdots\right] & n = 1, 3, 5, \ldots \end{cases}$$

that are valid for all arguments x. The infinite Fourier series

21:6:2
$$P_n(\cos(\theta)) = \frac{2}{\pi}\sum_{j=0}^{\infty}\frac{(j+1)_n}{(j+{}^1\!/_2)_{n+1}}\sin[(n+1+2j)\theta]$$

involves coefficients that are ratios of Pochhammer polynomials [Chapter 18]. A third expansion is

21:6:3
$$P_n(x) = \frac{1}{2}\sum_{j=0}^{n}\frac{(n+j)!}{(n-j)!(j!)^2}\left[\left(\frac{x-1}{2}\right)^j + (-1)^{n-j}\left(\frac{x+1}{2}\right)^j\right]$$

Other expansions can be found in Gradshteyn and Ryzhik [Section 8.911] and yet others may be derived from the formulas in Section 59:6 with $\nu = n$, or by combining equations 21:12:1–21:12:4 with the expansions given in Chapter 60.

21:7 PARTICULAR VALUES

	$P_n(-\infty)$	$P_n(-1)$	$P_n(0)$	$P_n(1)$	$P_n(\infty)$
$n = 0$	1	1	1	1	1
$n = 1, 3, 5, \ldots$	$-\infty$	-1	0	1	∞
$n = 2, 4, 6, \ldots$	∞	1	$\dfrac{(-1)^{n/2}(n-1)!!}{n!!}$	1	∞

21:8 NUMERICAL VALUES

For small n, values of $P_n(x)$ are easily found from the expressions in Section 21:4. The algorithm

```
Input n >>>>>>>
Input x >>>>>>>
          Set f = g = x
          If n = 1 go to (2)
          Set f = h = j = 1
          If n = 0 go to (2)
      (1) Replace j by j + 1
          Replace f by [(2j − 1)xg − (j − 1)h]/j
          If j = n go to (2)
          Set h = g
          Set g = f
          Go to (1)
      (2) Output f
f = Pn(x) <<<<<<<
```

Storage needed: n, x, f, g, h and j

Test values:
$P_0(\text{any } x) = 1$
$P_1(x) = x$
$P_{12}(\pm 3) = 251595969$
$P_5\left(\pm\sqrt{\dfrac{35 \pm \sqrt{280}}{63}}\right) = 0$
(but see text)

utilizes recurrence 21:5:2 and is exact although, of course, rounding may cause large relative errors close to the zeros of $P_n(x)$.

21:9 APPROXIMATE VALUES

For $0 \le n \le 6$ and $0 \le x \le 1$, approximate values of $P_n(x)$ may be read from Figure 21-1, and extension to the range $-1 \le x \le 0$ is accomplished using 21:5:1. For arguments of large magnitude Legendre polynomials are approximated by

21:9:1 $$P_n(x) \simeq \frac{(2n-1)!!x^n}{n!} \qquad \text{8-bit precision} \qquad |x| > 8\sqrt{n}$$

which follows from expansion 21:3:3.

21:10 OPERATIONS OF THE CALCULUS

Differentiation of a Legendre polynomial gives

21:10:1 $$\frac{d}{dx}P_n(x) = \frac{n}{1-x^2}[P_{n-1}(x) - xP_n(x)] \qquad n = 1, 2, 3, \ldots$$

while integration produces

21:10:2 $$\int_{-1}^{x} P_n(t)dt = \frac{P_{n+1}(x) - P_{n-1}(x)}{2n+1} \qquad n = 1, 2, 3, \ldots$$

From this last result it follows that

21:10:3 $$\int_{-1}^{1} P_n(t)dt = 0 \qquad n = 1, 2, 3, \ldots$$

in view of 21:7:4. Other definite integrals involving Legendre polynomials include

21:10:4 $$\int_{-1}^{1} \frac{P_n(t)dt}{\sqrt{1-t}} = \frac{\sqrt{8}}{2n+1}$$

21:10:5 $$\int_{-1}^{1} \frac{P_n(t)dt}{\sqrt{1-t^2}} = \begin{cases} \pi\left[\dfrac{(n-1)!!}{n!!}\right]^2 & n = 0, 2, 4, \ldots \\ 0 & n = 1, 3, 5, \ldots \end{cases}$$

21:10:6 $$\int_{-1}^{1} t^m P_n(t)dt = \begin{cases} 0 & m = n-1, n-2, n-3, \ldots \\ \dfrac{m!}{2^{m-1}\left(\dfrac{m-n}{2}\right)!(m+n+1)!!} & m = n, n+2, n+4, \ldots \\ 0 & m = n+1, n+3, n+5, \ldots \end{cases}$$

and

21:10:7 $$\int_{0}^{1} t^{\mu} P_n(t)dt = \frac{(\mu-n+2)(\mu-n+4)(\mu-n+6)\cdots\left(\mu-\frac{1}{2}\pm\frac{1}{2}\right)}{(\mu+n+1)(\mu+n-1)(\mu+n-3)\cdots\left(\mu+\frac{3}{2}\mp\frac{1}{2}\right)} \qquad \mu > -\frac{3}{2} \pm \frac{1}{2}$$

where the upper/lower signs apply according as n is even/odd. Gradshteyn and Ryzhik devote several pages [Sections 7.22–7.25] to a listing of yet other definite integrals involving Legendre polynomials.

The integral of a product of $P_n(x)$ with some other function $f(x)$ is frequently evaluable by utilizing Rodrigues' formula 21:3:2 and parts integration [Section 0:10]. When the integration limits are ± 1, the simple formula

21:10:8
$$\int_{-1}^{1} f(t)P_n(t)dt = \frac{1}{2^n n!}\int_{-1}^{1} \frac{d^n f(t)}{dt^n}(1-t^2)^n dt$$

emerges.

The integral

21:10:9
$$\int_{-1}^{1} P_m(t)P_n(t)dt = \begin{cases} 0 & m \neq n \\ \dfrac{2}{2n+1} & m = n \end{cases}$$

establishes the *orthogonality* of Legendre polynomials over the interval $-1 \leq x \leq 1$, a valuable property that is discussed in Section 21:14.

The Hilbert transform [Section 7:10] of a Legendre polynomial

21:10:10
$$\frac{1}{\pi}\int_{-1}^{1} \frac{P_n(t)dt}{t-s} = \frac{-2}{\pi}Q_n(s)$$

produces a *Legendre function of the second kind* [see Section 21:13], a result known as *Neumann's formula*.

21:11 COMPLEX ARGUMENT

The argument of Legendre polynomials is generally taken to be real. See Section 59:11, however, for a discussion of $P_\nu(x + iy)$.

21:12 GENERALIZATIONS

As explained in Section 21:1, Legendre polynomials are the special ν = integer cases of the Legendre functions of the first kind, dealt with in Chapter 59. However, Legendre polynomials may be generalized in a number of other directions.

Legendre polynomials generalize to give *associated Legendre functions of the first kind*, $P_\nu^{(\mu)}(x)$ [see Section 59:13], of which they are the ν = integer, $\mu = 0$ instances. Likewise, Legendre polynomials result as the $\nu = \mu = 0$ instances of the Jacobi polynomials $P_n^{(\nu;\mu)}(x)$ that are discussed in Section 22:12.

Legendre polynomials are also special cases of the Gauss function [Chapter 60], the simple identity

21:12:1
$$P_n(x) = F\left(-n, n+1; 1; \frac{1-x}{2}\right)$$

being known as *Murphy's formula*. Other relationships between Legendre polynomials and Gauss functions are

21:12:2
$$P_n(x) = \frac{(2n-1)!!}{n!}(x-1)^n F\left(-n, -n; -2n; \frac{2}{1-x}\right)$$

21:12:3
$$P_n(x) = \frac{(2n-1)!!}{n!}x^n F\left(\frac{-n}{2}, \frac{1-n}{2}; \frac{1-2n}{2}; \frac{1}{x^2}\right)$$

21:12:4
$$P_n(x) = \begin{cases} (-1)^{n/2}\dfrac{(n-1)!!}{n!!}F\left(\dfrac{-n}{2}, \dfrac{n+1}{2}; \dfrac{1}{2}; x^2\right) & n = 0, 2, 4, \ldots \\ (-1)^{(n-1)/2}\dfrac{n!!}{(n-1)!!}xF\left(\dfrac{1-n}{2}, \dfrac{n+2}{2}; \dfrac{3}{2}; x^2\right) & n = 1, 3, 5, \ldots \end{cases}$$

and still other formulas may be developed using the relationships of Chapter 60.

21:13 COGNATE FUNCTIONS

The polynomials sometimes denoted $P_n^*(x)$ and $P^n(x)$ are minor variants of $P_n(x)$, defined as follows:

21:13:1 $$P_n^*(x) = P_n(2x-1)$$

known as the *shifted Legendre polynomials* [they are orthogonal, see Section 21:14, on $0 \le x \le 1$], and

21:13:2 $$P^n(x) = \frac{n!}{(2n-1)!!} P_n(x)$$

Neither variant is used in this *Atlas*.

The functions $Q_n(x)$ have been cited as solutions to differential equation 21:3:7 and in connection with Neumann's formula 21:10:8. Their behavior is illustrated in Figure 21-4: note that $Q_n(x)$ is infinite at $x = 1$ (and at $x = -1$). They are the special $\nu = 0, 1, 2, \ldots$ cases of the functions $Q_\nu(x)$, *Legendre's functions of the second kind* [see Chapter 59]. Even for integer ν, the functions $Q_\nu(x)$ are not polynomials, the first few cases being

21:13:3 $$Q_0(x) = \operatorname{artanh}(x) \qquad -1 < x < 1$$

21:13:4 $$Q_1(x) = x \operatorname{artanh}(x) - 1 \qquad -1 < x < 1$$

21:13:5 $$Q_2(x) = \frac{3x^2-1}{2} \operatorname{artanh}(x) - \frac{3x}{2} \qquad -1 < x < 1$$

and

21:13:6 $$Q_3(x) = \frac{5x^3-3x}{2} \operatorname{artanh}(x) - \frac{15x^2-4}{6} \qquad -1 < x < 1$$

where artanh is the inverse hyperbolic tangent function [see Chapter 31]. The general case is

21:13:7 $$Q_n(x) = P_n(x) \operatorname{artanh}(x) - 2\sum_j \frac{2n-2j+1}{j(2n-j-1)} P_{n-j}(x) \qquad j = 1, 3, 5, \ldots, n - \tfrac{1}{2} \mp \tfrac{1}{2}$$

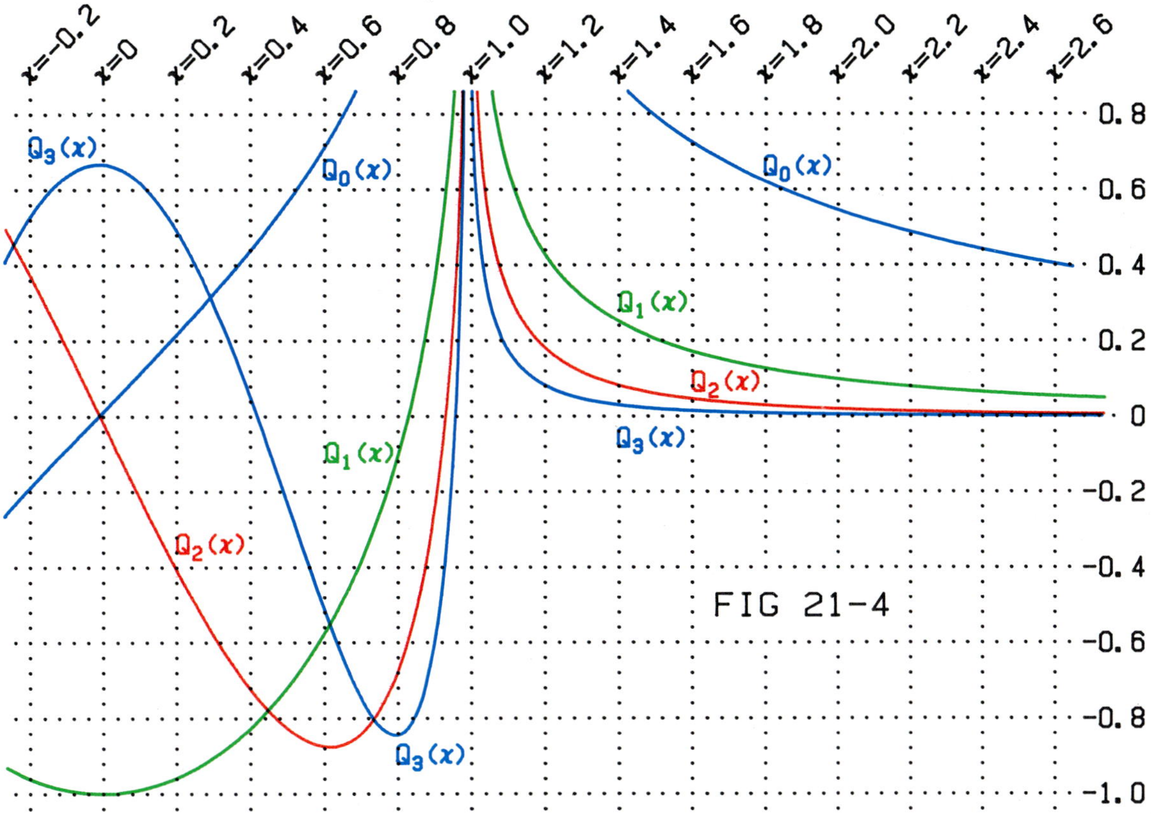

for $-1 < x < 1$. For $|x| > 1$, the artanh function is to be replaced by arcoth. Many of the properties of $P_n(x)$ are mimicked by $Q_n(x)$; for example, the recurrence formula

21:13:8 $$Q_n(x) = \frac{2n-1}{n} x Q_{n-1}(x) - \frac{n-1}{n} Q_{n-2}(x) \qquad n = 2, 3, 4, \ldots$$

is obeyed. However, the symmetry properties of $Q_n(x)$ differ from those in 21:5:1 because $Q_n(x)$ is an even function when n is odd, and vice versa.

21:14 RELATED TOPICS

A family of functions $\Psi_0(x)$, $\Psi_1(x)$, $\Psi_2(x)$, ... is said to be *orthogonal* on an interval $x_0 \le x \le x_1$ when its family members satisfy the relationship

21:14:1 $$\int_{x_0}^{x_1} w(t)\Psi_m(t)\Psi_n(t)dt = 0 \qquad m \neq n$$

but

21:14:2 $$\int_{x_0}^{x_1} w(t)[\Psi_n(t)]^2 dt = \Omega_n \neq 0$$

The family is said to be *orthonormal* if 21:14:1 and 21:14:2 hold and if $\Omega_n = 1$, $n = 0, 1, 2, \ldots$. The function $w(x)$ is known as the *weight function*; it is required to be nonnegative on the interval $x_0 \le x \le x_1$. For example, integral 21:10:7 shows that the Legendre polynomials constitute an orthogonal family on the interval $-1 \le x \le 1$ with a weight function of unity.

The orthogonality property permits the functions $\Psi_0(x)$, $\Psi_1(x)$, $\Psi_2(x)$, ... to be used as a *basis set* for the expansion

21:14:3 $$f(x) = c_0\Psi_0(x) + c_1\Psi_1(x) + c_2\Psi_2(x) + \cdots$$

of a wide range of functions $f(x)$. The constants c_0, c_1, c_2, ... in this so-called *orthogonal series* or *generalized Fourier expansion* may be evaluated by the integral

21:14:4 $$c_n = \frac{1}{\Omega_n}\int_{x_0}^{x_1} f(t)w(t)\Psi_n(t)dt$$

For example, the function $1/\sqrt{1-x^2}$ may be expanded as a series

21:14:5 $$\frac{1}{\sqrt{1-x^2}} = c_0P_0(x) + c_1P_1(x) + c_2P_2(x) + \cdots \qquad |x| < 1$$

of Legendre polynomials, the constants being evaluable, via 21:10:5, as

21:14:6 $$c_n = \left(n + \frac{1}{2}\right)\pi\left[\frac{(n-1)!!}{n!!}\right]^2 \qquad n = 0, 2, 4, \ldots$$

with $c_1 = c_3 = c_5 = \cdots = 0$.

CHAPTER 22

THE CHEBYSHEV POLYNOMIALS $T_n(x)$ AND $U_n(x)$

Chebyshev polynomials of the first kind, $T_n(x)$, constitute an orthogonal family [Section 21:14] on the interval $-1 \le x \le 1$ with a weight function $1/\sqrt{1-x^2}$. Similarly, the second kind of Chebyshev polynomials, $U_n(x)$, are orthogonal on the same interval with weight function $\sqrt{1-x^2}$. Chebyshev polynomials are extensively used in "fitting" techniques, such as those discussed in Section 22:14.

22:1 NOTATION

Alternative transliteration from the Cyrillic alphabet leads to spellings ranging from Chebyshev to Tschebischeff for the names of these polynomials.

Unfortunately, mathematicians differ in their definitions of Chebyshev polynomials, and there is not even unanimity on which family constitutes the first and which the second kind. The notations $T_n(x)$ and $U_n(x)$ may be encountered with meanings equivalent to the $T_n(x)/2^{n-1}$ and $nU_{n-1}(x)$ of this *Atlas*. Confusingly, the symbol $U_n(x)$ and the name *Chebyshev polynomials of the second kind* are commonly applied to functions defined by our $\sqrt{1-x^2}\,U_{n-1}(x)$ for $n = 1, 2, 3, \ldots$ and by $\arcsin(x)$ for $n = 0$, despite these not being polynomial functions at all.

Several supplementary notations are encountered. $T_n^*(x)$ and $U_n^*(x)$ are so-called *shifted Chebyshev polynomials* equal to our $T_n(2x-1)$ and $U_n(2x-1)$. The symbols $C_n(x)$ and $S_n(x)$ have been used for $2T_n(x/2)$ and $U_n(x/2)$, respectively. *Chebyshev functions*, $\overline{T}_n(x)$ and $\overline{U}_n(x)$ are defined identically to the corresponding polynomials, but n is not restricted to be an integer. None of these supplementary notations is employed in this *Atlas*.

22:2 BEHAVIOR

The degree n of a Chebyshev polynomial may take any nonnegative integer value, although $U_0(x)$ is not a member of the orthogonal family [see Section 22:10]. Interest in Chebyshev polynomials is confined to the $-1 \le x \le 1$ range, as portrayed in Figures 22-1 and 22-2, although several of the definitions remain valid outside this argument range.

Chebyshev polynomials of each kind, $T_n(x)$ and $U_n(x)$, have exactly n zeros, $\mathrm{Int}(n/2)$ minima and $\mathrm{Int}[(n-1)/2]$ maxima, all of which lie within the $-1 < x < 1$ range. The $T_n(x)$ polynomial takes the value -1 at each minimum and $+1$ at each maximum, but no comparable rule applies to the Chebyshev polynomials of the second kind. The polynomials are confined to the ranges

22:2:1 $$-1 \le T_n(x) \le 1 \qquad -1 \le x \le 1$$

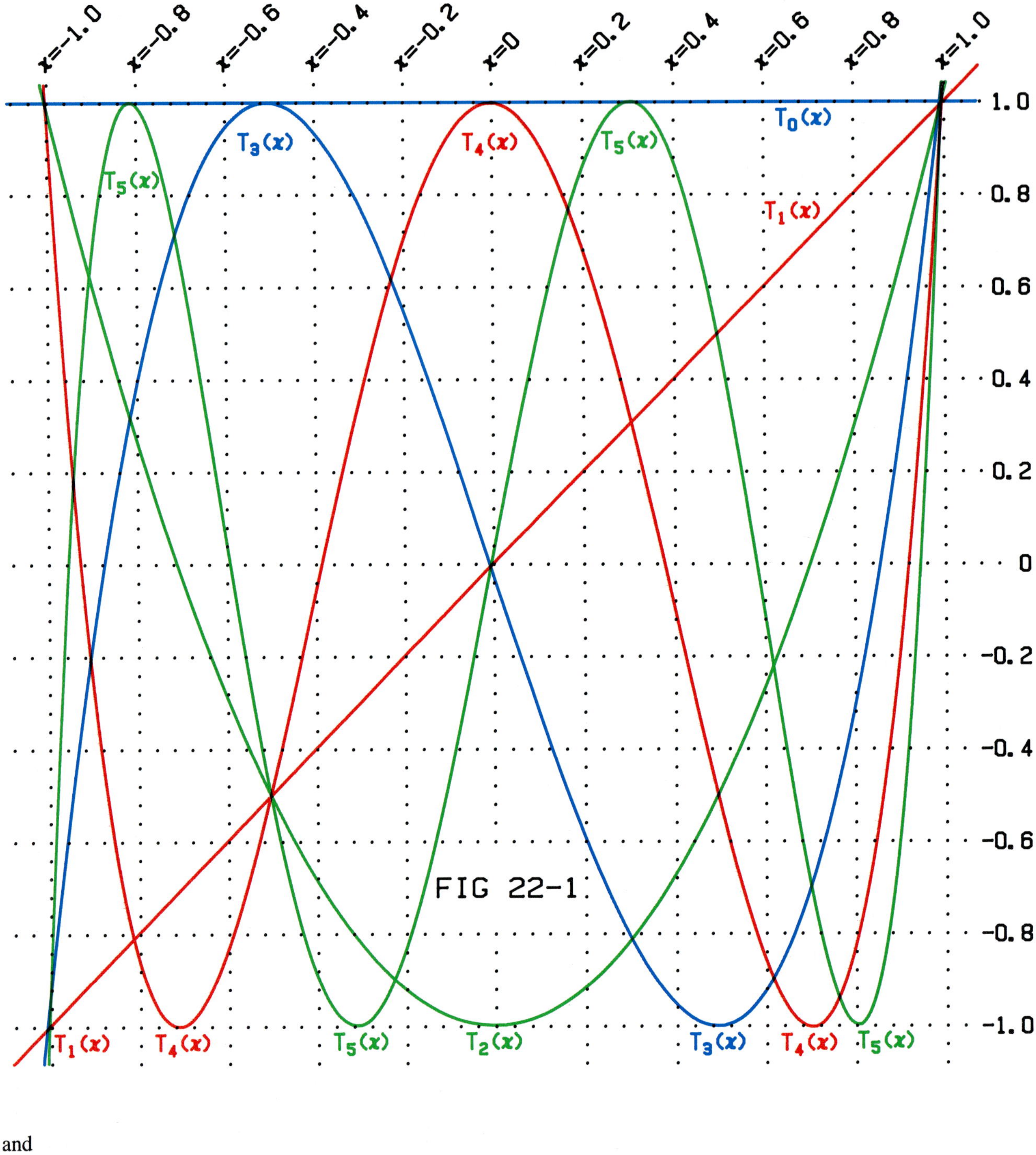

and

22:2:2 $$-n - 1 \leq U_n(x) \leq n + 1 \qquad -1 \leq x \leq 1$$

22:3 DEFINITIONS

Chebyshev polynomials are given by the trigonometric definitions

22:3:1 $$T_n(x) = \cos(n\theta)$$

22:3:2 $$U_n(x) = \csc(\theta)\sin[(n + 1)\theta]$$

$$\left.\begin{matrix} \\ \\ \end{matrix}\right\} \quad \theta = \arccos(x) \qquad -1 \leq x \leq 1$$

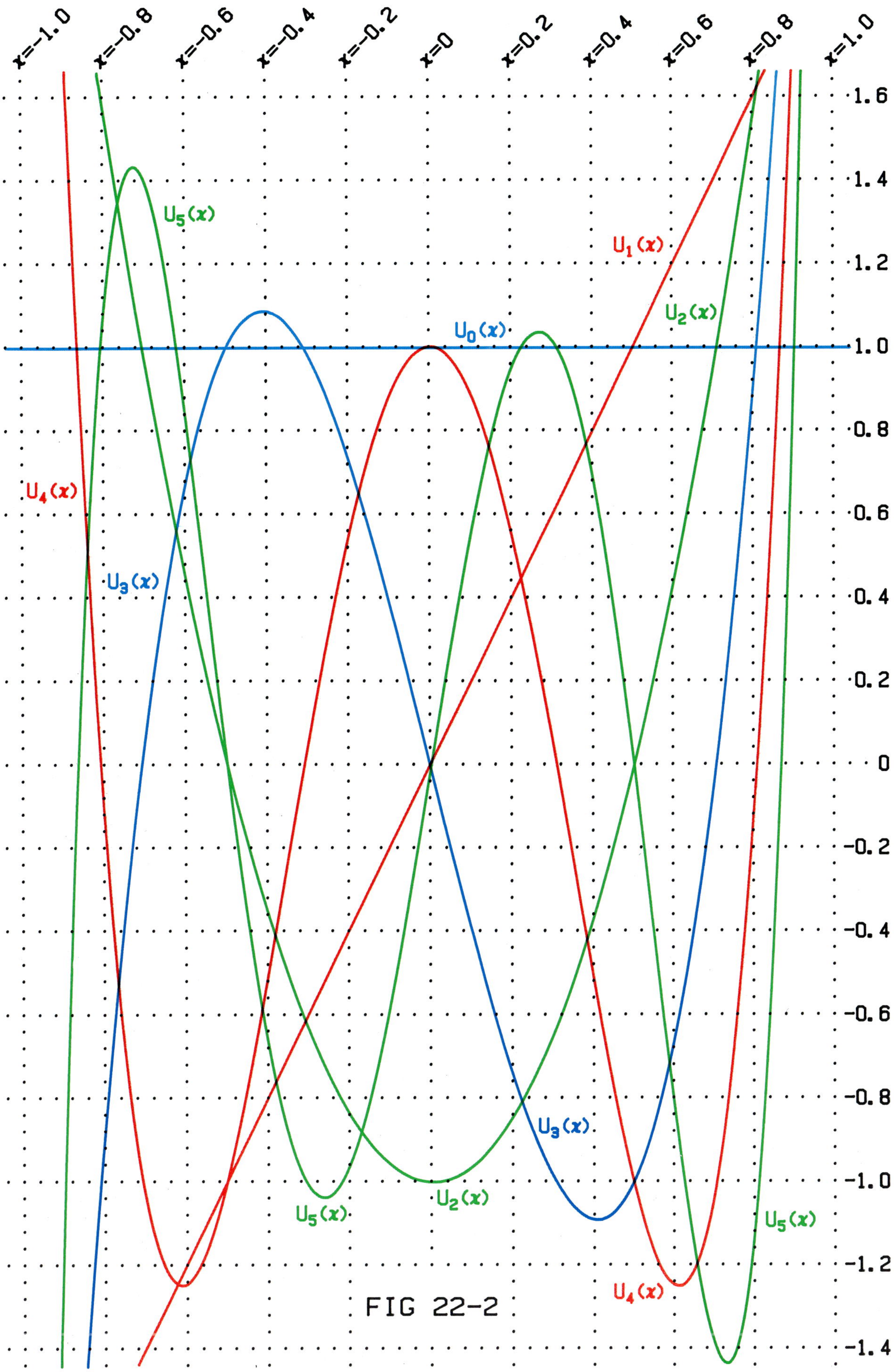
x=-1.0
x=-0.8
x=-0.6
x=-0.4
x=-0.2
x=0
x=0.2
x=0.4
x=0.6
x=0.8
x=1.0
1.6
1.4
1.2
1.0
0.8
0.6
0.4
0.2
0
-0.2
-0.4
-0.6
-0.8
-1.0
-1.2
-1.4
U5(x)
U1(x)
U2(x)
U0(x)
U4(x)
U3(x)
U3(x)
U5(x)
U2(x)
U5(x)
U4(x)

FIG 22-2

The purely algebraic definition

22:3:3
$$T_n(x) = \frac{1}{2}(x + \sqrt{x^2 - 1})^n + \frac{1}{2}(x - \sqrt{x^2 - 1})^n$$

holds for Chebyshev polynomials of the first kind.

These polynomials may be defined by the generating functions

22:3:4
$$\frac{1 - tx}{1 - 2\,tx + t^2} = \sum_{n=0}^{\infty} T_n(x)t^n \qquad -1 < t < 1$$

22:3:5
$$\frac{1}{1 - 2tx + t^2} = \sum_{n=0}^{\infty} U_n(x)t^n \qquad -1 < t < 1$$

or by the *Rodrigues formulas*

22:3:6
$$T_n(x) = \frac{(-1)^n \sqrt{1 - x^2}}{(2n - 1)!!} \frac{d^n}{dx^n}(1 - x^2)^{n-1/2}$$

22:3:7
$$U_n(x) = \frac{(-1)^n(n + 1)}{(2n + 1)!!\sqrt{1 - x^2}} \frac{d^n}{dx^n}(1 - x^2)^{n+1/2}$$

For $n = 1, 2, 3, \ldots$ the most general solution of *Chebyshev's differential equation*

22:3:8
$$(1 - x^2)\frac{d^2 f}{dx^2} - x\frac{df}{dx} + n^2 f = 0$$

is $f = c_1 T_n(x) + c_2\sqrt{1 - x^2}\, U_{n-1}(x)$, where c_1 and c_2 are arbitrary constants. For $n = 0$ the solution is $f = c_1 + c_2 \arcsin(x)$.

22:4 SPECIAL CASES

The first eight polynomials of each kind are

22:4:1 $$T_0(x) = U_0(x) = 1$$

22:4:2 $$T_1(x) = x$$

22:4:3 $$U_1(x) = 2x$$

22:4:4 $$T_2(x) = 2x^2 - 1$$

22:4:5 $$U_2(x) = 4x^2 - 1$$

22:4:6 $$T_3(x) = 4x^3 - 3x$$

22:4:7 $$U_3(x) = 8x^3 - 4x$$

22:4:8 $$T_4(x) = 8x^4 + 8x^2 + 1$$

22:4:9 $$U_4(x) = 16x^4 - 12x^2 + 1$$

22:4:10 $$T_5(x) = 16x^5 - 20x^3 + 5x$$

22:4:11 $$U_5(x) = 32x^5 - 32x^3 + 6x$$

22:4:12 $$T_6(x) = 32x^6 - 48x^4 - 18x^2 - 1$$

22:4:13 $$U_6(x) = 64x^6 - 80x^4 + 24x^2 - 1$$

22:4:14 $$T_7(x) = 64x^7 - 112x^5 + 56x^3 - 7x$$

22:4:15 $$U_7(x) = 128x^7 - 192x^7 + 80x^3 - 8x$$

22:5 INTRARELATIONSHIPS

Chebyshev polynomials are even or odd

22:5:1 $$f_n(-x) = (-1)^n f_n(x) \qquad f_n = T_n \text{ or } U_n$$

according to the parity of the degree n. The recurrence formula

22:5:2 $$f_n(x) = 2xf_{n-1}(x) - f_{n-2}(x) \qquad n = 2, 3, 4, \ldots$$

also applies equally to both kinds of Chebyshev polynomials.

The formulas

22:5:3 $$T_n(x) = U_n(x) - xU_{n-1}(x) \qquad n = 1, 2, 3, \ldots$$

and

22:5:4 $$U_n(x) = \frac{xT_{n-1}(x) - T_n(x)}{1 - x^2} \qquad n = 1, 2, 3, \ldots$$

permit one kind of Chebyshev polynomial to be expressed in terms of the other.

Positive integer powers can be expressed as finite series of Chebyshev polynomials of either kind. For example:

22:5:5 $$x^2 = \frac{1}{2}T_0(x) + \frac{1}{2}T_2(x) = \frac{1}{4}U_0(x) + \frac{1}{4}U_2(x)$$

22:5:6 $$x^3 = \frac{3}{4}T_1(x) + \frac{1}{4}T_3(x) = \frac{1}{4}U_1(x) + \frac{1}{8}U_3(x)$$

and generally

22:5:7 $$x^n = \gamma_n^{(n)}T_n(x) + \gamma_{n-2}^{(n)}T_{n-2}(x) + \gamma_{n-4}^{(n)}T_{n-4}(x) + \cdots + \left\{\begin{matrix} \gamma_0^{(n)}T_0(x) & n = 0, 2, 4, \ldots \\ \gamma_1^{(n)}T_1(x) & n = 1, 3, 5, \ldots \end{matrix}\right\} = \sum_{j=0}^{n} \gamma_j^{(n)}T_j(x)$$

where $\gamma_j^{(n)} = 0$ if $j > n$ or if j and n have unlike parities. The simplest coefficient $\gamma_0^{(0)}$ equals unity, and others can be calculated by means of the recursions $\gamma_j^{(n)} = \frac{1}{2}\gamma_{j-1}^{(n-1)} + \frac{1}{2}\gamma_{j+1}^{(n-1)}$ except $\gamma_1^{(n)} = \gamma_0^{(n-1)} + \frac{1}{2}\gamma_2^{(n-1)}$ and $\gamma_0^{(n)} = \frac{1}{2}\gamma_1^{(n-1)}$.

There are formulas relating the product of two Chebyshev polynomials to (usually the sum of two) other Chebyshev polynomials:

22:5:8 $$T_n(x)T_m(x) = \frac{1}{2}T_{n+m}(x) + \frac{1}{2}T_{n-m}(x) \qquad n \ge m$$

22:5:9 $$U_n(x)U_m(x) = \frac{T_{n-m}(x) - T_{n+m+2}(x)}{2(1 - x^2)} \qquad n \ge m$$

22:5:10 $$T_n(x)U_m(x) = \begin{cases} \frac{1}{2}U_{n+m}(x) + \frac{1}{2}U_{n-m}(x) & n \le m \\ \frac{1}{2}U_{n+m}(x) & n = m + 1 \\ \frac{1}{2}U_{n+m}(x) - \frac{1}{2}U_{n-m-2}(x) & n \ge m + 2 \end{cases}$$

as well as the so-called *Christoffel-Darboux formulas* for sums of products of Chebyshev polynomials of different arguments:

22:5:11 $$\frac{1}{2} + \sum_{j=1}^{n} T_j(x)T_j(y) = \frac{T_{n+1}(x)T_n(y) - T_n(x)T_{n+1}(y)}{x - y}$$

22:5:12 $$1 + \sum_{j=1}^{n} U_j(x)U_j(y) = \frac{U_{n+1}(x)U_n(y) - U_n(x)U_{n+1}(y)}{x - y}$$

Series of Chebyshev polynomials of even degree have the sums

22:5:13 $$T_0(x) + T_2(x) + T_4(x) + \cdots + T_n(x) = \frac{1}{2} + \frac{1}{2} U_n(x) \qquad n = 0, 2, 4, \ldots$$

22:5:14 $$U_0(x) + U_2(x) + U_4(x) + \cdots + U_n(x) = \frac{1 - T_{n+2}(x)}{2(1 - x^2)} \qquad n = 0, 2, 4, \ldots$$

while the corresponding sums for odd degree are

22:5:15 $$T_1(x) + T_3(x) + T_5(x) + \cdots + T_n(x) = \frac{1}{2} U_n(x) \qquad n = 1, 3, 5, \ldots$$

22:5:16 $$U_1(x) + U_3(x) + U_5(x) + \cdots + U_n(x) = \frac{x - T_{n+2}(x)}{2(1 - x^2)} \qquad n = 1, 3, 5, \ldots$$

22:6 EXPANSIONS

Explicit expressions for the two Chebyshev polynomials are

22:6:1 $$T_n(x) = \frac{n}{2} \sum_{j=0}^{J} \frac{(-1)^j}{n-j} \binom{n-j}{j} (2x)^{n-2j} \qquad n = 1, 2, 3, \ldots$$

and

22:6:2 $$U_n(x) = \sum_{j=0}^{J} (-1)^j \binom{n-j}{j} (2x)^{n-2j} \qquad n = 0, 1, 2, \ldots$$

where $J = \mathrm{Int}(n/2)$, but these expansions may also be written

22:6:3 $$T_n(x) = x^n - \binom{n}{2} x^{n-2}(1 - x^2) + \binom{n}{4} x^{n-4}(1 - x^2)^2 - \cdots$$

and

22:6:4 $$U_n(x) = \binom{n+1}{1} x^n - \binom{n+1}{3} x^{n-2}(1 - x^2) + \binom{n+1}{5} x^{n-4}(1 - x^2)^2 - \cdots$$

If $t_n^{(j)}$ represents the coefficient of x^j in the expansion of $T_n(x)$ so that

22:6:5 $$T_n(x) = \sum_{j=0}^{n} t_n^{(j)} x^j$$

then $t_n^{(j)} = 0$ when j and n are of unlike parity. All nonzero values of $t_n^{(j)}$ are integers given by

22:6:6 $$t_n^{(j)} = (-1)^{(n-j)/2} n \left(\frac{n+j}{2} - 1\right)! \, 2^{j-1} \Big/ \left[\left(\frac{n-j}{2}\right)! \, j!\right] \qquad j = 1, 2, 3, \ldots, n$$

except that $t_0^{(0)}$ equals unity.

22:7 PARTICULAR VALUES

	$T_n(-1)$	$T_n(0)$	$T_n(1)$	$U_n(-1)$	$U_n(0)$	$U_n(1)$
$n = 0$	1	1	1	1	1	1
$n = 1, 3, 5, \ldots$	-1	0	1	$-n - 1$	0	$n + 1$
$n = 2, 4, 6, \ldots$	1	$(-1)^{n/2}$	1	$n + 1$	$(-1)^{n/2}$	$n + 1$

The zeros of the Chebyshev polynomials occur at arguments given by

22:7:1 $$T_n(r_j) = 0 \qquad r_j = \cos\left[\frac{(2j-1)\pi}{2n}\right] \qquad j = 1, 2, 3, \ldots, n$$

and

22:7:2 $$U_n(r_j) = 0 \qquad r_j = \cos\left(\frac{j\pi}{n+1}\right) \qquad j = 1, 2, 3, \ldots, n$$

22:8 NUMERICAL VALUES

For small degree, values of the Chebyshev polynomials are readily calculated from the expressions in Section 22:4.

The following algorithm employs recurrence formula 22:5:2 and provides exact values of the Chebyshev polynomials. $T_n(x)$ is generated if the command in black is executed; $U_n(x)$ is produced by the green alternative (the two algorithms differ only by a factor of 2 in the command that sets up the value of $f_1(x)$).

For $-1 \le x < 3$, values of the Chebyshev polynomials may also be found using the universal hypergeometric algorithm [Section 18:14].

```
Input n >>>>>>>
Input x >>>>>>>
          Set f = g = { x
                      { 2x
          If n = 1 go to (2)
          Set f = h = j = 1
          If n = 0 go to (2)
      (1) Replace j by j + 1
          Replace f by 2xg − h
          If j = n go to (2)
          Replace h by g
          Replace g by f
          Go to (1)
      (2) Output f
f = { Tn(x) } <<<<
    { Un(x) }
```

Storage needed: n, x, f, g, h and j

Input restrictions: n must be a nonnegative integer. x may adopt any value.

Test values:
$T_0(\text{any } x) = 1$
$U_1(x) = 2x$
$T_{12}(0.2) = -0.748302037$
$U_5(0.5) = 0$
$U_{11}(-0.6) = 1.239131013$

22:9 APPROXIMATIONS

Figures 22-1 and 22-2 can provide approximate values of $T_n(x)$ and $U_n(x)$ for degrees up to 5. For large values of the argument, the following approximations are valid:

22:9:1 $$T_n(x) \simeq \frac{1}{2}\left(2x - \frac{1}{2x}\right)^n \qquad \text{8-bit precision} \qquad x > (18n)^{1/4} \qquad n \ge 2$$

22:9:2 $$U_n(x) \simeq 2x\left(2x - \frac{1}{2x}\right)^{n-1} \qquad \text{8-bit precision} \qquad x > 2(n-1)^{1/4} \qquad n \ge 2$$

22:10 OPERATIONS OF THE CALCULUS

Differentiation gives

22:10:1 $$\frac{d}{dx} T_n(x) = \frac{n}{1-x^2}[xT_n(x) - T_{n+1}(x)] = nU_{n-1}(x)$$

22:10:2 $$\frac{d}{dx}U_n(x) = \frac{1}{1-x^2}[(n+1)U_{n-1}(x) - nxU_n(x)]$$

The Chebyshev polynomial of the second kind may be integrated indefinitely

22:10:3 $$\int_x^1 U_n(t)dt = \frac{1 - T_{n+1}(x)}{n+1}$$

but we know of no corresponding integral for $T_n(x)$.

The definite integral

22:10:4 $$\int_{-1}^1 \frac{T_m(t)T_n(t)}{\sqrt{1-t^2}}dt = \begin{cases} 0 & m \neq n \\ \pi & m = n = 0 \\ \dfrac{\pi}{2} & m = n = 1, 2, 3, \ldots \end{cases}$$

establishes that the polynomials $T_0(x)$, $T_1(x)$, $T_2(x)$, ... are orthogonal [see Section 21:14] on the interval $-1 \le x \le 1$ with respect to the weight function $1/\sqrt{1-x^2}$. Likewise, the integrals

22:10:5 $$\int_{-1}^1 U_m(t)U_n(t)\sqrt{1-t^2}\,dt = \begin{cases} 0 & m \neq n \quad \text{or} \quad m = n = 0 \\ \dfrac{\pi}{2} & m = n = 1, 2, 3, \ldots \end{cases}$$

establish the orthogonality of the polynomials $U_1(x)$, $U_2(x)$, $U_3(x)$, ... on the same interval with respect to the weight function $\sqrt{1-x^2}$.

Many other definite integrals are listed by Gradshteyn and Ryzhik [Sections 7.34 to 7.36], including

22:10:6 $$\int_{-1}^1 T_n^2(t)dt = \frac{4n^2-2}{4n^2-1}$$

22:10:7 $$\int_0^1 \frac{t^\nu T_n(t)}{\sqrt{1-t^2}}dt = \frac{\pi\Gamma(\nu+1)}{2^{\nu+1}\Gamma\left(\dfrac{\nu+n}{2}+1\right)\Gamma\left(\dfrac{\nu-n}{2}+1\right)} \qquad \nu > -1$$

22:10:8 $$\int_0^1 \frac{\cos(bt)T_n(t)}{\sqrt{1-t^2}}dt = \frac{(-1)^{n/2}\pi}{2}J_n(b) \qquad b > 0 \qquad n = 0, 2, 4, \ldots$$

and

22:10:9 $$\int_0^1 \frac{\sin(bt)T_n(t)}{\sqrt{1-t^2}}dt = \frac{(-1)^{(n-1)/2}\pi}{2}J_n(b) \qquad b > 0 \qquad n = 1, 3, 5, \ldots$$

where Γ denotes the gamma function [Chapter 43] and J_n the Bessel coefficient [Chapter 52] of order n.

On Hilbert transformation [Section 7:10] the product of one kind of Chebyshev polynomial and its weight function gives the other kind:

22:10:10 $$\frac{1}{\pi}\int_{-1}^1 \frac{T_n(t)dt}{\sqrt{1-t^2}\,(t-s)} = U_{n-1}(s) \qquad -1 < s < 1$$

22:10:11 $$\frac{1}{\pi}\int_{-1}^1 U_n(t)\sqrt{1-t^2}\frac{dt}{(t-s)} = -T_{n+1}(s)$$

22:11 COMPLEX ARGUMENTS

Applications of Chebyshev polynomials are usually restricted to real arguments.

22:12 GENERALIZATIONS

Chebyshev polynomials may be generalized to Gegenbauer polynomials, which are themselves special cases of Jacobi polynomials. This section briefly discusses both of these generalizations.

The *Jacobi polynomial* or *hypergeometric polynomial* $P_n^{(\nu;\mu)}(x)$ is a quadrivariate function, its value being dependent on: the argument x, which may take any real value with interest being concentrated on the $-1 \le x \le 1$ range; the degree n, which takes nonnegative integer values; and two distinct parameters, ν and μ, both of which exceed -1. These polynomials may be defined by the Rodrigues expression

22:12:1 $$P_n^{(\nu;\mu)}(x) = \frac{1}{(-2)^n n!(1-x)^\nu(1+x)^\mu}\frac{d^n}{dx^n}[(1-x)^{\nu+n}(1+x)^{\mu+n}]$$

They obey the reflection and recurrence formulas

22:12:2 $$P_n^{(\nu;\mu)}(-x) = (-1)^n P_n^{(\nu;\mu)}(x)$$

22:12:3 $$P_n^{(\nu;\mu)}(x) = \frac{(2n+\nu+\mu-1)[(2n+\nu+\mu-2)(2n+\nu+\mu)x+\nu^2-\mu^2]}{2n(n+\nu+\mu)(2n+\nu+\mu-2)}P_{n-1}^{(\nu;\mu)}(x) - \frac{(n+\nu-1)(n+\mu-1)(2n+\nu+\mu)}{n(n+\nu+\mu)(2n+\nu+\mu-2)}P_{n-2}^{(\nu;\mu)}(x)$$

with the first members being

22:12:4 $$P_0^{(\nu;\mu)}(x) = 1$$

and

22:12:5 $$P_1^{(\nu;\mu)}(x) = \frac{(\nu+\mu+2)x}{2} + \frac{\nu-\mu}{2}$$

A binomial coefficient [Chapter 6] expresses the value of the Jacobi polynomial at unity argument

22:12:6 $$P_n^{(\nu;\mu)}(1) = \binom{n+\nu}{n}$$

Jacobi polynomials are orthogonal [Section 21:14] on the interval $-1 \le x \le 1$ with weight function $(1-x)^\nu(1+x)^\mu$

22:12:7 $$\int_{-1}^{1}(1-t)^\nu(1+t)^\mu P_m^{(\nu;\mu)}(t)P_n^{(\nu;\mu)}(t)dt = \begin{cases} 0 & m \ne n \\ \dfrac{2^{\nu+\mu+1}\Gamma(n+\nu+1)\Gamma(n+\mu+1)}{(2n+\nu+\mu+1)n!\Gamma(n+\nu+\mu+1)} & m = n \end{cases}$$

For more information about Jacobi polynomials, the reader is referred to more advanced sources, particularly Szegö. Jacobi polynomials can be expressed in several ways as Gauss functions [Chapter 60]. For example:

22:12:8 $$P_n^{(\nu;\mu)}(x) = \binom{n+\nu}{n}F\left(-n, n+\nu+\mu+1; \nu+1; \frac{1-x}{2}\right)$$

and numerical values are therefore calculable using the algorithm of Section 18:14.

Gegenbauer polynomials, or *ultraspherical polynomials*, are related to those Jacobi polynomials in which the two parameters are equal: $\nu = \mu$. They may be defined by

22:12:9 $$C_n^{(\lambda)}(x) = \frac{(2\lambda)_n}{(\lambda+{}^1/_2)_n}P_n^{(\lambda-1/2;\lambda-1/2)}(x) \qquad -\frac{1}{2} < \lambda \ne 0$$

supplemented by the definition

22:12:10 $$C_n^{(0)}(x) = \lim_{\lambda\to 0}\frac{C_n^{(\lambda)}(x)}{\lambda} = \frac{4(2n-2)!!}{(2n-1)!!}P_n^{(-1/2;-1/2)}(x)$$

Most of the properties of Gegenbauer polynomials may be deduced as special cases of those of Jacobi polynomials.

Legendre polynomials [Chapter 21] and each kind of Chebyshev polynomial are special cases of Gegenbauer polynomials

22:12:11 $$P_n(x) = C_n^{(1/2)}(x) = P_n^{(0;0)}(x)$$

22:12:12 $$T_n(x) = \frac{n}{2} C_n^{(0)}(x) = \frac{(2n)!!}{(2n-1)!!} P_n^{(-1/2;-1/2)}(x)$$

22:12:13 $$U_n(x) = C_n^{(1)}(x) = \frac{(2n+2)!!}{2(2n+1)!!} P_n^{(1/2;1/2)}(x)$$

while Hermite polynomials [Chapter 24] are generated from Gegenbauer polynomials by the limiting process

22:12:14 $$H_n(x) = \lim_{\lambda\to\infty} \frac{n}{\lambda^{n/2}} C_n^{(\lambda)}\left(\frac{x}{\sqrt{\lambda}}\right)$$

22:13 COGNATE FUNCTIONS

Orthogonality defined in terms of integral 21:14:1 is generally satisfied by families of polynomials, each orthogonal family containing an infinite number of members. However, another variety of orthogonality, defined by

22:13:1 $$\sum_y w(y)\Psi_m(y)\Psi_n(y) = \begin{cases} 0 & m \neq n \\ \Omega_n \neq 0 & m = n \end{cases} \quad y = y_0, y_1, y_2, \ldots, y_J$$

may be satisfied by a finite set of polynomials $\Psi_0(y)$, $\Psi_1(y)$, $\Psi_2(y)$, ..., $\Psi_J(y)$, known as *discrete orthogonal polynomials*.

The simplest of these are the *discrete Chebyshev polynomials* $t_0(y)$, $t_1(y)$, $t_2(y)$, ..., $t_J(y)$, which correspond to weights $w(y)$ in 22:13:1 equal to unity and should not be confused with the coefficients discussed in Section 22:6. With the variable y confined to the range $-1 \le y \le 1$, the orthogonality condition is

22:13:2 $$\sum_{j=0}^{J} t_m(y_j)t_n(y_j) = \begin{cases} 0 & m \neq n \\ \Omega_n & m = n \end{cases}$$

where

22:13:3 $$\Omega_n = \frac{(J+n+1)!(J-n)!}{(2n+1)(J!)^2} = \frac{1}{2n+1}\frac{(J+1)_{n+1}}{(J-n+1)_n}$$

Values of the discrete Chebyshev polynomials generally depend on J as well as on the argument y. The first few are

22:13:4 $$t_0(y) = 1$$

22:13:5 $$t_1(y) = y$$

22:13:6 $$t_2(y) = \frac{3Jy^2 - (J+2)}{2(J-1)}$$

22:13:7 $$t_3(y) = \frac{5J^2y^3 - (3J^2 + 6J - 4)y}{2(J-1)(J-2)}$$

and others may be calculated via the recursion formula

22:13:8 $$t_n(y) = \frac{(2n-1)}{n(J-n+1)}[yt_{n-1}(y) - t_{n-2}(y)] + t_{n-2}(y)$$

The practical importance of discrete Chebyshev polynomials depends on the role that they play in the fitting of low-degree polynomials to equally spaced data points [see Section 17:14]. Let a K-degree polynomial $p_K(y)$ be sought that passes close to the points (y_0,f_0), (y_1,f_1), (y_2,f_2), ..., (y_J,f_J), as shown in Figure 22-3, so as to satisfy

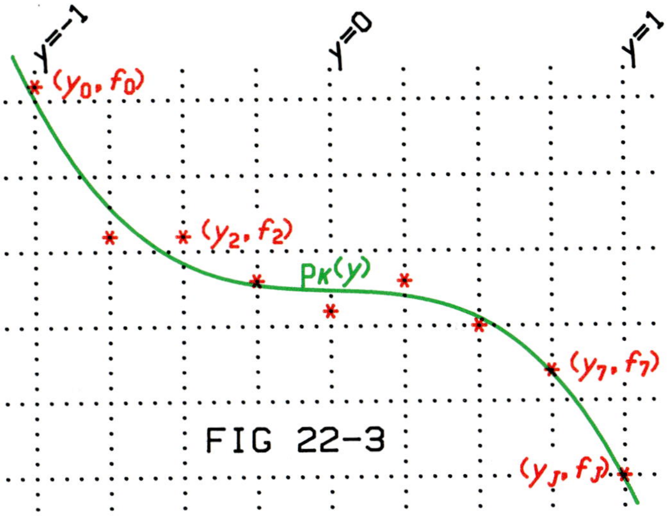

FIG 22-3

the *least squares condition*, that is, to minimize the quantity

22:13:9 $$\sum_{j=0}^{J} [f_j - p_K(y_j)]^2 \qquad y_j = 1 - \frac{2j}{J}$$

Then this "best" polynomial is given by

22:13:10 $$p_K(y) = \sum_{k=0}^{K} a_k t_k(y)$$

where

22:13:11 $$a_k = \frac{1}{\Omega_k} \sum_{j=0}^{J} f_j t_k(y_j)$$

These relationships are exploited in the algorithm presented in Section 17:14.

Despite their name, discrete Chebyshev polynomials have less in common with Chebyshev polynomials than they do with Legendre polynomials, to which they reduce as J approaches infinity:

22:13:12 $$\lim_{J\to\infty} t_n(y) = P_n(y)$$

22:14 RELATED TOPICS

As inspection of Section 6 of most of the chapters of this *Atlas* will attest, very many functions $f(x)$ can be represented by infinite (or less frequently by finite) power series of the form $\Sigma\, a_j x^j$. If only the summands $a_0 + a_1 x + a_2 x^2 + \cdots + a_n x^n$ are used, and n is large, an extremely accurate approximation to the true value of the function $f(x)$ may be obtained:

22:14:1 $$\left| f(x) - \sum_{j=0}^{n} a_j x^j \right| = \delta$$

where δ is very small. For computational purposes, however, it may be undesirable to require as many as $n + 1$ terms when n is large. Indeed, it may be unnecessary to use more than a few terms, especially if interest in the function $f(x)$ is restricted to a small range $x_0 \le x \le x_1$ of argument. *Economization of series* is a procedure that replaces a very accurate (or even exact) polynomial approximation $\Sigma\, a_j x^j$ of degree n by an "economized" polynomial $\Sigma\, e_j x^j$ of a smaller degree m such that, in the range of interest, the absolute error introduced by the replacement is less than some acceptable value, ε:

22:14:2 $$\left| \sum_{j=0}^{n} a_j x^j - \sum_{j=0}^{m} e_j x^j \right| < \varepsilon \qquad x_0 \le x \le x_1$$

The procedure of economization, or *telescoping* as it is sometimes called, is accomplished by utilizing the properties of Chebyshev polynomials of the first kind. A general algorithm follows by which any polynomial may be economized to any extent. The algorithm is lengthy because it has seven distinct phases.

The first phase merely allows the input parameters $x_0, x_1, n, a_0, a_1, a_2, \ldots, a_n$ to be loaded. Note that x_0 and x_1 are not stored as such but as $(x_0 + x_1)/2 = u$ and $(x_1 - x_0)/2 = h$.

The second phase calculates the coefficients $b_0, b_1, b_2, \ldots, b_n$ of a second polynomial of degree n

22:14:3
$$\sum_{j=0}^{n} b_j y^j = \sum_{j=0}^{n} a_j x^j \qquad \text{where} \qquad x = u + hy$$

whose argument has been defined so that the region of interest is $-1 \le y \le 1$. The second phase uses the binomial relationship

22:14:4
$$b_j = h^j \sum_{k=j}^{n} \binom{k}{j} a_k u^{k-j} = \frac{h^j}{j!}\left[a_j j! + \sum_{k=j+1}^{n} a_k k! u^{k-j}/(k-j)!\right]$$

to calculate the new coefficients and stores them in place of the original set.

The third phase computes the coefficients $c_0, c_1, c_2, \ldots, c_n$ such that

22:14:5
$$\sum_{j=0}^{n} b_j y^j = \sum_{j=0}^{n} c_j T_j(y) = \sum_{j=0}^{n} c_j \sum_{k=0}^{j} t_j^{(k)} y^k$$

by first exploiting the fact that, because only the $T_n(y)$ polynomial contains a term in y^n, $c_n = b_n/t_n^{(n)} = b_n/2^{n-1}$ (if $n \neq 0$). Using expression 22:6:6, $c_n t_n^{(j)}$ is now subtracted from each stored b_j other than b_n so that the degree of the $\Sigma\, b_j y^j$ polynomial is reduced to $n - 1$. The procedure is then repeated until all c_j coefficients have been calculated and their values used to replace the now-redundant b_j coefficients.

To start the fourth phase of the algorithm, the maximum permissible error ε is input. The smallest integer m that satisfies

22:14:6
$$\sum_{j=m+1}^{n} |c_j| \le \varepsilon$$

is then calculated by the algorithm, output, and used to replace n in storage. Thereafter, no further use is made of coefficients $c_{m+1}, c_{m+2}, c_{m+3}, \ldots, c_n$.

The fifth phase reverses the process of the third phase, replacing $\Sigma\, c_j T_j(y)$ by the polynomial $\Sigma\, d_j y^j$ of degree m. The properties of $t_j^{(k)}$ [see Section 22:6] are used in the identity

22:14:7
$$\sum_{j=0}^{m} d_j y^j = \sum_{j=0}^{m} c_j \sum_{k=0 \text{ or } 1}^{j} t_j^{(k)} y^k = \sum_{j=0}^{m} y^j \left[c_j t_j^{(j)} + \sum_{k} c_k t_k^{(j)}\right]$$

where in the final summation k takes the values $j + 2, j + 4, j + 6, \ldots, (m$ or $m - 1)$ to establish that

22:14:8
$$d_j = 2^{j-1} c_j + \sum_{k} (-1)^{(k-j)/2} k c_k \left(\frac{k+j}{2} - 1\right)! 2^{j-1} \Big/ \left[\left(\frac{k-j}{2}\right)! j!\right]$$

except that, when $j = 0$, the $2^{j-1}c_j$ term is to be doubled. The algorithm replaces $c_0, c_1, c_2, \ldots, c_m$ in storage by $d_0, d_1, d_2, \ldots, d_m$.

The sixth phase of the algorithm reverses the effect of the second phase in that it converts $\Sigma\, d_j y^j$ to $\Sigma\, e_j x^j$ and replaces the $m + 1$ stored values of d_j by the corresponding e_j coefficients. The binomial identity

22:14:9
$$\sum_{j=0}^{m} e_j x^j = \sum_{j=0}^{m} d_j \left(\frac{x}{h} - \frac{u}{h}\right)^j = \sum_{j=0}^{m} \frac{x^j}{j!} \sum_{k=j}^{m} \frac{(-u)^{k-j} d_k k!}{(k-j)! h^k}$$

shows that

22:14:10
$$e_j = \frac{1}{h^j j!}\left[j! d_j + \sum_{k=j+1}^{m} \frac{d_k k!}{(k-j)!}\left(\frac{-u}{h}\right)^{k-j}\right]$$

and this is the relationship used by the algorithm.

The seventh phase simply outputs the economized coefficients in the order $e_0, e_1, e_2, \ldots, e_m$.

---

first phase

Input x_0 >>>>>>>

Set $u = x_0$

Input x_1 >>>>>>>

Set $h = (x_1 - u)/2$

Replace u by $u + h$

Input n >>>>>>>

Set $j = 0$

(1) Output j

j (as cue) <<<<<<

Input a_j >>>>>>>

Replace j by $j + 1$

If $j \leq n$ go to (1)

---

second phase

Set $j = 0$

(2) Replace a_j by $a_j j!$

If $j = n$ go to (4)

Set $k = j + 1$

(3) Replace a_j by $a_j + a_k k! u^{k-j}/(k - j)!$

Replace k by $k + 1$

If $k \leq n$ go to (3)

(4) Replace a_j by $a_j h^j / j!$

Replace j by $j + 1$

If $j \leq n$ go to (2)

---

third phase

Set $k = n$

(5) Set $j = k$

Replace a_j by $a_j / 2^{j-1}$

(6) Replace j by $j - 2$

If $j < 0$ go to (7)

Set $\varepsilon = (k - j)/2$

Replace a_j by $a_j - (-1)^{\varepsilon} k a_k (\varepsilon + j - 1)! 2^{j-1}/(\varepsilon! j!)$

Go to (6)

(7) Replace k by $k - 1$

If $k \neq 1$ go to (5)

---

fourth phase

Set $j = n$

Input ε >>>>>>>

(8) Replace ε by $\varepsilon - |a_j|$

Replace j by $j - 1$

If $\varepsilon > 0$ go to (8)

Set $m = j + 1$

Output m

m <<<<<<<<<<

---

fifth phase

Replace a_0 by $2a_0$

Set $j = -1$

(9) Replace j by $j + 1$

If $j > m$ go to (11)

Replace a_j by $a_j 2^{j-1}$

Set $k = j$

(10) Replace k by $k + 2$

If $k > m$ go to (9)

Set $\varepsilon = (k - j)/2$

Replace a_j by $a_j + (-1)^{\varepsilon} k a_k (\varepsilon + j - 1)! 2^{j-1}/(\varepsilon! j!)$

Go to (10)

Storage needed: $n + 7$ registers are required for u, h, n, j, k, ε, m and for the coefficients $a_0, a_1, a_2, \ldots, a_n$. Note that the n register may be used to store m and that the ε register is used for temporary storage twice by the algorithm. The coefficients denoted $b_0, b_1, b_2, \ldots, b_n$; $c_0, c_1, c_2, \ldots, c_n$; $d_0, d_1, d_2, \ldots, d_m$ and $e_0, e_1, e_2, \ldots, e_m$ in the text are stored in the same registers as $a_0, a_1, a_2, \ldots, (a_n$ or $a_m)$ and are denoted a_j in the algorithm.

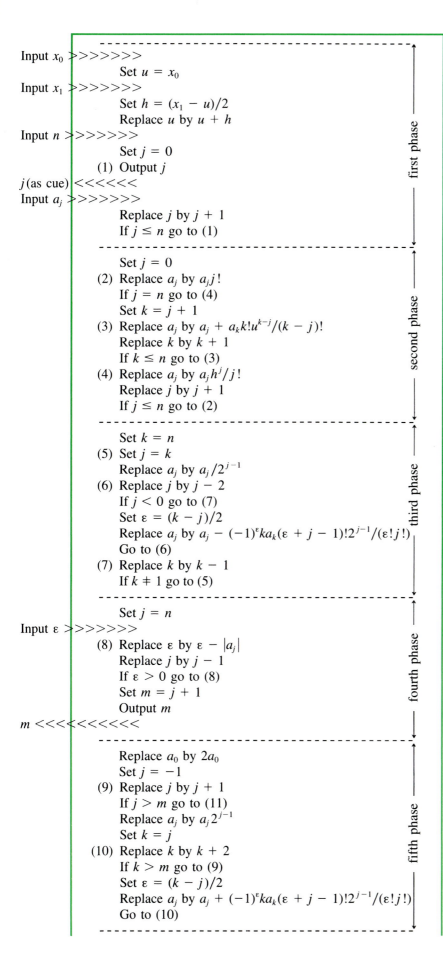

(11) Set $j = 0$
(12) Replace a_j by $a_j j!$
If $j = m$ go to (14)
Set $k = j + 1$
(13) Replace a_j by $a_j + a_k k!(-u/h)^{k-j}/(k-j)!$
Replace k by $k + 1$
If $k \leq m$ go to (13)
(14) Replace a_j by $a_j/(h^j!)$
Replace j by $j + 1$
If $j \leq m$ go to (12)
(sixth phase)

Set $j = 0$
(15) Output a_j
$e_0, e_1, \ldots, e_m$ <<<<
Replace j by $j + 1$
If $j \leq m$ go to (15)
(seventh phase)

Test values: $x_0 = 0$, $x_1 = 1$, $n = 9$, $a_0 = 1$, $a_1 = -1$, $a_2 = 1/2$, $a_3 = -1/6$, $a_4 = 1/24$, $a_5 = 1/120$, $a_6 = 1/720$, $a_7 = -1/5040$, $a_8 = 1/40320$, $a_9 = -1/362880$ [i.e., $f(x) = \exp(-x)$], $\varepsilon = 1/256$
Output: $m = 2$, $e_0 = 0.99658$, $e_1 = -0.93532$, $e_2 = 0.30963$

Because of the very large number of arithmetic operations executed by the algorithm, rounding errors may accumulate and the values of the economized coefficients $e_0, e_1, e_2, \ldots, e_m$ generated by a program may not be quite the "best." To evaluate this effect, the magnitude of which will depend on the characteristics of the particular computing device employed, one may repeat the calculation with $\varepsilon = 0$. Then m will equal n and the coefficients $e_0, e_1, e_2, \ldots, e_m$ should, but will probably not exactly, equal $a_0, a_1, a_2, \ldots, a_n$.

Apart from these rounding errors, the economized series $\Sigma\, e_j x^j$ is *guaranteed* to differ from the original $\Sigma\, a_j x^j$ series by no more than ε over the $x_0 \leq x \leq x_1$ range. However, the actual absolute error $|\Sigma\, a_j x^j - \Sigma\, e_j x^j|$ will often be less than ε. This actual economization error E is found in the following algorithm by calculating the maximum value of

22:14:11 $$\left|\sum_{j=0}^{n} A_j x^j\right|$$

at one hundred points in the $x_0 \leq x \leq x_1$ range. The input coefficients A_j are given by

22:14:12 $$A_j = \begin{cases} a_j - e_j & j = 0, 1, 2, \ldots, m \\ a_j & j = m+1, m+2, m+3, \ldots, n \end{cases}$$

Input x_0 >>>>>>>
Set $x = x_0$
Input x_1 >>>>>>>
Set $h = (x_1 - x)/99$
Set $j = k = E = 0$
Input n >>>>>>>
(1) Output j
j(as cue) <<<<<<
Input A_j >>>>>>>
Replace j by $j + 1$
If $j \leq n$ go to (1)
(2) Set $f = 0$
Replace j by $n + 1$
(3) Replace j by $j - 1$
Replace f by $fx + A_j$
If $j \neq 0$ go to (3)
If $E > |f|$ go to (4)

Storage needed: $n + 8$ registers are required for $A_0, A_1, A_2, \ldots, A_n, x, h, j, k, n, E$ and f.

Replace E by $|f|$
(4) Replace x by $x + h$
Replace k by $k + 1$
If $k \neq 100$ go to (2)
Output E
E <<<<<<<<<<

Test values (taken from the preceding algorithm): $x_0 = 0$, $x_1 = 1$, $n = 9$, $A_0 = 0.00342$, $A_1 = -0.06468$, $A_2 = 0.19037$, $A_3 = -0.16667$, $A_4 = 0.04167$, $A_5 = -0.00833$, $A_6 = 0.00139$, $A_7 = -0.00020$, $A_8 = 0.00002$, $A_9 = -0.00000$
Output: $E = 0.00342$

CHAPTER
23

THE LAGUERRE POLYNOMIALS $L_n(x)$

The Laguerre polynomials are orthogonal [see Section 21:14] on the interval $0 \le x \le \infty$ with a weight function $\exp(-x)$ [see Chapter 26]. They arise in a variety of scientific applications, for example, in solutions to the wave equation of quantum physics.

23:1 NOTATION

Though the $L_n(x)$ symbolism is used for both, Laguerre polynomials are defined in two different ways. Some authorities use definitions that differ by a multiplicative factor of $n!$ from those adopted in this *Atlas*. The terminology *Laguerre polynomial* is sometimes applied to the more general class of polynomials that we discuss in Section 23:12 under the name *associated Laguerre polynomials*.

Although this *Atlas* avoids the ambiguity, the symbols $L_n(x)$ and $L_\nu(x)$ are sometimes used to denote the hyperbolic Struve function [see Section 57:13].

23:2 BEHAVIOR

While interest is usually confined to positive arguments, the Laguerre polynomial $L_n(x)$ is defined for all values of the argument x and all nonnegative integer degrees n. Laguerre polynomials of odd degree adopt all values, whereas those of even degree have a restricted range. Figure 23-1 shows the behavior of the first few members of this family.

The Laguerre polynomial $L_n(x)$ has exactly n zeros, $\text{Int}(n/2)$ minima and $\text{Int}[(n-1)/2]$ maxima, all of which are located in the range $0 < x < 2n + 1 + \sqrt{4n^2 + 4n + (5/4)}$. For $x > 2n + 1 + \sqrt{4n^2 + 4n + (5/4)}$, $L_n(x)$ increases or decreases monotonically according to whether n is even or odd, remaining bounded by

23:2:1 $$-\exp\left(\frac{x}{2}\right) \le L_n(x) \le \exp\left(\frac{x}{2}\right) \qquad x \ge 0$$

23:3 DEFINITIONS

Laguerre polynomials are defined by the generating function

23:3:1 $$\frac{1}{1-t}\exp\left(\frac{-xt}{1-t}\right) = \sum_{n=0}^{\infty} L_n(x)t^n \qquad -1 < t < 1$$

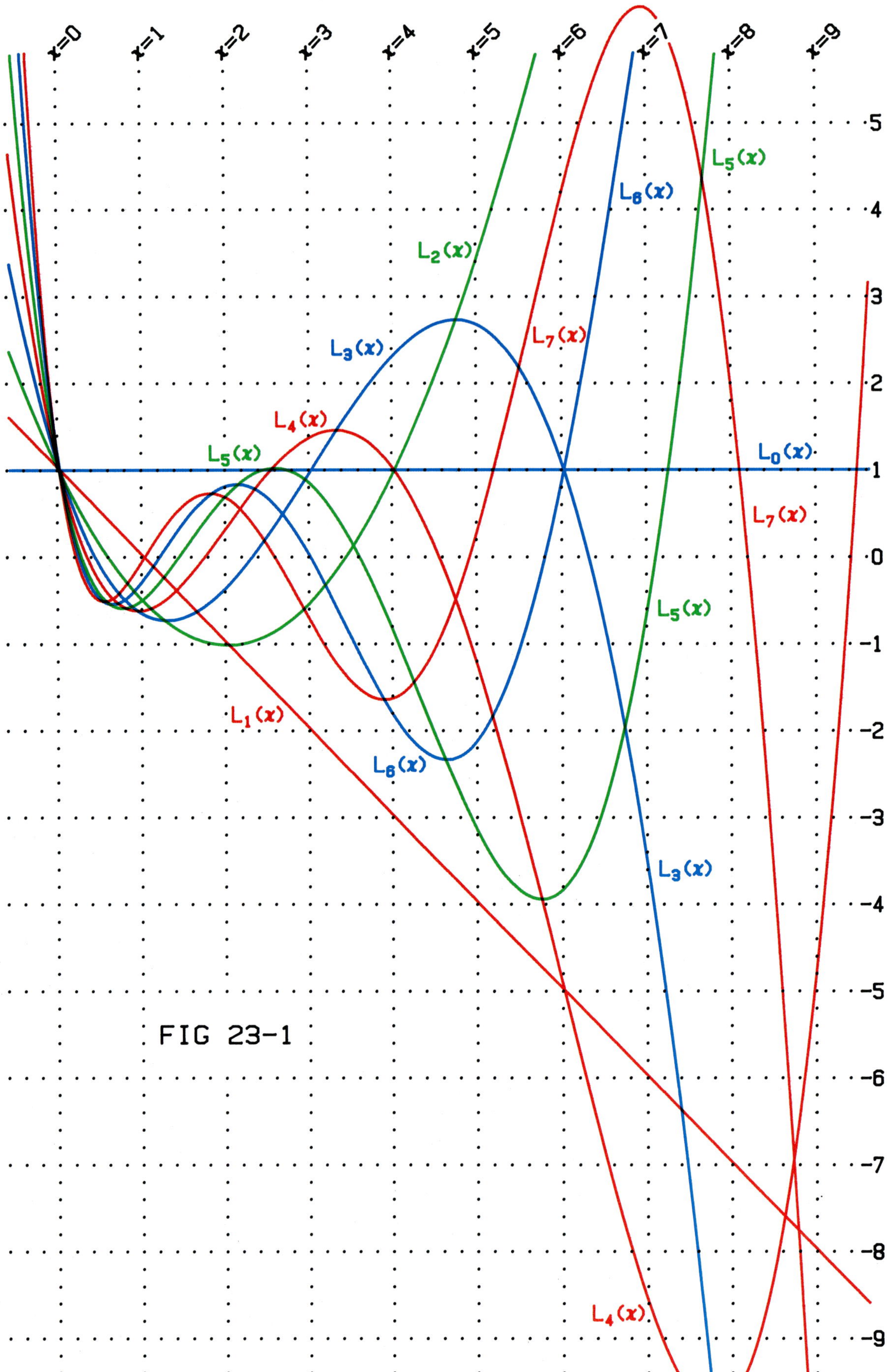
x=0
x=1
x=2
x=3
x=4
x=5
x=6
x=7
x=8
x=9
5
4
3
2
1
0
-1
-2
-3
-4
-5
-6
-7
-8
-9
L5(x)
L6(x)
L2(x)
L7(x)
L3(x)
L4(x)
L5(x)
L0(x)
L7(x)
L5(x)
L1(x)
L6(x)
L3(x)
L4(x)

FIG 23-1

or by *Rodrigues' formula*

23:3:2 $$L_n(x) = \frac{\exp(x)}{n!}\frac{d^n}{dx^n}\left[x^n \exp(-x)\right] \qquad n = 0, 1, 2, \ldots$$

A third definition is provided by expansion 23:6:1. The limiting operation

23:3:3 $$L_n(x) = \lim_{\mu\to\infty} P_n^{(0;\mu)}\left(1 - \frac{2x}{\mu}\right)$$

links the Laguerre polynomials to Jacobi polynomials [Section 22:12], while the integral representation

23:3:4 $$L_n(x) = \frac{\exp(x)}{n!}\int_0^\infty t^n \exp(-t) J_0(2\sqrt{tx})dt$$

establishes a connection with the zero-order Bessel coefficient of Chapter 52.

One solution of *Laguerre's differential equation*

23:3:5 $$x\frac{d^2f}{dx^2} + (1 - x)\frac{df}{dx} + nf = 0$$

is $f = cL_n(x)$, c being an arbitrary constant.

23:4 SPECIAL CASES

The first eight members of the family of Laguerre polynomials are

23:4:1 $$L_0(x) = 1$$

23:4:2 $$L_1(x) = -x + 1$$

23:4:3 $$L_2(x) = \frac{x^2}{2} - 2x + 1$$

23:4:4 $$L_3(x) = -\frac{x^3}{6} + \frac{3x^2}{2} - 3x + 1$$

23:4:5 $$L_4(x) = \frac{x^4}{24} - \frac{2x^3}{3} + 3x^2 - 4x + 1$$

23:4:6 $$L_5(x) = -\frac{x^5}{120} + \frac{5x^4}{24} - \frac{5x^3}{3} + 5x^2 - 5x + 1$$

23:4:7 $$L_6(x) = \frac{x^6}{720} - \frac{x^5}{20} + \frac{5x^4}{8} - \frac{10x^3}{3} + \frac{15x^2}{2} - 6x + 1$$

23:4:8 $$L_7(x) = -\frac{x^7}{5040} + \frac{7x^6}{720} - \frac{7x^5}{40} + \frac{35x^4}{24} - \frac{35x^3}{6} + \frac{21x^2}{2} - 7x + 1$$

23:5 INTRARELATIONSHIPS

We know of no reflection or recursion formulas that express $L_n(-x)$ or $L_n(x + 1)$ in terms of $L_n(x)$, but there exists the recurrence relation

23:5:1 $$L_n(x) = \frac{2n - 1 - x}{n} L_{n-1}(x) - \frac{n - 1}{n} L_{n-2}(x) \qquad n = 2, 3, 4, \ldots$$

relating a Laguerre polynomial to its congenitors of lower degree but identical argument.

Integer powers of x can be expressed as sums of Laguerre polynomials. The identity

23:5:2 $$x^6 = 720[L_0(x) + L_6(x)] - 4320[L_1(x) + L_5(x)] + 10800[L_2(x) + L_4(x)] - 14400\, L_3(x)$$

provides an example. The reader is referred to Abramowitz and Stegun [Table 22.10] for a tabulation of the coefficients that occur in such expressions. Other series of Laguerre polynomials include

23:5:3 $$\sum_{j=0}^{n} L_j(x) = \frac{x-n}{x} L_n(x) + \frac{n}{x} L_{n-1}(x) = L_n^{(1)}(x)$$

where $L_n^{(1)}$ is an associated Laguerre polynomial [Section 23:12]. A series of Laguerre polynomials also occurs in the multiplication formula

23:5:4 $$L_n(bx) = (1-b)^n \sum_{j=0}^{n} \binom{n}{j} \left(\frac{b}{1-b}\right)^j L_j(x)$$

while the addition formula

23:5:5 $$L_n(x+y) = \frac{1}{n!}\left(\frac{-1}{4}\right)^n \sum_{j=0}^{n} \binom{n}{j} H_{2j}(\sqrt{x})H_{2n-2j}(\sqrt{y})$$

involves a sum of products of Hermite polynomials [Chapter 24]. The *Christoffel-Darboux formula* for Laguerre polynomials is

23:5:6 $$\sum_{j=0}^{n} L_j(x)L_j(y) = \frac{n+1}{x-y}[L_n(x)L_{n+1}(y) - L_{n+1}(x)L_n(y)]$$

23:6 EXPANSIONS

An explicit expression for the Laguerre polynomial is

23:6:1 $$L_n(x) = (-1)^n \left[\frac{x^n}{n!} - \frac{nx^{n-1}}{1!(n-1)!} + \frac{n(n-1)x^{n-2}}{2!(n-2)!} - \cdots + (-1)^n\right] = \sum_{j=0}^{n} \binom{n}{n-j} \frac{(-x)^j}{j!}$$

23:7 PARTICULAR VALUES

	$L_n(-\infty)$	$L_n(0)$	$L_n(\infty)$
$n = 0$	1	1	1
$n = 1, 2, 3, \ldots$	∞	1	$(-\infty)^n$

We know of no exact expressions for the arguments of the zeros, minima or maxima of the Laguerre polynomials, beyond the first few.

23:8 NUMERICAL VALUES

When the portions of the algorithm that are shown in green are omitted, exact values of $L_n(x)$ are generated, via formula 23:5:1, when n and x are input. Laguerre polynomials, and their associated counterparts [see Section 23:12], may also be evaluated through the universal hypergeometric algorithm that is presented in Section 18:14.

Set $h = f = j = 1$

Input n >>>>>>>

Storage needed: h, f, j, n, m, x and g

```
                  If n = 0 go to (2)
Input m >>>>>>>>
Input x >>>>>>>>
                  Replace f by 1 + m - x
                  If n = 1 go to (2)
              (1) Replace j by j + 1
                  Set g = f
                  Replace f by [(2j + m - 1 - x)g - (j + m - 1)h]/j
                  Replace h by g
                  If j < n go to (1)
              (2) Output f
f = L_n^(m)(x) <<<<<<
```

Input restrictions: n must be a nonnegative integer. x and m are unrestricted.

Test values:
$L_0(\text{any } x) = 1$, $L_1(9) = -8$
$L_6(8.5) = 12.6229384$
$L_3^{(2)}(1) = 2.33333333$

23:9 APPROXIMATIONS

For large x one can approximate

23:9:1 $$L_n(x) \simeq \frac{(n-x)^n}{n!} \qquad \text{8-bit precision} \qquad x \geq (6n-2)^{3/2}$$

while for small x and large n

23:9:2 $$L_n(x) \simeq \exp\left(\frac{x}{2}\right)\left[J_0(\sqrt{Nx}) + \frac{x}{2N}J_2(\sqrt{Nx}) - \frac{x^{3/2}}{6\sqrt{N}}J_3(\sqrt{Nx}) + \cdots\right]$$

where $N = 4n + 2$. For further details of the asymptotic expansion 23:9:2 and for approximations to $L_n(x)$ as $n \to \infty$ for other ranges of x, see Erdélyi, Magnus, Oberhettinger and Tricomi [*Higher Transcendental Functions*, Section 10.15].

23:10 OPERATIONS OF THE CALCULUS

Derivatives and indefinite integrals of $L_n(x)$ can be expressed in terms of other Laguerre polynomials or, more succinctly, as associated Laguerre polynomials [Section 23:12] of unity order

23:10:1 $$\frac{d}{dx}L_n(x) = \frac{n}{x}[L_n(x) - L_{n-1}(x)] = -L_{n-1}^{(1)}(x)$$

23:10:2 $$\int_0^x L_n(t)dt = L_n(x) - L_{n+1}(x) = \frac{x}{n+1}L_n^{(1)}(x)$$

Other important integrals are

23:10:3 $$\int_x^\infty \exp(-t)L_n(t)dt = \exp(-x)[L_n(x) - L_{n-1}(x)] = \frac{-n\exp(-x)}{x}L_{n-1}^{(1)}(x)$$

and

23:10:4 $$\int_0^x L_n(t)L_m(x-t)dt = \int_0^x L_{n+m}(t)dt = \frac{x}{n+m+1}L_{n+m}^{(1)}(x)$$

and many more are listed by Gradshteyn and Ryzhik [Sections 7.41, 7.42]. The orthogonality of the Laguerre polynomials is established by the definite integral

23:10:5 $$\int_0^\infty \exp(-t)L_n(t)L_m(t)dt = \begin{cases} 0 & m \neq n \\ 1 & m = n \end{cases}$$

23:11 COMPLEX ARGUMENT

Laguerre polynomials are usually encountered with real arguments only.

23:12 GENERALIZATIONS

In previous sections of this chapter we refer to the family of *associated Laguerre polynomials* $L_n^{(m)}(x)$ of which the Laguerre polynomial is the $m = 0$ case

23:12:1 $$L_n^{(0)}(x) = L_n(x)$$

Each of the definitions 23:3:1–23:3:5 may be generalized as follows:

23:12:2 $$\frac{1}{(1-t)^{m+1}}\exp\left(\frac{-xt}{1-t}\right) = \sum_{n=0}^{\infty} L_n^{(m)}(x)t^n \qquad -1 < t < 1$$

23:12:3 $$L_n^{(m)}(x) = \frac{\exp(x)}{n!x^m}\frac{d^n}{dx^n}[x^{n+m}\exp(-x)]$$

23:12:4 $$L_n^{(m)}(x) = \lim_{\mu\to\infty} P_n^{(m;\mu)}\left(1 - \frac{2x}{\mu}\right)$$

23:12:5 $$L_n^{(m)}(x) = \frac{\exp(x)}{n!x^{m/2}}\int_0^{\infty} t^{(2n+m)/2}J_m(2\sqrt{tx})dt$$

23:12:6 $$x\frac{d^2f}{dx^2} + (m + 1 - x)\frac{df}{dx} + nf = 0 \qquad f = cL_n^{(m)}(x) + \cdots$$

The associated Laguerre polynomials may be expanded via

23:12:7 $$L_n^{(m)}(x) = \sum_{j=0}^{n}\binom{n+m}{n-j}\frac{(-x)^j}{j!}$$

and the early members are listed in Table 23.12.1. Numerical values of $L_n^{(m)}(x)$ are given by the algorithm in Section 23:8 if the portions printed in green are included. Analogs exist of most of the formulas of Sections 23:5, 23:9 and 23:10, including the following recursion, differentiation and orthogonality formulas:

23:12:8 $$L_n^{(m)}(x) = \frac{2n+m-1-x}{n}L_{n-1}^{(m)}(x) - \frac{n+m-1}{n}L_{n-2}^{(m)}(x) = \frac{x-n}{x}L_n^{(m-1)}(x) + \frac{n+m-1}{x}L_{n-1}^{(m-1)}(x)$$

Table 23.12.1

	$L_0^{(m)}(x)$	$L_1^{(m)}(x)$	$L_2^{(m)}(x)$	$L_3^{(m)}(x)$	$L_4^{(m)}(x)$	
$m = 0$	1	$-x + 1$	$\frac{x^2}{2} - 2x + 1$	$-\frac{x^3}{6} + \frac{3x^2}{2} - 3x + 1$	$\frac{x^4}{24} - \frac{2x^3}{3} + 3x^2 - 4x + 1$	$m = 0$
$m = 1$	1	$-x + 2$	$\frac{x^2}{2} - 3x + 3$	$-\frac{x^3}{6} + 2x^2 - 6x + 4$	$\frac{x^4}{24} - \frac{5x^3}{6} + 5x^2 - 10x + 5$	$m = 1$
$m = 2$	1	$-x + 3$	$\frac{x^2}{2} - 4x + 6$	$-\frac{x^3}{6} + \frac{5x^2}{2} - 10x + 10$	$\frac{x^4}{24} - x^3 + \frac{15x^2}{2} - 20x + 15$	$m = 2$
$m = 3$	1	$-x + 4$	$\frac{x^2}{2} - 5x + 10$	$-\frac{x^3}{6} + 3x^2 - 15x + 20$	$\frac{x^4}{24} - \frac{7x^3}{6} + \frac{21x^2}{2} - 35x + 35$	$m = 3$
$m = 4$	1	$-x + 5$	$\frac{x^2}{2} - 6x + 15$	$-\frac{x^3}{6} + \frac{7x^2}{2} - 21x + 35$	$\frac{x^4}{24} - \frac{4x^3}{6} + 14x^2 - 56x + 70$	$m = 4$

23:12:9 $$\frac{d}{dx}L_n^{(m)}(x) = \frac{n}{x}L_n^{(m)}(x) - \frac{n+m}{x}L_{n-1}^{(m)}(x) = -L_{n-1}^{(m+1)}(x)$$

23:12:10 $$\int_0^\infty t^m \exp(-t)\,L_n^{(m)}(t)L_N^{(m)}(t)dt = \begin{cases} 0 & n \neq N \\ 1 & n = N \end{cases}$$

Caution is necessary because the associated Laguerre polynomials are sometimes defined as

23:12:11 $$n!\frac{d^m}{dx^m}L_n(x)$$

which definition leads to properties radically different from those of the polynomials defined by 23:12:2–23:12:6. According to Sneddon [Section 44], the definitions adopted in this *Atlas* are generally favored by pure mathematicians, whereas 23:12:11 is preferred by applied mathematicians. Although interest concentrates on nonnegative integer values of the order m, definitions 23:12:2–23:12:7 may remain valid for negative integer m, and even for noninteger values provided $m > -1$. Polynomials so defined are often named *generalized Laguerre polynomials;* the simplest instances are

23:12:12 $$L_n^{(1/2)}(x) = \frac{1}{2}\left(\frac{-1}{4}\right)^n \frac{H_{2n+1}(\sqrt{x})}{n!\sqrt{x})}$$

and

23:12:13 $$L_n^{(-1/2)}(x) = \left(\frac{-1}{4}\right)^n \frac{H_{2n}(\sqrt{x})}{n!}$$

where H is the Hermite polynomial [Chapter 24].

The Kummer function [Chapter 47] is a broader generalization of the Laguerre polynomial and of its associated counterparts. The relationship is

23:12:14 $$L_n^{(m)}(x) = \binom{n+m}{n} M(-n;m+1;x)$$

23:13 COGNATE FUNCTIONS

Several formulas are presented in previous sections that relate Laguerre polynomials to Jacobi polynomials [Section 22:12], to Hermite polynomials [Chapter 24] and to Bessel coefficients [Chapter 52]. To the last category the generating function

23:13:1 $$\frac{\exp(t)}{(xt)^{m/2}}J_m(2\sqrt{xt}) = \sum_{n=0}^{\infty} L_n^{(m)}(x)\frac{t^n}{(m+n)!}$$

should be added.

23:14 RELATED TOPICS

The term *Laguerre function* is encountered with two distinct meanings. It is used to signify the $L_n(x)$ function in which n is not restricted to integer values, being then an example of the Kummer function [Chapter 47]

23:14:1 $$L_\nu(x) = M(-\nu;1;x) = \sum_{j=0}^{\infty}\frac{(-\nu)_j x^j}{(j!)^2}$$

The generalized Laguerre function is addressed in Section 47:1. Alternatively, the expression

23:14:2 $$(n+1)!\exp\left(\frac{-x}{2}\right)x^l\frac{d^{2l+1}}{dx^{2l+1}}L_{n+1}(x)$$

is defined as the *Laguerre function*. It is a solution to the radial portion of the Schrödinger equation for a hydrogen-like atom

23:14:3
$$\frac{d^2f}{dx^2} + \frac{2}{x}\frac{df}{dx} - \left[\frac{1}{4} - \frac{n}{x} + \frac{l(l+1)}{x^2}\right] f = 0 \qquad n \geq l + 1$$

where n and l are the appropriate quantum numbers [see Sneddon, Section 44].

CHAPTER 24

THE HERMITE POLYNOMIALS $H_n(x)$

The Hermite polynomials are orthogonal [see Section 21:14] on the interval $-\infty < x < \infty$ with a weight function $\exp(-x^2)$ [see Chapter 27]. They occur in several applications to physical problems, for example, in the solution of the Schrödinger equation [see Section 24:14].

24:1 NOTATION

The notation $H_n(x)$ is in general use for the Hermite polynomial of degree n and argument x. An alternative Hermite polynomial, symbolized $He_n(x)$, is discussed in Section 24:13; unfortunately, it also is sometimes denoted $H_n(x)$, especially in texts on mathematical statistics.

Occasionally, Hermite polynomials of odd degree are defined with the opposite sign to that adopted in this *Atlas*.

Although this *Atlas* avoids the ambiguity, the symbol $H_n(x)$ is sometimes used for the Struve function [see Chapter 57].

24:2 BEHAVIOR

The Hermite polynomial $H_n(x)$ has exactly n zeros, $\mathrm{Int}(n/2)$ minima and $\mathrm{Int}((n-1)/2)$ maxima. These features are symmetrically disposed about $x = 0$, and all occur in the region $-\sqrt{2n} < x < \sqrt{2n}$. Outside this range $|H_n(x)|$ increases monotonically, except for $H_0(x)$, remaining bounded globally by

24:2:1 $$|H_n(x)| < 1.09\sqrt{2^n\, n!\, \exp(x^2)}$$

The range spanned by the oscillations of $H_n(x)$ increases so dramatically with increasing n that Figure 24-1 portrays the behavior of $H_n(x)/n!$, rather than that of the Hermite polynomials themselves.

24:3 DEFINITIONS

Hermite polynomials are defined by the generating function

24:3:1 $$\exp(2tx - t^2) = \sum_{n=0}^{\infty} H_n(x)\frac{t^n}{n!}$$

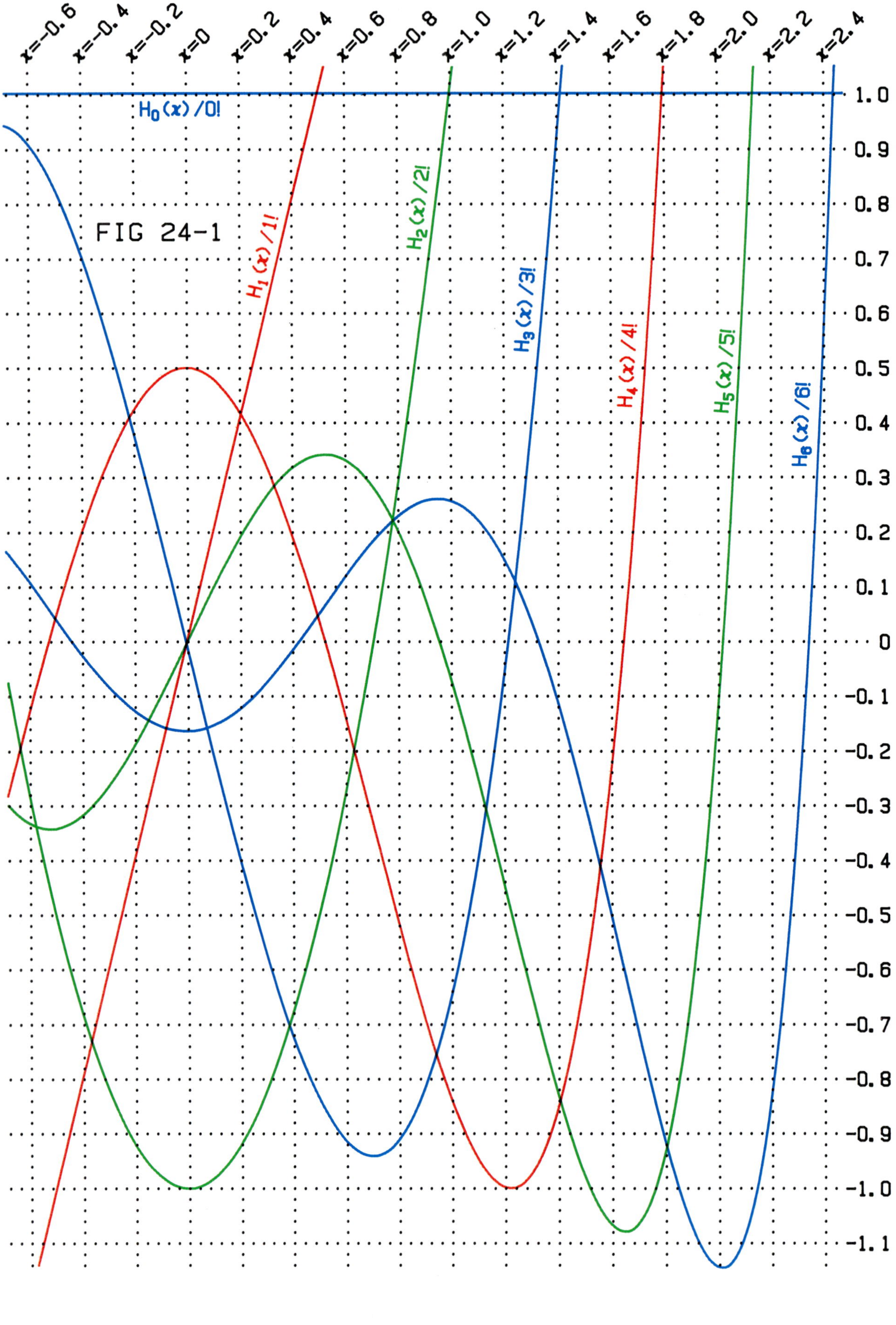
x=−0.6
x=−0.4
x=−0.2
x=0
x=0.2
x=0.4
x=0.6
x=0.8
x=1.0
x=1.2
x=1.4
x=1.6
x=1.8
x=2.0
x=2.2
x=2.4
FIG 24-1
H0(x)/0!
H1(x)/1!
H2(x)/2!
H3(x)/3!
H4(x)/4!
H5(x)/5!
H6(x)/6!
1.0
0.9
0.8
0.7
0.6
0.5
0.4
0.3
0.2
0.1
0
−0.1
−0.2
−0.3
−0.4
−0.5
−0.6
−0.7
−0.8
−0.9
−1.0
−1.1

or by the *Rodrigues formula*

24:3:2 $$H_n(x) = (-1)^n \exp(x^2) \frac{d^n}{dx^n} \exp(-x^2)$$

As well, the integral representation

24:3:3 $$H_n(x) = \frac{2^{n+1}}{\sqrt{\pi}} \exp(x^2) \int_0^\infty t^n \exp(-t^2) \cos\left(2xt - \frac{n\pi}{2}\right) dt$$

may be regarded as defining $H_n(x)$, as may expansion 24:6:1 or the limiting operation

24:3:4 $$H_n(x) = \lim_{\lambda\to\infty} \frac{n!}{\lambda^{n/2}} C_n^{(\lambda)}\left(\frac{x}{\sqrt{\lambda}}\right)$$

applied to the Gegenbauer polynomials [Section 22:13].

One solution of *Hermite's differential equation*

24:3:5 $$\frac{d^2f}{dx^2} - 2x\frac{df}{dx} + 2nf = 0$$

is $f = cH_n(x)$, where c is an arbitrary constant; a second solution is a closely related infinite series [see Murphy, page 321, but note that his definition of $H_n(x)$ differs from ours]. Another important differential equation

24:3:6 $$\frac{d^2f}{dx^2} = (x^2 - 2n - 1)f$$

is solved by $f = c \exp(-x^2/2)H_n(x)$ [see Section 24:14].

24:4 SPECIAL CASES

The first eight members of the family of Hermite polynomials are

24:4:1 $$H_0(x) = 1$$

24:4:2 $$H_1(x) = 2x$$

24:4:3 $$H_2(x) = 4x^2 - 2$$

24:4:4 $$H_3(x) = 8x^3 - 12x$$

24:4:5 $$H_4(x) = 16x^4 - 48x^2 + 12$$

24:4:6 $$H_5(x) = 32x^5 - 160x^3 + 120x$$

24:4:7 $$H_6(x) = 64x^6 - 480x^4 + 720x^2 - 120$$

24:4:8 $$H_7(x) = 128x^7 - 1344x^5 + 3360x^3 - 1680x$$

24:5 INTRARELATIONSHIPS

The *reflection formula*

24:5:1 $$H_n(-x) = (-1)^n H_n(x)$$

shows that a Hermite polynomial is an even or odd function according to whether its degree is even or odd. The addition formula

24:5:2 $$H_n(x + y) = 2^{-n/2} \sum_{j=0}^{n} \binom{n}{j} H_j(x\sqrt{2})H_{n-j}(y\sqrt{2})$$

and the *Christoffel-Darboux formula*

24:5:3 $$\sum_{j=0}^{n} \frac{H_j(x)H_j(y)}{2^j j!} = \frac{H_{n+1}(x)H_n(y) - H_n(x)H_{n+1}(y)}{2^{n+1}n!(x-y)}$$

are obeyed by Hermite polynomials. Other summation formulas include the infinite series

24:5:4 $$H_0(x) - \frac{H_2(x)}{2!} + \frac{H_4(x)}{4!} - \cdots = e\cos(2x)$$

and

24:5:5 $$H_1(x) - \frac{H_3(x)}{3!} + \frac{H_5(x)}{5!} - \cdots = e\sin(2x)$$

where e is the base of natural logarithms [Chapter 1]. Series akin to 24:5:4 and 24:5:5 in which the signs do not alternate sum to $[\cosh(2x)]/e$, and $[\sinh(2x)]/e$, respectively.

The recursion formula

24:5:6 $$H_n(x) = 2xH_{n-1}(x) - 2(n-1)H_{n-2}(x) \qquad n = 2, 3, 4, \ldots$$

together with $H_0(x) = 1$ and $H_1(x) = 2x$, permits the polynomial form of any Hermite polynomial function to be evaluated.

24:6 EXPANSIONS

The explicit expansion of the Hermite polynomial may be written in several equivalent ways; for example:

24:6:1 $$H_n(x) = (2x)^n - \frac{n(n-1)}{1!}(2x)^{n-2} + \frac{n(n-1)(n-2)(n-3)}{2!}(2x)^{n-4} - \cdots + \begin{cases} (-2)^{n/2}(n-1)!! & n = 0, 2, 4, \ldots \\ -(-2)^{(n+1)/2}n!!x & n = 1, 3, 5, \ldots \end{cases}$$
$$= n! \sum_{j=0}^{J} \frac{(-1)^j(2x)^{n-2j}}{j!(n-2j)!} \qquad J = \mathrm{Int}(n/2)$$

24:7 PARTICULAR VALUES

	$H_n(-\infty)$	$H_n(0)$	$H_n(\infty)$
$n = 0$	1	1	1
$n = 1, 3, 5, \ldots$	$-\infty$	0	∞
$n = 2, 4, 6, \ldots$	∞	$(-2)^{n/2}(n-1)!!$	∞

Because of relationship 24:10:1, certain characteristic values of the argument correspond to features in a number of consecutive Hermite polynomials. For example, the four values $\pm\sqrt{(3 \pm \sqrt{6})/2}$ are the zeros of $H_4(x)$, generate maxima or minima of $H_5(x)$ and correspond to inflection points of $H_6(x)$.

The asterisks in Figure 24-2 show the exact locations of the zeros for $H_1(x)$ through $H_{12}(x)$. The colored lines in this diagram are explained in Section 24:9.

24:8 NUMERICAL VALUES

The algorithm

Set $h = f = j = 1$

Storage needed: h, f, j, n, x and g

```
Input n >>>>>>>
            If n = 0 go to (2)
Input x >>>>>>>
            Set f = 2x
            If n = 1 go to (2)
        (1) Replace j by j + 1
            Set g = f
            Set f = 2[xg + (1 − j)h]
            Set h = g
            If j < n go to (1)
        (2) Output f
f = Hn(x) <<<<<<<
```

Input restrictions: n must be a nonnegative integer; x is unrestricted.

Test values:
$H_0(\text{any } x) = 1$
$H_1(-12) = -24$
$H_7(4.2) = 1436292.99$

utilizes equation 24:5:6 to generate exact values of the Hermite polynomial. Hermite polynomials may also be enumerated via the universal hypergeometric algorithm of Section 18:14.

24:9 APPROXIMATE VALUES

For large (positive or negative) values of x, the approximation

24:9:1 $$H_n(x) \simeq \left[2x - \frac{n-1}{2}\right]^n \qquad \text{8-bit precision} \qquad |x| > 2.3(n-1)^{3/4}$$

holds. For large n, the Hermite polynomial is approximated by the oscillatory function

24:9:2 $$H_n(x) \simeq \begin{cases} (-2)^{n/2}(n-1)!!\,\exp(x^2/2)\cos(x\sqrt{2n+1}) & \text{even } n \to \infty \\ (-2)^{(n-1)/2}\sqrt{\dfrac{2}{n}}\,n!!\,\exp(x^2/2)\sin(x\sqrt{2n+1}) & \text{odd } n \to \infty \end{cases}$$

This approximation is best for small $|x|$ and fails hopelessly for $|x| > \sqrt{2n}$, that is, when the Hermite polynomial leaves its oscillatory region [see discussion in Section 24:2].

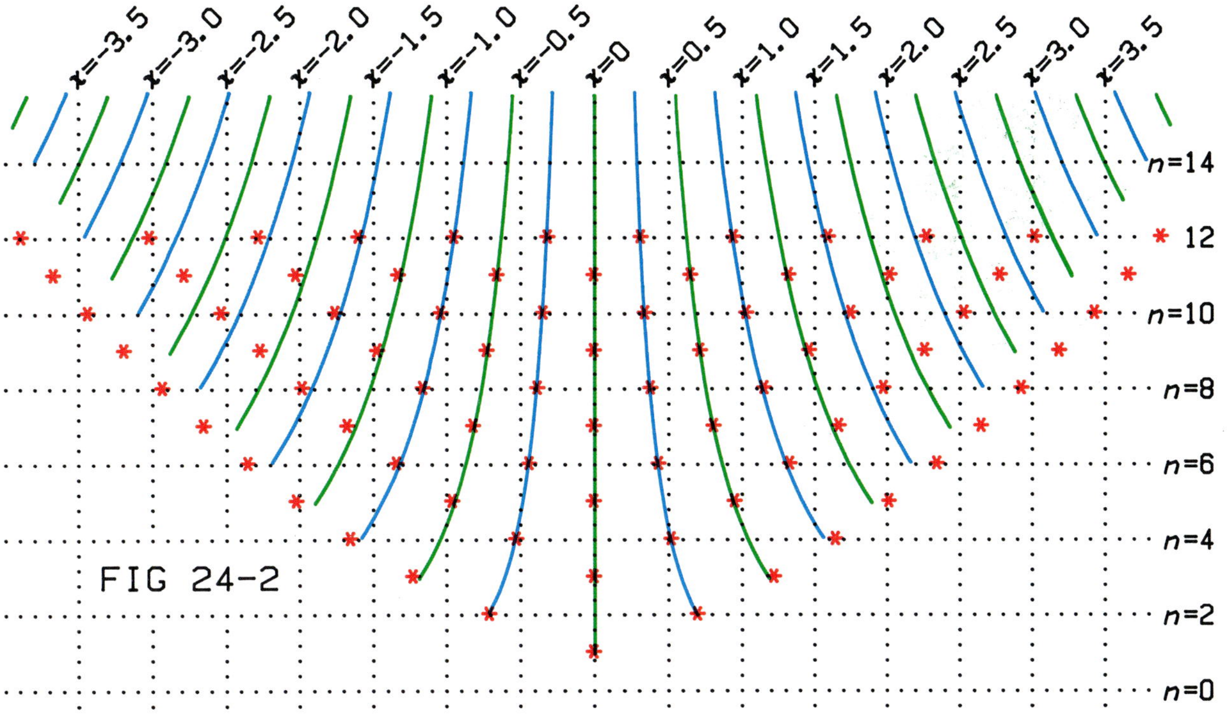

FIG 24-2

From 24:9:2 it follows that the n zeros of $H_n(x)$ are approximated by

24:9:3
$$r \simeq \frac{j\pi}{2\sqrt{2n+1}} \begin{cases} j = \pm 1, \pm 3, \pm 5, \ldots, \pm(n-1) & n \text{ even} \\ j = 0, \pm 2, \pm 4, \ldots, \pm(n-1) & n \text{ odd} \end{cases}$$

The lines drawn on Figure 24-2 connect the zeros predicted by this approximation, blue for even n and green for odd n. Evidently, the approximation is good for the inner zeros but worsens the further the zero is from $x = 0$.

24:10 OPERATIONS OF THE CALCULUS

We have the following simple results for differentiation and indefinite integration:

24:10:1
$$\frac{d}{dx} H_n(x) = 2nH_{n-1}(x) \qquad n = 1, 2, 3, \ldots$$

24:10:2
$$\frac{d}{dx} [\exp(-x^2)H_n(x)] = -\exp(-x^2)H_{n+1}(x)$$

24:10:3
$$\int_0^x H_n(t)dt = \begin{cases} \dfrac{H_{n+1}(x)}{2(n+1)} & n = 0, 2, 4, \ldots \\ \dfrac{H_{n+1}(x) - (-2)^{(n+1)/2} n!!}{2(n+1)} & n = 1, 3, 5, \ldots \end{cases}$$

and

24:10:4
$$\int_0^x \exp(-t^2)H_n(t)dt = \begin{cases} -\exp(-x^2)H_{n-1}(x) & n = 2, 4, 6, \ldots \\ (-2)^{(n-1)/2}(n-2)!! - \exp(-x^2)H_{n-1}(x) & n = 1, 3, 5, \ldots \end{cases}$$

Three important definite integrals are

24:10:5
$$\int_{-\infty}^{\infty} \exp(-t^2)H_n(bt)dt = \begin{cases} \sqrt{\pi}\, n!(b^2 - 1)^{n/2}/(n/2)! & n = 0, 2, 4, \ldots \\ 0 & n = 1, 3, 5, \ldots \end{cases}$$

24:10:6
$$\int_{-\infty}^{\infty} t \exp(-t^2)H_n(bt)dt = \begin{cases} 0 & n = 0, 2, 4, \ldots \\ \sqrt{\pi} n!! b(2b^2 - 2)^{(n-1)/2} & n = 1, 3, 5, \ldots \end{cases}$$

and

24:10:7
$$\int_{-\infty}^{\infty} t^n \exp(-t^2)H_n(bt)dt = \sqrt{\pi} n! P_n(b)$$

where P_n is the Legendre polynomial of degree n [Chapter 21], and many others are listed by Gradshteyn and Ryzhik [Sections 7.37 and 7.38].

The orthogonality of Hermite polynomials is established by

24:10:8
$$\int_{-\infty}^{\infty} \exp(-t^2)H_n(t)H_m(t)dt = \begin{cases} 0 & m \neq n \\ \sqrt{\pi}\, 2^n n! & m = n \end{cases}$$

24:11 COMPLEX ARGUMENT

As for other orthogonal polynomials, the argument of the Hermite polynomials is generally encountered as a real variable.

24:12 GENERALIZATIONS

Hermite polynomials are special cases of the parabolic cylinder function [Chapter 46] in which the order is a nonnegative integer:

24:12:1 $$H_n(x) = 2^{n/2}\exp\left(\frac{x^2}{2}\right) D_n(x\sqrt{2}) \qquad n = 0, 1, 2, \ldots$$

Because the parabolic cylinder function itself generalizes to the Tricomi function [Chapter 48] or the Kummer function [Chapter 47] these latter functions may also be regarded as generalizations of Hermite polynomials

24:12:2 $$H_n(x) = 2^n U\left(\frac{1-n}{2}; \frac{3}{2}; x^2\right)$$

24:12:3 $$H_n(x) = \begin{cases} (-2)^{n/2}(n-1)!!\, M\left(\frac{-n}{2}; \frac{1}{2}; x^2\right) & n = 0, 2, 4, \ldots \\ -(-2)^{(n+1)/2} n!!\, M\left(\frac{1-n}{2}; \frac{3}{2}; x^2\right) & n = 1, 3, 5, \ldots \end{cases}$$

24:13 COGNATE FUNCTIONS

The *alternative Hermite polynomial* $He_n(x)$ is related to $H_n(x)$ by a simple variable change

24:13:1 $$He_n(x) = 2^{-n/2} H_n(x/\sqrt{2})$$

24:14 RELATED TOPICS

Equation 24:3:6 arises in the application of Schrödinger's equation to simple harmonic oscillators. The solution

24:14:1 $$f = \exp\left(\frac{-x^2}{2}\right) H_n(x) = f_n(x)$$

is known as a *Hermite function*. In addition to satisfying a typical orthogonality relationship

24:14:2 $$\int_{-\infty}^{\infty} f_m(t)\, f_n(t) dt = \begin{cases} 0 & m \neq n \\ \sqrt{\pi}\, n!\, 2^n & m = n \end{cases}$$

Hermite functions also satisfy such related expressions as

24:14:3 $$\int_{-\infty}^{\infty} t f_m(t)\, f_n(t) dt = \begin{cases} 0 & m \neq n \pm 1 \\ \sqrt{\pi}\, n!\, 2^{n-1} & m = n - 1 \end{cases}$$

CHAPTER 25

THE LOGARITHMIC FUNCTION ln(x)

The logarithmic function may be regarded as the simplest *transcendental function,* that is, ln(x) is the simplest function of x that cannot be expressed as a finite combination of algebraic terms. Prior to the advent of electronic calculators, logarithms [especially decadic logarithms, see Section 25:14] were extensively used as aids to arithmetic computation, both in the form of tables and in the design of devices such as slide rules.

In addition to its treatment of the logarithmic function, this chapter includes brief discussions of the logarithmic integral li(x) and the dilogarithm diln(x).

25:1 NOTATION

The name *logarithm* of x is used synonymously with logarithmic function of x to describe ln(x). Alternative notations are log(x) and $\log_e(x)$, although the former is also used to describe a decadic logarithm. The initial letter is sometimes written in script type to avoid possible confusion with the numeral "one." See Section 25:11 for the significance of Ln(z).

To emphasize the distinction from logarithms to other bases [Section 25:14], ln(x) is variously referred to as the *natural logarithm,* the *Naperian logarithm,* the *hyperbolic logarithm* or the *logarithm to base e.*

25:2 BEHAVIOR

The logarithm ln(x) is not defined as a real-valued function for negative argument [but see Section 25:11] and takes the value $-\infty$ when x is zero. As shown in Figure 25-1, ln(x) is negative for $0 \le x < 1$ and positive for $x > 1$.

The slope $\mathrm{dln}(x)/\mathrm{d}x$ continuously decreases as x increases so that, even though ln(x) increases without limit as x tends to infinity, its rate of increase becomes ever more leisurely. In fact, ln(x) increases more slowly than any positive power of x:

25:2:1 $$\frac{\ln(x)}{x^{\nu}} \to 0 \qquad x \to \infty \qquad \nu > 0$$

25:3 DEFINITIONS

As illustrated in Figure 25-2, the logarithmic function is defined through the integral

25:3:1 $$\ln(x) = \int_1^x \frac{1}{t}\,\mathrm{d}t \qquad x > 0$$

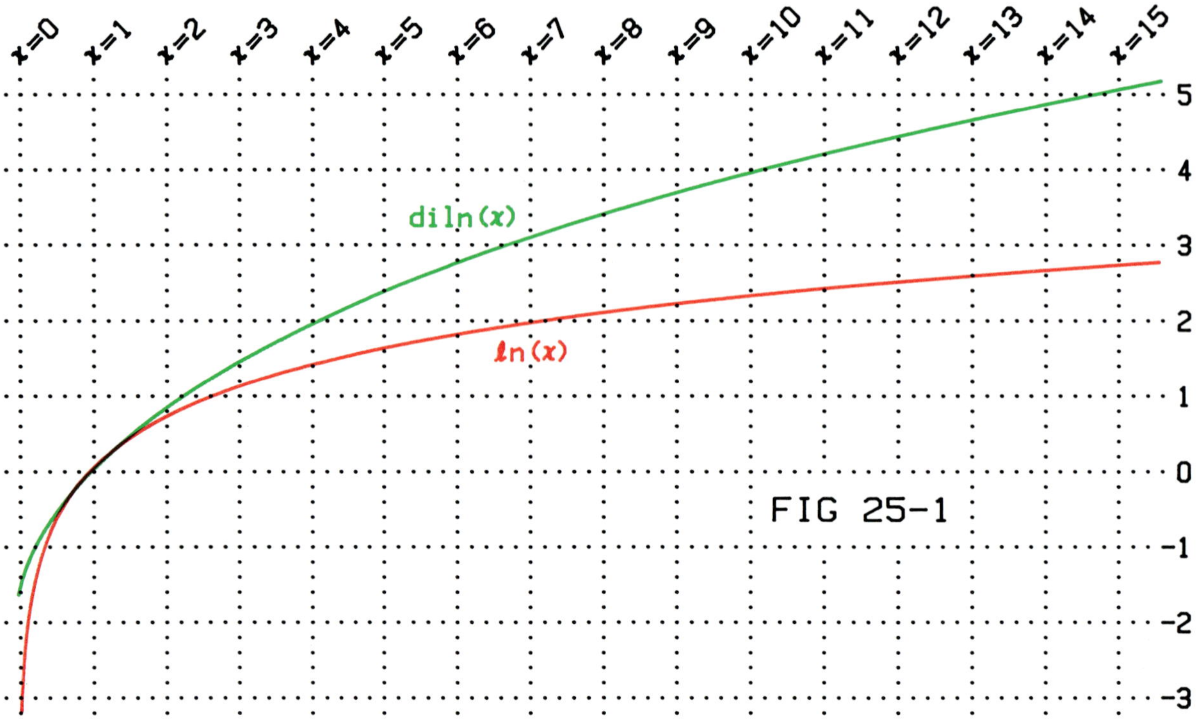

FIG 25-1

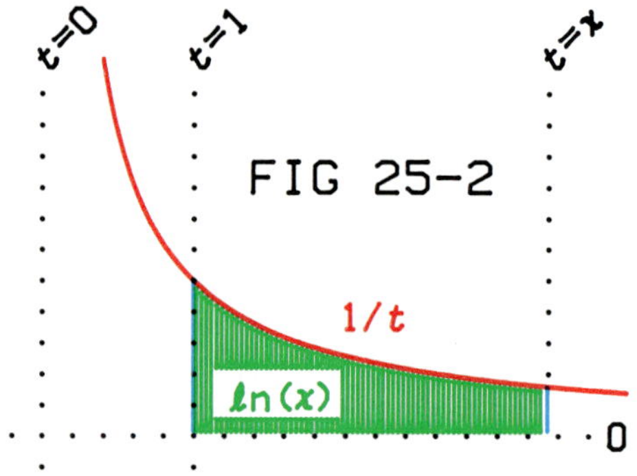

FIG 25-2

It may also be defined as the limit

25:3:2 $$\ln(x) = \lim_{v\to\infty}[v(x^{1/v} - 1)] \qquad x > 0 \qquad v > 0$$

or by the rule

25:3:3 $$\ln(x) = f \qquad \text{if} \qquad x = e^f$$

where e is the number 2.1718281828 defined in Chapter 1.

25:4 SPECIAL CASES

There are none.

25:5 INTRARELATIONSHIPS

The logarithmic function of any argument in the range zero to unity may be related to ln(x) where $1 < x < \infty$ by the reciprocation formula

25:5:1 $$\ln\left(\frac{1}{x}\right) = -\ln(x) \qquad x > 0$$

The logarithmic functions of products, quotients and powers may be expressed by the formulas

25:5:2 $$\ln(xy) = \ln(x) + \ln(y) \qquad x > 0 \qquad y > 0$$

25:5:3 $$\ln(x/y) = \ln(x) - \ln(y) \qquad x > 0 \qquad y > 0$$

25:5:4 $$\ln(x^\nu) = \nu\ln(x) \qquad x > 0$$

If a function $f(x)$ is expressible as the finite (or infinite) product $\Pi f_j(x)$, its logarithm is the finite (or infinite) sum

25:5:5 $$\ln(f(x)) = \sum \ln(f_j(x)) \qquad \text{all } f_j(x) > 0$$

provided, in the infinite case, that the series in question converges.

25:6 EXPANSIONS

The logarithmic function may be expanded as a power series in a variety of ways, of which the following are representative:

25:6:1 $$\ln(1+x) = x - \frac{x^2}{2} + \frac{x^3}{3} - \cdots = -\sum_{j=1}^{\infty} \frac{(-x)^j}{j} \qquad -1 < x \leq 1$$

25:6:2 $$\ln(x) = \frac{x-1}{x} + \frac{(x-1)^2}{2x^2} + \frac{(x-1)^3}{3x^3} + \cdots = \sum_{j=1}^{\infty} \frac{(x-1)^j}{jx^j} \qquad x \geq \frac{1}{2}$$

25:6:3 $$\ln(x) = \frac{x^2-1}{x^2+1} + \frac{(x^2-1)^3}{3(x^2+1)^3} + \frac{(x^2-1)^5}{5(x^2+1)^5} + \cdots = \sum_{j=0}^{\infty} \frac{1}{2j+1}\left(\frac{x^2-1}{x^2+1}\right)^{2j+1} \qquad x > 0$$

As well, the logarithm is expansible through the continued fractions

25:6:4 $$\ln(1+x) = 1 - \frac{1}{1+}\,\frac{x}{1-x+}\,\frac{x}{2-x+}\,\frac{4x}{3-2x+}\,\frac{9x}{4-3x+}\,\frac{16x}{5-4x+}\cdots$$

and

25:6:5 $$\frac{\ln(1+x)}{x} = \frac{1}{1+}\,\frac{x}{2+}\,\frac{x}{3+}\,\frac{4x}{4+}\,\frac{4x}{5+}\,\frac{9x}{6+}\,\frac{9x}{7+}\,\frac{16x}{8+}\cdots$$

25:7 PARTICULAR VALUES

$\ln(0)$	$\ln(1)$	$\ln(e)$	$\ln(e^n)$	$\ln(\infty)$
$-\infty$	0	1	n	∞

25:8 NUMERICAL VALUES

Computer languages invariably permit logarithms to be calculated by a single command. Likewise, scientific calculators mostly incorporate a key that generates the value of $\ln(x)$ from an x value in the calculator's register. Sometimes a "log" key must be struck and the result multiplied by 2.302585093 [see Section 25:14]. Because of the widespread availability of such devices, we include no algorithm for the logarithmic function, although the universal hypergeometric algorithm [Section 18:14] does permit logarithms to be evaluated to any sought precision.

25:9 APPROXIMATIONS

A number of approximations, including

25:9:1 $$\ln(x) \simeq \frac{x-1}{\sqrt{x}} \qquad \text{8-bit precision} \qquad \frac{3}{4} \le x \le \frac{4}{3}$$

and

25:9:2 $$\ln(x) \simeq (x-1)\left(\frac{6}{1+5x}\right)^{3/5} \qquad \text{8-bit precision} \qquad \frac{1}{2} \le x \le 2$$

are available when x is close to unity.

Based on the limiting expression

25:9:3 $$\ln(n) \to -\gamma + \sum_{j=1}^{n} \frac{1}{j} \qquad n \to \infty \qquad \gamma = 0.5772156649$$

where γ is Euler's constant [Chapter 1], we have the approximation

25:9:4 $$\ln(x) \simeq \frac{\text{frac}(x)}{x} - 0.58 + \sum_{j=1}^{\text{Int}(x)} \frac{1}{j} \qquad \text{8-bit precision} \qquad x \ge 31$$

valid for large arguments.

25:10 OPERATIONS OF THE CALCULUS

Single and multiple differentiation give

25:10:1 $$\frac{d}{dx}\ln(bx+c) = \frac{b}{bx+c}$$

25:10:2 $$\frac{d^n}{dx^n}\ln(bx+c) = -(n-1)!\left(\frac{-b}{bx+c}\right)^n \qquad n = 1, 2, 3, \ldots$$

while indefinite integration yields the following results:

25:10:3 $$\int_{(1-c)/b}^{x} \ln(bt+c)dt = \left(x + \frac{c}{b}\right)[\ln(bx+c) - 1]$$

25:10:4 $$\int_{1}^{x} \ln^n(t)dt = (-1)^n n! x \sum_{j=0}^{n} \frac{[-\ln(x)]^j}{j!} \qquad n = 1, 2, 3, \ldots$$

25:10:5 $$\int_{1}^{x} t^\nu \ln(t)dt = \begin{cases} \dfrac{x^{\nu+1}}{\nu+1}\left[\ln(x) - \dfrac{1}{\nu+1}\right] & \nu \neq -1 \\ \dfrac{\ln^2(x)}{2} & \nu = -1 \end{cases}$$

25:10:6 $$\int_{-c/b}^{x} \frac{dt}{\ln(bt+c)} = \frac{\text{li}(bx+c)}{b}$$

25:10:7 $$\int_{0}^{x} \frac{t^\nu dt}{\ln(t)} = \begin{cases} \text{li}(x^{\nu+1}) & \nu \neq -1 \\ \ln(\ln(x)) & \nu = -1 \end{cases}$$

The li function in the last two formulas is the logarithmic integral function [see Section 25:13]. In addition to definite integrals that can be evaluated as special cases of the last five formulas, the following are of interest:

25:10:8 $$\int_0^1 \ln(-\ln(t))\mathrm{d}t = -\gamma$$

25:10:9 $$\int_0^1 \frac{\ln(t)\mathrm{d}t}{1 \pm t} = \begin{cases} -\pi^2/12 \\ -\pi^2/6 \end{cases}$$

25:10:10 $$\int_0^1 \frac{\ln(t)\mathrm{d}t}{1 \pm t^2} = \begin{cases} -G \\ -\pi^2/8 \end{cases}$$

25:10:11 $$\int_0^1 \ln(t)\ln(1 \pm t)\mathrm{d}t = \begin{cases} 2 - (\pi^2/12) - \ln(4) \\ 2 - (\pi^2/6) \end{cases}$$

Others are given by Gradshteyn and Ryzhik [Section 4.2–4.4]. The constants appearing in these expressions are defined in Section 1:7.

Semidifferentiation and semiintegration, with a lower limit of zero, yield

25:10:12 $$\frac{\mathrm{d}^{1/2}}{\mathrm{d}x^{1/2}} \ln(bx) = \frac{\ln(4bx)}{\sqrt{\pi x}}$$

and

25:10:13 $$\frac{\mathrm{d}^{-1/2}}{\mathrm{d}x^{1/2}} \ln(bx) = 2\sqrt{\frac{x}{\pi}}\,[\ln(4bx) - 2]$$

25:11 COMPLEX ARGUMENT

When the argument of the logarithmic function is replaced by the complex number $x + iy$, the result is a function that adopts infinitely many complex values for each pair of x,y values. To emphasize this many-valued property, the first letter of the function's symbol is capitalized in writing:

25:11:1 $$\mathrm{Ln}(x + iy) = \ln(\sqrt{x^2 + y^2}) + i[2k\pi + \theta] \qquad -\pi < \theta \le \pi \qquad x^2 + y^2 \neq 0 \qquad k = 0, \pm 1, \pm 2, \ldots$$

where θ is the polar angle corresponding to the Cartesian coordinates x, y. In selecting a single value, the so-called *principal value,* to represent the logarithm of a complex argument, it is usual to take $k = 0$ in 25:11:1. The determination of θ uniquely in terms of x and y leads to

25:11:2 $$\ln(x + iy) = \frac{1}{2}\ln(x^2 + y^2) + i\,\mathrm{sgn}(y)\,\mathrm{arccot}(x/|y|) \qquad y \neq 0$$

in the general case of a complex argument and to

25:11:3 $$\ln(x) = \ln(|x|) + \frac{i\pi}{2}[1 - \mathrm{sgn}(x)] \qquad x \neq 0$$

when $y = 0$ (for example, $\ln(-1) = i\pi$).

Setting $x = 0$ in result 25:11:2 yields the formula

25:11:4 $$\ln(iy) = \ln(|y|) + \frac{i\pi}{2}\mathrm{sgn}(y)$$

for the logarithm of an imaginary argument (for example, $\ln(\pm i) = \pm i\pi/2$).

25:12 GENERALIZATIONS

We defer to Section 25:14 a discussion of the generalization of $\ln(x)$ "to other bases." Polylogarithms are addressed in Section 25:13.

The logarithmic function is the special $\nu = 1$ case of the function defined by

25:12:1 $$\sum_{j=0}^{\infty} \frac{1}{j+\nu}\left(\frac{x-1}{x}\right)^{j+\nu} \qquad x \geq \frac{1}{2}$$

which has been termed the *generalized logarithmic function* and is denoted $\ln_\nu(x)$ [see Oldham and Spanier, Section 10.5]. This generalized function is also representable as

25:12:2 $$\ln_\nu(x) = \mathrm{B}\left(\nu; 0; \frac{x-1}{x}\right)$$

in terms of an incomplete beta function [Chapter 58].

25:13 COGNATE FUNCTIONS

The logarithmic function is closely related to its inverse, the exponential function [Chapter 26], and to the inverse hyperbolic functions [Chapter 31].

The *logarithmic integral*, defined by

25:13:1 $$\mathrm{li}(x) = \int_0^x \frac{dt}{\ln(t)} = x\int_0^1 \frac{dt}{\ln(x) + \ln(t)}$$

is a related function with importance in number theory. Figure 25-3 illustrates its definition, and its behavior is portrayed in Figure 37-1. Because of the identity

25:13:2 $$\mathrm{li}(x) = \mathrm{Ei}(\ln(x)) \qquad x > 0$$

the properties of the logarithmic integral follow from those of the exponential integral of Chapter 37. Numerical values of li(x) may be generated via the algorithm of that chapter.

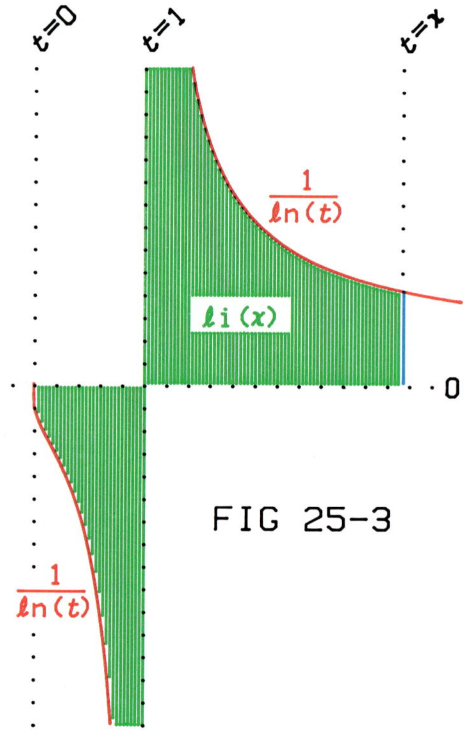

FIG 25-3

Yet another related function is the *dilogarithm* (or *Spence's integral*)

25:13:3 $$\mathrm{diln}(x) = \int_1^x \frac{\ln(t)}{t-1}\,dt = \int_0^x \frac{\ln(t)}{t-1}\,dt - \frac{\pi^2}{6}$$

which occurs in radiation theory, among other applications. Its definition is illustrated in Figure 25-4 and its behavior in Figure 25-1. The dilogarithm is also expressible as the sum

25:13:4 $$\mathrm{diln}(x) = -\sum_{j=1}^{\infty} \frac{(1-x)^j}{j^2} \qquad 0 \le x \le 2$$

from which the particular values $\mathrm{diln}(0) = -\zeta(2) = -\pi^2/6$ and $\mathrm{diln}(2) = \eta(2) = \pi^2/12$ follow. Here ζ and η signify zeta and eta numbers [Chapter 3]. Confusingly, there are a number of alternative definitions and notations for the dilogarithm. Series 25:13:4 is rapidly convergent for $1/2 \le x < 1$ and advantage of this fact is taken in the algorithm

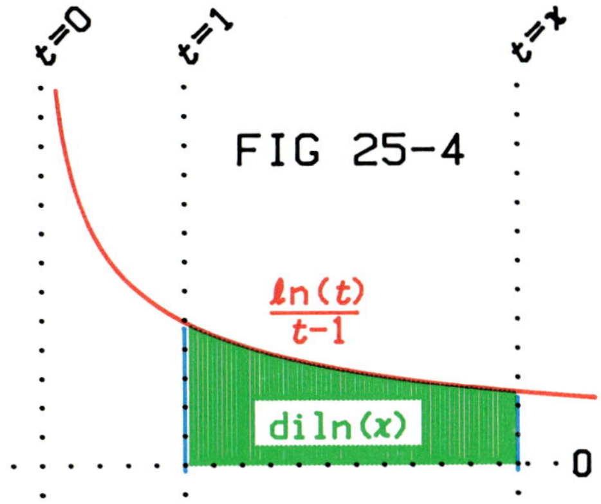

Set $f = g = 0$
Set $s = 1$
Input x >>>>>>>
If $x \le 1$ go to (1)
Set $g = -[\ln^2(x)]/2$
Set $s = -1$
Replace x by $1/x$
(1) If $x \ge 1/2$ go to (2)
Replace g by $g + s[(\pi^2/6) - \ln(x)\ln(1-x)]$
Replace s by $-s$
Replace x by $1 - x$
(2) Set $j = 2 + \mathrm{Int}(4/x^2)$
(3) Replace f by $[f+(1/j^2)](1-x)$
Replace j by $j - 1$
If $j \ne 0$ go to (3)
Replace f by $-g - sf$
Output f
$f \simeq \mathrm{diln}(x)$ <<<<<

Storage needed: f, g, s, x and j

Input restriction: x must be positive.

Test values:
diln(0.7) = −0.326129511
diln(0.2) = −1.07479460
diln(10) = +3.95066378

by replacing the infinite upper limit in 25:13:4 by $2 + \mathrm{Int}(4/x^2)$. The algorithm calculates dilogarithms of arguments in the $0 < x < \frac{1}{2}$ range by exploiting the reflection formula

25:13:5 $$\mathrm{diln}(1-x) = \ln(x)\ln(1-x) - \mathrm{diln}(x) - \frac{\pi^2}{6} \qquad 0 \le x \le 1$$

and those in the range $x > 1$ are accessed via the reciprocation formula

25:13:6 $$\mathrm{diln}(1/x) = \frac{\ln^2(x)}{2} - \mathrm{diln}(x) \qquad x > 0$$

The algorithm generates values with 24-bit precision (relative error $\leq 6 \times 10^{-8}$) for all positive arguments. The universal hypergeometric algorithm [Section 18:14] may also be used to evaluate the dilogarithm in the range $0 \leq x \leq 2$.

One may similarly define a *trilogarithm*

25:13:7 $$\mathrm{triln}(x) = \int_1^x \frac{\mathrm{diln}(t)}{t-1}\,dt = -\sum_{j=1}^{\infty} \frac{(1-x)^j}{j^3}$$

It occurs in distribution theory [Section 27:14] and is also evaluable, for $0 \leq x \leq 2$, by the universal hypergeometric algorithm of Section 18:14. The logarithm, dilogarithm and trilogarithm generalize to the *polylogarithm*

25:13:8 $$\mathrm{poln}_\nu(x) = -\sum_{j=1}^{\infty} \frac{(1-x)^j}{j^\nu}$$

where ν is not necessarily an integer. The polylogarithm is encountered also in Section 64:12 and, in the alternative notation $-\mathrm{F}(1 - x, \nu)$, some of its properties were reported by Bateman [see Erdélyi, Magnus, Oberhettinger and Tricomi, *Higher Transcendental Functions,* Volume 1, pages 30 and 31].

25:14 RELATED TOPICS

If β is a positive number other than unity and $x = \beta^f$, then f is termed the *logarithm* of x *to the base* β and is denoted $\log_\beta(x)$

25:14:1 $$\log_\beta(x) = \log_\beta(\beta^f) = f \qquad 0 < \beta \neq 1$$

A logarithm to any base is proportional to the logarithmic function $\ln(x)$

25:14:2 $$\log_\beta(x) = \log_\beta(e)\ln(x) = \frac{\ln(x)}{\ln(\beta)} \qquad 0 < \beta \neq 1$$

Other than e, the most commonly encountered logarithmic bases are 2 and 10.

Logarithms to base 2 are called *binary logarithms*

25:14:3 $$\log_2(x) = (1.44269504)\ln(x) = \frac{\ln(x)}{0.6931471806}$$

Logarithms to base 10 are called *common logarithms*, *Briggsian logarithms* or *decadic logarithms*. The subscript 10 is often omitted so that $\log_{10}(x)$ is written $\log(x)$ or even $1g(x)$. The identities

25:14:4 $$\log_{10}(x) = (0.4342944819)\ln(x) = \frac{\ln(x)}{2.302585093}$$

obtain. The decadic logarithm of a number in scientific notation [Section 9:14] is easily found because

25:14:5 $$\log_{10}(n_0 \cdot n_{-1}n_{-2}n_{-3} \cdots \times 10^N) = N + \log_{10}(n_0 \cdot n_{-1}n_{-2}n_{-3} \cdots)$$

The final term in 25:14:5, which necessarily lies in the range .000... through .999..., is called the *mantissa* of the logarithm, while N is its *characteristic*. Chemists use the p notation

25:14:6 $$\mathrm{p}x = \log_{10}(1/x) = -\log_{10}(x)$$

For example, $\mathrm{p}H = -\log_{10}(H)$, where H is the activity (or concentration) of hydrogen ions. The name *cologarithm* of x is sometimes applied to $\mathrm{p}x$.

CHAPTER 26

THE EXPONENTIAL FUNCTION exp($bx + c$)

Many natural and manmade assemblies (populations of bacteria, bank balances, collections of radionuclei, arrays of lightbulbs) increase or decrease at a rate that is proportional to their size. This characteristic, described by equation 26:3:4, leads to *exponential growth* or *exponential decay,* whose functional dependences are expressed by $\exp(bx + c)$ with b positive for growth and negative for decay.

The function $\exp(bx + c)$, or sometimes the special cases $\exp(\pm x)$, are considered in this chapter. The exponential functions of more complicated arguments are deferred to the next chapter.

26:1 NOTATION

The symbolism e^x, where e is the base of natural logarithms [Section 1:4], often replaces $\exp(x)$. The name "natural antilogarithm" and the symbol $\ln^{-1}(x)$ are occasionally encountered. Colloquially, exponential functions are often called "exponentials."

26:2 BEHAVIOR

The exponential function accepts arguments of either sign and any magnitude, but itself adopts only positive values. As illustrated in Figure 26-1, $\exp(x)$ rapidly increases while $\exp(-x)$ decreases towards zero, as the magnitude of x increases.

The exponential function shares with the unity function the remarkable property $f(x)\, f(-x) = 1$.

Figure 26-1 also displays a map of the *self-exponential function* x^x. This exhibits a minimum value of $\exp(-1/e) = 0.69220 \cdots$ at $x = \exp(-1) = 0.36787 \cdots$.

26:3 DEFINITIONS

The exponential function is defined as the inverse of the logarithmic function; that is:

26:3:1 $$f = \exp(x) \qquad \text{where} \qquad x = \ln(f)$$

Alternatively, the definition as a power of the number e [see Section 1:7]

26:3:2 $$\exp(x) = e^x = (2.7182818284 \cdots)^x$$

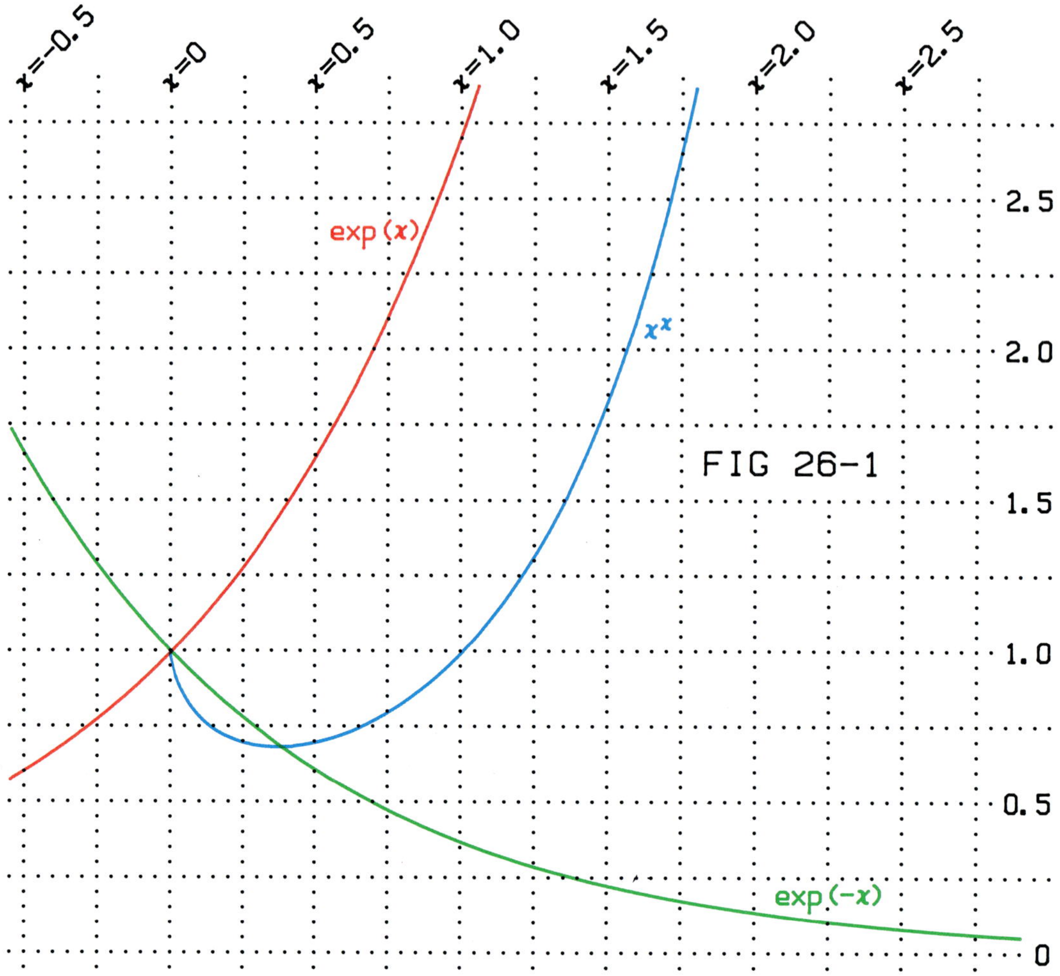

may be used, or expansion 26:6:1 may serve instead. The limiting operation

26:3:3 $$\exp(x) = \lim_{n\to\infty}\left(1 + \frac{x}{n}\right)^n$$

is yet another definition of the exponential function, albeit not a very useful one in practice.

The differential equation

26:3:4 $$\frac{df}{dx} = bf$$

is solved by $f = \exp(bx + c)$ where c is an arbitrary constant. Thus, $\exp(x + c)$ is the only function (other than the zero function) that remains unchanged on differentiation.

26:4 SPECIAL CASES

There are none.

26:5 INTRARELATIONSHIPS

The remarkably simple reflection, addition, subtraction and involution formulas

26:5:1 $$\exp(-x) = \frac{1}{\exp(x)}$$

26:5:2 $$\exp(x + y) = \exp(x)\exp(y)$$

26:5:3 $$\exp(x - y) = \frac{\exp(x)}{\exp(y)}$$

26:5:4 $$\exp^{\nu}(x) = \exp(\nu x)$$

partly explain the widespread utility of the exponential function.

Infinite series of exponential functions of negative argument that may be summed geometrically [see equation 1:6:5] or as hyperbolic functions [Chapters 29 and 30] include

26:5:5 $$\exp(-x) + \exp(-2x) + \exp(-3x) + \cdots = \frac{1}{\exp(x) - 1} = \frac{1}{2}\coth\left(\frac{x}{2}\right) - \frac{1}{2} \qquad x > 0$$

25:5:6 $$\exp(-x) - \exp(-2x) + \exp(-3x) - \cdots = \frac{1}{\exp(x) + 1} = \frac{1}{2} - \frac{1}{2}\tanh\left(\frac{x}{2}\right) \qquad x > 0$$

26:5:7 $$\exp(-x) + \exp(-3x) + \exp(-5x) + \cdots = \frac{\exp(x)}{\exp(2x) - 1} = \frac{1}{2}\operatorname{csch}(x) \qquad x > 0$$

26:5:8 $$\exp(-x) - \exp(-3x) + \exp(-5x) - \cdots = \frac{\exp(x)}{\exp(2x) + 1} = \frac{1}{2}\operatorname{sech}(x) \qquad x > 0$$

Similar sums in which the exponential arguments involve the squares of the natural (or the odd) numbers lead to theta functions [see Section 27:13]

26:5:9 $$\exp(-x) + \exp(-4x) + \exp(-9x) + \cdots = \sum_{j=1}^{\infty} \exp(-j^2 x) = \frac{1}{2}\theta_3\left(0; \frac{x}{\pi^2}\right) - \frac{1}{2} \qquad x > 0$$

26:5:10 $$\exp(-x) - \exp(-4x) + \exp(-9x) - \cdots = -\sum_{j=1}^{\infty} (-1)^j \exp(-j^2 x) = \frac{1}{2} - \frac{1}{2}\theta_4\left(0; \frac{x}{\pi^2}\right) \qquad x > 0$$

26:5:11 $$\exp(-x) + \exp(-9x) + \exp(-25x) + \cdots = \sum_{j=1}^{\infty} \exp[-(2j-1)^2 x] = \frac{1}{2}\theta_2\left(0; \frac{4x}{\pi^2}\right) \qquad x > 0$$

26:5:12 $$\exp(-x) - \exp(-9x) + \exp(-25x) - \cdots = -\sum_{j=1}^{\infty} (-1)^j \exp[-(2j-1)^2 x] = \frac{1}{2}\theta_2\left(\frac{1}{2}; \frac{x}{\pi^2}\right) \qquad x > 0$$

26:6 EXPANSIONS

The expansion

26:6:1 $$\exp(bx + c) = 1 + \frac{bx + c}{1!} + \frac{(bx + c)^2}{2!} + \cdots = \sum_{j=0}^{\infty} \frac{(bx + c)^j}{j!}$$

is valid for all arguments. Another expansion is in terms of hyperbolic Bessel functions [Chapter 49]:

26:6:2 $$\exp(\pm x) = \mathrm{I}_0(x) \pm 2\mathrm{I}_1(x) + 2\mathrm{I}_2(x) \pm 2\mathrm{I}_3(x) + \cdots = \sum_{j=-\infty}^{\infty} (\pm 1)^j \mathrm{I}_j(x)$$

Among continued fraction expansions of the exponential function are

26:6:3 $$\exp(x) = \frac{1}{1-}\,\frac{x}{1+}\,\frac{x}{2-}\,\frac{x}{3+}\,\frac{x}{2-}\,\frac{x}{5+}\cdots$$

and

26:6:4 $$\exp(x) = 1 + \frac{x}{1-}\,\frac{x}{2+}\,\frac{x}{3-}\,\frac{x}{2+}\,\frac{x}{5-}\cdots$$

26:7 PARTICULAR VALUES

exp(−∞)	exp(−1)	exp(0)	exp(1)	exp(∞)
0	$1/e$	1	e	∞

26:8 NUMERICAL VALUES

Most calculators and computers incorporate means by which exp(x) can be calculated directly. Therefore, no algorithm for the exponential function is included in this chapter although the universal hypergeometric algorithm [Section 18:14] may be employed to calculate exp(x) for any finite value of the argument x.

26:9 APPROXIMATIONS

Relaxing the $n \to \infty$ condition in definition 26:3:3 leads to

26:9:1 $$\exp(x) \simeq \left(1 + \frac{x}{130}\right)^{130} \qquad \text{8-bit precision} \qquad -1 \le x \le 1$$

See the test values for an "economized polynomial" approximation $e_0 + e_1x + e_2x^2$ to $\exp(-x)$ in the range $0 \le x \le 1$. This approximation has a precision superior to 8 bit, and is presented in Section 22:14. Some rational function approximations to exp(x) are given in Section 17:13.

26:10 OPERATIONS OF THE CALCULUS

Single differentiation, multiple differentiation and indefinite integration of the function exp($bx + c$) give

26:10:1 $$\frac{d}{dx}\exp(bx+c) = b\exp(bx+c)$$

26:10:2 $$\frac{d^n}{dx^n}\exp(bx+c) = b^n\exp(bx+c)$$

and

26:10:3 $$\int_{-\infty}^{x}\exp(bt+c)dt = \frac{\exp(bx+c)}{b} \qquad b>0$$

These three results may be generalized to the rule

26:10:4 $$\frac{d^\nu}{[d(x+\infty)]^\nu}\exp(bx+c) = b^\nu\exp(bx+c) \qquad b>0$$

for differintegration [Section 0:10] with a lower limit of $-\infty$. The generalized differintegral with a lower limit of zero is a function containing an incomplete gamma function [Chapter 45]

26:10:5 $$\frac{d^\nu}{dx^\nu}\exp(bx+c) = b^\nu \exp(bx+c)\frac{\gamma(-\nu;bx)}{\Gamma(-\nu)} = x^{-\nu}\exp(bx+c)\gamma^*(-\nu;bx)$$

Differentiation of the self-exponential function gives

26:10:6 $$\frac{d}{dx}x^x = x^x[1+\ln(x)]$$

For positive b, indefinite integrals of $t^\nu\exp(bt+c)$ are generally evaluable as Kummer functions [Chapter 47]

26:10:7 $$\int_0^x t^\nu \exp(bt+c)dt = \frac{x^{\nu+1}\exp(c)}{\nu+1} M(\nu+1;\nu+2;bx) \qquad \nu > -1$$

and the following special cases apply:

26:10:8 $$\int_0^x \exp(bt+c)dt = \frac{\exp(bx+c)-\exp(c)}{b}$$

26:10:9 $$\int_0^x t^n \exp(bt+c)dt = \frac{n!\exp(bx+c)}{(-b)^{n+1}}[\exp(-bx)-e_n(-bx)] \qquad n = 0, 1, 2, \ldots$$

26:10:10 $$\int_0^x \frac{\exp(bt+c)}{\sqrt{t}}dt = \frac{2}{\sqrt{b}}\exp(bx+c)\,\mathrm{daw}\sqrt{bx} \qquad b > 0$$

26:10:11 $$\int_{-\infty}^x \frac{\exp(bt+c)}{t}dt = \exp(c)\,\mathrm{Ei}(bx) \qquad b > 0$$

26:10:12 $$\int_{-\infty}^x \frac{\exp(bt+c)}{t^n}dt = \frac{b^{n-1}}{(n-1)!}\exp(bx+c)\left[\exp(-bx)\,\mathrm{Ei}(bx) - \frac{0!}{bx} - \frac{1!}{(bx)^2} - \cdots - \frac{(n-2)!}{(bx)^{n-1}}\right] \qquad n = 2, 3, 4, \ldots \qquad b > 0$$

the e_n, daw and Ei functions being explained in Section 26:13, Chapter 42, and Chapter 37. See Table 37.14.1 for a compilation of indefinite integrals of $x^\nu\exp(x)$ for commonly encountered values of ν.

Equations 26:10:7–26:10:9 apply also when b is negative, but the general expressions

26:10:13 $$\int_0^x t^\nu \exp(bt+c)dt = \frac{\exp(c)}{(-b)^{\nu+1}}\gamma(\nu+1;-bx) \qquad \nu > -1 \qquad b < 0$$

26:10:14 $$\int_x^\infty t^\nu \exp(bt+c)dt = \frac{\exp(c)}{(-b)^{\nu+1}}\Gamma(\nu+1;-bx) \qquad \nu \le -1 \qquad b < 0$$

in terms of incomplete gamma functions [Chapter 42] are then more useful. Special cases include

26:10:15 $$\int_0^x \frac{\exp(bt+c)}{\sqrt{t}}dt = \sqrt{\frac{\pi}{-b}}\exp(c)\,\mathrm{erf}\sqrt{-bx} \qquad b < 0$$

26:10:16 $$\int_x^\infty \frac{\exp(bt+c)}{t}dt = -\exp(c)\,\mathrm{Ei}(bx) \qquad b < 0$$

26:10:17 $$\int_x^\infty \frac{\exp(bt+c)}{t^n}dt = \frac{b^{n-1}\exp(bx+c)}{(n-1)!}\left[\frac{0!}{bx} + \frac{1!}{(bx)^2} + \cdots + \frac{(n-2)!}{(bx)^{n-1}} - \exp(-bx)\,\mathrm{Ei}(bx)\right] \qquad n = 2, 3, 4, \ldots \qquad b < 0$$

and specific expressions for $\int_x^\infty t^\nu \exp(-t)dt$ will be found in Table 37.14.1 for a variety of ν values.

Some other important indefinite integrals include

26:10:18
$$\int_0^x \frac{dt}{a + \exp(bt + c)} = \frac{x}{a} - \frac{1}{ab} \ln\left[\frac{a + \exp(bx + c)}{a + \exp(c)}\right]$$

and

26:10:19
$$\int_{-\infty}^x \frac{dt}{\exp(bt) + a\exp(-bt)} = \begin{cases} \dfrac{1}{b\sqrt{a}} \arctan\left(\dfrac{\exp(bx)}{\sqrt{a}}\right) & a > 0 \\ \dfrac{1}{b\sqrt{-a}} \operatorname{artanh}\left(\dfrac{\exp(bx)}{\sqrt{-a}}\right) & a < 0 \end{cases}$$

Some thirty pages are devoted by Gradshteyn and Ryzhik [Sections 3.3 and 3.4] to definite integrals of exponential functions. Here we cite only four general examples:

26:10:20
$$\int_1^\infty t^n \exp(-xt)dt = \frac{n!}{x^{n+1}} \exp(-x)\, e_n(x) \qquad n = 0, 1, 2, \ldots \qquad x > 0$$

26:10:21
$$\int_{-1}^1 t^n \exp(-xt)dt = \frac{n!}{x^{n+1}} [\exp(x)e_n(-x) - \exp(-x)e_n(x)] \qquad n = 0, 1, 2, \ldots \qquad x > 0$$

26:10:22
$$\int_1^\infty \frac{\exp(-xt)}{t^n} dt = -\frac{(-x)^{n-1}}{(n-1)!}\left[\operatorname{Ei}(-x) + \exp(-x)\left\{\frac{0!}{x} - \frac{1!}{x^2} + \cdots + \frac{(-1)^n(n-2)!}{x^{n-1}}\right\}\right] \qquad n = 2, 3, 4, \ldots \qquad x > 0$$

and

26:10:23
$$\int_0^\infty \frac{t^n dt}{\exp(t) \pm 1} dt = \begin{Bmatrix} n!\, \eta(n+1) \\ n!\, \zeta(n+1) \end{Bmatrix} \qquad n = 0, 1, 2, \ldots$$

The η and ζ functions are those discussed in Chapter 3.

Definite integrals of the form

26:10:24
$$\int_0^\infty f(t) \exp(bt + c)dt$$

which generally exist only for a restricted range of b values, can be evaluated via the *Laplace transforms* that are listed in Section 14 of this chapter. A Laplace transform is defined by

26:10:25
$$\bar{f}_L(s) = \int_0^\infty f(t) \exp(-st)dt$$

and integral 26:10:24 can therefore be evaluated via the identity

26:10:26
$$\int_0^\infty f(t) \exp(bt + c)dt = \exp(c)\, \bar{f}_L(-b)$$

26:11 COMPLEX ARGUMENT

When the argument of the exponential function is the complex number $x + iy$, the relationship

26:11:1
$$\exp(x + iy) = \exp(x)[\cos(y) + i \sin(y)]$$

known as *Euler's formula*, is obeyed. The special cases

26:11:2 $$\exp(\pm i\pi) = -1$$

26:11:3 $$\exp\left(\pm i\frac{\pi}{2}\right) = \pm i$$

are noteworthy.

26:12 GENERALIZATIONS

The function β^x, where β is a positive constant other than unity or e, is known as the *general exponential function* or *antilogarithm to base* β. It is equivalent to an exponential function with changed argument

26:12:1 $$\beta^x = \exp(bx) \qquad b = \ln(\beta)$$

This formula enables the properties of β^x to be deduced simply from those of $\exp(bx + c)$. The function 10^x is frequently encountered and is the *common antilogarithm* of x.

Generalization in a different direction is provided by the concept of hypergeometric functions introduced in Section 18:14. In the nomenclature of that section, $\exp(x)$ is the $K = 0$, $L = 1$ hypergeometric function in which the sole parameter equals unity. Generalizing this parameter to ν gives the function $\Gamma(\nu)\exp(x)\gamma^*(\nu - 1;x)$ [see Chapter 45 for the γ^* function], which could therefore claim to be a generalized exponential.

26:13 COGNATE FUNCTIONS

The *exponential polynomial* $e_n(x)$ may be derived from the 26:6:1 expansion by truncation after the $x^n/n!$ term:

26:13:1 $$e_n(x) = 1 + \frac{x}{1!} + \frac{x^2}{2!} + \cdots + \frac{x^n}{n!} = \sum_{j=0}^{n} \frac{x^j}{j!}$$

Thus, $\exp(x) = e_\infty(x)$. The concatenation

26:13:2 $$e_n(x) = \left(\left(\cdots\left(\left(\frac{x}{n} + 1\right)\frac{x}{n-1} + 1\right)\frac{x}{n-2} + \cdots + 1\right)\frac{x}{2} + 1\right)\frac{x}{1} + 1$$

provides a convenient method of evaluating the exponential polynomial if its degree n is not too large. Section 18:14 shows how the universal hypergeometric algorithm may be employed to calculate $n!x^{-n}e_n(x)$. For large n the asymptotic approximation

26:13:3 $$e_{n-2}(x) \sim \exp(x) - \frac{n^2x^{n-1}}{n!(n-x)} \qquad n \to \infty$$

is useful.

The self-exponential function x^x was mentioned in Section 26:2 and is graphed in Figure 26-1. The related function $x^{1/x}$ displays a maximum value at $x = e$. The inverse function [Section 0:3] of the latter function is the limit of the sequence

26:13:4 $$x,\ x^x,\ x^{(x^x)},\ x^{(x^{(x^x)})},\ x^{(x^{(x^{(x^x)})})},\ \ldots$$

corresponding to infinitely repeated exponentiation. Figure 26-2 maps these interesting functions.

Those Basset functions [see Chapter 51] $K_\nu(x)$, in which ν is an odd multiple of $\frac{1}{2}$, are related in a very straightforward way to the exponential function $\exp(-x)$. Thus, we have

26:13:5 $$K_{1/2}(x) = \sqrt{\frac{\pi}{2x}}\exp(-x)$$

26:13:6 $$K_{3/2}(x) = \sqrt{\frac{\pi}{2x}}\exp(-x)\left[1 + \frac{1}{x}\right]$$

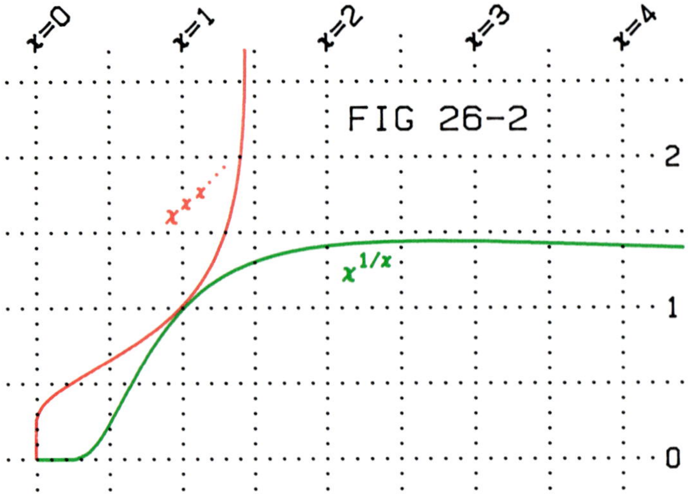

26:13:7 $$K_{5/2}(x) = \sqrt{\frac{\pi}{2x}}\exp(-x)\left[1 + \frac{3}{x} + \frac{3}{x^2}\right]$$

and further members of the sequence may be constructed via the recurrence formula

26:13:8 $$K_\nu(x) = K_{\nu-2}(x) + \frac{2\nu - 2}{x} K_{\nu-1}$$

Maps of the functions $K_\nu(x)$ for $\nu = \frac{1}{2}, \frac{3}{2}, \frac{5}{2}$ and $\frac{7}{2}$ are displayed in Figure 26-3. The symbol $k_n(x)$ and the name *modified spherical Bessel function of the third kind* are sometimes applied to the function $\sqrt{\pi/2x}\,K_{n+1/2}(x)$.

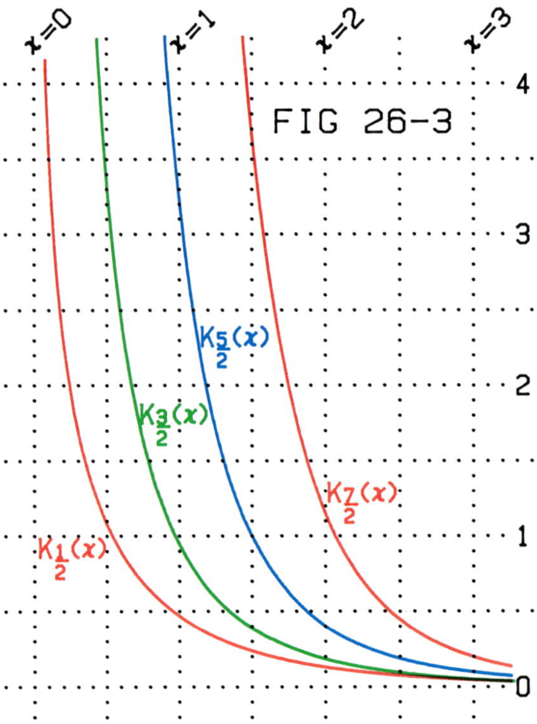

26:14 RELATED TOPICS

Laplace transformation is one of the most valuable techniques available for the solving of practical problems in engineering and the physical sciences. Particularly susceptible to this approach are situations characterized by a set of simultaneous ordinary differential equations, or by a partial differential equation with attendant boundary conditions. The overall strategy and motive for the use of Laplace transformation are illustrated in the following scheme:

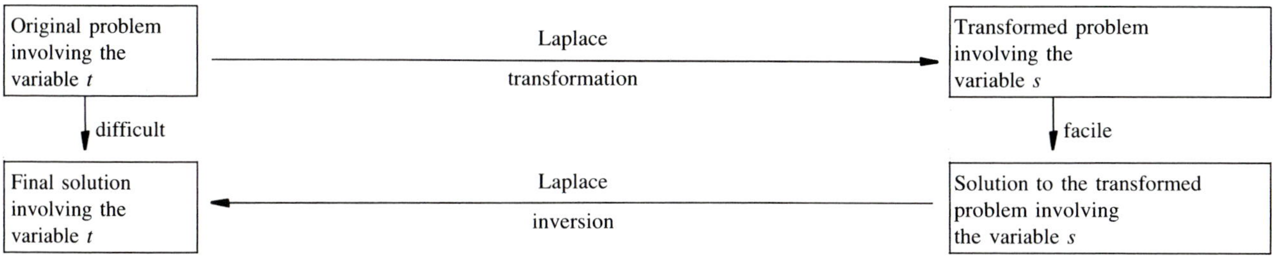

In this section we present some general rules for Laplace transformation and inversion; the classic book by Churchill may be consulted for further information. A short table of transform pairs is also included.

In the following we use f and g to represent functions of t whose Laplace transforms $\bar{f}_L$ and $\bar{g}_L$ are functions of the variable s, which is sometimes called the *dummy variable*. Equation 26:10:25 shows how $\bar{f}_L(s)$ is related to $f(t)$, but it is often more economical to use the operator notation

26:14:1 $$L\{f(t)\} = \int_0^\infty f(t)\exp(-st)dt = \bar{f}_L(s)$$

to represent this relationship. In most applications t is a real variable whereas s is often complex. Here, however, we shall not pursue the complex nature of the dummy variable.

It should be recognized that not all functions have Laplace transforms. To be transformable, a function $f(t)$ must at least be defined as a real function over the entire range $0 \le t \le \infty$. Any function that is finite throughout this semiinfinite range has a Laplace transform, although $\bar{f}_L(s)$ is not necessarily a named function. Moreover, many functions that encounter infinities in the $0 \le t \le \infty$ range may nevertheless be transformed. For example, an infinity at $t = 0$ will not impede Laplace transformation provided that $tf(t)$ approaches zero as t tends to zero from positive values.

Later in this section a table [Table 26.14.1] of Laplace transforms is presented. Its utility may be greatly enhanced by making use of certain general properties that obtain under Laplace transformation. Some of these general rules are explored in the next few paragraphs.

The weighted sum or difference of two (or more) functions may be Laplace transformed by using the *linearity property*

26:14:2 $$L\{c_1 f(t) \pm c_2 g(t)\} = c_1 \bar{f}_L(s) \pm c_2 \bar{g}_L(s)$$

Provided that both series converge, this rule may be extended to the infinite summation $\Sigma c_j f_j(t)$, which thereby gives $\Sigma c_j \bar{f}_{jL}(s)$ on transformation.

The *scaling property*

26:14:3 $$L\{f(bt)\} = \frac{1}{b}\bar{f}_L\left(\frac{s}{b}\right) \qquad b > 0$$

of Laplace transformation is easily implemented but the *linear shift property*

26:14:4 $$L\{f(bt + c)\} = \frac{1}{b}\exp\left(\frac{cs}{b}\right)\left[\bar{f}_L\left(\frac{s}{b}\right) - \int_0^c f(t)\exp\left(\frac{-st}{b}\right)dt\right] \qquad b > 0$$

is generally less useful than the corresponding rule [equation 26:14:23] for Laplace inversion. Of course, if f is a function (such as sin or exp) for which an argument-addition formula [Section 0:5] exists, transformation of $f(bt + c)$ may be accomplished by that route.

With $a > 0$ and u representing the unit-step function [Chapter 8], the function $u(t-a)f(t-a)$ may be obtained by nullifying the $t < 0$ portion of f(t) and then translating the residue along the t-axis as portrayed in Figure 26-4. Such a function may be Laplace transformed by the *delay property*

26:14:5 $$L\{u(t-a)f(t-a)\} = \exp(-as)\bar{f}_L(s) \qquad a > 0$$

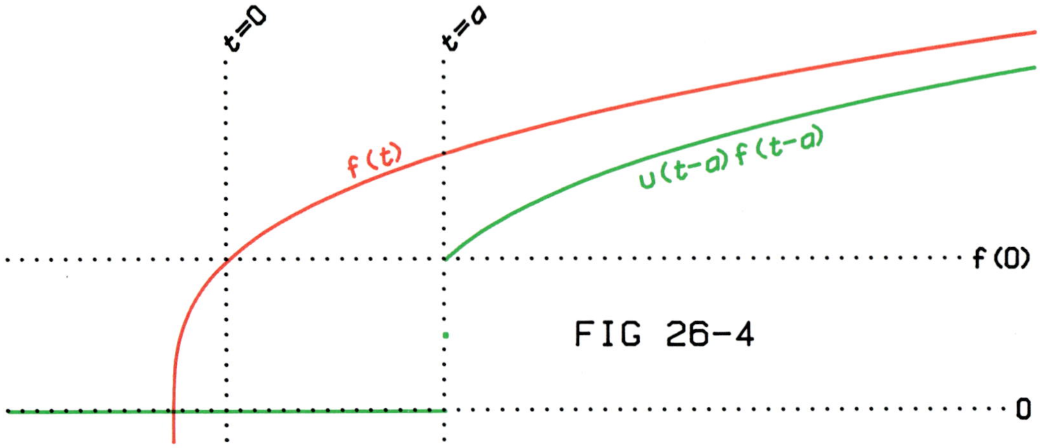

FIG 26-4

For certain simple functions g(t), the *composite function* f{g(t)} may be Laplace transformed as the definite integral

26:14:6 $$L\{f\{g(t)\}\} = \int_0^\infty G(s;t)\bar{f}_L(t)dt$$

where G(s;t) is a bivariate function, some examples of which are shown in Table 26.14.1. The most general *chain rule* for Laplace transformation is

26:14:7 $$L\{f\{g(t)\}\} = \int_{g(0)}^{g(\infty)} G(s;t)\bar{f}_L(t)dt$$

where G(u;t) is the function resulting from Laplace inversion of $\exp\{-uG(s)\}dG/ds$, G being the inverse function of g [that is, $G\{g(x)\} = x$; see Section 0:3].

There is no simple *product rule* for Laplace transformation; that is, the transform of f(t)g(t) cannot be simply written in terms of $\bar{f}_L(s)$ and $\bar{g}_L(s)$. For certain instances of f(t), however, simple relationships do exist. Thus:

26:14:8 $$L\{\exp(bt+c)g(t)\} = \exp(c)\bar{g}_L(s-b)$$

Table 26.14.1

g(t)	G(s;t)
$\frac{1}{t}$	$\sqrt{\frac{t}{s}}J_1(2\sqrt{st})$
t^2	$\frac{1}{2\sqrt{\pi t}}\exp\left(\frac{-s^2}{4t}\right)$
$\exp(t)-1$	$\frac{t^s\exp(-t)}{\Gamma(1+s)}$
$\sinh(t)$	$J_s(t)$

26:14:9 $$L\{\sinh(bt)g(t)\} = \frac{1}{2}[\bar{g}_L(s-b) - \bar{g}_L(s+b)]$$

and similarly the Laplace transform of $\cosh(bt)g(t)$ is $[\bar{g}_L(s-b) + \bar{g}_L(s+b)]/2$.

When $f(t)$ is a power of t, Laplace transformation of the product $f(t)g(t)$ corresponds to performing operations of the calculus on $\bar{g}_L(s)$. The simplest cases

26:14:10 $$L\{tg(t)\} = -\frac{d}{ds}\bar{g}_L(s)$$

and

26:14:11 $$L\left\{\frac{g(t)}{t}\right\} = \int_s^\infty \bar{g}_L(s)ds$$

iterate to

26:14:12 $$L\{t^n g(t)\} = (-1)^n \frac{d^n}{ds^n}\bar{g}_L(s) \qquad n = 0, 1, 2, \ldots$$

and

26:14:13 $$L\left\{\frac{g(t)}{t^n}\right\} = \int_s^\infty \cdots \int_s^\infty \bar{g}_L(s)(ds)^n \qquad n = 2, 3, 4, \ldots$$

and all these rules are subsumed in the general *power multiplication property*

26:14:14 $$L\{t^\nu g(t)\} = \frac{d^\nu}{[d(-s+\infty)]^\nu}\bar{g}_L(s)$$

which corresponds to differintegration [Section 0:10] with respect to $-s$, the lower limit being $-\infty$.

If $f(t)$ is the derivative of a function $F(t)$, then $L\{f(t)\} = s\bar{F}_L(s) - F(0)$ and if $f(t)$ is an indefinite integral of $\phi(t)$, then $L\{f(t)\} = [\bar{\phi}_L(s) + f(0)]/s$. Both these rules of Laplace transformation are special cases of the general *differintegral transformation property* [see Oldham and Spanier, Section 8.1]

26:14:15 $$L\left\{\frac{d^\nu f}{dt^\nu}(t)\right\} = s^\nu \bar{f}_L(s) - \sum_{j=1}^{J} s^{j-1}\frac{d^{\nu-j}f}{dt^{\nu-j}}(0) \qquad J = -\mathrm{Int}(-\nu)$$

This rule applies for all ν, but the summation is empty if $\nu \leq 0$.

The *periodicity property* permits the Laplace transform of a periodic function [Chapter 36] to be replaced by an integral over a single period

26:14:16 $$L\{\mathrm{per}(t)\} = \frac{1}{1-\exp(-Ps)}\int_0^P \mathrm{per}(t)\exp(-st)dt = \left[\frac{1}{2} - \frac{1}{2}\coth\left(\frac{Ps}{2}\right)\right]\int_0^P \mathrm{per}(t)\exp(-st)dt$$

Many important periodic functions are symmetrical in the sense that translation by a half-period changes the sign; that is, if $\mathrm{qer}(t)$ is such a *symmetrical periodic function* of period Q, $\mathrm{qer}[t + (Q/2)] = -\mathrm{qer}(t)$. In such cases Laplace transformation may be accomplished by the formula

26:14:17 $$L\{\mathrm{qer}(t)\} = \overline{\mathrm{qer}}_L(s) = \frac{1}{1+\exp\left(\frac{-Qs}{2}\right)}\int_0^{Q/2}\mathrm{qer}(t)\exp(-st)dt$$
$$= \left[\frac{1}{2} + \frac{1}{2}\tanh\left(\frac{Qs}{4}\right)\right]\int_0^{Q/2}\mathrm{qer}(t)\exp(-st)dt$$

in which integration is over a half-period. The half- and fully rectified analogs [Section 36:13] of such a symmetrical periodic function have transforms equal to $[\frac{1}{2} - \frac{1}{2}\coth(Qs/4)]\,\overline{\mathrm{qer}}_L(s)$ and $\coth(Qs/4)\,\overline{\mathrm{qer}}_L(s)$, respectively, where $\overline{\mathrm{qer}}_L(s)$ is given by 26:14:17.

The rules above, in conjunction with Table 26.14.2, permit a very large number of Laplace transforms to be

Table 26.14.2

Chapter or section no.	f(t)		$\bar{f}_L(s)$		Chapter or section no.
1	c		$\frac{c}{s}$		7
10	$c\delta(t)$		c		1
1:13	$p(c;h;t-a)$		$\frac{2c}{s}\exp(-as)\sinh\left(\frac{hs}{2}\right)$		28
8	$\Sigma u\{t - \ln(j)\}$ $\quad j = 0, 1, 2, \ldots$	steepening staircase	$\frac{1}{s}\zeta(s)$		3
7	$bt + c$		$\frac{cs + b}{s^2}$		11
7	$\frac{1}{bt + c}$		$\frac{-1}{b}\exp\left(\frac{cs}{b}\right)\operatorname{Ei}\left(\frac{-cs}{b}\right) = \frac{1}{cs}\operatorname{E}\left(\frac{b}{cs}\right)$	Euler's function	37
10:12	$b\delta'(t) + c\delta(t)$		$bs + c$		7
26	$\exp\left(\frac{-ct}{b}\right)$		$\frac{b}{bs + c}$		7
8	$u(t - a) \quad a > 0$		$\frac{1}{s}\exp(-as)$		26
8	$2\Sigma(-1)^j u(t - b - 2jb) \quad j = 0, 1, 2, \ldots$	square pulses	$\frac{1}{s}\operatorname{sech}(bs)$		29
8	$2\Sigma u(t - (j + \frac{1}{2})^2\pi^2) \quad j = 0, 1, 2, \ldots$	Staircases with	$\frac{1}{s}\theta_2(0;s)$		27:13
8	$1 + 2\Sigma u(t - j^2\pi^2) \quad j = 1, 2, 3, \ldots$	lengthening treads	$\frac{1}{s}\theta_3(0;s)$		27:13
8	$\lvert c - bt\rvert \quad b > 0$		$\frac{2b}{s^2}\exp\left(\frac{-cs}{b}\right) + \frac{cs - b}{s^2}$		26
9	$\operatorname{Int}(bt)$	staircase	$\frac{b}{2s}\left[\coth\left(\frac{s}{2b}\right) - 1\right]$		30
9	$\operatorname{frac}\left(\frac{t}{P}\right)$	sawtooth	$\frac{1}{2Ps}\left[\frac{2}{s} + 1 - \coth\left(\frac{Ps}{2}\right)\right]$		30
9	$\operatorname{Int}(bt + \frac{1}{2})$	staircase	$\frac{1}{2s}\operatorname{csch}\left(\frac{s}{2b}\right)$		29
9	$\frac{1}{2} + (-1)^{\operatorname{Int}(2t/P)}\left[\operatorname{frac}\left(\frac{2t}{P}\right) - \frac{1}{2}\right]$	triangular wave	$\frac{2}{Ps^2}\tanh\left(\frac{Ps}{4}\right)$		30
10	$\delta(t - a) \quad a > 0$		$\exp(-as)$		26
10:12	$\delta'(t - a) \quad a > 0$		$s\exp(-as)$		26
11	$t^n \quad n = 0, 1, 2, \ldots$		$n!/s^{1+n}$		11
11	$(bt + c)^n \quad n = 2, 3, 4, \ldots$		$(n!)^2 s\sum\frac{1}{j!(n-j)!}\left(\frac{b}{cs}\right)^j$	$j = 0, 1, 2, \ldots, n$	11
11	$bc^{n-1}(bt + c)^{-n} \quad n = 2, 3, 4, \ldots$		$\exp\left(\frac{cs}{b}\right)\operatorname{E}_n\left(\frac{cs}{b}\right)$	Schlömilch functions	37:13
26	$t^{n-1}\exp\left(\frac{-ct}{b}\right)$		$(n-1)!b^n(bs + c)^{-n}$	$n = 2, 3, 4, \ldots$	11
16	$at^2 + bt + c$		$\frac{c}{s} + \frac{b}{s^2} + \frac{2a}{s^3}$		11

Table 26.14.2 (Continued)

Chapter or section no.	f(t)	$\bar{f}_L(s)$	Chapter or section no.
23	$L_n(bt) \quad n = 0, 1, 2, \ldots$	$\frac{1}{s}\left(\frac{s-b}{s}\right)^n$	11
12	$\frac{1}{\sqrt{t}}$	$\sqrt{\frac{\pi}{s}}$	12
12	$2\sqrt{\frac{t}{\pi}}$	$s^{-3/2}$	13
12	$\sqrt{bt + c}$	$\frac{\sqrt{bc}}{s} + \frac{b}{2}\sqrt{\frac{\pi}{s^3}}\exp\left(\frac{cs}{b}\right)\operatorname{erfc}\left(\sqrt{\frac{cs}{b}}\right)$	41
12	$\frac{1}{\sqrt{bt + c}}$	$\sqrt{\frac{\pi}{bs}}\exp\left(\frac{cs}{b}\right)\operatorname{erfc}\left(\sqrt{\frac{cs}{b}}\right)$	41
12	$\sqrt{\frac{bc}{t}}\frac{1}{bt + c}$	$\pi\exp\left(\frac{cs}{b}\right)\operatorname{erfc}\left(\sqrt{\frac{cs}{b}}\right)$	41
12	$\frac{1}{\sqrt{t}}[1 - u(t - b)]$	$\sqrt{\frac{\pi}{s}}\operatorname{erf}(\sqrt{bs})$	40
12	$\frac{1}{t}\sqrt{\frac{b}{t - b}}u(t - b)$	$\pi\operatorname{erfc}(\sqrt{bs})$	40
26	$\frac{1}{\sqrt{t}}\exp\left(\frac{-ct}{b}\right)$	$\sqrt{\frac{\pi b}{bs + c}}$	12
41, 42	$\exp(-bt)\operatorname{erf}(\sqrt{-bt})$ or $\frac{2}{\sqrt{\pi}}\operatorname{daw}(\sqrt{bt})$	$\sqrt{\frac{\lvert b\rvert}{s}\frac{1}{s + b}} \quad b \lessgtr 0$	12
41	$\frac{1}{\sqrt{\pi t}} - a\exp(a^2t)\operatorname{erfc}(a\sqrt{t})$	$\frac{1}{\sqrt{s} + a}$	12
41	$\exp(a^2t)\operatorname{erfc}(a\sqrt{t})$	$\frac{1}{\sqrt{s}(\sqrt{s} + a)}$	12
40	$\operatorname{erf}(\sqrt{bt})$	$\frac{1}{s}\sqrt{\frac{b}{s + b}}$	12
13	$t^{\nu-1} \quad \nu > 0$	$\Gamma(\nu)/s^\nu$	13
13	$(bt + c)^\nu \quad \nu > -1$	$\frac{1}{s}\left(\frac{b}{s}\right)^\nu\exp\left(\frac{cs}{b}\right)\Gamma\left(1 + \nu; \frac{cs}{b}\right)$	45
13	$(t - a)^{\nu-1}u(t - a) \quad \nu > 0$	$\frac{\Gamma(\nu)}{s^\nu}\exp(-as)$	26
13	$\frac{b^{1+\nu}t^\nu}{bt + c} \quad \nu > -1$	$\Gamma(1 + \nu)c^\nu\exp\left(\frac{cs}{b}\right)\Gamma\left(-\nu; \frac{cs}{b}\right)$	45
13	$\frac{1}{t}\left(\frac{t - b}{b}\right)^\nu u(t - b) \quad \nu > -1$	$\Gamma(1 + \nu)\Gamma(-\nu; bs)$	45
13	$t^\nu[1 - u(t - b)] \quad \nu > -1$	$\Gamma(\nu)b^{1+\nu}\gamma^*(1 + \nu; bs)$	45
13	$b^\nu t^{\nu-1}(bt + c)^{\mu-1} \quad \nu > 0 < \mu$	$\Gamma(\nu)c^{\nu+\mu-1}U\left(\nu; \nu + \mu; \frac{cs}{b}\right)$	48

Table 26.14.2 (Continued)

Chapter or section no.	f(t)	$\bar{f}_L(s)$	Chapter or section no.
26	$t^{\nu-1}\exp(bt+c) \quad \nu > 0$	$\Gamma(\nu)\dfrac{\exp(c)}{(s-b)^\nu}$	13
45	$\exp(bt)\gamma(\nu;bt) \quad \nu > 0$	$\left(\dfrac{c}{bs}\right)^\nu \dfrac{b\Gamma(\nu)}{bs-c}$	13
47	$t^{\nu+\mu-1}\exp(-at)\mathrm{M}(\nu;\nu+\mu;at) \quad \nu+\mu > 0$	$\dfrac{\Gamma(\nu+\mu)}{s^\nu(s+a)^\mu}$	13
46	$\dfrac{\nu}{t}t^{\nu/2}\exp\left(\dfrac{bt}{4}\right)\mathrm{D}_{-\nu-1}(\sqrt{bt}) \quad \nu > 0$	$\dfrac{\sqrt{2\pi}}{(\sqrt{2s}+\sqrt{b})^\nu}$	13
14	$\dfrac{2[1-\mathrm{u}(t-a)]}{\pi\sqrt{a^2-t^2}}$	$\mathrm{I}_0(as) - \ell_0(as)$	57
15	$\dfrac{2}{\pi\sqrt{t^2+a}} \quad a > 0$	$\mathrm{h}_0(s\sqrt{a}) - \mathrm{Y}_0(s\sqrt{a})$	57
15	$\sqrt{t^2+a} \quad a > 0$	$\dfrac{\sqrt{a}}{s} + \dfrac{1}{2}\sqrt{\dfrac{\pi}{s^3}}\exp(as)\,\mathrm{erfc}(\sqrt{as})$	41
15	$\dfrac{\mathrm{u}(t-b)}{\sqrt{t^2-b^2}}$	$\mathrm{K}_0(bs)$	51
49, 52	$\mathrm{I}_0(t\sqrt{-a})$ or $\mathrm{J}_0(t\sqrt{a})$	$\dfrac{1}{\sqrt{s^2+a}} \quad a \lessgtr 0$	15
49, 52	$\mathrm{I}_1(t\sqrt{-a})$ or $\mathrm{J}_1(t\sqrt{a})$	$\dfrac{\sqrt{\lvert a\rvert}}{a}\left(1 - \dfrac{s}{\sqrt{s^2+a}}\right) \quad a \lessgtr 0$	15
50, 53	$\mathrm{I}_\nu(t\sqrt{-a})$ or $\mathrm{J}_\nu(t\sqrt{a})$	$\left[\dfrac{\sqrt{\lvert a\rvert}(\sqrt{s^2+a}-s)}{a}\right]^\nu \Big/ \sqrt{s^2+a} \quad a \lessgtr 0$	15
39	$\mathrm{S}(bt)$ or $\mathrm{C}(bt)$	$\dfrac{1}{2s}\sqrt{b(\sqrt{s^2+b^2} \mp s)/(s^2+b^2)}$	15
55	$\sqrt{2}\,\mathrm{ber}(bt)$ or $\sqrt{2}\,\mathrm{bei}(bt)$	$\sqrt{\dfrac{1}{\sqrt{s^4+b^4}} \pm \dfrac{s^2}{s^4+b^4}}$	15
51	$2b^\nu\mathrm{K}_\nu(bt) \quad \nu \neq 0, 1, 2, \ldots$	$\pi\csc(\nu\pi)[(s+\sqrt{s^2-b^2})^\nu - (s-\sqrt{s^2-b^2})^\nu]/\sqrt{s^2-b^2}$	15
26	$\exp\left\{\dfrac{(\sqrt{-\Delta}-b)t}{2a}\right\} - \exp\left\{\dfrac{-(\sqrt{-\Delta}+b)t}{2a}\right\}$	$\dfrac{\sqrt{-\Delta}}{as^2+bs+c} \quad 4ac-b^2 = \Delta < 0$	16
32	$2\exp\left(\dfrac{-bt}{2a}\right)\sin\left(\dfrac{t\sqrt{\Delta}}{2a}\right)$	$\dfrac{\sqrt{\Delta}}{as^2+bs+c} \quad 4ac-b^2 = \Delta > 0$	16
28, 32	$\sinh(t\sqrt{-a})$ or $\sin(t\sqrt{a})$	$\dfrac{\sqrt{\lvert a\rvert}}{s^2+a} \quad a \lessgtr 0$	16
28, 32	$\cosh(t\sqrt{-a})$ or $\cos(t\sqrt{a})$	$\dfrac{s}{s^2+a} \quad a \lessgtr 0$	16
16:13	$\dfrac{\mathrm{u}(t-b)-\mathrm{u}(t-3b)}{\sqrt{-t^2+4bt-3b^2}}$	$\mathrm{I}_0(bs)$	49

Table 26.14.2 (Continued)

Chapter or section no.	f(t)	$\bar{f}_L(s)$	Chapter or section no.
49, 52	$\exp\left(\frac{-bt}{a}\right)\left[I_0\left(\frac{t\sqrt{-\Delta}}{2a}\right) \quad \text{or} \quad J_0\left(\frac{t\sqrt{\Delta}}{2a}\right)\right]$	$\sqrt{\frac{a}{as^2 + bs + c}} \quad 4ac - b^2 = \Delta \lesseqgtr 0$	16:13
17–24	$\Sigma a_j t^j \quad a_j = b_j/j!$	$\frac{1}{s}\sum b_j\left(\frac{1}{s}\right)^j \quad b_j = j!a_j$	17–24
52	$t^\nu J_0(\sqrt{bt}) \quad \nu > -1$	$\frac{\Gamma(1+\nu)}{s^{1+\nu}}\exp\left(\frac{-b}{4s}\right)L_\nu\left(\frac{b}{4s}\right)$	23:14
25	$\gamma + \ln(t) = \ln(at) \quad a = 1.78107\cdots$	$\frac{1}{s}\ln\left(\frac{1}{s}\right) = \frac{-\ln(s)}{s}$	25
25	$\ln(bt + c)$	$\frac{1}{s}\left[\ln(c) - \ln\left(\frac{cs}{b}\right)\exp\left(\frac{cs}{b}\right)\mathrm{Ei}\left(\frac{-cs}{b}\right)\right]$	37
37	$\ln(c) - \mathrm{Ei}\left(\frac{-ct}{b}\right)$	$\frac{1}{s}\ln(bs + c)$	25
38	$\mathrm{Ci}(bt) - \ln(b)$	$\frac{1}{s}\ln(\sqrt{s^2 + b^2})$	25
27	$t^\nu\exp(-a^2/t)$	$2\left(\frac{a}{\sqrt{s}}\right)^{1+\nu}K_{1+\nu}(2a\sqrt{s})$	51
27	$(2t)^\nu\exp(-a\sqrt{t}) \quad a > 0$	$\frac{\Gamma(2+2\nu)}{s^{1+\nu}}\exp\left(\frac{a^2}{8s}\right)D_{-2\nu-2}\left(\frac{a}{\sqrt{2s}}\right) \quad \nu > -1$	46
27	$b^\nu t^{\nu-1}\exp\left(\frac{-b^2t^2}{2}\right) \quad \nu > 0$	$\Gamma(\nu)\exp\left(\frac{s^2}{4b^2}\right)D_{-\nu}\left(\frac{s}{b}\right)$	46
50, 53	$t^{\nu/2}[I_\nu(2\sqrt{bt}) \quad \text{or} \quad J_\nu(2\sqrt{-bt})] \quad \nu > -1$	$\frac{\lvert b\rvert^{\nu/2}}{s^{1+\nu}}\exp\left(\frac{b}{s}\right) \quad b \gtrless 0$	27
40:13	$t^{n/2}i^n\mathrm{erfc}(\sqrt{b/t}) \quad n = -1, 0, 1, 2, \ldots$	$\frac{1}{s}\left(\frac{1}{2\sqrt{s}}\right)^n\exp(-2\sqrt{bs})$	27
41	$\exp(bt + c)\,\mathrm{erfc}\left(\sqrt{bt} + \frac{c}{2\sqrt{bt}}\right)$	$\frac{1}{\sqrt{s}(\sqrt{s} + \sqrt{b})}\exp\left(-c\sqrt{\frac{s}{b}}\right)$	27
28, 32	$[\cosh(2\sqrt{bt}) \quad \text{or} \quad \cos(2\sqrt{-bt})]/\sqrt{t}$	$\sqrt{\frac{\pi}{s}}\exp\left(\frac{b}{s}\right) \quad b \gtrless 0$	27
28, 32	$\sinh(2\sqrt{bt}) \quad \text{or} \quad \sin(2\sqrt{-bt})$	$\sqrt{\frac{\pi\lvert b\rvert}{s^3}}\exp\left(\frac{b}{s}\right) \quad b \gtrless 0$	27
46	$\left(\frac{1}{\sqrt{t}}\right)^{1+\nu}\exp\left(\frac{-b}{4t}\right)D_\nu\left(\sqrt{\frac{b}{t}}\right)$	$\sqrt{\frac{\pi}{s}}(2s)^{\nu/2}\exp(-\sqrt{2bs})$	27
27:13	$\theta_1(\nu;t) \quad \lvert\nu\rvert \le \frac{1}{2}$	$\frac{-1}{\sqrt{s}}\mathrm{sech}(\sqrt{s})\sinh(2\nu\sqrt{s})$	28, 29
27:13	$\theta_2(\nu;t) \quad \lvert\nu\rvert \le \frac{1}{2}$	$\frac{1}{\sqrt{s}}\mathrm{sech}(\sqrt{s})\sinh(\sqrt{s} - 2\nu\sqrt{s})$	28, 29
27:13	$\theta_3(\nu;t) \quad 0 \le \nu \le 1$	$\frac{1}{\sqrt{s}}\mathrm{csch}(\sqrt{s})\cosh(\sqrt{s} - 2\nu\sqrt{s})$	28, 29

Table 26.14.2 (Continued)

Chapter or section no.	f(t)	$\bar{f}_L(s)$	Chapter or section no.
27:13	$\theta_4(\nu;t) \quad \lvert\nu\rvert \le \frac{1}{2}$	$\frac{1}{\sqrt{s}}\operatorname{csch}(\sqrt{s})\cosh(2\nu\sqrt{s})$	28, 29
32	$\operatorname{sgn}\{\sin(\sqrt{t})\}$ lengthening squarewave	$\frac{1}{s}\theta_4(0;s)$	27:13
29	$-2b(-bt)^n\exp\left(\frac{bt}{2}\right)\operatorname{csch}\left(\frac{bt}{2}\right)$	$\psi^{(n)}\left(\frac{s}{b}\right) \quad n = 0, 1, 2, \ldots$	44:12
29	$\operatorname{sech}(bt)$	$\frac{1}{2b}G\left(\frac{s+b}{2b}\right)$	44:13
29	$(2bt)^\nu\operatorname{csch}(bt) \quad \nu > 0$	$\frac{\Gamma(1+\nu)}{b}\zeta\left(1+\nu;\frac{s+b}{2b}\right)$	64
29, 30	$b^\nu t^{\nu-1}[1 + \operatorname{csch}(bt) + \coth(bt)]$	$2\Gamma(\nu)\zeta\left(\nu;\frac{s}{b}\right)$	64
29, 30	$b^\nu t^{\nu-1}[1 - \operatorname{csch}(bt) + \coth(bt)]$	$2\Gamma(\nu)\eta\left(\nu;\frac{s}{b}\right)$	64:13
30	$\tanh(bt)$	$\frac{1}{2b}G\left(\frac{s}{2b}\right) - \frac{1}{s}$	44:13
32	$\operatorname{sgn}\left\{\sin\left(\frac{\pi t}{2b}\right)\right\}$ squarewave	$\frac{1}{s}\tanh(bs)$	30
30:13	$\coth(bt) - bt$	$\frac{1}{b}\ln\left(\frac{s}{2b}\right) - \frac{1}{b}\psi\left(\frac{s}{2b}\right) - \frac{1}{s}$	44
31	$2\operatorname{arsinh}(bt)$	$\frac{\pi}{s}\left[h_0\left(\frac{s}{b}\right) - Y_0\left(\frac{s}{b}\right)\right]$	57
31	$\operatorname{arcosh}(1 + bt)$	$\frac{1}{s}\exp\left(\frac{s}{b}\right)K_0\left(\frac{s}{b}\right)$	51
51	$K_0(bt)$	$\frac{1}{\sqrt{s^2 - b^2}}\operatorname{arcosh}\left(\frac{s}{b}\right)$	31
54, 57	$\pi Y_0(bt)$ or $\pi h_0(bt)$	$\frac{2}{\sqrt{s^2 + b^2}}\left[\operatorname{arsinh}\left(\frac{-s}{b}\right) \text{ or } \operatorname{arsinh}\left(\frac{b}{s}\right)\right]$	31
32	$\frac{1}{\sqrt{t}}\sin\left(\frac{b}{2t}\right)$ or $\frac{1}{\sqrt{t}}\cos\left(\frac{b}{2t}\right)$	$\sqrt{\frac{\pi}{s}}\exp(-\sqrt{bs})\,[\sin(\sqrt{bs}) \text{ or } \cos(\sqrt{bs})]$	32
32	$t^{\nu-1}\sin(bt) \quad 0 \ne \nu > -1$	$\frac{\Gamma(\nu)}{(s^2 + b^2)^{\nu/2}}\sin\left\{\nu\operatorname{arccot}\left(\frac{s}{b}\right)\right\}$	35
32	$\frac{1}{t}\sin(\sqrt{bt})$	$\pi\operatorname{erf}(\sqrt{b/4s})$	40
55	$\operatorname{ber}(2\sqrt{bt})$ or $\operatorname{bei}(2\sqrt{bt})$	$\frac{1}{s}\cos\left(\frac{b}{s}\right)$ or $\frac{1}{s}\sin\left(\frac{b}{s}\right)$	32
32:13	$\frac{1}{b}\operatorname{sinc}\left(\frac{bt}{\pi}\right) = \frac{1}{t}\sin(bt)$	$\operatorname{arccot}\left(\frac{s}{b}\right) = \arctan\left(\frac{b}{s}\right)$	35
35	$2\arctan(\sqrt{bt})$	$\frac{\pi}{s}\exp\left(\frac{s}{b}\right)\operatorname{erfc}\left(\sqrt{\frac{s}{b}}\right)$	41

Table 26.14.2 (Continued)

Chapter or section no.	f(t)	$\bar{f}_L(s)$	Chapter or section no.
35	$\arctan(bt)$	$\frac{\cos(s/b)}{s}\left[\frac{\pi}{2} - \mathrm{Si}\left(\frac{s}{b}\right)\right] - \frac{\sin(s/b)}{s}\mathrm{Ci}\left(\frac{s}{b}\right)$	38
57:13	$\pi\ell_0(bt)$	$\frac{2}{\sqrt{s^2 - b^2}}\arcsin\left(\frac{b}{s}\right)$	35
38	$\mathrm{Si}(bt)$	$\frac{1}{s}\operatorname{arccot}\left(\frac{s}{b}\right) = \frac{1}{s}\arctan\left(\frac{b}{s}\right)$	35
42	$2\,\mathrm{daw}\left(\frac{bt}{2}\right)$	$\frac{1}{b}\exp\left(\frac{s^2}{b^2}\right)\mathrm{Ei}\left(\frac{-s^2}{b^2}\right)$	37
40	$\mathrm{erf}(bt)$	$\frac{1}{s}\exp\left(\frac{s^2}{4b^2}\right)\mathrm{erfc}\left(\frac{s}{2b}\right)$	41
45	$2^{\nu}\gamma\left(\nu;\frac{b^2t^2}{2}\right)$ $\nu > 0$	$\frac{\Gamma(2\nu)}{2s}\exp\left(\frac{s^2}{4b^2}\right)\mathrm{D}_{-2\nu}\left(\frac{s}{b}\right)$	46
45	$\Gamma\left(2\nu;\frac{b}{t}\right)$	$\frac{2(bs)^{\nu}}{s}\mathrm{K}_{2\nu}(2\sqrt{bs})$	51
56	$2\,\mathrm{Ai}\{(9bt)^{1/3}\}$	$\Gamma\left(\frac{4}{3}\right)\left(\frac{3b}{s^4}\right)^{1/3}\exp\left(\frac{b}{s}\right)\Gamma\left(\frac{-1}{3};\frac{b}{s}\right)$	45
47	$t^{c}\mathrm{M}(a;2c;bt)$ $c > -1$	$\frac{\Gamma(1+c)}{s^{1+c}}\mathrm{F}\left(a,1+c;2c;\frac{b}{s}\right)$	60
50	$\frac{\Gamma(c)\sqrt{bt}^{\,a-1/2}}{(bt)^{c/2}}\mathrm{I}_{c-1}(2\sqrt{bt})$	$\frac{\Gamma(a)}{s^{a}}\mathrm{M}\left(a;c;\frac{b}{s}\right)$	47
48	$t^{a-1}\exp(-Bt)\mathrm{U}(c-b;1+a-b;Bt)$	$B^{2-a}\,\mathrm{F}\left(a,b;c;\frac{-s}{B}\right)$	60
48:4	$\mathrm{k}_{2\nu}(bt)$ Bateman's k function	$2b\,\mathrm{sinc}(\nu)(b-s)^{\nu-1}(b+s)^{-\nu-1}$ $\mathrm{B}\left(1-\nu;\,1+\nu;\,\frac{b-s}{2b}\right)$	58
49	$\pi\mathrm{I}_0^{\,2}(bt)$	$\frac{2}{s}\mathrm{K}\left(\frac{2b}{s}\right)$	61
49	$b\mathrm{I}_0(bt)\mathrm{I}_1(bt)$	$s\left[\frac{1}{2} - \frac{1}{\pi}\mathrm{E}\left(\frac{2b}{s}\right)\right]$	61
50	$\sqrt{\frac{\pi b}{2t}}\,\mathrm{I}_{1/2+\nu}(bt)$ $\nu > -1$	$\mathrm{Q}_{\nu}\left(\frac{s}{b}\right)$	59
51	$\frac{2}{\sqrt{t}}\mathrm{K}_0(\sqrt{bt})$	$\sqrt{\frac{\pi}{s}}\exp\left(\frac{b}{8s}\right)\mathrm{K}_0\left(\frac{b}{8s}\right)$	51
51	$\sqrt{\frac{2b}{\pi t}}\,\mathrm{K}_{1/2+\nu}(bt)$ $-1 < \nu < 0$	$-\pi\csc(\nu\pi)\mathrm{P}_{\nu}\left(\frac{s}{b}\right)$	59
59	$\sqrt{\pi b}\,\mathrm{P}_{\nu}(1+bt)$	$\sqrt{\frac{2}{s}}\exp\left(\frac{s}{b}\right)\mathrm{K}_{1/2+\nu}\left(\frac{s}{b}\right)$	51

determined. The table lists pairs $f(t)$, $\bar{f}_L(s)$ of functions and their Laplace transforms. To use this table for transformation, first locate the chapter or section of the *Atlas* in which the f function appears. Then scan the numerical index that constitutes the first column of the table to find entries containing the f function. Of course, there exist many Laplace transform pairs that are not listed in Table 26.14.2 and that cannot be conveniently deduced via rules 26:14:2–26:14:18. We recommend the comprehensive tabulation by Roberts and Kaufman as a source of several thousand varied transform pairs.

A useful check on the accuracy of an $f(t)$, $\bar{f}_L(s)$ pair is provided by their limiting forms as the arguments approach zero and infinity. These *limiting properties* are

26:14:18 $$\mathrm{L}\left\{\lim_{t\to 0}\{f(t)\}\right\} = \lim_{s\to\infty}\{s\bar{f}_L(s)\} \qquad \text{and} \qquad \mathrm{L}\left\{\lim_{t\to\infty}\{f(t)\}\right\} = \lim_{s\to 0}\{s\bar{f}_L(s)\}$$

An even simpler check is provided by the requirement that the inverse Laplace transform of any continuous function $f(t)$ must tend to zero as $s \to \infty$.

Section 18:14 demonstrates that the majority of the functions in this *Atlas* may be expressed as hypergeometric functions. Let $f(t)$ be such a function and let us adopt $(a_{1\to K})_j$ as an abbreviation for the product of K Pochhammer polynomials

26:14:19 $$f(t) = \sum_{j=0}^{\infty} \frac{(a_1)_j(a_2)_j(a_3)_j \cdots (a_K)_j}{(c_1)_j(c_2)_j(c_3)_j \cdots (c_L)_j} t^j = \sum_{j=0}^{\infty} \frac{(a_{1\to K})_j}{(c_{1\to L})_j} t^j$$

Then $\bar{f}_L(s)$ is also a hypergeometric function, namely that given by expression 18:14:7. Moreover, many functions closely related to $f(t)$ may also be Laplace transformed to hypergeometric functions with one or more extra numeratorial parameters. Thus:

26:14:20 $$\mathrm{L}\{t^\nu f(bt)\} = \frac{\Gamma(1+\nu)}{s^{1+\nu}} \sum_{j=0}^{\infty} \frac{(1+\nu)_j (a_{1\to K})_j}{(c_{1\to L})_j} \left(\frac{b}{s}\right)^j \qquad \nu > -1$$

and

26:14:21 $$\mathrm{L}\{t^\nu f(\pm b^2t^2)\} = \frac{\Gamma(1+\nu)}{s^{1+\nu}} \sum_{j=0}^{\infty} \frac{\left(\frac{1+\nu}{2}\right)_j \left(1+\frac{\nu}{2}\right)_j (a_{1\to K})_j}{(c_{1\to L})_j} \left(\frac{\pm 4b^2}{s^2}\right)^j \qquad \nu > -1$$

while the transform of $t^\nu f(\pm\sqrt{bt})$ is the sum

26:14:22 $$\frac{\Gamma(1+\nu)}{s^{1+\nu}} \sum_{j=1}^{\infty} \frac{(1+\nu)_j(\frac{1}{2}a_{1\to K})_j(\frac{1}{2}+\frac{1}{2}a_{1\to K})_j}{(\frac{1}{2}c_{1\to L})_j(\frac{1}{2}+\frac{1}{2}c_{1\to L})_j} \left(\frac{4^{K-L}b}{s}\right)^j$$
$$\pm \frac{\Gamma(\frac{3}{2}+\nu)a_{1\to K}\sqrt{b}}{s^{3/2+\nu}c_{1\to L}} \sum_{j=0}^{\infty} \frac{(\frac{3}{2}+\nu)_j(\frac{1}{2}+\frac{1}{2}a_{1\to K})_j(1+\frac{1}{2}a_{1\to K})_j}{(\frac{1}{2}+\frac{1}{2}c_{1\to L})_j(1+\frac{1}{2}c_{1\to L})_j} \left(\frac{4^{K-L}b}{s}\right)^j$$

of two hypergeometric functions, each with $2K + 1$ numeratorial and $2L$ denominatorial parameters.

The table of transform pairs 26:14:2 is also valuable for Laplace inversion. To use it for that purpose, first scan the right-most column to locate the chapter or section number of the *Atlas* where the function $\bar{f}_L$ is discussed. Then the second column lists the inverse transform of the function in the third column. As for transformation, there are a number of general properties of Laplace inversion that permit the table to be used to invert many more than the 120 entries that are displayed. Some of these general rules are presented in the following paragraphs.

The *linearity property* of Laplace inversion follows immediately from equation 26:14:2. The *linear shift property*

26:14:23 $$\bar{f}_L(bs+c) = \mathrm{L}\left\{\frac{1}{b}\exp\left(\frac{-ct}{b}\right)f\left(\frac{t}{b}\right)\right\}$$

of Laplace inversion incorporates the *scaling property*, which is the $c = 0$ instance of 26:14:23.

The most general *chain rule* for Laplace inversion gives the inverse transform of the *composite function* $\bar{f}_L\{g(s)\}$ as the definite integral of the product $F(t;u)f(u)$, where $f(t)$ and $F(t;u)$ are respectively the inverse transforms of $\bar{f}_L(s)$ and $\exp\{-ug(s)\}$. That is:

26:14:24 $$\bar{f}_L\{g(s)\} = L\left\{\int_0^\infty F(t;u)f(u)du\right\} \qquad \text{where} \qquad \exp\{-ug(s)\} = L\{F(t;u)\}$$

Unfortunately, there exist rather few g(s), F(t;u) pairs that can be used to exploit relationship 26:14:24. Table 26.14.3 lists three.

Whereas there is no such simple rule for Laplace transformation, there does exist a general *product rule* for Laplace inversion. This is the so-called *convolution property*

26:14:25 $$\bar{f}_L(s)\bar{g}_L(s) = L\left\{\int_0^t f(t-u)g(u)du\right\} = L\{f(t)*g(t)\}$$

Two straightforward applications of this property lead to the inversion rules

26:14:26 $$\exp(-bs)\bar{g}_L(s) = L\{u(t-b)g(t-b)\} \qquad b > 0$$

and

26:14:27 $$\coth\left(\frac{bs}{2}\right)\bar{g}_L(s) = L\left\{g(t) + 2\sum_{j=1}^{J} u(t-jb)g(t-jb)\right\} \qquad J = \text{Int}\left(\frac{t}{b}\right)$$

To Laplace invert the product $s^\nu\bar{g}_L(s)$ involves operations of the calculus applied to the inverse transform. The simplest cases

26:14:28 $$\frac{1}{s}\,\bar{g}_L(s) = L\left\{\int_0^t g(t)dt\right\}$$

and

26:14:29 $$s\bar{g}_L(s) = L\left\{\frac{dg}{dt}(t) + g(0)\delta(t)\right\}$$

iterate to

26:14:30 $$\frac{\bar{g}_L(s)}{s^n} = L\left\{\int_0^t \cdots \int_0^t g(t)(dt)^n\right\} \qquad n = 2, 3, 4, \ldots$$

and

26:14:31 $$s^n\bar{g}_L(s) = L\left\{\frac{d^n g}{dt^n}(t) + \sum_{j=0}^{n-1}\frac{d^j g}{dt^j}(0)\delta^{(j)}(t)\right\} \qquad n = 1, 2, 3, \ldots$$

Here $\delta^{(n)}(t)$ is the nth derivative of the Dirac delta function [Section 10:13] at $t = 0$; away from the immediate vicinity of $t = 0$ all such functions are zero. Ignoring the contributions from these delta derivatives, the inverse Laplace transform of $s^\nu\bar{g}_L(s)$ is the differintegral $d^\nu g/dt^\nu$ for all values of ν.

A number of useful rules exist for the Laplace inversion of the quotients $\bar{f}_L\{g(s)\}/g(s)$ and $\bar{f}_L\{g(s)\}/h(s)$ where g(s) and h(s) are such simple functions as $\sqrt{s}$, $1/s$, $\sqrt{s^2 \pm a^2}$, etc. Inversion usually leads to definite or indefinite

Table 26.14.3

g(s)	F(t;u)
ln(s)	$\dfrac{t^{u-1}}{\Gamma(u)}$
$\sqrt{s}$	$\dfrac{u}{2\sqrt{\pi t^3}}\exp\left(\dfrac{-u^2}{4t}\right)$
$s^{1/3}$	$\dfrac{u}{(3t^4)^{1/3}}\text{Ai}\left(\dfrac{u}{(3t)^{1/3}}\right)$

integrals of the product of f(t), the inverse transform of $\bar{f}_L(s)$, with other functions. Examples include

26:14:32 $$\frac{\bar{f}_L\left(\frac{1}{s}\right)}{\sqrt{s}} = L\left\{\frac{1}{\sqrt{\pi t}}\int_0^\infty \cos(2\sqrt{tu})f(u)du\right\}$$

26:14:33 $$\frac{\bar{f}_L(\sqrt{s^2+a})}{\sqrt{s^2+a}} = L\left\{\int_0^t J_0(\sqrt{at^2-au^2})f(u)du\right\} \qquad a \ge 0$$

with $I_0(\sqrt{au^2-at^2})$ replacing $J_0(\sqrt{at^2-au^2})$ if a is negative; Roberts and Kaufman [pages 171–174] may be consulted for others.

If $\bar{f}_L(s)$ is the hypergeometric function

26:14:34 $$\bar{f}_L(s) = \frac{1}{s}\sum_{j=0}^\infty \frac{(a_1)_j(a_2)_j\cdots(a_K)_j}{(c_1)_j(c_2)_j\cdots(c_L)_j}\left(\frac{1}{s}\right)^j = \frac{1}{s}\sum_{j=0}^\infty \frac{(a_{1\to K})_j}{(c_{1\to L})_j}\left(\frac{1}{s}\right)^j$$

then inverse transformation of $\bar{f}_L(s)$, and a number of related functions, give hypergeometric functions of t. The results

26:14:35 $$\frac{1}{s^\nu}\bar{f}_L\left(\frac{s}{b}\right) = L\left\{\frac{t^\nu}{\Gamma(1+\nu)}\sum_{j=0}^\infty \frac{(a_{1\to K})_j(bt)^j}{(1+\nu)_j(c_{1\to L})_j}\right\} \qquad \nu > -1$$

26:14:36 $$\frac{1}{s^\nu}\bar{f}_L\left(\frac{\pm s^2}{b^2}\right) = L\left\{\frac{t^\nu}{\Gamma(1+\nu)}\sum_{j=0}^\infty \frac{(a_{1\to K})_j(\pm b^2t^2/4)^j}{\left(\frac{1+\nu}{2}\right)_j\left(1+\frac{\nu}{2}\right)_j(c_{1\to L})_j}\right\} \qquad \nu > -1$$

26:14:37 $$\frac{1}{\nu}\bar{f}_L\left(\pm\sqrt{\frac{s}{b}}\right) = L\left\{\frac{t^\nu}{\Gamma(1+\nu)}\sum_{j=0}^\infty \frac{(\frac{1}{2}a_{1\to K})_j(\frac{1}{2}+\frac{1}{2}a_{1\to K})_j(bt/4^{K-L})^j}{(1+\nu)_j(\frac{1}{2}c_{1\to L})_j(\frac{1}{2}+\frac{1}{2}c_{1\to L})_j}\right\}$$

$$\pm L\left\{\frac{t^{1/2+\nu}a_{1\to K}}{\Gamma(\frac{3}{2}+\nu)\sqrt{b}\,c_{1\to L}}\sum_{j=0}^\infty \frac{(\frac{1}{2}+\frac{1}{2}a_{1\to K})_j(1+\frac{1}{2}a_{1\to K})_j(bt/4^{K-L})^j}{(\frac{3}{2}+\nu)_j(\frac{1}{2}+\frac{1}{2}c_{1\to L})_j(1+\frac{1}{2}c_{1\to L})_j}\right\}$$

are seen to be strict counterparts of the transforms 26:14:20–26.14:22.

We conclude this section by drawing to the attention of the reader the *Heaviside expansion theorem* that is discussed in Section 17:10. This theorem provides a procedure for the Laplace inversion of $1/p_n(s)$ where $p_n(s)$ is a polynomial function of degree n. A similar procedure may be employed to invert the rational function $p_m(s)/p_n(s)$, provided $m < n$, via the decomposition detailed in Section 17:13. The Heaviside theorem may even be applied for infinite values of m and n, provided that the denominatorial degree exceeds that of the numerator, and hence may be employed to invert the quotient of two transcendental functions that may be written, for example, as $(A_0 + A_1s + A_2s^2 + \cdots)/(a_1s + a_2s^2 + a_3s^3 + \cdots)$.

CHAPTER 27

EXPONENTIALS OF POWERS $\exp(-\alpha x^\nu)$

Many of the results of this chapter follow by combining the properties of the functions discussed in Chapters 12, 13 and 26. It is appropriate to devote a chapter to the functions $\exp(-\alpha x^\nu)$, however, because they are of such widespread importance, particularly when α is positive. For example, the temperature dependence of many physical properties obeys this functionality with $\nu = -1$. Random events often involve the $\nu = 2$ instance and lead to important distributions, as discussed in Section 27:14.

27:1 NOTATION

No special notation or symbolism need be considered beyond that addressed in Sections 12:1 and 26:1.

27:2 BEHAVIOR

As with the functions discussed in Chapter 13, the range of $\exp(-\alpha x^\nu)$ depends in a rather detailed way on the characteristics of the number ν. In this chapter, however, attention will be confined to the range $x \geq 0$, except when ν is an integer.

Figures 27-1 and 27-2 are maps of the functions $\exp(x^\nu)$ and $\exp(-x^\nu)$ for assorted values of ν. The important graph of $\exp(-x^2)$ versus x, known as a *Gauss curve*, displays a maximum value of unity at $x = 0$ and points of inflection [see Section 0:7] at $x = \pm 1/\sqrt{2}$. Similarly, $\exp(-1/x)$ has an inflection at $x = \frac{1}{2}$.

27:3 DEFINITIONS

With x replaced by $-\alpha x^\nu$, any of the definitions in Section 26:3 can serve to define $\exp(-\alpha x^\nu)$.

27:4 SPECIAL CASES

When $\nu = 0$ or 1, the $\exp(-\alpha x^\nu)$ function reduces respectively to a constant [Chapter 1] or a simple exponential [Chapter 26].

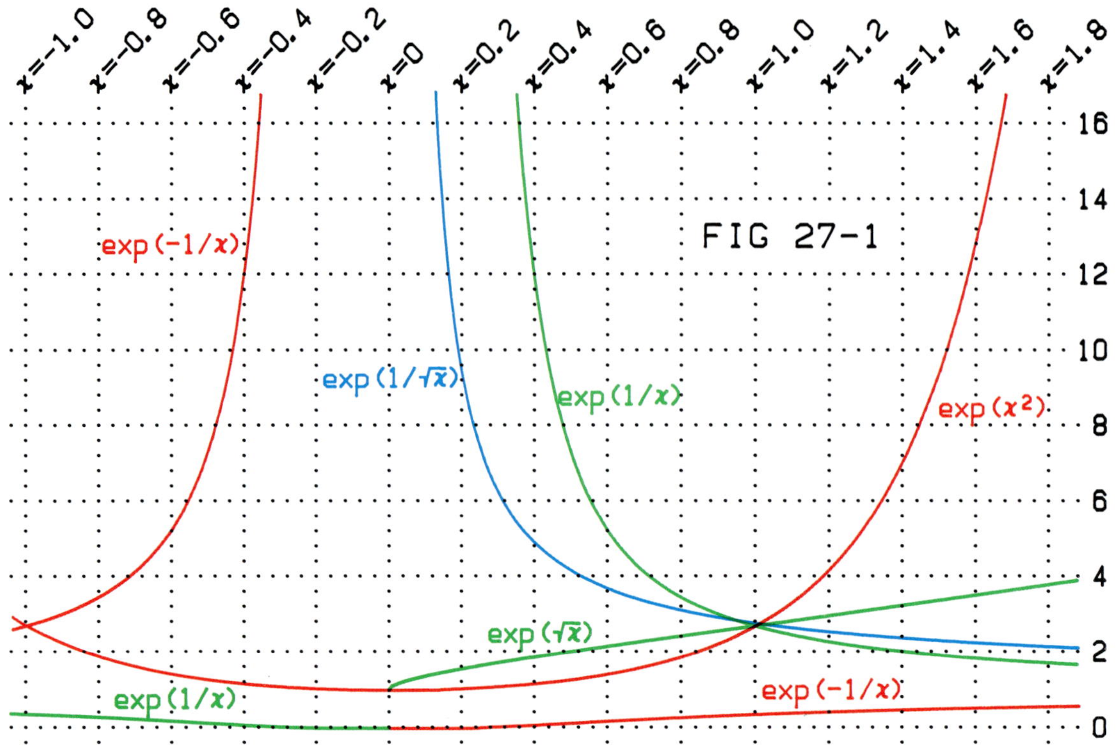

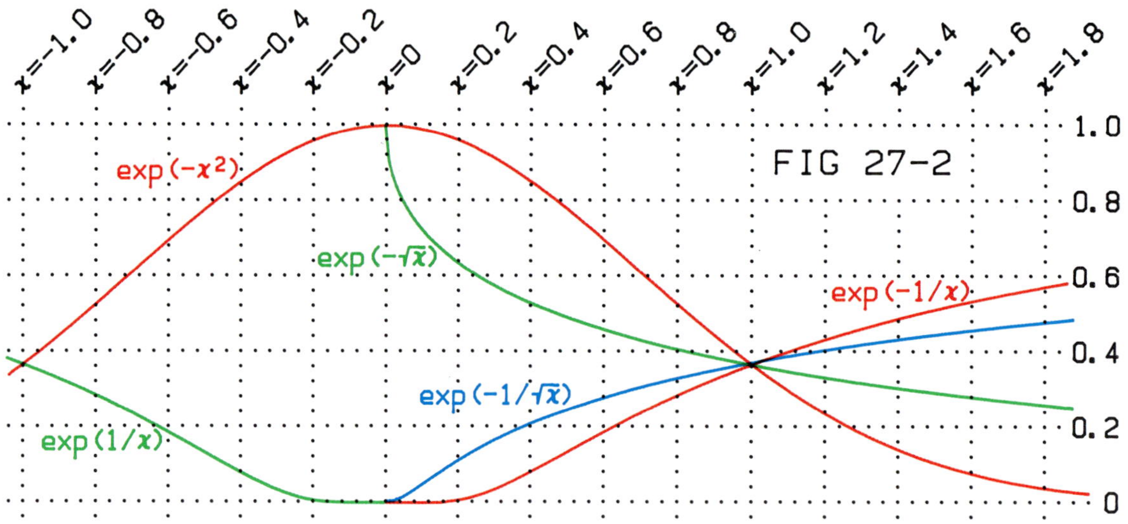

27:5 INTRARELATIONSHIPS

The relationships of Section 26:5 hold when x is replaced by $-\alpha x^{\nu}$.

Summation formulas for series of exponentials of $1/x$ include

27:5:1 $$\exp\left(\frac{-1}{x}\right) + \exp\left(\frac{-4}{x}\right) + \exp\left(\frac{-9}{x}\right) + \cdots = \frac{\sqrt{\pi x}}{2}\,\theta_3(0;x) - \frac{1}{2} \simeq \frac{\sqrt{\pi x}-1}{2} + \sqrt{\pi x}\exp(-\pi^2 x) \qquad x>0$$

27:5:2 $$\exp\left(\frac{-1}{x}\right) - \exp\left(\frac{-4}{x}\right) + \exp\left(\frac{-9}{x}\right) - \cdots = \frac{1}{2} - \frac{\sqrt{\pi x}}{2}\,\theta_2(0;x) \simeq \frac{1}{2} - \sqrt{\pi x}\exp\left(\frac{-\pi^2 x}{4}\right) \qquad x>0$$

and

27:5:3 $$\exp\left(\frac{-1}{x}\right)+\exp\left(\frac{-9}{x}\right)+\exp\left(\frac{-25}{x}\right)+\cdots=\frac{\sqrt{\pi x}}{4}\theta_4\left(0;\frac{x}{4}\right)\simeq\frac{\sqrt{\pi x}}{4}-\frac{\sqrt{\pi x}}{2}\exp\left(\frac{-\pi^2 x}{4}\right)\qquad x>0$$

in addition to variants of equations 26:5:5–26:5:12. The θ functions are explained in Section 27:13. The final approximations improve as x increases but are valid to better than 1 part in 10^9 for $x \geq 1$.

27:6 EXPANSIONS

The series expansion

27:6:1 $$\exp(-\alpha x^\nu)=1-\frac{\alpha x^\nu}{1!}+\frac{\alpha^2 x^{2\nu}}{2!}-\cdots=\sum_{j=0}^{\infty}\frac{(-\alpha x^\nu)^j}{j!}$$

holds for all α, for all ν and for all x for which x^ν is defined.

The continued fraction expansion

27:6:2 $$\exp(\sqrt{x})=1+\frac{2\sqrt{x}}{2-\sqrt{x}+}\,\frac{x/3}{2+}\,\frac{x/15}{2+}\,\frac{x/35}{2+}\,\frac{x/(4j^2-1)}{2+}\cdots$$

is rapidly convergent.

27:7 PARTICULAR VALUES

Choosing $\alpha = 1$ or $\alpha = -1$ and taking care that x^ν is defined we find

	$x=0$	$x=1$	$x=\infty$
$\exp(x^\nu)$	$\begin{cases}\infty & \nu<0\\ 1 & \nu>0\end{cases}$	e	$\begin{cases}1 & \nu<0\\ \infty & \nu>0\end{cases}$
$\exp(-x^\nu)$	$\begin{cases}0 & \nu<0\\ 1 & \nu>0\end{cases}$	$\frac{1}{e}$	$\begin{cases}1 & \nu<0\\ 0 & \nu>0\end{cases}$

27:8 NUMERICAL VALUES

Values of $\exp(-\alpha x^\nu)$ are calculable by first evaluating $-\alpha x^\nu$, followed by exponentiation.

27:9 APPROXIMATIONS

Though crude, the approximation

27:9:1 $$\exp(-x^2)\simeq\begin{cases}1-\dfrac{|x|}{\sqrt{\pi}} & -\sqrt{\pi}\leq x\leq\sqrt{\pi}\\ 0 & |x|\geq\sqrt{\pi}\end{cases}$$

is never in error by more than 0.09. This triangular approximation to the Gauss curve has an area of $\sqrt{\pi}$, equal to that under the curve.

27:10 OPERATIONS OF THE CALCULUS

Differentiation gives

27:10:1
$$\frac{d}{dx}\exp(-\alpha x^\nu) = -\alpha\nu x^{\nu-1}\exp(-\alpha x^\nu)$$

Indefinite integration of the general expressions $\exp(-\alpha t^\nu)$ and $t^w\exp(-\alpha t^\nu)$ can be accomplished by making the substitution $-\alpha t^\nu = \pm y$ and utilizing the general formulas contained in equations 26:10:7, 26:10:13 and 26:10:14. Below are listed some important indefinite integrals:

27:10:2
$$\int_0^x \exp(t^2)dt = \exp(x^2)\,\mathrm{daw}(x)$$

27:10:3
$$\int_1^x \exp(\sqrt{t})dt = 2(\sqrt{x}-1)\exp(\sqrt{x})$$

27:10:4
$$\int_0^x \exp\left(\frac{1}{\sqrt{t}}\right)dt = (x+\sqrt{x})\exp\left(\frac{1}{\sqrt{x}}\right) - \mathrm{Ei}\left(\frac{1}{\sqrt{x}}\right) \qquad x>0$$

27:10:5
$$\int_0^x \exp\left(\frac{1}{t}\right)dt = x\exp\left(\frac{1}{x}\right) - \mathrm{Ei}\left(\frac{1}{x}\right) \qquad x>0$$

27:10:6
$$\int_0^x \exp(-t^2)dt = \frac{\sqrt{\pi}}{2}\mathrm{erf}(x)$$

27:10:7
$$\int_x^\infty \exp(-\sqrt{t})dt = 2(\sqrt{x}+1)\exp(-\sqrt{x})$$

27:10:8
$$\int_0^x \exp\left(\frac{-1}{\sqrt{t}}\right)dt = (x-\sqrt{x})\exp\left(\frac{-1}{\sqrt{x}}\right) - \mathrm{Ei}\left(\frac{-1}{\sqrt{x}}\right) \qquad x>0$$

27:10:9
$$\int_0^x \exp\left(\frac{-1}{t}\right)dt = x\exp\left(\frac{-1}{x}\right) + \mathrm{Ei}\left(\frac{-1}{x}\right) \qquad x>0$$

27:10:10
$$\int_x^\infty \exp(-\alpha t^2-\beta t-\gamma)dt = \frac{1}{2}\sqrt{\frac{\pi}{\alpha}}\exp\left(\frac{\beta^2-4\alpha\gamma}{4\alpha}\right)\mathrm{erfc}\left(\frac{-2\alpha x+\beta}{2\sqrt{\alpha}}\right) \qquad \alpha>0$$

Chapters 42, 37 and 40 are devoted to the daw, Ei and erfc functions.

Some important definite integrals include

27:10:11
$$\int_0^\infty t^\omega\exp(-\alpha t^\nu)dt = \frac{\Gamma\left(\frac{\omega+1}{\nu}\right)}{\alpha^{(\omega+1)/\nu}} \qquad \frac{\omega+1}{\nu}>0 \qquad \alpha>0$$

and

27:10:12
$$\int_0^\infty \exp\left(-\alpha t^2-\frac{\beta}{t^2}\right)dt = \frac{1}{2}\sqrt{\frac{\pi}{\alpha}}\exp(-2\sqrt{\alpha\beta}) \qquad \alpha>0<\beta$$

where Γ is the gamma function [see Chapter 43].

27:11 COMPLEX ARGUMENT

When the arguments of the two most important functions of this chapter are replaced by $x+iy$, we find

27:11:1
$$\exp[-\alpha(x+iy)^2] = \exp(\alpha y^2-\alpha x^2)[\cos(2\alpha xy) - i\sin(2\alpha xy)]$$

and

27:11:2 $$\exp\left(\frac{-\alpha}{x+iy}\right) = \exp\left(\frac{-\alpha x}{x^2+y^2}\right)\left[\cos\left(\frac{\alpha y}{x^2+y^2}\right) + i\sin\left(\frac{\alpha y}{x^2+y^2}\right)\right]$$

27:12 GENERALIZATIONS

The function $\exp(-\alpha x^\nu)$ may be generalized "to other bases" as described in Section 26:12 so that evaluation of $\beta^{(-\alpha x^\nu)}$ is possible.

27:13 COGNATE FUNCTIONS

The four *theta functions* are bivariate, depending on the parameter ν as well as on the argument x. We define

27:13:1 $$\theta_1(\nu;x) = \frac{1}{\sqrt{\pi x}}\sum_{j=-\infty}^{\infty}(-1)^j \exp[-(\nu - {}^1\!/_2 + j)^2/x]$$

27:13:2 $$\theta_2(\nu;x) = \frac{1}{\sqrt{\pi x}}\sum_{j=-\infty}^{\infty}(-1)^j \exp[-(\nu + j)^2/x] = \theta_1(\nu + {}^1\!/_2;x)$$

27:13:3 $$\theta_3(\nu;x) = \frac{1}{\sqrt{\pi x}}\sum_{j=-\infty}^{\infty}\exp[-(\nu + j)^2/x]$$

27:13:4 $$\theta_4(\nu;x) = \frac{1}{\sqrt{\pi x}}\sum_{j=-\infty}^{\infty}\exp[-(\nu + {}^1\!/_2 + j)^2/x] = \theta_3(\nu + {}^1\!/_2;x)$$

but the reader should be alert to the wide variety of symbolisms in use in the literature. Thus, the quantity defined in equation 27:13:1 would be denoted $\theta_1(\pi\nu,\exp(-\pi^2 x))$ by Abramowitz and Stegun [Sections 16.27–16.30].

It is evident that these four functions are dependent periodically on ν so that $\theta(\nu + 2;x) = -\theta(\nu + 1;x) = \theta(\nu;x)$ for θ_1 and θ_2, whereas $\theta(\nu + 1;x) = \theta(\nu;x)$ for θ_3 and θ_4. This periodicity is brought out more clearly in the alternative representations

27:13:5 $$\theta_1(\nu;x) = 2\sum_{j=0}^{\infty}(-1)^j \exp[-(j + {}^1\!/_2)^2\pi^2 x]\sin[2(j + {}^1\!/_2)\nu\pi]$$

27:13:6 $$\theta_2(\nu;x) = 2\sum_{j=0}^{\infty}\exp[-(j + {}^1\!/_2)^2\pi^2 x]\cos[2(j + {}^1\!/_2)\nu\pi]$$

27:13:7 $$\theta_3(\nu;x) = 1 + 2\sum_{j=1}^{\infty}\exp[-j^2\pi^2 x]\cos[2j\nu\pi]$$

27:13:8 $$\theta_4(\nu;x) = 1 + 2\sum_{j=1}^{\infty}(-1)^j\exp[-j^2\pi^2 x]\cos[2j\nu\pi]$$

of the theta functions.

The theta functions satisfy the intriguing *quadruplication formulas*

27:13:9 $$\theta_1(\nu;4x) = \frac{1}{2}\theta_3\left(\frac{\nu}{2} - \frac{1}{4};x\right) - \frac{1}{2}\theta_4\left(\frac{\nu}{2} - \frac{1}{4};x\right)$$

27:13:10 $$\theta_2(\nu;4x) = \frac{1}{2}\theta_3\left(\frac{\nu}{2};x\right) - \frac{1}{2}\theta_4\left(\frac{\nu}{2};x\right)$$

27:13:11 $$\theta_3(\nu;4x) = \frac{1}{2}\theta_3\left(\frac{\nu}{2};x\right) + \frac{1}{2}\theta_4\left(\frac{\nu}{2};x\right)$$

27:13:12 $$\theta_4(\nu;4x) = \frac{1}{2}\theta_3\left(\frac{\nu}{2} - \frac{1}{4}; x\right) + \frac{1}{2}\theta_4\left(\frac{\nu}{2} - \frac{1}{4}; x\right)$$

that may be rephrased in several alternative ways.

The series in 27:13:1–27:13:4 converge so rapidly if $x \le 1/\pi$ that it is rarely necessary to use $|j| > 3$ to compute very accurate values of the theta functions. On the other hand, if $x \ge 1/\pi$, the series in 27:13:5–27:13:8 converge even more rapidly. This means that accurate computation of theta functions is *always* a simple matter, making these functions useful in a variety of applications. Their utility is aided by the widespread occurrence of theta functions in Laplace transformation [see Section 26:14]. As examples

27:13:13 $$\int_0^\infty \theta_3(\nu;t)\exp(-st)\mathrm{d}t = \frac{\operatorname{csch}(\sqrt{s})}{\sqrt{s}}\cosh[(2\nu - 1)\sqrt{s}] \qquad 0 \le \nu \le 1 \qquad s > 0$$

27:13:14 $$\frac{1}{s}\theta_3(0;s) = \int_0^\infty \left[1 + 2\sum_{j=1}^{\infty} \mathrm{u}(t - j^2\pi^2)\right]\exp(-st)\mathrm{d}t \qquad s > 0$$

but for a more comprehensive listing see Roberts and Kaufman, in which the notation for theta functions is almost identical with that adopted in this *Atlas*.

The identity of expressions 27:13:5–27:13:8 with definitions 27:13:1–27:13:4 permits a reformulation of series of exponential functions, as follows:

27:13:15 $$\frac{1}{\sqrt{\pi x}}\left[1 + 2\sum_{j=1}^{\infty}(-1)^j\exp\left(\frac{-j^2}{x}\right)\right] = \theta_2(0;x) = 2\sum_{j=0}^{\infty}\exp\left(\frac{-(2j+1)^2\pi^2 x}{4}\right) \qquad x > 0$$

27:13:16 $$\frac{1}{\sqrt{\pi x}}\left[1 + 2\sum_{j=1}^{\infty}\exp\left(\frac{-j^2}{x}\right)\right] = \theta_3(0;x) = 1 + 2\sum_{j=1}^{\infty}\exp(-j^2\pi^2 x) \qquad x > 0$$

27:13:17 $$\frac{2}{\sqrt{\pi x}}\sum_{j=0}^{\infty}\exp\left(\frac{-(2j+1)^2}{4x}\right) = \theta_4(0;x) = 1 + 2\sum_{j=1}^{\infty}(-1)^j\exp(-j^2\pi^2 x) \qquad x > 0$$

Theta functions with $\nu = 0$ are important in their own right. Several of the representations coalesce when $x = {}^1/_\pi$, leading to the particular values

27:13:18 $$2^{1/4}\theta_2\left(0;\frac{1}{\pi}\right) = \theta_3\left(0;\frac{1}{\pi}\right) = 2^{1/4}\theta_4\left(0;\frac{1}{\pi}\right) = 1.086434811 = \frac{1}{\sqrt{U}}$$

where U is the ubiquitous constant discussed in Section 1:7.

Theta functions play an important role in the theory of elliptic functions [Chapter 63]. As well as the theta functions discussed above, there are other versions known as Neville's theta functions and Jacobi's theta functions [see Sections 63:8 and 63:13].

27:14 RELATED TOPICS

When a measurement or observation is repeated a large number N of times, it frequently happens that the values found are not identical. We say that the measured quantity x has a *distribution*. Sometimes (as in rolling dice) only a finite set of values is available for x. Here, however, we consider the continuous case (exemplified by sizes of raindrops) in which possible x values are limited in proximity only by the discrimination of the measuring device.

Let the measurements be arranged in order of size along the line $-\infty < x < \infty$. Then, as $N \to \infty$, it often becomes possible to delineate a *density function* or *frequency function* $\mathrm{f}(x)$ with the property that $2\mathrm{f}(x)\mathrm{d}x$ gives the approximate probability that any single measurement of x will lie in the range $x - \mathrm{d}x$ to $x + \mathrm{d}x$. Associated with each distribution is a *mean* defined by

27:14:1 $$\mu = \int_{x_0}^{x} t\,\mathrm{f}(t)\mathrm{d}t$$

and a *variance* defined by

27:14:2 $$\sigma^2 = \int_{x_0}^{x_1} (t-\mu)^2 f(t)dt = -\mu^2 + \int_{x_0}^{x_1} t^2 f(t)dt \geq 0$$

The limits, x_0 to x_1, on the integrals demarcate the range that is accessible to x, often $-\infty$ to ∞ or 0 to ∞. The mean of a distribution provides a measure of the average value of the property x, while the variance describes the dispersion of the distribution about the mean. If the distribution has too great a dispersion, no finite variance exists. The Lorentz distribution [see Table 27.14.1] is a case in point: for this particular distribution, depicted in Figure 27-3 for $\alpha = 8$, the integral in 27:14:2 diverges. Even integral 27:14:1 diverges for the Lorentz distribution, but the so-called *Cauchy principal value* of the integral, namely

27:14:3 $$\lim_{L\to\infty} \int_{-L}^{L} t f(t)dt$$

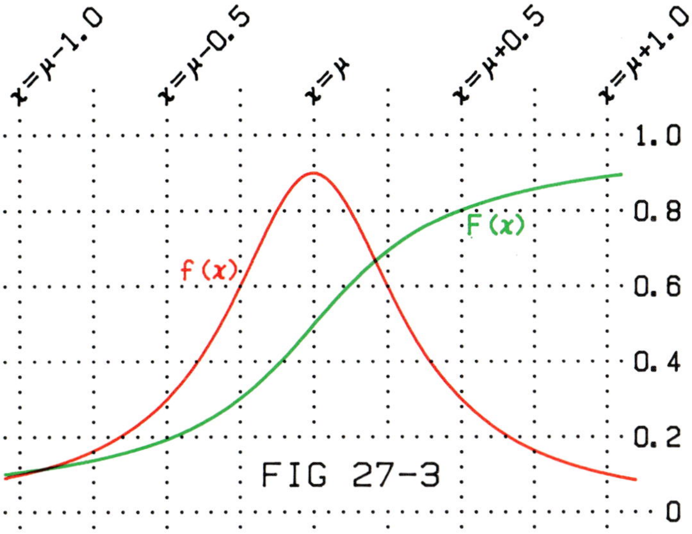

is as tabled.

Accompanying each distribution is a *cumulative function* or *distribution function*, given by

27:14:4 $$F(x) = \int_{x_0}^{x} f(t)dt$$

where, by definition of $f(x)$, $F(x_1) = 1$. The value of $F(x)$ necessarily lies in the range $0 \leq F(x) \leq 1$ and expresses the fraction of the measurements that (for $N \to \infty$) will lie in the interval between x_0 and x. Statisticians speak of *percentiles* or *percentage points*: the pth *percentile* is the value $x_{p/100}$ of x such that

27:14:5 $$F(x_{p/100}) = \frac{p}{100}$$

The fiftieth percentile $x_{1/2}$ is also called the *median* of the distribution and satisfies the condition

27:14:6 $$\int_{x_0}^{x_{1/2}} f(t)dt = \int_{x_{1/2}}^{x_1} f(t)dt = \frac{1}{2}$$

Though with symmetrical distributions they are the same, the mean and the median are not, in general, identical, nor does either necessarily correspond to the peak that often exists in a graph of $f(x)$ versus x. Where such a peak exists, its x-coordinate is known as the *mode* or *most probable value* of the distribution.

Distributions occur widely, especially in statistics and physics. Examples from both of these fields have been assembled into Table 27.14.1, and a selection is displayed graphically. The archetypal distribution is the *normal distribution* shown in Figure 27-4 for $\alpha = 3$. The *Gauss distribution* is the $\mu = 0$ instance of this. Notwithstanding

Table 27.14.1

Distribution	Limits x_0 to x_1	Density function f(x)	Mean μ	Variance σ^2	Cumulative function F(x)
Uniform (or rectangular)	x_0 to x_1	$\frac{1}{x_1 - x_0}$	$\frac{x_1 + x_0}{2}$	$\frac{(x_1 - x_0)^2}{12}$	$\frac{x - x_0}{x_1 - x_0}$
Gauss	$-\infty$ to ∞	$\sqrt{\frac{\alpha}{\pi}}\exp(-\alpha x^2)$	0	$\frac{1}{2\alpha}$	$\frac{1}{2} + \frac{1}{2}\mathrm{erf}(\sqrt{\alpha}x)$
Normal (or shifted Gauss)	$-\infty$ to ∞	$\sqrt{\frac{\alpha}{\pi}}\exp[-\alpha(x-\mu)^2]$	μ	$\frac{1}{2\alpha}$	$\frac{1}{2} + \frac{1}{2}\mathrm{erf}[\sqrt{\alpha}(x-\mu)]$
Log-normal	0 to ∞	$\sqrt{\frac{\alpha\mu}{\pi x^3}}\exp\left[-\alpha\ln^2\left(\frac{x}{\mu}\right) - \frac{1}{16\alpha}\right]$	μ	$\mu^2\left[\exp\left(\frac{1}{2\alpha}\right) - 1\right]$	$\frac{1}{2} + \frac{1}{2}\mathrm{erf}\left[\sqrt{\alpha}\ln\left(\frac{x}{\mu}\right) + \frac{1}{4\sqrt{\alpha}}\right]$
Rayleigh (or circular normal)	0 to ∞	$2\alpha x\exp(-\alpha x^2)$	$\frac{1}{2}\sqrt{\frac{\pi}{\alpha}}$	$\frac{1}{\alpha} - \frac{\pi}{4\alpha}$	$1 - \exp(-\alpha x^2)$
Maxwell	0 to ∞	$4\sqrt{\frac{\alpha^3}{\pi}}x^2\exp(-\alpha x^2)$	$\frac{2}{\sqrt{\pi\alpha}}$	$\frac{3}{2\alpha} - \frac{4}{\pi\alpha}$	$\mathrm{erf}(\sqrt{\alpha}x) - 2x\sqrt{\frac{\alpha}{\pi}}\exp(-\alpha x^2)$
Weibull	0 to ∞	$\frac{\sigma x^{\alpha-1}}{u^{\alpha}}\exp\left[-\left(\frac{x}{u}\right)^{\alpha}\right]$	$\frac{u}{\alpha}\Gamma\left(\frac{1}{\alpha}\right)$	$\frac{2u^2}{\alpha}\Gamma\left(\frac{2}{\alpha}\right) - \mu^2$	$1 - \exp\left[-\left(\frac{x}{u}\right)^{\alpha}\right]$
Sech-square	$-\infty$ to ∞	$\frac{\alpha}{2}\mathrm{sech}^2[\alpha(x-\mu)]$	μ	$\frac{\pi^2}{12\alpha^2}$	$\frac{1}{2} + \frac{1}{2}\tanh[\alpha(x-\mu)]$
Laplace	$-\infty$ to ∞	$\frac{\alpha}{2}\exp[-\alpha\lvert x-\mu\rvert]$	μ	$\frac{2}{\alpha^2}$	$\frac{1}{2} + \frac{\mathrm{sgn}(x-\mu)}{2}\{1 - \exp[-\alpha\lvert x-\mu\rvert]\}$
Boltzmann (or exponential)	0 to ∞	$\alpha\exp(-\alpha x)$	$\frac{1}{\alpha}$	$\frac{1}{\alpha^2}$	$1 - \exp(-\alpha x)$
Fermi-Dirac	0 to ∞	$\frac{\alpha\beta}{\ln(1+\beta)[\beta + \exp(\alpha x)]}$	$\frac{\mathrm{diln}(1+\beta)}{\alpha\ln(1+\beta)}$	$\frac{2\,\mathrm{triln}(1+\beta)}{\alpha^2\ln(1+\beta)} - \mu^2$	$1 + \frac{\alpha x - \ln[\beta + \exp(\alpha x)]}{\ln(1+\beta)}$
Bose-Einstein	0 to ∞	$\frac{-\alpha\beta}{\ln(1-\beta)[\exp(\alpha x) - \beta]}$	$\frac{\mathrm{diln}(1-\beta)}{\alpha\ln(1-\beta)}$	$\frac{2\,\mathrm{triln}(1-\beta)}{\alpha^2\ln(1-\beta)} - \mu^2$	$1 + \frac{\alpha x - \ln[\exp(\alpha x) - \beta]}{\ln(1-\beta)}$
Chi-square	0 to ∞	$\left(\frac{x}{2}\right)^{n/2}\frac{\exp(-x/2)}{x\Gamma(n/2)}$	n	$2n$	$\gamma\left(\frac{n}{2};\frac{x}{2}\right)/\Gamma\left(\frac{n}{2}\right)$
Gamma	0 to ∞	$\frac{(\alpha x/\mu)^2}{\Gamma(\alpha)x}\exp\left(\frac{-\alpha x}{\mu}\right)$	μ	$\frac{\mu^2}{2}$	$\gamma\left(\alpha;\frac{\alpha x}{\mu}\right)/\Gamma(\alpha)$
Beta	0 to ∞	$\frac{x^{\alpha-1}(1-x)^{\beta-1}}{\mathrm{B}(\alpha,\beta)}$	$\frac{\alpha}{\alpha+\beta}$	$\frac{\alpha\beta}{(\alpha+\beta)^2(\alpha+\beta+1)}$	$\mathrm{B}(\alpha;\beta;x)/\mathrm{B}(\alpha,\beta)$
Snedecor's-F ($n > 4$)	0 to ∞	$\frac{1}{\mathrm{B}\left(\frac{n}{2},\frac{m}{2}\right)x}\left(\frac{n}{n+mx}\right)^{n/2}\left(\frac{mx}{n+mx}\right)^{m/2}$	$\frac{n}{n-2}$	$\frac{2\mu^2(n+m-2)}{m(n-4)}$	$\mathrm{B}\left(\frac{n}{2};\frac{m}{2};\frac{mx}{n+mx}\right)/\mathrm{B}\left(\frac{n}{2},\frac{m}{2}\right)$
Lorentz (or Cauchy)	$-\infty$ to ∞	$\frac{\sqrt{\alpha}}{\pi[1 + \alpha(x-\mu)^2]}$	none exists, although μ = Cauchy principal value	none exists	$\frac{1}{2} + \frac{1}{\pi}\arctan[\sqrt{\alpha}(x-\mu)]$
Delta (or causal)	$-\infty$ to ∞	$\delta(x-\mu)$	μ	0	$\mathrm{u}(x-\mu)$
Student's-t n = 4, 6, 8, ...	$-\infty$ to ∞	$\frac{(n-1)!!}{2\sqrt{n}(n-2)!!}\left(\frac{n}{n+x^2}\right)^{(n+1)/2}$	0	$\frac{n}{n-2}$	$\frac{1}{2} + \frac{x}{2\sqrt{n+x^2}}\sum_{j=0}^{n/2-1}\frac{(2j-1)!!}{(2j)!!}\left(\frac{n}{n+x^2}\right)^{j}$
Student's-t n = 3, 5, 7, ...	$-\infty$ to ∞	$\frac{(n-1)!!}{\pi\sqrt{n}(n-2)!!}\left(\frac{n}{n+x^2}\right)^{(n+1)/2}$	0	$\frac{n}{n-2}$	$\frac{1}{2} + \frac{1}{\pi}\arctan(x/\sqrt{n}) + \frac{x}{\pi\sqrt{n}}\sum_{j=0}^{n/2-3/2}\frac{(2j)!!}{(2j+1)!!}\left(\frac{n}{n+x^2}\right)^{j+1}$

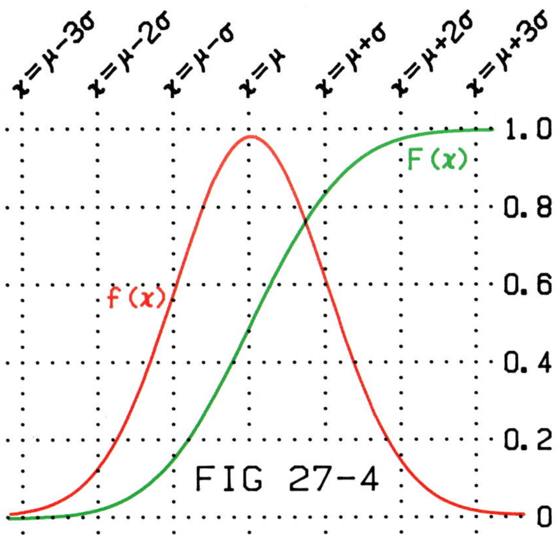
x=μ-3σ
x=μ-2σ
x=μ-σ
x=μ
x=μ+σ
x=μ+2σ
x=μ+3σ
1.0
0.8
0.6
0.4
0.2
0
F(x)
f(x)
FIG 27-4

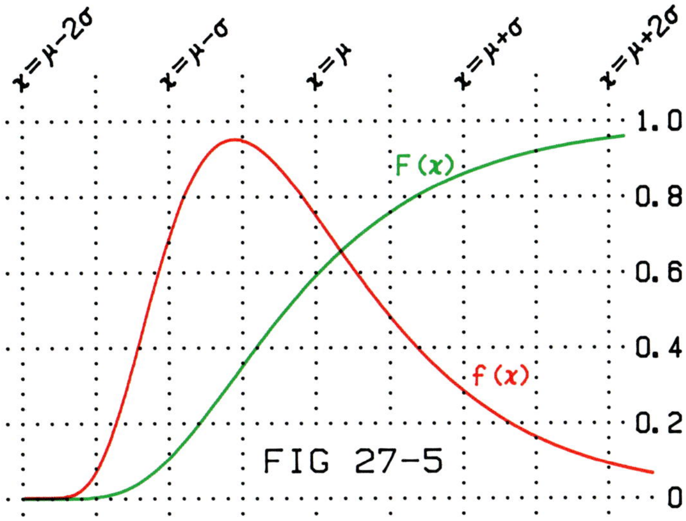
x=μ-2σ
x=μ-σ
x=μ
x=μ+σ
x=μ+2σ
1.0
0.8
0.6
0.4
0.2
0
F(x)
f(x)
FIG 27-5

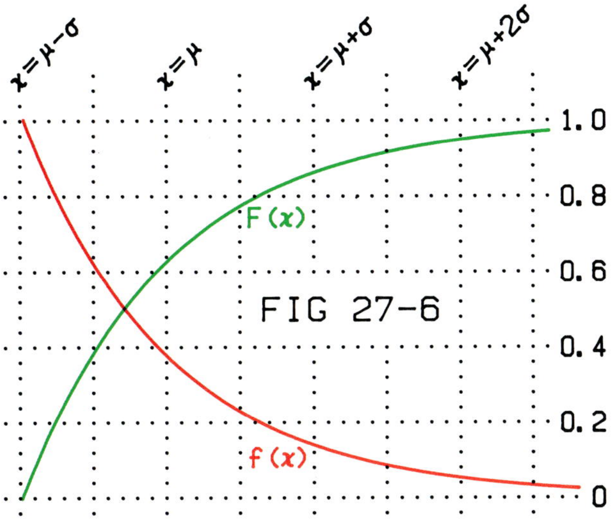
x=μ-σ
x=μ
x=μ+σ
x=μ+2σ
1.0
0.8
0.6
0.4
0.2
0
F(x)
f(x)
FIG 27-6

its name, the normal distribution is encountered in practice rather seldom. The asymmetrical *log-normal distribution*, in contrast, finds more frequent application. An example of this latter distribution in which $\mu = 2\sigma$ is illustrated in Figure 27-5. Even less symmetrical is the *Boltzmann distribution* depicted in Figure 27-6: such a distribution describes, for example, the variation of the numbers of molecules with height in the earth's atmosphere.

The shape of the *Weibull distribution* is a very strong function of α (which must be positive) and this distribution therefore finds widespread empirical application in failure analysis.

CHAPTER 28

THE HYPERBOLIC SINE sinh(x) AND COSINE cosh(x) FUNCTIONS

This chapter and the next two chapters address the six so-called *hyperbolic functions*. The present chapter deals with the two most important of the six: the hyperbolic sine and the hyperbolic cosine. These two functions are interrelated by

28:0:1 $$\cosh^2(x) - \sinh^2(x) = 1$$

and by each being the derivative of the other [see equations 28:10:1 and 28:10:2].

28:1 NOTATION

The names of these functions arise because of their complex algebraic relationship [see Section 32:11] to the sine and cosine functions. Their association with the hyperbola is explained in Section 28:3.

The notations sh(x) and ch(x) sometimes replace sinh(x) and cosh(x). Although they cause confusion, the symbolisms Sin(x) and Cos(x) are occasionally encountered.

28:2 BEHAVIOR

Both functions are defined for all arguments but, whereas the hyperbolic sine adopts all values, the hyperbolic cosine is restricted in range to $\cosh(x) \geq 1$. Figure 28-1 shows the behavior of the functions for rather small arguments. For arguments of large absolute magnitude, both functions tend exponentially towards infinite values.

28:3 DEFINITIONS

The hyperbolic sine and cosine functions are defined in terms of the exponential function of Chapter 26 by

28:3:1 $$\sinh(x) = \frac{\exp(x) - \exp(-x)}{2}$$

28:3:2 $$\cosh(x) = \frac{\exp(x) + \exp(-x)}{2}$$

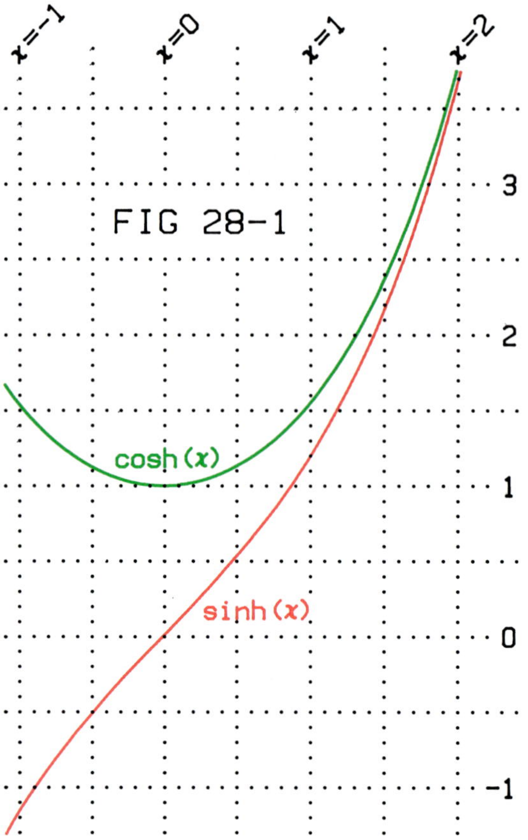

To provide a geometric definition of the hyperbolic functions, consider the positive branch of the rectangular hyperbola [Section 15:4] $\pm\sqrt{x^2 - 1}$ depicted in Figure 28-2. The green area is bounded by the hyperbola and by a pair of straight lines *OP* and *OP'* through the origin with slopes that are equal in magnitude but opposite in sign. Let *a* denote this shaded area: it can take values between zero (corresponding to points *P* and *P'* coinciding with *A*) and infinity (corresponding to lines *OP* and *OP'* having slopes of +1 and −1 and constituting the asymptotes of the hyperbola). The lengths *PQ* and *OQ* may then be regarded as functions of *a* and are, in fact, the hyperbolic sine and cosine of *a*

28:3:3 $$PQ = \sinh(a) = \sinh\text{ (green area)}$$

28:3:4 $$OQ = \cosh(a) = \cosh\text{ (green area)}$$

The second-order differential equation

28:3:5 $$\frac{d^2f}{dx^2} = b^2x$$

has the general solution $f = c_1 \sinh(bx) + c_2 \cosh(bx)$, where b, c_1 and c_2 are constants.

28:4 SPECIAL CASES

There are none.

28:5 INTRARELATIONSHIPS

The hyperbolic cosine function is even

28:5:1 $$\cosh(-x) = \cosh(x)$$

whereas the hyperbolic sine is an odd function

28:5:2 $$\sinh(-x) = -\sinh(x)$$

The duplication and triplication formulas

28:5:3 $$\cosh(2x) = \cosh^2(x) + \sinh^2(x) = 2\cosh^2(x) - 1 = 1 + 2\sinh^2(x)$$

28:5:4 $$\sinh(2x) = 2\sinh(x)\cosh(x) = 2\sinh(x)\sqrt{1 + \sinh^2(x)}$$

28:5:5 $$\cosh(3x) = 4\cosh^3(x) - 3\cosh(x)$$

28:5:6 $$\sinh(3x) = 4\sinh^3(x) + 3\sinh(x) = \sinh(x)[4\cosh^2(x) - 1]$$

generalize to

28:5:7 $$\cosh(nx) = \mathrm{T}_n(\cosh(x)) = \sum_{j=0}^{n} \mathrm{t}_n^{(j)} \cosh^j(x)$$

and

28:5:8 $$\sinh(nx) = \sinh(x)\mathrm{U}_{n-1}(\cosh(x)) = \sum_{j=0}^{n} \frac{[\mathrm{t}_n^{(j)}\cosh(x) - \mathrm{t}_{n-1}^{(j)}]\cosh^j(x)}{\sinh(x)}$$

where the T_n and U_n Chebyshev polynomials are discussed in Chapter 22, as are the Chebyshev coefficients $\mathrm{t}_n^{(j)}$. *De Moivre's theorem*

28:5:9 $$\cosh(nx) \pm \sinh(nx) = [\cosh(x) \pm \sinh(x)]^n = \exp(\pm nx)$$

is also useful.

Equations 28:5:3 and 28:5:4 may be regarded as special cases of the argument-addition formulas

28:5:10 $$\cosh(x \pm y) = \cosh(x)\cosh(y) \pm \sinh(x)\sinh(y)$$

28:5:11 $$\sinh(x \pm y) = \sinh(x)\cosh(y) \pm \cosh(x)\sinh(y)$$

From 28:5:3 one may derive the expressions

28:5:12 $$\cosh\left(\frac{x}{2}\right) = \sqrt{\frac{\cosh(x) + 1}{2}}$$

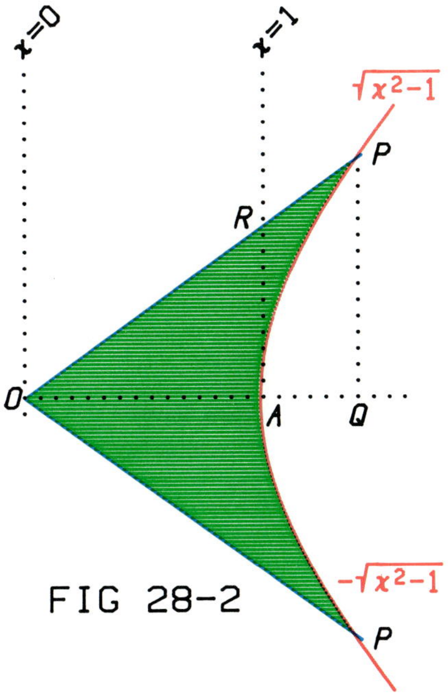

FIG 28-2

and

28:5:13 $$\sinh\left(\frac{x}{2}\right) = \mathrm{sgn}(x)\sqrt{\frac{\cosh(x) - 1}{2}}$$

for the hyperbolic functions of half argument, as well as the formulas

28:5:14 $$\cosh^2(x) = \frac{\cosh(2x) + 1}{2}$$

and

28:5:15 $$\sinh^2(x) = \frac{\cosh(2x) - 1}{2}$$

for the squares. These latter may be generalized to the expressions

28:5:16 $$\cosh^n(x) = \frac{1}{2^n}\sum_{j=0}^{n}\binom{n}{j}\cosh[(n-2j)x] = \begin{cases} \dfrac{1}{2^{n-1}}\displaystyle\sum_{j=0}^{(n-1)/2}\binom{n}{j}\cosh[(n-2j)x] & n = 1, 3, 5, \ldots \\ \dfrac{(n-1)!!}{n!!} + \dfrac{1}{2^{n-1}}\displaystyle\sum_{j=0}^{(n/2)-1}\binom{n}{j}\cosh[(n-2j)x] & n = 2, 4, 6, \ldots \end{cases}$$

and

28:5:17 $$\sinh^n(x) = \begin{cases} \dfrac{1}{2^n}\displaystyle\sum_{j=0}^{n}(-1)^j\binom{n}{j}\sinh[(n-2j)x] = \dfrac{1}{2^{n-1}}\sum_{j=0}^{(n-1)/2}(-1)^j\binom{n}{j}\sinh[(n-2j)x] & n = 1, 3, 5, \ldots \\ \dfrac{1}{2^n}\displaystyle\sum_{j=0}^{n}(-1)^j\binom{n}{j}\cosh[(n-2j)x] = \dfrac{(-1)^{n/2}(n-1)!!}{n!!} + \dfrac{1}{2^{n-1}}\sum_{j=0}^{(n/2)-1}(-1)^j\binom{n}{j}\cosh[(n-2j)x] & n = 2, 4, 6, \ldots \end{cases}$$

for any positive integer power of the hyperbolic cosine or sine.

The function-addition formulas

28:5:18 $$\cosh(x) \pm \sinh(x) = \exp(\pm x)$$

28:5:19 $$\cosh(x) + \cosh(y) = 2\cosh\left(\frac{x+y}{2}\right)\cosh\left(\frac{x-y}{2}\right)$$

28:5:20 $$\cosh(x) - \cosh(y) = 2\sinh\left(\frac{x+y}{2}\right)\sinh\left(\frac{x-y}{2}\right)$$

28:5:21 $$\sinh(x) \pm \sinh(y) = 2\sinh\left(\frac{x \pm y}{2}\right)\cosh\left(\frac{x \mp y}{2}\right)$$

and the function-multiplication formulas

28:5:22 $$\sinh(x)\sinh(y) = \frac{1}{2}\cosh(x+y) - \frac{1}{2}\cosh(x-y)$$

28:5:23 $$\sinh(x)\cosh(y) = \frac{1}{2}\sinh(x+y) + \frac{1}{2}\sinh(x-y)$$

28:5:24 $$\cosh(x)\cosh(y) = \frac{1}{2}\cosh(x+y) + \frac{1}{2}\cosh(x-y)$$

complete our listing of intrarelationships between these most maleable functions.

28:6 EXPANSIONS

The hyperbolic sine and cosine functions may be expanded as infinite series

28:6:1 $$\sinh(x) = x + \frac{x^3}{3!} + \frac{x^5}{5!} + \cdots = \sum_{j=0}^{\infty} \frac{x^{2j+1}}{(2j+1)!}$$

28:6:2 $$\cosh(x) = 1 + \frac{x^2}{2!} + \frac{x^4}{4!} + \cdots = \sum_{j=0}^{\infty} \frac{x^{2j}}{(2j)!}$$

or as infinite products

28:6:3 $$\sinh(x) = x\left(1 + \frac{x^2}{\pi^2}\right)\left(1 + \frac{x^2}{4\pi^2}\right) \cdots = x\prod_{j=1}^{\infty} 1 + \frac{x^2}{j^2\pi^2}$$

28:6:4 $$\cosh(x) = \left(1 + \frac{4x^2}{\pi^2}\right)\left(1 + \frac{4x^2}{9\pi^2}\right)\left(1 + \frac{4x^2}{25\pi^2}\right) \cdots = \prod_{j=1}^{\infty} 1 + \frac{x^2}{(j+\frac{1}{2})^2\pi^2}$$

28:7 PARTICULAR VALUES

	$x = -\infty$	$x = -1$	$x = 0$	$x = 1$	$x = \infty$
sinh(x)	$-\infty$	$\frac{1 - e^2}{2e}$	0	$\frac{e^2 - 1}{2e}$	∞
cosh(x)	∞	$\frac{1 + e^2}{2e}$	1	$\frac{e^2 + 1}{2e}$	∞

28:8 NUMERICAL VALUES

These are easily calculated via equations 28:3:1 and 28:3:2. There is an algorithm in Section 29:8 that enables any one of the six hyperbolic functions, including sinh(x) or cosh(x), to be evaluated. As well, the universal hypergeometric algorithm [Section 18:14] permits values of sinh(x)/x and of cosh(x) to be found.

28:9 APPROXIMATIONS

The hyperbolic sine and cosine may be approximated by polynomials; for example,

28:9:1 $$\sinh(x) \simeq x + \frac{x^3}{6} \qquad \text{8-bit precision} \qquad |x| < 0.84$$

28:9:2 $$\cosh(x) \simeq \left(1 + \frac{x^2}{4}\right)^2 \qquad \text{8-bit precision} \qquad |x| < 0.70$$

at small arguments and by exponential functions

28:9:3 $$\sinh(x) \simeq \operatorname{sgn}(x)\frac{\exp(|x|)}{2} \qquad \text{8-bit precision} \qquad |x| > 2.78$$

28:9:4 $$\cosh(x) \simeq \frac{\exp(|x|)}{2} \qquad \text{8-bit precision} \qquad |x| > 2.78$$

when the argument is large.

28:10 OPERATIONS OF THE CALCULUS

Differentiation and indefinite integration of sinh(bx) and cosh(bx) give

28:10:1
$$\frac{d}{dx}\sinh(bx) = b\cosh(bx)$$

28:10:2
$$\frac{d}{dx}\cosh(bx) = b\sinh(bx)$$

28:10:3
$$\int_0^x \sinh(bt)dt = \frac{\cosh(bx) - 1}{b}$$

28:10:4
$$\int_0^x \cosh(bt)dt = \frac{\sinh(bx)}{b}$$

The general formulas

28:10:5
$$\int_0^x \cosh^n(t)dt = \begin{cases} \dfrac{(n-1)!!\sinh(x)}{n!!}\displaystyle\sum_{j=0}^{(n-1)/2}\frac{(2j-1)!!}{(2j)!!}\cosh^{2j}(x) & n = 1, 3, 5, \ldots \\ \dfrac{(n-1)!!}{n!!}\left[x + \sinh(x)\displaystyle\sum_{j=0}^{(n/2)-1}\frac{(2j)!!}{(2j+1)!!}\cosh^{2j+1}(x)\right] & n = 2, 4, 6, \ldots \end{cases}$$

28:10:6
$$\int_0^x \sinh^n(t)dt = \begin{cases} \dfrac{(-1)^{(n-1)/2}(n-1)!!\cosh(x)}{n!!}\displaystyle\sum_{j=0}^{(n-1)/2}(-1)^j\frac{(2j-1)!!}{(2j)!!}\sinh^{2j}(x) & n = 1, 3, 5, \ldots \\ \dfrac{(-1)^{n/2}(n-1)!!}{n!!}\left[x + \cosh(x)\displaystyle\sum_{j=0}^{(n/2)-1}\frac{(2j)!!}{(2j+1)!!}(-1)^j\sinh^{2j+1}(x)\right] & n = 2, 4, 6, \ldots \end{cases}$$

permit the indefinite integration of integer powers of the hyperbolic sine and cosine functions. Alternative expressions may be derived by integration of equations 28:5:16 and 28:5:17. Noninteger powers are treated in Section 58:14. Other important classes of indefinite integral include

28:10:7
$$\int_0^x t^n\cosh(bt)dt = \begin{cases} \dfrac{n!\sinh(bx)}{b^{n+1}}\left[\dfrac{(bx)^n}{n!} + \dfrac{(bx)^{n-2}}{(n-2)!)} + \cdots + 1\right] - \dfrac{n!\cosh(bx)}{b^{n+1}}\left[\dfrac{(bx)^{n-1}}{(n-1)!} + \dfrac{(bx)^{n-3}}{(n-3)!} + \cdots + bx\right] & n = 2, 4, 6, \ldots \\ \dfrac{n!\sinh(bx)}{b^{n+1}}\left[\dfrac{(bx)^n}{n!} + \dfrac{(bx)^{n-2}}{(n-2)!} + \cdots + bx\right] - \dfrac{n!\cosh(bx)}{b^{n+1}}\left[\dfrac{(bx)^{n-1}}{(n-1)!} + \dfrac{(bx)^{n-3}}{(n-3)!} + \cdots + 1\right] + \dfrac{n!}{b^{n+1}} & n = 1, 3, 5, \ldots \end{cases}$$

28:10:8
$$\int_0^x t^n\sinh(bt)dt = \begin{cases} \dfrac{n!\cosh(bx)}{b^{n+1}}\left[\dfrac{(bx)^n}{n!} + \dfrac{(bx)^{n-2}}{(n-2)!} + \cdots + 1\right] - \dfrac{n!\cosh(bx)}{b^{n+1}}\left[\dfrac{(bx)^{n-1}}{(n-1)!} + \dfrac{(bx)^{n-3}}{(n-3)!} + \cdots + bx\right] - \dfrac{n!}{b^{n+1}} & n = 2, 4, 6, \ldots \\ \dfrac{n!\cosh(bx)}{b^{n+1}}\left[\dfrac{(bx)^n}{n!} + \dfrac{(bx)^{n-2}}{(n-2)!} + \cdots + bx\right] - \dfrac{n!\sinh(bx)}{b^{n+1}}\left[\dfrac{(bx)^{n-1}}{(n-1)!} + \dfrac{(bx)^{n-3}}{(n-3)!} + \cdots + 1\right] & n = 1, 3, 5, \ldots \end{cases}$$

The bracketed series in the above integrals may be expressed as $[e_n(bx) \pm e_n(-bx)]/2$, where the e_n function is discussed in Section 26:13. Similar indefinite integrals for $n = -1, -2, -3, \ldots$ are listed by Gradshteyn and Ryzhik [Section 2.475]; they involve the chi and shi functions defined in Chapter 38.

A large number of indefinite and definite integrals involving the hyperbolic sine and cosine functions exist. The reader is referred to Chapters 2.4 and 3.5 of Gradshteyn and Ryzhik.

28:11 COMPLEX ARGUMENT

When the argument x of sinh(x) or cosh(x) is replaced by $x + iy$, we have

28:11:1 $$\sinh(x + iy) = \sinh(x)\cos(y) + i\cosh(x)\sin(y)$$

28:11:2 $$\cosh(x + iy) = \cosh(x)\cos(y) + i\sinh(x)\sin(y)$$

For a purely imaginary argument

28:11:3 $$\sinh(iy) = i\sin(y)$$

28:11:4 $$\cosh(iy) = \cos(y)$$

28:12 GENERALIZATIONS

The Jacobian elliptic functions nc(x;p) and nd(x;p) may be regarded as generalizations of cosh(x), to which they reduce when $p = 1$. Likewise, sc(x;p) and sd(x;p) reduce to sinh(x) when $p = 1$ and therefore generalize the hyperbolic sine. See Chapter 63 for all these Jacobian elliptic functions.

28:13 COGNATE FUNCTIONS

The expressions

28:13:1 $$\sinh(x) = \mathrm{sgn}(x)\sqrt{1 - \cosh^2(x)} = \frac{\mathrm{sgn}(x)\sqrt{1 - \mathrm{sech}^2(x)}}{\mathrm{sech}(x)} = \frac{1}{\mathrm{csch}(x)} = \frac{\tanh(x)}{\sqrt{1 + \tanh^2(x)}} = \frac{\mathrm{sgn}(x)}{\sqrt{\coth^2(x) - 1}}$$

28:13:2 $$\cosh(x) = \sqrt{1 + \sinh^2(x)} = \frac{1}{\mathrm{sech}(x)} = \frac{\sqrt{1 + \mathrm{csch}^2(x)}}{|\mathrm{csch}(x)|} = \frac{1}{\sqrt{1 - \tanh^2(x)}} = \frac{|\coth(x)|}{\sqrt{\coth^2(x) - 1}}$$

relate the hyperbolic sine and cosine to the other hyperbolic functions [see Chapters 29 and 30].

The hyperbolic sine and cosine functions are closely related to those hyperbolic Bessel functions $I_\nu(x)$ in which ν is an odd multiple of $\pm\frac{1}{2}$. Examples are

28:13:3 $$I_{1/2}(x) = \sqrt{\frac{2}{\pi x}}\sinh(x)$$

28:13:4 $$I_{-1/2}(x) = \sqrt{\frac{2}{\pi x}}\cosh(x)$$

28:13:5 $$I_{3/2}(x) = \sqrt{\frac{2}{\pi x}}\left[\cosh(x) - \frac{\sinh(x)}{x}\right]$$

and others may be constructed by use of the recursion formula

28:13:6 $$I_{\nu+1}(x) + \frac{2\nu}{x}I_\nu(x) - I_{\nu-1}(x) = 0$$

These functions, some of which are graphed in Figure 28-3, share the properties of all hyperbolic Bessel functions as discussed in Chapter 50. The name *modified spherical Bessel function* and the symbol $i_n(x)$ is sometimes given to the function $\sqrt{\pi/2x}\, I_{n+1/2}(x)$.

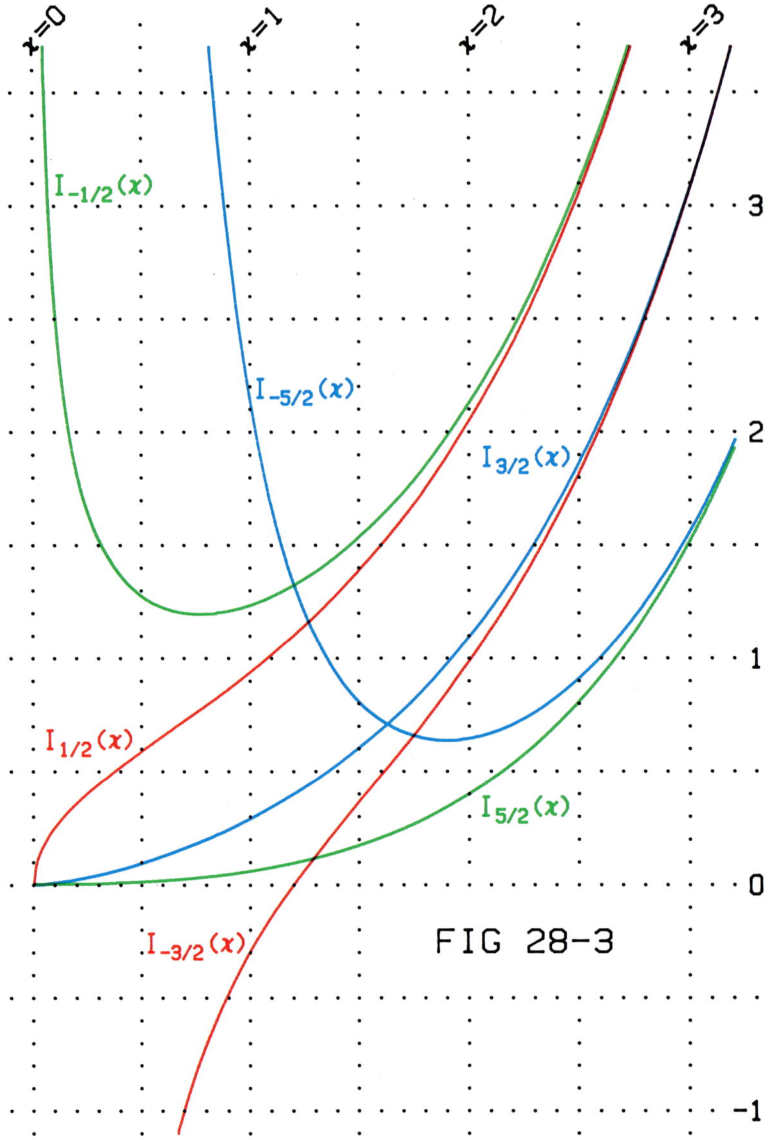

FIG 28-3

28:14 RELATED TOPICS

If a heavy rope or flexible chain of length $2L$ is freely suspended from two points, separated by a horizontal distance $2h$, but at the same level, then the rope adopts a characteristic shape known as a *catenary*. The equation of the catenary is

28:14:1 $$f(x) = \frac{\cosh(bx) - \cosh(bh)}{b}$$

and its shape is illustrated in Figure 28-4. The coefficient b is related to the lengths L and h by the implicit definition

28:14:2 $$bL = \sinh(bh)$$

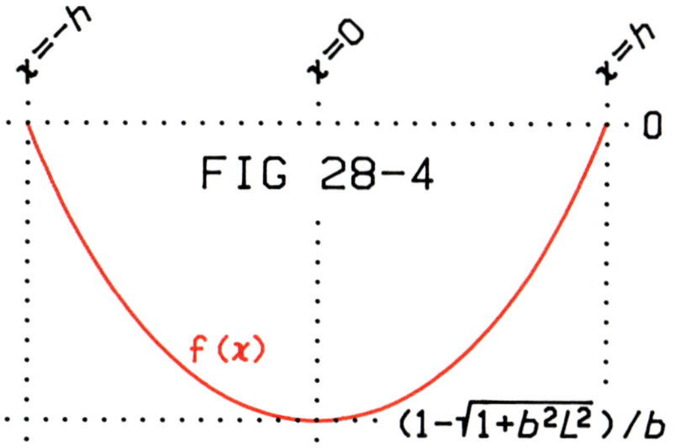

FIG 28-4

CHAPTER 29

THE HYPERBOLIC SECANT sech(x) AND COSECANT csch(x) FUNCTIONS

Of the six hyperbolic functions, the two treated in this chapter are perhaps the least frequently encountered. The property

29:0:1 $$(1 - \mathrm{sech}^2(x))(1 + \mathrm{csch}^2(x)) = 1$$

interrelates the two. Two features of this chapter—see Sections 29:3 and 29:8—deal with all six hyperbolic functions.

29:1 NOTATION

The notation cosech(x) is sometimes used for the hyperbolic cosecant. Some authors admit only four hyperbolic functions, using $1/\cosh(x)$ and $1/\sinh(x)$ to represent the secant and cosecant.

29:2 BEHAVIOR

Figure 29-1 shows the behavior of the two functions. Both approach zero as their arguments tend to $\pm\infty$. Both functions accept any argument but, whereas csch(x) adopts all values, the hyperbolic secant is restricted in range to $0 \le \mathrm{sech}(x) \le 1$.

29:3 DEFINITIONS

The relationships

29:3:1 $$\mathrm{sech}(x) = \frac{2}{\exp(x) + \exp(-x)} = \frac{1}{\cosh(x)}$$

and

29:3:2 $$\mathrm{csch}(x) = \frac{2}{\exp(x) - \exp(-x)} = \frac{1}{\sinh(x)}$$

are the usual definitions of the hyperbolic secant and cosecant.

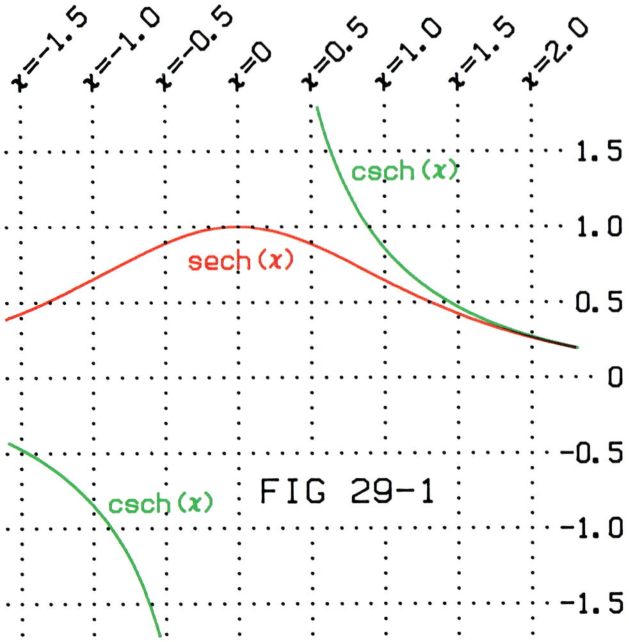

FIG 29-1

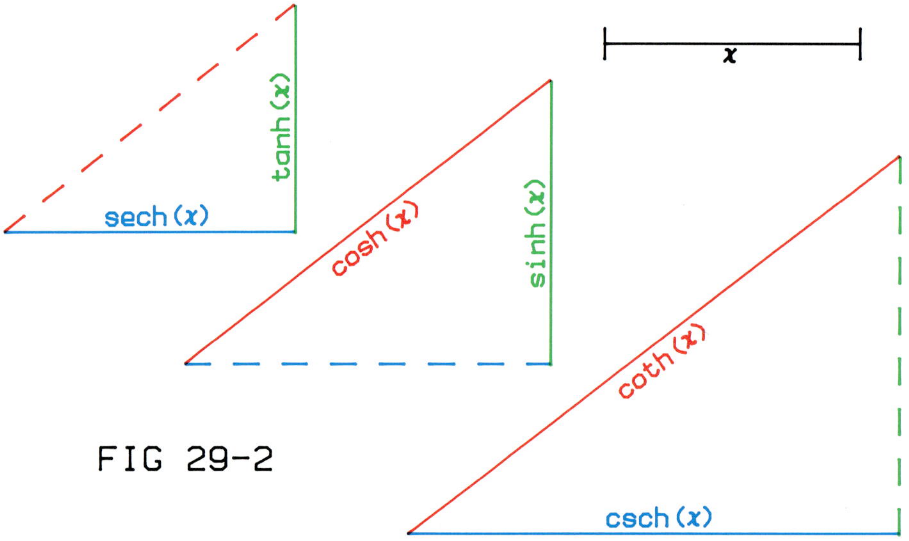

FIG 29-2

Figure 29-2 depicts three similar right-angled triangles. If the sides that are drawn as dashed lines are of unit length, then each of the other six sides equals one of the hyperbolic functions as labeled. Pythagorean and similarity properties thus enable the interrelationship between any two hyperbolic functions to be deduced and may serve as the definition of any one function in terms of any other. Note that the argument x is shown in the diagram, but it is not related to the triangles by any simple construction. It is, in fact, related to the angles in the triangles via the gudermannian function [see Section 33:14].

The differential equations

29:3:3
$$\frac{df}{dx} + f\sqrt{a^2 \pm f^2} = 0$$

are satisfied respectively by $f = a\ \mathrm{csch}(ax)$ and $f = a\ \mathrm{sech}(ax)$.

29:4 SPECIAL CASES

There are none.

29:5 INTRARELATIONSHIPS

The hyperbolic secant and cosecant obey the reflection formulas

29:5:1 $$\operatorname{sech}(-x) = \operatorname{sech}(x)$$

29:5:2 $$\operatorname{csch}(-x) = -\operatorname{csch}(x)$$

and the duplication formulas

29:5:3 $$\operatorname{sech}(2x) = \frac{\operatorname{sech}^2(x)}{2 - \operatorname{sech}^2(x)}$$

29:5:4 $$\operatorname{csch}(2x) = \frac{\operatorname{sech}(x)\operatorname{csch}(x)}{2}$$

Other relationships may be derived via the equations of Section 28:5, but these are generally more complicated and less useful than are the intrarelationships of the hyperbolic sine or cosine.

29:6 EXPANSIONS

The functions sech(x) and x csch(x) may be expanded as power series

29:6:1 $$\operatorname{sech}(x) = 1 - \frac{x^2}{2} + \frac{5x^4}{24} - \frac{61x^6}{720} + \cdots = -2\sum_{j=0}^{\infty} \beta(2j+1)\left(\frac{-4x^2}{\pi^2}\right)^j = \sum_{j=0}^{\infty} \frac{E_{2j}x^{2j}}{(2j)!} \qquad -\frac{\pi}{2} < x < \frac{\pi}{2}$$

and

29:6:2 $$\operatorname{csch}(x) = \frac{1}{x} - \frac{x}{6} + \frac{7x^3}{360} - \frac{31x^5}{15120} + \cdots = \frac{1}{x} + \frac{2}{x}\sum_{j=1}^{\infty} \eta(2j)\left(\frac{-x^2}{\pi^2}\right)^j = \sum_{j=0}^{\infty} \frac{(2-4^j)}{(2j)!} B_{2j}x^{2j-1} \qquad -\pi < x < \pi$$

The β and η coefficients are defined in Chapter 3 and the Euler E and Bernoulli B numbers in Chapters 5 and 4, respectively.

Expansions as exponentials take the forms

29:6:3 $$\operatorname{sech}(x) = 2\exp(-|x|) - 2\exp(-3|x|) + 2\exp(-5|x|) - \cdots = 2\sum_{j=0}^{\infty} (-1)^j \exp[-(2j+1)|x|] \qquad x \neq 0$$

and

29:6:4 $$\operatorname{csch}(x) = 2\operatorname{sgn}(x)[\exp(-|x|) + \exp(-3|x|) + \exp(-5|x|) + \cdots] = 2\operatorname{sgn}(x)\sum_{j=1}^{\infty} \exp[(1-2j)|x|] \qquad x \neq 0$$

As well, the hyperbolic secant and cosecant can be expanded as partial fractions

29:6:5 $$\operatorname{sech}(x) = \frac{4\pi}{\pi^2 + 4x^2} - \frac{12\pi}{9\pi^2 + 4x^2} + \frac{20\pi}{25\pi^2 + 4x^2} - \cdots = \sum_{j=0}^{\infty} (-1)^j \frac{(2j+1)\pi}{(j+\frac{1}{2})^2\pi^2 + x^2}$$

29:6:6 $$\operatorname{csch}(x) = \frac{1}{x} - \frac{2x}{\pi^2 + x^2} + \frac{2x}{4\pi^2 + x^2} - \frac{2x}{9\pi^2 + x^2} + \cdots = \sum_{j=-\infty}^{\infty} \frac{(-1)^j x}{j^2\pi^2 + x^2}$$

29:7 PARTICULAR VALUES

	$x = -\infty$	$x = -1$	$x = 0$	$x = 1$	$x = \infty$
csch(x)	0	$\dfrac{-2e}{e^2 - 1}$	$\mp\infty$	$\dfrac{2e}{e^2 - 1}$	0
sech(x)	0	$\dfrac{2e}{e^2 + 1}$	1	$\dfrac{2e}{e^2 + 1}$	0

29:8 NUMERICAL VALUES

Calculation via definitions 29:3:1 and 29:3:2 is usually a simple matter.

If many values of the hyperbolic functions are needed, the following algorithm may be useful. It generates exact values of any one of the six hyperbolic functions according to the value (0, 1, 2, 3, 4 or 5) of a code that is input in addition to the argument x.

Input x >>>>>>>
Set $f = \exp(x)$
Set $g = 1/f$
Replace f by $(f - g)/2$
Input code c >>>>
If $c \leq 1$ go to (2)
Replace g by $f + g$
If $c > 3$ go to (1)
Replace f by g
Go to (2)
(1) Replace f by f/g
(2) If frac$(c/2) = 0$ go to (3)
Replace f by $1/f$
(3) Output f
<<<<

$f_0 = \sinh(x)$, $f_1 = \text{csch}(x)$, $f_2 = \cosh(x)$, $f_3 = \text{sech}(x)$, $f_4 = \tanh(x)$, $f_5 = \coth(x)$

Storage needed: x, f, g and c

Input restriction: $x \neq 0$

code, c	function
0	sinh
1	csch
2	cosh
3	sech
4	tanh
5	coth

Test value: $x = \pi$
$f_0 = 11.54873936$
$f_2 = 11.59195328$
$f_5 = 1.003741873$

29:9 APPROXIMATIONS

For large values of x, one can approximate

29:9:1 $\quad \text{sech}(x) \simeq 2\exp(-|x|) \quad$ 8-bit precision $\quad |x| \geq 2.8$

29:9:2 $\quad \text{csch}(x) \simeq 2\,\text{sgn}(x)\exp(-|x|) \quad$ 8-bit precision $\quad |x| \geq 2.8$

29:10 OPERATIONS OF THE CALCULUS

Differentiation of the hyperbolic secant or cosecant gives

29:10:1 $$\frac{d}{dx}\text{sech}(x) = -\text{sech}(x)\tanh(x) = -\text{sech}(x)\sqrt{1 - \text{sech}^2(x)}$$

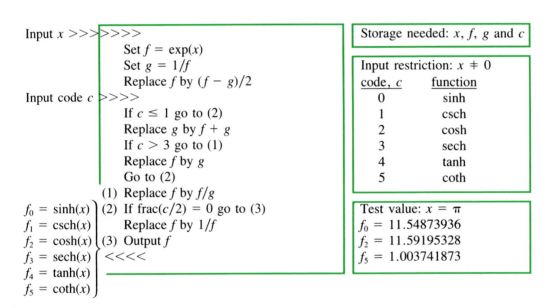

29:10:2 $$\frac{d}{dx}\operatorname{csch}(x) = -\operatorname{csch}(x)\coth(x) = -\operatorname{csch}(x)\sqrt{1+\operatorname{csch}^2(x)}$$

The indefinite integration of these functions yields rather complicated results:

29:10:3 $$\int_0^x \operatorname{sech}(t)dt = \arctan(\sinh(x)) = \operatorname{gd}(x)$$

where gd is the gudermannian function discussed in Section 33:14, and

29:10:4 $$\int_x^\infty \operatorname{csch}(t)dt = \ln\left(\coth\left(\frac{x}{2}\right)\right) \qquad x > 0$$

but their squares integrate more simply

29:10:5 $$\int_0^x \operatorname{sech}^2(t)dt = \tanh(x)$$

29:10:6 $$\int_x^\infty \operatorname{csch}^2(t)dt = -\coth(x) \qquad x > 0$$

Indefinite integration of the square roots of the hyperbolic secant and cosecant functions generates special cases of the incomplete elliptic integral [see Chapter 62] in which the parameter equals $1/\sqrt{2}$.

29:10:7 $$\int_0^x \sqrt{\operatorname{sech}(t)}\, dt = \sqrt{2}\, F\left(\frac{1}{\sqrt{2}}; \theta\right) \qquad \sin(\theta) = \sqrt{1-\operatorname{sech}(x)}$$

29:10:8 $$\int_x^\infty \sqrt{\operatorname{csch}(t)}\, dt = F\left(\frac{1}{\sqrt{2}}; \phi\right) \qquad \cos(\phi) = \frac{\operatorname{csch}(x)-1}{\operatorname{csch}(x)+1}$$

For a generalization to arbitrary power, see Section 58:14.

Useful definite integrals include

29:10:9 $$\int_0^\infty t^n \operatorname{sech}(t)dt = 2n!\,\beta(n+1) \qquad n = 0, 1, 2, \ldots$$

29:10:10 $$\int_0^\infty t^n \operatorname{csch}(t)dt = 2n!\,\lambda(n+1) \qquad n = 1, 2, 3, \ldots$$

where Chapter 3 describes the beta and lambda numbers.

29:11 COMPLEX ARGUMENT

Equations 28:11:1 and 28:11:2 may be employed to evaluate $\operatorname{csch}(x+iy)$ and $\operatorname{sech}(x+iy)$, respectively. For purely imaginary argument

29:11:1 $$\operatorname{sech}(iy) = \sec(y)$$

29:11:2 $$\operatorname{csch}(iy) = -i\csc(y)$$

where sec and csc are the functions to which Chapter 33 is devoted.

29:12 GENERALIZATIONS

The Jacobian elliptic functions $\operatorname{cn}(x;p)$ and $\operatorname{dn}(x;p)$, discussed in Chapter 63, are generalizations of $\operatorname{sech}(x)$, while $\operatorname{cs}(x;p)$ and $\operatorname{ds}(x;p)$ similarly generalize $\operatorname{csch}(x)$.

29:13 COGNATE FUNCTIONS

The expressions

29:13:1 $$\mathrm{sech}(x) = \frac{1}{\sqrt{1+\sinh^2(x)}} = \frac{1}{\cosh(x)} = \frac{|\mathrm{csch}(x)|}{\sqrt{1+\mathrm{csch}^2(x)}} = \sqrt{1-\tanh^2(x)} = \frac{\sqrt{\coth^2(x)-1}}{|\coth(x)|}$$

29:13:2 $$\mathrm{csch}(x) = \frac{1}{\sinh(x)} = \frac{\mathrm{sgn}(x)}{\sqrt{\cosh^2(x)-1}} = \frac{\mathrm{sgn}(x)\,\mathrm{sech}(x)}{\sqrt{1-\mathrm{sech}^2(x)}} = \frac{\sqrt{1-\tanh^2(x)}}{\tanh(x)} = \mathrm{sgn}(x)\sqrt{\coth^2(x)-1}$$

relate the hyperbolic secant and cosecant to the other hyperbolic functions [Chapters 28 and 30].

CHAPTER 30

THE HYPERBOLIC TANGENT tanh(x) AND COTANGENT coth(x) FUNCTIONS

The two functions of this chapter are the reciprocals of each other

30:0:1 $$\tanh(x)\coth(x) = 1$$

and are closely related to the other hyperbolic functions [Chapters 28 and 29].

30:1 NOTATION

The symbolism th(x) sometimes replaces tanh(x); cth(x) or ctnh(x) sometimes replaces coth(x). The misleading notations Tan(x) and Cot(x) are occasionally encountered.

30:2 BEHAVIOR

These functions are defined for all values of the argument x, but both are restricted in range. The hyperbolic tangent takes values only in the range -1 to $+1$, whereas coth(x) assumes all values except those between -1 and $+1$.

As shown in Figure 30-1, both functions lie exclusively in the first and third quadrants [see Section 0:2], and both approach $+1$ as $x \to \infty$ and -1 as $x \to -\infty$. The diagram also shows the $\coth(x) - 1/x$ function, which also approaches ± 1 at its limits, and which is cited in Section 30:13.

30:3 DEFINITIONS

The hyperbolic tangent and cotangent may be defined in terms of the exponential function

30:3:1 $$\tanh(x) = \frac{\exp(2x) - 1}{\exp(2x) + 1} = \frac{1 - \exp(-2x)}{1 + \exp(-2x)} = \frac{\exp(x) - \exp(-x)}{\exp(x) + \exp(-x)}$$

30:3:2 $$\coth(x) = \frac{\exp(2x) + 1}{\exp(2x) - 1} = \frac{1 + \exp(-2x)}{1 - \exp(-2x)} = \frac{\exp(x) + \exp(-x)}{\exp(x) - \exp(-x)}$$

or in terms of the other hyperbolic functions

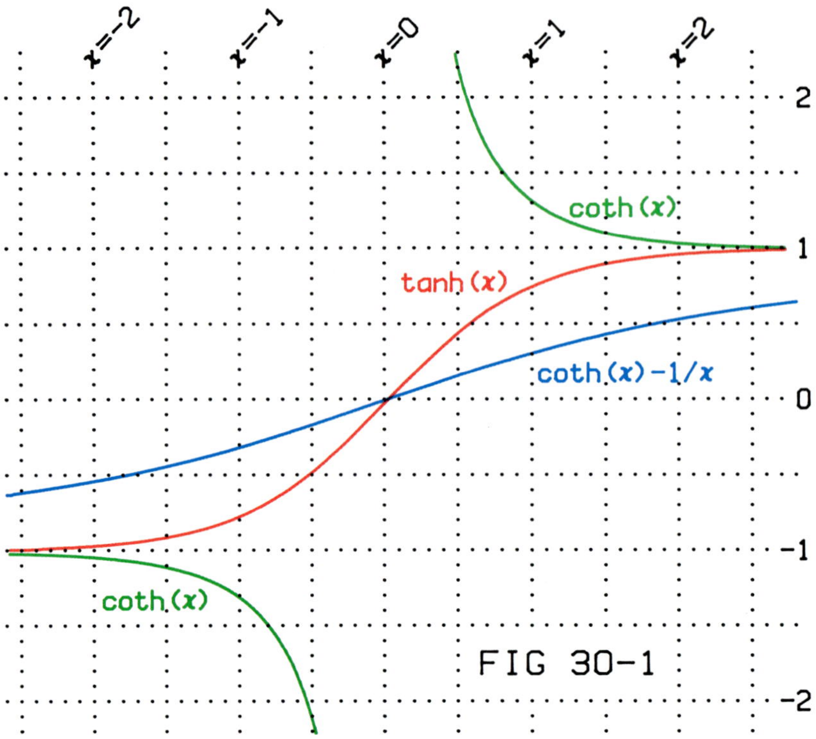

FIG 30-1

30:3:3 $$\tanh(x) = \frac{\sinh(x)}{\cosh(x)} = \frac{\operatorname{sech}(x)}{\operatorname{csch}(x)} = \frac{1}{\coth(x)}$$

30:3:4 $$\coth(x) = \frac{\cosh(x)}{\sinh(x)} = \frac{\operatorname{csch}(x)}{\operatorname{sech}(x)} = \frac{1}{\tanh(x)}$$

Figure 28-2 may be used to provide a geometric definition of the hyperbolic tangent, namely

30:3:5 $$RA = \tanh(a) = \tanh\text{ (shaded area)}$$

The hyperbolic tangent and cotangent each satisfy the differential equation

30:3:6 $$\frac{df}{dx} = 1 - f^2$$

30:4 SPECIAL CASES

There are none.

30:5 INTRARELATIONSHIPS

Both functions are odd:

30:5:1 $$f(-x) = -f(x) \qquad f = \tanh \text{ or } \coth$$

The duplication formulas

30:5:2 $$\tanh(2x) = \frac{2\tanh(x)}{1 + \tanh^2(x)} = \frac{2}{\tanh(x) + \coth(x)}$$

and

30:5:3 $$\coth(2x) = \frac{\coth^2(x) + 1}{2\coth(x)} = \frac{\coth(x) + \tanh(x)}{2}$$

are special cases of the argument-addition expressions

30:5:4 $$\tanh(x \pm y) = \frac{\tanh(x) \pm \tanh(y)}{1 \pm \tanh(x)\tanh(y)}$$

and

30:5:5 $$\coth(x \pm y) = \frac{1 \pm \coth(x)\coth(y)}{\coth(x) \pm \coth(y)}$$

The equations in this paragraph may be used to build formulas for tanh(3x), coth(4x), etc.

The half-argument formulas

30:5:6 $$\tanh\left(\frac{x}{2}\right) = \frac{1 - \sqrt{1 - \tanh^2(x)}}{\tanh(x)} = \coth(x) - \operatorname{sgn}(x)\sqrt{\coth^2(x) - 1} = \coth(x) - \operatorname{csch}(x)$$

and

30:5:7 $$\coth\left(\frac{x}{2}\right) = \coth(x) + \operatorname{sgn}(x)\sqrt{\coth^2(x) - 1} = \frac{1 + \sqrt{1 - \tanh^2(x)}}{\tanh(x)} = \coth(x) + \operatorname{csch}(x)$$

are useful.

The sums and differences of the functions of this chapter may be expressed in terms of hyperbolic sines and cosines:

30:5:8 $$\tanh(x) \pm \tanh(y) = \frac{\sinh(x \pm y)}{\cosh(x)\cosh(y)}$$

30:5:9 $$\coth(x) \pm \tanh(y) = \frac{\cosh(x \pm y)}{\sinh(x)\cosh(y)}$$

30:5:10 $$\coth(x) \pm \coth(y) = \frac{\sinh(x \pm y)}{\sinh(x)\sinh(y)}$$

30:6 EXPANSIONS

The functions $\tanh(x)/x$ and $x\coth(x)$ may be expanded as power series in x^2:

30:6:1 $$\tanh(x) = x - \frac{x^3}{3} + \frac{2x^5}{15} - \frac{17x^7}{315} + \cdots = \frac{-2}{x}\sum_{j=1}^{\infty}\lambda(2j)\left(\frac{-4x^2}{\pi^2}\right)^j = \frac{1}{x}\sum_{j=1}^{\infty}\frac{(4^j - 1)B_{2j}(4x^2)^j}{(2j)!} \qquad -\frac{\pi}{2} < x < \frac{\pi}{2}$$

30:6:2 $$\coth(x) = \frac{1}{x} + \frac{x}{3} - \frac{x^3}{45} + \frac{2x^5}{945} - \cdots = \frac{1}{x} - \frac{2}{x}\sum_{j=1}^{\infty}\zeta(2j)\left(\frac{-x^2}{\pi^2}\right)^j = \frac{1}{x}\sum_{j=0}^{\infty}\frac{B_{2j}(4x^2)^j}{(2j)!} \qquad -\pi < x < \pi$$

where the lambda and zeta functions are discussed in Chapter 3 and the Bernoulli numbers in Chapter 4.

The hyperbolic tangent and cotangent may alternatively be expanded as series of exponentials. The tanh(x) expression

30:6:3 $$\tanh(x) = \operatorname{sgn}(x)[1 - 2\exp(-2|x|) + 2\exp(-4|x|) - 2\exp(-6|x|) + \cdots] = \operatorname{sgn}(x)\sum_{j=-\infty}^{\infty}(-1)^j\exp(-2j|x|)$$

has alternating signs; the expression for coth(x) is similar but has uniformly positive signs:

30:6:4 $$\coth(x) = \operatorname{sgn}(x)[1 + 2\exp(-2|x|) + 2\exp(-4|x|) + 2\exp(-6|x|) + \cdots]$$
$$= \operatorname{sgn}(x)\sum_{j=-\infty}^{\infty} \exp(-2j|x|) \qquad x \neq 0$$

Partial fraction expansions exist for both functions

30:6:5 $$\tanh(x) = \frac{8x}{\pi^2 + 4x^2} + \frac{8x}{9\pi^2 + 4x^2} + \frac{8x}{25\pi^2 + 4x^2} + \cdots = \sum_{j=0}^{\infty} \frac{8x}{(2j+1)^2\pi^2 + 4x^2}$$

30:6:6 $$\coth(x) = \frac{1}{x} + \frac{2x}{\pi^2 + x^2} + \frac{2x}{4\pi^2 + x^2} + \frac{2x}{9\pi^2 + x^2} + \cdots = \sum_{j=-\infty}^{\infty} \frac{x}{j^2\pi^2 + x^2}$$

and the continued fraction expansion

30:6:7 $$\tanh(x) = \frac{x}{1+}\,\frac{x^2}{3+}\,\frac{x^2}{5+}\,\frac{x^2}{7+}\cdots$$

holds for the hyperbolic tangent.

30:7 PARTICULAR VALUES

	$x = -\infty$	$x = -1$	$x = -\frac{1}{2}$	$x = 0$	$x = \frac{1}{2}$	$x = 1$	$x = \infty$
tanh(x)	-1	$\frac{1-e^2}{1+e^2}$	$\frac{1-e}{1+e}$	0	$\frac{e-1}{e+1}$	$\frac{e^2-1}{e^2+1}$	1
coth(x)	-1	$\frac{1+e^2}{1-e^2}$	$\frac{1+e}{1-e}$	$\mp\infty$	$\frac{e+1}{e-1}$	$\frac{e^2+1}{e^2-1}$	1

30:8 NUMERICAL VALUES

The hyperbolic tangent and cotangent functions may be evaluated via formulas 30:3:1 and 30:3:2 or by the algorithm in Section 29:8.

30:9 APPROXIMATIONS

For small arguments, the approximations

30:9:1 $$\tanh(x) \simeq x\left(1 - \frac{x^2}{3}\right) \qquad \text{8-bit precision} \qquad |x| \le 0.41$$

30:9:2 $$\coth(x) \simeq \frac{1}{x}\left(1 + \frac{x^2}{3}\right) \qquad \text{8-bit precision} \qquad |x| \le 0.65$$

are valid. For large arguments, one may use

30:9:3 $$\tanh(x) \simeq \operatorname{sgn}(x)[1 - 2\exp(-2|x|)] \qquad \text{8-bit precision} \qquad |x| \ge 1.6$$

30:9:4 $$\coth(x) \simeq \operatorname{sgn}(x)[1 + 2\exp(-2|x|)] \qquad \text{8-bit precision} \qquad |x| \ge 1.6$$

30:10 OPERATIONS OF THE CALCULUS

Differentiation and indefinite integration give

30:10:1 $$\frac{d}{dx}\tanh(bx) = b\,\mathrm{sech}^2(bx) = b[1 - \tanh^2(bx)]$$

30:10:2 $$\frac{d}{dx}\coth(bx) = -b\,\mathrm{csch}^2(bx) = b[1 - \coth^2(bx)]$$

30:10:3 $$\int_0^x \tanh(bt)dt = \frac{1}{b}\ln(\cosh(bx))$$

30:10:4 $$\int_{x_0}^x \coth(t)dt = \ln(\sinh(x)) \qquad x > 0 \qquad x_0 = \ln(1 + \sqrt{2}) = 0.88137\ldots$$

30:10:5 $$\int_0^x \left[\coth(t) - \frac{1}{t}\right]dt = \ln\left(\frac{\sinh(x)}{x}\right)$$

30:10:6 $$\int_x^\infty [\coth(t) - 1]dt = \ln[1 - \exp(-2x)] \qquad x > 0$$

30:10:7 $$\int_x^\infty [1 - \tanh(t)]dt = \ln[1 + \exp(-2x)]$$

30:10:8 $$\int_0^x \tanh^2(bt)dt = x - \frac{\tanh(bx)}{b}$$

30:10:9 $$\int_{x_0}^x \coth^2(t)dt = x - \coth(x) \qquad x_0 = 1.19967864$$

30:10:10 $$\int_0^x \tanh^n(t)dt = \begin{Bmatrix} \ln(\cosh(x)) - \sigma & n = 1, 3, 5, \ldots \\ x - \sigma & n = 2, 4, 6, \ldots \end{Bmatrix} \qquad \sigma = \sum_{j=1}^{\mathrm{Int}(n/2)} \frac{\tanh^{n-2j+1}(x)}{n - 2j + 1}$$

The indefinite integrals of $\tanh^\lambda(x)$ and $\coth^\lambda(x)$, where λ is an arbitrary power, are discussed in Section 58:14.

The semiderivative, with lower limit $-\infty$, of $[1 + \tanh(x/2)]/2$ is an important function, known as the *Randles-Sevcik function* in electrochemistry

30:10:11 $$\mathrm{rsf}(x) = \frac{d^{1/2}}{[d(x + \infty)]^{1/2}} \frac{1 + \tanh\left(\frac{x}{2}\right)}{2} = -\sum_{j=1}^\infty (-1)^j \sqrt{j}\exp(jx) = \sqrt{\frac{\pi}{2}}\sum_{j=1}^\infty \frac{\sqrt{X - x}(X + 2x)}{X^3}$$

where $X = \sqrt{(2j - 1)^2\pi^2 + x^2}$. It is related to Lerch's function [Section 64:12] by the identity $\mathrm{rsf}\{\ln(x)\} = x\Phi(-x; -1/2; 1)$.

30:11 COMPLEX ARGUMENT

When the argument of the hyperbolic tangent and cotangent functions becomes $x + iy$, we have

30:11:1 $$\tanh(x + iy) = \frac{\sinh(2x) + i\sin(2y)}{\cosh(2x) + \cos(2y)}$$

30:11:2 $$\coth(x + iy) = \frac{\sinh(2x) - i\sin(2y)}{\cosh(2x) - \cos(2y)}$$

For purely imaginary argument

30:11:3 $$\tanh(iy) = i\tan(y)$$

30:11:4 $$\coth(iy) = -i\cot(y)$$

30:12 GENERALIZATIONS

The Jacobian elliptic functions sn($x;p$) and ns($x;p$) [see Chapter 63] may be regarded as generalizations of tanh(x) and coth(x), respectively. As $p \to 1$, sn($x;p$) $\to$ tanh(x) and ns($x;p$) $\to$ coth(x).

30:13 COGNATE FUNCTIONS

The expressions

30:13:1 $$\tanh(x) = \frac{\sinh(x)}{\sqrt{1+\sinh^2(x)}} = \frac{\sqrt{\cosh^2(x)-1}}{\mathrm{sgn}(x)\cosh(x)} = \mathrm{sgn}(x)\sqrt{1-\mathrm{sech}^2(x)} = \frac{\mathrm{sgn}(x)}{\sqrt{1+\mathrm{csch}^2(x)}} = \frac{1}{\coth(x)}$$

and

30:13:2 $$\coth(x) = \frac{\sqrt{1+\sinh^2(x)}}{\sinh(x)} = \frac{\mathrm{sgn}(x)\cosh(x)}{\sqrt{\cosh^2(x)-1}} = \frac{\mathrm{sgn}(x)}{\sqrt{1-\mathrm{sech}^2(x)}} = \mathrm{sgn}(x)\sqrt{1+\mathrm{csch}^2(x)} = \frac{1}{\tanh(x)}$$

relate the tangent and cotangent to other members of the hyperbolic family. Figure 29-2 is useful in expressing these relationships.

The function coth(x) $-$ ($1/x$), which occurs in the theory of dielectrics, is known as the *Langevin function*. It is mapped in Figure 30-1 and its integral is given in 30:10:5. The Langevin function can be expanded via 30:6:6 and its reciprocal as

30:13:3 $$\frac{1}{\coth(x)-(1/x)} = \frac{3}{x} + 2x\sum_{j=1}^{\infty}\frac{1}{x^2+r_j^2(1)}$$

where $r_j(1)$ denotes the jth positive root of the equation $\tan(y) = y$ [see Section 34:7].

CHAPTER 31

THE INVERSE HYPERBOLIC FUNCTIONS

The six functions of this chapter are interrelated by the following permutations of argument:

31:0:1 $$\operatorname{arsinh}(x) = \operatorname{sgn}(x)\operatorname{arcosh}(\sqrt{1+x^2}) = \operatorname{sgn}(x)\operatorname{arsech}\left(\frac{1}{\sqrt{1+x^2}}\right) = \operatorname{arcsch}\left(\frac{1}{x}\right) = \operatorname{artanh}\left(\frac{x}{\sqrt{1+x^2}}\right) = \operatorname{arcoth}\left(\frac{\sqrt{1+x^2}}{x}\right)$$

31:0:2 $$\operatorname{arcosh}(x) = \operatorname{arsinh}(\sqrt{x^2-1}) = \operatorname{arsech}\left(\frac{1}{x}\right) = \operatorname{arcsch}\left(\frac{1}{\sqrt{x^2-1}}\right) = \operatorname{artanh}\left(\frac{\sqrt{x^2-1}}{x}\right) = \operatorname{arcoth}\left(\frac{x}{\sqrt{x^2-1}}\right) \qquad x \ge 1$$

31:0:3 $$\operatorname{arsech}(x) = \operatorname{arsinh}\left(\frac{\sqrt{1-x^2}}{x}\right) = \operatorname{arcosh}\left(\frac{1}{x}\right) = \operatorname{arcsch}\left(\frac{x}{\sqrt{1-x^2}}\right) = \operatorname{artanh}(\sqrt{1-x^2}) = \operatorname{arcoth}\left(\frac{1}{\sqrt{1-x^2}}\right) \qquad 0 \le x \le 1$$

31:0:4 $$\operatorname{arcsch}(x) = \operatorname{arsinh}\left(\frac{1}{x}\right) = \operatorname{sgn}(x)\operatorname{arcosh}(\sqrt{1+1/x^2}) = \operatorname{sgn}(x)\operatorname{arsech}(\sqrt{x^2/(1+x^2)}) = \operatorname{sgn}(x)\operatorname{artanh}\left(\frac{1}{\sqrt{1+x^2}}\right) = \operatorname{sgn}(x)\operatorname{arcoth}(\sqrt{1+x^2}) \qquad x \ne 0$$

31:0:5 $$\operatorname{artanh}(x) = \operatorname{arsinh}\left(\frac{x}{\sqrt{1-x^2}}\right) = \operatorname{sgn}(x)\operatorname{arcosh}\left(\frac{1}{\sqrt{1-x^2}}\right) = \operatorname{sgn}(x)\operatorname{arsech}(\sqrt{1-x^2}) = \operatorname{arcsch}\left(\frac{\sqrt{1-x^2}}{x}\right) = \operatorname{arcoth}\left(\frac{1}{x}\right) \qquad -1 \le x \le 1$$

31:0:6 $$\operatorname{arcoth}(x) = \operatorname{sgn}(x)\operatorname{arsinh}\left(\frac{1}{\sqrt{x^2-1}}\right) = \operatorname{sgn}(x)\operatorname{arcosh}(\sqrt{x^2/(x^2-1)}) = \operatorname{sgn}(x)\operatorname{arsech}(\sqrt{1-1/x^2})$$
$$= \operatorname{sgn}(x)\operatorname{arcsch}(\sqrt{x^2-1}) = \operatorname{artanh}\left(\frac{1}{x}\right) \qquad |x| > 1$$

31:1 NOTATION

The prefix "ar" means "area," and its pertinence can be appreciated by reference to Figure 28-2. The inverse hyperbolic sine of x, for example, denotes that area which, in the Figure 28-2 construct, has a hyperbolic sine of a. That is:

31:1:1 $$\operatorname{arsinh}(x) = a = \text{shaded area} \qquad \text{if } x = \sinh(a) = PQ$$

The symbolisms arcsinh(x), argsinh(x), arsh(x) and $\sinh^{-1}(x)$ are all used in place of arsinh(x). Corresponding variants are encountered for the other inverse hyperbolic functions. The same symbolisms with a capitalized initial letter—Arcsinh(x), Arsh(x), $\mathrm{Sinh}^{-1}(x)$, etc.—are sometimes used synonymously with arsinh(x), but more often they denote the multiple-valued functions that we discuss in Section 31:12.

31:2 BEHAVIOR

The behaviors of the six inverse hyperbolic functions are shown in Figure 31-1. Notice that all functions exist in the first and third quadrants [see Section 0:2] only.

The inverse hyperbolic sine has a behavior that is simpler than the other five. It accepts all arguments and adopts the sign of its argument.

The inverse hyperbolic cosine is normally defined only for arguments $x \geq 1$ although some authors extend its domain of definition to $|x| \geq 1$ via $\operatorname{arcosh}(-x) = \operatorname{arcosh}(x)$. It adopts only positive values.

The inverse hyperbolic secant is likewise normally defined only for $0 \leq x \leq 1$ although its definition may be extended to $0 \leq |x| \leq 1$ by means of $\operatorname{arsech}(-x) = \operatorname{arsech}(x)$. The values of arsech($x$) are always positive.

The inverse hyperbolic cosecant has two branches, as mapped in Figure 31-1. It adopts the sign of its argument and exhibits a discontinuity at $x = 0$.

The inverse hyperbolic tangent is defined only for $-1 \leq x \leq 1$ and approaches $\pm\infty$ as $x \to \pm 1$.

The inverse hyperbolic cotangent has two branches. For $1 \leq x \leq \infty$, arcoth(x) is positive; for $-\infty \leq x \leq -1$, arcoth(x) is negative. The function is not defined in the $-1 < x < 1$ gap.

31:3 DEFINITIONS

Indefinite integrals define each of the six inverse hyperbolic functions. Small diagrams, Figures 31-2 through 31-7, illustrate these definitions.

31:3:1 $$\operatorname{arsinh}(x) = \int_0^x \frac{dt}{\sqrt{1+t^2}}$$

31:3:2 $$\operatorname{arcosh}(x) = \int_0^x \frac{dt}{\sqrt{t^2-1}} \qquad x \geq 1$$

31:3:3 $$\operatorname{arsech}(x) = \int_x^1 \frac{dt}{t\sqrt{1-t^2}} \qquad 0 \leq x \leq 1$$

31:3:4 $$\operatorname{arcsch}(x) = \begin{cases} \displaystyle\int_x^\infty \frac{dt}{t\sqrt{1+t^2}} & 0 \leq x \leq \infty \\ \displaystyle\int_{-\infty}^x \frac{dt}{t\sqrt{1+t^2}} & -\infty \leq x \leq 0 \end{cases}$$

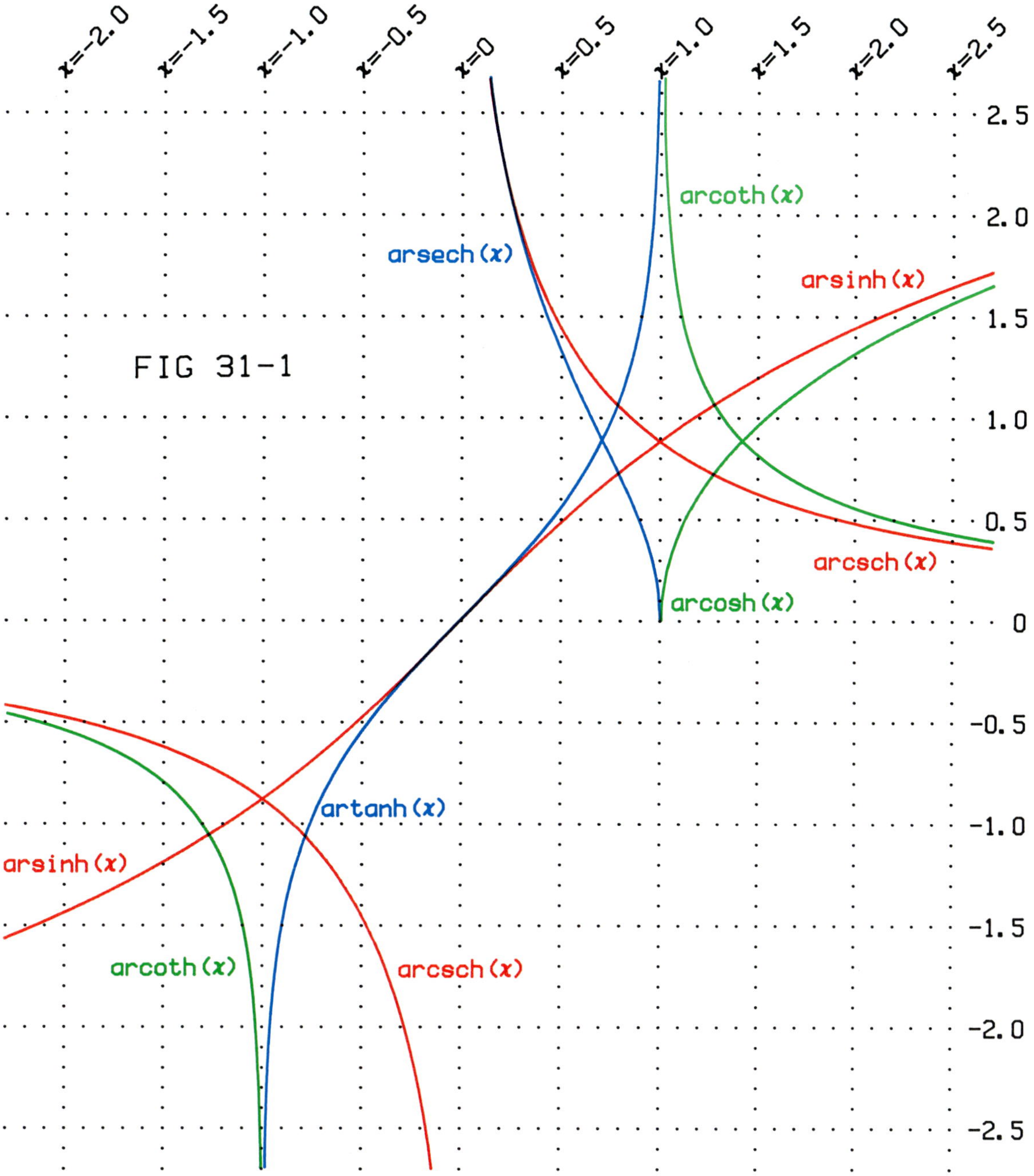

FIG 31-1

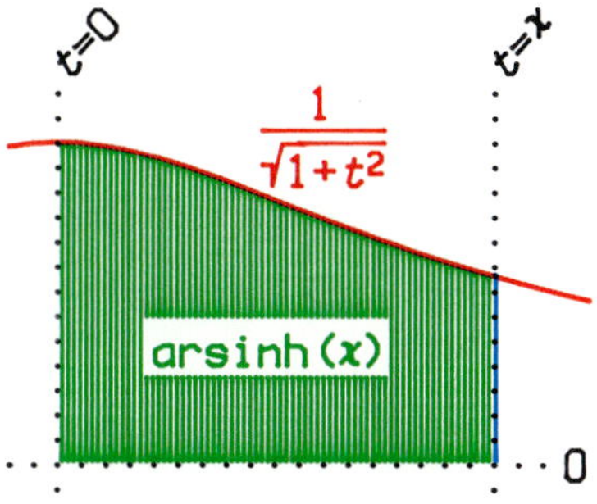

FIG 31-2

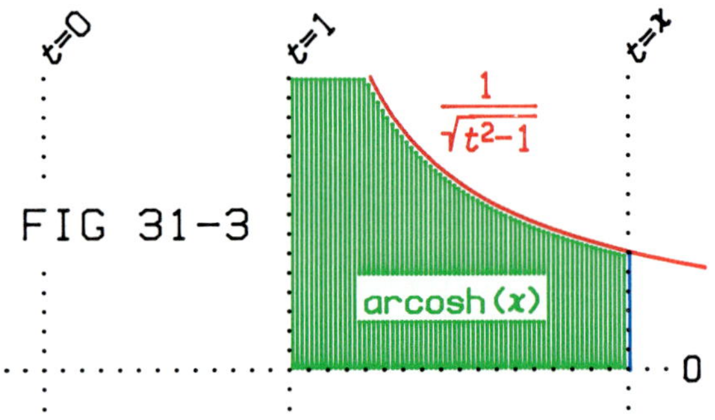

FIG 31-3

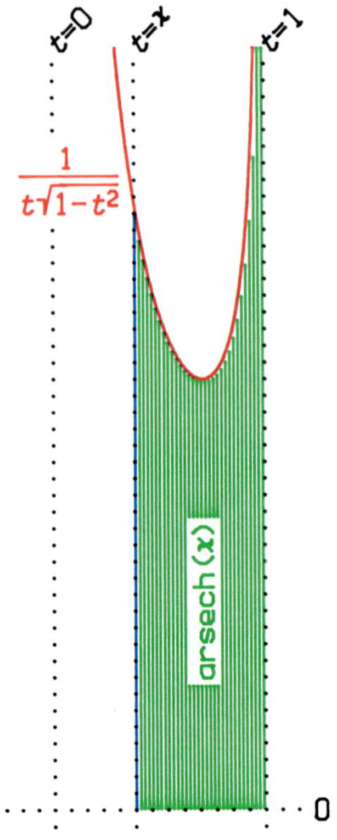

FIG 31-4

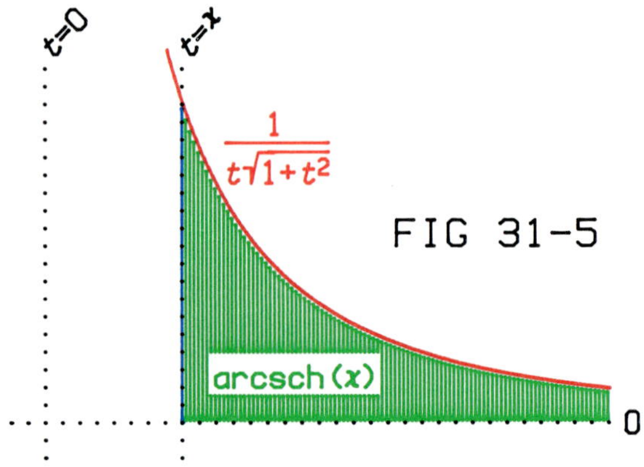

FIG 31-5

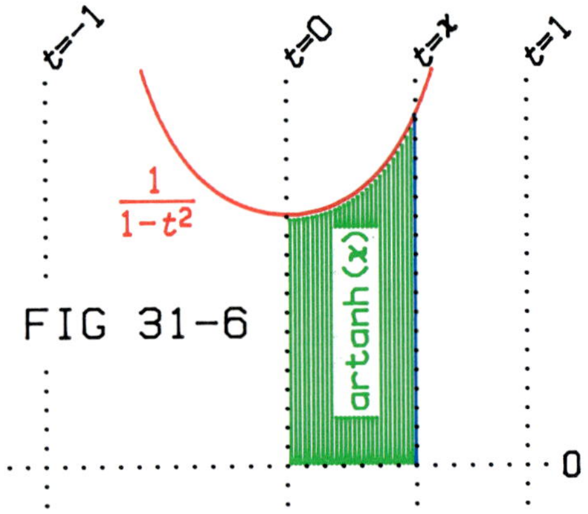

FIG 31-6

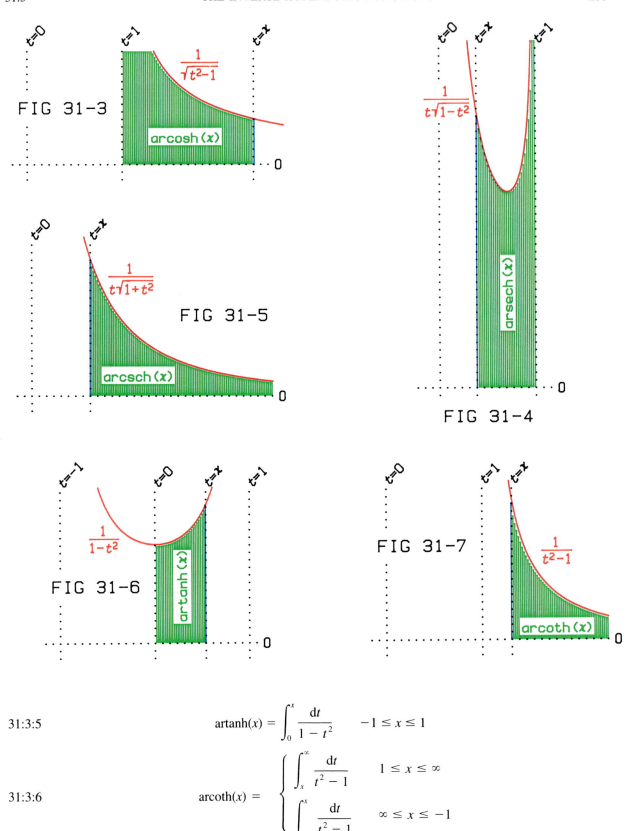

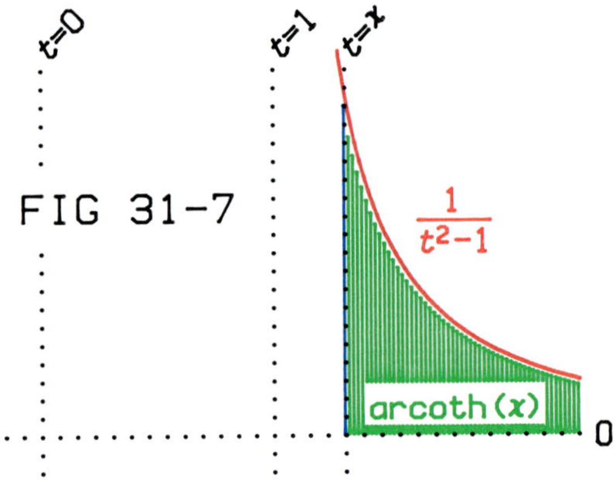

FIG 31-7

31:3:5
$$\operatorname{artanh}(x) = \int_0^x \frac{dt}{1-t^2} \qquad -1 \le x \le 1$$

31:3:6
$$\operatorname{arcoth}(x) = \begin{cases} \displaystyle\int_x^{\infty} \frac{dt}{t^2-1} & 1 \le x \le \infty \\ \displaystyle\int_{-\infty}^{x} \frac{dt}{t^2-1} & \infty \le x \le -1 \end{cases}$$

Another approach to inverse hyperbolic functions is to use definitions as the logarithms of algebraic functions of the argument x. Thus:

31:3:7 $$\mathrm{arsinh}(x) = \ln(x + \sqrt{x^2 + 1})$$

31:3:8 $$\mathrm{arcosh}(x) = \ln(x + \sqrt{x^2 - 1}) \qquad x \geq 1$$

31:3:9 $$\mathrm{arsech}(x) = \ln\left(\frac{1 + \sqrt{1 - x^2}}{x}\right) \qquad 0 \leq x \leq 1$$

31:3:10 $$\mathrm{arcsch}(x) = \ln\left(\frac{1}{x} + \sqrt{1 + \frac{1}{x^2}}\right)$$

31:3:11 $$\mathrm{artanh}(x) = \ln\left(\sqrt{\frac{1 + x}{1 - x}}\right) \qquad -1 \leq x \leq 1$$

31:3:12 $$\mathrm{arcoth}(x) = \ln\left(\sqrt{\frac{x + 1}{x - 1}}\right) \qquad |x| \geq 1$$

The definition from which the names of these functions arise must be treated with caution. The arsinh, arcsch, artanh, and arcoth functions are straightforward inverses of the corresponding hyperbolic functions

31:3:13 $$f = \mathrm{arsinh}(x) \qquad \text{where} \qquad x = \sinh(f)$$

31:3:14 $$f = \mathrm{arcsch}(x) \qquad \text{where} \qquad x = \mathrm{csch}(f)$$

31:3:15 $$f = \mathrm{artanh}(x) \qquad \text{where} \qquad x = \tanh(f)$$

31:3:16 $$f = \mathrm{arcoth}(x) \qquad \text{where} \qquad x = \coth(f)$$

For the arcosh and arsech functions, however, a proviso is needed:

31:3:17 $$f = \mathrm{arcosh}(x) \qquad \text{where} \qquad x = \cosh(f) \text{ and } f \geq 0$$

31:3:18 $$f = \mathrm{arsech}(x) \qquad \text{where} \qquad x = \mathrm{sech}(f) \text{ and } f \geq 0$$

31:4 SPECIAL CASES

There are none.

31:5 INTRARELATIONSHIPS

No reflection formulas exist for the arcosh or arsech functions [although see Section 31:2], but the other four inverse hyperbolic functions are odd:

31:5:1 $$\mathrm{f}(-x) = -\mathrm{f}(x) \qquad \mathrm{f} = \text{arsinh, arcsch, artanh, arcoth}$$

A number of function-addition formulas exist for the inverse hyperbolic functions. These include

31:5:2 $$\mathrm{arsinh}(x) \pm \mathrm{arsinh}(y) = \mathrm{arsinh}(x\sqrt{1 + y^2} \pm y\sqrt{1 + x^2})$$

31:5:3 $$\mathrm{arsinh}(x) \pm \mathrm{arcosh}(y) = \mathrm{arsinh}(xy \pm \sqrt{(x^2 + 1)(y^2 - 1)})$$

31:5:4 $$\mathrm{arcosh}(x) \pm \mathrm{arcosh}(y) = \mathrm{arcosh}(xy \pm \sqrt{(x^2 - 1)(y^2 - 1)})$$

31:5:5 $$\mathrm{artanh}(x) \pm \mathrm{artanh}(y) = \mathrm{artanh}\left(\frac{x \pm y}{1 \pm xy}\right)$$

31:5:6 $$\mathrm{artanh}(x) \pm \mathrm{arcoth}(y) = \mathrm{artanh}\left(\frac{1 \pm xy}{x \pm y}\right)$$

31:5:7 $$\mathrm{arcoth}(x) \pm \mathrm{arcoth}(y) = \mathrm{arcoth}\left(\frac{1 \pm xy}{x \pm y}\right)$$

and many others may be constructed using the equivalences listed in formulas 31:0:1–31:0:7. Use of these formulas requires, of course, that each argument lie within the domain of its function.

Special cases of the relationships of the last paragraph may be converted into such formulas as

31:5:8 $$\mathrm{arsinh}(x) = 2\,\mathrm{sgn}(x)\,\mathrm{arsinh}\left(\sqrt{\frac{\sqrt{1+x^2}-1}{2}}\right)$$

31:6 EXPANSIONS

Series expansions exist for the inverse hyperbolic sine and tangent

31:6:1 $$\mathrm{arsinh}(x) = x - \frac{x^3}{6} + \frac{3x^5}{40} - \frac{5x^7}{112} + \cdots = x\sum_{j=0}^{\infty}\frac{(2j-1)!!}{(2j)!!}\frac{(-x^2)^j}{2j+1} \qquad -1 < x < 1$$

31:6:2 $$\mathrm{artanh}(x) = x + \frac{x^3}{3} + \frac{x^5}{5} + \frac{x^7}{7} + \cdots = \sum_{j=0}^{\infty}\frac{x^{2j+1}}{2j+1} \qquad -1 < x < 1$$

Replacing x by $1/x$ and restricting $|x| > 1$, these same series depict arcsch(x) and arcoth(x), respectively. We also have

31:6:3 $$\ln\left(\frac{2}{x}\right) - \mathrm{arsech}(x) = \frac{x^2}{4} + \frac{3x^4}{16} + \frac{5x^6}{96} + \cdots = \sum_{j=1}^{\infty}\frac{(2j-1)!!}{(2j)!!}\frac{x^{2j}}{2j} \qquad 0 < x < 1$$

Replacing x by $1/x$ and restricting x to exceed unity converts this series into an expression for $\ln(2x) - \mathrm{arcosh}(x)$. The similar series with alternating signs is

31:6:4 $$\mathrm{sgn}(x)\,\mathrm{arsinh}(x) - \ln(2|x|) = \frac{1}{4x^2} - \frac{3}{16x^4} + \frac{5}{96x^6} - \cdots = -\sum_{j=1}^{\infty}\frac{(2j-1)!!}{2j(2j)!!(-x^2)^j} \qquad |x| \geq 1$$

The inverse hyperbolic sine and tangent may each be expressed as continued fractions

31:6:5 $$\frac{\mathrm{arsinh}(\sqrt{x})}{\sqrt{x}\sqrt{1+x}} = \frac{1}{1+}\,\frac{1\times 2x}{3+}\,\frac{1\times 2x}{5+}\,\frac{3\times 4x}{7+}\,\frac{3\times 4x}{9+}\,\frac{5\times 6x}{11+}\cdots$$

and

31:6:6 $$\frac{\mathrm{artanh}(x)}{x} = \frac{1}{1-}\,\frac{x^2}{3-}\,\frac{(2x)^2}{5-}\,\frac{(3x)^2}{7-}\,\frac{(4x)^2}{9-}\cdots$$

31:7 PARTICULAR VALUES

In Table 31.7.1 "undef" means that the function is not defined for the argument in question.

Table 31.7.1

	$x=-\infty$	$x=-1$	$x=0$	$x=1$	$x=\infty$
arsinh(x)	$-\infty$	$\ln(\sqrt{2}-1)$	0	$\ln(\sqrt{2}+1)$	∞
arcosh(x)	undef	undef	undef	0	∞
arsech(x)	undef	undef	$+\infty$	0	undef
arcsch(x)	0	$\ln(\sqrt{2}-1)$	$\pm\infty$	$\ln(\sqrt{2}+1)$	0
artanh(x)	undef	$-\infty$	0	$+\infty$	undef
arcoth(x)	0	$-\infty$	undef	$+\infty$	0

31:8 NUMERICAL VALUES

The algorithm below, which is exact, uses definitions 31:3:7–31:3:12 to calculate values of the inverse hyperbolic functions. The universal hypergeometric algorithm [Section 18:14] may also be used to evaluate these functions for suitable arguments.

```
Input x >>>>>>>>>>>
Input code >>>>>>>>
          If frac(c/2) = 0 go to (1)
          Replace x by 1/x
      (1) Set f = √(x² + 1)
          If c ≤ 1 go to (3)
          If c ≤ 3 go to (2)
          Set f = √((1 + x)/(1 − x))
          Go to (4)
      (2) Set f = √(x² − 1)
      (3) Replace f by f + x
      (4) Replace f by ln(f)
          Output f
f0 = arsinh(x)
f1 = arcsch(x)
f2 = arcosh(x)   <<<<<
f3 = arsech(x)
f4 = artanh(x)
f5 = arcoth(x)
```

Storage needed:
x, c and f

Input restrictions: Argument must lie in the range:

code, c	function	argument
0	arsinh	any
1	arcsch	$x \neq 0$
2	arcosh	$x \geq 1$
3	arsech	$0 < x \leq 1$
4	artanh	$-1 < x < 1$
5	arcoth	$\|x\| > 1$

Test values:
$\text{arsinh}(-\pi) = -1.862295743$
$\text{arcosh}(\pi) = 1.811526272$
$\text{arcoth}(\pi) = 0.3297653150$

31:9 APPROXIMATIONS

For large x we can use equations 31:6:4 and 31:6:3 to approximate

31:9:1 $$\text{arsinh}(x) \simeq \text{sgn}(x)\left[\ln(2|x|) + \frac{1}{4x^2}\right] \qquad \text{8-bit precision} \qquad |x| \geq 2$$

31:9:2 $$\text{arcosh}(x) \simeq \ln(2x) - \frac{1}{4x^2} \qquad \text{8-bit precision} \qquad |x| \geq 2.2$$

For small x the approximation

31:9:3 $$\text{artanh}(x) \simeq \frac{x}{\sqrt{1 - \dfrac{2x^2}{3}}} \qquad \text{8-bit precision} \qquad -\frac{1}{2} \leq x \leq \frac{1}{2}$$

is useful.

31:10 OPERATIONS OF THE CALCULUS

Differentiation of the six inverse hyperbolic functions gives

31:10:1 $$\frac{d}{dx}\text{arsinh}(x) = \frac{1}{\sqrt{1 + x^2}}$$

31:10:2 $$\frac{d}{dx}\text{arcosh}(x) = \frac{1}{\sqrt{x^2 - 1}} \qquad x \geq 1$$

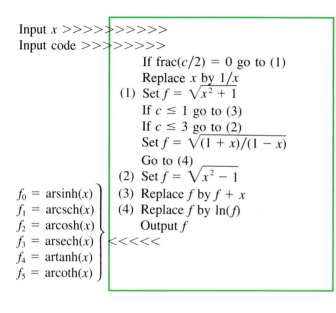

31:10:3 $$\frac{d}{dx}\operatorname{arsech}(x) = \frac{-1}{x\sqrt{1-x^2}} \qquad 0 \le x \le 1$$

31:10:4 $$\frac{d}{dx}\operatorname{arcsch}(x) = \frac{-1}{|x|\sqrt{1+x^2}}$$

31:10:5 $$\frac{d}{dx}\operatorname{artanh}(x) = \frac{1}{1-x^2} \qquad -1 \le x \le 1$$

31:10:6 $$\frac{d}{dx}\operatorname{arcoth}(x) = \frac{1}{1-x^2} \qquad |x| \ge 1$$

The corresponding indefinite integrals are

31:10:7 $$\int_0^x \operatorname{arsinh}(t)dt = x\operatorname{arsinh}(x) - \sqrt{1+x^2} + 1$$

31:10:8 $$\int_1^x \operatorname{arcosh}(t)dt = x\operatorname{arcosh}(x) - \sqrt{x^2-1} \qquad x \ge 1$$

31:10:9 $$\int_0^x \operatorname{arsech}(t)dt = x\operatorname{arsech}(x) + \arcsin(x) \qquad 0 \le x \le 1$$

31:10:10 $$\int_0^x \operatorname{arcsch}(t)dt = x\operatorname{arcsch}(x) + |\operatorname{arsinh}(x)|$$

31:10:11 $$\int_0^x \operatorname{artanh}(t)dt = x\operatorname{artanh}(x) + \ln(\sqrt{1-x^2}) \qquad -1 < x < 1$$

31:10:12 $$\int_2^x \operatorname{arcoth}(t)dt = x\operatorname{arcoth}(x) + \ln\left(\sqrt{\frac{x^2-1}{27}}\right) \qquad 1 < x < \infty$$

The semiderivative

31:10:13 $$\frac{d^{1/2}}{dx^{1/2}}\operatorname{artanh}(\sqrt{x}) = \frac{1}{2}\sqrt{\frac{\pi}{1-x}}$$

links the inverse hyperbolic tangent to a simple algebraic function.

31:11 COMPLEX ARGUMENT

Either the definitions 31:3:1–31:3:6 or 31:3:7–31:3:12 may be used to extend the inverse hyperbolic functions to complex arguments. For example, using 31:3:7, 12:11:1 and 25:11:1, one obtains the multivalued function

31:11:1 $$\operatorname{Arsinh}(x+iy) = \frac{1}{2}\ln(X^2+Y^2) + i(2k\pi + \theta)$$

where k is any integer:

31:11:2 $$\theta = \begin{cases} \operatorname{sgn}(Y)\operatorname{arccot}(X/|Y|) & Y \ne 0 \\ \frac{\pi}{2}[1-\operatorname{sgn}(X)] & Y = 0 \end{cases}$$

31:11:3 $$X = x + \sqrt{\frac{x^2 - y^2 + 1 + \sqrt{(1+x^2+y^2)^2 - 4y^2}}{2}}$$

and

31:11:4 $$Y = y + \mathrm{sgn}(xy)\sqrt{\frac{y^2 - x^2 - 1 + \sqrt{(1 + x^2 + y^2)^2 - 4y^2}}{2}}$$

The resulting explicit expressions in terms of x and y are rather complicated. However, several useful special cases of such formulas arise upon selecting $k = 0$ and by restricting the argument to be purely imaginary. This produces

31:11:5 $$\mathrm{arsinh}(iy) = i\,\mathrm{arcsin}(y)$$

31:11:6 $$\mathrm{arcsch}(iy) = -i\,\mathrm{arccsc}(y)$$

31:11:7 $$\mathrm{artanh}(iy) = i\,\mathrm{arctan}(y)$$

31:11:8 $$\mathrm{arcoth}(iy) = -\,i\,\mathrm{arccot}(y)$$

where the functions appearing on the right-hand sides are inverse trigonometric functions [see Chapter 35].

31:12 GENERALIZATIONS

The inverse hyperbolic tangent is a special case of the generalized logarithmic function [Section 25:12]

31:12:1 $$\mathrm{artanh}(x) = \frac{\mathrm{sgn}(x)}{2}\,\ln_{1/2}\left(\frac{1}{1 - x^2}\right)$$

and of the incomplete beta function [Chapter 58]

31:12:2 $$\mathrm{artanh}(x) = \frac{\mathrm{sgn}(x)}{2}\,\mathrm{B}(1/2;0;x^2) \qquad -1 < x < 1$$

while the inverse hyperbolic sine is an instance of the Gauss function [Chapter 60]

31:12:3 $$\mathrm{arsinh}(x) = x\,\mathrm{F}\left(\frac{1}{2}, \frac{1}{2}; \frac{3}{2}; -x^2\right) \qquad -1 < x < 1$$

The generalization of arsinh to the multivalued Arsinh function is discussed in Section 31:11. The multivalued Arcosh, Arsech, Arcsch, Artanh and Arcoth functions may be constructed analogously.

31:13 COGNATE FUNCTIONS

The six inverse hyperbolic functions are closely related to the logarithmic function and to the six inverse trigonometric functions that are discussed in Chapter 34.

CHAPTER 32

THE SINE sin(*x*) AND COSINE cos(*x*) FUNCTIONS

The functions of this chapter are of paramount importance: they are the units from which all periodic functions [Chapter 36] can be built. The sine and cosine functions are interrelated by

32:0:1 $$\sin^2(x) + \cos^2(x) = 1$$

and by

32:0:2 $$\sin\left(x + \frac{\pi}{2}\right) = \cos(x)$$

32:1 NOTATION

The symbolism sin(*x*) and cos(*x*) is universal. The notations Sin(*x*) and Cos(*x*) refer to the hyperbolic functions of Chapter 28, not to those presently under consideration. Sometimes the names *circular sine* and *circular cosine* are employed to emphasize the distinction from the corresponding hyperbolic functions.

In this *Atlas* we generally use *x* to represent the argument of a function, and this convention is retained in the present chapter. As explained in Section 32:2, however, the arguments of the sine and cosine functions [and of the functions addressed in Chapters 33 and 34] are often regarded as angles, rather than simply as numbers to which no special geometric significance is to be attached. We shall write sin(θ) and cos(θ) [as well as sec(θ), tan(θ), etc. in Chapters 33 and 34] when we particularly want to emphasize the angular interpretation of the argument.

The sine and cosine functions, and any "mixture" of them $c_1 \sin(x) + c_2 \cos(x)$, are known as *sinusoidal functions* or *sinusoids*. Via equation 32:5:26, such a "mixture" may be expressed as $\sqrt{c_1^2 + c_2^2}\sin(x + \phi)$ where the angle ϕ is known as the *phase* of the sinusoid and $\sqrt{c_1^2 + c_2^2}$ as its *amplitude*.

Collectively, the functions of Chapters 32, 33 and 34 are often known as the *trigonometric functions* [see Section 34:14].

32:2 BEHAVIOR

The sinusoids are periodic functions with period equal to 2π, that is, their values at $x \pm 2\pi$, $x \pm 4\pi$, $x \pm 6\pi$, ... exactly equal their values at argument *x*. This is evident from Figure 32-1.

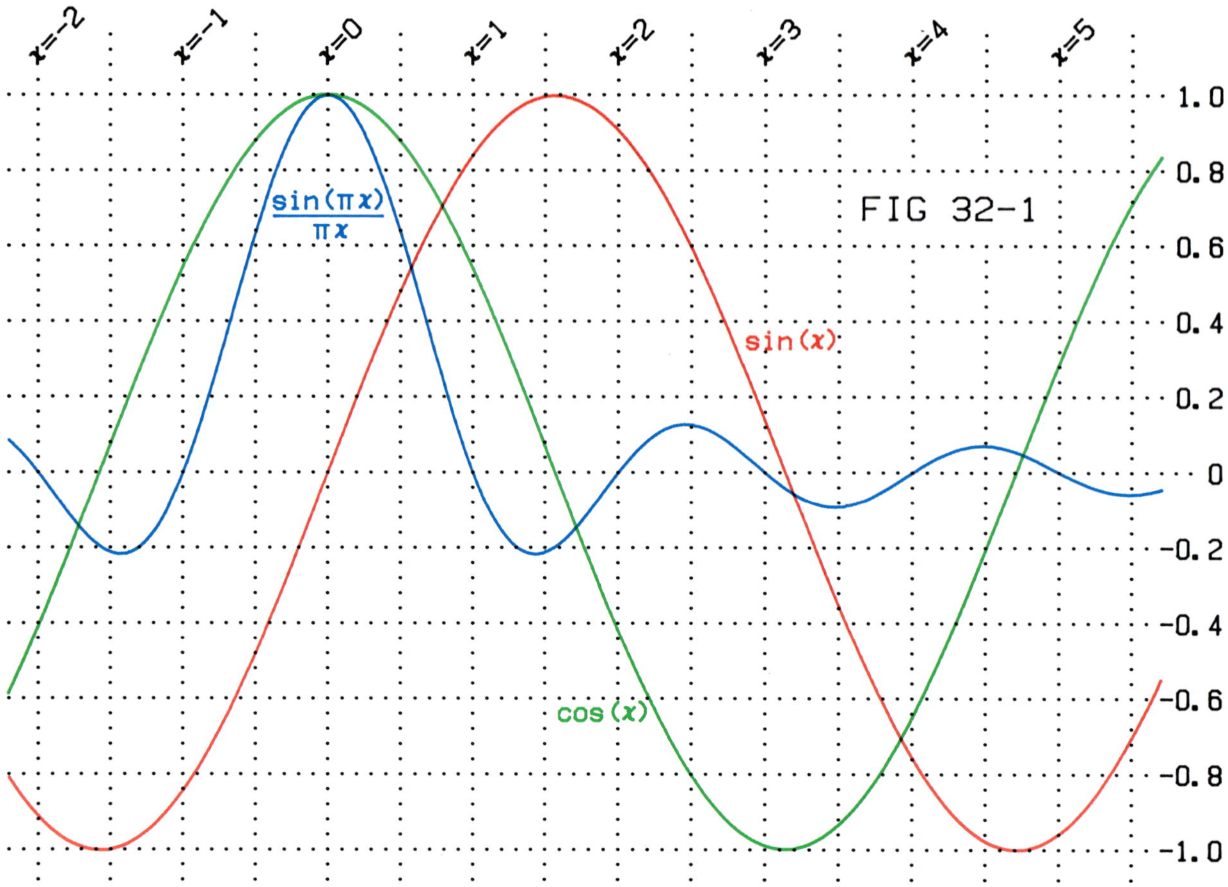

In each period the sinusoidal functions display two zeros, one maximum and one minimum. The sine function has maxima of $\sin(x) = 1$ at $x = \pi/2, -3\pi/2, 5\pi/2, -7\pi/2, \ldots$ and minima of $\sin(x) = -1$ at $x = -\pi/2, 3\pi/2, -5\pi/2, 7\pi/2, \ldots$, a zero being located midway between each adjacent maximum-minimum pair (i.e., at $x = 0, \pm\pi, \pm 2\pi, \ldots$). The behavior of the cosine function is similar: $\cos(x) = 1$ maxima occur at $x = 0, \pm 2\pi, \pm 4\pi, \ldots$; $\cos(x) = -1$ minima at $x = \pm\pi, \pm 3\pi, \pm 5\pi, \ldots$; and zeros at $x = \pm\pi/2, \pm 3\pi/2, \pm 5\pi/2, \ldots$.

The periodicity of the sine and cosine functions can be appreciated perhaps more easily when the argument is regarded as an angle. Since the angle θ is coincident with the angles $\theta \pm 2\pi$, $\theta \pm 4\pi$, $\theta \pm 6\pi$, ..., it comes as no surprise that

32:2:1 $$\sin(\theta \pm 2\pi) = \sin(\theta \pm 4\pi) = \sin(\theta \pm 6\pi) = \cdots = \sin(\theta)$$

and similarly for the cosine [and, in fact, for the other four circular functions discussed in Chapters 33 and 34]. Angles in the ranges

32:2:2 $$2k\pi < \theta < \frac{\pi}{2} + 2k\pi \qquad k = 0, \pm 1, \pm 2, \ldots$$

are said to be in the *first quadrant*. Similarly the second, third and fourth quadrants encompass the angles

32:2:3 $$(\pi/2) + 2k\pi < \theta < \pi + 2k \qquad \textit{second quadrant} \qquad k = 0, \pm 1, \pm 2, \ldots$$

32:2:4 $$\pi + 2k\pi < \theta < (3\pi/2) + 2k\pi \qquad \textit{third quadrant} \qquad k = 0, \pm 1, \pm 2, \ldots$$

and

32:2:5 $$(3\pi/2) + 2k\pi < \theta < 2(k+1)\pi \qquad \textit{fourth quadrant} \qquad k = 0, \pm 1, \pm 2, \ldots$$

This usage of the term "quadrant" is related to, but differs from, that described in Section 0:2. Some simple rules respecting the signs of the trigonometric functions in the four quadrants are assembled in Section 33:2.

32:3 DEFINITIONS

The trigonometric definitions of the sine and cosine function are illustrated in Figure 32-2. Consider the point P to have arrived at its present location by having moved from A along the circular path AP such that the distance from point O has remained constant and equal to unity. Then the sine and cosine functions are defined as the lengths

32:3:1 $$\sin(x) = PQ \qquad [\text{negative if } P \text{ lies below line } OA]$$

32:3:2 $$\cos(x) = OQ \qquad [\text{negative if } Q \text{ lies to the left of } O]$$

PQO being a right angle. The argument x may be interpreted in three distinct ways: first, as the length of the arc AP, this being assigned a negative value if the rotation of OP was clockwise; second, as the angle POA, measured in radians, this again being considered negative if P rotated clockwise to arrive at its present location; and third, as the area [shaded in Figure 32-2] enclosed by the arc AP, the line OP and a mirror-symmetrical construct below the OA line. It is this third interpretation of x that parallels the one used in Section 28:2 as a geometric definition of the hyperbolic sine and cosine.

FIG 32-2

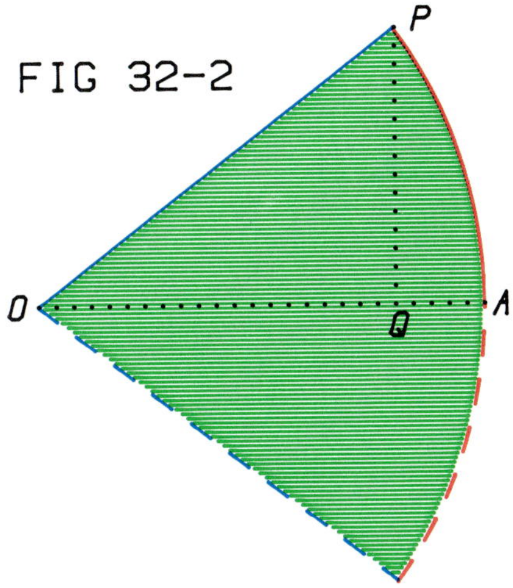

Expansions 32:6:1 and 32:6:2 may serve as definitions of the cosine and sine functions. Alternatively, these functions may be defined via the formulas

32:3:3 $$\sin(x) = \frac{\exp(ix) - \exp(-ix)}{2i}$$

32:3:4 $$\cos(x) = \frac{\exp(ix) + \exp(-ix)}{2}$$

which utilize exponential functions of imaginary arguments [see Section 26:11].

Several polynomial functions exhibit sinusoidal behavior in the limit of large order; see equation 18:9:4 for the example of Pochhammer polynomials. Similarly, for Euler polynomials [Chapter 20]

32:3:5 $$\sin(x) = \frac{\pi}{4} \lim_{n\to\infty}\left\{\frac{(-\pi^2)^n}{(2n)!} E_{2n}\left(\frac{x}{\pi}\right)\right\}$$

and

32:3:6 $$\cos(x) = \frac{1}{2} \lim_{n\to\infty}\left\{\frac{n(-\pi^2)^n}{(2n)!} E_{2n-1}\left(\frac{x}{\pi}\right)\right\}$$

For negative b, the differential equation

32:3:7 $$\frac{d^2f}{dx^2} = bf + c \qquad b < 0$$

is solved by the general sinusoid $f = -(c/b) + c_1 \sin(x\sqrt{-b}) + c_2 \cos(x\sqrt{-b})$, where c_1 and c_2 are arbitrary constants. Similarly, for suitable values of the coefficients, the differential equation

32:3:8 $$\frac{df}{dx} = \sqrt{af^2 + bf + c} \qquad a < 0 \qquad b^2 > 4ac$$

is solved by $f = -(b/2a) + [\sqrt{b^2 - 4ac}/2a][\sqrt{\lambda}\,\sin(x\sqrt{-a}) - \sqrt{1-\lambda}\,\cos(x\sqrt{-a})]$ where λ is an arbitrary constant in the range $0 \le \lambda \le 1$. Hence, equation 32:3:7 or 32:3:8 can be considered to define the sinusoidal function.

32:4 SPECIAL CASES

There are none.

32:5 INTRARELATIONSHIPS

The cosine function is even

32:5:1 $$\cos(-x) = \cos(x)$$

whereas the sine is an odd function

32:5:2 $$\sin(-x) = -\sin(x)$$

The duplication and triplication formulas

32:5:3 $$\cos(2x) = \cos^2(x) - \sin^2(x) = 2\cos^2(x) - 1 = 1 - 2\sin^2(x)$$

32:5:4 $$\sin(2x) = 2\sin(x)\cos(x) = 2\sin(x)\sqrt{1 - \sin^2(x)}$$

32:5:5 $$\cos(3x) = 4\cos^3(x) - 3\cos(x)$$

32:5:6 $$\sin(3x) = 3\sin(x) - 4\sin^3(x) = \sin(x)[4\cos^2(x) - 1]$$

generalize to

32:5:7 $$\cos(nx) = \mathrm{T}_n(\cos(x)) = \sum_{j=0}^{n} \mathrm{t}_n^{(j)} \cos^j(x)$$

and

32:5:8 $$\sin(nx) = \sin(x)\mathrm{U}_{n-1}(\cos(x)) = \csc(x) \sum_{j=0}^{n-1} [\mathrm{t}_{n-2}^{(j)} \cos(x) - \mathrm{t}_{n-1}^{(j)} \cos^j(x)]$$

where T_n and U_n are Chebyshev polynomials discussed in Chapter 22, as are the Chebyshev coefficients $\mathrm{t}_n^{(j)}$.

Equations 32:5:3 and 32:5:4 are special cases of the argument-addition formulas

32:5:9 $$\cos(x \pm y) = \cos(x)\cos(y) \mp \sin(x)\sin(y)$$

32:5:10 $$\sin(x \pm y) = \sin(x)\cos(y) \pm \cos(x)\sin(y)$$

which, in turn, have the important special cases

32:5:11 $$\cos\left(x \pm \frac{n\pi}{2}\right) = \begin{cases} \cos(x) & n = 0, 4, 8, \ldots \\ \mp\sin(x) & n = 1, 5, 9, \ldots \\ -\cos(x) & n = 2, 6, 10, \ldots \\ \pm\sin(x) & n = 3, 7, 11, \ldots \end{cases}$$

32:5:12
$$\sin\left(x \pm \frac{n\pi}{2}\right) = \begin{cases} \sin(x) & n = 0, 4, 8, \ldots \\ \pm\cos(x) & n = 1, 5, 9, \ldots \\ -\sin(x) & n = 2, 6, 10, \ldots \\ \mp\cos(x) & n = 3, 7, 11, \ldots \end{cases}$$

constituting recursion formulas.

From equation 32:5:3 one may derive the expressions

32:5:13
$$\cos\left(\frac{x}{2}\right) = (-1)^m\sqrt{\frac{1+\cos(x)}{2}} \qquad m = \mathrm{Int}\left(\frac{\pi + |x|}{2\pi}\right)$$

32:5:14
$$\sin\left(\frac{x}{2}\right) = (-1)^m\sqrt{\frac{1-\cos(x)}{2}} \qquad m = \mathrm{Int}\left(\frac{|x|}{2\pi}\right)$$

for the cosine and sine of half argument, as well as the formulas

32:5:15
$$\cos^2(x) = \frac{1+\cos(2x)}{2}$$

32:5:16
$$\sin^2(x) = \frac{1-\cos(2x)}{2}$$

for the squares. These latter may be generalized to the expressions

32:5:17
$$\cos^n(x) = \frac{1}{2^n}\sum_{j=0}^{n}\binom{n}{j}\cos[(n-2j)x] = \begin{cases} \dfrac{1}{2^{n-1}}\displaystyle\sum_{j=0}^{N}\binom{n}{j}\cos[(n-2j)x] & n = 2N+1 = 1, 3, 5, \ldots \\ \dfrac{(n-1)!!}{n!!} + \dfrac{1}{2^{n-1}}\displaystyle\sum_{j=0}^{N-1}\binom{n}{j}\cos[(n-2j)x] & n = 2N = 2, 4, 6, \ldots \end{cases}$$

and

32:5:18
$$\sin^n(x) = \begin{cases} \dfrac{(-1)^N}{2^n}\displaystyle\sum_{j=0}^{n}(-1)^j\binom{n}{j}\sin[(n-2j)x] = \dfrac{(-1)^N}{2^{n-1}}\displaystyle\sum_{j=0}^{N}(-1)^j\binom{n}{j}\sin[(n-2j)x] & n = 2N+1 = 1, 3, 5, \ldots \\ \dfrac{(-1)^N}{2^n}\displaystyle\sum_{j=0}^{n}(-1)^j\binom{n}{j}\cos[(n-2j)x] = \dfrac{(n-1)!!}{n!!} + \dfrac{(-1)^N}{2^{n-1}}\displaystyle\sum_{j=0}^{N-1}(-1)^j\binom{n}{j}\cos[(n-2j)x] & n = 2N = 2, 4, 6, \ldots \end{cases}$$

for positive integer powers of the cosine and sine functions.

The cosine and sine functions satisfy the function-addition formulas

32:5:19
$$\cos(x) \pm \sin(x) = \sqrt{2}\sin\left(x \pm \frac{\pi}{4}\right) = \sqrt{2}\cos\left(x \mp \frac{\pi}{4}\right)$$

32:5:20
$$\cos(x) + \cos(y) = 2\cos\left(\frac{x+y}{2}\right)\cos\left(\frac{x-y}{2}\right)$$

32:5:21
$$\cos(x) - \cos(y) = -2\sin\left(\frac{x+y}{2}\right)\sin\left(\frac{x-y}{2}\right)$$

32:5:22
$$\sin(x) \pm \sin(y) = 2\sin\left(\frac{x\pm y}{2}\right)\cos\left(\frac{x\mp y}{2}\right)$$

as well as the function-multiplication expressions

32:5:23 $$\cos(x)\cos(y) = \frac{1}{2}\cos(x+y) + \frac{1}{2}\cos(x-y)$$

32:5:24 $$\cos(x)\sin(y) = \frac{1}{2}\sin(x+y) - \frac{1}{2}\sin(x-y)$$

32:5:25 $$\sin(x)\sin(y) = \frac{1}{2}\cos(x-y) - \frac{1}{2}\cos(x+y)$$

Equation 32:5:19 is a special case of the important formula

32:5:26 $$c_1\sin(x) + c_2\cos(x) = \sqrt{c_1^2+c_2^2}\,\sin(x+\arctan(c_2/c_1)) = \sqrt{c_1^2+c_2^2}\,\cos(x-\operatorname{arccot}(c_2/c_1))$$

by which any sinusoid may be expressed as a sine or as a cosine. The arctan and arccot functions occurring in 32:5:26 are discussed in Chapter 35.

Infinite series of the forms

32:5:27 $$\frac{c_0}{2} + c_1\cos(x) + c_2\cos(2x) + c_3\cos(3x) + \cdots = \frac{c_0}{2} + \sum_{j=1}^{\infty} c_j\cos(jx)$$

or

32:5:28 $$s_1\sin(x) + s_2\sin(2x) + s_3\sin(3x) + \cdots = \sum_{j=1}^{\infty} s_j\sin(jx)$$

or sometimes of the combination $\frac{1}{2}c_0 + \Sigma\, c_j\cos(jx) + s_j\sin(jx)$, with the c_j and s_j being specified coefficients, are termed *Fourier series* and are discussed in Chapter 36. Here we merely quote two examples with relevance to the present chapter. If ν is any number other than an integer, then

32:5:29 $$\cos(\nu x) = \frac{2\nu}{\pi}\sin(\nu\pi)\left[\frac{1}{2\nu^2} + \frac{\cos(x)}{1-\nu^2} - \frac{\cos(2x)}{4-\nu^2} + \frac{\cos(3x)}{9-\nu^2} - \cdots\right]$$
$$= \frac{2\nu}{\pi}\sin(\nu\pi)\left[\frac{1}{2\nu^2} - \sum_{j=1}^{\infty}\frac{(-1)^j\cos(jx)}{j^2-\nu^2}\right] \qquad -\pi < x < \pi$$

32:5:30 $$\sin(\nu x) = \frac{2}{\pi}\sin(\nu\pi)\left[\frac{\sin(x)}{1-\nu^2} - \frac{2\sin(2x)}{4-\nu^2} + \frac{3\sin(3x)}{9-\nu^2} - \cdots\right]$$
$$= \frac{-2}{\pi}\sin(\nu\pi)\sum_{j=1}^{\infty}\frac{(-1)^j j\sin(jx)}{j^2-\nu^2} \qquad -\pi < x < \pi$$

The infinite series

32:5:31 $$\sum_{j=1}^{\infty}\frac{\sin(jx)}{j^n} \quad n = 1, 3, 5, \ldots \quad \text{and} \quad \sum_{j=1}^{\infty}\frac{\cos(jx)}{j^n} \quad n = 2, 4, 6, \ldots$$

may be summed in terms of the Hurwitz function $\zeta(1-n;x/2\pi)$ [see equation 64:6:2]. The sum $\Sigma\cos(jx)/j$ equals $\ln\{\csc(x/2)/2\}$, while $\Sigma\sin(jx)/j^2$ is the integral of this function, known as *Clausen's integral* [see Abramowitz and Stegun, Section 27.8].

32:6 EXPANSIONS

Taylor series [see equation 0:5:1] exist for the cosine and sine functions and for their logarithms. Those for the functions

32:6:1 $$\cos(x) = 1 - \frac{x^2}{2!} + \frac{x^4}{4!} - \frac{x^6}{6!} + \cdots = \sum_{j=0}^{\infty}\frac{(-x^2)^j}{(2j)!}$$

32:6:2
$$\sin(x) = x - \frac{x^3}{3!} + \frac{x^5}{5!} - \frac{x^7}{7!} + \cdots = x \sum_{j=0}^{\infty} \frac{(-x^2)^j}{(2j+1)!}$$

converge for all x, but more rapidly for small arguments. The logarithmic series have limited domains of convergence:

32:6:3
$$\ln(\cos(x)) = -\frac{x^2}{2} - \frac{x^4}{12} - \frac{x^6}{45} - \cdots = -\sum_{j=1}^{\infty} \frac{\lambda(2j)}{j}\left(\frac{2x}{\pi}\right)^j = -\sum_{j=1}^{\infty} \frac{(4^j - 1)|B_{2j}|}{2j(2j)!}(2x)^{2j} \qquad -\frac{\pi}{2} < x < \frac{\pi}{2}$$

32:6:4
$$\ln\left(\frac{\sin(x)}{x}\right) = -\frac{x^2}{6} - \frac{x^4}{180} - \frac{x^6}{2835} - \cdots = -\sum_{j=1}^{\infty} \frac{\zeta(2j)}{j}\left(\frac{x}{\pi}\right)^j = -\sum_{j=1}^{\infty} \frac{|B_{2j}|}{2j(2j)!}(2x)^{2j} \qquad -\pi < x < \pi$$

and the general terms involve eta, zeta or Bernoulli numbers [see Chapters 3 and 4].

The sine and cosine functions are also expansible as infinite products:

32:6:5
$$\sin(x) = x\left(1 - \frac{x^2}{\pi^2}\right)\left(1 - \frac{x^2}{4\pi^2}\right)\left(1 - \frac{x^2}{9\pi^2}\right)\cdots = x\prod_{j=1}^{\infty}\left(1 - \frac{x^2}{j^2\pi^2}\right)$$

32:6:6
$$\cos(x) = \left(1 - \frac{4x^2}{\pi^2}\right)\left(1 - \frac{4x^2}{9\pi^2}\right)\left(1 - \frac{4x^2}{25\pi^2}\right) = \prod_{j=1}^{\infty}\left(1 - \frac{4x^2}{(2j-1)^2\pi^2}\right)$$

both of which are encompassed by the general formula

32:6:7
$$f(x) = \prod_j \left(1 - \frac{x}{r_j}\right) \qquad f(x) = \frac{\sin(x)}{x}, \cos(x)$$

where the r values are the zeros of $f(x)$. Yet another expansion as an infinite product is

32:6:8
$$\frac{\sin(x)}{x} = \cos\left(\frac{x}{2}\right)\cos\left(\frac{x}{4}\right)\cos\left(\frac{x}{8}\right)\cdots = \prod_{j=1}^{\infty}\cos\left(\frac{x}{2^j}\right)$$

The cosine and sine functions may be expressed as infinite sums of Bessel functions [Chapter 52]

32:6:9
$$\cos(x) = J_0(x) - 2J_2(x) + 2J_4(x) - \cdots = J_0(x) + 2\sum_{j=1}^{\infty}(-1)^j J_{2j}(x)$$

32:6:10
$$\sin(x) = 2J_1(x) - 2J_3(x) + 2J_5(x) - \cdots = 2\sum_{j=0}^{\infty}(-1)^j J_{2j+1}(x)$$

32:7 PARTICULAR VALUES

Table 32.7.1 evaluates the cosine and sine functions for special values of the argument x in the range $-\pi/2 \le x \le \pi/2$ (angles θ in the range $-90°$ to $90°$).

For particular values outside this range, use the recursion formulas 32:5:11 and 32:5:12.

32:8 NUMERICAL VALUES

Almost all calculators have keys by which $\sin(x)$ and $\cos(x)$ may be computed, and almost all computers incorporate sine and cosine functions; therefore, no algorithms are presented here. Sometimes the computer generates $\sin(\theta)$ or $\cos(\theta)$ where θ is an angle in degrees; in this case $\sin(x)$ can be calculated as $\sin(\theta) = \sin(180x/\pi)$. Computing

Table 32.7.1

θ	0°	±15°	±18°	$\pm 22\frac{1}{2}°$	±30°	±36°	±45°	±54°	±60°	$\pm 67\frac{1}{2}°$	±72°	±75°	±90°
x	0	$\frac{\pm\pi}{12}$	$\frac{\pm\pi}{10}$	$\frac{\pm\pi}{8}$	$\frac{\pm\pi}{6}$	$\frac{\pm\pi}{5}$	$\frac{\pm\pi}{4}$	$\frac{\pm 3\pi}{10}$	$\frac{\pm\pi}{3}$	$\frac{\pm 3\pi}{8}$	$\frac{\pm 2\pi}{5}$	$\frac{\pm 5\pi}{12}$	$\frac{\pm\pi}{2}$
cos	1	$\frac{\sqrt{3}+1}{\sqrt{8}}$	$\sqrt{\frac{5+\sqrt{5}}{8}}$	$\frac{\sqrt{2+\sqrt{2}}}{2}$	$\frac{\sqrt{3}}{2}$	$\sqrt{\frac{3+\sqrt{5}}{8}}$	$\frac{1}{\sqrt{2}}$	$\sqrt{\frac{5-\sqrt{5}}{8}}$	$\frac{1}{2}$	$\frac{\sqrt{2-\sqrt{2}}}{2}$	$\sqrt{\frac{3-\sqrt{5}}{8}}$	$\frac{\sqrt{3}-1}{\sqrt{8}}$	0
sin	0	$\frac{\pm 1}{\sqrt{6}+\sqrt{2}}$	$\pm\sqrt{\frac{3-\sqrt{5}}{8}}$	$\frac{\pm 1}{\sqrt{4+\sqrt{8}}}$	$\frac{\pm 1}{2}$	$\pm\sqrt{5-\frac{\sqrt{5}}{8}}$	$\frac{\pm 1}{\sqrt{2}}$	$\pm\sqrt{\frac{3+\sqrt{5}}{8}}$	$\frac{\pm\sqrt{3}}{2}$	$\frac{\pm 1}{\sqrt{4-\sqrt{8}}}$	$\pm\sqrt{\frac{5+\sqrt{5}}{8}}$	$\frac{\pm 1}{\sqrt{6}-\sqrt{2}}$	±1

devices often offer the user a choice of "radian mode" or "degree mode" when evaluating trigonometric or inverse trigonometric functions; be aware that, because of rounding errors inherent in the internal routines of the device, the "degree mode" is usually the more precise of the two modes. (For example, sin(1440°) will be given as zero, whereas $\sin(8\pi)$ will often fail to be calculated as zero.)

The universal hypergeometric algorithm [Section 18:14] permits $\cos(x)$ and $\sin(x)/x$ to be evaluated for any x.

32:9 APPROXIMATIONS

For small values of the argument, the approximations

32:9:1
$$\sin(x) \simeq x\left(1 - \frac{x^2}{2\pi}\right) \qquad \text{8-bit precision} \qquad -1.1 \le x \le 1.1$$

and

32:9:2
$$\cos(x) \simeq \sqrt{\left(1 - \frac{x^2}{3}\right)^3} \qquad \text{8-bit precision} \qquad -0.9 \le x \le 0.9$$

are useful.

32:10 OPERATIONS OF THE CALCULUS

The differentiation formulas

32:10:1
$$\frac{d}{dx}\sin(bx) = b\cos(bx) = b\sin\left(bx + \frac{\pi}{2}\right)$$

32:10:2
$$\frac{d}{dx}\cos(bx) = -b\sin(bx) = b\cos\left(bx + \frac{\pi}{2}\right)$$

can be generalized to

32:10:3
$$\frac{d^n}{dx^n}f(bx) = b^n f\left(bx + \frac{n\pi}{2}\right) \qquad n = 0, 1, 2, \ldots \qquad f = \sin, \cos$$

Indefinite integration gives

32:10:4
$$\int_0^x \sin(bt)dt = \frac{1 - \cos(bx)}{b}$$

32:10:5
$$\int_0^x \cos(bt)dt = \frac{\sin(bx)}{b}$$

which are special cases of the more general results

32:10:6 $$\int_0^x \sin^n(t)dt = \begin{cases} \dfrac{(n-1)!!}{n!!}\left[1 - \cos(x)\displaystyle\sum_{j=0}^{N}\dfrac{(2j-1)!!}{(2j)!!}\sin^{2j}(x)\right] & n = 2N+1 = 1, 3, 5, \ldots \\ \dfrac{(n-1)!!}{n!!}\left[x - \cos(x)\displaystyle\sum_{j=0}^{N-1}\dfrac{(2j)!!}{(2j+1)!!}\sin^{2j+1}(x)\right] & n = 2N = 2, 4, 6, \ldots \end{cases}$$

32:10:7 $$\int_0^x \cos^n(t)dt = \begin{cases} \dfrac{(n-1)!!}{n!!}\sin(x)\displaystyle\sum_{j=0}^{N}\dfrac{(2j-1)!!}{(2j)!!}\cos^{2j}(x) & n = 2N+1 = 1, 3, 5, \ldots \\ \dfrac{(n-1)!!}{n!!}\left[x + \sin(x)\displaystyle\sum_{j=0}^{N-1}\dfrac{(2j)!!}{(2j+1)!!}\cos^{2j+1}(x)\right] & n = 2N = 2, 4, 6, \ldots \end{cases}$$

Other important indefinite integrals include

32:10:8 $$\int_0^x \frac{dt}{1 + a\sin(bt)} = \begin{cases} \dfrac{2}{b\sqrt{a^2-1}}\left[\operatorname{arcosh}(a) - \operatorname{arcoth}\left(\dfrac{a + \tan(bx/2)}{\sqrt{a^2-1}}\right)\right] & |a| > 1 \\ \dfrac{1}{b}\left[\tan\left(\dfrac{bx}{2} \mp \dfrac{\pi}{4}\right) \pm 1\right] & a = \pm 1 \\ \dfrac{2}{b\sqrt{1-a^2}}\left[\arctan\left(\dfrac{a + \tan(bx/2)}{\sqrt{1-a^2}}\right) - \arcsin(a)\right] & -1 < a < 1 \end{cases}$$

32:10:9 $$\int_0^x \frac{dt}{1 + a\cos(bt)} = \begin{cases} \dfrac{2}{b\sqrt{a^2-1}}\operatorname{artanh}\left(\sqrt{\dfrac{a+1}{a-1}}\cot\left(\dfrac{bx}{2}\right)\right) & |a| > 1 \\ \dfrac{1}{b}\tan\left(\dfrac{bx}{2}\right) & a = 1 \\ \dfrac{2}{b\sqrt{1-a^2}}\arctan\left(\sqrt{\dfrac{1-a}{1+a}}\tan\left(\dfrac{bx}{2}\right)\right) & -1 < a < 1 \end{cases}$$

Indefinite integrals of the sine or cosine of the quadratic function $ax^2 + bx + c$ may be expressed in terms of Fresnel integrals [Chapter 39]

32:10:10 $$\int_{-b/2a}^x \sin(at^2 + bt + c)dt = \sqrt{\frac{\pi}{2a}}\left[\cos\left(\frac{b^2-4ac}{4a}\right)S\left(\frac{2ax+b}{2\sqrt{a}}\right) - \sin\left(\frac{b^2-4ac}{4a}\right)C\left(\frac{2ax+b}{2\sqrt{a}}\right)\right] \qquad a > 0$$

32:10:11 $$\int_{-b/2a}^x \cos(at^2 + bt + c)dt = \sqrt{\frac{\pi}{2a}}\left[\cos\left(\frac{b^2-4ac}{4a}\right)C\left(\frac{2ax+b}{2\sqrt{a}}\right) + \sin\left(\frac{b^2-4ac}{4a}\right)S\left(\frac{2ax+b}{2\sqrt{a}}\right)\right] \qquad a > 0$$

as may

32:10:12 $$\int_0^x \frac{\sin(bt)dt}{\sqrt{t}} = \sqrt{\frac{2\pi}{b}}\,S(\sqrt{bx})$$

and

32:10:13 $$\int_0^x \frac{\cos(bt)\mathrm{d}t}{\sqrt{t}} = \sqrt{\frac{2\pi}{b}}\,\mathrm{C}(\sqrt{bx})$$

With Si and Ci denoting the sine and cosine integrals [Chapter 38], we also have

32:10:14 $$\int_0^x \frac{\sin(bt)}{t}\,\mathrm{d}t = \mathrm{Si}(bx)$$

and

32:10:15 $$\int_x^\infty \frac{\cos(bt)}{t}\,\mathrm{d}t = -\mathrm{Ci}(bx)$$

The technique exemplified in 34:10:19 is useful for evaluating many integrals of sine and cosine functions.

Definite integrals involving integrands of the forms f(x) cos(sx) and f(x) sin(sx) are called Fourier transforms and are discussed in Section 32:14. One definition of the so-called *cosine transformation* is

32:10:16 $$\bar{\mathrm{f}}_{\mathrm{C}}(s) = \sqrt{\frac{2}{\pi}}\int_0^\infty \mathrm{f}(t)\cos(st)\mathrm{d}t = \frac{1}{\sqrt{2\pi}}\int_0^\infty \mathrm{f}(t)\exp(-ist)\mathrm{d}t + \frac{1}{\sqrt{2\pi}}\int_0^\infty \mathrm{f}(t)\exp(ist)\mathrm{d}t$$

while the corresponding *sine transformation* is defined by

32:10:17 $$\bar{\mathrm{f}}_{\mathrm{S}}(s) = \sqrt{\frac{2}{\pi}}\int_0^\infty \mathrm{f}(t)\sin(st)\mathrm{d}t = \frac{i}{\sqrt{2\pi}}\int_0^\infty \mathrm{f}(t)\exp(-ist)\mathrm{d}t - \frac{i}{\sqrt{2\pi}}\int_0^\infty \mathrm{f}(t)\exp(ist)\mathrm{d}t$$

Erdélyi, Magnus, Oberhettinger and Tricomi [*Tables of Integral Transforms,* Volume 1, Chapters I and II] list many such transforms, although their definitions omit the $\sqrt{2/\pi}$ multiplier. The final equalities in 32:10:16 and 32:10:17 show how cosine transforms and sine transforms are related to the Laplace transforms of Section 26:14. Thereby the tabulation in that section may be used to evaluate many cosine and sine transforms. Consider, for example, the function $\mathrm{f}(x) = 1/\sqrt{x}$ for which the Laplace transform is $\bar{\mathrm{f}}_{\mathrm{L}}(s) = \sqrt{\pi/s}$. It follows then from 32:10:16 that $\bar{\mathrm{f}}_{\mathrm{C}}(s) = \sqrt{\pi/is}/\sqrt{2\pi} + \sqrt{\pi/-is}/\sqrt{2\pi} = 1/\sqrt{s}$.

The general formula

32:10:18 $$\int_0^\infty \mathrm{f}(bt^\nu)\mathrm{d}t = \Gamma\left(\frac{1+\nu}{\nu}\right)b^{-1/\nu}\mathrm{f}\left(\frac{\pi}{2\nu}\right) \qquad b>0 \qquad \nu>1 \qquad \mathrm{f} = \sin, \cos$$

also holds.

With a lower limit of $-\infty$, differintegration to order ν generally causes the addition of $\nu\pi/2$ to the argument

32:10:19 $$\frac{\mathrm{d}^\nu}{[\mathrm{d}(x+\infty)]^\nu}\mathrm{f}(bx) = b^\nu \mathrm{f}\left(bx + \frac{\nu\pi}{2}\right) \qquad \nu > -1 \qquad \mathrm{f} = \sin, \cos$$

and multiplication by the νth power of the argument multiplier. Equations 32:10:1–32:10:3 are instances of this rule; another example is

32:10:20 $$\frac{\mathrm{d}^{-1/2}}{[\mathrm{d}(x+\infty)]^{-1/2}}\sin(bx) = \frac{1}{\sqrt{b}}\sin\left(bx - \frac{\pi}{4}\right) = \frac{\sin(bx) - \cos(bx)}{\sqrt{2b}} \qquad b>0$$

With a lower limit of zero, semidifferentiation and semiintegration generate auxiliary Fresnel integrals [see Chapter 39]

32:10:21 $$\frac{\mathrm{d}^{1/2}}{\mathrm{d}x^{1/2}}\sin(bx) = \sqrt{b}\sin\left(bx + \frac{\pi}{4}\right) - \sqrt{2b}\,\mathrm{Gres}(\sqrt{bx})$$

32:10:22 $$\frac{\mathrm{d}^{1/2}}{\mathrm{d}x^{1/2}}\cos(bx) = \frac{1}{\sqrt{\pi x}} + \sqrt{b}\cos\left(bx + \frac{\pi}{4}\right) - \sqrt{2b}\,\mathrm{Fres}(\sqrt{bx})$$

32:10:23 $$\frac{d^{-1/2}}{dx^{-1/2}}\sin(bx) = \frac{1}{\sqrt{b}}\sin\left(bx - \frac{\pi}{4}\right) + \sqrt{\frac{2}{b}}\,\text{Fres}(\sqrt{bx})$$

32:10:24 $$\frac{d^{-1/2}}{dx^{-1/2}}\cos(bx) = \frac{1}{\sqrt{b}}\cos\left(bx - \frac{\pi}{4}\right) - \sqrt{\frac{2}{b}}\,\text{Gres}(\sqrt{bx})$$

That the functions $1 = \cos(0x)$, $\sin(x)$, $\cos(x)$, $\sin(2x)$, $\cos(2x)$, $\sin(3x)$, ... form an orthogonal family [see Section 21:14] on the interval $0 \le x \le 2\pi$ (or alternatively on $-\pi \le x \le \pi$) is established by the integrals

32:10:25 $$\int_0^{2\pi} \cos(nt)\sin(mt)dt = 0 \qquad n = 0, 1, 2, \ldots \qquad m = 1, 2, 3, \ldots$$

32:10:26 $$\int_0^{2\pi} \cos(nt)\cos(mt)dt = \begin{Bmatrix} 0 & m \neq n \\ 2\pi & m = n = 0 \\ \pi & m = n \neq 0 \end{Bmatrix} \qquad n, m = 0, 1, 2, \ldots$$

32:10:27 $$\int_0^{2\pi} \sin(nt)\sin(mt)dt = \begin{Bmatrix} 0 & m \neq n \\ \pi & m = n \end{Bmatrix} \qquad n, m = 1, 2, 3, \ldots$$

On the interval $0 \le x \le \pi$, the functions $\cos(0x)$, $\cos(x)$, $\cos(2x)$, ... *or* the functions $\sin(x)$, $\sin(2x)$, $\sin(3x)$, ... are orthogonal, but the conjoined families are not.

32:11 COMPLEX ARGUMENT

For argument $(x + iy)$ we have

32:11:1 $$\sin(x + iy) = \sin(x)\cosh(y) + i\cos(x)\sinh(y)$$

32:11:2 $$\cos(x + iy) = \cos(x)\cosh(y) - i\sin(x)\sinh(y)$$

and therefore for purely imaginary argument

32:11:3 $$\sin(iy) = i\sinh(y)$$

32:11:4 $$\cos(iy) = \cosh(y)$$

Useful relationships are embodied in *De Moivre's theorem:*

32:11:5 $$[\cos(z) + i\sin(z)]^{\nu} = \exp(i\nu z) = \cos(\nu z) + i\sin(\nu z)$$

where z itself may be complex. Unless ν is an integer, the real part of z must lie between $-\pi$ and π.

32:12 GENERALIZATIONS

As explained in Chapter 36, any periodic function can be regarded as a generalized sinusoid.

The Jacobian elliptic functions $\text{sn}(p;x)$ and $\text{sd}(p;x)$ generalize the sine function, to which they reduce when p is zero:

32:12:1 $$\text{sn}(0;x) = \text{sd}(0;x) = \sin(x)$$

Likewise, because

32:12:2 $$\text{cn}(0;x) = \text{cd}(0;x) = \cos(x)$$

the functions $\text{cn}(p;x)$ and $\text{cd}(p;x)$ are generalizations of the cosine function. The Jacobian elliptic functions are the subject of Chapter 63. Finally, the sine function is a special case of the incomplete elliptic integral of the second kind [Chapter 62]:

32:12:3 $$\sin(x) = \text{E}(x;1)$$

32:13 COGNATE FUNCTIONS

The sine and cosine functions are related to the other *circular functions* [those treated in Chapters 32, 33 and 34] by

32:13:1 $$\sin(x) = \sigma_2\sqrt{1-\cos^2(x)} = \frac{\sigma_3\sqrt{\sec^2(x)-1}}{\sec(x)} = \frac{1}{\csc(x)} = \frac{\sigma_4\tan(x)}{\sqrt{1+\tan^2(x)}} = \frac{\sigma_2}{\sqrt{\cot^2(x)+1}}$$

32:13:2 $$\cos(x) = \sigma_4\sqrt{1-\sin^2(x)} = \frac{1}{\sec(x)} = \frac{\sigma_3\sqrt{\csc^2(x)-1}}{\csc(x)} = \frac{\sigma_4}{\sqrt{1+\tan^2(x)}} = \frac{\sigma_2\cot(x)}{\sqrt{\cot^2(x)+1}}$$

where the multipliers σ_2, σ_3 and σ_4 take the values $+1$ or -1 according to the quadrant [see Section 33:2] in which x (interpreted as an angle) lies. All multipliers are positive in the first quadrant, σ_2 also in the second, σ_3 in the first and third, σ_4 in the first and fourth. The explicit formulas

32:13:3 $$\sigma_2 = (-1)^{\mathrm{Int}(x/\pi)}$$

32:13:4 $$\sigma_3 = (-1)^{\mathrm{Int}(2x/\pi)}$$

32:13:5 $$\sigma_4 = (-1)^{\mathrm{Int}[(2x+\pi)/2\pi]} = \sigma_2\sigma_3$$

permit the multipliers to be calculated for any x value. Figure 33-2 provides a geometric interpretation of the interrelationships among the circular function.

The *versine function* vers(x), *coversine function* covers(x), and *haversine function* hav(x), defined by

32:13:6 $$\mathrm{vers}(x) = 1 - \cos(x)$$

32:13:7 $$\mathrm{covers}(x) = 1 - \sin(x)$$

and

32:13:8 $$\mathrm{hav}(x) = \frac{1}{2}[1-\cos(x)] = \sin^2\left(\frac{x}{2}\right)$$

are rather archaic functions seldom encountered nowadays. The function sinc(x), sometimes known as the *sampling function*, is important in spectral theory. It is usually defined by

32:13:9 $$\mathrm{sinc}(x) = \frac{\sin(\pi x)}{\pi x}$$

although the definition $\sin(x)/x$ is also encountered. Figure 32-1 includes a graph of sinc(x), and equations 32:6:2, 32:6:4, 32:6:5, 32:6:7 and 32:6:8 are readily adapted to provide information on this function.

Bessel functions [Chapter 53] and Neumann functions [Chapter 54] of orders equal to one-half of a (positive or negative) odd integer are related in a very simple way to sines and cosines. The simplest cases are

32:13:10 $$\mathrm{J}_{1/2}(x) = \mathrm{Y}_{-1/2}(x) = \sqrt{\frac{2}{\pi x}}\sin(x)$$

32:13:11 $$\mathrm{Y}_{1/2}(x) = -\mathrm{J}_{-1/2}(x) = -\sqrt{\frac{2}{\pi x}}\cos(x)$$

and the general expressions for $n = 1, 2, 3, \ldots$ are

32:13:12 $$\mathrm{J}_{n+1/2}(x) = (-1)^n\mathrm{Y}_{-n-1/2}(x) = \sqrt{\frac{2}{\pi x}}\sin\left(x-\frac{n\pi}{2}\right)\mathrm{W}_n(x) + \frac{1}{\sqrt{2\pi x^3}}\cos\left(x-\frac{n\pi}{2}\right)\mathrm{V}_n(x)$$

32:13:13 $$\mathrm{Y}_{n+1/2}(x) = (-1)^{n+1}\mathrm{J}_{-n-1/2}(x) = \frac{1}{\sqrt{2\pi x^3}}\sin\left(x-\frac{n\pi}{2}\right)\mathrm{V}_n(x) - \sqrt{\frac{2}{\pi x}}\cos\left(x-\frac{n\pi}{2}\right)\mathrm{W}_n(x)$$

where $W_n(x)$ and $V_n(x)$ are the finite sums

32:13:14 $$W_n(x) = \sum_{j=0}^{J} \frac{(n+2j)!}{(2j)!(n-2j)!}\left(\frac{-1}{4x^2}\right)^j \qquad J = \text{Int}\left(\frac{n}{2}\right)$$

32:13:15 $$V_n(x) = \sum_{j=0}^{J} \frac{(n+2j+1)!}{(2j+1)!(n-2j-1)!}\left(\frac{-1}{4x^2}\right)^j \qquad J = \text{Int}\left(\frac{n-1}{2}\right)$$

For example, $W_1(x) = 1$, $V_1(x) = 2$, $W_2(x) = 1 - 3x^{-2}$ and $V_2(x) = 6$. Of course, all the general properties of the Bessel and Neumann functions [Chapter 53 and 54] apply to these spherical functions, the first few of which are mapped in Figure 32-3. The symbols $j_n(x)$ and $y_n(x)$ and the names *spherical Bessel function of the first kind* and *spherical Bessel function of the second kind*, respectively, are sometimes applied to the composite functions $\sqrt{\pi/2x}\,J_{n+1/2}(x)$ and $\sqrt{\pi/2x}\,Y_{n+1/2}(x)$. The significance of the adjective "spherical" will be evident from Section 59:14.

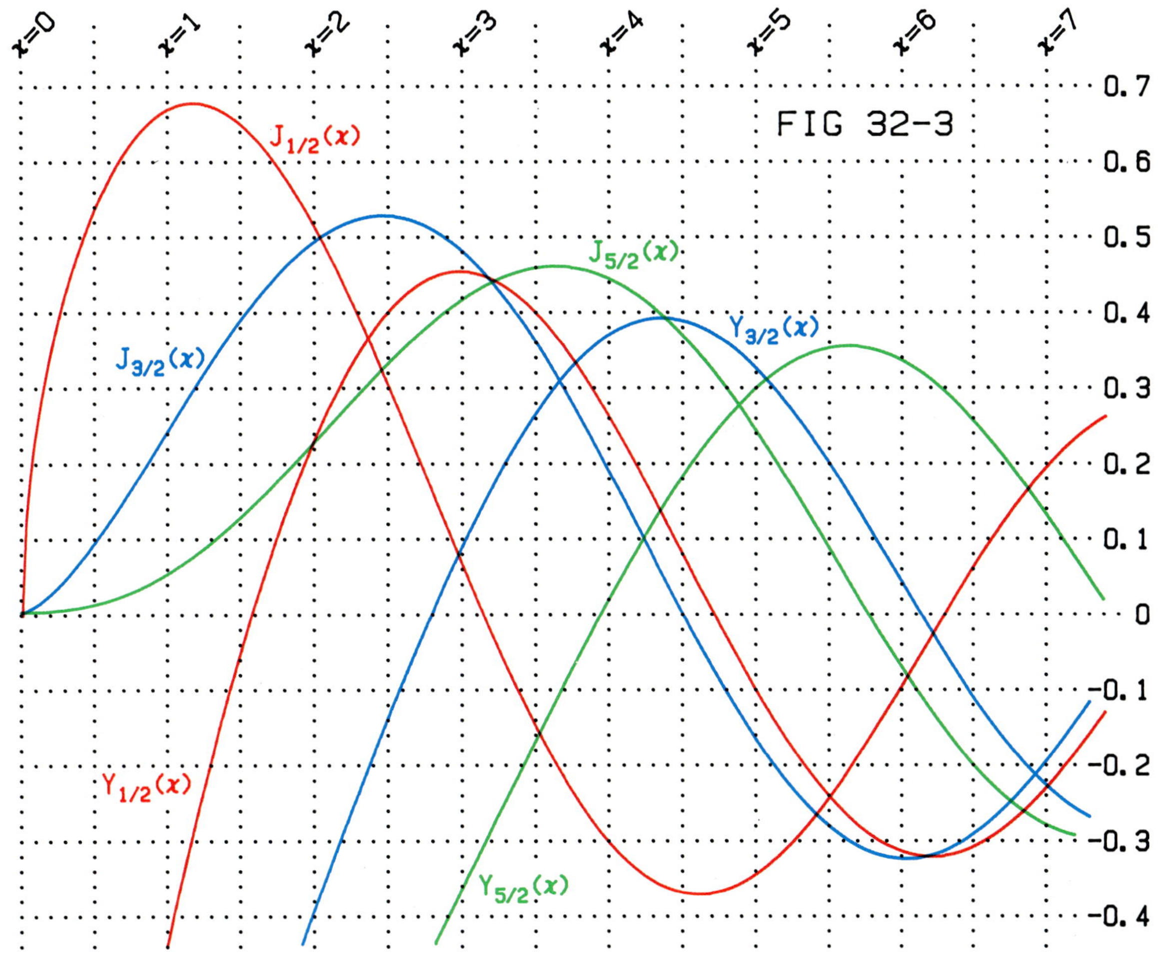

32:14 RELATED TOPICS

In this section we are concerned with the transform

32:14:1 $$\int_{-\infty}^{\infty} f(t)\exp(-2i\pi st)dt = \int_{-\infty}^{\infty} f(t)[\cos(2\pi st) - i\sin(2\pi st)]dt = F(s)$$

which converts a function f of t into a function F of s and is known as the *exponential Fourier transform*. In general, f and F may be complex, but t and s are usually real. Properties of the transform and examples for specific f(t) functions are tabulated [for example, by Beyer, *Handbook of Mathematical Sciences,* page 597, who, however, uses a definition slightly different from 32:14:1]. The function f(t) may be regenerated by the inversion formula

32:14:2 $$\int_{-\infty}^{\infty} \mathrm{F}(s)\exp(2i\pi st)\mathrm{d}s = \int_{-\infty}^{\infty} \mathrm{F}(s)[\cos(2\pi st) + i\sin(2\pi st)]\mathrm{d}s = \mathrm{f}(t)$$

which differs from 32:14:1 only in signs.

The operations known as *Fourier transformation* and *Fourier inversion* that are of such practical importance in science and engineering are discrete-and-finite analogs of the continuous-and-infinite formulas 32:14:1 and 32:14:2. Most applications are in the fields of spectroscopy and acoustics where f and F describe the intensity of a wave motion: F in the space of frequency, s; and f in the space of time, t. Usually, measurements of f consist of recording $f_0, f_1, f_2, \ldots, f_j, \ldots, f_{N-1}$ at equally spaced values of t, namely $t = 0, T/N, 2T/N, \ldots, jT/N, \ldots, (N-1)T/N$, each value possibly having a real component r_j and an imaginary component i_j. The equations

32:14:3 $$R_k = \frac{1}{N}\sum_{j=0}^{N-1}\left[r_j\cos\left(\frac{2\pi jk}{N}\right) + i_j\sin\left(\frac{2\pi jk}{N}\right)\right]$$

32:14:4 $$I_k = \frac{1}{N}\sum_{j=0}^{N-1}\left[i_j\cos\left(\frac{2\pi jk}{N}\right) - r_j\sin\left(\frac{2\pi jk}{N}\right)\right]$$

are used to transform these data into a set of real and imaginary components of the variable F. The subscript k is intimately related to the frequency variable s; in fact, $R_k + iI_k = F_k$ is the value of F sampled at $s = k/T$. The highest k value that is accessible corresponds to the *Nyquist frequency* $N/2T$; see Hamming for further discussions on this topic.

Equations 32:14:3 and 32:14:4 are the fundamental formulas used for numerical Fourier transformation. When N is large, these equations call for vast computational resources, and to alleviate this demand the *fast fourier transform* has been developed. This invention [for full details, Section 2.3.2 of Stoer and Bulirsch, or the original work of Cooley and Tukey, may be consulted] takes advantage of the symmetry properties of the sine and cosine functions to abbreviate the volume of arithmetic. Such abbreviation is especially pronounced when N, the number of data pairs, is a power of 2 (e.g., $N = 2^7 = 128$ or $N = 2^{10} = 1024$) and this is the condition incorporated into the algorithm below, which performs fast fourier transformation.

The algorithm first requires the input of the integer p, equal to $\log_2(N)$, followed by the sampled data $r_0, i_0, r_1, i_1, r_2, \ldots, i_{N-2}, r_{N-1}, i_{N-1}$. The ultimate output is in the order $R_0, I_0, R_1, I_1, R_2, \ldots, I_{N-2}, R_{N-1}, I_{N-1}$.

The two portions of the algorithm shown in green are virtually identical and could therefore be programmed as a subroutine. They are used to set register M equal to the *bit-reversed complement* of the integer K. This means that the bit pattern of the number K written in binary notation (e.g., 0101001 if $K = 41$ and $p = 7$) is reversed in M (e.g., 1001010 = 74 = M).

Even though the algorithm makes provision for imaginary components of the sampled input function, many applications of Fourier transformation utilize real data only. Then, of course, one enters the values $i_0 = i_1 = i_2 = \cdots = i_{N-1} = 0$ into the fast fourier transform algorithm. In such applications, only the first half of the output has useful information content because the second half merely duplicates the first in magnitude: $R_{N-k} = R_k$, $I_{N-k} = -I_k$.

Input p >>>>>>>
    Set $k = N = 2^{|p|}$
    Set $j = 0$
(1) Output j
j(as cue) <<<<<<
Input r_j >>>>>>>
Input i_j >>>>>>>
    Replace j by $j + 1$
    If $j \neq k$ go to (1)
(2) Replace k by $k/2$
    Set $m = -2k$

Storage needed: $2N + 9$ registers are required to store: $p, k, N, j, r_0, i_0, r_1, i_1, r_2, \ldots, i_{N-2}, r_{N-1}, i_{N-1}, m, K, M, I$ and R.

Use degree mode or replace 360 by 2π.

(3) Replace m by $m + 2k$
Set $K = m/k$
Set $M = 0$
Set $j = |p|$
(4) Replace j by $j - 1$
Replace M by $M + K2^j$
Replace K by $\text{Int}(K/2)$
Replace M by $M - 2K2^j$
If $j \neq 0$ go to (4)
Set $I = 360M/N$
Set $R = \cos(I)$
Replace I by $(p/|p|)\sin(I)$
Set $j = k + m - 1$
(5) Replace j by $j + 1$
Set $K = r_jR + i_jI$
Set $M = i_jR - r_jI$
Replace r_{j-k} by $r_{j-k} + K$
Set $r_j = r_{j-k} - 2K$
Replace i_{j-k} by $i_{j-k} + M$
Set $i_j = i_{j-k} - 2M$
If $j + 1 < 2k + m$ go to (5)
If $m + 2k < N$ go to (3)
If $k > 1$ go to (2)
Set $m = -1$
(6) Replace m by $m + 1$
Set $K = m$
Set $M = 0$
Set $j = |p|$
(7) Replace j by $j - 1$
Replace M by $M + K2^j$
Replace K by $\text{Int}(K/2)$
Replace M by $M - 2K2^j$
If $j \neq 0$ go to (7)
If $M \leq m$ go to (8)
Set $R = r_m$
Set $r_m = r_M$
Set $r_M = R$
Set $I = i_m$
Set $i_m = i_M$
Set $i_M = I$
(8) If $m < N - 1$ go to (6)
Set $k = -1$
If $p > 0$ go to (9)
Set $p = 0$
Set $N = 1$
(9) Replace k by $k + 1$
Output k
k <<<<<<<<<<
Set $R = r_k/N$
Output R
$R = R_k$ <<<<<<<
Set $I = i_k/N$
Output I

Input restriction: The parameter p must be an integer.

Test values for transformation:
$p = 4$
$r_0 = r_8 = 2,\ i_0 = i_8 = 3$
$r_1 = r_9 = \sqrt{2},\ i_1 = i_9 = 3$
$r_2 = r_{10} = 0,\ i_2 = i_{10} = 3$
$r_3 = r_{11} = -\sqrt{2},\ i_3 = i_{11} = 3$
$r_4 = r_{12} = -2,\ i_4 = i_{12} = 3$
$r_5 = r_{13} = -\sqrt{2},\ i_5 = i_{13} = 3$
$r_6 = r_{14} = 0,\ i_6 = i_{14} = 3$
$r_7 = r_{15} = \sqrt{2},\ i_7 = i_{15} = 3$
Output:
$I_0 = P_0 = 3$
$R_2 = P_2 = 1$
$R_{14} = P_{14} = 1$
All other outputs are zero.

$I = I_k$ <<<<<<<<<

If $p = 0$ go to (10)
Set $M = \sqrt{R^2 + I^2}$
Output M

$M = P_k$ <<<<<<<<

(10) If $k < m$ go to (9)

To test inversion, input $p = -4$ followed by the R_0, I_0, R_1, ..., I_{15} values listed above. Output values: those listed above as r_0, i_0, r_1, ..., i_{15}.

The physically significant result of the transformation is usually the real quantity $\sqrt{R_k^2 + I_k^2}$ which, as a function of the frequency k/t, is known as the *power spectrum* of the input data. The algorithm also outputs this quantity, P_k.

Many applications of Fourier transformation require that the transformed data be processed in some way in the frequency domain and then inverted back to the time domain. The equations describing the inversion are

32:14:5 $$r_j = \sum_{k=0}^{N-1} \left[R_k \cos\left(\frac{2\pi kj}{N}\right) - I_k \sin\left(\frac{2\pi kj}{N}\right) \right]$$

32:14:6 $$i_j = \sum_{k=0}^{N-1} \left[I_k \cos\left(\frac{2\pi kj}{N}\right) + R_k \sin\left(\frac{2\pi kj}{N}\right) \right]$$

and are so similar to equations 32:14:3 and 32:14:4 that the same algorithm may be used to effect inversion as transformation. Thus, the algorithm presented above is actually a fast fourier transformation/inversion algorithm. To select inversion, one merely inputs $-p$ instead of p, followed by R_0, I_0, R_1, I_1, R_2, ..., I_{N-2}, R_{N-1}, I_{N-1}. The output is r_0, i_0, r_1, i_1, r_2, ... i_{N-2}, r_{N-1}, i_{N-1}.

CHAPTER 33

THE SECANT sec(x) AND COSECANT csc(x) FUNCTIONS

These functions are the reciprocals of those discussed in Chapter 32. The secant and cosecant functions are interrelated by

33:0:1 $$[\sec^2(x) - 1][\csc^2(x) - 1] = 1$$

and by

33:0:2 $$\csc\left(x + \frac{\pi}{2}\right) = \sec(x)$$

33:1 NOTATION

The symbol cosec(x) sometimes replaces csc(x). To avoid possible confusion with the functions of Chapter 29, the names *circular secant* and *circular cosecant* may be used. Because of their applicability to triangles [see Section 34:14], the functions of Chapters 32–34 are known collectively as *trigonometric functions*.

33:2 BEHAVIOR

Figure 33-1 shows that sec(x) and csc(x) adopt all values except those between -1 and $+1$. The two functions may receive any argument, but sec(x) is ill defined when $x = \pm\pi/2, \pm3\pi/2, \pm5\pi/2, \ldots$ and csc(x) is similarly indefinite at $x = 0, \pm\pi, \pm2\pi, \ldots$. Both functions are periodic with period 2π. With the argument interpreted as an angle, the signs acquired by the sec and csc functions depend on the quadrant [Section 32:2] in which the argument lies, as illustrated in Table 33.2.1, which includes the other four trigonometric functions.

33:3 DEFINITIONS

The secant and cosecant functions may be defined as the reciprocals of the functions of Chapter 32:

33:3:1 $$\sec(x) = \frac{1}{\cos(x)}$$

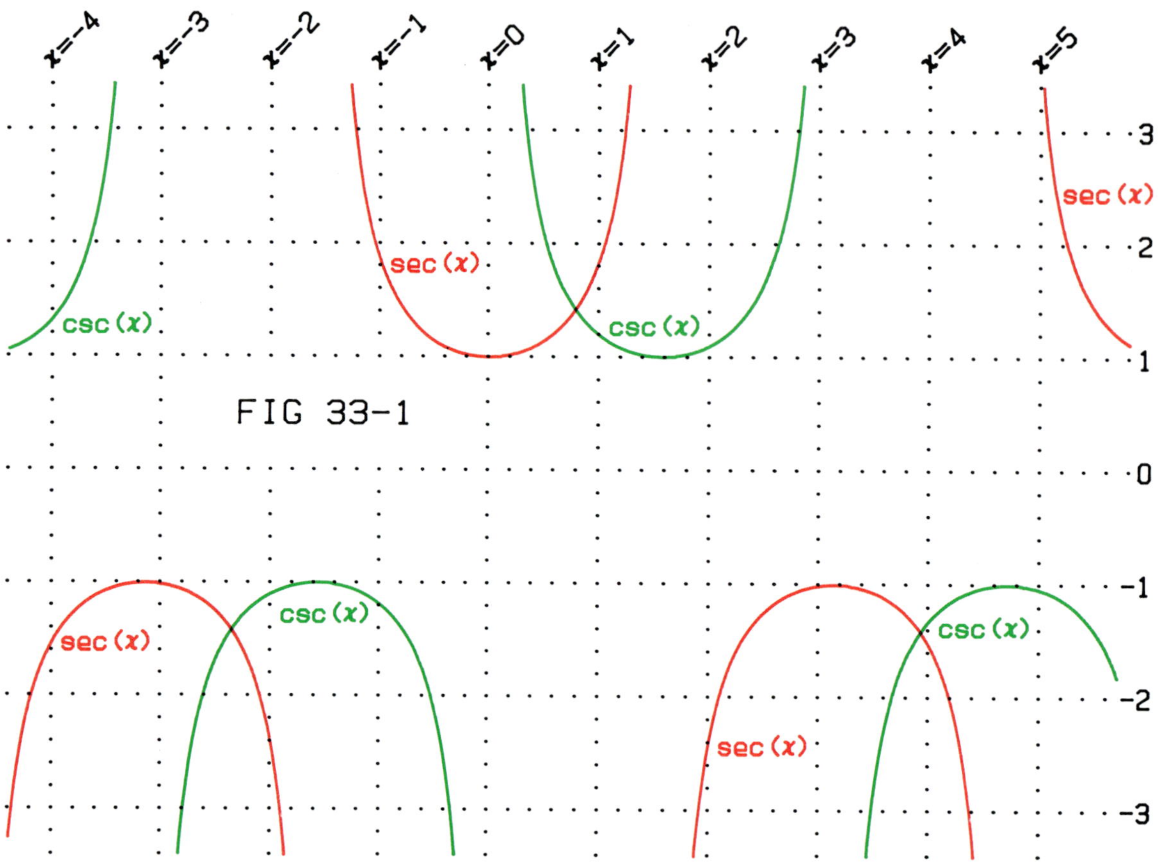

33:3:2 $$\csc(x) = \frac{1}{\sin(x)}$$

Three similar right-angled triangles are shown in Figure 33-2. If the sides represented by the dashed lines are all of unity length, then the other six sides have lengths that define the six trigonometric functions as depicted in the diagram. The argument x of these functions may be interpreted either as the marked angle or as the length of the arc of a unity-radius circle subtended by the angle, as diagrammed. Besides serving to define the six functions, many interrelations between them may be derived by applying similarity and Pythagorean relationships to the triangles. For example:

33:3:3 $$\frac{1}{\cos(x)} = \frac{\sec(x)}{1} = \frac{\csc(x)}{\cot(x)}$$

follows by equating the ratio of the red to the blue sides in the three triangles, while

33:3:4 $$\csc^2(x) = 1 + \cot^2(x)$$

is a consequence of applying Pythagoras' theorem to the third triangle.

Table 33.2.1

	First quadrant	Second quadrant	Third quadrant	Fourth quadrant
cos and sec	+	−	−	+
sin and csc	+	+	−	−
tan and cot	+	−	+	−

As well, the secant and cosecant functions may be defined in terms of exponential functions of imaginary argument

33:3:5 $$\sec(x) = \frac{2\exp(ix)}{\exp(2ix) + 1}$$

33:3:6 $$\csc(x) = \frac{2i\exp(ix)}{\exp(2ix) - 1}$$

or via the integral transforms

33:3:7 $$\sec(x) = \frac{2}{\pi}\int_0^\infty \frac{t^{2x/\pi}dt}{t^2 + 1} \qquad \frac{-\pi}{2} < x < \frac{\pi}{2}$$

33:3:8 $$\csc(x) = \frac{1}{\pi}\int_0^\infty \frac{t^{x/\pi}dt}{t^2 + t} \qquad 0 < x < \pi$$

33:4 SPECIAL CASES

There are none.

33:5 INTRARELATIONSHIPS

Whereas the secant function is even:

33:5:1 $$\sec(-x) = \sec(x)$$

the cosecant is an odd function:

33:5:2 $$\csc(-x) = -\csc(x)$$

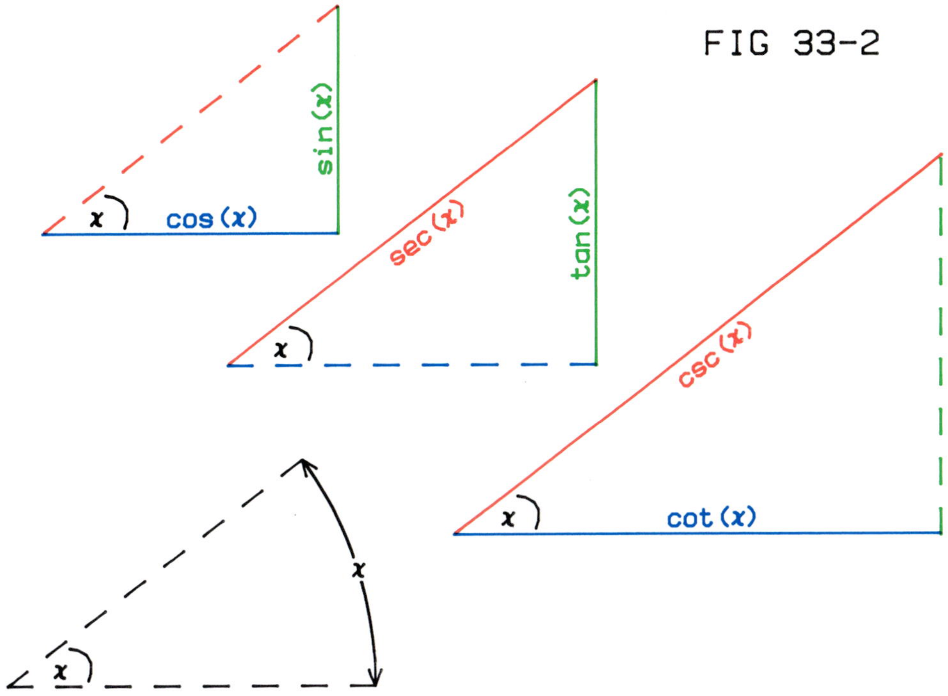

FIG 33-2

The recursion formulas

33:5:3
$$\sec\left(x \pm \frac{n\pi}{2}\right) = \begin{cases} \sec(x) & n = 0, 4, 8, \dots \\ \mp\csc(x) & n = 1, 5, 9, \dots \\ -\sec(x) & n = 2, 6, 10, \dots \\ \pm\csc(x) & n = 3, 7, 11, \dots \end{cases}$$

and

33:5:4
$$\csc\left(x + \frac{n\pi}{2}\right) = \begin{cases} \csc(x) & n = 0, 4, 8, \dots \\ \pm\sec(x) & n = 1, 5, 9, \dots \\ -\csc(x) & n = 2, 6, 10, \dots \\ \mp\sec(x) & n = 3, 7, 11, \dots \end{cases}$$

parallel those of the cosine and sine functions, but there are no simple formulas, akin to 32:5:9 or 32:5:10, to express $\sec(x \pm y)$ or $\csc(x \pm y)$. Such expressions—as well as those for $\sec^n(x)$, $\csc(x) \pm \csc(y)$, etc.—are best evaluated via the formulas of Section 32:5, making use of identities 33:3:1 and 33:3:2.

The secants and cosecants of double argument and half argument may be evaluated through the formulas

33:5:5
$$\sec(2x) = \frac{\sec^2(x)}{2 - \sec^2(x)}$$

33:5:6
$$\csc(2x) = \frac{\sec(x)\csc(x)}{2} = \frac{\csc^2(x)}{2\sqrt{\csc^2(x) - 1}}$$

33:5:7
$$\sec\left(\frac{x}{2}\right) = (-1)^m \sqrt{\frac{2\sec(x)}{1 + \sec(x)}} \qquad m = \mathrm{Int}\left(\frac{\pi + |x|}{2\pi}\right)$$

and

33:5:8
$$\csc\left(\frac{x}{2}\right) = (-1)^m \sqrt{\frac{\sec(x)}{\sec(x) - 1}} \qquad m = \mathrm{Int}\left(\frac{|x|}{2\pi}\right)$$

33:6 EXPANSIONS

The power series expansion of the secant

33:6:1
$$\left.\begin{aligned} \sec(x) &= 1 + \frac{x^2}{2} + \frac{5x^4}{24} + \frac{61x^6}{720} + \cdots = \sum_{j=0}^{\infty} \frac{|\mathrm{E}_{2j}|}{(2j)!} x^{2j} \\ &= \frac{2}{x} \sum_k \beta(k)(2x/\pi)^k \qquad k = 1, 3, 5, \dots \end{aligned}\right\} \quad \frac{-\pi}{2} < x < \frac{\pi}{2}$$

can be written in terms of Euler numbers [Chapter 5] or beta numbers [Chapter 3], whereas that of the cosecant

33:6:2
$$\left.\begin{aligned} \csc(x) &= \frac{1}{x} + \frac{x}{6} + \frac{7x^3}{360} + \frac{31x^5}{15120} + \cdots = \sum_{j=0}^{\infty} \frac{|(4^j - 2)\mathrm{B}_{2j}|x^{2j-1}}{(2j)!} \\ &= \frac{1}{x} + \frac{2}{x} \sum_{j=1}^{\infty} \eta(2j)(x/\pi)^{2j} \end{aligned}\right\} \quad -\pi < x < \pi$$

involves the Bernoulli numbers [Chapter 4] or the eta numbers [Chapter 3]. Series for the logarithms of sec(x) and of x csc(x) invoke the lambda and zeta numbers [Chapter 3]

33:6:3
$$\ln(\sec(x)) = \sum_{j=1}^{\infty} \frac{\lambda(2j)}{j} \left(\frac{2x}{\pi}\right)^{2j} \qquad \frac{-\pi}{2} < x < \frac{\pi}{2}$$

33:6:4 $$\ln(x\csc(x)) = \sum_{j=1}^{\infty} \frac{\zeta(2j)}{j}\left(\frac{x}{\pi}\right)^{2j} \qquad -\pi < x < \pi$$

The secant and cosecant functions may be expanded as partial fractions

33:6:5 $$\sec(x) = \frac{4\pi}{\pi^2 - 4x^2} - \frac{12\pi}{9\pi^2 - 4x^2} + \frac{20\pi}{25\pi^2 - 4x^2} - \cdots = \pi \sum_{j=0}^{\infty} \frac{(-1)^j(2j+1)}{(j+\frac{1}{2})^2\pi^2 - x^2}$$

33:6:6 $$\csc(x) = \frac{1}{x} + \frac{2x}{\pi^2 - x^2} - \frac{2x}{4\pi^2 - x^2} + \frac{2x}{9\pi^2 - x^2} - \cdots = \frac{1}{x} - 2x\sum_{j=1}^{\infty} \frac{(-1)^j}{j^2\pi^2 - x^2}$$

as may their squares

33:6:7 $$\sec^2(x) = \frac{4}{(\pi - 2x)^2} + \frac{4}{(\pi + 2x)^2} + \frac{4}{(3\pi - 2x)^2} + \frac{4}{(3\pi + 2x)^2} + \cdots = \sum_{j=-\infty}^{\infty} \frac{1}{[x + (j+\frac{1}{2})\pi]^2}$$

33:6:8 $$\csc^2(x) = \frac{1}{x^2} + \frac{1}{(\pi - x)^2} + \frac{1}{(\pi + x)^2} + \frac{1}{(2\pi - x)^2} + \frac{1}{(2\pi + x)^2} + \cdots = \sum_{j=-\infty}^{\infty} \frac{1}{[x + j\pi]^2}$$

33:7 PARTICULAR VALUES

Table 33.7.1 lists values of the secant and cosecant functions for special values of the argument x in the range $-\pi/2 \le x \le \pi/2$ (angles θ in the range $-90° \le \theta \le 90°$). For particular values outside this range, use the recursion formulas 33:5:3 and 33:5:4.

33:8 NUMERICAL VALUES

These are easily found by taking the reciprocals of the values of the cosine or sine [see Section 32:8].

33:9 APPROXIMATIONS

For small values of the argument, the approximations

33:9:1 $$\sec(x) \simeq \left(1 - \frac{x^2}{3}\right)^{-3/2} \qquad \text{8-bit precision} \qquad -0.9 \le x \le 0.9$$

33:9:2 $$\csc(x) \simeq \frac{1}{x} + \frac{x}{6} \qquad \text{8-bit precision} \qquad -\frac{2}{3} \le x \le \frac{2}{3}$$

are valid.

Table 33.7.1

θ	0°	±15°	±18°	$\pm 22\frac{1}{2}°$	±30°	±36°	±45°	±54°	±60°	$\pm 67\frac{1}{2}°$	±72°	±75°	±90°
x	0	$\frac{\pm\pi}{12}$	$\frac{\pm\pi}{10}$	$\frac{\pm\pi}{8}$	$\frac{\pm\pi}{6}$	$\frac{\pm\pi}{5}$	$\frac{\pm\pi}{4}$	$\frac{\pm 3\pi}{10}$	$\frac{\pm\pi}{3}$	$\frac{\pm 3\pi}{8}$	$\frac{\pm 2\pi}{5}$	$\frac{\pm 5\pi}{12}$	$\frac{\pm\pi}{2}$
sec	1	$\sqrt{6} - \sqrt{2}$	$\sqrt{2 - \frac{2}{\sqrt{5}}}$	$\sqrt{4 - \sqrt{8}}$	$\frac{2}{\sqrt{3}}$	$\sqrt{6 - \sqrt{20}}$	$\sqrt{2}$	$\sqrt{2 + \frac{2}{\sqrt{5}}}$	2	$\sqrt{4 + \sqrt{8}}$	$\sqrt{6 + \sqrt{20}}$	$\sqrt{6} + \sqrt{2}$	$\pm\infty$
csc	$\pm\infty$	$\pm\sqrt{6} \pm \sqrt{2}$	$\pm\sqrt{6 + \sqrt{20}}$	$\pm\sqrt{4 + \sqrt{8}}$	± 2	$\pm\sqrt{2 + \frac{2}{\sqrt{5}}}$	$\pm\sqrt{2}$	$\pm\sqrt{6 - \sqrt{20}}$	$\frac{\pm 2}{\sqrt{3}}$	$\pm\sqrt{4 - \sqrt{8}}$	$\pm\sqrt{2 - \frac{2}{\sqrt{5}}}$	$\pm\sqrt{6} \mp \sqrt{2}$	1

33:10 OPERATIONS OF THE CALCULUS

Differentiation of the secant and cosecant functions gives

33:10:1 $$\frac{d}{dx}\sec(x) = \sec(x)\tan(x) = \frac{\sec^2(x)}{\csc(x)}$$

and

33:10:2 $$\frac{d}{dx}\csc(x) = -\csc(x)\cot(x) = \frac{-\csc^2(x)}{\sec(x)}$$

whereas integration leads to

33:10:3 $$\int_{\pi/2}^{x}\csc(t)dt = \ln\left[\tan\left(\frac{x}{2}\right)\right] = \ln[\csc(x) - \cot(x)]$$

and

33:10:4 $$\int_{0}^{x}\sec(t)dt = \ln\left[\tan\left(\frac{x}{2} + \frac{\pi}{4}\right)\right] = \ln[\sec(x) + \tan(x)] = \mathrm{invgd}(x)$$

where invgd is the inverse gudermannian function discussed in Section 33:14. Indefinite integrals of the squares of the secant and cosecant functions have simpler formulations:

33:10:5 $$\int_{0}^{x}\sec^2(t)dt = \tan(x)$$

and

33:10:6 $$\int_{x}^{\pi/2}\csc^2(t)dt = \cot(x)$$

33:11 COMPLEX ARGUMENT

The formulas

33:11:1 $$\sec(x + iy) = \frac{\coth(y) + i\tan(x)}{\sec(x)\sinh(y) + \cos(x)\operatorname{csch}(y)}$$

33:11:2 $$\csc(x + iy) = \frac{\coth(y) - i\cot(x)}{\sin(x)\operatorname{csch}(y) + \csc(x)\sinh(y)}$$

reduce to

33:11:3 $$\sec(iy) = \operatorname{sech}(y)$$

33:11:4 $$\csc(iy) = -i\operatorname{csch}(y)$$

when the argument is purely imaginary.

33:12 GENERALIZATIONS

The Jacobian elliptic functions [see Chapter 63] nc(p;x) and dc(p;x) are generalizations of sec(x), to which they reduce as $p \to 0$. Similarly, csc(x) is the $p = 0$ limit of the ns(p;x) and ds(p;x) functions.

33:13 COGNATE FUNCTIONS

The secant and cosecant functions are related to the other circular functions [Chapters 32, 33 and 34] by

33:13:1 $$\sec(x) = \frac{\sigma_4}{\sqrt{1-\sin^2(x)}} = \frac{1}{\cos(x)} = \frac{\sigma_3 \csc(x)}{\sqrt{\csc^2(x)-1}} = \sigma_4\sqrt{1+\tan^2(x)} = \frac{\sigma_2\sqrt{\cot^2(x)+1}}{\cot(x)}$$

33:13:2 $$\csc(x) = \frac{1}{\sin(x)} = \frac{\sigma_2}{\sqrt{1-\cos^2(x)}} = \frac{\sigma_3 \sec(x)}{\sqrt{\sec^2(x)-1}} = \frac{\sigma_4\sqrt{1+\tan^2(x)}}{\tan(x)} = \sigma_2\sqrt{\cot^2(x)+1}$$

The multipliers σ_2, σ_3 and σ_4 equal ± 1 according to the magnitude of x, as explained in Section 32:13; for $0 \le x \le \pi/2$ (i.e., in the first quadrant), they are all plus unity.

The secant and cosecant functions are related to hyperbolic functions in two distinct ways: through the imaginary-argument formulas 33:11:3 and 33:11:4, and via the gudermannian function of Section 33:14.

Of course, sec and csc are closely related to their inverses, the arcsec and arccsc functions that are the subject of Chapter 35. Occasionally encountered is the *exsecant* function

33:13:3 $$\text{exsec}(x) = \sec(x) - 1 = \tan(x)\tan\left(\frac{x}{2}\right) \qquad \frac{-\pi}{2} < x < \frac{\pi}{2}$$

33:14 RELATED TOPICS

Comparison of Figures 29-2 and 33-2 suggests that the six hyperbolic functions and the six circular functions are closely interrelated. In fact, one has

33:14:1 $$\sin(x) = \tanh(y)$$

33:14:2 $$\cos(x) = \text{sech}(y)$$

33:14:3 $$\sec(x) = \cosh(y)$$

33:14:4 $$\csc(x) = \coth(y)$$

33:14:5 $$\tan(x) = \sinh(y)$$

33:14:6 $$\cot(x) = \text{csch}(y)$$

$$-\frac{\pi}{2} < x < \frac{\pi}{2}$$

provided that the arguments x and y are suitably related. The relationship involves the *gudermannian function*

33:14:7 $$x = \text{gd}(y)$$

or the *inverse gudermannian function*

33:14:8 $$y = \text{invgd}(x)$$

also denoted $\text{gd}^{-1}(x)$.

The gudermannian function may be defined in a variety of ways, including

33:14:9 $$\text{gd}(y) = 2\arctan(\exp(y)) - \frac{\pi}{2} = 2\arctan\left(\tanh\left(\frac{y}{2}\right)\right)$$

and

33:14:10 $$\text{gd}(y) = \int_0^y \text{sech}(t)\,dt$$

Similarly, the inverse gudermannian function may be defined by

33:14:11 $$\text{invgd}(x) = \ln\left(\tan\left(\frac{x}{2} + \frac{\pi}{4}\right)\right) = \ln(\sec(x) + \tan(x)) \qquad -\frac{\pi}{2} < x < \frac{\pi}{2}$$

or by the integral

33:14:12 $$\mathrm{invgd}(x) = \int_0^x \sec(t)\,dt \qquad -\frac{\pi}{2} < x < \frac{\pi}{2}$$

The same function may be realized as a special case of the incomplete elliptic integral of the first kind [Chapter 62]:

33:14:13 $$\mathrm{invgd}(x) = \mathrm{F}(1;x)$$

As Figure 33-3 demonstrates, both the gudermannian and its inverse are odd functions. The gudermannian function approaches $\pm\pi/2$ as its argument approaches $\pm\infty$. Conversely, invgd(x) approaches $\pm\infty$ as $x \to \pm\pi/2$. The derivatives of gd(x) and invgd(x) are sech(x) and sec(x), respectively.

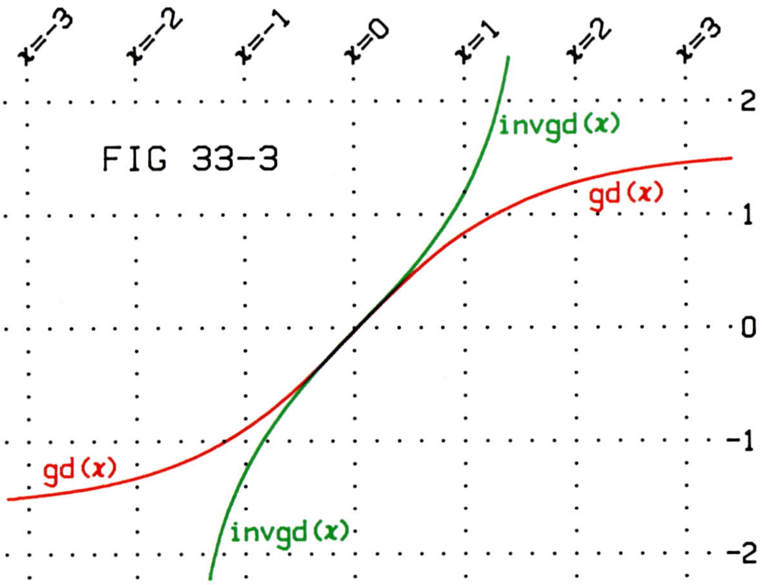

The power series of gd(x) and invgd(x) involve Euler numbers [Chapter 5] and are remarkably similar:

33:14:14 $$\mathrm{gd}(x) = x - \frac{x^3}{6} + \frac{x^5}{24} - \frac{61x^7}{5040} + \cdots = \sum_{j=0}^{\infty} \frac{\mathrm{E}_{2j}x^{2j+1}}{(2j+1)!} \qquad -1 < x < 1$$

33:14:15 $$\mathrm{invgd}(x) = x + \frac{x^3}{6} + \frac{x^5}{24} + \frac{61x^7}{5040} + \cdots = \sum_{j=0}^{\infty} \frac{|\mathrm{E}_{2j}|x^{2j+1}}{(2j+1)!} \qquad -\frac{\pi}{2} < x < \frac{\pi}{2}$$

Several other expansions exist for the gudermannian function and its inverse [see Beyer, *Handbook of Mathematical Sciences,* pages 323–325].

CHAPTER 34

THE TANGENT tan(x) AND COTANGENT cot(x) FUNCTIONS

The functions of this chapter are the reciprocals of each other:

34:0:1 $$\tan(x)\cot(x) = 1$$

Together with the sine, cosine, secant and cosecant functions [Chapters 32 and 33] they constitute the six trigonometric functions that play important roles in the mensuration of triangles [see Section 34:14].

34:1 NOTATION

The alternative notation tg(x) is occasionally encountered for the tangent function, while cotan(x) or ctg(x) sometimes replaces cot(x). To emphasize the distinction from the functions addressed in Chapter 30, the names *circular tangent* and *circular cotangent* are often applied to tan(x) and cot(x). As in Chapters 32 and 33, we use tan(θ) and cot(θ) as alternatives to tan(x) and cot(x) whenever we wish to stress the angular interpretation of the argument [A, B and C are used as angular arguments in Section 34:14].

34:2 BEHAVIOR

Like the other four trigonometric functions, the tangent and cotangent are periodic functions, but unlike the other four the period is π, not 2π. This is evident in Figure 34-1. The signs acquired by tan(θ) and cot(θ) in the four quadrants [as defined in Section 32:2] are tabulated in Section 33:2.

The tangent function has zeros at $x = 0, \pm\pi, \pm2\pi, \ldots$ but is undefined at $x = \pm\pi/2, \pm3\pi/2, \pm5\pi/2, \ldots$. Conversely, cot($x$) has zeros at $x = \pm\pi/2, \pm3\pi/2, \pm5\pi/2, \ldots$ but is undefined at $x = 0, \pm\pi, \pm2\pi, \ldots$.

34:3 DEFINITIONS

Geometric definitions of the tangent and cotangent functions are possible with the aid of Figure 32-2, namely

34:3:1 $$\tan(x) = \frac{PQ}{OQ}$$

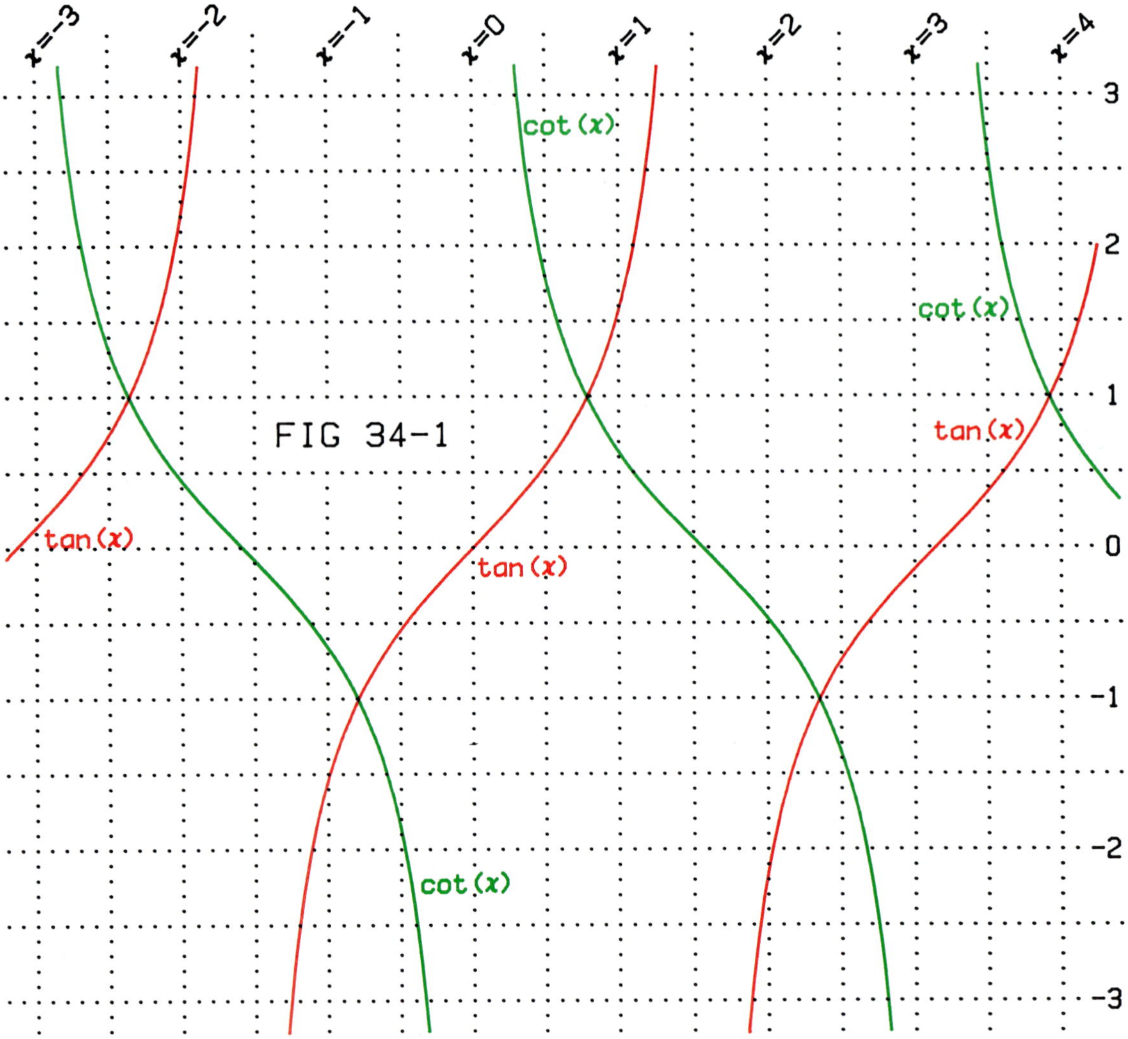

34:3:2
$$\cot(x) = \frac{OQ}{PQ}$$

The functions may also be defined in terms of the sine and cosine functions:

34:3:3
$$\tan(x) = \frac{\sin(x)}{\cos(x)}$$

34:3:4
$$\cot(x) = \frac{\cos(x)}{\sin(x)}$$

or via exponentials of imaginary argument:

34:3:5
$$\tan(x) = \frac{2i}{\exp(2ix) + 1} - i$$

34:3:6
$$\cot(x) = \frac{2i}{\exp(2ix) - 1} + i$$

or trigonometrically as in Figure 33-2.

The indefinite integrals

34:3:7 $$\tan(x) = \int_0^x \sec^2(t)\mathrm{d}t$$

34:3:8 $$\cot(x) = \int_x^{\pi/2} \csc^2(t)\mathrm{d}t$$

also define the tangent and cotangent functions, as do the definite integrals

34:3:9 $$\tan(x) = \frac{2}{\pi}\int_0^\infty \frac{t^{2x/\pi} - 1}{t^2 - 1}\mathrm{d}t \qquad x < \frac{\pi}{2}$$

34:3:10 $$\cot(x) = \frac{2}{\pi}\int_0^\infty \frac{t^{2x/\pi}\mathrm{d}t}{t - t^3} \qquad x < \pi$$

For appropriate values of a, b and c, the differential equation

34:3:11 $$\frac{\mathrm{d}f}{\mathrm{d}x} = af^2 + bf + c \qquad b^2 < 4ac$$

is solved by

34:3:12 $$f = \frac{\sqrt{4ac - b^2}}{2a}\tan\left(\frac{x\sqrt{4ac - b^2}}{2}\right) - \frac{b}{2a} \qquad \text{or} \qquad -\frac{\sqrt{4ac - b^2}}{2a}\cot\left(\frac{x\sqrt{4ac - b^2}}{2}\right) - \frac{b}{2a}$$

When $b^2 > 4ac$, a solution is $f = -(b/2a) - (\sqrt{b^2 - 4ac}/2a)\tanh(x\sqrt{b^2 - 4ac}/2)$ or $-(b/2a) - (\sqrt{b^2 - 4ac}/2a)\coth(x\sqrt{b^2 - 4ac}/2)$ [see equation 30:3:6 for a special case]. When $b^2 = 4ac$, a solution is $f = -(1 + bx)/4ax$.

34:4 SPECIAL CASES

There are none.

34:5 INTRARELATIONSHIPS

The tangent and cotangent functions are both odd:

34:5:1 $$\mathrm{f}(-x) = -\mathrm{f}(x) \qquad \mathrm{f} = \tan \text{ or } \cot$$

The argument-addition formulas

34:5:2 $$\tan(x \pm y) = \frac{\tan(x) \pm \tan(y)}{1 \mp \tan(x)\tan(y)}$$

and

34:5:3 $$\cot(x \pm y) = \frac{\cot(x)\cot(y) \mp 1}{\cot(y) \pm \cot(x)}$$

have the special cases

34:5:4 $$\tan(2x) = \frac{2\tan(x)}{1 - \tan^2(x)} = \frac{2\cot(x)}{\cot^2(x) - 1} = \frac{1}{\cot(2x)}$$

and

34:5:5 $$\tan(3x) = \frac{3\tan(x) - \tan^3(x)}{1 - 3\tan^2(x)} = \frac{3\cot^2(x) - 1}{\cot^3(x) - 3\cot(x)} = \frac{1}{\cot(3x)}$$

and also generate the recursion formulas

34:5:6
$$\tan\left(x \pm \frac{n\pi}{4}\right) = \begin{cases} \tan(x) & n = 0, 4, 8, \ldots \\ \dfrac{\tan(x) \pm 1}{1 \mp \tan(x)} & n = 1, 5, 9, \ldots \\ -\cot(x) & n = 2, 6, 10, \ldots \\ \dfrac{\tan(x) \mp 1}{1 \pm \tan(x)} & n = 3, 7, 11, \ldots \end{cases}$$

and

34:5:7
$$\cot\left(x \pm \frac{n\pi}{4}\right) = \begin{cases} \cot(x) & n = 0, 4, 8, \ldots \\ \dfrac{\cot(x) \mp 1}{1 \pm \cot(x)} & n = 1, 5, 9, \ldots \\ -\tan(x) & n = 2, 6, 10, \ldots \\ \dfrac{\cot(x) \pm 1}{1 \mp \cot(x)} & n = 3, 7, 11, \ldots \end{cases}$$

The tangent of half argument, which we represent here by

34:5:8
$$\tau = \tan\left(\frac{x}{2}\right) = (-1)^m \sqrt{\frac{1 - \cos(x)}{1 + \cos(x)}} \qquad m = \mathrm{Int}\left(\frac{2x}{\pi}\right)$$

is a useful quantity because it is related in a very simple fashion to the following trigonometric functions of x:

34:5:9
$$\sin(x) = \frac{2\tau}{1 + \tau^2}$$

34:5:10
$$\cos(x) = \frac{1 - \tau^2}{1 + \tau^2}$$

34:5:11
$$\sec(x) = \frac{1 + \tau^2}{1 - \tau^2}$$

34:5:12
$$\csc(x) = \frac{1 + \tau^2}{2\tau}$$

34:5:13
$$\tan(x) = \frac{1}{1 - \tau} - \frac{1}{1 + \tau}$$

34:5:14
$$\cot(x) = \frac{1 - \tau^2}{2\tau}$$

34:5:15
$$1 + \cos(x) = \frac{2}{1 + \tau^2}$$

34:5:16
$$\sec(x) + 1 = \frac{1}{1 - \tau} + \frac{1}{1 + \tau}$$

34:5:17
$$\sec(x) + \tan(x) = \frac{1 + \tau}{1 - \tau}$$

34:5:18
$$\sec(x) - \tan(x) = \frac{1 - \tau}{1 + \tau}$$

34:5:19 $$\csc(x) + \cot(x) = \frac{1}{\tau}$$

34:5:20 $$\csc(x) - \cot(x) = \tau$$

These equations apply irrespective of the magnitude of x, that is, in all four quadrants, as defined in Section 32:2. The series

34:5:21 $$\tau + \frac{\tau^3}{3} + \frac{\tau^5}{5} + \cdots = \sum_{j=0}^{\infty} \frac{\tau^{2j+1}}{2j+1} = \frac{\mathrm{invgd}(x)}{2}$$

sums to the inverse gudermannian function [Section 33:14].

The function-addition formulas

34:5:22 $$\tan(x) \pm \tan(y) = \sin(x \pm y)\sec(x)\sec(y)$$

34:5:23 $$\cot(x) \pm \cot(y) = \sin(y \pm x)\csc(x)\csc(y)$$

34:5:24 $$\cot(x) + \tan(x) = 2\csc(2x)$$

34:5:25 $$\cot(x) - \tan(x) = 2\cot(2x)$$

and function-multiplication formulas

34:5:26 $$\tan(x)\tan(y) = \frac{\cos(x-y) - \cos(x+y)}{\cos(x-y) + \cos(x+y)}$$

34:5:27 $$\cot(x)\cot(y) = \frac{\cos(x-y) + \cos(x+y)}{\cos(x-y) - \cos(x+y)}$$

34:5:28 $$\tan(x)\cot(y) = \frac{\sin(x+y) + \sin(x-y)}{\sin(x+y) - \sin(x-y)}$$

complete our list of intrarelationships.

34:6 EXPANSIONS

Power series corresponding to the tangent and cotangent functions and their logarithms are

34:6:1 $$\tan(x) = x + \frac{x^3}{3} + \frac{2x^5}{15} + \frac{17x^7}{315} + \cdots = \frac{2}{x}\sum_{j=1}^{\infty} \lambda(2j)\left(\frac{4x^2}{\pi^2}\right)^j$$
$$= \sum_{j=1}^{\infty} \frac{4^j(4^j-1)|B_{2j}|}{(2j)!} x^{2j-1} \qquad -\frac{\pi}{2} < x < \frac{\pi}{2}$$

34:6:2 $$\cot(x) = \frac{1}{x} - \frac{x}{3} - \frac{x^3}{45} - \frac{2x^5}{945} - \cdots = \frac{1}{x} - \frac{2}{x}\sum_{j=1}^{\infty} \zeta(2j)\frac{x^{2j}}{\pi^{2j}}$$
$$= \frac{1}{x} - \sum_{j=1}^{\infty} \frac{4^j|B_{2j}|}{(2j)!} x^{2j-1} \qquad -\pi < x < \pi$$

and

34:6:3 $$\ln\left[\frac{\tan(x)}{x}\right] = -\ln[x\cot(x)] = \frac{x^2}{3} + \frac{7x^4}{90} + \frac{62x^6}{2835} + \cdots = \sum_{j=1}^{\infty} \frac{4^j(4^j-2)|B_{2j}|}{2j(2j)!} x^{2j}$$
$$= \sum_{j=1}^{\infty} \frac{\eta(2j)}{j}\left(\frac{4x^2}{\pi^2}\right)^j \qquad -\frac{\pi}{2} < x < \frac{\pi}{2}$$

where the λ, ζ and η numbers are considered in Chapter 3 and the Bernoulli numbers in Chapter 4.

The partial fraction expansions of the tangent and cotangent functions

34:6:4 $$\tan(x) = \frac{8x}{\pi^2 - 4x^2} + \frac{8x}{9\pi^2 - 4x^2} + \frac{8x}{25\pi^2 - 4x^2} + \cdots$$

34:6:5 $$\cot(x) = \frac{1}{x} - \frac{2x}{\pi^2 - x^2} - \frac{2x}{4\pi^2 - x^2} - \frac{2x}{9\pi^2 - x^2} - \cdots$$

can be concisely written

34:6:6 $$f(x) = \sum_j \frac{1}{x - x_j} \qquad f(x) = \cot(x) \text{ or } -\tan(x)$$

where x_j are the values of x that make f(x) infinite; for example, for cot(x): $x_0 = 0$, $x_1 = \pi$, $x_2 = -\pi$, $x_3 = 2\pi$, ...

The continued fraction expansion

34:6:7 $$\tan x = \frac{x}{1-}\,\frac{x^2}{3-}\,\frac{x^2}{5-}\,\frac{x^2}{7-}\cdots \qquad x \neq \pm\frac{\pi}{2}, \pm\frac{3\pi}{2}, \pm\frac{5\pi}{2}, \ldots$$

holds for the tangent function.

34:7 PARTICULAR VALUES

Values of the tangent and cotangent functions for particular values of the argument x in the range $-\pi/2 \le x \le \pi/2$ (angles θ between $-90°$ and $90°$) are listed in Table 34.7.1, and particular values outside this range are calculable via the recursion formulas 34:5:6 and 34:5:7.

The values of x that satisfy the equation

34:7:1 $$\tan(x) = bx \qquad -\infty < b < \infty$$

arise in the solutions to certain problems in applied mathematics. These values, the so-called *roots* of equation 34:7:1, depend on b and will be denoted $r_j(b)$. They form an infinite set, symmetrically disposed about $x = 0$, viz

34:7:2 $$x = 0, \pm r_0(b), \pm r_1(b), \pm r_2(b), \ldots, \pm r_j(b), \ldots \qquad \text{where} \qquad j\pi - \frac{\pi}{2} < r_j < j\pi + \frac{\pi}{2}$$

except that the roots $\pm r_0(b)$ exist only if $b \ge 1$. The positive members of this set may be found by utilizing the following simple algorithm, which is based on inversion of equation 34:7:1 to

34:7:3 $$x = \mathrm{Arctan}(bx) = j\pi + \arctan(bx) \qquad x = r_j(b)$$

where the Arctan and arctan functions are discussed in Chapter 35. Starting with a crude estimate of $r_j(b)$, namely $j\pi$ (or $\pi/4$ for $j = 0$), repeated application of 34:7:3 converges to the exact value of the jth positive root. The algorithm halts when two successive estimates of $r_j(b)$ differ by less than 10^{-9}.

Table 34.7.1

θ	0°	±15°	±18°	$\pm 22\frac{1}{2}°$	±30°	±36°	±45°	±54°	±60°	$\pm 67\frac{1}{2}°$	±72°	±75°	±90°
x	0	$\frac{\pm\pi}{12}$	$\frac{\pm\pi}{10}$	$\frac{\pm\pi}{8}$	$\frac{\pm\pi}{6}$	$\frac{\pm\pi}{5}$	$\frac{\pm\pi}{4}$	$\frac{\pm 3\pi}{10}$	$\frac{\pm\pi}{3}$	$\frac{\pm 3\pi}{8}$	$\frac{\pm 2\pi}{5}$	$\frac{\pm 5\pi}{12}$	$\pm\frac{\pi}{2}$
tan	0	$\pm(2 - \sqrt{3})$	$\pm\sqrt{1 - \frac{2}{\sqrt{5}}}$	$\pm(\sqrt{2} - 1)$	$\frac{\pm 1}{\sqrt{3}}$	$\pm\sqrt{5 - \sqrt{20}}$	± 1	$\pm\sqrt{1 + \frac{2}{\sqrt{5}}}$	$\pm\sqrt{3}$	$\pm(\sqrt{2} + 1)$	$\pm\sqrt{5 + \sqrt{20}}$	$\pm(2 + \sqrt{3})$	$\pm\infty$
cot	$\pm\infty$	$\pm(2 + \sqrt{3})$	$\pm\sqrt{5 + \sqrt{20}}$	$\pm(\sqrt{2} + 1)$	$\pm\sqrt{3}$	$\pm\sqrt{1 + \frac{2}{\sqrt{5}}}$	± 1	$\pm\sqrt{5 - \sqrt{20}}$	$\frac{\pm 1}{\sqrt{3}}$	$\pm(\sqrt{2} - 1)$	$\pm\sqrt{1 - \frac{2}{\sqrt{5}}}$	$\pm(2 - \sqrt{3})$	0

```
Input b >>>>>>>
Input j >>>>>>>
          Set r = jπ
          If r ≠ 0 go to (1)
          If b ≤ 1 go to (1)
          Set r = π/4
      (1) Set q = r
          Replace r by arctan(br) + jπ
          If |r − q| ≥ 10^-9 go to (1)
          Output r
r ≃ r_j(b) <<<<<<
```

Storage needed: b, j, r and q

Use radian mode or change arctan(br) to (π/180) arctan(br).

Test values:
$r_9(-1) = 26.7409160$
$r_0(2) = 1.16556118$

The case $b = 1$ is especially important and these roots are given by

34:7:4 $$r_j(1) = J - \frac{1}{J} - \frac{2}{3J^3} - \frac{13}{15J^5} - \frac{146}{105J^7} - \frac{781}{315J^9} - \cdots \qquad \text{where} \qquad J = (j + {}^1\!/_2)\pi$$

With j taking values from 1 through ∞, the sums $\Sigma r_j^{-n}(1)$ occur in certain problems and have the values $\frac{1}{10}$, $\frac{1}{350}$ and $\frac{1}{7875}$ for $n = 2$, 4 and 6. The values of $r_j(1)$ occur in the expansion of the Langevin function [Section 30:13] and correspond to the zeros of the spherical Bessel functions of the first kind [see Section 53:7].

Similarly, the roots of the equation

34:7:5 $$\cot(x) = bx \qquad -\infty < b < \infty$$

which we denote

34:7:6 $$x = \pm\rho_1(b),\ \pm\rho_2(b),\ \pm\rho_3(b),\ \ldots,\ \pm\rho_j(b),\ \ldots \qquad \text{where} \qquad j\pi - \pi < \rho_j(b) < j\pi$$

are determinable via the recursive algorithm

```
Input b >>>>>>>
Input j >>>>>>>
          Set ρ = (j − 1/2)π
          If b = 0 go to (2)
          Replace ρ by ρ − π|b|/2b
      (1) Set q = ρ
          Replace ρ by ρ + arctan(1/bρ)
          If |ρ − q| ≥ 10^-9 go to (1)
      (2) Output ρ
ρ ≃ ρ_j(b) <<<<<<
```

Storage needed: b, j, ρ and q

Use radian mode or change arctan($1/b\rho$) to (π/180) arctan($1/b\rho$).

Test values:
$\rho_5(1) = 12.6452872$
$\rho_2(-1) = 6.12125047$

34:8 NUMERICAL VALUES

Computer languages and programmable calculators invariably permit the tangent function to be evaluated either straightforwardly or via definition 34:3:3. Accordingly, we present no algorithms here.

34:9 APPROXIMATIONS

For small values of the argument, the tangent and cotangent functions can be approximated by

34:9:1 $$\tan(x) \simeq \frac{x}{1 - \dfrac{x^2}{3}} \qquad \text{8-bit precision} \qquad -0.6 \le x \le 0.6$$

and

34:9:2 $$\cot(x) \simeq \frac{\left(1 - \frac{x^2}{6}\right)^2}{x} \qquad \text{8-bit precision} \qquad -0.5 \le x \le 0.5$$

From expansion 34:6:6 it is evident that $1/(x - x_j)$ is an approximation to cot(x) or −tan(x) whenever x is close to one of the arguments, x_j, at which the function is infinite. In fact:

34:9:3 $$\tan(x) \simeq \frac{1}{x_j - x} \qquad x_j = \pm\pi/2, \pm 3\pi/2, \pm 5\pi/2, \ldots \qquad \text{8-bit precision} \qquad -0.1 \le x_j - x \le 0.1$$

34:9:4 $$\cot(x) \simeq \frac{1}{x - x_j} \qquad x_j = 0, \pm\pi, \pm 2\pi, \ldots \qquad \text{8-bit precision} \qquad -0.1 \le x - x_j \le 0.1$$

34:10 OPERATIONS OF THE CALCULUS

Differentiation and indefinite integration give

34:10:1 $$\frac{d}{dx}\tan(x) = \sec^2(x) = 1 + \tan^2(x)$$

34:10:2 $$\frac{d}{dx}\cot(x) = -\csc^2(x) = -1 - \cot^2(x)$$

34:10:3 $$\int_0^x \tan(t)dt = \ln(\sec(x)) = \frac{1}{2}\ln(1 + \tan^2(x))$$

34:10:4 $$\int_x^{\pi/2} \cot(t)dt = \ln(\csc(x)) = \frac{1}{2}\ln(1 + \cot^2(x))$$

Integer powers of the tangent function integrate to give

34:10:5 $$\int_0^x \tan^n(t)dt = \frac{\tan^{n-1}(x)}{n-1} - \frac{\tan^{n-3}(x)}{n-3} + \frac{\tan^{n-5}(x)}{n-5} - \cdots \begin{cases} -\frac{1}{2}\tan^2(x) + \ln(\sec(x)) & n = 1, 5, 9, \ldots \\ +\tan(x) - x & n = 2, 6, 10, \ldots \\ +\frac{1}{2}\tan^2(x) - \ln(\sec(x)) & n = 3, 7, 11, \ldots \\ -\tan(x) + x & n = 4, 8, 12, \ldots \end{cases}$$

The integral of $\cot^n(t)$ is similar, but without alternating signs. Section 58:14 discusses the indefinite integrals of an arbitrary power of the tangent and cotangent functions.

Some important definite integrals include

34:10:6 $$\int_0^{\pi/2} \tan^\nu(t)dt = \int_0^{\pi/2} \cot^\nu(t)dt = \frac{\pi}{2}\sec\left(\frac{\nu\pi}{2}\right) \qquad -1 < \nu < 1$$

34:10:7 $$\int_0^{\pi/4} \tan^\nu(t)dt = \int_{\pi/4}^{\pi/2} \cot^\nu(t)dt = \frac{1}{2}\beta\left(\frac{1+\nu}{2}\right) \qquad \nu > -1$$

and

34:10:8 $$-\int_0^{\pi/4} \ln(\tan(t))dt = \int_0^{\pi/4} \ln(\cot(t))dt = G$$

and involve the secant function [Chapter 33], the beta function [Chapter 3] and Catalan's constant [Section 1:5].

The use of formulas 34:5:9–34:5:20, together with the differential identity

34:10:9 $$dt = \frac{2d\tau}{1+\tau^2} \qquad \tau = \tan\left(\frac{t}{2}\right)$$

can often aid the integration of expressions involving trigonometric functions by converting the integrand to a purely algebraic function. As a simple example

34:10:10 $$\int_0^x \frac{dt}{[1+\cos(t)]^2} = \frac{1}{2}\int_0^{\tan(x/2)} (1+\tau^2)d\tau = \frac{1}{2}\tan\left(\frac{x}{2}\right) + \frac{1}{6}\tan^3\left(\frac{x}{2}\right)$$

34:11 COMPLEX ARGUMENT

With argument $x + iy$ the tangent and cotangent functions become

34:11:1 $$\tan(x+iy) = \frac{\sin(2x) + i\sinh(2y)}{\cos(2x) + \cosh(2y)}$$

34:11:2 $$\cot(x+iy) = \frac{\sin(2x) - i\sinh(2y)}{\cosh(2y) - \cos(2x)}$$

For purely imaginary arguments, these results reduce to

34:11:3 $$\tan(iy) = i\tanh(y)$$

and

34:11:4 $$\cot(iy) = -i\coth(y)$$

34:12 GENERALIZATIONS

As periodic functions, the tangent and cotangent are special cases of the functions of Chapter 36.

Respectively, tan(x) and cot(x) generalize to the Jacobian elliptic functions sc(p;x) and cs(p;x) that are discussed in Chapter 63. When $p = 0$, we have

34:12:1 $$\mathrm{sc}(0;x) = \tan(x)$$

34:12:2 $$\mathrm{cs}(0;x) = \cot(x)$$

34:13 COGNATE FUNCTIONS

The tangent and cotangent functions are related to the other circular functions [those addressed in Chapters 32 and 33] by

34:13:1 $$\tan(x) = \frac{\sigma_4\sin(x)}{\sqrt{1-\sin^2(x)}} = \frac{\sigma_2\sqrt{1-\cos^2(x)}}{\cos(x)} = \sigma_3\sqrt{\sec^2(x)-1} = \frac{\sigma_3}{\sqrt{\csc^2(x)-1}} = \frac{1}{\cot(x)}$$

34:13:2 $$\cot(x) = \frac{\sigma_4\sqrt{1-\sin^2(x)}}{\sin(x)} = \frac{\sigma_2\cos(x)}{\sqrt{1-\cos^2(x)}} = \frac{\sigma_3}{\sqrt{\sec^2(x)-1}} = \sigma_3\sqrt{\csc^2(x)-1} = \frac{1}{\tan(x)}$$

where σ_2, σ_3 and σ_4 take the values +1 or −1 according to the magnitude of x, as discussed in Section 32:13.

The inverse functions of the tangent and cotangent are among the functions that are the subject of the next chapter.

34:14 RELATED TOPICS

Trigonometric functions and their inverses [Chapter 35] play an indispensable role in the mensuration of triangles. Historically, it was the need to determine the angles and sides of triangles that led to the invention of the trigonometric functions.

Figure 34-2 shows a triangle in which angle C is a right angle. Note that angle A is the angle opposite side a, and similarly for angles B and C. The side c, opposite the right angle, is known as the *hypotenuse*. There are five undefined parameters: the sides a, b and c, and the angles A and B. If any two of these five parameters are specified, then the other three are calculable via Table 34.14.1, provided that at least one of the two known parameters is the length of a side. Checkmarks indicate the given parameters in the table, which also includes a column giving the area of the triangle. Of course, the most important relationship applicable to a right-angled triangle is the *theorem of Pythagoras*, $a^2 + b^2 = c^2$. Sets of integers that satisfy this relationship are known as *Pythagorean trios;* examples are (3,4;5), (5,12;13), (8,15;17), (7,24;25), (20,21;29), (12,35;37), (9,40;41), (28,45;53), (11,60;61), (16,63;65), (33,56;65), (48,55;73), (13,84;85), (36,77;85), (39,80;89), (65,72;97), (20,99;101), (60,91;109), (15,112;113), (44,117;125), (88,105;137) and (119,120;169).

FIG 34-2

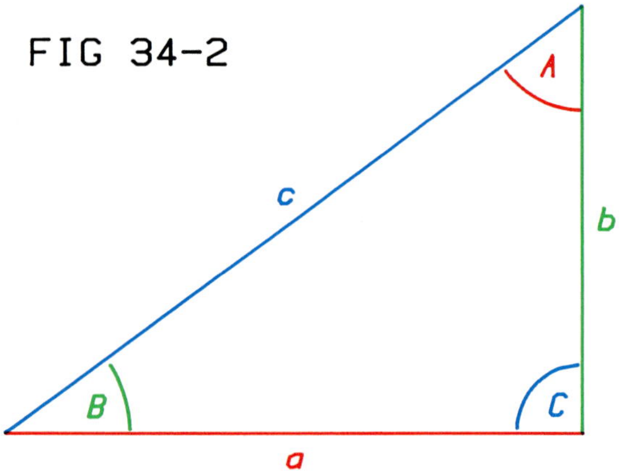

FIG 34-3

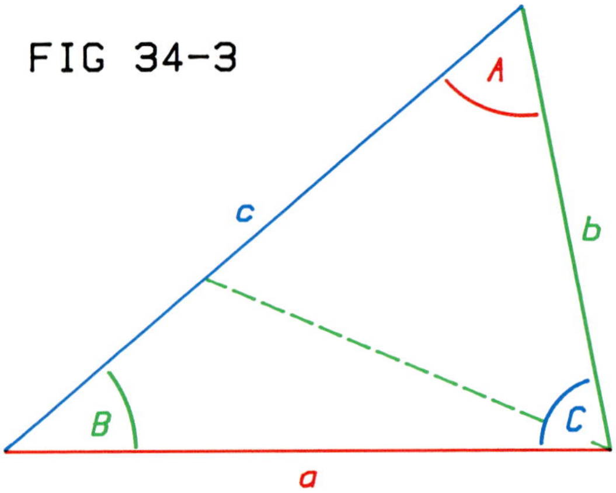

Table 34.14.1

a	b	c	A	B	Area
√	√	$\sqrt{a^2+b^2}$	arctan(a/b)	arctan(b/a)	$\frac{ab}{2}$
√	$\sqrt{c^2-a^2}$	√	arcsin(a/c)	arcsin(c/a)	$\frac{a}{2}\sqrt{c^2-a^2}$
√	a cot(A)	a csc(A)	√	$\frac{\pi}{2}-A$	$\frac{a^2}{2}\cot(A)$
√	a tan(B)	a sec(B)	$\frac{\pi}{2}-B$	√	$\frac{a^2}{2}\tan(B)$
$\sqrt{c^2-b^2}$	√	√	arccos(b/c)	arcsin(b/c)	$\frac{b}{2}\sqrt{c^2-b^2}$
b tan(A)	√	b sec(A)	√	$\frac{\pi}{2}-A$	$\frac{b^2}{2}\tan(A)$
b cot(B)	√	b csc(B)	$\frac{\pi}{2}-B$	√	$\frac{b^2}{2}\cot(B)$
c sin(A)	c cos(A)	√	√	$\frac{\pi}{2}-A$	$\frac{c^2}{4}\sin(2A)$
c cos(B)	c sin(B)	√	$\frac{\pi}{2}-B$	√	$\frac{c^2}{4}\sin(2B)$

Table 34.14.2

Known parameters	Formula for unknown parameters	Area
The three sides a, b and c; $-c < a - b < c < a + b$	$A = 2\arctan\sqrt{\frac{a^2-(b-c)^2}{(b+c)^2-a^2}}$ $B = 2\arctan\sqrt{\frac{b^2-(c-a)^2}{(c+a)^2-b^2}}$ $C = 2\arctan\sqrt{\frac{c^2-(a-b)^2}{(a+b)^2-c^2}}$	$\sqrt{\frac{(b+c)^2-a^2}{4}}\sqrt{\frac{a^2-(b-c)^2}{4}}$
Two sides and the angle opposite the longer of those two sides, e.g., a, b, A where $a \geq b$	$c = \sqrt{a^2 + b^2\cos(2A) - 2b\cos(A)\sqrt{a^2 - b^2\sin^2(A)}}$ $B = \arcsin\left[\frac{b\sin(A)}{a}\right]$ $C = \arccos\left[\frac{b}{a}\sin^2(A) + \cos(A)\sqrt{1 - \frac{b^2}{a^2}\sin^2(A)}\right]$	$\frac{b^2\sin(A)}{2}\left[\cos(A) + \sqrt{\frac{a^2}{b^2} - \sin^2(A)}\right]$
Two sides and the angle opposite the shorter of those two sides, e.g., a, b, B where $a \geq b \geq a\sin(B)$	$c = \sqrt{a^2\cos(2B) + b^2 \pm 2a\cos(B)\sqrt{b^2 - a^2\sin^2(B)}}$ $A = \frac{\pi}{2} \pm \arccos\left[\frac{a\sin(B)}{b}\right]$ $C = \arccos\left[\frac{a}{b}\sin^2(B) \mp \cos(B)\sqrt{1 - \frac{a^2}{b^2}\sin^2(B)}\right]$	$\frac{a^2\sin(B)}{2}\left[\cos(B) \mp \sqrt{\frac{b^2}{a^2} - \sin^2(B)}\right]$
Two sides and the angle between them, e.g., a, b, C	$c = \sqrt{a^2 + b^2 - 2ab\cos(C)}$ $A = \arcsin\left[\frac{a\sin(C)}{\sqrt{a^2+b^2-2ab\cos(C)}}\right]$ $B = \arcsin\left[\frac{b\sin(C)}{\sqrt{a^2+b^2-2ab\cos(C)}}\right]$	$\frac{ab}{2}\sin(C)$
Two angles and the side opposite one of them, e.g., a, A, B	$b = a\csc(A)\sin(B)$ $c = a[\cot(A)\sin(B) + \cos(B)]$ $C = \pi - A - B$	$\frac{a^2}{2}\sin^2(B)[\cot(A) + \cot(B)]$
Two angles and the side between them, e.g., a, B, C	$b = \frac{a\csc(C)}{\cot(B)+\cot(C)}$ $c = \frac{a\csc(B)}{\cot(B)+\cot(C)}$ $A = \pi - B - C$	$\frac{\frac{1}{2}a^2}{\cot(B)+\cot(C)}$

For a triangle that is not necessarily right angled, there are six parameters, and three of these (including at least one side length) must be known in order for the triangle to be determined. For example, if the side a is of known length and the angles B and C are also known [see Figure 34-3], sides b and c are calculable by the formulas given in Table 34.14.2, as is the angle A. Notice that prescribing the magnitudes of a, b and B does not fully delineate the triangle if a exceeds b. The line of length b has two alternative positions, in that case, as illustrated in Figure 34-3 by the full green line and the dashed green line, and consequently the magnitudes of c, A and C each have two alternative values. The alternative magnitudes of these quantities are given in Table 34.14.2, as are the two possible areas. The formulas in this table are based on the *cosine law*

34:14:1 $$2bc\cos(A) = b^2 + c^2 - a^2$$

the *sine law*

34:14:2 $$\frac{\sin(A)}{a} = \frac{\sin(B)}{b} = \frac{\sin(C)}{c}$$

and the area identities

34:14:3 $$\text{area} = \frac{bc}{2}\sin(A) = \sqrt{s(s-a)(s-b)(s-c)}$$

where s is the *semiperimeter*, $s = (a + b + c)/2$. Another useful rule in the mensuration of triangles is the *tangent rule*

34:14:4 $$\frac{\tan[(A-B)/2]}{\tan[(A+B)/2]} = \frac{a-b}{a+b}$$

Three constructions that are important in trigonometry are shown in Figure 34-4. The *altitude AO*, that is the line through A perpendicular to BC, has a length

34:14:5 $$AO = b\sin(C) = c\sin(B) \qquad (\text{angle } BOA) = \pi/2$$

A line through a vertex of the triangle that bisects the angle there is known as a *bisector;* the bisector of angle A in Figure 34-4 has a length

34:14:6 $$AN = \frac{2bc\cos(A/2)}{b+c} \qquad (\text{angle } BAN) = (\text{angle } NAC) = \frac{A}{2}$$

Line AM, which bisects the line BC, is termed a *median* and has the length

34:14:7 $$AM = \frac{1}{2}\sqrt{b^2 + c^2 + 2bc\cos(A)} \qquad BM = MC = \frac{a}{2}$$

FIG 34-4

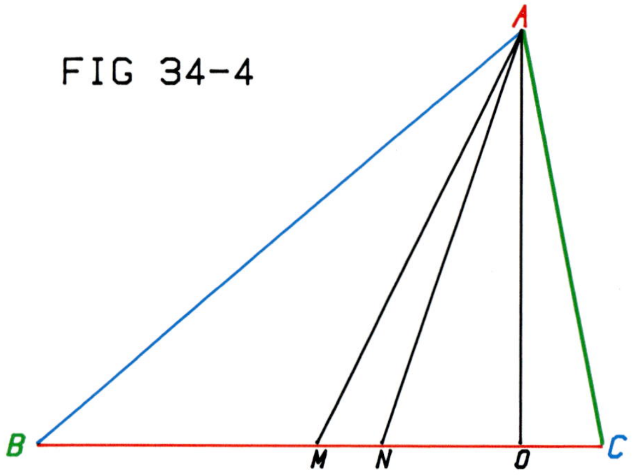

CHAPTER 35

THE INVERSE TRIGONOMETRIC FUNCTIONS

The fundamental interrelationships between the six subjects of this chapter are

35:0:1 $$\arctan(x) = \arcsin\left(\frac{x}{\sqrt{1+x^2}}\right) = \operatorname{arccsc}\left(\frac{\sqrt{x^2+1}}{x}\right)$$

and

35:0:2 $$\arctan(x) + \operatorname{arccot}(x) = \arcsin(x) + \arccos(x) = \operatorname{arccsc}(x) + \operatorname{arcsec}(x) = \frac{\pi}{2}$$

but some of the many alternative links are presented below:

35:0:3 $$\arctan(x) = \arcsin\left(\frac{x}{\sqrt{1+x^2}}\right) = \operatorname{arccsc}\left(\frac{\sqrt{x^2+1}}{x}\right)$$
$$= \begin{cases} \operatorname{arccot}\left(\dfrac{1}{x}\right) = \arccos\left(\dfrac{1}{\sqrt{1+x^2}}\right) = \operatorname{arcsec}(\sqrt{x^2+1}) & x > 0 \\ \operatorname{arccot}\left(\dfrac{1}{x}\right) - \pi = -\arccos\left(\dfrac{1}{\sqrt{1+x^2}}\right) = -\operatorname{arcsec}(\sqrt{x^2+1}) & x < 0 \end{cases}$$

35:0:4 $$\arcsin(x) = \arctan\left(\frac{x}{\sqrt{1-x^2}}\right) = \operatorname{arccsc}\left(\frac{1}{x}\right) = \frac{\operatorname{sgn}(x)}{2}\arccos(1-2x^2) \qquad -1 \le x \le 1$$
$$= \begin{cases} \operatorname{arccot}\left(\dfrac{\sqrt{1-x^2}}{x}\right) = \arccos(\sqrt{1-x^2}) = \operatorname{arcsec}\left(\dfrac{1}{\sqrt{1-x^2}}\right) & 0 < x < 1 \\ \operatorname{arccot}\left(\dfrac{\sqrt{1-x^2}}{x}\right) - \pi = -\arccos(\sqrt{1-x^2}) = -\operatorname{arcsec}\left(\dfrac{1}{\sqrt{1-x^2}}\right) & -1 \le x \le 0 \end{cases}$$

35:0:5
$$\operatorname{arccsc}(x) = \arctan\left(\frac{\operatorname{sgn}(x)}{\sqrt{x^2-1}}\right) = \arcsin\left(\frac{1}{x}\right) \qquad |x| \geq 1$$

$$= \begin{cases} \operatorname{arccot}(\sqrt{x^2-1}) = \arccos\left(\dfrac{\sqrt{x^2-1}}{x}\right) = \operatorname{arcsec}\left(\dfrac{x}{\sqrt{x^2-1}}\right) & x > 1 \\ \operatorname{arccot}(\sqrt{x^2-1}) - \pi = \arccos\left(\dfrac{\sqrt{x^2-1}}{x}\right) - \pi = \operatorname{arcsec}\left(\dfrac{x}{\sqrt{x^2-1}}\right) - \pi & x < -1 \end{cases}$$

35:0:6
$$\operatorname{arccot}(x) = \arccos\left(\frac{x}{\sqrt{1+x^2}}\right) = \operatorname{arcsec}\left(\frac{\sqrt{1+x^2}}{x}\right)$$

$$= \begin{cases} \arctan\left(\dfrac{1}{x}\right) = \arcsin\left(\dfrac{1}{\sqrt{1+x^2}}\right) = \operatorname{arccsc}(\sqrt{1+x^2}) & x > 0 \\ \pi + \arctan\left(\dfrac{1}{x}\right) = \pi - \arcsin\left(\dfrac{1}{\sqrt{1+x^2}}\right) = \pi - \operatorname{arccsc}(\sqrt{1+x^2}) & x < 0 \end{cases}$$

35:0:7
$$\arccos(x) = \operatorname{arccot}\left(\frac{x}{\sqrt{1-x^2}}\right) = \operatorname{arcsec}\left(\frac{1}{x}\right) = 2\arcsin\left(\sqrt{\frac{1-x}{2}}\right) \qquad -1 \leq x \leq 1$$

$$= \begin{cases} \arctan\left(\dfrac{\sqrt{1-x^2}}{x}\right) = \arcsin(\sqrt{1-x^2}) = \operatorname{arccsc}\left(\dfrac{1}{\sqrt{1-x^2}}\right) & 0 < x < 1 \\ \pi + \arctan\left(\dfrac{\sqrt{1-x^2}}{x}\right) = \pi - \arcsin(\sqrt{1-x^2}) = \pi - \operatorname{arccsc}\left(\dfrac{1}{\sqrt{1-x^2}}\right) & -1 < x < 0 \end{cases}$$

35:0:8
$$\operatorname{arcsec}(x) = \operatorname{arccot}\left(\frac{\operatorname{sgn}(x)}{\sqrt{x^2-1}}\right) = \arccos\left(\frac{1}{x}\right) \qquad |x| \geq 1$$

$$= \begin{cases} \arctan(\sqrt{x^2-1}) = \arcsin\left(\dfrac{\sqrt{x^2-1}}{x}\right) = \operatorname{arccsc}\left(\dfrac{x}{\sqrt{x^2-1}}\right) & x > 1 \\ \pi - \arctan(\sqrt{x^2-1}) = \pi + \arcsin\left(\dfrac{\sqrt{x^2-1}}{x}\right) = \pi + \operatorname{arccsc}\left(\dfrac{x}{\sqrt{x^2-1}}\right) & x < -1 \end{cases}$$

Many more interrelationships may be constructed by combining equations 35:0:3–35:0:8 with those in 35:0:2. For example:

35:0:9
$$\arctan(x) = \frac{\pi}{2} - \arccos\left(\frac{x}{\sqrt{1+x^2}}\right) = \frac{\pi}{2} - \operatorname{arcsec}\left(\frac{\sqrt{x^2+1}}{x}\right)$$

$$= \operatorname{sgn}(x)\left[\frac{\pi}{2} - \arcsin\left(\frac{1}{\sqrt{1+x^2}}\right)\right] = \operatorname{sgn}(x)\left[\frac{\pi}{2} - \operatorname{arccsc}(\sqrt{x^2+1})\right]$$

follows from combining 35:0:2 with the right-hand members of equation 35:0:3.

35:1 NOTATION

An alternative collective title for the functions of this chapter—the inverse tangent, the inverse sine, the inverse cosecant, the inverse cotangent, the inverse cosine, and the inverse secant—is the *inverse circular functions*. The symbolism $\tan^{-1}(x)$, $\sin^{-1}(x)$, etc., often replaces arctan(x), arcsin(x), etc. Variants such as arctg(x), arcctg(x) and arccosec(x) are occasionally encountered. The origin of the "arc" prefix is made evident in Section 35:3.

The notation arccot(x) is sometimes used to denote a function defined, for negative x, somewhat differently than here, being equal to our arccot(x) $- \pi$. Similar discrepancies may be encountered for other inverse trigonometric functions.

The multivalued functions Arctan(x), Arcsin(x), etc., are discussed in Section 35:12, but, confusingly, these are often also denoted arctan(x), arcsin(x), etc.

We use arctrig(x) to represent any one of the six inverse trigonometric functions.

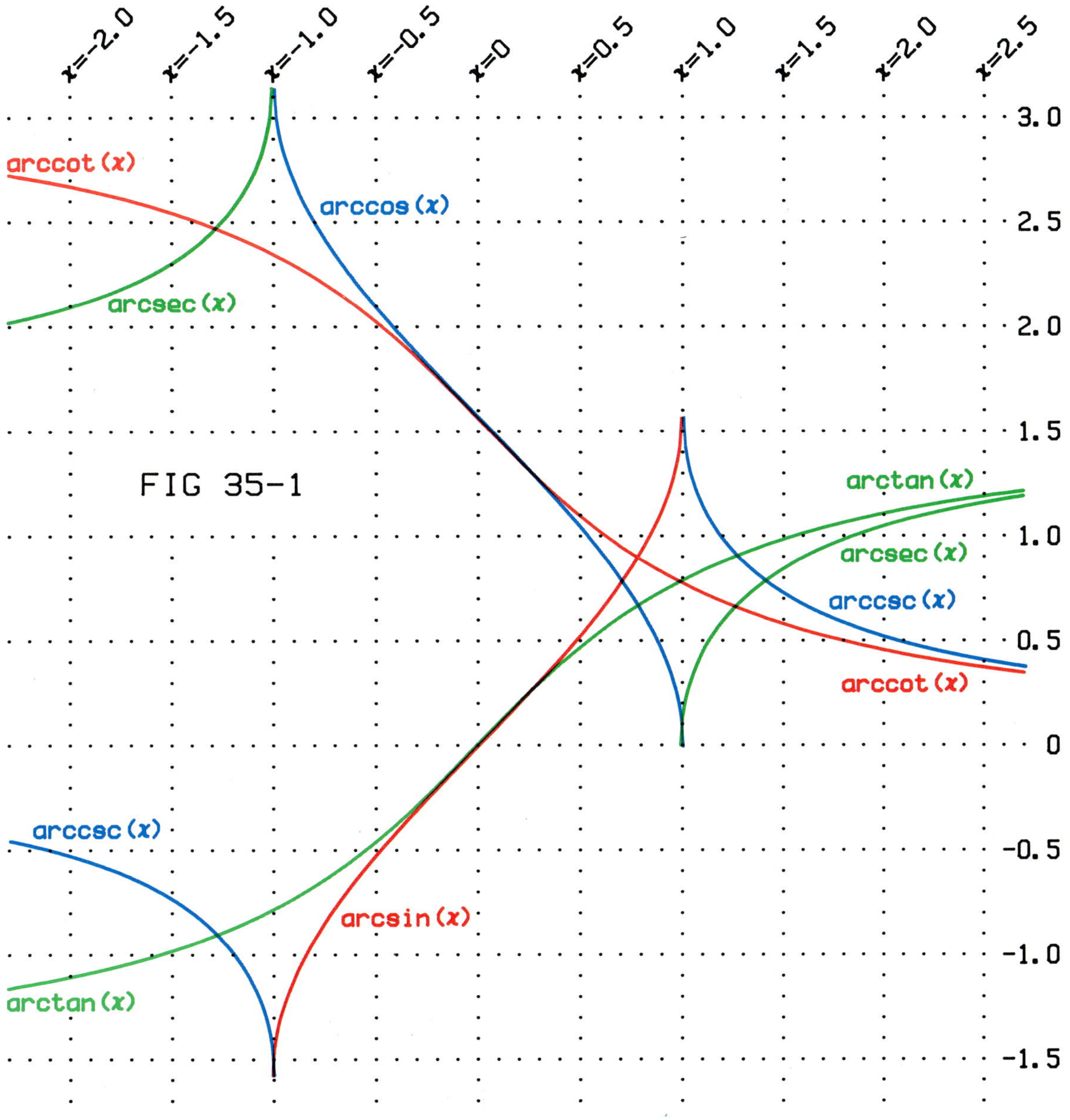

35:2 BEHAVIOR

Figure 35-1 depicts the behavior of the six functions. Note that both the inverse cosecant and inverse secant functions have two branches. The domains and ranges of all the functions are summarized in Table 35.2.1.

Table 35.2.1

f(x)	Domain of f(x)	Range of f(x)
arctan(x)	$-\infty < x < \infty$	$-\pi/2 < \text{arctan}(x) < \pi/2$
arcsin(x)	$-1 \le x \le 1$	$-\pi/2 \le \text{arcsin}(x) \le \pi/2$
arccsc(x)	$\begin{cases} -\infty < x \le -1 \\ 1 \le x < \infty \end{cases}$	$\begin{Bmatrix} -\pi/2 \le \text{arccsc}(x) \le 0 \\ 0 < \text{arccsc}(x) \le \pi/2 \end{Bmatrix}$
arccot(x)	$-\infty < x < \infty$	$0 < \text{arccot}(x) < \pi$
arccos(x)	$-1 \le x \le 1$	$0 \le \text{arccos}(x) \le \pi$
arcsec(x)	$\begin{cases} -\infty < x \le -1 \\ 1 \le x < \infty \end{cases}$	$\begin{Bmatrix} \pi/2 < \text{arcsec}(x) \le \pi \\ 0 \le \text{arcsec}(x) < \pi/2 \end{Bmatrix}$

35:3 DEFINITIONS

Indefinite integrals define each of the six inverse trigonometric functions

35:3:1
$$\text{arctan}(x) = \int_0^x \frac{dt}{1+t^2}$$

35:3:2
$$\text{arcsin}(x) = \int_0^x \frac{dt}{\sqrt{1-t^2}} \qquad -1 \le x \le 1$$

35:3:3
$$\text{arccsc}(x) = \begin{cases} \displaystyle\int_{-\infty}^x \frac{dt}{t\sqrt{t^2-1}} & -\infty < x \le -1 \\ \displaystyle\int_x^{\infty} \frac{dt}{t\sqrt{t^2-1}} & 1 \le x < \infty \end{cases}$$

35:3:4
$$\text{arccot}(x) = \int_x^{\infty} \frac{dt}{1+t^2}$$

35:3:5
$$\text{arccos}(x) = \int_x^1 \frac{dt}{\sqrt{1-t^2}} \qquad -1 \le x \le 1$$

35:3:6
$$\text{arcsec}(x) = \int_1^x \frac{dt}{t\sqrt{t^2-1}} \qquad x \ge 1$$

Diagrams illustrating these definitions, for $x > 0$, are included as Figures 35-2, 35-3 and 35-4.

Geometric definitions of the inverse trigonometric functions may be illustrated by reference to Figure 35-5 for arctan(x). A line of length x is made part of a right-angled triangle, the side shown dashed being of unity length. Thus, x is the tangent of the marked angle. The arc of a unity-radius circle subtended by this marked angle, shown in black on the diagram, is then defined as the inverse tangent of x. Alternatively, the angle itself may be identified with arctan(x).

The names "inverse trigonometric functions" imply that these functions are the inverses of the functions of Chapters 32, 33 and 34. This is true, however, only if appropriate restrictions, namely

35:3:7 $f = \text{arctan}(x)$ where $x = \tan(f)$

35:3:8 $f = \text{arcsin}(x)$ where $x = \sin(f)$

35:3:9 $f = \text{arccsc}(x)$ where $x = \csc(f)$

$$-\frac{\pi}{2} \le f \le \frac{\pi}{2}$$

35:3:10 $\quad f = \mathrm{arccot}(x) \quad$ where $\quad x = \cot(f)$

35:3:11 $\quad f = \arccos(x) \quad$ where $\quad x = \cos(f) \qquad 0 \le f \le \pi$

35:3:12 $\quad f = \mathrm{arcsec}(x) \quad$ where $\quad x = \sec(f)$

are placed upon the magnitudes of the arguments of the trigonometric functions.

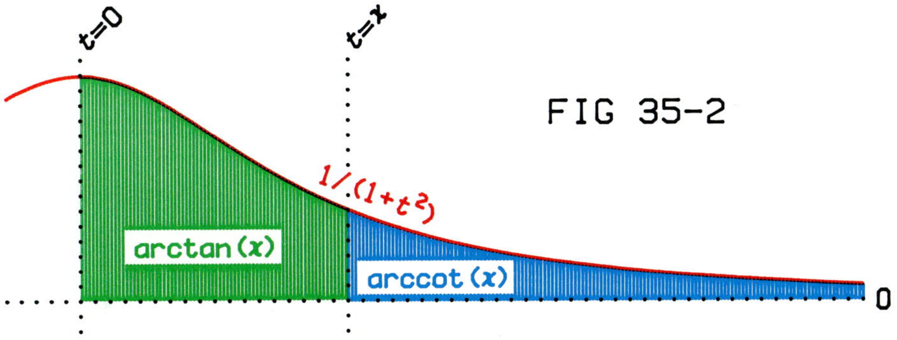

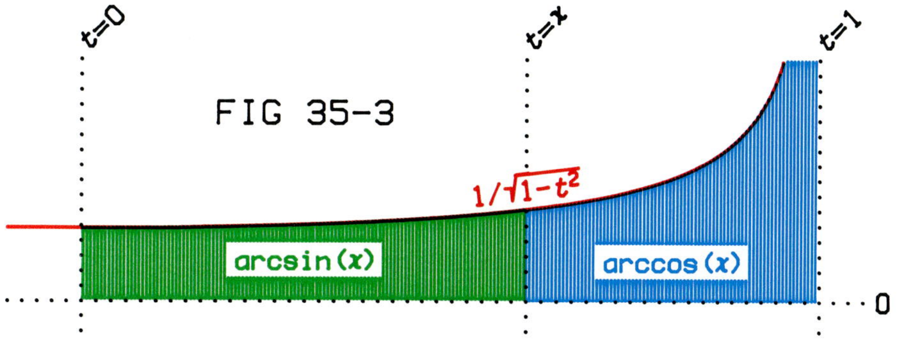

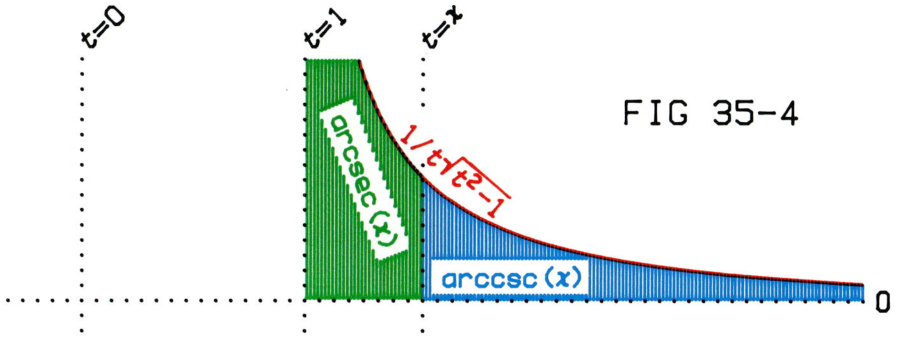

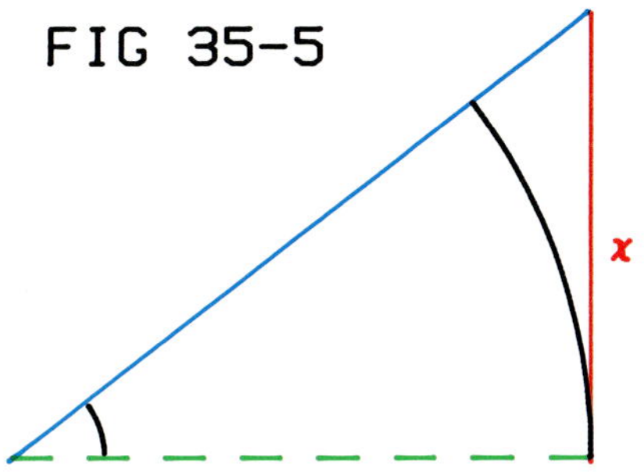

35:4 SPECIAL CASES

There are none.

35:5 INTRARELATIONSHIPS

The inverse tangent, sine and cosecant are odd functions:

35:5:1 $$\mathrm{f}(-x) = -\mathrm{f}(x) \qquad \mathrm{f} = \mathrm{arctan,\ arcsin\ or\ arccsc}$$

whereas the other three inverse trigonometric functions obey the reflection formula

35:5:2 $$\mathrm{f}(-x) = \pi - \mathrm{f}(x) \qquad \mathrm{f} = \mathrm{arccot,\ arccos\ or\ arcsec}$$

The inverse tangent and cotangent functions obey the simple reciprocation formulas

35:5:3 $$\arctan\left(\frac{1}{x}\right) = -\arctan(x) + \frac{\pi\,\mathrm{sgn}(x)}{2} \qquad x \neq 0$$

35:5:4 $$\mathrm{arccot}\left(\frac{1}{x}\right) = -\mathrm{arccot}(x) + \pi - \frac{\pi\,\mathrm{sgn}(x)}{2} \qquad x \neq 0$$

For the other inverse trigonometric functions, formulas for arctrig$(1/x)$ will be found among the equations in Section 35:0. Other formulas relating two values of a single inverse trigonometric function include

35:5:5 $$\arcsin(x) = \frac{1}{2}\arcsin(2x\sqrt{1-x^2}) \qquad \frac{-1}{\sqrt{2}} \leq x \leq \frac{1}{\sqrt{2}}$$

and

35:5:6 $$\arccos(x) = 2\arccos\left(\sqrt{\frac{1+x}{2}}\right) \qquad -1 \leq x \leq 1$$

If f is an inverse trigonometric function, then $\mathrm{f}(x) \pm \mathrm{f}(y)$ may lie within the range of f, or above it, or below it. These three possibilities are incorporated into the following function-addition formulas by allowing the integer k to adopt one of three values:

35:5:7 $$\arctan(x) \pm \arctan(y) = k\pi + \arctan\left(\frac{x \pm y}{1 \mp xy}\right)$$

35:5:8 $$\arcsin(x) \pm \arcsin(y) = k\pi + \arcsin(x\sqrt{1-y^2} \pm y\sqrt{1-x^2})$$

35:5:9 $$\mathrm{arccsc}(x) \pm \mathrm{arccsc}(y) = k\pi + \mathrm{arccsc}\left(\frac{xy}{\sqrt{y^2-1} \pm \sqrt{x^2-1}}\right)$$

$$k = 0, \pm 1$$

The appropriate k value may be found from the formula

35:5:10 $$k = \mathrm{Int}\left(\frac{1}{2} + \frac{\mathrm{f}(x) \pm \mathrm{f}(y)}{\pi}\right) \qquad \mathrm{f} = \mathrm{arctan,\ arcsin\ or\ arccsc}$$

that utilizes the integer-value function [Chapter 9]. Similarly:

35:5:11 $$\mathrm{arccot}(x) \pm \mathrm{arccot}(y) = \left(\frac{1}{2} - k\right)\pi + \mathrm{arccot}\left(\frac{x \pm y}{1 \mp xy}\right)$$

35:5:12 $$\arccos(x) \pm \arccos(y) = \left(\frac{1}{2} - k\right)\pi + \arccos(xy \mp \sqrt{1-x^2}\sqrt{1-y^2})$$

35:5:13 $$\mathrm{arcsec}(x) \pm \mathrm{arcsec}(y) = \left(\frac{1}{2} - k\right)\pi + \mathrm{arcsec}\left(\frac{xy}{1 \mp \sqrt{x^2-1}\sqrt{y^2-1}}\right)$$

$$k = 0, \pm 1$$

where

35:5:14 $$k = \mathrm{Int}\left(\frac{\mathrm{f}(x) \pm \mathrm{f}(y)}{\pi}\right) \qquad \mathrm{f} = \text{arccot, arccos or arccsc}$$

Application of the formulas in Section 35:0 enables many more intrarelationships to be constructed.

35:6 EXPANSIONS

There are three basic power series expansions for inverse trigonometric functions:

35:6:1 $$\arctan(x) = x - \frac{x^3}{3} + \frac{x^5}{5} - \frac{x^7}{7} + \cdots = x\sum_{j=0}^{\infty}\frac{(-x^2)^j}{2j+1} \qquad -1 \le x \le 1$$

35:6:2 $$\arctan(x) = \frac{\pi\,\mathrm{sgn}(x)}{2} - \frac{1}{x} + \frac{1}{3x^3} - \frac{1}{5x^5} + \cdots = \frac{\pi\,\mathrm{sgn}(x)}{2} - \frac{1}{x}\sum_{j=0}^{\infty}\frac{(-x^2)^{-j}}{2j+1} \qquad |x| \ge 1$$

35:6:3 $$\arcsin(x) = x + \frac{1}{2}\frac{x^3}{3} + \frac{1\times 3}{2\times 4}\frac{x^5}{5} + \frac{1\times 3\times 5}{2\times 4\times 6}\frac{x^7}{7} + \cdots = \sum_{j=0}^{\infty}\frac{(2j-1)!!}{(2j)!!}\frac{x^{2j+1}}{2j+1} \qquad -1 \le x \le 1$$

but the multifarious interrelationships between the six functions allow easy adaptation of these series to other inverse trigonometric functions. For example, in light of equation 35:0:7, replacement of x on the right side of 35:6:3 by $\sqrt{(1-x)/2}$ provides an expansion for 2 arccos(x).

Continued fraction expansions of the inverse trigonometric functions include

35:6:4 $$\arctan(x) = \frac{x}{1+}\,\frac{x^2}{3+}\,\frac{4x^2}{5+}\,\frac{9x^2}{7+}\,\frac{16x^2}{9+}\cdots$$

and, for $-1 < x < 1$

35:6:5 $$\arcsin(x) = \frac{x\sqrt{1-x^2}}{1-}\,\frac{(1\times 2)x}{3-}\,\frac{(1\times 2)x}{5-}\,\frac{(3\times 4)x}{7-}\,\frac{(3\times 4)x}{9-}\,\frac{(5\times 6)x}{11-}\cdots$$

35:7 PARTICULAR VALUES

The entry "undef" in Table 35.7.1 indicates that the function is undefined at the argument in question.

35:8 NUMERICAL VALUES

Calculators usually incorporate keys for evaluating the arctan, arcsin and arccos functions, but computer languages rarely incorporate any inverse trigonometric function other than arctan. Accordingly, we present an algorithm that

Table 35.7.1

	$x = -\infty$	$x = -\sqrt{2}$	$x = -1$	$x = \frac{-1}{\sqrt{2}}$	$x = 0$	$x = \frac{1}{\sqrt{2}}$	$x = 1$	$x = \sqrt{2}$	$x = \infty$
arctan(x)	$-\pi/2$	$-0.955\ldots$	$-\pi/4$	$-0.615\ldots$	0	$0.615\ldots$	$\pi/4$	$0.955\ldots$	$\pi/2$
arcsin(x)	undef	undef	$-\pi/2$	$-\pi/4$	0	$\pi/4$	$\pi/2$	undef	undef
arccsc(x)	0	$-\pi/4$	$-\pi/2$	undef	undef	undef	$\pi/2$	$\pi/4$	0
arccot(x)	π	$2.526\ldots$	$3\pi/4$	$2.186\ldots$	$\pi/2$	$0.955\ldots$	$\pi/4$	$0.615\ldots$	0
arccos(x)	undef	undef	π	$3\pi/4$	$\pi/2$	$\pi/4$	0	undef	undef
arcsec(x)	$\pi/2$	$3\pi/4$	π	undef	undef	undef	0	$\pi/4$	$\pi/2$

computes any one of the six inverse trigonometric functions by utilizing a built-in arctan function. The algorithm, which is exact, uses relationships 35:0:1 and 35:0:2.

As well, the universal hypergeometric algorithm [Section 18:14] may be used to calculate inverse trigonometric functions.

Input x >>>>>>>
Set $f = 1$
Input code c >>>>
If $c \leq 1$ go to (2)
Set $f = 1 - x^2$
If $f \neq 0$ go to (1)
Set $f = \pi x/2$
Go to (3)
(1) If $c \leq 3$ go to (2)
Replace f by $-x^2 f$
(2) Replace f by $\arctan(x/\sqrt{f})$
(3) If $\mathrm{frac}(c/2) = 0$ go to (4)
Replace f by $(\pi/2) - f$
(4) Output f
$f_0 = \arctan(x)$
$f_1 = \mathrm{arccot}(x)$
$f_2 = \arcsin(x)$
$f_3 = \arccos(x)$
$f_4 = \mathrm{arccsc}(x)$
$f_5 = \mathrm{arcsec}(x)$ <<<

Storage needed:
x, c (the code) and f

Use radian mode

Input restrictions: Argument must lie in the range:

code	function	argument
0	arctan	any
1	arccot	any
2	arcsin	$-1 \leq x \leq 1$
3	arccos	$-1 \leq x \leq 1$
4	arccsc	$\lvert x \rvert \geq 1$
5	arcsec	$\lvert x \rvert \geq 1$

Test values:
$\mathrm{arccot}(\pi) = 0.30816907$
$\arcsin(-0.7) = -0.77439750$
$\mathrm{arccsc}(4) = 0.25268026$
$\mathrm{arcsec}(-1) = 3.14159265$

35:9 APPROXIMATIONS

Among the approximations to inverse trigonometric functions are

35:9:1 $$\arctan(x) \simeq \frac{3x}{3 + x^2} \qquad \text{8-bit precision} \qquad -0.45 \leq x \leq 0.45$$

35:9:2 $$\mathrm{arccot}(x) \simeq \frac{3x}{3x^2 + 1} \qquad \text{8-bit precision} \qquad x \geq 1.8$$

35:9:3 $$\arcsin(x) \simeq x\sqrt{\frac{3}{3 - x^2}} \qquad \text{8-bit precision} \qquad 0.5 \leq x \leq 0.5$$

35:10 OPERATIONS OF THE CALCULUS

Derivatives of the six inverse trigonometric functions are

35:10:1 $$\frac{\mathrm{d}}{\mathrm{d}x}\arctan(bx + c) = -\frac{\mathrm{d}}{\mathrm{d}x}\mathrm{arccot}(bx + c) = \frac{b}{1 + (bx + c)^2}$$

35:10:2 $$\frac{\mathrm{d}}{\mathrm{d}x}\arcsin(bx + c) = -\frac{\mathrm{d}}{\mathrm{d}x}\arccos(bx + c) = \frac{b}{\sqrt{1 - (bx + c)^2}}$$

35:10:3 $$\frac{\mathrm{d}}{\mathrm{d}x}\mathrm{arccsc}(bx + c) = -\frac{\mathrm{d}}{\mathrm{d}x}\mathrm{arcsec}(bx + c) = \frac{-b}{\lvert bx + c\rvert\sqrt{(bx + c)^2 - 1}}$$

while the corresponding indefinite integrals are

35:10:4 $$\int_{-c/b}^{x} \operatorname{arctan}(bt + c)dt = \left(x + \frac{c}{b}\right)\operatorname{arctan}(bx + c) - \frac{1}{b}\ln\sqrt{1 + (bx + c)^2}$$

35:10:5 $$\int_{-c/b}^{x} \operatorname{arccot}(bt + c)dt = \left(x + \frac{c}{b}\right)\operatorname{arccot}(bx + c) + \frac{1}{b}\ln\sqrt{1 + (bx + c)^2}$$

35:10:6 $$\int_{-c/b}^{x} \arcsin(bt + c)dt = \left(x + \frac{c}{b}\right)\arcsin(bx + c) - \frac{1}{b} + \sqrt{\frac{1}{b^2} - \left(x + \frac{c}{b}\right)^2} \qquad -c - 1 \le bx \le 1 - c$$

35:10:7 $$\int_{-c/b}^{x} \arccos(bt + c)dt = \left(x + \frac{c}{b}\right)\arccos(bx + c) + \frac{1}{b} - \sqrt{\frac{1}{b^2} - \left(x + \frac{c}{b}\right)^2} \qquad -c - 1 \le bx \le 1 - c$$

35:10:8 $$\int_{(1-c)/b}^{x} \operatorname{arccsc}(bt + c)dt = \left(x + \frac{c}{b}\right)\operatorname{arccsc}(bx + c) + \frac{1}{b}\operatorname{arcosh}(bx + c) \qquad bx > 1 - c$$

35:10:9 $$\int_{(1-c)/b}^{x} \operatorname{arcsec}(bt + c)dt = \left(x + \frac{c}{b}\right)\operatorname{arcsec}(bx + c) - \frac{1}{b}\operatorname{arcosh}(bx + c) \qquad bx > 1 - c$$

Indefinite integrals of the form $\int t\,\mathrm{arctrig}(bt + c)dt$ include

35:10:10 $$\int_{-c/b}^{x} t\operatorname{arctan}(bt + c)dt = \frac{1 + b^2x^2 - c^2}{2b^2}\operatorname{arctan}(bx + c) - \frac{bx + c}{2b^2} + \frac{c}{b^2}\ln\sqrt{1 + (bx + c)^2}$$

35:10:11 $$\int_{-c/b}^{x} t\operatorname{arccot}(bt + c)dt = \frac{1 + b^2x^2 - c^2}{2b^2}\operatorname{arccot}(bx + c) + \frac{2bx + 2c - \pi}{4b^2} - \frac{c}{b^2}\ln\sqrt{1 + (bx + c)^2}$$

35:10:12 $$\int_{-c/b}^{x} t\arcsin(bt + c)dt = \frac{2b^2x^2 - 2c^2 - 1}{4b^2}\arcsin(bx + c) + \frac{c}{b^2} + \frac{bx - 3c}{4b^2}\sqrt{1 - (bx + c)^2} \qquad -c - 1 \le bx \le 1 - c$$

35:10:13 $$\int_{-c/b}^{x} t\arccos(bt + c)dt = \frac{2b^2x^2 - 2c^2 - 1}{4b^2}\arccos(bx + c) + \frac{\pi - 8c}{8b^2} - \frac{bx - 3c}{4b^2}\sqrt{1 - (bx + c)^2} \qquad -c - 1 \le bx \le 1 - c$$

but none of $\int t^{-1}\mathrm{arctrig}(bt + c)dt$ has been evaluated as a finite number of terms. See Spiegel [pages 82–84] for a long list of indefinite integrals of the form $\int t^{\pm n}\mathrm{arctrig}(t)dt$ for integer n. Gradshteyn and Ryzhik [Section 2.8] list similar integrals and include additional entries such as $\int \mathrm{arctrig}^n(t)dt$ and $\int (bt + c)^{-\nu}\mathrm{arctrig}(t)dt$.

Gradshteyn and Ryzhik [Section 4.5] also present over 100 definite integrals involving inverse trigonometric functions;

35:10:14 $$\int_0^1 \frac{\operatorname{arctan}(t)}{t}dt = \int_0^\infty \frac{\operatorname{arccot}(t)}{t}dt = G$$

and

35:10:15 $$\int_0^1 \frac{\arcsin(t)}{t}dt = \frac{\pi}{2}\ln(2)$$

are typical examples. G is Catalan's constant (see Section 1:7).

The inverse trigonometric functions play a role in the fractional calculus [see Oldham and Spanier]. For example:

35:10:16 $$\frac{d^{1/2}}{dx^{1/2}}\sqrt{1+x} = \frac{1}{\sqrt{\pi x}} + \frac{\arctan(\sqrt{x})}{\sqrt{\pi}}$$

35:10:17 $$\frac{d^{-1/2}}{dx^{-1/2}}\operatorname{arccot}(\sqrt{x}) = \sqrt{\pi}(1 + \sqrt{x} - \sqrt{1+x})$$

35:10:18 $$\frac{d^{1/2}}{dx^{1/2}}\frac{\arcsin(\sqrt{x})}{\sqrt{1-x}} = \frac{\sqrt{\pi}}{2(1-x)}$$

35:11 COMPLEX ARGUMENT

A variety of ways may be used to extend the inverse trigonometric functions to complex argument. One might couple the definitions 35:3:1–35:3:6 with integration in the complex plane or, alternatively, permit power series such as 35:6:1 and 35:6:3 to accept a complex argument. Another route makes use of the relationships 35:13:1–35:13:6, which are valid for complex argument as well. The results, however one proceeds, are rather formidable looking. The most useful of the resulting multivalued functions are

35:11:1 $$\operatorname{Arcsin}(x+iy) = k\pi + (-1)^k\arcsin(Y) + (-1)^k i\ln[X + \sqrt{X^2-1}]$$

35:11:2 $$\operatorname{Arccos}(x+iy) = 2k\pi \pm \{\arccos(Y) - i\ln[X + \sqrt{X^2-1}]\}$$

35:11:3 $$\operatorname{Arctan}(x+iy) = k\pi + \frac{1}{2}\arctan\left(\frac{2x}{1-x^2-y^2}\right) + \frac{i}{4}\ln\left[\frac{x^2+(y+1)^2}{x^2+(y-1)^2}\right] \qquad x^2+(y-1)^2 \neq 0$$

where k is any integer and

35:11:4 $$X = \frac{1}{2}\sqrt{(x+1)^2+y^2} + \frac{1}{2}\sqrt{(x-1)^2+y^2} \qquad Y = \frac{1}{2}\sqrt{(x+1)^2+y^2} - \frac{1}{2}\sqrt{(x-1)^2+y^2}$$

The logarithmic function ln occurring in the above equation is the single-valued logarithm defined in Chapter 25.

35:12 GENERALIZATIONS

Because the trigonometric functions are periodic, their unrestricted inverses have infinitely many values for each acceptable argument. These multivalued inverse functions are distinguished from the single-valued functions treated elsewhere in this chapter by having their initial letter capitalized. The definitions are encompassed by

35:12:1 $$\operatorname{Arctrig}(x) = y \qquad \text{where} \qquad \operatorname{trig}(y) = x$$

With $k = 0, \pm1, \pm2, \ldots,$ the relationships

35:12:2 $$\operatorname{Arctrig}(x) = \operatorname{arctrig}(x) + 2k\pi \qquad \operatorname{arctrig} = \arcsin, \arccos, \operatorname{arccsc}, \operatorname{arcsec}$$

35:12:3 $$\operatorname{Arctrig}(x) = \operatorname{arctrig}(x) + k\pi \qquad \operatorname{arctrig} = \arctan, \operatorname{arccot}$$

hold.

The inverse tangent and inverse sine functions are each special cases of the Gauss F function [see Chapter 60]:

35:12:4 $$\arctan(x) = xF\left(\frac{1}{2},1;\frac{3}{2};-x^2\right) \qquad -1 < x < 1$$

35:12:5 $$\arcsin(x) = F\left(\frac{1}{2},\frac{1}{2};\frac{3}{2};x^2\right) \qquad -1 < x < 1$$

Also, the inverse sine and cosine are special cases of the incomplete beta function of Chapter 58:

35:12:6 $$\arcsin(x) = \frac{1}{2} \mathrm{B}\left(\frac{1}{2}; \frac{1}{2}; x^2\right)$$

35:12:7 $$\arccos(x) = \frac{1}{2} \mathrm{B}\left(\frac{1}{2}; \frac{1}{2}; 1 - x^2\right)$$

35:13 COGNATE FUNCTIONS

The inverse trigonometric functions are closely related to the inverse hyperbolic functions that are considered in Chapter 30. One has

35:13:1 $$\arctan(x) = -i \operatorname{artanh}(ix)$$

35:13:2 $$\operatorname{arccot}(x) = i \operatorname{arcoth}(ix)$$

35:13:3 $$\arcsin(x) = -i \operatorname{arsinh}(ix)$$

35:13:4 $$\arccos(x) = \pm i \operatorname{arcosh}(x)$$

35:13:5 $$\operatorname{arccsc}(x) = i \operatorname{arcsch}(ix)$$

35:13:6 $$\operatorname{arcsec}(x) = \pm i \operatorname{arsech}(x)$$

CHAPTER 36

PERIODIC FUNCTIONS

Periodic functions play an important role in the solutions of many difficult problems in applied mathematics. Also, periodic functions, or nearly periodic functions, form the medium by which a good many information transfers take place. For example, beams of light, sound waves, and a variety of telecommunications signals are instances of periodic functions, often "modulated" in some way.

36:1 NOTATION

We shall use per(x) to represent any periodic function, and occasionally qer(x) to represent a second periodic function. Throughout this chapter P will denote the period of per(x). The quantity $2\pi/P$ is sometimes known as the *frequency* of the periodic function and is often denoted by ω.

36:2 BEHAVIOR

Apart from their characteristic of indefinitely repeating, periodic functions share no common behavior. Periodic functions may be continuous or discontinuous, simple or complicated. Examples of periodic functions are shown in Figures 36-1 through 36-7, exhibited later in this chapter.

36:3 DEFINITIONS

A function of argument x that satisfies the condition

36:3:1 $$f(x) = f(x + kP) \qquad k = 0, \pm1, \pm2, \ldots$$

for all x is a periodic function of period equal to the smallest positive value of P that satisfies equality 36:3:1.

A periodic function may be "naturally" periodic, as are each of the functions cited in Section 36:4, or it may be "synthesized" from an aperiodic function. An example of a "synthetic" periodic function, in this case created from the square function [Chapter 11], is

36:3:2 $$\mathrm{per}(x) = (x - 2k)^2 \qquad 2k \le x < 2 + 2k \qquad k = 0, \pm1, \pm2, \ldots$$

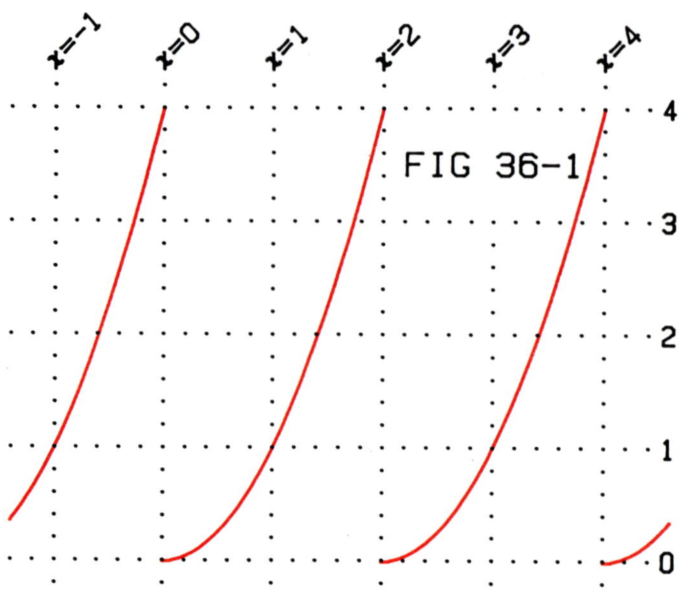

This particular periodic function, which is mapped in Figure 36-1, has a period of 2. As here, synthetic periodic functions often exhibit a discontinuity at one point, at least, in the $0 \le x < P$ range.

36:4 SPECIAL CASES

A number of periodic functions are considered in previous chapters. The fractional value function frac(x) is treated in Chapter 9, as is f(frac(x)); these have periods of unity. The functions sin(x), cos(x), csc(x) and sec(x), discussed in Chapters 32 and 33, have periods of 2π. In Chapter 34 the tan(x) and cot(x) functions, which are periodic functions of period π, are considered. The theta functions [Section 27:13] are periodic, as are the functions of Chapter 63.

The constant function [Chapter 1] is a degenerate case of a periodic function for which no smallest nonzero value of P exists.

36:5 INTRARELATIONSHIPS

The fundamental intrarelationship of periodic functions is their periodicity, which constitutes a recurrence formula

36:5:1 $$\mathrm{per}(x + P) = \mathrm{per}(x)$$

If two periodic functions per(x) and qer(x), with periods P and Q, are added, subtracted, multiplied or divided, the resultant function will be periodic only if the quotient P/Q is rational. If P/Q is rational, the period of per(x) $\pm$ qer(x), per(x)qer(x) and per(x)/qer(x) will normally be the least common multiple of P and Q. There are, however, frequent exceptions to this rule; for example, sin(x) and cos(x), each of period 2π, have products and quotients of period π, not 2π as predicted by the rule.

If the argument x of a periodic function per(x) is replaced by a linear function of x, to give per($bx + c$), periodicity is maintained but the period is altered from P to P/b. Replacement of the argument by any other aperiodic function of x destroys the periodicity; the functions per($ax^2 + bx + c$) and per($1/x$), for example, are not periodic. Multiplication or division by, or addition or subtraction of, any aperiodic function other than a constant likewise destroys the periodicity.

On the other hand, composites of periodic functions, such as $a\,\mathrm{per}^2(x) + b\,\mathrm{per}(x) + c$, exp(per($x$)) or $1/[c + \mathrm{per}(x)]$ are themselves periodic. Such functions have the same period P as per(x) or, occasionally, a submultiple of it such as $P/2$.

36:6 EXPANSIONS

In a process known as *harmonic analysis*, any periodic function per(x) of period P may be represented, exactly or approximately, as the *Fourier series*

36:6:1
$$\mathrm{F}(x) = \frac{c_0}{2} + \sum_{j=1}^{\infty} c_j \cos\left(\frac{2j\pi x}{P}\right) + s_j \sin\left(\frac{2j\pi x}{P}\right) \simeq \mathrm{per}(x)$$

The c and s coefficients in this series, known as *Fourier coefficients*, are given by the so-called *Euler formulas*

36:6:2
$$c_j = \frac{2}{P}\int_{x_0}^{x_0+P} \mathrm{per}(t)\cos\left(\frac{2j\pi t}{P}\right)\mathrm{d}t \qquad j = 0, 1, 2, \ldots$$

36:6:3
$$s_j = \frac{2}{P}\int_{x_0}^{x_0+P} \mathrm{per}(t)\sin\left(\frac{2j\pi t}{P}\right)\mathrm{d}t \qquad j = 1, 2, 3, \ldots$$

where x_0 is arbitrary. The Fourier coefficients satisfy *Parseval's relation*

36:6:4
$$\frac{c_0^2}{2} + \sum_{j=1}^{\infty} c_j^2 + s_j^2 = \frac{2}{P}\int_{x_0}^{x_0+P} \mathrm{per}^2(t)\mathrm{d}t$$

provided that per(x) is everywhere finite. A brief tabulation of Fourier coefficients for a variety of periodic functions is included in Section 36:14.

The Fourier series 36:6:1 may be written in the alternative form

36:6:5
$$\mathrm{F}(x) = \frac{c_0}{2} + \sum_{j=1}^{\infty} \sqrt{c_j^2 + s_j^2}\,\sin\left(\frac{2j\pi x}{P} + \mathrm{arccot}(s_j/c_j)\right)$$

or, in terms of exponential functions of imaginary argument, as

36:6:6
$$\mathrm{F}(x) = \sum_{j=-\infty}^{\infty} a_j \exp\left(\frac{2ij\pi x}{P}\right)$$

where

36:6:7
$$a_j = \frac{1}{P}\int_0^P \mathrm{per}(t)\,\exp\left(\frac{-2ij\pi t}{P}\right)\mathrm{d}t = \frac{c_{|j|} - \mathrm{sgn}(j) i s_{|j|}}{2}$$

We shall not discuss the convergence of Fourier series in detail [see Hamming, Chapter 32]. Suffice it to state that the Fourier series of a finite continuous periodic function converges to the function, whereas the Fourier series of a finite discontinuous periodic function converges to the function except at the points of discontinuity. If the periodic function per(x) has a discontinuity at $x = d$ (as well as at $x = d \pm P$, $x = d \pm 2P$, etc., of course), such that

36:6:8
$$\lim_{\varepsilon \to 0} \mathrm{per}(d - \varepsilon) = \mathrm{per}(d-) \qquad \varepsilon > 0$$

and

36:6:9
$$\lim_{\varepsilon \to 0} \mathrm{per}(d + \varepsilon) = \mathrm{per}(d+) \qquad \varepsilon > 0$$

then the Fourier series converges to $[\mathrm{per}(d-)+\mathrm{per}(d+)]/2$ at $x = d$. For example, the periodic function defined in 36:3:2 and mapped in Figure 36-1 has the Fourier series

36:6:10
$$\frac{4}{3} + \sum_{j=1}^{\infty} \frac{4}{j^2\pi^2}\cos(j\pi x) - \frac{4}{j\pi}\sin(j\pi x)$$

This evaluates, for $x = 0$, to $\frac{4}{3} + 4\zeta(2)/\pi^2 = 2$, which is the mean of the values, 4 and 0, on either side of the discontinuity at per(0).

36:7 PARTICULAR VALUES

The particular values of per(x) depend on the identity of the periodic function. In terms of the Fourier coefficients, however, it is evident that

36:7:1 $$\mathrm{per}(0) = \mathrm{per}(P) = \frac{c_0}{2} + c_1 + c_2 + c_3 + c_4 + c_5 + c_6 + \cdots$$

36:7:2 $$\mathrm{per}\left(\frac{P}{4}\right) = \frac{c_0}{2} + s_1 - c_2 - s_3 + c_4 + s_5 - c_6 - \cdots$$

36:7:3 $$\mathrm{per}\left(\frac{P}{2}\right) = \frac{c_0}{2} - c_1 + c_2 - c_3 + c_4 - c_5 + c_6 - \cdots$$

36:7:4 $$\mathrm{per}\left(\frac{3P}{4}\right) = \frac{c_0}{2} - s_1 - c_2 + s_3 + c_4 - s_5 - c_6 + \cdots$$

36:8 NUMERICAL VALUES

The calculation of values of per(x) generally presents no special problems.

36:9 APPROXIMATIONS

A truncated Fourier series

36:9:1 $$\mathrm{F}_J(x) = \frac{c_0}{2} + \sum_{j=1}^{J} c_j \cos\left(\frac{2j\pi x}{P}\right) + s_j \sin\left(\frac{2j\pi x}{P}\right) \simeq \mathrm{per}(x)$$

produces an approximation to per(x) that is best in the least-squares sense [see Section 7:14].

36:10 OPERATIONS OF THE CALCULUS

The derivative of a periodic function per(x) is another periodic function of identical period. Discontinuities in per(x) will generate Dirac delta functions [Chapter 10] in $\mathrm{d\,per}(x)/\mathrm{d}x$.

The $c_0/2$ Fourier coefficient represents the average value of per(x) over the period, that is:

36:10:1 $$\frac{1}{P}\int_{x_0}^{x_0+P} \mathrm{per}(t)\mathrm{d}t = \frac{c_0}{2} \qquad x_0 \text{ arbitrary}$$

Only if this coefficient is zero will indefinite integration of per(x) give rise to another periodic function.

After subtraction of $c_0/2$, differintegration with lower limit minus infinity converts per(x) into another periodic function of identical period. In terms of its Fourier series, the new function is

36:10:2 $$\frac{\mathrm{d}^\nu}{[\mathrm{d}(x+\infty)]^\nu}\left[\mathrm{per}(x) - \frac{c_0}{2}\right] = \sum_{j=1}^{\infty}\left(\frac{2j\pi}{P}\right)^\nu\left[c_j\cos\left(\frac{2j\pi x}{P} + \frac{\nu\pi}{2}\right) + s_j\sin\left(\frac{2j\pi x}{P} + \frac{\nu\pi}{2}\right)\right]$$

$$= \left(\frac{2\pi}{P}\right)^\nu \sum_{j=1}^{\infty} j^\nu(c_jC + s_jS)\cos\left(\frac{2j\pi x}{P}\right) + j^\nu(s_jC - c_jS)\sin\left(\frac{2j\pi x}{P}\right)$$

where $C = \cos(\nu\pi/2)$ and $S = \sin(\nu\pi/2)$. This formulation is valid for $\nu > -1$.

36:11 COMPLEX ARGUMENT

For constant y, the periodic function per($x + iy$) remains periodic in x. The Fourier coefficients of such functions are then complex numbers that can be evaluated with the help of equations 32:11:1 and 32:11:2.

Certain functions, such as ln and sinh, which are aperiodic for real argument, become periodic when the argument is imaginary or complex.

The Jacobian elliptic functions [Chapter 63] are *doubly periodic* when their argument is complex. That is, they satisfy the recurrence relations

36:11:1 $$\mathrm{per}(x + iy) = \mathrm{per}(x + P + iy) = \mathrm{per}(x + iy + iQ) = \mathrm{per}(x + P + iy + iQ)$$

where P and Q are the real and imaginary periods.

36:12 GENERALIZATIONS

Apart from the double periodicity cited in Section 36:11, no generalization of the concept of a periodic function has been made.

36:13 COGNATE FUNCTIONS

The terms *full-rectification* and *half-rectification*, which have their origin in the technology of alternating electrical currents, are sometimes encountered in connection with periodic functions. Using the notation of the absolute-value function [see Chapter 8], full- and half-rectification converts the function per(x) into

36:13:1 $$|\mathrm{per}(x)| \qquad \text{full-rectification}$$

and

36:13:2 $$\frac{1}{2}\{|\mathrm{per}(x)| + \mathrm{per}(x)\} \qquad \text{half-rectification}$$

Figure 36-2 shows an example of each. Full- or half-rectification of a periodic function maintains its periodicity and, generally, the period remains unchanged. Sometimes, as in the full-rectification example shown in Figure 36-2, rectification decreases the period (increases the frequency) by a factor of 2 (or more).

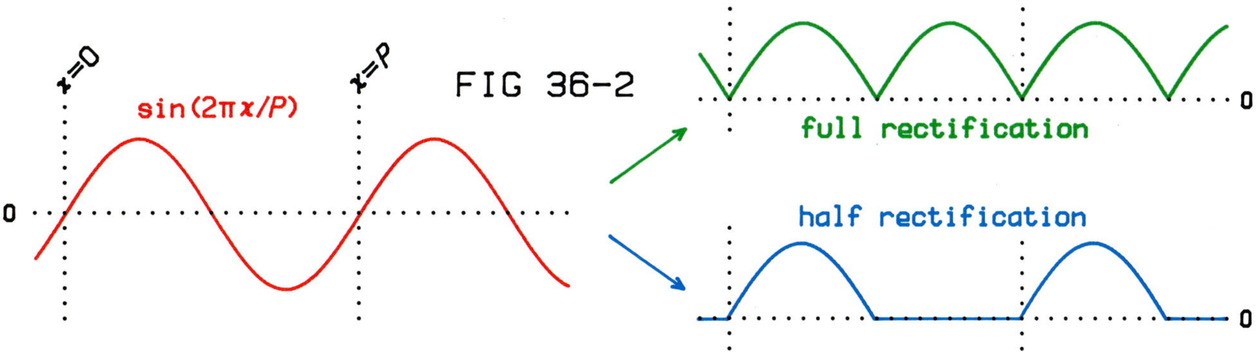

FIG 36-2

36:14 RELATED TOPICS

Table 36.14.1 lists some frequently encountered periodic functions, together with their Fourier coefficients [see equations 36:6:1–36:6:3]. For further examples of Fourier coefficients, see equations 32:5:29, 32:5:30, 20:6:4–20:6:6 and 19:6:5–19:6:7.

Table 36.14.1

Name of function	Figure	per(x) =	$\frac{c_0}{2}$	c_j: $j = 1, 3, 5, \ldots$	c_j: $j = 2, 4, 6, \ldots$	s_j: $j = 1, 3, 5, \ldots$	s_j: $j = 2, 4, 6, \ldots$
Square wave (odd)	36-3	$(-1)^{\mathrm{Int}(2x/P)}$	0	0	0	$\frac{4}{j\pi}$	0
Square wave (even)	36-4	$(-1)^n \; n = \mathrm{Int}\left(\frac{4x+P}{2P}\right)$	0	$\frac{(-1)^{(j-1)/2}4}{j\pi}$	0	0	0
Sawtooth wave	36-5	$\mathrm{frac}\left(\frac{1}{2}+\frac{x}{P}\right)$	$\frac{1}{2}$	0	0	$\frac{1}{j\pi}$	$\frac{-1}{j\pi}$
Triangular wave	36-6	$(-1)^{\mathrm{Int}(2x/P)}\left[2\,\mathrm{frac}\left(\frac{1}{2}+\frac{x}{P}\right)-1\right]$	$\frac{1}{2}$	$\frac{-4}{j^2\pi^2}$	0	0	0
Pulse train	36-7	$\sum_{j=-\infty}^{\infty} p(c;h;t-\frac{P}{2}-jP)$	$\frac{ch}{P}$	$\frac{-4c}{h}\sin\left(\frac{j\pi h}{P}\right)$	$\frac{4c}{h}\sin\left(\frac{j\pi h}{P}\right)$	0	0
Full-rectified sine wave	36-2	$\left\lvert\sin\left(\frac{2\pi x}{P}\right)\right\rvert$	$\frac{2}{\pi}$	0	$\frac{-4}{(j^2-1)\pi}$	0	0
Half-rectified sine wave	36-2	$\frac{1}{2}\sin\left(\frac{2\pi x}{P}\right)+\frac{1}{2}\left\lvert\sin\left(\frac{2\pi x}{P}\right)\right\rvert$	$\frac{1}{\pi}$	0	$\frac{-2}{(j^2-1)\pi}$	$\frac{\mathrm{u}(2-j)}{2}$	0

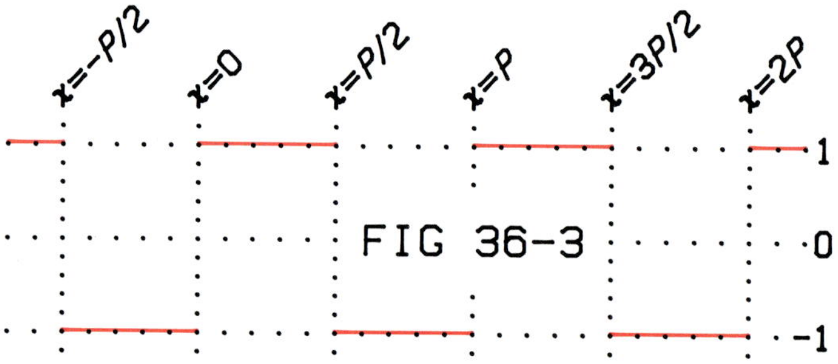

FIG 36-3

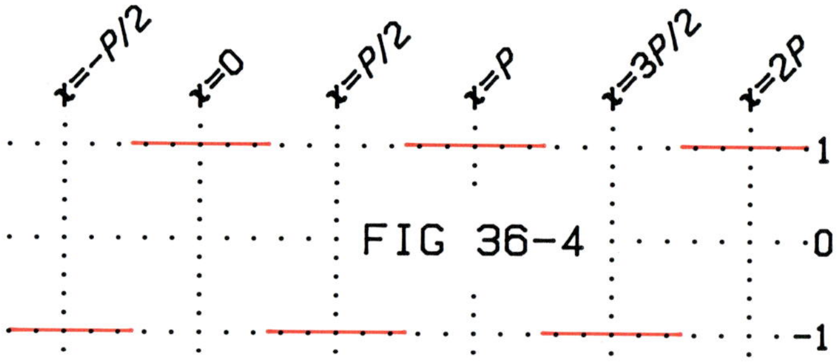

FIG 36-4

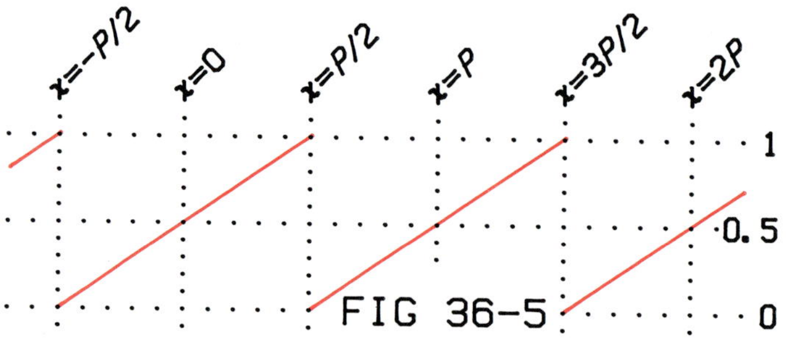

FIG 36-5

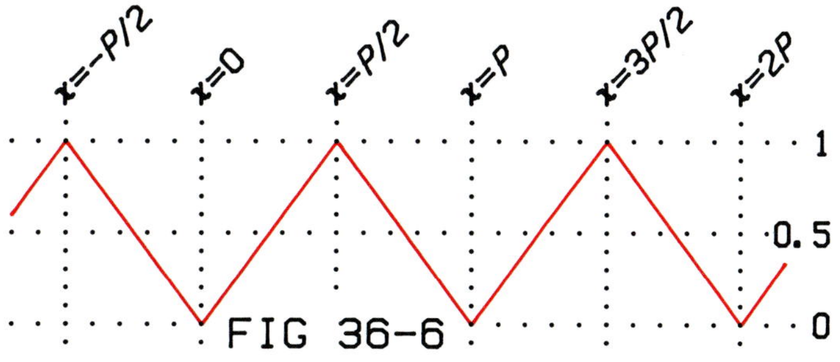

FIG 36-6

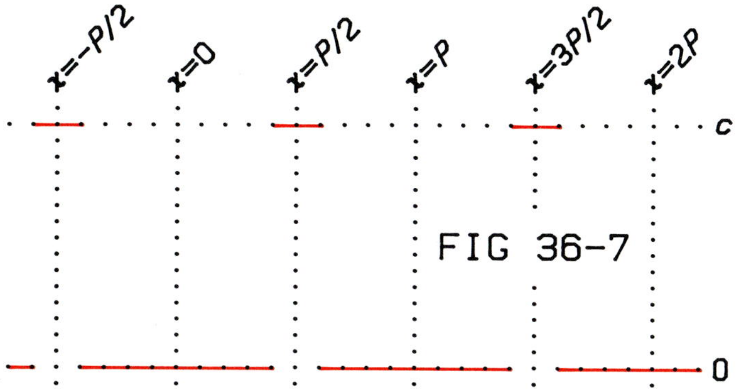

FIG 36-7

CHAPTER 37

THE EXPONENTIAL INTEGRAL Ei(x) AND RELATED FUNCTIONS

Indefinite integrals of the form $\int t^{-n}\exp(\pm t)dt$ cannot be expressed in terms of elementary functions for $n = 1, 2, 3, \ldots$. The exponential integral function Ei(x) fills this deficiency. A closely related function, the *entire exponential integral*, is also discussed in this chapter; it is related to Ei(x) by an identity

37:0:1 $$\mathrm{Ein}(x) = \gamma + \ln(|x|) - \mathrm{Ei}(-x) \qquad \gamma = 0.5772156649$$

involving Euler's constant γ [Section 1:7] and the logarithm of the absolute value of the argument x.

The *logarithmic integral* li(x) is discussed in Chapter 25 but, because of its simple relationship

37:0:2 $$\mathrm{li}(x) = \mathrm{Ei}(\ln(x))$$

to the exponential integral, many of its properties are given in this chapter.

37:1 NOTATION

Notations abound for the exponential integral function: Ei*(x), E*(x), $\bar{\mathrm{E}}\mathrm{i}(x)$, $\overline{\mathrm{Ei}}(x)$, $\mathrm{E}^+(x)$ and $\mathrm{E}^-(x)$ have all been used to denote Ei(x) or some very similar function. When one of these symbols is encountered, care is needed to ascertain whether its definition differs from that which is employed for Ei(x) in this *Atlas*.

When x is negative, the expression $-\mathrm{E}_1(-x)$ is often used to replace Ei(x) in view of relationship 37:13:7. The $\mathrm{E}_1(x)$ symbol here represents a Schlömilch function [see Section 37:13] and should not be confused with the identical symbol used for Euler polynomials [Chapter 20]. To add to the confusion, the notation ei(x) is sometimes encountered for the Schlömilch $\mathrm{E}_1(x)$ function.

37:2 BEHAVIOR

Figure 37-1 maps the behaviors of the exponential integral Ei, entire exponential integral Ein and logarithmic integral li functions. All three functions increase without limit as $x \to \infty$ but, following an inflection [Section 0:7] at $x = 1$, the increase is dramatic for the exponential integral function (e.g., $\mathrm{Ei}(10) \simeq 2 \times 10^3$ and $\mathrm{Ei}(100) \simeq 3 \times 10^{41}$). Note also that Ei($x$) rapidly approaches zero as $x \to -\infty$ and exhibits a discontinuity at $x = 0$. Observe also that li(x) is discontinuous at $x = 1$ and is defined only for $x \geq 0$.

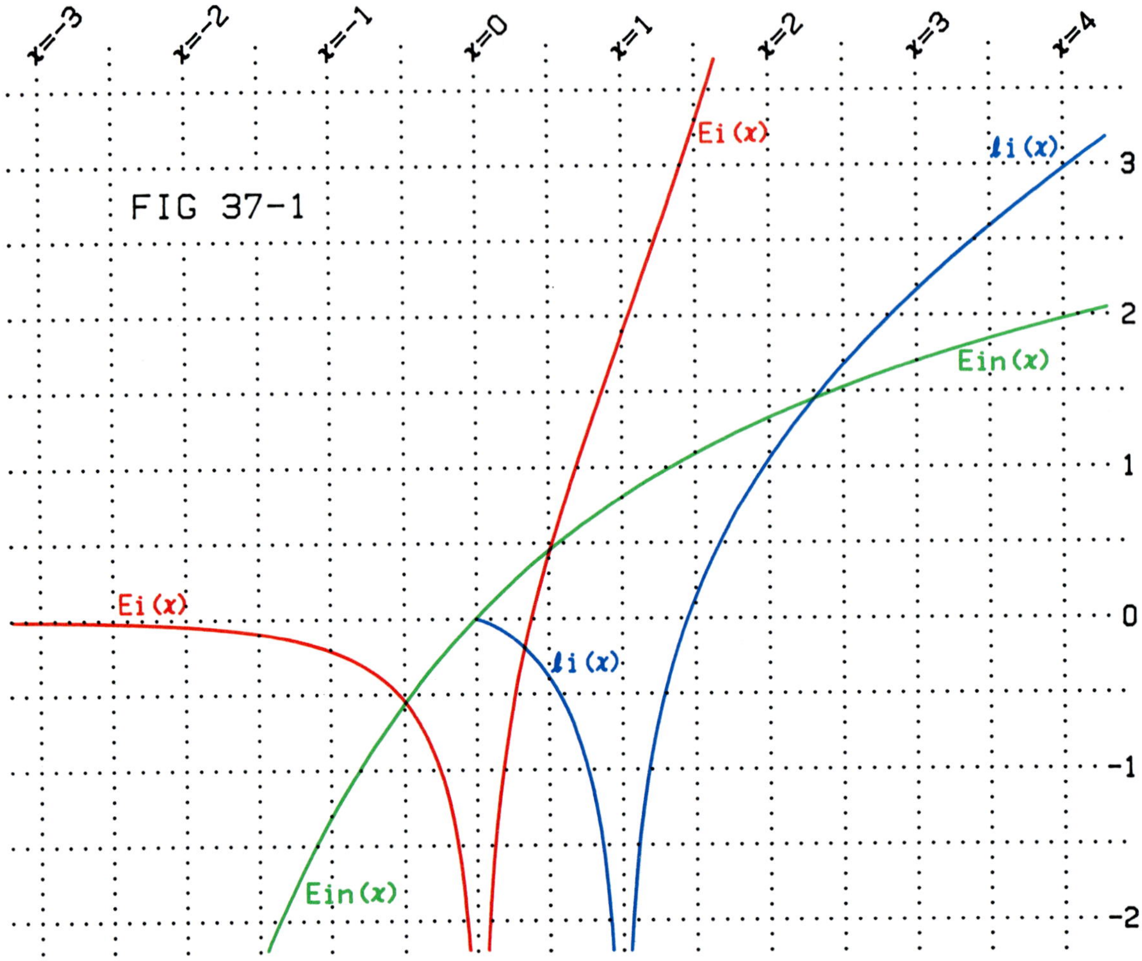

37:3 DEFINITIONS

The exponential integral function is defined by the indefinite integral

37:3:1
$$\mathrm{Ei}(x) = \int_{-\infty}^{x} \frac{\exp(t)}{t}\,\mathrm{d}t$$

for all x, as illustrated in Figure 37-2. At zero argument the integrand in 37:3:1 encounters an infinity so that for $x > 0$ the integral is to be interpreted as the *Cauchy limit*

37:3:2
$$\lim_{\varepsilon\to 0}\left\{\mathrm{Ei}(-\varepsilon) + \int_{\varepsilon}^{x} \frac{\exp(t)}{t}\,\mathrm{d}t\right\} \qquad x > 0 \qquad \varepsilon > 0$$

The exponential integral may also be expressed as a definite integral in a number of ways, including

37:3:3
$$\mathrm{Ei}(\pm x) = \exp(\pm x)\int_{0}^{1} \frac{\mathrm{d}t}{\ln(t) \pm x} \qquad x > 0$$

Gradshteyn and Ryzhik [Section 8.212] list many other integral representations of Ei(x).

The definition of the entire exponential integral

37:3:4
$$\mathrm{Ein}(x) = \int_{0}^{x} \frac{1 - \exp(-t)}{t}\,\mathrm{d}t$$

FIG 37-2

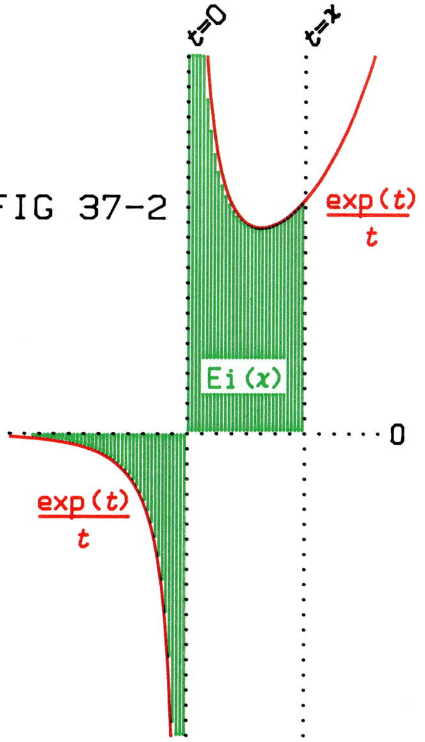

is illustrated in Figure 37-3. Its relationship, equation 37:0:1, to the exponential integral Ei permits several alternative definitions of the Ein function.

Definitions of the logarithmic integral function were presented in Section 25:13. Some authorities regard the logarithmic integral as defined only for arguments exceeding unity, but in this *Atlas* li(x) exists for $0 \le x < 1$ with a Cauchy limit interpretation, similar to 37:3:2, serving to extend the definition to $x > 1$.

37:4 SPECIAL CASES

There are none.

FIG 37-3

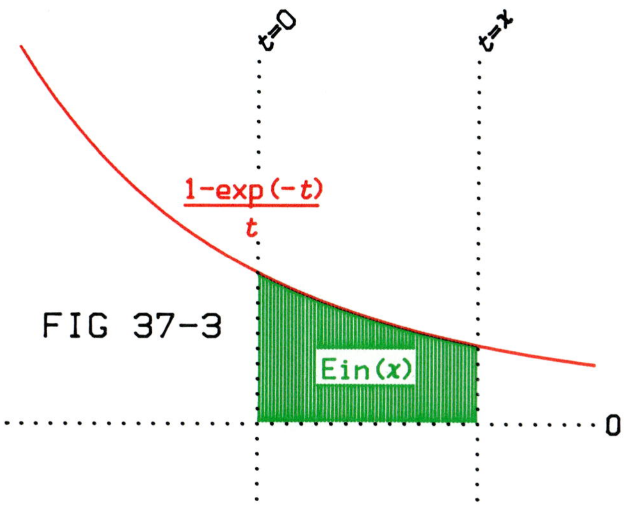

37:5 INTRARELATIONSHIPS

Relating the exponential integral and its entire analog is the identity

37:5:1 $$\mathrm{Ei}(x) + \mathrm{Ein}(-x) = \mathrm{Ei}(-x) + \mathrm{Ein}(x)$$

that follows from equation 37:0:1, or through definitions 37:3:1 and 37:3:4.

We know of no reflection, recurrence, addition or multiplication formulas for the exponential integral functions that involve a finite number of elementary functions. However the reflection formulas

37:5:2 $$\mathrm{Ei}(-x) = \mathrm{Ei}(x) - 2\,\mathrm{Shi}(x)$$

and

37:5:3 $$\mathrm{Ein}(-x) = \mathrm{Ein}(x) - 2\,\mathrm{Shi}(x) = -\mathrm{Ein}(x) - 2\,\mathrm{Chin}(x)$$

may be written in terms of the functions of Chapter 38. Also, the argument-addition formula

37:5:4 $$\mathrm{Ei}(x + y) = \mathrm{Ei}(x) + \exp(x) \sum_{j=0}^{\infty} \frac{j![\exp(y)\mathrm{e}_j(-y) - 1]}{x^{j+1}} \qquad |x| > |y|$$

for the exponential integral exists and utilizes the exponential polynomial [see Section 26:13].

37:6 EXPANSIONS

Two alternative power series:

37:6:1 $$\mathrm{Ein}(x) = x - \frac{x^2}{4} + \frac{x^3}{18} - \frac{x^4}{96} + \cdots = -\sum_{j=1}^{\infty} \frac{(-x)^j}{j!j}$$

and

37:6:2 $$\exp(x)\,\mathrm{Ein}(x) = x + \frac{3x^2}{4} + \frac{11x^3}{36} + \frac{25x^4}{288} + \cdots = \sum_{j=1}^{\infty}\left(1 + \frac{1}{2} + \frac{1}{3} + \cdots + \frac{1}{j}\right)\frac{x^j}{j!}$$
$$= \sum_{j=1}^{\infty} [\gamma + \psi(j+1)] \frac{x^j}{j!}$$

are available to express the entire exponential integral function. In 37:6:2, ψ is the digamma function described in Chapter 44.

The exponential integral function is expansible in terms of Laguerre polynomials [Chapter 23]:

37:6:3 $$\mathrm{Ei}(x) = -\exp(x) \sum_{j=0}^{\infty} \frac{\mathrm{L}_j(-x)}{j+1} \qquad x < 0$$

as the continued fraction:

37:6:4 $$\mathrm{Ei}(x) = \frac{\exp(x)}{x-}\;\frac{1}{1-}\;\frac{1}{x-}\;\frac{2}{1-}\;\frac{2}{x-}\;\frac{3}{1-}\cdots$$

or as the asymptotic series

37:6:5 $$\mathrm{Ei}(x) \sim \exp(x)\left[\frac{1}{x} + \frac{1}{x^2} + \frac{2}{x^3} + \frac{6}{x^4} + \cdots + \frac{j!}{x^{j+1}} + \cdots\right] \qquad x \to \infty$$

valid for large argument x. For small arguments, combination of expansion 37:6:1 with relation 37:0:1 leads to

37:6:6 $$\mathrm{Ei}(x) = \ln(|x|) + \gamma + x + \frac{x^2}{4} + \frac{x^3}{18} + \cdots$$

which describes the behavior of the exponential integral close to $x = 0$.

Expansions for li(x) arise by replacing x in 37:6:3–37:6:6 by ln(x).

Table 37.7.1

	$x = -\infty$	$x = 0$	$x = 1$	$x = \infty$
Ei(x)	0	$-\infty$	1.895117816	∞
Ein(x)	$-\infty$	0	0.7965995993	∞
li(x)	undef	0	$-\infty$	∞

37:7 PARTICULAR VALUES

As well as the values given in Table 37.7.1, note that Ei(0.3725074) = li(1.4513692) = 0.

37:8 NUMERICAL VALUES

The algorithm below is designed to generate values of Ei(x), for x values of any magnitude and either sign, with 24-bit precision. However, less precise values may be encountered in the immediate vicinity of the zero of the Ei(x) function, that is, between $x = 0.370$ and 0.375. For $-2 < x < 22$, the algorithm uses series 37:6:1 in the truncated and concatenated form

37:8:1 $$\mathrm{Ein}(-x) \simeq \left(\left(\left(\cdots\left(\left(\frac{Jx}{(J+1)^2}+1\right)\frac{(J-1)x}{J^2}+1\right)\frac{(J-2)x}{(J-1)^2}+\cdots+1\right)\frac{2x}{9}+1\right)\frac{x}{4}+1\right)\frac{(-x)}{1}$$

Set $f = -10^{99}$
Input x >>>>>>>
Replace x by $\ln(x)$
If $|x - 10| \geq 12$ go to (2)
If $x = 0$ go to (4)
Set $j = \mathrm{Int}(10 + 2|x|)$
Set $f = 1/(j+1)^2$
(1) Replace f by $(fjx + 1)/j^2$
Replace j by $j - 1$
If $j \neq 0$ go to (1)
Replace f by $fx + \ln(1.781072418|x|)$
Go to (4)
(2) Set $j = \mathrm{Int}(5 + 20/|x|)$
Set $f = x$
(3) Replace f by $[1/(1/f - 1/j)] + x$
Replace j by $j - 1$
If $j \neq 0$ go to (3)
Replace f by $[\exp(x)]/f$
(4) Output f
$f \simeq \left\{\begin{matrix}\mathrm{Ei}(x)\\ \mathrm{li}(x)\end{matrix}\right\}$ <<<<

Storage needed: x, j and f

Input restrictions: For the calculation of li(x), x must be positive.

Test values:
Ei(25) = 3.00595092×10^9
Ei(−1.5) = −0.100019585
li(7) = 4.75705176

with J assigned the empirical value Int($10 + 2|x|$), and then employs interrelation 37:0:1 to generate Ei(x). For $x \leq -2$ or $x \geq 22$, the algorithm employs the continued fraction 37:6:4 terminated at $(J-2)/(1-(J-1)/(x-J/(1-J/x)))$ where $J = \mathrm{Int}(5 + 20)/|x|)$. The algorithm will generate li(x) when the additional instruction shown in green is included. Because Ei(0) = li(1) = $-\infty$, the algorithm returns -10^{99} at these arguments.

The universal hypergeometric algorithm [Section 18:14] also permits Ei(x), Ein(x) and li(x) to be determined for suitable values of the argument x.

37:9 APPROXIMATION

Based on the identity 37:0:1 and expansion 37:6:1, the approximation

37:9:1 $$\mathrm{Ei}(x) \simeq \frac{36x + x^2}{36 - 8x} + \ln(1.78|x|) \qquad \text{8-bit precision} \qquad \begin{cases} -1.05 \le x < 0 \\ 0 < x \le 0.36 \\ 0.38 \le x \le 1.25 \end{cases}$$

is valid for most small values of x, while the approximation

37:9:2 $$\mathrm{Ei}(x) \simeq \frac{\exp(x)}{\sqrt{x^2 - 2x}} \qquad \text{8-bit precision} \qquad \begin{cases} x \le -9 \\ x \ge 15 \end{cases}$$

based on expansion 37:6:5, is useful for arguments of large magnitude. The very simple approximation to the entire exponential integral

37:9:3 $$\mathrm{Ein}(x) \simeq \frac{36x - x^2}{36 + 8x} \qquad \text{8-bit precision} \qquad -1.1 \le x \le 1.4$$

may be used near the origin.

37:10 OPERATIONS OF THE CALCULUS

The following rules apply for differentiation and indefinite integration:

37:10:1 $$\frac{\mathrm{d}}{\mathrm{d}x}\mathrm{Ei}(bx + c) = \frac{\exp(bx + c)}{x + (c/b)}$$

37:10:2 $$\frac{\mathrm{d}}{\mathrm{d}x}\mathrm{Ein}(bx + c) = \frac{1 - \exp(-bx - c)}{x + (c/b)}$$

37:10:3 $$\frac{\mathrm{d}}{\mathrm{d}x}\mathrm{li}(x^\nu) = \frac{x^{\nu-1}}{\ln(x)} \qquad \nu \neq 0$$

37:10:4 $$\int_{-c/b}^{x} \mathrm{Ei}(bt + c)\mathrm{d}t = \left(x + \frac{c}{b}\right)\mathrm{Ei}(bx + c) + \frac{1 - \exp(bx + c)}{b}$$

37:10:5 $$\int_{-c/b}^{x} \mathrm{Ein}(bt + c)\mathrm{d}t = \left(x + \frac{c}{b}\right)[\mathrm{Ein}(bx + c) - 1] + \frac{1 - \exp(-bx - c)}{b}$$

37:10:6 $$\int_{-c/b}^{x} \mathrm{li}(bt + c)\mathrm{d}t = \left(x + \frac{c}{b}\right)\mathrm{li}(bx + c) - \frac{1}{b}\mathrm{li}(b^2x^2 + 2bcx + c^2)$$

37:10:7 $$\int_{-c/b}^{x} \exp(Bt + C)\,\mathrm{Ei}(bt + c)\mathrm{d}t = \begin{cases} \dfrac{\exp\left(C - \dfrac{Bc}{b}\right)}{b}\left[\exp\left(Bx + \dfrac{Bc}{b}\right)\mathrm{Ei}(bx + c) - \mathrm{Ei}\left((B + b)\left(x + \dfrac{c}{b}\right)\right) + \ln\left|\dfrac{B + b}{b}\right|\right] & B + b \neq 0 \\ \dfrac{\exp(C + c)}{b}\left[\ln\left(x + \dfrac{c}{b}\right) + \gamma - \exp(-bx - c)\mathrm{Ei}(bx + c)\right] & B + b = 0 \end{cases}$$

When $n = -1$, the indefinite integral $\int t^n \mathrm{Ei}(bt + c)\mathrm{d}t$ cannot be evaluated in a finite number of terms, but for $n = 0, 1, 2, 3, \ldots$ we have

37:10:8 $$\int_{-c/b}^{x} t^n\,\mathrm{Ei}(bt+c)\mathrm{d}t = \frac{\mathrm{Ei}(bx+c)}{n+1}\left[x^{n+1} - \left(\frac{-c}{b}\right)^{n+1}\right] + \frac{n!}{(-b)^{n+1}}\sum_{j=0}^{n}\frac{c^{n-j}[\exp(bx+c)\mathrm{e}_j(-bx-c)-1]}{(j+1)(n-j)!}$$

while the $n = -2, -3, -4, \ldots$ cases can be evaluated via the lengthy expression

37:10:9 $$\int_{-c/b}^{x}\frac{\mathrm{Ei}(bt+c)}{t^{n+1}}\mathrm{d}t = \frac{\mathrm{Ei}(bx+c)}{n(-c/b)^n}\left[1-\left(\frac{-c}{bx}\right)^n\right] + \frac{\exp(c)\mathrm{e}_{n-1}(-c)}{n(-c/b)^n}\left\{\mathrm{Ei}(-c) - \mathrm{Ei}(bx) + \sum_{j=0}^{n-2} j!\left[\frac{\exp(bx)}{(bx)^{j+1}} - \frac{\exp(-c)}{(-c)^{j+1}}\right]\left[1 - \frac{\mathrm{e}_j(-c)}{\mathrm{e}_{n-1}(-c)}\right]\right\}$$

in which $\mathrm{e}_j(x)$ denotes the exponential polynomial [Section 26:13].

Included among the definite integrals tabulated by Gradshteyn and Ryzhik [Section 6.21–6.23] are

37:10:10 $$\int_{-\infty}^{0}\mathrm{Ei}(bt)\mathrm{d}t = \frac{-1}{b} \qquad b>0$$

37:10:11 $$\int_{-\infty}^{0}\mathrm{Ei}(bt)\,\mathrm{Ei}(Bt)\mathrm{d}t = \frac{1}{bB}\ln\left(\frac{(b+B)^{b+B}}{b^bB^B}\right) \qquad b>0 \qquad B>0$$

37:10:12 $$\int_{0}^{\infty}\ln(Bt)\,\mathrm{Ei}(-bt)\mathrm{d}t = \frac{1+\gamma+\ln(b/B)}{b} \qquad b>0 \qquad B>0$$

37:10:13 $$\int_{0}^{\infty}t^{\nu}\,\mathrm{Ei}(-bt)\mathrm{d}t = \frac{-\Gamma(1+\nu)}{(1+\nu)b^{1+\nu}} \qquad b\geq 0 \qquad \nu>-1$$

37:10:14 $$\int_{0}^{\infty}t^{\nu-1}\exp(-Bt)\,\mathrm{Ei}(-bt)\mathrm{d}t = \frac{-\Gamma(\nu)}{B^{\nu}}\mathrm{B}\left(0;\nu;\frac{B}{b+B}\right) \qquad b+B>0 \qquad \nu>0$$

37:10:15 $$\int_{0}^{1}t^{\nu}\,\mathrm{li}(t)\mathrm{d}t = \frac{-\ln(2+\nu)}{1+\nu} \qquad \nu>-2$$

37:10:16 $$\int_{1}^{\infty}\frac{\mathrm{li}(t)}{t^{\nu}}\mathrm{d}t = \frac{-\ln(\nu-2)}{\nu-1} \qquad \nu>2$$

The Γ and B functions occurring in integrals 37:10:13 and 37:10:14 are, respectively, the gamma function [Chapter 43] and the incomplete beta function [Chapter 58].

37:11 COMPLEX ARGUMENT

Even with a complex argument, the Ein function is single valued and continuous. For small complex arguments, the series

37:11:1 $$\mathrm{Ein}(x+iy) = \left[x - \frac{x^2-y^2}{2!2} + \frac{x^3-3xy^3}{3!3} - \frac{x^4-6x^2y^2+y^4}{4!4} + \cdots\right] + i\left[y - \frac{2xy}{2!2} + \frac{3xy^2-y^3}{3!3} - \frac{4x^3y-4xy^3}{4!4} + \cdots\right]$$

converges rapidly and can be used to find values of $\mathrm{Ein}(x+iy)$. For purely imaginary argument, one has

37:11:2 $$\mathrm{Ein}(iy) = \mathrm{Cin}(y) + i\mathrm{Si}(y)$$

where the Cin and Si functions are discussed in Chapter 38.

The exponential integral Ei is a many-valued function, when its argument is complex, because equation 37:0:1 generalizes to

37:11:3 $$\mathrm{Ein}(x+iy) + \mathrm{Ei}(-x-iy) = \gamma + \mathrm{Ln}(x+iy)$$

where Ln is the multivalued logarithmic function discussed in Section 25:11. However, as explained in that section, one usually selects a principal value of the Ln function, which leads to the single value

37:11:4 $$\mathrm{Ei}(x+iy) = \gamma + \frac{1}{2}\ln(x^2+y^2) - \mathrm{Ein}(-x-iy) + i\theta$$

where θ is the angle defined in Section 25:11. Notice that, according to 37:11:4, the exponential integral may have a complex value even when its argument is real. In fact:

37:11:5 $$\mathrm{Ei}(x+0\mathrm{i}) = \gamma + \ln(x) - \mathrm{Ein}(-x) + i\pi \qquad x>0$$

The imaginary term in this expression is ignored in the rest of the chapter. Abramowitz and Stegun [Tables 5.6 and 5.7] provide tables from which the real and imaginary parts of $\mathrm{Ei}(x+iy)$ may be determined. For purely imaginary argument, one has

37:11:6 $$\mathrm{Ei}(iy) = \mathrm{Ci}(y) + i\left[\mathrm{Si}(y) - \frac{\pi}{2}\mathrm{sgn}(y)\right]$$

in terms of the functions of Chapter 38.

37:12 GENERALIZATIONS

The exponential integral function is a special case:

37:12:1 $$\mathrm{Ei}(x) = -\Gamma(0;-x)$$

of the complementary incomplete gamma function [Chapter 45] and of the Tricomi function [Chapter 48]

37:12:2 $$\mathrm{Ei}(x) = -\exp(x)\,\mathrm{U}(1;1;-x)$$

The entire exponential integral function is a hypergeometric function [Section 18:14]:

37:12:3 $$\mathrm{Ein}(x) = x\sum_{j=0}^{\infty}\frac{(1)_j(-x)^j}{(2)_j(2)_j}$$

37:13 COGNATE FUNCTIONS

In addition to the *Schlömilch functions* [discussed later in this section], two other function families, related to exponential integrals, are sometimes encountered.

The *alpha exponential integral* function of order n is defined by

37:13:1 $$\alpha_n(x) = \int_1^{\infty} t^n \exp(-xt)\mathrm{d}t \qquad n = 0, 1, 2, \ldots$$

The first member $\alpha_0(x)$ equals $[\exp(-x)]/x$ and because of the recurrence relation

37:13:2 $$\alpha_n(x) = \frac{n}{x}\alpha_{n-1}(x) + \alpha_0(x) \qquad n = 1, 2, 3, \ldots$$

all of these functions can be reduced to the elementary functions

37:13:3 $$\alpha_n(x) = \frac{n!\exp(-x)}{x^{n+1}}\mathrm{e}_n(x) \qquad n = 0, 1, 2, \ldots$$

where e_n is the exponential polynomial [Section 26:13].

The *beta exponential integral* function family is defined by a similar integral, but with changed limits:

37:13:4 $$\beta_n(x) = \int_{-1}^{1} t^n \exp(-xt)\mathrm{d}t \qquad n = 0, 1, 2, \ldots$$

The first member $\beta_0(x) = 2[\sinh(x)]/x$ and all members are reducible, via the expression

37:13:5 $$\beta_n(x) = \frac{n!}{x^{n+1}}[\exp(x)e_n(-x) - \exp(-x)e_n(x)] \qquad n = 0, 1, 2, \ldots$$

to more elementary functions.

A family of functions defined by

37:13:6 $$E_n(x) = \int_1^\infty \frac{\exp(-xt)}{t^n}dt \qquad n = 0, 1, 2, \ldots$$

was introduced by Schlömilch. The first member $E_0(x) = [\exp(-x)]/x$ is elementary, while the second

37:13:7 $$E_1(x) = -Ei(-x)$$

is related in a simple way to the exponential integral function. These identities, coupled with the recurrence relationship

37:13:8 $$E_n(x) = \frac{x}{n-1}[E_0(x) - E_{n-1}(x)] \qquad n = 2, 3, 4, \ldots$$

enable the properties of $E_n(x)$ to be deduced from those of $Ei(-x)$. The general relationship is

37:13:9 $$E_n(x) = \frac{-(-x)^{n-1}}{(n-1)!}\left[Ei(-x) + \frac{\exp(-x)}{x}\left\{1 - \frac{1}{x} + \frac{2!}{x^2} - \cdots + \frac{(n-2)!}{(-x)^{n-2}}\right\}\right] \qquad n = 2, 3, 4, \ldots$$

Some familial properties of Schlömilch functions are

37:13:10 $$E_n(0) = \frac{1}{n-1} \qquad n = 2, 3, 4, \ldots$$

37:13:11 $$\frac{d}{dx}E_n(x) = -E_{n-1}(x) \qquad n = 1, 2, 3, \ldots$$

37:13:12 $$E_n(x) = x^{n-1}\Gamma(1-n;x) \qquad n = 0, 1, 2, \ldots$$

and the asymptotic expansion

37:13:13 $$E_n(x) \sim \exp(-x)\left[\frac{1}{x} - \frac{n}{x^2} + \frac{n(n+1)}{x^3} - \cdots - \frac{(n)_j}{(-x)^{j+1}} + \cdots\right] \qquad x \to \infty$$

The function

37:13:14 $$\frac{-1}{x}\exp\left(\frac{1}{x}\right)Ei\left(\frac{-1}{x}\right) \sim 1 - x + 2x^2 - 6x^3 + \cdots + j!(-x)^j + \cdots$$

has been named *Euler's function* and symbolized E(x). It should not be confused with the complete elliptic integral of the second kind [Chapter 61] for which the same notation is adopted.

The avoid an unnecessary proliferation of functions, none of the functions discussed in this section is used elsewhere in the *Atlas*.

37:14 RELATED TOPICS

Very general expressions for indefinite integrals of the form $\int t^\nu \exp(bt)dt$ were presented in Section 26:10. Nevertheless, because of the very widespread occurrence of the integrals

37:14:1 $$\int t^\nu \exp(\pm t)dt \qquad \nu = 0, \pm\frac{1}{2}, \pm 1, \pm\frac{3}{2}, \ldots$$

in applied mathematics, Table 37.14.1, containing such indefinite integrals, may be useful. Integrals of the forms

37:14:2 $$\int u^{\nu} \exp\left(\frac{\pm 1}{u}\right) du \qquad \nu = 0, \pm\frac{1}{2}, +1, \pm\frac{3}{2}, \ldots$$

and

37:14:3 $$\int v^{n} \exp(\pm v^2) dv \qquad n = 0, \pm 1, \pm 2, \pm 3, \ldots$$

are similarly ubiquitous. The substitutions $u = 1/t$ or $v = \sqrt{t}$ convert these into the form of 37:14:1 so that Table 37.14.1 may also be used to evaluate integrals of the 37:14:2 and 37:14:3 families.

The erfc and daw functions occurring in Table 37.14.1 are the error function complement [Chapter 40] and Dawson's integral [Chapter 42]. The lower limits x_M and x_i are, respectively, the argument values corresponding to the maximum and the point of inflection of a graph of daw($\sqrt{x}$) [see Section 42:7].

Table 37.14.1

ν	x_0	$\int_{x_0}^{x} t^{\nu} \exp(t) dt$	$\int_{x}^{\infty} t^{\nu} \exp(-t) dt$
-3	$-\infty$	$\frac{\mathrm{Ei}(x)}{2} - \frac{(x+1)\exp(x)}{2x^2}$	$\frac{(1-x)\exp(-x)}{2x^2} - \frac{\mathrm{Ei}(-x)}{2}$
$\frac{-5}{2}$	x_i	$\frac{2}{3}\exp(x)\left[\mathrm{daw}(\sqrt{x}) - \frac{2x+1}{\sqrt{x^3}}\right]$	$\frac{2\exp(-x)}{3\sqrt{x}}\left[\frac{1}{x} - 2\right] + \frac{4\sqrt{\pi}}{3}\mathrm{erfc}(\sqrt{x})$
-2	$-\infty$	$\mathrm{Ei}(x) - \frac{\exp(x)}{x}$	$\mathrm{Ei}(-x) + \frac{\exp(-x)}{x}$
$\frac{-3}{2}$	x_M	$2\exp(x)\left[2\,\mathrm{daw}(\sqrt{x}) - \frac{1}{\sqrt{x}}\right]$	$\frac{2\exp(-x)}{\sqrt{x}} - 2\sqrt{\pi}\,\mathrm{erfc}(\sqrt{x})$
-1	$-\infty$	$\mathrm{Ei}(x)$	$-\mathrm{Ei}(-x)$
$\frac{-1}{2}$	0	$2\exp(x)\,\mathrm{daw}(\sqrt{x})$	$\sqrt{\pi}\,\mathrm{erfc}(\sqrt{x})$
0	$-\infty$	$\exp(x)$	$\exp(-x)$
$\frac{1}{2}$	0	$\exp(x)[\sqrt{x} - \mathrm{daw}(\sqrt{x})]$	$\frac{\sqrt{\pi}}{2}\mathrm{erfc}(\sqrt{x}) + \sqrt{x}\exp(-x)$
1	$-\infty$	$(x-1)\exp(x)$	$(x+1)\exp(-x)$
$\frac{3}{2}$	0	$\exp(x)\left[\sqrt{x^3} - \frac{3\sqrt{x}}{2} + \frac{3}{2}\mathrm{daw}(\sqrt{x})\right]$	$\frac{3\sqrt{\pi}}{4}\mathrm{erfc}(\sqrt{x}) + \sqrt{x}\left(x + \frac{3}{2}\right)\exp(-x)$
2	$-\infty$	$(x^2 - 2x + 2)\exp(x)$	$(x^2 + 2x + 2)\exp(-x)$

CHAPTER 38

SINE AND COSINE INTEGRALS

This chapter concerns functions defined in terms of indefinite integrals of $\sin(x)/x$, $\cos(x)/x$ and their hyperbolic analogs. Whereas the definitions of the *hyperbolic sine integral* $\mathrm{Shi}(x)$ and the *sine integral* $\mathrm{Si}(x)$ present no difficulty, the corresponding integrals of cosines diverge at zero argument. Accordingly, in addition to the *hyperbolic cosine integral* $\mathrm{Chi}(x)$ and the *cosine integral* $\mathrm{Ci}(x)$, it is useful also to define an *entire hyperbolic cosine integral* $\mathrm{Chin}(x)$ and an *entire cosine integral* $\mathrm{Cin}(x)$. These entire integrals are related by

38:0:1 $$\mathrm{Chin}(x) = \mathrm{Chi}(x) - \ln(|x|) - \gamma = \mathrm{Chi}(x) - \ln(|x|) - 0.5772156649$$

and

38:0:2 $$\mathrm{Cin}(x) = \gamma + \ln(|x|) - \mathrm{Ci}(x) = \ln(1.781072418|x|) - \mathrm{Ci}(x)$$

to the Chi and Ci functions.

For large arguments, certain auxiliary functions discussed in Section 38:13 are more convenient than Ci and Si.

Because of their wider applicability, this chapter places more emphasis on the Si and Ci functions than on their hyperbolic counterparts.

38:1 NOTATION

The initial letter of Shi and Chi is not always capitalized. The "h" that is used to identify the hyperbolic integrals may occur elsewhere in the function's symbol. For example, $\mathrm{Sih}(x)$ sometimes replaces $\mathrm{Shi}(x)$, and the anagram $\mathrm{Cinh}(x)$ is often used instead of $\mathrm{Chin}(x)$.

Some authors use $\mathrm{ci}(x)$ synonymously with $\mathrm{Ci}(x)$, but others employ it to denote $-\mathrm{Ci}(x)$. The notation $\mathrm{si}(x)$ is usually encountered with the meaning

38:1:1 $$\mathrm{si}(x) = \mathrm{Si}(x) - \frac{\pi}{2}$$

Neither ci nor si is utilized in this *Atlas*.

38:2 BEHAVIOR

Figure 38-1 maps the Shi, Chi and Chin functions. Note that for large positive arguments Shi(x) and Chi(x) converge and approach the value $\frac{1}{2}$ Ei(x), where Ei is the exponential integral [Chapter 37].

The damped oscillatory behavior of the Si and Ci functions is evident in Figure 38-2. As $x \to \pm\infty$, Si(x) approaches $\pm(\pi/2)$ and Ci(x) approaches zero, whereas Cin(x) approaches infinity logarithmically via a series of plateaus.

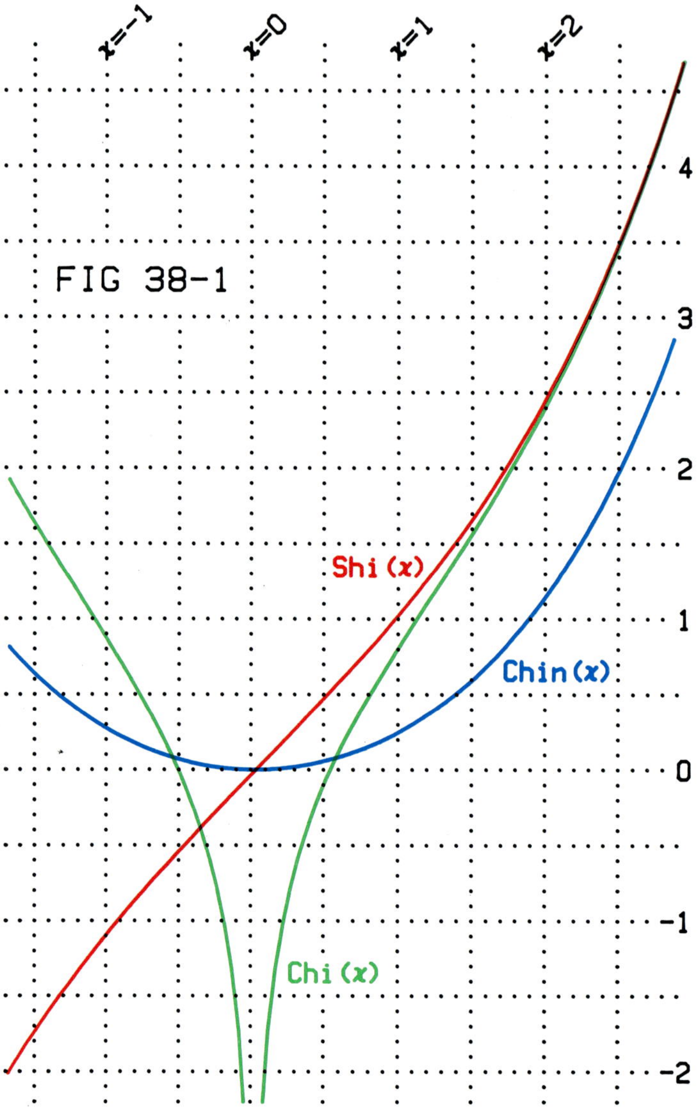

38:3 DEFINITIONS

In addition to their definitions as integrals, the hyperbolic sine and cosine integrals may be defined in terms of the functions of Chapter 37:

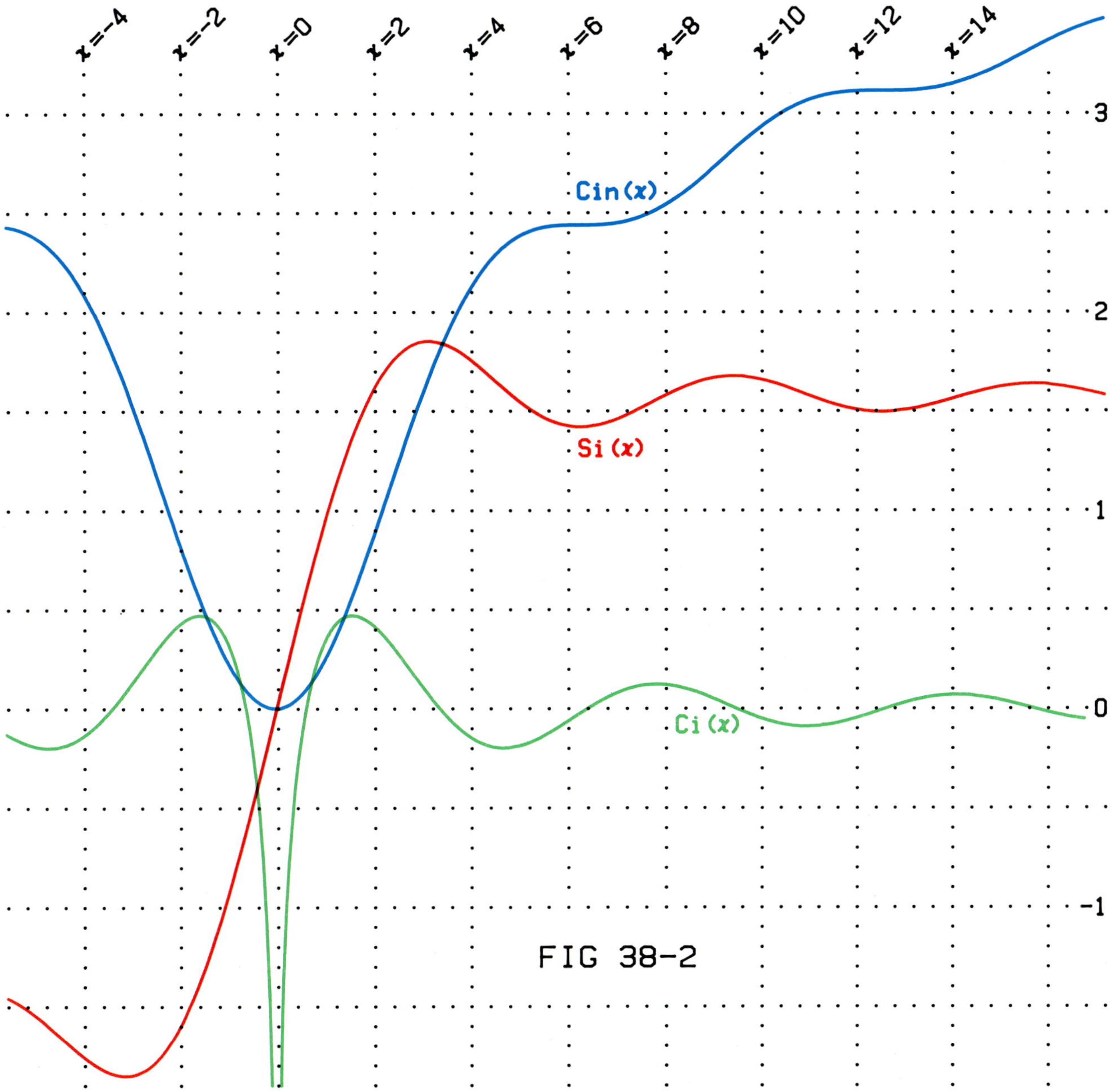

FIG 38-2

38:3:1 $$\mathrm{Shi}(x) = \frac{\mathrm{Ei}(x) - \mathrm{Ei}(-x)}{2} = \int_0^x \frac{\sinh(t)}{t}\,\mathrm{d}t$$

38:3:2 $$\mathrm{Chi}(x) = \frac{\mathrm{Ei}(x) + \mathrm{Ei}(-x)}{2} = \int_{x_0}^x \frac{\cosh(t)}{t}\,\mathrm{d}t \qquad x_0 = 0.52382257$$

38:3:3 $$\mathrm{Chin}(x) = -\frac{\mathrm{Ein}(x) + \mathrm{Ein}(-x)}{2} = \int_0^x \frac{\cosh(t) - 1}{t}\,\mathrm{d}t$$

The sine integral, cosine integral and entire cosine integral are defined by

38:3:4 $$\mathrm{Si}(x) = \int_0^x \frac{\sin(t)}{t}\,\mathrm{d}t = \frac{1}{\pi}\int_0^{\pi x} \mathrm{sinc}(t)\mathrm{d}t = \frac{\pi}{2} - \int_x^\infty \frac{\sin(t)}{t}\,\mathrm{d}t$$

38:3:5 $$\mathrm{Ci}(x) = -\int_x^\infty \frac{\cos(t)}{t}\,\mathrm{d}t$$

and

38:3:6 $$\mathrm{Cin}(x) = \int_0^x \frac{1-\cos(t)}{t}\,\mathrm{d}t$$

Some other definitions may be constructed by employing relationships 38:0:1 and 38:0:2.

38:4 SPECIAL CASES

There are none.

38:5 INTRARELATIONSHIPS

All cosine integrals, like the cosine itself, are even functions:

38:5:1 $$\mathrm{f}(-x) = \mathrm{f}(x) \qquad \mathrm{f} = \mathrm{Chi, Chin, Ci, Cin}$$

whereas the sine integrals are odd:

38:5:2 $$\mathrm{f}(-x) = -\mathrm{f}(x) \qquad \mathrm{f} = \mathrm{Shi, Si}$$

as is the sine function.

Expressions for the Ci and Si functions in terms of the auxiliary functions of Section 38:13 are

38:5:3 $$\mathrm{Ci}(x) = \sin(x)\,\mathrm{fi}(x) - \cos(x)\,\mathrm{gi}(x)$$

38:5:4 $$\mathrm{Si}(x) = \frac{\pi}{2} - \cos(x)\,\mathrm{fi}(x) - \sin(x)\,\mathrm{gi}(x)$$

38:6 EXPANSIONS

The sine integral may be expanded as the power series

38:6:1 $$\mathrm{Si}(x) = x - \frac{x^3}{18} + \frac{x^5}{600} - \frac{x^7}{35280} + \cdots = x\sum_{j=0}^{\infty} \frac{(-x^2)^j}{(2j+1)(2j+1)!}$$

The similar series without alternating signs, namely $\Sigma x^{2j+1}/(2j+1)(2j+1)!$, represents Shi($x$). Likewise, replacement of all the $-$ signs by $+$ in the expansion

38:6:2 $$\mathrm{Cin}(x) = \frac{x^2}{4} - \frac{x^4}{96} + \frac{x^6}{4320} - \frac{x^8}{322560} + \cdots = -\sum_{j=0}^{\infty} \frac{(-x^2)^j}{2j(2j)!}$$

of the entire cosine integral produces the series $\Sigma x^{2j}/(2j)(2j)!$, which represents Chin(x).

The sine integral is also expansible in terms of spherical Bessel functions of the first kind [Section 32:13]:

38:6:3 $$\mathrm{Si}(x) = \pi\sum_{j=0}^{\infty} J_{j+1/2}^2\left(\frac{x}{2}\right) = \frac{4}{x}\sin^2\left(\frac{x}{2}\right) + \frac{16}{x^3}\left[\sin\left(\frac{x}{2}\right) - \frac{x}{2}\cos\left(\frac{x}{2}\right)\right]^2 + \frac{576}{x^5}\left[\left(1-\frac{x^2}{12}\right)\sin\left(\frac{x}{2}\right) - \frac{x^2}{2}\cos\left(\frac{x}{2}\right)\right]^2 + \cdots$$

More rapidly convergent even than 38:6:1 and 38:6:2 are the composite power series

38:6:4 $$\mathrm{Si}(x) - \frac{2-\cos(x)}{x} + \frac{\sin(x)}{x^2} = \frac{x}{3} - \frac{x^3}{180} + \frac{x^5}{12600} - \cdots = 2x\sum_{j=0}^{\infty}\frac{(-x^2)^j}{(2j+1)(2j+3)!}$$

and

38:6:5 $$\mathrm{Cin}(x) + \frac{\sin(x)}{x} + \frac{1-\cos(x)}{x^2} - \frac{3}{2} = \frac{x^2}{24} - \frac{x^4}{1440} + \frac{x^6}{120960} - \cdots = -\sum_{j=1}^{\infty}\frac{(-x^2)^j}{j(2j+2)!}$$

Asymptotic expansions of Ci(x) and of Si(x) − ($\pi/2$) may be constructed by combining expressions 38:5:3 and 38:5:4 with the expansions 38:13:5 and 38:13:6.

38:7 PARTICULAR VALUES

With $n = 1, 2, 3, \ldots$, Table 38.7.1 lists particular values and features of the six functions. Included in Table 38.7.1 are the arguments at which Si(x) and Ci(x) acquire values that are locally maximal or minimal. These extrema also correspond to special values of the auxiliary integrals discussed in Section 38:13:

38:7:1 $$\mathrm{Si}(\pm 2n\pi) = \mp\frac{\pi}{2} \mp \mathrm{fi}(2n\pi)$$ minimum / maximum of Si(x), $n = 1, 2, 3, \ldots$

38:7:2 $$\mathrm{Si}(\pm(2n-1)\pi) = \pm\frac{\pi}{2} \pm \mathrm{fi}((2n-1)\pi)$$ maximum / minimum of Si(x), $n = 1, 2, 3, \ldots$

38:7:3 $$\mathrm{Ci}\left(\pm\left(2n-\frac{1}{2}\right)\pi\right) = \mp\mathrm{gi}\left(\left(2n-\frac{1}{2}\right)\pi\right)$$ minimum / maximum of Ci(x), $n = 1, 2, 3, \ldots$

38:7:4 $$\mathrm{Ci}\left(\pm\left(2n-\frac{3}{2}\right)\pi\right) = \pm\mathrm{gi}\left(\left(2n-\frac{3}{2}\right)\pi\right)$$ maximum / minimum of Ci(x), $n = 1, 2, 3, \ldots$

At $x = \pm 2\pi, \pm 4\pi, \pm 6\pi, \ldots$, a horizontal inflection is displayed by Cin(x), that is, d Cin(x)/dx and d^2 Cin(x)/dx^2 are both zero at these values of the argument [see Section 0:7].

The zeros of the Chi(x) function occur at $x = \pm 0.52382257$. The first zeros of the Ci(x) function are found at ± 0.61650549, and others occur close to the points of inflection of the function, which are given by

38:7:5 $$\frac{\mathrm{d}^2}{\mathrm{d}x^2}\mathrm{Ci}(x) = 0 \qquad x = \pm\rho_j(-1) \qquad j = 1, 2, 3, \ldots$$

Similarly, the sine integral inflects at

Table 38.7.1

	$x=-\infty$	$x=-2n\pi$	$x = (\frac{1}{2}-2n)\pi$	$x = (1-2n)\pi$	$x = (\frac{3}{2}-2n)\pi$	$x=0$	$x = (2n-\frac{3}{2})\pi$	$x = (2n-1)\pi$	$x = (2n-\frac{1}{2})\pi$	$x=2n\pi$	$x=\infty$
Shi(x)	$-\infty$					0					∞
Chi(x)	∞					$-\infty$					∞
Chin(x)	∞					0					∞
Si(x)	$\frac{-\pi}{2}$	max		min		0		max		min	$\frac{\pi}{2}$
Ci(x)	0		min		max	$-\infty$	max		min		0
Cin(x)	∞	horiz. inflc.				0				horiz. inflc.	∞

38:7:6 $$\frac{d^2}{dx^2}\mathrm{Si}(x) = 0 \qquad x = 0, \pm r_j(1) \qquad j = 1, 2, 3, \ldots$$

Here $\rho_j(-1)$ and $r_j(1)$ are, respectively, the roots of the equations $\cot(x) = -x$ and $\tan(x) = x$ [see Section 34:7].

38:8 NUMERICAL VALUES

The hyperbolic integrals Shi(x) and Chi(x) are conveniently evaluated by using the algorithm in Section 37:8 to generate values of Ei($\pm x$) and then employing the identities given in equations 38:3:1 and 38:3:2.

Input x >>>>>>>

If $14 < |x|$ go to (3)

If $|x| < 0.14$ go to (2)

Set $J = 2\mathrm{Int}\left(3 + \frac{5|x|}{4}\right)$

Set $f = -x^2/(J + 2)^3$

(1) Replace f by $\left(f + \frac{1}{J-1}\right)\left(\frac{-x^2}{J(J+1)}\right)$

Replace J by $J - 2$

If $J \neq 0$ go to (1)

Replace f by $2(1 - f) - \cos(x) - \frac{\sin(x)}{x}$

Go to (5)

(2) Set $f = x\exp(-x^2/18)$

Go to (6)

(3) Set $J = 2\ \mathrm{Int}\left(2 + \frac{56}{|x|}\right)$

Set $f = g = [(J + 2)/x]^2$

(4) Replace g by $1 - g(J + 1)J/x^2$

Replace f by $1 - fJ(J - 1)/x^2$

Replace J by $J - 2$

If $J \neq 0$ go to (4)

Replace f by $\frac{\pi|x|}{2} - f\cos(x) - \frac{g\sin(x)}{x}$

(5) Replace f by f/x

(6) Output f

$f \simeq \mathrm{Si}(x)$ <<<<<<<

Storage needed: x, J, f and g

Use radian mode.

Test values:
Si(π) = 1.85193705
Si(0.1) = 0.0999444641
Si(-20) = -1.54824170

The algorithm above permits Si(x) to be calculated to 24-bit precision (i.e., the relative error never exceeds 6×10^{-8}) for any argument. For values of x in the range $-0.14 < x < 0.14$, the approximation formula 38:9:1 is used. When $|x|$ lies in the $0.14 \le |x| \le 14$ range, the algorithm uses expansion 38:6:4 in the concatenated and truncated form

38:8:1 $$\mathrm{Si}(x) \simeq \left(\left(\left(\left(\cdots\left(\left(\frac{-x^2}{(J+2)^3} + \frac{1}{J-1}\right)\frac{-x^2}{J(J+1)} + \frac{1}{J-3}\right)\frac{-x^2}{(J-2)(J-1)} + \cdots + \frac{1}{3}\right)\frac{-x^2}{4\times 5} + \frac{1}{1}\right)\frac{x^2}{2\times 3} + 1\right)2 - \cos(x) - \frac{\sin(x)}{x}\right)\frac{1}{x}$$

with $J = 2\ \mathrm{Int}(3 + 5|x|/4)$. For $|x| > 14$ the expression

38:8:2
$$\mathrm{Si}(x) \simeq \frac{\pi|x|}{2x} - \left(\left(\left(\cdots\left(\left(\frac{(J+2)^2}{-x^2}+1\right)\frac{J(J-1)}{-x^2}+1\right)\frac{(J-2)(J-3)}{-x^2} + \cdots + 1\right)\frac{4\times 3}{-x^2}+1\right)\frac{2\times 1}{-x^2}+1\right)\frac{\cos(x)}{x}$$
$$- \left(\left(\left(\cdots\left(\left(\frac{(J+2)^2}{-x^2}+1\right)\frac{(J+1)J}{-x^2}+1\right)\frac{(J-1)(J-2)}{-x^2} + \cdots + 1\right)\frac{5\times 4}{-x^2}+1\right)\frac{3\times 2}{-x^2}+1\right)\frac{\sin(x)}{x}$$

(which follows from equations 38:5:4, 38:13:5 and 38:13:6) is used with $J = 2\,\mathrm{Int}(2 + 56/|x|)$.

The algorithm below is designed to compute values of $\mathrm{Cin}(x)$ to 24-bit precision. It is very similar to the algorithm for $\mathrm{Si}(x)$, being based on equations 38:9:2, 38:6:5, 38:13:5, 38:13:6, 38:5:3 and 38:0:2.

Input x >>>>>>>

- If $16 < |x|$ go to (3)
- If $|x| < 0.17$ go to (2)
- Set $J = 2\,\mathrm{Int}(3 + 4|x|/\pi)$
- Set $f = -x^2/(J+3)^3$
- (1) Replace f by $\left(f + \frac{1}{J}\right)\left(\frac{-x^2}{(J+1)(J+2)}\right)$
- Replace J by $J - 2$
- If $J \neq 0$ go to (1)
- Replace f by $\frac{3}{2} - f - \frac{\sin(x)}{x} - \frac{1-\cos(x)}{x^2}$
- Go to (6 or 5)
- (2) Set $f = [x^2 \exp(-x^2/24)]/4$
- Go to (6 or 5)
- (3) Set $J = 2\,\mathrm{Int}\left(2 + \frac{96}{|x|}\right)$
- Set $f = g = [(J+2)/x]^2$
- (4) Replace g by $1 - g(J+1)J/x^2$
- Replace f by $1 - fJ(J-1)/x^2$
- Replace J by $J - 2$
- If $J \neq 0$ go to (4)
- Replace f by $f\frac{\sin(x)}{x} - g\frac{\cos(x)}{x^2}$
- Go to (5 or 6)
- (5) Replace f by $\ln(1.781072418|x|) - f$
- (6) Output f

$f \simeq \mathrm{Cin}(x)$ <<<<<
or $\mathrm{Ci}(x)$

Storage needed: x, J, f and g

Use radian mode.

Test values:
$\mathrm{Cin}(\pi) = 1.64827764$
$\mathrm{Cin}(0.1) = 0.00249895856$
$\mathrm{Cin}(20) = 3.5285212$
$\mathrm{Ci}(2) = 0.422980829$
$\mathrm{Ci}(-0.15) = -1.32552405$
$\mathrm{Ci}(20) = 0.0444198210$

When the alternate commands shown in green are substituted, the algorithm generates values of $\mathrm{Ci}(x)$ instead of $\mathrm{Cin}(x)$. However, the fractional error in the calculated values of $\mathrm{Ci}(x)$ may exceed 6×10^{-8}, especially near the zeros of the cosine integral.

Also, values of $\mathrm{Shi}(x)$, $\mathrm{Chin}(x)$, $\mathrm{Si}(x)$ and $\mathrm{Cin}(x)$ are calculable via the universal hypergeometric algorithm of Section 18:14.

38:9 APPROXIMATIONS

Close to $x = 0$, the formulas

38:9:1 $$\mathrm{Si}(x) \simeq x \exp\left(\frac{-x^2}{18}\right) \qquad \text{8-bit precision} \qquad |x| \le 2.2$$

and

38:9:2 $$\mathrm{Cin}(x) \simeq \frac{x^2}{4} \exp\left(\frac{-x^2}{14}\right) \qquad \text{8-bit precision} \qquad |x| \le 2.5$$

are useful, while for large x the approximations

38:9:3 $$\mathrm{Si}(x) \simeq \frac{\pi}{2} - \frac{x\cos(x)}{x^2+2} - \frac{\sin(x)}{x^2+6} \qquad \text{large } x$$

and

38:9:4 $$\mathrm{Ci}(x) \simeq \frac{x\sin(x)}{x^2+2} - \frac{\cos(x)}{x^2+6} \qquad \text{large } x$$

may be used.

38:10 OPERATIONS OF THE CALCULUS

The differentiation and indefinite integration operators, applied to the six functions of this chapter, give

38:10:1 $$\frac{d}{dx}\mathrm{Shi}(bx+c) = \frac{\sinh(bx+c)}{x+(c/b)}$$

38:10:2 $$\frac{d}{dx}\mathrm{Chi}(bx+c) = \frac{\cosh(bx+c)}{x+(c/b)}$$

38:10:3 $$\frac{d}{dx}\mathrm{Chin}(bx+c) = \frac{\cosh(bx+c)-1}{x+(c/b)}$$

38:10:4 $$\frac{d}{dx}\mathrm{Si}(bx+c) = \frac{\sin(bx+c)}{x+(c/b)} = \frac{b}{\pi}\mathrm{sinc}(\pi bx + \pi c)$$

38:10:5 $$\frac{d}{dx}\mathrm{Ci}(bx+c) = \frac{\cos(bx+c)}{x+(c/b)}$$

38:10:6 $$\frac{d}{dx}\mathrm{Cin}(bx+c) = \frac{1-\cos(bx+c)}{x+(c/b)}$$

38:10:7 $$\int_{-c/b}^{x} \mathrm{Shi}(bt+c)dt = \left(x+\frac{c}{b}\right)\mathrm{Shi}(bx+c) - \frac{\cosh(bx+c)-1}{b}$$

38:10:8 $$\int_{-c/b}^{x} \mathrm{Chi}(bt+c)dt = \left(x+\frac{c}{b}\right)\mathrm{Chi}(bx+c) - \frac{\sinh(bx+c)}{b}$$

38:10:9 $$\int_{-c/b}^{x} \mathrm{Chin}(bt+c)dt = \left(x+\frac{c}{b}\right)[1+\mathrm{Chin}(bx+c)] - \frac{\sinh(bx+c)}{b}$$

38:10:10 $$\int_{-c/b}^{x} \mathrm{Si}(bt+c)dt = \left(x+\frac{c}{b}\right)\mathrm{Si}(bx+c) - \frac{1-\cos(bx+c)}{b}$$

38:10:11 $$\int_{-c/b}^{x} \mathrm{Ci}(bt + c)dt = \left(x + \frac{c}{b}\right) \mathrm{Ci}(bx + c) - \frac{\sin(bx + c)}{b}$$

38:10:12 $$\int_{-c/b}^{x} \mathrm{Cin}(bt + c)dt = \left(x + \frac{c}{b}\right) [\mathrm{Cin}(bx + c) - 1] + \frac{\sin(bx + c)}{b}$$

Other indefinite integrals are listed by Gradshteyn and Ryzhik [Section 5.3]. The most useful of these are probably the integrals of products of sinusoidal functions with the sine and cosine integrals. We have the results

38:10:13 $$\int_{x}^{\infty} \sin(Bt + C)\, \mathrm{Si}(bt + c)dt = \frac{1}{B} \cos(Bx + C)\left[\mathrm{Si}(bx + c) - \frac{\pi}{2}\,\mathrm{sgn}(b)\right] - \cos\left(C - \frac{Bc}{b}\right) \frac{\mathrm{Si}(+) - \mathrm{Si}(-)}{2B} - \sin\left(C - \frac{Bc}{b}\right) \frac{\mathrm{Ci}(+) - \mathrm{Ci}(-)}{2B}$$

38:10:14 $$\int_{x}^{\infty} \cos(Bt + C)\, \mathrm{Si}(bt + c)dt = \frac{1}{B} \sin(Bx + C)\left[\frac{\pi}{2}\,\mathrm{sgn}(b) - \mathrm{Si}(bx + c)\right] + \sin\left(C - \frac{Bc}{b}\right) \frac{\mathrm{Si}(+) - \mathrm{Si}(-)}{2B} - \cos\left(C - \frac{Bc}{b}\right) \frac{\mathrm{Ci}(+) - \mathrm{Ci}(-)}{2B}$$

38:10:15 $$\int_{x}^{\infty} \sin(Bt + C)\, \mathrm{Ci}(bt + c)dt = \frac{1}{B} \cos(Bx + C)\, \mathrm{Ci}(bx + c) + \sin\left(C - \frac{Bc}{b}\right) \frac{\mathrm{Si}(+) + \mathrm{Si}(-)}{2B} - \cos\left(C - \frac{Bc}{b}\right) \frac{\mathrm{Ci}(+) + \mathrm{Ci}(-)}{2B}$$

38:10:16 $$\int_{x}^{\infty} \cos(Bt + C)\, \mathrm{Ci}(bt + c)dt = \frac{-1}{B} \sin(Bx + C)\, \mathrm{Ci}(bx + c) + \cos\left(C - \frac{Bc}{b}\right) \frac{\mathrm{Si}(+) + \mathrm{Si}(-)}{2B} - \sin\left(C - \frac{Bc}{b}\right) \frac{\mathrm{Ci}(+) + \mathrm{Ci}(-)}{2B}$$

where $\mathrm{Ci}(\pm)$ is an abbreviation for $\mathrm{Ci}[(B \pm b)(x + c/b)]$ and $\mathrm{Si}(\pm)$ similarly abbreviates $\mathrm{Si}[(B \pm b)(x + c/b)] - (\pi/2)\,\mathrm{sgn}(B \pm b)$.

Definite integrals of the sine and cosine integral functions include

38:10:17 $$\int_{0}^{\infty} \mathrm{Ci}(Bt)\, \mathrm{Ci}(bt)dt = \begin{cases} \pi/2B & B \geq b \\ \pi/2b & b \geq B \end{cases}$$

and many others are listed by Gradshteyn and Ryzhik [Sections 6.26 and 6.27].

38:11 COMPLEX ARGUMENTS

The six functions of this chapter acquire complex values when their arguments are complex and even, in the case of the Chi function, for example, for real values:

38:11:1 $$\mathrm{Chi}(x + 0i) = \gamma + \ln(|x|) + \mathrm{Chin}(x) + \frac{i\pi}{2}$$

With argument $x + iy$, values of the functions may be calculated from those of the Chapter 37 functions via the identities

38:11:2 $$\mathrm{Shi}(x + iy) = \frac{1}{2} [\mathrm{Ei}(x + iy) - \mathrm{Ei}(-x - iy)]$$

38:11:3 $$\mathrm{Chi}(x + iy) = \frac{1}{2}\,[\mathrm{Ei}(x + iy) + \mathrm{Ei}(-x - iy)]$$

38:11:4 $$\mathrm{Chin}(x + iy) = \frac{-1}{2}\,[\mathrm{Ein}(x + iy) + \mathrm{Ein}(-x - iy)]$$

38:11:5 $$\mathrm{Si}(x + iy) = \frac{\pi}{2} + \frac{i}{2}\,[\mathrm{Ei}(y - ix) - \mathrm{Ei}(-y + ix)]$$

38:11:6 $$\mathrm{Ci}(x + iy) = \frac{1}{2}\,[\mathrm{Ei}(y - ix) + \mathrm{Ei}(-y + ix)]$$

38:11:7 $$\mathrm{Cin}(x + iy) = \frac{1}{2}\,[\mathrm{Ein}(y - ix) - \mathrm{Ein}(-y + ix)]$$

For purely imaginary arguments, one has

38:11:8 $$\mathrm{Shi}(iy) = i\left[\mathrm{Si}(y) - \frac{\pi}{2}\right]$$

38:11:9 $$\mathrm{Chi}(iy) = \mathrm{Ci}(y)$$

38:11:10 $$\mathrm{Si}(iy) = i\,\mathrm{Shi}(y) + \frac{\pi}{2}$$

38:11:11 $$\mathrm{Ci}(iy) = \mathrm{Chi}(y) + \frac{i\pi}{2}$$

38:12 GENERALIZATIONS

The sine integral is the special $a = 0$ instance of the more general function

38:12:1 $$\int_0^x \frac{\sin(\sqrt{a^2 + t^2})\mathrm{d}t}{\sqrt{a^2 + t^2}} = \int_a^{\sqrt{a^2+x^2}} \frac{\sin(t)\mathrm{d}t}{\sqrt{t^2 - a^2}}$$

[see Erdélyi, Magnus, Oberhettinger and Tricomi, *Higher Transcendental Functions,* Volume 2, page 147].

The Shi, Chin, Si and Cin functions are all instances of hypergeometric functions, as discussed in Section 18:14.

Boehmer integrals [Section 39:12] also generalize the sine and cosine integrals. We have

38:12:2 $$\mathrm{Ci}(x) = -\mathrm{C}(x;0)$$

and

38:12:3 $$\mathrm{Si}(x) = \frac{\pi}{2} - \mathrm{S}(x;0)$$

38:13 COGNATE FUNCTIONS

The so-called *auxiliary sine integral* and *auxiliary cosine integral* functions are defined by the definite integrals

38:13:1 $$\mathrm{fi}(x) = \int_0^\infty \frac{\sin(t)}{t + x}\,\mathrm{d}t = \int_0^\infty \frac{\exp(-xt)}{t^2 + 1}\,\mathrm{d}t$$

and

38:13:2 $$\mathrm{gi}(x) = \int_0^\infty \frac{\cos(t)}{(t+x)}\,dt = \int_0^\infty \frac{t\exp(-xt)}{t^2+1}\,dt$$

There appears to be no standard notation. Abramowitz and Stegun [Section 5.2] use the f(x) and g(x) symbolism. Figure 38-3 maps the two functions, which are defined for positive real argument only.

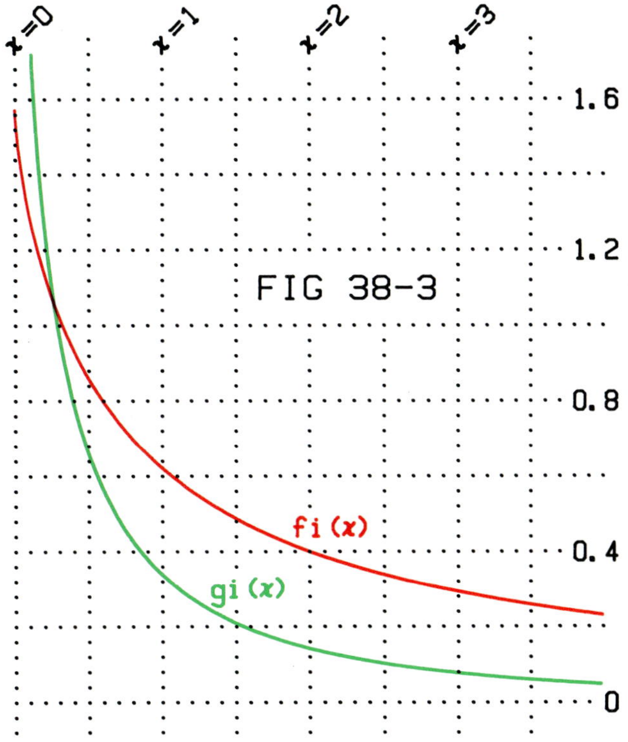

The functions are related by

38:13:3 $$\mathrm{fi}(x) = \sin(x)\,\mathrm{Ci}(x) + \cos(x)\left[\frac{\pi}{2} - \mathrm{Si}(x)\right]$$

and

38:13:4 $$\mathrm{gi}(x) = \sin(x)\left[\frac{\pi}{2} - \mathrm{Si}(x)\right] - \cos(x)\,\mathrm{Ci}(x)$$

to the sine and cosine integrals. The simple asymptotic formulas, valid for large argument

38:13:5 $$\mathrm{fi}(x) \sim \frac{1}{x} - \frac{2}{x^3} + \frac{24}{x^5} - \frac{720}{x^7} + \cdots + \frac{(2j)!}{x(-x^2)^j} + \cdots \qquad x \to \infty$$

and

38:13:6 $$\mathrm{gi}(x) \sim \frac{1}{x^2} - \frac{6}{x^4} + \frac{120}{x^6} - \frac{5040}{x^8} + \cdots + \frac{(2j+1)!}{x^2(-x^2)^j} + \cdots \qquad x \to \infty$$

permit convenient calculation of the auxiliary functions and are incorporated into the hypergeometric representation [Section 18:14].

The differential identities

38:13:7 $$\frac{d}{dx}\,\mathrm{fi}(x) = -\mathrm{gi}(x)$$

and

38:13:8 $$\frac{d}{dx}\,\mathrm{gi}(x) = \mathrm{fi}(x) - \frac{1}{x}$$

relate the two auxiliary integrals, which therefore satisfy the differential equations

38:13:9 $$\frac{d^2f}{dx^2} + f = \frac{1}{x} \qquad f = \mathrm{fi}(x) + c_1\sin(x) + c_2\cos(x)$$

and

38:13:10 $$\frac{d^2f}{dx^2} + f = \frac{1}{x^2} \qquad f = \mathrm{gi}(x) + c_1\sin(x) + c_2\cos(x)$$

respectively, where c_1 and c_2 are arbitrary constants. One also has

38:13:11 $$\int_0^\infty \mathrm{gi}(t)dt = \frac{\pi}{2}$$

but the corresponding integral for fi is infinite.

CHAPTER 39

THE FRESNEL INTEGRALS S(x) AND C(x)

Among other applications, Fresnel integrals occur in theories of physical optics [see Section 39:14]. The so-called auxiliary Fresnel integrals, discussed in Section 39:13, are also of some importance.

39:1 NOTATION

The symbols S(x) and C(x) are almost universally adopted for the *Fresnel sine-integral* and *Fresnel cosine-integral,* respectively. However, there is a wide variation in the definitions chosen for these functions by various authors. Thus, in addition to our definition 39:3:1, each of the following integrals is cited as the "Fresnel sine-integral" by at least one authority:

39:1:1 $$\int_0^x \sin(t^2)\mathrm{d}t = \sqrt{\frac{\pi}{2}}\,\mathrm{S}(x)$$

39:1:2 $$\frac{1}{\sqrt{2\pi}}\int_0^x \frac{\sin(t)}{\sqrt{t}}\,\mathrm{d}t = \mathrm{S}(\sqrt{x}) \qquad x \geq 0$$

39:1:3 $$\int_0^x \sin\left(\frac{\pi t^2}{2}\right)\mathrm{d}t = \mathrm{S}\left(\sqrt{\frac{\pi}{2}}\,x\right)$$

39:1:4 $$\frac{1}{2\sqrt{x}}\int_0^x \frac{\sin(t)}{\sqrt{t}}\,\mathrm{d}t = \sqrt{\frac{\pi}{2x}}\,\mathrm{S}(\sqrt{x}) \qquad x \geq 0$$

The right-hand member of each of the above equations is the expression for the left-hand integral in the notation of this *Atlas*. In all cases the definitions adopted for the Fresnel cosine-integral involves straightforward replacement of sin in the integrand by cos. Some authors adopt multiple definitions of these functions and utilize subscripts (as in $\mathrm{S}_1(x)$, $\mathrm{C}_2(x)$, etc.) to distinguish the options. There is, however, no greater unanimity in this secondary notation than in the primary.

The reader will note that the definitions of S(x) and C(x) adopted in this *Atlas* agree with those used by Gradshteyn and Ryzhik, and by Spiegel, but differ from those selected by Abramowitz and Stegun, and by a number of other authors.

39:2 BEHAVIOR

Figure 39-1 maps the behavior of the two functions. Both S(x) and C(x) display damped oscillations that decrease in both period and amplitude as $x \to \pm\infty$ approaching the values $\pm\frac{1}{2}$ in these limits.

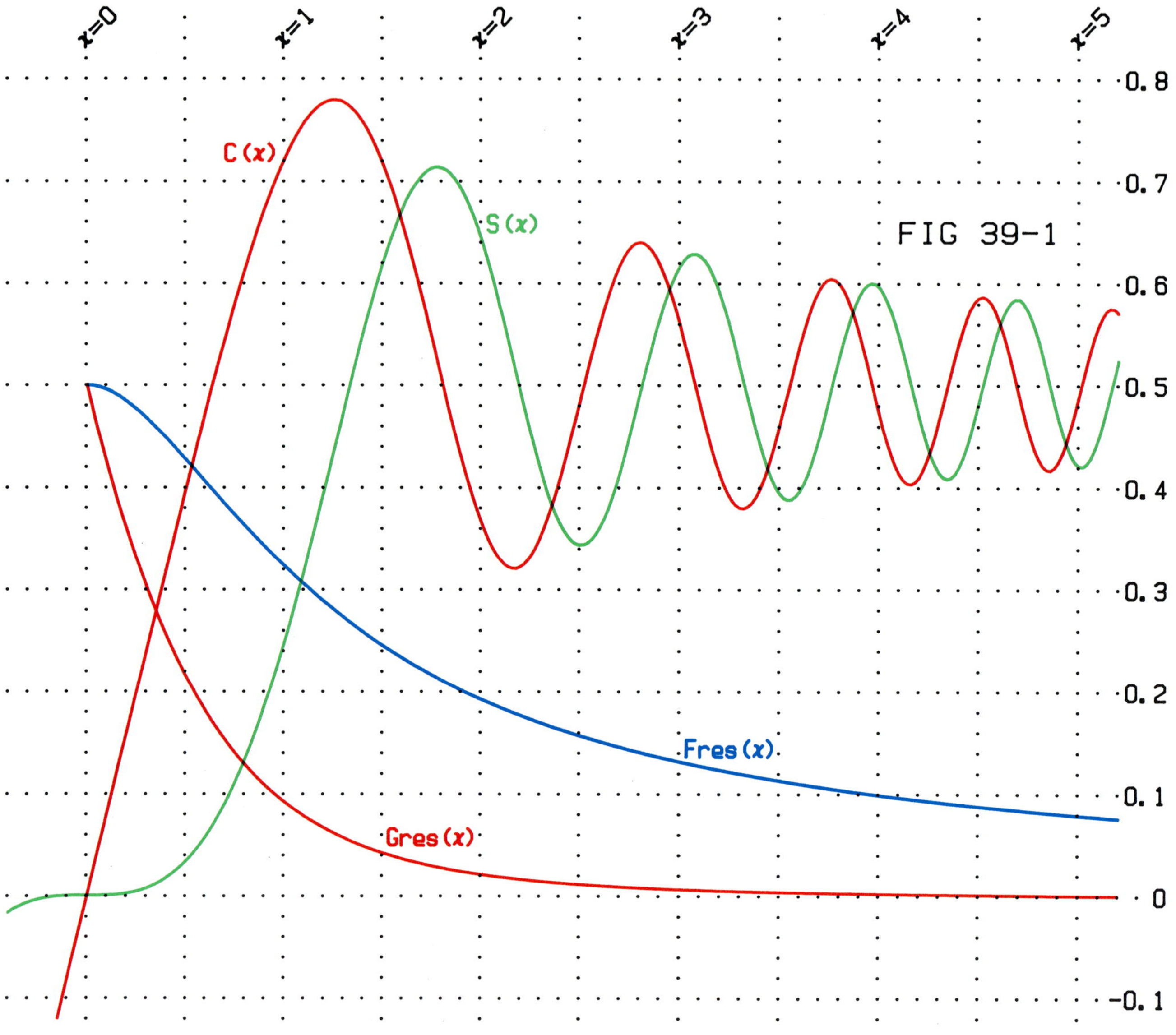

For $x \geq 0$, the S(x) function lies entirely between the curves $\frac{1}{2}$ − Fres(x) − Gres(x) and $\frac{1}{2}$ + Fres(x) + Gres(x) where Fres and Gres are the functions discussed in Section 39:13. The same bounds apply for C(x), and a similar relationship holds for $x \leq 0$. For all arguments

39:2:1 $$\frac{\text{sgn}(x)}{2} - \text{Fres}(|x|) - \text{Gres}(|x|) \leq \text{f}(x) \leq \frac{\text{sgn}(x)}{2} + \text{Fres}(|x|) + \text{Gres}(|x|) \qquad \text{f} = \text{S or C}$$

39:3 DEFINITIONS

The Fresnel sine-integral is defined by either of the equivalent representations

39:3:1 $$\text{S}(x) = \sqrt{\frac{2}{\pi}} \int_0^x \sin(t^2)\text{d}t = \frac{\text{sgn}(x)}{\sqrt{2\pi}} \int_0^{x^2} \frac{\sin(t)}{\sqrt{t}}\,\text{d}t$$

and similarly for the Fresnel cosine-integral

39:3:2 $$C(x) = \sqrt{\frac{2}{\pi}}\int_0^x \cos(t^2)dt = \frac{\text{sgn}(x)}{\sqrt{2\pi}}\int_0^{x^2}\frac{\cos(t)}{\sqrt{t}}dt$$

The integrands may be regarded as spherical Bessel functions [Section 32:13]

39:3:3 $$S(x) = \frac{\text{sgn}(x)}{2}\int_0^{x^2} J_{1/2}(t)dt = \frac{\text{sgn}(x)}{2}\int_0^{x^2} Y_{-1/2}(t)dt$$

39:3:4 $$C(x) = \frac{\text{sgn}(x)}{2}\int_0^{x^2} J_{-1/2}(t)dt = \frac{-\text{sgn}(x)}{2}\int_0^{x^2} Y_{1/2}(t)dt$$

39:4 SPECIAL CASES

There are none.

39:5 INTRARELATIONSHIPS

Both Fresnel integrals are odd functions

39:5:1 $$f(-x) = -f(x) \qquad f = S \text{ or } C$$

They may be expressed in terms of the auxiliary Fresnel integrals [Section 39:13] by

39:5:2 $$S(x) = \text{sgn}(x)\left[\frac{1}{2} - \cos(x^2)\,\text{Fres}(|x|) - \sin(x^2)\,\text{Gres}(|x|)\right]$$

and

39:5:3 $$C(x) = \text{sgn}(x)\left[\frac{1}{2} + \sin(x^2)\,\text{Fres}(|x|) - \cos(x^2)\,\text{Gres}(|x|)\right]$$

39:6 EXPANSIONS

Power series expansions of the Fresnel integrals are

39:6:1 $$S(x) = \sqrt{\frac{2}{\pi}}\left(\frac{x^3}{3} - \frac{x^7}{42} + \frac{x^{11}}{1320} - \cdots\right) = x^3\sqrt{\frac{2}{\pi}}\sum_{j=0}^{\infty}\frac{(-x^4)^j}{(4j+3)(2j+1)!}$$

and

39:6:2 $$C(x) = \sqrt{\frac{2}{\pi}}\left(x - \frac{x^5}{10} + \frac{x^9}{216} - \cdots\right) = x\sqrt{\frac{2}{\pi}}\sum_{j=0}^{\infty}\frac{(-x^4)^j}{(4j+1)(2j)!}$$

More rapidly convergent than these are the composite expansions

39:6:3 $$\sqrt{\frac{\pi}{2}}S(x) + \frac{\sin(x^2) + 2x^2\cos(x^2)}{4x^3} = \frac{3}{4x} + \frac{x^3}{24} - \frac{x^7}{1120} + \frac{x^{11}}{73920} - \cdots = \frac{-3}{4x}\sum_{j=0}^{\infty}\frac{(-x^4)^j}{(4j-1)(2j+1)!}$$

and

39:6:4 $$\sqrt{\frac{\pi}{2}}C(x) + \frac{\cos(x^2) - 2x^2\sin(x^2)}{4x^3} = \frac{1}{4x^3} + \frac{3x}{8} - \frac{x^5}{160} + \frac{x^9}{8640} - \cdots = \frac{-3}{4x^3}\sum_{j=0}^{\infty}\frac{(-x^4)^j}{(4j-3)(2j)!}$$

Via relationships 39:5:2 and 39:5:3, one can utilize the various expansions of the auxiliary Fresnel integrals [Section 39:13] to produce the summations

39:6:5
$$S(x) = x\sqrt{\frac{2}{\pi}}\left[\sin(x^2)\sum_{j=0}^{\infty}\frac{(-4x^4)^j}{(4j+1)!!} - 2x^2\cos(x^2)\sum_{j=0}^{\infty}\frac{(-4x^4)^j}{(4j+3)!!}\right]$$

and

39:6:6
$$C(x) = x\sqrt{\frac{2}{\pi}}\left[\cos(x^2)\sum_{j=0}^{\infty}\frac{(-4x^4)^j}{(4j+1)!!} + x^2\sin(x^2)\sum_{j=0}^{\infty}\frac{(-4x^4)^j}{(4j+3)!!}\right]$$

as well as the asymptotic representations

39:6:7
$$S(x) \sim \frac{\operatorname{sgn}(x)}{2} - \frac{\cos(x^2)}{x\sqrt{2\pi}}\left[1 - \frac{3}{4x^4} + \frac{105}{16x^8} - \cdots + \frac{(4j-1)!!}{(-4x^4)^j} + \cdots\right]$$
$$- \frac{\sin(x^2)}{x^3\sqrt{8\pi}}\left[1 - \frac{15}{4x^4} + \frac{945}{16x^8} - \cdots + \frac{(4j+1)!!}{(-4x^4)^j} + \cdots\right] \qquad x \to \infty$$

and

39:6:8
$$C(x) \sim \frac{\operatorname{sgn}(x)}{2} + \frac{\sin(x^2)}{x\sqrt{2\pi}}\left[1 - \frac{3}{4x^4} + \frac{105}{16x^8} - \cdots + \frac{(4j-1)!!}{(-4x^4)^j} + \cdots\right]$$
$$- \frac{\cos(x^2)}{x^3\sqrt{8\pi}}\left[1 - \frac{15}{4x^4} + \frac{945}{16x^8} - \cdots + \frac{(4j+1)!!}{(-4x^4)^j} + \cdots\right] \qquad x \to \infty$$

The Fresnel integrals are expansible in terms of spherical Bessel functions [see Section 32:13]

39:6:9
$$S(x) = \operatorname{sgn}(x)[J_{3/2}(x^2) + J_{7/2}(x^2) + J_{11/2}(x^2) + \cdots] = \operatorname{sgn}(x)\sum_{j=0}^{\infty} J_{2j+3/2}(x^2)$$

and

39:6:10
$$C(x) = \operatorname{sgn}(x)[J_{1/2}(x^2) + J_{5/2}(x^2) + J_{9/2}(x^2) + \cdots] = \operatorname{sgn}(x)\sum_{j=0}^{\infty} J_{2j+1/2}(x^2)$$

39:7 PARTICULAR VALUES

Included in the Table 39.7.1 are the values at which the Fresnel integrals exhibit maxima, minima and points of inflection. The functions also inflect at the origin $x = 0$. In this table n represents any positive integer, $n = 1, 2, 3, \ldots$.

Table 39.7.1

x	$-\infty$	$-\sqrt{2n\pi}$	$-\sqrt{2n\pi - \frac{\pi}{2}}$	$-\sqrt{2n\pi - \pi}$	$-\sqrt{2n\pi - \frac{3\pi}{2}}$	0	$\sqrt{2n\pi - \frac{3\pi}{2}}$	$\sqrt{2n\pi - \pi}$	$\sqrt{2n\pi - \frac{\pi}{2}}$	$\sqrt{2n\pi}$	∞
S(x)	$-\frac{1}{2}$	max	infl	min	infl	0	infl	max	infl	min	$\frac{1}{2}$
C(x)	$-\frac{1}{2}$	infl	max	infl	min	0	max	infl	min	infl	$\frac{1}{2}$

As $x \to \pm\infty$, the values of the maxima and minima converge towards $\pm\frac{1}{2}$ according to the formulas

39:7:1 $$S(\pm\sqrt{2n\pi - \pi}) = \pm\frac{1}{2} \pm \mathrm{Fres}(\sqrt{2n\pi - \pi}) \quad \begin{matrix}\text{maximum}\\ \text{minimum}\end{matrix}$$

39:7:2 $$S(\pm\sqrt{2n\pi}) = \pm\frac{1}{2} \pm \mathrm{Fres}(\sqrt{2n\pi}) \quad \begin{matrix}\text{minimum}\\ \text{maximum}\end{matrix}$$

(39:7:1 and 39:7:2: of S(x))

39:7:3 $$C\left(\pm\sqrt{2n\pi - \frac{3\pi}{2}}\right) = \pm\frac{1}{2} \pm \mathrm{Fres}\left(\sqrt{2n\pi - \frac{3\pi}{2}}\right) \quad \begin{matrix}\text{maximum}\\ \text{minimum}\end{matrix}$$

39:7:4 $$C\left(\pm\sqrt{2n\pi - \frac{3\pi}{2}}\right) = \pm\frac{1}{2} \pm \mathrm{Fres}\left(\sqrt{2n\pi - \frac{\pi}{2}}\right) \quad \begin{matrix}\text{minimum}\\ \text{maximum}\end{matrix}$$

(39:7:3 and 39:7:4: of C(x))

$n = 1, 2, 3, \ldots$

where the Fres function is the auxiliary Fresnel cosine-integral discussed in Section 39:13. The values at the points of inflection converge more rapidly towards $\pm\frac{1}{2}$ and are easily calculated from

39:7:5 $$S\left(\pm\sqrt{m\pi - \frac{\pi}{2}}\right) = \pm\left[\frac{1}{2} + (-1)^m \mathrm{Gres}\left(\sqrt{m\pi - \frac{\pi}{2}}\right)\right] \quad \text{inflection}$$

39:7:6 $$C(\pm\sqrt{m\pi}) = \pm\left[\frac{1}{2} + (-1)^m \mathrm{Gres}(\sqrt{m\pi})\right] \quad \text{inflection}$$

$m = 1, 2, 3, \ldots$

where Gres is the auxiliary Fresnel sine-integral [Section 39:13].

39:8 NUMERICAL VALUES

In Section 18:14 we show how the universal hypergeometric algorithm may be utilized to generate values of the Fresnel integrals (as well as those of the auxiliary Fresnel integrals) in several different ways.

Because applications of Fresnel integrals usually require values of both S(x) and C(x), the algorithm presented below generates numerical values of the Fresnel sine-integral and the Fresnel cosine-integral simultaneously. The algorithm has 24-bit precision (i.e., the relative error never exceeds 6×10^{-8}). For very small absolute values of x, approximations 39:9:1 and 39:9:2 are used by the algorithm. For intermediate values of $|x|$, the composite expansions 39:6:3 and 39:6:4 are used in the truncated and concatenated forms.

39:8:1 $$S(x) \simeq \left(\left(\left(\left(\cdots\left(\left(\frac{-x^4}{(2J-1)(J+1)J} + \frac{1}{2J-5}\right)\frac{-x^4}{(J-1)(J-2)} + \frac{1}{2J-9}\right)\frac{-x^4}{(J-3)(J-4)} + \cdots + \frac{1}{7}\right)\frac{-x^4}{5\times 4} + \frac{1}{3}\right)\frac{-x^4}{3\times 2} + \frac{1}{-1}\right)\frac{-3}{2} - \frac{\sin(x^2)}{2x^2} - \cos(x^2)\right)\frac{1}{x\sqrt{2\pi}}$$

39:8:2 $$C(x) \simeq \left(\left(\left(\left(\cdots\left(\left(\frac{-x^4}{(2J-3)J(J-1)} + \frac{1}{2J-7}\right)\frac{-x^4}{(J-2)(J-3)} + \frac{1}{2J-11}\right)\frac{-x^4}{(J-4)(J-5)} + \cdots + \frac{1}{5}\right)\frac{-x^4}{4\times 3} + \frac{1}{1}\right)\frac{-x^4}{2\times 1} + \frac{1}{-3}\right)\frac{-3}{1} - 2x^2\sin(x^2) - \cos(x^2)\right)\frac{1}{x^3\sqrt{8\pi}}$$

with J given the empirical value 2 Int$(5 + 6x^2/5)$. Similarly, the concatenated expansions

39:8:3 $$\left(\left(\cdots\left(\left(\frac{(J-\frac{1}{2})(J-\frac{3}{2})}{-x^4}+1\right)\frac{(J-\frac{5}{2})(J-\frac{7}{2})}{-x^4}+1\right)\frac{(J-\frac{9}{2})(J-\frac{11}{2})}{-x^4} + \cdots + 1\right)\frac{\frac{7}{2}\times\frac{5}{2}}{-x^4}+1\right)\frac{\frac{3}{2}\times\frac{1}{2}}{-x^4}+1$$

39:8:4 $$\left(\left(\cdots\left(\left(\frac{(J+\frac{1}{2})(J-\frac{1}{2})}{-x^4}+1\right)\frac{(J-\frac{3}{2})(J-\frac{5}{2})}{-x^4}+1\right)\frac{(J-\frac{7}{2})(J-\frac{9}{2})}{-x^4} + \cdots + 1\right)\frac{\frac{9}{2}\times\frac{7}{2}}{-x^4}+1\right)\frac{\frac{5}{2}\times\frac{3}{2}}{-x^4}+1$$

are used to evaluate the two asymptotic series in 39:6:7 and 39:6:8, and these latter are then employed to evaluate S(x) and C(x) for large values of $|x|$.

Input x >>>>>>>

Storage needed: x, f, g and J

Use radian mode.

If $0.1 < x^2$ go to (1)
Set $f = (2x^3/3)\exp(-x^4/14)$
Set $g = 2x\exp(-x^4/10)$
Go to (5)

(1) If $x^2 > 15$ go to (3)
Set $J = 2\,\mathrm{Int}(5 + 1.2x^2)$
Set $f = 1/(2J - 1)$
Set $g = 1/(2J - 3)$

(2) Replace f by $\dfrac{1}{2J-5} - \dfrac{x^4 f}{J(J+1)}$
Replace g by $\dfrac{1}{2J-7} - \dfrac{x^4 g}{(J-1)J}$
Replace J by $J - 2$
If $J \neq 0$ go to (2)
Replace f by $\left[-\dfrac{\sin(x^2)}{2x^2} - \cos(x^2) - \dfrac{3f}{2}\right]\Big/ x$
Replace g by $[2x^2\sin(x^2) - \cos(x^2) - 3g]/2x^3$
Go to (5)

(3) Set $J = 2\,\mathrm{Int}(4 + 60/x^2)$
Set $f = g = 1$

(4) Replace f by $1 - f\left(J - \dfrac{1}{2}\right)\left(J - \dfrac{3}{2}\right)\Big/ x^4$
Replace g by $1 - g\left(J^2 - \dfrac{1}{4}\right)\Big/ x^4$
Replace J by $J - 2$
If $J \neq 0$ go to (4)
Set $J = f\sin(x^2)$
Replace f by $\left[-f\cos(x^2) - \dfrac{g\sin(x^2)}{2x^2} + \sqrt{\dfrac{\pi x^2}{2}}\right]\Big/ x$

Replace g by $\left[-\dfrac{g\cos(x^2)}{2x^2} + J + \sqrt{\dfrac{\pi x^2}{2}}\right] \Big/ x$

(5) Replace f by $f/\sqrt{2\pi}$
Output f

$f \simeq S(x)$ <<<<<<

Replace g by $g/\sqrt{2\pi}$
Output g

$g \simeq C(x)$ <<<<<<

Test values:
S(0.2) = 0.00212744901
C(0.2) = 0.159551382
S(−π) = −0.616486872
C(−π) = −0.451358120
S(5) = 0.421217048
C(5) = 0.487879892

This algorithm is readily adapted to generate values of the auxiliary integrals Fres(x) and Gres(x) [see Section 39:13].

39:9 APPROXIMATIONS

For arguments of small magnitude one has

39:9:1 $$S(x) \simeq \sqrt{\frac{2}{\pi}}\frac{x^3}{3}\exp\left(\frac{-x^4}{14}\right) \qquad \text{8-bit precision} \qquad |x| \le 1.3$$

39:9:2 $$C(x) \simeq \sqrt{\frac{2}{\pi}}\,x\exp\left(\frac{-x^4}{10}\right) \qquad \text{8-bit precision} \qquad |x| \le 1.3$$

while for large arguments

39:9:3 $$S(x) \simeq \frac{1}{2} - \frac{\cos(x^2)}{x\sqrt{2\pi}} - \frac{\sin(x^2)}{x^3\sqrt{8\pi}} \qquad \text{8-bit precision} \qquad |x| \ge \pi$$

39:9:4 $$C(x) \simeq \frac{1}{2} + \frac{\sin(x^2)}{x\sqrt{2\pi}} - \frac{\cos(x^2)}{x^3\sqrt{8\pi}} \qquad \text{8-bit precision} \qquad |x| \ge \pi$$

The approximation

39:9:5 $$\left[S(x) - \frac{1}{2}\right]^2 + \left[C(x) - \frac{1}{2}\right]^2 \simeq \frac{1}{2\pi x^2} \qquad \text{8-bit precision} \qquad x \ge 3$$

shows that, for a small range of x values greater than 3, a cartesian graph of the Fresnel sine-integral versus the Fresnel cosine-integral is roughly a circle, centered at the point $(\frac{1}{2},\frac{1}{2})$ and with radius $1/x\sqrt{2\pi}$. The precise shape of such a graph is discussed in Section 39:14.

39:10 OPERATIONS OF THE CALCULUS

The derivatives and indefinite integrals of the Fresnel integrals are

39:10:1 $$\frac{d}{dx}S(bx) = b\sqrt{\frac{2}{\pi}}\sin(b^2x^2)$$

39:10:2 $$\frac{d}{dx}C(bx) = b\sqrt{\frac{2}{\pi}}\cos(b^2x^2)$$

39:10:3 $$\int_0^x S(bt)dt = x\,S(bx) - \frac{1-\cos(b^2x^2)}{b\sqrt{2\pi}}$$

39:10:4 $$\int_0^x C(bt)dt = x\,C(bx) - \frac{\sin(b^2x^2)}{b\sqrt{2\pi}}$$

Among the definite integrals listed by Gradshteyn and Ryzhik [Section 6.32] are

39:10:5 $$\int_0^\infty t^\nu\left[\frac{1}{2} - \mathrm{S}(bt)\right]\mathrm{d}t = \frac{\Gamma\left(1+\frac{\nu}{2}\right)\cos\left(\frac{\nu\pi}{4}\right)}{\sqrt{2\pi}(1+\nu)b^{1+\nu}} \qquad -1<\nu<2 \qquad b>0$$

39:10:6 $$\int_0^\infty t^\nu\left[\mathrm{C}(bt) - \frac{1}{2}\right]\mathrm{d}t = \frac{\Gamma\left(1+\frac{\nu}{2}\right)\sin\left(\frac{\nu\pi}{4}\right)}{\sqrt{2\pi}(1+\nu)b^{1+\nu}} \qquad -1<\nu<2 \qquad b>0$$

and

39:10:7 $$\int_0^\infty \sin(Bt^2)\,\mathrm{S}(bt)\mathrm{d}t = \int_0^\infty \cos(Bt^2)\,\mathrm{C}(bt)\mathrm{d}t = \begin{cases}\sqrt{\dfrac{\pi}{32B}} & 0<B<b^2\\ 0 & B>b^2\end{cases}$$

39:11 COMPLEX ARGUMENT

Fresnel integrals of complex argument may be related via the formulas

39:11:1 $$\mathrm{S}(x+iy) = \frac{1+i}{4}\mathrm{erf}\left(\frac{x-y}{\sqrt{2}} + i\frac{x+y}{\sqrt{2}}\right) + \frac{1-i}{4}\mathrm{erf}\left(\frac{x+y}{\sqrt{2}} - i\frac{x-y}{\sqrt{2}}\right)$$

and

39:11:2 $$\mathrm{C}(x+iy) = \frac{1-i}{4}\mathrm{erf}\left(\frac{x-y}{\sqrt{2}} + i\frac{x+y}{\sqrt{2}}\right) + \frac{1+i}{4}\mathrm{erf}\left(\frac{x+y}{\sqrt{2}} - i\frac{x-y}{\sqrt{2}}\right)$$

to the error function of complex argument [Section 40:11].

39:12 GENERALIZATIONS

Fresnel integrals may be generalized to *Boehmer integrals,* also known as *generalized Fresnel integrals,* and defined by

39:12:1 $$\mathrm{S}(x;\nu) = \int_x^\infty t^{\nu-1}\sin(t)\mathrm{d}t \qquad x\geq 0 \qquad \nu<1$$

39:12:2 $$\mathrm{C}(x;\nu) = \int_x^\infty t^{\nu-1}\cos(t)\mathrm{d}t \qquad x\geq 0 \qquad \nu<1$$

When the parameter ν is zero, these integrals reduce to sine and cosine integrals [Chapter 38]

39:12:3 $$\mathrm{S}(x;0) = \frac{\pi}{2} - \mathrm{Si}(x)$$

39:12:4 $$\mathrm{C}(x;0) = -\mathrm{Ci}(x)$$

and reduction to the Fresnel integrals occurs when the parameter is a moiety:

39:12:5 $$\mathrm{S}(x;\tfrac{1}{2}) = \sqrt{2\pi}\left[\frac{1}{2} - \mathrm{S}(\sqrt{x})\right]$$

39:12:6 $$\mathrm{C}(x;\tfrac{1}{2}) = \sqrt{2\pi}\left[\frac{1}{2} - \mathrm{C}(\sqrt{x})\right]$$

The recurrence relationships

39:12:7
$$S(x;\nu) = -\frac{C(x;\nu+1)}{\nu} - \frac{x^\nu \sin(x)}{\nu}$$

39:12:8
$$C(x;\nu) = \frac{S(x;\nu+1)}{\nu} - \frac{x^\nu \cos(x)}{\nu}$$

permit any indefinite integral of $t^{-n/2}\sin(t)$ or $t^{-n/2}\cos(t)$ to be evaluated for $n = 1, 2, 3, \ldots$ in terms of the special cases 39:12:3–39:12:6. Boehmer integrals have the particular values

39:12:9
$$C(0;\nu) = \Gamma(\nu)\cos\left(\frac{\nu\pi}{2}\right)$$

39:12:10
$$S(0;\nu) = \Gamma(\nu)\sin\left(\frac{\nu\pi}{2}\right)$$

at zero argument.

Convergent expansions in power series

39:12:11
$$S(x;\nu) = S(0;\nu) - x^{1+\nu}\sum_{j=0}^{\infty}\frac{(-x^2)^j}{(2j+1)!(2j+1+\nu)}$$

39:12:12
$$C(x;\nu) = C(0;\nu) - x^{\nu}\sum_{j=0}^{\infty}\frac{(-x^2)^j}{(2j)!(2j+\nu)}$$

as well the asymptotic series

39:12:13
$$S(x;\nu) \sim x^{\nu-1}\cos(x)\left[1 - \frac{(1-\nu)(2-\nu)}{x^2} + \frac{(1-\nu)(2-\nu)(3-\nu)(4-\nu)}{x^4} - \cdots + \frac{(1-\nu)_{2j}}{(-x^2)^j} + \cdots\right] + (1-\nu)x^{\nu-2}\sin(x)\left[1 - \frac{(2-\nu)(3-\nu)}{x^2} + \frac{(2-\nu)(3-\nu)(4-\nu)(5-\nu)}{x^4} - \cdots + \frac{(2-\nu)_{2j}}{(-x^2)^j} + \cdots\right]$$

39:12:14
$$C(x;\nu) \sim x^{\nu-1}\sin(x)\left[1 - \frac{(1-\nu)(2-\nu)}{x^2} + \frac{(1-\nu)(2-\nu)(3-\nu)(4-\nu)}{x^4} - \cdots + \frac{(1-\nu)_{2j}}{(-x^2)^j} + \cdots\right] + (1-\nu)x^{\nu-2}\cos(x)\left[1 - \frac{(2-\nu)(3-\nu)}{x^2} + \frac{(2-\nu)(3-\nu)(4-\nu)(5-\nu)}{x^4} - \cdots + \frac{(2-\nu)_{2j}}{(-x^2)^j} + \cdots\right]$$

valid as $x \to \infty$, hold for the Boehmer integrals. These expansions permit the numerical evaluation of $S(x;\nu)$ and $C(x;\nu)$: for example, via the hypergeometric algorithm of Section 18:14.

39:13 COGNATE FUNCTIONS

There appears to be no definitive symbolism for the *auxiliary Fresnel cosine-integral* or the *auxiliary Fresnel sine-integral*, for which this *Atlas* uses the notations Fres(x) and Gres(x), respectively. Abramowitz and Stegun [Chapter 7] use $f(\sqrt{2/\pi x})$ and $g(\sqrt{2/\pi x})$ for the same functions. Maps of these functions, which we define only for positive argument, are included in Figure 39-1.

The auxiliary Fresnel integrals are defined by the transforms

39:13:1 $$\mathrm{Fres}(x) = \sqrt{\frac{2}{\pi}}\int_0^\infty \exp(-2xt)\cos(t^2)dt \qquad x \geq 0$$

39:13:2 $$\mathrm{Gres}(x) = \sqrt{\frac{2}{\pi}}\int_0^\infty \exp(-2xt)\sin(t^2)dt \qquad x \geq 0$$

or as the nonperiodic component of the semiintegral and semiderivative, respectively, of the sine function

39:13:3 $$\frac{d^{-1/2}}{dx^{-1/2}}\sin(x) = \sin\left(x - \frac{\pi}{4}\right) + \sqrt{2}\,\mathrm{Fres}(\sqrt{x})$$

39:13:4 $$\frac{d^{1/2}}{dx^{1/2}}\sin(x) = \sin\left(x + \frac{\pi}{4}\right) - \sqrt{2}\,\mathrm{Gres}(\sqrt{x})$$

with zero lower limit. Their relationships to the ordinary Fresnel integrals

39:13:5 $$\mathrm{Fres}(x) = \cos(x^2)\left[\frac{1}{2} - \mathrm{S}(x)\right] - \sin(x^2)\left[\frac{1}{2} - \mathrm{C}(x)\right] \qquad x \geq 0$$

39:13:6 $$\mathrm{Gres}(x) = \sin(x^2)\left[\frac{1}{2} - \mathrm{S}(x)\right] + \cos(x^2)\left[\frac{1}{2} - \mathrm{C}(x)\right] \qquad x \geq 0$$

may also serve as definitions.

The auxiliary Fresnel integrals may be expanded as the convergent power series

39:13:7 $$\mathrm{Fres}(x) = \frac{1}{2}\left[\cos(x^2) - \sin(x^2)\right] = \sqrt{\frac{2}{\pi}}\left[\frac{2x^3}{3} - \frac{8x^7}{105} + \frac{32x^{11}}{10395} - \cdots\right] = x^3\sqrt{\frac{8}{\pi}}\sum_{j=0}^{\infty}\frac{(-4x^4)^j}{(4j+3)!!}$$

39:13:8 $$\frac{1}{2}\left[\sin(x^2) + \cos(x^2)\right] - \mathrm{Gres}(x) = \sqrt{\frac{2}{\pi}}\left[x - \frac{4x^5}{15} + \frac{16x^9}{945} - \cdots\right] = x\sqrt{\frac{2}{\pi}}\sum_{j=0}^{\infty}\frac{(-4x^4)^j}{(4j+1)!!}$$

or as asymptotic inverse-power series

39:13:9 $$\mathrm{Fres}(x) \sim \frac{1}{x\sqrt{2\pi}}\left[1 - \frac{3}{4x^4} + \frac{105}{16x^8} - \cdots + \frac{(4j-1)!!}{(-4x^4)^j} + \cdots\right] \qquad x \to \infty$$

39:13:10 $$\mathrm{Gres}(x) \sim \frac{1}{x^3\sqrt{8\pi}}\left[1 - \frac{15}{4x^4} + \frac{945}{16x^8} - \cdots + \frac{(4j+1)!!}{(-4x^4)^j} + \cdots\right] \qquad x \to \infty$$

the latter being valid for large x.

39:14 RELATED TOPICS

Fresnel integrals play a paramount role in the theory of the diffraction of light. Indeed, it was his studies in physical optics that led Augustin Fresnel to construct the integrals that now bear his name.

A very useful construct in diffraction theory is *Cornu's spiral,* also known as the *clotoid* curve. This double spiral is defined parametrically by

39:14:1 $$f = \mathrm{S}(w) \qquad \text{where} \qquad \mathrm{C}(w) = x$$

and is shown, in part, in Figure 39-2. The two points $x = \pm\frac{1}{2}, f = \pm\frac{1}{2}$ (marked A and B on the figure) are approached asymptotically by the ever-tightening spirals. Cornu's spiral has the interesting property that its *curvature,* that is, the quantity

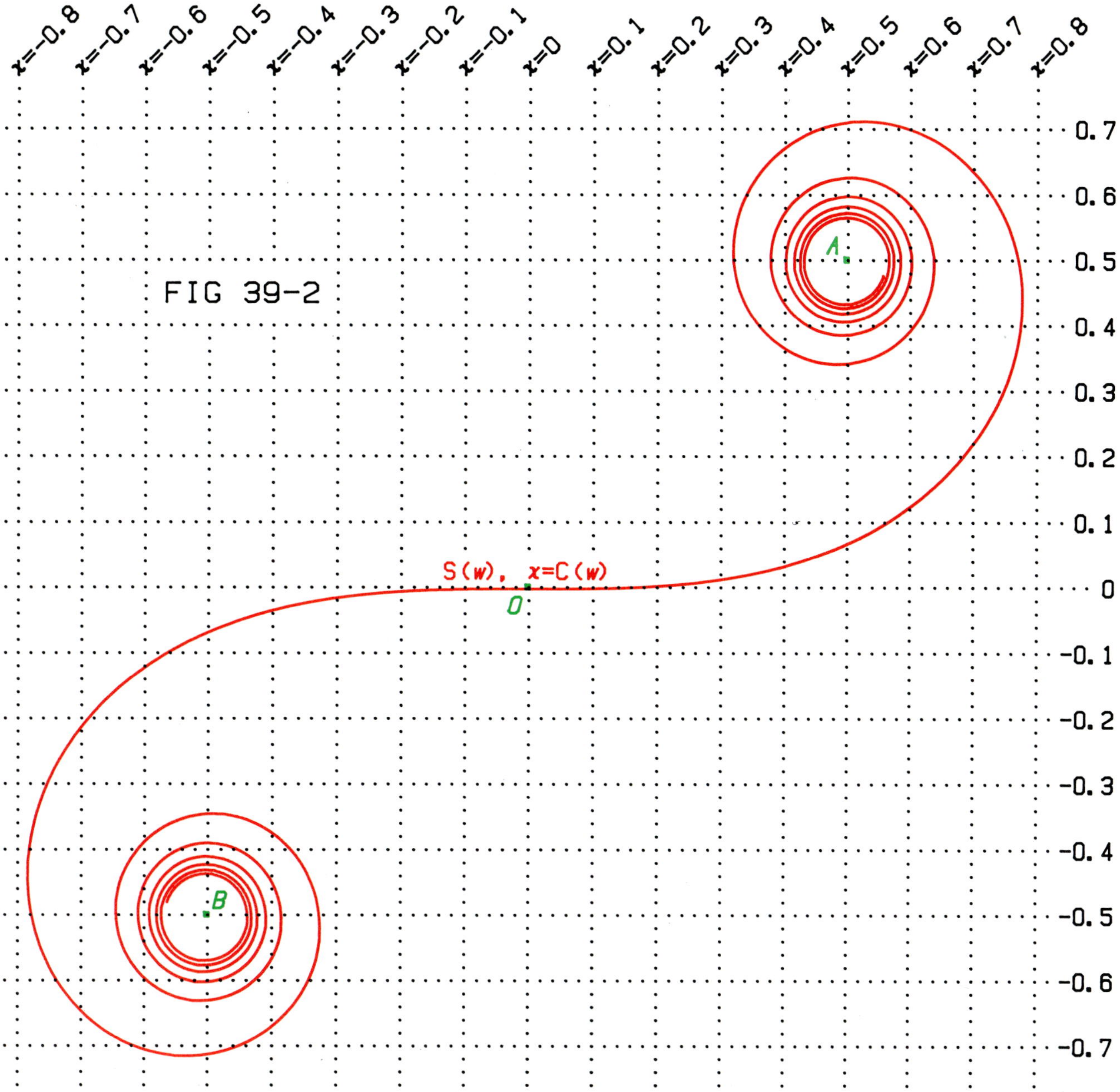

39:14:2
$$\frac{\dfrac{d^2f}{dx^2}}{\left[1+\left(\dfrac{df}{dx}\right)^2\right]^{3/2}}$$

at any point *P* is directly proportional to the *arc length*

39:14:3
$$\int_0^x \sqrt{1+\left(\frac{df}{dt}\right)^2}\,dt$$

between the origin *O* and point *P*. These two quantities are, in fact, $w\sqrt{2\pi}$ and $w\sqrt{2/\pi}$, respectively, where w is the parameter defined in 39:14:1.

CHAPTER 40

THE ERROR FUNCTION erf(x) AND ITS COMPLEMENT erfc(x)

The functions of this chapter are interrelated by

40:0:1 $$\mathrm{erf}(x) + \mathrm{erfc}(x) = 1$$

and occur widely in problems of heat conduction and similar instances of the diffusion of matter or energy. The name of the function arises from its importance in probability theory [see Section 40:14].

40:1 NOTATION

The function erf(x) is also known as the *probability integral* and is sometimes denoted H(x) or $\Phi(x)$. The related notations $\Phi_1(x)$, $\Phi_2(x)$, etc., then denote *successive derivatives of the error function*

40:1:1 $$\Phi_n(x) = \frac{\mathrm{d}^n}{\mathrm{d}x^n}\,\mathrm{erf}(x)$$

Sometimes the initial letter of the erf and erfc notation is capitalized without change of meaning, but Erf(x) and Erfc(x) may also denote $(\sqrt{\pi}/2)\,\mathrm{erf}(x)$ and $(\sqrt{\pi}/2)\,\mathrm{erfc}(x)$, respectively. Changed arguments are common, the name "probability integral" or the *Gauss probability integral* often being given to $\mathrm{erf}(x/\sqrt{2})$ or $[\mathrm{erf}(x/\sqrt{2})]/2$.

40:2 BEHAVIOR

Figure 40-1 includes maps of erf(x) and erfc(x), both of which are sigmoidal functions that rapidly approach limits as $x \to \pm\infty$. These limits are ± 1 for erf(x) and 0 or 2 for erfc(x).

40:3 DEFINITIONS

The most useful definitions of these functions are as the indefinite integrals

40:3:1 $$\mathrm{erf}(x) = \frac{2}{\sqrt{\pi}}\int_0^x \exp(-t^2)\mathrm{d}t = \frac{\mathrm{sgn}(x)}{\sqrt{\pi}}\int_0^{x^2} \frac{\exp(-t)}{\sqrt{t}}\,\mathrm{d}t$$

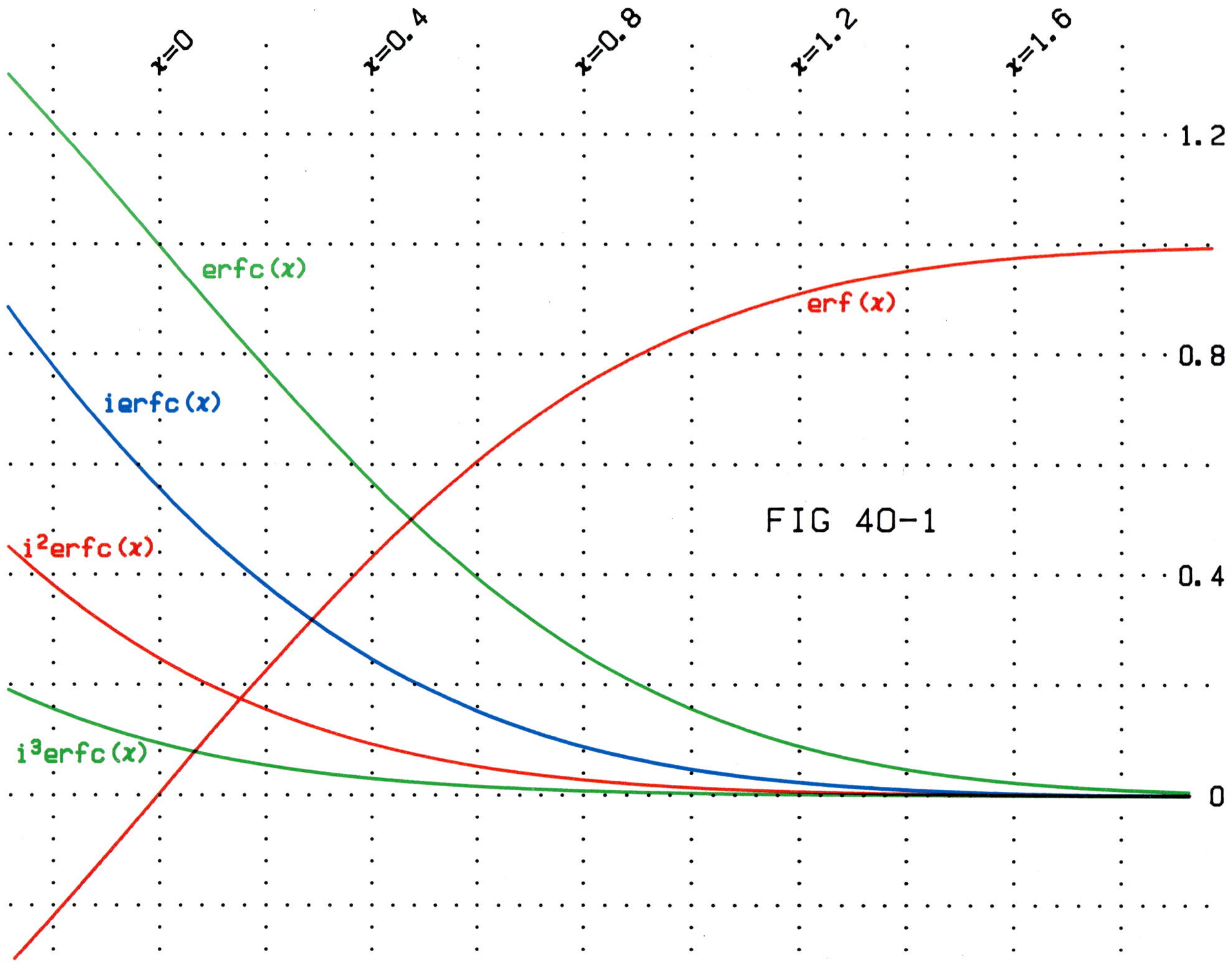

and

40:3:2 $$\operatorname{erfc}(x) = \frac{2}{\sqrt{\pi}} \int_x^\infty \exp(-t^2)\mathrm{d}t = 1 - \operatorname{erf}(x)$$

but definitions as the definite integrals

40:3:3 $$\operatorname{erf}(x) = \frac{2}{\pi} \int_0^\infty \frac{\exp(-t^2)\sin(2xt)}{t}\,\mathrm{d}t$$

40:3:4 $$\operatorname{erf}(x) = \frac{2x}{\sqrt{\pi}} \int_0^1 \exp(-x^2t^2)\mathrm{d}t$$

as well as others, also apply.

40:4 SPECIAL CASES

There are none.

40:5 INTRARELATIONSHIPS

The error function is an odd function

40:5:1 $$\mathrm{erf}(-x) = -\mathrm{erf}(x)$$

while its complement obeys the reflection formula

40:5:2 $$\mathrm{erfc}(-x) = 2 - \mathrm{erfc}(x)$$

40:6 EXPANSIONS

The error function may be expanded in many ways, including

40:6:1 $$\mathrm{erf}(x) = \frac{2}{\sqrt{\pi}}\left[x - \frac{x^3}{3} + \frac{x^5}{10} - \cdots\right] = \frac{x}{\sqrt{\pi}}\sum_{j=0}^{\infty}\frac{(-x^2)^j}{j!(j+\frac{1}{2})}$$

40:6:2 $$\mathrm{erf}(x) = \frac{2}{\sqrt{\pi}}\exp(-x^2)\left[x + \frac{2x^3}{3} + \frac{4x^5}{15} + \cdots\right] = \exp(-x^2)\sum_{j=0}^{\infty}\frac{x^{2j+1}}{\Gamma(j+\frac{3}{2})}$$

40:6:3 $$\mathrm{erf}(x) = \sqrt{2}[\mathrm{I}_{1/2}(x^2) - \mathrm{I}_{3/2}(x^2) - \mathrm{I}_{5/2}(x^2) + \mathrm{I}_{7/2}(x^2) + \cdots] = \sqrt{2}\sum_{j=0}^{\infty}(-1)^J\mathrm{I}_{j+1/2}(x^2) \qquad J = \mathrm{Int}\left(\frac{2+j}{4}\right)$$

while its complement is expansible asymptotically as

40:6:4 $$\mathrm{erfc}(x) \sim \frac{\exp(-x^2)}{x\sqrt{\pi}}\left[1 - \frac{1}{2x^2} + \frac{3}{4x^4} - \frac{15}{8x^6} + \cdots + \frac{(2j-1)!!}{(-2x^2)^j} + \cdots\right] \qquad x \to \infty$$

The functions Γ and $\mathrm{I}_{j+1/2}$ are discussed in Chapter 43 and Section 28:13, respectively. See Section 41:6 for other related expansions.

40:7 PARTICULAR VALUES

Included in Table 40.7.1 are values that have a relevance to probability theory [Section 40:14].

40:8 NUMERICAL VALUES

The following algorithm calculates numerical values of the error function or its complement with 24-bit precision (i.e., the relative error in erf(x) or erfc(x) never exceeds 6×10^{-8}) for any value of the argument. The erf(x) function is computed if the code c is set equal to 0; erfc(x) is evaluated when $c = 1$ is input. For $-1.5 \le x \le 1.5$, the algorithm uses a concatenated form of expansion 40:6:1, truncated at $j = 3 + \mathrm{Int}(9|x|)$. Note that this truncation actually generates more precision in erf(x) than is necessary but that this gratuitous accuracy is needed to preserve sufficient precision on complementation to give erfc(x) when x is close to 1.5. For $x > 1.5$, the algorithm utilizes the continued fraction expression 41:6:3 in the truncated form

Table 40.7.1

	$x = -\infty$	$x = 0$	$x = 0.476936276$	$x = 1/\sqrt{2}$	$x = \infty$
erf(x)	−1	0	$\frac{1}{2}$	0.682689492	1
erfc(x)	2	1	$\frac{1}{2}$	0.317310508	0

40:8:1
$$\operatorname{erfc}(x) = \frac{\sqrt{2/\pi}\,\exp(-x^2)}{x\sqrt{2}+}\;\frac{1}{x\sqrt{2}+}\;\frac{2}{x\sqrt{2}+}\;\frac{3}{x\sqrt{2}+}\cdots\frac{2+\operatorname{Int}(32/x)}{x\sqrt{2}}$$

The same continued fraction is employed for $x < -1.5$, with $-x$ replacing x, and the reflection formula 40:5:2 is then utilized.

Some care is needed in programming this algorithm. For instance, the final parentheses in the penultimate command should not be omitted. If they are, the altered order of operation may cause insufficient precision in an output value of the error function complement. For example, $\operatorname{erfc}(3.5) = 7.43098372 \times 10^{-7}$ but, on a computing device that retains only 10 significant digits, the result of the calculation $\operatorname{erfc}(3.5) + 1 - 1$ is $7.43000000 \times 10^{-7}$.

Numerical values of $[\sqrt{\pi}\,\operatorname{erf}(x)]/2x$ are also available via the universal hypergeometric algorithm of Section 18:14.

Input code c >>>>
Input x >>>>>>>
If $1.5 < |x|$ go to (2)
Set $j = 3 + \operatorname{Int}(9|x|)$
Set $f = 1$
(1) Replace f by $1 + fx^2(\frac{1}{2} - j)/[j(\frac{1}{2} + j)]$
Replace j by $j - 1$
If $j \neq 0$ go to (1)
Replace f by $c + fx(2 - 4c)/\sqrt{\pi}$
Go to (4)
(2) Replace c by $c|x|/x$
Set $j = 3 + \operatorname{Int}(32/|x|)$
Set $f = 0$
(3) Replace f by $1/[fj + \sqrt{2x^2}]$
Replace j by $j - 1$
If $j \neq 0$ go to (3)
Replace f by $f(c^2 + c - 1)(\sqrt{2/\pi})\exp(-x^2) + (1 - c)$
(4) Output f
$f_0 \simeq \operatorname{erf}(x)$, $f_1 \simeq \operatorname{erfc}(x)$ <<<<

Storage needed: c (the code), x, j and f

code	output
0	erf(x)
1	erfc(x)

Test values:
erf(1.45) = 0.959695026
erfc(1.45) = 0.0403049740
erf(−0.5) = −0.520499879
erfc(−0.5) = 1.52049988
erfc(5) = $1.53745979 \times 10^{-12}$

40:9 APPROXIMATIONS

For small arguments, one can use

40:9:1
$$\operatorname{erf}(x) \simeq \frac{3.372x}{3 + x^2} \qquad \text{8-bit precision} \qquad -0.96 \le x \le 0.96$$

whereas for large values of x

40:9:2
$$\operatorname{erfc}(x) \simeq \frac{1.132x\exp(-x^2)}{2x^2 + 1} \qquad \text{8-bit precision} \qquad x \ge 2.7$$

40:10 OPERATIONS OF THE CALCULUS

Differentiation gives

40:10:1
$$\frac{d}{dx}\operatorname{erf}(bx + c) = -\frac{d}{dx}\operatorname{erfc}(bx + c) = \frac{2b}{\sqrt{\pi}}\exp(-b^2x^2 - 2bcx - c^2)$$

while successive differentiation gives rise to Hermite polynomials [Chapter 24]

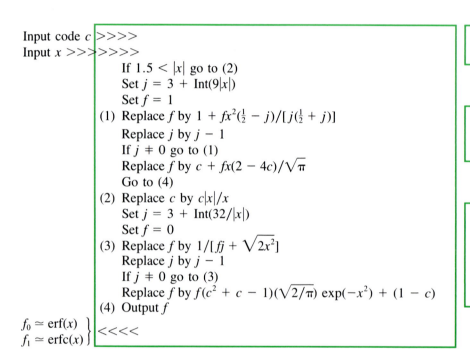

40:10:2 $$-\frac{d^n}{dx^n}\operatorname{erf}(bx) = \frac{d^n}{dx^n}\operatorname{erfc}(bx) = \frac{2}{\sqrt{\pi}}(-b)^n \exp(-b^2x^2)\mathrm{H}_{n-1}(bx) \qquad n = 1, 2, 3, \ldots$$

Indefinite integration gives

40:10:3 $$\int_0^x \operatorname{erf}(bt)dt = x\operatorname{erf}(bx) - \frac{1-\exp(-b^2x^2)}{b\sqrt{\pi}}$$

and

40:10:4 $$\int_x^\infty \operatorname{erfc}(bt)dt = \frac{\operatorname{ierfc}(bx)}{b} = \frac{\exp(-b^2x^2)}{b\sqrt{\pi}} - x\operatorname{erfc}(bx)$$

where the ierfc function is discussed in Section 40:13. The same section discusses repeated integrals of the error function complement.

Of widespread occurrence are indefinite and definite integrals of the product of an exponential function with an error function or complementary error function. Such integrals will be found in Section 41:10 or may be evaluated via the table of Laplace transforms in Section 26:14. Other definite integrals are listed by Gradshteyn and Ryzhik [Sections 6.28–6.31] and include

40:10:5 $$\int_0^\infty t^\nu \operatorname{erfc}(bt)dt = \frac{\Gamma\left(1+\frac{\nu}{2}\right)}{(1+\nu)\sqrt{\pi}b^{1+\nu}} \qquad b > 0 \qquad \nu > -1$$

40:10:6 $$\int_0^\infty \sin(Bt)\operatorname{erfc}(bt)dt = \frac{1-\exp(-B^2/4b^2)}{B} \qquad B > 0 \qquad b > 0$$

40:11 COMPLEX ARGUMENT

If the argument in definition 40:3:1 is replaced by the complex variable $x + iy$, one obtains a complex function given by

40:11:1 $$\operatorname{erf}(x+iy) = \frac{2}{\sqrt{\pi}}\left[\exp(y^2)\int_0^x \exp(-t^2)\cos(2yt)dt + \exp(-x^2)\int_0^y \exp(t^2)\sin(2xt)dt\right]$$
$$- \frac{2i}{\sqrt{\pi}}\left[\exp(y^2)\int_0^x \exp(-t^2)\sin(2yt)dt - \exp(-x^2)\int_0^y \exp(t^2)\cos(2xt)dt\right]$$

in terms of four integrals that are not evaluable in simpler terms. When the real part of the argument is either zero or infinity, however, reduction occurs to the simple results

40:11:2 $$\operatorname{erf}(iy) = \frac{2i}{\sqrt{\pi}}\exp(y^2)\operatorname{daw}(y)$$

40:11:3 $$\operatorname{erf}(\infty + iy) = 1 - i\operatorname{erf}(y)$$

where daw is Dawson's integral, discussed in Chapter 42.

Confusingly, the function generally known as the *error function of complex argument*, and denoted by the symbolism $\mathrm{W}(x + iy)$, is not that given by 40:11:1 but is instead defined by equation 41:11:1 of the next chapter.

40:12 GENERALIZATIONS

The error function and its complement generalize to the incomplete gamma functions of Chapter 45, of which they are the $\nu = \frac{1}{2}$ cases

40:12:1 $$\sqrt{\pi}\,\mathrm{erf}(x) = \mathrm{sgn}(x)\,\gamma(\tfrac{1}{2};x^2)$$

40:12:2 $$\sqrt{\pi}\,\mathrm{erfc}(x) = \Gamma(\tfrac{1}{2};x^2) \qquad x \ge 0$$

The error function complement may also be considered as a special case of the parabolic cylinder function discussed in Chapter 46:

40:12:3 $$\mathrm{erfc}(x) = \sqrt{\frac{2}{\pi}}\exp\left(\frac{-x^2}{2}\right)D_{-1}(x\sqrt{2})$$

Because the incomplete gamma functions and the parabolic cylinder function themselves generalize to the Kummer function [Chapter 47] and Tricomi function [Chapter 48], the error function and its complement may be regarded as special cases of those functions. The relationships are

40:12:4 $$\frac{\sqrt{\pi}}{2x}\,\mathrm{erf}(x) = M(\tfrac{1}{2};\tfrac{3}{2};-x^2) = \exp(-x^2)\,M(1;\tfrac{3}{2};x^2)$$

40:12:5 $$\sqrt{\pi}\,\mathrm{erfc}(x) = \exp(-x^2)\,U(\tfrac{1}{2};\tfrac{1}{2};x^2) = \frac{\exp(-x^2)}{\sqrt{2}}\,U(1;\tfrac{3}{2};x^2) \qquad x > 0$$

40:13 COGNATE FUNCTIONS

A set of functions is defined by the integral

40:13:1 $$i^n\mathrm{erfc}(x) = \frac{2}{\sqrt{\pi}}\int_x^\infty \frac{(t-x)^n}{n!}\exp(-t^2)dt \qquad n = 2, 3, 4, \ldots$$

The notation is sometimes extended to embrace

40:13:2 $$i^1\mathrm{erfc}(x) = \mathrm{ierfc}(x) = \frac{2}{\sqrt{\pi}}\int_x^\infty (t-x)\exp(-t^2)dt = \int_x^\infty \mathrm{erfc}(t)dt$$

40:13:3 $$i^0\mathrm{erfc}(x) = \frac{2}{\sqrt{\pi}}\int_x^\infty \exp(-t^2)dt = \mathrm{erfc}(x)$$

and even

40:13:4 $$i^{-1}\mathrm{erfc}(x) = \frac{2}{\sqrt{\pi}}\exp(-x^2)$$

The ierfc and i^nerfc functions are known as the *complementary error function integral* and the *repeated integrals of the error function complement*, respectively. The behavior of some of these functions is shown in Figure 40-1. Because values for $x < 0$ are seldom of importance in practical applications, these are excluded from the diagram.

The power series expansion

40:13:5 $$i^n\mathrm{erfc}(x) = \frac{1}{2^n}\sum_{j=0}^{\infty}\frac{(-2x)^j}{j!\,\Gamma\left(1+\frac{n-j}{2}\right)}$$

shows that the functions have the particular values

40:13:6 $$i^n\mathrm{erfc}(0) = \frac{1}{2^n\Gamma\left(1+\frac{n}{2}\right)}$$

of which the first few are shown in Table 40.13.1. For $n \ge 1$ i^nerfc(x) approaches ∞ as $x \to -\infty$ and approaches zero asymptotically as $x \to \infty$ according to the asymptotic expansion

Table 40.13.1

n	$\mathrm{i}^n\mathrm{erfc}(0)$
-1	$2/\sqrt{\pi}$
0	1
1	$1/\sqrt{\pi}$
2	$1/4$
3	$1/6\sqrt{\pi}$
4	$1/32$

40:13:7
$$\mathrm{i}^n\mathrm{erfc}(x) \sim \frac{2\exp(-x^2)}{\sqrt{\pi}(2x)^{n+1}}\left[1 - \frac{(n+1)(n+2)}{4x^2} + \frac{(n+1)(n+2)(n+3)(n+4)}{32x^4} - \cdots + \frac{(n+2j)!}{n!j!(-4x^2)^j} + \cdots\right] \qquad x \to \infty$$

The functions obey the recurrence relationship

40:13:8
$$\mathrm{i}^n\mathrm{erfc}(x) = \frac{-x}{n}\,\mathrm{i}^{n-1}\mathrm{erfc}(x) + \frac{1}{2n}\,\mathrm{i}^{n-2}\mathrm{erfc}(x) \qquad n = 1, 2, 3, \ldots$$

and sufficient applications of this formula permit any of the $\mathrm{i}^n\mathrm{erfc}(x)$ functions to be expressed in terms of erfc(x) and $\exp(-x^2)$, and hence evaluated. Early examples are

40:13:9
$$\mathrm{ierfc}(x) = \frac{\exp(-x^2)}{\sqrt{\pi}} - x\,\mathrm{erfc}(x)$$

40:13:10
$$\mathrm{i}^2\mathrm{erfc}(x) = \frac{1+2x^2}{4}\,\mathrm{erfc}(x) - \frac{x}{2\sqrt{\pi}}\exp(-x^2)$$

and

40:13:11
$$\mathrm{i}^3\mathrm{erfc}(x) = \frac{1+x^2}{6\sqrt{\pi}}\exp(-x^2) - \frac{3x+2x^3}{12}\,\mathrm{erfc}(x)$$

Alternatively, these functions are evaluable via the universal hypergeometric algorithm of Section 18:14.

The repeated integrals of the error function complement satisfy the following operations of the calculus:

40:13:12
$$\frac{\mathrm{d}}{\mathrm{d}x}\,\mathrm{i}^n\mathrm{erfc}(x) = -\mathrm{i}^{n-1}\mathrm{erfc}(x)$$

40:13:13
$$\int_x^\infty \mathrm{i}^n\mathrm{erfc}(t)\mathrm{d}t = \mathrm{i}^{n+1}\mathrm{erfc}(x)$$

it being the latter property that explains the name of the function family. It is possible to generalize the family to $\mathrm{i}^\nu\mathrm{erfc}(x)$, where ν is not necessarily an integer, by using procedures of the fractional calculus [see Oldham and Spanier, Section 10.5]. The differential equation

40:13:14
$$a\frac{\mathrm{d}^2f}{\mathrm{d}x^2} + x\frac{\mathrm{d}f}{\mathrm{d}x} - nf = 0 \qquad a > 0$$

is solved by $f = c_1\,\mathrm{i}^n\mathrm{erfc}(x/\sqrt{2a}) + c_2\,\mathrm{i}^n\mathrm{erfc}(-x/\sqrt{2a})$, where c_1 and c_2 are arbitrary constants.

A useful definite integral is

40:13:15
$$\int_0^\infty \exp(Bt)\mathrm{i}^n\mathrm{erfc}(bt+c)\mathrm{d}t = \frac{-1}{B}\,\mathrm{i}^n\mathrm{erfc}(c) + \frac{b}{B}\int_0^\infty \exp(Bt)\mathrm{i}^{n-1}\mathrm{erfc}(bt+c)\mathrm{d}t$$

40:14 RELATED TOPICS

As discussed in Section 27:14, the so-called *normal distribution function* is given by

40:14:1
$$f(x) = \frac{1}{\sigma\sqrt{2\pi}}\exp\left(\frac{-(x-\mu)^2}{2\sigma^2}\right)$$

in terms of the mean μ and variance σ^2 of the distribution, the corresponding cumulative function being

40:14:2
$$F(x) = \int_{-\infty}^{x} f(t)dt = \tfrac{1}{2} + \tfrac{1}{2}\,\text{erf}\left(\frac{x-\mu}{\sigma\sqrt{2}}\right)$$

Random events often obey, or are assumed to obey, distribution 40:14:1, and the quantity $|x - \mu|$ is termed the *error* of a single measurement or observation of x. Accordingly, σ is known as the *standard error*: about 68% of all normally distributed samples lie in the range $\mu - \sigma \le x \le \mu + \sigma$. A similar concept is the *probable error*, equal to 0.6745σ. Because $\text{erf}(0.6745/\sqrt{2}) = \text{erf}(0.4769) = \frac{1}{2}$, it follows that the probability of a normally distributed measurement x lying in the range $\mu - 0.6745\sigma \le x \le \mu + 0.6745\sigma$ is exactly one-half so that as many events lie outside this interval as within it.

The functions

40:14:3
$$f(x) = \frac{1}{\sqrt{2\pi}}\exp\left(\frac{-x^2}{2}\right)$$

and

40:14:4
$$F(x) = \int_{-\infty}^{x} f(t)dt = \tfrac{1}{2} + \tfrac{1}{2}\,\text{erf}\left(\frac{x}{\sqrt{2}}\right) = 1 - \tfrac{1}{2}\,\text{erfc}\left(\frac{x}{\sqrt{2}}\right)$$

are normalized versions of 40:14:1 and 40:14:2 known as the *standard normal density function* and the *cumulative standard normal probability function*, respectively.

A common need in simulation studies is to create a set of numbers that mimic random errors, having a specified mean μ and variance σ^2. Such a set may be constructed by the formula

40:14:5
$$x_j = \mu + \sqrt{2\sigma^2 \ln\left(\frac{N}{n_j}\right)}\cos\left(\frac{2\pi}{N}n_{j+1}\right)$$

where N is a large integer and n_j and n_{j+1} are each random integers in the range $0 < n \le N$. Knuth recommends $N = 199017$ and the formula

40:14:6
$$n_{j+1} = 1 + (24298n_j + 99991)\,\text{mod}(199017) \qquad j = 0, 1, 2, \ldots$$

as a means of creating successive *pseudorandom integers*. The "seed" n_0 is selected arbitrarily in the range $0 < n_0 \le 199017$, and the integers n_j are confined to the same range.

An algorithm follows for generating a sequence of J successive numbers that are normally distributed about a mean μ with a variance σ^2. It uses formulas 40:14:5 and 40:14:6, taking $2J$ as the seed.

Input μ >>>>>>>

Set $N = 199017$

Input σ >>>>>>>

Set $S = 2\sigma^2$

Input J >>>>>>>

Set $j = n = 2J$

(1) Replace n by $24298n + 99991$

Replace n by $1 + n - N\,\text{Int}(n/N)$

Storage needed: μ, N, S, j, n and x

Use radian mode.

Replace j by $j - 1$

If frac($j/2$) = 0 go to (2)

Set $x = \sqrt{S \ln(N/n)}$

Go to (1)

(2) Replace x by $\mu + x \cos(2\pi n/N)$

Output x

$x = x_j$ <<<<<<<

If $j \neq 0$ go to (1)

Test values:

$\mu = 5$, $\sigma = 2$, $J = 6$

$x_1 = 5.46780187$

$x_2 = 2.27940439$

$x_3 = 4.56323336$

$x_4 = 3.85012900$

$x_5 = 5.10798656$

$x_6 = 6.18719458$

CHAPTER 41

THE $\exp(x)\ \mathrm{erfc}(\sqrt{x})$ AND RELATED FUNCTIONS

Although the functions considered in this chapter are composites of those discussed in Chapters 12, 26 and 40, their practical importance warrants a separate treatment. Moreover, many of the properties of these functions transcend those of their component factors.

Primarily, this chapter is devoted to the following functions, each of which involves the product of an exponential function and an error function complement:

41:0:1 $$\exp(x)\ \mathrm{erfc}(\sqrt{x})$$

41:0:2 $$\exp(x)\ \mathrm{erfc}(-\sqrt{x}) = 2\exp(x) - \exp(x)\ \mathrm{erfc}(\sqrt{x})$$

41:0:3 $$\sqrt{\pi x}\ \exp(x)\ \mathrm{erfc}(\sqrt{x})$$

41:0:4 $$\frac{1}{\sqrt{\pi x}} + \exp(x)\ \mathrm{erfc}(-\sqrt{x}) = 2\exp(x) + \frac{1}{\sqrt{\pi x}} - \exp(x)\ \mathrm{erfc}(\sqrt{x})$$

and

41:0:5 $$\exp(x^2)\ \mathrm{erfc}(x)$$

All these functions arise in physical applications, such as problems in diffusion. The similar set of functions with the error function replacing its complement are also important; these are addressed more briefly in Section 41:13.

41:1 NOTATION

Symbol variants such as $(2/\sqrt{\pi})e^x\ \mathrm{Erfc}(x^{1/2})$ and $[1 - \Phi(\sqrt{x})]\exp(x)$ occur, as explained in Sections 12:1, 26:1 and 40:1. In addition, a unitary notation is sometimes adopted; for example, $\mathrm{erc}(x)$ and $\mathrm{eerfc}(x)$ have both been used as abbreviations for $\exp(x^2)\ \mathrm{erfc}(x)$.

41:2 BEHAVIOR

Most of the functions of this chapter are defined as real functions only for arguments $x \geq 0$, and this is the only range shown in Figure 41-1.

The function $\exp(x)$ increases rapidly in magnitude as x increases, whereas $\mathrm{erfc}(\sqrt{x})$ suffers a dramatic decrease. The competition between these two effects causes $\exp(x)\ \mathrm{erfc}(\sqrt{x})$ to diminish in a very leisurely fashion toward

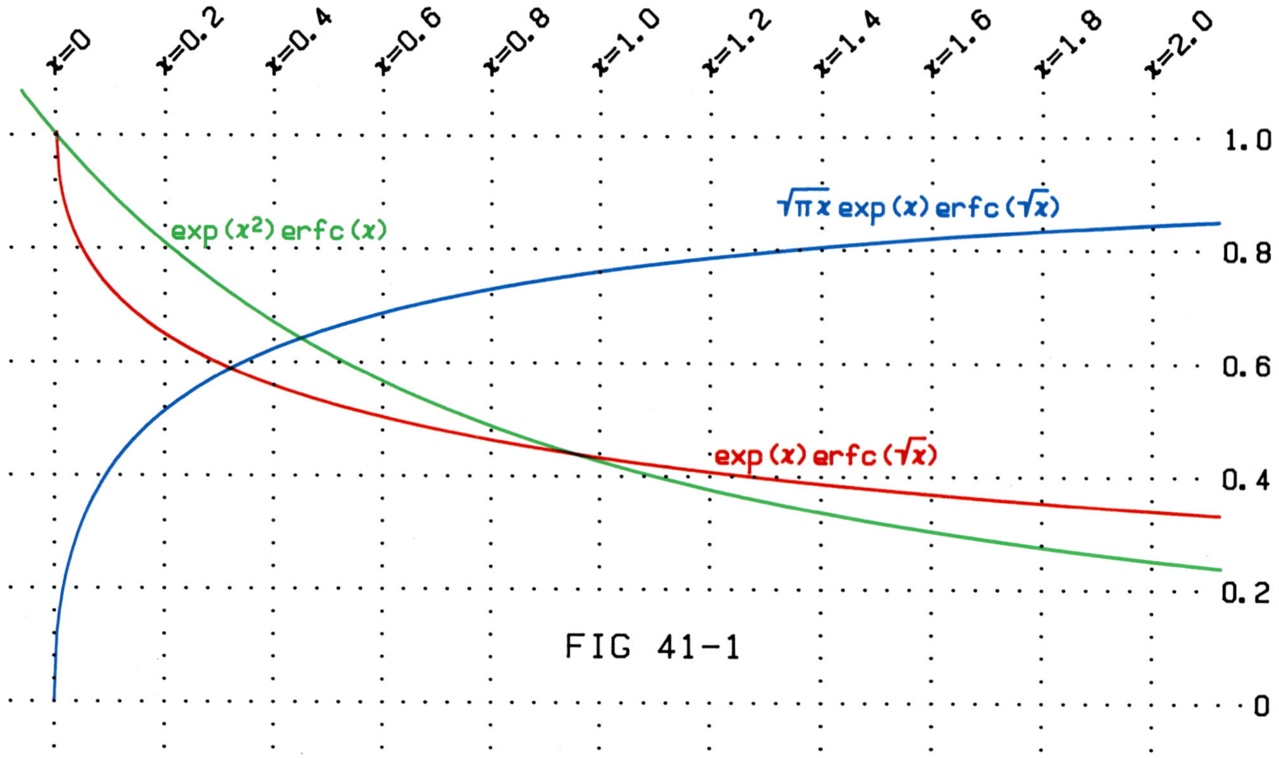

FIG 41-1

zero as $x \to \infty$, whereas $\sqrt{\pi x}\exp(x)\,\mathrm{erfc}(\sqrt{x})$ slowly approaches a unity asymptote in the same limit. These trends are mapped in Figure 41-1.

No such compensation influences the functions $\exp(x)\,\mathrm{erfc}(-\sqrt{x})$ or $(1/\sqrt{\pi x}) + \exp(x)\,\mathrm{erfc}(-\sqrt{x})$, both of which increase exponentially as $x \to \infty$ as depicted in Figure 41-2. The function $(1/\sqrt{\pi x}) + \exp(x)\,\mathrm{erfc}(-\sqrt{x})$ displays a minimum value of about 3.1 close to $x = 0.20$. The same argument corresponds to a point of inflection [Section 0:7] in $\exp(x)\,\mathrm{erfc}(-\sqrt{x})$, as reported in Section 41:7.

41:3 DEFINITIONS

The $\exp(x)\,\mathrm{erfc}(\sqrt{x})$ function may be defined as the indefinite integral

41:3:1
$$\exp(x)\,\mathrm{erfc}(\sqrt{x}) = \frac{1}{\sqrt{\pi}}\int_x^\infty \frac{\exp(x-t)\mathrm{d}t}{\sqrt{t}} = \frac{2}{\sqrt{\pi}}\int_{\sqrt{x}}^\infty \exp(x-t^2)\mathrm{d}t$$

or via the definite integrals

41:3:2
$$\exp(x)\,\mathrm{erfc}(\sqrt{x}) = \frac{1}{\sqrt{\pi}}\int_0^\infty \frac{\exp(-t)\mathrm{d}t}{\sqrt{t+x}} = \frac{2\sqrt{x}}{\pi}\int_0^\infty \frac{\exp(-t^2)\mathrm{d}t}{t^2+x}$$

41:3:3
$$\exp(x)\,\mathrm{erfc}(\sqrt{x}) = \frac{2}{\sqrt{\pi}}\int_0^\infty \exp(-t^2-2t\sqrt{x})\mathrm{d}t$$

and

41:3:4
$$\exp(x)\,\mathrm{erfc}(\sqrt{x}) = \sqrt{\frac{x}{\pi}}\int_0^\infty \frac{\exp(-xt)\mathrm{d}t}{\sqrt{t+1}} = \frac{1}{\pi}\int_0^\infty \frac{\exp(-xt)}{(t+1)\sqrt{t}}\mathrm{d}t$$

The integrals in 41:3:4 may be regarded as Laplace transforms [Section 26:14], in applications of which the functions of this chapter play a distinguished role.

Functions 41:0:1–41:0:5 all satisfy rather simple first order differential equations, namely

41:3:5 $$\frac{df}{dx} - f = \frac{\mp 1}{\sqrt{\pi x}} \qquad f = \exp(x)[\mathrm{erfc}(\pm\sqrt{x}) + c]$$

41:3:6 $$\frac{df}{dx} - \left(1 + \frac{1}{2x}\right)f = -1 \qquad f = \sqrt{\pi x}\,\exp(x)[\mathrm{erfc}(\sqrt{x}) + c]$$

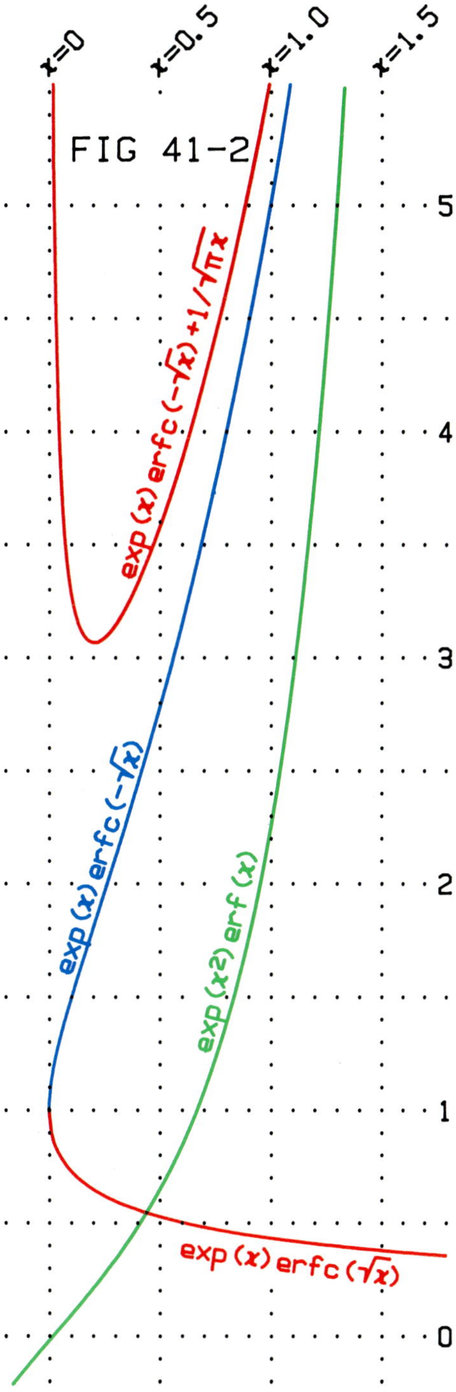

41:3:7 $$\frac{df}{dx} - f = \frac{-1}{2\sqrt{\pi x^3}} \qquad f = \frac{1}{\sqrt{\pi x}} + \exp(x)[\mathrm{erfc}(-\sqrt{x}) + c]$$

41:3:8 $$\frac{df}{dx} - 2xf = \frac{-2}{\sqrt{\pi}} \qquad f = \exp(x^2)[\mathrm{erfc}(x) + c]$$

where c is an arbitrary constant. Thus, the various functions may be defined as the special $c = 0$ cases of the solutions to these differential equations.

The fractional calculus may be employed to define some of the functions of this chapter. Thus, with lower limit zero, semiintegration of the exponential function gives

41:3:9 $$\frac{d^{-1/2}}{dx^{-1/2}} \exp(x) = \exp(x)\,\mathrm{erf}(\sqrt{x}) = \exp(x) - \exp(x)\,\mathrm{erfc}(\sqrt{x})$$

Function 41:0:4 is preserved on semidifferentiation; that is:

41:3:10 $$\frac{d^{1/2}f}{dx^{1/2}} = f \qquad f = \frac{1}{\sqrt{\pi x}} + \exp(x)\,\mathrm{erfc}(-\sqrt{x})$$

so that this function plays a role in the semicalculus analogous to that played by the exponential function in the classical calculus.

41:4 SPECIAL CASES

There are none.

41:5 INTRARELATIONSHIPS

The reflection formula

41:5:1 $$\exp(x)\,\mathrm{erfc}(-\sqrt{x}) = 2\exp(x) - \exp(x)\,\mathrm{erfc}(\sqrt{x})$$

applies. The addition formula

41:5:2 $$\exp(x+y)\,\mathrm{erfc}(\sqrt{x+y}) = \exp(x+y)\,\mathrm{erfc}(\sqrt{x}) - \frac{\exp(y)}{\sqrt{\pi x}} \sum_{j=0}^{\infty} \frac{(2j-1)!!}{(2j)!!} \frac{\gamma(j+1;y)}{(-x)^j}$$

involves a series of incomplete gamma functions [Chapter 45].

41:6 EXPANSIONS

A power series for $\exp(x)\,\mathrm{erfc}(\pm\sqrt{x})$

41:6:1 $$\exp(x)\,\mathrm{erfc}(\pm\sqrt{x}) = 1 \mp 2\sqrt{\frac{x}{\pi}} + x \mp \frac{4}{3}\sqrt{\frac{x^3}{\pi}} + \frac{x^2}{2} \mp \cdots = \sum_{j=0}^{\infty} \frac{(\mp\sqrt{x})^j}{\Gamma\left(1 + \frac{j}{2}\right)}$$

involves the gamma function [Chapter 43]. Alternate terms may be summed separately:

41:6:2 $$\exp(x)\,\mathrm{erfc}(\pm\sqrt{x}) = 1 + x + \frac{x^2}{2} + \cdots \mp 2\sqrt{\frac{x}{\pi}}\left[1 + \frac{2x}{3} + \frac{4x^2}{15} + \cdots\right]$$
$$= \sum_{j=0}^{\infty} \frac{x^j}{j!} \mp 2\sqrt{\frac{x}{\pi}} \sum_{j=0}^{\infty} \frac{(2x)^j}{(2j+1)!!} = \exp(x) \mp \exp(x)\,\mathrm{erf}(\sqrt{x})$$

Of course, replacement of $\sqrt{x}$ by x in 41:6:1 gives the power series expansion of $\exp(x^2)\,\mathrm{erfc}(\pm x)$.

The $\exp(x)\,\mathrm{erfc}(\sqrt{x})$ function may also be expanded as a continued fraction:

41:6:3 $$\exp(x)\,\mathrm{erfc}(\sqrt{x}) = \frac{\sqrt{2/\pi}}{\sqrt{2x}+}\,\frac{1}{\sqrt{2x}+}\,\frac{2}{\sqrt{2x}+}\,\frac{3}{\sqrt{2x}+}\,\frac{4}{\sqrt{2x}+}\cdots$$

Truncated versions of the asymptotic series

41:6:4 $$\sqrt{\pi x}\,\exp(x)\,\mathrm{erfc}(\sqrt{x}) \sim 1 - \frac{1}{2x} + \frac{3}{4x^2} - \cdots + \frac{(2j-1)!!}{(-2x)^j} + \cdots \qquad x \to \infty$$

are useful for large x. The truncation error is less in absolute value than the first neglected term and of the same sign.

It follows from 41:6:2 that the function $(1/\sqrt{\pi x}) + \exp(x)\,\mathrm{erfc}(-\sqrt{x})$ is expansible as the series

41:6:5 $$\frac{1}{\sqrt{\pi x}} + \exp(x)\,\mathrm{erfc}(-\sqrt{x}) = \frac{1}{\sqrt{\pi x}} + 1 + 2\sqrt{\frac{x}{\pi}} + x + \cdots = \sum_{j=0}^{\infty} \frac{x^{(j-1)/2}}{\Gamma\left(\frac{j+1}{2}\right)}$$
$$= \exp(x) + \frac{1}{\sqrt{\pi x}} \sum_{j=0}^{\infty} \frac{(2x)^j}{(2j-1)!!}$$

with uniformly positive terms.

41:7 PARTICULAR VALUES

	$x = -\infty$	$x = 0$	$x = 0.204053939$	$x = \infty$
$\exp(x)\,\mathrm{erfc}(\sqrt{x})$	undef	1	0.641304446	0
$\exp(x)\,\mathrm{erfc}(-\sqrt{x})$	undef	1	1.81142415 = inflection	∞
$\sqrt{\pi x}\,\exp(x)\,\mathrm{erfc}(\sqrt{x})$	undef	0	0.513465985	1
$\frac{1}{\sqrt{\pi x}} + \exp(x)\,\mathrm{erfc}(-\sqrt{x})$	undef	∞	3.06039578 = minimum	∞
$\exp(x^2)\,\mathrm{erfc}(x)$	∞	1	0.805767647	0

41:8 NUMERICAL VALUES

The algorithm presented below generates values of $\exp(x)\,\mathrm{erfc}(\sqrt{x})$ to 24-bit precision (i.e., the relative error in the output does not exceed 6×10^{-8}) for any input $x \ge 0$. Because $\exp(x)\,\mathrm{erfc}(\sqrt{x})$ does not exist as a real number for $x < 0$, the input of a negative number $-x$ is treated by the algorithm as an instruction to calculate the alternative function $\exp(x)\,\mathrm{erfc}(-\sqrt{x})$, using identity 41:0:2. If the facility to calculate this alternative function is not needed, the portion of the algorithm shown in green may be omitted.

The algorithm resembles that in Section 40:8. For $x \le 2.25$, it uses the concatenation

41:8:1 $$\mathrm{erf}(\sqrt{x}) \simeq \left(\left(\left(\cdots\left(\left(\frac{x(\frac{1}{2}-J)}{J(J+\frac{1}{2})}+1\right)\frac{x(\frac{3}{2}-J)}{(J-1)(J-\frac{1}{2})}+1\right)\frac{x(\frac{5}{2}-J)}{(J-2)(J-\frac{3}{2})}\right.\right.\right.$$
$$\left.\left.\left.+\cdots+1\right)\frac{-\frac{3}{2}x}{2\times\frac{5}{2}}+1\right)\frac{-\frac{1}{2}x}{1\times\frac{3}{2}}+1\right)2\sqrt{\frac{x}{\pi}}$$

which is based on expansion 40:6:1, truncated with $J = 3 + \mathrm{Int}(9\sqrt{x})$. For $x > 2.25$ a continued fraction expansion analogous to 40:8:1 is utilized.

Set $f = g = 0$	Storage needed: f, g, x and j
Input $\pm x$ >>>	
If $x \ge 0$ go to (1)	

Replace x by $-x$
Set $g = 2\exp(x)$
(1) If $1.5 < \sqrt{x}$ go to (3)
Set $j = 3 + \mathrm{Int}(9\sqrt{x})$
Set $f = 1$
(2) Replace f by $1 + fx(\tfrac{1}{2} - j)/[j(\tfrac{1}{2} + j)]$
Replace j by $j - 1$
If $j \neq 0$ go to (2)
Replace f by $\exp(x)[1 - 2f\sqrt{x/\pi}]$
Go to (5)
(3) Set $j = 3 + \mathrm{Int}(32/\sqrt{x})$
(4) Replace f by $1/(fj + \sqrt{2x})$
Replace j by $j - 1$
If $j \neq 0$ go to (4)
Replace f by $f\sqrt{2/\pi}$
(5) If $g = 0$ go to (6)
Replace f by $g - f$
(6) Output f
<<<<
$f \simeq \exp(x)\,\mathrm{erfc}(\pm\sqrt{x})$

Input restrictions: When $x \geq 0$ is input, $\exp(x)\,\mathrm{erfc}(\sqrt{x})$ is output. When $-x \leq 0$ is input, $\exp(x)\,\mathrm{erfc}(-\sqrt{x})$ is output.

Test values:
$\exp(1)\,\mathrm{erfc}(1) = 0.427583576$
$\exp(\pi)\,\mathrm{erfc}(\sqrt{\pi}) = 0.282059176$
$\exp(1)\,\mathrm{erfc}(-1) = 5.00898008$
$\exp(\pi)\,\mathrm{erfc}(-\sqrt{\pi}) = 45.9993261$

41:9 APPROXIMATIONS

The approximation 40:9:2 is easily adapted to provide an estimate of the $\exp(x)\,\mathrm{erfc}(\sqrt{x})$ function for large x. A more accurate approximation, valid for all arguments, is

41:9:1
$$\sqrt{\pi x}\,\exp(x)\,\mathrm{erfc}(\sqrt{x}) \simeq \frac{2}{1 + \sqrt{1 + \dfrac{2}{x}\left(1 - \dfrac{1 - (2/\pi)}{\exp(\sqrt{5x/7})}\right)}} \qquad \text{10-bit precision}$$

This approximation exploits the inequality

41:9:2
$$\frac{2}{1 + \sqrt{1 + \dfrac{2}{x}}} < \sqrt{\pi x}\,\exp(x)\,\mathrm{erfc}(\sqrt{x}) \leq \frac{2}{1 + \sqrt{1 + \dfrac{4}{\pi x}}} \qquad \infty > x \geq 0$$

41:10 OPERATIONS OF THE CALCULUS

Differentiation yields

41:10:1
$$\frac{d}{dx}\exp(bx)\,\mathrm{erfc}(\pm\sqrt{bx}) = b\exp(bx)\,\mathrm{erfc}(\pm\sqrt{bx}) \mp \sqrt{\frac{b}{\pi x}}$$

41:10:2
$$\frac{d}{dx}\sqrt{\pi x}\,\exp(bx)\,\mathrm{erfc}(\sqrt{bx}) = \sqrt{\frac{\pi}{x}}\,(bx + \tfrac{1}{2})\exp(bx)\,\mathrm{erfc}(\sqrt{bx}) - \sqrt{b}$$

41:10:3
$$\frac{d}{dx}\left(\frac{1}{\sqrt{\pi bx}} + \exp(bx)\,\mathrm{erfc}(-\sqrt{bx})\right) = \sqrt{\frac{b}{\pi x}} + b\exp(bx)\,\mathrm{erfc}(-\sqrt{bx}) - \frac{1}{2\sqrt{bx^3}}$$

41:10:4
$$\frac{d}{dx}\exp(ax^2)\,\mathrm{erfc}(bx) = 2\exp(ax^2)\left[ax\,\mathrm{erfc}(bx) - \frac{b}{\sqrt{\pi}}\exp(-b^2x^2)\right]$$

Repeated differentiation of $\exp(x)\,\mathrm{erfc}(\pm\sqrt{x})$ regenerates the original function together with a finite series of algebraic terms

41:10:5
$$\frac{\mathrm{d}^n}{\mathrm{d}x^n}\exp(x)\,\mathrm{erfc}(\pm\sqrt{x}) = \exp(x)\,\mathrm{erfc}(\pm\sqrt{x}) \mp \frac{1}{\sqrt{\pi x}} \pm \frac{1}{2\sqrt{\pi x^3}} \mp \frac{3}{4\sqrt{\pi x^5}} \pm \cdots$$
$$= \exp(x)\,\mathrm{erfc}(\pm\sqrt{x}) \mp \frac{1}{\sqrt{\pi x}}\sum_{j=0}^{n-1}\frac{(2j-1)!!}{(-2x)^j}$$

The series is identical with the leading members of the asymptotic expansion 41:6:4.

Indefinite integrals include

41:10:6
$$\int_0^x \exp(-b^2t^2)\,\mathrm{erfc}(\pm bt)\mathrm{d}t = \frac{\pm\sqrt{\pi}}{4b}[1 - \mathrm{erfc}^2(\pm bx)]$$

41:10:7
$$\int_0^x \exp(Bt)\,\mathrm{erfc}(bt + c)\mathrm{d}t = \frac{1}{B}\exp(Bx)[\mathrm{erfc}(bx + c) - \mathrm{erfc}(c)] - \frac{1}{B}\exp\left(\frac{B^2}{4b^2} - \frac{Bc}{b}\right)\left[\mathrm{erfc}\left(bx + c - \frac{B}{2b}\right) - \mathrm{erfc}\left(c - \frac{B}{2b}\right)\right]$$

41:10:8
$$\int_0^x \exp(-Bt)\,\mathrm{erfc}\left(\sqrt{\frac{a}{t}}\right)\mathrm{d}t = \frac{\exp(2\sqrt{aB})}{2B}\mathrm{erfc}\left(\sqrt{\frac{a}{x}} + \sqrt{Bx}\right) + \frac{\exp(-2\sqrt{aB})}{2B}\mathrm{erfc}\left(\sqrt{\frac{a}{x}} - \sqrt{Bx}\right) - \frac{\exp(-Bx)}{B}\mathrm{erfc}\left(\sqrt{\frac{a}{x}}\right) \qquad B > 0 \quad a > 0 \quad x > 0$$

and

41:10:9
$$\int_0^x \exp(Bt)\,\mathrm{erfc}(\sqrt{bt + c})\mathrm{d}t = \frac{1}{B}[\exp(Bx)\,\mathrm{erfc}(\sqrt{bx + c}) - \mathrm{erfc}(\sqrt{c}) + \phi(x)]$$

where

41:10:10
$$\phi(x) = \begin{cases} -\sqrt{\dfrac{b}{b-B}}\exp\left(\dfrac{-Bc}{b}\right)\left[\mathrm{erfc}\left(\sqrt{b-B}\sqrt{x + \dfrac{c}{b}}\right) - \mathrm{erfc}\left(\sqrt{c - \dfrac{cB}{b}}\right)\right] & B < b \\ \dfrac{2}{\sqrt{\pi}}\exp(-c)[\sqrt{bx + c} - \sqrt{c}] & B = b \\ \sqrt{\dfrac{4b}{\pi(B-b)}}\exp(-c)\left[\exp(Bx - bx)\,\mathrm{daw}\left(\sqrt{B-b}\sqrt{x + \dfrac{c}{b}}\right) - \mathrm{daw}\left(\sqrt{\dfrac{Bc}{b} - c}\right)\right] & B > b \end{cases}$$

Representing the error function by Φ, Gradshteyn and Ryzhik [Sections 6.28–6.31] list over thirty definite integrals relevant to this chapter. Some of the more useful are

41:10:11
$$\int_0^\infty \exp(at^2)\,\mathrm{erfc}(bt)\mathrm{d}t = \begin{cases} \dfrac{\mathrm{artanh}(\sqrt{a}/b)}{\sqrt{\pi a}} & a > 0 \\ \dfrac{1}{b\sqrt{\pi}} & a = 0 \\ \dfrac{\arctan(\sqrt{-a}/b)}{\sqrt{-\pi a}} & a < 0 \end{cases}$$

41:10:12 $$\int_0^\infty t^\nu \exp(at^2)\,\mathrm{erfc}(bt)\mathrm{d}t = \frac{\Gamma\left(1+\frac{\nu}{2}\right)}{\sqrt{\pi}(1+\nu)b^{1+\nu}} \mathrm{F}\left(\frac{\nu+1}{2},\frac{\nu+2}{2};\frac{\nu+3}{2};\frac{a}{b^2}\right) \qquad b^2 > a \qquad \nu > -1$$

41:10:13 $$\int_0^\infty \exp(Bt)\,\mathrm{erfc}(\sqrt{bt})\mathrm{d}t = \begin{cases} 1/(b - B + \sqrt{b^2 - bB}) & B < 0 \\ [\sqrt{b/(b-B)} - 1]/B & 0 < B < b \end{cases}$$

41:10:14 $$\int_0^\infty \frac{\exp(bt)\,\mathrm{erfc}\sqrt{bt}}{t^\nu}\mathrm{d}t = \frac{\sec(\nu\pi)\,\Gamma(\nu)}{b^\nu} \qquad 0 < \nu < {}^1\!/_2$$

With zero lower limit, the operations of semidifferentiation and semiintegration give

41:10:15 $$\frac{\mathrm{d}^{1/2}}{\mathrm{d}x^{1/2}} \exp(bx)\,\mathrm{erfc}(\pm\sqrt{bx}) = \frac{1}{\sqrt{\pi x}} \mp \sqrt{b}\exp(bx)\,\mathrm{erfc}(\pm\sqrt{bx})$$

and

41:10:16 $$\frac{\mathrm{d}^{-1/2}}{\mathrm{d}x^{-1/2}} \exp(bx)\,\mathrm{erfc}(\pm\sqrt{bx}) = \frac{\pm 1}{\sqrt{b}}[1 - \exp(bx)\,\mathrm{erfc}(\pm\sqrt{bx})]$$

41:11 COMPLEX ARGUMENT

The so-called *error function of complex argument* is, in fact, the function $\exp(x)\,\mathrm{erfc}(-\sqrt{x})$ with x replaced by $-(x + iy)^2$. It is defined by

41:11:1 $$\mathrm{W}(z) = \exp(-z^2)\left[1 + \frac{2i}{\sqrt{\pi}}\int_0^z \exp(t^2)\mathrm{d}t\right] = \exp(-z^2)\,\mathrm{erfc}(-iz) \qquad z = x + iy$$

This is an important complex-valued function, the real and imaginary portions of which arise in practical problems. Accordingly, its values have been tabulated [for example by Abramowitz and Stegun, Table 7.9], or, alternatively, numerical values may be calculated via expansions 41:6:2 and 41:6:5, which remain valid when x is replaced by $-(x + iy)^2$.

The real and imaginary parts of $\mathrm{W}(x + iy)$, defined by

41:11:2 $$\mathrm{W}(x+iy) = \mathrm{Re}\{\mathrm{W}(x+iy)\} + i\,\mathrm{Im}\{\mathrm{W}(x+iy)\}$$

are expressible as the integrals

41:11:3 $$\mathrm{Re}\{\mathrm{W}(x+iy)\} = \frac{y}{\pi}\int_{-\infty}^{\infty} \frac{\exp(-t^2)\mathrm{d}t}{(x-t)^2 + y^2}$$

and

41:11:4 $$\mathrm{Im}\{\mathrm{W}(x+iy)\} = \frac{1}{\pi}\int_{-\infty}^{\infty} \frac{(x-t)\exp(-t^2)\mathrm{d}t}{(x-t)^2 + y^2}$$

When the argument of the W function is real (i.e., $y = 0$), the function has real and imaginary parts given by

41:11:5 $$\mathrm{W}(x) = \exp(-x^2) + \frac{2i}{\sqrt{\pi}}\,\mathrm{daw}(x)$$

but when the argument is purely imaginary (i.e., $x = 0$), the function is real

41:11:6 $$\mathrm{W}(iy) = \exp(y^2)\,\mathrm{erfc}(y)$$

41:12 GENERALIZATIONS

The $\exp(x)\,\mathrm{erf}(\sqrt{x})$ function is a special case of the Kummer function [Chapter 47]

41:12:1 $$M(1;\tfrac{3}{2};x) = \frac{1}{2}\sqrt{\frac{\pi}{x}}\exp(x)\,\mathrm{erf}(\sqrt{x}) = \frac{1}{2}\sqrt{\frac{\pi}{x}}\,[\exp(x) - \exp(x)\,\mathrm{erfc}(\sqrt{x})]$$

The asymptotic representation 41:6:4 of the $\sqrt{\pi x}\exp(x)\,\mathrm{erfc}(\sqrt{x})$ function generalizes to the simple hypergeometric function [Section 18:14]

41:12:2 $$\sum_{j=0}^{\infty}\frac{(a)_j}{(-x)^j}$$

of which it is the $a = \frac{1}{2}$ special case.

41:13 COGNATE FUNCTIONS

Functions formed by replacement of erfc in formulas 41:0:1–41:0:5 by erf are almost as important as the originals. We do not explicitly treat such cognate functions in this *Atlas* because their properties are so easily deduced from those of the primary functions of this chapter via the identities

41:13:1 $$\exp(x)\,\mathrm{erf}(\sqrt{x}) = \exp(x) - \exp(x)\,\mathrm{erfc}(\sqrt{x})$$

and

41:13:2 $$\exp(x)\,\mathrm{erf}(\sqrt{x}) = \frac{\exp(x)\,\mathrm{erfc}(-\sqrt{x}) - \exp(x)\,\mathrm{erfc}(\sqrt{x})}{2}$$

In addition, explicit reference to the $\exp(x)\,\mathrm{erf}(\sqrt{x})$ function is made in equations 41:3:9, 41:6:2 and 41:13:1. Figure 17-2 includes a graph of $\exp(x^2)\,\mathrm{erf}(x)$.

An expansion in terms of hyperbolic Bessel functions [Section 28:13] and particular values of the exponential polynomial [Section 26:13] is

41:13:3 $$\exp(x)\,\mathrm{erf}(\sqrt{x}) = \sum_{j=0}^{\infty} e_j(-1)(\sqrt{x})^{j+1/2}\,I_{j+1/2}(2\sqrt{x})$$

while a more conventional expansion is

41:13:4 $$\exp(x)\,\mathrm{erf}(\sqrt{x}) = 2\sqrt{\frac{x}{\pi}}\left[1 + \frac{2x}{3} + \frac{4x^2}{15} + \cdots\right] = 2\sqrt{\frac{x}{\pi}}\sum_{j=0}^{\infty}\frac{(2x)^j}{(2j+1)!!}$$

The latter is useful for calculating values of $\exp(x)\,\mathrm{erf}(\sqrt{x})$; for example via the universal hypergeometric algorithm of Section 18:14.

CHAPTER 42

DAWSON'S INTEGRAL

The functions of this chapter have much in common with those of the previous chapter. Dawson's integral daw(x) is related to $(\sqrt{\pi}/2)\exp(-x^2)\,\text{erf}(x)$ via the identity

42:0:1
$$\text{daw}(x) = \frac{-i\sqrt{\pi}}{2}\exp(-x^2)\,\text{erf}(ix)$$

in much the same way that sin(x) is related to sinh(x) [see 28:11:3].

42:1 NOTATION

Because no standard notation exists, this *Atlas* uses the symbol daw(x) for Dawson's integral of argument x. Abramowitz and Stegun [Chapter 7] use F(x) and, in the light of relationship 42:0:1, Gradshteyn and Ryzhik adopt a symbolism equivalent to $-i(\sqrt{\pi}/2)\exp(-x^2)\,\text{erf}(ix)$.

Sometimes the name "Dawson's integral" is given to $\exp(x^2)\text{daw}(x)$ and Erfi(x) or erfi(x) is employed to denote this function. A rescaling of the argument often leads to simpler formulas; the function $\sqrt{2x}\,\text{daw}(\sqrt{x/2})$ has been symbolized D(x).

42:2 BEHAVIOR

The functions daw($\sqrt{x}$), daw(x) and x daw(x) are all important and all three are mapped in Figure 42-1. Each of these functions is zero at $x = 0$, displays a maximum in the vicinity of $x = 1$ and then slowly declines, approaching $\frac{1}{2}$ asymptotically in the case of x daw(x) and zero in the other two cases.

42:3 DEFINITIONS

The usual definition of Dawson's integral is through the indefinite integral

42:3:1
$$\text{daw}(x) = \int_0^x \exp(t^2 - x^2)dt = \frac{\text{sgn}(x)}{2}\int_0^{x^2}\frac{\exp(t - x^2)dt}{\sqrt{t}}$$

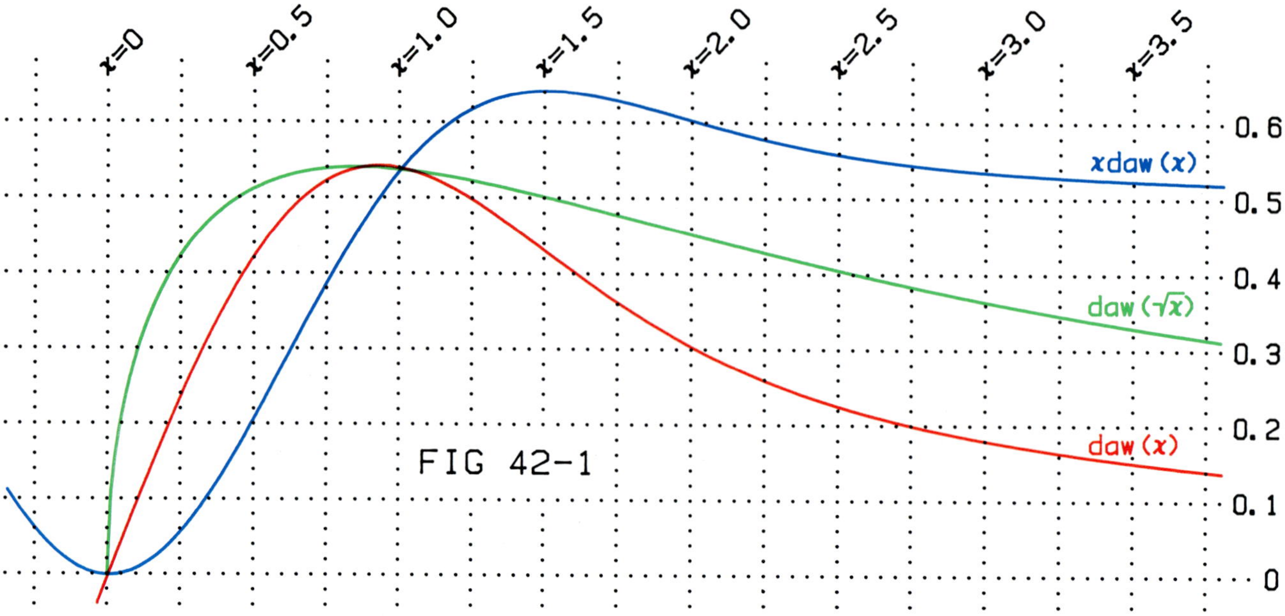

FIG 42-1

but it may also be defined as an inverse Laplace transform [see integral 42:10:7], in terms of the error function of complex argument [Section 41:11] or as the semiintegral

42:3:2
$$\mathrm{daw}(\sqrt{x}) = \frac{\sqrt{\pi}}{2}\frac{\mathrm{d}^{-1/2}}{\mathrm{d}x^{-1/2}}\exp(-x)$$

with lower limit zero.

Dawson's integral may be considered to arise as the *complementary function* [see Murphy, page 82] in the solution of the very general first-order differential equation

42:3:3
$$\frac{\mathrm{d}f}{\mathrm{d}x} + bxf = c$$

where b and c are constants. The general solution is

42:3:4
$$f = \frac{2c}{b}\,\mathrm{daw}\left(\frac{bx}{2}\right) + (\text{constant})\exp\left(\frac{-bx^2}{2}\right)$$

42:4 SPECIAL CASES

There are none.

42:5 INTRARELATIONSHIPS

Dawson's integral is an odd function:

42:5:1
$$\mathrm{daw}(-x) = -\mathrm{daw}(x)$$

42:6 EXPANSIONS

The power series expansion

42:6:1
$$\mathrm{daw}(x) = x - \frac{2x^3}{3} + \frac{4x^5}{15} - \cdots = x\sum_{j=0}^{\infty}\frac{(-2x^2)^j}{(2j+1)!!}$$

may be written in the alternative forms

42:6:2 $$\mathrm{daw}(\sqrt{x}) = \frac{\sqrt{\pi x}}{2}\sum_{j=0}^{\infty}\frac{(-x)^j}{\Gamma(j+\frac{3}{2})} = \sqrt{x}\sum_{j=0}^{\infty}\frac{(-x)^j}{(\frac{3}{2})_j}$$

or as the continued fraction

42:6:3 $$\mathrm{daw}\left(\sqrt{\frac{x}{2}}\right) = \frac{\sqrt{x/2}}{1+}\ \frac{x}{3-x+}\ \frac{3x}{5-x+}\ \frac{5x}{7-x+}\cdots$$

The expansion

42:6:4 $$\mathrm{daw}(\sqrt{x}) = \exp(-x)\left[\sqrt{x} + \frac{\sqrt{x^3}}{3} + \frac{\sqrt{x^5}}{10} + \cdots\right] = \sqrt{x}\exp(-x)\sum_{j=0}^{\infty}\frac{x^j}{j!(2j+1)}$$

may be developed into

42:6:5 $$\mathrm{daw}(\sqrt{x}) = \frac{1}{2\sqrt{x}}\sum_{j=0}^{n-1}\frac{(2j-1)!!}{(2x)^j} + \frac{(2n-1)!!\sqrt{x}\exp(-x)}{(2x)^n}\sum_{j=0}^{\infty}\frac{x^j}{j!(2j-2n+1)}$$

for $n = 1, 2, 3, \ldots$. As $n \to \infty$, this becomes the asymptotic series

42:6:6 $$\mathrm{daw}(x) \sim \frac{1}{2x} + \frac{1}{4x^3} + \frac{3}{8x^5} + \cdots + \frac{(2j-1)!!}{(2x^2)^j\,2x} + \cdots \qquad x \to \infty$$

valid for large x.

42:7 PARTICULAR VALUES

All three functions acquire the value daw(1) = 0.538079507 at unity argument:

	$x = -\infty$	$x = -1$	$x = 0$	$x = 1$	$x = \infty$
$\mathrm{daw}(\sqrt{x})$	undef	undef	0	daw(1)	0
$\mathrm{daw}(x)$	0	−daw(1)	0	daw(1)	0
$x\,\mathrm{daw}(x)$	$\frac{1}{2}$	daw(1)	0	daw(1)	$\frac{1}{2}$

The values of the maxima and the inflection of each of the functions are related in a simple way to the corresponding arguments:

	Maximum		Inflection(s)	
	Value of x	Value of f	Value of x	Value of f
$f = \mathrm{daw}(\sqrt{x})$	$0.854032657 = x_M$	$\dfrac{1}{2\sqrt{x_M}}$	$1.84365320 = x_i$	$\dfrac{2x_i+1}{4\sqrt{x_i^3}}$
$f = \mathrm{daw}(x)$	$0.924138873 = \sqrt{x_M}$	$\dfrac{1}{2\sqrt{x_M}}$	$1.50197527 = x_c$	$\dfrac{x_c}{2x_c^2-1}$
$f = x\,\mathrm{daw}(x)$	$1.50197527 = x_c$	$\dfrac{x_c^2}{2x_c^2-1}$	$\begin{Bmatrix}0.595677278\\1.95562141\end{Bmatrix}$	$\dfrac{x^2-1}{2x^2-3}$

The particular arguments x_M and x_i, which correspond respectively to the maximum and inflection point of the $\mathrm{daw}(\sqrt{x})$ function, are cited in Section 37:14 as the lower limits of important indefinite integrals.

42:8 NUMERICAL VALUES

The algorithm below evaluates daw(x) to 24-bit precision (i.e., the relative error never exceeds 6×10^{-8}) for any input value of x.

For arguments in the range $0 < |x| < 5$, the algorithm uses a double concatenation formula, based on expansion 42:6:5. With J set to $1 + \mathrm{Int}(12\sqrt{|x|})$ and n to $\mathrm{Int}(x^2)$, the expression

42:8:1
$$f = \left(\left(\cdots\left(\left(\frac{x^2}{J(J+\frac{1}{2}-n)} + \frac{1}{J-\frac{1}{2}-n}\right)\frac{x^2}{J-1} + \frac{1}{J-\frac{3}{2}-n}\right)\frac{x^2}{J-2} + \cdots + \frac{1}{\frac{3}{2}-n}\right)\frac{x^2}{1} + \frac{1}{\frac{1}{2}-n}\right) x^2 \exp(-x^2)$$

is first evaluated. This value then initiates the second concatenation

42:8:2
$$\mathrm{daw}(x) = \left(\left(\left(\cdots\left(\left(\frac{(\frac{1}{2}-n)f}{-x^2} + 1\right)\frac{\frac{3}{2}-n}{-x^2} + 1\right)\frac{\frac{5}{2}-n}{-x^2} + \cdots + 1\right)\frac{-\frac{3}{2}}{-x^2} + 1\right)\frac{-\frac{1}{2}}{-x^2} + 1\right)\frac{1}{2x}$$

For $|x| \ge 5$, the asymptotic expansion 42:6:6 is employed, being suitably terminated and evaluated by the same set of commands that are used by 42:8:2.

Input x >>>>>>>

If $5 \le |x|$ go to (3)
Set $J = 1 + \mathrm{Int}(12\sqrt{|x|})$
Set $p = J + \frac{1}{2} - \mathrm{Int}(x^2)$
Set $f = x^2/p$
If $f = 0$ go to (5)
(1) Replace p by $p - 1$
Replace f by $[(f/J) + (1/p)]x^2$
Replace J by $J - 1$
If $J \neq 0$ go to (1)
Replace f by $f\exp(-x^2)$
(2) If $p > 0$ go to (4)
Replace f by $1 - (pf/x^2)$
Replace p by $p + 1$
Go to (2)
(3) Set $p = -\frac{3}{2} - \mathrm{Int}(30/|x|)$
Set $f = 1$
Go to (2)
(4) Replace f by $f/2x$
(5) Output f

$f \simeq \mathrm{daw}(x)$ <<<<<

Storage needed: x, J, p and f

Test values:
daw(0) = 0
daw(0.7) = 0.510504*058*
daw(−2) = −0.301340*389*
daw(10) = 0.0502538*471*

Alternatively, the universal hypergeometric algorithm of Section 18:14 may be utilized to evaluate Dawson's integral.

42:9 APPROXIMATIONS

For small and large arguments, one may use the approximations

42:9:1
$$\mathrm{daw}(x) \simeq \frac{x}{3}(3 - 2x^2) \qquad \text{8-bit precision} \qquad -0.34 \le x \le 0.34$$

and

42:9:2 $$\mathrm{daw}(x) \simeq \frac{2x^2+1}{4x^3} \qquad \text{8-bit precision} \qquad |x| \geq 3.9$$

respectively.

42:10 OPERATIONS OF THE CALCULUS

Differentiation gives

42:10:1 $$\frac{d}{dx}\mathrm{daw}(bx) = b - 2bx\,\mathrm{daw}(bx)$$

42:10:2 $$\frac{d}{dx}\mathrm{daw}(\sqrt{bx}) = \frac{1}{2}\sqrt{\frac{b}{x}} - b\,\mathrm{daw}(\sqrt{bx})$$

Repeated derivatives involve Hermite polynomials [Chapter 24]

42:10:3 $$\frac{d^n}{dx^n}\mathrm{daw}(x) = (-1)^n\,\mathrm{H}_n(x)\,\mathrm{daw}(x) - \begin{cases} 2^J \displaystyle\sum_{j=0}^{J} \frac{(J+j)!\mathrm{H}_{2j}(x)}{(-2)^j(2j)!} & n = 2J+1 = 1, 3, 5, \ldots \\ 2^J \displaystyle\sum_{j=0}^{J} \frac{(J+j)!\mathrm{H}_{2j+1}(x)}{(-2)^j(2j+1)!} & n = 2J = 2, 4, 6, \ldots \end{cases}$$

Indefinite integrals include

42:10:4 $$\int_0^x \mathrm{daw}(\sqrt{bt})dt = \sqrt{\frac{x}{b}} - \frac{\mathrm{daw}(\sqrt{bx})}{b}$$

and

42:10:5 $$\int_0^x t\,\mathrm{daw}(bt)dt = \frac{bx - \mathrm{daw}(bx)}{2b^2}$$

The indefinite integral $\int \mathrm{daw}(t)dt$ is not expressible in terms of any named function. It is, however, quite a simple hypergeometric function [see Section 18:14]

42:10:6 $$\int_0^x \mathrm{daw}(t)dt = \frac{x^2}{2}\sum_{j=0}^{\infty}\frac{(1)_j}{(\frac{3}{2})_j(2)_j}(-x^2)^j$$

The first three of the definite integrals listed below may be regarded as Laplace transforms [Section 26:14] but are included here because of the importance of these particular integrals:

42:10:7 $$\int_0^\infty \mathrm{daw}(\sqrt{bt})\exp(-st)dt = \frac{\sqrt{\pi b}}{2\sqrt{s(s+b)}}$$

42:10:8 $$\int_0^\infty \mathrm{daw}(bt)\exp(-st)dt = \frac{1}{4b}\exp\left(\frac{s^2}{4b^2}\right)\mathrm{Ei}\left(\frac{-s^2}{4b^2}\right) \qquad b>0 \qquad s>0$$

42:10:9 $$\int_0^\infty t\,\mathrm{daw}(bt)\exp(-st)dt = \frac{1}{2bs} + \frac{s}{8b^2}\mathrm{Ei}\left(\frac{-s^2}{4b^2}\right) \qquad b>0 \qquad s>0$$

42:10:10 $$\int_0^\infty \exp(-at^2)\,\mathrm{daw}(bt)dt = \sqrt{\frac{\pi}{a}}\frac{b}{4(b^2+a)} \qquad a>0 \qquad b>0$$

42:10:11 $$\int_0^\infty \sin(Bt)\,\mathrm{daw}(bt)dt = \frac{\sqrt{\pi^3}}{8b}\exp\left(\frac{-B^2}{4b}\right) \qquad B>0$$

With lower limit zero, the semiderivative and semiintegral of $\text{daw}(\sqrt{x})$ are

42:10:12
$$\frac{d^{1/2}}{dx^{1/2}}\text{daw}(\sqrt{x}) = \frac{\sqrt{\pi}}{2}\exp(-x)$$

and

42:10:13
$$\frac{d^{-1/2}}{dx^{-1/2}}\text{daw}(\sqrt{x}) = \frac{\sqrt{\pi}}{2}[1 - \exp(-x)]$$

42:11 COMPLEX ARGUMENT

The relationship of Dawson's integral to the error function of complex argument $W(x + iy)$ is discussed in Section 41:11. With a purely imaginary argument one finds

42:11:1
$$\text{daw}(iy) = \frac{i\sqrt{\pi}}{2}\exp(y^2)\,\text{erf}(y)$$

42:12 GENERALIZATIONS

Dawson's integral is a special case of the incomplete gamma function [Chapter 45]

42:12:1
$$\text{daw}(x) = \frac{\sqrt{\pi}}{2}x\exp(-x^2)\,\gamma^*(\tfrac{1}{2};-x^2)$$

and of the Kummer function [Chapter 47]

42:12:2
$$\text{daw}(x) = x\,M(1;\tfrac{3}{2};-x^2)$$

42:13 COGNATE FUNCTIONS

Dawson's integral is the $n = 2$ instance of functions defined by

42:13:1
$$\int_0^x \exp(t^n - x^n)dt \qquad n = 2, 3, 4, \ldots$$

The other members also have some importance. All display a peak in the vicinity of $x = 0.7$.

CHAPTER 43

THE GAMMA FUNCTION $\Gamma(x)$

The gamma function is unusual in the simplicity of its recurrence properties. It is because of this that the gamma function (and its special case, the factorial) plays such an important role in the theory of other functions. The reciprocal $1/\Gamma(x)$ and the logarithm $\ln(\Gamma(x))$ are also important and are discussed in this chapter, as is the related complete beta function $B(x,y)$, which is addressed in Section 43:13.

Formulas involving the gamma function often become simpler when written for argument $1 + x$ rather than x, and we have sometimes taken advantage of this fact. Because $\Gamma(x) = \Gamma(1 + x)/x$, a change of argument is readily achieved.

43:1 NOTATION

The gamma function is also known as *Euler's integral of the second kind.* $\Gamma(1 + x)$ is sometimes symbolized $x!$ or $\Pi(x)$ and termed the *factorial function* or *pi function*, respectively. To avoid possible confusion with the functions of Chapter 45, Γ is distinguished as the *complete gamma function*.

43:2 BEHAVIOR

The behavior of $\Gamma(x)$ for $-5 < x < 6$ is shown on the accompanying map, Figure 43-1; it is complicated. For positive argument, the gamma function passes through a shallow minimum between $x = 1$ and $x = 2$ and increases steeply as $x = 0$ or $x = \infty$ is approached. On the negative side, $\Gamma(x)$ is segmented: it has positive values for $-2 < x < -1$, $-4 < x < -3$, $-6 < x < -5$, ... but negative values for $-1 < x < 0$, $-3 < x < -2$, $-5 < x < -4$, ..., with discontinuities at $x = 0, -1, -2, \ldots$. The gamma function never takes the value zero, but it comes very close between consecutive large negative integers.

The reciprocal $1/\Gamma(x)$ has no discontinuities. It rapidly approaches zero as $x \to \infty$ and equals zero at $x = 0, -1, -2, \ldots$. As shown on the map, its oscillations become increasingly violent as the argument becomes more and more negative.

The logarithm $\ln(\Gamma(x))$ is usually only considered for $x > 0$. This is the convention adopted in drawing the map, which shows that $\ln(\Gamma(x))$ is a positive function except between $x = 1$ and $x = 2$, where it is briefly and slightly negative.

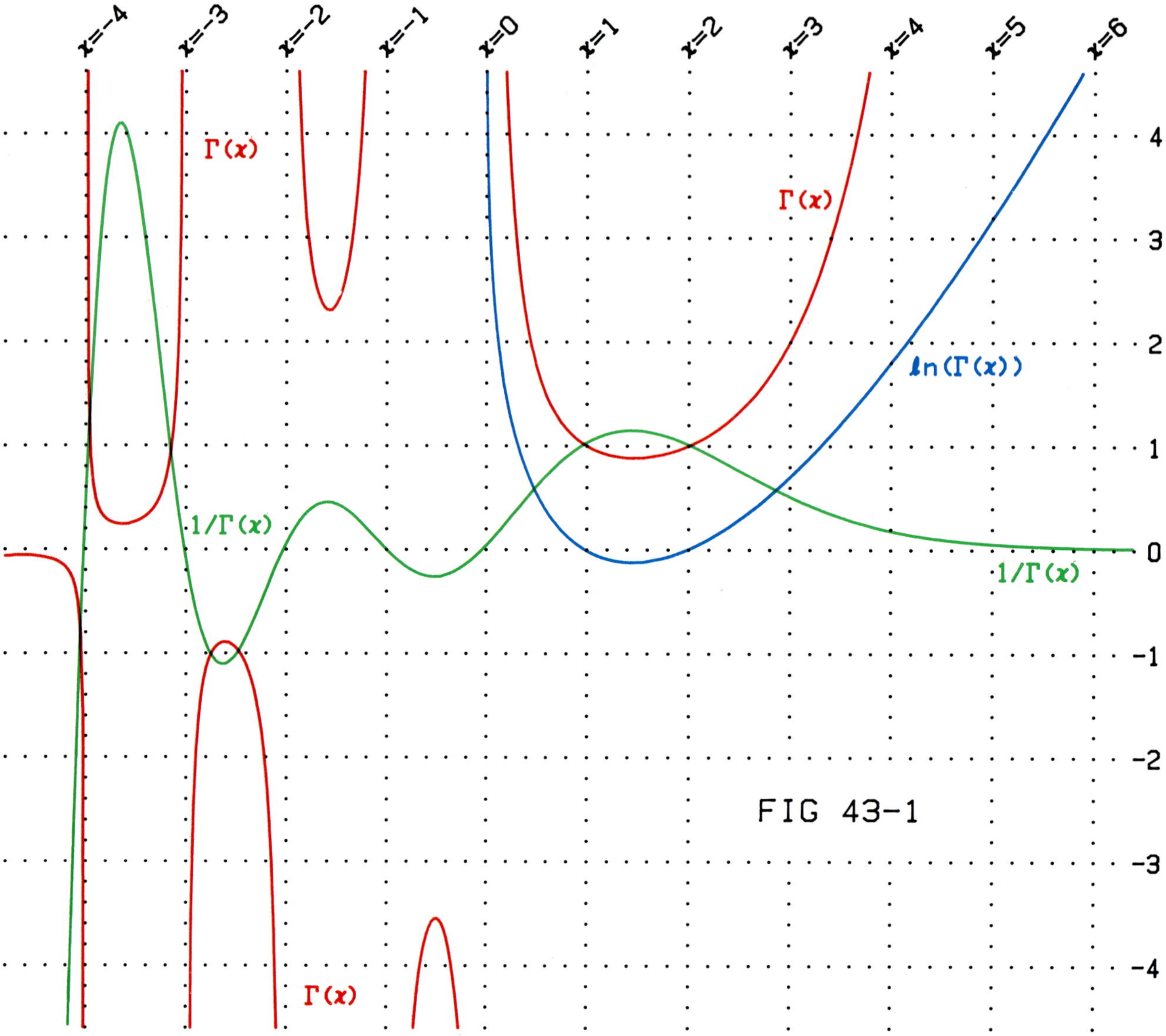

FIG 43-1

43:3 DEFINITIONS

Although it is restricted to positive arguments, the most useful definition of the gamma function is the *Euler integral*

43:3:1 $$\Gamma(x) = \int_0^\infty t^{x-1}\exp(-t)\mathrm{d}t \qquad x > 0$$

More comprehensive are the Gauss limit definition

43:3:2 $$\Gamma(x) = \lim_{n\to\infty}\left\{\frac{n^x}{x(1+x)\left(1+\frac{x}{2}\right)\cdots\left(1+\frac{x}{n-1}\right)\left(1+\frac{x}{n}\right)}\right\}$$

and the infinite product definition of Weierstrass

43:3:3 $$\frac{1}{\Gamma(x)} = x\exp(\gamma x)\prod_{j=1}^{\infty}\left(1+\frac{x}{j}\right)\exp\left(\frac{-x}{j}\right)$$

where γ is Euler's constant [Chapter 1].

The gamma function may be expressed as a definite integral in many ways apart from the one given above. Gradshteyn and Ryzhik [Section 8.31] give a long list of which the following are representative:

43:3:4 $$\Gamma(1 + x) = \int_0^1 \ln^x\left(\frac{1}{t}\right)dt \qquad x > -1$$

43:3:5 $$\Gamma(x) = s^x \int_0^\infty t^{x-1} \exp(-st)dt \qquad x > 0 \qquad s > 0$$

43:3:6 $$\Gamma(x) = s^x \sec\left(\frac{\pi x}{2}\right) \int_0^\infty t^{x-1} \cos(st)dt \qquad 0 < x < 1 \qquad s > 0$$

Alternatively, in 43:3:6 sec and cos may be replaced by csc and sin, respectively.

Likewise, there are several ways of representing the logarithm of the gamma function by means of integrals. One is

43:3:7 $$\ln(\Gamma(1 + x)) = \int_0^1 \left(\frac{t^x - 1}{t - 1} - x\right) \frac{dt}{\ln(t)} \qquad x > -1$$

and others may be found in Gradshteyn and Ryzhik [Section 8.34].

43:4 SPECIAL CASES

The gamma function reduces to the factorial function [Chapter 2] when its argument is a positive integer:

43:4:1 $$\Gamma(n) = (n - 1)! \qquad n = 1, 2, 3, \ldots$$

The gamma function of $\frac{1}{2}$ is $\sqrt{\pi}$, and the gamma function of an odd multiple of $\frac{1}{2}$ involves a double factorial function [Section 2:13]. Comprehensive formulas are

43:4:2 $$\Gamma(n + \tfrac{1}{2}) = (2n - 1)!!\,\Gamma(\tfrac{1}{2})/2^n = (2n - 1)!!\sqrt{\pi}/2^n \qquad n = 0, 1, 2, \ldots \qquad \sqrt{\pi} = 1.772453851$$

and

43:4:3 $$\Gamma(\tfrac{1}{2} - n) = (-2)^n\,\Gamma(\tfrac{1}{2})/(2n - 1)!! = (-2)^n\sqrt{\pi}/(2n - 1)!! \qquad n = 0, 1, 2, \ldots$$

Similar to 43:4:2 and 43:4:3 are the formulas

43:4:4 $$\Gamma(n + \tfrac{1}{3}) = (3n - 2)!!!\,\Gamma(\tfrac{1}{3})/3^n \qquad n = 0, 1, 2, \ldots$$

43:4:5 $$\Gamma(\tfrac{1}{3} - n) = (-3)^n\,\Gamma(\tfrac{1}{3})/(3n - 1)!!! \qquad n = 0, 1, 2, \ldots$$

43:4:6 $$\Gamma(n + \tfrac{2}{3}) = (3n - 1)!!!\,\Gamma(\tfrac{2}{3})/3^n = 2\pi(3n - 1)!!!/3^{n+1/2}\,\Gamma(\tfrac{1}{3}) \qquad n = 0, 1, 2, \ldots$$

and

43:4:7 $$\Gamma(\tfrac{2}{3} - n) = (-3)^n\,\Gamma(\tfrac{2}{3})/(3n - 2)!!! = 2\pi(-3)^n/\sqrt{3}(3n - 2)!!!\,\Gamma(\tfrac{1}{3}) \qquad n = 0, 1, 2, \ldots$$

involving the triple factorial function [Section 2:13] and where

43:4:8 $$\Gamma(\tfrac{1}{3}) = 2.678938535 \qquad \text{and} \qquad \Gamma(\tfrac{2}{3}) = \frac{2\pi}{\sqrt{3}\,\Gamma(\tfrac{1}{3})} = 1.354117939$$

Likewise, for arguments that are odd multiples of $\frac{1}{4}$, the gamma function takes the values

43:4:9 $$\Gamma(\tfrac{1}{4} + n) = (4n - 3)!!!!\,\Gamma(\tfrac{1}{4})/4^n \qquad n = 0, 1, 2, \ldots$$

43:4:10 $$\Gamma(\tfrac{3}{4} + n) = (4n - 1)!!!!\Gamma(\tfrac{3}{4})/4^n = \sqrt{2}\pi(4n - 1)!!!!/4^n\,\Gamma(\tfrac{1}{4}) \qquad n = 0, 1, 2, \ldots$$

43:4:11 $$\Gamma(\tfrac{1}{4} - n) = (-4)^n\,\Gamma(\tfrac{1}{4})/(4n - 1)!!!! \qquad n = 0, 1, 2, \ldots$$

or

43:4:12 $$\Gamma(\tfrac{3}{4} - n) = (-4)^n\Gamma(\tfrac{3}{4})/(4n - 3)!!!! = \sqrt{2}\pi(-4)^n/(4n - 3)!!!!\,\Gamma(\tfrac{1}{4}) \qquad n = 0, 1, 2, \ldots$$

involving the quadruple factorial function [Section 2:13] and where

43:4:13 $$\Gamma(\tfrac{1}{4}) = \pi^{3/4}\sqrt{\frac{2}{U}} = 3.625609908$$

or

43:4:14 $$\Gamma(\tfrac{3}{4}) = \frac{\sqrt{2}\pi}{\Gamma(\frac{1}{4})} = \pi^{1/4}\sqrt{U} = 1.225416702$$

U being the ubiquitous constant [see Section 1:7].

43:5 INTRARELATIONSHIPS

The gamma function obeys the reflection formulas

43:5:1 $$\Gamma(-x) = \frac{-\pi\csc(\pi x)}{x\Gamma(x)} = \frac{\pi\csc(-\pi x)}{\Gamma(1+x)}$$

and

43:5:2 $$\Gamma(\tfrac{1}{2} - x) = \frac{\pi\sec(\pi x)}{\Gamma(\frac{1}{2} + x)}$$

The recurrence formulas

43:5:3 $$\Gamma(1 + x) = x\Gamma(x)$$

and

43:5:4 $$\Gamma(x - 1) = \frac{\Gamma(x)}{x - 1}$$

generalize to

43:5:5 $$\Gamma(n + x) = x(1 + x)(2 + x)\cdots(n - 1 + x)\,\Gamma(x) = (x)_n\,\Gamma(x) \qquad n = 0, 1, 2, \ldots$$

and

43:5:6 $$\Gamma(x - n) = \frac{\Gamma(x)}{(x-1)(x-2)(x-3)\cdots(x-n)} = \frac{(-1)^n\,\Gamma(x)}{(1-x)_n} \qquad n = 0, 1, 2, \ldots$$

in terms of the Pochhammer polynomial [Chapter 18].

The duplication and triplication formulas

43:5:7 $$\Gamma(2x) = \frac{4^x}{2\sqrt{\pi}}\Gamma(x)\,\Gamma(\tfrac{1}{2} + x)$$

and

43:5:8 $$\Gamma(3x) = \frac{(27)^x}{2\pi\sqrt{3}}\Gamma(x)\,\Gamma(\tfrac{1}{3} + x)\Gamma(\tfrac{2}{3} + x)$$

are the $n = 2$ and 3 cases of the general *Gauss-Legendre formula*

43:5:9 $$\Gamma(nx) = \sqrt{\frac{2\pi}{n}\frac{n^{nx}}{(2\pi)^{n/2}}}\prod_{j=0}^{n-1}\Gamma\left(\frac{j}{n} + x\right) \qquad n = 2, 3, 4, \ldots$$

which applies for any positive integer multiplier n.

From 43:5:5 and 43:5:6, one may derive the expressions

43:5:10 $$\frac{\Gamma(n+x)}{\Gamma(x)} = (x)_n \qquad n = 1, 2, 3, \ldots$$

and

43:5:11 $$\frac{\Gamma(x-n)}{\Gamma(x)} = \frac{(-1)^n}{(1-x)_n} \qquad n = 1, 2, 3, \ldots$$

for the ratio of the gamma functions of two arguments that differ by an integer. These formulas may be used even when the individual gamma functions are infinite (e.g., $\Gamma(x-3)/\Gamma(x) \to -\frac{1}{6}$ as $x \to 0$).

Because of the frequent occurrence of the reciprocal gamma function in power series expansions of transcendental functions, particular values of the latter functions often serve as sums of infinite series of reciprocal gamma functions. For example, on account of expansion 41:6:5, we have

43:5:12 $$\frac{1}{\Gamma(\frac{1}{2})} + \frac{1}{\Gamma(1)} + \frac{1}{\Gamma(\frac{3}{2})} + \cdots = \sum_{j=1}^{\infty} \frac{1}{\Gamma(j/2)} = \frac{1}{\sqrt{\pi}} + e\,\mathrm{erfc}(-1) = 5.573169664$$

while, from the expansions in Section 45:6

43:5:13 $$\frac{1}{\Gamma(x)} - \frac{1}{\Gamma(x+1)} + \frac{1}{\Gamma(x+2)} - \cdots = \sum_{j=0}^{\infty} \frac{(-1)^j}{\Gamma(x+j)} = \frac{\gamma^*(x-1;-1)}{e}$$

43:6 EXPANSIONS

The power series expansions for the gamma function and its reciprocal are

43:6:1 $$[\Gamma(x)]^{\pm 1} = x^{\mp 1}\left\{1 \mp \gamma x + \left(\frac{\gamma^2}{2} \pm \frac{\pi^2}{12}\right)x^2 \cdots\right\} = \sum_{j=0}^{\infty} a_{\pm j} x^{j \mp 1} \qquad x \neq -1, -2, -3, \ldots$$

where $a_0 = 1$ and

43:6:2 $$a_{\pm j} = \frac{\mp \gamma a_{\pm j-1}}{j} \pm \frac{(-1)^j}{j} \sum_{k=0}^{j-2} (-1)^k \zeta(j-k) a_{\pm k} \qquad j = 1, 2, 3, \ldots \qquad \gamma = 0.5772156649$$

Here γ denotes Euler's constant [Chapter 1] and $\zeta(n)$ is the nth zeta number [Chapter 3]. Numerical values of a_0, $a_{-1}, a_{-2}, \ldots, a_{-25}$ are listed by Abramowitz and Stegun [page 256]. The power series expansion of the logarithm of the gamma function is less complicated:

43:6:3 $$\ln(\Gamma(1+x)) = -\gamma x + \frac{\pi^2}{12}x^2 - \cdots = -\gamma x + \sum_{j=2}^{\infty} \frac{\zeta(j)}{j}(-x)^j \qquad -1 < x \le 1$$

More rapidly convergent is the similar series

43:6:4 $$\ln(\Gamma(1+x)) = (1-\gamma)x + \frac{1}{2}\ln\left(\frac{\pi x(1-x)}{(1+x)\sin(\pi x)}\right) - \sum_{j} \frac{\zeta(j)-1}{j} x^j \qquad j = 3, 5, 7, \ldots \qquad -1 < x < 1$$

The gamma function may also be expanded as the infinite product

43:6:5 $$\Gamma(1+x) = \prod_{j=1}^{\infty} \frac{\left(1 + \frac{1}{j}\right)^x}{1 + \frac{x}{j}}$$

An asymptotic expansion of the gamma function is provided by *Stirling's formula*:

43:6:6 $$\Gamma(x) \sim \sqrt{\frac{2\pi}{x}}\exp(-x)\,x^x\left(1 + \frac{1}{12x} + \frac{1}{288x^2} - \frac{139}{51840x^3} - \cdots\right) \qquad x \to \infty$$

The corresponding asymptotic expansion for the gamma function's logarithm is

43:6:7 $$\ln(\Gamma(x)) \sim \ln(\sqrt{2\pi}) - x + \left(x - \frac{1}{2}\right)\ln(x) + \frac{1}{12x} - \frac{1}{360x^3} + \frac{1}{1260x^5} - \cdots + \frac{B_{2j}}{2j(2j-1)x^{2j-1}} + \cdots \qquad x \to \infty$$

where B_{2j} denotes a Bernoulli number [Chapter 4]. Though exact only in the $x \to \infty$ limit, these expansions are remarkably accurate for modest values of the argument [see 43:9:1, for example]. Relationship 43:6:7 is a special case of *Barnes' asymptotic expansion*

43:6:8 $$\ln(\Gamma(x+c)) \sim (x+c)\ln(x) - x + \frac{1}{2}\ln\left(\frac{2\pi}{x}\right) + \frac{6c^2 - 6c + 1}{12x} - \frac{2c^3 - 3c^2 + c}{12x^2} + \cdots - \frac{B_{j+1}(c)}{j(j+1)(-x)^j} + \cdots \qquad x \to \infty$$

where $B_{j+1}(c)$ denotes a Bernoulli polynomial [Chapter 19]. From this expansion one may also derive the useful expansion

43:6:9 $$\frac{\Gamma(x+c)}{\Gamma(x)} \sim x^c\left[1 + \frac{c(c-1)}{2x} + \frac{c(c-1)(c-2)(3c-1)}{24x^2} + \frac{c^2(c-1)^2(c-2)(c-3)}{48x^3} + \cdots\right] \qquad x \to \infty$$

for the ratio of two gamma functions of large, but not very different, arguments. For example:

43:6:10 $$\frac{\Gamma(x+\frac{1}{2})}{\sqrt{x}\,\Gamma(x)} \sim 1 - X + \frac{X^2}{2} + \frac{5X^3}{2} - \frac{21X^4}{8} - \frac{399X^5}{8} + \cdots \qquad X = \frac{1}{8x} \to 0$$

43:7 PARTICULAR VALUES

In addition to those reported in 43:4:14 and 43:4:15, there are the following:

x	-3	$\frac{-5}{2}$	-2	$\frac{-3}{2}$	-1	$\frac{-1}{2}$	0	$\frac{1}{2}$	1	$\frac{3}{2}$	2	$\frac{5}{2}$	3	$\frac{7}{2}$	4	∞
$\Gamma(x)$	$\pm\infty$	$\frac{-8\sqrt{\pi}}{15}$	$\pm\infty$	$\frac{4\sqrt{\pi}}{3}$	$\pm\infty$	$-2\sqrt{\pi}$	$\pm\infty$	$\sqrt{\pi}$	1	$\frac{\sqrt{\pi}}{2}$	1	$\frac{3\sqrt{\pi}}{4}$	2	$\frac{15\sqrt{\pi}}{8}$	6	∞

The arguments yielding local maxima or minima of $\Gamma(x)$ correspond to the zeros of the digamma function [see Section 44:7].

43:8 NUMERICAL VALUES

For $x \geq 3$, the following algorithm uses a truncated version of expansion 43:6:6 to calculate a sufficiently precise value of $\Gamma(x)$. For smaller arguments, the recurrence 43:5:4 is used repeatedly until x is brought within the $3 \leq x < 4$ range. Since the gamma functions of 0, -1, -2, ... are infinite, the algorithm returns a value of 10^{99} at these arguments. Apart from these cases, the algorithm has a precision of 24 bits (i.e., the relative error in $\Gamma(x)$ never exceeds 6×10^{-8}).

Set $f = 10^{99}$
Set $g = 1$

Storage needed: f, g and x

Input x >>>>>>>

(1) If $x = 0$ go to (2)
  Replace g by gx
  Replace x by $x + 1$
  If $x < 3$ go to (1)
  Set $f = \left[1 - \dfrac{2}{7x^2}\left(1 - \dfrac{2}{3x^2}\right)\right] \Big/ 30x^2$
  Replace f by $\dfrac{1-f}{12x} + x[\ln(x) - 1]$
  Replace f by $\dfrac{\exp(f)}{g}\sqrt{\dfrac{2\pi}{x}}$
(2) Output f

$f \simeq \Gamma(x)$ <<<<<<<

Test values:
$\Gamma(0.6) = 1.48919225$
$\Gamma(7.4) = 1541.33619$
$\Gamma(-4.2) = -0.164061051$

43:9 APPROXIMATIONS

From 43:6:5, the *Stirling approximation*

43:9:1 $$\Gamma(x) = \sqrt{\frac{2\pi}{x}}\exp(-x)x^x\left(1 + \frac{1}{12x}\right) \qquad \text{8-bit precision} \qquad x \geq 0.4$$

may be derived by truncation. By splitting the argument x into its integer-value $I = \text{Int}(x)$ and fractional-value $f = \text{frac}(x)$ components [Chapter 9], the gamma function may be approximated by

43:9:2 $$\Gamma(x) = \frac{I!}{I^{1-f}}\left[1 - \frac{f(1-f)}{2I}\right] \qquad \text{8-bit precision} \qquad x \geq 2$$

an expression derived from 43:6:9 that permits the gamma function to be expressed in terms of the factorial function [Chapter 2].

Close to its zeros, the reciprocal gamma function is approximated by

43:9:3 $$\frac{1}{\Gamma(x)} = (-1)^n n!(x+n)[1 - (x+n)\psi(n+1)] \qquad x \simeq -n = 0, -1, -2, \ldots$$

where ψ denotes the digamma function [Chapter 44].

43:10 OPERATIONS OF THE CALCULUS

Differentiation gives

43:10:1 $$\frac{d}{dx}\Gamma(x) = \psi(x)\,\Gamma(x)$$

43:10:2 $$\frac{d}{dx}\ln(\Gamma(x)) = \psi(x)$$

43:10:3 $$\frac{d^n}{dx^n}\ln(\Gamma(x)) = \psi^{(n-1)}(x) \qquad n = 1, 2, 3, \ldots$$

where ψ and $\psi^{(n)}$ are the digamma and polygamma functions [Chapter 44].

Few simple integrals involving the gamma function have been established, although there are numerous integrals of products and quotients of gamma functions [see Gradshteyn and Ryzhik, Sections 6.41 and 6.42]. A number of integrals involving the logarithm of the gamma function, and including

43:10:4 $$\int_x^{x+1} \ln(\Gamma(t))dt = \ln(\sqrt{2\pi}) + x\ln(x) - x \qquad x > 0$$

are also listed by Gradsheyn and Ryzhik [Section 6.44]. The latter formula leads to

43:10:5 $$\int_1^n \ln(\Gamma(t))dt = \frac{n-1}{2}[\ln(2\pi) - n] + \sum_{j=2}^{n-1} j\ln(j) \qquad n = 2, 3, 4, \ldots$$

43:11 COMPLEX ARGUMENT

The gamma function with complex argument has real and imaginary components:

43:11:1 $$\Gamma(x + iy) = [\cos(\theta) + i\sin(\theta)]|\Gamma(x)| \prod_{j=0}^{\infty} \frac{|j + x|}{\sqrt{y^2 + (j + x)^2}}$$

$$\text{where} \qquad \theta = y\psi(x) + \sum_{j=0}^{\infty} \frac{y}{j + x} - \arctan\left(\frac{y}{j + x}\right)$$

Here ψ is the digamma function [Chapter 44]. Tables from which $\Gamma(x + iy)$ may be evaluated are given by Abramowitz and Stegun [pages 277–287]. It follows from 43:11:1 that the product $\Gamma(x + iy)\Gamma(x - iy)$ is real; special cases are

43:11:2 $$\Gamma(1 + iy)\Gamma(1 - iy) = \pi y\,\mathrm{csch}(\pi y)$$

and

43:11:3 $$\Gamma(\tfrac{1}{2} + iy)\Gamma(\tfrac{1}{2} - iy) = \pi\,\mathrm{sech}(\pi y)$$

For purely imaginary argument

43:11:4 $$\Gamma(iy) = \sqrt{\frac{\pi}{y}\,\mathrm{csch}(\pi y)}\,[\sin(\theta) - i\cos(\theta)] \qquad \theta = -\gamma y + \sum_{j=1}^{\infty} \frac{y}{j} - \arctan\left(\frac{y}{j}\right)$$

where γ is Euler's constant [Section 1:7]

43:12 GENERALIZATIONS

The gamma function is a special case of the incomplete gamma function $\gamma(\nu;x)$

43:12:1 $$\Gamma(\nu) = \gamma(\nu;\infty)$$

and of its complement $\Gamma(\nu;x)$

43:12:2 $$\Gamma(\nu) = \Gamma(\nu;0)$$

both of which are discussed in Chapter 45. Moreover, these two functions sum to give the gamma function

43:12:3 $$\gamma(\nu;x) + \Gamma(\nu;x) = \Gamma(\nu)$$

for all values of x.

43:13 COGNATE FUNCTIONS

The functions discussed in Chapters 2, 6, 18, 44 and 45 are all closely related to the gamma function.

Another important function that is intimately related to the gamma function is the *complete beta function* $\mathrm{B}(x,y)$. Also known as *Euler's integral of the first kind* or simply as the *beta function*, it is defined by the Euler integral

43:13:1 $$B(x,y) = \int_0^1 t^{x-1}(1-t)^{y-1}dt \qquad x>0 \qquad y>0$$

and is related to the gamma function through

43:13:2 $$B(x,y) = \frac{\Gamma(x)\Gamma(y)}{\Gamma(x+y)} = B(y,x)$$

As with the gamma function, the complete beta function may be expressed as a definite integral in many ways other than 43:13:1. These include

43:13:3 $$B(x,y) = \int_0^\infty \frac{t^{x-1}+t^{y-1}}{(1+t)^{x-y}}dt \qquad x<0 \qquad y>0$$

and

43:13:4 $$B(x,y) = 2\int_0^1 t^{2x-1}(1-t^2)^{y-1}dt = 2\int_0^{\pi/2} \sin^{2x-1}(t)\cos^{2y-1}(t)dt \qquad x>0 \qquad y>0$$

and extensive lists may be found in Gradshteyn and Ryzhik [Section 8.38] and in Magnus, Oberhettinger and Soni [page 7]. The integral representations of $B(x,y)$ apply only when both arguments are positive, but relationship 43:13:2 extends the definition to any pair of real arguments, as does the infinite-product expression

43:13:5 $$B(x,y) = \prod_{j=0}^{\infty} \frac{(j+1)(x+y+j)}{(x+j)(y+j)}$$

If the two arguments of the complete beta function are equal in magnitude, or sum to a moiety, we have the special cases

43:13:6 $$\frac{4^x}{2}B(x,x) = \cos(\pi x)B\left(x,\frac{1}{2}-x\right) = B\left(\frac{1}{2},x\right) = \sum_{j=0}^{\infty}\frac{(2j-1)!!}{(2j)!!(j+x)}$$

and

43:13:7 $$B(x,-x) = \begin{cases} (-1)^x/x & x = 0, \pm1, \pm2, \ldots \\ 0 & \text{otherwise} \end{cases}$$

Important special cases of 43:13:6 are $B(\frac{1}{4},\frac{1}{4}) = 2\pi/U$, $B(\frac{1}{2},\frac{1}{2}) = \pi$ and $B(\frac{3}{4},\frac{3}{4}) = 2U$, where U is the ubiquitous constant [Section 1:7]. When one (or both) of the arguments is a positive integer, the reciprocal of the complete beta function reduces to a binomial coefficient [Chapter 6]

43:13:8 $$\frac{1}{B(x,n)} = x\binom{n+x-1}{n-1}$$

Intrarelationships of complete beta functions, such as

43:13:9 $$B(x+1,y) = \frac{x}{x+y}B(x,y) = \frac{x}{y}B(x,y+1)$$

as well as expansions, such as

43:13:10 $$B(x,y) = \frac{1}{x} + \frac{1-y}{1+x} + \frac{(1-y)(2-y)}{2(2+x)} + \cdots = \sum_{j=0}^{\infty}\frac{(1-y)_j}{j!(x+j)} \qquad y>0$$

and infinite sums, such as

43:13:11 $$B(x,y) + B(x+1,y) + B(x+2,y) + \cdots = \sum_{j=0}^{\infty} B(x+j,y) = B(x,y-1)$$

may be established via the identity 43:13:2. The same expression may be used to calculate numerical values of $B(x,y)$. The argument values that lead to $B(x,y) = 0$ and $B(x,y) = \pm\infty$ are indicated by green and black lines,

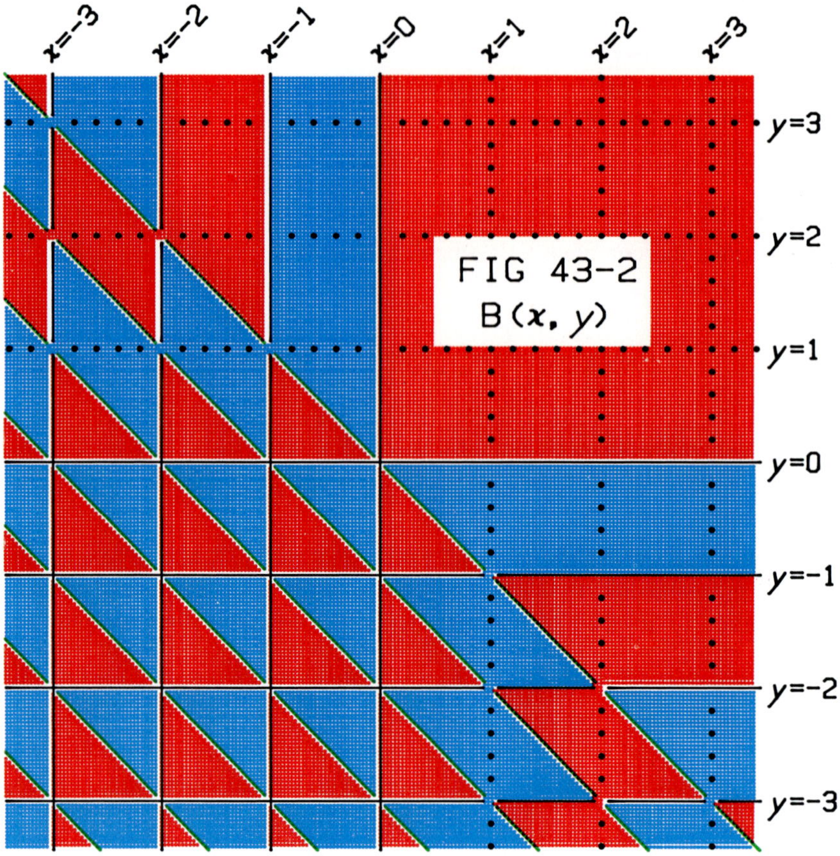

respectively, in Figure 43-2. The areas shaded in red on this map show where $B(x,y)$ is positive; blue shading indicates that the complete beta function is negative.

43:14 RELATED TOPICS

Together with very few others, the functions

43:14:1
$$\sum_{j=0}^{\infty} \frac{X^j}{(1)_j^2} = \begin{cases} I_0(2\sqrt{X}) & X \geq 0 \\ J_0(2\sqrt{-X}) & X \leq 0 \end{cases}$$

43:14:2
$$\sum_{j=0}^{\infty} \frac{X^j}{(1)_j} = \exp(X)$$

43:14:3
$$\sum_{j=0}^{\infty} X^j = \frac{1}{1-X} \qquad -1 < X < 1$$

and the asymptotic representation

43:14:4
$$1 + X + 2X^2 + 6X^3 + \cdots + (1)_j X^j + \cdots \sim \frac{1}{X}\exp\left(\frac{-1}{X}\right) \mathrm{Ei}\left(\frac{1}{X}\right) \qquad |X| \text{ small}$$

can claim a distinguished role in the theory of transcendental functions. The Bessel functions I_0 and J_0 occurring in 43:14:1 are treated in Chapters 49 and 52, respectively, while the exponential integral Ei is the subject of Chapter 37. These four have been termed *basis hypergeometric functions* because they can be employed to generate almost *any* hypergeometric function (i.e., any of the functions tabulated in Section 18:14), as will now be demonstrated.

Functions 43:14:1–43:14:4 have been expressed in terms of $(1)_j$, the Pochhammer polynomial [Chapter 18] of unity argument, but since $(1)_j = j! = \Gamma(1+j)$, each of these functions can be written alternatively as $\Sigma X^j/\Gamma^n(1+j)$ with $n = 2, 1, 0$ or -1. Here X represents the argument x of interest, or some simple variant of it such as $-2x$ or $x^2/4$. Now, let $\underset{\mu}{\overset{\nu}{\rightarrow}}$ represent the following sequence of operations: (a) division by $X^\mu\Gamma(1-\mu)$, (b) differintegration [see Section 0:10] to order $\nu - \mu$ with respect to X using lower limit zero and (c) multiplication by $X^\nu\Gamma(1-\nu)$. Then, making use of equations 13:10:11 and 43:5:10, we find for the $\nu = 0$ case

43:14:5
$$\sum_{j=0}^{\infty}\frac{X^j}{(1)_j^n}\underset{\mu}{\overset{0}{\rightarrow}}\sum_{j=0}^{\infty}\frac{(1-\mu)_jX^j}{(1)_j^{n+1}}$$

while for the $\mu = 0$ case

43:14:6
$$\sum_{j=0}^{\infty}\frac{X^j}{(1)_j^n}\underset{0}{\overset{\nu}{\rightarrow}}\sum_{j=0}^{\infty}\frac{X^j}{(1)_j^{n-1}(1-\nu)_j}$$

In the general case when neither ν nor μ is necessarily zero

43:14:7
$$\sum_{j=0}^{\infty}\frac{X^j}{(1)_j^n}\underset{\mu}{\overset{\nu}{\rightarrow}}\sum_{j=0}^{\infty}\frac{(1-\mu)_jX^j}{(1)_j^n(1-\nu)_j}$$

In the terminology of Section 18:14, the $\underset{\mu}{\overset{\nu}{\rightarrow}}$ operation is seen to preserve the difference $L - K$ between the numbers of denominatorial and numeratorial parameters but to alter the values of some of those parameters. Hence, starting with the appropriate basis hypergeometric function, one may "synthesize" a sought hypergeometric function by one or more $\underset{\mu}{\overset{\nu}{\rightarrow}}$ operations. Examples are

43:14:8
$$I_0(x) = \sum_{j=0}^{\infty}\frac{(x^2/4)^j}{(1)_j^2}\underset{0}{\overset{\frac{1}{2}}{\rightarrow}}\sum_{j=0}^{\infty}\frac{(x^2/4)^j}{(1)_j(\frac{1}{2})_j} = \cosh(x) \qquad x = 2\sqrt{X}$$

43:14:9
$$\exp(2x) = \sum_{j=0}^{\infty}\frac{(2x)^j}{(1)_j}\underset{-\frac{1}{2}}{\overset{-2}{\longrightarrow}}\sum_{j=0}^{\infty}\frac{(\frac{3}{2})_j(2x)^j}{(1)_j(3)_j} = \frac{2}{x}\exp(x)\,I_1(x) \qquad x = \frac{X}{2}$$

43:14:10
$$\frac{1}{1-x^2} = \sum_{j=0}^{\infty}(x^2)^j\underset{\frac{1}{2}}{\overset{0}{\rightarrow}}\sum_{j=0}^{\infty}\frac{(\frac{1}{2})_j(x^2)^j}{(1)_j}\underset{\frac{3}{2}}{\overset{0}{\rightarrow}}\sum_{j=0}^{\infty}\frac{(\frac{1}{2})_j(\frac{-1}{2})_j(x^2)^j}{(1)_j^2} = \frac{2}{\pi}E(x) \qquad x = \sqrt{X}$$

and

43:14:11
$$x\exp(-x)\,\mathrm{Ei}(x) \sim 1 + \frac{1}{x} + \frac{2}{x^2} + \cdots + \frac{(1)_j}{x^j} + \cdots \underset{\frac{1}{2}}{\overset{0}{\rightarrow}} 1 + \frac{1}{2x} + \frac{3}{4x^2}$$
$$+ \cdots + \frac{(\frac{1}{2})_j}{x^j} + \cdots \sim 2\sqrt{x}\,\mathrm{daw}(\sqrt{x}) \qquad x = \frac{1}{X}$$

where cosh, I_1, E and daw are the functions of Chapters 28, 49, 61 and 42, respectively. Further information on this topic will be found in Oldham and Spanier [Chapter 9].

CHAPTER 44

THE DIGAMMA FUNCTION $\psi(x)$

The digamma function is derived by differentiation of the gamma function of Chapter 43. Multiple differentiation yields the *polygamma functions* that are addressed in Section 44:12. Digamma and polygamma functions are useful in summing certain algebraic series, a subject discussed in Section 44:14.

44:1 NOTATION

The digamma function is also known as the *psi function*. Some authors employ a translated argument and denote our $\psi(x)$ by $\psi(x-1)$.

44:2 BEHAVIOR

Figure 44-1 includes a map showing the global behavior of the digamma function. The function $\psi(x)$ has an infinite discontinuity at $x = 0$ and at each negative integer argument. Otherwise, $\mathrm{d}\psi(x)/\mathrm{d}x$ is always positive, this gradient decreasing steadily as x increases through positive values. As $x \to \infty$, $\psi(x)$ approaches infinity logarithmically. Figure 44-2 shows the detailed behavior of $\psi(x)$, and some other functions, for $0.5 \le x \le 5.5$.

44:3 DEFINITIONS

The digamma function is the derivative of the logarithm of the gamma function [Chapter 43]

44:3:1 $$\psi(x) = \frac{\mathrm{d}}{\mathrm{d}x}\ln(\Gamma(x)) = \frac{1}{\Gamma(x)}\frac{\mathrm{d}}{\mathrm{d}x}\Gamma(x)$$

A second definition is as the limit

44:3:2 $$\psi(x) = \lim_{n\to\infty}\left\{\ln(n) - \sum_{j=0}^{n}\frac{1}{j+x}\right\}$$

Representations as definite integrals are numerous and include

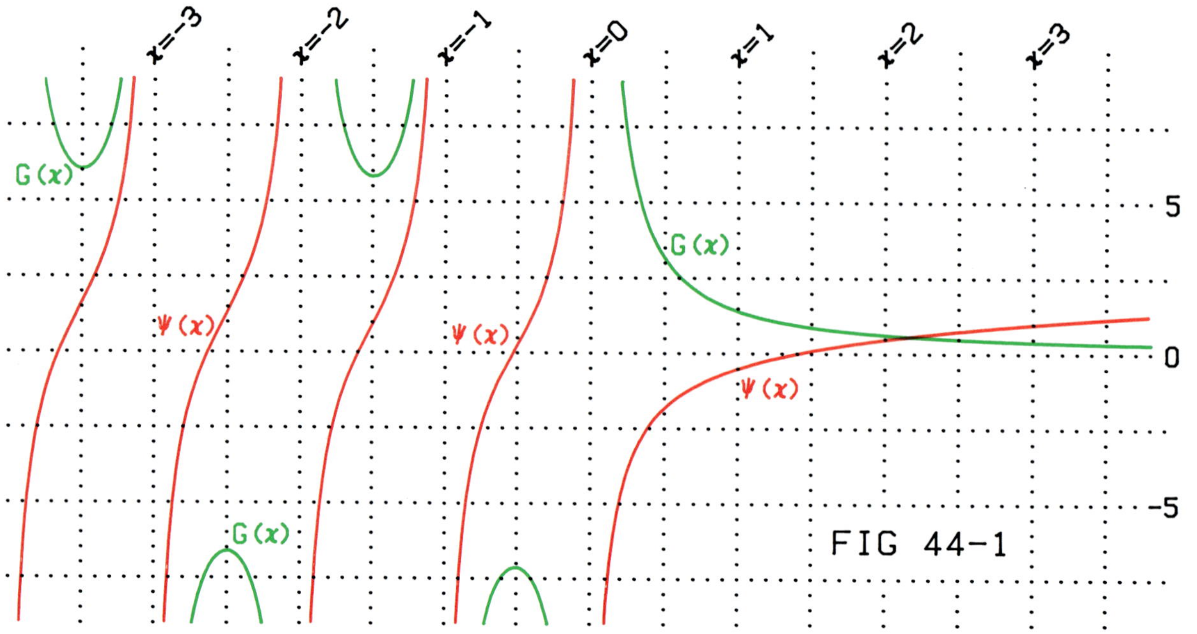

FIG 44-1

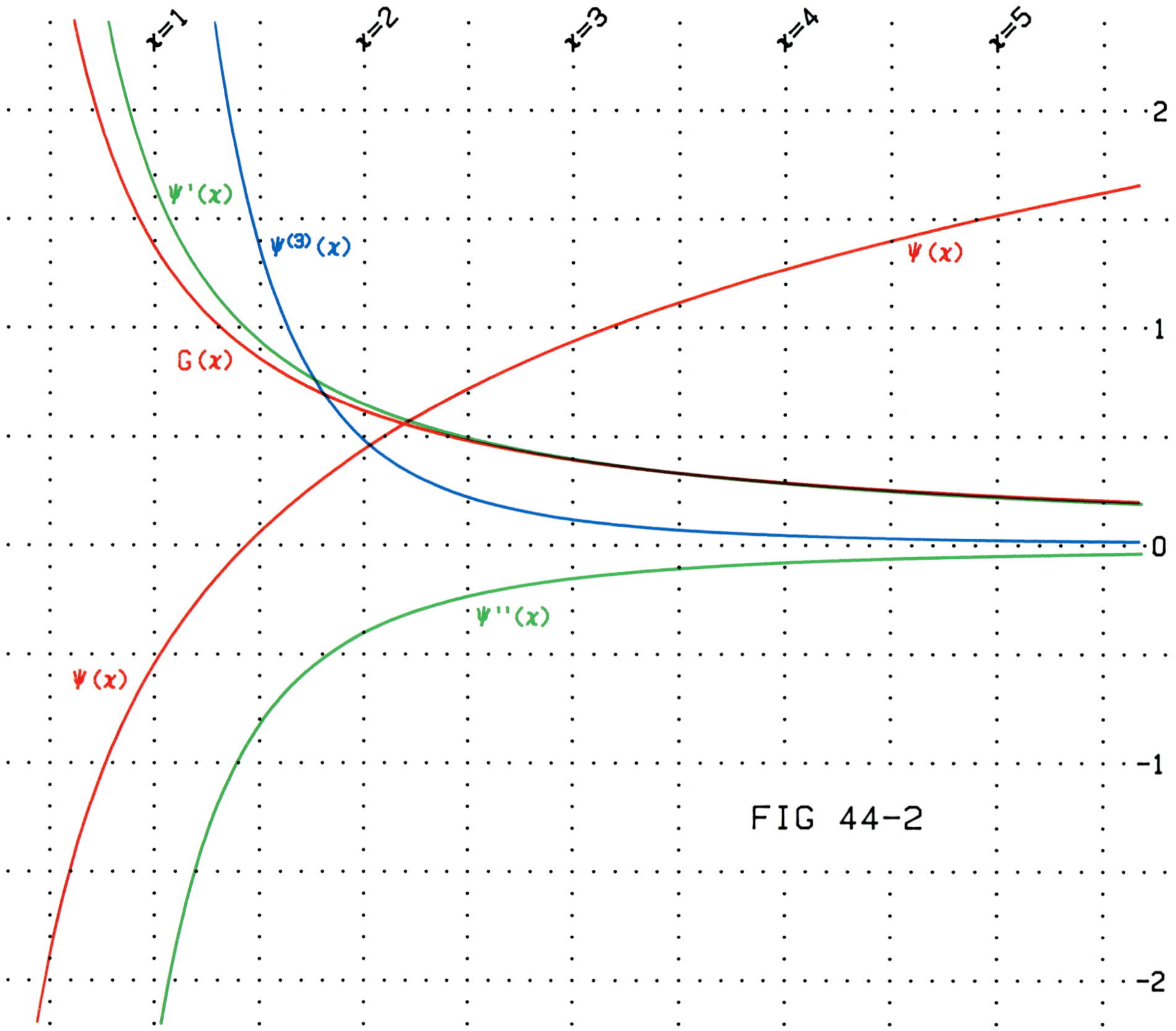

FIG 44-2

44:3:3 $$\psi(x) = -\gamma + \int_0^1 \frac{1 - t^{x-1}}{1 - t}\,dt = -\gamma + \int_0^\infty \frac{\exp(-t) - \exp(-xt)}{1 - \exp(-t)}\,dt \qquad x > 0$$

where γ is Euler's constant [Section 1:7]. The integrals

44:3:4 $$\psi(x) = \int_0^\infty \left[\frac{\exp(-t)}{t} - \frac{\exp(-xt)}{1 - \exp(-t)}\right] dt = \int_0^\infty \frac{\exp(-t) - (1 + t)^{-x}}{t}\,dt \qquad x > 0$$

and

44:3:5 $$\psi(x) = \ln(x) - \frac{1}{2x} - 2\int_0^\infty \frac{t\,dt}{(t^2 + x^2)[\exp(2\pi t) - 1]} \qquad x > 0$$

are attributed respectively to Gauss, Dirichlet and Binet. For other definitions of $\psi(x)$ as a definite integral, see Erdélyi, Magnus, Oberhettinger and Tricomi [*Higher Transcendental Functions*, Volume 1, Section 1.7.2.].

44:4 SPECIAL CASES

When its argument is a positive integer, the digamma function is given by

44:4:1 $$\psi(n) = -\gamma + 1 + \tfrac{1}{2} + \tfrac{1}{3} + \cdots + \frac{1}{n-1} = -\gamma + \sum_{j=1}^{n-1} \frac{1}{j} \qquad n = 2, 3, 4, \ldots \qquad \gamma = 0.5772156649$$

an expression involving Euler's constant γ and the first $n - 1$ terms of the harmonic series. The function $\psi(n + 1) + \gamma$ is sometimes denoted $\Phi(n)$.

If x is a rational number (i.e., capable of being expressed as a ratio m/n of two integers) then $\psi(x)$ is expressible as a finite sum of terms involving simpler functions (logarithmic and trigonometric) and constants (Archimedes' number π and Euler's constant γ). For $0 < x = m/n < 1$, where at least one of m and n is odd, the *theorem of Gauss* states

44:4:2 $$\psi\left(\frac{m}{n}\right) = -\gamma - \frac{\pi}{2}\cot\left(\frac{m\pi}{n}\right) + 2\sum_{j=1}^{J}\left[\cos\left(\frac{2jm\pi}{n}\right)\ln\left(2\sin\left(\frac{j\pi}{n}\right)\right)\right] - \begin{cases} \ln(2n) & n = 2J + 2 = 2, 4, 6, \ldots \\ \ln(n) & n = 2J + 1 = 3, 5, 7, \ldots \end{cases}$$

If $m > n$, recursion 44:5:5 should be used before 44:4:2.

44:5 INTRARELATIONSHIPS

The reflection formulas

44:5:1 $$\psi(1 - x) = \psi(x) + \pi\cot(\pi x)$$

and

44:5:2 $$\psi\left(\frac{1}{2} - x\right) = \psi\left(\frac{1}{2} + x\right) - \pi\tan(\pi x)$$

apply for any value of x, whereas the formula

44:5:3 $$\psi\left(\frac{1}{2} - J\right) = \psi\left(\frac{1}{2} + J\right) = -\gamma - \ln(4) + \sum_{j=1}^{J} \frac{2}{2j - 1} \qquad J = 0, 1, 2, \ldots$$

is restricted to nonnegative integer J.

The recurrence relationship

44:5:4
$$\psi(1 + x) = \psi(x) + \frac{1}{x}$$

may be generalized to

44:5:5
$$\psi(n + x) = \psi(x) + \sum_{j=0}^{n-1} \frac{1}{j + x} \qquad n = 1, 2, 3, \ldots$$

Similarly, the duplication formula

44:5:6
$$\psi(2x) = \ln(2) + \frac{1}{2}\psi(x) + \frac{1}{2}\psi\left(x + \frac{1}{2}\right)$$

generalizes to

44:5:7
$$\psi(nx) = \ln(n) + \frac{1}{n}\sum_{j=0}^{n-1} \psi\left(x + \frac{j}{n}\right) \qquad n = 2, 3, 4, \ldots$$

The difference between two digamma functions may be expressed as

44:5:8
$$\psi(x) - \psi(y) = (x - y)\sum_{j=0}^{\infty} \frac{1}{(j + x)(j + y)}$$

or as a hypergeometric function [Section 18:14] of unit argument:

44:5:9
$$\psi(x) - \psi(y) = \frac{x - y}{x}\sum_{j=0}^{\infty} \frac{(1)_j(x - y + 1)_j}{(2)_j(x + 1)_j} \qquad y > 0$$

Infinite sums of quotients of digamma functions by gamma functions include

44:5:10
$$-\frac{\psi(1)}{\Gamma(1)} + \frac{\psi(2)}{\Gamma(2)} - \frac{\psi(3)}{\Gamma(3)} + \cdots = \sum_{j=1}^{\infty} \frac{(-1)^j\psi(j)}{\Gamma(j)} = \frac{\mathrm{Ei}(1)}{e} = 0.697174883$$

and

44:5:11
$$-\frac{\psi(1)}{\Gamma^2(1)} + \frac{\psi(2)}{\Gamma^2(2)} - \frac{\psi(3)}{\Gamma^2(3)} + \cdots = \sum_{j=1}^{\infty} \frac{(-1)^j\psi(j)}{\Gamma^2(j)} = \frac{\pi Y_0(2)}{2} = 0.801696703$$

where Ei is the exponential integral function [Chapter 37] and Y_0 is the zero-order Neumann function [Chapter 54]. The corresponding series without alternating signs $\Sigma\psi(j)/\Gamma^n(j)$ sum to $-e\mathrm{Ei}(-1) = 0.596347362$ and $K_0(2) = 0.113893873$ for $n = 1$ and 2, respectively, K_0 being the zero-order Basset function [Chapter 51].

44:6 EXPANSIONS

The Taylor expansion

44:6:1
$$\psi(x + 1) = -\gamma - \sum_{j=1}^{\infty} \zeta(j + 1)(-x)^j \qquad -1 < x < 1$$

involves the zeta numbers [Chapter 3] and is valid only for arguments between 0 and 2. Of wider validity are the expansions

44:6:2
$$\psi(x) = -\gamma + \sum_{j=0}^{\infty}\left[\frac{1}{j + 1} - \frac{1}{j + x}\right] = -\gamma + \sum_{j=0}^{\infty} \frac{x - 1}{(j + 1)(j + x)} = -\gamma - \frac{1}{x} + x\sum_{j=1}^{\infty} \frac{1}{j(j + x)}$$

44:6:3
$$\psi(x) = \ln(x) - \sum_{j=0}^{\infty}\left[\frac{1}{j + x} - \ln\left(1 + \frac{1}{j + x}\right)\right] \qquad x > 0$$

and

44:6:4 $$\psi(x) = -\gamma - \frac{x^2}{1-x^2} - \frac{\pi}{2}\cot(\pi x) - \frac{1}{x}\left[\frac{1}{2} + \sum_k (\zeta(k) - 1)x^k\right] \qquad -2 < x < 2 \qquad k = 3, 5, 7, \ldots$$

An asymptotic expansion for the digamma function, valid for large argument, is

44:6:5 $$\psi(x) \sim \ln(x) - \frac{1}{2x} - \frac{1}{12x^2} + \frac{1}{120x^4} - \frac{1}{252x^6} + \cdots - \frac{B_j}{jx^j} + \cdots \qquad x \to \infty$$

where B_j is a Bernoulli number [Chapter 4].

44:7 PARTICULAR VALUES

Euler's constant, $\gamma = 0.5772156649$, is a component of most particular values of the digamma function. Table 44.7.1 includes particular values of the trigamma function [Section 44:12] and of Bateman's G function [Section 44:13], as well as those of the digamma function $\psi(x)$.

By exploiting formulas 44:4:1 and 44:5:4, it is possible to evaluate the digamma function of *any* rational argument $x = m/n$. For small values of n, rather simple expressions result. In addition to those incorporated into equation 44:5:3, we have the following examples:

44:7:1 $$\psi\left(\frac{1}{3} + J\right) + \frac{\pi}{2\sqrt{3}} = \psi\left(\frac{2}{3} - J\right) - \frac{\pi}{2\sqrt{3}} = -\gamma - \ln\sqrt{27} + \sum_{j=1}^{J} \frac{3}{3j-2} \qquad J = 0, 1, 2, \ldots$$

44:7:2 $$\psi\left(\frac{2}{3} + J\right) - \frac{\pi}{2\sqrt{3}} = \psi\left(\frac{1}{3} - J\right) + \frac{\pi}{2\sqrt{3}} = -\gamma - \ln\sqrt{27} + \sum_{j=1}^{J} \frac{3}{3j-1} \qquad J = 0, 1, 2, \ldots$$

44:7:3 $$\psi\left(\frac{1}{4} + J\right) + \frac{\pi}{2} = \psi\left(\frac{3}{4} - J\right) - \frac{\pi}{2} = -\gamma - \ln(8) + \sum_{j=1}^{J} \frac{4}{4j-3} \qquad J = 0, 1, 2, \ldots$$

44:7:4 $$\psi\left(\frac{3}{4} + J\right) - \frac{\pi}{2} = \psi\left(\frac{1}{4} - J\right) + \frac{\pi}{2} = -\gamma - \ln(8) + \sum_{j=1}^{J} \frac{4}{4j-1} \qquad J = 0, 1, 2, \ldots$$

The digamma function encounters zeros at argument values of $+1.461632145$, -0.504083008, -1.57349847, -2.61072087, ... Denoting these zeros by r_0, r_1, r_2, r_3, ... then the following approximation holds:

44:7:5 $$r_j \simeq -j + \frac{1}{\pi}\arctan\left(\frac{\pi}{\ln(j)}\right) \qquad \text{large } j$$

44:8 NUMERICAL VALUES

The algorithm below has an *absolute* accuracy of better than 1×10^{-8}, except when x is a nonpositive integer, in which case the algorithm returns 10^{99} instead of $\pm\infty$. For $x \geq 5$, the algorithm uses the expansion 44:6:5 in a

Table 44.7.1

	$x = -2$	$x = \frac{-3}{2}$	$x = -1$	$x = \frac{-1}{2}$	$x = 0$	$x = \frac{1}{2}$	$x = 1$	$x = \frac{3}{2}$	$x = 2$	$x = \frac{5}{2}$	$x = 3$	$x = \infty$
$\psi(x)$	$\pm\infty$	$\frac{8}{3} - \gamma - \ln(4)$	$\pm\infty$	$2 - \gamma - \ln(4)$	$\pm\infty$	$-\gamma - \ln(4)$	$-\gamma$	$2 - \gamma - \ln(4)$	$1 - \gamma$	$\frac{8}{3} - \gamma - \ln(4)$	$\frac{3}{2} - \gamma$	∞
$\psi'(x)$	∞	$\frac{\pi^2}{2} + \frac{40}{9}$	∞	$\frac{\pi^2}{2} + 4$	∞	$\frac{\pi^2}{2}$	$\frac{\pi^2}{6}$	$\frac{\pi^2}{2} - 4$	$\frac{\pi^2}{6} - 1$	$\frac{\pi^2}{2} - \frac{40}{9}$	$\frac{\pi^2}{6} - \frac{5}{4}$	0
$G(x)$	$\pm\infty$	$\pi + \frac{8}{3}$	$\pm\infty$	$-4 - \pi$	$\pm\infty$	π	$\ln(4)$	$4 - \pi$	$2 - \ln(4)$	$\pi - \frac{8}{3}$	$\ln(4) - 1$	0

truncated and compensated version. For smaller arguments, the algorithm uses the recursion $\psi(x) = \psi(x+1) - (1/x)$ a sufficient number of times until the argument is brought to 5 or above.

Set $f = 10^{99}$
Set $g = 0$
Input x >>>>>>>
(1) If $x = 0$ go to (2)
Replace g by $g + \dfrac{1}{x}$
Replace x by $x + 1$
If $x < 5$ go to (1)
Set $f = 1 + \left[\left(\dfrac{0.46}{x^2} - 1\right) \Big/ 10x^2\right]$
Replace f by $-\left[\left(\dfrac{f}{6x} + 1\right) \Big/ 2x\right] + \ln(x) - g$
(2) Output f
$f = \psi(x)$ <<<<<<

Storage needed: f, g and x

Test values:
$\psi(7) = 1.87278434$
$\psi(1.461632145) = 0$
$\psi(-1.05) = 20.2897713$

44:9 APPROXIMATIONS

For large x the approximations

44:9:1 $$\psi(-x) \simeq \ln\left(x + \frac{1}{2}\right) + \pi \cot(\pi x) \qquad \text{large } x$$

and

44:9:2 $$\psi(x) \simeq \ln\left(x - \frac{1}{2}\right) \qquad \text{8-bit precision} \qquad x \geq 4$$

are useful.

44:10 OPERATIONS OF THE CALCULUS

One or more differentiations of the digamma function give the polygamma functions that are discussed in Section 44:12:

44:10:1 $$\frac{d^n}{dx^n}\psi(x) = \psi^{(n)}(x) \qquad n = 1, 2, 3, \ldots$$

Formulas for indefinite integration include

44:10:2 $$\int_1^x \psi(t)dt = \ln(\Gamma(x)) \qquad x > 0$$

and

44:10:3 $$\int_x^{x+1} \psi(t)dt = \ln(x) \qquad x > 0$$

Many definite integrals, including

44:10:4 $$\int_0^\infty \frac{\gamma + \psi(1+t)}{t^\nu}\,dt = -\pi \csc(\nu\pi)\zeta(\nu) \qquad 1 < \nu < 2$$

and

44:10:5 $$\int_0^1 \psi(t)\sin(n\pi t)\mathrm{d}t = \begin{cases} 0 & n = 1 \\ \dfrac{-\pi}{2} & n = 2, 4, 6, \ldots \end{cases}$$

are listed by Gradshteyn and Ryzhik [Sections 6.46 and 6.47].

44:11 COMPLEX ARGUMENT

The digamma function of argument $x + iy$ is given by

44:11:1 $$\psi(x + iy) = -\gamma - \frac{x}{x^2 + y^2} + \sum_{j=1}^{\infty} \frac{jx + x^2 + y^2}{j[(j + x)^2 + y^2]} + i\sum_{j=0}^{\infty} \frac{y}{(j + x)^2 + y^2}$$

Tables permitting the evaluation of $\psi(x + iy)$ are presented by Abramowitz and Stegun [Table 6.8].

For purely imaginary argument

44:11:2 $$\psi(iy) = -\gamma + \sum_{j=1}^{\infty} \frac{y^2}{j(j^2 + y^2)} + i\left[\frac{1}{2y} + \frac{\pi}{2}\coth(\pi y)\right]$$

For the related special case

44:11:3 $$\psi(\tfrac{1}{2} + iy) = -\gamma - \frac{2}{1 + 4y^2} + \frac{\pi}{4y}\tanh(\pi y) + \sum_{j=1}^{\infty} \frac{1 + 4y^2}{j[(2j + 1)^2 + y^2]} + \frac{i\pi}{2}\tanh(\pi y)$$

the imaginary part of the digamma function is seen to contain a hyperbolic function [see Chapter 30].

44:12 GENERALIZATIONS

One differentiation of $\ln(\Gamma(x))$ gives the digamma function; multiple differentiations generate the family of functions known as the *polygamma functions* and denoted $\psi^{(n)}(x)$. One has

44:12:1 $$\psi^{(n)}(x) = \frac{\mathrm{d}^n}{\mathrm{d}x^n}\psi(x) \qquad n = 1, 2, 3, \ldots$$

Another definition is

44:12:2 $$\psi^{(n)}(x) = (-1)^n \int_0^{\infty} \frac{t^n \exp(-xt)}{t - 1}\mathrm{d}t \qquad n = 1, 2, 3, \ldots$$

The names *trigamma function* and *tetragamma function*, respectively, are given to $\psi^{(1)}$ and $\psi^{(2)}$: these particular polygamma functions are more commonly denoted ψ' and ψ''. The functions $\psi^{(3)}$ and $\psi^{(4)}$ are known as the *pentagamma function* and *hexagamma function*.

For odd n the polygamma functions $\psi^{(n)}$ are restricted to positive values, but ψ'', $\psi^{(4)}$, $\psi^{(6)}$, ... take values of both signs. As illustrated in Figures 44-3 through 44-5, the polygamma functions are infinite at $x = 0, -1, -2, \ldots$. By combining the recurrence formula

44:12:3 $$\psi^{(n)}(x + 1) = \psi^{(n)}(x) - \frac{n!}{(-x)^{n+1}} \qquad n = 1, 2, 3, \ldots$$

with the particular values

44:12:4 $$\psi^{(n)}(1) = (-1)^{n+1} n!\, \zeta(n + 1) \qquad n = 1, 2, 3, \ldots$$

and

44:12:5 $$\psi^{(n)}\left(\frac{1}{2}\right) = (-2)^{n+1} n!\, \lambda(n + 1) \qquad n = 1, 2, 3, \ldots$$

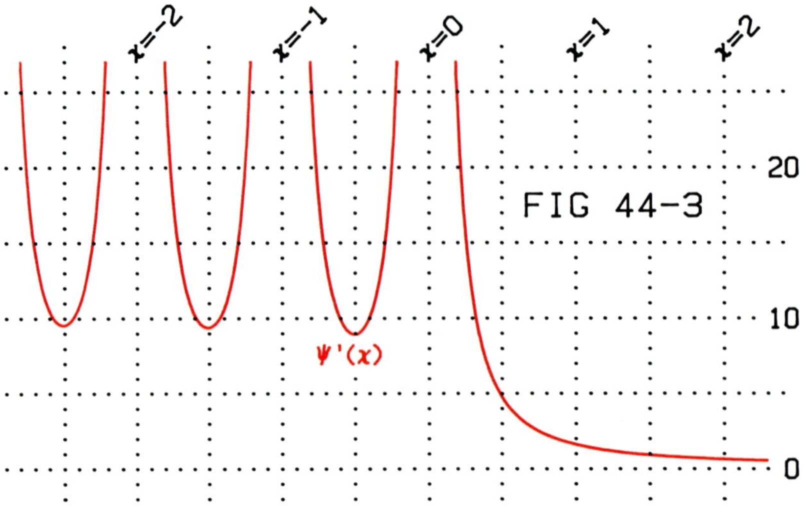
x=-2
x=-1
x=0
x=1
x=2
20
FIG 44-3
10
ψ'(x)
0

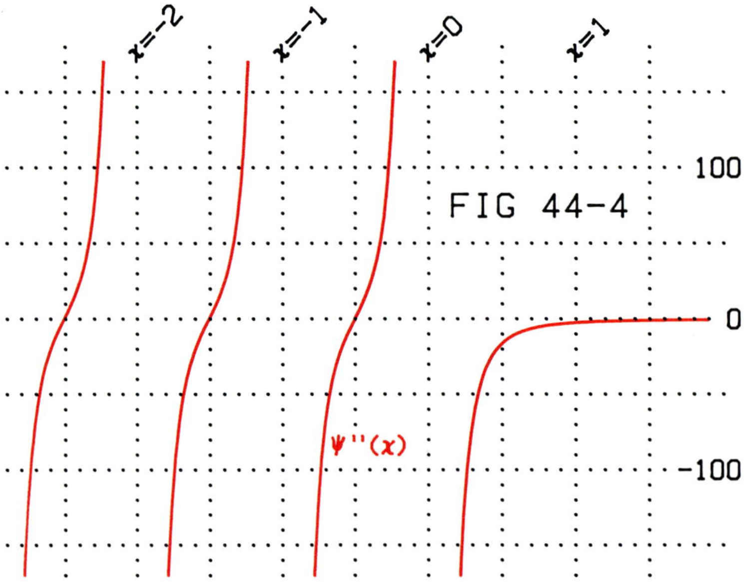
x=-2
x=-1
x=0
x=1
100
FIG 44-4
0
ψ''(x)
-100

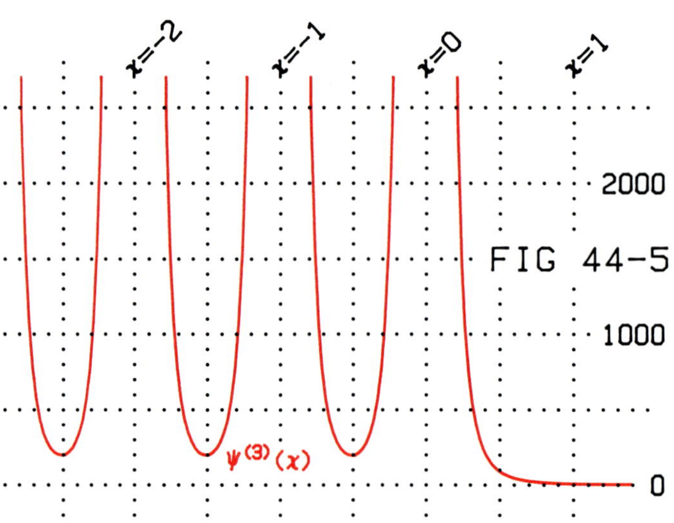
x=-2
x=-1
x=0
x=1
2000
FIG 44-5
1000
ψ(3)(x)
0

involving zeta ζ and lambda λ numbers [Chapter 3], one may derive the expressions

$$\left.\begin{aligned} &\text{44:12:6} \quad \psi^{(n)}(J) = (-1)^{n+1}n!\left[\zeta(n+1) - \sum_{j=1}^{J-1} j^{-n-1}\right] = n!\sum_{j=J}^{\infty}\left(\frac{-1}{j}\right)^{n+1} \\ &\text{44:12:7} \quad \psi^{(n)}\left(J-\frac{1}{2}\right) = (-2)^{n+1}n!\left[\lambda(n+1) - \sum_{j=1}^{J-1}(2j-1)^{-n-1}\right] = n!\sum_{j=J}^{\infty}\left(\frac{-2}{2j-1}\right)^{n+1} \end{aligned}\right\} \quad \begin{aligned} n &= 1, 2, 3, \ldots \\ J &= 1, 2, 3, \ldots \end{aligned}$$

for the polygamma functions of positive integers and half-integers. Some such values are included in Table 44.7.1 for the $n = 1$ case.

For large positive values of their arguments, the polygamma functions are well approximated by

44:12:8 $$\psi^{(n)}(x) \simeq \frac{-(n-1)!}{(\frac{1}{2}-x)^n} \qquad \text{large } x$$

Yet more refined approximations are provided by 44:12:12 for large x. As is evident in Figures 44-3–44-5, for negative x values of any magnitude, the values of $\psi^{(n)}(x)$ are largely determined by the fractional value [Chapter 9] of x and hardly at all by its integer value. This is a consequence of the fact that, in the reflection formula

44:12:9 $$\psi^{(n)}(-x) = (-1)^n\pi\frac{d^n}{dx^n}\cot(\pi x) + (-1)^n\psi^{(n)}(x+1)$$

the final term for $x > 0$ is generally negligible compared with the periodic derivative term.

Among the relationships for the polygamma functions that can be derived by differentiation of the corresponding formulas for the digamma function are the following expansions:

44:12:10 $$\psi^{(n)}(x) = \frac{n!}{(-x)^{n+1}} - \sum_{j=0}^{\infty}(-1)^{n+j}\frac{(n+j)!}{j!}\zeta(n+j+1)x^j \qquad n = 1, 2, 3, \ldots \qquad -1 < x < 1$$

44:12:11 $$\psi^{(n)}(x) = n!\sum_{j=0}^{\infty}\left(\frac{-1}{j+x}\right)^{n+1} \qquad n = 1, 2, 3, \ldots$$

44:12:12 $$\psi^{n}(x) \sim \frac{-(n-1)!}{(-x)^n}\left[1 + \frac{n}{2x} + \frac{n(n+1)}{12x^2} - \frac{n(n+1)(n+2)(n+3)}{720x^4} + \cdots + \binom{j+n-1}{j}\frac{B_j}{x^j} + \cdots\right] \qquad x \to \infty$$

the latter asymptotic expansion being valid for large x.

The algorithm below is designed to calculate values of $\psi^{(n)}(x)$ for small values of n (e.g., $n = 1, 2, 3$). Since the polygamma functions take infinite values for $x = 0, -1, -2, \ldots,$ the algorithm returns 10^{99} under these conditions. Otherwise, the algorithm has high accuracy, although it becomes increasingly imprecise as n increases. Note that n must be a positive integer.

Input n >>>>>>>
Set $f = 10^{99}$
Set $g = 0$
Input x >>>>>>>
(1) If $x = 0$ go to (4)
Set $h = -1/x$
Set $j = n$
(2) Replace h by $-hj/x$
Replace j by $j - 1$
If $j \neq 0$ go to (2)
Replace g by $g + h$
Replace x by $x + 1$

Storage needed: n, f, g, x, h and j

Input restriction: n must be a positive integer.

If $x < 5$ go to (1)
Set $f = [1 - (n + 4)(n + 5)/45x^2]/60x^2$
Replace f by $(n + 1)[1 - f(n + 2)(n + 3)]/6x$
Replace f by $-\left[\frac{1}{n} + \frac{f+1}{2x}\right]$
Set $j = n$
(3) Replace f by $-fj/x$
Replace j by $j - 1$
If $j \neq 0$ go to (3)
Replace f by $f + g$
(4) Output f
$f = \psi^{(n)}(x)$ <<<<<

Test values:
$\psi'(5.5) = 0.199342388$
$\psi''(-0.5) = -0.828796644$
$\psi^{(3)}(9) = 0.00323439609$

The algorithm does *not* generate $\psi(x)$ when $n = 0$ is input. The algorithm utilizes equation 44:12:3 together with a truncated, compensated and concatenated version of 44:12:10; it parallels the algorithm of Section 44:8 in its manner of operation. Some approximate values of $\psi^{(n)}(x)$ may also be read from maps in this chapter.

44:13 COGNATE FUNCTIONS

In addition to its definition as the integral

44:13:1 $$G(x) = 2\int_0^1 \frac{t^{x-1}}{1+t}\,dt = 2\int_0^\infty \frac{\exp(-xt)}{1+\exp(-t)}\,dt$$

Bateman's G function is defined in terms of the digamma function as

44:13:2 $$G(x) = \psi\left(\frac{x+1}{2}\right) - \psi\left(\frac{x}{2}\right)$$

The notation $2\beta(x)$ is also used for this function, but confusion must be avoided with the dissimilar function denoted by β in Chapter 3. In fact it is to the *bivariate eta function* of Section 64:13 that $G(x)$ is related:

44:13:3 $$G(x) = 2\sum_{j=0}^{\infty} \frac{(-1)^j}{x+j} = 2\eta(1;x)$$

Some properties of Bateman's G function are

44:13:4 $$G(1+x) = \frac{2}{x} - G(x)$$

and

44:13:5 $$G(1-x) = 2\pi\csc(\pi x) - G(x)$$

and, in addition to 44:13:3, we have the expansions

44:13:6 $$G(x) = \sum_{j=0}^{\infty} \frac{2}{(2j+x)(2j+x+1)} = \sum_{j=0}^{\infty} \frac{j!}{2^j (x)_{j+1}} = 2\sum_{j=0}^{\infty} \eta(j+1)(1-x)^j$$

and

44:13:7 $$G(x) = \frac{1}{x} - \frac{2}{1-x^2} + \csc(\pi x) + 2\sum_{j=0}^{\infty} [1 - \eta(2j+1)]x^{2j}$$

where $\eta(j)$ is the eta number of Chapter 3. The behavior of $G(x)$ is mapped in Figures 44-1 and 44-2; Section 18:14 shows how the universal hypergeometric algorithm may be used to calculate its values. Some particular values of $G(x)$ will be found in Section 44:7.

Many properties of the *successive derivatives* $G^{(n)}(x)$ of *Bateman's G function* are given by Erdélyi, Magnus, Oberhettinger and Tricomi [*Higher Transcendental Functions*, Volume 1, Section 1.16]. We mention only

44:13:8
$$G^{(n)}(x) = 2n! \sum_{j=0}^{\infty} \frac{(-1)^{n+j}}{(j+x)^{n+1}} = 2(-1)^n n! \eta(n+1;x)$$

where η is the bivariate eta function of Section 64:13.

44:14 RELATED TOPICS

The digamma function and Bateman's G function are useful for summing certain algebraic series involving reciprocal linear functions. Thus:

44:14:1
$$\frac{1}{c} + \frac{1}{b+c} + \frac{1}{2b+c} + \frac{1}{3b+c} + \cdots + \frac{1}{Jb-b+c} = \sum_{j=0}^{J-1} \frac{1}{jb+c} = \frac{1}{b}\left[\psi\left(\frac{Jb+c}{b}\right) - \psi\left(\frac{c}{b}\right)\right]$$

and

44:14:2
$$\frac{1}{c} - \frac{1}{b+c} + \frac{1}{2b+c} - \frac{1}{3b+c} + \cdots \pm \frac{1}{Jb-b+c} = \sum_{j=0}^{J-1} \frac{(-1)^j}{bj+c}$$
$$= \frac{1}{2b}\left[G\left(\frac{b+c}{b}\right) - (-1)^J G\left(\frac{Jb+c}{b}\right)\right]$$

Because the G function approaches zero as its argument approaches infinity, the sum to $j = \infty$ of $(-1)^j/(bj+c)$ is simply $G[(b+c)/b]/2b$. Special cases of these results are presented as equations 1:14:5 and 1:14:10.

Similarly, infinite series of reciprocal powers have the sums

44:14:3
$$\frac{1}{c^n} + \frac{1}{(b+c)^n} + \frac{1}{(2b+c)^n} + \frac{1}{(3b+c)^n} + \cdots = \sum_{j=0}^{\infty} \frac{1}{(jb+c)^n} = \frac{\psi^{(n-1)}(c/b)}{(-b)^n(n-1)!} \qquad n = 2, 3, 4, \ldots$$

and

44:14:4
$$\frac{1}{c^n} - \frac{1}{(b+c)^n} + \frac{1}{(2b+c)^n} - \frac{1}{(3b+c)^n} + \cdots = \sum_{j=0}^{\infty} \frac{(-1)^j}{(jb+c)^n} = \frac{\psi^{(n-1)}\left(\frac{c}{2b}\right) - \psi^{(n-1)}\left(\frac{b+c}{2b}\right)}{(-2b)^n(n-1)!}$$
$$n = 1, 2, 3, \ldots$$

in terms of polygamma functions. A corresponding finite series may be expressed as the difference between two infinite series and hence may be summed via equation 44:14:3 or 44:14:4.

Consider the summation of a series of rational functions [Section 17:13] of the summation index k

44:14:5
$$S = \sum_{k=0}^{\infty} \frac{p_m(k)}{p_n(k)} = \sum_{k=0}^{\infty} \frac{A_m k^m + A_{m-1}k^{m-1} + \cdots + A_1 k + A_0}{a_n k^n + a_{n-1}k^{n-1} + \cdots + a_1 k + a_0}$$

where p_m and p_n are polynomial functions of degrees m and n. If $n \geq m+2$, the series will generally converge, and, in these circumstances, S may be expressed in terms of digamma and/or polygamma functions, as we now demonstrate.

In Section 17:6 it is shown that $1/p_n(k)$ may be expressed as the sum of n partial fractions, each of form $C_j/[a_n(k-\rho_j)]$ where ρ_j is the jth zero of p_n and C_j is a constant. A similar fractionation of $p_m(k)/p_n(k)$ is possible so that

44:14:6
$$S = \frac{1}{a_n} \sum_{j=1}^{n} \sum_{k=0}^{\infty} \frac{C_j}{k-\rho_j} = \frac{-1}{a_n} \sum_{j=1}^{n} C_j \psi(-\rho_j)$$

Because $m \le n - 2$, it follows that $\Sigma C_j = 0$, and this fact is used, together with the first of the 44:6:2 expansions, to establish 44:14:6. Some (or all or none) of the zeros ρ_j may be complex, in which case the corresponding C_j constants will also be complex.

The foregoing presupposes that the zeros $\rho_1, \rho_2, \ldots \rho_\nu$ are all distinct. When this is not the case, one follows the procedures outlined in Section 17:13. For example, if the zeros ρ_1, ρ_2 and ρ_3 (but no others) were coincident, one would replace 44:14:6 by

44:14:7
$$S = \frac{1}{a_n}\sum_{k=0}^{\infty}\left[\frac{C_1}{k-\rho_1} + \frac{C_2}{(k-\rho_1)^2} + \frac{C_3}{(k-\rho_1)^3}\right] + \frac{1}{a_n}\sum_{j=4}^{n}\sum_{k=0}^{\infty}\frac{C_j}{k-\rho_j}$$
$$= \frac{1}{a_n}\left[-C_1\psi(-\rho_1) + \frac{C_2}{2!}\psi'(-\rho_1) - \frac{C_3}{3!}\psi''(-\rho_1) - \sum_{j=4}^{n} C_j\psi(-\rho_j)\right]$$

equation 44:12:9 having been used.

CHAPTER 45

THE INCOMPLETE GAMMA $\gamma(\nu;x)$ AND RELATED FUNCTIONS

Many of the functions discussed in preceding chapters are special cases of the functions considered in this chapter. Besides the incomplete gamma function $\gamma(\nu;x)$ itself, we shall make frequent use in this chapter of two closely related functions: the *complementary incomplete gamma function* $\Gamma(\nu;x)$ and the *entire incomplete gamma function* $\gamma^*(\nu;x)$. The interrelationships

45:0:1 $$x^{\nu}\gamma^*(\nu;x) = \frac{\gamma(\nu;x)}{\Gamma(\nu)} = 1 - \frac{\Gamma(\nu;x)}{\Gamma(\nu)}$$

link the three, where $\Gamma(\nu)$ denotes the (complete) gamma function [Chapter 43]. The entire incomplete gamma function remains useful for negative x, whereas $\gamma(\nu;x)$ and $\Gamma(\nu;x)$ are then often ill defined.

45:1 NOTATION

The variables ν and x are termed, respectively, the parameter and the argument of the functions. The adjective "incomplete" reflects the restricted ranges of integration in definitions 45:3:1 and 45:3:2 compared with that in Euler's integral of the second kind, 43:3:1.

Alternative symbolisms abound. $\Gamma_x(\nu)$ has been used for $\gamma(\nu;x)$; $P(\nu,x)$ and $Q(\nu,x)$ have been used for $\gamma(\nu;x)/\Gamma(\nu)$ and $\Gamma(\nu;x)/\Gamma(\nu)$, respectively. The notation $E_\nu(x)$ or $K_\nu(x)$ is sometimes adopted for $x^{\nu-1}\Gamma(1-\nu;x)$.

45:2 BEHAVIOR

Because, for negative arguments, the functions $\gamma(\nu;x)$ and $\Gamma(\nu;x)$ are often complex and/or multivalued, we restrict our discussion of these functions to $x \geq 0$. Figure 45-1 maps typical behaviors of these two functions for a variety of values of the parameter.

For $x > 0$, the $\gamma(\nu;x)$ function is well defined except for $\nu = 0, -1, -2, \ldots$ when, like the *complete* gamma function [Chapter 43], it encounters infinite discontinuities viewed as a function of its parameter. The $\gamma(\nu;x)$ function is always a monotonically increasing function of x. For $\nu > 0$, $\gamma(\nu;x)$ is zero at $x = 0$ and positive elsewhere. For $-1 < \nu < 0$, $-3 < \nu < -2$, $-5 < \nu < -4$, etc., the function is invariably negative, infinitely so at $x = 0$. For $-2 < \nu < -1$, $-4 < \nu < -3$, etc., $\gamma(\nu;x)$ equals $-\infty$ at zero argument, then increases monotonically, crossing zero to reach the positive value $\Gamma(\nu)$ at $x = \infty$.

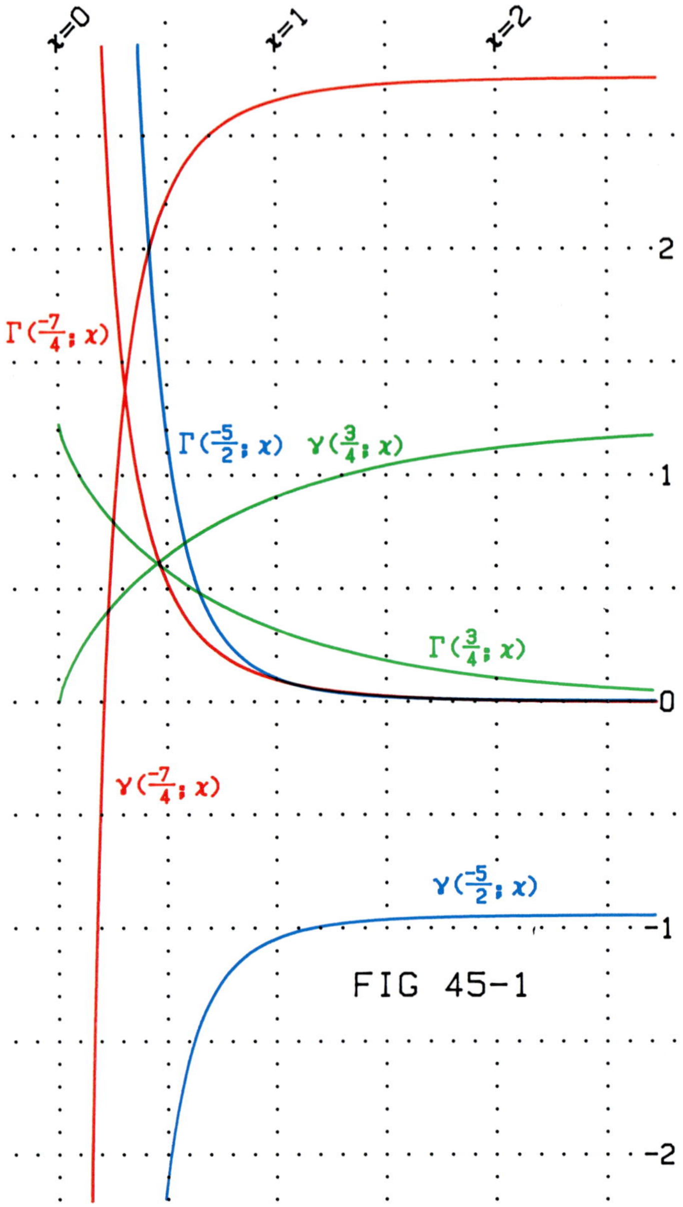

FIG 45-1

The complementary incomplete gamma function is invariably positive and decreases monotonically with x, approaching zero asymptotically as $x \to \infty$.

The entire incomplete gamma function $\gamma^*(\nu;x)$ is single valued, real and finite at all values of ν and x that are real and finite. Its behavior is comprehensively described by the contour map Figure 45-2. Notice that γ^* has no zeros for $\nu > 0$, has one zero (at a negative value of x) for $-1 < \nu < 0$, $-3 < \nu < -2$, etc., and two zeros (one at positive and one at negative arguments) for $-2 < \nu < -1$, $-4 < \nu < -3$, etc.

45:3 DEFINITIONS

The $\gamma(\nu;x)$ and $\Gamma(\nu;x)$ incomplete gamma functions are usually defined via the integrals

45:3:1 $$\gamma(\nu;x) = \int_0^x t^{\nu-1} \exp(-t)dt \qquad x \geq 0 \qquad \nu > 0$$

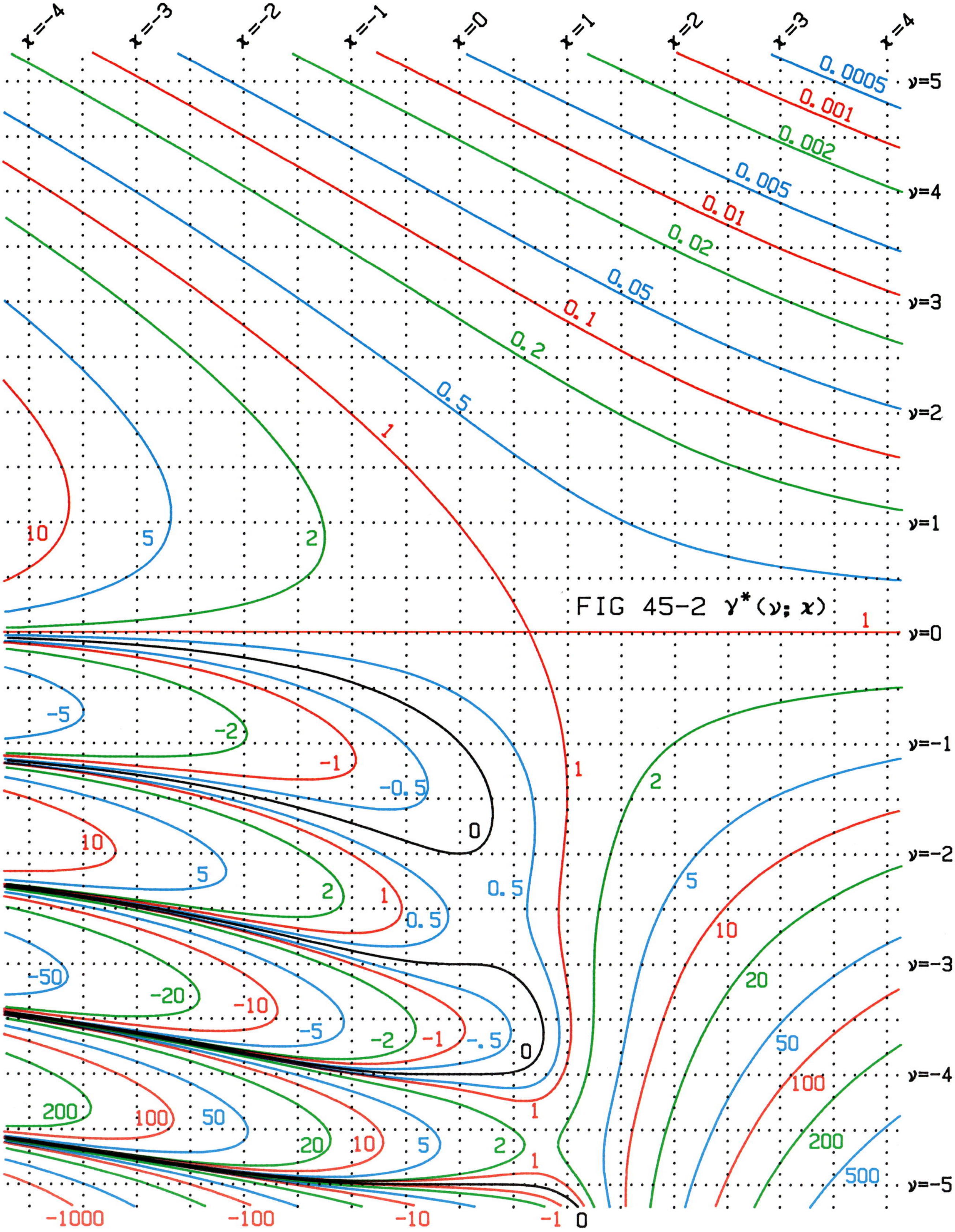
FIG 45-2 γ*(ν; x)
x=−4
x=−3
x=−2
x=−1
x=0
x=1
x=2
x=3
x=4
ν=5
ν=4
ν=3
ν=2
ν=1
ν=0
ν=−1
ν=−2
ν=−3
ν=−4
ν=−5
0.0005
0.001
0.002
0.005
0.01
0.02
0.05
0.1
0.2
0.5
1
10
5
2
−5
−2
−1
−0.5
0
0.5
−50
−20
−10
−.5
20
50
100
200
500
−1000
−100
−10

45:3:2 $$\Gamma(\nu;x) = \int_x^\infty t^{\nu-1}\exp(-t)\mathrm{d}t \qquad x > 0$$

Definition 45:3:1 may be extended to negative ν by utilizing the recursion formula 45:5:2. The definite integrals

45:3:3 $$\Gamma(\nu;x) = \frac{x^\nu \exp(-x)}{\Gamma(1-\nu)} \int_0^\infty \frac{t^{-\nu}\exp(-t)}{t+x}\mathrm{d}t \qquad x > 0 \qquad \nu < 1$$

and

45:3:4 $$\gamma^*(\nu;x) = \frac{1}{\Gamma(\nu)} \int_0^1 t^{\nu-1}\exp(-xt)\mathrm{d}t \qquad \nu > 0$$

open up other ranges of the parameter and argument. The more complicated integral representations

45:3:5 $$\gamma(\nu;x) = x^{\nu/2} \int_0^\infty t^{(\nu-2)/2}\exp(-t)\,\mathrm{J}_\nu(2\sqrt{xt})\mathrm{d}t \qquad \nu > 0$$

45:3:6 $$\Gamma(\nu;x) = \frac{2x^{\nu/2}\exp(-x)}{\Gamma(1-\nu)} \int_0^\infty t^{-\nu/2}\exp(-t)\,\mathrm{K}_\nu(2\sqrt{xt})\mathrm{d}t \qquad \nu < 1$$

link the incomplete gamma functions to Bessel [Chapter 53] and Basset [Chapter 51] functions.

Often the series 45:6:1 is taken as the definition of the incomplete gamma function. As well, the entire incomplete gamma function may be defined succinctly by the differintegration [Section 0:10] formulation

45:3:7 $$\frac{\mathrm{d}^\nu}{\mathrm{d}x^\nu}\exp(\pm x) = x^{-\nu}\exp(\pm x)\,\gamma^*(-\nu;\pm x)$$

45:4 SPECIAL CASES

Certain values of the parameter convert the incomplete gamma function into simpler functions treated elsewhere in this *Atlas*.

When the parameter is a positive integer:

45:4:1 $$\Gamma(n;x) = (n-1)!\exp(-x)\,\mathrm{e}_{n-1}(x) \qquad n = 1, 2, 3, \ldots$$

where $\mathrm{e}_n(x)$ denotes the exponential polynomial function [Section 26:13]. For example, $\Gamma(1;x) = \exp(-x)$, so that $\gamma(1;x) = 1 - \exp(-x)$ and $\gamma^*(1;x) = [1 - \exp(-x)]/x$. When $\nu = 0$, $\gamma(\nu;x)$ is undefined but $\Gamma(0;x) = -\mathrm{Ei}(-x)$ and $\gamma^*(0;x) = 1$, where Ei is the exponential integral of Chapter 37.

For a parameter equal to a negative integer:

45:4:2 $$\gamma^*(-n,x) = x^n \qquad n = 0, 1, 2, \ldots$$

a simple result that provides a link to Chapter 11. The complementary incomplete gamma function for a negative integer parameter may be related to a Schlömilch function [Section 37:13], but an equivalent expression is

45:4:3 $$\Gamma(-n;x) = \frac{(-1)^{n+1}}{n!}\left[\frac{\exp(-x)}{x}\sum_{j=0}^{n-1}\frac{j!}{(-x)^j} + \mathrm{Ei}(-x)\right] \qquad n = 0, 1, 2, \ldots$$

in terms of the exponential integral function [Chapter 37].

Incomplete gamma functions of moiety parameter reduce to the functions of Chapters 40 and 42:

45:4:4 $$\gamma(\tfrac{1}{2};x) = \sqrt{\pi}\,\mathrm{erf}(\sqrt{x})$$

45:4:5 $$\Gamma(\tfrac{1}{2};x) = \sqrt{\pi}\,\mathrm{erfc}(\sqrt{x})$$

45:4:6 $$\gamma^*(\tfrac{1}{2};-x) = \frac{2}{\sqrt{\pi x}}\exp(x)\,\mathrm{daw}(\sqrt{x})$$

Notice also that by 45:3:2

45:4:7 $$\Gamma(\nu + 1;x) = \int_x^\infty t^\nu \exp(-t)dt \qquad x > 0$$

Special cases of such integrals are tabulated in Table 37.14.1 for $\nu = -3, -\frac{5}{2}, -2, \ldots, 2$. Thus, this table serves to provide a listing of the special cases $\Gamma(\nu;x)$ for $\nu = -2, \frac{-3}{2}, -1, \ldots, 3$.

45:5 INTRARELATIONSHIPS

The incomplete gamma function and its complementary cohort sum to the (complete) gamma function [Chapter 43]

45:5:1 $$\gamma(\nu;x) + \Gamma(\nu;x) = \Gamma(\nu)$$

Recurrence formulas for the three functions are

45:5:2 $$\gamma(\nu + 1;x) = \nu\gamma(\nu;x) - x^\nu \exp(-x)$$

45:5:3 $$\Gamma(\nu + 1;x) = \nu\Gamma(\nu;x) + x^\nu \exp(-x)$$

and

45:5:4 $$\gamma^*(\nu + 1;x) = \frac{\gamma^*(\nu;x)}{x} - \frac{\exp(-x)}{x\Gamma(1 + \nu)} \qquad x \neq 0$$

The first of these may be generalized to

45:5:5 $$\gamma(\nu + n;x) = (\nu)_n\left[\gamma(\nu;x) - x^\nu \exp(-x)\sum_{j=0}^{n-1}\frac{x^j}{(\nu)_{j+1}}\right] = (\nu)_n\gamma(\nu;x) = x^\nu \mathrm{p}_{n-1}(x)\exp(-x)$$

where $\mathrm{p}_{n-1}(x)$ is the polynomial defined in 45:8:2.

Nielsen's expansion provides an addition formula for the argument:

45:5:6 $$\gamma(\nu;x + y) = \gamma(\nu;x) + \frac{\exp(-x)}{x^{1-\nu}}\sum_{j=0}^{\infty}\frac{(1 - \nu)_j}{(-x)^j}[1 - \exp(-y)\,\mathrm{e}_j(y)] \qquad |y| < |x|$$

45:6 EXPANSIONS

A large number of expansions exist for the incomplete gamma functions. The most important of these are

45:6:1 $$\gamma(\nu;x) = \frac{x^\nu}{\nu} - \frac{x^{\nu+1}}{1 + \nu} + \frac{x^{\nu+2}}{2(2 + \nu)} - \frac{x^{\nu+3}}{6(3 + \nu)} + \cdots = x^\nu \sum_{j=0}^{\infty}\frac{(-x)^j}{j!(j + \nu)}$$

and

45:6:2 $$\gamma(\nu;x) = \exp(-x)\left[\frac{x^\nu}{\nu} + \frac{x^{\nu+1}}{\nu(\nu + 1)} + \frac{x^{\nu+2}}{\nu(\nu + 1)(\nu + 2)} + \cdots\right] = \exp(-x)\sum_{j=0}^{\infty}\frac{x^{j+\nu}}{(\nu)_{j+1}} = \frac{\exp(-x)}{\nu}\sum_{j=0}^{\infty}\frac{x^{j+\nu}}{(\nu + 1)_j}$$

which lend themselves to expression as hypergeometric series [Section 18:14]. Involving the associated Laguerre polynomials [Section 23:12] and hyperbolic Bessel functions [Chapter 50], respectively, are the expansions

45:6:3 $$\Gamma(\nu;x) = x^\nu \exp(-x)\sum_{j=0}^{\infty}\frac{\mathrm{L}^\nu j(x)}{j + 1}$$

and

45:6:4 $$\gamma^*(\nu;x) = \frac{\exp(-x)}{x^{\nu/2}}\sum_{j=0}^{\infty} x^{j/2}\mathrm{e}_j(-1)\mathrm{I}_{j+\nu}(2\sqrt{x})$$

The complementary incomplete gamma function is expansible as the continued fraction

45:6:5
$$\Gamma(\nu;x) = \frac{x^\nu \exp(-x)}{x+} \; \frac{1-\nu}{1+} \; \frac{1}{x+} \; \frac{2-\nu}{1+} \; \frac{2}{x+} \; \frac{3-\nu}{1+} \cdots$$

or as the important asymptotic expansion

45:6:6
$$\Gamma(\nu;x) \sim \frac{\exp(-x)}{x^{1-\nu}}\left[1 - \frac{1-\nu}{x} + \frac{(1-\nu)(2-\nu)}{x^2} - \frac{(1-\nu)(2-\nu)(3-\nu)}{x^3} + \cdots + \frac{(1-\nu)_j}{(-x)^j} + \cdots\right] \qquad x \to \infty$$

valid for large x.

Expansion 45:6:6 applies when x, but not ν, is large. Conversely, there is an asymptotic expansion

45:6:7
$$\gamma(\nu;x) \sim x^\nu\left[\frac{1}{\nu} - \frac{x}{\nu+1} + \frac{x^2}{2(\nu+2)} - \frac{x^3}{6(\nu+3)} + \cdots + \frac{(-x)^j}{j!(\nu+j)} + \cdots\right] \qquad \nu \to \infty$$

that is valid when the parameter, but not the argument, is large.

45:7 PARTICULAR VALUES

The values of the three functions for arguments of zero and infinity sometimes depend on the sign of the parameter. In Table 45.7.1 "undef" means undefined.

Table 45.7.1

	$x = 0$		$x = \infty$	
	$\nu < 0$	$\nu > 0$	$\nu < 0$	$\nu > 0$
$\gamma(\nu;x)$	$-\infty$	0	$\Gamma(\nu)$	$\Gamma(\nu)$
$\Gamma(\nu;x)$	undef	$\Gamma(\nu)$	0	0
$\gamma^*(\nu;x)$	$\dfrac{1}{\Gamma(1+\nu)}$	$\dfrac{1}{\Gamma(1+\nu)}$	$+\infty$	0

45:8 NUMERICAL VALUES

By repeated applications of recursion formula 45:5:4, and using expansion 45:6:2, one can write

45:8:1
$$\gamma^*(\nu;x) = x^n\gamma^*(\nu+n;x) + \mathrm{p}_{n-1}(x)\frac{\exp(-x)}{\Gamma(n+\nu)} = \frac{\exp(-x)}{\Gamma(n+\nu)}\left[\mathrm{p}_{n-1}(x) + x^n\sum_{j=0}^{\infty}\frac{x^j}{(n+\nu)_{j+1}}\right]$$

where $\mathrm{p}_{n-1}(x)$ is a polynomial function of x that may be written

45:8:2
$$\mathrm{p}_{n-1}(x) = ((\cdots(((1+\nu)+x)(2+\nu)+x^2)(3+\nu)+\cdots+x^{n-3})(n-2+\nu) + x^{n-2})(n-1+\nu) + x^{n-1}$$

These are the formulas used in the algorithm for $\gamma^*(\nu;x)$ that is presented below. The integer n is chosen to be at least 1 and such that $n + \nu > 2$. The infinite series in 45:8:1 is evaluated via the truncated concatenation

45:8:3
$$\sum_{j=0}^{\infty}\frac{x^j}{(n+\nu)_{j+1}} \simeq \left(\left(\left(\cdots\left(\left(\frac{x}{n+\nu+J-x}+1\right)\frac{x}{n+\nu+J-1}+1\right)\frac{x}{n+\nu+J-2}\right.\right.\right.$$

$$+\cdots+1\Bigg)\frac{x}{n+\nu+2}+1\Bigg)\frac{x}{n+\nu+1}+1\Bigg)\frac{1}{n+\nu}$$

By use of the empirical assignment $J = \text{Int}(5 + (3 + |x|)/2)$ one ensures that approximation 45:8:3 has 24-bit precision (the relative error is less than 6×10^{-8}). The final segment of the algorithm is concerned with calculating the multiplier $\exp(-x)/\Gamma(n + \nu)$, using a procedure similar to that employed in Section 43:8.

Input ν >>>>>>>
 Set $g = p = 1$
Input x >>>>>>>
(1) Replace ν by $1 + \nu$
 If $\nu > 2$ go to (2)
 Replace g by gx
 Replace p by $p\nu + g$
 Go to (1)
(2) Set $j = \text{Int}(5(3 + |x|)/2)$
 Set $f = 1/(j + \nu - x)$
(3) Replace j by $j - 1$
 Replace f by $(fx + 1)/(j + \nu)$
 If $j \neq 0$ go to (3)
 Replace p by $p + fgx$
 Set $g = \left[1 - \dfrac{2}{7\nu^2}\left(1 - \dfrac{2}{3\nu^2}\right)\right] \Big/ 30\nu^2$
 Replace g by $\dfrac{g-1}{12\nu} - \nu[\ln(\nu) - 1]$
 Set $f = p \exp(g - x)\sqrt{\nu/2\pi}$
 Output f
$f \simeq \gamma^*(\nu;x)$ <<<<<

Storage needed: ν, g, p, x, j and f

Test values:
$\gamma^*(\frac{1}{2};-0.49) = 1.34326134$
$\gamma^*(-3.9;0) = -0.521258841$
$\gamma^*(1;\pi) = 0.304554469$

Of course, numerical values of $\gamma(\nu;x)$ and $\Gamma(\nu;x)$ may be computed from $\gamma^*(\nu;x)$ by using equations 45:0:1. Alternatively, the universal hypergeometric algorithm of Section 18:14 may be used in several versions.

45:9 APPROXIMATIONS

Whereas it is exact only if the parameter is a nonpositive integer, the expression

45:9:1 $$\gamma^*(\nu;x) \simeq x^{-\nu}$$

is a useful approximation for all ν if x is positive and sufficiently large.

45:10 OPERATIONS OF THE CALCULUS

Differentiation of the incomplete gamma function and its complementary analog give

45:10:1 $$\frac{d}{dx}\gamma(\nu;x) = -\frac{d}{dx}\Gamma(\nu;x) = x^{\nu-1}\exp(-x) = \gamma(\nu;x) - (\nu - 1)\gamma(\nu - 1;x)$$

The differintegration [Section 0:10] formula for the entire incomplete gamma function

45:10:2 $$\frac{d^\mu}{dx^\mu}[x^\nu \exp(x)\gamma^*(\nu;x)] = x^{\nu-\mu}\exp(x)\gamma^*(\nu - \mu;x)$$

holds for any value of μ, positive or negative, integer or noninteger.

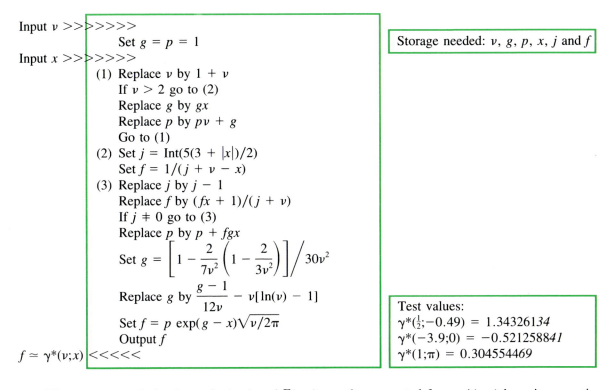

Two definite integrals are

45:10:3 $$\int_0^\infty \gamma(\nu;bt)\exp(-st)\mathrm{d}t = \frac{\gamma(\nu)}{s}\left(\frac{b}{b+s}\right)^\nu \qquad \nu > 0$$

and

45:10:4 $$\int_0^\infty t^{\mu-1}\Gamma(\nu;bt)\mathrm{d}t = \frac{\Gamma(\nu+\mu)}{\mu b^\mu} \qquad \nu+\mu > 0 \qquad \mu > 0$$

and many others are listed by Gradshteyn and Ryzhik [Section 6.45].

45:11 COMPLEX ARGUMENT

Here we shall cite the form acquired by the complementary incomplete gamma function when its argument is imaginary and omit the more general case of complex argument. The result:

45:11:1 $$\Gamma(\nu;iy) = \sin\left(\frac{\pi\nu}{2}\right)\mathrm{S}(y;\nu) + \cos\left(\frac{\pi\nu}{2}\right)\mathrm{C}(y;\nu) + i\left[\sin\left(\frac{\pi\nu}{2}\right)\mathrm{C}(y;\nu) - \cos\left(\frac{\pi\nu}{2}\right)\mathrm{S}(y;\nu)\right]$$

involves Boehmer integrals [see Section 39:12].

45:12 GENERALIZATIONS

The three incomplete gamma functions are special cases either of the Kummer function [Chapter 47]

45:12:1 $$\gamma(\nu;x) = \frac{x^\nu}{\nu}\exp(-x)\,\mathrm{M}(1;1+\nu;x) = \frac{x^\nu}{\nu}\mathrm{M}(\nu;1+\nu;-x)$$

45:12:2 $$\gamma^*(\nu;x) = \frac{\exp(-x)}{\Gamma(1+\nu)}\mathrm{M}(1;1+\nu;x) = \frac{\mathrm{M}(\nu;1+\nu;-x)}{\Gamma(1+\nu)}$$

or of the Tricomi function [Chapter 48]

45:12:3 $$\Gamma(\nu;x) = x^\nu\exp(-x)\,\mathrm{U}(1;1+\nu;x) = \exp(-x)\,\mathrm{U}(1-\nu;1-\nu;x)$$

45:13 COGNATE FUNCTIONS

Just as the incomplete gamma function derives from Euler's integral of the second kind [the complete gamma function, Chapter 43] by allowing the upper limit to become indefinite, so the incomplete beta function [Chapter 58] derives from Euler's integral of the first kind [the complete beta function, Section 43:13] by similarly making the upper limit indefinite.

45:14 RELATED TOPICS

Several functions are arranged below in a hierarchical chart. The higher placed functions are the more general: arrows represent the effect of restricting one variable to a specific value. Numbers in brackets refer to chapters or sections of this *Atlas*. This chart is not, of course, exhaustive; thus, the Basset function [$\mathrm{K}_\nu(x)$, Chapter 51] is another special case of the Tricomi function, while both $\gamma^*(\nu,x)$ and $\mathrm{K}_\nu(x)$ include the exponential function [Chapter 26] among their special cases.

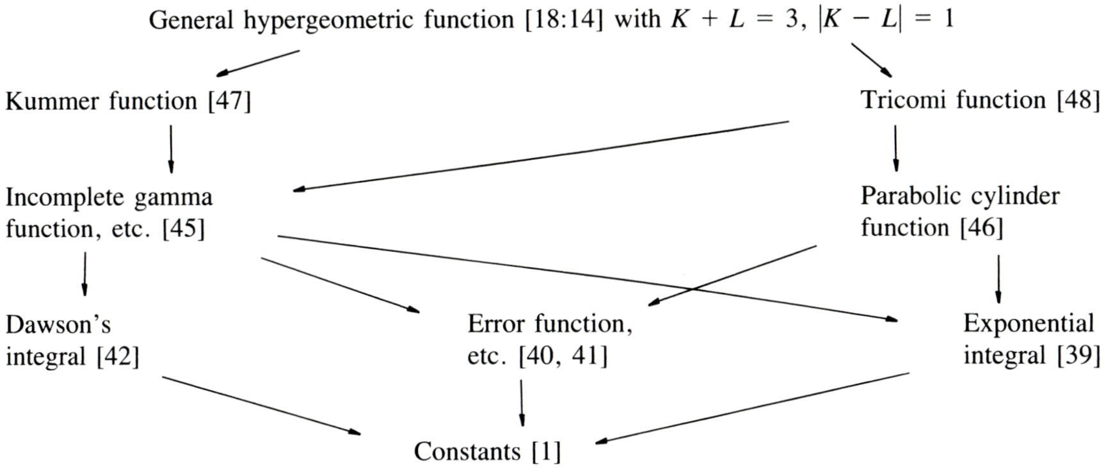
General hypergeometric function [18:14] with $K + L = 3$, $|K - L| = 1$
Kummer function [47]
Tricomi function [48]
Incomplete gamma function, etc. [45]
Parabolic cylinder function [46]
Dawson's integral [42]
Error function, etc. [40, 41]
Exponential integral [39]
Constants [1]

CHAPTER 46

THE PARABOLIC CYLINDER FUNCTION $D_\nu(x)$

This function arises in the solution to many practical problems that are conveniently expressed in parabolic cylindrical coordinates. This coordinate system, and others, are discussed in Section 46:14.

46:1 NOTATION

The parabolic cylinder function is also known as the *Weber function* or the *Weber-Hermite function*. The name "Whittaker's function" is also encountered, but confusion should be avoided with the identically named functions of Section 48:13. An alternative to the usual $D_\nu(x)$ notation is $U(-\nu - \frac{1}{2}, x)$.

The variables ν and x are known, respectively, as the order of the parabolic cylinder function and its argument. The origin of the name "parabolic cylinder" function is made evident in Section 46:14.

46:2 BEHAVIOR

The $D_\nu(x)$ function is defined for all real values of ν and x. Figure 46-1 is a contour map showing some values of the parabolic cylinder function.

For ν less than about -0.20494, $D_\nu(x)$ is a monotonically decreasing positive function of x. For larger orders, the parabolic cylinder function displays a number of zeros, maxima and minima in the $-\infty < x < \infty$ range, as shown in Table 46.2.1.

Table 46.2.1

	Number of zeros	Number of maxima	Number of minima
$-0.20494 < \nu \le 0$	0	1	0
$0 < \nu \le 1$	1	1	0
$1 < \nu \le 2$	2	1	1
$2 < \nu \le 3$	3	2	1
etc.			
$n-1 < \nu \le n$	n	$n - \text{Int}\left(\frac{n}{2}\right)$	$\text{Int}\left(\frac{n}{2}\right)$

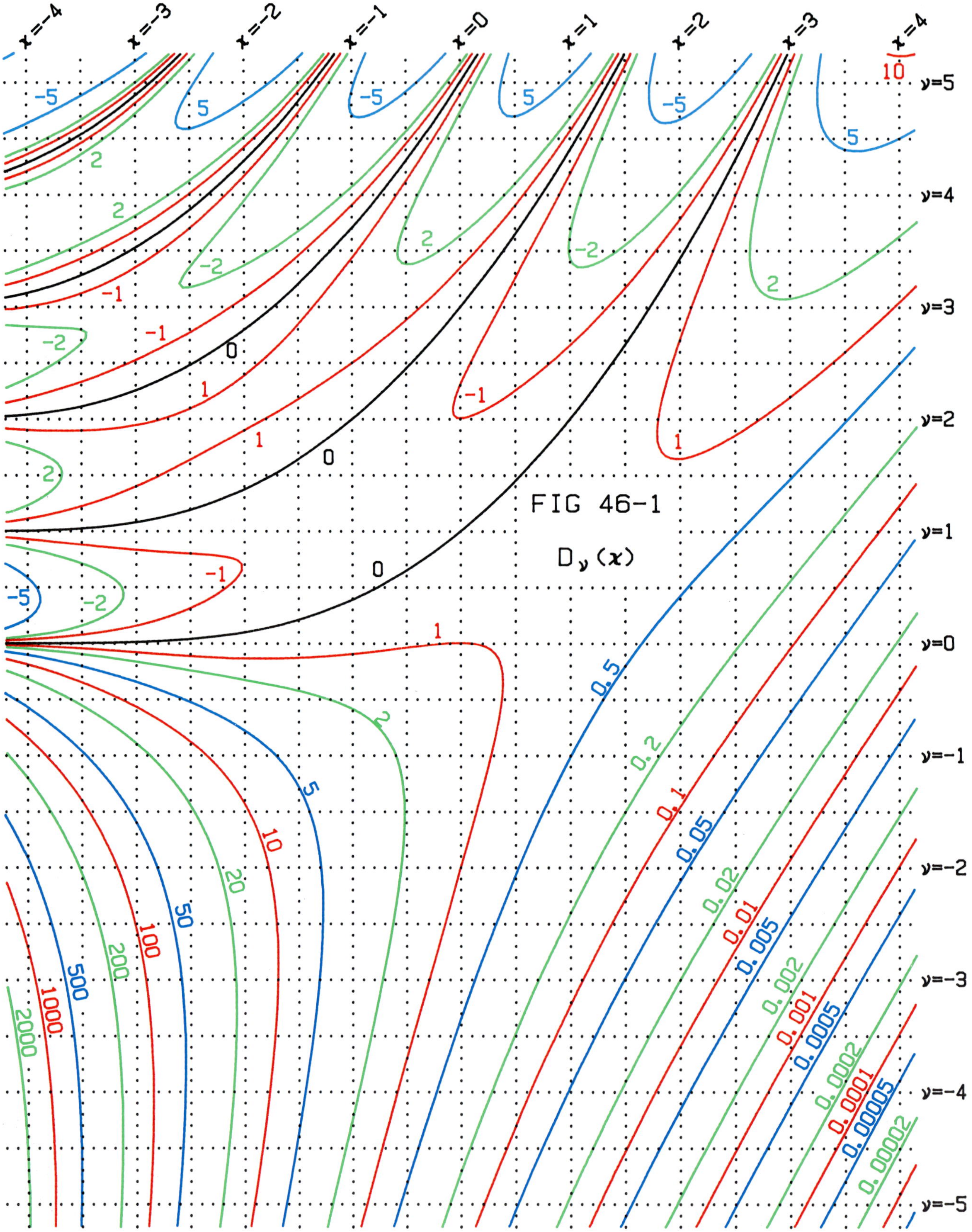
x=-4
x=-3
x=-2
x=-1
x=0
x=1
x=2
x=3
x=4
ν=5
ν=4
ν=3
ν=2
ν=1
ν=0
ν=-1
ν=-2
ν=-3
ν=-4
ν=-5
FIG 46-1
Dν(x)
0.5
0.2
0.1
0.05
0.02
0.01
0.005
0.002
0.001
0.0005
0.0002
0.0001
0.00005
0.00002
2000
1000
500
200
100
50
20
10
5
2
1
0
-1
-2
-5
10

When ν is an even/odd nonnegative integer, the function $D_\nu(x)$ is even/odd:

46.2.1 $$D_n(-x) = (-1)^n D_n(x) \qquad n = 0, 1, 2, \ldots$$

but otherwise the parabolic cylinder function is neither even nor odd.

46:3 DEFINITIONS

The parabolic cylinder function is defined by a number of integrals, including

46:3:1 $$D_\nu(x) = \sqrt{\frac{2}{\pi}}\exp\left(\frac{x^2}{4}\right)\int_0^\infty t^\nu \exp\left(\frac{-t^2}{2}\right)\cos\left(xt - \frac{\nu\pi}{2}\right)dt \qquad \nu > -1$$

and

46:3:2 $$D_\nu(x) = \frac{1}{\Gamma(-\nu)}\exp\left(\frac{-x^2}{4}\right)\int_0^\infty \exp\left(\frac{-t^2}{2} - xt\right)\frac{dt}{t^{\nu+1}} \qquad \nu < 0$$

Many other integral representations are listed by Erdélyi, Magnus, Oberhettinger and Tricomi [*Higher Transcendental Functions*, Volume 2, Section 8.3].

The differential equation known as *Weber's equation*:

46:3:3 $$\frac{d^2f}{dx^2} = \left(\frac{x^2}{4} - \nu - \frac{1}{2}\right)f$$

is satisfied by $D_\nu(x)$ and, if ν is not an integer, independently by $D_\nu(-x)$.

The equations of Section 46:12 sometimes serve as the definition of the parabolic cylinder function.

46:4 SPECIAL CASES

When ν is a nonnegative integer, the parabolic cylinder function involves a Hermite polynomial [Chapter 24] or an alternative Hermite polynomial [Section 24:13]:

46:4:1 $$D_n(x) = \frac{1}{2^{n/2}}\exp\left(\frac{-x^2}{4}\right)H_n\left(\frac{x}{\sqrt{2}}\right) = \exp\left(\frac{-x^2}{4}\right)He_n(x) \qquad n = 0, 1, 2, \ldots$$

and hence $D_0(x) = \exp(-x^2/4)$, $D_1(x) = x\exp(-x^2/4)$, etc.

With $\nu = -1$ we have reduction to the product of an exponential function and an error function complement [but note that the product is not an instance of the functions of Chapter 41]:

46:4:2 $$D_{-1}(x) = \sqrt{\frac{\pi}{2}}\exp\left(\frac{x^2}{4}\right)\operatorname{erfc}\left(\frac{x}{\sqrt{2}}\right)$$

This is the $n = 0$ case of a general expression for a parabolic cylinder function of negative integer order, which may be written as either

46:4:3 $$D_{-n-1}(x) = \frac{(-1)^n}{n!}\sqrt{\frac{\pi}{2}}\exp\left(\frac{-x^2}{4}\right)\frac{d^n}{dx^n}\left[\exp\left(\frac{x^2}{2}\right)\operatorname{erfc}\left(\frac{x}{\sqrt{2}}\right)\right] \qquad n = 0, 1, 2, \ldots$$

or

46:4:4 $$D_{-n-1}(x) = \sqrt{\frac{\pi}{2}}\,2^{n/2}\exp\left(\frac{x^2}{4}\right)i^n\operatorname{erfc}\left(\frac{x}{\sqrt{2}}\right) \qquad n = 0, 1, 2, \ldots$$

in terms of either successive derivatives or repeated integrals [see Section 40:13] of the error function complement.

When the argument of a parabolic cylinder function is positive and its order is an odd multiple of $\frac{1}{2}$, it may be expressed in terms of one or more Basset functions [Chapter 51] of orders that are odd multiples of $\frac{1}{4}$. The simplest cases are

46:4:5 $$D_{-1/2}(x) = \sqrt{\frac{x}{2\pi}}\,K_{1/4}\left(\frac{x^2}{4}\right) \qquad x>0$$

and

46:4:6 $$D_{1/2}(x) = \sqrt{\frac{x^3}{8\pi}}\left[K_{1/4}\left(\frac{x^2}{4}\right) + K_{3/4}\left(\frac{x^2}{4}\right)\right] \qquad x>0$$

and others may be evaluated via recurrence 46:5:1.

46:5 INTRARELATIONSHIPS

The parabolic cylinder function satisfies the recursion formula

46:5:1 $$D_{\nu+1}(x) = xD_\nu(x) - \nu D_{\nu-1}(x)$$

and the argument-addition formulas

46:5:2 $$D_\nu(x+y) = \exp\left(\frac{2xy+y^2}{4}\right)\sum_{j=0}^{\infty}\frac{(-y)^j}{j!}D_{\nu+j}(x) = \exp\left(\frac{-2xy-y^2}{4}\right)\sum_{j=0}^{\infty}\binom{\nu}{j}y^jD_{\nu-j}(x)$$

The sum or difference $D_\nu(x) \pm D_\nu(-x)$ can be expressed in terms of Kummer functions [Chapter 47]

46:5:3 $$D_\nu(x) + D_\nu(-x) = \frac{\sqrt{2^{\nu+2}\pi}}{\Gamma\left(\frac{1-\nu}{2}\right)}\exp\left(\frac{-x^2}{4}\right)M\left(\frac{-\nu}{2};\frac{1}{2};\frac{x^2}{2}\right) = \frac{\sqrt{2^{\nu+2}\pi}}{\Gamma\left(\frac{1-\nu}{2}\right)}\exp\left(\frac{x^2}{4}\right)M\left(\frac{1+\nu}{2};\frac{1}{2};\frac{-x^2}{2}\right)$$

46:5:4 $$D_\nu(x) - D_\nu(-x) = \frac{-\sqrt{2^{\nu+3}\pi}}{\Gamma\left(\frac{-\nu}{2}\right)}x\exp\left(\frac{-x^2}{4}\right)M\left(\frac{1-\nu}{2};\frac{3}{2};\frac{x^2}{2}\right) = \frac{-\sqrt{2^{\nu+3}\pi}}{\Gamma\left(\frac{-\nu}{2}\right)}x\exp\left(\frac{x^2}{4}\right)M\left(1+\frac{\nu}{2};\frac{3}{2};\frac{-x^2}{2}\right)$$

46:6 EXPANSIONS

A number of power series expansions exist for $\exp(\pm x^2/4)\,D_\nu(x)$. Thus, one has

46:6:1 $$D_\nu(x) = \frac{\exp(-x^2/4)}{2^{(2+\nu)/2}\Gamma(-\nu)}\sum_{j=0}^{\infty}\frac{\Gamma\left(\frac{j-\nu}{2}\right)(-x\sqrt{2})^j}{j!}$$

46:6:2 $$D_\nu(x) = \sqrt{2^\nu\pi}\exp\left(\frac{x^2}{4}\right)\sum_{j=0}^{\infty}\frac{(-x\sqrt{2})^j}{j!\,\Gamma\left(\frac{1-\nu-j}{2}\right)}$$

46:6:3 $$D_\nu(x) = \sqrt{\frac{2^\nu}{\pi}}\exp\left(\frac{x^2}{4}\right)\sum_{j=0}^{\infty}\cos\left(\frac{\nu\pi+j\pi}{2}\right)\frac{\Gamma\left(\frac{1+\nu+j}{2}\right)(x\sqrt{2})^j}{j!}$$

Alternatively, one may add equations 46:5:3 and 46:5:4 and then use expansion 47:6:1 so as to express the parabolic cylinder function as the sum of two power series.

Expansion in terms of Hermite polynomials [Chapter 24] is possible in two distinct ways:

46:6:4 $$D_\nu(x) = \frac{2^{\nu/2}}{\Gamma(-\nu/2)} \exp\left(\frac{-x^2}{4}\right) \sum_{j=0}^{\infty} \frac{H_{2j}(x/\sqrt{2})}{j!\left(j - \dfrac{\nu}{2}\right)} \left(\frac{-1}{4}\right)^j \qquad x > 0$$

and

46:6:5 $$D_\nu(x) = \frac{2^{(\nu-2)/2}}{\Gamma\left(\dfrac{1-\nu}{2}\right)} \exp\left(\frac{-x^2}{4}\right) \sum_{j=0}^{\infty} \frac{H_{2j+1}(x/\sqrt{2})}{j!\left(j + \dfrac{1-\nu}{2}\right)} \left(\frac{-1}{4}\right)^j \qquad x > 0$$

The asymptotic expansion of the parabolic cylinder function

46:6:6 $$D_\nu(x) \sim x^\nu \exp\left(\frac{-x^2}{4}\right)\left[1 - \frac{(-\nu)(1-\nu)}{2x^2} + \frac{(-\nu)(1-\nu)(2-\nu)(3-\nu)}{2!(2x^2)^2} - \cdots + \frac{(-\nu)_{2j}}{j!(-2x^2)^j} + \cdots\right] \qquad x \to \infty$$

valid for large x, is a consequence of relationship 46:12:2 or 46:12:3, together with 48:6:1.

46:7 PARTICULAR VALUES

	$D_\nu(-\infty)$	$D_\nu(0)$	$D_\nu(\infty)$
$1 < \nu < 2,\ 3 < \nu < 4,\ 5 < \nu < 6, \ldots$	$+\infty$	$\dfrac{\sqrt{2^\nu \pi}}{\Gamma\left(\dfrac{1-\nu}{2}\right)}$	0
$0 < \nu < 1,\ 2 < \nu < 3,\ 4 < \nu < 5, \ldots$	$-\infty$		
$\nu = 0, 1, 2, \ldots$	0		
$\nu < 0$	$+\infty$		

46:8 NUMERICAL VALUES

From equation 46:6:1 one can develop the equation

46:8:1 $$D_\nu(x) = \sqrt{2^\nu}\left\{(F_0 + G_0)\exp\left(\frac{x^2}{4}\right) + \exp\left(\frac{-x^2}{4}\right) \sum_{j=1}^{\infty} (F_j - f_j) + (G_j - g_j)\right\}$$

where $F_0 = f_0 = \sqrt{\pi}/\Gamma[(1-\nu)/2]$, $G_0 = g_0 = -x\sqrt{2\pi}/\Gamma(-\nu/2)$, $F_j/F_{j-1} = [j - 1 - (\nu/2)]x^2/j(2j-1)$, $G_j/G_{j-1} = [j - \frac{1}{2} - (\nu/2)]x^2/j(2j+1)$ and $f_j/f_{j-1} = g_j/g_{j-1} = x^2/2j$. This expression is more convenient than any of the expansions in Section 46:6 and is the basis of the algorithm presented below. In the algorithm the upper infinite limit in equation 46:8:1 is replaced by an empirically determined integer dependent upon x and ν. Generally, the precision of the algorithm exceeds 24 bits (i.e., the relative accuracy of $D_\nu(x)$ is 6×10^{-8} or better) but this may be degraded near the zeros of $D_\nu(x)$ or when $D_\nu(x)$ has a very small magnitude (e.g., when x and $-\nu$ are both large and positive).

The first part of the algorithm calculates $\sqrt{\pi}/\Gamma[(1-\nu)/2]$ and $\sqrt{\pi}/\Gamma(-\nu/2)$ using essentially the same procedure as in Section 43:8. Because these values are stored permanently (in registers a and b, respectively), there is no need to access the first part of the algorithm if one wants to calculate another parabolic cylinder function $D_\nu(x)$ without altering the order ν.

Set $a = 1$
Input ν >>>>>>>
(1) Set $x = (a - \nu)/2$
Set $g = 1$

Storage needed:
a, ν, x, f, g, b, w, d
G, J, F, h and j

(2) If $x = 0$ go to (3)
Replace g by gx
Replace x by $x + 1$
If $x \le 3$ go to (2)
Set $b = 2/x^2$
Set $f = b\left[1 + \frac{b}{7}\left(\frac{b}{3} - 1\right)\right] \Big/ 60$
Replace f by $\frac{f-1}{12x} - x[\ln(x) - 1]$
Replace f by $g\sqrt{x/2}\,\exp(f)$
(3) If $a = 0$ go to (4)
Set $a = 0$
Set $w = f$
Go to (1)
(4) Set $a = w$
Set $b = f$
Input x >>>>>>>
Set $G = -bx\sqrt{2}$
Set $J = 3 + 6|x| + (|(x + 2)(2 - \nu)|/5) + (x^2/2)$
Replace x by $x^2/2$
Set $w = \exp\left(\nu \ln\sqrt{2} - \frac{x}{2}\right)$
Replace G by Gw
Set $f = F = aw$
Set $g = G$
Set $d = (F + G)\exp(x)$
Set $w = (\nu + 1)/2$
Set $j = 0$
(5) Replace j by $j + 1$
Replace F by $\left(j - \frac{1}{2} - w\right) Fx/j\left(j - \frac{1}{2}\right)$
Replace f by fx/j
Replace g by gx/j
Replace G by $(j - w)\,Gx/j\left(j + \frac{1}{2}\right)$
Replace d by $d + (F - f) + (G - g)$
If $j \le J$ go to (5)
Output d
$d \simeq D_\nu(x)$ <<<<<<<

Test values:
$D_3(0) = D_2(\pm 1) = 0$
$D_{-2}(1.7) = 0.0944123874$
$D_{-3.4}(-2.8) = 85.6541111$

For large arguments, the parabolic cylinder function is best determined via expansion 46:6:6 [see also Section 48:8].

46:9 APPROXIMATIONS

If x is small and of either sign

46:9:1
$$D_\nu(x) \simeq \sqrt{2^\nu \pi}\left[\frac{1 - \left(\frac{1}{4} + \frac{\nu}{2}\right)x^2}{\Gamma\left(\frac{1-\nu}{2}\right)} - \frac{x\sqrt{2}}{\Gamma\left(\frac{-\nu}{2}\right)}\right]$$

whereas for large positive arguments

46:9:2 $$D_\nu(x) \simeq x^\nu \exp\left(\frac{-x^2}{4}\right)\left[1 + \frac{\nu - \nu^2}{2x^2}\right]$$

The latter approximation is valid even for moderate positive x if ν is close to one of the integers 0, 1, 2 or 3.

46:10 OPERATIONS OF THE CALCULUS

Rules for single and double differentiation are

46:10:1 $$\frac{d}{dx} D_\nu(x) = \frac{x}{2} D_\nu(x) - D_{\nu+1}(x) = \nu D_{\nu-1}(x) - \frac{x}{2} D_\nu(x)$$

46:10:2 $$\frac{d^2}{dx^2} D_\nu(x) = \left(\frac{x^2}{4} - \frac{1}{2} - \nu\right) D_\nu(x)$$

and we also have the elegant relationship

46:10:3 $$\frac{d^n}{dx^n} \exp\left(\frac{-x^2}{4}\right) D_\nu(x) = (-1)^n \exp\left(\frac{-x^2}{4}\right) D_{\nu+n}(x) \qquad n = 0, 1, 2, \ldots$$

The very general definite integral

46:10:4 $$\int_0^\infty t^p \exp\left(\frac{-\alpha t^2}{4}\right) D_\nu(t)dt = \sqrt{\frac{2^\nu \pi}{(\alpha+1)^{p+1}}} \frac{\Gamma(p+1)}{\Gamma\left(\frac{p-\nu}{2}+1\right)} F\left(\frac{p+1}{2}, \frac{-\nu}{2}; 1 + \frac{p-\nu}{2}; \frac{\alpha-1}{\alpha+1}\right) \qquad p > -1 \qquad \alpha > 0$$

yields a Gauss function [Chapter 60].

46:11 COMPLEX ARGUMENT

The parabolic cylinder function is complex when its argument is complex. Here we present only the formula

46:11:1 $$D_\nu(iy) = \frac{\Gamma(1+\nu)}{\sqrt{2\pi}}\left\{\cos\left(\frac{\pi\nu}{2}\right)[D_{-\nu-1}(y) + D_{-\nu-1}(-y)] - i \sin\left(\frac{\pi\nu}{2}\right)[D_{-\nu-1}(y) - D_{-\nu-1}(-y)]\right\}$$

for the case of imaginary argument.

46:12 GENERALIZATIONS

The Kummer function [Chapter 47] and the Tricomi function [Chapter 48] may be regarded as generalizations of the parabolic cylinder function. The relationships are

46:12:1 $$D_\nu(x) = \sqrt{2^\nu \pi} \exp\left(\frac{-x^2}{4}\right)\left[\frac{M\left(\frac{-\nu}{2}; \frac{1}{2}; \frac{x^2}{2}\right)}{\Gamma\left(\frac{1-\nu}{2}\right)} - \sqrt{2}x \frac{M\left(\frac{1-\nu}{2}; \frac{3}{2}; \frac{x^2}{2}\right)}{\Gamma\left(\frac{-\nu}{2}\right)}\right]$$

$$= \sqrt{2^\nu \pi} \exp\left(\frac{x^2}{4}\right)\left[\frac{M\left(\frac{1+\nu}{2}; \frac{1}{2}; \frac{-x^2}{2}\right)}{\Gamma\left(\frac{1-\nu}{2}\right)} - \sqrt{2}x \frac{M\left(1 + \frac{\nu}{2}; \frac{3}{2}; \frac{-x^2}{2}\right)}{\Gamma\left(\frac{-\nu}{2}\right)}\right] \qquad x \geq 0$$

46:12:2 $$D_\nu(x) = 2^{\nu/2}\exp\left(\frac{-x^2}{4}\right)U\left(-\frac{\nu}{2};\frac{1}{2};\frac{x^2}{2}\right) \qquad x \geq 0$$

46:12:3 $$D_\nu(x) = 2^{(\nu-1)/2}\exp\left(\frac{-x^2}{4}\right)U\left(\frac{1-\nu}{2};\frac{3}{2};\frac{x^2}{2}\right) \qquad x \geq 0$$

46:13 COGNATE FUNCTIONS

The function

46:13:1 $$\frac{\Gamma(-\nu)}{\pi}[D_\nu(-x) - \cos(\pi\nu)D_\nu(x)]$$

is sometimes encountered and is symbolized $V(-\nu - \frac{1}{2}, x)$.

46:14 RELATED TOPICS

In many contexts there arises a need to map space using an origin and three coordinates. In this section we discuss a number of *orthogonal coordinate systems*, which constitute the most useful ways of performing such mappings.

If we use r, q and p to denote three general coordinates, specifying the triplet (r,q,p) of numbers locates a unique point in space. Specifying two of the coordinates, say q and p, but allowing the third to adopt any permissible value, defines a line (generally a space curve) that we can denote (q,p). Specifying only one coordinate, say r, defines a surface (r). It is a characteristic of an *orthogonal* coordinate system that, at any point (r,q,p), the three surfaces (r), (q) and (p) are mutually perpendicular. Likewise, the three lines (q,p), (r,p) and (r,q) are mutually perpendicular at (r,q,p).

In simple physical applications, each (r) surface, defined by allocating a specific value to the r coordinate, may correspond to a particular value of some scalar quantity F (temperature, energy, concentration, electric potential, etc.). Such surfaces are sometimes called *equipotentials*. Some physical body, known as the *generator* of the coordinate system, may occupy the $r = 0$ surface. The line (p,q) corresponding to specified values of the p and q coordinates is known by a variety of names such as "line of force," "flux line," "field vector," "line of steepest descent," "streamline," etc., depending on the field of application. Here we use the name *streamline*.

Of course, the most familiar set of orthogonal coordinates is the *cartesian coordinate system*, (x,y,z). We may think of this as arising from a generator corresponding to the infinite plane $y = 0$ with uniformly spaced equipotentials $y = \pm 1$, $y = \pm 2$, $y = \pm 3$, ... The streamlines are straight lines. A cartesian coordinate system is depicted in Figure 46-2. As in the other diagrams of this section, streamlines are shown in red, equipotentials in green and the generator in blue; the z-coordinate is perpendicular to the plane of the paper. The *rectangular coordinate system* may equally well be represented by Figure 46-2; it is the two-dimensional equivalent of the cartesian system.

The *cylindrical coordinate system* is illustrated in Figure 46-3. The coordinates consist of two lengths, r and z, and one angle, θ. The relationship to cartesian coordinates is

46:14:1 $$x = r\cos(\theta) \qquad y = r\sin(\theta) \qquad 0 \leq r < \infty \qquad -\pi < \theta \leq \pi$$

The generator is the line ($r = 0$, θ arbitrary); streamlines are straight lines radiating from the generator; equipotentials are circular cylinders having the generator as their common axis. The *polar coordinate system* is the two-dimensional analog of the cylindrical coordinate system: equations 46:14:1 and Figure 46-3 apply equally to both.

The generator of the *parabolic cylindrical coordinate system* is the half-plane ($y = 0$, $x < 0$) as depicted in Figure 46-4. The coordinates are r, q and z, the first two of these being related to cartesian coordinates by

46:14:2 $$x = \frac{1}{2}(r^2 - q^2) \qquad y = rq \qquad 0 \leq r < \infty \qquad -\infty < q < \infty$$

The equipotentials are parabolic cylinders with a common focal axis ($x = 0$, $y = 0$) [or ($r = 0$, $q = 0$)], while the streamlines are semiparabolas [see Chapter 12] whose foci lie on the same axis.

In the *elliptic cylindrical coordinate system* the generator is the strip ($y = 0$, $-1 < x < 1$). The r and θ

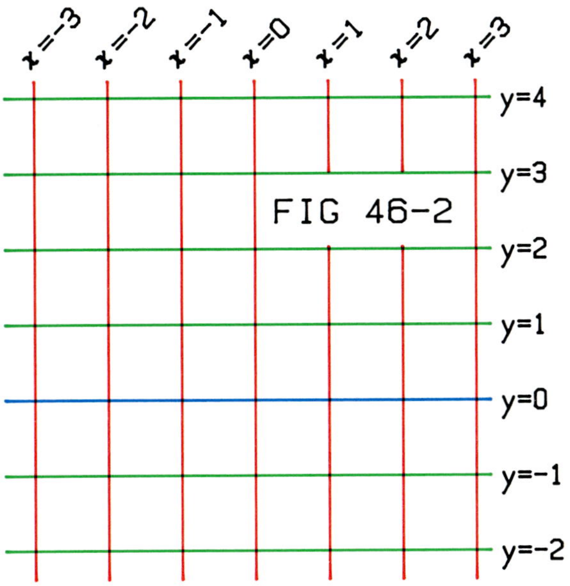

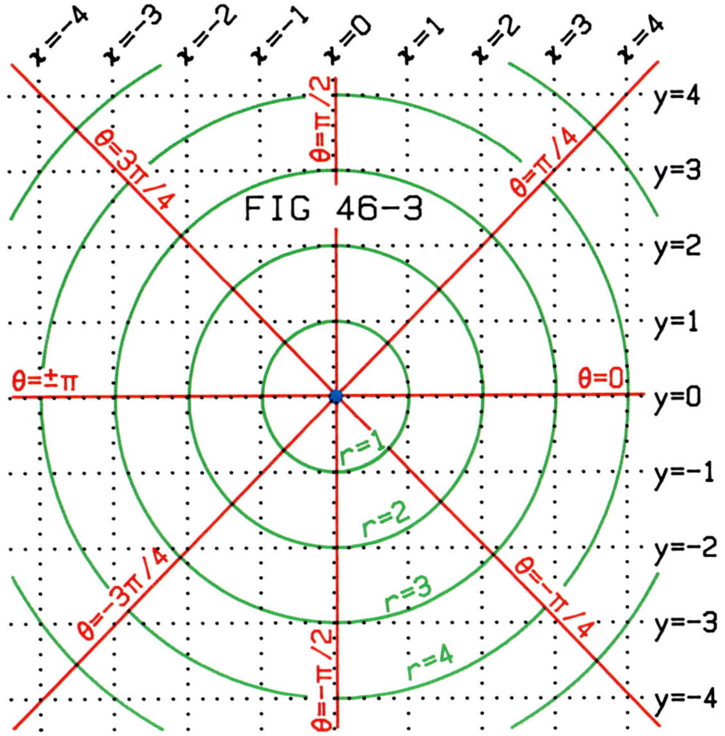

coordinates are related by

46:14:3 $x = \cosh(r)\cos(\theta) \qquad y = \sinh(r)\sin(\theta) \qquad 0 \le r < \infty \qquad -\pi < \theta \le \pi$

to cartesian coordinates, while the third coordinate is again z. As shown in Figure 46-5, the equipotentials are elliptic cylinders and the streamlines are semihyperbolas [see Chapter 15].

All the coordinate systems we have considered up to this point have had a single generator at $r = 0$, although one might postulate a second generator at $r = \infty$. The *bipolar coordinate system* has two generators, corresponding to $r = +\infty$ and $r = -\infty$, located on the lines ($x = 0$, $y = 1$) and ($x = 0$, $y = -1$), respectively. Representing the three coordinates by r, θ and z, we have the relationships

46:14:4 $$x = \frac{\sinh(r)}{\cosh(r) - \cos(\theta)} \qquad y = \frac{\sin(\theta)}{\cosh(r) - \cos(\theta)} \qquad -\infty < r < \infty \qquad -\pi < \theta \le \pi$$

to cartesian coordinates. As Figure 46-6 shows, the equipotentials are mostly circular cylinders; their axes are the straight lines ($y = 0$, $x = \coth(r)$) and their radii are $|\mathrm{csch}(r)|$. Streamlines are arcs of circles whose centers lie on the line ($x = 0$, z) at $y = \cot(\theta)$ with radii equal to $|\csc(\theta)|$.

The *spherical coordinate system* employs one length r and two angles θ and ϕ as its three coordinates. Because it is inherently three dimensional, we do not include a diagrammatic representation of this system, which is related to cartesian coordinates by the equations

46:14:5 $$x = r\sin(\theta)\cos(\phi) \quad y = r\sin(\theta)\sin(\phi) \quad z = r\cos(\theta) \quad 0 \le r < \infty \quad -\frac{\pi}{2} \le \theta < \frac{\pi}{2} \quad -\pi < \phi \le \pi$$

The generator is a single point at the origin; equipotentials are spheres centered on the origin, and streamlines are straight lines radiating in all directions from the origin.

There are a number of other orthogonal coordinate systems that we shall not discuss in detail. The generator of the *paraboloidal coordinate system* is the half-line ($x = 0$, $y = 0$, $z > 0$). The generator of the *prolate spheroidal coordinate system* is the line segment ($x = 0$, $y = 0$, $-1 < z < 1$). The generator of the *oblate spheroidal coordinate system* is the disc ($x^2 + y^2 < 1$, $z = 0$). The generator of the *toroidal coordinate system* is the hoop ($x^2 + y^2 = 1$, $z = 0$). For details of these systems, and three others, see Spiegel [pages 126–130].

With k sometimes positive, sometimes negative and sometimes zero, the equation

46:14:6 $$\nabla^2 F = kF$$

often known as the *Helmholtz equation*, is ubiquitous throughout all of science. Special instances are the *wave equation* for the propagation of vibrations, *Laplace's equation* describing electric fields, the *Fourier equation* of

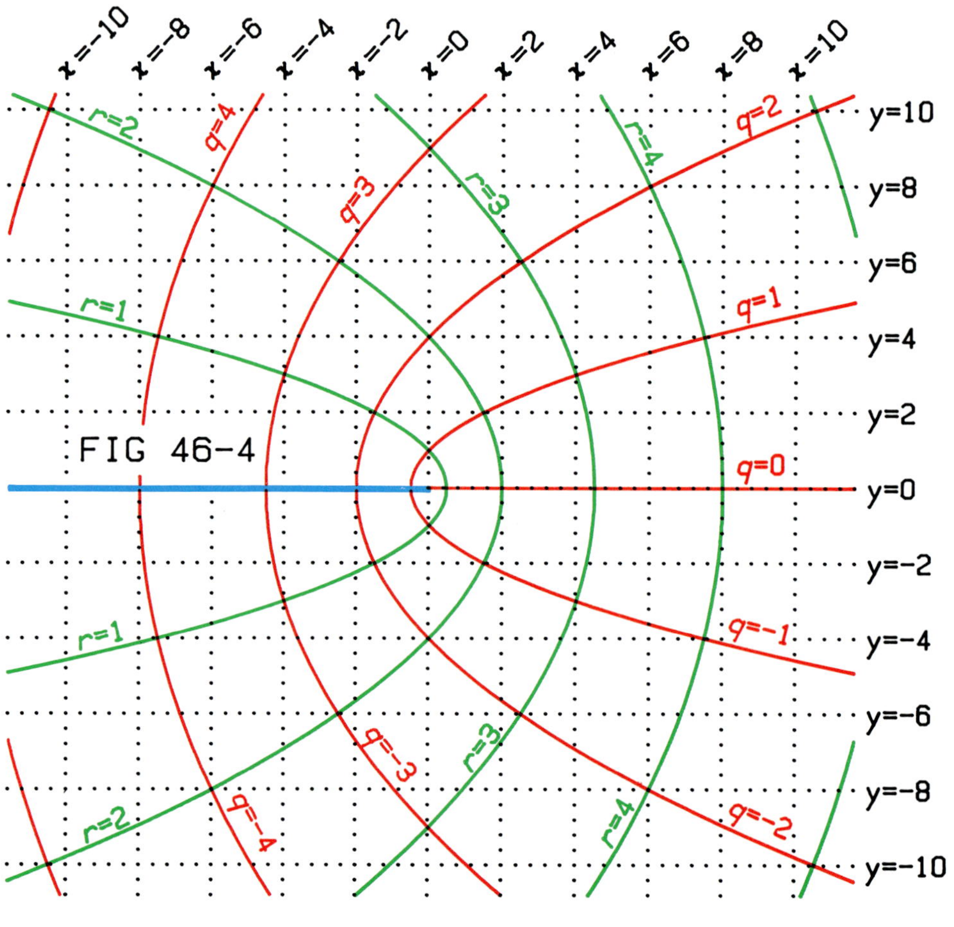

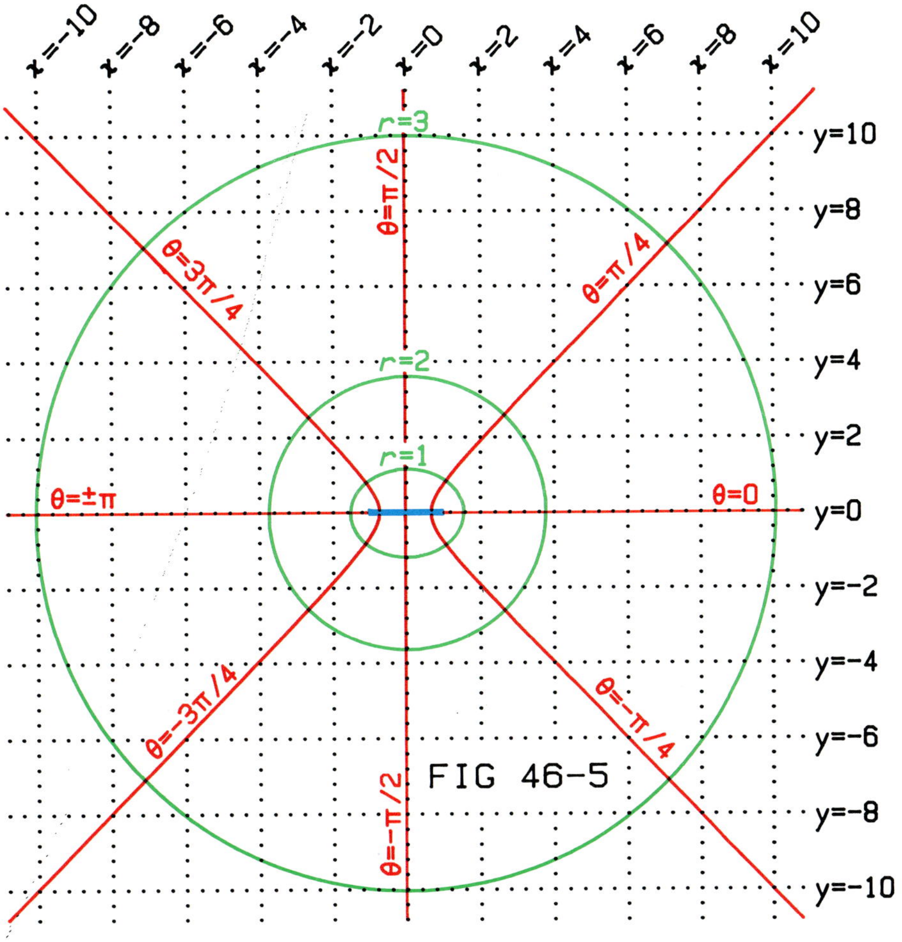

heat conduction, the *Schrödinger equation* of quantum physics and *Fick's second law* of diffusion. In all these cases F is some scalar property whose value depends on the spatial coordinates (and often on time also, although here we shall ignore the temporal variation of F). The so-called *Laplacian operator* ∇^2 performs double differentiation with respect to the spatial coordinates, the precise form of the operator depending on the coordinate system adopted. Table 46.14.1 lists the forms of ∇^2F for a variety of orthogonal coordinate systems.

Table 46.14.1

Coordinate system	∇^2F
Cartesian	$\frac{\partial^2 F}{\partial x^2} + \frac{\partial^2 F}{\partial y^2} + \frac{\partial^2 F}{\partial z^2}$
Cylindrical	$\frac{\partial^2 F}{\partial r^2} + \frac{1}{r}\frac{\partial F}{\partial r} + \frac{1}{r^2}\frac{\partial^2 F}{\partial \theta^2} + \frac{\partial^2 F}{\partial z^2}$
Parabolic cylindrical	$\frac{1}{r^2 + q^2}\left(\frac{\partial^2 F}{\partial r^2} + \frac{\partial^2 F}{\partial q^2}\right) + \frac{\partial^2 F}{\partial z^2}$
Elliptic cylindrical	$\frac{1}{\sinh^2(r) + \sin^2(\theta)}\left(\frac{\partial^2 F}{\partial r^2} + \frac{\partial^2 F}{\partial \theta^2}\right) + \frac{\partial^2 F}{\partial z^2}$
Bipolar	$[\cosh(r) - \cos(\theta)]^2\left(\frac{\partial^2 F}{\partial r^2} + \frac{\partial^2 F}{\partial \theta^2}\right) + \frac{\partial^2 F}{\partial z^2}$
Spherical	$\frac{\partial^2 F}{\partial r^2} + \frac{2}{r}\frac{\partial F}{\partial r} + \frac{\cot(\theta)}{r^2}\frac{\partial F}{\partial \theta} + \frac{1}{r^2}\frac{\partial^2 F}{\partial \theta^2} + \frac{\csc^2(\theta)}{r^2}\frac{\partial^2 F}{\partial \phi^2}$

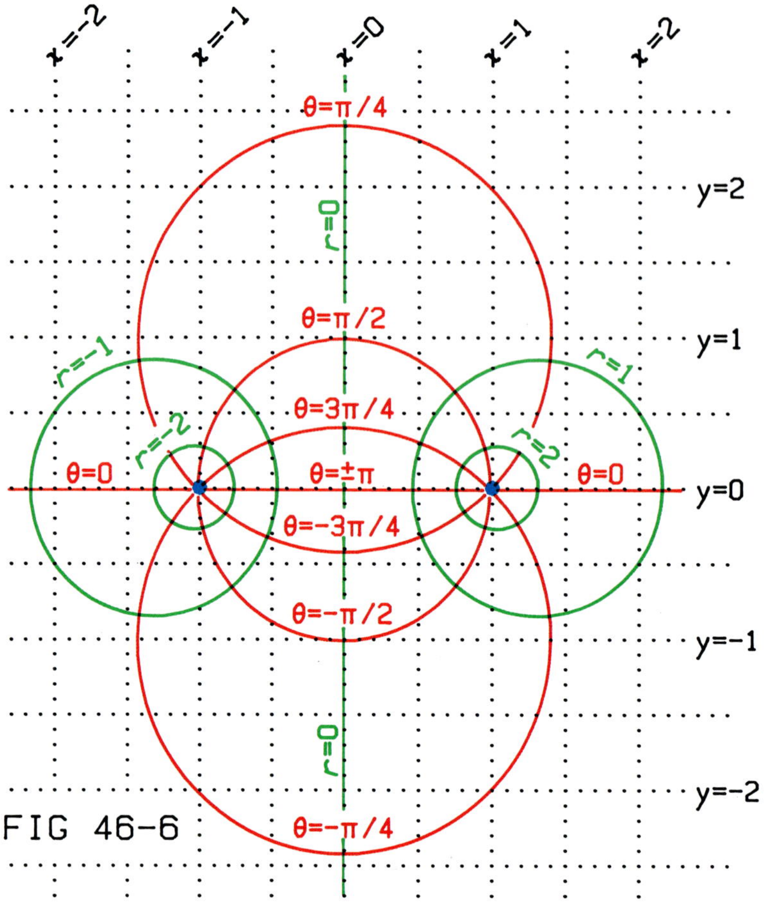

FIG 46-6

Because the cartesian coordinate system is undoubtedly the simplest, both conceptually and in the form of its Laplacian, one might wonder why other systems are ever used. One answer is that, for a physical body that matches the corresponding generator, the Helmholtz equation is most easily solved in the appropriate orthogonal coordinate system. We shall illustrate the method of solution by reference to the parabolic cylindrical coordinate system.

The function F that is a solution of the Helmholtz equation

46:14:7 $$\nabla^2 F = \frac{1}{r^2+q^2}\left(\frac{\partial^2 F}{\partial r^2} + \frac{\partial^2 F}{\partial q^2}\right) + \frac{\partial^2 F}{\partial z^2} = kF$$

will generally be a function of all three coordinates: r, q and z. Assume, however, that $F = R(r)\,Q(q)\,Z(z)$ where R is a function of r, but not of q or z, and similarly for Q and Z. Then equation 46:14:7 is easily transformed to

46:14:8 $$\frac{1}{r^2+q^2}\left(\frac{1}{R}\frac{d^2R}{dr^2} + \frac{1}{Q}\frac{d^2Q}{dq^2}\right) = k - \frac{1}{Z}\frac{d^2Z}{dz^2}$$

A consequence of our assumption is that the left-hand side of this equation is a function of r and q, but not of z, whereas the right-hand side is a function of z, but of neither r nor q. It follows that each side must equal the same constant, a so-called *separation constant*, say c^4, so that

46:14:9 $$\frac{d^2Z}{dz^2} = (k - c^4)\,Z$$

and

46:14:10 $$\frac{1}{R}\frac{d^2R}{dr^2} - c^4r^2 = c^4q^2 - \frac{1}{Q}\frac{d^2Q}{dq^2}$$

Once again, we can argue that, because the left-hand side of 46:14:10 does not depend on q and the right-hand side does not depend on r, each side must equal another separation constant. It is convenient to represent this new constant by $c^2(2\nu + 1)$, whence

46:14:11 $$\frac{1}{c^2}\frac{d^2Q}{dq^2} = (c^2q^2 - 2\nu - 1)Q$$

and

46:14:12 $$\frac{1}{c^2}\frac{d^2R}{dr^2} = (c^2r^2 + 2\nu + 1)R$$

Thus, the Helmholtz equation has been decomposed into three ordinary differential equations. It requires only a redefinition of the independent variable to convert equation 46:14:11 to 46:3:3, showing that a parabolic cylinder function can be a solution of the Helmholtz equation expressed in parabolic cylindrical coordinates.

Section 59:14 presents a second example of the solution of the Helmholtz equation.

CHAPTER 47

THE KUMMER FUNCTION M(a;c;x)

The Kummer function is one of the most important instances of the hypergeometric function [Section 18:14] and is closely related to the Tricomi function of Chapter 48.

47:1 NOTATION

The symbol $\Phi(a;c;x)$ often replaces $M(a;c;x)$. The variables a and c are termed the numeratorial parameter and denominatorial parameter, respectively; x is the argument.

Alternative notations for the Kummer function make recourse to a generalization of the Laguerre function [Section 23:14]

47:1:1 $$L_\nu^{(\mu)}(x) = \frac{M(-\nu;\mu+1;x)}{\Gamma(\nu+1)}$$

or to a simple instance of the generalized hypergeometric function [Section 60:12]

47:1:2 $${}_1F_1(a;c;x) = M(a;c;x)$$

Collectively, the Kummer function and the Tricomi function [Chapter 48] are known as *confluent hypergeometric functions* or as *degenerate hypergeometric functions*. These puzzling names have their origins in definitions 47:3:2 and 48:3:5.

47:2 BEHAVIOR

The Kummer function is defined for all real values of a and x and for all values of c except $c = 0, -1, -2, \ldots$

Being trivariate, $M(a;c;x)$ is incapable of having its behavior comprehensively described in a two-dimensional map. Some examples of the wide variety of behaviors are depicted in Figure 47-1. In its dependence upon x, the Kummer function may, or may not, encounter zeros, maxima or minima, depending on the values of the two parameters. Figure 47-2 may be used to find, for specific values of a and c, the number of negative zeros of the Kummer function, that is, the number of times that $M(a;c;x)$ acquires the value zero in the argument range $-\infty < x < 0$. Similarly, Figure 47-3 can be used to find the number of positive zeros of $M(a;c;x)$. In the regions in which the Kummer function lacks zeros, it is invariably positive.

For the behavior of the Kummer function as its argument approaches infinity in magnitude, see Section 47:7.

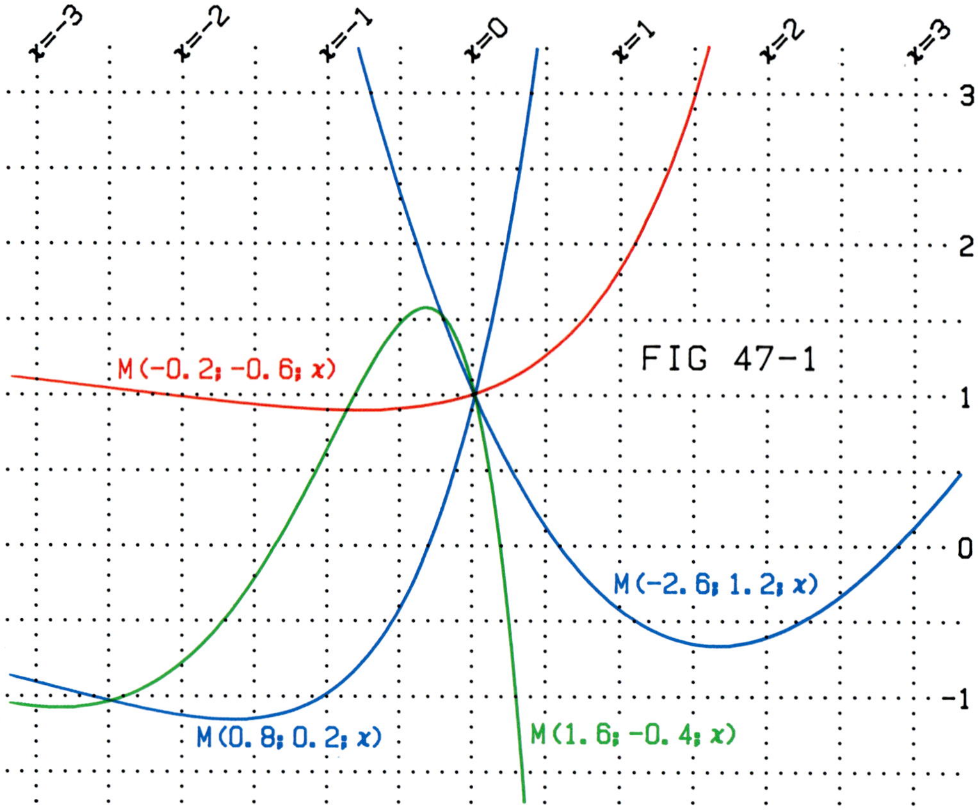

x=−3
x=−2
x=−1
x=0
x=1
x=2
x=3
3
2
1
0
−1
FIG 47-1
M(−0.2; −0.6; x)
M(−2.6; 1.2; x)
M(0.8; 0.2; x)
M(1.6; −0.4; x)

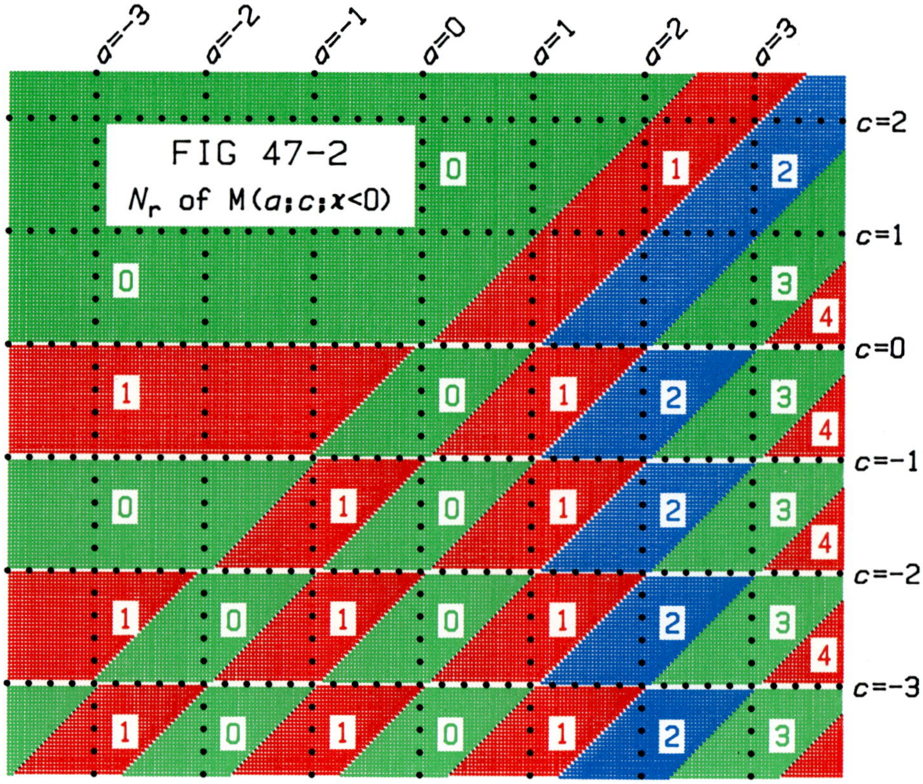

a=−3
a=−2
a=−1
a=0
a=1
a=2
a=3
FIG 47-2
N_r of M(a;c;x<0)
c=2
c=1
c=0
c=−1
c=−2
c=−3

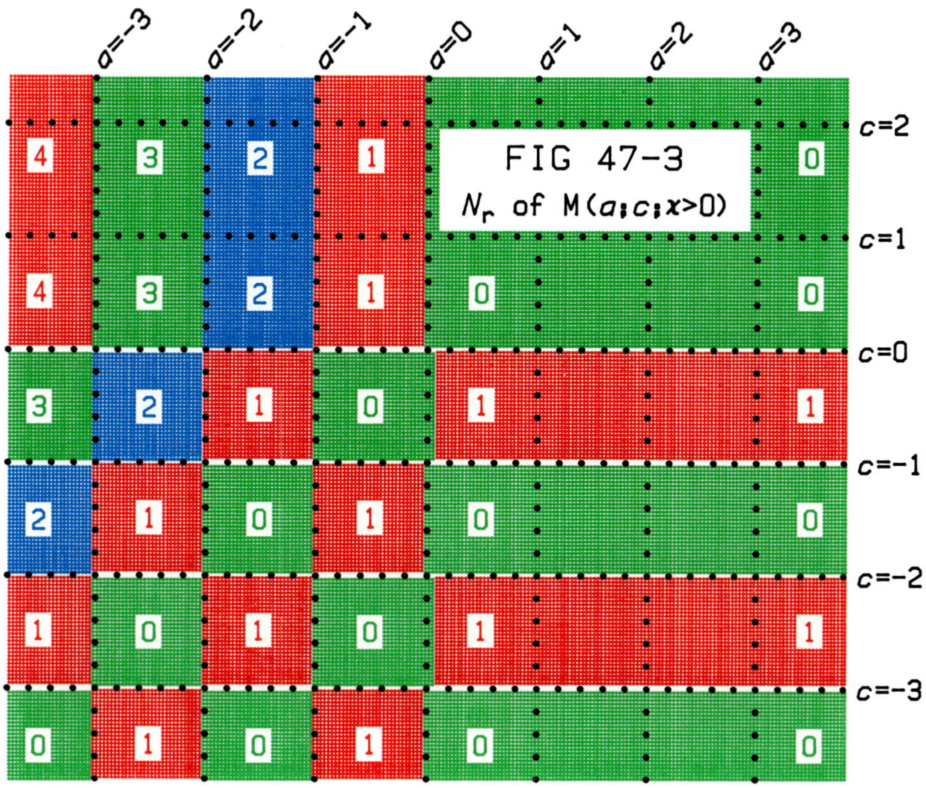

47:3 DEFINITIONS

The Kummer function may be defined as a hypergeometric function [Section 18:14] with one numeratorial parameter and two denominatorial parameters, one of the latter being equal to unity:

47:3:1 $$\mathrm{M}(a;c;x) = \sum_{j=0}^{\infty} \frac{(a)_j}{(c)_j(1)_j} x^j$$

It may also be defined in terms of the Gauss function [Chapter 60] by means of the limiting operation

47:3:2 $$\mathrm{M}(a;c;x) = \lim_{b\to\infty} \mathrm{F}\left(a,b;c;\frac{x}{b}\right)$$

The indefinite integral

47:3:3 $$\mathrm{M}(a;c;x) = \frac{\Gamma(c)x^{1-c}}{\Gamma(c-a)\,\Gamma(a)} \int_0^x \frac{t^{a-1}\exp(t)}{(x-t)^{1+a-c}}\,\mathrm{d}t \qquad 0 < a < c$$

is a representation of the Kummer function, as are the definite integrals

47:3:4 $$\mathrm{M}(a;c;x) = \frac{\Gamma(c)}{\Gamma(c-a)\,\Gamma(a)} \int_0^1 \frac{t^{a-1}\exp(xt)}{(1-t)^{1+a-c}}\,\mathrm{d}t \qquad 0 < a < c$$

and

47:3:5 $$\mathrm{M}(a;c;x) = \frac{2^{1-c}\exp(x/2)\,\Gamma(c)}{\Gamma(c-a)\,\Gamma(a)} \int_{-1}^1 \frac{(1+t)^{a-1}\exp(xt/2)}{(1-t)^{1+a-c}}\,\mathrm{d}t \qquad 0 < a < c$$

The so-called *confluent hypergeometric equation*

47:3:6
$$x\frac{d^2f}{dx^2} + (c - x)\frac{df}{dx} - af = 0$$

has the solution $f = c_1M(a;c;x) + c_2x^{1-c}M(a - c + 1;2 - c;x)$ provided that c is not an integer; c_1 and c_2 are arbitrary constants.

The Kummer function is obtained by differintegrating the function $x^{a-1}\exp(x)$ with respect to x, using a lower limit of zero:

47:3:7
$$M(a;c;x) = \frac{\Gamma(c)}{\Gamma(a)} x^{1-c} \frac{d^{a-c}}{dx^{a-c}} [x^{a-1}\exp(x)]$$

Thus, in the notation of Section 43:14, the synthesis

47:3:8
$$\exp(x) \xrightarrow[1-a]{1-c} M(a;c;x)$$

gives rise to the Kummer function.

47:4 SPECIAL CASES

The exponential function [Chapter 26] is the special case of the Kummer function in which the parameters are identical:

47:4:1
$$M(a;a;x) = \exp(x)$$

If the denominatorial parameter exceeds the numeratorial parameter by unity, the Kummer function reduces to an incomplete gamma function [Chapter 45]:

47:4:2
$$M(a;1 + a;x) = \frac{a}{(-x)^a}\gamma(a;-x) = \Gamma(1 + a)\gamma^*(a;-x)$$

The Kummer function $M(a;a + n;x)$ in which the denominatorial parameter exceeds the numeratorial parameter by a positive integer may also, by sufficient applications of recurrence formula 47:5:3, be expressed as an entire incomplete gamma function; for example:

47:4:3
$$M(a;2 + a;x) = \frac{a + x}{x}\Gamma(2 + a)\gamma^*(a;-x) - \frac{a(1 + a)}{x}\exp(x)$$

When the denominatorial parameter is twice the numeratorial parameter, $c = 2a$, the Kummer function specializes to the product of the exponential function and the hyperbolic Bessel function [Chapter 50] of order $a - \frac{1}{2}$:

47:4:4
$$M(a;2a;x) = \frac{\Gamma(\frac{1}{2} + a)\exp\left(\frac{x}{2}\right)}{(x/4)^{a-1/2}} I_{a-1/2}\left(\frac{x}{2}\right)$$

With the numeratorial parameter equal to a negative integer, the Kummer function becomes a generalized Laguerre polynomial [Section 23:12]:

47:4:5
$$M(-n;c;x) = L_n^{(c-1)}(x) \Big/ \binom{n + c - 1}{n} \qquad n = 1, 2, 3, \ldots \qquad c > 0$$

If $a = 0$, the Kummer function equals unity, and if $a = 1$, specialization to the incomplete gamma function occurs again:

47:4:6
$$M(1;c;x) = 1 + x^{1-c}\exp(x)\,\gamma(c;x) = 1 + \Gamma(c)x\exp(x)\,\gamma^*(c;x)$$

Coupled with recurrence relationship 47:5:2, this last result also enables $M(2;c;x)$, $M(3;c;x)$, etc., to be expressed in terms of the entire incomplete gamma function or, alternatively, if x is positive, in terms of the incomplete gamma function.

Except perhaps when a is also a negative integer, setting $c = 0, -1, -2, \ldots$ causes $\mathrm{M}(a;c;x)$ to be undefined [but see Section 47:12]. When $c = 1$, reduction occurs to a Laguerre function [Section 23:14].

47:4:7 $$\mathrm{M}(a;1;x) = \mathrm{L}_{-a}(x)$$

Use of recurrence formula 47:5:3 then permits the special cases $\mathrm{M}(a;2;x)$, $\mathrm{M}(a;3;x)$, etc., to be expressed as Laguerre functions also.

With a denominatorial parameter equal to $\frac{1}{2}$, the Kummer function becomes the sum of two parabolic cylinder functions [Chapter 46]

47:4:8 $$\mathrm{M}\left(a;\frac{1}{2};x\right) = \begin{cases} \dfrac{2^{a-1}\Gamma(a+\frac{1}{2})}{\sqrt{\pi}} \exp\left(\dfrac{x}{2}\right)[\mathrm{D}_{-2a}(\sqrt{2x}) + \mathrm{D}_{-2a}(-\sqrt{2x})] & x \geq 0 \\ \dfrac{\Gamma(1-a)}{2^a\sqrt{2\pi}} \exp\left(\dfrac{x}{2}\right)[\mathrm{D}_{2a-1}(\sqrt{-2x}) + \mathrm{D}_{2a-1}(-\sqrt{-2x})] & x \leq 0 \end{cases}$$

The difference of two other parabolic cylinder functions is the special $c = \frac{3}{2}$ case of the Kummer function:

47:4:9 $$\mathrm{M}\left(a;\frac{3}{2};x\right) = \begin{cases} \dfrac{-2^a\Gamma(a-\frac{1}{2})}{\sqrt{32\pi x}} \exp\left(\dfrac{x}{2}\right)[\mathrm{D}_{1-2a}(\sqrt{2x}) - \mathrm{D}_{1-2a}(-\sqrt{2x})] & x > 0 \\ \dfrac{-\Gamma(1-a)}{2^{a+1}\sqrt{-\pi x}} \exp\left(\dfrac{x}{2}\right)[\mathrm{D}_{2a-2}(\sqrt{-2x}) - \mathrm{D}_{2a-2}(-\sqrt{-2x})] & x < 0 \end{cases}$$

Bessel functions [Chapter 53] and their hyperbolic counterparts [Chapter 51] arise from the limiting operations

47:4:10 $$\lim_{a\to\infty} \mathrm{M}\left(a;c+1;\frac{-x}{a}\right) = \frac{\Gamma(1+c)}{x^{c/2}} \mathrm{J}_c(2\sqrt{x}) \qquad x \geq 0$$

47:4:11 $$\lim_{a\to\infty} \mathrm{M}\left(a;c+1;\frac{x}{a}\right) = \frac{\Gamma(1+c)}{x^{c/2}} \mathrm{I}_c(2\sqrt{x}) \qquad x \geq 0$$

Further specialization of incomplete gamma functions [Section 45:4], parabolic cylinder functions [Section 46:4], hyperbolic Bessel functions [Section 51:4] and Bessel functions [Section 53:4] may also occur to yield still more elementary functions.

47:5 INTRARELATIONSHIPS

Known as *Kummer's transformation*, the important identity

47:5:1 $$\mathrm{M}(a;c;-x) = \exp(-x)\,\mathrm{M}(c-a;c;x)$$

constitutes a reflection formula for the Kummer function.

Recurrence formulas may be written interrelating three Kummer functions whose parameters differ by unity, as follows:

47:5:2 $$\mathrm{M}(a+1;c;x) = \frac{2a-c+x}{a}\mathrm{M}(a;c;x) + \frac{c-a}{a}\mathrm{M}(a-1;c;x)$$
$$= \frac{a-c+1}{a}\mathrm{M}(a;c;x) + \frac{c-1}{a}\mathrm{M}(a;c-1;x)$$

47:5:3 $$\mathrm{M}(a;c+1;x) = \frac{c(c-1+x)}{(c-a)x}\mathrm{M}(a;c;x) - \frac{c(c-1)}{(c-a)x}\mathrm{M}(a;c-1;x) = \frac{c}{x}\mathrm{M}(a;c;x) - \frac{c}{x}\mathrm{M}(a-1;c;x)$$

47:5:4
$$\mathrm{M}(a;c;x) = \frac{a}{a+x}\mathrm{M}(a+1;c;x) + \frac{(c-a)x}{c(a+x)}\mathrm{M}(a;c+1;x)$$
$$= \frac{c-1}{a-1+x}\mathrm{M}(a;c-1;x) - \frac{c-a}{a-1+x}\mathrm{M}(a-1;c;x)$$

These relationships may be developed into formulas expressing, for example, $\mathrm{M}(a+n;c;x)$ where n is an integer.

The Kummer function obeys a number of argument-addition relationships. One is

47:5:5
$$\mathrm{M}(a;c;x+y) = \sum_{j=0}^{\infty} \frac{(a)_j y^j}{(c)_j j!} \mathrm{M}(a+j;c+j;x)$$

and others are given by Erdélyi, Magnus, Oberhettinger and Tricomi [*Higher Transcendental Functions*, Volume 1, Section 6:14]. By setting $y = (\nu - 1)x$, 47:5:5 becomes an argument-multiplication formula.

A special linear combination of Kummer functions:

47:5:6
$$\frac{\Gamma(1-c)}{\Gamma(a-c+1)}\mathrm{M}(a;c;x) + \frac{\Gamma(c-1)}{\Gamma(a)x^{c-1}}\mathrm{M}(a-c+1;2-c;x) = \mathrm{U}(a;c;x)$$

generates the Tricomi function [Chapter 48]. Among the summable infinite series of Kummer functions is

47:5:7
$$\mathrm{M}(a;c;x) + \mathrm{M}(a-1;c;x) + \cdots = \sum_{j=0}^{\infty} \mathrm{M}(a-j;c;x) = \frac{c-1}{x}\mathrm{M}(a;c-1;x) \qquad c > \frac{5}{2}$$

and several others are reported by Erdélyi, Magnus, Oberhettinger and Tricomi [*Higher Transcendental Functions*, Volume 1, Section 6.15.1].

47:6 EXPANSIONS

The power series expansion

47:6:1
$$\mathrm{M}(a;c;x) = 1 + \frac{ax}{c} + \frac{a(a+1)x^2}{2!c(c+1)} + \frac{a(a+1)(a+2)x^3}{3!c(c+1)(c+2)} + \cdots = \sum_{j=0}^{\infty} \frac{(a)_j x^j}{j!(c)j}$$

known as *Kummer's series*, is convergent for all arguments and for all parameter values except $c = 0, -1, 2, \ldots$. Using ν as an abbreviation for $c - a - \frac{1}{2}$, a series expansion in hyperbolic Bessel functions [Chapter 50] is

47:6:2
$$\mathrm{M}(a;c;x) = \Gamma(\nu)\left(\frac{4}{x}\right)^{\nu} \exp\left(\frac{x}{2}\right) \sum_{j=0}^{\infty} \frac{(-1)^j (2\nu)_j (c-2a)_j}{j!(c)_j} \mathrm{I}_{\nu+j}\left(\frac{x}{2}\right)$$

Expansions may also be made in Bessel functions of order $c - 1 + j$: see Abramowitz and Stegun [Section 13.3] for the coefficients involved. Another expansion is in terms of generalized Laguerre polynomials [Section 23:12]:

47:6:3
$$\mathrm{M}(a;c;-x) = \sum_{j=0}^{\infty} \frac{(a)_j}{2^{j+a}(c)_j} \mathrm{L}_j^{(c-1)}(x) \qquad x > 0$$

The Kummer function may be written as the sum of two terms involving Tricomi functions [Chapter 48]

47:6:4
$$\mathrm{M}(a;c;x) = \frac{(-1)^a\,\Gamma(c)}{\Gamma(c-a)}\mathrm{U}(a;c;x) + \frac{(-1)^{a-c}\,\Gamma(c)}{\Gamma(a)}\exp(x)\,\mathrm{U}(c-a;c;-x)$$

where powers of -1 are to be interpreted, as explained in Section 13:11, as the (generally complex) number

47:6:5
$$(-1)^{\nu} = \cos(\nu\pi) + i\sin(\nu\pi)$$

Because the Tricomi function has a simple asymptotic expansion [see 48:6:1], an expansion of the Kummer function valid for large x may be constructed using 47:6:4.

47:7 PARTICULAR VALUES

The Kummer function equals unity at zero argument, irrespective of the parameter values

47:7:1 $$M(a;c;0) = 1$$

As $x \to -\infty$, the Kummer function acquires one of four values: $-\infty$, 0, 1 or $+\infty$ depending on the values of the a and c parameters. From Figure 47-4 one can identify the limiting value acquired in any particular case. Note that $M(0;c;-\infty) = 1$ is illustrated in black on the map and that we have $M(a;c;-\infty) = 0$ if $a - c$ is a nonnegative integer, as illustrated in green.

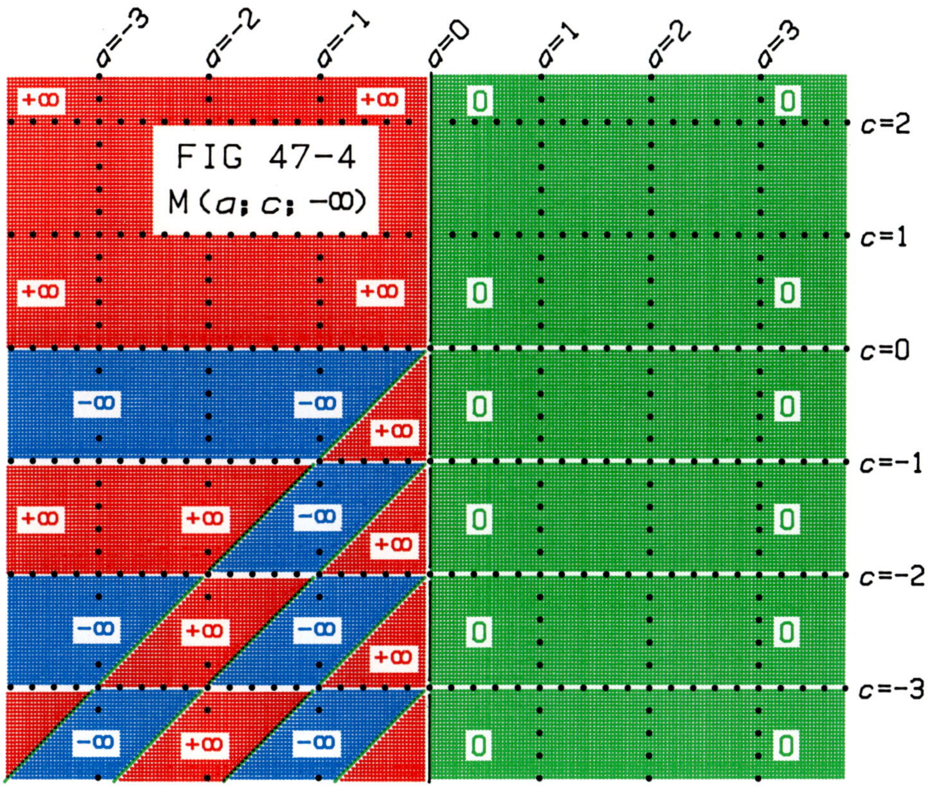

FIG 47-4
M(a; c; −∞)

Figure 47-5 similarly depicts the values of $M(a;c;+\infty)$. Notice the exclusion from both diagrams of $c = 0, -1, -2, \ldots$ where $M(a;c;x)$ is undefined.

47:8 NUMERICAL VALUES

Because the Kummer series, equation 47:6:1, converges rapidly, it constitutes a convenient method for calculating values of $M(a;c;x)$. If x is negative, the terms in the expansion

47:8:1 $$M(a;c;x) = \sum_{j=0}^{\infty} \frac{(a)_j x^j}{(c)_j j!} = \sum_{j=0}^{\infty} T_j$$

ultimately constitute an alternating series [see Section 0:6]. One can show that, for all values of j greater than some integer J, the series $T_J + T_{J+1} + T_{J+2} + \cdots + T_j + \cdots$ is alternating provided that $|T_J| < |T_{J-1}|$ and that $J > 2|a| + |c| + 1$. These considerations form the basis of the algorithm presented below, which exploits the unique advantage, namely

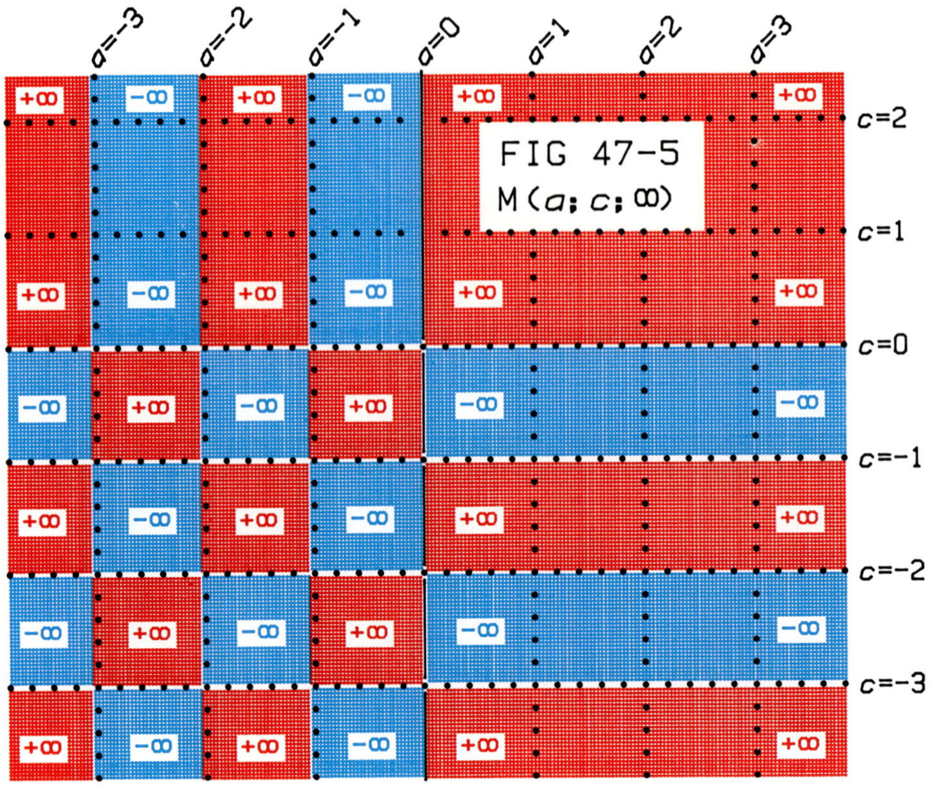

47:8:2
$$M(a;c;x) \simeq \frac{1}{2}T_J + \sum_{j=0}^{J-1} T_j \qquad |\text{error}| \le \frac{1}{2}|T_J|$$

that alternating series possess for the placing of bounds on the error introduced by truncation. The algorithm tests each T_j until a value T_J is reached that satisfies $|T_J| < 10^{-7} \mid T_0 + T_1 + T_2 + \cdots T_{J-1}|$, thus ensuring a precision of better than 6×10^{-8}. The algorithm also checks to ascertain that the necessary conditions $J > 2|a| + |c| + 1$ and $|T_J| < |T_{J-1}|$ are met. If not, j is incremented until all three conditions are satisfied.

Input a >>>>>>>
Set $f = t = 1$
Input c >>>>>>>
Set $j = g = 0$
Input x >>>>>>>
If $x < 0$ go to (1)
Set $f = \exp(x)$
Replace x by $-x$
Replace a by $c - a$
(1) Replace g by $g + t$
Set $p = (a + j)x/(c + j)(j + 1)$
Replace j by $j + 1$
Replace t by tp
If $|t| \ge 10^{-7}|g|$ go to (1)
If $|p| > 1$ go to (1)
If $j \le 2|a| + |c| + 2$ go to (1)
Replace f by $f\left(g + \frac{t}{2}\right)$

Storage needed: a, f, t, c, j, g, x and p

Input restrictions: The c parameter cannot take values $0, -1, -2, \ldots$.

Test values:
$M(0.7;0.6;2) = 8.94061153$
$M(-4;2;-1) = 4.175$
$M(0.5;1;-2) = 0.465759608$

Output f

$f \simeq \mathrm{M}(a;c;x)$ <<<<

If x is positive, the Kummer transformation, equation 47:5:1, is invoked by the algorithm.

For large argument, numerical values of $\mathrm{M}(a;c;x)$ may be calculated via expansion 47:6:4 and the algorithm in Section 48:8.

47:9 APPROXIMATIONS

If either a or x, but not c, is small:

47:9:1 $$\mathrm{M}(a;c;x) \simeq 1 + \frac{ax}{c}$$

If one of a, c or x is large, the other two remaining modest in magnitude, the following approximations hold:

47:9:2 $$\mathrm{M}(a;c;x) \simeq \frac{\Gamma(c)}{\sqrt{\pi}}\left(\frac{cx}{2} - ax\right)^{(1-2c)/4} \exp\left(\frac{x}{2}\right)\cos\left(\frac{\pi}{4} - \frac{c\pi}{2} + \sqrt{2cx - 4ax}\right) \qquad a \text{ large and negative}$$

47:9:3 $$\mathrm{M}(a;c;x) \simeq \Gamma(c)\left(ax - \frac{cx}{2}\right)^{(1-c)/2} \exp\left(\frac{x}{2}\right)\mathrm{I}_{c-1}(\sqrt{4ax - 2cx}) \qquad a \text{ large and positive}$$

47:9:4 $$\mathrm{M}(a;c;x) \simeq \left(1 - \frac{x}{c}\right)^{-a} \qquad c \text{ large, either sign}$$

47:9:5 $$\mathrm{M}(a;c;x) \simeq \frac{\Gamma(c)}{\Gamma(c-a)}(-x)^a \qquad x \text{ large and negative} \qquad c - a \neq 0, -1, -2, \ldots$$

47:9:6 $$\mathrm{M}(a;c;x) \simeq \frac{\Gamma(c)}{\Gamma(a)} x^{a-c}\exp(x) \qquad x \text{ large and positive} \qquad a \neq 0, -1, -2, \ldots$$

47:10 OPERATIONS OF THE CALCULUS

Single and multiple differentiation of the Kummer function satisfy the rules

47:10:1 $$\frac{\mathrm{d}}{\mathrm{d}x}\mathrm{M}(a;c;x) = \frac{a}{c}\mathrm{M}(1 + a;1 + c;x)$$

and

47:10:2 $$\frac{\mathrm{d}^n}{\mathrm{d}x^n}\mathrm{M}(a;c;x) = \frac{(a)_n}{(c)_n}\mathrm{M}(n + a;n + c;x)$$

Formulas for indefinite integration include

47:10:3 $$\int_0^x \mathrm{M}(a;c;t)\mathrm{d}t = \frac{c-1}{a-1}[\mathrm{M}(a - 1;c - 1;x) - 1] \qquad a \neq 1 \qquad c \neq 1$$

47:10:4 $$\int_0^x \frac{t^{c-1}\mathrm{M}(a;c;t)}{(x-t)^{1+c-C}}\mathrm{d}t = \frac{\Gamma(c)\,\Gamma(C-c)}{\Gamma(C)}\mathrm{M}(a;C;x) \qquad c > 0 \qquad C > 0$$

and we also cite the important definite integral

47:10:5 $$\int_0^\infty t^{\nu-1}\mathrm{M}(a;c;-t)\mathrm{d}t = \frac{\Gamma(\nu)\,\Gamma(c)\,\Gamma(a-\nu)}{\Gamma(a)\,\Gamma(c-\nu)} \qquad 0 < \nu < a$$

With a lower limit of zero, differintegration of the Kummer function gives a hypergeometric function [Section 18:14]; the formula

47:10:6
$$\frac{d^\nu}{dx^\nu}M(a;c;x) = \frac{x^{-\nu}}{\Gamma(1-\nu)}\sum_{j=0}^{\infty}\frac{(a)_j x^j}{(c)_j(1-\nu)_j}$$

represents a generalization of 47:10:2. The differintegration formula

47:10:7
$$\frac{d^\nu}{dx^\nu}x^{c-1}M(a;c;x) = \frac{x^{c-1-\nu}\Gamma(c)}{\Gamma(c-\nu)}M(a;c-\nu;x)$$

is equivalent to the integral transform 47:10:4. Hence, in the notation of Section 43:14, one Kummer function can be synthesized from another by

47:10:8
$$M(a;c;x) \xrightarrow[1-c]{1-C} M(a;C;x)$$

Similarly:

47:10:9
$$M(a;c;x) \xrightarrow[1-A]{1-a} M(A;c;x)$$

47:11 COMPLEX ARGUMENT

The replacement of x in $M(a;c;x)$ by $x + iy$ produces a complex-valued quadrivariate function, the properties of which are not pursued in this *Atlas*.

When its argument is purely imaginary, the Kummer function has real and imaginary parts that are higher-order [$K = 2$, $L = 4$; see Section 18:14] hypergeometric functions

47:11:1
$$M(a;c;iy) = \sum_{j=0}^{\infty}\frac{\left(\frac{a}{2}\right)_j\left(\frac{a+1}{2}\right)_j\left(\frac{-y^2}{4}\right)^j}{\left(\frac{c}{2}\right)_j\left(\frac{c+1}{2}\right)_j\left(\frac{1}{2}\right)_j(1)_j} + \frac{iay}{c}\sum_{j=0}^{\infty}\frac{\left(\frac{a+1}{2}\right)_j\left(\frac{a+2}{2}\right)_j\left(\frac{-y^2}{4}\right)^j}{\left(\frac{c+1}{2}\right)_j\left(\frac{c+2}{2}\right)_j(1)_j\left(\frac{3}{2}\right)_j}$$

47:12 GENERALIZATIONS

The *entire Kummer function*, defined by

47:12:1
$$\frac{M(a;c;x)}{\Gamma(c)}$$

is a modest generalization of the Kummer function designed to embrace those values of the denominatorial parameter ($c = 0,-1,-2,\ldots$) that are excluded from the definition of the Kummer function itself. Unlike $M(a;c;x)$, the entire Kummer function encounters no discontinuities as $c \to 1 - n = 0, -1, -2, \ldots$, having the finite values

47:12:2
$$\frac{M(a;1-n;x)}{\Gamma(1-n)} = \frac{(a)_n x^n}{n!}M(a+n;1+n;x) \qquad n = 1, 2, 3, \ldots$$

at these points. It is unfortunate that the entire Kummer function is not widely used because its properties are significantly simpler than those of M(a;c;x).

As well, all hypergeometric functions with $L = K + 1$, $K \geq 1$ [see Section 18:14] may be regarded as generalizations of the Kummer function.

47:13 COGNATE FUNCTIONS

Closely related to the Kummer function are the Tricomi function [Chapter 48] and the Whittaker function $M_{\nu;\mu}(x)$, which is discussed in Section 48:13.

CHAPTER 48

THE TRICOMI FUNCTION U(a;c;x)

Together with the Kummer function, the Tricomi function is a solution of an important differential equation. These two functions constitute the so-called *confluent hypergeometric functions*. The closely related Whittaker functions are discussed in Section 48:13.

48:1 NOTATION

The symbol $\psi(a;c;x)$ often replaces $\mathrm{U}(a;c;x)$, and Tricomi used the G symbol. On account of asymptotic expansion 48:6:1, the notation $x^{-a}{}_2\mathrm{F}_0(a,1+a-c;-1/x)$ is sometimes used for the Tricomi function.

The variable x is the argument of the Tricomi function. The variables a and c are the parameters but, in contrast to their roles in the Kummer function, they cannot be separately assigned as numeratorial and denominatorial parameters. The composite variable $1 + a - c$ is important in determining the properties of $\mathrm{U}(a;c;x)$ and will sometimes be denoted by b.

48:2 BEHAVIOR

The Tricomi function is defined for all values of a, c and x. However, when x is negative, $\mathrm{U}(a;c;x)$ is generally complex and we therefore exclude the $x < 0$ range from most of the discussion of the Tricomi function.

As a function of a and c, Figure 48-1 shows the number of positive zeros of $\mathrm{U}(a;c;x)$, that is, the number of times the Tricomi function acquires the value zero in the range $0 < x < \infty$. Where zeros are absent, $\mathrm{U}(a;c;x)$ is invariably positive, being a monotonically increasing function if $a < 0$ and monotonically decreasing for $a > 0$.

The behavior of the Tricomi function when its positive argument is close to zero or infinity can be deduced from the approximations given in Section 48:9.

48:3 DEFINITIONS

In terms of the Kummer function [Chapter 47], the Tricomi function is defined as

48:3:1
$$\mathrm{U}(a;c;x) = \frac{\Gamma(1-c)}{\Gamma(1+a-c)}\mathrm{M}(a;c;x) + \frac{\Gamma(c-1)}{\Gamma(a)x^{c-1}}\mathrm{M}(1+a-c;2-c;x)$$

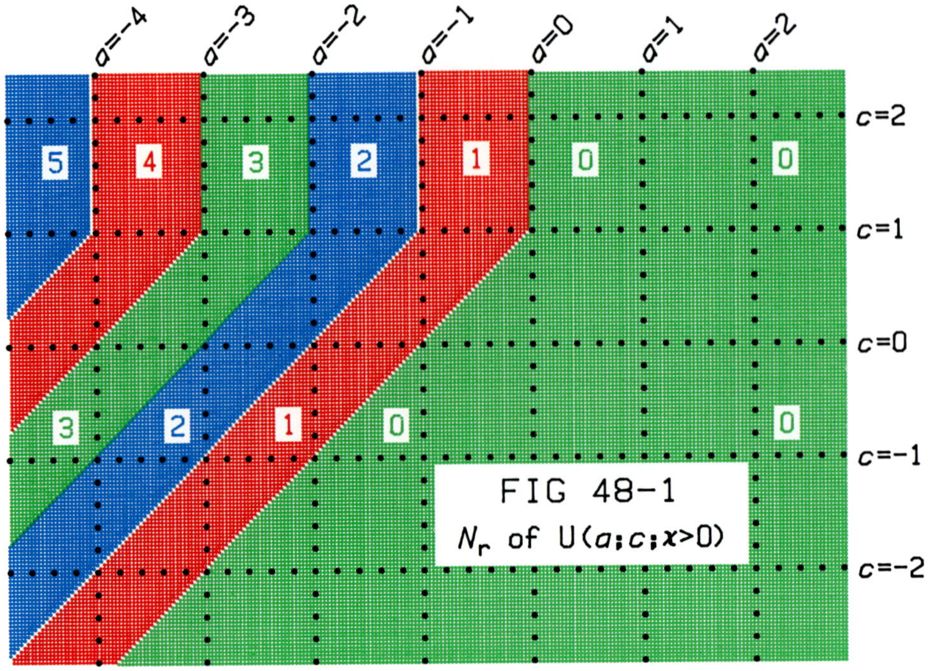

FIG 48-1
N_r of U(a;c;x>0)

This definition may be invalid if c is an integer. The complicated equation

48:3:2 $$\mathrm{U}(a;n+1;x) = \frac{(n-1)!}{\Gamma(a)x^n}\sum_{j=0}^{n-1}\frac{(a-n)_j x^j}{(1-n)_j j!} - \frac{(-1)^n}{n!\Gamma(a-n)}\left\{\ln(x)\,\mathrm{M}(a;n+1;x) + \sum_{j=0}^{\infty}\frac{(a)_j x^j}{(1+n)_j j!}[\psi(a+j)-\psi(1+j)-\psi(1+n+j)]\right\} \qquad n=0,1,2,\ldots$$

provides a supplementary definition to cover the cases $c = 1, 2, 3, \ldots$, while the $c = 0, -1, -2, \ldots$ instances may be accessed via transformation 48:5:1. Here ψ is the digamma function [Chapter 44]. When the argument is negative, definition 48:3:1 is to be replaced by the complex expression

48:3:3 $$\mathrm{U}(a;c;-x) = \frac{\Gamma(1-c)}{\Gamma(b)}\mathrm{M}(a;c;-x) - \frac{(-1)^c\Gamma(c-1)}{\Gamma(a)x^{c-1}}\mathrm{M}(b;2-c;-x) \qquad x>0$$

where $b = 1 + a - c$ and $(-1)^c = \cos(\pi c) + i\sin(\pi c)$. This formula may be rewritten in a large number of equivalent forms. Thus, either or both of the Kummer functions may be transformed via 47:5:1. Again, any of the gamma functions may be subjected to the reflection formula 43:5:1, and the resulting cosecant function may be combined with the trigonometric terms in $(-1)^c$.

The Tricomi function is also defined by the limiting operation

48:3:4 $$\mathrm{U}(a;c;x) = x^{-a}\lim_{\gamma\to\infty}\mathrm{F}\left(a,1+a-c;\gamma;1-\frac{\gamma}{x}\right)$$

performed on the Gauss function [Chapter 60] or as the Laplace transform [Section 26:14]

48:3:5 $$\mathrm{U}(a;c;x) = \frac{1}{\Gamma(a)}\int_0^{\infty}\frac{t^{a-1}\exp(-xt)}{(1+t)^{1+a-c}}\,\mathrm{d}t \qquad c>a>0$$

The confluent hypergeometric differential equation, 47:3:6, is solved by $f = c_1\mathrm{U}(a;c;x) + c_2\exp(x)\,\mathrm{U}(c-a;c;-x)$ for all values of the variables, where c_1 and c_2 are arbitrary constants.

48:4 SPECIAL CASES

When the two parameters of the Tricomi function are identical, reduction occurs to the complementary incomplete gamma function [Chapter 45]

48:4:1 $$\mathrm{U}(a;a;x) = \exp(x)\,\Gamma(1-a;x)$$

If the c parameter exceeds a by unity, the result is a simple power function [Chapter 13]:

48:4:2 $$\mathrm{U}(a;a+1;x) = x^{-a}$$

Using these two formulas, and the recursion formulas in Section 48:5, one may deduce expressions for U(a;$a \pm n$;x) with $n = 1, 2, 3, \ldots$.

When the c parameter is twice the a parameter, the Tricomi function specializes to a function involving the exponential and a Basset function [Chapter 51]:

48:4:3 $$\mathrm{U}(a;2a;x) = \frac{\exp(x/2)}{\sqrt{\pi x^{a-1/2}}}\mathrm{K}_{a-1/2}\left(\frac{x}{2}\right)$$

Further specialization may occur [see Section 51:4].

With the a parameter equal to a negative integer, the Tricomi function becomes a generalized Laguerre polynomial [Section 23:12]

48:4:4 $$\mathrm{U}(-n;c;x) = (-1)^n n!\,\mathrm{L}_n^{(c-1)}(x)$$

Simplifications of the Tricomi function for $a = 0$ or 1 are

48:4:5 $$\mathrm{U}(0;c;x) = 1$$

and

48:4:6 $$\mathrm{U}(1;c;x) = x^{1-c}\exp(x)\,\Gamma(c-1;x)$$

The latter function has an asymptotic expansion

48:4:7 $$\mathrm{U}(1;c;x) \sim \frac{1}{x}\sum_{j=0}^{\infty}\frac{(2-c)_j}{(-x)^j} = \frac{1}{c-1}\sum_{j=1}^{\infty}\frac{(1-c)_j}{(-x)^j} \qquad x \to \infty$$

that represents the simplest of the $K = L + 1$ family of hypergeometric functions [Section 18:14]. With the aid of equations 48:4:5 and 48:4:6, recursion 48:5:2 permits expressions for U(2;c;x), U(3,c;x), etc., to be derived.

Definition 48:3:2 can be simplified somewhat when $c = 2$ and, employing 48:5:1, the result can be transformed to

48:4:8 $$\mathrm{U}(a;0;x) = \frac{1}{\Gamma(1+a)}\sum_{j=0}^{\infty}\frac{(a)_j x^j}{(j!)^2}[1 + j\ln(x) + j\psi(a+j) - 2j\psi(j)]$$

This special case of the Tricomi function is related to *Bateman's k function*:

48:4:9 $$\mathrm{U}(a;0;x) = x\mathrm{U}(1+a;2;x) = \Gamma(1-a)\exp\left(\frac{x}{2}\right)\mathrm{k}_{-2a}\left(\frac{x}{2}\right) \qquad x > 0$$

which is defined by the integral

48:4:10 $$\mathrm{k}_\nu(x) = \frac{2}{\pi}\int_0^{\pi/2}\cos(x\tan(t) - \nu t)\,\mathrm{d}t$$

With the c parameter equal to $\frac{1}{2}$ or $\frac{3}{2}$, the Tricomi function is a parabolic cylinder function [Chapter 46]:

48:4:11 $$\mathrm{U}(a;\tfrac{1}{2};x) = 2^a\exp\left(\frac{x}{2}\right)\mathrm{D}_{-2a}(2\sqrt{x})$$

48:4:12 $$\mathrm{U}(a;\tfrac{3}{2};x) = 2^a\exp\left(\frac{x}{2}\right)\mathrm{D}_{1-2a}(2\sqrt{x})$$

Clearly, these last two expressions can be used in conjunction with recursion 48:5:3 to express $U(a;n+\frac{1}{2};x)$ where n is any integer.

The limit operation

48:4:13 $$\lim_{a\to\infty}\left\{\Gamma(1+a-c)U\left(a;c;\frac{x}{a}\right)\right\} = 2x^{(1-c)/2}K_{c-1}(2\sqrt{x})$$

produces a Basset function [Chapter 51] of order $c-1$.

In this section we have addressed the effect of specializing one of the two parameters. As will be clear from Sections 4 of Chapters 13, 23, 45, 46 and 51, still simpler functions are generated when *both* a and c are specialized.

48:5 INTRARELATIONSHIPS

The important transformation

48:5:1 $$U(a;c;x) = x^{1-c}\,U(1+a-c;2-c;x)$$

relates two Tricomi functions of common argument.

Recurrence formulas may be written interrelating three Tricomi functions whose parameters differ by unity. Examples are

48:5:2 $$U(a+1;c;x) = \frac{2a-c+x}{a(1+a-c)}U(a;c;x) - \frac{1}{a(1+a-c)}U(a-1;c;x) = \frac{1}{a}U(a;c;x) - \frac{1}{a}U(a;c-1;x)$$

48:5:3 $$U(a;c+1;x) = \frac{c-1+x}{x}U(a;c;x) + \frac{1+a-c}{x}U(a;c-1;x) = \frac{c-a}{x}U(a;c;x) + \frac{1}{x}U(a-1;c;x)$$

and

48:5:4 $$U(a;c;x) = \frac{a(1+a-c)}{a+x}U(a+1;c;x) + \frac{x}{a+x}U(a;c+1;x)$$
$$= \frac{1}{a-1+x}U(a-1;c;x) - \frac{1+a-c}{a-1+x}U(a;c-1;x)$$

Analogous to equation 47:5:5 is the argument-addition formula

48:5:5 $$U(a;c;x+y) = \sum_{j=0}^{\infty}\frac{(a)_j(-y)^j}{j!}U(a+j;c+j;x) \qquad |y|<|x|$$

and it may be converted to an argument-multiplication formula in a similar manner.

48:6 EXPANSIONS

The function $U(a;c;x)$ may be written as a power series in x by combining definition 48:3:1 with expansion 47:6:1.

The Tricomi function may be expanded asymptotically as

48:6:1 $$U(a;c;x) \sim x^{-a}\left[1 - \frac{ab}{x} + \frac{a(a+1)b(b+1)}{2x^2} - \cdots + \frac{(a)_j(b)_j}{j!(-x)^j} + \cdots\right] \qquad x\to\infty$$

if x is large, where $b = 1+a-c$.

48:7 PARTICULAR VALUES

When its argument equals zero or $+\infty$, the Tricomi function takes the values

48:7:1 $$U(a;c;0) = \begin{cases}\Gamma(1-c)/\Gamma(1+a-c) & c<1\\ \infty & c\geq 1\end{cases}$$

48:7:2
$$\mathrm{U}(a;c;\infty) = \begin{cases} \infty & a < 0 \\ 1 & a = 0 \\ 0 & a > 0 \end{cases}$$

48:8 NUMERICAL VALUES

For small values of x, U(a;c;x) is best evaluated via definition 48:3:1 and the algorithm of Section 47:8.

For large positive x the following algorithm may be employed.

Input a >>>>>>>
  Set $t = 1$
Input c >>>>>>>
  Set $b = a + 1 - c$
Input x >>>>>>>
  Set $f = x^a$
  Set $j = g = 0$
(1) Replace g by $g + t$
  Set $p = (a + j)(b + j)/x(1 + j)$
  Replace j by $j + 1$
  Replace t by $-tp$
  If $|p| < 1$ go to (1)
  Replace f by $\left[g - \left(\frac{5}{2} - a - b - x - j\right)t/4px\right]\Big/f$
  Output f
$f \simeq \mathrm{U}(a;c;x)$ <<<<

Storage needed: a, t, b, x, f, j, g, t and p

Test values:
$\mathrm{U}(\frac{1}{2}; \frac{1}{2}; 3\pi) = 0.310662341$
$\mathrm{U}(\frac{3}{2};3;5) = 0.101573807$
$\mathrm{U}(1;1;20) = 0.0477185455$

The algorithm utilizes expansion 48:6:1 with an appropriate convergence factor:

48:8:1
$$x^a\mathrm{U}(a;c;x) \simeq 1 - T_1 + T_2 - T_3 + \cdots \mp T_{J-1} \pm T_J\left(\frac{3}{4} + \frac{3 - 2a - 2b - 2J}{8x}\right) \qquad b = 1 + a - c$$

where $T_j = (a)_j(b)_j/j!x^j$ and J is the positive integer such that T_J is smaller in magnitude than any other T_j. No particular accuracy is claimed for this algorithm, but the precision will increase with x, rarely being acceptable for $x < 5$.

48:9 APPROXIMATIONS

For small positive values of its argument, the behavior of the Tricomi function depends dramatically on the c parameter. This leads to a multiplicity of approximation formulas:

48:9:1
$$\mathrm{U}(a;c;x) \simeq \frac{\Gamma(1-c)}{\Gamma(1+a-c)} - \frac{a\Gamma(-c)x}{\Gamma(1+a-c)} \qquad c < 0$$

48:9:2
$$\mathrm{U}(a;0;x) \simeq \frac{1}{\Gamma(1+a)} - \frac{x\ln(x)}{\Gamma(-a)} \qquad c = 0$$

48:9:3
$$\mathrm{U}(a;c;x) \simeq \frac{\Gamma(1-c)}{\Gamma(1+a-c)} - \frac{\Gamma(c)x^{1-c}}{(1-c)\Gamma(a)} \qquad 0 < c < 1$$

48:9:4
$$\mathrm{U}(a;1;x) \simeq \frac{-\ln(x)}{\Gamma(a)} - \frac{2\gamma - \psi(a)}{\Gamma(a)} \qquad c = 1$$

48:9:5
$$\mathrm{U}(a;c;x) \simeq \frac{\Gamma(c-1)}{\Gamma(a)x^{c-1}} - \frac{\Gamma(2-c)}{(c-1)\,\Gamma(1+a-c)} \qquad 1 < c < 2$$

48:9:6 $$\mathrm{U}(a;2;x) \simeq \frac{1}{\Gamma(a)x} + \frac{1}{\Gamma(a-1)}\ln(x) \qquad c = 2$$

48:9:7 $$\mathrm{U}(a;c;x) = \frac{\Gamma(c-1)}{\Gamma(a)x^{c-1}} + \frac{(c-1-a)\,\Gamma(c-2)}{\Gamma(a)x^{c-2}} \qquad c > 2$$

all of which are valid for small x.

The crude approximation x^{-a} may be refined to

48:9:8 $$\mathrm{U}(a;c;x) \simeq \left[\frac{(x-\nu)^{\nu-1-a}}{x^{2\nu-1-a}}\right]^{a/\nu} \qquad \nu = 2 + 2a - c$$

for large positive argument. Abramowitz and Stegun [Section 13.5] give formulas valid when a (but not c or x) is of large magnitude, or when all three variables are large.

48:10 OPERATIONS OF THE CALCULUS

Single and multiple differentiation of the Tricomi function lead to

48:10:1 $$\frac{\mathrm{d}}{\mathrm{d}x}\mathrm{U}(a;c;x) = -a\mathrm{U}(a+1;c+1;x)$$

and

48:10:2 $$\frac{\mathrm{d}^n}{\mathrm{d}x^n}\mathrm{U}(a;c;x) = (-1)^n(a)_n\,\mathrm{U}(a+n;c+n;x)$$

Formulas for indefinite and definite integration include

48:10:3 $$\int_x^{\infty}\mathrm{U}(a;c;t)\mathrm{d}t = \frac{\mathrm{U}(a-1;c-1;x)}{a-1} \qquad a < 1$$

48:10:4 $$\int_0^{\infty}\mathrm{U}(a;c;t)\mathrm{d}t = \frac{\Gamma(2-c)}{(a-1)\,\Gamma(1+a-c)} \qquad a > 1 \qquad c < 2$$

and

48:10:5 $$\int_0^{\infty} t^{\nu}\,\mathrm{U}(a;c;bt)\mathrm{d}t = \frac{\Gamma(1+\nu)\,\Gamma(a-1-\nu)\,\Gamma(2+\nu-c)}{b^{1+\nu}\,\Gamma(a)\,\Gamma(1+a-c)} \qquad b > 0 \qquad 0 < 1+\nu < a \qquad c < 2+\nu$$

48:11 COMPLEX ARGUMENT

The Tricomi function generally is complex valued when its argument is complex or negative. The latter case was discussed in connection with definition 48:3:3. The case of complex argument is not treated in this *Atlas*.

48:12 GENERALIZATIONS

Via its asymptotic representation, equation 48:6:1, the Tricomi function generalizes to the unconstrained $K = 2$, $L = 1$ hypergeometric function [see Section 18:14]

48:12:1 $$\sum_{j=0}^{\infty}\frac{(a)_j(b)_j(x)^j}{(\gamma)_j} \qquad b = 1 + a - c$$

48:13 COGNATE FUNCTIONS

In this section we discuss the *Whittaker functions* $M_{\nu;\mu}(x)$ and $W_{\nu;\mu}(x)$. These are closely related, respectively, to the Kummer and Tricomi functions

48:13:1 $$M_{\nu;\mu}(x) = x^{1/2+\mu}\exp\left(\frac{-x}{2}\right)M\left(\frac{1}{2}+\mu-\nu;1+2\mu;x\right)$$

48:13:2 $$W_{\nu;\mu}(x) = x^{1/2+\mu}\exp\left(\frac{-x}{2}\right)U\left(\frac{1}{2}+\mu-\nu;1+2\mu;x\right)$$

Thus, the Whittaker functions have parameters related to those of their equivalent confluent hypergeometric functions by $\nu = (c/2) - a$, $\mu = (c/2) - 1$. This choice of parameter definitions makes the Whittaker functions more symmetrical under transformations equivalent to 47:5:1 and 48:5:1:

48:13:3 $$M_{\nu;\mu}(-x) = (-1)^{1/2+\mu}M_{-\nu;\mu}(x) = [-\sin(\mu\pi) + i\cos(\mu\pi)]\,M_{-\nu;\mu}(x)$$

48:13:4 $$W_{\nu;\mu}(x) = W_{\nu;-\mu}(x)$$

Most of the formulas of this, and the preceding, chapter may be redrafted in terms of Whittaker functions; sometimes the resulting expressions are simpler. For example, when $\nu = 0$:

48:13:5 $$M_{0;\mu}(x) = 4^{\mu}\Gamma(1+\mu)\sqrt{x}\,I_{\mu}\left(\frac{x}{2}\right)$$

48:13:6 $$W_{0;\mu}(x) = \sqrt{\frac{x}{\pi}}\,K_{\mu}\left(\frac{x}{2}\right)$$

CHAPTER 49

THE HYPERBOLIC BESSEL FUNCTIONS $I_0(x)$ AND $I_1(x)$

Although they are simply the $\nu = 0$ and $\nu = 1$ cases of the general function addressed in Chapter 50, the *zero-order hyperbolic Bessel function* $I_0(x)$ and the *first-order hyperbolic Bessel function* $I_1(x)$ are sufficiently important in their own right to warrant separate consideration. The present chapter also contains some results that relate to $I_n(x)$, the hyperbolic Bessel function of arbitrary integer order.

49:1 NOTATION

The names *modified Bessel function of the first kind of order zero* and *modified Bessel function of the first kind of order one* are also applied to $I_0(x)$ and $I_1(x)$, respectively.

The adjective "hyperbolic" indicates that $I_0(x)$ and $I_1(x)$ are related to the Bessel functions $J_0(x)$ and $J_1(x)$ [see Chapter 52] in the same way that the functions of Chapters 28, 29 and 30 are related to those of Chapters 32, 33 and 34.

49:2 BEHAVIOR

Figure 49-1 shows that both functions increase rapidly in magnitude as the argument x increases in magnitude. That the rate of increase is somewhat less than exponential is evident from Figure 49-2. The latter map also includes a graph of $1/\sqrt{2\pi x}$, the common asymptote to which the product of $\exp(-x)$ with *any* hyperbolic Bessel function (including those of noninteger order [see Chapter 50]) tends as $x \to \infty$.

49:3 DEFINITIONS

Hyperbolic Bessel functions of integer order are defined by the generating functions

49:3:1 $$\exp\left(\frac{x + t^2x}{2t}\right) = \sum_{j=-\infty}^{\infty} t^j I_j(x) = I_0(x) + \left(t + \frac{1}{t}\right) I_1(t) + \left(t^2 + \frac{1}{t^2}\right) I_2(t) + \cdots$$

49:3:2 $$\exp(x\cos(t)) = I_0(x) + 2\cos(t)\,I_1(x) + 2\cos(2t)\,I_2(x) + \cdots = I_0(x) + 2\sum_{j=1}^{\infty}\cos(jt)I_j(x)$$

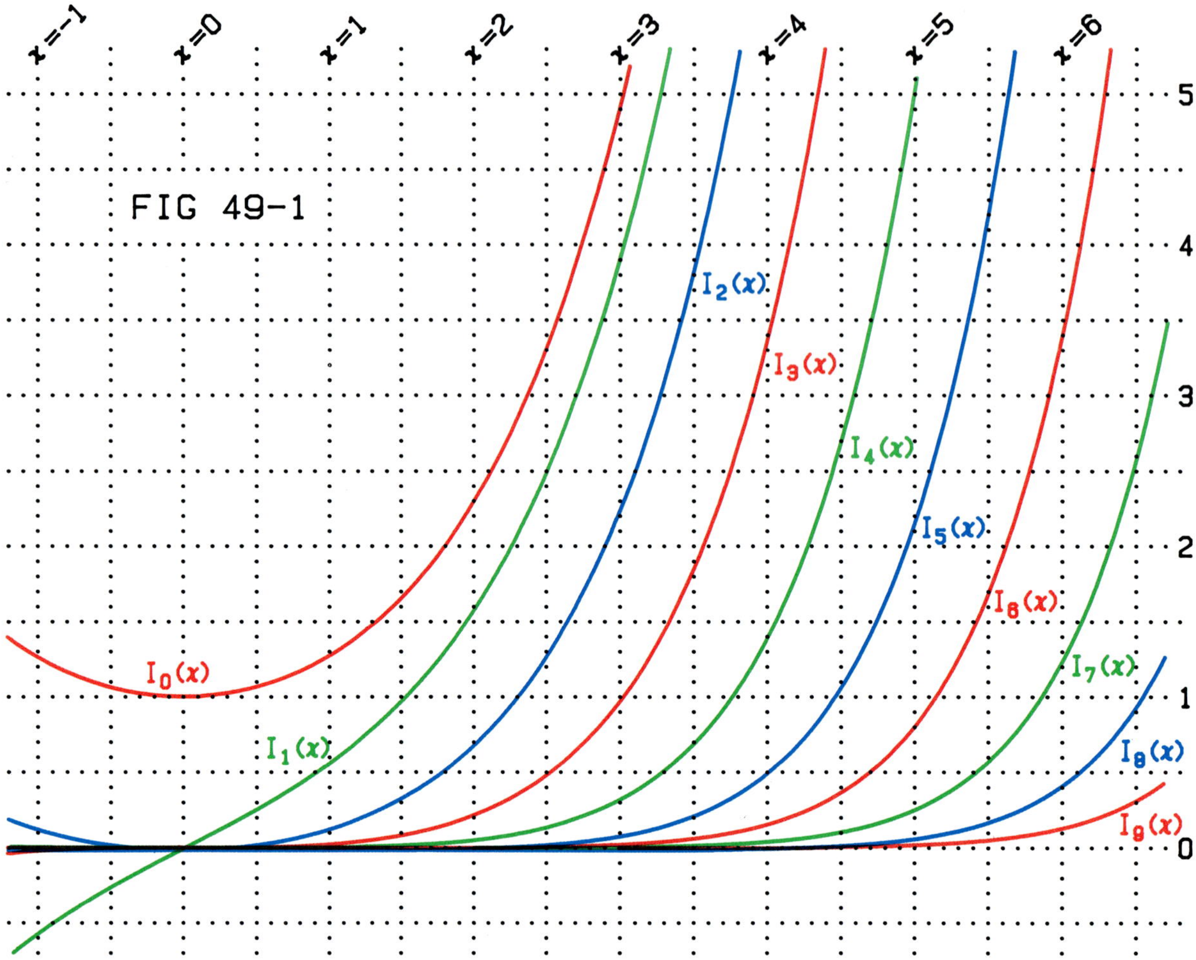

49:3:3 $$\exp(x\sin(t)) = I_0(x) + 2\sin(t)\,I_1(x) + 2\sum_{j=1}^{\infty}(-1)^j\{\cos(2jt)\,I_{2j}(x) + \sin(t+2jt)\,I_{2j+1}(x)\}$$

A number of indefinite integrals, such as

49:3:4 $$I_1(x) = \frac{1}{\pi x}\int_0^{2x}\frac{(x-t)\exp(x-t)dt}{\sqrt{2xt-t^2}}$$

and definite integrals, exemplified by

49:3:5 $$I_0(x) = \frac{1}{\pi}\int_0^{\pi}\exp(\pm x\cos(t))dt = \frac{1}{\pi}\int_0^{\pi}\cosh(x\cos(t))dt$$

$$= \frac{1}{2\pi}\int_0^{2\pi}\exp(\pm x\sin(t))dt = \frac{1}{\pi}\int_0^{\pi}\cosh(x\sin(t))dt$$

49:3:6 $$I_0(x) = \frac{2}{\pi}\int_0^1\frac{\cosh(xt)dt}{\sqrt{1-t^2}}$$

and

49:3:7 $$I_1(x) = \frac{2x}{\pi}\int_0^1\sqrt{1-t^2}\cosh(xt)dt$$

may be employed to define the hyperbolic Bessel functions of orders zero and unity. Some of these latter definitions may be generalized:

49:3:8 $$I_n(x) = \frac{1}{\pi}\int_0^{\pi} \cos(nt)\exp(x\cos(t))dt \qquad n = 0, \pm1, \pm2, \ldots$$

49:3:9 $$I_n(x) = \frac{x^n}{(2n-1)!!\pi}\int_0^{\pi} \sin^{2n}(t)\exp(\pm x\cos(t))dt \qquad n = 0, 1, 2, \ldots$$

to arbitrary integer order.

The hyperbolic Bessel functions of orders zero and unity satisfy some simple second order differential equations, as follows:

49:3:10 $$x\frac{d^2f}{dx^2} + \frac{df}{dx} - xf = 0 \qquad f = c_1I_0(x) + c_2K_0(x)$$

49:3:11 $$x\frac{d^2f}{dx^2} + \frac{df}{dx} - f = 0 \qquad f = c_1I_0(2\sqrt{x}) + c_2K_0(2\sqrt{x})$$

49:3:12 $$x\frac{d^2f}{dx^2} - \frac{df}{dx} - xf = 0 \qquad f = c_1xI_1(x) + c_2xK_1(x)$$

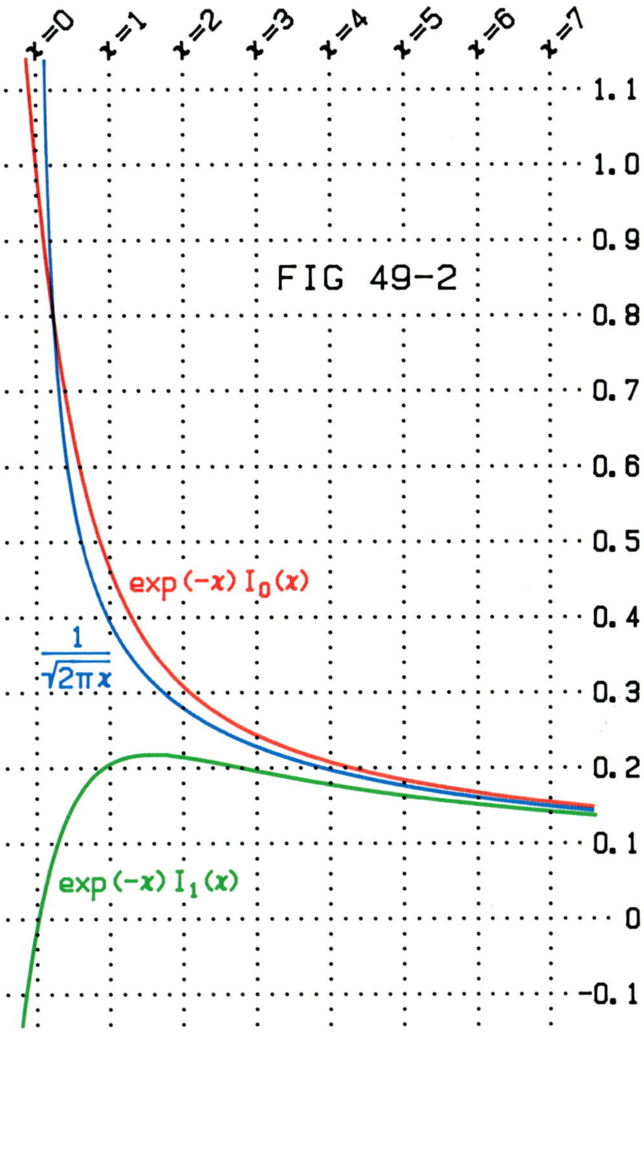

FIG 49-2

49:3:13 $$x\frac{d^2f}{dx^2} - f = 0 \qquad f = c_1\sqrt{x}\,I_1(2\sqrt{x}) + c_2\sqrt{x}\,K_1(2\sqrt{x})$$

where c_1 and c_2 are arbitrary constants and K_0 and K_1 denote the Basset functions [Chapter 52] of orders zero and unity.

Hyperbolic Bessel functions of orders zero and unity may be generated by the operations of semiintegration or semidifferentiation acting upon functions involving exponentials [see Chapter 26]:

49:3:14 $$I_0(x) = \exp(-x)\frac{d^{-1/2}}{dx^{-1/2}}\frac{\exp(2x)}{\sqrt{\pi x}}$$

49:3:15 $$I_1(x) = \frac{x\exp(x)}{2}\frac{d^{1/2}}{dx^{1/2}}\frac{\exp(-2x)}{\sqrt{\pi x^3}}$$

error functions [Chapter 40]:

49:3:16 $$I_0(x) = \frac{\exp(x)}{\sqrt{2}}\frac{d^{1/2}}{dx^{1/2}}\operatorname{erf}(\sqrt{2x})$$

49:3:17 $$I_1(x) = \frac{\exp(x)}{x}\frac{d^{-1/2}}{dx^{-1/2}}\left\{\frac{\operatorname{erf}(\sqrt{2x})}{\sqrt{8}} - \sqrt{\frac{x}{\pi}}\exp(-2x)\right\}$$

or hyperbolic functions [Chapter 28]:

49:3:18 $$I_0(\sqrt{x}) = \frac{2}{\sqrt{\pi}}\frac{d^{1/2}}{dx^{1/2}}\sinh(\sqrt{x}) = \frac{d^{-1/2}}{dx^{-1/2}}\frac{\cosh(\sqrt{x})}{\sqrt{\pi x}}$$

49:3:19 $$I_1(\sqrt{x}) = \frac{1}{\sqrt{\pi x}}\frac{d^{-1/2}}{dx^{-1/2}}\sinh(\sqrt{x}) = 2\sqrt{\frac{x}{\pi}}\frac{d^{1/2}}{dx^{1/2}}\frac{\cosh(\sqrt{x})}{\sqrt{x}}$$

49:4 SPECIAL CASES

There are no special cases, but the important role played by $I_0(2\sqrt{x})$ as a basis hypergeometric function [see Section 43:14] should be noted.

49:5 INTRARELATIONSHIPS

The reflection formula

49:5:1 $$I_n(-x) = (-1)^n\,I_n(x) \qquad n = 0, \pm1, \pm2, \ldots$$

shows that $I_0(x)$ is an even function, whereas $I_1(x)$ is odd. Hyperbolic Bessel functions of negative integer order are identical with their counterparts of positive order:

49:5:2 $$I_{-n}(x) = I_n(x) \qquad n = 1, 2, 3, \ldots$$

but this is not true for noninteger orders [see Chapter 50].

The multiplication formulas

49:5:3 $$I_0(bx) = \sum_{j=0}^{\infty}\frac{1}{j!}\left(\frac{b^2x - x}{2}\right)^j I_j(x)$$

and

49:5:4 $$I_1(bx) = b\sum_{j=0}^{\infty}\frac{1}{j!}\left(\frac{b^2x - x)}{2}\right)^j I_{j+1}(x)$$

generate series of hyperbolic Bessel functions of integer order. Other series of such functions have the sums

49:5:5 $$\frac{1}{2}I_0(x) + I_1(x) + I_2(x) + I_3(x) + \cdots = \frac{1}{2}\sum_{j=-\infty}^{\infty} I_j(x) = \frac{\exp(x)}{2}$$

49:5:6 $$\frac{1}{2}I_0(x) - I_1(x) + I_2(x) - I_3(x) + \cdots = \frac{1}{2}\sum_{j=-\infty}^{\infty} (-1)^j I_j(x) = \frac{\exp(-x)}{2}$$

49:5:7 $$\frac{1}{2}I_0(x) + I_2(x) + I_4(x) + \cdots = \frac{\cosh(x)}{2}$$

49:5:8 $$I_1(x) + I_3(x) + I_5(x) + \cdots = \frac{\sinh(x)}{2}$$

49:5:9 $$\frac{1}{2}I_0(x) - I_2(x) + I_4(x) - I_6(x) + \cdots = \frac{1}{2}$$

49:5:10 $$\frac{I_2(x)}{2} + \frac{I_4(x)}{4} + \frac{I_6(x)}{6} + \cdots = \frac{1}{4}\left\{\left[\gamma + \ln\left(\frac{x}{2}\right)\right] I_0(x) + K_0(x)\right\}$$

and yet other similar expressions are derivable via the generating functions 49:3:1–49:3:3. Here K_0 is the zero-order Basset function of Chapter 51.

By sufficient applications of the recursion formula

49:5:11 $$I_{n+1}(x) = I_{n-1}(x) - \frac{2n}{x} I_n(x)$$

any hyperbolic Bessel function $I_n(x)$ of integer order may be expressed in terms of $I_0(x)$ and $I_1(x)$. The first few examples are

49:5:12 $$I_2(x) = I_0(x) - \frac{2}{x} I_1(x)$$

49:5:13 $$I_3(x) = \frac{-4}{x} I_0(x) + \left(1 + \frac{8}{x^2}\right) I_1(x)$$

49:5:14 $$I_4(x) = \left(1 + \frac{24}{x^2}\right) I_0(x) - \left(\frac{8}{x} + \frac{48}{x^3}\right) I_1(x)$$

49:5:15 $$I_5(x) = -\left(\frac{12}{x} + \frac{192}{x^3}\right) I_0(x) + \left(1 + \frac{72}{x^2} + \frac{384}{x^4}\right) I_1(x)$$

49:5:16 $$I_6(x) = \left(1 + \frac{144}{x^2} + \frac{1920}{x^4}\right) I_0(x) - \left(\frac{18}{x} + \frac{768}{x^3} + \frac{3840}{x^5}\right) I_1(x)$$

49:6 EXPANSIONS

The hyperbolic Bessel functions of orders zero and one may be expanded as

49:6:1 $$I_0(x) = 1 + \frac{x^2}{4} + \frac{x^4}{64} + \frac{x^6}{2304} + \cdots = \sum_{j=0}^{\infty} \frac{(x^2/4)^j}{(j!)^2}$$

and

49:6:2 $$I_1(x) = \frac{x}{2} + \frac{x^3}{16} + \frac{x^5}{384} + \frac{x^7}{18432} + \cdots = \frac{x}{2}\sum_{j=0}^{\infty} \frac{(x^2/4)^j}{j!(j+1)!}$$

each of which is a special case of

49:6:3 $$I_n(x) = \left(\frac{x}{2}\right)^n \sum_{j=0}^{\infty} \frac{(x^2/4)^j}{j!(j+n)!}$$

Asymptotic expansions of $I_0(x)$ and $I_1(x)$, valid for large argument, are

49:6:4 $$I_0(x) \sim \frac{\exp(x)}{\sqrt{2\pi x}}\left[1 + \frac{1}{8x} + \frac{9}{128x^2} + \frac{225}{3072x^3} + \cdots + \frac{[(2j)!]^3}{(j!)^3(32x)^j} + \cdots\right] \qquad x \to \infty$$

and

49:6:5 $$I_1(x) \sim \frac{\exp(x)}{\sqrt{2\pi x}}\left[1 - \frac{3}{8x} - \frac{15}{128x^2} - \frac{315}{3072x^3} - \frac{(2j-3)!!(2j+1)!!}{j!(8x)^j} - \cdots\right] \qquad x \to \infty$$

The asymptotic expansion of the general hyperbolic Bessel function $I_n(x)$ of integer order is given by equation 50:6:4 on replacement of ν by n.

49:7 PARTICULAR VALUES

	$x = -\infty$	$x = 0$	$x = \infty$
$I_0(x)$	∞	1	∞
$I_1(x), I_3(x), I_5(x), \ldots$	$-\infty$	0	∞
$I_2(x), I_4(x), I_6(x), \ldots$	∞	0	∞

49:8 NUMERICAL VALUES

If, for a given x, R_j denotes the ratio $I_{j+1}(x)/I_j(x)$ of two successive hyperbolic Bessel functions of integer order, then from equation 49:5:5

49:8:1 $$I_0(x) = \frac{\exp(x)}{1 + 2R_0(1 + R_1(1 + R_2(1 + R_3(1 + \cdots))))}$$

A truncated version of this relationship is employed by the following algorithm to calculate $I_0(x)$. The approximation is made that $R_j = 0$ for $j \geq J = 11 + \mathrm{Int}(|x|)$, which is sufficiently accurate to ensure 24-bit precision in the generated values of $I_0(x)$ and $I_1(x)$. The recurrence

49:8:2 $$R_{j-1} = \frac{1}{R_j + \dfrac{2j}{x}} \qquad j = J, J-1, \ldots, 3, 2, 1$$

Input x >>>>>>>

Set $s = 1$
If $x \geq 0$ go to (1)
Set $s = -1$
Replace x by $-x$
(1) Set $r = f = 0$
Set $j = 10 + \mathrm{Int}(x)$
(2) Replace f by $1 + fr$
Replace r by $x/(xr + 2j)$

Storage needed: x, s, r, f and j

Replace j by $j - 1$
If $j \neq 0$ go to (2)
Replace f by $\exp(x)/[1 + 2rf]$
Output f
$f \simeq I_0(x)$ <<<<<<<
Replace f by frs
Output f
$f \simeq I_1(x)$ <<<<<<<

Test values:
$I_0(6.9) = 153.698996$
$I_1(6.9) = 142.079028$
$I_0(-3) = 4.88079259$
$I_1(-3) = -3.95337022$

[which follows from 49:5:10] is used to calculate R_j values for successfully smaller values of j.

The algorithm is designed to produce values of $I_0(x)$ and $I_1(x)$ with a relative error of less than 6×10^{-8} for any nonnegative argument x. When those portions of the algorithm shown in green are included, negative arguments are also admissible.

There are algorithms in Sections 50:8 and 51:8 that may also be used to generate values of $I_0(x)$ and $I_1(x)$. As well, the universal hypergeometric algorithm of Section 18:14 can perform these tasks.

49:9 APPROXIMATIONS

For small arguments, the algebraic approximations

49:9:1 $$I_0(x) \simeq \left(1 + \frac{x^2}{8}\right)^2 \qquad \text{8-bit precision} \qquad -1.5 \le x \le 1.5$$

49:9:2 $$I_1(x) \simeq \frac{x}{2}\left(1 + \frac{x^2}{24}\right)^3 \qquad \text{8-bit precision} \qquad -2.3 \le x \le 2.3$$

are valid. For sufficiently large positive argument, the approximation

49:9:3 $$I_n(x) \simeq \frac{\exp(x)}{\sqrt{2\pi x}} \qquad x \to \infty$$

is valid irrespective of the order of the hyperbolic Bessel function.

49:10 OPERATIONS OF THE CALCULUS

The differentiation formulas

49:10:1 $$\frac{d}{dx} I_0(x) = I_1(x)$$

49:10:2 $$\frac{d}{dx} I_1(x) = I_0(x) - \frac{1}{x} I_1(x)$$

are special cases of the general expressions

49:10:3 $$\frac{d}{dx} I_n(x) = \frac{I_{n-1}(x) + I_{n+1}(x)}{2} = I_{n-1}(x) - \frac{n}{x} I_n(x) = I_{n+1}(x) + \frac{n}{x} I_n(x)$$

The formulas

49:10:4 $$\frac{d}{dx}\{x^{\pm n} I_n(x)\} = x^{\pm n} I_{n\mp 1} x$$

are attractively symmetrical.

Formulas for indefinite integration include

49:10:5 $$\int_0^x I_0(t)dt = \frac{\pi x}{2}[I_0(x)\,\ell_{-1}(x) - I_1(x)\,\ell_0(x)]$$

49:10:6 $$\int_0^x I_1(t)dt = I_0(x)$$

49:10:7 $$\int_0^x tI_0(t)dt = xI_1(x)$$

49:10:8 $$\int_0^x tI_1(t)dt = \frac{\pi x}{2}[I_1(x)\,\ell_0(x) - I_0(x)\,\ell_1(x)]$$

Here ℓ_{-1}, ℓ_0 and ℓ_1 denote hyperbolic Struve functions [Section 57:13]. The indefinite integrals

49:10:9 $$\int_0^x \exp(-t)\,I_0(t)dt = x\exp(-x)[I_0(x) + I_1(x)]$$

and

49:10:10 $$\int_0^x \exp(-t)\,I_1(t)dt = (x+1)\exp(-x)\,I_0(x) + x\exp(-x)\,I_1(x) - 1$$

of products of the exponential function and the hyperbolic Bessel functions $I_0(x)$ and $I_1(x)$ generalize to

49:10:11 $$\int_0^x \exp(-t)\,I_n(t)dt = \exp(-x)\left[(x+n)\,I_0(x) + xI_1(x) + 2\sum_{j=1}^{n-1}(n-j)\,I_j(x)\right] - n$$

Formulas for definite integrals involving the I_0 and I_1 functions may be obtained by specialization of those of Section 50:10.

With lower limit zero, the following are included among differintegration formulas:

49:10:12 $$\frac{d^{1/2}}{dx^{1/2}} I_0(\sqrt{x}) = \frac{\cosh(\sqrt{x})}{\sqrt{\pi x}}$$

49:10:13 $$\frac{d^{-1/2}}{dx^{-1/2}} I_0(\sqrt{x}) = \frac{2}{\sqrt{\pi}}\sinh(\sqrt{x})$$

For others, see Oldham and Spanier.

49:11 COMPLEX ARGUMENT

When their arguments are complex, the hyperbolic Bessel functions of orders zero and unity are expressible as the series

49:11:1 $$I_0(x+iy) = \sum_{j=0}^{\infty} \frac{\left(\dfrac{x^2-y^2}{4} + \dfrac{ixy}{2}\right)^j}{(j!)^2} = \left(1 + \frac{x^2-y^2}{4} + \frac{x^4 - 6x^2y^2 + y^4}{64} + \cdots\right) + i\left(\frac{xy}{2} + \frac{x^3y - xy^3}{16} + \cdots\right)$$

49:11:2 $$I_1(x+iy) = \frac{x+iy}{2}\sum_{j=0}^{\infty} \frac{\left(\dfrac{x^2-y^2}{4} + \dfrac{ixy}{2}\right)^j}{j!(j+1)!} = \left(\frac{x}{2} + \frac{x^3 - 3xy^2}{16} + \frac{x^5 - 10x^3y^2 + 5xy^4}{384} + \cdots\right)$$

$$+ i\left(\frac{y}{2} + \frac{3x^2y - y^3}{16} + \frac{5x^4y - 10x^2y^3 + y^5}{384} + \cdots\right)$$

For purely imaginary arguments, these formulas reduce to

49:11:3 $$I_0(iy) = J_0(y)$$

49:11:4 $$I_1(iy) = iJ_1(y)$$

which generalize to

49:11:5 $$I_n(iy) = i^n J_n(y)$$

and show that the hyperbolic Bessel functions of imaginary argument and integer order n are real or imaginary according as n is even or odd. The functions J_n are those discussed in Chapter 52.

Equations 49:11:1 and 49:11:2 simplify when the magnitudes of the real and imaginary components of the arguments are equal. Then one has

49:11:6 $$I_0(x \pm ix) = \sum_{j=0}^{\infty} \frac{(\pm ix^2/2)^j}{(j!)^2} = \mathrm{ber}_0(x\sqrt{2}) \pm i\,\mathrm{bei}_0(x\sqrt{2})$$

49:11:7 $$I_1(x \pm ix) = \frac{x \pm ix}{2} \sum_{j=0}^{\infty} \frac{(\pm ix^2/2)^j}{j!(j+1)!} = \mathrm{bei}_1(x\sqrt{2}) \mp i\,\mathrm{ber}_1(x\sqrt{2})$$

where ber_n and bei_n denote Kelvin functions, as discussed in Chapter 57.

49:12 GENERALIZATIONS

The next chapter discusses hyperbolic Bessel functions of arbitrary order.

49:13 COGNATE FUNCTIONS

The functions I_0 and I_1 are closely related to the Bessel coefficients J_0 and J_1, [Chapter 52], to the Basset functions K_0 and K_1 [Chapter 51] and to the hyperbolic Struve functions ℓ_{-1}, ℓ_0 and ℓ_1 [Section 57:13].

CHAPTER 50

THE GENERAL HYPERBOLIC BESSEL FUNCTION $I_\nu(x)$

The function $I_\nu(x)$ is defined for all values of its order ν. Many physically important functions arise when ν acquires special values.

50:1 NOTATION

The names *modified Bessel function of the first kind* and *Bessel function of imaginary argument* are commonly applied to $I_\nu(x)$, but these titles are less informative than the name *hyperbolic Bessel function* that we adopt. The adjective "hyperbolic" indicates that the relationship [see 50:11:2] between $I_\nu(x)$ and the Bessel function itself [$J_\nu(x)$, see Chapter 53] resembles that between hyperbolic functions [Chapters 28–30] and circular functions [Chapters 32–34]. See Section 50:4 for the $i_n(x)$ symbol.

The parameters ν and x are named the order and argument, respectively, of the function.

50:2 BEHAVIOR

Unless ν is an integer, the hyperbolic Bessel function $I_\nu(x)$ is complex when its argument is negative. Accordingly, this chapter concentrates on positive values of x, and this is the only range depicted in the contour map, Figure 50-1.

When the order ν is an integer, the functions $I_\nu(x)$ and $I_{-\nu}(x)$ are identical, but these two functions are also close to each other in value whenever x exceeds $|\nu|$. Moreover, for large enough argument, the hyperbolic Bessel function $I_\nu(x)$ becomes almost independent, not only of the sign of its order, but also of the magnitude of ν, being well approximated by $\exp(x)/\sqrt{2\pi x}$.

Whereas $I_\nu(x)$ is generally complex for negative x, the composite function $x^{-\nu}I_\nu(x)$ remains real for all real arguments and is, in fact, even so that

50:2:1 $$(-x)^{-\nu}\, I_\nu(-x) = x^{-\nu} I_\nu(x)$$

The evenness is evident on writing 50:2:1 as $2^\nu C_\nu(-x^2/4)$, in terms of *Clifford's notation* [see Section 53:2].

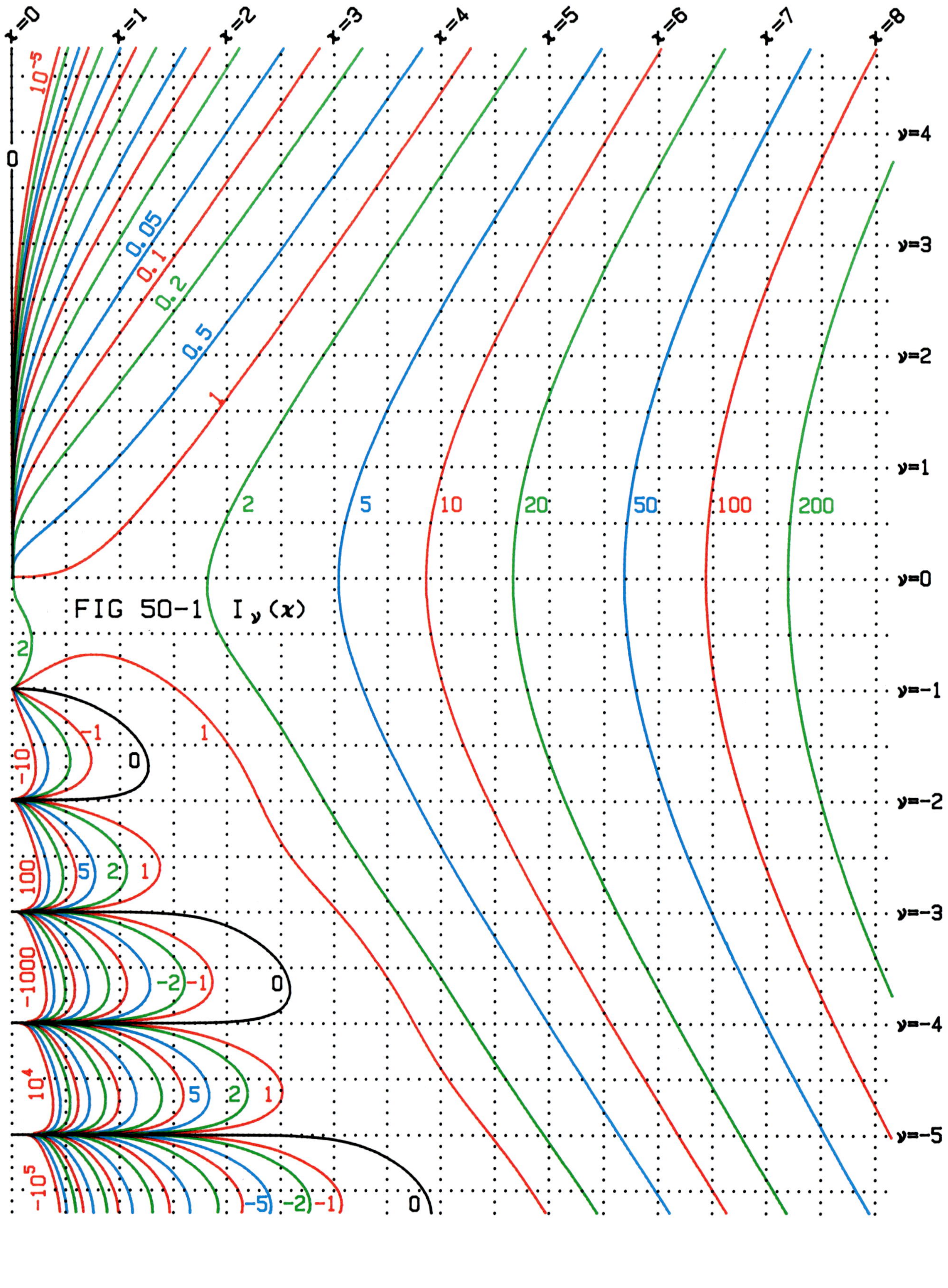
FIG 50-1 $I_\nu(x)$
x=0
x=1
x=2
x=3
x=4
x=5
x=6
x=7
x=8
ν=4
ν=3
ν=2
ν=1
ν=0
ν=−1
ν=−2
ν=−3
ν=−4
ν=−5
10^{-5}
0
0.05
0.1
0.2
0.5
1
2
5
10
20
50
100
200
2
−1
0
1
−10
100
5
2
1
−1000
−2
−1
0
10^4
5
2
1
-10^5
−5
−2
−1
0

50:3 DEFINITIONS

The hyperbolic Bessel function is defined by a number of definite integrals, including

50:3:1 $$I_\nu(x) = \frac{(x/2)^\nu}{\sqrt{\pi}\,\Gamma(\frac{1}{2}+\nu)} \int_{-1}^{1} (1-t^2)^{\nu-1/2} \exp(\pm xt)\mathrm{d}t \qquad \nu > -\tfrac{1}{2}$$

Substituting $t = \cos(\theta)$ in 50:3:1 gives a second definition and some others are listed by Gradshteyn and Ryzhik [Section 8.431].

A Kummer function [Chapter 47] whose denominatorial and numeratorial parameters are in a two-to-one ratio is related to a hyperbolic Bessel function by

50:3:2 $$I_\nu(x) = \frac{(x/2)^\nu \exp(-x)}{\Gamma(1+\nu)} M(\tfrac{1}{2}+\nu; 1+2\nu; 2x)$$

The operations discussed in Section 43:14 can generate any hyperbolic Bessel function from any other hyperbolic Bessel function and hence from any of the functions discussed in Chapter 49. For example, in the notation of Section 43:14

50:3:3 $$I_0(x) \xrightarrow[0]{-\nu} \frac{\Gamma(1+\nu)}{(x/2)^\nu} I_\nu(x) \qquad x = 2\sqrt{X}$$

A solution of the *modified Bessel equation*

50:3:4 $$x^2\frac{\mathrm{d}^2f}{\mathrm{d}x^2} + x\frac{\mathrm{d}f}{\mathrm{d}x} - (x^2+\nu^2)f = 0$$

is $c_1 I_\nu(x) + c_2 I_{-\nu}(x)$ unless ν is an integer [see Section 51:3 for that case]. The terms c_1 and c_2 are arbitrary constants. Other differential equations that are satisfied by hyperbolic Bessel functions are

50:3:5 $$x\frac{\mathrm{d}^2f}{\mathrm{d}x^2} + (2\nu+1)\frac{\mathrm{d}f}{\mathrm{d}x} - xf = 0 \qquad f = c_1 x^{-\nu} I_\nu(x) + c_2 x^\nu I_{-\nu}(x)$$

and

50:3:6 $$x\frac{\mathrm{d}^2f}{\mathrm{d}x^2} + (\nu+1)\frac{\mathrm{d}f}{\mathrm{d}x} - f = 0 \qquad f = c_1(2\sqrt{x})^{-\nu} I_\nu(2\sqrt{x}) + c_2(2\sqrt{x})^\nu I_{-\nu}(2\sqrt{x})$$

unless ν is an integer.

50:4 SPECIAL CASES

The hyperbolic Bessel functions of integer order are the subject of Chapter 49.

When the order of the hyperbolic Bessel function is an odd multiple of $\pm\frac{1}{2}$, reduction occurs to the simpler functions discussed in Section 28:13. The simplest cases are

50:4:1 $$I_{-3/2}(x) = \sqrt{\frac{2}{\pi x}}\left[\sinh(x) - \frac{\cosh(x)}{x}\right] = \left(\frac{x-1}{x}\right)\frac{\exp(x)}{\sqrt{2\pi x}} - \left(\frac{x+1}{x}\right)\frac{\exp(-x)}{\sqrt{2\pi x}}$$

50:4:2 $$I_{-1/2}(x) = \sqrt{\frac{2}{\pi x}}\cosh(x) = \frac{\exp(x)}{\sqrt{2\pi x}} + \frac{\exp(-x)}{\sqrt{2\pi x}}$$

50:4:3 $$I_{1/2}(x) = \sqrt{\frac{2}{\pi x}}\sinh(x) = \frac{\exp(x)}{\sqrt{2\pi x}} - \frac{\exp(-x)}{\sqrt{2\pi x}}$$

50:4:4 $$I_{3/2}(x) = \sqrt{\frac{2}{\pi x}}\left[\cosh(x) - \frac{\sinh(x)}{x}\right] = \left(\frac{x-1}{x}\right)\frac{\exp(x)}{\sqrt{2\pi x}} + \left(\frac{x+1}{x}\right)\frac{\exp(-x)}{\sqrt{2\pi x}}$$

and others may be constructed via recursion formula 50:5:1. Some of these functions are mapped in Figure 28-3. The symbol $i_n(x)$ and the name *modified spherical Bessel function of the first kind* are sometimes applied to the $\sqrt{\pi/2x}\, I_{n+1/2}(x)$ function.

Hyperbolic Bessel functions of order $\pm\frac{1}{3}$ are closely related to Airy functions [Chapter 56]. One has

50:4:5 $$I_{\pm 1/3}(x) = \sqrt{\frac{3}{4X}}\,[\mathrm{Bi}(X) \mp \sqrt{3}\mathrm{Ai}(X)] \qquad X = \left(\frac{3x}{2}\right)^{2/3}$$

Hyperbolic Bessel functions of order $\pm\frac{1}{4}$ are expressible as parabolic cylinder functions [Chapter 46] of order $-\frac{1}{2}$. The relationships are

50:4:6 $$I_{\pm 1/4}(x) = \frac{D_{-1/2}(-2\sqrt{x}) \mp D_{-1/2}(2\sqrt{x})}{\sqrt{2\pi}x^{1/4}}$$

Similarly, $I_{\pm 3/4}(x)$ may be expressed in terms of parabolic cylinder functions by

50:4:7 $$I_{\pm 3/4}(x) = \frac{\pm\sqrt{x}D_{-1/2}(2\sqrt{x}) - \sqrt{x}D_{-1/2}(-2\sqrt{x}) \mp D_{1/2}(2\sqrt{x}) - D_{1/2}(-2\sqrt{x})}{\sqrt{2\pi}x^{3/4}}$$

and, hence, with the help of recursion formula 50:5:1, so may $I_{\pm 5/4}(x)$, etc.

50:5 INTRARELATIONSHIPS

Hyperbolic Bessel functions satisfy the recursion relationship

50:5:1 $$I_{\nu+1}(x) = I_{\nu-1}(x) - \frac{2\nu}{x} I_\nu(x)$$

and the argument-multiplication formula

50:5:2 $$I_\nu(bx) = b^\nu \sum_{j=0}^{\infty} \left(\frac{b^2x - x}{2}\right)^j \frac{1}{j!} I_{\nu+j}(x)$$

Setting $b = 1 + (y/x)$ converts 50:5:2 into an argument-addition formula, while setting $b = i = \sqrt{-1}$ generates the summation formula

50:5:3 $$I_\nu(x) - xI_{\nu+1}(x) + \frac{x^2}{2!} I_{\nu+2}(x) - \cdots = \sum_{j=0}^{\infty} \frac{(-x)^j}{j!} I_{\nu+j}(x) = i^{-\nu} I_\nu(ix) = J_\nu(x)$$

The hyperbolic Bessel functions $I_\nu(x)$ and $I_{-\nu}(x)$ are identical if ν is an integer; otherwise:

50:5:4 $$I_{-\nu}(x) = I_\nu(x) + \frac{2}{\pi}\sin(\nu\pi)\, K_\nu(x)$$

where K_ν is the Basset function [Chapter 51]. This equation may be regarded as an order-reflection formula, interrelating $I_{-\nu}(x)$ and $I_\nu(x)$. There are similar order-reflection formulas for the product of two hyperbolic Bessel functions whose orders sum to ± 1, namely

50:5:5 $$I_{-\nu}(x)\, I_{\nu\pm 1}(x) = I_\nu(x)\, I_{-\nu\mp 1}(x) + \frac{2}{\pi x}\sin(\nu\pi)$$

and

50:5:6 $$I_{-\nu-1/2}(x)\, I_{\nu-1/2}(x) = I_{1/2+\nu}(x)\, I_{1/2-\nu}(x) + \frac{2}{\pi x}\cos(\nu\pi)$$

50:6 EXPANSIONS

The hyperbolic Bessel functions may be expanded in the convergent series

50:6:1 $$I_\nu(x) = \frac{(x/2)^\nu}{\Gamma(1+\nu)} + \frac{(x/2)^{2+\nu}}{1!\Gamma(2+\nu)} + \frac{(x/2)^{4+\nu}}{2!\Gamma(3+\nu)} + \cdots = \sum_{j=0}^{\infty} \frac{(x/2)^{2j+\nu}}{j!\Gamma(j+\nu+1)}$$

That this expansion involves a hypergeometric series [Section 18:14] becomes more evident when it is rewritten in the form

50:6:2 $$I_\nu(x) = \frac{(x/2)^\nu}{\Gamma(1+\nu)}\left[1 + \frac{x^2}{4(1+\nu)} + \frac{x^4}{32(1+\nu)(2+\nu)} + \frac{x^6}{384(1+\nu)(2+\nu)(3+\nu)} + \cdots\right]$$
$$= \frac{(x/2)^\nu}{\Gamma(1+\nu)} \sum_{j=0}^{\infty} \frac{(x^2/4)^j}{(1)_j(1+\nu)_j}$$

where $(1+\nu)_j$ denotes a Pochhammer polynomial [Chapter 18]. The expansion

50:6:3 $$I_\nu(x) = J_\nu(x) + xJ_{\nu+1}(x) + \frac{x^2}{2!}J_{\nu+2}(x) + \cdots = \sum_{j=0}^{\infty} \frac{x^j}{j!} J_{j+\nu}(x)$$

in terms of Bessel functions [Chapter 53] is also convergent.

An asymptotic expansion is provided by the series

50:6:4 $$I_\nu(x) \sim \frac{\exp(x)}{\sqrt{2\pi x}}\left\{1 + \frac{\frac{1}{4}-\nu^2}{2x} + \frac{(\frac{9}{4}-\nu^2)(\frac{1}{4}-\nu^2)}{8x^2} + \frac{(\frac{25}{4}-\nu^2)(\frac{9}{4}-\nu^2)(\frac{1}{4}-\nu^2)}{48x^3} + \cdots + \frac{(\frac{1}{2}-\nu)_j(\frac{1}{2}+\nu)_j}{j!(2x)^j} + \cdots\right\} \qquad x \to \infty$$

which is valid if x, but not $|\nu|$, is large. This series terminates when $\nu = \pm\frac{1}{2}, \pm\frac{3}{2}, \pm\frac{5}{2}, \ldots$ but the expansion is not exact even under those circumstances.

50:7 PARTICULAR VALUES

		$x = 0$	$x = \infty$
$I_\nu(x)$	$\nu > 0$	0	∞
$I_0(x)$		1	∞
$I_\nu(x)$	$-1 < \nu < 0,\ -3 < \nu < -2,\ -5 < \nu < -4, \ldots$	∞	∞
$I_\nu(x)$	$\nu = -1, -2, -3, \ldots$	0	∞
$I_\nu(x)$	$-2 < \nu < -1,\ -4 < \nu < -3,\ -6 < \nu < -5, \ldots$	$-\infty$	∞

As is evident from the contour map, Figure 50-1, the hyperbolic Bessel function $I_\nu(x)$ encounters a zero, for $x > 0$, only if the order is negative and in one of the ranges $-2 < \nu < -1$, $-4 < \nu < -3$, $-6 < \nu < -5$, etc. For other negative values of ν, $I_\nu(x)$ generally displays a minimum.

50:8 NUMERICAL VALUES

The following algorithm for $I_\nu(x)$ utilizes expansion 50:6:2 in the truncated and concatenated version

50:8:1 $$I_\nu(x) \simeq \left(\left(\left(\cdots\left(\left(\frac{x^2}{4J(J+\nu)}+1\right)\frac{x^2}{4(J-1)(J-1+\nu)}+1\right)\frac{x^2}{4(J-2)(J-2+\nu)}\right.\right.\right.$$

$$+\cdots+1\Bigg)\frac{x^2}{8(2+\nu)}+1\Bigg)\frac{x^2}{4(1+\nu)}+1\Bigg)\frac{(x/2)^\nu}{\Gamma(1+\nu)}$$

when x is less than $8 + (\nu^2/12)$. The $1/\Gamma(1 + \nu)$ multiplier is calculated by a routine that is essentially the one used in Section 43:8, but special measures are taken if $\nu = 0$ or if ν is a negative integer. The integer J in 50:8:1 is calculated by an empirical expression to ensure 24-bit precision (the fractional error in the hyperbolic Bessel function does not exceed 6×10^{-8}). If x exceeds $8 + (\nu^2/12)$, the hyperbolic Bessel function is calculated via the formula

50:8:2 $$I_\nu(x) \simeq \Bigg(\Bigg(\Bigg(\cdots\Bigg(\Bigg(\frac{(J-\frac{1}{2})^2-\nu^2}{2Jx}+1\Bigg)\frac{(J-\frac{3}{2})^2-\nu^2}{2(J-1)x}+1\Bigg)\frac{(J-\frac{5}{2})^2-\nu^2}{2(J-2)x}$$

$$+\cdots+1\Bigg)\frac{(\frac{3}{2})^2-\nu^2}{4x}+1\Bigg)\frac{(\frac{1}{2})^2-\nu^2}{2x}+1\Bigg)\frac{\exp(x)}{\sqrt{2\pi x}}$$

which is a concatenated version of the asymptotic expansion 50:6:4, truncated by an empirical choice of J, again to ensure 24-bit precision.

Storage needed: ν, f, x and j

Input restrictions: Any value of ν may be input, but x must exceed zero.

ν >>>>>>>>>>

Set $f = 1$

x >>>>>>>>>>

If $x \geq 8 + (\nu^2/12)$ go to (4)

If $\nu + \text{Int}(|\nu|) \neq 0$ go to (1)

Replace ν by $-\nu$

(1) Set $j = \text{Int}(x + 3)$

If $(\nu - 5)\left(\nu + 10 + \frac{3x}{2}\right) > 0$ go to (2)

Replace j by $j + \text{Int}\left(4 - \frac{2\nu}{3}\right)$

(2) Replace f by $1 + fx^2/[4j(j + \nu)]$

Replace j by $j - 1$

If $j \neq 0$ go to (2)

If $\nu = 0$ go to (6)

Replace f by f/ν

(3) Replace f by $2f\nu/x$

Replace ν by $\nu + 1$

If $\nu \leq 3$ go to (3)

Set $j = \left[1 + \frac{2}{7\nu^2}\left(\frac{2}{3\nu^2} - 1\right)\right]\Big/ 30\nu^2$

Replace j by $\frac{j-1}{12\nu} + \nu[1 - \ln(2\nu/x)]$

Replace f by $f\exp(j)\sqrt{\nu/2\pi}$

Go to (6)

(4) Set $j = \text{Int}\left(5 + \frac{640}{x^2} + \frac{7|\nu|}{10}\right)$

If $x < 100$ go to (5)

Replace j by $j - \text{Int}\left[\left(\frac{1}{2} - \sqrt{\frac{247}{x\sqrt{x}}}\right)|\nu|\right]$

(5) Replace f by $1 + f\left[\left(j - \frac{1}{2}\right)^2 - \nu^2\right]\Big/ 2jx$

Replace j by $j - 1$

Test values:
$I_0(5) = 27.2398718$
$I_{1/2}(\pi) = 5.19875924$
$I_{-3}(1) = 0.0221684248$

If $j \neq 0$ go to (5)
(6) Replace f by $f\exp(x)/\sqrt{2\pi x}$
Output f
$f \simeq I_\nu(x)$ <<<<<<

$I_{-7/2}(2) = -0.628009049$
$I_2(10) = 2281.51897$
$I_{100}(100) = 4.64153494 \times 10^{21}$
$I_{\pm 1}(125) = 6.88584377 \times 10^{52}$

A simpler alternative is available in Section 49:8 for calculating hyperbolic Bessel functions of integer order. Several versions of the universal hypergeometric algorithm [see Section 18:14] are also useful for determining numerical values of $I_\nu(x)$.

50:9 APPROXIMATIONS

If ν is not a negative integer, the hyperbolic Bessel function of small positive argument is approximated crudely by

50:9:1 $$I_\nu(x) \simeq \frac{(x/2)^\nu}{\Gamma(1+\nu)} \qquad x \to 0 \qquad \nu \neq -1, -2, -3, \ldots$$

and more accurately (especially if ν is positive) by

50:9:2 $$I_\nu(x) \simeq \frac{(x/2)^\nu}{\Gamma(1+\nu)}\left[1 + \frac{x^2}{4(1+\nu)(2+\nu)}\right]^{2+\nu} \qquad \text{small } x$$

For large argument, the approximation

50:9:3 $$I_\nu(x) \simeq \frac{\exp(x)}{\sqrt{2\pi x}}\left(1 - \frac{1}{x}\right)^\mu \qquad x >> \mu = \frac{\nu^2}{2} - \frac{1}{8}$$

is valid and shows that increasing its argument makes the hyperbolic Bessel function increasingly independent of its order and increasingly closer to $\exp(x)/\sqrt{\pi x}$.

Making use of the reflection property [Section 43:5] of the gamma function, approximation 50:9:1 shows that

50:9:4 $$I_\nu(x)\,I_{-\nu}(x) \simeq \frac{1}{\Gamma(1+\nu)\,\Gamma(1-\nu)} = \frac{\sin(\nu\pi)}{\nu\pi} \qquad x \to 0$$

a result that remains valid for all orders. It may be combined with equation 50:5:4 to produce the approximation

50:9:5 $$\frac{1}{I_\nu(x)} - \frac{1}{I_{-\nu}(x)} \simeq 2\nu K_\nu(x) \qquad x \to 0 \qquad \nu \neq 0, \pm 1, \pm 2, \ldots$$

50:10 OPERATIONS OF THE CALCULUS

Differentiation of a hyperbolic Bessel function gives the alternative formulations

50:10:1 $$\frac{d}{dx} I_\nu(x) = \frac{I_{\nu-1}(x) + I_{\nu+1}(x)}{2} = I_{\nu-1}(x) - \frac{\nu}{x} I_\nu(x) = I_{\nu+1}(x) + \frac{\nu}{x} I_\nu(x)$$

but the indefinite integral of the general $I_\nu(x)$ function is expressible only as the summation

50:10:2 $$\int_0^x I_\nu(t)dt = 2\sum_{j=0}^{\infty} (-1)^j I_{\nu+1+2j}(x) \qquad \nu > -1$$

and not in closed form. On the other hand, the indefinite integrals of the following products:

50:10:3 $$\int_0^x t^\nu I_\nu(t)dt = \frac{\sqrt{\pi}\,\Gamma(\frac{1}{2}+\nu)x}{2^{1-\nu}}\,[I_\nu(x)\,\ell_{\nu-1}(x) - I_{\nu-1}(x)\ell_\nu(x)] \qquad \nu > -\frac{1}{2}$$

50:10:4 $$\int_0^x t^\nu I_{\nu-1}(t)dt = x^\nu I_\nu(x) \qquad \nu > 0$$

50:10:5 $$\int_0^x t^{-\nu} I_{\nu+1}(t)dt = x^{-\nu} I_\nu(x) - \frac{2^{-\nu}}{\Gamma(1+\nu)}$$

50:10:6 $$\int_0^x t^{-\nu}\exp(\pm t)\, I_\nu(t)dt = \frac{x^{1-\nu}\exp(\pm x)}{2\nu - 1}\left[\pm I_{\nu-1}(x) - I_\nu(x)\right] \mp \frac{2^{1-\nu}}{(2\nu-1)\,\Gamma(\nu)} \qquad \nu \neq \frac{1}{2}$$

are straightforward. The ℓ functions in 50:10:3 are instances of the hyperbolic Struve function discussed in Section 57:13, while Γ denotes the gamma function [Chapter 43].

Among formulas for definite integration is

50:10:7 $$\int_0^\infty \exp(-B^2t^2)\, I_\nu(bt)dt = \frac{\sqrt{\pi}}{2B}\exp\left(\frac{b^2}{8B^2}\right) I_{\nu/2}\left(\frac{b^2}{8B^2}\right) \qquad \nu > -1$$

Replacing the argument x by $2\sqrt{x}$ in the first of the two *Rayleigh formulas*

50:10:8 $$\left\{\frac{1}{x}\frac{d}{dx}\right\}^n [x^\nu I_\nu(x)] = x^{\nu-n} I_{\nu-n}(x)$$

50:10:9 $$\left\{\frac{1}{x}\frac{d}{dx}\right\}^n [x^{-\nu} I_\nu(x)] = x^{-\nu-n} I_{\nu+n}(x)$$

for multiple differentiation leads to a relationship that generalizes to the very simple expression

50:10:10 $$\frac{d^\mu}{dx^\mu}[x^{\nu/2} I_\nu(2\sqrt{x})] = x^{(\nu-\mu)/2} I_{\nu-\mu}(2\sqrt{x})$$

Here the operator d^μ/dx^μ signifies differintegration with lower limit zero to an arbitrary (positive or negative, integer or noninteger) order [see Section 0:10].

50:11 COMPLEX ARGUMENT

Replacing x in expansion 50:6:1 by $x + iy$ leads to

50:11:1 $$I_\nu(x+iy) = \sum_{j=0}^\infty \frac{\cos(\phi) + i\sin(\phi)}{j!\Gamma(j+1+\nu)}\left\{\frac{x^2+y^2}{4}\right\}^{j+(\nu/2)}$$

where $\phi = (2j + \nu)(\theta + 2k\pi)$, θ and k having retained their significances from Section 13:11.

When the argument is purely imaginary, the hyperbolic Bessel function becomes a Bessel function [Chapter 53] of real argument:

50:11:2 $$I_\nu(iy) = i^\nu J_\nu(y) = \left[\cos\left(\frac{\nu\pi}{2}\right) + i\sin\left(\frac{\nu\pi}{2}\right)\right] J_\nu(y)$$

Unless its order is an integer, the hyperbolic Bessel function is complex when its argument is negative. We have

50:11:3 $$I_\nu(-x) = (-1)^\nu I_\nu(x) = [\cos(\nu\pi) + i\sin(\nu\pi)]\, I_\nu(x)$$

50:12 GENERALIZATIONS

The function

50:12:1 $$\left(\frac{x}{2}\right)^{\mu+\nu}\sum_{j=0}^\infty \frac{(x^2/4)^j}{\Gamma(1+\mu+j)\Gamma(1+\nu+j)}$$

closely related to the Kummer function [Chapter 47], is a generalization of the hyperbolic Bessel function, which is the $\mu = 0$ instance of 50:12:1. The hyperbolic Struve ℓ function [Section 57:13] is also a particular instance of the general function 50:12:1.

50:13 COGNATE FUNCTIONS

In the next chapter the cognate Basset function is discussed. Also closely related to hyperbolic Bessel functions are the Kelvin functions [Chapter 55] and, of course, the Bessel functions [Chapter 53] themselves.

CHAPTER

51

THE BASSET FUNCTION $K_\nu(x)$

Because the Basset function of noninteger order is related so simply [see 51:3:5] to the hyperbolic Bessel function of the previous chapter, this chapter concentrates on the Basset functions $K_0(x)$, $K_1(x)$, $K_2(x)$, ... of integer order.

51:1 NOTATION

Alternative names for the Basset function are the *modified Bessel function of the third kind, Bessel's function of the second kind of imaginary argument, Macdonald's function* and the *modified Hankel function.*

We use ν generally to represent the order of a Basset function but replace this symbol by n to specify integer order.

51:2 BEHAVIOR

The Basset function $K_\nu(x)$ is infinite for $x = 0$ and complex for $x < 0$ [see equation 51:11:3]. Accordingly, we restrict attention to $x > 0$ here and generally throughout this chapter.

For all ν, $K_\nu(x)$ is a positive and monotonically decreasing function of its argument x, approaching zero as $x \to \infty$, in accordance with expression 51:9:6, in a manner increasingly independent of the order ν. However, the approach to infinity as $x \to 0$ is a strong function of ν as detailed in Section 51:9.

For a constant positive argument, $K_\nu(x)$ is an even function of its order ν, as evidenced by equation 51:5:1. Moreover, for constant positive argument, $K_\nu(x)$ invariably increases as $|\nu|$ increases.

Figure 51-1 shows maps of $K_n(x)$ for $n = 0, 1, 2, 3, 4, 5$ and 6. The curves for noninteger order Basset functions smoothly interpolate between these mapped curves; for example, $K_{9/2}(x)$ lies intermediate between $K_4(x)$ and $K_5(x)$.

51:3 DEFINITIONS

Basset functions may be defined via those Tricomi functions [Chapter 48] in which the a parameter is a moiety of the c parameter:

51:3:1 $$K_\nu(x) = \sqrt{\pi}(2x)^\nu \exp(-x)\, U(\tfrac{1}{2} + \nu; 1 + 2\nu; 2x) \qquad \nu \geq 0$$

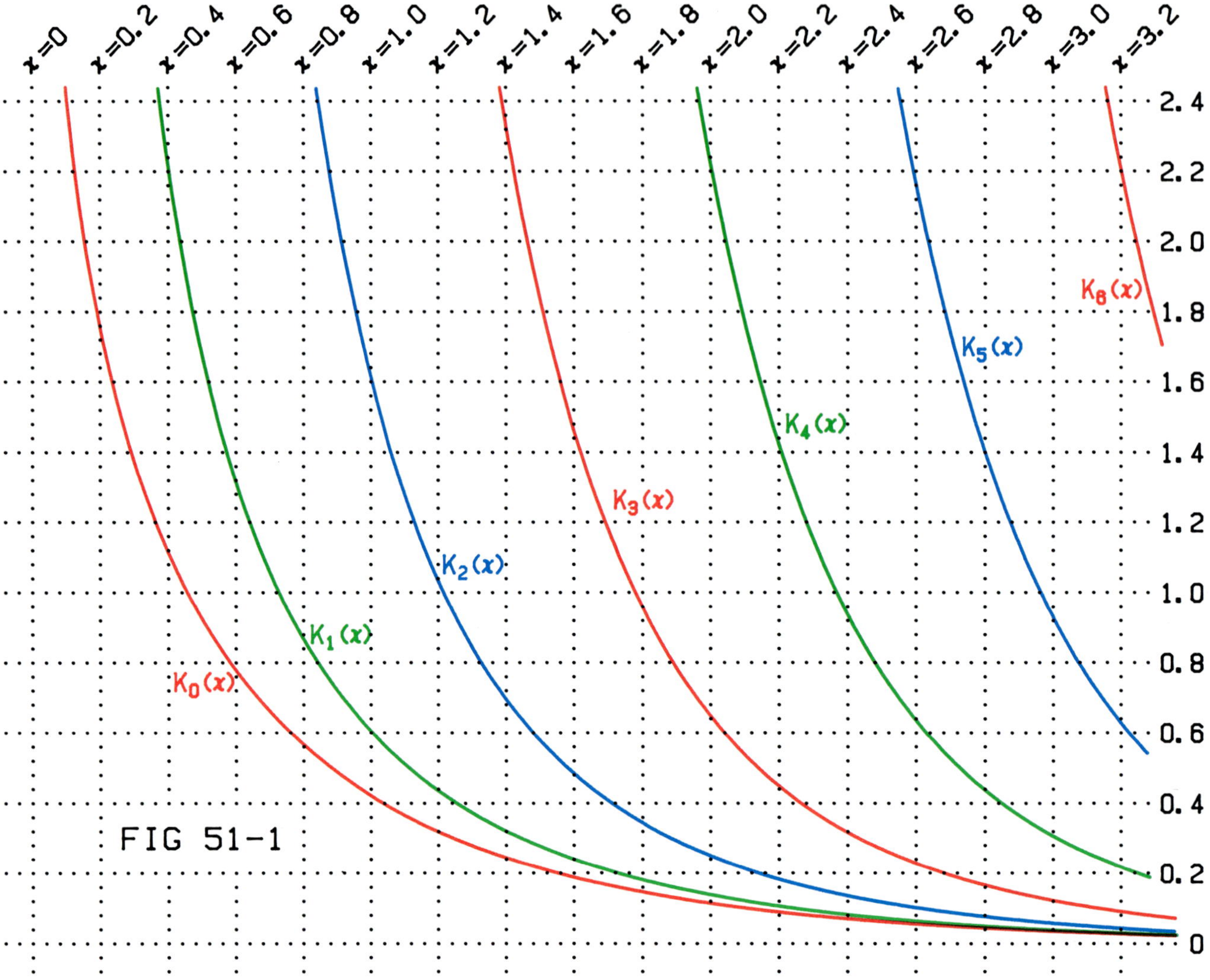

FIG 51-1

or approaches infinity:

51:3:2 $$K_\nu(x) = \frac{1}{2}\left(\frac{x}{2}\right)^\nu \lim_{a\to\infty}\left\{\Gamma(a-\nu)\,U\left(a;1+\nu;\frac{x^2}{4a}\right)\right\} \qquad \nu \geq 0$$

Equivalent to 51:3:1, but somewhat simpler, is the definition of a Basset function as a special case of Whittaker's W function [see 48:13:6].

Among definite integrals that define the Basset function are

51:3:3 $$K_\nu(x) = \frac{\sqrt{\pi}}{\Gamma(\frac{1}{2}+\nu)}\left(\frac{x}{2}\right)^\nu \int_1^\infty \frac{\exp(-xt)\mathrm{d}t}{(t^2-1)^{1/2-\nu}} \qquad \nu > -\frac{1}{2}$$

and

51:3:4 $$K_\nu(x) = \frac{\Gamma(\frac{1}{2}+\nu)}{\sqrt{\pi}}(2x)^\nu \int_0^\infty \frac{\cos(t)\mathrm{d}t}{(t^2+x^2)^{1/2+\nu}} \qquad \nu \geq -\frac{1}{2}$$

and a large number of alternatives were assembled in the Bateman manuscript [see Erdélyi, Magnus, Oberhettinger and Tricomi, *Higher Transcendental Functions*, Volume 2, pages 82 and 83]. Because of relationship 51:5:1, the sign of ν may be changed in definitions 51:3:3 and 51:3:4.

Basset functions of noninteger order may be defined in terms of hyperbolic Bessel functions by

51:3:5 $$K_\nu(x) = \frac{\pi[I_{-\nu}(x) - I_\nu(x)]}{2\sin(\nu\pi)} \qquad \nu \neq 0, \pm 1, \pm 2, \ldots$$

but this expression must be written as a limit:

51:3:6 $$K_n(x) = \frac{\pi}{2}\lim_{\nu\to n}\left\{\frac{I_{-\nu}(x) - I_\nu(x)}{\sin(\nu\pi)}\right\} \qquad n = 0, \pm 1, \pm 2, \ldots$$

when the order is an integer.

With c_1 and c_2 representing arbitrary constants, the differential equation

51:3:7 $$x^2\frac{d^2f}{dx^2} + x\frac{df}{dx} - (x^2 + n^2)f = 0 \qquad n = 0, \pm 1, \pm 2, \ldots$$

is solved by $f = c_1 I_n(x) + c_2 K_n(x)$. Other differential equations whose solutions involve the K_0 or K_1 functions are listed as 49:3:10–49:3:13.

Semiintegration with lower limit zero is a powerful method of generating Basset functions. Examples include

51:3:8 $$K_0\left(\frac{1}{2x}\right) = \sqrt{\pi x}\exp\left(\frac{1}{2x}\right)\frac{d^{-1/2}}{dx^{-1/2}}\left\{\frac{1}{x}\exp\left(\frac{-1}{x}\right)\right\}$$

51:3:9 $$K_1\left(\frac{1}{\sqrt{x}}\right) = \frac{x\sqrt{\pi}}{2}\frac{d^{-1/2}}{dx^{-1/2}}\left\{\frac{1}{\sqrt{x^3}}\exp\left(\frac{-1}{\sqrt{x}}\right)\right\}$$

51:3:10 $$K_{1/4}\left(\frac{1}{2x^2}\right) = x\sqrt{2\pi}\exp\left(\frac{1}{2x^2}\right)\frac{d^{-1/2}}{dx^{-1/2}}\left\{\frac{1}{\sqrt{x^3}}\exp\left(\frac{-1}{x^2}\right)\right\}$$

51:4 SPECIAL CASES

When its order is an odd multiple of $\frac{1}{2}$, the Basset function reduces to a simple function involving an exponential. The simplest cases are

51:4:1 $$K_{1/2}(x) = K_{-1/2}(x) = \sqrt{\frac{\pi}{2x}}\exp(-x)$$

51:4:2 $$K_{3/2}(x) = K_{-3/2}(x) = \sqrt{\frac{\pi}{2x}}\left[1 + \frac{1}{x}\right]\exp(-x)$$

[see Section 26:13 and Figure 26-2] and others may be constructed via recurrence formula 51:5:2. The symbol $k_n(x)$ and the name *modified spherical Bessel function of the third kind* are sometimes applied to the $\sqrt{\pi/2x}\,K_{n+1/2}(x)$ function.

A Basset function of order $\frac{1}{3}$ is related to an Airy function [Chapter 56]:

51:4:3 $$K_{1/3}(x) = K_{-1/3}(x) = \pi\sqrt{\frac{3}{X}}\,\mathrm{Ai}(X) \qquad X = \left(\frac{3x}{2}\right)^{2/3}$$

while that of order $\frac{1}{4}$ is related to a parabolic cylinder function [Chapter 46]:

51:4:4 $$K_{1/4}(x) = \frac{\sqrt{\pi}}{x^{1/4}}D_{-1/2}(2\sqrt{x})$$

51:5 INTRARELATIONSHIPS

With respect to its order, the Basset function is even

51:5:1 $$K_{-\nu}(x) = K_\nu(x)$$

so that ν may be replaced by $|\nu|$ in most of the formulas of this chapter.

The Basset function obeys the recurrence relations

51:5:2 $$K_{\nu+1}(x) = K_{\nu-1}(x) + \frac{2\nu}{x} K_\nu(x)$$

and

51:5:3 $$K_{\nu+1}(x)\, I_\nu(x) = \frac{1}{x} - K_\nu(x)\, I_{\nu+1}(x)$$

The zero-order Basset function satisfies the argument-addition and -subtraction formulas

51:5:4 $$K_0(x \pm y) = I_0(y)\, K_0(y) + 2\sum_{j=1}^{\infty} (\mp 1)^j I_j(y)\, K_j(x) \qquad x > y > 0$$

By sufficient applications of formula 51:5:2, any Basset function $K_n(x)$ of integer order may be expressed in terms of $K_0(x)$ and $K_1(x)$. The first two examples are

51:5:5 $$K_2(x) = K_0(x) + \frac{2}{x} K_1(x)$$

51:5:6 $$K_3(x) = \frac{4}{x} K_0(x) + \left(1 + \frac{8}{x^2}\right) K_1(x)$$

and expressions for $K_4(x)$, $K_5(x)$ and $K_6(x)$ are analogous to 49:5:13–49:5:15, but with uniformly positive signs.

51:6 EXPANSIONS

The Basset function of noninteger order is expansible as the sum of two convergent series

51:6:1 $$K_\nu(x) = \frac{\Gamma(\nu)}{2}\left(\frac{x}{2}\right)^{-\nu} \sum_{j=0}^{\infty} \frac{(x^2/4)^j}{j!(1-\nu)_j} + \frac{\Gamma(-\nu)}{2}\left(\frac{x}{2}\right)^{\nu} \sum_{j=0}^{\infty} \frac{(x^2/4)^j}{j!(1+\nu)_j} \qquad \nu \neq 0, \pm 1, \pm 2, \ldots$$

that coalesce only if ν is an odd multiple of $\frac{1}{2}$. For integer order the series are

51:6:2 $$K_0(x) = \left[-\gamma - \ln\left(\frac{x}{2}\right)\right] + \left[-\gamma + 1 - \ln\left(\frac{x}{2}\right)\right]\frac{(x^2/4)}{(1!)^2} + \left[-\gamma + 1 + \frac{1}{2} - \ln\left(\frac{x}{2}\right)\right]\frac{(x^2/4)^2}{(2!)^2} + \left[-\gamma + 1 + \frac{1}{2} + \frac{1}{3} - \ln\left(\frac{x}{2}\right)\right]\frac{(x^2/4)^3}{(3!)^2} + \cdots$$

51:6:3 $$K_1(x) = \frac{1}{x} - \left[-\gamma + \frac{1}{2} - \ln\left(\frac{x}{2}\right)\right]\frac{(x^2/4)^{1/2}}{0!1!} - \left[-\gamma + 1 + \frac{1}{4} - \ln\left(\frac{x}{2}\right)\right]\frac{(x^2/4)^{3/2}}{1!2!} - \left[-\gamma + 1 + \frac{1}{2} + \frac{1}{6} - \ln\left(\frac{x}{2}\right)\right]\frac{(x^2/4)^{5/2}}{2!3!} - \cdots$$

and generally

51:6:4 $$K_n(x) = \frac{1}{2}\left(\frac{2}{x}\right)^n \sum_{j=0}^{n-1} \frac{(n-j-1)!}{j!}\left(\frac{-x^2}{4}\right)^j + (-1)^n \sum_{j=0}^{\infty}\left[\frac{\psi(1+j)}{2} + \frac{\psi(1+n+j)}{2} - \ln\left(\frac{x}{2}\right)\right]\frac{(x/2)^{2j+n}}{j!(n+j)!}$$

In these formulas Γ denotes the gamma function [Chapter 43], ψ the digamma function [Chapter 44], $(1-\nu)_j$ is a Pochhammer polynomial [Chapter 18] and γ is Euler's constant [Section 1:7]. Alternatively, Basset functions of integer order may be expanded as *Neumann series* in terms of hyperbolic Bessel functions:

51:6:5 $$K_0(x) = \left[-\gamma - \ln\left(\frac{x}{2}\right)\right] I_0(x) + 2\sum_{j=1}^{\infty} \frac{I_{2j}(x)}{j} \qquad \gamma = 0.5772156649$$

51:6:6 $$K_1(x) = -\left[-\gamma + 1 - \ln\left(\frac{x}{2}\right)\right] I_1(x) + \frac{1}{x} I_0(x) - \sum_{j=1}^{\infty} \frac{(1+2j)}{j(1+j)} I_{1+2j}(x)$$

and generally

51:6:7 $$K_n(x) = (-1)^n \left[\psi(1+n) - \ln\left(\frac{x}{2}\right)\right] I_n(x) + \frac{1}{2}\left(\frac{2}{x}\right)^n \sum_{k=0}^{n-1} \frac{n!}{k!(n-k)} \left(\frac{-x}{2}\right)^k I_k(x) + (-1)^n \sum_{j=1}^{\infty} \frac{(n+2j)}{j(n+j)} I_{n+2j}(x)$$

The series

51:6:8 $$K_\nu(x) \sim \sqrt{\frac{\pi}{2x}} \exp(-x) \left\{1 - \frac{\frac{1}{4} - \nu^2}{2x} + \frac{(\frac{9}{4} - \nu^2)(\frac{1}{4} - \nu^2)}{8x^2} - \frac{(\frac{25}{4} - \nu^2)(\frac{9}{4} - \nu^2)(\frac{1}{4} - \nu^2)}{48x^3} + \cdots + \frac{(\frac{1}{2} - \nu)_j (\frac{1}{2} + \nu)_j}{j!(-2x)^j} + \cdots\right\} \qquad x \to \infty$$

is generally asymptotic but it does terminate when $|\nu|$ is an odd multiple of $\frac{1}{2}$, and under these circumstances the expansion is exact. If $|\nu|$ is not an odd multiple of $\frac{1}{2}$ and lies between the numbers $n - \frac{1}{2}$ and $n + \frac{1}{2}$, where n is an integer, then the terms in series 51:6:8 corresponding to $j = 0, j = 1, j = 2, \ldots, j = n$ are uniformly positive (we are treating only positive x), whereas the $j = (n+1)^{\text{th}}$ term is negative; thereafter, the terms alternate in sign. Accordingly, it follows from the properties of alternating series [see Section 0:5] that if the series 51:6:8 is truncated, terms after $j = J$ being ignored, then the partial sum is related to the true value of $K_\nu(x)$ by one of the inequalities

51:6:9 $$(\text{series of } J \text{ terms}) \lesseqgtr K_\nu(x) \lesseqgtr (\text{series of } J+1 \text{ terms}) \qquad J \geq n = \text{Int}\left(|\nu| + \frac{1}{2}\right)$$

for large enough J.

51:7 PARTICULAR VALUES

	$x = 0$	$x = \infty$
$K_\nu(x)$	∞	0

51:8 NUMERICAL VALUES

Two algorithms are presented in this section. The first generates values of $K_0(x)$ and $K_1(x)$ for sufficiently small arguments, say $x \leq 3$. From these, recurrence 51:5:2 shows how it is possible to compute the $K_n(x)$ function for any positive integer n. The second algorithm produces values of the Basset function for arguments exceeding 2.5 and for orders (integer or noninteger) less than about $[24x/\ln(x)]^{2/3}$. Neither algorithm is suitable for evaluating $K_\nu(x)$ for small x and noninteger ν, but definition 51:3:5 provides easy access to these values via the algorithm of Section 50:8.

With R_j having the significance accorded to it in Section 49:8, the Neumann series 51:6:5 may be rewritten as the concatenation

51:8:1 $$K_0(x) = I_0(x)\left[4R_0R_1\left(\frac{1}{2} + R_2R_3\left(\frac{1}{4} + R_4R_5\left(\frac{1}{6} + \cdots\right)\right)\right) - \gamma - \ln\left(\frac{x}{2}\right)\right]$$

This is the formula used by the first algorithm to calculate $K_0(x)$, the required value of $I_0(x)$ being determined via the similar concatenation

51:8:2 $$\cosh(x) = I_0(x)[1 + 2R_0R_1(1 + R_2R_3(1 + R_4R_5(1 + \cdots)))]$$

which follows from series 49:5:7. In practice, both concatenations are terminated by setting $R_j = 0$ for j greater than an empirically chosen integer J. Recurrence 49:8:2 then serves to calculate all other R_j values. The hyperbolic Bessel function $I_1(x)$ is calculated by the algorithm as $R_0 I_0(x)$ and thence the Basset function $K_1(x)$ as $R_0\{[1/xI_1(x)] - K_0(x)\}$, an identity that follows from 51:5:3. Many problems in applied mathematics require values of the functions $I_0(x)$ and $I_1(x)$ as well as those of $K_0(x)$ and $K_1(x)$; the algorithm below delivers all four functions if the portions shown in green are included.

Input x >>>>>>>

Set $j = 8 + 2\,\mathrm{Int}(x)$

Set $I = K = r = R = 0$

(1) Replace I by $1 + IrR$

Replace K by $(1/j) + KrR$

Set $r = 1 \bigg/ \left(R + \frac{2j}{x}\right)$

Replace j by $j - 1$

Set $R = 1 \bigg/ \left(r + \frac{2j}{x}\right)$

Replace j by $j - 1$

If $j \neq 0$ go to (1)

Replace I by $[\exp(x) + \exp(-x)]/[2 + 4IrR]$

Output I

$I \simeq I_0(x)$ <<<<<<

Replace K by $I[4KrR - \ln(.890536209x)]$

Output K

$K \simeq K_0(x)$ <<<<<

Replace I by IR

Output I

$I \simeq I_1(x)$ <<<<<<<

Replace K by $R[(1/xI) - K]$

Output K

$K \simeq K_1(x)$ <<<<<

Storage needed: x, j, I, K, r and R

Input restrictions: x must exceed zero. If x exceeds about 3 (the precise limit depends on the computing device), accuracy may be impaired.

Test values:
$I_0(1) = 1.26606588$
$K_0(1) = 0.421024438$
$I_1(1) = 0.565159104$
$K_1(1) = 0.601907230$
$K_0(3) = 0.0347395044$
$K_1(3) = 0.0401564311$
$K_0(5) = 0.00369109833$
$K_1(5) = 0.00404461344$

Although the accuracy of this algorithm is inherently high, it may not deliver precise values of $K_0(x)$ and $K_1(x)$ unless x is small. The difficulty arises because equation 51:8:1 calls for the differencing of two terms whose values become increasingly close as x increases; therefore, significance is lost in their subtraction. The seriousness of this effect depends on the number of significant bits utilized by the computing device. Changes to the algorithm cannot ameliorate the problem.

The second algorithm utilizes expansion 51:6:8 and overcomes its asymptoticity by invoking the Padé technique as described in Section 17:13. An empirical formula is used to determine the order J, up to a maximum of 30, of the diagonal Padé approximant R_J^J that is generated. Within the permitted ranges of x and ν, $K_\nu(x)$ is generally returned with a precision better than 24 bits.

ν >>>>>>>>>>

Set $j = t = h_0 = 1$

x >>>>>>>>>>

Set $J = 2\,\mathrm{Int}\{6 + [|\nu|^{3/2}\ln(x)/x]\}$

(1) Replace t by $t\left[\nu^2 - \left(j - \frac{1}{2}\right)^2\right] \bigg/ (2jx)$

Set $h_j = t + h_{j-1}$

Set $q = 0$

Set $k = j$

Storage needed: ν, j, t, x, J, k, p, q, h_0, h_1, $h_2, \ldots, h_{59}$, h_{60}

(2) Set $p = h_{k-1}$
If $p \neq h_k$ go to (3)
Set $h_{k-1} = 10^{99}$
If $p \neq 10^{99}$ go to (4)
Set $h_{k-1} = q$
Go to (4)
(3) Set $h_{k-1} = q + [1/(h_k - p)]$
(4) Set $q = p$
Replace k by $k - 1$
If $k \neq 0$ go to (2)
Replace j by $j + 1$
If $j \leq J$ go to (1)
Set $p = h_0\sqrt{\pi/2x}\exp(-x)$
Output p
$p \simeq K_\nu(x)$ <<<<<<

Input restrictions:
$x \geq 2.5$
$|\nu| < [24x/\ln(x)]^{2/3}$

Test values:
$K_2(5) = 0.00530894371$
$K_{1/2}(\pi) = 0.0305568546$
$K_{-1}(10) = 1.86487735 \times 10^{-5}$
$K_{50}(100) = 9.27452265 \times 10^{-40}$
$K_{16}(5) = 186233.583$

51:9 APPROXIMATIONS

From expansions 51:6:1–51:6:4, the following two-term approximations, valid as $x \to 0$, may be derived:

51:9:1 $$K_0(x) \simeq \ln\left(\frac{2}{x}\right) - \gamma = 0.11593 - \ln(x) \qquad \nu = 0 \qquad \text{small } x$$

51:9:2 $$K_\nu(x) \simeq \frac{\Gamma(\nu)x^{-\nu}}{2^{1-\nu}} + \frac{\Gamma(-\nu)x^{\nu}}{2^{1+\nu}} \qquad 0 < \nu < 1 \qquad \text{small } x$$

51:9:3 $$K_1(x) \simeq \frac{1}{x} + \frac{x}{2}\ln\left(\frac{x}{2}\right) \qquad \nu = 1 \qquad \text{small } x$$

51:9:4 $$K_\nu(x) \simeq \Gamma(|\nu|)\left(\frac{2}{x}\right)^{|\nu|}\left[\frac{1}{2} - \frac{x^2}{8|\nu|-8}\right] \qquad |\nu| > 1 \qquad \text{small } x$$

More accurate than 51:9:4, for $|\nu| > 2$, is the approximation

51:9:5 $$K_\nu(x) \simeq \frac{\Gamma(|\nu|)}{2}\left(\frac{2}{x}\right)^{|\nu|}\left[1 - \frac{x^2}{4(|\nu| - 1)(|\nu| - 2)}\right]^{|\nu|-2} \qquad |\nu| > 2 \qquad \text{small } x$$

For large x, the approximation

51:9:6 $$K_\nu(x) \simeq \sqrt{\frac{\pi}{2x}}\exp(-x)\left(1 + \frac{1}{x}\right)^{\mu} \qquad x >> \mu = \frac{\nu^2}{2} - \frac{1}{8}$$

follows from the asymptotic expansion 51:6:8. It is exact only for $\nu = \pm\frac{1}{2}, \pm\frac{3}{2}$.

51:10 OPERATIONS OF THE CALCULUS

The general differentiation formulas

51:10:1 $$\frac{d}{dx}K_\nu(x) = \frac{-1}{2}\left[K_{\nu-1}(x) + K_{\nu+1}(x)\right] = \frac{\nu}{x}K_\nu(x) - K_{\nu+1}(x) = -K_{\nu-1}(x) - \frac{\nu}{x}K_\nu(x)$$

have the special cases

51:10:2 $$\frac{d}{dx}K_0(x) = -K_1(x)$$

and

51:10:3 $$\frac{d}{dx}K_1(x) = -K_0(x) - \frac{K_1(x)}{x}$$

and lead to the derivatives

51:10:4 $$\frac{d}{dx}x^{\pm\nu}K_\nu(x) = -x^{\pm\nu}K_{\nu\mp1}(x)$$

The formulas

51:10:5 $$\int_0^x K_0(t)dt = \frac{\pi x}{2}[K_0(x)\,\ell_{-1}(x) + K_1(x)\,\ell_0(x)]$$

and

51:10:6 $$\int_x^\infty K_1(t)dt = K_0(x)$$

for the indefinite integrals of the zero-order and unity-order Basset functions may be obtained as special cases of the general expressions 51:10:7 and 51:10:9 below. In equations 51:10:5 and 51:10:7 ℓ is the hyperbolic Struve function discussed in Section 57:13. Closed-form expressions are available for the following indefinite integrals of the Basset function multiplied by a power:

51:10:7 $$\int_0^x t^\nu K_\nu(t)dt = 2^{\nu-1}\sqrt{\pi}\,\Gamma(\nu+\tfrac{1}{2})x[K_\nu(x)\,\ell_{\nu-1}(x) + K_{\nu-1}(x)\,\ell_\nu(x)] \qquad \nu > -\frac{1}{2}$$

51:10:8 $$\int_0^x t^\nu K_{\nu-1}(t)dt = 2^{\nu-1}\Gamma(\nu) - x^\nu K_\nu(x) \qquad \nu > 0$$

51:10:9 $$\int_x^\infty t^{-\nu}K_{\nu+1}(t)dt = x^{-\nu}K_\nu(x)$$

as well as for the following expressions involving exponential functions:

51:10:10 $$\int_0^x t^\nu \exp(\pm t)\,K_\nu(t)dt = \frac{x^{\nu+1}\exp(\pm x)}{2\nu+1}[K_\nu(x) \pm K_{\nu+1}(x)] \mp \frac{2^\nu\Gamma(\nu+1)}{2\nu+1} \qquad \nu > -\frac{1}{2}$$

and

51:10:11 $$\int_x^\infty t^{-\nu}\exp(t)\,K_\nu(t)dt = \frac{x^{1-\nu}\exp(x)}{2\nu-1}[K_\nu(x) + K_{\nu-1}(x)] \qquad \nu > \frac{1}{2}$$

The important *King's integral* is the $\nu = 0$ instance of 51:10:10 with upper signs selected.

For $|\nu| > 1$ the definite integral $\int K_\nu(t)dt$ between limits of $t = 0$ and $t = \infty$ diverges. The convergent cases are the $\mu = 0$ instances of the general formula

51:10:12 $$\int_0^\infty t^\mu K_\nu(t)dt = 2^{\mu-1}\Gamma\left(\frac{\mu+\nu+1}{2}\right)\Gamma\left(\frac{\mu-\nu+1}{2}\right) \qquad 0 \le \nu < 1+\mu$$

namely

51:10:13 $$\int_0^\infty K_\nu(t)dt = \frac{\pi}{2}\sec\left(\frac{\nu\pi}{2}\right) \qquad |\nu| < 1$$

51:11 COMPLEX ARGUMENT

Definition 51:3:5 may be combined with equation 50:11:1 to produce an expression for the Basset function of complex argument when ν is noninteger.

For purely imaginary argument, the Basset function is related to the functions of Chapters 53 and 54 by

51:11:1 $$K_\nu(iy) = \frac{-\pi}{2i^\nu}[Y_\nu(y) + iJ_\nu(y)] = \frac{-\pi}{2}\left[\cos\left(\frac{\nu\pi}{2}\right)Y_\nu(y) + \sin\left(\frac{\nu\pi}{2}\right)J_\nu(y)\right] + \frac{i\pi}{2}\left[\sin\left(\frac{\nu\pi}{2}\right)Y_\nu(y) - \cos\left(\frac{\nu\pi}{2}\right)J_\nu(y)\right]$$

When the real and imaginary components of the argument of a Basset function are equal in magnitude, we have

51:11:2 $$K_\nu(x \pm ix) = i^{\pm\nu}[\mathrm{ker}_\nu(x\sqrt{2}) \pm i\,\mathrm{kei}_\nu(x\sqrt{2})]$$

where ker and kei are Kelvin functions [Chapter 55] and $i^{\pm\nu}$ is to be interpreted as $\cos(\nu\pi/2) \pm i\sin(\nu\pi/2)$.

The Basset function is complex for negative real argument. The real and imaginary parts are given by

51:11:3 $$K_\nu(-x) = \cos(\nu\pi)K_\nu(x) - \frac{i\pi}{2}[I_\nu(x) + I_{-\nu}(x)] \qquad \nu \neq 0, \pm 1, \pm 2, \ldots$$

51:12 GENERALIZATIONS

The Basset function may be generalized to Whittaker's function [Section 48:13]

51:12:1 $$K_\nu\left(\frac{x}{2}\right) = \sqrt{\frac{\pi}{x}}\,W_{0;\nu}(x)$$

and thence to the Tricomi function [see Chapter 48].

51:13 COGNATE FUNCTIONS

The Basset function is related to all the functions discussed in Chapters 49–57.

CHAPTER 52

THE BESSEL COEFFICIENTS $J_0(x)$ AND $J_1(x)$

Those Bessel functions [Chapter 53] in which the order is a nonnegative integer are known as "Bessel coefficients" and are the subject of this chapter. Emphasis is placed on the most important of the $J_n(x)$ functions: those with $n = 0$ and 1.

52:1 NOTATION

Bessel coefficients are also known as *Bessel functions of the first kind of nonnegative integer order*.

52:2 BEHAVIOR

All Bessel coefficients are oscillatory functions whose oscillations become increasingly damped as their arguments approach large values of either sign. As Figure 52-1 shows, all Bessel coefficients except $J_0(x)$ are zero at $x = 0$. Moreover, as n increases, $J_n(x)$ retains near-zero values over an increasing range in the vicinity of $x = 0$; for example, $J_9(x)$ does not exceed 0.01 in magnitude in the range $-5.4 \le x \le 5.4$.

Some regularities are apparent in Figure 52-1. Notice that each local maximum or minimum of J_0 corresponds to a zero of J_1, but this rule does not extend to larger orders. On the other hand, for $n \ge 1$, each local maximum or minimum of J_n corresponds to a point of intersection of the J_{n-1} and J_{n+1} curves. Moreover, at each zero of J_n, J_{n-1} and J_{n+1} have equal magnitudes but opposite signs.

For $x \ge 0$, each Bessel coefficient encounters its first (and largest) maximum at an argument that, for small n, is close to $n\pi/2$. This rule fails for larger n, however, and for *very* large order, the Bessel coefficient $J_n(x)$ attains its first maximum close to $x = n$. Subsequent, and ever-smaller, maxima are encountered with a spacing of approximately 2π. Local minima occur approximately midway between consecutive maxima. For $x > 0$ the first minimum is the largest and subsequent minima decrease progressively in size. A zero is encountered between each minimum and its adjacent maxima so that the spacing of the zeros is approximately π.

52:3 DEFINITIONS

Bessel coefficients are defined by a number of generating functions. Those of even order arise from

52:3:1 $$\cos[x\cos(t)] = J_0(x) - 2J_2(x)\cos(2t) + 2J_4(x)\cos(4t) - \cdots = J_0(x) + 2\sum_{j=1}^{\infty}(-1)^j J_{2j}(x)\cos(2jt)$$

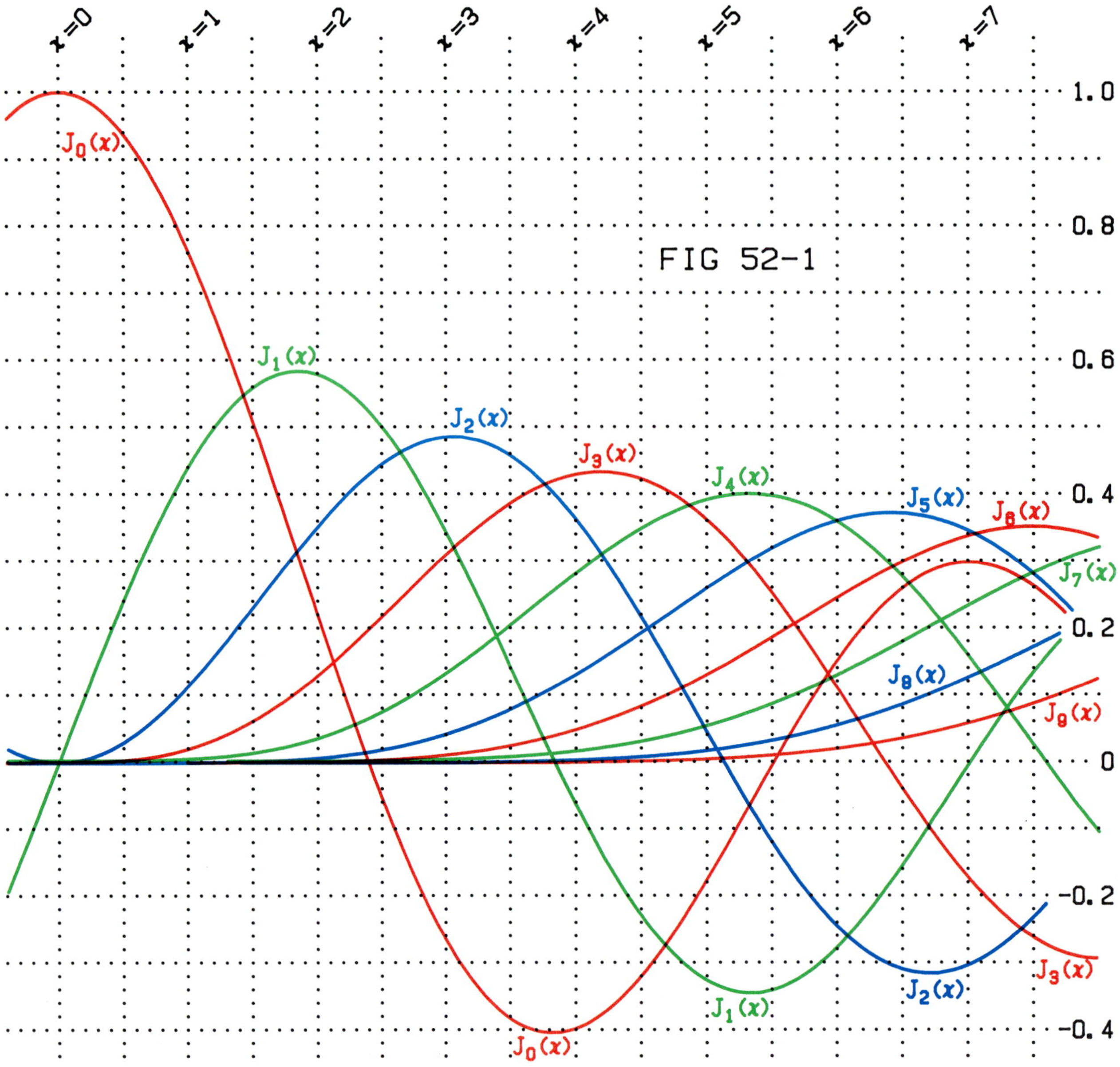

FIG 52-1

and if the signs in this series are made uniformly positive, the terms are generated by $\cos(x\sin(t))$. Bessel coefficients of odd order arise from

52:3:2 $$\sin[x\cos(t)] = 2J_1(x)\cos(t) - 2J_3(x)\cos(3t) + 2J_5(x)\cos(5t) - \cdots = 2\sum_{j=0}^{\infty}(-1)^j J_{2j+1}(x)\cos(t+2jt)$$

or from the corresponding series for $\sin[x\sin(t)]$, which differs from 52:3:2 in that sines replace cosines, and all terms are positive. The generating function

52:3:3 $$\exp\left(\frac{xt^2-x}{2t}\right) = J_0(x) + \left(t-\frac{1}{t}\right)J_1(x) + \left(t^2+\frac{1}{t^2}\right)J_2(x) + \left(t^3-\frac{1}{t^3}\right)J_3(x) + \cdots = \sum_{j=-\infty}^{\infty} t^j J_j(x)$$

gives rise to Bessel coefficients of both parities.

The Bessel coefficient J_0 may be represented as a definite integral in a number of ways, including

52:3:4 $$J_0(x) = \frac{1}{\pi}\int_0^{\pi}\cos[x\cos(t)]dt = \frac{2}{\pi}\int_0^{\infty}\sin[x\cosh(t)]dt$$

The following are among the integral definitions of the general Bessel coefficient

52:3:5 $$J_n(x) = \frac{1}{\pi}\int_0^\pi \cos[x\sin(t) - nt]dt \qquad n = 0, 1, 2, \ldots$$

52:3:6 $$J_n(x) = \frac{2}{\pi}\int_0^{\pi/2} f[x\sin(t)]\, f(nt)dt \begin{cases} n = 0, 2, 4, \ldots & f = \cos \\ n = 1, 3, 5, \ldots & f = \sin \end{cases}$$

and others may be obtained by specializing the formulas of Section 53:3.

The Bessel coefficients J_0 and J_1 arise in a large number of physical contexts because they satisfy certain very simple second-order differential equations, as follows:

52:3:7 $$x\frac{d^2f}{dx^2} + \frac{df}{dx} + xf = 0 \qquad f = c_1J_0(x) + c_2Y_0(x)$$

52:3:8 $$x\frac{d^2f}{dx^2} + \frac{df}{dx} + f = 0 \qquad f = c_1J_0(2\sqrt{x}) + c_2Y_0(2\sqrt{x})$$

52:3:9 $$x\frac{d^2f}{dx^2} - \frac{df}{dx} + xf = 0 \qquad f = c_1xJ_1(x) + c_2xY_1(x)$$

52:3:10 $$x\frac{d^2f}{dx^2} + f = 0 \qquad f = c_1\sqrt{x}\,J_0(2\sqrt{x}) + c_2\sqrt{x}\,Y_0(2\sqrt{x})$$

Here c_1 and c_2 are arbitrary constants and Y_0 and Y_1 are Neumann functions [Chapter 54] of zero and unity orders.

The Bessel coefficients J_0 and J_1 are generated by applying the operations of semiintegration or semidifferentiation, with lower limit zero, upon certain functions involving sinusoids. Examples include

52:3:11 $$J_0(x) = \frac{\sec(x)}{\sqrt{\pi}}\frac{d^{-1/2}}{dx^{-1/2}}\left(\frac{\cos(2x)}{\sqrt{x}}\right) = \frac{\csc(x)}{\sqrt{\pi}}\frac{d^{-1/2}}{dx^{-1/2}}\left(\frac{\sin(2x)}{\sqrt{x}}\right)$$

52:3:12 $$J_0(\sqrt{x}) = \frac{2}{\sqrt{\pi}}\frac{d^{1/2}}{dx^{1/2}}(\sin(\sqrt{x}))$$

52:3:13 $$J_1(\sqrt{x}) = \frac{1}{\sqrt{\pi x}}\frac{d^{-1/2}}{dx^{-1/2}}(\sin(\sqrt{x}))$$

and

52:3:14 $$J_0\left(\frac{1}{\sqrt{x}}\right) = \sqrt{\frac{x}{\pi}}\frac{d^{-1/2}}{dx^{-1/2}}\left(\frac{\sin(1/\sqrt{x})}{x}\right)$$

52:4 SPECIAL CASES

There are none.

52:5 INTRARELATIONSHIPS

Bessel coefficients are even or odd

52:5:1 $$J_n(-x) = (-1)^n J_n(x) \qquad n = 0, 1, 2, \ldots$$

according to the parity of n. A similar reflection formula

52:5:2 $$J_{-n}(x) = (-1)^n J_n(x) \qquad n = 0, 1, 2, \ldots$$

exists for the order of Bessel functions of integer order.

Many formulas exist for infinite sums of Bessel coefficients. Some of these, such as

52:5:3 $$J_1(x) - J_3(x) + J_5(x) - J_7(x) + \cdots = \frac{1}{2}\sin(x)$$

52:5:4 $$\frac{1}{2}J_0(x) - J_2(x) + J_4(x) - J_6(x) + \cdots = \frac{1}{2}\cos(x)$$

and

52:5:5 $$\frac{1}{2}J_0(x) + J_2(x) + J_4(x) + J_6(x) + \cdots = \frac{1}{2}\sum_{j=-\infty}^{\infty} J_j(x) = \frac{1}{2}$$

may be derived as special cases of formulas 52:3:1–52:3:3. Others, for example:

52:5:6 $$J_1(x) + 3J_3(x) + 5J_5(x) + \cdots = \sum_{j=0}^{\infty}(2j+1)\,J_{2j+1}(x) = \frac{x}{2}$$

52:5:7 $$2J_2(x) - 4J_4(x) + 6J_6(x) - \cdots = -2\sum_{j=1}^{\infty}(-1)^j jJ_{2j}(x) = \frac{x}{2}J_1(x)$$

52:5:8 $$4J_2(x) + 16J_4(x) + 36J_6(x) + \cdots = 4\sum_{j=1}^{\infty} j^2 J_{2j}(x) = \frac{x^2}{2}$$

and, for $n = 1, 2, 3, \ldots$:

52:5:9 $$(n-1)\,!nJ_n(x) + n!(n+2)\,J_{n+2}(x) + \frac{(n+1)!(n+4)}{2!}J_{n+4}(x) + \cdots$$
$$= \sum_{j=0}^{\infty}\frac{(n+j-1)!(n+2j)}{j!}J_{n+2j}(x) = \left(\frac{x}{2}\right)^n$$

follow from the properties of Neumann series [see Section 53:14]. Yet others, such as

52:5:10 $$\frac{1}{2}J_0^2(x) + J_1^2(x) + J_2^2(x) + \cdots = \frac{1}{2}J_0^2(x) + \sum_{j=1}^{\infty} J_j^2(x) = \frac{1}{2}$$

and

52:5:11 $$J_0(x)\,J_1(x) - J_1(x)\,J_2(x) + J_2(x)\,J_3(x) - \cdots = \sum_{j=0}^{\infty}(-1)^j\,J_j(x)\,J_{j+1}(x) = \frac{1}{2}J_1(2x)$$

may be derived from *Neumann's addition formula*

52:5:12 $$J_n(x+y) = \sum_{j=-\infty}^{\infty} J_{n-j}(x)\,J_j(y) \qquad n = 0, \pm 1, \pm 2, \ldots$$

Included among still other types of expansions are

52:5:13 $$J_1(x) + J_2(2x) + J_3(3x) + \cdots = \sum_{j=1}^{\infty} J_j(jx) = \frac{x}{2(1-x)}$$

and

52:5:14 $$J_1^2(x) + J_2^2(2x) + J_3^2(3x) + \cdots = \sum_{j=1}^{\infty} J_j^2(jx) = \frac{1}{2}\left[\frac{1}{\sqrt{1-x^2}} - 1\right]$$

that are examples of *Kapteyn series* [see Erdélyi, Magnus, Oberhettinger and Tricomi, *Higher Transcendental Functions*, Volume 2, pages 66–68].

By applying the recursion formula

52:5:15 $$J_{n+1}(x) = \frac{2n}{x} J_n(x) - J_{n-1}(x)$$

for a sufficient number of times, any Bessel coefficient may be expressed in terms of J_0 and J_1. Examples are

52:5:16 $$J_2(x) = \frac{2}{x} J_1(x) - J_0(x)$$

52:5:17 $$J_3(x) = \left(\frac{8}{x^2} - 1\right) J_1(x) - \frac{4}{x} J_0(x)$$

52:5:18 $$J_4(x) = \left(\frac{48}{x^3} - \frac{8}{x}\right) J_1(x) - \left(\frac{24}{x^2} - 1\right) J_0(x)$$

52:5:19 $$J_5(x) = \left(\frac{384}{x^4} - \frac{72}{x^2} + 1\right) J_1(x) - \left(\frac{192}{x^3} - \frac{12}{x}\right) J_0(x)$$

52:5:20 $$J_6(x) = \left(\frac{3840}{x^5} - \frac{768}{x^3} + \frac{18}{x}\right) J_1(x) - \left(\frac{1920}{x^4} - \frac{144}{x^2} + 1\right) J_0(x)$$

The ratio R_n of two consecutive Bessel coefficients of common argument obeys the simple recursion formula

52:5:21 $$\frac{J_{n+1}(x)}{J_n(x)} = R_n = \frac{2n}{x} - \frac{1}{R_{n-1}} \qquad n = 1, 2, 3, \ldots$$

which makes these ratios computationally useful [see Section 52:8].

Of course, all formulas in Section 53:5 can be applied to Bessel coefficients by setting $\nu = n$, a nonnegative integer. For example, the argument-multiplication formula yields

52:5:22 $$J_n(bx) = b^n \sum_{j=0}^{\infty} \frac{1}{j!} \left(\frac{x - b^2 x}{2}\right)^j J_{n+j}(x)$$

52:6 EXPANSIONS

Any Bessel coefficient may be expanded as the power series

52:6:1 $$J_n(x) = \left(\frac{x}{2}\right)^n \left[\frac{1}{n!} - \frac{x^2/4}{(n+1)!} + \frac{(x^2/4)^2}{2!(n+2)!} - \frac{(x^2/4)^3}{3!(n+3)!} + \cdots\right] = \left(\frac{x}{2}\right)^n \sum_{j=0}^{\infty} \frac{(-x^2/4)^j}{j!(n+j)!}$$

of which the first two instances are

52:6:2 $$J_0(x) = 1 - \frac{x^2}{4} + \frac{x^4}{64} - \frac{x^6}{2304} + \frac{x^8}{147456} - \cdots = \sum_{j=0}^{\infty} \frac{(-x^2/4)^j}{(j!)^2}$$

and

52:6:3 $$J_1(x) = \frac{x}{2} - \frac{x^3}{16} + \frac{x^5}{384} - \frac{x^7}{18432} + \cdots = \frac{x}{2} \sum_{j=0}^{\infty} \frac{(-x^2/4)^j}{j!(j+1)!}$$

Expansions as infinite products take the form

52:6:4 $$J_0(x) = \left(1 - \frac{x^2}{j_{0;1}^2}\right)\left(1 - \frac{x^2}{j_{0;2}^2}\right)\left(\frac{1 - x^2}{j_{0;3}^2}\right) \cdots = \prod_{k=1}^{\infty} 1 - \frac{x^2}{j_{0;k}^2}$$

52:6:5 $$J_1(x) = x\left(1 - \frac{x^2}{j_{1;1}^2}\right)\left(1 - \frac{x^2}{j_{1;2}^2}\right)\left(1 - \frac{x^2}{j_{1;3}^2}\right) \cdots = x \prod_{k=1}^{\infty} 1 - \frac{x^2}{j_{1;k}^2}$$

where $j_{n;k}$ denotes the kth zero of the Bessel coefficient $J_n(x)$ [see Section 52:7].

By making use of formula 52:5:22, it is possible to expand a Bessel coefficient $J_n(x)$ in terms of the Bessel coefficients J_n, J_{n+1}, J_{n+2}, ... at any fixed nonzero argument. For example, choosing 2 for this fixed argument gives

52:6:6 $$J_n(x) = \left(\frac{x}{2}\right)^n \sum_{j=0}^{\infty} \frac{1}{j!}\left(1 - \frac{x^2}{4}\right)^j J_{n+j}(2) \qquad n = 0, 1, 2, \ldots$$

Such an expansion converges extremely rapidly.

The Bessel coefficient ratio defined in equation 52:5:21 may be expanded in partial fractions

52:6:7 $$R_n = \frac{2x}{j_{n;1}^2 - x^2} + \frac{2x}{j_{n;2}^2 - x^2} + \frac{2x}{j_{n;3}^2 - x^2} + \cdots = 2x \sum_{k=1}^{\infty} \frac{1}{j_{n;k}^2 - x^2}$$

or as the infinite continued fraction

52:6:8 $$R_n = \frac{J_{n+1}(x)}{J_n(x)} = \frac{x/2}{1+n-}\,\frac{x^2/4}{2+n-}\,\frac{x^2/4}{3+n-}\,\frac{x^2/4}{4+n-}\cdots$$

The effect of curtailing this continued fraction at any point may be represented as another Bessel coefficient ratio. For example:

52:6:9 $$R_n = \frac{x/2}{1+n-}\,\frac{x^2/4}{2+n-}\,\frac{x^2/4}{3+n-}\,\frac{xR_{n+3}}{2}$$

For large arguments, the asymptotic expansions

52:6:10 $$J_0(x) \sim \frac{\cos(x)}{\sqrt{\pi x}}\left[1 - \frac{1}{8x} - \frac{9}{128x^2} + \frac{225}{3072x^3} + \cdots\right] + \frac{\sin(x)}{\sqrt{\pi x}}\left[1 + \frac{1}{8x} - \frac{9}{128x^2} - \frac{225}{3072x^3} + \cdots\right] \qquad x \to \infty$$

52:6:11 $$J_1(x) \sim \frac{\sin(x)}{\sqrt{\pi x}}\left[1 + \frac{3}{8x} + \frac{15}{128x^2} - \frac{315}{3072x^3} - \cdots\right] - \frac{\cos(x)}{\sqrt{\pi x}}\left[1 - \frac{3}{8x} + \frac{15}{128x^2} + \frac{315}{3072x^3} - \cdots\right] \qquad x \to \infty$$

hold. See equations 53:6:6–53:6:8 for the general formulation.

52:7 PARTICULAR VALUES

	$x = -\infty$	$x = 0$	$x = \infty$
$J_0(x)$	0	1	0
$J_1(x)$, $J_2(x)$, $J_3(x)$, ...	0	0	0

Those positive argument values that cause $J_n(x)$ to acquire the value zero are denoted $j_{n;1}$, $j_{n;2}$, $j_{n;3}$, ..., and $j_{n;k}$ is called the kth *zero of the* n^{th} *Bessel coefficient*. The value of the derivative $dJ_n(x)/dx$ at the zero is a quantity whose value is needed in certain practical problems [see Section 52:14]. Such a derivative is known as the *associated value of the zero*, and the usual notation is $J_n'(j_{n;k})$, although two alternative notations follow from the identity

52:7:1 $$J_n'(j_{n;k}) = J_{n-1}(j_{n;k}) = -J_{n+1}(j_{n;k})$$

Table 52.7.1

$j_{0;k}$	$J'_0(j_{0;k})$	$j_{1;k}$	$J'_1(j_{1;k})$		$j'_{0;k}$	$J_0(j'_{0;k})$	$j'_{1;k}$	$J_1(j'_{1;k})$
2.4048	−0.5191	3.8317	−0.4028	$k = 1$	0.0000	+1.0000	1.8412	+0.5819
5.5201	+0.3403	7.0156	+0.3001	$k = 2$	3.8317	−0.4028	5.3314	−0.3461
8.6537	−0.2715	10.1735	−0.2497	$k = 3$	7.0156	+0.3001	8.5363	+0.2733
11.7915	+0.2325	13.3237	+0.2184	$k = 4$	10.1735	−0.2497	11.7060	−0.2333
14.9309	−0.2065	16.4706	−0.1965	$k = 5$	13.3237	+0.2184	14.8636	+0.2070
18.0711	+0.1877	19.6159	+0.1801	$k = 6$	16.4706	−0.1965	18.0155	−0.1880
21.2116	−0.1733	22.7601	−0.1672	$k = 7$	19.6159	+0.1801	21.1644	+0.1735

which is a consequence of 52:10:1. Approximate values of the first seven zeros, and their associated values, for $n = 0$ and 1, constitute the left-hand side of Table 52.7.1. Other approximate values of $j_{n;k}$ may be read from the map in Section 53:7 (Figure 53-2). More precise values of $j_{n;k}$ and $J'_n(j_{n;k})$ are available from the algorithm that follows.

The algorithm operates in two modes according to the relative values of n and k. If $k \geq n - 3$, the algorithm adopts $x_1 = \pi(2k + n - 1)/2$ as an approximate value of $j_{n;k}$ and then uses Newton's method [Section 17:7] to generate the improved approximation

52:7:2
$$x_2 = x_1 - \frac{J_n(x_1)}{J'_n(x_1)} = x_1 - \frac{x_1 J_n(x_1)}{nJ_n(x_1) - x_1 J_{n+1}(x_1)} = x_1 - \frac{x_1}{n - x_1 R_n}$$

and this procedure is repeated until two successive approximants differ by less than 10^{-7}. The method used to compute R_n, the Bessel coefficient ratio, is described in Section 52:8. When $k < n - 3$, the algorithm first calculates $n + 2n^{1/3}$ [see 53:7] as a crude approximation to $j_{n;1}$ and refines this approximation by a single application of 52:7:2. A crude approximation to $j_{n;2}$ is then calculated by adding 4 to the $j_{n;1}$ approximant and this, in turn, is refined by a single application of Newton's method. The procedure of adding 4 and refining is repeated until a crude value of $j_{n;k}$ is reached. This is then improved by repeated applications of formula 52:7:2. The second half of the algorithm, which calculates $-J_{n+1}(j_{n;k})$, is essentially the same as the algorithm in Section 52:8.

Input n >>>>>>>>>
Set $J = 0$
Input k >>>>>>>>>
Set $x = \pi\left(k + \frac{n-1}{2}\right)$ $\left[\text{or} = \pi\left(k - 1 + \frac{n}{2}\right)\right]$
If $x = 0$ go to (8)
If $k > n - 4$ go to (4)
Set $x = n - 4 + 2n^{1/3}$ [or $= n - 4 + 0.8n^{1/3}$]
(1) Replace x by $x + 4$
Set $l = \text{Int}\sqrt{n^2 + 2(x + 6)^2}$
Replace J by $J + 1$
Go to (6)
(2) If $J \neq 1$ go to (3)
Replace x by $x + (n/14)$
(3) If $J < k$ go to (1)
Set $J = 0$
(4) Set $L = 2 + 2\,\text{Int}\sqrt{[n^2/4] + [(x + 7)^2/2]}$
(5) Set $l = L$
(6) Set $R = 0$
(7) If $\frac{2l}{x} = R$ go to (8)

Storage needed: n, k, x, J, l, R, L and D

Input restrictions:
$n = 0, 1, 2, \ldots$
$k = 1, 2, 3, \ldots$

Replace R by $1\Big/\left(\frac{2l}{x} - R\right)$
Replace l by $l - 1$
If $l > n$ go to (7)
If $\frac{n}{x} = R$ go to (8)
Set $D = 1\Big/\left(\frac{n}{x} - R\right)$
Replace D by $x^2/[D(n^2 - n - x^2 + xR)]$
Replace x by $x - D$
If $J \neq 0$ go to (3) [or to (2)]
If $|D| > 10^{-7}$ go to (5)
(8) Output x

$x \simeq \left\{\begin{matrix} j_{n;k} \\ \text{or} \\ j'_{n;k} \end{matrix}\right\}$ <<<<<<<

Set $D = J = 1$
If $x = 0$ go to (11)
(9) Replace R by $1\Big/\left(\frac{2L}{x} - R + 10^{-15}\right)$
If $L - n > 1$ [or > 0] go to (10)
Replace J by JR
(10) Replace D by $DR + 2\,\text{frac}\left(\frac{L}{2}\right)$
Replace L by $L - 1$
If $L \neq 0$ go to (9)
Replace J by $J/(1 - 2D)$ [or by $J/(2D - 1)$]
(11) Output J

$J \simeq \left\{\begin{matrix} J'_n(j_{n;k}) \\ \text{or} \\ J_n(j'_{n;k}) \end{matrix}\right\}$ <<<<<

Test values:
$j_{2;2} = 8.41724414$
$J'_2(j_{2;2}) = 0.271382590$
$j_{10;5} = 28.8873751$
$J'_{10}(j_{10;5}) = 0.143812670$
$j'_{0;1} = 0.000000000$
$J_0(j'_{0;1}) = 1.000000000$
$j'_{2;2} = 6.70613319$
$J_2(j'_{2;2}) = -0.313530445$
$j'_{10;5} = 27.1820215$
$J_{10}(j'_{10;5}) = 0.158631021$

Also important in a number of physical applications are those nonnegative values of the argument that cause $J_n(x)$ to attain an extremum (a local maximum or minimum): we call these values *extrema of the* n^{th} *Bessel coefficient* and denote the kth such value by $j'_{n;k}$. Some approximate values of the extrema of J_0 and J_1 occupy the right-hand side of the table above, together with the corresponding *associated values of the extrema of the* n^{th} *Bessel coefficient* $J_n(j'_{n;k})$. These associated values obey the relationships

52:7:3 $$J_n(x) = \frac{x}{n} J_{n+1}(x) = \frac{x}{n} J_{n-1}(x) \qquad x = j'_{n;k}$$

Note also that $j'_{0;k+1} = j_{1;k}$ and that $J_0(j'_{0;k+1}) = J'_1(j_{1;k})$. Figure 53-2 is useful for finding approximate values of the extrema, and accurate values of both $j'_{n;k}$ and $J_n(j'_{n;k})$ are given by the algorithm above if the additions and modifications shown in green are instituted. The procedures adopted by the modified algorithm generally resemble its unmodified counterpart, but the Newtonian improvement formula is

52:7:4 $$x_2 = x_1 - \frac{J'_n(x_1)}{J''_n(x_1)} = x_1 - \frac{nx_1J_n(x_1) - x_1^2J_{n+1}(x_1)}{(n^2 - n - x^2)J_n(x_1) + x_1J_{n+1}(x_1)} = x_1 - \frac{x_1(n - x_1R_n)}{n^2 - n - x_1^2 + x_1R_n}$$

in this case.

Whether used for finding zeros or extrema, the algorithm of this section is slow, particularly for large n and/or k.

52:8 NUMERICAL VALUES

Recursion 52:5:19 may be rewritten as

52:8:1
$$R_{l-1} = \frac{1}{\dfrac{2l}{x} - R_l} \qquad l = 1, 2, 3, \ldots$$

and this formula has the useful property that an error in R_l generally induces a smaller error in R_{l-1}. This means that the concatenation

52:8:2
$$1 = J_0(x)(1 + 2R_0R_1(1 + R_2R_3(1 + R_4R_5(1 + R_6R_7(1 + \cdots)))))$$

that follows from 52:5:5, may be curtailed by setting $R_{L-1} = 1$ and $R_L = 0$ with little error, provided that L is even and large enough. Because $J_n(x) = R_{n-1}\,R_{n-2}\,R_{n-3} \cdots R_1R_0J_0(x)$, the formula

52:8:3
$$J_n(x) \simeq \frac{R_{n-1}R_{n-2}R_{n-3} \cdots R_1R_0}{((\cdots((R_{L-1}R_{L-2} + 1)R_{L-3}R_{R-4} + 1)R_{L-5}R_{L-6} + \cdots + 1)R_3R_2 + 1)2R_1R_0 + 1}$$

can therefore be used to calculate numerical values of Bessel coefficients.

The algorithm that follows utilizes expressions 52:8:1 and 52:8:3. An empirical formula selects a value of L, depending on n and x, that is large enough to ensure that the error in $J_n(x)$ does not exceed 6×10^{-8}.

n >>>>>>>>>>>
    Set $J = D = R = 1$
x >>>>>>>>>>>
    Set $l = 2 + 2\,\mathrm{Int}\sqrt{[(|x| + 7)^2/2] + [n^2/4]}$
(1) Replace R by $x/(2l - Rx)$
    If $l > n$ go to (1)
    Replace J by JR
(2) Replace D by $DR + 2\,\mathrm{frac}(l/2)$
    Replace l by $l - 1$
    If $l \neq 0$ go to (2)
    Replace J by $J/(2D - 1)$
    Output J
$J \simeq J_n(x)$ <<<<<<<

Storage needed: n, J, D, R, x and l

Test values:
$J_0(8) = 0.171650807$
$J_{100}(-10) = 6.59731606 \times 10^{-89}$

The algorithm of Section 53:8 may also be used to calculate $J_n(x)$ for $x > 0$ and the universal hypergeometric algorithm [Section 18:14] can perform this calculation as well.

52:9 APPROXIMATIONS

For small arguments, the first two Bessel coefficients are well approximated by

52:9:1
$$J_0(x) \simeq \left(1 - \frac{x^2}{8}\right)^2 \qquad \text{8-bit precision} \qquad -1.3 \le x \le 1.3$$

52:9:2
$$J_1(x) \simeq \frac{x}{2}\left(1 - \frac{x^2}{24}\right)^3 \qquad \text{8-bit precision} \qquad -2 \le x \le 2$$

while the corresponding approximations for large x are

52:9:3
$$J_0(x) \simeq \frac{\sin(x + {}^{\pi}/_{4})}{\sqrt{\pi x/2}} - \frac{\cos(x + {}^{\pi}/_{4})}{4\sqrt{2\pi x^3}}$$

52:9:4
$$J_1(x) \simeq \frac{\sin(x - {}^{\pi}/_{4})}{\sqrt{\pi x/2}} + \frac{3\cos(x - {}^{\pi}/_{4})}{4\sqrt{2\pi x^3}}$$

For large k, the zeros and extrema of the zeroth and first Bessel coefficients may be conveniently computed from the asymptotic formulas

52:9:5
$$j_{0;k} \sim K + \frac{1}{8K} - \frac{31}{384K^3} + \frac{3779}{15360K^5} - \cdots \qquad K = \left(k - \frac{1}{4}\right)\pi$$

52:9:6 and 52:9:7
$$\left.\begin{matrix} j_{1;k} \\ \\ j'_{0;k} \end{matrix}\right\} \sim K - \frac{3}{8K} + \frac{3}{128K^3} - \frac{1179}{5120K^5} + \cdots \qquad \left\{\begin{matrix} K = \left(k + \frac{1}{4}\right)\pi \\ \\ K = \left(k - \frac{3}{4}\right)\pi \end{matrix}\right.$$

52:9:8
$$j'_{1;k} \sim K - \frac{7}{8K} - \frac{431}{384K^3} - \frac{29893}{15360K^5} - \cdots \qquad K = \left(k - \frac{1}{4}\right)\pi$$

52:10 OPERATIONS OF THE CALCULUS

The derivative of a Bessel coefficient may be expressed as three equivalent formulations

52:10:1
$$\frac{d}{dx} J_n(x) = \frac{J_{n-1}(x) - J_{n+1}(x)}{2} = J_{n-1}(x) - \frac{n}{x} J_n(x) = \frac{n}{x} J_n(x) - J_{n+1}(x)$$

of which $d[J_0(x)]dx = -J_1(x)$ and $d[J_1(x)]/dx = J_0(x) - [J_1(x)/x]$ are special cases. The coefficients occurring in the formulas

52:10:2
$$\frac{d^2}{dx^2} J_n(x) = \frac{1}{2^2} [J_{n-2}(x) - 2J_n(x) + J_{n+2}(x)]$$

52:10:3
$$\frac{d^3}{dx^3} J_n(x) = \frac{1}{2^3} [J_{n-3}(x) - 3J_{n-1}(x) + 3J_{n+1}(x) - J_{n+3}(x)]$$

for multiple differentiation will be recognized as binomial coefficients [Chapter 6] of alternating sign.

Indefinite integrals of Bessel coefficients include

52:10:4
$$\int_0^x J_0(t)dt = xJ_0(x) + \frac{\pi x}{2} [J_1(x)\, h_0(x) - J_0(x)\, h_1(x)] \qquad x > 0$$

52:10:5
$$\int_0^x J_1(t)dt = 1 - J_0(x)$$

and generally

52:10:6
$$\int_0^x J_n(t)dt = -2\sum_k J_k(x) + \begin{cases} \int_0^x J_0(t)dt & n = 0, 2, 4, \ldots \quad k = 1, 3, 5, \ldots, n-1 \\ 1 - J_0(x) & n = 1, 3, 5, \ldots \quad k = 2, 4, 6, \ldots, n-1 \end{cases}$$

and some others may be obtained by specialization of formulas in Section 53:10. The h_0 and h_1 functions in formula 52:10:4 are discussed in Chapter 57.

Among important definite integrals are

52:10:7 $$\int_0^\infty \frac{J_n(bt)}{t}\,dt = \frac{1}{n} \qquad n = 1, 2, 3, \ldots$$

52:10:8 $$\int_0^\infty \cos(Bt)\, J_0(bt)dt = \begin{cases} 0 & B < b \\ \dfrac{1}{\sqrt{B^2 - b^2}} & B > b \end{cases}$$

52:10:9 $$\int_0^\infty \exp(-Bt)\, J_0(b\sqrt{t})dt = \frac{1}{B}\exp\left(\frac{-b^2}{4B}\right)$$

along with many others that follow from formulas in Section 53:10.

With $n = 0$ and 1, Erdélyi, Magnus, Oberhettinger and Tricomi [*Tables of Integral Transforms*, Volume 2, pages 9–21] list many definite integrals of the form

52:10:10 $$\int_0^\infty \sqrt{st}\, J_n(st)\, f(t)dt$$

these being examples of Hankel transforms [see Section 53:10].

The operations of semidifferentiation and semiintegration frequently convert Bessel coefficients into functions involving sinusoids [Chapter 32]. Examples are

52:10:11 $$\frac{d^{1/2}}{dx^{1/2}} J_0(\sqrt{x}) = \frac{\cos(\sqrt{x})}{\sqrt{\pi x}}$$

and

52:10:12 $$\frac{d^{-1/2}}{dx^{-1/2}} \frac{J_0(\sqrt{x})}{2} = \frac{d^{1/2}}{dx^{1/2}} \sqrt{x}\, J_1(\sqrt{x}) = \frac{\sin(\sqrt{x})}{\sqrt{\pi}}$$

52:11 COMPLEX ARGUMENT

We discuss specific instances of $J_n(x + iy)$, but not the general case.

When the argument of the J_n Bessel coefficient is purely imaginary, one has a function that is real or imaginary according to the parity of n:

52:11:1 $$J_n(iy) = \begin{cases} I_n(y) & n = 0, 4, 8, \ldots \\ iI_n(y) & n = 1, 5, 9, \ldots \\ -I_n(y) & n = 2, 6, 10, \ldots \\ -iI_n(y) & n = 3, 7, 11, \ldots \end{cases}$$

Here I_n is the hyperbolic Bessel function of Chapter 49.

When the real and imaginary components of its argument are equal in magnitude, the real and imaginary components of a Bessel coefficient correspond to Kelvin functions [Chapter 55]

52:11:2 $$J_n(x \pm ix) = (-1)^n[\mathrm{ber}_n(x\sqrt{2}) \mp i\,\mathrm{bei}_n(x\sqrt{2})]$$

52:12 GENERALIZATIONS

The generalization of Bessel coefficients to arbitrary order is the subject of the next chapter.

52:13 COGNATE FUNCTIONS

Similarities exist among all the functions discussed in Chapters 49–57.

52:14 RELATED TOPICS

The definite integral

52:14:1
$$\int_0^1 tJ_n(\alpha t)\,J_n(\beta t)dt = \begin{cases} \dfrac{\alpha J_n(\beta)\,J_n'(\alpha) - \beta J_n(\alpha)\,J_n'(\beta)}{\beta^2 - \alpha^2} & \beta \neq \alpha \\ \dfrac{[\alpha J_n'(\alpha)]^2 + (\alpha^2 - n^2)[J_n(\alpha)]^2}{2\alpha^2} & \beta = \alpha \end{cases}$$

shows that the functions $J_n(j_{n;1}x)$, $J_n(j_{n;2}x)$, $J_n(j_{n;3}x)$, ... are orthogonal [Section 21:14] on the interval $0 \le x \le 1$ with weight function x. For, if we abbreviate $J_n(j_{n;k}x) = \Psi_k(x)$, then

52:14:2
$$\int_0^1 t\Psi_k(t)\,\Psi_l(t)dt = \begin{cases} 0 & l \neq k \\ \dfrac{[J_n'(j_{n;k})]^2}{2} & l = k \end{cases}$$

Hence, any function $f(x)$ which exists in the $0 \le x \le 1$ range can be expanded in the orthogonal series

52:14:3
$$f(x) = c_1\Psi_1(x) + c_2\Psi_2(x) + c_3\Psi_3(x) + \cdots = \sum_{k=1}^{\infty} c_k J_n(j_{n;k}x)$$

where

52:14:4
$$c_k = \frac{2}{[J_n'(j_{n;k})]^2}\int_0^1 tf(t)\,J_n(j_{n;k}t)dt$$

Note that the orthogonal series may be based on *any* Bessel coefficient; in practice, J_0 is often the most suitable.

It is equally evident from 52:14:1 that

52:14:5
$$\int_0^1 t\,\Psi_k'(t)\,\Psi_l'(t)dt = \begin{cases} 0 & l \neq k \\ \left[1 - \left(\dfrac{n}{j_{n;k}'}\right)^2\right]\dfrac{[J_n(j_{n;k}')]^2}{2} & l = k \end{cases}$$

where $\Psi_k'(x)$ abbreviates $J_n(j_{n;k}'x)$. Hence, the functions $\Psi_1'(x)$, $\Psi_2'(x)$, $\Psi_3'(x)$, ... are also orthogonal. In fact, one may generalize the orthogonality of Bessel coefficients more widely, as discussed by Spiegel [page 144].

CHAPTER 53

THE BESSEL FUNCTION $J_\nu(x)$

Some authors classify transcendental functions (those that cannot be expressed by a finite number of algebraic operations) as either "elementary" or "higher." The Bessel function is generally regarded as the most important member of the latter category, the "higher transcendental functions." This function was invented in the context of planetary astronomy but finds application in a wide variety of practical problems.

53:1 NOTATION

The symbolism $J_\nu(x)$ is almost universal, ν being termed the order and x the argument of the Bessel function. The name *cylinder function* is sometimes applied to $J_\nu(x)$, but this name is also used to describe collectively the broad class of functions that this *Atlas* treats in Chapters 49–57.

A change of argument to $2\sqrt{x}$ often simplifies formulas, and *Clifford's notation*

53:1:1 $$C_\nu(x) = x^{-\nu/2}J_\nu(2\sqrt{x})$$

is sometimes useful in this context.

53:2 BEHAVIOR

Except when ν is an integer, the Bessel function $J_\nu(x)$ is complex for negative argument. Therefore, in this chapter we deal exclusively with $x \geq 0$, and this is the only range depicted in the contour map, Figure 53-1.

The map suggests that an infinite number of bands of alternating positive and negative values cross the x, ν plane, and this is, in fact, the case. This means that the Bessel function, viewed either as a function of x or of ν, displays an infinite number of zeros, separated by local maxima and minima.

For noninteger ν less than zero, $J_\nu(x)$ approaches $\pm\infty$ as $x \to 0$, while for positive ν, $J_\nu(x)$ remains bounded. For $\nu \geq 0$, $J_\nu(x)$ never exceeds unity.

At constant small ν, $J_\nu(x)$ is oscillatory throughout the entire $x > 0$ domain, whereas if the order takes large values of either sign there is a range of the argument, approximately $0 < x < \nu$, during which $J_\nu(x)$ increases (if $\nu > 0$) or decreases (if $\nu < 0$) in magnitude prior to breaking into oscillations. Once they are established, the oscillations steadily attenuate, as x increases, remaining centered about zero and eventually conforming to the asymptotic expression

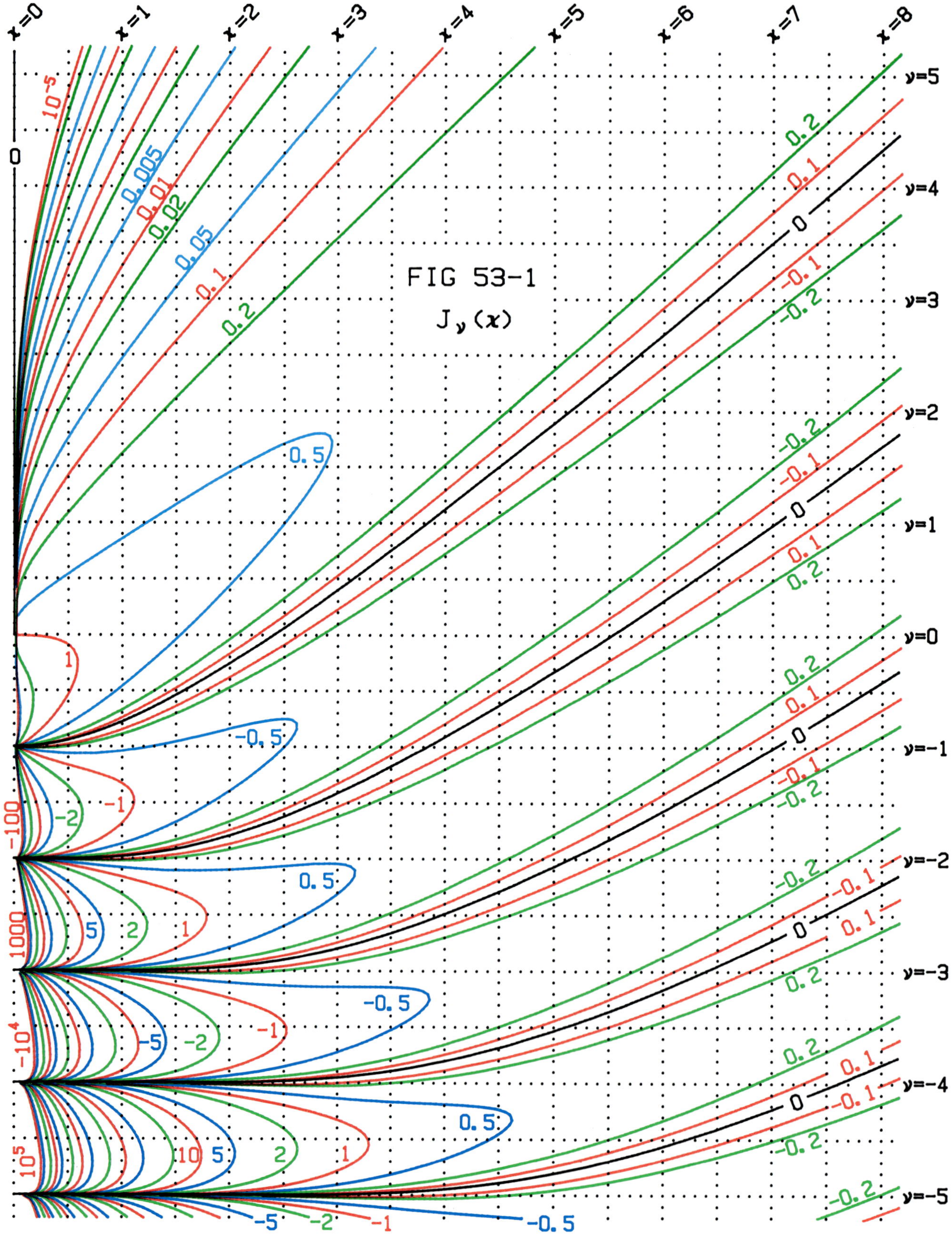
FIG 53-1
Jν(x)
x=0
x=1
x=2
x=3
x=4
x=5
x=6
x=7
x=8
ν=5
ν=4
ν=3
ν=2
ν=1
ν=0
ν=−1
ν=−2
ν=−3
ν=−4
ν=−5
10⁻⁵
0.005
0.01
0.02
0.05
0.1
0.2
0.5
−0.5
1
−1
−2
−100
1000
5
2
−10⁴
10⁵
10
0
−0.1
−0.2

53:2:1 $$J_\nu(x) \sim \sqrt{\frac{2}{\pi x}}\cos\left(x - \frac{\nu\pi}{2} - \frac{\pi}{4}\right) \qquad x \to \infty$$

If the argument x takes some constant value greater than zero, Figure 53-1 also illustrates how $J_\nu(x)$ varies with ν. As ν takes increasingly positive values, the oscillations of $J_\nu(x)$ sooner or later cease and the Bessel function becomes a monotonically decreasing positive function that eventually obeys the asymptotic expression

53:2:2 $$J_\nu(x) \sim \frac{1}{\sqrt{2\pi\nu}}\left(\frac{ex}{2\nu}\right)^\nu \qquad \nu \to \infty \qquad x = \text{constant} > 0$$

On the other hand, as ν acquires ever-more-negative values, $J_\nu(x)$ remains oscillatory, and the oscillations become increasingly amplified and eventually satisfy

53:2:3 $$J_\nu(x) \sim -\sqrt{\frac{-2}{\pi\nu}}\left(\frac{-ex}{2\nu}\right)^\nu \sin(\nu\pi) \qquad \nu \to -\infty \qquad x = \text{constant} > 0$$

An advantage of Clifford's notation [equation 53:1:1] is that whereas $J_\nu(x)$ is generally defined as a real function only for $x \geq 0$, $C_\nu(x)$ is real for all values of its argument. In fact, for negative x, $C_\nu(x)$ becomes related to the hyperbolic Bessel function of Chapter 50:

53:2:4 $$C_\nu(x) = (-x)^{-\nu/2}\, I_\nu(2\sqrt{-x}) \qquad x \leq 0$$

53:3 DEFINITIONS

The function $(x/2)^\nu$ is a generating function for the Bessel functions $J_\nu(x)$, $J_{\nu+1}(x)$, $J_{\nu+2}(x)$, ... Thus:

53:3:1 $$\frac{(x/2)^\nu}{\Gamma(1+\nu)} = \sum_{k=0}^{\infty} \frac{(x/2)^k}{k!} J_{\nu+k}(x)$$

Bessel functions are defined by a number of definite integrals, including

53:3:2 $$J_\nu(x) = \frac{(x/2)^\nu}{\sqrt{\pi}\,\Gamma(\nu+\frac{1}{2})}\int_0^\pi \cos(x\cos(t))\sin^{2\nu}(t)dt$$

53:3:3 $$J_\nu(x) = \frac{2(x/2)^\nu}{\sqrt{\pi}\,\Gamma(\nu+\frac{1}{2})}\int_0^1 (1-t^2)^{\nu-1/2}\cos(xt)dt \qquad \nu > -\frac{1}{2}$$

and

53:3:4 $$J_\nu(x) = \frac{2(2/x)^\nu}{\sqrt{\pi}\,\Gamma(\frac{1}{2}-\nu)}\int_1^\infty \frac{\sin(xt)}{(t^2-1)^{\nu+1/2}}dt \qquad -\frac{1}{2} < \nu < \frac{1}{2}$$

The general solution to the differential equation

53:3:5 $$x\frac{d}{dx}\left(x\frac{df}{dx}\right) = x^2\frac{d^2f}{dx^2} + x\frac{df}{dx} = (\nu^2 - x^2)f$$

is $f = c_1J_\nu(x) + c_2J_{-\nu}(x)$ if ν is not an integer, c_1 and c_2 being arbitrary constants. Equation 53:3:5 is known as *Bessel's equation* but Bessel functions also satisfy the simpler *Bessel-Clifford equation*

53:3:6 $$x\frac{d^2f}{dx^2} + (1+\nu)\frac{df}{dx} + f = 0 \qquad f = x^{-\nu/2}[c_1J_\nu(2\sqrt{x}) + c_2J_{-\nu}(2\sqrt{x})]$$

as well as many other differential equations [see Murphy]. See Section 52:3 for the corresponding solutions when ν is an integer.

The differintegration formula

53:3:7 $$J_\nu(\sqrt{x}) = \frac{2}{\sqrt{\pi}(2\sqrt{x})^\nu}\frac{d^{1/2-\nu}}{dx^{1/2-\nu}}\sin(\sqrt{x})$$

is valid for all ν and illustrates the close relationship of the Bessel function to the sine. Differintegration can also be employed [see 53:10:10] to convert a Bessel function of any order to a Bessel function of any other order. Hence, a Bessel function of arbitrary order may be regarded as derived from the J_0 Bessel coefficient [Chapter 52] by the operation

53:3:8
$$J_\nu(2\sqrt{x}) = x^{-\nu/2}\frac{d^{-\nu}}{dx^{-\nu}}J_0(2\sqrt{x})$$

or, in Clifford's notation [see 53:1:1]:

53:3:9
$$C_\nu(x) = x^{-\nu}\frac{d^{-\nu}}{dx^{-\nu}}C_0(x)$$

53:4 SPECIAL CASES

A Bessel function of nonnegative integer order is known as a Bessel coefficient [Chapter 52]. When the order is a negative integer, we have

53:4:1
$$J_{-n}(x) = (-1)^n J_n(x) \qquad n = 1, 2, 3, \ldots$$

Those Bessel functions of order equal to one-half of an odd integer are termed *spherical Bessel functions of the first kind*, and the notation

53:4:2
$$j_n(x) = \sqrt{\frac{\pi}{2x}}\,J_{n+1/2}(x) \qquad n = 0, \pm1, \pm2, \ldots$$

is sometimes adopted. These functions may be expressed either as an n-fold derivative:

53:4:3
$$j_n(\sqrt{x}) = (-2\sqrt{x})^n\frac{d^n}{dx^n}\frac{\sin(\sqrt{x})}{\sqrt{x}} \qquad n = 0, \pm1, \pm2, \ldots$$

or as an n-fold integral:

53:4:4
$$j_n(\sqrt{x}) = \frac{2}{(2\sqrt{x})^{n+1}}\frac{d^{-n}}{dx^{-n}}\sin(\sqrt{x}) \qquad n = 0, \pm1, \pm2, \ldots$$

involving $\sin(\sqrt{x})$. Because of this intimate relation to the sine, spherical Bessel functions are addressed in Section 32:13.

Bessel functions of orders $\pm\frac{1}{3}$ are related to Airy functions [Chapter 56]

53:4:5
$$\sqrt{2x}\,J_{\pm1/3}(x) = X^{1/4}[\sqrt{3}\,\mathrm{Ai}(-X) \mp \mathrm{Bi}(-X)] \qquad X = \left(\frac{3x}{2}\right)^{2/3}$$

53:5 INTRARELATIONSHIPS

Bessel functions satisfy the recursion relationship

53:5:1
$$J_{\nu+1}(x) = \frac{2\nu}{x}J_\nu(x) - J_{\nu-1}(x)$$

and the argument-multiplication formula

53:5:2
$$J_\nu(bx) = b^\nu\sum_{j=0}^{\infty}\left(\frac{x-b^2x}{2}\right)^j\frac{1}{j!}J_{\nu+j}(x)$$

Setting $b = 1 + (y/x)$ converts 53:5:2 into an argument-addition formula; another is *Neumann's addition theorem*:

53:5:3 $$J_\nu(x \pm y) = \sum_{j=-\infty}^{\infty} J_{\nu\mp j}(x)\,J_j(y) \qquad |y| < |x|$$

Setting $b = i = \sqrt{-1}$ in relationship 53:5:2 generates an infinite series

53:5:4 $$J_\nu(x) + x\,J_{\nu+1}(x) + \frac{x^2}{2!}J_{\nu+2}(x) + \cdots = \sum_{j=0}^{\infty}\frac{x^j}{j!}J_{\nu+j}(x) = i^{-\nu}J_\nu(ix) = I_\nu(x)$$

that sums to give a hyperbolic Bessel function [Chapter 50]. Expression 53:5:4 is a special case of a Neumann series [see Section 53:14]; another similar example is

53:5:5 $$J_\nu(2x) + xJ_{\nu+1}(2x) + \frac{x^2}{2!}J_{\nu+2}(2x) + \cdots = \sum_{j=0}^{\infty}\frac{x^j}{j!}J_{\nu+j}(2x) = \frac{x^\nu}{\Gamma(1+\nu)}$$

Unless ν is an integer (in which case 53:4:1 applies), there is no straightforward order-reflection formula, although $J_{-\nu}(x)$ and $J_\nu(x)$ are related in terms of the Neumann function [Chapter 54]

53:5:6 $$J_{-\nu}(x) = \cos(\nu\pi)\,J_\nu(x) - \sin(\nu\pi)\,Y_\nu(x)$$

and via the relationship

53:5:7 $$J_{-\nu}(x)\,J_{\nu+1}(x) + J_{-1-\nu}(x)\,J_\nu(x) = \frac{-2\sin(\nu\pi)}{\pi x}$$

53:6 EXPANSIONS

As a power series, the Bessel function is expressible by

53:6:1 $$J_\nu(x) = \frac{(x/2)^\nu}{\Gamma(1+\nu)} - \frac{(x/2)^{2+\nu}}{1!\Gamma(2+\nu)} + \frac{(x/2)^{4+\nu}}{2!\Gamma(3+\nu)} - \cdots = \left(\frac{x}{2}\right)^\nu \sum_{j=0}^{\infty}\frac{(-x^2/4)^j}{j!\Gamma(1+j+\nu)}$$

which is convergent for all values of ν, although the leading terms vanish if ν is a negative integer. That $J_\nu(x)$ may be expressed in terms of a hypergeometric series [Section 18:14] is evident by rewriting 53:6:1 as $[(x/2)^\nu/\Gamma(1+\nu)]\,\Sigma\,(-x^2/4)^j/[(1)_j(1+\nu)_j]$ or, in Clifford's notation [see 53:1:1]:

53:6:2 $$C_\nu(x) = \frac{1}{\Gamma(1+\nu)} - \frac{x}{1!\Gamma(2+\nu)} + \frac{x^2}{2!\Gamma(3+\nu)} - \cdots = \frac{1}{\Gamma(1+\nu)}\sum_{j=0}^{\infty}\frac{(-x)^j}{(1)_j(1+\nu)_j}$$

The product of two Bessel functions of common argument is expressible as a hypergeometric series with two numeratorial and four denominatorial parameters:

53:6:3 $$J_\nu(x)\,J_\mu(x) = \frac{(x/2)^{\nu+\mu}}{\Gamma(1+\nu)\,\Gamma(1+\mu)}\sum_{j=0}^{\infty}\frac{\left(\frac{1+\nu+\mu}{2}\right)_j\left(1+\frac{\nu+\mu}{2}\right)_j(-x^2)^j}{(1)_j(1+\nu)_j(1+\mu)_j(1+\nu+\mu)_j}$$

From this it follows by setting $\mu = \pm\nu$ that $J_\nu^2(x)$ or $J_\nu(x)\,J_{-\nu}(x)$ may each be expressed as a $K = 1$, $L = 3$ [see Section 18:14] hypergeometric series.

The Bessel function may be expanded as the infinite product

53:6:4 $$J_\nu(x) = \frac{(x/2)^\nu}{\Gamma(1+\nu)}\left(1-\frac{x^2}{j_{\nu;1}^2}\right)\left(1-\frac{x^2}{j_{\nu;2}^2}\right)\left(1-\frac{x^2}{j_{\nu;3}^2}\right)\cdots = \frac{(x/2)^\nu}{\Gamma(1+\nu)}\prod_{k=1}^{\infty}1-\frac{x^2}{j_{\nu;k}^2}$$

where $j_{\nu;k}$ denotes the kth zero of J_ν [see Section 53:7]. A similar infinite product holds for the derivative $J_\nu'(x) = dJ_\nu(x)/dx$

53:6:5 $$\frac{J_{\nu-1}(x) - J_{\nu+1}(x)}{2} = J_\nu'(x) = \frac{(x/2)^{\nu-1}}{2\Gamma(\nu)}\prod_{k=1}^{\infty}1-\left(\frac{x}{j_{\nu;k}'}\right)^2 \qquad \nu > 0$$

where $j_{\nu;k}'$ is the kth extremum of J_ν [also discussed in Section 53:7].

Although they are of greatest utility when the order is an integer, equations 52:6:6–52:6:9 apply to Bessel functions generally, on replacement of n by ν. Also, see 50:5:3 for an expansion of $J_\nu(x)$ in terms of hyperbolic Bessel functions.

When x, but not $|\nu|$, is large, the asymptotic expansion

53:6:6
$$J_\nu(x) = \sqrt{\frac{2}{\pi x}}\left[P(\nu;x)\cos\left(x - \frac{\nu\pi}{2} - \frac{\pi}{4}\right) - Q(\nu;x)\sin\left(x - \frac{\nu\pi}{2} - \frac{\pi}{4}\right)\right]$$

holds, where, as $x \to \infty$:

53:6:7
$$P(\nu;x) \sim 1 - \frac{(\frac{9}{4} - \nu^2)(\frac{1}{4} - \nu^2)}{2!(2x)^2} + \frac{(\frac{49}{4} - \nu^2)(\frac{25}{4} - \nu^2)(\frac{9}{4} - \nu^2)(\frac{1}{4} - \nu^2)}{4!(2x)^4} - \cdots + \frac{(\frac{1}{2} - \nu)_{2j}(\frac{1}{2} + \nu)_{2j}}{(2j)!(-4x^2)^j} + \cdots$$

and

53:6:8
$$Q(\nu;x) \sim -\frac{(\frac{1}{4} - \nu^2)}{1!(2x)} + \frac{(\frac{25}{4} - \nu^2)(\frac{9}{4} - \nu^2)(\frac{1}{4} - \nu^2)}{3!(2x)^3} - \frac{(\frac{81}{4} - \nu^2)(\frac{49}{4} - \nu^2)(\frac{25}{4} - \nu^2)(\frac{9}{4} - \nu^2)(\frac{1}{4} - \nu^2)}{5!(2x)^5} + \cdots$$
$$- \frac{(\frac{1}{2} - \nu)_{2j+1}(\frac{1}{2} + \nu)_{2j+1}}{2(2j+1)!x(-4x^2)^j} \cdots$$

Each of the functions $P(\nu;x)$ and $Q(\nu;x)$ is expressible as a $K = 4$, $L = 2$ hypergeometric function, as tabulated in Section 18:14. Although they are generally valid only for sufficiently large x, the series 53:6:7 and 53:6:8 terminate when ν is an odd multiple of $\pm\frac{1}{2}$, and in this circumstance an exact expression for $J_\nu(x)$ is provided for all x.

53:7 PARTICULAR VALUES

		$x = 0$	$x = \infty$
$J_\nu(x)$	$\nu > 0$	0	0
$J_0(x)$		1	0
$J_\nu(x)$	$-1 < \nu < 0,\ -3 < \nu < -2,\ -5 < \nu < -4, \ldots$	∞	0
$J_{-n}(x)$	$n = 1, 2, 3, \ldots$	0	0
$J_\nu(x)$	$-2 < \nu < -1,\ -4 < \nu < -3,\ -6 < \nu < -5, \ldots$	$-\infty$	0

The values of x that cause $J_\nu(x)$ to acquire the value zero are denoted by $j_{\nu;1}$, $j_{\nu;2}$, $j_{\nu;3}$, $\ldots$, and the name k^{th} *zero of the Bessel function* $J_\nu(x)$ is given to $j_{\nu;k}$. This name is appropriate for $\nu > -1$ but may be misleading for $\nu < -1$ because not all these zeros then exist (i.e., as real numbers). As an example, for $-4 < \nu < -3$ there exists the zeros $j_{\nu;4}$, $j_{\nu;5}$, $j_{\nu;6}$, $\ldots$ but there is no real $j_{\nu;1}$, $j_{\nu;2}$ or $j_{\nu;3}$. For nonnegative integer ν, values of $j_{\nu;k}$ are discussed in Section 52:7, and an algorithm is presented there for their calculation. For order -1 we have $j_{-1;2} = j_{1;1}$, $j_{-1;3} = j_{1;2}$, etc., and, generally:

53:7:1
$$j_{-n;k} = j_{n;k-n} \qquad k = n+1, n+2, n+3, \ldots$$

For orders $\pm\frac{1}{2}$ we have the simple rules $j_{1/2;k} = k\pi$ and $j_{-1/2;k} = (k - \frac{1}{2})\pi$. The zeros of order $\frac{3}{2}$ are given by $j_{3/2;k} = r_k(1)$ where the latter is discussed in Section 34:7. The same section discusses, and provides an algorithm for, $\rho_k(b)$, which is related to the zeros of order $-\frac{3}{2}$ by $j_{-3/2;k} = \rho_{k-1}(-1)$. For small ν of either sign $j_{\nu;k}$ is given by the approximation formula

53:7:2
$$\frac{2}{\pi}j_{\nu;k} - \nu \simeq 2k - \frac{1}{2} \qquad \text{small } |\nu|$$

as is evident from Figure 53-2, in which the red lines display values of $(2/\pi)j_{\nu;k} - \nu$ as a function of ν for $1 \le k \le 9$, permitting approximate values of the zeros to be found easily. For large positive orders, the first zero of $J_\nu(x)$ is given approximately by

53:7:3 $$j_{\nu;1} \simeq \nu + 1.8558\nu^{1/3} + 1.033\nu^{-1/3} \qquad \text{large } \nu$$

The symbols $j'_{\nu;1}$, $j'_{\nu;2}$, $j'_{\nu;3}$, ... are used to denote those values of the argument x that cause $J_\nu(x)$ to exhibit a local maximum or minimum, and $j'_{\nu;k}$ is known as the kth *extremum of the Bessel function* $J_\nu(x)$. Again, this name is not always appropriate because, for example, the "first" extremum of $J_{-3}(x)$ is $j'_{-3;4}$. Values of $j'_{n;k}$ were discussed in Section 52:7 for $n = 0, 1, 2, \ldots$, while for negative integer order we have the rule

53:7:4 $$j'_{-n;k} = j'_{n;k-n} \qquad k = n + 1, n + 2, n + 3, \ldots$$

For moiety orders $j'_{1/2;k} = r_{k-1}(2)$ and $j'_{-1/2;k} = \rho_{k-1}(-2)$, where r and ρ are the functions discussed in Section 34:7. For small orders of either sign, the approximation

53:7:5 $$\frac{2}{\pi} j'_{\nu;k} - \nu \simeq 2k - \frac{3}{2} \qquad k = 2, 3, 4, \ldots \qquad \text{small } |\nu|$$

holds except for $k = 1$. Figure 53-2 contains graphs of $(2/\pi)\, j'_{\nu;k} - \nu$ versus ν for the first nine extrema. Green lines on this graph represent maxima, blue lines minima. The approximation

53:7:6 $$j'_{\nu;1} \simeq \nu + 0.8086\nu^{1/3} + 0.072\nu^{-1/3}$$

is valid for large ν.

53:8 NUMERICAL VALUES

For arguments not exceeding $7.5 + 0.3\,|\nu|$, the algorithm below utilizes the formula

53:8:1 $$J_\nu(x) \simeq \frac{(x/2)^\nu}{\Gamma(1+\nu)} \left[t_0 + t_1 + t_2 + \cdots + t_k + \cdots + t_{K-1} + \frac{t_K}{2} \right]$$

where $t_0 = 1$ and $t_k = -x^2 t_{k-1}/4k(k+\nu)$, which is a truncated version of expansion 53:6:1. Provided $K > -\nu$, series 53:8:1 eventually alternates and the properties of alternating series [see Section 0:6] then guarantee that the fractional error in approximation 53:8:1 is less than $\varepsilon/2$ if $|t_K(2/x)^\nu\, \Gamma(1+\nu)\, J_\nu(x)| < \varepsilon$. The algorithm increments K until it exceeds $-\nu$ and until $|t_K/(t_0 + t_1 + \cdots + t_{K-1} + \frac{1}{2}t_K)| < 10^{-7}$, thus ensuring 24-bit precision in $J_\nu(x)$ except in the immediate vicinity of a zero of the Bessel function. The multiplier $(x/2)^\nu/\Gamma(1+\nu)$ is calculated by a modified version of the algorithm in Section 43:8, but special measures are adopted if ν is a negative integer or zero.

Input ν >>>>>>>
Set $f = h = \frac{1}{2}$
Set $k = 0$
Input x >>>>>>>
If $x > 7.5 + .3|\nu|$ go to (3)
If $\nu + \text{Int}(|\nu|) \neq 0$ go to (1)
Replace ν by $-\nu$
Set $f = h = (-1)^\nu/2$
(1) Replace k by $k + 1$
Replace f by $f + h$
Replace h by $-\,hx^2/4k(k + \nu)$
Replace f by $f + h$
If $|f/h| < 2 \times 10^7$ go to (1)
If $\nu \leq -k$ go to (1)
If $\nu = 0$ go to (5)
(2) Replace ν by $\nu + 1$
Replace f by $2f\nu/x$
If $\nu \leq 3$ go to (2)

Storage needed: ν, f, h, k, x, t and θ

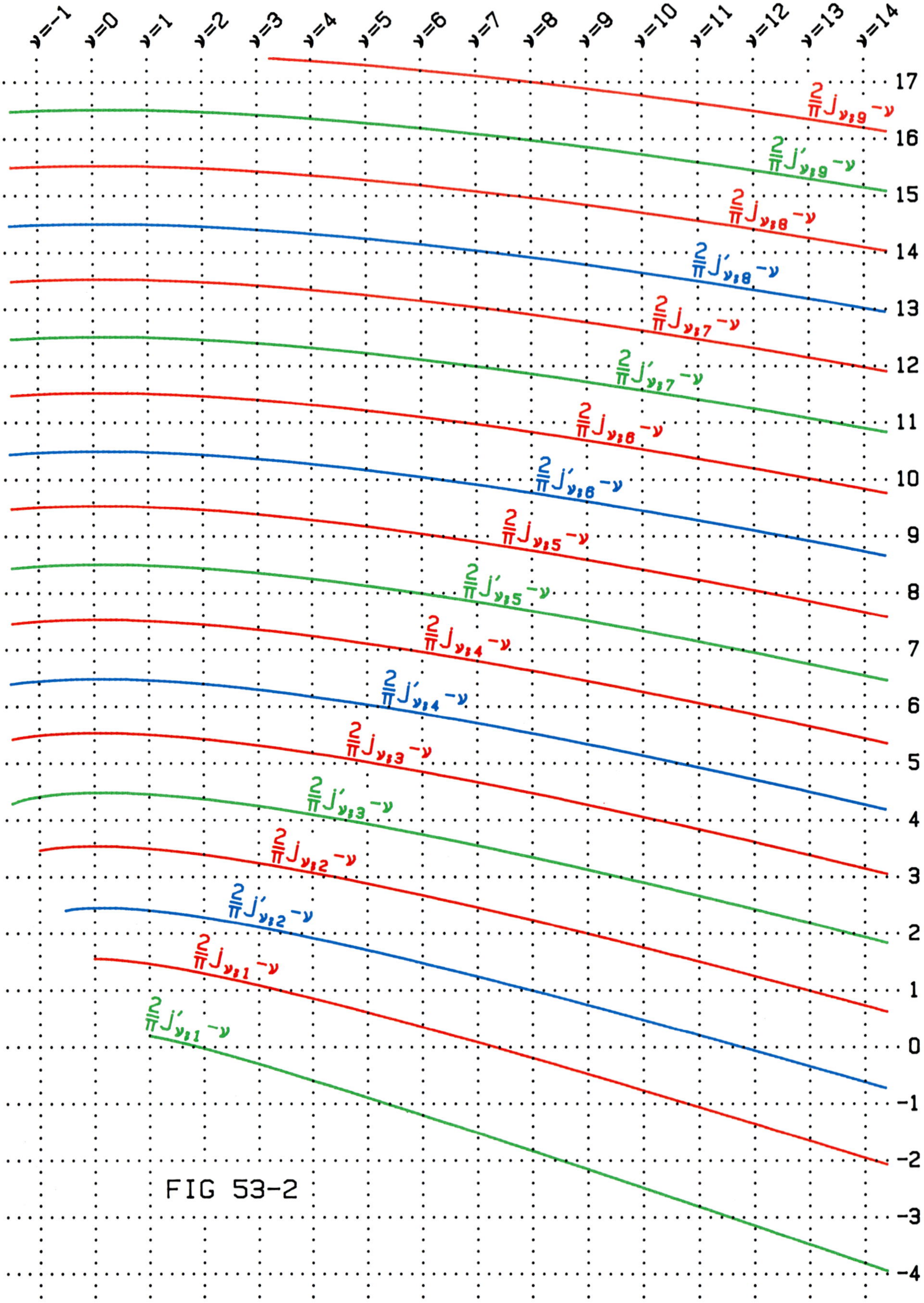

ν=−1
ν=0
ν=1
ν=2
ν=3
ν=4
ν=5
ν=6
ν=7
ν=8
ν=9
ν=10
ν=11
ν=12
ν=13
ν=14
17
16
15
14
13
12
11
10
9
8
7
6
5
4
3
2
1
0
−1
−2
−3
−4
$\frac{2}{\pi}j_{\nu;9}-\nu$
$\frac{2}{\pi}j'_{\nu;9}-\nu$
$\frac{2}{\pi}j_{\nu;8}-\nu$
$\frac{2}{\pi}j'_{\nu;8}-\nu$
$\frac{2}{\pi}j_{\nu;7}-\nu$
$\frac{2}{\pi}j'_{\nu;7}-\nu$
$\frac{2}{\pi}j_{\nu;6}-\nu$
$\frac{2}{\pi}j'_{\nu;6}-\nu$
$\frac{2}{\pi}j_{\nu;5}-\nu$
$\frac{2}{\pi}j'_{\nu;5}-\nu$
$\frac{2}{\pi}j_{\nu;4}-\nu$
$\frac{2}{\pi}j'_{\nu;4}-\nu$
$\frac{2}{\pi}j_{\nu;3}-\nu$
$\frac{2}{\pi}j'_{\nu;3}-\nu$
$\frac{2}{\pi}j_{\nu;2}-\nu$
$\frac{2}{\pi}j'_{\nu;2}-\nu$
$\frac{2}{\pi}j_{\nu;1}-\nu$
$\frac{2}{\pi}j'_{\nu;1}-\nu$

FIG 53-2

Set $h = \left[1 + \dfrac{2}{7\nu^2}\left(\dfrac{2}{3\nu^2} - 1\right)\right] \Big/ 30\nu^3$

Replace h by $\dfrac{h-1}{12\nu} + \nu\left[1 + \ln\left(\dfrac{x}{2\nu}\right)\right]$

Replace f by $f \exp(h)/\sqrt{2\pi\nu}$

Go to (5)

(3) Set $t = \sqrt{2/\pi x}$

Set $\theta = 90(\nu + h - t^2x^2)$

Set $f = 0$

(4) Replace f by $f + t\cos(\theta)$

Replace t by $t(h^2 - \nu^2)/[(2h + 1)x]$

Replace θ by $\theta + 90$

Replace h by $h + 1$

If $h < |\nu|$ go to (4)

If $|f| > 2 \times 10^7|t|$ go to (5)

If $h^2 < (2h + 1)x + \nu^2$ go to (4)

(5) Output f

$f \simeq J_\nu(x)$ <<<<<<

Use degree mode or replace 90 by $\pi/2$.

Test values:
$J_2(10) = 0.254630314$
$J_0(7) = 0.300079271$
$J_{-1}(4) = 0.0660433280$
$J_{5/2}(\pi) = 0.429869376$
$J_{-3.6}(0.5) = -169.507814$

For arguments greater than $7.5 + 0.3|\nu|$, equations 53:6:6–53:6:8 are used in the form

53:8:2 $$J_\nu(x) \sim \sum_{k=0}^{K} t_k \cos\left(\frac{k\pi}{2} - x + \frac{\nu\pi}{2} + \frac{\pi}{4}\right)$$

where $t_0 = \sqrt{2/\pi x}$ and other t_k values obey the recurrence $t_k = [(k - \frac{1}{2})^2 - \nu^2]t_{k-1}/2kx$. The number of terms included in this asymptotic series is determined by three criteria. First, to ensure that the terms in the $P(\nu;x)$ and $Q(\nu;x)$ series alternate in sign, we include terms in 53:8:2 up to at least $k = |\nu| - \frac{1}{2}$. Second, we cease adding terms to the cumulative sum f when the magnitude of t_k becomes less than $(5 \times 10^{-8})|f|$. Third, irrespective of size of the contribution of t_k to f, we never allow k to reach the critical value $x - \frac{1}{2} + \sqrt{x^2 + x + \nu^2}$ after which t_k increases in magnitude and the series diverges.

53:9 APPROXIMATIONS

The first three terms of expansion 53:6:1 are reproduced exactly by the approximation

53:9:1 $$J_\nu(x) \simeq \frac{4}{x^2\Gamma(1+\nu)}\left[\frac{x}{2} - \frac{x^3}{8(1+\nu)(2+\nu)}\right]^{2+\nu} \qquad \text{small } x$$

and subsequent terms are close when ν is large. An asymptotic approximation, valid for large x, is given by 53:2:1.

53:10 OPERATIONS OF THE CALCULUS

The derivative of the Bessel function is given by the equivalent formulas

53:10:1 $$\frac{d}{dx}J_\nu(x) = \frac{1}{2}J_{\nu-1}(x) - \frac{1}{2}J_{\nu+1}(x) = J_{\nu-1}(x) - \frac{\nu}{x}J_\nu(x) = \frac{\nu}{x}J_\nu(x) - J_{\nu+1}(x)$$

The expressions

53:10:2 $$\frac{d}{dx}x^{\pm\nu}J_\nu(x) = \pm x^{\pm\nu}J_{\nu\mp1}(x)$$

also hold.

Corresponding to 53:10:2 are the indefinite integrals

53:10:3 $$\int_0^x t^{1+\nu} J_\nu(t)dt = x^{1+\nu} J_{1+\nu}(x) \qquad \nu > -1$$

and

53:10:4 $$\int_0^x t^{1-\nu} J_\nu(t)dt = \frac{2}{2^\nu \Gamma(\nu)} - x^{1-\nu} J_{\nu-1}(x)$$

Important definite integrals include

53:10:5 $$\int_0^\infty J_\nu(t)dt = 1 \qquad \nu > -1$$

53:10:6 $$\int_0^\infty \exp(-bt)\, J_\nu(t)dt = \frac{(\sqrt{1+b^2} - b)^\nu}{\sqrt{1+b^2}} \qquad \nu > -1$$

53:10:7 $$\int_0^\infty \cos(bt)\, J_\nu(t)dt = \left\{\begin{array}{ll} \cos(\nu \arcsin(b))/\sqrt{1-b^2} & 0 \le b < 1 \\ \dfrac{-\sin(\pi\nu/2)}{\sqrt{b^2-1}\,(b+\sqrt{b^2-1})^\nu} & b > 1 \end{array}\right\} \quad \nu > -1$$

53:10:8 $$\int_0^\infty \sin(bt)\, J_\nu(t)dt = \left\{\begin{array}{ll} \sin(\nu \arcsin(b))/\sqrt{1-b^2} & 0 < b < 1 \\ \dfrac{\cos(\pi\nu/2)}{\sqrt{b^2-1}\,[b+\sqrt{b^2-1}]^\nu} & b > 1 \end{array}\right\} \quad \nu > -2$$

and many others are listed by Gradshteyn and Ryzhik (Sections 6.5–6.7). Definite integrals of the form

53:10:9 $$\int_0^\infty f(t)\, J_\nu(st)dt$$

are known as *Hankel transforms* and tables of $\int \sqrt{st}\, f(t)\, J_\nu(st)dt$ are given by Erdélyi, Magnus, Oberhettinger and Tricomi [*Tables of Integral Transforms*, Volume 2, Chapter 8].

The operations of differintegration are particularly simple when Clifford's notation [see 53:1:1] is adopted. Thus:

53:10:10 $$\frac{d^\mu}{dx^\mu} x^\nu C_\nu(x) = x^{\nu-\mu} C_{\nu-\mu}(x)$$

and, in terms of the function discussed in Section 53:12:

53:10:11 $$\frac{d^\mu}{dx^\mu} C_\nu(x) = \frac{x^{-\mu}}{\Gamma(1+\nu)\,\Gamma(1-\mu)} \sum_{j=0}^\infty \frac{(-x)^j}{(1+\nu)_j(1-\mu)_j}$$

53:11 COMPLEX ARGUMENT

The Bessel function adopts complex values when its argument is complex and, unless the order is an integer, even when the argument is negative. We report the expressions

53:11:1 $$J_\nu(-x) = (-1)^\nu J_\nu(x) = [\cos(\nu\pi) + i \sin(\nu\pi)]\, J_\nu(x)$$

and

53:11:2 $$J_\nu(iy) = i^\nu I_\nu(y) = \left[\cos\left(\frac{\nu\pi}{2}\right) + i \sin\left(\frac{\nu\pi}{2}\right)\right] I_\nu(y)$$

for negative and imaginary argument, but exclude the case of general complex argument. I_ν is the hyperbolic Bessel function of Chapter 50. With n replaced by ν, equation 52:11:2 applies for the special case of complex arguments in which the real and imaginary parts have equal magnitudes.

53:12 GENERALIZATIONS

As a hypergeometric function [Section 18:14], the Bessel function may be written

53:12:1 $$J_\nu(x) = \frac{(x/2)^\nu}{\Gamma(1)\,\Gamma(1+\nu)} \sum_{j=0}^{\infty} \frac{(-x^2/4)^j}{(1)_j(1+\nu)_j}$$

so that $K = 0$, $L = 2$ and one of the denominatorial parameters is constrained to be unity. Removal of this constraint generates a function

53:12:2 $$\frac{(x/2)^\nu}{\Gamma(1+\mu)\,\Gamma(1+\nu-\mu)} \sum_{j=0}^{\infty} \frac{(-x^2/4)^j}{(1+\mu)_j(1+\nu-\mu)_j}$$

that thus represents a generalization of the Bessel function. The $\mu = \frac{1}{2}$ instance of this generalized function is the Struve function $h_{\nu-1}(x)$ of Chapter 57.

53:13 COGNATE FUNCTIONS

The functions addressed in Chapters 49–57 all have common features.

53:14 RELATED TOPICS

If, for some restricted or unrestricted range of its arguments, a function $f(x)$ is expansible as a power series [see Section 11:14]:

53:14:1 $$f(x) = \sum_{j=0}^{\infty} a_j x^j \qquad a_j = \frac{1}{j!}\frac{d^j f}{dx^j}(0)$$

then it may also be expanded as the *Neumann series*

53:14:2 $$f(x) = \left(\frac{2}{x}\right)^\nu \sum_{k=0}^{\infty} (\nu+k)\, b_k J_{\nu+k}(x)$$

Here ν is arbitrary except that it may not be a negative integer. The coefficient b_k is given by

53:14:3 $$b_k = \sum_{j=0}^{J} \frac{4^j\Gamma(J+j+\nu)}{(J-j)!} a_{2j} \qquad 2J = k = 0, 2, 4, \ldots$$

when k is even, and by

53:14:4 $$b_k = 2\sum_{j=0}^{J} \frac{4^j\Gamma(J+j+1+\nu)}{(J-j)!} a_{2j+1} \qquad 2J+1 = k = 1, 3, 5, \ldots$$

for odd k. Equations 52:5:3–52:5:7 are all examples of Neumann series.

Replacing $f(x)$ in 53:14:2 by the unity function leads to the expression

53:14:5 $$x^\nu = 2^\nu \sum_{k=0}^{\infty} (2k+\nu)\frac{\Gamma(k+\nu)}{k!} J_{2k+\nu}(x) \qquad \nu \neq -1, -2, -3, \ldots$$

as an expansion of an arbitrary power. Equations 52:5:6, 52:5:8 and 52:5:9 are special cases.

As an alternative to the 53:14:2 expansion, the function f(x) may be expanded as the *modified Neumann series*

53:14:6 $$f(x) = \sum_{k=0}^{\infty} c_k \left(\frac{\sqrt{x}}{2}\right)^{k-\nu} J_{k+\nu}(\sqrt{x}) \qquad \nu \neq -1, -2, -3, \ldots$$

where the coefficients are given by

53:14:7 $$c_k = \sum_{j=0}^{k} 4^j \frac{\Gamma(1+j+\nu)}{(k-j)!} a_j$$

Equations 53:5:2 and 53:5:4 are examples of modified Neumann series. Replacing f(x) in 53:14:6 by unity leads to expression 53:3:1.

CHAPTER 54

THE NEUMANN FUNCTION $Y_\nu(x)$

In linear combination with a Bessel function [Chapter 53], the Neumann function satisfies Bessel's equation, and, accordingly, it appears in the solution to many physical problems, especially those involving cylindrical symmetry.

54:1 NOTATION

The symbol $N_\nu(x)$ is a common alternative to $Y_\nu(x)$ to denote the Neumann function of order ν and argument x. The names *Bessel function of the second kind* and *Weber's function* are also in use.

54:2 BEHAVIOR

Because the Neumann function is complex for negative argument, we restrict attention to $x \geq 0$. Behavior in this range can be appreciated best by consideration of the contour map, Figure 54-1.

Viewed as a function either of x, or of ν, the $Y_\nu(x)$ function displays an infinite number of oscillations. For a constant value of ν not very different from zero, the oscillations commence close to $x = 0$, and their amplitudes steadily diminish as the argument increases, eventually conforming to the asymptotic expression

54:2:1 $$Y_\nu(x) \sim \sqrt{\frac{2}{\pi x}}\sin\left(x - \frac{\nu\pi}{2} - \frac{\pi}{4}\right) \qquad x \to \infty$$

As ν departs further from zero in either direction, the oscillatory behavior persists, but its onset becomes increasingly delayed. This is very clearly illustrated in Figure 54-2, which depicts Neumann functions of nonnegative integer orders.

At any constant positive value of x, the Neumann function is an oscillatory function of ν in the vicinity of $\nu = 0$, and the oscillations become increasingly amplified as ν becomes more negative until the asymptotic formulation

54:2:2 $$Y_\nu(x) \sim -\sqrt{\frac{-2}{\pi\nu}}\left(\frac{-ex}{2\nu}\right)^\nu \cos(\nu\pi) \qquad \nu \to -\infty \qquad x = \text{constant} > 0$$

comes to be obeyed. On the other hand, as ν becomes increasingly positive, the oscillations of $Y_\nu(x)$ eventually cease, after which the Neumann function acquires negative values of increasing magnitude, given by

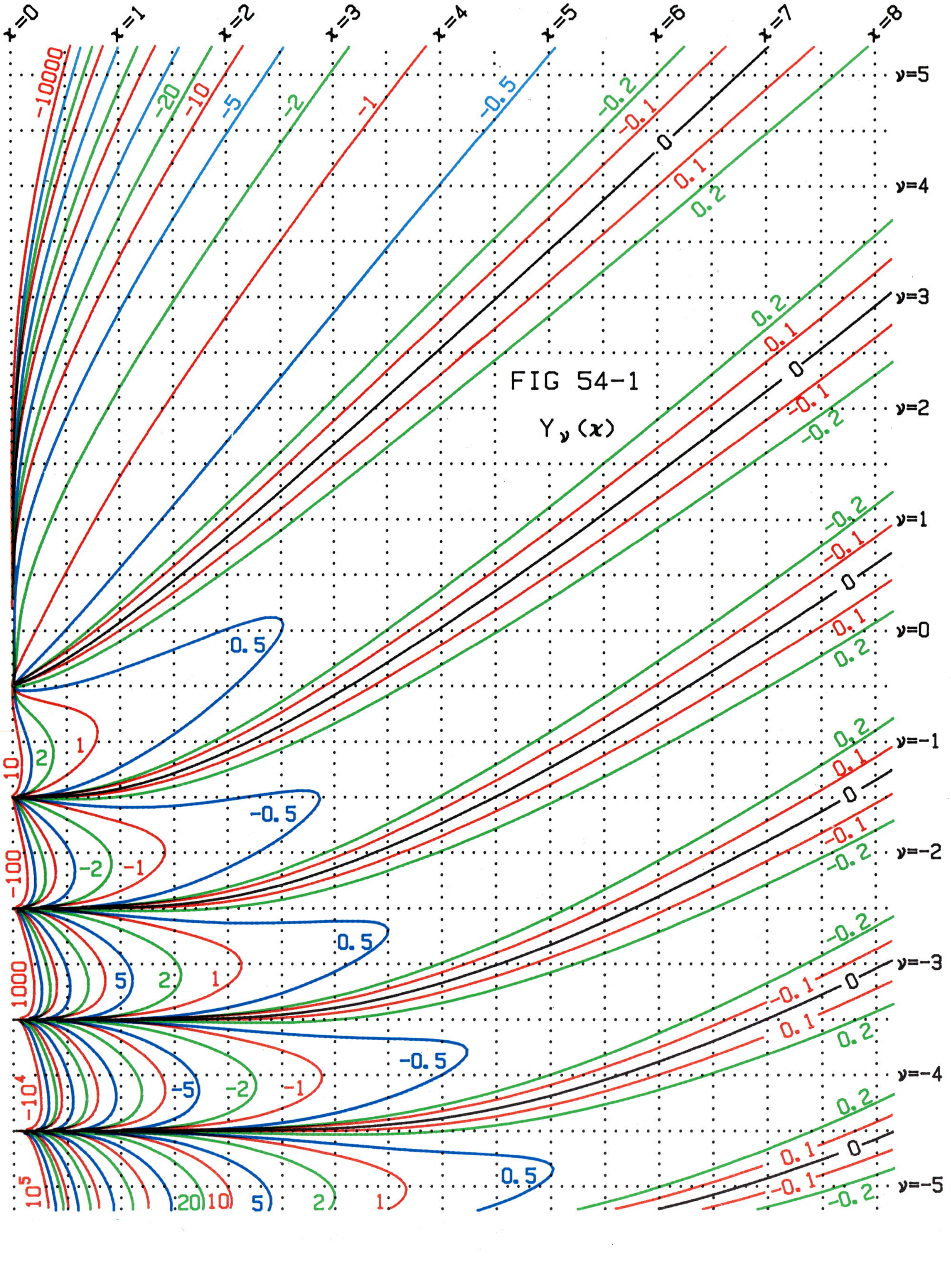
x=0
x=1
x=2
x=3
x=4
x=5
x=6
x=7
x=8
ν=5
ν=4
ν=3
ν=2
ν=1
ν=0
ν=−1
ν=−2
ν=−3
ν=−4
ν=−5
−10000
−20
−10
−5
−2
−1
−0.5
−0.2
−0.1
0
0.1
0.2
0.5
10
2
1
−100
−2
−1
−0.5
1000
5
2
1
0.5
−10⁴
−5
−2
−1
−0.5
10⁵
20
10
5
2
1
0.5

FIG 54-1

$Y_\nu(x)$

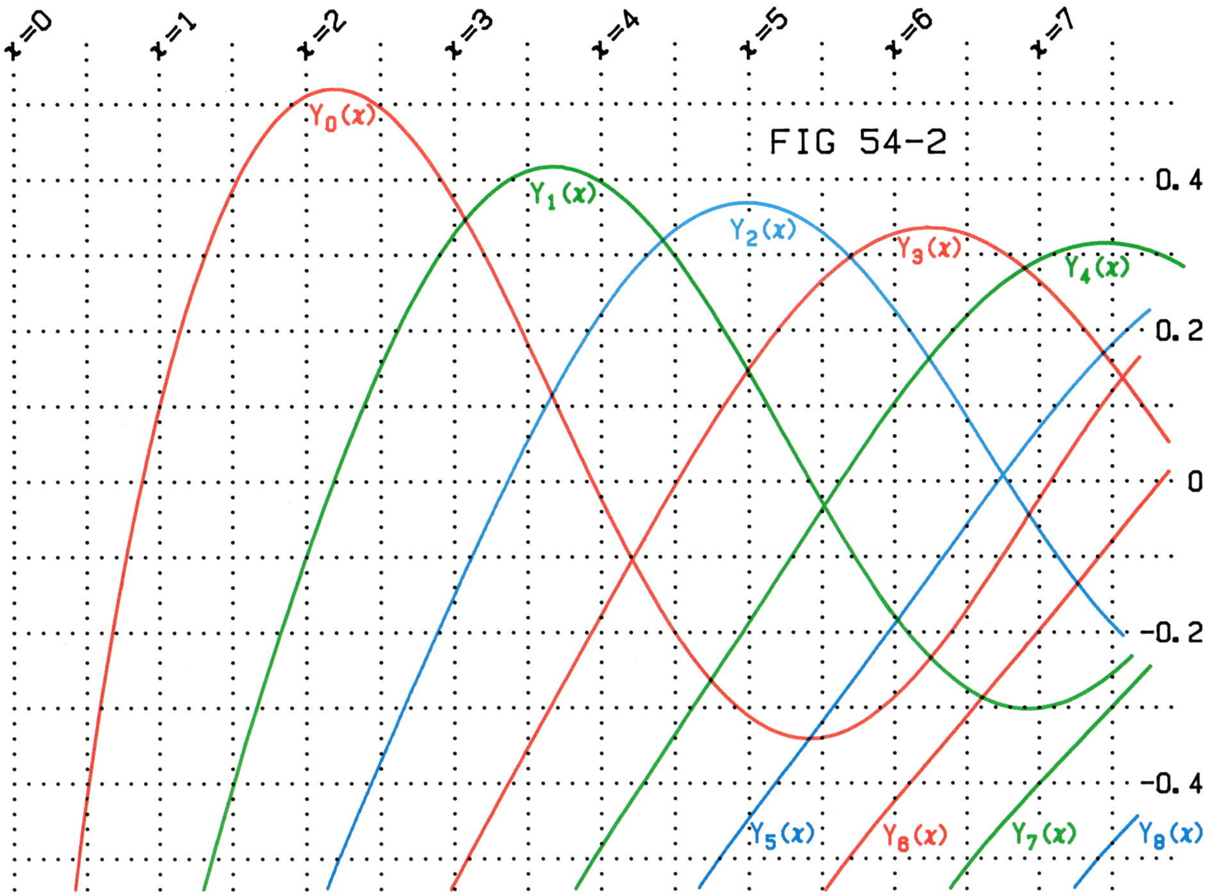

54:2:3 $$Y_\nu(x) \sim -\sqrt{\frac{2}{\pi\nu}}\left(\frac{2\nu}{ex}\right)^\nu \qquad \nu \to \infty \qquad x = \text{constant} > 0$$

Comparison of Figures 53-1 and 54-1 reveals considerable similarities. The major point of distinction occurs for positive ν as $x \to 0$. Whereas the Bessel function approaches zero in this limit, the Neumann function approaches $-\infty$. Another difference concerns the phase of the oscillations, which generally differ by $\pi/2$ or thereabouts. This is evident on comparison of equation 53:2:1 with 54:2:1 and of equation 53:2:3 with 54:2:2.

54:3 DEFINITIONS

For noninteger order, the Neumann function is defined in terms of Bessel functions

54:3:1 $$Y_\nu(x) = \frac{\cos(\nu\pi)\, J_\nu(x) - J_{-\nu}(x)}{\sin(\nu\pi)} \qquad \nu \neq 0, \pm1, \pm2, \ldots$$

For integer order, this definition is to be replaced by its limit

54:3:2 $$Y_n(x) = \lim_{\nu\to n} \{\cot(\nu\pi)\, J_\nu(x) - \csc(\nu\pi)\, J_{-\nu}(x)\} \qquad n = 0, \pm1, \pm2, \ldots$$

The combination $c_1 J_\nu(x) + c_2 Y_\nu(x)$ provides a solution to Bessel's equation 53:3:5, for integer or noninteger ν. Similarly, $c_1 J_\nu(2\sqrt{x}) + c_2 Y_\nu(2\sqrt{x})$ satisfies the Bessel-Clifford equation 53:3:6 for all ν. The coefficients c_1 and c_2 are arbitrary constants.

Some integral representations of the Neumann function are given by Gradshteyn and Rhyzik [Section 8.415].

The close relationship of the Neumann function of zero order to the cosine function is evidenced by the semiintegration formula

54:3:3 $$\frac{d^{-1/2}}{dx^{-1/2}}\frac{1}{x}\cos\left(\frac{1}{\sqrt{x}}\right) = -\sqrt{\frac{\pi}{x}}\,Y_0\left(\frac{1}{\sqrt{x}}\right)$$

54:4 SPECIAL CASES

Known as *spherical Bessel functions of the second kind*, the functions

54:4:1 $$y_n(x) = \sqrt{\frac{\pi}{2x}}\,Y_{n+1/2}(x) \qquad n = 0, \pm1, \pm2, \ldots$$

are related to those of the first kind by the simple equivalence

54:4:2 $$y_{n+1}(x) = (-1)^n\, j_{-n}(x) \qquad n = 0, \pm1, \pm2, \ldots$$

and the relationships given in Section 53:4 are therefore readily adapted to $y_n(x)$.

Neumann functions of order $\pm\frac{1}{3}$ are related to Airy functions [Chapter 56] via the expressions

54:4:3 $$Y_{\pm1/3}(x) = \frac{-3\mathrm{Bi}(-X) \mp \sqrt{3}\,\mathrm{Ai}(-X)}{2\sqrt{X}} \qquad X = \left(\frac{3x}{2}\right)^{2/3}$$

Similarly, Neumann functions of order $\pm\frac{2}{3}$ are related to the derivatives of the Airy functions.

54:5 INTRARELATIONSHIPS

The order-reflection formula

54:5:1 $$Y_{-n}(x) = (-1)^n\, Y_n(x) \qquad n = 0, \pm1, \pm2, \ldots$$

applies only for integer order, but the recurrence formula

54:5:2 $$Y_{\nu+1}(x) = \frac{2\nu}{x}\,Y_\nu(x) - Y_{\nu-1}(x)$$

applies for all ν. Because the form of 54:5:2 is identical to that of equation 52:5:15, expressions analogous to 52:5:16–52:5:20 relate $Y_2(x)$, $Y_3(x)$, $Y_4(x)$, $Y_5(x)$ and $Y_6(x)$ to $Y_0(x)$ and $Y_1(x)$.

A number of relationships interconnect Neumann and Bessel functions:

54:5:3 $$Y_\nu(x) + Y_{-\nu}(x) = \cot\left(\frac{\nu\pi}{2}\right)[J_\nu(x) - J_{-\nu}(x)]$$

54:5:4 $$Y_\nu(x) - Y_{-\nu}(x) = -\tan\left(\frac{\nu\pi}{2}\right)[J_\nu(x) + J_{-\nu}(x)]$$

54:5:5 $$Y_\nu^2(x) + J_\nu^2(x) = Y_{-\nu}^2(x) + J_{-\nu}^2(x)$$

54:5:6 $$J_{\nu+1}(x)\,Y_\nu(x) - J_\nu(x)\,Y_{\nu+1}(x) = \frac{2}{\pi x}$$

54:6 EXPANSIONS

Using definition 54:3:1 and expansion 53:6:1, $Y_\nu(x)$ may be expressed as the sum of two infinite series. Alternatively, by exploiting the properties of the gamma function [Section 43:5] and of the trigonometric functions, the two infinite series may be conjoined into a polynomial function and a single infinite series:

54:6:1 $$Y_\nu(x) = As(|\nu|)\sum_{j=0}^{n-1}\frac{(-x^2/4)^j}{j!(1-|\nu|)_j}$$

$$+\sum_{j=0}^{\infty}\left|\frac{Bs(-|\nu|)}{j!(1+|\nu|)_j}+\frac{Cs(|\nu|-n)(x/2)^n}{(n+j)!(1+n-|\nu|)_j}\right|\left(\frac{-x^2}{4}\right)^j \qquad \nu \neq 0, \pm1, \pm2, \ldots$$

where we use the abbreviation $s(p) = -(2/x)^p\,\Gamma(p)/\pi$. When ν is positive, the coefficients A and C are unity and B is $\cos(\nu\pi)$; when $\nu < 0$, $A = C = \cos(\nu\pi)$ and $B = 1$. Here n may take any positive integer value, but the expansion is most useful when n is the integer closest to $|\nu|$, for then the B and C terms in 54:6:1 have similar magnitudes, permitting the summation to be evaluated with little loss of precision.

Both of the expansions discussed above become indeterminate when the order of the Neumann function is an integer. For nonnegative integer order

54:6:2
$$Y_n(x) = \frac{2}{\pi}\ln\left(\frac{x}{2}\right)J_n(x) - \frac{1}{\pi}\left(\frac{2}{x}\right)^n\sum_{j=0}^{n-1}\frac{(n-j-1)!}{j!}\left(\frac{x^2}{4}\right)^j$$
$$-\frac{1}{\pi}\left(\frac{x}{2}\right)^n\sum_{j=0}^{\infty}\frac{\psi(j+1)+\psi(j+n+1)}{j!(j+n)!}\left(\frac{-x^2}{4}\right)^j \qquad n = 0, 1, 2, \ldots$$

where J_n and ψ are the functions of Chapters 52 and 44. The polynomial term in 54:6:2 is absent when $n = 0$; under these conditions one has

54:6:3
$$\frac{\pi}{2}Y_0(x) = \left[\ln\left(\frac{x}{2}\right)+\gamma\right]J_0(x) + \frac{x^2}{4} - \frac{3x^4}{128} + \frac{11x^6}{13824} - \cdots - \left(1+\frac{1}{2}+\frac{1}{3}+\cdots+\frac{1}{j}\right)\frac{(-x^2/4)^j}{(j!)^2} + \cdots$$

where γ is Euler's constant, 0.5772156649. Neumann functions of negative integer order can be expanded as in 54:6:2 by taking advantage of relationship 54:5:1.

Neumann functions of integer order are also expansible as an infinite series of Bessel coefficients [Chapter 52]. The general case is

54:6:4
$$Y_n(x) = \frac{-n!}{\pi}\left(\frac{2}{x}\right)^n\sum_{j=0}^{n-1}\frac{(x/2)^j}{(n-j)j!}J_j(x) + \frac{2}{\pi}\left[\ln\left(\frac{x}{2}\right) - \psi(n+1)\right]J_n(x)$$
$$-\frac{2}{\pi}\sum_{j=1}^{\infty}\frac{(-1)^j(n+2j)}{j(n+j)}J_{n+2j}(x) \qquad n = 0, 1, 2, \ldots$$

and the $n = 0$ case reduces to

54:6:5
$$\frac{\pi}{2}Y_0(x) = \left[\ln\left(\frac{x}{2}\right)+\gamma\right]J_0(x) - 2\sum_{j=1}^{\infty}\frac{(-1)^j}{j}J_{2j}(x)$$

When x, but not $|\nu|$, is large, the Neumann function obeys the asymptotic expansion

54:6:6
$$Y_\nu(x) \sim \sqrt{\frac{2}{\pi x}}\left[P(\nu;x)\sin\left(x-\frac{\nu\pi}{2}-\frac{\pi}{4}\right) + Q(\nu;x)\cos\left(x-\frac{\nu\pi}{2}-\frac{\pi}{4}\right)\right] \qquad x \to \infty$$

where $P(\nu;x)$ and $Q(\nu;x)$ are the functions discussed in Section 53:6.

54:7 PARTICULAR VALUES

		$x = 0$	$x = \infty$
$Y_\nu(x)$	$\nu > \frac{-1}{2}$	$-\infty$	0
$Y_{-1/2}(x)$, $Y_{-3/2}(x)$, $Y_{-5/2}(x)$, ...		0	0
$Y_\nu(x)$	$\frac{-3}{2} < \nu < \frac{-1}{2}, \frac{-7}{2} < \nu < \frac{-5}{2}, \frac{-11}{2} < \nu < \frac{-9}{2}, \ldots$	$+\infty$	0
$Y_\nu(x)$	$\frac{-5}{2} < \nu < \frac{-3}{2}, \frac{-9}{2} < \nu < \frac{-7}{2}, \frac{-13}{2} < \nu < \frac{-11}{2}, \ldots$	$-\infty$	0

The values $y_{\nu;1}$, $y_{\nu;2}$, $y_{\nu;3}$, ... of the argument x that cause $Y_\nu(x)$ to acquire the value zero are known as the *zeros of the Neumann function*, whereas the values $y'_{\nu;1}$, $y'_{\nu;2}$, $y'_{\nu;3}$, ... of the argument that cause $Y_\nu(x)$ to exhibit a local maximum or minimum are called *extrema of the Neumann function*. Qualitatively, $y_{\nu;k}$ and $y'_{\nu;k}$ are very similar to $j_{\nu;k}$ and $j'_{\nu;k}$ so that Section 53:7 contains useful information on these zeros and extrema. Numerically, the relationships

54:7:1 $$\nu \le j'_{\nu;1} < y_{\nu;1} < y'_{\nu;1} < j_{\nu;1} < j'_{\nu;2} < y_{\nu;2} < y'_{\nu;2} < j_{\nu;2} < j'_{\nu;3} < \cdots$$

hold, and for the more remote zeros and extrema:

54:7:2 $$y_{\nu;k} \simeq j'_{\nu;k} \simeq \left(\frac{\nu}{2} + k - \frac{3}{4}\right)\pi \qquad \text{large } k$$

54:7:3 $$y'_{\nu;k} \simeq j_{\nu;k} \simeq \left(\frac{\nu}{2} + k - \frac{1}{4}\right)\pi \qquad \text{large } k$$

For large positive orders, the first zero of the Neumann function is approximated by

54:7:4 $$y_{\nu;1} \simeq \nu + 0.9316\nu^{1/3} + 0.260\nu^{-1/3} \qquad \text{large } \nu$$

while the first extremum, a maximum, obeys the approximation

54:7:5 $$y'_{\nu;1} \simeq \nu + 1.8211\nu^{1/3} + 0.940\nu^{-1/3} \qquad \text{large } \nu$$

54:8 NUMERICAL VALUES

In this section we present a lengthy algorithm for calculating $Y_\nu(x)$ for all orders ν and for any nonnegative argument x. The precision is generally 24 bits (i.e., the relative error does not exceed 6×10^{-8}) except in the immediate vicinity of a zero of the Neumann function. Except when $\nu = \frac{-1}{2}, \frac{-3}{2}, \frac{-5}{2}, \ldots$, $Y_\nu(0)$ equals $-\infty$ or $+\infty$; the algorithm returns 0, -10^{99} or 10^{99}, as appropriate, when $x = 0$ is input. The algorithm is long because it employs three distinct procedures for calculating $Y_\nu(x)$, depending on the values of ν and x.

When ν is the nonnegative integer n, the algorithm employs expansion 54:6:2, which may be rewritten as

54:8:1 $$\pi Y_n(x) = h^n \sum_{j=0}^{\infty} \frac{2\ln(h) + 2\gamma - \phi(j) - \phi(j+n)}{j!(n+j)!}(-h^2)^j - h^{n-2}\sum_{j=0}^{n-1}\frac{j!}{(n-j-1)!}\left(\frac{1}{h^2}\right)^j$$

where $h = x/2$, $\phi(0) = 0$ and $\phi(k) = 1 + \frac{1}{2} + \frac{1}{3} + \cdots + 1/k$. A satisfactory approximation to the right-hand side of this equation is given by the finite sum $S_J = 2t_0 + 2t_1 + 2t_2 + \cdots + 2t_{n-1} + 2t_n + \cdots + 2t_{J-1} + t_J$ for some suitably large J. The general term, t_j, in this summation may be expressed as $A_{n;j}B_{n;j} - C_{n;j}$ where $A_{0;0} = 2\ln(h) + 2\gamma$, $B_{0;0} = \frac{1}{2}$, $C_{0;0} = 0$ and $C_{1;0} = 1/(2h)$; for $m = 1, 2, 3, \ldots, n$, $A_{m;0} = A_{m-1;0} - (1/m)$, $B_{m;0} = hB_{m-1;0}/m$ and $C_{m+1;0} = hC_{m;0}/m$; and for $k = 1, 2, 3, \ldots, j$, $A_{n;k} = A_{n;k-1} - (1/k) - [1/(n+k)]$, $B_{n;k} = -h^2B_{n;k-1}/k(n+k)$ and $C_{n;k} = k(n-k)C_{n;k-1}/h^2$. For a sufficiently large J, the terms $2t_J$, $2t_{J+1}$, $2t_{J+2}$, ... will alternate in sign and progressively diminish in magnitude. It then follows from the properties of convergent alternating series [Section 0:6] that the error in approximating $\pi Y_n(x)$ by S_J cannot exceed $|t_J|$ in magnitude. The algorithm therefore tests, for $j = 1, 2, 3, \ldots$, until $|t_j|/|s_j| \le 5 \times 10^{-8}$. Additionally, three other tests must be passed before J can be identified with certitude. First, j must be at least equal to $n - 1$, to ensure that all terms in the polynomial component of 54:8:1 have been included. Second, alternation of sign will not occur consistently until $\phi(j) + \phi(n+j)$ exceeds $2\ln(h) + 2$: this is guaranteed if the sign of t_j is negative for even j. Third, it is necessary to confirm that the t terms are, indeed, alternating and convergent.

Relationship 54:5:1 is invoked when the order is a negative integer.

The asymptotic expansion 54:6:6 is employed whenever x exceeds both 8 and $5.6 + 0.7|\nu|$, or for all x when ν is an odd multiple of $\pm\frac{1}{2}$. Provided $(j + \frac{1}{2})^2 > \nu^2$ the terms in the series for $P(\nu;x)$ and $Q(\nu;x)$ alternate after the jth term, t_j. Moreover, t_j progressively diminishes in magnitude provided $j < J = x + \frac{1}{2} + \sqrt{x^2 + x + \nu^2}$. Incorporating these two provisos, the algorithm appends terms to the series for $P(\nu;x)\sqrt{2/\pi x}$ and $Q(\nu;x)\sqrt{2/\pi x}$ until the jth term satisfies the inequality $|t_j| < 5 \times 10^{-8}\{P(\nu;x)\sin[x - \nu\pi/2 - \pi/4] + Q(\nu;x)\cos[x - \nu\pi/2 - \pi/4]\}\sqrt{2/\pi x}$.

Storage needed: ν, f, h, A, B, C, t, p, n, q, s, j and J

Input restrictions: Order ν may take any value but argument x may not be negative.

Use degree mode or change 180 to π and 90 to $\pi/2$.

Input ν >>>>>>>
Set $f = -10^{99}$
Input x >>>>>>>
Set $h = x/2$
If $\mathrm{frac}\left(\nu + \dfrac{1}{2}\right) = 0$ go to (16)
If $h \neq 0$ go to (1)
If $\nu > 0$ go to (19)
Replace f by $f \cos\left[180\,\mathrm{Int}\left(\dfrac{1}{2} - \nu\right)\right]$
Go to (19)
(1) If $h \le 4$ go to (2)
If $h > 0.35(8 + |\nu|)$ go to (17)
(2) If $\mathrm{frac}(\nu) = 0$ go to (11)
Set $B = \cos(180\nu)$
Set $A = C = f = t = 1$
If $\nu > 0$ go to (3)
Set $A = C = B$
Set $B = 1$
Replace ν by $-\nu$
(3) Set $p = \nu$
Set $n = \mathrm{Int}\left(\dfrac{1}{2} + \nu\right)$
(4) Set $q = h^{-p}$
(5) Replace q by q/p
Replace p by $p + 1$
If $p < 3$ go to (5)
Set $s = \left[1 + \dfrac{2}{7p^2}\left(\dfrac{2}{3p^2} - 1\right)\right] \Big/ 30p^2$
Replace s by $\dfrac{1 - s}{12p} + p[\ln(p) - 1]$
Replace s by $-q \exp(s)\sqrt{2/\pi p}$
If $f < 0$ go to (6)
If $f = 0$ go to (7)
Replace A by As
Set $f = -1$
Set $p = -\nu$
Go to (4)
(6) Replace B by Bs
Set $f = j = 0$
Set $p = \nu - n$
Go to (4)
(7) Replace C by Cs
If $n = 0$ go to (9)
(8) Replace j by $j + 1$
Replace f by $f + t$
Replace C by Ch/j
Replace t by $-th^2/j(j - \nu)$
If $j \neq n$ go to (8)
(9) Replace A by Af
Set $j = f = 0$
(10) Replace f by $f + (B + C)$

Replace j by $j + 1$
Replace B by $-Bh^2/j(j + \nu)$
Replace C by $-Ch^2/(j + n)(j + n - \nu)$
If $j < 4 + 4h$ go to (10)
Replace f by $f + A + \dfrac{B + C}{2}$
Go to (19)
(11) Set $f = q = 1$
If $\nu \geq 0$ go to (12)
Set $f = \cos(180\nu)$
Replace ν by $-\nu$
(12) Set $A = 2\ln(1.78107242h)$
Set $B = 1/2$
Set $C = 0$
If $\nu = 0$ go to (14)
Replace A by $A - 1$
Replace B by Bh
Set $C = 1/(2h)$
If $\nu = 1$ go to (14)
Set $j = 1$
(13) Replace C by Ch/j
Replace j by $j + 1$
Replace B by Bh/j
Replace A by $A - \dfrac{1}{j}$
If $j \neq \nu$ go to (13)
(14) Set $s = t = AB - C$
Set $j = 0$
(15) Replace j by $j + 1$
Replace q by $-q$
Replace s by $s + t$
Replace A by $A - \dfrac{1}{j} - \dfrac{1}{j + \nu}$
Replace B by $-h^2B/j(\nu + j)$
Replace C by $j(\nu - j)C/h^2$
Set $p = t$
Set $t = AB - C$
Replace s by $s + t$
If $|s| \leq 2 \times 10^7|t|$ go to (15)
If $j < \nu$ go to (15)
If $qt > 0$ go to (15)
If $qt < -qp$ go to (15)
Replace f by fs/π
Go to (19)
(16) If $h \neq 0$ go to (17)
If $\nu > 0$ go to (19)
Set $f = 0$
Go to (19)
(17) Set $A = \nu - \dfrac{1}{2}$
Set $B = -1 - A$
Set $C = 2h + \dfrac{1}{2}$

Set $J = C + \sqrt{C^2 - AB}$
Set $s = 90\left(B + \frac{4h}{\pi}\right)$
Set $t = 1/\sqrt{\pi h}$
Set $p = q = j = 0$
(18) Replace p by $p + t\cos(90j)$
Replace q by $q - t\sin(90j)$
Replace j by $j + 1$
Replace A by $A + 1$
Replace B by $B + 1$
Replace t by $tAB/4hj$
If $AB < 0$ go to (18)
Set $f = p\sin(s) + q\cos(s)$
If $|f| > 2 \times 10^7\,|t|$ go to (19)
If $j < J$ go to (18)
(19) Output f
$f \simeq Y_\nu(x)$ <<<<<<

Test values:
$Y_{1/3}(4.93837580) = -0.163661435$
$Y_6(0) = -\infty$
$Y_3(1) = -5.82151761$
$Y_{-1}(0) = \infty$
$Y_{50}(1) = -2.19114281 \times 10^{77}$
$Y_{-1}(5) = -0.147863143$
$Y_{-9/2}(0) = 0$
$Y_{5/2}(2.6) = 0.530168510$
$Y_7(50) = 0.0959120278$

The third routine is used by the algorithm when $x \le 8$ or when $x \le 5.6 + 0.7|\nu|$, except when ν is an odd multiple of $\pm\frac{1}{2}$. It employs expansion 54:6:1 with the infinite series terminated empirically at $j = \text{Int}(4 + 2x)$. The required gamma functions are calculated by a procedure that is essentially that utilized in Section 43:8.

54:9 APPROXIMATIONS

For a range of integer orders and moderate arguments, the polynomial term in expansion 54:6:2 is dominant. This leads to the approximation

54:9:1 $$Y_n(x) \simeq \frac{-1}{\pi}\sum_{j=0}^{n-1}\frac{(n-j-1)!}{j!}\left(\frac{x}{2}\right)^{2j-n} \qquad \text{8-bit precision} \qquad x < 0.73(n-1) \qquad n = 5, 6, 7, \ldots$$

54:10 OPERATIONS OF THE CALCULUS

The derivative of the Neumann function may be expressed in the alternative forms

54:10:1 $$\frac{d}{dx}Y_\nu(x) = \frac{Y_{\nu-1}(x) - Y_{\nu+1}(x)}{2} = Y_{\nu-1}(x) - \frac{\nu}{x}Y_\nu(x) = \frac{\nu}{x}Y_\nu(x) - Y_{\nu+1}(x)$$

and we have the simpler relationships

54:10:2 $$\frac{d}{dx}[x^{\pm\nu}Y_\nu(x)] = \pm x^{\pm\nu}Y_{\nu\mp1}(x)$$

These results are identical to the corresponding equations for Bessel functions [Section 53:10].

Indefinite integrals include

54:10:3 $$\int_0^x t^\nu Y_{\nu-1}(t)dt = x^\nu Y_\nu(x) + \frac{2^\nu\Gamma(\nu)}{\pi} \qquad \nu > 0$$

The definite integral

54:10:4 $$\int_0^\infty Y_\nu(t)dt = -\tan(\nu\pi/2) \qquad -1 < \nu < 1$$

shows that the infinite integral of Y_0 is zero. A more general definite integral is

54:10:5 $$\int_0^\infty t^\mu Y_\nu(t)dt = \frac{2^\mu}{\pi}\Gamma\left(\frac{1+\mu+\nu}{2}\right)\Gamma\left(\frac{1+\mu-\nu}{2}\right)\sin\left(\frac{\pi\mu-\pi\nu}{2}\right) \qquad \mu<\frac{1}{2} \qquad \mu\pm\nu>-1$$

and other definite integrals are listed by Gradshteyn and Ryzhik [Sections 6.5–6.7].

54:11 COMPLEX ARGUMENT

Because the Neumann function is complex even for real negative argument

54:11:1 $$Y_\nu(-x) = \cos(\nu\pi)\,Y_\nu(x) + i[\cos(\nu\pi)\,J_\nu(x) + J_{-\nu}(x)]$$

we omit the general formula for $Y_\nu(x+iy)$ and present only that for the Neumann function of imaginary argument

54:11:2 $$Y_\nu(iy) = \frac{2}{\pi}\cos\left(\frac{\nu\pi}{2}\right)K_\nu(y) - \sin\left(\frac{\nu\pi}{2}\right)I_\nu(y) + i\left[\cos\left(\frac{\nu\pi}{2}\right)I_\nu(y) - \frac{2}{\pi}\sin\left(\frac{\nu\pi}{2}\right)K_\nu(y)\right]$$

In these equations J_ν, K_ν and I_ν are the functions addressed in Chapters 53, 51 and 50.

54:12 GENERALIZATIONS

In light of the limiting operation

54:12:1 $$\lim_{\lambda\to 0}\{\lambda^\nu Q_{1/2}^{(\nu)}(\cos(\lambda x))\} = \frac{-\pi}{2}Y_{-\nu}(x) \qquad \nu>0 \qquad x>0$$

the Neumann function can be regarded as a special limiting case of the second Legendre function [Chapter 59].

54:13 COGNATE FUNCTIONS

The two *Hankel functions*, also known as *Bessel functions of the third kind*, are generally complex functions defined as

54:13:1 $$J_\nu(x) \pm i\,Y_\nu(x)$$

They satisfy Bessel's equation and are sometimes preferable to J_ν or Y_ν in such solutions.

CHAPTER 55

THE KELVIN FUNCTIONS

Of the four Kelvin functions, $\mathrm{ber}_\nu(x)$, $\mathrm{bei}_\nu(x)$, $\mathrm{ker}_\nu(x)$ and $\mathrm{kei}_\nu(x)$, those with zero order, $\nu = 0$, are most important and are emphasized in this chapter.

55:1 NOTATION

These functions are named after Lord Kelvin, the title adopted by William Thomson (1824–1907) on his elevation to the peerage. Accordingly, the name *Thomson function* is given to these functions by those with distaste for the aristocracy.

The initial letter, b or k (after Bessel and Kelvin), is sometimes capitalized. The terminal letter, r or i, refers to "real" or "imaginary," as in definitions 55:3:3–55:3:6. The absence of a subscript, as in $\mathrm{ber}(x)$ and $\mathrm{kei}(x)$, indicates an order of zero:

55:1:1 $$\mathrm{ker}(x) = \mathrm{ker}_0(x), \ \mathrm{bei}(x) = \mathrm{bei}_0(x), \text{ etc.}$$

55:2 BEHAVIOR

For unrestricted values of ν and x, the functions $\mathrm{ber}_\nu(x)$, $\mathrm{bei}_\nu(x)$, $\mathrm{ker}_\nu(x)$ and $\mathrm{kei}_\nu(x)$ display rather complicated behaviors, and a comprehensive verbal description wil not be attempted. Moreover, we shall restrict attention to arguments $x \geq 0$, although ber_ν and bei_ν are real for negative arguments if ν is an integer.

The four Kelvin functions are oscillatory functions of x when ν is in the neighborhood of zero, but the onset of oscillations becomes increasingly delayed as $|\nu|$ increases. Once established, the periods of the oscillations approach $\sqrt{8}\pi$ for all four functions, but while the amplitudes increase rapidly for $\mathrm{ber}_\nu(x)$ and $\mathrm{bei}_\nu(x)$, those of $\mathrm{ker}_\nu(x)$ and $\mathrm{kei}_\nu(x)$ decline as $x \to \infty$.

The magnitudes acquired by $\mathrm{ber}(x)$, $\mathrm{bei}(x)$, $\mathrm{ker}(x)$ and $\mathrm{kei}(x)$ are so diverse that it is impractical to plot the four zero-order Kelvin functions on the same map, except for arguments in the rather narrow range displayed in Figure 55-1. For larger arguments we make use of the asymptotic properties of the Kelvin functions by mapping, in Figure 55-2, the curves labeled $\mathrm{A} = \sqrt{2\pi x}\exp(-x/\sqrt{2})\,\mathrm{ber}(x)$, $\mathrm{B} = \sqrt{2\pi x}\exp(-x/\sqrt{2})\,\mathrm{bei}(x)$, $\mathrm{C} = \sqrt{2x/\pi}\exp(x/\sqrt{2})\,\mathrm{ker}(x)$ and $\mathrm{D} = \sqrt{2x/\pi}\exp(x/\sqrt{2})\,\mathrm{kei}(x)$. As is evident from this diagram, or from equations 55:9:15–55:9:18, these four products approach pure sinusoids [Chapter 32] at large arguments.

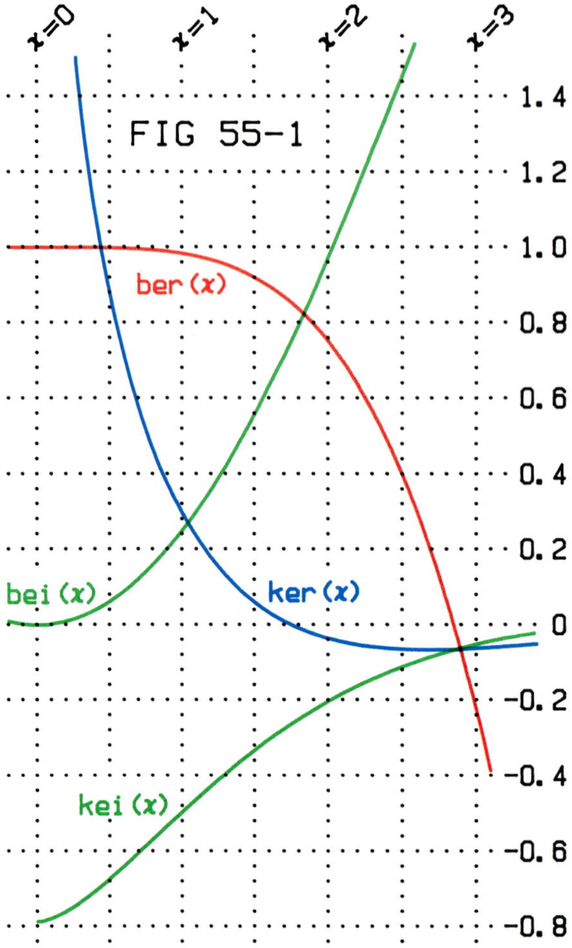

55:3 DEFINITIONS

The general ber_ν and bei_ν functions are defined in terms of Bessel functions of comlex argument, as

55:3:1 $$\mathrm{ber}_\nu(x) = \frac{1}{2}\mathrm{J}_\nu\left(\frac{-x}{\sqrt{2}} - \frac{ix}{\sqrt{2}}\right) + \frac{1}{2}\mathrm{J}_\nu\left(\frac{-x}{\sqrt{2}} + \frac{ix}{\sqrt{2}}\right)$$

55:3:2 $$\mathrm{bei}_\nu(x) = \frac{i}{2}\mathrm{J}_\nu\left(\frac{-x}{\sqrt{2}} - \frac{ix}{\sqrt{2}}\right) - \frac{i}{2}\mathrm{J}_\nu\left(\frac{-x}{\sqrt{2}} + \frac{ix}{\sqrt{2}}\right)$$

Although it is not immediately apparent, each of these formulas defines a function that is wholly real for nonnegative real x.

If we use the symbols Re{ } and Im{ } to signify "the real part of" and "the imaginary part of," so that $x = \mathrm{Re}\{x + iy\}$ and $y = \mathrm{Im}\{x + iy\}$, we may define

55:3:3 $$\mathrm{ber}_\nu(x) = \mathrm{Re}\left\{\mathrm{J}_\nu\left(\frac{-x}{\sqrt{2}} + \frac{ix}{\sqrt{2}}\right)\right\} = \mathrm{Re}\left\{\left[\cos\left(\frac{\nu\pi}{2}\right) + i\sin\left(\frac{\nu\pi}{2}\right)\right]\mathrm{I}_\nu\left(\frac{x}{\sqrt{2}} + \frac{ix}{\sqrt{2}}\right)\right\}$$

55:3:4 $$\mathrm{bei}_\nu(x) = \mathrm{Im}\left\{\mathrm{J}_\nu\left(\frac{-x}{\sqrt{2}} + \frac{ix}{\sqrt{2}}\right)\right\} = \mathrm{Im}\left\{\left[\cos\left(\frac{\nu\pi}{2}\right) + i\sin\left(\frac{\nu\pi}{2}\right)\right]\mathrm{I}_\nu\left(\frac{x}{\sqrt{2}} + \frac{ix}{\sqrt{2}}\right)\right\}$$

in terms of the Bessel function [Chapter 53] or hyperbolic Bessel function [Chapter 50] of argument whose real and imaginary parts are of equal magnitude. Similar formulas involving the Basset function [Chapter 51] define the ker_ν and kei_ν functions

55:3:5 $$\mathrm{ker}_\nu(x) = \mathrm{Re}\left\{\left[\cos\left(\frac{\nu\pi}{2}\right) - i\sin\left(\frac{\nu\pi}{2}\right)\right]\mathrm{K}_\nu\left(\frac{x}{\sqrt{2}} + \frac{ix}{\sqrt{2}}\right)\right\}$$

55:3:6 $$\mathrm{kei}_\nu(x) = \mathrm{Im}\left\{\left[\cos\left(\frac{\nu\pi}{2}\right) - i\sin\left(\frac{\nu\pi}{2}\right)\right]\mathrm{K}_\nu\left(\frac{x}{\sqrt{2}} + \frac{ix}{\sqrt{2}}\right)\right\}$$

The ker and kei Kelvin functions of zero order may be defined as the following definite integrals:

55:3:7 $$\mathrm{ker}(x) = -\frac{x^2}{8}\int_0^\infty \mathrm{Ci}\left(\frac{1}{t}\right)\exp\left(\frac{-x^2t}{4}\right)dt = \int_0^\infty \frac{t^3\mathrm{J}_0(xt)}{1+t^4}\,dt = \frac{x}{4}\int_0^\infty \ln(1+t^4)\,\mathrm{J}_1(xt)dt$$

55:3:8 $$\mathrm{kei}(x) = \frac{x^2}{8}\int_0^\infty \left[\mathrm{Si}\left(\frac{1}{t}\right) - \frac{\pi}{2}\right]\exp\left(\frac{-x^2t}{4}\right)dt = -\int_0^\infty \frac{t\mathrm{J}_0(xt)}{1+t^4}\,dt = \frac{-x}{2}\int_0^\infty \arctan(t^2)\,\mathrm{J}_1(xt)dt$$

in terms of functions discussed in Chapters 38 and 52.

55:4 SPECIAL CASES

Kelvin functions of orders equal to odd multiples of $\pm 1/2$ may be expressed as elementary transcendental functions. For example:

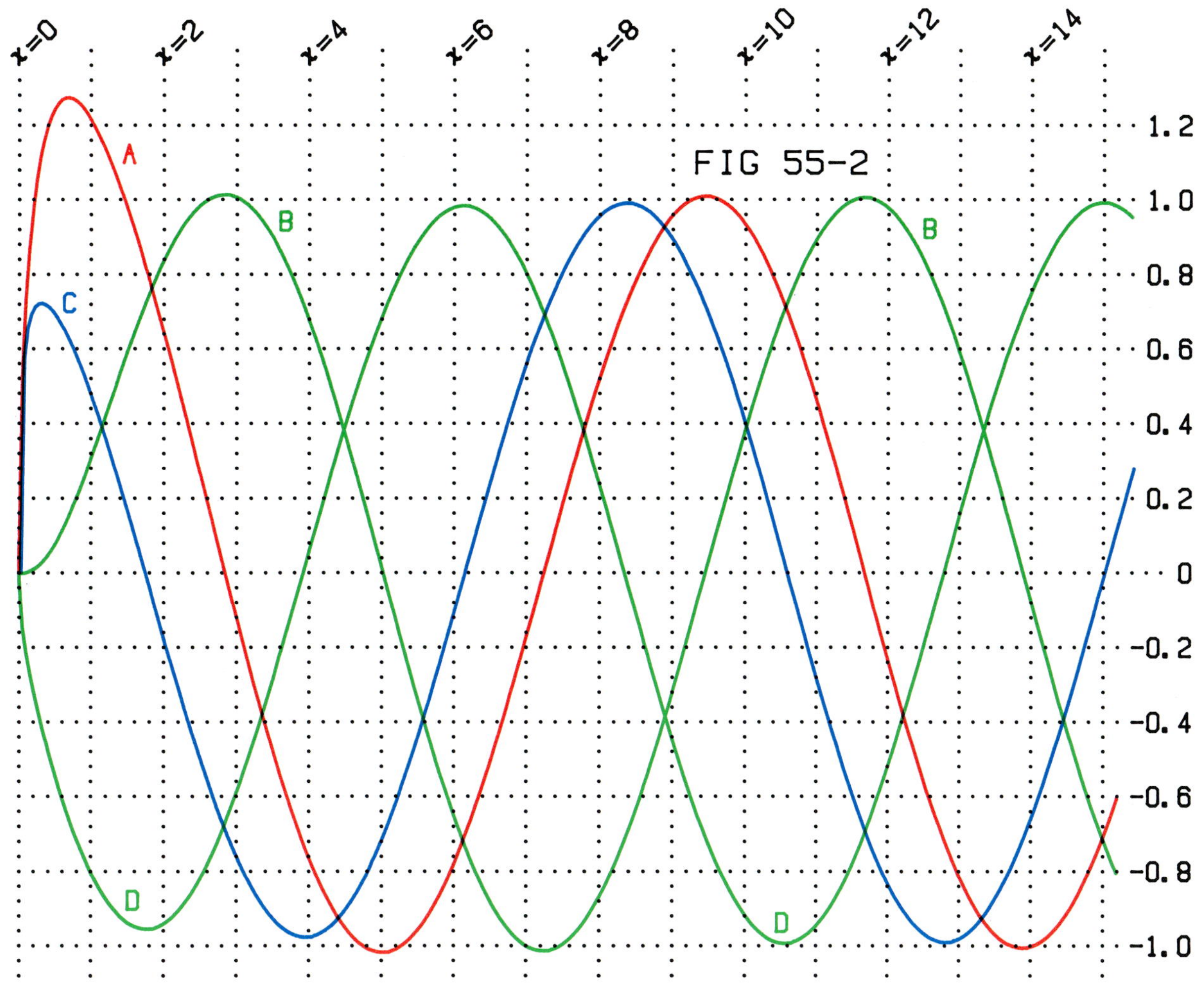

FIG 55-2

55:4:1 $$\mathrm{ker}_{1/2}(x) = -\sqrt{\frac{\pi}{2x}}\exp\left(\frac{-x}{\sqrt{2}}\right)\sin\left(\frac{x}{\sqrt{2}} - \frac{\pi}{8}\right)$$

55:4:2 $$\mathrm{kei}_{1/2}(x) = -\sqrt{\frac{\pi}{2x}}\exp\left(\frac{-x}{\sqrt{2}}\right)\cos\left(\frac{x}{\sqrt{2}} - \frac{\pi}{8}\right)$$

55:4:3 $$\mathrm{ber}_{1/2}(x) = \frac{1}{\sqrt{2\pi x}}\exp\left(\frac{x}{\sqrt{2}}\right)\cos\left(\frac{x}{\sqrt{2}} + \frac{\pi}{8}\right) + \frac{1}{\pi}\mathrm{kei}_{1/2}(x)$$

55:4:4 $$\mathrm{bei}_{1/2}(x) = \frac{1}{\sqrt{2\pi x}}\exp\left(\frac{x}{\sqrt{2}}\right)\sin\left(\frac{x}{\sqrt{2}} + \frac{\pi}{8}\right) + \frac{1}{\pi}\mathrm{ker}_{1/2}(x)$$

The Kelvin functions ber_ν and bei_ν become somewhat simpler when ν is a multiple of $\frac{1}{3}$. Only for these orders is $\mathrm{ber}_\nu(x)$ or $\mathrm{bei}_\nu(x)$ expressible as a simple power series. When the order is an even multiple of $\frac{1}{3}$ (i.e., a multiple of $\frac{2}{3}$), the power series contains terms of alternating signs $\cdots + - + - + - \cdots$; when the order is an odd multiple of $\frac{1}{3}$, signs occur in the sequence $\cdots + + - - + + - - \cdots$. Equations 55:6:1–55:6:4 provide details.

55:5 INTRARELATIONSHIPS

The order-reflection formulas

55:5:1 $$\mathrm{ber}_{-\nu}(x) = \cos(\nu\pi)\,\mathrm{ber}_\nu(x) + \sin(\nu\pi)\,\mathrm{bei}_\nu(x) + \frac{2}{\pi}\sin(\nu\pi)\,\mathrm{ker}_\nu(x)$$

55:5:2 $$\mathrm{bei}_{-\nu}(x) = -\sin(\nu\pi)\,\mathrm{ber}_\nu(x) + \cos(\nu\pi)\,\mathrm{bei}_\nu(x) + \frac{2}{\pi}\sin(\nu\pi)\,\mathrm{kei}_\nu(x)$$

55:5:3 $$\mathrm{ker}_{-\nu}(x) = \cos(\nu\pi)\,\mathrm{ker}_\nu(x) - \sin(\nu\pi)\,\mathrm{kei}_\nu(x)$$

55:5:4 $$\mathrm{kei}_{-\nu}(x) = \sin(\nu\pi)\,\mathrm{ker}_\nu(x) + \cos(\nu\pi)\,\mathrm{kei}_\nu(x)$$

all reduce to

55:5:5 $$\mathrm{f}_{-n}(x) = (-1)^n\,\mathrm{f}_n(x) \qquad \mathrm{f} = \mathrm{ber, bei, ker\ or\ kei} \qquad n = 0, \pm1, \pm2, \ldots$$

when the order is an integer.

Negative arguments generally render the Kelvin functions complex. Exceptions are ber_ν and bei_ν for integer order, to which the argument-reflection formula

55:5:6 $$\mathrm{f}_n(-x) = (-1)^n\,\mathrm{f}_n(x) \qquad \mathrm{f} = \mathrm{ber\ or\ bei} \qquad n = 0, \pm1, \pm2, \ldots$$

applies.

The recurrence relationships may be written

55:5:7 $$\mathrm{fer}_{\nu+1}(x) = \frac{-\nu\sqrt{2}}{x}[\mathrm{fer}_\nu(x) - \mathrm{fei}_\nu(x)] - \mathrm{fer}_{\nu-1}(x) \qquad \mathrm{fer} = \mathrm{ber\ or\ ker}$$

55:5:8 $$\mathrm{fei}_{\nu+1}(x) = \frac{-\nu\sqrt{2}}{x}[\mathrm{fei}_\nu(x) + \mathrm{fer}_\nu(x)] - \mathrm{fei}_{\nu-1}(x) \qquad \mathrm{fei} = \mathrm{bei\ or\ kei}$$

55:6 EXPANSIONS

All four Kelvin functions are expansible in terms of two series for which we shall use the abbreviations

55:6:1 $$\mathrm{Fe}_\nu(x) = \left(\frac{x}{2}\right)^\nu \sum_{j=0}^{\infty} \frac{(-x^4/16)^j}{(2j)!\Gamma(1+2j+\nu)} = \frac{\sqrt{\pi}(x/4)^\nu}{\Gamma\left(\frac{1}{2}+\frac{\nu}{2}\right)\Gamma\left(1+\frac{\nu}{2}\right)} \sum_{j=0}^{\infty} \frac{(-x^4/256)^j}{\left(\frac{1}{2}\right)_j (1)_j \left(\frac{1}{2}+\frac{\nu}{2}\right)_j \left(1+\frac{\nu}{2}\right)_j}$$

55:6:2 $$\mathrm{Ge}_\nu(x) = \left(\frac{x}{2}\right)^{2+\nu} \sum_{j=0}^{\infty} \frac{(-x^4/16)^j}{(1+2j)!\,\Gamma(2+2j+\nu)} = \frac{2\sqrt{\pi}(x/4)^{2+\nu}}{\Gamma\left(1+\frac{\nu}{2}\right)\Gamma\left(\frac{3}{2}+\frac{\nu}{2}\right)} \sum_{j=0}^{\infty} \frac{(-x^4/256)^j}{(1)_j \left(\frac{3}{2}\right)_j \left(1+\frac{\nu}{2}\right)_j \left(\frac{3}{2}+\frac{\nu}{2}\right)_j}$$

The second formulations above show that Fe_ν and Ge_ν are instances of $K = 0$, $L = 4$ hypergeometric functions [see Section 18:14]. When ν is a negative integer, formulas 55:6:1 and 55:6:2 are inconvenient and the identities $\mathrm{Fe}_{-2n}(x) = (-1)^n \mathrm{Fe}_{2n}(x)$, $\mathrm{Fe}_{1-2n}(x) = (-1)^n \mathrm{Ge}_{2n-1}(x)$, $\mathrm{Ge}_{-2n}(x) = (-1)^n \mathrm{Ge}_{2n}(x)$ and $\mathrm{Ge}_{1-2n}(x) = -(-1)^n \mathrm{Fe}_{2n-1}(x)$ may be used instead, where $n = 1, 2, 3, \ldots$. The expansions

55:6:3 $$\mathrm{ber}_\nu(x) = \left(\frac{x}{2}\right)^\nu \sum_{j=0}^{\infty} \frac{\cos\left(\frac{3\nu\pi}{4}+\frac{j\pi}{2}\right)}{j!\Gamma(1+j+\nu)}\left(\frac{x^2}{4}\right)^j = \cos\left(\frac{3\nu\pi}{4}\right)\mathrm{Fe}_\nu(x) - \sin\left(\frac{3\nu\pi}{4}\right)\mathrm{Ge}_\nu(x)$$

55:6:4 $$\mathrm{bei}_\nu(x) = \left(\frac{x}{2}\right)^\nu \sum_{j=0}^{\infty} \frac{\sin\left(\frac{3\nu\pi}{4}+\frac{j\pi}{2}\right)}{j!\Gamma(1+j+\nu)}\left(\frac{x^2}{4}\right)^j = \sin\left(\frac{3\nu\pi}{4}\right)\mathrm{Fe}_\nu(x) + \cos\left(\frac{3\nu\pi}{4}\right)\mathrm{Ge}_\nu(x)$$

are valid for all ν, but a restriction to noninteger ν is required to validate the expansions

55:6:5 $$\mathrm{ker}_\nu(x) = \frac{\pi\csc(\nu\pi)}{2}\left[\cos\left(\frac{3\nu\pi}{4}\right)\mathrm{Fe}_{-\nu}(x) + \sin\left(\frac{3\nu\pi}{4}\right)\mathrm{Ge}_{-\nu}(x) - \cos\left(\frac{\nu\pi}{4}\right)\mathrm{Fe}_\nu(x) - \sin\left(\frac{\nu\pi}{4}\right)\mathrm{Ge}_\nu(x)\right] \qquad \nu \neq 0, \pm1, \pm2, \ldots$$

55:6:6 $$\mathrm{kei}_\nu(x) = \frac{\pi\csc(\nu\pi)}{2}\left[-\sin\left(\frac{3\nu\pi}{4}\right)\mathrm{Fe}_{-\nu}(x) + \cos\left(\frac{3\nu\pi}{4}\right)\mathrm{Ge}_{-\nu}(x) + \sin\left(\frac{\nu\pi}{4}\right)\mathrm{Fe}_\nu(x) - \cos\left(\frac{\nu\pi}{4}\right)\mathrm{Ge}_\nu(x)\right] \qquad \nu \neq 0, \pm1, \pm2, \ldots$$

For nonnegative integer order, $\mathrm{ker}_\nu(x)$ and $\mathrm{kei}_\nu(x)$ are given by the expansions

55:6:7 $$\mathrm{ker}_n(x) = \ln\left(\frac{2}{x}\right)\mathrm{ber}_n(x) + \frac{\pi}{4}\mathrm{bei}_n(x) + \frac{1}{2}\left(\frac{2}{x}\right)^n \sum_{j=0}^{n-1} \frac{(n-j-1)!}{j!}\cos\left(\frac{3n\pi}{4}+\frac{j\pi}{2}\right)\left(\frac{x^2}{4}\right)^j$$
$$+ \frac{1}{2}\left(\frac{x}{2}\right)^n \sum_{j=0}^{\infty} \frac{\psi(1+j)+\psi(1+n+j)}{j!(n+j)!}\cos\left(\frac{3n\pi}{4}+\frac{j\pi}{2}\right)\left(\frac{x^2}{4}\right)^j \qquad n = 0, 1, 2, \ldots$$

55:6:8 $$\mathrm{kei}_n(x) = \ln\left(\frac{2}{x}\right)\mathrm{bei}_n(x) - \frac{\pi}{4}\mathrm{ber}_n(x) - \frac{1}{2}\left(\frac{2}{x}\right)^n \sum_{j=0}^{n-1} \frac{(n-j-1)!}{j!}\sin\left(\frac{3n\pi}{4}+\frac{j\pi}{2}\right)\left(\frac{x^2}{4}\right)^j$$
$$+ \frac{1}{2}\left(\frac{x}{2}\right)^n \sum_{j=0}^{\infty} \frac{\psi(1+j)+\psi(1+n+j)}{j!(n+j)!}\sin\left(\frac{3n\pi}{4}+\frac{j\pi}{4}\right)\left(\frac{x^2}{4}\right)^j \qquad n = 0, 1, 2, \ldots$$

where ψ is the digamma function of Chapter 44.

Asymptotic expansions of the Kelvin functions are provided by the formulas

55:6:9 $$\mathrm{ber}_\nu(x) \sim \frac{\exp(x/\sqrt{2})}{\sqrt{2\pi x}} \sum_{j=0} \frac{(\frac{1}{2}-\nu)_j(\frac{1}{2}+\nu)_j}{j!(2x)^j} \cos\left(\frac{x}{\sqrt{2}} + \frac{\nu\pi}{2} - \frac{j\pi}{4} - \frac{\pi}{8}\right) \qquad x \to \infty$$

55:6:10 $$\mathrm{bei}_\nu(x) \sim \frac{\exp(x/\sqrt{2})}{\sqrt{2\pi x}} \sum_{j=0} \frac{(\frac{1}{2}-\nu)_j(\frac{1}{2}+\nu)_j}{j!(2x)^j} \sin\left(\frac{x}{\sqrt{2}} + \frac{\nu\pi}{2} - \frac{j\pi}{4} - \frac{\pi}{8}\right) \qquad x \to \infty$$

55:6:11 $$\mathrm{ker}_\nu(x) \sim \sqrt{\frac{\pi}{2x}} \exp\left(\frac{-x}{\sqrt{2}}\right) \sum_{j=0} \frac{(\frac{1}{2}-\nu)_j(\frac{1}{2}+\nu)_j}{j!(-2x)^j} \cos\left(\frac{x}{\sqrt{2}} + \frac{\nu\pi}{2} + \frac{j\pi}{4} + \frac{\pi}{8}\right) \qquad x \to \infty$$

55:6:12 $$\mathrm{kei}_\nu(x) \sim -\sqrt{\frac{\pi}{2x}} \exp\left(\frac{-x}{\sqrt{2}}\right) \sum_{j=0} \frac{(\frac{1}{2}-\nu)_j(\frac{1}{2}+\nu)_j}{j!(-2x)^j} \sin\left(\frac{x}{\sqrt{2}} + \frac{\nu\pi}{2} + \frac{j\pi}{4} + \frac{\pi}{8}\right) \qquad x \to \infty$$

with suitably chosen upper limits to the summation index j. Although the series in 55:6:9 does, indeed, asymptotically represent $\mathrm{ber}_\nu(x)$ for large x, it more closely represents $\mathrm{ber}_\nu(x) + [\sin(2\nu\pi)\,\mathrm{ker}_\nu(x) + \cos(2\nu\pi)\,\mathrm{kei}_\nu(x)]/\pi$. Similarly, a better approximation to the series in 55:6:10 is $\mathrm{bei}_\nu(x) - [\cos(2\nu\pi)\,\mathrm{ker}_\nu(x) - \sin(2\nu\pi)\,\mathrm{kei}_\nu(x)]/\pi$.

When $\nu = 0$, the various expansions simplify to

55:6:13 $$\mathrm{ber}(x) = 1 - \frac{x^4}{64} + \frac{x^8}{147456} - \cdots + \frac{(-x^4/16)^j}{[(2j)!]^2} + \cdots$$

55:6:14 $$\mathrm{bei}(x) = \frac{x^2}{4} - \frac{x^6}{2304} + \frac{x^{10}}{14745600} - \cdots + \frac{x^2(-x^4/16)^j}{4[(2j+1)!]^2} + \cdots$$

55:6:15 $$\mathrm{ker}(x) = \left[\ln\left(\frac{2}{x}\right) - \gamma\right]\mathrm{ber}(x) + \frac{\pi}{4}\mathrm{bei}(x) + \sum_{j=1}^{\infty}\left(1 + \frac{1}{2} + \frac{1}{3} + \cdots + \frac{1}{2j}\right)\frac{(-x^4/16)^j}{[(2j)!]^2}$$

55:6:16 $$\mathrm{kei}(x) = \left[\ln\left(\frac{2}{x}\right) - \gamma\right]\mathrm{bei}(x) - \frac{\pi}{4}\mathrm{ber}(x) + \frac{x^2}{4}\sum_{j=0}^{\infty}\left(1 + \frac{1}{2} + \frac{1}{3} + \cdots + \frac{1}{2j+1}\right)\frac{(-x^4/16)^j}{[(2j+1)!]^2}$$

55:6:17 $$\begin{matrix}\mathrm{ber}\\ \mathrm{bei}\end{matrix}(x) \sim \frac{\exp(x/\sqrt{2})}{\sqrt{2\pi x}} \sum_{j=0}^{J} \frac{[(2j)!]^2}{(j!)^3(32x)^j} \begin{matrix}\cos\\ \sin\end{matrix}\left(\frac{x}{\sqrt{2}} - \frac{j\pi}{4} - \frac{\pi}{8}\right) \qquad J = \mathrm{Int}(2x) \qquad x \to \infty$$

55:6:18 $$\begin{matrix}\mathrm{ker}\\ \mathrm{kei}\end{matrix}(x) \sim \sqrt{\frac{\pi}{2x}} \exp\left(\frac{-x}{\sqrt{2}}\right) \sum_{j=0}^{J} \frac{[(2j)!]^2}{(j!)^3(32x)^j} \begin{matrix}\cos\\ \sin\end{matrix}\left(\frac{-x}{\sqrt{2}} + \frac{3j\pi}{4} - \frac{\pi}{8}\right) \qquad J = \mathrm{Int}(2x) \qquad x \to \infty$$

As well, $\mathrm{ber}(x)$ and $\mathrm{bei}(x)$ are expansible in terms of the functions of Chapters 49 and 52:

55:6:19 $$\mathrm{ber}(x) = \mathrm{I}_0\left(\frac{x}{\sqrt{2}}\right)\mathrm{J}_0\left(\frac{x}{\sqrt{2}}\right) - 2\mathrm{I}_2\left(\frac{x}{\sqrt{2}}\right)\mathrm{J}_2\left(\frac{x}{\sqrt{2}}\right) + 2\mathrm{I}_4\left(\frac{x}{\sqrt{2}}\right)\mathrm{J}_4\left(\frac{x}{\sqrt{2}}\right) - \cdots$$

$$= \sum_{j=-\infty}^{\infty} (-1)^j\, \mathrm{I}_{2j}\left(\frac{x}{\sqrt{2}}\right)\mathrm{J}_{2j}\left(\frac{x}{\sqrt{2}}\right)$$

55:6:20
$$\mathrm{bei}(x) = 2\mathrm{I}_1\left(\frac{x}{\sqrt{2}}\right)\mathrm{J}_1\left(\frac{x}{\sqrt{2}}\right) - 2\mathrm{I}_3\left(\frac{x}{\sqrt{2}}\right)\mathrm{J}_3\left(\frac{x}{\sqrt{2}}\right) + 2\mathrm{I}_5\left(\frac{x}{\sqrt{2}}\right)\mathrm{J}_5\left(\frac{x}{\sqrt{2}}\right) - \cdots$$
$$= \sum_{j=-\infty}^{\infty} (-1)^j \mathrm{I}_{2j+1}\left(\frac{x}{\sqrt{2}}\right)\mathrm{J}_{2j+1}\left(\frac{x}{\sqrt{2}}\right)$$

The sum of squares of the ber and bei functions is the simple series

55:6:21
$$\mathrm{ber}^2(x) + \mathrm{bei}^2(x) = 1 + \frac{x^4}{32} + \frac{x^8}{24576} + \cdots = \sum_{j=0}^{\infty} \frac{(x^4/16)^j}{(j!)^3\Gamma(j+\frac{1}{2})}$$

which is rapidly convergent for small arguments. Asymptotic expansions exist for $\mathrm{ber}^2(x) + \mathrm{bei}^2(x)$, for $\mathrm{ker}^2(x) + \mathrm{kei}^2(x)$ and for a number of similar functions; see Abramowitz and Stegun [Section 9.10] for these.

55:7 PARTICULAR VALUES

Kelvin functions display diverse behaviors as $x \to 0$, and this leads to very varied values for $\mathrm{ber}_\nu(0)$, $\mathrm{bei}_\nu(0)$, $\mathrm{ker}_\nu(0)$ and $\mathrm{kei}_\nu(0)$ as functions of ν. For the most part these particular values are $+\infty$, 0 or $-\infty$, but the following finite nonzero values are also realized:

55:7:1
$$\mathrm{ber}(0) = 1$$

55:7:2
$$\mathrm{ker}_{\pm 2}(0) = \frac{1}{2}$$

55:7:3
$$\mathrm{kei}(0) = -\frac{\pi}{4}$$

Figure 55-3 shows what value represents the limit of the four Kelvin functions of order ν as $x \to 0$. The color coding used in this diagram is that green represents zero, red represents $+\infty$, blue represents $-\infty$ and black represents one of the values given in equations 55:7:1–55:7:3. Notice the very elaborate behavior of $\mathrm{ber}_\nu(0)$ and $\mathrm{bei}_\nu(0)$ for negative ν. In this region these functions have a periodicity of 8 in their orders; that is:

55:7:4
$$\mathrm{f}_\nu(0) = \mathrm{f}_{\nu-8}(0) = \mathrm{f}_{\nu-16}(0) = \cdots \qquad \mathrm{f} = \mathrm{ber\ or\ bei} \qquad \nu < 0$$

As x approaches infinity, $\mathrm{ber}_\nu(x)$ and $\mathrm{bei}_\nu(x)$ oscillate with ever-increasing amplitude, but $\mathrm{ker}_\nu(x)$ and $\mathrm{kei}_\nu(x)$ converge towards zero:

55:7:5
$$\mathrm{ker}_\nu(\infty) = \mathrm{kei}_\nu(\infty) = 0$$

All four Kelvin functions have an infinite number of zeros. For a sufficiently large x, these are given by

55:7:6
$$\mathrm{ber}_\nu(r) = 0 \qquad r \simeq \sqrt{2}\pi\left(k - \frac{\nu}{2} - \frac{3}{8}\right) \qquad k = \text{large positive integer}$$

55:7:7
$$\mathrm{bei}_\nu(r) = 0 \qquad r \simeq \sqrt{2}\pi\left(k - \frac{\nu}{2} + \frac{1}{8}\right) \qquad k = \text{large positive integer}$$

55:7:8
$$\mathrm{ker}_\nu(r) = 0 \qquad r \simeq \sqrt{2}\pi\left(k - \frac{\nu}{2} - \frac{5}{8}\right) \qquad k = \text{large positive integer}$$

55:7:9
$$\mathrm{kei}_\nu(r) = 0 \qquad r \simeq \sqrt{2}\pi\left(k - \frac{\nu}{2} - \frac{1}{8}\right) \qquad k = \text{large positive integer}$$

The first few zeros for $\nu = 0$ are included in Table 55.7.1 together with argument and function values of early extrema, as well as the function values at the $x = 0$ origin. The final line in Table 55.7.1 provides an approximation formula for calculating the kth zero of each Kelvin function; it uses the abbreviations $a = \sqrt{2}(8k - 3)\pi$, $b = \sqrt{2}(8k + 1)\pi$, $c = \sqrt{2}(8k - 5)\pi$ and $d = \sqrt{2}(8k - 1)\pi$. Since the approximations are excellent, even for k as small as 4, the table permits *all* the zeros of the zero-order Kelvin functions to be estimated.

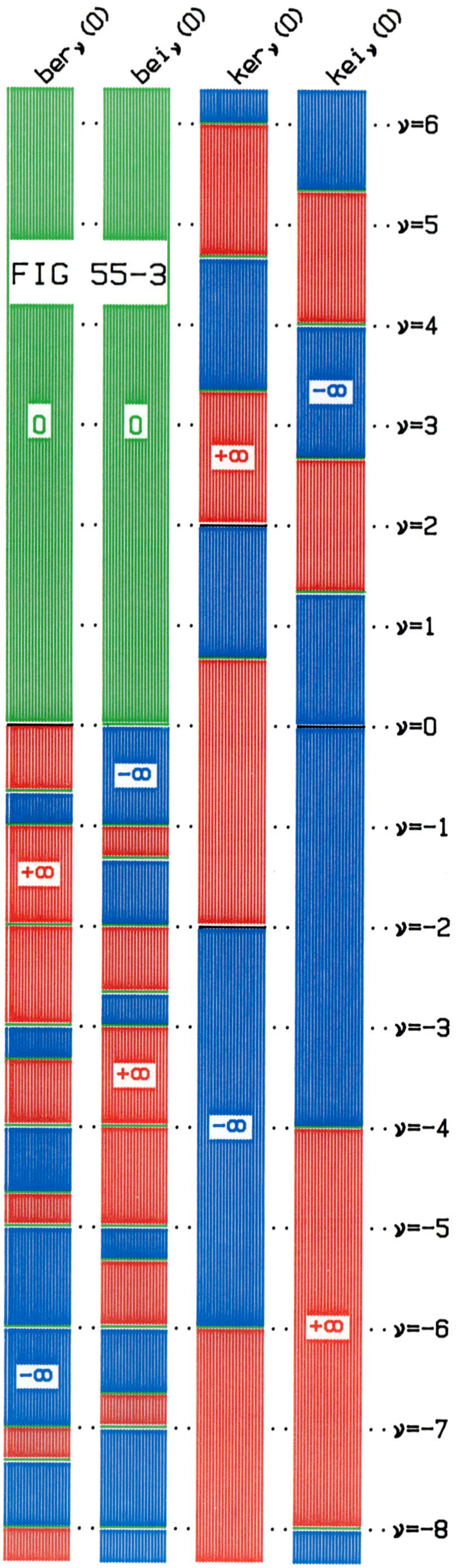
ber$_\nu$(0)
bei$_\nu$(0)
ker$_\nu$(0)
kei$_\nu$(0)
FIG 55-3
0
0
+∞
−∞
−∞
+∞
−∞
+∞
−∞
+∞
ν=6
ν=5
ν=4
ν=3
ν=2
ν=1
ν=0
ν=−1
ν=−2
ν=−3
ν=−4
ν=−5
ν=−6
ν=−7
ν=−8

Table 55.7.1

x	ber(x)	x	bei(x)	x	ker(x)	x	kei(x)
0	1.000000	0	0	0	∞	0	−0.7853982
2.848918	0	3.772674	2.346147	1.718543	0	3.914668	0
6.038711	−8.864036	5.026224	0	2.665845	−0.07102369	4.931811	0.01121607
7.238829	0	8.280989	−36.16540	6.127279	0	8.344225	0
10.51364	153.7818	9.455406	0	7.172120	0.001956681	9.404051	0.0003574233
11.67396	0	12.74215	670.1602	10.56294	0	12.78256	0
14.96845	−2968.681	13.89349	0	11.63219	-6.705969×10^{-5}	13.85827	1.280427×10^{-5}
16.11356	0	17.19343	−13305.52	15.00269	0	17.22314	0
$\frac{a}{8}+\frac{1}{a}+\frac{\sqrt{32}}{a^2}$	0	$\frac{b}{8}+\frac{1}{b}+\frac{\sqrt{32}}{b^2}$	0	$\frac{c}{8}+\frac{1}{c}-\frac{\sqrt{32}}{c^2}$	0	$\frac{d}{8}+\frac{1}{d}-\frac{\sqrt{32}}{d^2}$	0

55:8 NUMERICAL VALUES

The algorithm below calculates values of the zero order Kelvin function for all arguments $x \geq 0$. Its precision is 24 bits (relative error less than 6×10^{-8}), but some loss of significance may be encountered on computing devices that carry insufficient digits. The algorithm is based on expansions 55:6:13–55:6:18.

Storage needed: B, K, b, k, j, r, x, s and t

Set $B = K = b = k = j = r = 0$
Input x >>>>>>>
If $x < 8$ go to (1)
Set $r = 90\left(\frac{x\sqrt{2}}{\pi} - \frac{1}{4}\right)$
Set $s = -r - 45$
Set $t = 1/\sqrt{2x}$
Go to (4)
(1) Set $B = t = 1$
Set $s = 10^{99}$
If $x = 0$ go to (2)
Set $s = 0.115931516 - \ln(x)$
(2) Replace j by $j + 1$
Replace t by $t(x/2j)^2$
Replace r by $r + (1/j)$
Replace B by $B + t\cos(90j)$
Replace K by $K + rt\cos(90j)$
Replace b by $b + t\sin(90j)$
Replace k by $k + rt\sin(90j)$
If $j < 2x + 4$ go to (2)
Replace K by $K + sB + \pi b/4$
Replace k by $k + sb - \pi B/4$
Go to (5)
(3) Replace t by $\frac{t}{8x}\left(4j + \frac{1}{j+1}\right)$
Replace j by $j + 1$
Replace r by $r - 45$
Replace s by $s + 135$
(4) Replace B by $B + t\cos(r)$
Replace b by $b + t\sin(r)$

Use degree mode or alter 90 to $\pi/2$, 45 to $\pi/4$ and 135 to $3\pi/4$.

Input restriction: $x \geq 0$

Replace K by $K + t\cos(s)$
Replace k by $k + t\sin(s)$
If $j < 3 + \dfrac{70}{x}$ go to (3)
Set $t = (1/\sqrt{\pi})\exp(x/\sqrt{2})$
Replace K by K/t
Replace k by k/t
Replace B by $Bt - \dfrac{k}{\pi}$
Replace b by $bt + \dfrac{K}{\pi}$
(5) Output B
$B \simeq \mathrm{ber}(x)$ <<<<<
Output b
$b \simeq \mathrm{bei}(x)$ <<<<<
Output K
$K \simeq \mathrm{ker}(x)$ <<<<<
Output k
$k \simeq \mathrm{kei}(x)$ <<<<<

Test values:
ber(2) = 0.751734*183*
bei(2) = 0.972291*627*
ker(2) = −0.0416645*140*
kei(2) = −0.202400*068*
ber(10) = 138.840*467*
bei(10) = 56.3704*592*
ker(10) = 0.000129466*329*
kei(10) = −0.000307524*569*

55:9 APPROXIMATIONS

For small arguments, the following approximations generally hold:

55:9:1 $$\mathrm{ber}_\nu(x) \simeq \frac{\cos(3\nu\pi/4)}{\Gamma(1+\nu)}\left(\frac{x}{2}\right)^{\nu} - \frac{\sin(3\nu\pi/4)}{\Gamma(2+\nu)}\left(\frac{x}{2}\right)^{\nu+2} \qquad \nu \neq -1, -2, -3, \ldots$$

55:9:2 $$\mathrm{bei}_\nu(x) \simeq \frac{\sin(3\nu\pi/4)}{\Gamma(1+\nu)}\left(\frac{x}{2}\right)^{\nu} + \frac{\cos(3\nu\pi/4)}{\Gamma(2+\nu)}\left(\frac{x}{2}\right)^{\nu+2} \qquad \nu \neq -1, -2, -3, \ldots$$

55:9:3 $$\mathrm{ker}_\nu(x) \simeq \frac{\Gamma(\nu)\cos(3\nu\pi/4)}{2}\left(\frac{2}{x}\right)^{\nu} - \frac{\Gamma(\nu-1)\sin(3\nu\pi/4)}{2}\left(\frac{2}{x}\right)^{\nu-2} \qquad \nu > 2$$

55:9:4 $$\mathrm{ker}_\nu(x) \simeq \frac{1}{2} - \frac{\pi x^2}{16} \qquad \nu = \pm 2$$

55:9:5 $$\mathrm{ker}_\nu(x) \simeq \frac{\Gamma(\nu)\cos(3\nu\pi/4)}{2}\left(\frac{2}{x}\right)^{\nu} + \frac{\Gamma(-\nu)\cos(\nu\pi/4)}{2}\left(\frac{x}{2}\right)^{\nu} \qquad -2 < \nu < 2 \qquad \nu \neq 0, \pm 1$$

55:9:6 $$\mathrm{ker}_\nu(x) \simeq \frac{\mp 1}{x\sqrt{2}} \mp \frac{x}{\sqrt{8}}\left[\ln\left(\frac{2}{x}\right) + \frac{\pi}{4} + \gamma - \frac{1}{2}\right] \qquad \nu = \pm 1$$

55:9:7 $$\mathrm{ker}(x) \simeq \ln\left(\frac{2}{x}\right) - \gamma \qquad \nu = 0$$

55:9:8 $$\mathrm{ker}_\nu(x) \simeq \frac{\Gamma(-\nu)\cos(\nu\pi/4)}{2}\left(\frac{x}{2}\right)^{\nu} + \frac{\Gamma(-\nu-1)\sin(\nu\pi/4)}{2}\left(\frac{x}{2}\right)^{\nu+2} \qquad \nu < -2$$

55:9:9 $$\mathrm{kei}_\nu(x) = \frac{-\Gamma(\nu)\sin(3\nu\pi/4)}{2}\left(\frac{2}{x}\right)^{\nu} - \frac{\Gamma(\nu-1)\cos(3\nu\pi/4)}{2}\left(\frac{2}{x}\right)^{\nu-2} \qquad \nu > 2$$

55:9:10 $$\mathrm{kei}_\nu(x) \simeq \frac{2}{x^2} + \frac{x^2}{4}\left[\ln\left(\frac{2}{x}\right) - \gamma + \frac{1}{2}\right] \qquad \nu = \pm 2$$

55:9:11 $$\mathrm{kei}_\nu(x) \simeq \frac{-\Gamma(\nu)\sin(3\nu\pi/4)}{2}\left(\frac{2}{x}\right)^\nu - \frac{\Gamma(-\nu)\sin(\nu\pi/4)}{2}\left(\frac{x}{2}\right)^\nu \qquad -2<\nu<2 \qquad \nu \neq 0, \pm 1$$

55:9:12 $$\mathrm{kei}_\nu(x) \simeq \frac{\mp 1}{x\sqrt{2}} \pm \frac{x}{\sqrt{8}}\left[\ln\left(\frac{2}{x}\right) + \frac{\pi}{4} - \gamma + \frac{1}{2}\right] \qquad \nu = \pm 1$$

55:9:13 $$\mathrm{kei}(x) \simeq \frac{-\pi}{4} + \frac{x^2}{4}\left[\ln\left(\frac{2}{x}\right) - \gamma + \frac{1}{2}\right] \qquad \nu = 0$$

55:9:14 $$\mathrm{kei}_\nu(x) \simeq \frac{-\Gamma(-\nu)\sin(\nu\pi/4)}{2}\left(\frac{x}{2}\right)^\nu + \frac{\Gamma(-\nu-1)\cos(\nu\pi/4)}{2}\left(\frac{x}{2}\right)^{\nu+2} \qquad \nu < -2$$

and these may be supplemented by using equation 55:5:5 for cases of negative integer order. Generally, equations 55:9:1–55:9:14 give the first two terms in expansions of the Kelvin functions in ascending powers of the argument x. For certain special values of ν, however, some of these terms may vanish. The leading terms in equations 55:9:1, 55:9:3 and 55:9:5, for example, become zero when the order adopts one of the values $\frac{2}{3}$, 2, $\frac{10}{3}$, $\frac{14}{3}$, 6, ...

For large arguments, the approximations

55:9:15 $$\mathrm{ber}_\nu(x) \simeq \frac{1}{\sqrt{2\pi x}}\exp\left(\frac{x}{\sqrt{2}}\right)\cos\left(\frac{x}{\sqrt{2}} + \frac{\nu\pi}{2} - \frac{\pi}{8}\right) \qquad \text{large } x$$

55:9:16 $$\mathrm{bei}_\nu(x) \simeq \frac{1}{\sqrt{2\pi x}}\exp\left(\frac{x}{\sqrt{2}}\right)\sin\left(\frac{x}{\sqrt{2}} + \frac{\nu\pi}{2} - \frac{\pi}{8}\right) \qquad \text{large } x$$

55:9:17 $$\mathrm{ker}_\nu(x) \simeq \sqrt{\frac{\pi}{2x}}\exp\left(\frac{-x}{\sqrt{2}}\right)\cos\left(\frac{x}{\sqrt{2}} + \frac{\nu\pi}{2} + \frac{\pi}{8}\right) \qquad \text{large } x$$

55:9:18 $$\mathrm{kei}_\nu(x) \simeq \sqrt{\frac{\pi}{2x}}\exp\left(\frac{-x}{\sqrt{2}}\right)\sin\left(\frac{x}{\sqrt{2}} + \frac{\nu\pi}{2} + \frac{\pi}{8}\right) \qquad \text{large } x$$

hold. From these, one may derive the order-independent relationships

55:9:19 $$[\mathrm{ber}_\nu^2(x) + \mathrm{bei}_\nu^2(x)][\mathrm{ker}_\nu^2(x) + \mathrm{kei}_\nu^2(x)] \simeq \frac{1}{4x^2} \qquad \text{large } x$$

and

55:9:20 $$\mathrm{ber}_\nu(x)\,\mathrm{ker}_\nu(x) + \mathrm{bei}_\nu(x)\,\mathrm{kei}_\nu(x) = \mathrm{ber}_\nu(x)\,\mathrm{kei}_\nu(x) - \mathrm{bei}_\nu(x)\,\mathrm{ker}_\nu(x) \simeq \frac{1}{x\sqrt{8}} \qquad \text{large } x$$

55:10 OPERATIONS OF THE CALCULUS

The differentiation formulas

55:10:1 $$\frac{d}{dx}\mathrm{fer}_\nu(x) = \frac{1}{\sqrt{8}}[\mathrm{fer}_{\nu+1}(x) + \mathrm{fei}_{\nu+1}(x) - \mathrm{fer}_{\nu-1}(x) - \mathrm{fei}_{\nu-1}(x)] \qquad \mathrm{fer} = \mathrm{ber} \text{ or } \mathrm{ker}$$

55:10:2 $$\frac{d}{dx}\mathrm{fei}_\nu(x) = \frac{1}{\sqrt{8}}[\mathrm{fei}_{\nu+1}(x) - \mathrm{fer}_{\nu+1}(x) - \mathrm{fei}_{\nu-1}(x) + \mathrm{fer}_{\nu-1}(x)] \qquad \mathrm{fei} = \mathrm{bei} \text{ or } \mathrm{kei}$$

may be combined with recursions 55:5:7 and 55:5:8 to produce a number of alternative formulations.

Among indefinite integrals are

55:10:3 $$\int_0^x t^{1\pm\nu}\,\mathrm{fer}_\nu(t)\,dt = \frac{\mp x^{1\pm\nu}}{\sqrt{2}}[\mathrm{fer}_{\nu\pm1}(x) - \mathrm{fei}_{\nu\pm1}(x)]$$

while the following definite integrals establish links with the functions of Chapters 32 and 38:

55:10:4 $$\int_0^\infty \exp(-t/x)\,\mathrm{ber}(2\sqrt{t})\mathrm{d}t = x\cos(x)$$

55:10:5 $$\int_0^\infty \exp(-t/x)\,\mathrm{bei}(2\sqrt{t})\mathrm{d}t = x\sin(x)$$

55:10:6 $$\int_0^\infty \exp(-t/x)\,\mathrm{ker}(2\sqrt{t})\mathrm{d}t = \frac{-x}{2}\left\{\cos(x)\,\mathrm{Ci}(x) + \sin(x)\left[\mathrm{Si}(x) - \frac{\pi}{2}\right]\right\}$$

55:10:7 $$\int_0^\infty \exp(-t/x)\,\mathrm{kei}(2\sqrt{t})\mathrm{d}t = \frac{-x}{2}\left\{\sin(x)\,\mathrm{Ci}(x) - \cos(x)\left[\mathrm{Si}(x) - \frac{\pi}{2}\right]\right\}$$

Other indefinite integrals are given by Abramowitz and Stegun [Section 9.9] and other definite integrals by Gradshteyn and Ryzhik [Section 6.87].

55:11 COMPLEX ARGUMENT

Kelvin functions are seldom encountered with complex arguments, and this *Atlas* does not address this circumstance.

55:12 GENERALIZATIONS

If the complex variable $z = x + iy$ is represented by the polar equivalent $z = r\exp(i\theta)$, one sees from definitions 55:3:3 and 55:3:4 that $\mathrm{ber}_\nu(r)$ and $\mathrm{bei}_\nu(r)$ are the real and imaginary parts of $J_\nu(z)$ with $\theta = 3\pi/4$. Thus, one may generalize these Kelvin functions by allowing θ to adopt an arbitrary angular value.

55:13 COGNATE FUNCTIONS

Just as the ber_ν and bei_ν functions are defined in terms of the Bessel J_ν function to satisfy

55:13:1 $$\mathrm{ber}_\nu(x) + i\,\mathrm{bei}_\nu(x) = J_\nu\left(\frac{-x}{\sqrt{2}} + \frac{ix}{\sqrt{2}}\right)$$

so *Kelvin functions of the third kind* are defined by a similar relationship but with the Hankel function [Section 54:13] replacing the Bessel function. Thus:

55:13:2 $$\mathrm{her}_\nu(x) + i\,\mathrm{hei}_\nu(x) = J_\nu\left(\frac{-x}{\sqrt{2}} + \frac{ix}{\sqrt{2}}\right) + i\,Y_\nu\left(\frac{-x}{\sqrt{2}} + \frac{ix}{\sqrt{2}}\right)$$

One has the identities $\mathrm{her}_\nu(x) = (2/\pi)\,\mathrm{kei}_\nu(x)$ and $\mathrm{hei}_\nu(x) = (-2/\pi)\,\mathrm{ker}_\nu(x)$.

CHAPTER 56

THE AIRY FUNCTIONS Ai(x) AND Bi(x)

The Airy functions Ai(x) and Bi(x) are related to Bessel functions of order $\frac{1}{3}$ and $-\frac{1}{3}$, with rescaled arguments. The two so-called *auxiliary Airy functions* are also important and are briefly discussed in this chapter.

56:1 NOTATION

In formulas involving the Airy functions, the *auxiliary argument*

56:1:1 $$X = \frac{2}{3}(|x|)^{3/2}$$

is frequently more convenient than the usual argument x. Note that $dX = \sqrt{|x|}\,dx$.

There appears to be no standard notation for the auxiliary Airy functions. This *Atlas* employs fai(x) and gai(x) for the functions that Abramowitz and Stegun [Section 10.4] denote by f(x) and g(x).

56:2 BEHAVIOR

The functions Ai(x) and Bi(x) exist for all real arguments, but their behaviors depend crucially on the sign of x, as illustrated in Figure 56-1.

For $x \geq 0$, both Airy functions are positive, but whereas Bi(x) increases rapidly as $x \to \infty$, Ai(x) steadily decays towards zero. For negative arguments, Ai(x) and Bi(x) are oscillatory functions with oscillations whose frequencies gradually increase and whose amplitudes gradually decrease as $x \to -\infty$.

The auxiliary Airy functions are mapped in Figure 56-2. For positive arguments, both increase exponentially as $x \to \infty$, while for negative arguments each function exhibits oscillations similar to those of the Airy functions themselves.

56:3 DEFINITIONS

The *Airy integral*

56:3:1 $$\mathrm{Ai}(x) = \frac{1}{\pi}\int_0^\infty \cos\left(xt + \frac{t^3}{3}\right)dt$$

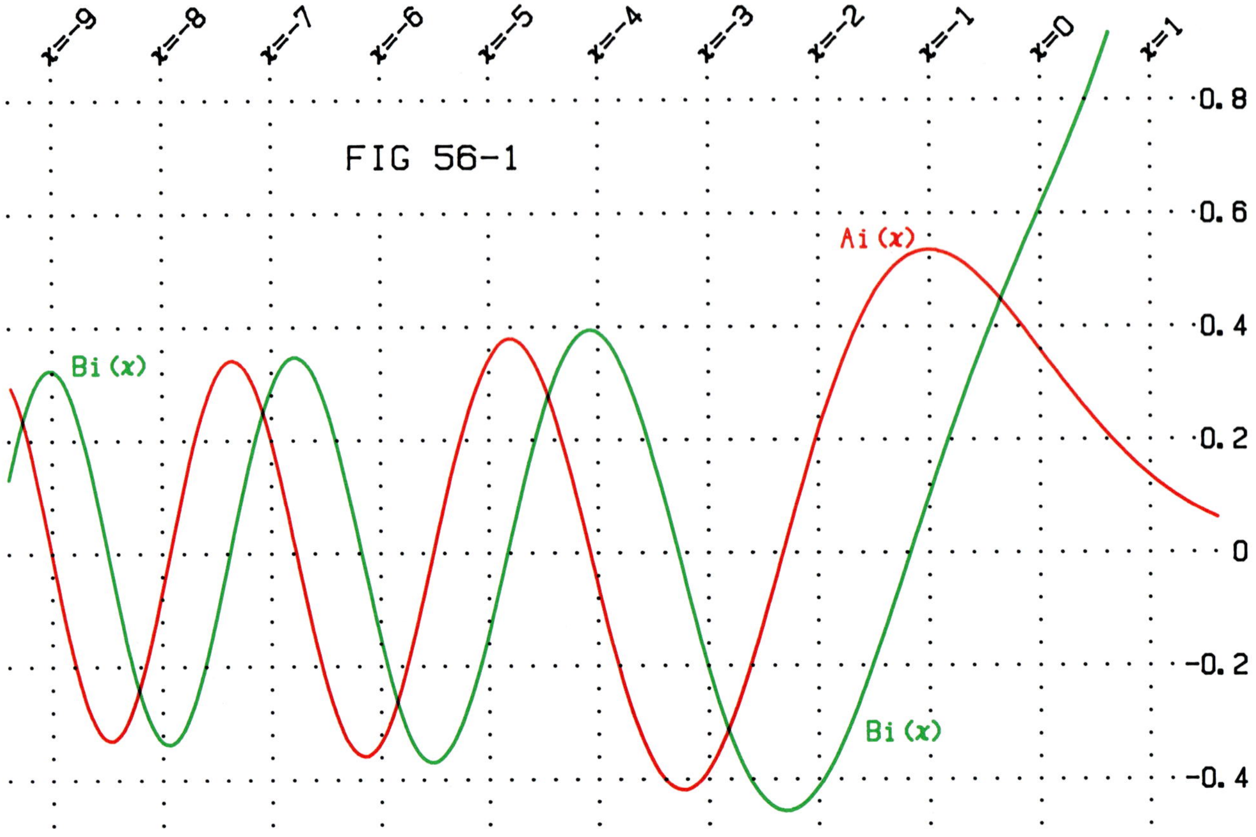

defines the Ai function for all arguments, while the similar definite integral sum

56:3:2 $$\mathrm{Bi}(x) = \frac{1}{\pi}\int_0^\infty \exp\left(xt - \frac{t^3}{3}\right)dt + \frac{1}{\pi}\int_0^\infty \sin\left(xt + \frac{t^3}{3}\right)dt$$

serves the same purpose for Bi.

Using the 56:1:1 definition of an auxiliary argument, the two Airy functions may be defined in terms of hyperbolic Bessel functions [Chapter 50], or the Basset function [Chapter 51], for positive arguments. The Bessel functions [Chapter 53], or Neumann functions [Chapter 54], provide the corresponding definitions when the argument is negative:

56:3:3 $$\mathrm{Ai}(x) = \begin{cases} \dfrac{\sqrt{x}}{3}\,[\mathrm{I}_{-1/3}(X) - \mathrm{I}_{1/3}(X)] = \dfrac{1}{\pi}\sqrt{\dfrac{x}{3}}\,\mathrm{K}_{1/3}(X) & x > 0 \\ \dfrac{\sqrt{-x}}{3}\,[\mathrm{J}_{-1/3}(X) + \mathrm{J}_{1/3}(X)] = \sqrt{\dfrac{-x}{3}}\,[\mathrm{Y}_{-1/3}(X) - \mathrm{Y}_{1/3}(X)] & x < 0 \end{cases}$$

56:3:4 $$\mathrm{Bi}(x) = \begin{cases} \sqrt{\dfrac{x}{3}}\,[\mathrm{I}_{-1/3}(X) + \mathrm{I}_{1/3}(X)] & x > 0 \\ \sqrt{\dfrac{-x}{3}}\,[\mathrm{J}_{-1/3}(X) - \mathrm{J}_{1/3}(X)] = \dfrac{-\sqrt{-x}}{3}\,[\mathrm{Y}_{-1/3}(X) + \mathrm{Y}_{1/3}(X)] & x < 0 \end{cases}$$

A linear combination of Ai(x) and Bi(x) functions satisfies the *Airy differential equation*

56:3:5 $$\frac{d^2f}{dx^2} = xf \qquad f = c_1\mathrm{Ai}(x) + c_2\mathrm{Bi}(x)$$

where c_1 and c_2 are arbitrary constants.

The auxiliary Airy functions are defined by

56:3:6 $$\mathrm{fai}(x) = \frac{3^{2/3}\Gamma(\frac{2}{3})}{2}\left[\frac{\mathrm{Bi}(x)}{\sqrt{3}} + \mathrm{Ai}(x)\right] = \begin{cases} \dfrac{\Gamma(\frac{2}{3})\sqrt{x}}{3^{1/3}}\,\mathrm{I}_{-1/3}(X) & x > 0 \\ \dfrac{\Gamma(\frac{2}{3})\sqrt{-x}}{3^{1/3}}\,\mathrm{J}_{-1/3}(X) & x < 0 \end{cases}$$

56:3:7 $$\mathrm{gai}(x) = \frac{3^{1/3}\Gamma(\frac{1}{3})}{2}\left[\frac{\mathrm{Bi}(x)}{\sqrt{3}} - \mathrm{Ai}(x)\right] = \begin{cases} \dfrac{\Gamma(\frac{1}{3})\sqrt{x}}{3^{2/3}}\,\mathrm{I}_{1/3}(X) & x > 0 \\ \dfrac{\Gamma(\frac{1}{3})\sqrt{-x}}{3^{2/3}}\,\mathrm{J}_{1/3}(X) & x < 0 \end{cases}$$

$$X = \tfrac{2}{3}(|x|)^{3/2}$$

where Γ denotes the gamma function [Chapter 43]. In the notation of Section 43:14, these auxiliary functions may

FIG 56-2

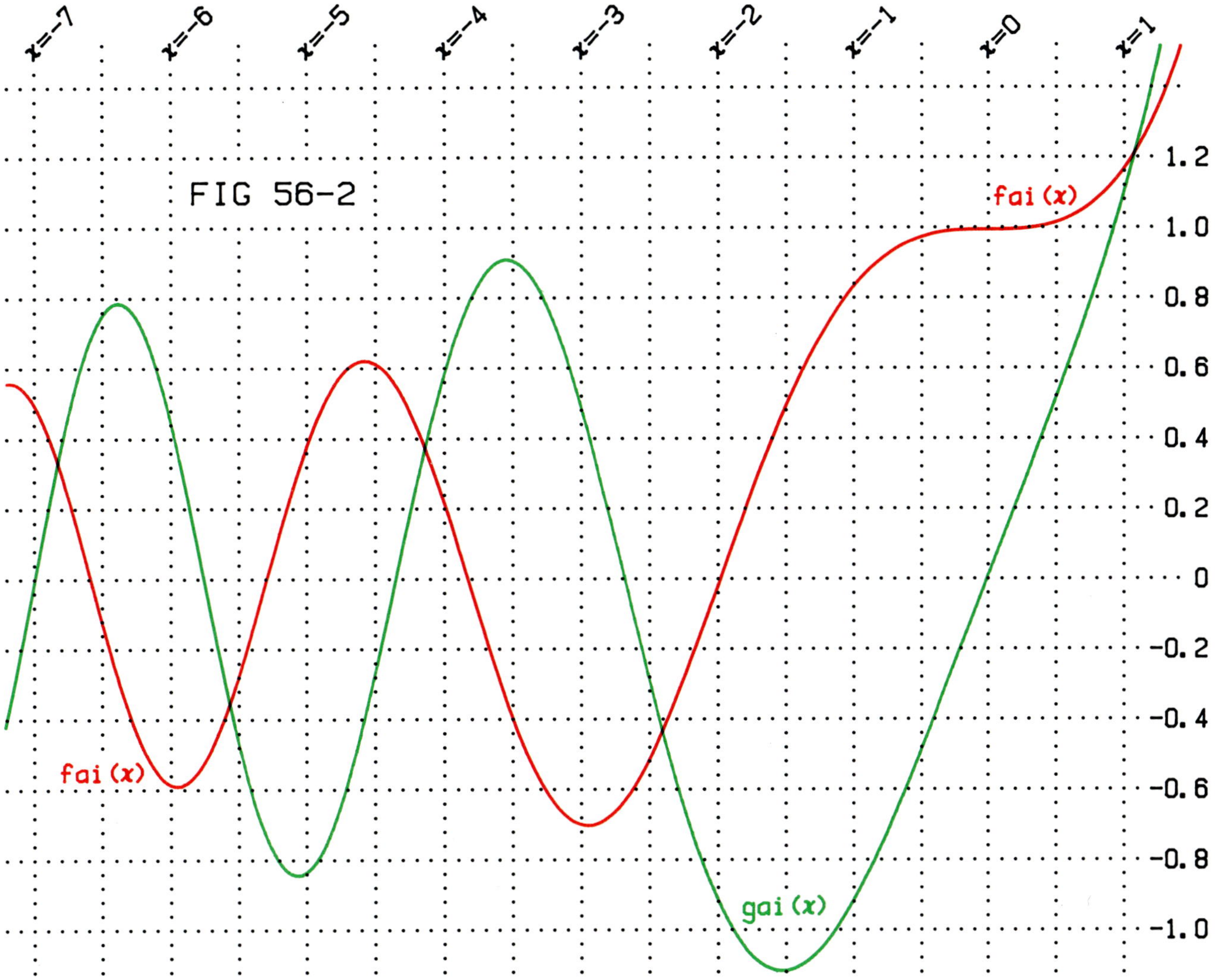

be synthesized from the zero-order hyperbolic Bessel function by the operations

56:3:8 $$I_0\left(\frac{2}{3}x^{3/2}\right) \xrightarrow[0]{\frac{1}{3}} \mathrm{fai}(x) \qquad (9X)^{1/3} = x \ge 0$$

56:3:9 $$I_0\left(\frac{2}{3}x^{3/2}\right) \xrightarrow[0]{\frac{-1}{3}} \frac{\mathrm{gai}(x)}{x} \qquad (9X)^{1/3} = x \ge 0$$

Syntheses similar to 56:3:8 and 56:3:9, but involving the Bessel J_0 coefficient instead of its hyperbolic counterpart, provide definitions of fai(x) and gai(x) for negative arguments.

56:4 SPECIAL CASES

There are none.

56:5 INTRARELATIONSHIPS

The formulas

56:5:1 $$\mathrm{Ai}(x) = \mathrm{Ai}(0)\,\mathrm{fai}(x) - \frac{\mathrm{gai}(x)}{2\pi \mathrm{Bi}(0)}$$

and

56:5:2 $$\mathrm{Bi}(x) = \mathrm{Bi}(0)\,\mathrm{fai}(x) + \frac{\mathrm{gai}(x)}{2\pi \mathrm{Ai}(0)}$$

relate Airy functions to their auxiliary counterparts. See Section 56:7 for values of Ai(0) and Bi(0).

56:6 EXPANSIONS

The auxiliary Airy functions are expansible as the convergent series

56:6:1 $$\mathrm{fai}(x) = 1 + \frac{x^3}{3!} + (1 \times 4)\frac{x^6}{6!} + (1 \times 4 \times 7)\frac{x^9}{9!} + \cdots = \sum_{j=0}^{\infty} \frac{(3j+1)!!!}{(3j+1)!} x^{3j}$$

56:6:2 $$\mathrm{gai}(x) = x + 2\frac{x^4}{4!} + (2 \times 5)\frac{x^7}{7!} + (2 \times 5 \times 8)\frac{x^{10}}{10!} + \cdots = \sum_{j=0}^{\infty} \frac{(3j+2)!!!}{(3j+2)!} x^{3j+1}$$

in which the triple factorial [see Section 2:13] occurs. Notice that fai and gai are two members of a trio of functions, the third of which

56:6:3 $$\mathrm{hai}(x) = \frac{x^2}{2!} + 3\frac{x^5}{5!} + (3 \times 6)\frac{x^8}{8!} + (3 \times 6 \times 9)\frac{x^{11}}{11!} + \cdots = \sum_{j=0}^{\infty} \frac{(3j)!!!x^{3j+2}}{(3j+2)!} = \sum_{j=0}^{\infty} \frac{(3j+3)!!!}{(3j+3)!} x^{3j+2}$$

is encountered in Section 56:13.

Expansions 56:6:1 and 56:6:2 may be used in conjunction with relationships 56:5:1 and 56:5:2 to generate power series expansions of the Airy functions Ai and Bi. Terms involving x^2, x^5, x^8, ... are absent from such expansions.

Asymptotic expansions of the Airy functions take different forms according to whether the large argument is positive or negative:

56:6:4
$$\mathrm{Ai}(x) \sim \begin{cases} \dfrac{\exp(-X)}{2\sqrt{\pi\sqrt{x}}}\left[a_0 - \dfrac{a_1}{X} + \dfrac{a_2}{X^2} - \dfrac{a_3}{X^3} + \cdots + \dfrac{(-1)^j a_j}{X^j} + \cdots\right] & x \to \infty \\ \dfrac{\sin\left(X + \dfrac{\pi}{4}\right)}{\sqrt{\pi\sqrt{-x}}}\left[a_0 - \dfrac{a_2}{X^2} + \dfrac{a_4}{X^4} - \cdots\right] - \dfrac{\cos\left(X + \dfrac{\pi}{4}\right)}{\sqrt{\pi\sqrt{-x}}}\left[\dfrac{a_1}{X} - \dfrac{a_3}{X^3} + \dfrac{a_5}{X^5} - \cdots\right] & x \to -\infty \end{cases}$$

56:6:5
$$\mathrm{Bi}(x) \sim \begin{cases} \dfrac{\exp(X)}{\sqrt{\pi\sqrt{x}}}\left[a_0 + \dfrac{a_1}{X} + \dfrac{a_2}{X^2} + \dfrac{a_3}{X^3} + \cdots + \dfrac{a_j}{X^j}\right] & x \to \infty \\ \dfrac{\cos\left(X + \dfrac{\pi}{4}\right)}{\sqrt{\pi\sqrt{-x}}}\left[a_0 - \dfrac{a_2}{X^2} + \dfrac{a_4}{X^4} - \cdots\right] + \dfrac{\sin\left(X + \dfrac{\pi}{4}\right)}{\sqrt{\pi\sqrt{-x}}}\left[\dfrac{a_1}{X} - \dfrac{a_3}{X^3} + \dfrac{a_5}{X^5} - \cdots\right] & x \to -\infty \end{cases}$$

where the auxiliary argument X is given by 56:1:1, $a_0 = 1$, and other coefficients obey the definition and recursion

56:6:6
$$a_j = \frac{(6j-1)!!}{j!(2j-1)!!(216)^j} = \left(j - 1 + \frac{5}{36j}\right)\frac{a_{j-1}}{2} \qquad j = 1, 2, 3, \ldots$$

An asymptotic expansion, namely

56:6:7
$$\mathrm{Ai}^2(x) + \mathrm{Bi}^2(x) \sim \frac{1}{\pi\sqrt{-x}}\left[1 + \frac{5}{32x^3} + \frac{1155}{2048x^6} + \cdots + \frac{(6j-1)!!}{j!(96x^3)^j} + \cdots\right] \qquad x \to -\infty$$

describes the sum of the squares of the Airy functions as the argument acquires very negative values.

56:7 PARTICULAR VALUES

	$x = -\infty$	$x = 0$	$x = \infty$
Ai(x)	0	$\dfrac{1}{3^{2/3}\Gamma(\frac{2}{3})} = \dfrac{\Gamma(\frac{1}{3})}{2\pi 3^{1/6}} = 0.3550280539$	0
Bi(x)	0	$\dfrac{1}{3^{1/6}\Gamma(\frac{2}{3})} = \dfrac{3^{1/3}\Gamma(\frac{1}{3})}{2\pi} = 0.6149266276$	∞
fai(x)	0	1	∞
gai(x)	0	0	∞

Table 56.7.1 records the first few zeros and extrema of Ai(x) and Bi(x), all of which occur at negative arguments. The final tabular entries, which use the abbreviations $a = [3\pi(4k-1)]^{2/3}$ and $b = [3\pi(4k-3)]^{2/3}$, provide excellent approximations for the kth zeros of the Ai and Bi functions, for $k \geq 5$.

56:8 NUMERICAL VALUES

The algorithm below calculates Ai(x) and Bi(x) with 24-bit precision, that is, the relative error does not exceed 6×10^{-8}, except when the value of the Airy function is close to zero. For arguments in the range $-5 \leq x \leq 5$, the algorithm uses expansions 56:6:1 and 56:6:2, together with relationships 56:5:1 and 56:5:2. When the argument exceeds 5 in absolute value, expansions 56:6:4 and 56:6:5 are utilized. When $x < -5$, these expansions are rewritten as

Table 56.7.1

x	Ai(x)	x	Bi(x)
−1.018793	0.5356567	−1.173712	0
−2.338107	0	−2.294440	−0.4549444
−3.248198	−0.4190155	−3.271093	0
−4.087949	0	−4.073155	0.3965228
−4.820099	0.3804065	−4.830738	0
−5.520560	0	−5.512396	−0.3679692
−6.163307	−0.3579079	−6.169852	0
−6.786708	0	−6.781294	0.3494992
$-\frac{a}{4}-\frac{5}{3a^2}$	0	$-\frac{b}{4}-\frac{5}{3b^2}$	0

56:8:1

$$\begin{matrix}\mathrm{Ai}(x)\\ \mathrm{Bi}(x)\end{matrix} \sim \frac{\cos(X)}{\sqrt{2\pi\sqrt{-x}}}\left[a_0 \mp \frac{a_1}{X} - \frac{a_2}{X^2} \pm \frac{a_3}{X^3} + \frac{a_4}{X^4} \mp \cdots\right]$$
$$+ \frac{\sin(X)}{\sqrt{2\pi\sqrt{-x}}}\left[\pm a_0 + \frac{a_1}{X} \mp \frac{a_2}{X^2} - \frac{a_3}{X^3} \pm \frac{a_4}{X^4} + \cdots\right]$$

Set $p = q = j = 1$
Input x >>>>>>>
If $5 < |x|$ go to (2)
Set $t = .614926628$
Set $u = x\sqrt{3}/2\pi t$
Set $A = t - u$
Set $B = t + u$
Set $X = x^3/3$
(1) Replace t by $tX/j(3j - 1)$
Replace u by $uX/j(3j + 1)$
Replace A by $A + t - u$
Replace B by $B + t + u$
Replace j by $j + 1$
If $j < 2 + 3|x|$ go to (1)
Replace A by $A/\sqrt{3}$
Go to (6)
(2) If $x > 0$ go to (3)
Set $q = -1$
Replace x by $-x$
(3) Set $A = t = B = u = 1/\sqrt{2\pi\sqrt{x}}$
Set $X = 2x^{3/2}/3$
(4) Replace B by $Bp\left(j - 1 + \frac{1}{7.2j}\right)\Big/2X$
Replace A by $-Ap\left(j - 1 + \frac{1}{7.2j}\right)\Big/2X$
Replace t by $t + A$
Replace u by $u + B$
Replace p by pq
Replace j by $j + 1$

Storage needed: p, q, j, x, t, u, A, B and X

Use radian mode.

Input restrictions: There are no restrictions on the magnitude of x.

If $j < 3 + \dfrac{90}{X}$ go to (4)
If $q < 0$ go to (5)
Set $B = u\sqrt{2}\exp(X)$
Set $A = t/[\sqrt{2}\exp(X)]$
Go to (6)
(5) Set $B = u\cos(X) - t\sin(X)$
Set $A = t\cos(X) + u\sin(X)$
(6) Output A
$A \simeq \mathrm{Ai}(x)$ <<<<<
Output B
$B \simeq \mathrm{Bi}(x)$ <<<<<

Test values:
$\mathrm{Ai}(4) = 9.516074 \times 10^{-4}$
$\mathrm{Bi}(4) = 83.8470714$
$\mathrm{Ai}(-4) = -0.0702655329$
$\mathrm{Bi}(-4) = 0.392234706$
$\mathrm{Ai}(6) = 9.94769437 \times 10^{-6}$
$\mathrm{Bi}(6) = 6536.44608$
$\mathrm{Ai}(-6) = -0.329145174$
$\mathrm{Bi}(-6) = -0.146698376$

The universal hypergeometric algorithm [Section 18:14] permits values of fai(x) and gai(x) to be determined.

56:9 APPROXIMATIONS

For small arguments, the approximations

56:9:1 $$\mathrm{Ai}(x) \simeq 0.355 - 0.259x + 0.059x^3 - 0.022x^4 \qquad \text{small } |x|$$

56:9:2 $$\mathrm{Bi}(x) \simeq 0.615 + 0.448x + 0.102x^3 + 0.037x^4 \qquad \text{small } |x|$$

hold, while for large positive arguments

56:9:3 $$\mathrm{Ai}(x) \simeq \frac{\exp(-2x^{3/2}/3)}{2\sqrt{\pi\sqrt{x}}} \qquad \text{large positive } x$$

56:9:4 $$\mathrm{Bi}(x) \simeq \frac{\exp(2x^{3/2}/3)}{\sqrt{\pi\sqrt{x}}} \qquad \text{large positive } x$$

so that the product $\mathrm{Ai}(x)\,\mathrm{Bi}(x) \simeq 1/2\pi\sqrt{x}$. For large negative argument

56:9:5 $$\mathrm{Ai}(x) \simeq \frac{\cos(X) + \sin(X)}{\sqrt{2\pi\sqrt{-x}}} \qquad X = \frac{2(-x)^{3/2}}{3} \qquad \text{large negative } x$$

56:9:6 $$\mathrm{Bi}(x) \simeq \frac{\cos(X) - \sin(X)}{\sqrt{2\pi\sqrt{-x}}} \qquad X = \frac{2(-x)^{3/2}}{3} \qquad \text{large negative } x$$

so that in this range $\mathrm{Ai}^2(x) + \mathrm{Bi}^2(x) \simeq 1/\pi\sqrt{-x}$, a result that is also a consequence of expansion 56:6:7.

56:10 OPERATIONS OF THE CALCULUS

Derivatives of the Airy functions involve Bessel and related functions of order $\frac{2}{3}$:

56:10:1 $$\frac{d}{dx}\mathrm{Ai}(x) = \begin{cases} \dfrac{x}{3}[\mathrm{I}_{-2/3}(X) - \mathrm{I}_{2/3}(X)] = \dfrac{x}{\pi\sqrt{3}}\mathrm{K}_{2/3}(X) & x > 0 \\ \dfrac{x}{3}[\mathrm{J}_{-2/3}(X) - \mathrm{J}_{2/3}(X)] & x < 0 \end{cases}$$

56:10:2 $$\frac{d}{dx}\mathrm{Bi}(x) = \begin{cases} \dfrac{x}{\sqrt{3}}[\mathrm{I}_{-2/3}(X) + \mathrm{I}_{2/3}(X)] & x > 0 \\ \dfrac{x}{\sqrt{3}}[\mathrm{J}_{-2/3}(X) + \mathrm{J}_{2/3}(X)] & x < 0 \end{cases}$$

The two derivatives are interrelated by

56:10:3 $$\mathrm{Ai}(x)\frac{d}{dx}\mathrm{Bi}(x) - \mathrm{Bi}(x)\frac{d}{dx}\mathrm{Ai}(x) = \frac{1}{\pi}$$

irrespective of sign of the argument. Second derivatives obey the simple relationship

56:10:4 $$\frac{d^2f}{dx^2} = xf \qquad f = \mathrm{Ai}(x) \text{ or } \mathrm{Bi}(x)$$

Indefinite integrals of the Airy functions are expressible in terms of their derivatives and that of the function Hi(x), discussed in Section 56:13:

56:10:5 $$\int_{-\infty}^{x}\mathrm{Ai}(t)dt = \frac{2}{3} + \int_0^x \mathrm{Ai}(t)dt = 1 - \int_x^{\infty}\mathrm{Ai}(t)dt = \pi\left[\mathrm{Ai}(x)\frac{d}{dx}\mathrm{Hi}(x) - \mathrm{Hi}(x)\frac{d}{dx}\mathrm{Ai}(x)\right]$$

56:10:6 $$\int_{-\infty}^{x}\mathrm{Bi}(t)dt = \int_0^x \mathrm{Bi}(t)dt = \pi\left[\mathrm{Bi}(x)\frac{d}{dx}\mathrm{Hi}(x) - \mathrm{Hi}(x)\frac{d}{dx}\mathrm{Bi}(x)\right]$$

These relationships are valid for either sign of the argument x. Three useful definite integrals result from setting $x = 0$ in 56:10:5 and 56:10:6.

56:11 COMPLEX ARGUMENT

By using definitions 56:3:3 and 56:3:4, the information in Sections 50:11 and 53:11 may be applied to Airy functions.

56:12 GENERALIZATIONS

Inasmuch as Airy functions are Bessel functions of order $\pm\frac{1}{3}$, they may be generalized to other fractional orders.

56:13 COGNATE FUNCTIONS

The two components of the integral definition, 56:3:2, of Bi(x) are sometimes regarded as distinct functions and denoted as follows:

56:13:1 $$\mathrm{Hi}(x) = \frac{1}{\pi}\int_0^{\infty}\exp\left(xt - \frac{t^3}{3}\right)dt = \frac{\Gamma(\frac{1}{3})}{3^{2/3}\pi}\sum_{j=0}^{\infty}\frac{(j-2)!!!}{j!}x^j = \frac{2\mathrm{Bi}(0)}{3}[\mathrm{fai}(x) + \mathrm{gai}(x) + \mathrm{hai}(x)]$$

56:13:2 $$\mathrm{Gi}(x) = \frac{1}{\pi}\int_0^{\infty}\sin\left(xt + \frac{t^3}{3}\right)dt = \frac{\mathrm{Bi}(0)}{3}\left[\mathrm{fai}(x) + \left(\frac{\sqrt{27}}{2\pi\mathrm{Bi}^2(0)} - 2\right)\mathrm{gai}(x) - 2\mathrm{hai}(x)\right]$$

Of course, Hi(x) + Gi(x) = Bi(x).

CHAPTER 57

THE STRUVE FUNCTION

The Struve function $h_\nu(x)$ has many analogies with the Bessel function of Chapter 53. Likewise, the hyperbolic Struve function $\ell_\nu(x)$, which is discussed briefly in Section 57:13, is analogous to the hyperbolic Bessel function of Chapter 50.

57:1 NOTATION

The usual notation for the Struve function is $H_\nu(x)$ with the capital H printed in bold type to avoid confusion with Hermite polynomials [Chapter 24]. However, because they are difficult to reproduce by hand or on an office machine, this *Atlas* avoids abnormal fonts. Thus, we adopt $h_\nu(x)$ for the Struve function and $\ell_\nu(x)$, instead of the more usual boldface $L_\nu(x)$, for its hyperbolic counterpart.

We use $h_\nu(x)$ to denote the Struve function of argument x and unrestricted order ν, but $h_n(x)$ is employed when the order is constrained to be an integer.

57:2 BEHAVIOR

Like the Bessel function, the Struve function is real for negative argument only when its order is an integer. Accordingly, the contour map Figure 57-1 is confined to $x \geq 0$.

Also like the Bessel function, $h_\nu(x)$ is an oscillatory function of its argument, the amplitude of the oscillations decreasing with increasing x. Unlike the Bessel function, however, the damped oscillations of the Struve function do not generally occur symmetrically about $x = 0$. For $\nu > \frac{1}{2}$, this asymmetry is so pronounced that $h_\nu(x)$ is uniformly positive when its argument is positive.

57:3 DEFINITIONS

The definite integral

57:3:1 $$h_\nu(x) = \frac{2(x/2)^\nu}{\sqrt{\pi}\,\Gamma(\nu + \frac{1}{2})} \int_0^1 (1 - t^2)^{\nu-1/2} \sin(xt)\mathrm{d}t \qquad \nu > -\frac{1}{2}$$

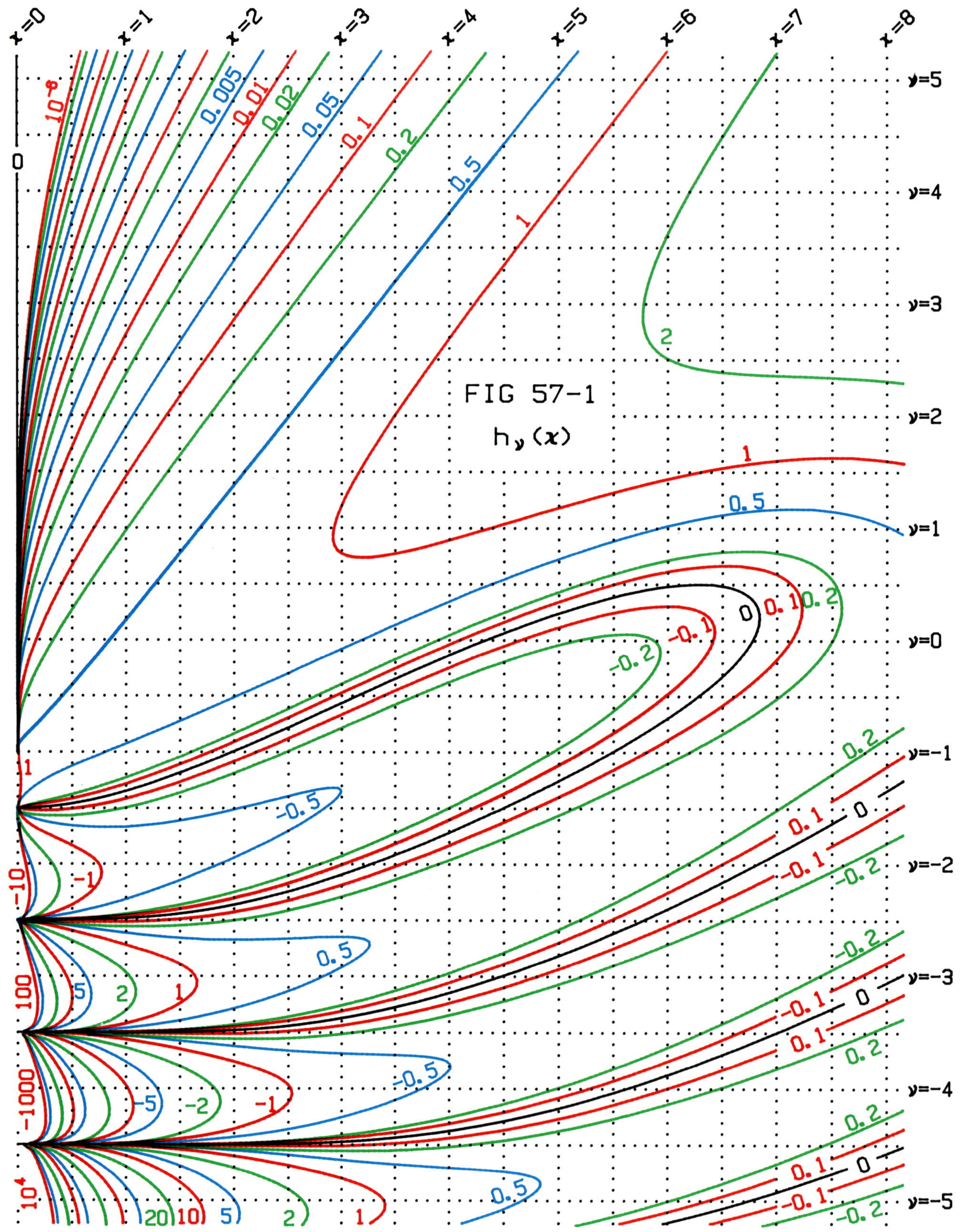
x=0
x=1
x=2
x=3
x=4
x=5
x=6
x=7
x=8
y=5
y=4
y=3
y=2
y=1
y=0
y=−1
y=−2
y=−3
y=−4
y=−5
FIG 57-1
$h_\nu(x)$

or its equivalent on replacement of t by $\sin(\theta)$ or $\cos(\theta)$, defines the Struve function for orders exceeding $-\frac{1}{2}$. Another integral

57:3:2 $$h_\nu(x) - Y_\nu(x) = \frac{2(x/2)^\nu}{\sqrt{\pi}\,\Gamma(\nu+\frac{1}{2})}\int_0^\infty (1+t^2)^{\nu-1/2}\exp(-xt)dt \qquad \nu > -\frac{1}{2}$$

provides a representation of the difference between the Struve and Neumann functions of common order and argument. This time $\tan(\theta)$ or $\sinh(u)$ may replace t, but the restriction to orders in excess of $-\frac{1}{2}$ again applies.

Using the notation of Section 43:14, the Struve function may be synthesized by

57:3:3 $$J_0(x) \xrightarrow[0]{-\frac{1}{2}} \xrightarrow[0]{-\nu-\frac{1}{2}} \frac{\sqrt{\pi}\,\Gamma(\frac{3}{2}+\nu)}{2(x/2)^{1+\nu}} h_\nu(x) \qquad X = \frac{-x^2}{4}$$

from the zero order Bessel coefficient. This, then, constitutes yet another way of defining $h_\nu(x)$.

57:4 SPECIAL CASES

When its order is half a negative odd integer, the Struve function becomes a spherical Bessel function [see Sections 32:13, 53:4 and 54:4]:

57:4:1 $$h_{-n-1/2}(x) = Y_{-n-1/2}(x) = (-1)^n J_{n+1/2}(x) = \sqrt{\frac{2x}{\pi}}\, y_{-n}(x) = (-1)^n \sqrt{\frac{2x}{\pi}}\, j_n(x) \qquad n = 0, 1, 2, \ldots$$

Reduction to elementary functions also occurs for orders equal to a moiety of a positive odd integer; thus:

57:4:2 $$h_{1/2}(x) = \sqrt{\frac{2}{\pi x}}(1-\cos(x))$$

57:4:3 $$h_{3/2}(x) = \sqrt{\frac{2}{\pi x}}\left(\frac{1-\cos(x)}{x} - \sin(x) + \frac{x}{2}\right)$$

and others may be derived from the recursion

57:4:4 $$h_{n+1/2}(x) = \frac{2n-1}{x} h_{n-1/2}(x) - h_{n-3/2}(x) + \sqrt{\frac{2}{\pi x}}\frac{(x/2)^n}{n!} \qquad n = 0, 1, 2, \ldots$$

In some respects the Struve function is also "special" when its order is an integer. Because of the preeminent interest in these cases, they are nevertheless treated in the main sections of this chapter.

57:5 INTRARELATIONSHIPS

The recurrence relationship

57:5:1 $$h_{\nu+1}(x) = \frac{(x/2)^\nu}{\sqrt{\pi}\,\Gamma(\nu+{}^3/_2)} + \frac{2\nu}{x} h_\nu(x) - h_{\nu-1}(x)$$

is obeyed for all orders, but it takes the simpler forms

57:5:2 $$h_{n+1}(x) + h_{n-1}(x) = \frac{2n}{x} h_n(x) \begin{cases} +\dfrac{2x^n}{\pi(2n+1)!!} & n = -1, 0, 1, 2, \ldots \\ -\dfrac{2(-2n-3)!!}{\pi(-x)^{-n}} & n = -1, -2, -3, \ldots \end{cases}$$

when the order of the Struve function is an integer. The double factorial symbol, which occurs frequently in this chapter, is explained in Section 2:13.

Reflection formulas exist only for integer orders. For argument-reflection, we have

57:5:3 $$h_n(-x) = -(-1)^n\, h_n(x) \qquad n = 0, \pm 1, \pm 2, \ldots$$

so that the Struve function of even order is odd, and vice versa. Order-reflection introduces an n-term polynomial function

57:5:4 $$(-1)^n\, h_{-n}(x) = h_n(x) - \frac{2}{\pi}\left(\frac{x^{n-1}}{(2n-1)!!} + \frac{1!!x^{n-3}}{(2n-3)!!} + \frac{3!!x^{n-5}}{(2n-5)!!} + \cdots + \frac{(2n-3)!!x^{1-n}}{1!!}\right) \qquad n = 1, 2, 3, \ldots$$

so that $h_{-1}(x) = (2/\pi) - h_1(x)$, $h_{-2}(x) = h_2(x) - (2/\pi x) - (2x/3\pi)$, etc.

57:6 EXPANSIONS

For all finite values of ν, the expansion

57:6:1 $$h_\nu(x) = \left(\frac{x}{2}\right)^{\nu+1} \sum_{j=0}^{\infty} \frac{(-x^2/4)^j}{\Gamma(j+\frac{3}{2})\,\Gamma(j+\nu+\frac{3}{2})} = \frac{2(x/2)^{1+\nu}}{\sqrt{\pi}\,\Gamma(\frac{3}{2}+\nu)} \sum_{j=0}^{\infty} \frac{(-x^2/4)^j}{(\frac{3}{2})_j(\frac{3}{2}+\nu)_j}$$

converges, provided x is real and positive; negative arguments are admissible if the order is an integer. The most important instances correspond to $\nu = -1$, 0 and 1, and are

57:6:2 $$h_{-1}(x) = \frac{2}{\pi}\left(1 - \frac{x^2}{3} + \frac{x^4}{45} - \frac{x^6}{1575} + \cdots + \frac{(-x^2)^j}{(2j-1)!!(2j+1)!!} + \cdots\right)$$

57:6:3 $$h_0(x) = \frac{2}{\pi}\left(x - \frac{x^3}{9} + \frac{x^5}{225} - \frac{x^7}{11025} + \cdots + \frac{x(-x^2)^j}{[(2j+1)!!]^2} + \cdots\right)$$

57:6:4 $$h_1(x) = \frac{2}{\pi}\left(\frac{x^2}{3} - \frac{x^4}{45} + \frac{x^6}{1575} - \frac{x^8}{99225} + \cdots + \frac{(-x^2)^{j+1}}{(2j+1)!!\,(2j+3)!!} + \cdots\right)$$

Integer order Struve functions are expansible in terms of Bessel coefficients [Chapter 52]; for example:

57:6:5 $$h_0(x) = \frac{4}{\pi}\left(J_1(x) + \frac{J_3(x)}{3} + \frac{J_5(x)}{5} + \cdots\right) = \frac{4}{\pi}\sum_{j=0}^{\infty}\frac{J_{2j+1}(x)}{2j+1}$$

57:6:6 $$h_1(x) = \frac{4}{\pi}\left(\frac{1-J_0(x)}{2} + \frac{J_2(x)}{3} + \frac{J_4(x)}{15} + \frac{J_6(x)}{35} + \cdots \frac{J_{2j}(x)}{4j^2-1} + \cdots\right)$$

An asymptotic expansion exists for the difference between Struve and Neumann functions of common argument and order:

57:6:7 $$h_\nu(x) - Y_\nu(x) \sim \frac{(x/2)^{\nu-1}}{\sqrt{\pi}\,\Gamma(\frac{1}{2}+\nu)}\left(1 - \frac{1-2\nu}{x^2} + \frac{3!!(1-2\nu)(3-2\nu)}{x^4} - \cdots + \frac{(2j-1)!!\,(\frac{1}{2}-\nu)_j}{(-x^2/2)^j} + \cdots\right) \qquad x \to \infty$$

See Section 54:6 for an asymptotic expansion of $Y_\nu(x)$ that may be used in conjunction with the 57:6:7 series. The important cases of orders -1, 0 and 1 are

57:6:8 $$h_{-1}(x) - Y_{-1}(x) \sim \frac{2}{\pi}\left(-\frac{1}{x^2} + \frac{3}{x^4} - \frac{45}{x^6} + \frac{1575}{x^8} - \cdots + \frac{(2j+1)!!(2j-1)!!}{(-x^2)^{j+1}} + \cdots\right) \qquad x \to \infty$$

57:6:9 $$h_0(x) - Y_0(x) \sim \frac{2}{\pi}\left(\frac{1}{x} - \frac{1}{x^3} + \frac{9}{x^5} - \frac{225}{x^7} + \cdots + \frac{[(2j-1)!!]^2}{x(-x^2)^j} + \cdots\right) \qquad x \to \infty$$

and

57:6:10 $$h_1(x) - Y_1(x) \sim \frac{2}{\pi}\left(1 + \frac{1}{x^2} - \frac{3}{x^4} + \frac{45}{x^6} - \cdots - \frac{(2j-1)!!(2j-3)!!}{(-x^2)^j} + \cdots\right) \qquad x \to \infty$$

57:7 PARTICULAR VALUES

		$x = -\infty$	$x = 0$	$x = \infty$
$h_n(x)$	$n = 3, 5, 7, \ldots$	∞	0	∞
$h_n(x)$	$n = 2, 4, 6, \ldots$	$-\infty$	0	∞
$h_\nu(x)$	$1 < \nu \neq 2, 3, 4, \ldots$	undef	0	∞
$h_1(x)$		$2/\pi$	0	$2/\pi$
$h_\nu(x)$	$1 > \nu > -1;\ \nu \neq 0$	undef	0	0
$h_0(x)$		0	0	0
$h_{-1}(x)$		0	$2/\pi$	0
$h_\nu(x)$	$-1 > \nu > \frac{-3}{2}$	undef	∞	0
$h_\nu(x)$	$\nu = \frac{-3}{2}, \frac{-5}{2}, \frac{-7}{2}, \ldots$	undef	0	0
$h_\nu(x)$	$\frac{-3}{2} > \nu > -2,\ -2 > \nu > \frac{-5}{2},\ \frac{-7}{2} > \nu > -4, \ldots$	undef	$-\infty$	0
$h_n(x)$	$n = -2, -4, -6, \ldots$	0	$\pm\infty$	0
$h_\nu(x)$	$\frac{-5}{2} > \nu > -3,\ -3 > \nu - \frac{-7}{2},\ \frac{-9}{2} > \nu > -5, \ldots$	undef	∞	0
$h_n(x)$	$n = -3, -5, -7, \ldots$	0	∞	0

Apart from its $x = 0$ value, the Struve function has no zeros for $\nu > \frac{1}{2}$. For $\nu \le \frac{1}{2}$, however, there is an infinite set of zeros that, for large arguments, correspond to the roots of the equation

57:7:1 $$\sin\left(r - \frac{\nu\pi}{2} + \frac{3\pi}{4}\right) \simeq \frac{(r/2)^{\nu-1/2}}{\Gamma(\nu + \frac{1}{2})} \qquad h_\nu(r) = 0 \qquad \text{large } r \qquad \nu \le \frac{1}{2}$$

and therefore ultimately occur close to $r \sim \pi(2k + \nu + \frac{1}{2})/2$ where k is a large positive integer.

57:8 NUMERICAL VALUES

The algorithm below is designed to calculate values of the Struve function in the approximate ranges $-15 \le \nu \le 15$, $0 < x \le 30$ to a 24-bit precision (i.e., the fractional error does not exceed 6×10^{-8}) except in the immediate vicinity of the function's zeros.

Input ν >>>>>>>

Set $a = \frac{1}{2}$

Input x >>>>>>>
[or $-x$]

Set $J = |x|\left(1.5 - \frac{\nu}{40}\right) + 5 - \frac{\nu}{3}$

Replace x by $(-x/2)(|x/2|)$

Set $t = 1/\pi x$

Storage needed: ν, a, x, J, t, p, f and j

(1) Replace t by tx/a
Replace a by $a + 1$
Set $p = a + \nu$
If $p + \mathrm{Int}(|p|) = 0$ go to (1)
(2) Replace t by tp
Replace p by $p + 1$
If $p < 3$ go to (2)
Set $f = \left(1 + \dfrac{2}{7p^2}\left(\dfrac{2}{3p^2} - 1\right)\right) \Big/ 30p^2$
Replace f by $\dfrac{f-1}{12p} + p[1 - \ln(p)] + \dfrac{\nu+1}{2}\ln(|x|)$
Replace t by $t\exp(f)\sqrt{p/2}$
Set $j = f = 0$
(3) Replace f by $f + t$
Replace t by $tx/[(a + j)(a + j + \nu)]$
Replace j by $j + 1$
If $j < J$ go to (3)
Output f

$f \simeq \mathrm{h}_\nu(x)$ <<<<<<
[or $\ell_\nu(x)$]

Input restrictions: $x > 0$
If $-x$, a negative number, is input, the algorithm computes $\ell_\nu(x)$ instead of $\mathrm{h}_\nu(x)$.

Test values:
$\mathrm{h}_1(5) = 0.807811945$
$\mathrm{h}_{-5/2}(\pi) = 0.429869378$
$\ell_1(5) = 23.7282158$
$\ell_{-5/2}(\pi) = 1.79596683$

The algorithm computes $\mathrm{h}_\nu(x)$ as the sum $t_0 + t_1 + t_2 + \cdots + t_j + \cdots + t_J$ where t_0, equal to $2(|x|/2)^{1+\nu}/\sqrt{\pi}\,\Gamma(\frac{3}{2} + \nu)$, is evaluated by a modified version of the algorithm of Section 43:8, t_j is calculated as $(-x^2/4)t_{j-1}/(\frac{3}{2} + j)(\frac{3}{2} + j + \nu)$ and J is given by an empirical function of x and ν designed to ensure that the series has converged adequately. Special measures are taken if ν has one of the values $-\frac{3}{2}, -\frac{5}{2}, -\frac{7}{2}, \ldots$. This algorithm will also compute the hyperbolic Struve function $\ell_\nu(x)$, discussed in Section 57:13, when $-x$ is input in place of x.

57:9 APPROXIMATIONS

Expansions 57:6:1–57:6:4 can readily provide approximations to the Struve function that are valid for small arguments. For large arguments, one may similarly use the asymptotic expressions 57:6:7 and 54:6:6. For the cases of orders equal to -1, 0 or 1, the leading asymptotic terms are

57:9:1 $$\mathrm{h}_{-1}(x) \simeq \sqrt{\frac{2}{\pi x}}\sin\left(x + \frac{\pi}{4}\right) + \frac{3}{\sqrt{32\pi x^3}}\cos\left(x + \frac{\pi}{4}\right) - \frac{2}{\pi x^2} \qquad \text{large } x$$

57:9:2 $$\mathrm{h}_0(x) \simeq \sqrt{\frac{2}{\pi x}}\sin\left(x - \frac{\pi}{4}\right) + \frac{2}{\pi x} - \frac{1}{\sqrt{32\pi x^3}}\cos\left(x - \frac{\pi}{4}\right) \qquad \text{large } x$$

57:9:3 $$\mathrm{h}_1(x) \simeq \frac{2}{\pi} + \sqrt{\frac{2}{\pi x}}\sin\left(x - \frac{3\pi}{4}\right) + \frac{3}{\sqrt{32\pi x^3}}\cos\left(x - \frac{3\pi}{4}\right) \qquad \text{large } x$$

57:10 OPERATIONS OF THE CALCULUS

The formula

57:10:1 $$\frac{d}{dx}\mathrm{h}_\nu(x) \simeq \frac{(x/2)^\nu}{2\sqrt{\pi}\,\Gamma(\nu + \frac{3}{2})} + \frac{\mathrm{h}_{\nu-1}(x) - \mathrm{h}_{\nu+1}(x)}{2}$$

for the derivative of the Struve function simplifies to $d\,\mathrm{h}_0(x)/dx = \mathrm{h}_{-1}(x)$ for zero order. One also has

57:10:2
$$\frac{d}{dx}(x^\nu h_\nu(x)) = x^\nu h_{\nu-1}(x)$$

The indefinite integral of $t^\mu h_\nu(t)$ is expressible in closed form only when $\mu = 1 \pm \nu$:

57:10:3
$$\int_0^x t^{1+\nu} h_\nu(t)dt = x^{1+\nu} h_{1+\nu}(x) \qquad \nu > -1$$

57:10:4
$$\int_x^\infty t^{1+\nu} h_\nu(t)dt = -x^{1+\nu} h_{1+\nu}(x) \qquad \nu < -\frac{1}{2}$$

57:10:5
$$\int_0^x t^{1-\nu} h_\nu(t)dt = \frac{2x}{2^\nu\sqrt{\pi}\,\Gamma(\frac{1}{2}+\nu)} - x^{1-\nu} h_{\nu-1}(x)$$

but the corresponding definite integral

57:10:6
$$\int_0^\infty t^\mu h_\nu(t)dt = \frac{-2^\mu \Gamma\left(\frac{\nu}{2}+\frac{\mu}{2}+\frac{1}{2}\right)}{\Gamma\left(\frac{\nu}{2}-\frac{\mu}{2}-\frac{1}{2}\right)} \cot\left(\frac{\nu\pi}{2}+\frac{\mu\pi}{2}\right) \qquad \mu < \frac{1}{2} \qquad -2 < \nu + \mu < 0$$

holds for a small range of μ values. Selected indefinite and definite integrals are

57:10:7
$$\frac{2x}{\pi} - \int_0^x h_1(t)dt = \int_0^x h_{-1}(t)dt = h_0(x)$$

57:10:8
$$\int_0^x t h_0(t)\,dt = x\left(\frac{2}{\pi} - h_{-1}(x)\right) = x h_1(x)$$

57:10:9
$$\int_0^\infty \frac{h_0(t)}{t}\,dt = \frac{\pi}{2}$$

With a redefinition of the argument, equations 57:10:2 and 57:10:3 are the special $\mu = 1$ and $\mu = -1$ cases of the general differintegration formula

57:10:10
$$\frac{d^\mu}{dx^\mu}\{x^{\nu/2} h_\nu(2\sqrt{x})\} = x^{(\nu-\mu)/2} h_{\nu-\mu}(2\sqrt{x}) \qquad \nu > -\frac{1}{2}$$

in which μ may adopt any value.

57:11 COMPLEX ARGUMENT

We present here only the case of the Struve function with a purely imaginary argument:

57:11:1
$$h_\nu(iy) = i^{1-\nu}\ell_\nu(y) = \left[\sin\left(\frac{\nu\pi}{2}\right) + i\cos\left(\frac{\nu\pi}{2}\right)\right]\ell_\nu(y)$$

Here ℓ_ν denotes the hyperbolic Struve function discussed in Section 57:13.

57:12 GENERALIZATIONS

The function introduced in Section 53:12 is a generalization of the Struve function.

57:13 COGNATE FUNCTIONS

The hyperbolic Struve function is to the Struve function as the hyperbolic Bessel function $I_\nu(x)$ is to the Bessel function $J_\nu(x)$. It may be defined by the definite integral

57:13:1 $$\ell_\nu(x) = \frac{2(x/2)^\nu}{\sqrt{\pi}\,\Gamma(\frac{1}{2}+\nu)} \int_0^1 (1-t^2)^{\nu-1/2}\sinh(xt)dt \qquad \nu > -\frac{1}{2}$$

or by its equivalent on replacement of t by $\sin(\theta)$ or $\cos(\theta)$.

A similar definite integral represents the difference between the hyperbolic Struve and the hyperbolic Bessel function of identical argument but order of opposite sign

57:13:2 $$\ell_\nu(x) - I_{-\nu}(x) = \frac{-2(x/2)^\nu}{\sqrt{\pi}\,\Gamma(\frac{1}{2}+\nu)} \int_0^\infty (1+t^2)^{\nu-1/2}\sin(xt)dt \qquad \nu < \frac{1}{2}$$

Replacement of t by $\tan(\theta)$ or $\sinh(u)$ leads to equivalent definitions.

The hyperbolic Struve function is defined (as a real function) for negative argument only if ν is an integer: the function $\ell_n(x)$ is odd if n is even and vice versa. Figure 57-2 shows maps of $\ell_n(x)$ for $n = 0, 1, 2, \ldots$. The corresponding hyperbolic Struve functions of negative integer order are related to those graphed by

57:13:3 $$\ell_{-n}(x) = \ell_n(x) + \frac{2}{\pi}\left(\frac{x^{n-1}}{(2n-1)!!} - \frac{1!!x^{n-3}}{(2n-3)!!} + \frac{3!!x^{n-5}}{(2n-5)!!} - \cdots - \frac{(-1)^n(2n-3)!!x^{1-n}}{1!!}\right)$$

$$n = 1, 2, 3, \ldots$$

for example, $\ell_{-1}(x) = \ell_1(x) + (2/\pi)$ and $\ell_{-2}(x) = \ell_2(x) + (2x/3\pi) - (2/\pi x)$.

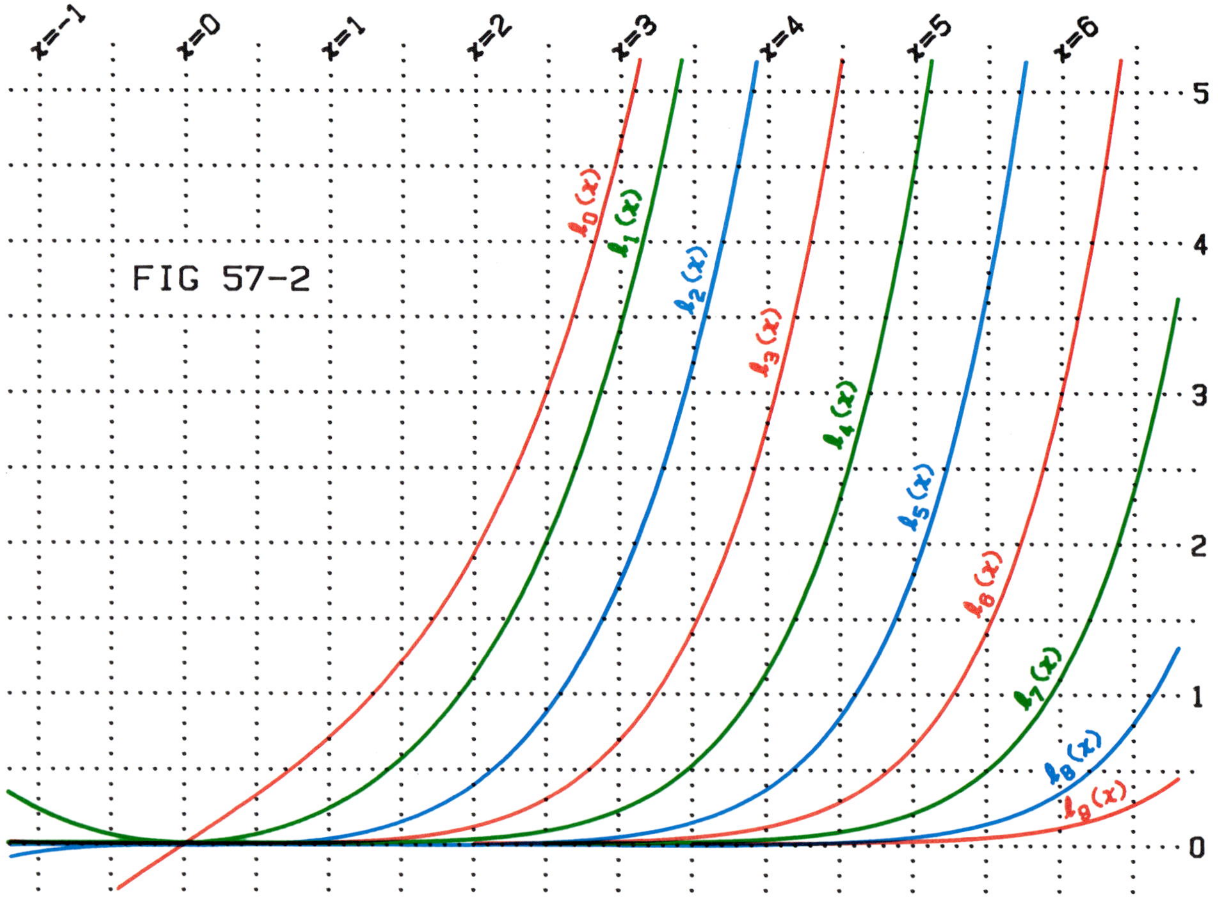

For orders of $-\frac{1}{2}$, $-\frac{3}{2}$, ..., we have the identity $\ell_{-n-1/2}(x) = \mathrm{I}_{n+1/2}(x)$ where the latter functions are discussed in Sections 50:4 and 28:13.

The expansion of $\ell_\nu(x)$ mirrors that given for $\mathrm{h}_\nu(x)$ in 57:6:1 except that the negative signs should be expunged. Similarly, if the alternating signs on the right-hand side of 57:6:7 are replaced by uniformly positive signs, the series is an asymptotic representation if $\mathrm{I}_{-\nu}(x) - \ell_\nu(x)$. Again, it is only by distinctions in sign that the recursion formula

57:13:4 $$\ell_{\nu+1}(x) = \ell_{\nu-1}(x) - \frac{2\nu}{x}\ell_\nu(x) - \frac{(x/2)^\nu}{\sqrt{\pi}\,\Gamma(\nu+\frac{3}{2})}$$

for the hyperbolic Struve function differs from 57:3:1 for $\mathrm{h}_{\nu+1}(x)$.

Because the Struve function and the hyperbolic Struve function are generally complex for negative argument, the algorithm in this chapter will accept only positive arguments. Accordingly, the input of a negative x into the Section 57:8 algorithm is interpreted as an instruction to compute $\ell_\nu(x)$, instead of the $\mathrm{h}_\nu(x)$ that is calculated when the input x is a positive number.

The differentiation formula

57:13:5 $$\frac{\mathrm{d}}{\mathrm{d}x}\ell_\nu(x) = \frac{\ell_{\nu-1}(x) + \ell_{\nu+1}(x)}{2} + \frac{(x/2)^\nu}{2\sqrt{\pi}\,\Gamma(\nu+\frac{3}{2})}$$

produces $\mathrm{d}\,\ell_0(x)/\mathrm{d}x = \ell_{-1}(x) = \ell_1(x) + (2/\pi)$ as a special case, and, apart from a sign, is analogous to the first equation in Section 57:10. Similarly, allowing for changes of sign, all the other relationships in that section have their analogs for the hyperbolic Struve function.

CHAPTER 58

THE INCOMPLETE BETA FUNCTION $B(\nu;\mu;x)$

Many simpler functions are special cases of the incomplete beta function. For example, as described in Section 58:14, the indefinite integral of any trigonometric or hyperbolic function, raised to an arbitrary power, is expressible as an incomplete beta function.

58:1 NOTATION

Alternative symbolisms for $B(\nu;\mu;x)$ are $B_x(\nu,\mu)$ and the product $I_x(\nu,\mu)\,B(\nu,\mu)$ where $B(\nu,\mu)$ is the complete beta function discussed in Section 43:13. The $I_x(\nu,\mu)$ function, which has statistical applications [see Section 27:14], is the incomplete beta function ratio, equal to the quotient

58:1:1
$$I_x(\nu,\mu) = \frac{B(\nu;\mu;x)}{B(\nu,\mu)} = \frac{\Gamma(\nu+\mu)\,B(\nu;\mu;x)}{\Gamma(\nu)\,\Gamma(\mu)}$$

of the incomplete beta function by the complete beta function.

The adjective *incomplete* reflects the fact that the upper limit of *Euler's integral of the first kind* [see definition 58:3:1] is generally less than the value of unity that is required to "complete," that is, symmetrize, the integral. This incompleteness prevents the interchangeability of the ν and μ parameters.

58:2 BEHAVIOR

Being trivariate, the incomplete beta function has a behavior sufficiently complicated that it cannot be adequately described either in words or planar maps.

While allowing μ to adopt any real value, we shall require ν to be positive. For otherwise unrestricted values of the ν and μ parameters, $B(\nu;\mu;x)$ is defined only for arguments in the range $0 \le x < 1$. However, if ν is a positive integer, the formula

58:2:1
$$B(n;\mu;-x) = (-1)^n\,B\left(n;1-\mu-n;\frac{x}{1+x}\right) \qquad x>0 \qquad n=1,2,3,\ldots$$

permits extension of the argument to general negative values. Similarly, extension to arguments exceeding unity is possible if the μ parameter is a positive integer:

58:2:2 $$B(\nu;m;x+1) = \frac{(m-1)!}{(\nu)_m} - (-1)^m B\left(m;1-\nu-m;\frac{x}{1+x}\right) \qquad x \geq 0 \qquad m = 1, 2, 3, \ldots$$

Here $(\nu)_m$ is a Pochhammer polynomial [Chapter 18].

58:3 DEFINITIONS

The incomplete beta function is defined by the indefinite integrals

58:3:1 $$B(\nu;\mu;x) = \int_0^x t^{\nu-1}(1-t)^{\mu-1}dt \qquad 0 \leq x < 1$$

58:3:2 $$B(\nu;\mu;x) = \int_0^T \frac{t^{\nu-1}dt}{(1+t)^{\nu+\mu}} \qquad 0 \leq T = \frac{x}{1-x} < \infty$$

58:3:3 $$B(\nu;\mu;x) = 2\int_0^T \sin^{2\nu-1}(t)\cos^{2\mu-1}(t)dt \qquad 0 \leq T = \arcsin(\sqrt{x}) < \frac{\pi}{2}$$

and

58:3:4 $$B(\nu;\mu;x) = 2\int_0^T \tanh^{2\nu-1}(t)\,\mathrm{sech}^{2\mu}(t)dt \qquad 0 \leq T = \mathrm{artanh}(\sqrt{x}) < \infty$$

As well, the incomplete beta function may be defined as a number of definite integrals, the simplest being

58:3:5 $$B(\nu;\mu;x) = x^\nu \int_0^1 t^{\nu-1}(1-xt)^{\mu-1}dt$$

Using the notation of Section 43:14, the incomplete beta function may be synthesized from a reciprocal linear function:

58:3:6 $$\frac{1}{1-x} \xrightarrow[1-\nu-\mu]{-\nu} \frac{\nu B(\nu;\mu;x)}{x^\nu(1-x)^\mu} \qquad x = X \qquad 0 \leq x < 1$$

58:4 SPECIAL CASES

When the ν parameter equals unity:

58:4:1 $$B(1:\mu;x) = \frac{1-(1-x)^\mu}{\mu}$$

By sufficient applications of the recurrence formula 58:5:2, this case can also provide expressions for $B(2;\mu;x)$, $B(3;\mu;x)$, etc. Formula 58:4:1 becomes indeterminate if μ is zero, but, by considering the $\mu \to 0$ limit, one concludes that $B(1;0;x) = -\ln(1-x)$.

When the μ parameter equals unity:

58:4:2 $$B(\nu;1;x) = \frac{x^\nu}{\nu}$$

This formula may be used in conjunction with recursion 58:5:3 to produce the polynomial expression

58:4:3 $$B(\nu;m;x) = x^\nu \sum_{j=0}^{m-1} \binom{m-1}{j} \frac{(-x)^j}{j+\nu} \qquad m = 1, 2, 3, \ldots$$

for the incomplete beta function having a positive integer as its μ parameter. When μ is zero, one has

58:4:4 $$B(\nu;0;x) = \sum_{j=0}^{\infty} \frac{x^{j+\nu}}{j+\nu} = \ln_\nu\left(\frac{1}{1-x}\right) = x^\nu \Phi(x;1;\nu)$$

where $\ln_\nu$ is the generalized logarithm discussed in Section 25:12 and Φ denotes Lerch's function [see Section 64:12]. Some important instances of this result are

58:4:5 $$B(n+1;0;x) = -\ln(1-x) - x - \frac{x^2}{2} - \frac{x^3}{3} - \cdots - \frac{x^n}{n} = \sum_{j=n}^{\infty} \frac{x^{j+1}}{j+1} \qquad n = 0, 1, 2, \ldots$$

58:4:6 $$B(n+\tfrac{1}{2};0;x) = 2\left[\operatorname{artanh}(\sqrt{x}) - \sqrt{x} - \frac{\sqrt{x^3}}{3} - \frac{\sqrt{x^5}}{5} - \cdots - \frac{\sqrt{x^{2n-1}}}{2n-1}\right]$$
$$= \sum_{j=n}^{\infty} \frac{x^{j+1/2}}{j+\frac{1}{2}} \qquad n = 0, 1, 2, \ldots$$

and

58:4:7 $$B(\tfrac{1}{2} \pm \tfrac{1}{4};0;x) = 2\operatorname{artanh}(|x^{1/4}|) \mp 2\arctan(|x^{1/4}|)$$

Negative integer μ cases can be deduced from 58:4:4 by one or more applications of the recurrence $(\mu - 1)\,B(\nu;\mu - 1;x) = (\nu - 1)\,B(\nu - 1;\mu{:}x) - x^{\nu-1}(1-x)^{\mu-1}$, which is a restatement of 58:5:2.

Whenever ν and μ are both positive multiples of $\frac{1}{2}$, the incomplete beta function reduces to an elementary function. There are four general cases. When both ν and μ are even multiples of $\frac{1}{2}$ (i.e., both are integers), the general formula is

58:4:8 $$B(n+1;m+1;x) = \int_0^x t^n(1-t)^m dt = x^{n+1}\sum_{j=0}^{m}\binom{m}{j}\frac{(-x)^j}{n+j+1} \qquad n, m = 0, 1, 2, \ldots$$

When ν is an odd but μ an even multiple of $\frac{1}{2}$, the general formula is

58:4:9 $$B(n+\tfrac{1}{2};m+1;x) = 2\int_0^{\sqrt{x}} t^{2n}(1-t^2)^m dt = x^{n+1/2}\sum_{j=0}^{m}\binom{m}{j}\frac{(-x)^j}{n+j+\frac{1}{2}} \qquad n, m = 0, 1, 2, \ldots$$

When ν is an even but μ an odd multiple of $\frac{1}{2}$, the general formula is

58:4:10 $$B(n+1;m+\tfrac{1}{2};x) = 2\int_{\sqrt{1-x}}^{1} t^{2m}(1-t^2)^n dt = \sum_{j=0}^{n}(-1)^j\binom{n}{j}\frac{1-(1-x)^{m+j+1/2}}{m+j+\frac{1}{2}} \qquad n, m = 0, 1, 2, \ldots$$

The final case, when both ν and μ are odd multiples of $\frac{1}{2}$, is more complicated and invariably leads to an $\arcsin(\sqrt{x})$ term. The general formula can be obtained by combining

58:4:11 $$B(n+\tfrac{1}{2};m+\tfrac{1}{2};x) = 2\int_0^T \sin^{2n}(t)\cos^{2m}(t)dt = 2\sum_{j=0}^{m}(-1)^j\binom{m}{j}\int_0^T \sin^{2n+2j}(t)dt \qquad n, m = 0, 1, 2, \ldots$$

where $T = \arcsin(\sqrt{x})$, with the indefinite integral 32:10:6. The simplest instance is $B(\frac{1}{2};\frac{1}{2};x) = 2\arcsin(\sqrt{x})$, and other simple cases can be obtained from this result and recursion formulas 58:5:2 and 58:5:3.

When the two parameters differ only in sign, there is the simple result

58:4:12 $$B(\nu;-\nu;x) = \frac{1}{\nu}\left(\frac{x}{1-x}\right)^\nu$$

but there is no comparable general formula for $B(\nu;\nu;x)$.

Many other special cases, while not evaluable as established functions, can be expressed as simple indefinite integrals. Some of these are cited in Section 58:14.

58:5 INTRARELATIONSHIPS

The formula

58:5:1 $$B(\mu;\nu;x) = \frac{\Gamma(\mu)\Gamma(\nu)}{\Gamma(\nu+\mu)} - B(\nu;\mu;1-x)$$

shows the effect of interchanging the parameters. As well, it provides an argument-reflection formula.

The recursion formulas

58:5:2 $$B(\nu+1;\mu;x) = \frac{\nu}{\mu}B(\nu;\mu+1;x) - \frac{x^\nu(1-x)^\mu}{\mu}$$

58:5:3 $$B(\nu;\mu+1;x) = \frac{\mu}{\nu}B(\nu+1;\mu;x) + \frac{x^\nu(1-x)^\mu}{\nu}$$

link two incomplete beta functions. There is a large number of interrelationships connecting three incomplete beta functions, of which

58:5:4 $$B(\nu;\mu;x) = B(\nu+1;\mu;x) + B(\nu;\mu+1;x)$$

is the simplest.

58:6 EXPANSIONS

The important power series

58:6:1 $$B(\nu;\mu;x) = x^\nu(1-x)^\mu\left[\frac{1}{\nu} + \frac{\nu+\mu}{\nu(\nu+1)}x + \frac{(\nu+\mu)(\nu+\mu+1)}{\nu(\nu+1)(\nu+2)}x^2 + \cdots\right] = \frac{(1-x)^\mu}{\nu}\sum_{j=0}^{\infty}\frac{(\nu+\mu)_j}{(1+\nu)_j}x^{j+\nu}$$

converges for $0 \le x < 1$. When x is close to unity, convergence may be slow and the alternative series

58:6:2 $$B(\nu;\mu;x) = \frac{\Gamma(\mu)\,\Gamma(\nu)}{\Gamma(\mu+\nu)} - \frac{x^\nu}{\mu}\sum_{j=0}^{\infty}\frac{(\nu+\mu)_j}{(1+\mu)_j}(1-x)^{j+\mu}$$

may then be preferable.

58:7 PARTICULAR VALUES

Generally, we have

	$x = 0$	$x = 1$
$B(\nu;\mu;x)$	0	$\dfrac{\Gamma(\nu)\,\Gamma(\mu)}{\Gamma(\nu+\mu)}$

but for certain interrelationships between the ν and μ parameters, the incomplete beta function of moiety argument also acquires particular values. For example:

58:7:1 $$B(\nu;\nu;\tfrac{1}{2}) = \frac{\Gamma^2(\nu)}{2\Gamma(2\nu)} = \frac{\sqrt{\pi}\,\Gamma(\nu)}{4^\nu\Gamma(\nu+\frac{1}{2})}$$

58:7:2 $$B(\nu;-\nu;\tfrac{1}{2}) = \frac{1}{\nu}$$

and

58:7:3 $$B(\nu;1-\nu;\tfrac{1}{2}) = \tfrac{1}{2}G(\nu)$$

where G is Bateman's function described in Section 44:13. The special cases $B(\frac{1}{4};\frac{1}{4};\frac{1}{2}) = \pi/U$, $B(\frac{1}{2};\frac{1}{2};\frac{1}{2}) = \pi/2$ and $B(\frac{3}{4};\frac{3}{4};\frac{1}{2}) = U$ of equation 58:7:1 should be noted, U being the ubiquitous constant cited in Section 1:7.

58:8 NUMERICAL VALUES

The algorithm below accepts any positive ν value and any argument in the range $0 \le x < 1$. The μ parameter is unrestricted except that if μ adopts any one of the values $-1, -2, -3, \ldots$, then $\nu + \mu$ must be positive.

If $x < 0.7$, the algorithm uses expansion 58:6:1 in the form

58:8:1 $$B(\nu;\mu;x) = f_0 + t_0 + t_1 + \cdots + t_j + \cdots + t_{J-1} + R_J$$

where $f_0 = 0$, $t_0 = x^\nu(1-x)^\mu/\nu$ and $t_{j+1} = t_j x(\nu + \mu + j)/(\nu + 1 + j)$. The R_J term is an approximation to the infinite sum of the terms $t_J + t_{J+1} + t_{J+2} + \cdots$ which is given exactly by

58:8:2 $$R_J = t_J\left\{1 + x\frac{1 + (\nu+\mu)/J}{1 + (\nu+1)/J} + x^2\frac{[1 + (\nu+\mu+1)/J]\,[1 + (\nu+\mu)/J]}{[1 + (\nu+2)/J]\quad[1 + (\nu+1)/J]} + \cdots\right\}$$

and approximately, for large J, by

58:8:3 $$R_J \simeq t_J\left\{[1 + x + x^2 + \cdots] + (\mu - 1)\left(\frac{x}{J} + \frac{2x^2}{J} + \frac{3x^3}{J}\right) + \text{smaller terms}\right\}$$
$$= t_J\left(\frac{1}{1-x} + \frac{(\mu-1)x}{J(1-x)^2} + \cdots\right)$$

If $\mu = 1$, this evaluation of R_J is exact, even for $J = 1$, and $B(\nu;\mu;x)$ is calculated as $f_0 + t_0 + t_1/(1-x)$. Otherwise, J is determined by adding terms into series 58:8:1 until $t_j(\mu - 1)x/[f_j j(1-x)^2]$ is less than 10^{-8}, with $f_j = f_0 + t_0 + t_1 + \cdots + t_{j-1}$, whereupon $J = j$ and the output value is $B(\nu;\mu;x) \simeq f_J + t_J[1 + (\mu - 1)x/J(1-x)]/(1-x)$.

If μ is zero or a negative integer, the procedure of the last paragraph is adopted irrespective of the magnitude of x. However, the recurrence $B(\nu;\mu;x) = -x^{\nu-1}(1-x)^\mu/\mu + [(\nu - 1)/\mu]B(\nu - 1;\mu + 1;x)$ is first applied as many times as necessary until $B(\nu + \mu; 0;x)$ is reached. The sought incomplete beta function is then evaluated as $f_0 + kB(\nu + \mu;0;x)$ where f_0 and k result from updating at each recursion stage.

If $x \ge 0.7$ and μ is not a negative integer, the reflection formula 58:5:1 is utilized. The incomplete beta function $B(\mu;\nu;1-x)$ is calculated as described above, and the three gamma functions are evaluated by a routine similar to that described in Section 43:8.

Generally, the algorithm is of 24-bit precision (i.e., the fractional error does not exceed 6×10^{-8}). Its implementation may be slow, especially when the argument is close to unity and the μ parameter equals zero or a negative integer.

Storage needed: ν, f, j, μ, g, t, x and a

Input restrictions: $0 \le x < 1$, $\nu > 0$, μ unrestricted but if $\mu = -1, -2, -3, \ldots$ then $(\nu + \mu) > 0$

```
Input ν >>>>>>>
        Set f = j = 0
Input μ >>>>>>>
        Set g = t = 1
Input x >>>>>>>
        If μ + Int(|μ|) ≠ 0 go to (2)
    (1) If μ = 0 go to (7)
        Replace f by f − x^(ν−1)(1 − x)^μ/μ
        Replace t by (ν − 1) t/μ
        Replace ν by ν − 1
        Replace μ by μ + 1
        Go to (1)
    (2) If x < 0.7 go to (7)
        Set a = ν + μ
        If a + Int(|a|) ≠ 0 go to (3)
        Set t = −1
        Go to (6)
    (3) Replace g by g/a
        Replace a by a + 1
```

If $a < 3$ go to (3)

Set $f = \left[1 + \dfrac{2}{7a^2}\left(\dfrac{2}{3a^2} - 1\right)\right] \Big/ 30a^2$

Replace f by $[(f-1)/12a] - a[\ln(a) - 1]$

Replace g by $g\exp(-f)\sqrt{2\pi/a}$

If $t < 0$ go to (5)

If $t = 0$ go to (4)

Replace g by $1/g$

Set $t = 0$

Set $a = \nu$

Go to (3)

(4) Set $t = -1$

Set $a = \mu$

Go to (3)

(5) Set $f = g$

(6) Set $a = \mu$

Set $\mu = \nu$

Set $\nu = a$

Replace x by $1 - x$

(7) Replace t by $tx^{\nu}(1-x)^{\mu}/\nu$

Set $a = 10^8(\mu - 1)x/(1-x)^2$

(8) Replace f by $f + t$

Replace t by $tx(\nu + \mu + j)/(\nu + 1 + j)$

Replace j by $j + 1$

If $j < |at|$ go to (8)

Replace f by $f + t\left[\dfrac{a}{10^8 j} + \dfrac{1}{1-x}\right]$

Output f

$f \simeq B(\nu;\mu;x)$ <<<<

Test values:
$B(3;3;\frac{1}{2}) = 0.0166666667$
$B(\frac{1}{4};0;0.8) = 5.09468017$
$B(2;-1;0.2) = 0.0268564487$
$B(\frac{1}{2};-1.5;0.9) = 24.0000000$
$B(\frac{1}{2};\frac{1}{2};0.75) = 2.09439510$

58:9 APPROXIMATIONS

For small enough values of the argument, the approximation

58:9:1
$$B(\nu;\mu;x) \simeq x^{\nu}\left[\frac{1}{\nu} + \left(\frac{1-\mu}{1+\nu}\right)x\right] \qquad \text{small } x$$

holds. The corresponding approximation when x is close to unity is

58:9:2
$$B(\nu;\mu;x) \simeq \frac{\Gamma(\nu)\,\Gamma(\mu)}{\Gamma(\nu+\mu)} - (1-x)\left[\frac{1}{\mu} + \left(\frac{1-\nu}{1+\mu}\right)(1-x)\right] \qquad \text{small } (1-x)$$

58:10 OPERATIONS OF THE CALCULUS

Differentiation of the incomplete beta function with respect to its argument gives

58:10:1
$$\frac{d}{dx}B(\nu;\mu;x) = x^{\nu-1}(1-x)^{\mu-1}$$

58:11 COMPLEX ARGUMENT

The incomplete beta function is seldom encountered with an argument that is imaginary or complex.

58:12 GENERALIZATIONS

The Gauss function of Chapter 60 represents a generalization of the incomplete beta function. The incomplete beta function may be expressed as a Gauss function with a unity numeratorial parameter

58:12:1 $$B(\nu;\mu;x) = \frac{1}{\nu} x^{\nu}(1-x)^{\mu}F(1,\nu+\mu;1+\nu;x)$$

or with a numeratorial parameter less by unity than the denominatorial parameter

58:12:2 $$B(\nu;\mu;x) = \frac{1}{\nu} x^{\nu}F(\nu,1-\mu;1+\nu;x)$$

58:13 COGNATE FUNCTIONS

The incomplete beta function has some similarities to the Legendre function of Chapter 59. Also, it is allied to the incomplete gamma function discussed in Chapter 45.

58:14 RELATED TOPICS

If $f(x)$ denotes any hyperbolic or trigonometric function (i.e., any one of the twelve functions addressed in Chapters 28–30 and 32–34) and $f^{\lambda}(x)$ represents such a function raised to an arbitrary power, then the indefinite integral $\int f^{\lambda}(t)dt$ is an incomplete beta function.

Table 58.14.1

f	ν	μ
sinh	$\frac{1+\lambda}{2}$	$\frac{-\lambda}{2}$
cosh	$\frac{1}{2}$	$\frac{-\lambda}{2}$
sech	$\frac{1}{2}$	$\frac{\lambda}{2}$
csch	$\frac{1-\lambda}{2}$	$\frac{\lambda}{2}$
tanh	$\frac{1+\lambda}{2}$	0
coth	$\frac{1-\lambda}{2}$	0
sin	$\frac{1+\lambda}{2}$	$\frac{1}{2}$
cos	$\frac{1}{2}$	$\frac{1+\lambda}{2}$
sec	$\frac{1}{2}$	$\frac{1-\lambda}{2}$
csc	$\frac{1-\lambda}{2}$	$\frac{1}{2}$
tan	$\frac{1+\lambda}{2}$	$\frac{1-\lambda}{2}$
cot	$\frac{1-\lambda}{2}$	$\frac{1+\lambda}{2}$

The general formula is

58:14:1
$$\int_0^x f^\lambda(t)dt = \frac{1}{2}B(\nu;\mu;\tanh^2(x)) \qquad x \ge 0$$

for hyperbolic functions and

58:14:2
$$\int_0^x f^\lambda(t)dt = \frac{1}{2}B(\nu;\mu;\sin^2(x)) \qquad 0 < x < \frac{\pi}{2}$$

for trigonometric functions. The parameters ν and μ of the incomplete beta functions depend on λ as shown in Table 58.14.1. For the integral to be finite, the ν parameter of the corresponding incomplete beta function must be positive: this limits the range of λ in certain cases; for example, it requires $\lambda < 1$ for convergence of $\int \text{csch}^\lambda(t)dt$ with zero lower limit.

Interchanging the roles of the ν and μ parameters produces expressions for certain complementary indefinite integrals. Thus, for all hyperbolic functions, except tanh and coth, one has

58:14:3
$$\int_x^\infty f^\lambda(t)dt = \frac{1}{2}B(\mu;\nu;\text{sech}^2(x))$$

while for any trigonometric function whatsoever

58:14:4
$$\int_x^{\pi/2} f^\lambda(t)dt = \frac{1}{2}B(\mu;\nu;\cos^2(x))$$

The same table still applies, but the restriction, if any, on λ is now governed by the requirement that μ be positive.

CHAPTER

59

THE LEGENDRE FUNCTIONS $P_\nu(x)$ AND $Q_\nu(x)$

The functions of this chapter arise in several physical contexts, notably to describe the properties of the surface of a sphere [see Section 59:14]. Most of the chapter is concerned with the bivariate $P_\nu(x)$ and $Q_\nu(x)$ functions, but attention is given in Section 59:13 to the *associated Legendre functions* $P_\nu^{(\mu)}(x)$ and $Q_\nu^{(\mu)}(x)$, which are trivariate.

59:1 NOTATION

$P_\nu(x)$ is known as the *Legendre function of the first kind* and $Q_\nu(x)$ as the *Legendre function of the second kind*. The variables ν and x are termed the degree and argument, respectively, of the Legendre functions. Frequently, the argument is expressed as a cosine so that the symbols $P_\nu(\cos(\theta))$ and $P_\nu(\cosh(X))$ are often encountered.

Sometimes the title "Legendre function" is taken to embrace the associated Legendre functions that we discuss separately in Section 59:13. The third variable, μ, in these functions is termed the order of $P_\nu^{(\mu)}(x)$ or $Q_\nu^{(\mu)}(x)$ and is zero for $P_\nu(x)$ and $Q_\nu(x)$.

As explained in Section 59:11, slightly different definitions for the Legendre functions may be adopted according to whether the argument is regarded as an unrestricted complex variable or as a real number confined to $-1 < x < 1$. Slightly different notations sometimes reflect this distinction. Other notational conventions are summarized by Abramowitz and Stegun [Chapter 8].

59:2 BEHAVIOR

Unless the degree is an integer, the Legendre functions adopt complex values when the argument is real and less than -1. Accordingly, we restrict consideration to the range $x \geq -1$. Perhaps it is better to consider two subranges, $-1 < x \leq 1$ and $1 \leq x < \infty$, inasmuch as the behaviors of the Legendre functions are substantially different in the two. In practical applications $-1 < x \leq 1$ is the more important subrange and accordingly it is emphasized in this chapter.

The contour maps Figures 59-1 and 59-2 depict the rather elaborate behaviors of the Legendre functions. Notice that these diagrams segment naturally into rectangular domains. The boundaries of these domains are $x = \pm 1$ and $\nu = 0, 1$ for the $P_\nu(x)$ function; for the $Q_\nu(x)$ function they are $x = \pm 1$ and $\nu = -1, -2, -3, \ldots$. Section 59:7 details the values adopted by the Legendre functions along these boundaries.

The number of zeros of the Legendre function of the first kind is given by

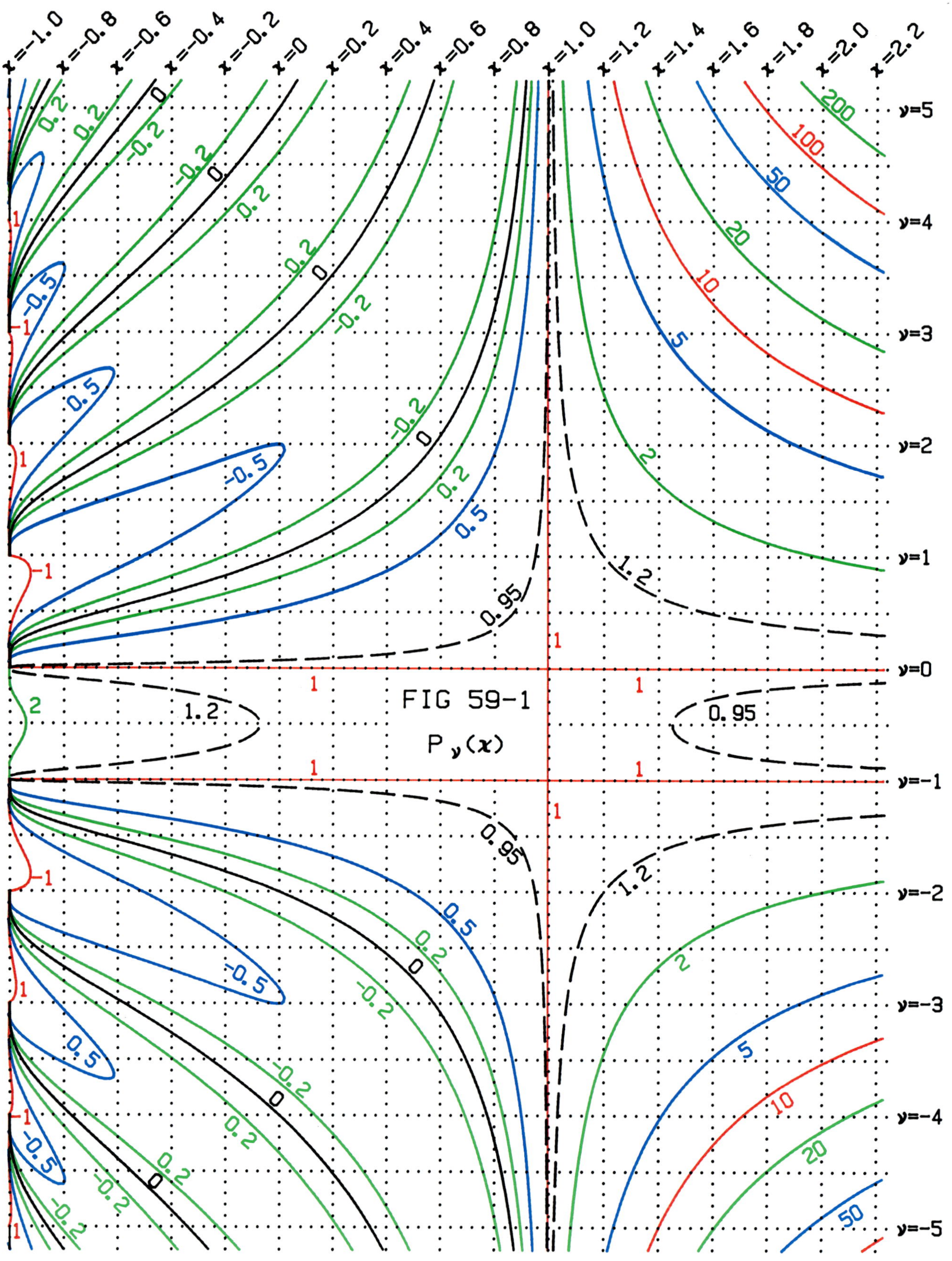
x=−1.0
x=−0.8
x=−0.6
x=−0.4
x=−0.2
x=0
x=0.2
x=0.4
x=0.6
x=0.8
x=1.0
x=1.2
x=1.4
x=1.6
x=1.8
x=2.0
x=2.2
ν=5
ν=4
ν=3
ν=2
ν=1
ν=0
ν=−1
ν=−2
ν=−3
ν=−4
ν=−5
FIG 59-1
P ν(x)
200
100
50
20
10
5
2
1.2
0.95
0.5
0.2
0
−0.2
−0.5
1
−1

59:2:1
$$N_r(P_\nu) = \begin{cases} n & n-1 < \nu \le n \\ 0 & -1 \le \nu \le 0 \\ n & -n-1 \le \nu < -n \end{cases} \qquad n = 1, 2, 3, \ldots$$

and all these zeros lie in the range $-1 < x < 1$. For the Legendre function of the second kind, the formula

59:2:2
$$N_r(Q_\nu) = \begin{cases} n & n - \frac{3}{2} < \nu \le n - \frac{1}{2} \\ 0 & -1 < \nu \le -\frac{1}{2} \\ n+1+1^* & -\frac{1}{2} - n < \nu < -n \\ n & -n-1 < \nu \le -\frac{1}{2} - n \end{cases} \qquad n = 1, 2, 3, \ldots$$

specifies the number of zeros as a function of the degree ν. Most of these zeros of Q_ν lie in the $-1 < x < 1$ range, but the asterisk in 59:2:2 indicates that sometimes a single zero is located in the $1 < x < \infty$ range of real arguments. Note that for $-\frac{3}{2} < \nu < -1$, $-\frac{5}{2} < \nu < -2, \ldots$ one zero of $Q_\nu(x)$ is located so close to $x = -1$ that it cannot be properly portrayed on Figure 59-2.

The values $\nu = -1, -2, -3, \ldots$ are omitted from formula 59:2:2 [and also from the table of particular values of $Q_\nu(x)$ in Section 59:7] because $Q_\nu(x)$ is poorly defined for these degrees. Consider $Q_{-3}(x)$ and refer to Figure 59-2. As ν approaches -3 from more positive values, $Q_\nu(x) \to -\infty$ for $-(1\sqrt{3}) < x < (1/\sqrt{3})$ whereas $Q_\nu(x) \to \infty$ for $-1 < x < -(1/\sqrt{3})$ and $(1/\sqrt{3}) < x < 1$. On the other hand, the approach of ν to -3 from more negative values leads $Q_\nu(x)$ toward $+\infty$ within the $\pm 1/\sqrt{3}$ range and toward $-\infty$ outside. At the values $x = \pm 1/\sqrt{3}$ (these are the zeros of P_{-3}, the *first* kind of Legendre function), there is a flip in the signs of the infinities so that all the contour lines crowd to these points and, as it were, penetrate the $Q_{-3}(x) = \pm\infty$ barrier there.

Behaviors for very large argument, and for x in the vicinities of ± 1, may be deduced from the formulas presented in Section 59:9.

59:3 DEFINITIONS

Several definite integrals, including

59:3:1
$$P_\nu(x) = \frac{1}{\pi}\int_0^\pi [x + \sqrt{x^2-1}\cos(t)]^\nu dt \qquad x \ge 1$$

59:3:2
$$Q_\nu(x) = \int_0^\infty [x + \sqrt{x^2-1}\cosh(t)]^{-\nu-1} dt \qquad x \ge 1 \qquad \nu > -1$$

59:3:3
$$P_\nu(x) = \frac{-\sin(\nu\pi)}{\pi}\int_0^\infty \frac{t^\nu dt}{\sqrt{1 + 2xt + t^2}} \qquad -1 < x < 1 \qquad -1 < \nu < 0$$

59:3:4
$$Q_\nu(x) = \frac{1}{2^{1+\nu}}\int_{-1}^1 \frac{(1-t^2)^\nu dt}{(x-t)^{1+\nu}} \qquad x > 1 \qquad \nu > -1$$

59:3:5
$$P_\nu(x) = \frac{1}{\pi}\int_1^\infty i\{[x + it\sqrt{1-x^2}]^{-1-\nu} - [x - it\sqrt{1-x^2}]^{-1-\nu}\}\frac{dt}{\sqrt{t^2-1}} \qquad -1 < x \le 1 \qquad \nu > -1$$

59:3:6
$$Q_\nu(x) = \frac{1}{2}\int_1^\infty \{[x + it\sqrt{1-x^2}]^{-1-\nu} + [x - it\sqrt{1-x^2}]^{-1-\nu}\}\frac{dt}{\sqrt{t^2-1}} \qquad -1 < x \le 1 \qquad \nu > -1$$

may be used to define Legendre functions. Despite appearances to the contrary, the integrands in the final two definitions above are wholly real: all imaginary terms cancel after binomial expansion.

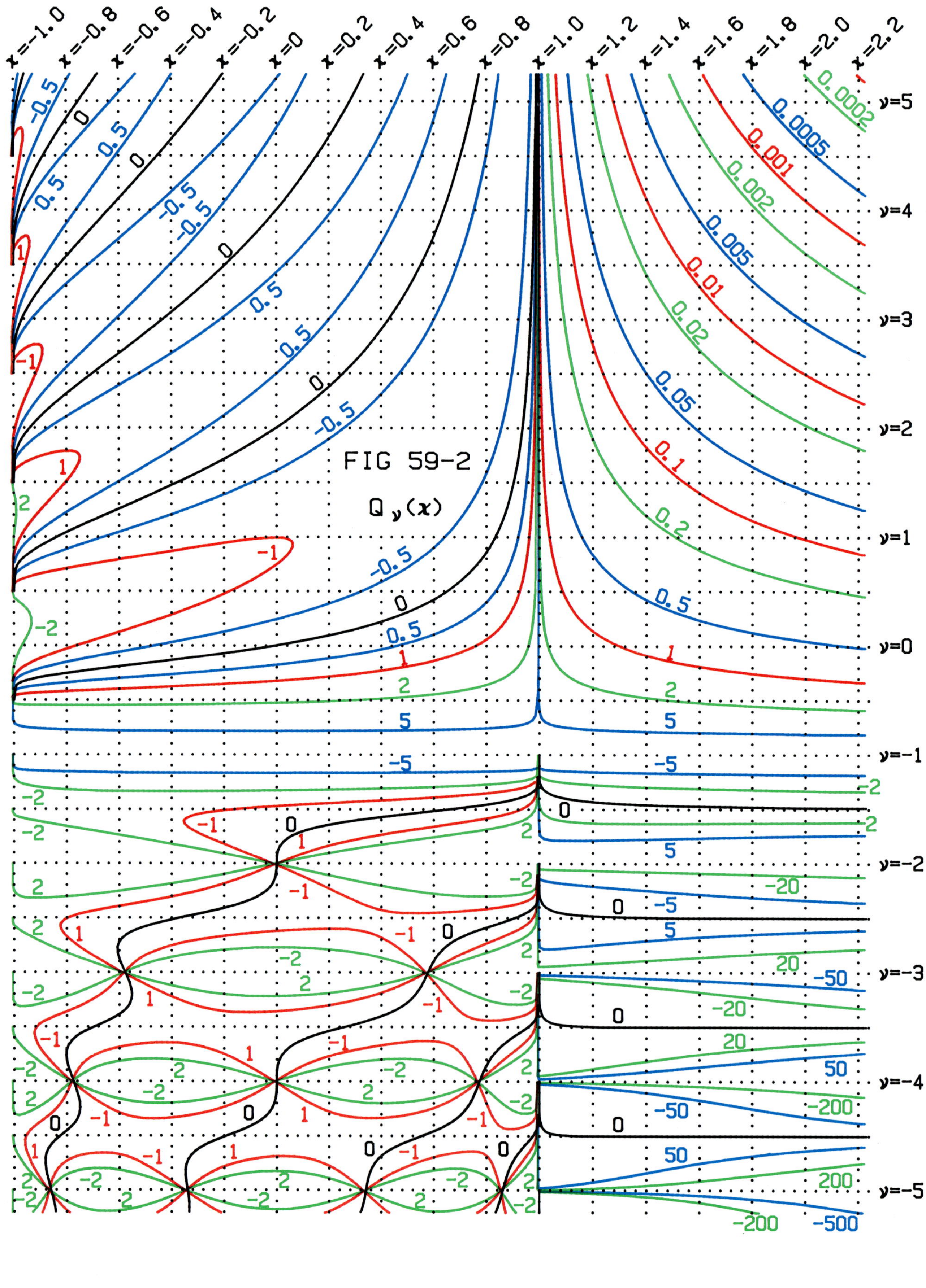
x=−1.0
x=−0.8
x=−0.6
x=−0.4
x=−0.2
x=0
x=0.2
x=0.4
x=0.6
x=0.8
x=1.0
x=1.2
x=1.4
x=1.6
x=1.8
x=2.0
x=2.2
ν=5
ν=4
ν=3
ν=2
ν=1
ν=0
ν=−1
ν=−2
ν=−3
ν=−4
ν=−5
FIG 59-2
Q ν (x)

The Legendre functions may also be represented by the indefinite integrals

59:3:7 $$P_\nu(\cosh(X)) = \frac{\sqrt{2}}{\pi}\int_0^X \frac{\cosh((\nu+\frac{1}{2})t)}{\sqrt{\cosh(t)-\cosh(X)}}\,dt \qquad \cosh(X) = x > 1$$

59:3:8 $$Q_\nu(\cosh(X)) = \frac{1}{\sqrt{2}}\int_X^\infty \frac{\exp(-(\nu+\frac{1}{2})t)}{\sqrt{\cosh(t)-\cosh(X)}}\,dt \qquad \cosh(X) = x > 1$$

and

59:3:9 $$P_\nu(\cos(\theta)) = \frac{1}{\pi\sqrt{2}}\int_0^\theta \frac{\cos((\nu+\frac{1}{2})t)}{\sqrt{\cos(t)-\cos(\theta)}}\,dt \qquad 0 < \theta < \pi$$

the last being known as the *Mehler-Dirichlet formula*.

The formulas of Section 59:12, and especially equations 59:6:7 and 59:6:8, often serve to define Legendre functions as special cases of the Gauss function.

With c_1 and c_2 as arbitrary constants, a linear combination of the two Legendre functions generally satisfies *Legendre's differential equation*:

59:3:10 $$(1-x^2)\frac{d^2f}{dx^2} - 2x\frac{df}{dx} + \nu(\nu+1)f = 0 \qquad f = c_1P_\nu(x) + c_2Q_\nu(x)$$

In the notation of Section 43:14, the Legendre function of the first kind may be synthesized from the binomial function $(1+x)^\nu$ by the procedure

59:3:11 $$\left(\frac{1+x}{2}\right)^\nu \xrightarrow[-\nu]{0} P_\nu(x) \qquad X = \frac{1-x}{2}$$

59:4 SPECIAL CASES

When ν is the nonnegative integer n, the Legendre function $P_\nu(x)$ reduces to the Legendre polynomial discussed in Chapter 21. The same polynomials result when n is a negative integer, on account of the identity

59:4:1 $$P_{-n-1}(x) = P_n(x) = \text{polynomial of degree } n \qquad n = 0, 1, 2, \ldots$$

The Legendre function of the second kind becomes an inverse hyperbolic function [Chapter 31] when its degree is zero:

59:4:2 $$Q_0(x) = \begin{cases} \operatorname{artanh}(x) & -1 < x < 1 \\ \operatorname{arcoth}(x) & |x| > 1 \end{cases}$$

Other integer degrees lead to the following special cases:

59:4:3 $$Q_n(x) = \begin{cases} P_n(x)\,Q_0(x) - \text{polynomial of degree } (n-1) & n = 1, 2, 3, \ldots \\ \pm\infty & n = -1, -2, -3, \ldots \end{cases}$$

See equation 59:5:9 for details and Section 21:13 for examples.

When the degree ν is an odd multiple of $\pm\frac{1}{2}$, Legendre functions of both kinds reduce to complete elliptic integrals [see Chapter 61]. The simplest cases are

59:4:4 $$P_{-1/2}(x) = \begin{cases} \dfrac{2}{\pi}K\left(\sqrt{\dfrac{1-x}{2}}\right) & -1 \le x \le 1 \\ \dfrac{2}{\pi}\sqrt{\dfrac{2}{x+1}}\,K\left(\sqrt{\dfrac{x-1}{x+1}}\right) & x > 1 \end{cases}$$

59:4:5 $$Q_{-1/2}(x) = \begin{cases} K\left(\sqrt{\dfrac{1+x}{2}}\right) & -1 \le x \le 1 \\ \sqrt{\dfrac{2}{x+1}}\, K\left(\sqrt{\dfrac{2}{x+1}}\right) & x > 1 \end{cases}$$

59:4:6 $$P_{1/2}(x) = P_{-3/2}(x) = \begin{cases} \dfrac{4}{\pi} E\left(\sqrt{\dfrac{1-x}{2}}\right) - P_{-1/2}(x) & -1 \le x \le 1 \\ \dfrac{2}{\pi}\sqrt{x+\sqrt{x^2-1}}\, E\left(\sqrt{\dfrac{2\sqrt{x^2-1}}{x+\sqrt{x^2-1}}}\right) & x > 1 \end{cases}$$

59:4:7 $$Q_{1/2}(x) = Q_{-3/2}(x) = \begin{cases} K\left(\sqrt{\dfrac{1+x}{2}}\right) - 2E\left(\sqrt{\dfrac{1+x}{2}}\right) & -1 \le x \le 1 \\ xQ_{-1/2}(x) - \sqrt{2x+2}\, E\left(\sqrt{\dfrac{2}{x+1}}\right) & x > 1 \end{cases}$$

and others may be deduced with the aid of recursion formula 59:5:5.

59:5 INTRARELATIONSHIPS

The argument-reflection formulas

59:5:1 $$P_n(-x) = (-1)^n P_n(x) \qquad n = 0, 1, 2, \ldots$$

59:5:2 $$Q_n(-x) = -(-1)^n Q_n(x) \qquad n = 0, 1, 2, \ldots$$

are valid only for integer degree, whereas the degree-reflection formulas

59:5:3 $$P_{-\nu-1}(x) = P_\nu(x)$$

59:5:4 $$Q_{-\nu-1}(x) = Q_\nu(x) - \pi \cot(\nu\pi)\, P_\nu(x)$$

have general validity for real arguments.

The same recursion formula

59:5:5 $$(\nu + 1)\, f_{\nu+1}(x) = (2\nu x + x)\, f_\nu(x) - \nu f_{\nu-1}(x) \qquad f = P \text{ or } Q$$

is obeyed by Legendre functions of both kinds.

The equations

59:5:6 $$Q_\nu(\pm x) = \frac{\pi}{2}[\cot(\nu\pi)P_\nu(\pm x) - \csc(\nu\pi)\, P_\nu(\mp x)] \qquad \nu \neq 0, \pm 1, \pm 2, \ldots \qquad -1 < x < 1$$

59:5:7 $$P_\nu(\pm x) = \frac{-2}{\pi}[\cot(\nu\pi)\, Q_\nu(\pm x) + \csc(\nu\pi)\, Q_\nu(\mp x)] \qquad \nu \neq 0, \pm 1, \pm 2, \ldots \qquad -1 < x < 1$$

59:5:8 $$Q_\nu(x) = P_\nu(x)[Q_0(x) - \psi(\nu + 1) - \gamma] + \sum_{j=1}^{\infty} \frac{(-\nu)_j (1+\nu)_j}{(j!)^2}[\gamma + \psi(j+1)]\left(\frac{1-x}{2}\right)^j$$

$$\nu \neq 0, \pm 1, \pm 2, \ldots$$

interrelate the two kinds of Legendre function when the degree is not an integer. See 59:4:2 for $Q_0(x)$, Chapter 44 for ψ and γ. When the degree is a positive integer:

59:5:9 $$Q_n(x) = P_n(x)\left[Q_0(x) - 1 - \frac{1}{2} - \frac{1}{3} - \cdots - \frac{1}{n}\right] + \sum_{j=1}^{n} \frac{(n+j)!}{(n-j)!(j!)^2}\left[1 + \frac{1}{2} + \frac{1}{3} + \cdots + \frac{1}{j}\right]\left(\frac{x-1}{2}\right)^j$$

$$n = 1, 2, 3, \ldots$$

59:6 EXPANSIONS

As detailed in Section 59:12, Legendre functions may be expressed as Gauss functions in a multitude of ways. Because each of these latter functions may be expanded as the power series 60:6:1, the Legendre functions have very many expansions. Here we present only those that are the most useful computationally.

If, for $-1 < x < 1$, we define the auxiliary Legendre functions

59:6:1 $$\mathrm{lef}_\nu(x) = F\left(\frac{-\nu}{2}, \frac{1+\nu}{2}; \frac{1}{2}; x^2\right) = \sum_{j=0}^{\infty} \left(\frac{-\nu}{2}\right)_j \left(\frac{1+\nu}{2}\right)_j \frac{(2x)^{2j}}{(2j)!}$$

and

59:6:2 $$\mathrm{leg}_\nu(x) = 2xF\left(\frac{1-\nu}{2}, 1+\frac{\nu}{2}; \frac{3}{2}; x^2\right) = \sum_{j=0}^{\infty} \left(\frac{1-\nu}{2}\right)_j \left(1+\frac{\nu}{2}\right)_j \frac{(2x)^{2j+1}}{(2j+1)!}$$

then

59:6:3 $$P_\nu(x) = \frac{1}{\sqrt{\pi}}\left[\frac{\Gamma\left(\frac{1+\nu}{2}\right)}{\Gamma\left(1+\frac{\nu}{2}\right)}\cos\left(\frac{\nu\pi}{2}\right)\mathrm{lef}_\nu(x) + \frac{\Gamma\left(1+\frac{\nu}{2}\right)}{\Gamma\left(\frac{1+\nu}{2}\right)}\sin\left(\frac{\nu\pi}{2}\right)\mathrm{leg}_\nu(x)\right]$$

whereas

59:6:4 $$Q_\nu(x) = \frac{\sqrt{\pi}}{2}\left[\frac{-\Gamma\left(\frac{1+\nu}{2}\right)}{\Gamma\left(1+\frac{\nu}{2}\right)}\sin\left(\frac{\nu\pi}{2}\right)\mathrm{lef}_\nu(x) + \frac{\Gamma\left(1+\frac{\nu}{2}\right)}{\Gamma\left(\frac{1+\nu}{2}\right)}\cos\left(\frac{\nu\pi}{2}\right)\mathrm{leg}_\nu(x)\right]$$

For the same argument range, one may use the trigonometric series

59:6:5 $$P_\nu(\cos(\theta)) = \frac{2\Gamma(1+\nu)}{\sqrt{\pi}\,\Gamma(\frac{3}{2}+\nu)} \sum_{j=0}^{\infty} \frac{(1+\nu)_j(\frac{1}{2})_j}{(\frac{3}{2}+\nu)_j j!} \sin[(1+\nu+2j)\theta] \qquad 0<\theta<\pi$$

59:6:6 $$Q_\nu(\cos(\theta)) = \frac{\sqrt{\pi}\,\Gamma(1+\nu)}{\Gamma(\frac{3}{2}+\nu)} \sum_{j=0}^{\infty} \frac{(1+\nu)_j(\frac{1}{2})_j}{(\frac{3}{2}+\nu)_j j!} \cos[(1+\nu+2j)\theta] \qquad 0<\theta<\pi$$

For arguments in the range $-1 < x < 3$, the series

59:6:7 $$P_\nu(x) = 1 - \frac{(\nu^2+\nu)(1-x)}{2} + \frac{(\nu^4+2\nu^3-\nu^2-2\nu)(1-x)^2}{16} - \cdots$$
$$= \sum_{j=0}^{\infty} \frac{(-\nu)_j(1+\nu)_j}{(j!)^2}\left(\frac{1-x}{2}\right)^j \qquad -1<x<3$$

is a convenient way of calculating values of the Legendre function of the first kind. Similarly, for $x > 1$, one may utilize the series

59:6:8 $$Q_\nu(x) = \frac{1}{2}\sum_{j=0}^{\infty} \frac{\Gamma\left(\frac{1}{2}+j+\frac{\nu}{2}\right)\Gamma\left(1+j+\frac{\nu}{2}\right)}{j!\Gamma(\frac{3}{2}+j+\nu)\,x^{2j+\nu+1}} = \frac{\sqrt{\pi}\Gamma(1+\nu)}{\Gamma(\frac{3}{2}+\nu)(2x)^{1+\nu}} \sum_{j=0}^{\infty} \frac{\left(\frac{1+\nu}{2}\right)_j\left(1+\frac{\nu}{2}\right)_j}{j!(\frac{3}{2}+\nu)_j x^{2j}} \qquad x>1$$

to calculate values of the Legendre function of the second kind, except for negative integer degree (for which $Q_\nu(x)$ is infinite) or when $\nu = \frac{-3}{2}, \frac{-5}{2}, \frac{-7}{2}, \ldots$. In the latter event, the first $(-\nu - \frac{1}{2})$ terms of series 59:6:8 become zero, permitting a redefinition of the summation index and leading to

59:6:9 $$Q_\nu(x) = \frac{\sqrt{\pi}\,\Gamma(-\nu)}{\Gamma(\frac{1}{2}-\nu)(2x)^{-\nu}} \sum_{j=0}^{\infty} \frac{\left(\frac{-\nu}{2}\right)_j \left(\frac{1-\nu}{2}\right)_j}{j!(\frac{1}{2}-\nu)_j x^{2j}} = Q_{-\nu-1}(x) \qquad \nu = \frac{-3}{2}, \frac{-5}{2}, \frac{-7}{2}, \ldots \qquad x > 1$$

in conformity with 59:5:4.

59:7 PARTICULAR VALUES

	$x = -1$	$x = 0$	$x = 1$	$x = \infty$
$P_\nu(x)$ $\left\{\begin{matrix} \nu = 2, 4, 6, \ldots = 2n \\ \nu = -3, -5, -7, \ldots = -1-2n \end{matrix}\right\}$	1	$\dfrac{(2n-1)!!}{(-2)^n n!}$	1	∞
$P_\nu(x)$ $\left\{\begin{matrix} 1<\nu<2,\ 3<\nu<4,\ 5<\nu<6, \ldots \\ -3<\nu<-2,\ -5<\nu<-4,\ -7<\nu<-6, \ldots \end{matrix}\right\}$	∞	$\dfrac{\Gamma\left(\frac{1+\nu}{2}\right)}{\sqrt{\pi}\,\Gamma\left(1+\frac{\nu}{2}\right)}\cos\left(\dfrac{\nu\pi}{2}\right)$	1	∞
$P_\nu(x)$ $\left\{\begin{matrix} \nu = 1, 3, 5, \ldots \\ \nu = -2, -4, -6, \ldots \end{matrix}\right\}$	-1	0	1	∞
$P_\nu(x)$ $\left\{\begin{matrix} 0<\nu<1,\ 2,<\nu<3,\ 4<\nu<5, \ldots \\ -2<\nu<-1,\ -4<\nu<-3,\ -6<\nu<-5, \ldots \end{matrix}\right\}$	$-\infty$	$\dfrac{\Gamma\left(\frac{1+\nu}{2}\right)}{\sqrt{\pi}\,\Gamma\left(1+\frac{\nu}{2}\right)}\cos\left(\dfrac{\nu\pi}{2}\right)$	1	∞
$P_0(x)$, $P_{-1}(x)$	1	1	1	1
$P_\nu(x)$ $\quad -1<\nu<0$	$-\infty$	$\dfrac{\Gamma\left(\frac{1+\nu}{2}\right)}{\sqrt{\pi}\,\Gamma\left(1+\frac{\nu}{2}\right)}\cos\left(\dfrac{\nu\pi}{2}\right)$	1	0
$Q_\nu(x)$ $\quad \nu = \frac{1}{2}, \frac{5}{2}, \frac{9}{2}, \ldots = 2n + \frac{1}{2}$	$\dfrac{-\pi}{2}$	$\dfrac{-(-1)^n U(4n-1)!!!!}{(4n+1)!!!!}$	∞	0
$Q_\nu(x)$ $\quad -\frac{1}{2}<\nu<\frac{1}{2}, \frac{3}{2}<\nu<\frac{5}{2}, \frac{7}{2}<\nu<\frac{9}{2}, \ldots$	$-\infty$	$\dfrac{-\sqrt{\pi}\,\Gamma\left(\frac{1+\nu}{2}\right)}{2\Gamma\left(1+\frac{\nu}{2}\right)}\sin\left(\dfrac{\nu\pi}{2}\right)$	∞	0
$Q_\nu(x)$ $\quad \nu = 0, 2, 4, \ldots$	$-\infty$	0	∞	0
$Q_\nu(x)$ $\quad \nu = \frac{-1}{2}, \frac{3}{2}, \frac{7}{2}, \frac{11}{2}, \ldots = 2n - \frac{1}{2}$	$\dfrac{\pi}{2}$	$\dfrac{(-1)^n \pi(4n-3)!!!!}{2U(4n-1)!!!!}$	∞	0
$Q_\nu(x)$ $\quad -1<\nu<\frac{-1}{2}, \frac{1}{2}<\nu<\frac{3}{2}, \frac{5}{2}<\nu<\frac{7}{2}, \ldots$	∞	$\dfrac{-\sqrt{\pi}\,\Gamma\left(\frac{1+\nu}{2}\right)}{2\Gamma\left(1+\frac{\nu}{2}\right)}\sin\left(\dfrac{\nu\pi}{2}\right)$	∞	0
$Q_\nu(x)$ $\quad \frac{-3}{2}<\nu<-1, \frac{-7}{2}<\nu<-3, \frac{-11}{2}<\nu<-5, \ldots$	∞	$\dfrac{\sqrt{\pi}\,\Gamma\left(\frac{-\nu}{2}\right)}{2\Gamma\left(\frac{1-\nu}{2}\right)}\cos\left(\dfrac{\nu\pi}{2}\right)$	∞	$-\infty$
$Q_\nu(x)$ $\quad \nu = \frac{-3}{2}, \frac{-7}{2}, \frac{-11}{2}, \ldots = \frac{-3}{2} - 2n$	$\dfrac{-\pi}{2}$	$\dfrac{-(-1)^n U(4n-1)!!!!}{(4n+1)!!!!}$	∞	0

		$x = -1$	$x = 0$	$x = 1$	$x = \infty$
$Q_\nu(x)$	$-2 < \nu < \frac{-3}{2},\ -4 < \nu < \frac{-7}{2},\ -6 < \nu < \frac{-11}{2}, \ldots$	$-\infty$	$\dfrac{\sqrt{\pi}\,\Gamma\left(\frac{-\nu}{2}\right)}{2\Gamma\left(\frac{1-\nu}{2}\right)} \cos\left(\frac{\nu\pi}{2}\right)$	∞	∞
$Q_\nu(x)$	$\frac{-5}{2} < \nu < -2,\ \frac{-9}{2} < \nu < -4,\ \frac{-13}{2} < \nu < -6, \ldots$	$-\infty$	$\dfrac{\sqrt{\pi}\,\Gamma\left(\frac{-\nu}{2}\right)}{2\Gamma\left(\frac{1-\nu}{2}\right)} \cos\left(\frac{\nu\pi}{2}\right)$	∞	$-\infty$
$Q_\nu(x)$	$\nu = \frac{-5}{2}, \frac{-9}{2}, \frac{-13}{2}, \ldots = \frac{-1}{2} - 2n$	$\frac{\pi}{2}$	$\dfrac{(-1)^n\,\pi(4n-3)!!!!}{2U(4n-1)!!!!}$	∞	0
$Q_\nu(x)$	$-3 < \nu < \frac{-5}{2},\ -5 < \nu < \frac{-9}{2},\ -7 < \nu < \frac{-13}{2}, \ldots$	∞	$\dfrac{\sqrt{\pi}\,\Gamma\left(\frac{-\nu}{2}\right)}{2\Gamma\left(\frac{1-\nu}{2}\right)} \cos\left(\frac{\nu\pi}{2}\right)$	∞	∞

The quadruple factorial $(4n - 1)!!!!$ that, with its analogs, appears in several of the tabular entries is defined in Section 2:13 and equals $3 \times 7 \times 11 \times \cdots \times (4n-5)(4n-1)$. The constant U [Section 1:7] equals 0.8472130848; note that $P_{1/2}(0) = 2U/\pi$, $P_{-1/2}(0) = 1/U$, $Q_{1/2}(0) = U$ and $Q_{-1/2}(0) = \pi/2U$.

59:8 NUMERICAL VALUES

Each auxiliary Legendre function, defined in equations 59:6:1 and 59:6:2, may be calculated as the sum of a series $t_0 + t_1 + t_2 + \cdots + t_j + \ldots$, where $t_0 = 1$ for the lef$_\nu$ function while $t_0 = 2x$ for the leg$_\nu$ function. The same recursion formula, $t_{j+1} = t_j x^2[k^2 - (k + \frac{1}{2})^2 - \frac{1}{4}]$, applies to both series; however, $k = 2j + \frac{1}{2}$ for lef$_\nu$, whereas $k = 2j + \frac{3}{2}$ for the leg$_\nu$ function. These relationships are used by the following algorithm, in conjunction with equations 59:6:3 and 59:6:4, to calculate $P_\nu(x)$ and $Q_\nu(x)$. The gamma function ratio, $R = \Gamma[1 + (\nu/2)]/\Gamma[(1+\nu)/2]$ is calculated with the aid of equation 43:6:10, if $\nu > 6$. If $\nu \le 6$, 43:6:10 is not used until the identity $\{\Gamma[l + 1 + (\nu/2)]/\Gamma[l + (1+\nu)/2]\} = \{[l + (1+\nu)/2]/[l + 1 + (\nu/2)]\}\{\Gamma[l + 2 + (\nu/2)]/\Gamma[l + (3+\nu)/2]\}$ has been applied $L + 1$ times with $l = 0, 1, 2, \ldots, L$ where $2 - (\nu/2) < L \le 3 - (\nu/2)$. Should ν be a negative integer, equation 59:5:3 is used to calculate $P_\nu(x)$ and a number of order 10^{99} is returned for $Q_\nu(x)$, in recognition of 59:4:3.

Set $a = R = 1$
Input ν >>>>>>>
Set $K = 4\sqrt{|\nu - \nu^2|}$
If $|1 + \nu| + \mathrm{Int}(1 + \nu) \neq 0$ go to (1)
Set $a = 10^{99}$
Replace ν by $-1 - \nu$
(1) Set $s = \sin(90\nu)$
Set $c = \cos(90\nu)$
Set $w = (^1/_2 + \nu)^2$
(2) If $\nu > 6$ go to (3)
Replace ν by $\nu + 2$
Replace R by $R(\nu - 1)/\nu$
Go to (2)
(3) Set $X = 1/(4 + 4\nu)$
Set $g = 1 + 5X(1 - 3X(.35 + 6.1X))$
Replace R by $R[1 - X(1 - gX/2)]/\sqrt{8X}$

Storage needed: a, R, ν, K, s, c, w, k, X, u, t, x, f and g

Use degree mode or change 90 to $\pi/2$.

Input restrictions: $-1 < x < 1$

To recalculate with unchanged ν, simply input new x.

Input x >>>>>>>>

Set $g = u = 2x$
Set $f = t = 1$
Set $k = \frac{1}{2}$
Set $X = 1 + [10^8/(1 - x^2)]$
(4) Replace t by $tx^2(k^2 - w)/[(k + 1)^2 - \frac{1}{4}]$
Replace k by $k + 1$
Replace f by $f + t$
Replace u by $ux^2(k^2 - w)/[(k + 1)^2 - \frac{1}{4}]$
Replace k by $k + 1$
Replace g by $g + u$
If $k < K$ go to (4)
If $|Xt| > |f|$ go to (4)
Replace f by $f + [x^2t/(1 - x^2)]$
Replace g by $g + [x^2u/(1 - x^2)]$
Set $p = [(sgR) + (cf/R)]/\sqrt{\pi}$
Output p

$p \simeq P_\nu(x)$<<<<<<

Set $q = a\sqrt{\pi}[(cgR) - (sf/R)]/2$
Output q

$q \simeq Q_\nu(x)$ <<<<<<

Test values:
$P_7(0.5) = 0.223144533$
$Q_7(0.5) = 0.343915293$
$P_{-2}(-1/\pi) = -0.318309886$
$Q_{-2}(-1/\pi) = \pm\infty$
$P_{3/2}(0) = -0.393446867$
$Q_{3/2}(0) = -0.618024892$
$P_1(8/9) = 0.888888889$
$Q_1(8/9) = 0.259205931$

For $J > 2\sqrt{|\nu^2 - \nu|}$ the infinite series $t_J + t_{J+1} + t_{J+2} + \cdots$ is well approximated by the geometric series $t_J(1 + x^2 + x^4 + x^6 + \cdots)$ and the sum is therefore approximately $t_J/(1 - x^2)$. Hence, the fractional error in truncating the infinite sum $t_0 + t_1 + t_2 + \cdots$ at the t_{J-1} term is approximately $[t_J/(1 - x^2)]/[t_0 + t_1 + \cdots + t_{J-1} + t_J(1 - x^2)^{-1}]$. The algorithm makes use of this principle in deciding when to truncate, thereby ensuring that lef_ν and leg_ν are computed to a precision of about 6×10^{-8}. Close to the zeros of $P_\nu(x)$ and $Q_\nu(x)$, this precision may not be carried over to the Legendre functions themselves.

The algorithm is valid only for arguments in the range $-1 < x < 1$ and is extremely slow close to the boundaries of this range.

59:9 APPROXIMATIONS

Close to $x = 1$, the Legendre function of the first kind is approximated by the linear function

59:9:1
$$P_\nu(x) \simeq \frac{1}{2}[(1 - \nu)(2 + \nu) + \nu(1 + \nu)x] \qquad |1 - x| \text{ small}$$

while that of the second kind obeys

59:9:2
$$Q_\nu(x) \simeq \ln\sqrt{\frac{2}{|1 - x|}} - \gamma - \psi(1 + \nu) \qquad |1 - x| \text{ small}$$

The corresponding approximations close to $x = -1$ are

59:9:3
$$P_\nu(x) \simeq \cos(\nu\pi) + \frac{\sin(\pi)}{\pi}\left[\ln\left(\frac{1 + x}{2}\right) + \gamma + 2\psi(1 + \nu)\right] \qquad (1 + x) \text{ small} \qquad \gamma = 0.5772156649$$

59:9:4
$$Q_\nu(x) \simeq \frac{\cos(\nu\pi)}{2}\left[\ln\left(\frac{1 + x}{2}\right) + \gamma + 2\psi(1 + \nu)\right] - \frac{\pi\sin(\nu\pi)}{2} \qquad (1 + x) \text{ small}$$

Here ψ is the digamma function [Chapter 44] and γ is Euler's constant [see Section 1:4].

As x acquires very large values, we have

59:9:5
$$P_\nu(x) \simeq \begin{cases} \dfrac{\Gamma(\frac{1}{2}+\nu)}{\sqrt{\pi}\,\Gamma(1+\nu)}(2x)^\nu & \nu > \dfrac{-1}{2} \\ \dfrac{1}{\pi}\sqrt{\dfrac{8}{x}}\ln(\sqrt{8x}) & \nu = \dfrac{-1}{2} \\ \dfrac{\Gamma(-\nu-\frac{1}{2})}{\sqrt{\pi}\,\Gamma(-\nu)}(2x)^{-\nu-1} & \nu < \dfrac{-1}{2} \end{cases} \quad x \to \infty$$

59:9:6
$$Q_\nu(x) \simeq \begin{cases} \dfrac{\sqrt{\pi}\,\Gamma(1+\nu)}{\Gamma(\frac{3}{2}+\nu)(2x)^{1+\nu}} & \nu > -1 \\ \dfrac{\pi}{\sqrt{2}}\dfrac{(-2\nu-2)!!}{(\frac{-1}{2}-\nu)!}(4x)^\nu & \nu = \dfrac{-3}{2}, \dfrac{-5}{2}, \dfrac{-7}{2}, \ldots \\ \dfrac{\sqrt{\pi}\,\Gamma(\frac{-1}{2}-\nu)}{\Gamma(-\nu)}\cot(\nu\pi)(2x)^{-\nu-1} & -\dfrac{1}{2} > \nu \neq \dfrac{-3}{2}, \dfrac{-5}{2}, \dfrac{-7}{2} \ldots \end{cases} \quad x \to \infty$$

59:10 OPERATIONS OF THE CALCULUS

Differentiation or integration of a Legendre function gives an associated Legendre function [Section 59:13]:

59:10:1
$$\frac{d}{dx}f_\nu(x) = \begin{cases} \dfrac{-1}{\sqrt{1-x^2}}f_\nu^{(1)}(x) & -1 < x < 1 \\ \dfrac{1}{\sqrt{x^2-1}}f_\nu^{(1)}(x) & x > 1 \end{cases} \quad f = P \text{ or } Q$$

59:10:2
$$\int_x^1 P_\nu(t)dt = \sqrt{1-x^2}\,P_\nu^{(-1)}(x) \qquad -1 < x < 1$$

59:10:3
$$\int_1^x P_\nu(t)dt = \sqrt{x^2-1}\,P_\nu^{(-1)}(x) \qquad x > 1$$

59:10:4
$$\int_x^\infty Q_\nu(t)(dt) = -\sqrt{x^2-1}\,Q_\nu^{(-1)}(x) \qquad x > 1$$

Alternatively, the derivatives may be expressed as

59:10:5
$$\frac{d}{dx}f_\nu(x) = \frac{\nu x + x}{1-x^2}f_\nu(x) - \frac{\nu+1}{1-x^2}f_{\nu+1}(x) = \frac{\nu}{1-x^2}f_{\nu-1}(x) - \frac{\nu x}{1-x^2}f_\nu(x)$$

a formulation that applies for either kind of Legendre function and for any x exceeding -1.

Definite integrals of products of Legendre functions include

59:10:6
$$\int_{-1}^1 P_\nu(t)P_\omega(t)dt = \frac{4\psi(1+\nu) - 4\psi(1+\omega) + 2\pi\cot(\nu\pi) - 2\pi\cot(\omega\pi)}{\pi^2(1+\nu+\omega)(\omega-\nu)\csc(\nu\pi)\csc(\omega\pi)} \qquad \nu+\omega \neq -1$$

59:10:7
$$\int_{-1}^1 P_\nu(t)Q_\omega(t)dt = \frac{\pi\cos(\omega\pi-\nu\pi) - \pi + \sin(2\nu\pi)[\psi(1+\nu) - \psi(1+\omega)]}{\pi(1+\nu+\omega)(\omega-\nu)} \qquad \nu > 0 \qquad \omega > 0$$

59:10:8
$$\int_{-1}^1 Q_\nu(t)\,Q_\omega(t)dt = \frac{[2+2\cos(\nu\pi)\cos(\omega\pi)][\psi(1+\nu) - \psi(1+\omega)] - \pi\sin(\nu\pi-\omega\pi)}{\pi(1+\nu+\omega)(\omega-\nu)} \qquad \nu+\omega \neq -1$$

when the integration limits are $-1 < t < 1$. These formulas are indeterminate when $\omega = \nu$; in that event the three integrals give $[1 - 2/\pi^2) \sin^2(\nu\pi)\, \psi'(1 + \nu)]/(\frac{1}{2} + \nu)$, $-\sin(2\nu\pi)\, \psi'(1 + \nu)/[(1 + 2\nu)\pi]$ and $\{(\pi^2/2) - [1 + \cos^2(\nu\pi)]\psi'(1 + \nu)\}/(1 + 2\nu)$, respectively. Similarly, the integral between limits $1 < t < \infty$:

59:10:9 $$\int_1^\infty Q_\nu(t)\, Q_\omega(t)dt = \frac{\psi(1 + \omega) - \psi(1 + \nu)}{(1 + \nu + \omega)(\omega - \nu)} \qquad \nu + \omega > -1$$

has $\psi'(1 + \nu)/(1 + 2\nu)$ as its special $\omega = \nu$ case, but

59:10:10 $$\int_1^\infty P_\nu(t)Q_\omega(t)dt = \frac{1}{(1 + \nu + \omega)(\omega - \nu)} \qquad \omega > \nu > 0$$

diverges if the degrees are equal.

Since $P_0(t) = 1$, $P_1(t) = t$, $\frac{2}{3}P_2(t) + \frac{1}{3}P_0(t) = t^2$, etc. [see Section 21:5], the above formulas may be adapted to give many definite integrals of the form $\int t^n f_\nu(t)dt$ for $f_\nu = P_\nu$ or Q_ν. Very many other definite integrals are listed by Gradshteyn and Ryzhik [Sections 7.1 and 7.2].

59:11 COMPLEX ARGUMENT

With the real argument x replaced by $x + iy$, equation 59:12:1 serves to define the complex-valued function $P_\nu(x + iy)$ as $F(-\nu,1 + \nu;1;(1 - x - iy)/2)$. The corresponding formula

59:11:1 $$Q_\nu(x + iy) = \frac{\pi}{2}[\cot(\nu\pi) \mp i]\, F\left(-\nu,1 + \nu;1;\frac{1 - x - iy}{2}\right) - \frac{\pi}{2}\csc(\nu\pi)\, F\left(-\nu,1 + \nu;1;\frac{1 + x + iy}{2}\right)$$

provides a valid definition of the Legendre function of the second kind for all combinations of x and y except when $y = 0$ and $|x| > 1$. The alternative signs in 59:11:1 apply accordingly as $y > 0$ (upper sign) or $y < 0$ (lower sign).

For $y = 0$ and $-1 < x < 1$, definition 59:11:1 yields a function that is multiple valued and complex. To avoid this difficulty, it is usual to redefine $Q_\nu(x)$ as the average value of the two limits $Q_\nu(x + iy)$ and $Q_\nu(x - iy)$ as $y \to 0$. This removes the i from 59:11:1, which then reduces to 59:5:6. This convention has been followed in all equations of this chapter (except 59:11:1), so that our $Q_\nu(x)$ is always real for real argument.

59:12 GENERALIZATIONS

Legendre functions may be generalized to the Gauss function of the next chapter. Included among the ways in which Legendre functions may be expressed as a single Gauss function are

59:12:1 $$P_\nu(x) = F\left(-\nu,1 + \nu;1;\frac{1 - x}{2}\right) \qquad -1 < x < 3$$

59:12:2 $$P_\nu(x) = \left(\frac{1 + x}{2}\right)^\nu F\left(-\nu,-\nu;1;\frac{x - 1}{x + 1}\right) = \left(\frac{2}{1 + x}\right)^{1+\nu} F\left(1 + \nu,1 + \nu;1;\frac{x - 1}{x + 1}\right) \qquad x > 0$$

59:12:3 $$P_\nu(x) = F\left(\frac{-\nu}{2},\frac{1 + \nu}{2};1;1 - x^2\right) = xF\left(\frac{1 - \nu}{2},1 + \frac{\nu}{2};1;1 - x^2\right) \qquad 0 < x < \sqrt{2}$$

59:12:4 $$P_\nu(x) = [x + \sqrt{x^2 - 1}]^\nu F\left(-\nu,\frac{1}{2};1;\frac{2\sqrt{x^2 - 1}}{x + \sqrt{x^2 - 1}}\right)$$

$$= \frac{[x - \sqrt{x^2 - 1}]^{1/2+\nu}}{\sqrt{x + \sqrt{x^2 - 1}}} F\left(\frac{1}{2},1 + \nu;1;\frac{2\sqrt{x^2 - 1}}{x + \sqrt{x^2 - 1}}\right) \qquad x > 1$$

59:12:5
$$P_\nu(x) = [x - \sqrt{x^2-1}]^\nu \, F\left(-\nu, \frac{1}{2}; 1; \frac{2\sqrt{x^2-1}}{\sqrt{x^2-1}-x}\right)$$
$$= \frac{[x+\sqrt{x^2-1}]^{1/2+\nu}}{\sqrt{x-\sqrt{x^2-1}}} F\left(\frac{1}{2}, 1+\nu; 1; \frac{2\sqrt{x^2-1}}{\sqrt{x^2-1}-x}\right) \qquad x > \frac{3}{\sqrt{8}}$$

59:12:6
$$P_\nu(x) = x^\nu \, F\left(\frac{-\nu}{2}, \frac{1-\nu}{2}; 1; 1 - \frac{1}{x^2}\right) = x^{-1-\nu} F\left(\frac{1+\nu}{2}, 1 + \frac{\nu}{2}; 1; 1 - \frac{1}{x^2}\right) \qquad x > \frac{1}{\sqrt{2}}$$

59:12:7
$$Q_\nu(x) = \frac{\sqrt{\pi}\,\Gamma(1+\nu)}{\Gamma(\frac{3}{2}+\nu)(2x \pm 2)^{1+\nu}} F\left(1+\nu, 1+\nu; 2+2\nu; \frac{2}{1 \pm x}\right) \qquad x > 2 \mp 1$$

59:12:8
$$Q_\nu(x) = \frac{\sqrt{\pi}\,\Gamma(1+\nu)}{\Gamma(\frac{3}{2}+\nu)(2x)^{1+\nu}} F\left(\frac{1+\nu}{2}, 1+\frac{\nu}{2}; \frac{3}{2} \pm \nu; \frac{1}{x^2}\right) \qquad x > 1$$

59:12:9
$$Q_\nu(x) = \frac{\sqrt{\pi}\,\Gamma(1+\nu)}{\Gamma(\frac{3}{2}+\nu)(2\sqrt{x^2-1})^{1+\nu}} F\left(\frac{1+\nu}{2}, \frac{1+\nu}{2}; \frac{3}{2}+\nu; \frac{1}{1-x^2}\right) \qquad x > \sqrt{2}$$

59:12:10
$$Q_\nu(x) = \frac{\Gamma(1+\nu)}{\Gamma(\frac{3}{2}+\nu)} \sqrt{\frac{\pi(x-\sqrt{x^2-1})^{1+2\nu}}{2\sqrt{x^2-1}}} F\left(\frac{1}{2}, \frac{1}{2}; \frac{3}{2}+\nu; \frac{\sqrt{x^2-1}-x}{2\sqrt{x^2-1}}\right)$$
$$= \frac{\sqrt{\pi}\,\Gamma(1+\nu)}{\Gamma(\frac{3}{2}+\nu)(2\sqrt{x^2-1})^{1+\nu}} F\left(1+\nu, 1+\nu, \frac{3}{2}+\nu; \frac{\sqrt{x^2-1}-x}{2\sqrt{x^2-1}}\right) \qquad x > \frac{3}{\sqrt{8}}$$

59:12:11
$$Q_\nu(x) = \frac{\sqrt{\pi}\,\Gamma(1+\nu)}{\Gamma(\frac{3}{2}+\nu)(x+\sqrt{x^2-1})^{1+\nu}} F\left(\frac{1}{2}, 1+\nu; \frac{3}{2}+\nu; \frac{x-\sqrt{x^2-1}}{x+\sqrt{x^2-1}}\right) \qquad x > 1$$

The restrictions on x attaching to the formulas above ensure that the Legendre functions are real and that the series expansion of the Gauss functions [see 60:6:1] converges. As well, there are many formulas that express the Legendre functions as the sum of two Gauss functions; some are

59:12:12
$$P_\nu(x) = \frac{\Gamma(-\nu-\frac{1}{2})}{\sqrt{\pi}\,\Gamma(-\nu)(2x \pm 2)^{1+\nu}} F\left(1+\nu, 1+\nu; 2+2\nu; \frac{2}{1 \pm x}\right)$$
$$+ \frac{\Gamma(\frac{1}{2}+\nu)(2x \pm 2)^\nu}{\sqrt{\pi}\,\Gamma(1+\nu)} F\left(-\nu, -\nu; -2\nu; \frac{2}{1 \pm x}\right) \qquad x > 2 \mp 1$$

59:12:13
$$P_\nu(x) = \frac{\Gamma(-\nu-\frac{1}{2})}{\sqrt{\pi}\,\Gamma(-\nu)(2\sqrt{x^2-1})^{1+\nu}} F\left(\frac{1+\nu}{2}, \frac{1+\nu}{2}; \frac{3}{2}+\nu; \frac{1}{1-x^2}\right)$$
$$+ \frac{\Gamma(\frac{1}{2}+\nu)(2\sqrt{x^2-1})^\nu}{\sqrt{\pi}\,\Gamma(1+\nu)} F\left(\frac{-\nu}{2}, \frac{-\nu}{2}; \frac{1}{2}-\nu; \frac{1}{1-x^2}\right) \qquad x > \sqrt{2}$$

59:12:14
$$P_\nu(x) = \frac{\sqrt{\pi}}{\Gamma\left(\frac{1-\nu}{2}\right)\Gamma\left(1+\frac{\nu}{2}\right)} F\left(\frac{-\nu}{2}, \frac{1+\nu}{2}; \frac{1}{2}; x^2\right)$$
$$- \frac{2\sqrt{\pi}\,x}{\Gamma\left(\frac{1+\nu}{2}\right)\Gamma\left(\frac{-\nu}{2}\right)} F\left(\frac{1-\nu}{2}, 1+\frac{\nu}{2}; \frac{3}{2}; x^2\right) \qquad -1 < x < 1$$

59:12:15
$$P_\nu(x) = \frac{\Gamma(-\nu-\frac{1}{2})}{\sqrt{\pi}\,\Gamma(-\nu)(2x)^{1+\nu}}F\left(\frac{1+\nu}{2},1+\frac{\nu}{2};\frac{3}{2}+\nu;\frac{1}{x^2}\right) + \frac{\Gamma(\frac{1}{2}+\nu)(2x)^\nu}{\sqrt{\pi}\,\Gamma(1+\nu)}F\left(\frac{-\nu}{2},\frac{1-\nu}{2};\frac{1}{2}-\nu;\frac{1}{x^2}\right) \qquad x>1$$

59:12:16
$$P_\nu(x) = \frac{\Gamma(-\nu-\frac{1}{2})}{\Gamma(-\nu)}\sqrt{\frac{(x-\sqrt{x^2-1})^{1+2\nu}}{2\pi\sqrt{x^2-1}}}\,F\left(\frac{1}{2},\frac{1}{2};\frac{3}{2}+\nu;\frac{1}{2}-\frac{x}{2\sqrt{x^2-1}}\right) + \frac{\Gamma(\frac{1}{2}+\nu)}{\Gamma(1+\nu)}\sqrt{\frac{(x-\sqrt{x^2-1})^{-2\nu-1}}{2\pi\sqrt{x^2-1}}}\,F\left(\frac{1}{2},\frac{1}{2};\frac{1}{2}-\nu;\frac{1}{2}-\frac{x}{2\sqrt{x^2-1}}\right) \qquad x>\frac{3}{\sqrt{8}}$$

59:12:17
$$P_\nu(x) = \frac{\Gamma(-\nu-\frac{1}{2})}{\sqrt{\pi}\,\Gamma(-\nu)(x+\sqrt{x^2-1})^{1+\nu}}F\left(\frac{1}{2},1+\nu;\frac{3}{2}+\nu;\frac{x-\sqrt{x^2-1}}{x+\sqrt{x^2-1}}\right) + \frac{\Gamma(\frac{1}{2}+\nu)(x+\sqrt{x^2-1})^\nu}{\sqrt{\pi}\,\Gamma(1+\nu)}F\left(\frac{1}{2},-\nu;\frac{1}{2}-\nu;\frac{x-\sqrt{x^2-1}}{x+\sqrt{x^2-1}}\right) \qquad x>1$$

59:12:18
$$Q_\nu(x) = \frac{-\sqrt{\pi}\,\Gamma\left(\frac{1+\nu}{2}\right)}{2\Gamma\left(1+\frac{\nu}{2}\right)}\sin\left(\frac{\nu\pi}{2}\right)F\left(\frac{-\nu}{2},\frac{1+\nu}{2};\frac{1}{2};x^2\right) - \frac{\sqrt{\pi}\,\Gamma\left(1+\frac{\nu}{2}\right)}{\Gamma\left(\frac{1+\nu}{2}\right)}\cos\left(\frac{\nu\pi}{2}\right)xF\left(\frac{1-\nu}{2},1+\frac{\nu}{2};\frac{3}{2};x^2\right) \qquad -1<x<1$$

Many of these formulas become invalid, or require modification, if ν is an integer. For $x > 1$ the substitution $x = \cosh(X)$, $\sqrt{x^2-1} = \sinh(X)$, $x \pm \sqrt{x^2-1} = \exp(\pm X/2)$ often yields simplifications.

59:13 COGNATE FUNCTIONS

The associated Legendre functions $P_\nu^{(\mu)}(x)$ and $Q_\nu^{(\mu)}(x)$ represent a generalization of the Legendre functions inasmuch as $P_\nu^{(0)}(x) = P_\nu(x)$ and $Q_\nu^{(0)}(x) = Q_\nu(x)$. They are sometimes named "spherical harmonics" but we reserve that title for the extended functions discussed in Section 59:14. In linear combination, they satisfy the associated Legendre equation

59:13:1
$$(1-x^2)\frac{d^2f}{dx^2} - 2x\frac{df}{dx} + \left[\nu(1+\nu) - \frac{\mu^2}{1-x^2}\right]f = 0 \qquad f = c_1P_\nu^{(\mu)}(x) + c_2Q_\nu^{(\mu)}(x)$$

[see 59:14:4 for a trigonometric equivalent of this equation]. Replacement of x by $1 - 2x$ and then f by $(x^2 - x)^{\mu/2}f$ leads to an example of the Gauss equation, 60:3:4, and accordingly associated Legendre functions are instances of the Gauss function of Chapter 60.

Generally, these functions are complex even when their arguments are real. However, it is conventional to adopt redefinitions similar to that discussed in Section 59:11 to ensure that $P_\nu^{(\mu)}(x)$ and $Q_\nu^{(\mu)}(x)$ are real for real x between -1 and $+1$. Here we discuss this range exclusively. Moreover, we shall emphasize cases in which ν and μ are nonnegative integers, using n and m to represent the degree and order, respectively, in these cases.

Calculation of values of the associated Legendre functions is possible via the definition

59:13:2
$$P_\nu^{(\mu)}(x) = \frac{2^\mu}{\sqrt{\pi}(1-x^2)^{\mu/2}}\left\{\frac{\Gamma\left(\frac{1+\nu+\mu}{2}\right)}{\Gamma\left(1+\frac{\nu-\mu}{2}\right)}\cos\left(\frac{\nu\pi+\mu\pi}{2}\right)\mathrm{lef}_\nu^{(\mu)}(x) + \frac{\Gamma\left(1+\frac{\nu+\mu}{2}\right)}{\Gamma\left(\frac{1+\nu-\mu}{2}\right)}\sin\left(\frac{\nu\pi+\mu\pi}{2}\right)\mathrm{leg}_\nu^{(\mu)}(x)\right\}$$

59:13:3
$$Q_\nu^{(\mu)}(x) = \frac{\sqrt{\pi}2^{\mu-1}}{(1-x^2)^{\mu/2}}\left\{\frac{-\Gamma\left(\frac{1+\nu+\mu}{2}\right)}{\Gamma\left(1+\frac{\nu-\mu}{2}\right)}\sin\left(\frac{\nu\pi+\mu\pi}{2}\right)\mathrm{lef}_\nu^{(\mu)}(x) + \frac{\Gamma\left(1+\frac{\nu-\mu}{2}\right)}{\Gamma\left(\frac{1+\nu-\mu}{2}\right)}\cos\left(\frac{\nu\pi+\mu\pi}{2}\right)\mathrm{leg}_\nu^{(\mu)}(x)\right\}$$

where the *auxiliary associated Legendre functions* are defined by

59:13:4
$$\mathrm{lef}_\nu^{(\mu)}(x) = F\left(\frac{-\nu-\mu}{2}, \frac{1+\nu-\mu}{2}; \frac{1}{2}; x^2\right) = \sum_{j=0}^{\infty}\left(\frac{-\nu-\mu}{2}\right)_j\left(\frac{1+\nu-\mu}{2}\right)_j\frac{(2x)^{2j}}{(2j)!}$$

59:13:5
$$\mathrm{leg}_\nu^{(\mu)}(x) = 2xF\left(\frac{1-\nu-\mu}{2}, 1+\frac{\nu-\mu}{2}; \frac{3}{2}; x^2\right) = \sum_{j=0}^{\infty}\left(\frac{1-\nu-\mu}{2}\right)_j\left(1+\frac{\nu-\mu}{2}\right)_j\frac{(2x)^{1+2j}}{(1+2j)!}$$

The algorithm of Section 59:8 may be modified to use these equations, although four distinct gamma function evaluations are required instead of the single gamma function ratio that sufficed with equations 59:6:1–59:6:4.

Some interrelationships between associated Legendre functions are

59:13:6
$$P_{-\nu-1}^{(\mu)}(x) = P_\nu^{(\mu)}(x)$$

59:13:7
$$[\tan(\nu\pi) - \tan(\mu\pi)]\, Q_{-\nu-1}^{(\mu)}(x) = [\tan(\nu\pi) + \tan(\mu\pi)]\, Q_\nu^{(\mu)}(x) - \pi P_\nu^{(\mu)}(x)$$

59:13:8
$$P_\nu^{(-\mu)}(x) = \frac{\Gamma(1+\nu-\mu)}{\Gamma(1+\nu+\mu)}\left[\cos(\mu\pi)\, P_\nu^{(\mu)}(x) - \frac{2\sin(\mu\pi)}{\pi} Q_\nu^{(\mu)}(x)\right]$$

59:13:9
$$Q_\nu^{(-\mu)}(x) = \frac{\Gamma(1+\nu-\mu)}{\Gamma(1+\nu+\mu)}\left[\cos(\mu\pi)\, Q_\nu^{(\mu)}(x) + \frac{\pi\sin(\mu\pi)}{2} P_\nu^{(\mu)}(x)\right]$$

59:13:10
$$P_\nu^{(\mu)}(-x) = \cos(\nu\pi + \mu\pi)P_\nu^{(\mu)}(x) - \frac{2\sin(\nu\pi+\mu\pi)}{\pi} Q_\nu^{(\mu)}(x)$$

59:13:11
$$Q_\nu^{(\mu)}(-x) = -\cos(\pi\nu + \pi\mu)\, Q_\nu^{(\mu)}(x) - \frac{\pi\sin(\pi\nu+\pi\mu)}{2} P_\nu^{(\mu)}(x)$$

constituting degree-, order- and argument-reflection formulas.

Some examples of associated Legendre functions are

59:13:12 $$P_1^{(1)}(x) = -\sqrt{1-x^2} \qquad Q_1^{(1)}(x) = \frac{-x}{\sqrt{1-x^2}} - \sqrt{1-x^2}\,\text{artanh}(x)$$

59:13:13 $$P_2^{(1)}(x) = -3x\sqrt{1-x^2} \qquad Q_2^{(1)}(x) = \frac{2-3x^2}{\sqrt{1-x^2}} - 3x\sqrt{1-x^2}\,\text{artanh}(x)$$

59:13:14 $$P_2^{(2)}(x) = 3(1-x^2) \qquad Q_2^{(2)}(x) = \frac{5x-3x^3}{1-x^2} + 3(1-x^2)\,\text{artanh}(x)$$

59:13:15 $$P_n^{(n)}(x) = (-1)^n(2n-1)!!\,(1-x^2)^{n/2} \qquad n = 0, 1, 2, \ldots$$

and others may be obtained via the differentiation

59:13:16 $$f_\nu^{(m)}(x) = (-1)^m(1-x^2)^{m/2}\frac{d^m}{dx^m}f_\nu(x) \qquad f = P \text{ or } Q$$

or recursion

59:13:17 $$f_{n+1}^{(m)}(x) = \frac{(1+2n)x}{1+n-m}f_n^{(m)}(x) + \frac{n+m}{1+n-m}f_{n-1}^{(m)}(x)$$

59:13:18 $$f_n^{(m+1)}(x) = \frac{-2mx}{\sqrt{1-x^2}}f_n^{(m)}(x) - (1+n-m)(n+m)\,f_n^{(m-1)}(x)$$

$$\left.\right\} \quad f = P \text{ or } Q$$

formulas. Formulas 59:13:12–59:13:15 demonstrate the inappropriateness of the name "associated Legendre polynomials" sometimes given to these functions. Be aware that the $(-1)^m$ factor in 59:13:16 is often omitted so that associated Legendre functions of odd order may be encountered with signs opposite to those employed in this *Atlas*.

The associated Legendre functions of the first kind satisfies the orthogonality relationships

59:13:19 $$\int_{-1}^{1} P_N^{(m)}(t)\,P_n^{(m)}(t)dt = \begin{cases} 0 & N \neq n \\ \dfrac{2}{2n+1}\dfrac{(n+m)!}{(n-m)!} & N = n \end{cases}$$

but no such relationship holds for the second kind.

59:14 RELATED TOPICS

In Section 46:14 we discuss the solution of the Helmholtz equation in various orthogonal coordinate systems. In the spherical system one finds

59:14:1 $$\frac{r^2}{R}\frac{d^2R}{dr^2} + \frac{2r}{R}\frac{dR}{dr} - kr^2 = h = -\frac{1}{Y}\frac{\partial^2 Y}{\partial\theta^2} - \frac{\cot(\theta)}{Y}\frac{\partial Y}{\partial\theta} - \frac{\csc^2(\theta)}{Y}\frac{\partial^2 Y}{\partial\theta^2}$$

where R is a function only of the radial coordinate r, Y is a function of the two angular coordinates, θ and ϕ, and h is a separation constant. With m^2 denoting a second separation constant, the second equality in 59:14:1 may be further decomposed into

59:14:2 $$\frac{\sin^2(\theta)}{\Theta}\frac{d^2\Theta}{d\theta^2} + \frac{\sin(2\theta)}{2\Theta}\frac{d\Theta}{d\theta} + h\sin^2(\theta) = m^2 = \frac{1}{\Phi}\frac{d^2\Phi}{d\phi^2}$$

where Θ is a function only of θ, and Φ is a function only of ϕ.

The second equality in 59:14:2 is satisfied by a sinusoidal function of the *longitude* ϕ

59:14:3 $$\Phi = C_m\cos(m\phi) + S_m\sin(m\phi) \qquad m = 0, 1, 2, \ldots$$

where C_m and S_m are constants and m is constrained to be an integer by the geometric constraint that $\Phi(-\pi) = \Phi(\pi)$.

The first equality in 59:14:2 may be rewritten as

59:14:4 $$\frac{d^2\Theta}{d\theta^2} + \cot(\theta)\frac{d\Theta}{d\theta} + [h - m^2\csc^2(\theta)]\Theta = 0$$

which, by the substitution $x = \cos(\theta)$, may be converted into the associated Legendre equation 59:13:1. Accordingly, the general solution is

59:14:5 $$\Theta = p_{\nu;m}P_\nu^{(m)}(\cos(\theta)) + q_{\nu;m}Q_\nu^{(m)}(\cos(\theta)) \qquad \nu = \sqrt{\tfrac{1}{4} + h} - \tfrac{1}{2}$$

where constant coefficients are denoted by $p_{\nu;m}$ and $q_{\nu;m}$. The latter is usually constrained to be zero by physical considerations: the prohibition on Y, and hence Θ, being infinite at $\theta = 0$. Because the *latitude* of the sphere is $(\pi/2) - \theta$, $\theta = 0$ represents its "north pole." At the "south pole," $\theta = \pi$, $\cos(\theta) = -1$ and, since $P_\nu^{(m)}(-1)$ is infinite unless ν is an integer, we are led to conclude that $\nu = n = 0, \pm1, \pm2, \ldots$. Because the $P_n^{(m)}$ and $P_{-n-1}^{(m)}$ functions are identical, we can ignore negative n values, leading to

59:14:6 $$\Theta = p_{n;m}P_n^{(m)}(\cos(\theta)) \qquad n = 0, 1, 2, \ldots \qquad h = n(n+1)$$

as the only physically significant solutions of 59:14:4. Moreover, since the associated Legendre function $P_n^{(m)}$ is zero if $|m| > n$, we can discount all values of m except $0, \pm1, \pm2, \ldots \pm n$. Where only certain values of separation constants are permitted, as n and m here, these values are named *eigenvalues* or *quantum numbers*.

Because the eigenvalue m enters into the solutions for both Φ and Θ, it is often counterproductive to factor the Y function into its longitudinal and latitudinal components. The general solution of the second equality in 59:14:1 has now been shown to be

59:14:7 $$Y = \sum_{n=0}^{\infty}\sum_{m=-n}^{n}[S_{n;m}\sin(m\phi) + C_{n;m}\cos(m\phi)]\,P_n^{(m)}(\cos(\theta))$$

where $S_{n;m}$ and $C_{n;m}$ are redefined constants. The components of this solution are known as *spherical harmonics* or *surface harmonics* and, in light of orthogonality relationships 59:13:19, 32:10:25 and 32:10:26, they are usually defined in the normalized versions

59:14:8 $$Y_{n;m}(\theta;\phi) = \sqrt{\left(\frac{2n+1}{2\pi}\right)\frac{(n-m)!}{(n+m)!}}\,f(m\phi)\,P_n^{(m)}(\cos(\theta))$$
$$f = \sin \text{ or } \cos \qquad n = 0, 1, 2, \ldots \qquad m = 0, \pm1, \ldots, \pm n$$

However, normalization conventions sometimes differ from author to author.

As their name suggests, spherical harmonics play a vital role in describing oscillatory behavior in systems of spherical symmetry: for example, in the quantum mechanics of atomic electrons. One distinguishes *zonal surface harmonics* when $m = 0$, *sectoral surface harmonics* when $m = n$ and *tesseral surface harmonics* when $m = 1, 2, 3, \ldots, n-1$.

Finally, let us return to the first equality in 59:14:1 with $n(n+1)$ now substituted for the separation constant h. Replacement of the radial coordinate r by $x/\sqrt{k}$ (or by $x/\sqrt{-k}$ for negative k) and the dependent variable R by $f/\sqrt{x}$ then leads to

59:14:9 $$x^2\frac{d^2f}{dx^2} + x\frac{df}{dx} - [(n+\tfrac{1}{2})^2 \pm x^2]f = 0 \qquad n = 0, 1, 2, \ldots$$

where the alternative upper/lower signs apply as k is positive/negative. This is the (hyperbolic Bessel)/(Bessel) equation 50:3:4/53:3:5, and the lower alternative is satisfied by

59:14:10 $$R = \frac{f}{\sqrt{x}} = \frac{A_n}{\sqrt{x}}J_{1/2+n}(x) + \frac{B_n}{\sqrt{x}}J_{-1/2-n}(x) = a_n j_n(x) + b_n y_n(x)$$

where J_ν represents the Bessel function [Chapter 53]; A_n, B_n, a_n and b_n are arbitrary constants; and j_n and y_n are spherical Bessel functions [see Sections 32:13, 53:4 and 54:4].

CHAPTER 60

THE GAUSS FUNCTION F(a,b;c;x)

A surprisingly large number of simple intrarelationships make the quadrivariate Gauss function unusually flexible. It embraces many of the functions discussed in previous chapters and may be further generalized as explained in Section 18:14.

60:1 NOTATION

The Gauss function is also known as the *hypergeometric function* or as the *Gauss hypergeometric function*. The subscripted symbolism ${}_2F_1(a,b;c;x)$ is sometimes encountered; Section 60:13 contains an explanation of the "2" and "1" numerals.

As usual, the x variable in the Gauss function is its argument. Because of their locations in expansion 60:6:1, the a and b variables are known as numeratorial parameters, while c is a denominatorial parameter. There is a second denominatorial parameter, equal to unity, whose presence is not explicitly displayed in the F(a,b;c;x) notation but is brought out in the first of the following alternative symbolisms:

60:1:1
$$\frac{\Gamma(c)}{\Gamma(a)\,\Gamma(b)}\left[x\,\frac{a-1,\,b-1}{0,\,c-1}\right] = {}_2F_1\left[\begin{matrix}a,\,b;\,x\\ c\end{matrix}\right] = F(a,b;c;x)$$

60:2 BEHAVIOR

Unless one of the four quantities a, b, $c - a$ or $c - b$ is a nonpositive integer [in which event, see 60:4:10, 60:4:11 or their analogs], the Gauss function is defined only for real values of its argument in the range $-\infty < x < 1$. The domain may be extended to embrace $x = 1$ provided that $c > a + b$.

When c is a nonpositive integer the Gauss function adopts infinite values [unless a or b is also an integer such that $a - c$ or $b - c$ equals 0, 1, 2, ..., so that 60:4:11 applies]. Nevertheless, significance can always be attributed to the ratio $F(a,b;c;x)/\Gamma(c)$, even when $c = 1 - n = 0, -1, -2, \ldots$, because of the limiting operation

60:2:1
$$\lim_{c\to 1-n}\left\{\frac{F(a,b;c;x)}{\Gamma(c)}\right\} = \frac{(a)_n(b)_n x^n}{n!}F(n+a,n+b;n+1;x) \qquad n = 1, 2, 3, \ldots$$

Because it is quadrivariate, the Gauss function displays such a wide variety of behaviors that it is impractical to depict graphically.

60:3 DEFINITIONS

Expansion 60:6:1 provides the usual definition of the Gauss function for $-1 < x < 1$. The transformation 60:5:3 permits extension to $-\infty < x < \frac{1}{2}$.

The *Euler hypergeometric integral*

60:3:1 $$F(a,b;c;x) = \frac{\Gamma(c)}{\Gamma(b)\,\Gamma(c-b)} \int_0^1 \frac{t^{b-1}\,dt}{(1-t)^{1+b-c}(1-xt)^a} \qquad c > b > 0 \qquad x < 1$$

serves as a definition, as do the equivalent definite integrals

60:3:2 $$F(a,b;c;x) = \frac{\Gamma(c)}{\Gamma(b)\,\Gamma(c-b)} \int_1^\infty \frac{(t-1)^{c-b-1}t^{a-c}\,dt}{(t-x)^a} \qquad 1 + a > c > b \qquad x < 1$$

and

60:3:3 $$F(a,b;c;x) = \frac{\Gamma(c)}{\Gamma(b)\,\Gamma(c-b)} \int_0^\infty \frac{t^{b-1}\,dt}{(1+t)^{c-a}(1+x-xt)^a} \qquad c > b > 0 \qquad x < 1$$

Because of identity 60:5:1, the two numeratorial parameters may be interchanged in any of these definitions.

A linear combination of two Gauss functions satisfies the *hypergeometric differential equation*

60:3:4 $$x(1-x)\frac{d^2f}{dx^2} + [1 - \gamma - (1 + \alpha + \beta)x]\frac{df}{dx} - \alpha\beta f = 0$$

$$f = c_1 F(\alpha,\beta;1-\gamma;x) + c_2 x^\gamma F(a+\gamma,\beta+\gamma;1+\gamma;x)$$

c_1 and c_2 being arbitrary constants.

In the notation of Section 43:14, the Gauss function may be synthesized from the binomial function [Section 6:14] by the operation

60:3:5 $$(1-x)^{-a} \xrightarrow[1-b]{1-c} F(a,b;c;x) \qquad -1 < x < 1$$

60:4 SPECIAL CASES

A substantial fraction of the functions treated in this *Atlas* are special cases of the Gauss function. Thus, any $K = L = 1$ hypergeometric function [see Section 18:14] is a Gauss function on account of the identity

60:4:1 $$\sum_{j=0}^{\infty} \frac{(a)_j}{(c)_j} x^j = F(1,a;c;x)$$

Similarly, any $K = L = 2$ hypergeometric function having unity as one of its denominatorial parameters is a Gauss function via

60:4:2 $$\sum_{j=0}^{\infty} \frac{(a_1)_j(a_2)_j}{(1)_j(c_2)_j} x^j = F(a_1,a_2;c_2;x)$$

A glance at Table 18.14.2 will reveal the large number of instances in which important functions conform to either 60:4:1 or 60:4:2.

If the two numeratorial parameters of the Gauss function are interrelated by any one of the three relationships

60:4:3 $$b = a - \tfrac{1}{2},\ a + \tfrac{1}{2} \text{ or } 1 - a$$

or if the denominatorial parameter is related by any one of

60:4:4 $$c = 2a,\ 2b,\ a + b - \tfrac{1}{2},\ a + b + \tfrac{1}{2},\ \frac{1+a+b}{2},\ 1 + a - b \text{ or } 1 + b - a$$

to the numeratorial parameters, then the Gauss function reduces to an associated Legendre function [Section 59:13] or to one of the special cases (a Legendre function, a Legendre polynomial or a complete elliptic integral, among others) of the latter. Of these 10 possibilities, seven are detailed in Table 60.4.1 by listing the appropriate values of "factor," μ, ν and X in the identity

60:4:5 $$F(a,b;c;x) = (\text{factor})P_\nu^{(\mu)}(X)$$

The three omissions are easily derived by interchanging a and b.

When the denominatorial parameter equals $\frac{1}{2}$ (or $\frac{3}{2}$), reduction occurs to the sum (or difference) of two associated Legendre functions; thus:

60:4:6 $$F(a,b;1 \mp \tfrac{1}{2};x) = (\text{factor})[P_\nu^{(\mu)}(X) \pm P_\nu^{(\mu)}(-X)]$$

These cases are also included in Table 60.4.1.

Another trivariate function that is a common special case of the Gauss function is the incomplete beta function. If either of the numeratorial parameters of the Gauss function equals unity, or is less by unity than the denominatorial parameter, we have reduction as follows:

60:4:7 $$F(a,1;c;x) = \frac{(c-1)\,B(c-1;1+a-c;x)}{x^{c-1}(1-x)^{1+a-c}}$$

60:4:8 $$F(a,c-1;c;x) = \frac{(c-1)\,B(c-1;1-a;x)}{x^{c-1}}$$

to instances of the function discussed in Chapter 58. Powers, polynomials, logarithms and inverse hyperbolic or trigonometric functions may arise by further specialization of the incomplete beta function, as explained in Section 58:4.

When the denominatorial parameter c equals one of the numeratorial parameters, say b, we have

60:4:9 $$F(a,b;b;x) = \sum_{j=0}^{\infty} \frac{(a)_j}{(1)_j} x^j = \frac{1}{(1-x)^a}$$

In effect, the two parameters have "cancelled."

Table 60.4.1

Parameter relation	Factor	μ	ν	X	Restriction
$b = a - \frac{1}{2}$	$2^{c-1}\Gamma(c)\sqrt{\lvert x\rvert^{1-c}(1-x)^{c-2a}}$	$1-c$	$2a-c-1$	$\frac{1}{\sqrt{1-x}}$	$x < 1$
$b = 1 - a$	$\Gamma(c)[\lvert x\rvert/(1-x)]^{(1-c)/2}$	$1-c$	$-a$	$1-2x$	$x < 1$
$c = 2a$	$2^{2a-1}\Gamma(\frac{1}{2}+a)\sqrt{x^{1-2a}(1-x)^{a-b-1/2}}$	$\frac{1}{2}-a$	$b-a-\frac{1}{2}$	$\frac{2-x}{2\sqrt{1-x}}$	$0 < x < 1$
$c = a + b - \frac{1}{2}$	$2^{a+b-3/2}\Gamma(a+b-\frac{1}{2})\sqrt{\lvert x\rvert^{3/2-a-b}/(1-x)}$	$\frac{3}{2}-a-b$	$b-a-\frac{1}{2}$	$\sqrt{1-x}$	$x < 1$
$c = a + b + \frac{1}{2}$	$\Gamma(\frac{1}{2}+a+b)(\sqrt{\lvert x\rvert}/2)^{1/2-a-b}$	$\frac{1}{2}-a-b$	$a-b-\frac{1}{2}$	$\sqrt{1-x}$	$x < 1$
$c = \frac{1+a+b}{2}$	$\Gamma\left(\frac{1+a+b}{2}\right)\lvert x-x^2\rvert^{(1-a-b)/4}$	$\frac{1-a-b}{2}$	$\frac{a-b-1}{2}$	$1-2x$	
$c = 1 + a - b$	$\Gamma(1+a-b)\lvert x\rvert^{(b-a)/2}/(1-x)^b$	$b-a$	$-b$	$\frac{1+x}{1-x}$	$x < 1$
$c = \frac{1}{2}$	$2^{a+b}\Gamma(\frac{1}{2}+a)\Gamma(\frac{1}{2}+b)\sqrt{\lvert 1-x\rvert^{1/2-a-b}/8\pi}$	$\frac{1}{2}-a-b$	$a-b-\frac{1}{2}$	$\sqrt{\lvert x\rvert}$ & $-\sqrt{\lvert x\rvert}$	
$c = \frac{3}{2}$	$-2^{a+b}\Gamma(a-\frac{1}{2})\Gamma(b-\frac{1}{2})\sqrt{(1-x)^{3/2-a-b}/128\pi x}$	$\frac{3}{2}-a-b$	$a-b-\frac{1}{2}$	$\sqrt{x}$ & $-\sqrt{x}$	$0 < x < 1$

If either of the numeratorial parameters is a nonpositive integer, for example, if $b = -n = 0, -1, -2, \ldots$, then the Gauss function reduces to a polynomial function [Chapter 17] of degree n:

60:4:10 $$\mathrm{F}(a,-n;c;x) = \sum_{j=0}^{n} \binom{n}{j} \frac{(a)_j}{(c)_j} (-x)^j = \frac{n!}{(c)_n} \mathrm{P}_n^{(c-1;a-n-c)}(1-2x)$$

where $\mathrm{P}_n^{(\nu;\mu)}(x)$ is a Jacobi polynomial [Section 22:12]. A similar polynomial combined with a binomial function [Section 6:14] is generated if $a - c$ or $b - c$ is a nonnegative integer; thus, for example:

60:4:11 $$\mathrm{F}(a,c+n;c;x) = (1-x)^{-a-n} \sum_{j=0}^{n} \binom{n}{j} \frac{(c-a)_j}{(c)_j} (-x)^j$$

$$= \frac{n!(1-x)^{-a-n}}{(c)_n} \mathrm{P}_n^{(c-1;-a-n)}(1-2x) \qquad n = 0, 1, 2, \ldots$$

The cases in which the denominatorial parameter c is zero or a negative integer are addressed in Section 60:2; however, 60:4:11 may still be valid if one of the numeratorial parameters is a less negative integer. Equations 22:12:11–22:12:14, combined with 60:4:11, show how a number of orthogonal polynomials are special cases of the Gauss function.

60:5 INTRARELATIONSHIPS

The Gauss function is symmetrical with respect to interchange of the two numeratorial parameters

60:5:1 $$\mathrm{F}(b,a;c;x) = \mathrm{F}(a,b;c;x)$$

The transformation

60:5:2 $$\mathrm{F}(a,b;c;x) = \frac{\mathrm{F}(c-a,c-b;c;x)}{(1-x)^{a+b-c}}$$

which may be regarded as a reflection formula for both numeratorial parameters, provides a relationship between two Gauss functions of common argument. On the other hand, the transformations

60:5:3 $$\mathrm{F}(a,b;c;x) = \frac{\mathrm{F}\left(a,c-b;c;\dfrac{x}{x-1}\right)}{(1-x)^a} = \frac{\mathrm{F}\left(c-a,b;c;\dfrac{x}{x-1}\right)}{(1-x)^b}$$

relate a Gauss function with argument in the range $0 \le x \le 1$ to ones with arguments in the $-\infty \le x \le 0$ range.

The six functions $\mathrm{F}(a \pm 1,b;c;x)$, $\mathrm{F}(a,b \pm 1;c;x)$ and $\mathrm{F}(a,b;c \pm 1;x)$ are said to be *contiguous* to the Gauss function $\mathrm{F}(a,b;c;x)$. The function $\mathrm{F}(a,b;c;x)$ is linearly related to *any pair* of its contiguous functions. Thus, there are 15 *contiguity relationships*. Three of these are

60:5:4 $$(2a - c + bx - ax)\,\mathrm{F}(a,b;c;x) = (a-c)\,\mathrm{F}(a-1,b;c;x) + (a-ax)\,\mathrm{F}(a+1,b;c;x)$$

60:5:5 $$(a-b)\,\mathrm{F}(a,b;c;x) = a\mathrm{F}(a+1,b;c;x) - b\mathrm{F}(a,b+1;c;x)$$

and

60:5:6 $$[1 - c - (1+a+b-2c)x]\,\mathrm{F}(a,b;c;x) = (c-1)(x-1)\,\mathrm{F}(a,b;c-1;x) + \frac{(c-a)(c-b)x}{c}\mathrm{F}(a,b;c+1;x)$$

and the remaining 12 may be found in Erdélyi, Magnus, Oberhettinger and Tricomi [*Higher Transcendental Functions*, Volume 1, Section 2.8, formulas (33)–(44)]. These contiguity relationships constitute recursion formulas, enabling $\mathrm{F}(a+1,b;c;x)$, for example, to be expressed in terms of $\mathrm{F}(a,b;c;x)$ and $\mathrm{F}(a-1,b;c;x)$.

The 15 contiguity relationships provide intrarelationships linking three Gauss functions of common argument, but there also exists a plethora of relationships between trios of Gauss functions with dissimilar arguments. Three typical relationships of this class are

60:5:7 $$\mathrm{F}(a,b;c;x) = \sum_{\lambda=a}^{b} \frac{\Gamma(c)\,\Gamma(a+b-2\lambda)}{\Gamma(a+b-\lambda)\,\Gamma(c-\lambda)} \frac{\mathrm{F}\left(\lambda, 1+\lambda-c; 1+2\lambda-a-b; \dfrac{1}{x}\right)}{(-x)^{\lambda}} \qquad x < 0$$

60:5:8 $$\mathrm{F}(a,b;c;x) = \sum_{\lambda=a}^{b} \frac{\Gamma(c)\,\Gamma(a+b-2\lambda)}{\Gamma(a+b-\lambda)\,\Gamma(c-\lambda)} \frac{\mathrm{F}\left(\lambda, c+\lambda-a-b; 1+2\lambda-a-b; \dfrac{1}{1-x}\right)}{(1-x)^{\lambda}} \qquad x < 1$$

and

60:5:9 $$\mathrm{F}(a,b;c,x) = \sum_{\lambda=c}^{a+b} \frac{\Gamma(c)\,\Gamma(a+b+c-2\lambda)}{\Gamma(a+c-\lambda)\,\Gamma(b+c-\lambda)} \frac{\mathrm{F}(\lambda-a, \lambda-b; 1+2\lambda-a-b-c; 1-x)}{(1-x)^{a+b-\lambda}} \qquad x < 1$$

The Σ notation in these three equations indicates that their right-hand members consist of two terms that differ only by virtue of λ adopting either of the two indicated values. These relationships may fail if the argument of one of the numeratorial gamma functions equals a nonpositive integer: see Abramowitz and Stegun [equations 15.3.10–15.3.14] for expansions that apply in these exceptional cases.

Because they interrelate Gauss functions whose arguments are linked by a linear relationship, formulas 60:5:2, 60:5:3 and 60:5:7–60:5:9 are examples of *linear transformations*; the Bateman manuscript [see Erdélyi, Magnus, Oberhettinger and Tricomi, *Higher Transcendental Equations*, Volume 1, Sections 2.9 and 2.11] contains a catalog of these transformations. Apart from the exceptional cases noted in the previous paragraph, linear transformations of Gauss functions are valid for all values of the numeratorial and denominatorial parameters. In contrast, the set of so-called *quadratic transformations*, in which a quadratic relationship exists between the Gauss function arguments, holds only if one of the following restrictive conditions apply:

60:5:10 $$\begin{cases} a+b = 1,\ c+\dfrac{1}{2},\ c-\dfrac{1}{2} \text{ or } 2c-1 \\ a-b = \dfrac{1}{2},\ -\dfrac{1}{2},\ c-1 \text{ or } 1-c \\ c = \dfrac{1}{2},\ \dfrac{3}{2},\ 2a \text{ or } 2b \end{cases}$$

These 12 conditions are precisely those detailed in Section 60:4 that cause reduction of the Gauss function to one or two associated Legendre functions. Accordingly, the table in that section (Table 60.4.1) may be used to deduce quadratic transformation formulas. For example, the first two tabular entries may be separately reformulated as

60:5:11 $$\mathrm{P}_{\nu}^{(\mu)}(X) = \left(\frac{2}{\sqrt{|1-X^2|}}\right)^{\mu} \frac{\mathrm{F}\left(\dfrac{1+\nu-\mu}{2}, 1+\dfrac{\nu-\mu}{2}; 1-\mu; 1-\dfrac{1}{X^2}\right)}{\Gamma(1-\mu)\,X^{1+\nu-\mu}} \qquad X = \frac{1}{\sqrt{1-x}} > 0$$

and

60:5:12 $$\mathrm{P}_{\nu}^{(\mu)}(X) = \left(\frac{1+X}{1-X}\right)^{\mu/2} \frac{\mathrm{F}\left(-\nu, 1+\nu; 1-\mu; \dfrac{1-X}{2}\right)}{\Gamma(1-\mu)} \qquad X = 1-2x > -1$$

The equality of the right-hand members of 60:5:11 and 60:5:12 constitutes a quadratic transformation, valid for $X > 0$. The Bateman manuscript [see reference above] includes a comprehensive listing of quadratic transformations and information on *cubic transformations*.

60:6 EXPANSIONS

The *Gauss series*

60:6:1
$$F(a,b;c;x) = 1 + \frac{abx}{c}\left(1 + \frac{(1+a)(1+b)x}{2(1+c)}\left(1 + \frac{(2+a)(2+b)x}{3(2+c)}\left(1 + \cdots + \frac{(j+a-1)(j+b-1)x}{j(j+c-1)}\left(1+\cdots\right)\cdots\right)\right)\right) = \sum_{j=0}^{\infty}\frac{(a)_j(b)_j}{(1)_j(c)_j}x^j \qquad -1 < x < 1$$

is the fundamental expansion of the Gauss function, here expressed in concatenated form and in terms of the Pochhammer polynomial [Chapter 18]. The restriction on the magnitude of the argument x may be discarded if the series terminates, a circumstance discussed in connection with equations 60:4:10 and 60:4:11.

For large negative argument, the expansion

60:6:2
$$F(a,b;c;x) = \sum_{\lambda=a}^{b}\frac{\Gamma(c)\,\Gamma(a+b-2\lambda)}{\Gamma(a+b-\lambda)\,\Gamma(c-\lambda)(-x)^{\lambda}}\left[1 - \frac{\lambda(1+\lambda-c)}{(1+2\lambda-a-b)x} + \cdots\right] \qquad x \to -\infty$$

valid unless a and b differ by an integer, is a consequence of linear transformation 60:5:7. Useful at the other end of the $-\infty$ to 1 range of argument is the similar expansion

60:6:3
$$F(a,b;c;1-x) = \sum_{\lambda=c}^{a+b}\frac{\Gamma(c)\,\Gamma(a+b+c-2\lambda)}{\Gamma(a+c-\lambda)\,\Gamma(b+c-\lambda)x^{a+b-\lambda}}\left[1 + \frac{(\lambda-a)(\lambda-b)x}{1+2\lambda-a-b-c} + \cdots\right] \qquad x \to 0$$

that derives from 60:5:9.

60:7 PARTICULAR VALUES

60:7:1
$$F(a,b;c;0) = 1$$

60:7:2
$$F(a,b;c;1) = \frac{\Gamma(c)\,\Gamma(c-a-b)}{\Gamma(c-a)\,\Gamma(c-b)} \qquad c > a+b$$

60:7:3
$$F(a,b;c;-\infty) = \begin{cases} 0 & a<0,\ b<0 \\ 1 & (a \text{ or } b) = 0,\ (b \text{ or } a) < 0 \\ \infty & (a \text{ or } b) > 0 \end{cases}$$

These are the only particular values that can be formulated without placing severe restrictions on the parameters. Some particular values under such restrictive conditions include

60:7:4
$$F(a,b;1+a-b;-1) = \frac{\sqrt{\pi}\,\Gamma(1+a-b)}{2^a\Gamma\left(1+\frac{a}{2}-b\right)\Gamma\left(\frac{1+a}{2}\right)} \qquad a-b \neq -1, -2, -3, \ldots$$

60:7:5
$$F\left(a,b;\frac{1+a+b}{2};\frac{1}{2}\right) = \frac{\sqrt{\pi}\,\Gamma\left(\frac{1+a+b}{2}\right)}{\Gamma\left(\frac{1+a}{2}\right)\Gamma\left(\frac{1+b}{2}\right)} \qquad a+b \neq -1, -3, -5, \ldots$$

and

60:7:6
$$F\left(a,1-a;c;\frac{1}{2}\right) = \frac{2\sqrt{\pi}\,\Gamma(c)}{2^c\Gamma\left(\frac{a+c}{2}\right)\left(\frac{1+c-a}{2}\right)} \qquad c \neq 0, -1, -2, \ldots$$

60:8 NUMERICAL VALUES

When $0 \le x < 1$, the following algorithm uses the Gauss series 60:6:1 in the form

60:8:1
$$F(a,b;c;x) = t_0 + t_1 + \cdots t_j + \cdots t_J + R_J$$

where $t_0 = 1$ and $t_j = (j + a - 1)(j + b - 1)\, t_{j-1}x/j(j + c - 1)$. An estimate of the remainder is provided by the geometric sum $R_J = t_Jx/(1 - x)$. The number $J + 1$ of summed terms is incremented until $J \ge |a| + |b| + |c|$ (ensuring that all t_j have uniform signs for $j > J$) and until $|R_J| \le 10^{-8}|t_0 + t_1 + \cdots + t_J|$. A similar technique is employed for negative x, but transformation 60:5:3 is first invoked.

Input a >>>>>>>
    Set $f = t = 1$
Input b >>>>>>>
    Set $j = 0$
Input c >>>>>>>
    Set $d = |a| + |b| + |c|$
Input x >>>>>>>
    If $x \ge 0$ go to (1)
    Set $f = t = (1 - x)^{-a}$
    Replace b by $c - b$
    Replace x by $x/(x - 1)$
(1) Replace t by $(a + j)(b + j)\, tx/(c + j)(1 + j)$
    Replace j by $j + 1$
    Replace f by $f + t$
    If $j < d$ go to (1)
    If $|f|/10^8 < x|t|/(1 - x)$ go to (1)
    Replace f by $f + [xt/(1 - x)]$
    Output f
$f \simeq F(a,b;c;x)$ <<<

Storage needed: a, f, t, b, j, c, d and x

Input restrictions: $x < 1$, $c \ne 0, -1, -2, \ldots$

Test values:
$F(\frac{1}{2},\frac{1}{2};\frac{1}{2};\frac{1}{2}) = 1.41421356$
$F(-0.1,0.4;0.5;-1) = 1.05857815$

No particular accuracy is claimed for this algorithm, but the precision will generally be better than 24 bits. The algorithm will often be intolerably slow if $x > 0.9$ or if $x < -9$. The linear transformations 60:5:9 or 60:5:8, respectively, are useful in these circumstances.

60:9 APPROXIMATIONS

The rational approximation

60:9:1
$$F(a,b;c;x) \simeq 1 + \frac{abx}{c}\left[1 + \cfrac{1}{\cfrac{2(c+1)}{(a+1)(b+1)x} - 1}\right] \qquad \text{small } x$$

is valid for sufficiently small argument. If one of the numeratorial parameters, say b, is large, then

60:9:2
$$F(a,b;c;x) \simeq M(a;c;bx) \qquad \text{large } b$$

where M denotes the Kummer function [Chapter 47]. Conversely, if the denominatorial parameter is large, we have an approximation as a Tricomi function [Chapter 48]:

60:9:3
$$F(a,b;c;x) = \left(\frac{c}{1-x}\right)^a U\left(a;1 + a - b;\frac{c}{1-x}\right) \qquad \text{large } c$$

60:10 OPERATIONS OF THE CALCULUS

The formulas for differentiation:

60:10:1 $$\frac{d}{dx}F(a,b;c;x) = \frac{ab}{c}F(a+1,b+1;c+1;x)$$

multiple differentiation:

60:10:2 $$\frac{d^n}{dx^n}F(a,b;c;x) = \frac{(a)_n(b)_n}{(c)_n}F(a+n,b+n;c+n;x)$$

and integration:

60:10:3 $$\int_0^x F(a,b;c;t)dt = \frac{c-1}{(a-1)(b-1)}[F(a-1,b-1;c-1;x)-1] \quad \boldsymbol{a,b,c \neq 1}$$

modify each of the three parameters equally. One may, however, selectively alter a single parameter of the Gauss function by such operations as

60:10:4 $$x^{1-b}\frac{d^n}{dx^n}\{x^{n-1+b}F(a,b;c;x)\} = (b)_n F(a,b+n;c;x)$$

60:10:5 $$\frac{x^{1-c+b}}{(1-x)^{a+b-c-n}}\frac{d^n}{dx^n}\left\{\frac{x^{n-1+c-b}}{(1-x)^{c-a-b}}F(a,b;c;x)\right\} = (c-b)_n F(a,b-n;c;x)$$

60:10:6 $$\frac{1}{(1-x)^{a+b-c-n}}\frac{d^n}{dx^n}\left\{\frac{F(a,b;c;x)}{(1-x)^{c-a-b}}\right\} = \frac{(c-a)_n(c-b)_n}{(c)_n}F(a,b;c+n;x)$$

60:10:7 $$x^{1-c+n}\frac{d^n}{dx^n}\{x^{c-1}F(a,b;c;x)\} = (c-n)_n F(a,b;c-n;x)$$

Differintegration of the product of a Gauss function and a power obeys the rule

60:10:8 $$\frac{d^\nu}{dx^\nu}x^\mu F(a,b;c;x) = \frac{\Gamma(1+\mu)}{\Gamma(1+\mu-\nu)}\sum_{j=0}^{\infty}\frac{(1+\mu)_j(a)_j(b)_j}{(1+\mu-\nu)_j(c)_j(1)_j}x^{\mu-\nu+j}$$

and generally creates a $K = L = 3$ hypergeometric function [Section 18:14]. By suitable choices of ν and μ, however, the generated function may be of the $K = L = 2$ or even the $K = L = 1$ class. The notation of Section 43:14 enables these conversions to be expressed succinctly; for example:

60:10:9 $$F(a,b;c;x) \xrightarrow[1-A]{1-a} F(A,b;c;x) \qquad X = x$$

60:10:10 $$F(a,b;c;x) \xrightarrow[1-c]{1-b} (1-x)^{-a} \qquad X = x$$

60:11 COMPLEX ARGUMENT

Generally, $F(a,b;c;x+iy)$ is complex valued, but we shall not pursue this topic.

60:12 GENERALIZATIONS

Because

60:12:1 $$F(a,b;c;x) = \sum_{j=0}^{\infty}\frac{(a)_j(b)_j}{(1)_j(c)_j}x^j$$

any of the following series:

60:12:2
$$\sum_{j=0}^{\infty}\frac{(a)_j(b)_j}{(c_1)_j(c_2)_j}x^j,\ \sum_{j=0}^{\infty}\frac{(a_1)_j(a_2)_j(a_3)_j}{(1)_j(c)_j}x^j,\ \sum_{j=0}^{\infty}\frac{(a)_j(b)_j}{(1)_j(c_2)_j(c_3)_j}x^j,\ \text{etc.}$$

represent functions that may be regarded as generalizations of the Gauss function. All of these are called *hypergeometric functions* in this *Atlas* and are discussed in Section 18:14.

There are several more profound generalizations of the Gauss function. These are the subject of a chapter by Erdélyi, Magnus, Oberhettinger and Tricomi [*Higher Transcendental Functions*, Volume 1, Chapter 4], but they will not be discussed here.

60:13 COGNATE FUNCTIONS

The misleading name *"generalized" hypergeometric function* is reserved for the class of hypergeometric functions in which at least one of the denominatorial parameters equals unity. These are symbolized by ${}_nF_d$ in which n and d are integers equalling the numbers of numeratorial and denominatorial parameters. The unity denominatorial parameter is excluded from this count so that, for example:

60:13:1
$${}_0F_0(x)=\sum_{j=0}^{\infty}\frac{x^j}{(1)_j}=\exp(x)$$

and

60:13:2
$${}_2F_1(a,b;c;x)=\sum_{j=0}^{\infty}\frac{(a)_j(b)_j}{(1)_j(c)_j}x^j=F(a,b;c;x)$$

In this *Atlas* the name "hypergeometric function" denotes *any* function that may be represented by a series of the form

60:13:3
$$\sum_{j=0}^{\infty}\frac{(a_1)_j(a_2)_j\cdots(a_K)_j}{(c_1)_j(c_2)_j\cdots(c_L)_j}x^j$$

irrespective of whether one of the denominatorial parameters happens to be unity. Moreover, the integers K and L, which we use to characterize the numbers of numeratorial and denominatorial parameters, include any unity parameters. Thus, the ${}_nF_d$ function generally has $K=n$ and $L=1+d$. In the ${}_nF_d$ notation, function 60:13:3 would be symbolized ${}_{K+1}F_L(1,a_1,a_2,\ldots a_K;c_1,c_2,\ldots c_L;x)$.

CHAPTER 61

THE COMPLETE ELLIPTIC INTEGRALS K(p) AND E(p)

The complete elliptic integrals of the first and second kinds are simple, although important, univariate functions. More complicated, but less important, is the complete elliptic integral of the third kind, discussed in Section 61:12. All these functions arise in the canonical representation of integrals of certain rational functions that do not reduce to simpler expressions.

61:1 NOTATION

Throughout this *Atlas* we normally use the name "argument" and the symbol x to represent the single variable of a univariate function. To achieve uniformity with Chapter 62, we here violate this rule and use the name *modulus* (*module* is also encountered) and the symbol p to represent the variable of K(p), the *complete elliptic integral of the first kind*, and of E(p), the *complete elliptic integral of the second kind*. Similarly, we denote the complete elliptic integral of the third kind [Section 61:12] by $\Pi(\nu;p)$.

An important auxiliary variable is the *complementary modulus*, defined by

61:1:1 $$q = \sqrt{1 - p^2}$$

The symbols k and k' frequently replace p and q. As well, the role of the modulus is often taken over by the *modular angle* α or the *parameter* m, where

61:1:2 $$p = k = \sin(\alpha) = \sqrt{m}$$

The complementary modulus then takes a symbol from the equivalences

61:1:3 $$\sqrt{1 - p^2} = q = k' = \cos(\alpha) = \sqrt{1 - m} = \sqrt{m_1}$$

Especially when interest is primarily in the functions of Chapters 62 and 63, the naked symbols K and E are frequently used to replace K(p) and E(p), a constant unspecified modulus p being understood. The notations K$'$ and E$'$ then mean K(q) and E(q), respectively.

61:2 BEHAVIOR

Figure 61-1 illustrates the simple behaviors of the complete elliptic integrals. They acquire values $\pi/2 \le \text{K}(p) \le \infty$ and $\pi/2 \ge \text{E}(p) \ge 1$ over their usual $0 \le p \le 1$ domain. The inequality

61:2:1 $$1.57 \simeq \frac{\pi}{2} \geq K(p) + \ln(q) \geq \ln(4) \simeq 1.39 \qquad 0 \leq p \leq 1$$

shows that K(p) + ln(q) varies little. The behaviors of K(p) and E(p) as $p \to 0$ or $p \to 1$ can be deduced from the expansions presented in Section 61:6.

In many applications the complete elliptic integrals of complementary modulus are almost as important as K(p) and E(p). For this reason, Figure 61-1 includes maps of K(q) and E(q).

Normally, interest in complete elliptic integrals is confined to the zero to unity range of modulus. When p exceeds unity, K(p) and E(p) are complex valued; this circumstance is discussed in Section 61:11. The evenness of the functions, which is evident from equations 61:3:1 and 61:3:2, permits extension to negative moduli. Both complete elliptic integrals acquire real values when their moduli are imaginary numbers of any magnitude; again, see Section 61:11.

61:3 DEFINITIONS

The complete elliptic integrals are defined by

61:3:1 $$K(p) = \int_0^{\pi/2} \frac{d\theta}{\sqrt{1 - p^2 \sin^2(\theta)}} = \int_0^1 \frac{dt}{\sqrt{(1 - t^2)(1 - p^2 t^2)}} = \int_0^\infty \frac{dt}{\sqrt{(1 + t^2)(1 + q^2 t^2)}}$$

and

61:3:2 $$E(p) = \int_0^{\pi/2} \sqrt{1 - p^2 \sin^2(\theta)}\, d\theta = \int_0^1 \sqrt{\frac{1 - p^2 t^2}{1 - t^2}}\, dt = \int_0^\infty \sqrt{\frac{1 + q^2 t^2}{(1 + t^2)^3}}\, dt$$

the two integrals that involve the angle θ being illustrated in Figures 62-2 and 62-3.

Alternative definitions are as special cases of the Gauss function [Chapter 60]:

61:3:3 $$K(p) = \frac{\pi}{2} F\left(\frac{1}{2}, \frac{1}{2}; 1; p^2\right) = \frac{\pi}{2q} F\left(\frac{1}{2}, \frac{1}{2}; 1; -p^2/q^2\right)$$

61:3:4 $$E(p) = \frac{\pi}{2} F\left(\frac{-1}{2}, \frac{1}{2}; 1; p^2\right) = \frac{\pi q^2}{2} F\left(\frac{1}{2}, \frac{3}{2}; 1; p^2\right)$$

or via Legendre functions of moiety order [see Section 59:4].

The perimeter of the ellipse [see Section 14:14] shown in Figure 61-2, which has semiaxes of q and 1, equals 4E(p). The common mean [Section 61:8] of the two semiaxes is $\pi/[2K(p)]$.

In the notation of Section 43:14, the complete elliptic integrals may be synthesized as follows:

61:3:5 $$K(p) \underset{\frac{1}{2}}{\overset{0}{\longleftarrow}} \frac{\pi}{2} \sqrt{1 - p^2} \underset{\frac{3}{2}}{\overset{0}{\longrightarrow}} E(p) \qquad X = p^2$$

61:4 SPECIAL CASES

There are none.

61:5 INTRARELATIONSHIPS

Both complete elliptic functions are even:

61:5:1 $$f(-p) = f(p) \qquad f = K \text{ or } E$$

The four complete elliptic integrals K(p), E(p), K(q) and E(q) are linked by the remarkable *Legendre relation*

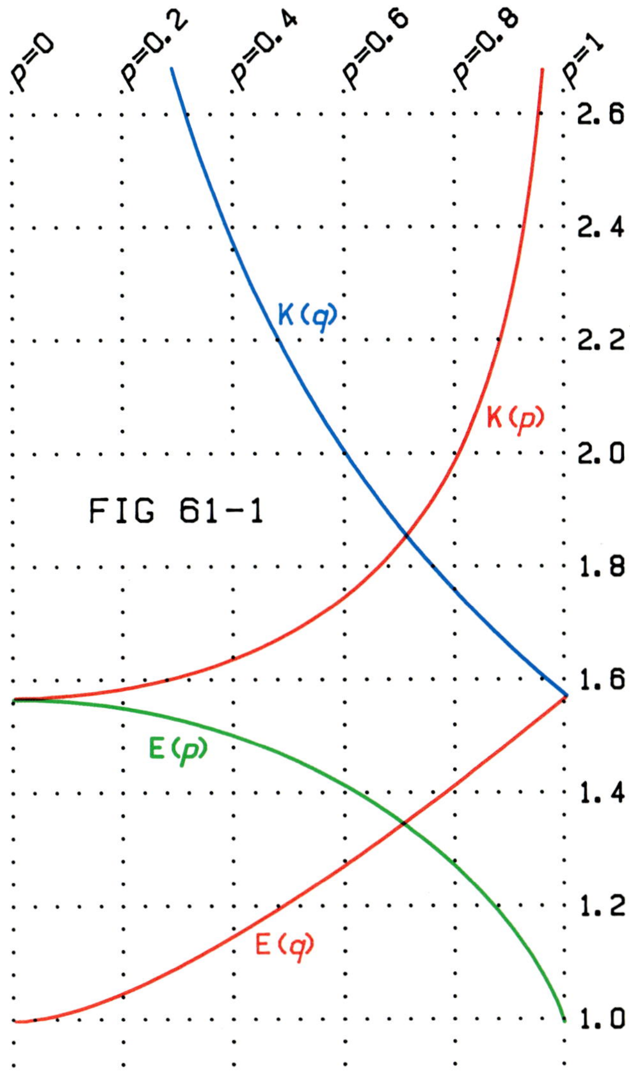

FIG 61-1

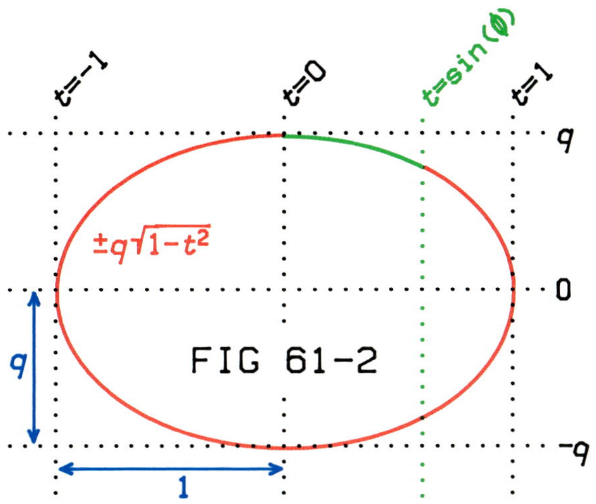

FIG 61-2

61:5:2 $$K(p)\,E(q) + K(q)\,E(p) = \frac{\pi}{2} + K(p)\,K(q)$$

Complete elliptic integrals of modulus p are related to those of modulus $2\sqrt{p}/(1+p)$ by the transformations

61:5:3 $$K\left(\frac{2\sqrt{p}}{1+p}\right) = (1+p)\,K(p) \qquad p \neq 1$$

61:5:4 $$E\left(\frac{2\sqrt{p}}{1+p}\right) = \frac{2E(p) - q^2K(p)}{1+p} \qquad q^2 = 1 - p^2$$

and to those of modulus $(1-q)/(1+q)$ by

61:5:5 $$K\left(\frac{1-q}{1+q}\right) = \frac{1+q}{2}\,K(p)$$

61:5:6 $$E\left(\frac{1-q}{1+q}\right) = \frac{E(p) + qK(p)}{1+q}$$

61:6 EXPANSIONS

Alternatives to the power series

61:6:1 $$K(p) = \frac{\pi}{2}\left\{1 + \left(\frac{1}{2}\right)^2 p^2 + \left(\frac{1\times 3}{2\times 4}\right)^2 p^4 + \left(\frac{1\times 3\times 5}{2\times 4\times 6}\right)^2 p^6 + \cdots\right\} = \frac{\pi}{2}\sum_{j=0}^{\infty}\left[\frac{(2j-1)!!}{(2j)!!}p^j\right]^2$$

61:6:2 $$E(p) = \frac{\pi}{2}\left\{1 - \left(\frac{1}{2}\right)^2 \frac{p^2}{1} - \left(\frac{1\times 3}{2\times 4}\right)^2 \frac{p^4}{3} - \left(\frac{1\times 3\times 5}{2\times 4\times 6}\right)^2 \frac{p^6}{5} - \cdots\right\} = \frac{-\pi}{2}\sum_{j=0}^{\infty}\frac{1}{2j-1}\left[\frac{(2j-1)!!}{(2j)!!}p^j\right]^2$$

may be constructed by use of transformations 61:5:2–61:5:5. Whereas expansions 61:6:1 and 61:6:2 are most useful for moduli close to zero, the pair

61:6:3 $$K(p) = \ln\left(\frac{4}{q}\right) + \left[\ln\left(\frac{4}{q}\right) - \frac{2}{1\times 2}\right]\left[\frac{1}{2}q\right]^2 + \left[\ln\left(\frac{4}{q}\right) - \frac{2}{1\times 2} - \frac{2}{3\times 4}\right]\left[\frac{1\times 3}{2\times 4}q^2\right]^2$$
$$\cdots + \left[\ln\left(\frac{4}{q}\right) - \frac{2}{1\times 2} - \frac{2}{3\times 4} - \cdots - \frac{2}{(2j-1)(2j)}\right]\left[\frac{(2j-1)!!}{(2j)!!}q^j\right]^2 + \cdots$$

and

61:6:4 $$E(p) = 1 + \frac{2}{1}\left[\ln\left(\frac{4}{q}\right) - \frac{1}{1\times 2}\right]\left[\frac{1}{2}q\right]^2 + \frac{4}{3}\left[\ln\left(\frac{4}{q}\right) - \frac{2}{1\times 2} - \frac{1}{3\times 4}\right]\left[\frac{1\times 3}{2\times 4}q^2\right]^2$$
$$\cdots + \frac{2j}{2j-1}\left[\ln\left(\frac{4}{q}\right) - \frac{2}{1\times 2} - \cdots - \frac{2}{(2j-3)(2j-2)} - \frac{1}{(2j-1)(2j)}\right]\left[\frac{(2j-1)!!}{(2j)!!}q^j\right]^2 + \cdots$$

is more valuable as $p \to 1$.

A trigonometric expansion is provided by

61:6:5 $$K(p) = \pi\left[\sin(\alpha) + \frac{1}{4}\sin(5\alpha) + \frac{9}{64}\sin(9\alpha) + \cdots\right] = \pi\sum_{j=0}^{\infty}\frac{[(2j)!]^2}{[2^j j!]^4}\sin(\alpha + 4j\alpha) \qquad \alpha = \arcsin(p)$$

For expansions involving the "nome," see Section 61:14.

61:7 PARTICULAR VALUES

In addition to listing values of the complete elliptic integrals at modular values of 0, $1/\sqrt{2}$ and 1, Table 61.7.1 includes p values at which the $K(q)/K(p)$ ratio acquires particular values. These ratios correspond to special values

Table 61.7.1

	$p = 0$	$p = 3 - \sqrt{8}$	$p = \dfrac{1}{\sqrt{6} + \sqrt{2}}$	$p = \sqrt{2} - 1$	$p = \dfrac{1}{\sqrt{2}}$	$p = \sqrt{\sqrt{8} - 2}$	$p = \dfrac{2 + \sqrt{3}}{4}$	$p = 2\sqrt{\sqrt{18} - 4}$	$p = 1$
K(p)	$\dfrac{\pi}{2}$				$\dfrac{\pi}{2U}$				∞
E(p)	$\dfrac{\pi}{2}$				$\dfrac{U}{2} + \dfrac{\pi}{4U}$				1
$\dfrac{K(q)}{K(p)}$	∞	2	$\sqrt{3}$	$\sqrt{2}$	1	$\dfrac{1}{\sqrt{2}}$	$\dfrac{1}{\sqrt{3}}$	$\dfrac{1}{2}$	0

of the "nome" [Section 61:14]. The constant U that appears in this table is the ubiquitous constant [Section 1:7], equal to 0.8472130848.

61:8 NUMERICAL VALUES

We start this section by describing the procedure for determining the so-called *common mean* of two numbers.

Let A_0 be the larger of the two positive numbers G_0 and A_0. The *geometric mean* of these numbers is $G_1 = \sqrt{G_0 A_0}$ while their *arithmetic mean* is $A_1 = (G_0 + A_0)/2$. From the properties of means [see, for example, Bronshtein and Semendyayev, Section 11.14] it follows that $G_0 < G_1 < A_1 < A_0$. If the procedure of forming geometric and arithmetic means is repeated indefinitely so that

61:8:1 $$G_{j+1} = \sqrt{G_j A_j} \qquad \text{and} \qquad A_{j+1} = (G_j + A_j)/2 \qquad j = 0, 1, 2, \ldots$$

one must have

61:8:2 $$G_0 < G_1 < G_2 < G_3 < \cdots < G_j < \cdots < A_j < \cdots < A_3 < A_2 < A_1 < A_0$$

and there is convergence towards a common mean, $G_\infty = A_\infty$, of G_0 and A_0.

The common mean of the complementary modulus q and unity is related to the complete elliptic integral of the first kind by

61:8:3 $$K(p) = \frac{\pi}{2G_\infty} \qquad G_\infty = \text{common mean of } q \text{ and } 1$$

This is the relationship that is exploited in the algorithm below. The complete elliptic integral of the second kind is evaluated via the expression

61:8:4 $$E(p) = \frac{K(p)}{2}\left[2 - p^2 - \sum_{j=1}^{\infty} 2^j (A_j^2 - G_j^2)\right] \qquad G_j(\text{or } A_j) = \text{successive geometric (or arithmetic) means of } q \text{ and } 1$$

Set $n = 0$
Input p >>>>>>
Set $g = \sqrt{1 - p^2}$
(1) Set $a = t = e = 1$
Replace e by $e + g^2$
(2) Set $s = ga$
Replace a by $(a + g)/2$
Set $g = \sqrt{s}$
Replace t by $2t$
Replace e by $e - t(a^2 - s)$
If $t(a^2 - s) \geq 6 \times 10^{-8}$ go to (2)
Set $k = \pi/2g$

Storage needed:
n, p, g, a, t, e, s and k

Input restrictions:
$0 \leq p < 1$
$0 < p < 1$

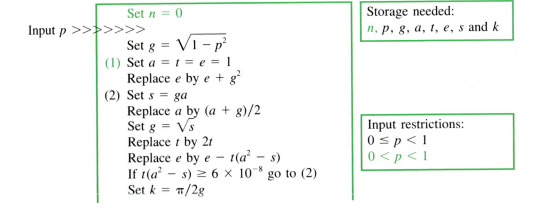

Replace e by $ek/2$
If $n \neq 0$ go to (3)
Output k
$k \simeq \mathrm{K}(p)$ <<<<<<
Output e
$e \simeq \mathrm{E}(p)$ <<<<<<
Set $n = k$
Set $g = p$
Go to (1)
(3) Replace n by $\exp(-\pi k/n)$
Output k
$k \simeq \mathrm{K}(q)$ <<<<<<
Output e
$e \simeq \mathrm{E}(q)$ <<<<<<
Output n
$n \simeq \mathrm{N}(p)$ <<<<<<

Test values:
$\mathrm{K}(1/2) = 1.68575035$
$\mathrm{E}(1/2) = 1.46746221$
$\mathrm{K}(\sqrt{3}/2) = 2.15651565$
$\mathrm{E}(\sqrt{3}/2) = 1.21105603$
$\mathrm{N}(1/2) = 0.0179723870$

The algorithm ceases to compute successive values of the geometric and arithmetic means when sufficient have been calculated to ensure 24-bit precision in K(p) and E(p). If the green portions of the algorithm are included, values of K(q), E(q) are also evaluated, as well as that of the nome [Section 61:14] N(p).

On account of the definitions 61:3:3 and 61:3:4, the universal hypergeometric algorithm of Section 18:14 may also be used to compute elliptic integrals of the first and second kinds. As well, the algorithm in Section 62:8 generates values of K(p) and E(p).

61:9 APPROXIMATIONS

For rather wide ranges of small values of the modulus, the approximations

61:9:1 $$\mathrm{K}(p) \simeq \frac{\pi}{2}\left(\frac{16-5p^2}{16-9p^2}\right) \qquad \text{8-bit precision} \qquad 0 \le p \le 0.67$$

61:9:2 $$\mathrm{E}(p) \simeq \frac{\pi}{2}\left(\frac{16-7p^2}{16-3p^2}\right) \qquad \text{8-bit precision} \qquad 0 \le p \le 0.71$$

61:9:3 $$\mathrm{K}(q) = \left[1+\frac{p^2}{4}\right]\ln\left(\frac{4}{p}\right) - \frac{p^2}{4} \qquad \text{8-bit precision} \qquad 0 < p \le 0.47$$

61:9:4 $$\mathrm{E}(q) \simeq 1 + \frac{p^2}{2}\ln\left(\frac{4}{p}\right) - \frac{p^2}{4} \qquad \text{8-bit precision} \qquad 0 < p \le 0.35$$

are valid, whereas for moduli close to unity we have

61:9:5 $$\mathrm{K}(p) \simeq \frac{5-p^2}{8}\ln\left(\frac{16}{1-p^2}\right) - \frac{1-p^2}{4} \qquad \text{8-bit precision} \qquad 0.88 \le p < 1$$

61:9:6 $$\mathrm{E}(p) \simeq \frac{3+p^2}{4} - \frac{1-p^2}{4}\ln\left(\frac{1-p^2}{16}\right) \qquad \text{8-bit precision} \qquad 0.94 \le p < 1$$

61:9:7 $$\mathrm{K}(q) \simeq \frac{\pi}{2}\left(\frac{11+5p^2}{7+9p^2}\right) \qquad \text{8-bit precision} \qquad 0.73 \le p \le 1$$

61:9:8 $$\mathrm{E}(q) \simeq \frac{\pi}{2}\left(\frac{9+7p^2}{13+3p^2}\right) \qquad \text{8-bit precision} \qquad 0.65 \le p \le 1$$

These results derive from expansions 61:6:1–61:6:4, from which more refined approximations are available. Because of their rapid convergence, the expansions in Section 61:14 give accurate approximations even when severely truncated.

61:10 OPERATIONS OF THE CALCULUS

Differentiation with respect to the modulus gives

61:10:1
$$\frac{\mathrm{d}}{\mathrm{d}p}\mathrm{K}(p) = \frac{\mathrm{E}(p)}{p(1-p^2)} - \frac{\mathrm{K}(p)}{p}$$

61:10:2
$$\frac{\mathrm{d}}{\mathrm{d}p}\mathrm{E}(p) = \frac{\mathrm{E}(p)-\mathrm{K}(p)}{p}$$

Indefinite integration of the complete elliptic integrals yields $K = L = 3$ hypergeometric functions [Section 18:14]

61:10:3
$$\int_0^p \mathrm{K}(t)\mathrm{d}t = \frac{\pi p}{2}\sum_{j=0}^{\infty}\frac{(\frac{1}{2})_j(\frac{1}{2})_j(\frac{1}{2})_j}{(1)_j(1)_j(\frac{3}{2})_j}p^{2j}$$

61:10:4
$$\int_0^p \mathrm{E}(t)\mathrm{d}t = \frac{\pi p}{2}\sum_{j=0}^{\infty}\frac{(\frac{-1}{2})_j(\frac{1}{2})_j(\frac{1}{2})_j}{(1)_j(1)_j(\frac{3}{2})_j}p^{2j}$$

but the results

61:10:5
$$\int_0^p t\mathrm{K}(t)\mathrm{d}t = \mathrm{E}(p) - (1-p^2)\,\mathrm{K}(p)$$

61:10:6
$$\int_0^p t\mathrm{E}(t)\mathrm{d}t = \frac{(1+p^2)\,\mathrm{E}(p) - (1-p^2)\,\mathrm{K}(p)}{3}$$

are simpler.

Some definite integrals lead to Catalan's constant [Section 1:7]

61:10:7
$$\frac{1}{2}\int_0^1 \mathrm{K}(t)\mathrm{d}t = \frac{-1}{2} + \int_0^1 \mathrm{E}(t)\mathrm{d}t = G$$

Interesting integrals involving the complementary modulus include

61:10:8
$$\frac{1}{2}\int_0^1 \mathrm{K}(\sqrt{1-t^2})\mathrm{d}t = \int_0^1 \mathrm{E}(\sqrt{1-t^2})\mathrm{d}t = \frac{\pi^2}{8}$$

61:10:9
$$\int_0^1 \frac{\mathrm{K}(t)\mathrm{d}t}{\sqrt{1-t^2}} = \mathrm{K}^2\left(\frac{1}{\sqrt{2}}\right) = \frac{\pi^2}{4U^2}$$

61:10:10
$$\int_0^1 \frac{\mathrm{E}(t)\mathrm{d}t}{\sqrt{1-t^2}} = 2\,\mathrm{E}^2\left(\frac{1}{\sqrt{2}}\right) - \frac{\pi}{2} = \frac{\pi^2}{8U^2} + \frac{U^2}{2}$$

and more are listed by Gradshteyn and Ryzhik [Sections 6.14 and 6.15].

The operations of semidifferentiation and semiintegration with respect to p^2, when applied to complete elliptic integrals, produce elementary functions. Examples include

61:10:11
$$\frac{\mathrm{d}^{1/2}}{(\mathrm{d}p^2)^{1/2}}\mathrm{E}(p) = \frac{\sqrt{\pi(1-p^2)}}{2p}$$

61:10:12
$$\frac{\mathrm{d}^{-1/2}}{(\mathrm{d}p^2)^{-1/2}}\mathrm{K}(p) = \sqrt{\pi}\,\arcsin(p)$$

61:11 COMPLEX ARGUMENT

The complete elliptic integrals of a complex modulus are themselves generally complex valued, but these functions have real values when their modulus is purely imaginary. The transformations

61:11:1
$$K(ip) = \frac{1}{\sqrt{1+p^2}} K\left(\frac{p}{\sqrt{1+p^2}}\right)$$

and

61:11:2
$$E(ip) = \sqrt{1+p^2}\, E\left(\frac{p}{\sqrt{1+p^2}}\right)$$

are valid for all real p. For example, $K(i) = (1/\sqrt{2})\, K(1/\sqrt{2})$ and $E(i) = \sqrt{2}\, E(1/\sqrt{2})$.

The complete elliptic integrals acquire complex values when their real moduli exceed unity. The transformation formulas

61:11:3
$$K(p) = \frac{1}{p} K\left(\frac{1}{p}\right) + \frac{i}{p} K\left(\frac{\sqrt{p^2-1}}{p}\right)$$

61:11:4
$$E(p) = \left[pE\left(\frac{1}{p}\right) - \left(p - \frac{1}{p}\right) K\left(\frac{1}{p}\right)\right] + i\left[pE\left(\frac{\sqrt{p^2-1}}{p}\right) - \frac{1}{p} K\left(\frac{\sqrt{p^2-1}}{p}\right)\right]$$

permit the evaluation of the real and imaginary parts in this circumstance. They show, for example, that

61:11:5
$$K(\sqrt{2}) = \left(\frac{1}{\sqrt{2}} + \frac{i}{\sqrt{2}}\right) K\left(\frac{1}{\sqrt{2}}\right) = (1+i)\frac{\pi}{\sqrt{8}\,U}$$

61:11:6
$$E(\sqrt{2}) = \left(\frac{1}{\sqrt{2}} + \frac{i}{\sqrt{2}}\right)\left[2E\left(\frac{1}{\sqrt{2}}\right) - K\left(\frac{1}{\sqrt{2}}\right)\right] = (1+i)\frac{U}{\sqrt{2}} = \frac{\pi i}{2K(\sqrt{2})}$$

indicating that the real and imaginary parts are equal for a modular value of $\sqrt{2}$.

61:12 GENERALIZATIONS

Allowing the upper limit to vary in the integral definitions 61:3:1 and 61:3:2 gives rise to the incomplete elliptic integrals of the next chapter.

In the remainder of this section we discuss the *complete elliptic integral of the third kind* $\Pi(\nu;p)$ defined by

61:12:1
$$\Pi(\nu;p) = \int_0^1 \frac{dt}{(1+\nu t^2)\sqrt{(1-t^2)(1-p^2t^2)}} = \int_0^{\pi/2} \frac{d\theta}{[1+\nu \sin^2(\theta)]\sqrt{1-p^2\sin^2(\theta)}}$$

It is a generalization of the first kind of complete elliptic integral inasmuch as $\Pi(0;p) = K(p)$. Be aware of the variety of definitions and notation. Thus, the integral 61:12:1 is variously denoted $\Pi_1(\nu,k)$, $\Pi(-\nu\backslash\arcsin(p))$, $\Pi(p,\nu,\pi/2)$, $\Pi(\pi/2,-\nu,p)$ or even $(-1/\nu)\Pi(-1/\nu,p,\pi/2)$ by different authors.

The *characteristic* ν may take real values in the range $-\infty < \nu < \infty$, although the integral is infinite for $\nu = -1$. Interest concentrates on the range $0 \le p \le 1$ of the *modulus*. Figure 61-3 shows the behavior of $\Pi(\nu;p)$ in its most important domain.

One may evaluate the complete elliptic integral of the third kind via the incomplete elliptic integrals of the first and second kinds [Chapter 62]. The formulas permitting this are as follows:

61:12:2
$$\Pi(\nu;p) = \frac{K(p)}{1+\nu} + \sqrt{\frac{\nu}{(1+\nu)(p^2+\nu)}}\left\{\frac{\pi}{2} + [K(p) - E(p)]\, F(q;\phi) - K(p)\, E(q;\phi)\right\}$$

$$\sin(\phi) = \frac{1}{\sqrt{1+\nu}} \qquad \nu > 0$$

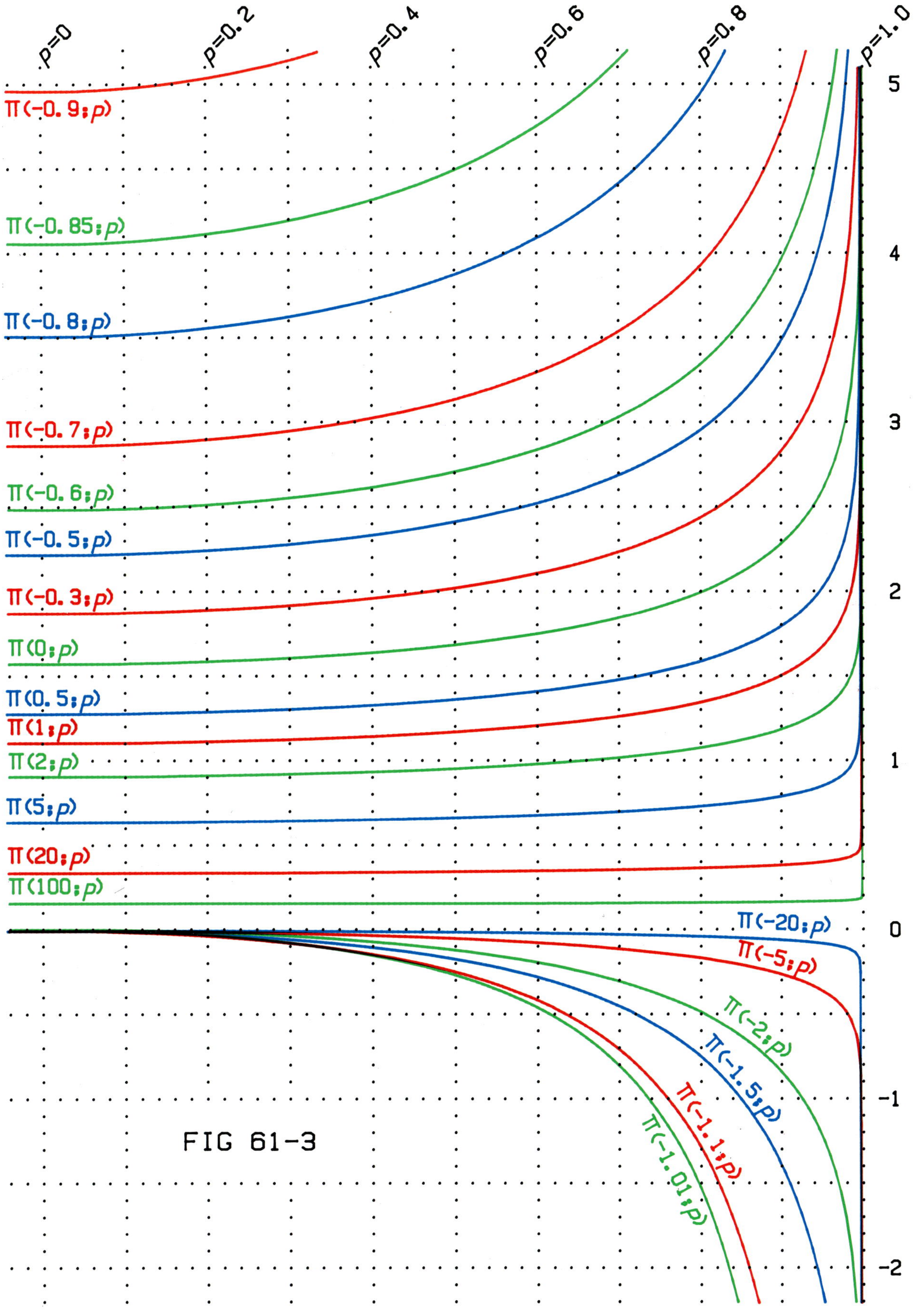

p=0
p=0.2
p=0.4
p=0.6
p=0.8
p=1.0
5
4
3
2
1
0
−1
−2
Π(−0.9;p)
Π(−0.85;p)
Π(−0.8;p)
Π(−0.7;p)
Π(−0.6;p)
Π(−0.5;p)
Π(−0.3;p)
Π(0;p)
Π(0.5;p)
Π(1;p)
Π(2;p)
Π(5;p)
Π(20;p)
Π(100;p)
Π(−20;p)
Π(−5;p)
Π(−2;p)
Π(−1.5;p)
Π(−1.1;p)
Π(−1.01;p)

FIG 61-3

61:12:3 $$\Pi(0;p) = \mathrm{K}(p) \qquad \nu = 0$$

61:12:4 $$\Pi(\nu;p) = \mathrm{K}(p) - \sqrt{\frac{-\nu}{(1+\nu)(p^2+\nu)}}\{\mathrm{E}(p)\,\mathrm{F}(p;\phi) - \mathrm{K}(p)\,\mathrm{E}(p;\phi)\} \qquad \sin(\phi) = \frac{\sqrt{-\nu}}{p} \qquad -p^2 < \nu < 0$$

61:12:5 $$\Pi(-p^2;p) = \frac{\mathrm{E}(p)}{1-p^2} = \frac{\mathrm{E}(\sqrt{-\nu})}{1+\nu} \qquad \nu = -p^2$$

61:12:6 $$\Pi(\nu;p) = \mathrm{K}(p) + \sqrt{\frac{\nu}{(1+\nu)(p^2+\nu)}}\left\{\frac{\pi}{2} + [\mathrm{K}(p) - \mathrm{E}(p)]\,\mathrm{F}(q;\phi) - \mathrm{K}(p)\,\mathrm{E}(q;\phi)\right\} \qquad \sin(\phi) = \frac{\sqrt{1+\nu}}{q} \qquad 1 < \nu < -p^2$$

61:12:7 $$\Pi(-1;p) = \pm\infty \qquad \nu = -1$$

61:12:8 $$\Pi(\nu;p) = \sqrt{\frac{\nu}{(1+\nu)(p^2+\nu)}}\{\mathrm{E}(p)\,\mathrm{F}(p;\phi) - \mathrm{K}(p)\,\mathrm{E}(p;\phi)\} \qquad \sin(\phi) = \frac{1}{\sqrt{-\nu}} \qquad \nu < -1$$

where $q = \sqrt{1-p^2}$ and $\mathrm{F}(p;\phi)$ and $\mathrm{E}(p;\phi)$ are incomplete elliptic integrals whose numerical values are calculable using the algorithm in Section 62:8.

For characteristics of magnitudes less than unity, the expansion

61:12:9 $$\Pi(\nu;p) = 1 - \frac{1}{2}\left(\nu - \frac{p^2}{2}\right) + \frac{3}{8}\left(\nu^2 - \frac{\nu p^2}{2} + \frac{3p^4}{8}\right) - \frac{5}{16}\left(\nu^3 - \frac{\nu^2 p^2}{2} + \frac{3\nu p^4}{8} - \frac{5p^6}{16}\right) + \cdots$$
$$= \sum_{j=0}^{\infty} \frac{(2j-1)!!}{(2j)!!}(-\nu)^j \sum_{k=0}^{j} \frac{(2k-1)!!}{(2k)!!}\left(\frac{-p^2}{\nu}\right)^k \qquad -1 < \nu < 1$$

may provide useful values of the complete elliptic integral of the third kind.

61:13 COGNATE FUNCTIONS

In addition to those of the first, second and third kinds, the following complete elliptic integrals may be encountered in some works

61:13:1 $$\mathrm{D}(p) = \int_0^{\pi/2} \frac{\sin^2(\theta)\mathrm{d}\theta}{\sqrt{1-p^2\sin^2(\theta)}} = \frac{\mathrm{K}(p) - \mathrm{E}(p)}{p^2}$$

61:13:2 $$\mathrm{B}(p) = \int_0^{\pi/2} \frac{\cos^2(\theta)\mathrm{d}\theta}{\sqrt{1-p^2\sin^2(\theta)}} = \frac{\mathrm{E}(p) - q^2\mathrm{K}(p)}{p^2}$$

61:13:3 $$\mathrm{C}(p) = \int_0^{\pi/2} \frac{\sin^2(\theta)\cos^2(\theta)\mathrm{d}\theta}{[1-p^2\sin^2(\theta)]^{3/2}} = \frac{2-p^2}{p^4}\mathrm{K}(p) - \frac{2}{p^4}\mathrm{E}(p)$$

although not in this *Atlas*.

61:14 RELATED TOPICS

Associated with the functions of Chapter 61, 62 and 63 is the elliptic *nome*. A common symbol is q, but to avoid confusion with the complementary modulus, and to emphasize that the nome is a function of the modulus, we shall use the N(p) notation. The nome is defined by

61:14:1 $$N(p) = \exp\left(\frac{-\pi K(q)}{K(p)}\right) \qquad q = \sqrt{1-p^2} \qquad 0 \le p \le 1$$

and its utility arises from the large number of functions that can be expressed as infinite sums or products of algebraic functions of N(p). Except in the immediate vicinity of $p = 1$, the nome is a small positive number [$0 \le N(p) \le \frac{1}{2}$ for $0 \le p \le 0.999995$; $\frac{1}{2} \le N(p) \le 1$ for $0.999995 \le p \le 1$], so these infinite sums and products generally converge very rapidly.

The functions that can be expressed in terms of the nome include the elliptic modulus:

61:14:2 $$p = 4\sqrt{N(p)}\left[\frac{1+N^2(p)}{1+N(p)}\right]^4\left[\frac{1+N^4(p)}{1+N^3(p)}\right]^4\left[\frac{1+N^6(p)}{1+N^5(p)}\right]^4\cdots$$

the complementary modulus:

61:14:3 $$q = \sqrt{1-p^2} = \left[\frac{1-N(p)}{1+N(p)}\right]^4\left[\frac{1-N^3(p)}{1+N^3(p)}\right]^4\left[\frac{1-N^5(p)}{1+N^5(p)}\right]^4\cdots$$

the complete elliptic integrals of the first:

61:14:4 $$\frac{K(p)}{2\pi} = \left[\frac{1}{2} + N(p) + N^4(p) + N^9(p) + \cdots\right]^2$$
$$= \sqrt{\frac{\sqrt{N(p)}}{4p}}\left[\frac{1-N^2(p)}{1-N(p)}\right]^2\left[\frac{1-N^4(p)}{1-N^3(p)}\right]^2\left[\frac{1-N^6(p)}{1-N^5(p)}\right]^2\cdots$$

and second:

61:14:5 $$E(p) = K(p) - \frac{\pi^2}{K(p)}\left[\frac{N(p) - 4N^4(p) + 9N^9(p) - \cdots}{\frac{1}{2} - N(p) + N^4(p) - N^9(p) + \cdots}\right]$$

kinds, the incomplete elliptic integral [Chapter 62] expression:

61:14:6 $$\frac{K(p)\,E(p;\phi) - xE(p)}{2\pi} = \frac{N(p)\sin[\pi x/K(p)]}{1-N^2(p)} + \frac{N^2(p)\sin[2\pi x/K(p)]}{1-N^4(p)} + \frac{N^3(p)\sin[3\pi x/K(p)]}{1-N^6(p)} + \cdots$$

where $x = F(p;\phi)$, the elliptic amplitude [Section 63:3]:

61:14:7 $$\frac{\mathrm{am}(p;x)}{2} - \frac{\pi x}{4K(p)} = \frac{N(p)\sin[\pi x/K(p)]}{1+N^2(p)} + \frac{N^2(p)\sin[2\pi x/K(p)]}{2[1+N^4(p)]} + \frac{N^3(p)\sin[3x/K(p)]}{3[1+N^6(p)]} + \cdots$$

all four of Neville's theta functions [Section 63:8]:

61:14:8 $$\theta_s(p;x) = \sqrt{\frac{2\pi\sqrt{N(p)}}{pqK(p)}}\left[\sin\left(\frac{\pi x}{2K(p)}\right) - N^2(p)\sin\left(\frac{3\pi x}{2K(p)}\right) + N^6(p)\sin\left(\frac{5\pi x}{2K(p)}\right) - \cdots\right]$$

61:14:9 $$\theta_c(p;x) = \sqrt{\frac{2\pi\sqrt{N(p)}}{pK(p)}}\left[\cos\left(\frac{\pi x}{2K(p)}\right) + N^2(p)\cos\left(\frac{3\pi x}{2K(p)}\right) + N^6(p)\cos\left(\frac{5\pi x}{2K(p)}\right) + \cdots\right]$$

61:14:10 $$\theta_d(p;x) = \sqrt{\frac{2\pi}{K(p)}}\left[\frac{1}{2} + N(p)\cos\left(\frac{\pi x}{K(p)}\right) + N^4(p)\cos\left(\frac{2\pi x}{K(p)}\right) + N^9(p)\cos\left(\frac{3\pi x}{K(p)}\right) + \cdots\right]$$

61:14:11 $$\theta_n(p;x) = \sqrt{\frac{2\pi}{qK(p)}}\left[\frac{1}{2} - N(p)\cos\left(\frac{\pi x}{K(p)}\right) + N^4(p)\cos\left(\frac{2\pi x}{K(p)}\right) - N^9(p)\cos\left(\frac{3\pi x}{K(p)}\right) + \cdots\right]$$

and all 12 Jacobian elliptic functions [Chapter 63], as well as products, quotients and logarithms of such functions. Gradshteyn and Ryzhik [Section 8.146] give a comprehensive listing of these, but the *Atlas* is content with the three most important:

61:14:12 $$\frac{pK(p)}{2\pi}\,\mathrm{sn}(p;x) = \frac{N^{1/2}(p)\sin[\pi x/2K(p)]}{1-N(p)} + \frac{N^{3/2}(p)\sin[3\pi x/2K(p)]}{1-N^3(p)} + \frac{N^{5/2}(p)\sin[5\pi x/2K(p)]}{1-N^5(p)} + \cdots$$

61:14:13 $$\frac{pK(p)}{2\pi}\,\mathrm{cn}(p;x) = \frac{N^{1/2}(p)\cos[\pi x/2K(p)]}{1+N(p)} + \frac{N^{3/2}(p)\cos[3\pi x/2K(p)]}{1+N^3(p)} + \frac{N^{5/2}(p)\cos[5\pi x/2K(p)]}{1+N^5(p)} + \cdots$$

61:14:14 $$\frac{K(p)}{2\pi}\,\mathrm{dn}(p;x) - \frac{1}{4} = \frac{N(p)\cos[\pi x/K(p)]}{1+N^2(p)} + \frac{N^2(p)\cos[2\pi x/K(p)]}{1+N^4(p)} + \frac{N^3(p)\cos[3\pi x/K(p)]}{1+N^6(p)} + \cdots$$

A deterrent to the full exploitation of the formulas above is the lack of a simple method of evaluating $N(p)$ from p. If p is sufficiently small, a truncated version of the series

61:14:15 $$N(p) = \lambda + 2\lambda^5 + 15\lambda^9 + 150\lambda^{13} + 1707\lambda^{17} + \cdots \qquad \lambda = \frac{1-(1-p^2)^{1/4}}{2+2(1-p^2)^{1/4}} \qquad 0 \le p \le 1$$

will serve, but otherwise one must resort to iterative or numerical methods; the extended algorithm of Section 61:8 provides one possibility. Some particular values of the nome are easily found by application of definition 61:14:1 to the values of $K(q)/K(p)$ listed in Table 61.7.1.

CHAPTER 62

THE INCOMPLETE ELLIPTIC INTEGRALS $F(p;\phi)$ AND $E(p;\phi)$

The incomplete elliptic integrals of the first and second kinds are important bivariate functions. Together with more elementary functions and the incomplete elliptic integral of the third kind [see Section 62:12], they may be used to express any indefinite integral of the form $\int R(t,\sqrt{p(t)})dt$ (where p is a polynomial of degree 3 or 4 and R is a rational function) [see Section 62:14].

62:1 NOTATION

These functions are sometimes named *Legendre's elliptic integrals*. The adjective "incomplete" is often omitted: it refers to the fact that the upper limits in the defining integrals 62:3:1 and 62:3:2 are usually smaller than those in the corresponding integrals 61:3:1 and 61:3:2 that define the "complete" elliptic integrals. The symbols F and E are in general use for the incomplete elliptic integrals of the first and second kinds, but there is no unanimity on the ways in which the variables are specified.

This *Atlas* uses the notation $F(p;\phi)$ for the *incomplete elliptic integral of the first kind* and we describe ϕ as the *amplitude* of the function and p as its *modulus* (k is often used). Other symbolisms are based on the *parameter* $m = p^2$ or the *modular angle* $\alpha = \arcsin(p)$, so the notations $F(\phi|m)$, $F(\phi\backslash\alpha)$, as well as $F(k,\phi)$, $F(\phi,k)$ and even $F(\phi)$ are encountered. Since ϕ is often interpreted as an angle, it may be expressed in degrees, and occasionally $\sin(\phi)$ is treated as the variable instead of ϕ itself.

The situation is even more confused for the *incomplete elliptic integral of the second kind* for which the *Atlas* uses the $E(p;\phi)$ symbol. In addition to analogs of the variants discussed in the last paragraph, one also encounters $E(x)$, where x represents the incomplete elliptic integral of the *first* kind, $F(p;\phi)$. As well, ambiguities can arise because the same E symbol is adopted for both the incomplete and complete [Chapter 61] elliptic integrals of the second kind.

62:2 BEHAVIOR

The incomplete elliptic integrals are sometimes discussed only for $0 \le p < 1$ and $0 \le \phi \le \pi/2$ but they may be defined for all moduli and amplitudes. They are real and finite, for all real p and ϕ, except that $F(1;\phi)$ is infinite when $|\phi| \ge \pi/2$. In fact, $F(p;\phi)$ and $E(p;\phi)$ may be real even when p is imaginary, as discussed in Section 62:11. Figure 62-1 shows graphs of $F(p;\phi)$ and $E(p;\phi)$ for $p = 0, 0.5, 0.7, 0.9$ and 1.

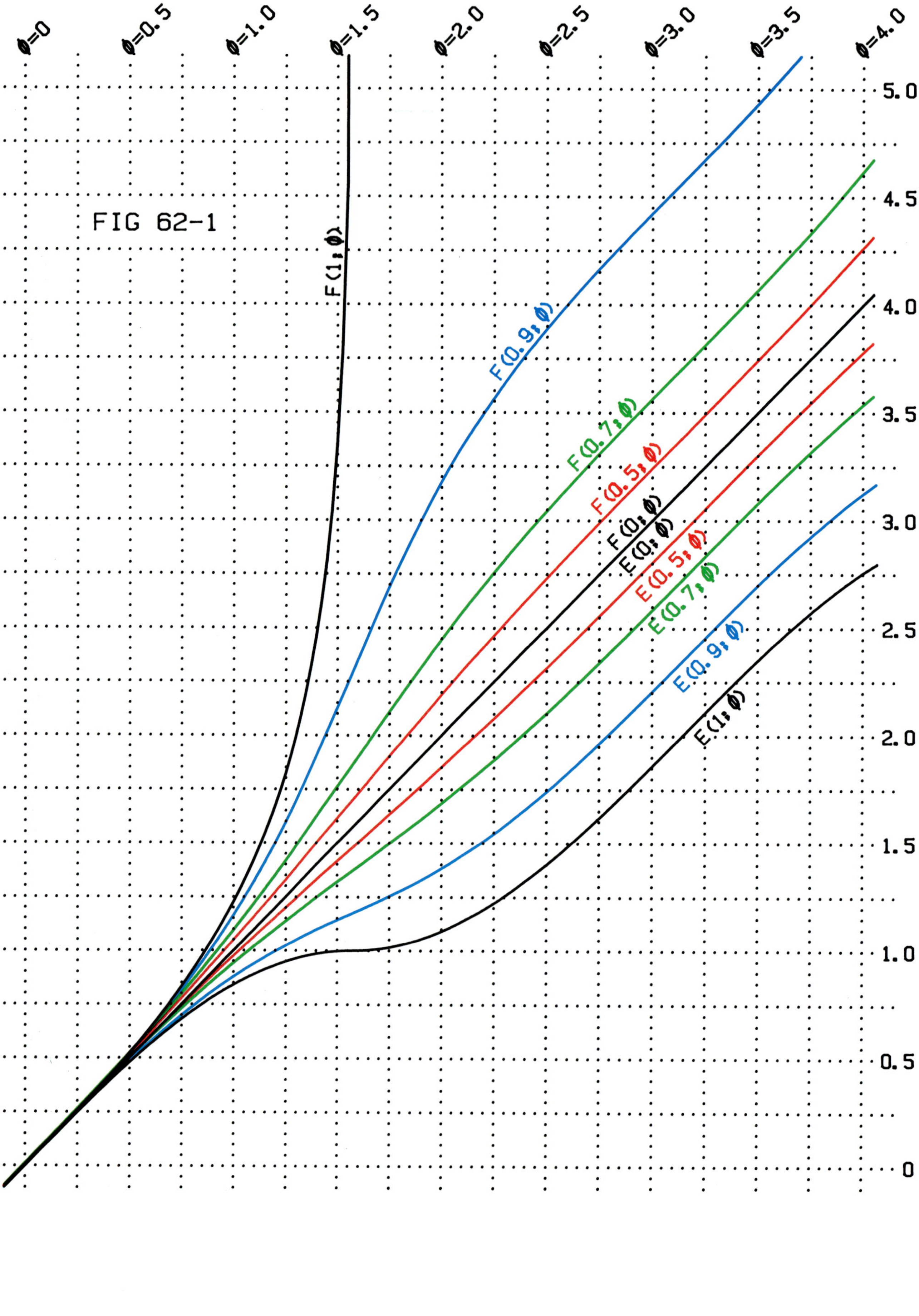
ϕ=0
ϕ=0.5
ϕ=1.0
ϕ=1.5
ϕ=2.0
ϕ=2.5
ϕ=3.0
ϕ=3.5
ϕ=4.0
5.0
4.5
4.0
3.5
3.0
2.5
2.0
1.5
1.0
0.5
0
FIG 62-1
F(1;ϕ)
F(0.9;ϕ)
F(0.7;ϕ)
F(0.5;ϕ)
F(0;ϕ)
E(0;ϕ)
E(0.5;ϕ)
E(0.7;ϕ)
E(0.9;ϕ)
E(1;ϕ)

The incomplete elliptic integrals possess a high degree of symmetry. They are even with respect to p and odd with respect to ϕ. The differences

62:2:1 $$F(p;\phi) - \frac{2\phi}{\pi}K(p) \quad\text{and}\quad E(p;\phi) - \frac{2\phi}{\pi}E(p)$$

are periodic (but *not* sinusoidal) with period π [see Chapter 36]. Moreover, each of these difference functions displays odd reflection symmetry about amplitude values of 0, $\pm\pi/2$, $\pm\pi$, $\pm3\pi/2$, ..., as evident in equation 62:5:2 and from the examples mapped in Figure 62-2.

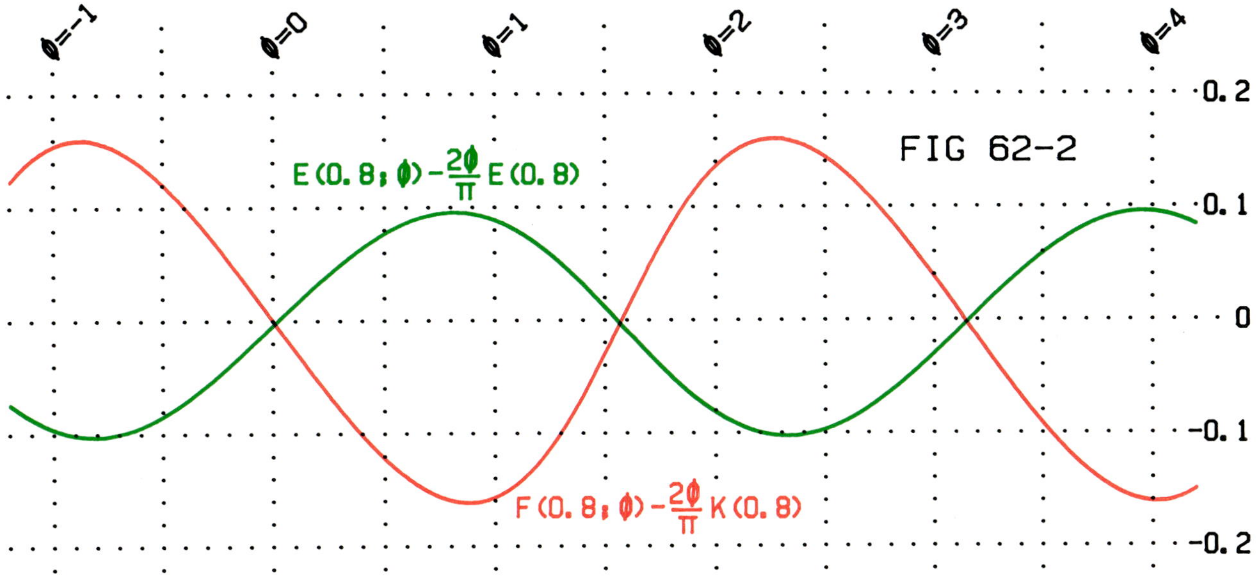

62:3 DEFINITIONS

The indefinite integrals

62:3:1 $$F(p;\phi) = \int_0^{\phi} \frac{d\theta}{\sqrt{1 - p^2\sin^2(\theta)}} = \int_0^{\sin(\phi)} \frac{dt}{\sqrt{1-t^2}\sqrt{1-p^2t^2}} = \int_0^{\tan(\phi)} \frac{dt}{\sqrt{1+t^2}\sqrt{1+q^2t^2}} \qquad q^2 = 1 - p^2$$

62:3:2 $$E(p;\phi) = \int_0^{\phi} \sqrt{1 - p^2\sin^2(\theta)}d\theta = \int_0^{\sin(\phi)} \sqrt{\frac{1-p^2t^2}{1-t^2}}dt = \int_0^{\tan(\phi)} \sqrt{\frac{1+q^2t^2}{(1+t^2)^3}}dt \qquad q^2 = 1 - p^2$$

define the incomplete elliptic integrals of the first and second kinds. The trigonometric integrals are illustrated by Figures 62-3 and 62-4, which also demonstrate the relationship to the complete elliptic integrals [Chapter 61].

The incomplete elliptic integrals may be expressed as inverse functions [Section 0:3] of the Jacobian elliptic functions [Chapter 63]. Thus, for example, if ϕ is the angle defined, in terms of variables p and x, by

62:3:3 $$\phi = \arcsin\{\mathrm{sn}(p;x)\} = \arccos\{\mathrm{cn}(p;x)\} = \arcsin\frac{\sqrt{1 - \mathrm{dn}^2(p;x)}}{p}$$

then $F(p;\phi) = x$ and

62:3:4 $$E(p;\phi) = \int_0^x \mathrm{dn}^2(p;t)dt = x - p^2\int_0^x \mathrm{sn}^2(p;t)dt$$

In discussions of elliptic functions, ϕ is often denoted am(p;x).

To define the incomplete elliptic integral of the first kind via a geometric construct, refer to Figure 62-5. Let point O be the $r = 0$ origin of a polar coordinate system [Section 46:14] (r,θ). In this system let points I and P

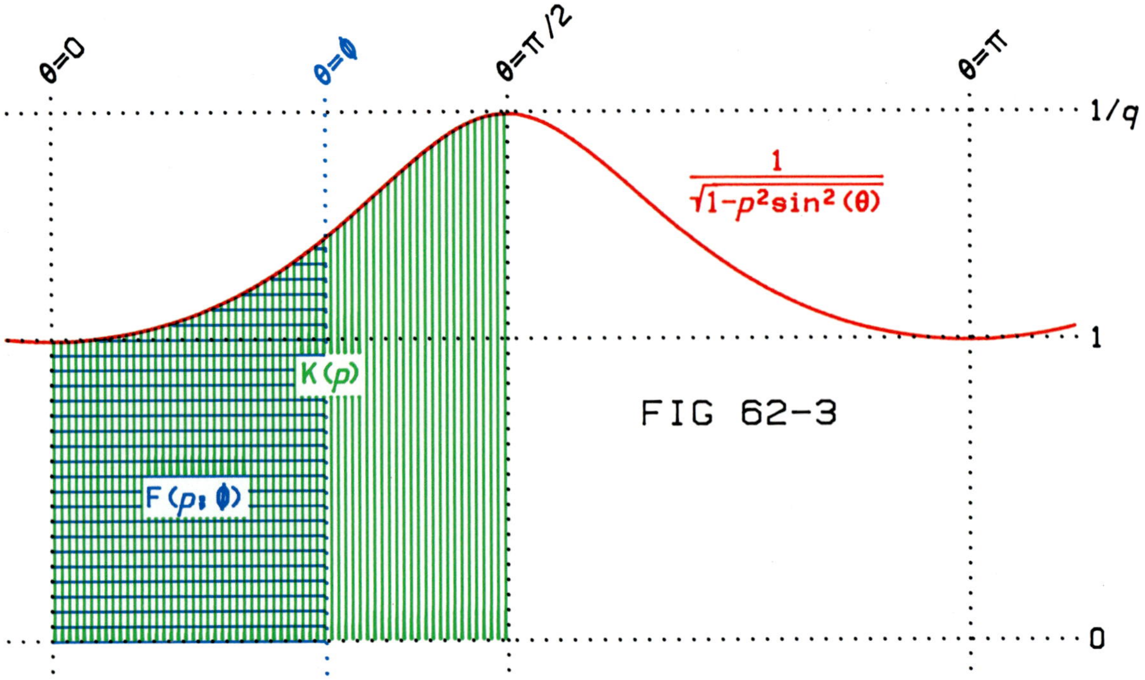

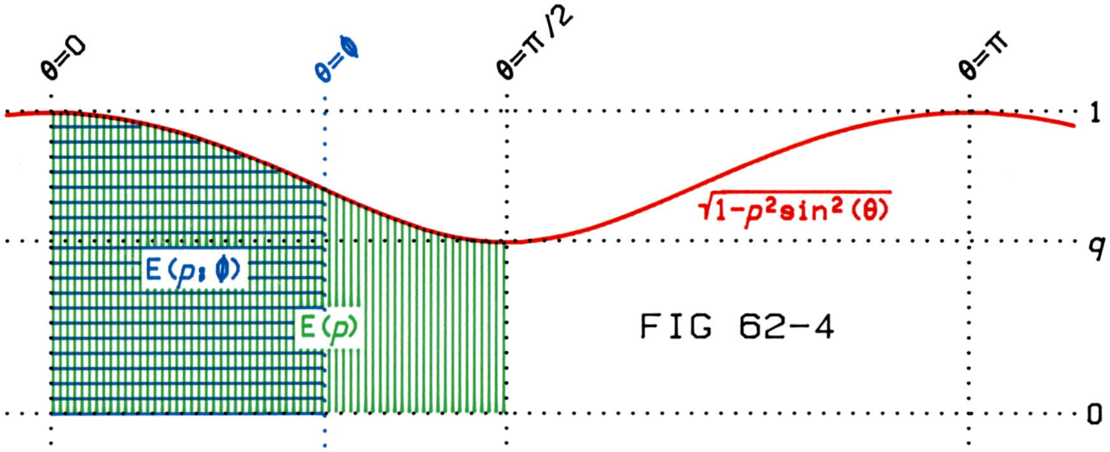

have coordinates (1,0) and $(1/q,\pi/2)$, respectively, where q is the complementary modulus of F(p;ϕ). Now construct an ellipse [Section 14:14] with OI and OP as its semiminor and semimajor axes. In polar coordinates the equation of this ellipse is $r = 1/\sqrt{1-p^2\sin^2(\theta)}$. Let ϕ be the amplitude and consider the average length $\bar{r}$ of the radius vector from the origin to the ellipse over the segment $0 \le \theta \le \phi$. Then the product $\bar{r}\phi$ is defined as the incomplete elliptic integral F(p;ϕ). That this definition is equivalent to 62:3:1 is evident by evaluating $\bar{r}$ as follows:

62:3:5 $$\bar{r} = \frac{1}{\phi}\int_0^\phi r\,d\theta = \frac{1}{\phi}\int_0^\phi \frac{d\theta}{\sqrt{1-p^2\sin^2(\theta)}} = \frac{F(p;\phi)}{\phi}$$

A geometric definition of the second kind of incomplete elliptic integral is illustrated in Figure 61-2 of the previous chapter. This shows a portion of the ellipse $\pm q\sqrt{1-t^2}$ marked in green and delineated by the ordinates $t = 0$ and $t = \sin(\phi)$. The length of this green arc defines the elliptic integral E(p;ϕ). The entire perimeter of the ellipse is 4E(p), four times the complete elliptic integral of the second kind.

62:4 SPECIAL CASES

The incomplete elliptic integral becomes equal to its amplitude

62:4:1 $$F(0;\phi) = E(0;\phi) = \phi$$

when the modulus is zero, to the inverse gudermannian [Section 33:14] or sine function [Chapter 32] when the modulus is unity:

62:4:2 $$F(1;\phi) = \mathrm{invgd}(\phi) = \ln\left(\tan\left(\frac{\pi}{4} + \frac{\phi}{2}\right)\right) \qquad |\phi| \le \pi/2$$

62:4:3 $$E(1;\phi) = \sin(\phi) \qquad |\phi| \le \pi/2$$

and to instances of the incomplete beta function [Chapter 58] when the modulus is $\sqrt{2}$ or $1/\sqrt{2}$:

62:4:4 $$F(\sqrt{2};\phi) = \frac{1}{4}\,B\left(\frac{1}{2};\frac{1}{4};\sin^2(2\phi)\right) \qquad 0 \le \phi \le \pi/4$$

62:4:5 $$E(\sqrt{2};\phi) = \frac{1}{4}\,B\left(\frac{1}{2};\frac{3}{4};\sin^2(2\phi)\right) \qquad 0 \le \phi \le \pi/4$$

62:4:6 $$F\left(\frac{1}{\sqrt{2}};\phi\right) = \frac{1}{\sqrt{8}}\,B\left(\frac{1}{2};\frac{1}{4};1-\cos^4(\phi)\right) \qquad 0 \le \phi \le \pi/2$$

62:4:7 $$E\left(\frac{1}{\sqrt{2}};\phi\right) = \frac{1}{4\sqrt{2}}\,B\left(\frac{1}{2};\frac{1}{4};1-\cos^4(\phi)\right) + \frac{1}{4\sqrt{2}}\,B\left(\frac{1}{2};\frac{3}{4};1-\cos^4(\phi)\right) \qquad 0 \le \phi < \pi/2$$

See Section 63:7 for F(p;ϕ) and E(p;ϕ) when ϕ takes particular values. When the amplitude and modulus are interrelated such that $\cot(\phi) = \sqrt{q} = (1-p^2)^{1/4}$, we have

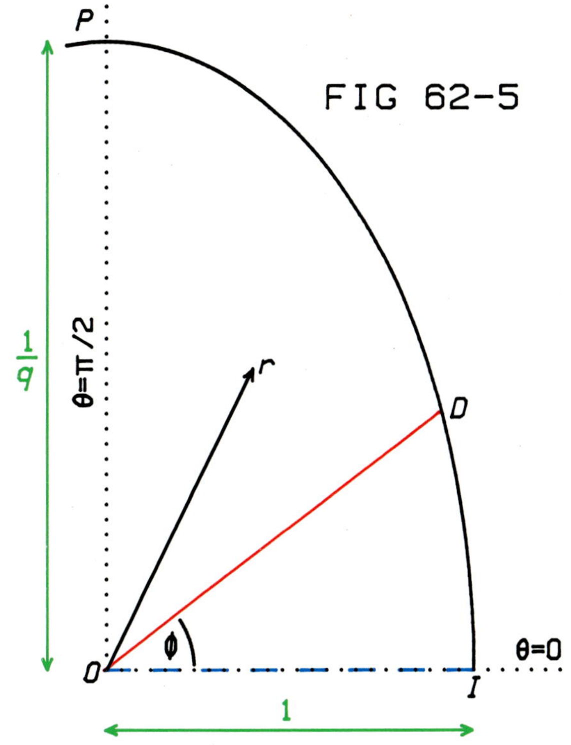

62:4:8 $$\mathrm{F}(p;\,\mathrm{arccot}(\sqrt{q})) = \mathrm{F}(\sqrt{1-\cot^4(\phi)};\phi) = \frac{1}{2}\mathrm{K}(p) \qquad \frac{\pi}{4} \le \phi \le \frac{\pi}{2}$$

62:4:9 $$\mathrm{E}(p;\,\mathrm{arccot}(\sqrt{q})) = \mathrm{E}(\sqrt{1-\cot^4(\phi)};\phi) = \frac{1}{2}[1 - q + \mathrm{E}(p)] \qquad \frac{\pi}{4} \le \phi \le \frac{\pi}{2}$$

62:5 INTRARELATIONSHIPS

The incomplete elliptic integrals are even with respect to their modulus but odd with respect to their amplitude:

62:5:1 $$\mathrm{f}(-p;\phi) = \mathrm{f}(p;\phi) = -\mathrm{f}(p;-\phi) \qquad \mathrm{f} = \mathrm{F} \text{ or } \mathrm{E}$$

The latter equality is a special case of the formula

62:5:2 $$\mathrm{f}\left(p;\frac{n\pi}{2} - \phi\right) = 2n\mathrm{f}\left(p;\frac{\pi}{2}\right) - \mathrm{f}\left(p;\frac{n\pi}{2} + \phi\right) \qquad n = 0, \pm1, \pm2, \ldots$$

for reflection of the amplitude across any multiple of $\pi/2$. In the equation above, and that below, $\mathrm{f}(p;\pi/2)$ is the complete elliptic integral $\mathrm{K}(p)$ or $\mathrm{E}(p)$. Both incomplete elliptic integrals obey the recurrence formula

62:5:3 $$\mathrm{f}(p;\phi + n\pi) = 2n\mathrm{f}\left(p;\frac{\pi}{2}\right) + \mathrm{f}(p;\phi) \qquad \mathrm{f} = \mathrm{F} \text{ or } \mathrm{E} \qquad n = \pm1, \pm2, \ldots$$

As a result, the difference functions 62:2:1 are periodic.

Let p_{-1}, p_0 and p_1 be three moduli interrelated by

62:5:4 $$p_{-1} = \frac{1 - \sqrt{1 - p_0^2}}{1 + \sqrt{1 - p_0^2}} \qquad \text{and} \qquad p_1 = \frac{2\sqrt{p_0}}{1 + p_0}$$

and ϕ_{-1}, ϕ_0 and ϕ_1 be three amplitudes interrelated by

62:5:5 $$\phi_{-1} = \phi_0 + \arctan\{\sqrt{1 - p_0^2}\tan(\phi_0)\} \qquad \text{and} \qquad \phi_1 = \frac{1}{2}[\phi_0 + \arcsin\{p_0 \sin(\phi_0)\}]$$

Then the corresponding incomplete elliptic integrals of the first kind are interrelated by

62:5:6 $$\mathrm{F}(p_{-1};\phi_{-1}) = [1 + \sqrt{1 - p_0^2}]\,\mathrm{F}(p_0;\phi_0) \qquad \text{and} \qquad \mathrm{F}(p_1;\phi_1) = \frac{1 + p_0}{2}\mathrm{F}(p_0;\phi_0)$$

and those of the second kind by

62:5:7 $$\begin{cases} \mathrm{E}(p_{-1};\phi_{-1}) = \dfrac{2}{1 + \sqrt{1 - p_0^2}}[\mathrm{E}(p_0;\phi_0) + \sqrt{1 - p_0^2}\,\mathrm{F}(p_0;\phi_0)] - \dfrac{(1 - \sqrt{1 - p_0^2})\cos(\phi_0)}{\sqrt{\csc^2(\phi_0) - p_0^2}} \\ \text{and} \\ \mathrm{E}(p_1;\phi_1) = \dfrac{\mathrm{E}(p_0;\phi_0)}{1 + p_0} - \dfrac{1 - p_0}{2}\mathrm{F}(p_0;\phi_0) \\ \qquad + \dfrac{p_0 \sin(\phi_0)[p_0\cos(\phi_0) + \sqrt{1 - p_0^2\sin^2(\phi_0)}]}{(1 + p_0^2)\sqrt{1 - p_0^2} + 2p_0\cos(\phi_0)[p_0\cos(\phi_0) + \sqrt{1 - p_0^2\sin^2(\phi_0)}]} \end{cases}$$

Going from elliptic integrals of variables p_0, ϕ_0 to those with variables p_{-1}, ϕ_{-1} is known as *descending Landen transformation*, while the conversion from variables p_0, ϕ_0 to p_1, ϕ_1 is called *ascending Landen transformation*. The adjectives "descending" and "ascending" recognize the fact that, for $0 < p_0 < 1$, the parameter p_{-1} is smaller than p_0, whereas p_1 is larger. Although it is not immediately apparent from 62:5:4, the numbers p_{-1}; p_0; p_1 form a *sequence*, that is, the rule by which p_0 is constructed from p_{-1} is identical with the rule by which p_1 is formed from p_0. Similarly, though more obscurely, ϕ_{-1}; ϕ_0; ϕ_1 form another sequence, as do $F(p_{-1};\phi_{-1})$; $F(p_0;\phi_0)$; $F(p_1;\phi_1)$ and $E(p_{-1};\phi_{-1})$; $E(p_0;\phi_0)$; $E(p_1;\phi_1)$. It is evident that each of these sequences may be extended indefinitely in both directions. In one popular method of computing numerical values of elliptic integrals, one or other of the Landen transformations is implemented repeatedly, starting with $p_0 = p$, $\phi_0 = \phi$, and generating a sequence of moduli (either $p_0; p_{-1}; p_{-2}; \ldots; p_{-n}$ or $p_0; p_1; p_2; \ldots; p_n$) that satisfy the inequalities

62:5:8 $$p_{-n} < p_{1-n} < \cdots < p_{-2} < p_{-1} < p_0 < p_1 < p_2 < \cdots < p_{n-1} < p_n$$

The limits as $n \to \infty$ of p_{-n} and p_n are 0 and 1, respectively; hence, if the transformation is carried out a sufficient number of times, it becomes possible to approximate $F(p_{\pm n};\phi_{\pm n})$ and $E(p_{\pm n};\phi_{\pm n})$ by use of equations 62:4:1–62:4:3. In this way F(p;ϕ) and E(p;ϕ), as well as K(p) and E(p) are calculable. The "common mean" technique used in Sections 61:8 and 62:8 is, in fact, an adaptation of the descending Landen transformation.

Incomplete elliptic integrals with moduli exceeding unity may be related to ones in the standard $0 < p \le 1$ range by the transformations

62:5:9 $$F\left(\frac{1}{p};\phi\right) = pF\left(p;\arcsin\left\{\frac{\sin(\phi)}{p}\right\}\right) \qquad p \le 1$$

62:5:10 $$E\left(\frac{1}{p};\phi\right) = \frac{1}{p}E\left(p;\arcsin\left\{\frac{\sin(\phi)}{p}\right\}\right) - \frac{1-p^2}{p}F\left(p;\arcsin\left\{\frac{\sin(\phi)}{p}\right\}\right) \qquad p \le 1$$

Yet another transformation is

62:5:11 $$F(p;\phi) = K(p) - F(p;\psi)$$

62:5:12 $$E(p;\phi) = E(p) - E(p;\psi) + p^2\sin(\phi)\sin(\psi)$$

where $\sin(\psi) = \cos(\phi)/\sqrt{1 - p^2\sin^2(\phi)}$.

62:6 EXPANSIONS

For small values of the amplitude and modulus, there exist expansions of which

62:6:1 $$F(p;\phi) = \frac{2K(p)}{\pi}\phi - \left[\frac{2K(p)}{\pi} - 1\right]\sin(\phi)\cos(\phi) - \left[\frac{4K(p)}{3\pi} - \frac{2}{3} - \frac{p^2}{6}\right]\sin^3(\phi)\cos(\phi) - \cdots$$

62:6:2 $$E(p;\phi) = \frac{2E(p)}{\pi}\phi + \left[1 - \frac{2E(p)}{\pi}\right]\sin(\phi)\cos(\phi) + \left[\frac{2}{3} - \frac{4E(p)}{3\pi} - \frac{p^2}{6}\right]\sin^3(\phi)\cos(\phi) + \ldots$$

are the leading terms. Similarly, when p is close to unity:

62:6:3 $$F(p;\phi) = \frac{2K(q)}{\pi}\operatorname{invgd}(\phi) - \left[\frac{2K(q)}{\pi} - 1\right]\frac{\sin(\phi)}{\cos^2(\phi)} + \left[\frac{4K(q)}{3\pi} - \frac{2}{3} - \frac{q^2}{6}\right]\frac{\sin^3(\phi)}{\cos^4(\phi)} - \cdots$$

62:6:4 $$E(p;\phi) = \frac{1 - \cos(\phi)\sqrt{1 - p^2\sin^2(\phi)}}{\sin(\phi)} + \frac{2K(q) - 2E(q)}{\pi}\left[\operatorname{invgd}(\phi) + \frac{\sin(\phi)}{\cos^2(\phi)}\right] - \frac{q^2\sin(\phi)}{2\cos^2(\phi)} + \cdots$$

where invgd is the inverse gudermannian function [Section 33:14]. Gradshteyn and Ryzhik [Sections 8.117 and 8.118] may be consulted for the general terms in these expansions.

62:7 PARTICULAR VALUES

Table 62.7.1 applies for $n = 1, 2, 3, \ldots$ and for any p in the range $0 < p < 1$. See Section 62:4 for cases in which p acquires special values.

Table 62.7.1

	$\phi = -\infty$	$\phi = \frac{-n\pi}{2}$	$\phi = 0$	$\phi = \frac{n\pi}{2}$	$\phi = \infty$
F(p; ϕ)	$-\infty$	$-n$K(p)	0	nK(p)	∞
E(p; ϕ)	$-\infty$	$-n$E(p)	0	nE(p)	∞

62:8 NUMERICAL VALUES

The algorithm of this section produces values of F(p;ϕ) and E(p;ϕ) using an extension of the "common mean" procedure that is described in Section 61:8. Because applications of incomplete elliptic integrals often require values of the complete integrals as well, the algorithm also generates K(p) and E(p).

Let A_j and G_j retain their significances from Section 61:8 and let a set of their values, for a given p, be generated by recursions 61:8:1. Further, let a set of angles ϕ_1, ϕ_2, ϕ_3,... be defined by the recurrence formula

62:8:1
$$\tan(2^{j+1}\phi_{j+1} - 2^j\phi_j) = \frac{G_j}{A_j}\tan(2^j\phi_j) \qquad j = 0, 1, 2, \ldots$$

with $\phi_0 = \phi$, the amplitude of the sought elliptic integrals. Then, in addition to the results $K(p) = \pi/(2G_\infty)$ and

62:8:2
$$E(p) = \frac{K(p)}{2}\left[2 - p^2 - \sum_{j=1}^{\infty} 2^j(A_j^2 - G_j^2)\right] = \frac{\pi}{8G_\infty}\left[4 - 2p^2 - \sum_{j=0}^{\infty} 2^j(A_j - G_j)^2\right]$$

which were given in the last chapter, we have

62:8:3
$$F(p;\phi) = 2K(p)\phi_\infty/\pi = \phi_\infty/G_\infty$$

and

62:8:4
$$E(p;\phi) = \frac{E(p)F(p;\phi)}{K(p)} + \sum_{j=1}^{\infty}\sqrt{A_j^2 - G_j^2}\,\sin(2^j\phi_j) = \frac{2\phi_\infty E(p)}{\pi} + \frac{1}{2}\sum_{j=0}^{\infty}(A_j - G_j)\sin(2^{j+1}\phi_{j+1})$$

These equations are the results of applying Landen's descending transformation an indefinite number of times. In practice, the angles are calculated via the formula

62:8:5
$$\phi_{j+1} = \phi_j - \frac{1}{2^{j+1}}\arctan\left\{\frac{(A_j - G_j)\tan(2^j\phi_j)}{A_j + G_j\tan^2(2^j\phi_j)}\right\}$$

and converge toward ϕ_∞ as rapidly as A_j and G_j converge toward their common mean G_∞.

```
              Set t = a = 1
              Set E = 0
Input p >>>>>>>
              Set g = √(1 − p²)
              Set e = 2(1 + g²)
Input φ >>>>>>>
          (1) Set T = tan(tφ)
              Replace φ by φ − {arctan[(a − g)T/(a + gT²)]}/(2t)
              Replace e by e − t(a − g)²
              Replace t by 2t
              Replace E by E + (a − g) sin(tφ)
              Replace g by √(ag)
              Replace a by [(g²/a) + a]/2
              If (a − g) ≥ 10⁻⁸ go to (1)
              Set k = π/(2g)
```

Storage needed: t, a, E, g, e, ϕ and T

Use radian mode.

Input restrictions: $-1 < p < 1$
ϕ must be in radians, not degrees

Output k
$k \simeq K(p)$ <<<<<<
Replace e by $\pi e/(8g)$
Output e
$e \simeq E(p)$ <<<<<<
Set $F = \phi/g$
Output F
$F \simeq F(p;\,\phi)$ <<<<
Replace E by $(E/2) + 2\phi e/\pi$)
Output E
$E \simeq E(p;\,\phi)$ <<<<

Test values:

$$K\left(\frac{1}{2}\right) = F\left(\frac{1}{2};\frac{\pi}{2}\right) = 1.68575\mathit{036}$$

$$E\left(\frac{1}{2}\right) = E\left(\frac{1}{2};\frac{\pi}{2}\right) = 1.46746\mathit{221}$$

$$F\left(\frac{1}{2};\frac{\pi}{4}\right) = 0.804366\mathit{101}$$

$$E\left(\frac{1}{2};\frac{\pi}{4}\right) = 0.767195\mathit{986}$$

The algorithm is designed to operate over the range $|p| < 1$ for all ϕ. The formulas of Section 62:5 may be employed for moduli outside this range. For $|\phi| \le \pi/2$, values of $E(p;\phi)$ and $F(p;\phi)$ generally are computed to 24-bit precision. However, $F(1;\pi/2) = \infty$, and when *both* input variables are very close to these critical values, less accuracy may be produced. In this circumstance, transformations 62:5:11 and 62:5:12 may be applied with advantage.

62:9 APPROXIMATIONS

Based on expansions 62:6:1 and 62:6:2, the approximations

62:9:1 $$F(p;\phi) \simeq \frac{K(p)}{\pi}[2\phi - \sin(2\phi)] + \frac{1}{2}\sin(2\phi) \qquad |p| \le 0.59 \ge |\phi|$$

62:9:2 $$E(p;\phi) \simeq \frac{E(p)}{\pi}[2\phi - \sin(2\phi)] + \frac{1}{2}\sin(2\phi) \qquad |p| \le 0.72 \ge |\phi|$$

are valid to 8-bit precision when both the modulus and amplitude fall within the specified range of small magnitudes. As p approaches unity, the approximations

62:9:3 $$F(p;\phi) \simeq \frac{2K(q)}{\pi}[\ln\{\tan(\phi) + \sec(\phi)\} - \tan(\phi)\sec(\phi)] + \tan(\phi)\sec(\phi) \qquad p \to 1$$

and

62:9:4 $$E(p;\phi) \simeq \frac{\sec(\phi) - \sqrt{1 - p^2\sin^2(\phi)}}{\tan(\phi)} + \frac{2}{\pi}[K(q) - E(q)][\ln\{\tan(\phi) + \sec(\phi)\} + \tan(\phi)\sec(\phi)] \qquad p \to 1$$

become increasingly accurate, but the precision depends on the amplitude ϕ.

62:10 OPERATIONS OF THE CALCULUS

Differentiation with respect to the amplitude gives

62:10:1 $$\frac{\partial}{\partial\phi}E(p;\phi) = \frac{1}{\dfrac{\partial}{\partial\phi}F(p;\phi)} = \sqrt{1 - p^2\sin^2(\phi)}$$

whereas, with respect to the modulus, we have

62:10:2 $$\frac{\partial}{\partial p}E(p;\phi) = \frac{E(p;\phi) - F(p;\phi)}{p}$$

and

62:10:3 $$\frac{\partial}{\partial p}\mathrm{F}(p;\phi) = \frac{\mathrm{E}(p;\phi)}{pq^2} - \frac{\mathrm{F}(p;\phi)}{p} - \frac{p\sin(\phi)\cos(\phi)}{q^2\sqrt{1-p^2\sin^2(\phi)}} \qquad q^2 = 1 - p^2$$

Indefinite integrals of incomplete elliptic integrals include

62:10:4 $$\int_0^\phi \sin(\theta)\,\mathrm{F}(p;\theta)d\theta = \frac{\arcsin(p\sin(\phi))}{p} - \cos(\phi)\,\mathrm{F}(p;\phi)$$

62:10:5 $$\int_0^\phi \sin(\theta)\mathrm{E}(p;\theta)d\theta = \frac{\arcsin(p\sin(\phi))}{2p} + \frac{\sin(\phi)}{2}\sqrt{1-p^2\sin^2(\phi)} - \cos(\phi)\,\mathrm{E}(p;\phi)$$

62:10:6 $$\int_0^\phi \frac{\mathrm{F}(p;\theta)d\theta}{\sqrt{1-p^2\sin^2(\theta)}} = \frac{\mathrm{F}^2(p;\phi)}{2}$$

62:10:7 $$\int_0^\phi \sqrt{1-p^2\sin^2(\theta)}\,\mathrm{E}(p;\theta)d\theta = \frac{\mathrm{E}^2(p;\phi)}{2}$$

See Gradshteyn and Ryzhik [Sections 5.12 and 6.11–6.13] for a few other indefinite integrals and many definite integrals.

62:11 COMPLEX ARGUMENT

The incomplete elliptic integrals of purely imaginary modulus are real:

62:11:1 $$\mathrm{F}(ip;\phi) = \frac{1}{\sqrt{1+p^2}}\left[\mathrm{K}\left(\frac{p}{\sqrt{1+p^2}}\right) - \mathrm{F}\left(\frac{p}{\sqrt{1+p^2}};\frac{\pi}{2}-\phi\right)\right] = \frac{1}{\sqrt{1+p^2}}\mathrm{F}\left(\frac{p}{\sqrt{1+p^2}};\psi\right)$$

62:11:2 $$\mathrm{E}(ip;\phi) = \sqrt{1+p^2}\left[\mathrm{E}\left(\frac{p}{\sqrt{1+p^2}}\right) - \mathrm{E}\left(\frac{p}{\sqrt{1+p^2}};\frac{\pi}{2}-\phi\right)\right]$$

$$= \sqrt{1+p^2}\,\mathrm{E}\left(\frac{p}{\sqrt{1+p^2}};\psi\right) - \frac{p^2\sin(\psi)\cos(\psi)}{\sqrt{1+p^2\cos^2(\psi)}}$$

where $\tan(\psi) = \sqrt{1+p^2}\tan(\phi)$. However, for imaginary amplitude, the incomplete elliptic integrals are themselves imaginary and involve the gudermannian function [Section 33:14] as well as hyperbolic functions [Chapters 30 and 28]:

62:11:3 $$\mathrm{F}(p;i\phi) = i\mathrm{F}(q;\mathrm{gd}(\phi)) \qquad q = \sqrt{1-p^2}$$

62:11:4 $$\mathrm{E}(p;i\phi) = i\,[\mathrm{F}(q;\mathrm{gd}(\phi)) - \mathrm{E}(q;\mathrm{gd}(\phi)) + \tanh(\phi)\sqrt{1+p^2\sinh^2(\phi)}]$$

For a complete list of similar transformations, see Gradshteyn and Ryzhik [Section 8.127].

62:12 GENERALIZATIONS

The incomplete elliptic integral of the third kind:

62:12:1 $$\Pi(\nu;p;\phi) = \int_0^\phi \frac{d\theta}{[1+\nu\sin^2(\theta)]\sqrt{1-p^2\sin^2(\theta)}}$$

is a generalization of the first kind because $\Pi(0;p;\phi) = \mathrm{F}(p;\phi)$. The notation is as confused as for the corresponding complete integral [Section 61:12]. Some other special cases include

62:12:2 $$\Pi(-1;p;\phi) = F(p;\phi) - \frac{1}{q}E(p;\phi) + \frac{1}{q}\tan(\phi)\sqrt{1-p^2\sin^2(\phi)} \qquad q=\sqrt{1-p^2}$$

62:12:3 $$\Pi(\pm p;p;\phi) = \frac{1}{2}F(p;\phi) + \frac{\arctan\{(1\pm p)\tan(\phi)/\sqrt{1-p^2\sin^2(\phi)}\}}{2\pm 2p}$$

62:12:4 $$\Pi(-p^2;p;\phi) = \frac{E(p;\phi)}{q^2} - \frac{p^2\sin(\phi)\cos(\phi)}{(1-p^2)\sqrt{1-p^2\sin^2(\phi)}}$$

62:12:5 $$\Pi(\nu;0;\phi) = \begin{cases} \dfrac{\arctan\{\sqrt{1+\nu}\tan(\phi)\}}{\sqrt{1+\nu}} & \nu > -1 \\ \tan(\phi) & \nu = -1 \\ \dfrac{\operatorname{artanh}\{\sqrt{-\nu-1}\tan(\phi)\}}{\sqrt{-\nu-1}} & \nu < -1 \end{cases}$$

62:12:6 $$\Pi(\nu;1;\phi) = \begin{cases} \dfrac{\operatorname{invgd}(\phi) - \sqrt{\nu}\operatorname{artanh}\{\sqrt{\nu}\sin(\phi)\}}{1+\nu} & \nu \ge 0 \\ \infty & \nu = -1 \\ \dfrac{\operatorname{invgd}(\phi) - \sqrt{-\nu}\operatorname{artanh}\{\sqrt{-\nu}\sin(\phi)\}}{1+\nu} & 0 \ge \nu \ne -1 \end{cases}$$

and

62:12:7 $$\Pi\left(\nu;p;\frac{\pi}{2}\right) = \Pi(\nu;p)$$

Because it is trivariate, numerical values of this third kind of elliptic integral are not easy to calculate. For small magnitudes of the characteristic, ν, the expansion

62:12:8 $$\Pi(\nu;p;\phi) = \sum_{j=0}^{\infty}\frac{(2j-1)!!}{(2j)!!}(-\nu)^j\sum_{k=0}^{j}\frac{(2k-1)!!}{(2k)!!}\left(\frac{-p^2}{\nu}\right)^k\left[\phi - \cos(\phi)\sum_{l=0}^{j-1}\frac{(2l)!!}{(2l+1)!!}\sin^{2l+1}(\phi)\right]$$

may be useful. The double factorial ratio [see Section 2:13] occurs repeatedly in this expression. When $|\nu| > 1$, 62:12:8 does not converge, and one must resort to rather complicated computational methods that are described by Milne-Thomson [see Abramowitz and Stegun, Section 17.7, but note that his n equals our $-\nu$].

62:13 COGNATE FUNCTIONS

Two functions that are closely related to the incomplete elliptic integrals are *Jacobi's zeta function*:

62:13:1 $$E(p;\phi) - \frac{E(p)}{K(p)}F(p;\phi)$$

and *Heuman's lambda function*:

62:13:2 $$\frac{2K(p)}{\pi}E(q;\phi) - \frac{2K(p)-2E(p)}{\pi}F(q;\phi) \qquad q=\sqrt{1-p^2}$$

Neither is used in this *Atlas*.

62:14 SPECIAL TOPICS

Historically, incomplete elliptic integrals arose as an attempt to codify the indefinite integrals $\int R(t,\sqrt{p(t)})dt$ where p(t) represents either a cubic function or a quartic function:

62:14:1 $$p(t) = \pm p_3(t) = \pm(t^3 + at^2 + bt + c) \qquad \text{or} \qquad p(t) = p_4(t) = a_4t^4 + a_3t^3 + a_2t^2 + a_1t + a_0$$

By R(x,y) we mean a rational function [Section 17:13] of x and y, that is, a quotient of polynomials in x and y. The name "elliptic integral" is sometimes given to this very general class of indefinite integrals.

By straightforward algebra [see Korn and Korn, Section 21.6, for details] *any* indefinite integral of the form $\int R(t,\sqrt{p(t)})dt$ may be reduced to a weighted sum of integrals of the following forms:

62:14:2 $$\int \frac{dt}{\sqrt{p(t)}}, \qquad \int \frac{t\,dt}{\sqrt{p(t)}}, \qquad \int \frac{dt}{(t - \text{constant})\sqrt{p(t)}}, \qquad \int R_n^m(t)dt$$

Here $R_n^m(t)$ is a rational function of t as discussed in Section 17:13, and its indefinite integral is easily found via the partial fraction decomposition discussed in that section.

The first three integral types in 62:14:2 may be expressed as incomplete elliptic integrals of the first and/or second and/or third kinds. We shall illustrate how this occurs, using the first member of 62:14:2, with $p(t) = \pm p_3(t)$ as our example.

As discussed in Section 17:7, the cubic function $p_3(t) = t^3 + at^2 + bt + c$ will have one, two or three distinct real zeros according to whether the discriminant D is positive, zero or negative. The $D = 0$ case is easily treated because the cubic function must then be of the form $p_3(t) = (t - r_1)(t - r_2)^2$, where r_1 and r_2 are the zeros, leading to

62:14:3 $$\int_x^\infty \frac{dt}{\sqrt{p_3(t)}} = \int_x^\infty \frac{dt}{(t - r_2)\sqrt{t - r_1}} = \frac{2}{\sqrt{r_2 - r_1}} \operatorname{arcoth}\sqrt{\frac{x - r_1}{r_2 - r_1}}$$

if $x > r_2 > r_1$, or to some equally elementary result otherwise.

If the cubic has a single (real) zero, r, then the sought integral involves the constants

62:14:4 $$h^2 = \frac{1}{\sqrt{3r^2 + 2ar + b}} \qquad \text{and} \qquad p^2 = 1 - q^2 = \frac{2 - (3r + a)h^2}{4}$$

and takes one of the following forms:

62:14:5 $$\int_x^\infty \frac{dt}{\sqrt{p_3(t)}} = hF(p; 2\operatorname{arccot}(h\sqrt{x - r}))$$

62:14:6 $$\int_r^x \frac{dt}{\sqrt{p_3(t)}} = hF(p; 2\arctan(h\sqrt{x - r}))$$

($r \le x$ for 62:14:5 and 62:14:6)

62:14:7 $$\int_x^r \frac{dt}{\sqrt{-p_3(t)}} = hF(q; 2\arctan(h\sqrt{r - x}))$$

62:14:8 $$\int_{-\infty}^x \frac{dt}{\sqrt{-p_3(t)}} = hF(q; 2\operatorname{arccot}(h\sqrt{r - x}))$$

($x \le r$ for 62:14:7 and 62:14:8)

according to the relative values of x and r.

When the cubic function has three zeros, denote them r_1, r_2 and r_3, where $r_1 < r_2 < r_3$, and define the constants

62:14:9 $$h^2 = \frac{4}{r_3 - r_1} \qquad \text{and} \qquad p^2 = 1 - q^2 = \frac{h^2(r_2 - r_1)}{4} = \frac{r_2 - r_1}{r_3 - r_1}$$

Then the following integrals are valid:

62:14:10 $$\int_x^\infty \frac{dt}{\sqrt{p_3(t)}} = hF\left(p;\arccos\left(\sqrt{\frac{x-r_3}{x-r_1}}\right)\right)$$

62:14:11 $$\int_{r_3}^x \frac{dt}{\sqrt{p_3(t)}} = hF\left(p;\arcsin\left(\sqrt{\frac{x-r_3}{x-r_2}}\right)\right)$$

$r_3 \le x$

62:14:12 $$\int_x^{r_3} \frac{dt}{\sqrt{-p_3(t)}} = hF\left(q;\arccos\left(\sqrt{\frac{x-r_2}{r_3-r_2}}\right)\right)$$

62:14:13 $$\int_{r_2}^x \frac{dt}{\sqrt{-p_3(t)}} = hF\left(q,\arcsin\left(\frac{1}{q}\sqrt{\frac{x-r_2}{x-r_1}}\right)\right)$$

$r_2 \le x \le r_3$

62:14:14 $$\int_x^{r_2} \frac{dt}{\sqrt{p_3(t)}} = hF\left(p;\arccos\left(\frac{q}{p}\sqrt{\frac{x-r_1}{r_3-x}}\right)\right)$$

62:14:15 $$\int_{r_1}^x \frac{dt}{\sqrt{p_3(t)}} = hF\left(p;\arcsin\left(\sqrt{\frac{x-r_1}{r_2-r_1}}\right)\right)$$

$r_1 \le x \le r_2$

62:14:16 $$\int_x^{r_1} \frac{dt}{\sqrt{-p_3(t)}} = hF\left(q;\arccos\left(\sqrt{\frac{r_2-r_1}{r_2-x}}\right)\right)$$

62:14:17 $$\int_{-\infty}^x \frac{dt}{\sqrt{-p_3(t)}} = hF\left(q;\arcsin\left(\sqrt{\frac{r_3-r_1}{r_3-x}}\right)\right)$$

$x \le r_1$

We refer the reader elsewhere [for example to Erdélyi, Oberhettinger, Magnus and Tricomi, *Higher Transcendental Functions*, Volume 2, Section 13.5] for a discussion of $\int dt/\sqrt{p_4(t)}$ and for treatments of the other forms in 62:14:2.

CHAPTER 63

THE JACOBIAN ELLIPTIC FUNCTIONS

These 12 functions have several interesting properties. One is their ability to bridge the gap between circular functions [Chapters 32–34] and hyperbolic functions [Chapters 28–30]. Another, discussed in Section 63:11, is their double periodicity. The close interrelationship of the twelve elliptic functions is detailed in Table 63.0.1.

The σ terms in this table take the values $+1$ or -1 according to the value of the quantity

63:0:1 $$w = \text{frac}\left(\frac{x}{4\text{K}(p)}\right)$$

as shown in Table 63.0.2. Explicit expressions for the σ terms are

63:0:2 $$\sigma_2 = (-1)^{\text{Int}(2w)} \qquad \sigma_3 = (-1)^{\text{Int}(4w)} \qquad \sigma_4 = \sigma_2\sigma_3$$

Notice the analogy to the concept [Section 32:13] of the sign of the circular functions being dependent on the "quadrant" in which argument falls. This is one manifestation of the fact that $\text{K}(p)$ plays the same role in elliptic trigonometry that $\pi/2$ plays in circular trigonometry.

63:1 NOTATION

Jacobi's name is not always attached to these functions. Alternatively, only the sn, cn and dn functions may be associated with Jacobi, the other nine being attributed to Glaisher, who invented their notation. Individual elliptic functions do not usually carry a name, although *sinus amplitudinis* has been applied to sn, *cosinus amplitudinis* to cn and *delta amplitudinis* to dn.

The Jacobian elliptic functions are bivariate, but one frequently encounters notations, such as $\text{sn}(x)$, that suggest only a single variable. The second variable is then implied and is regarded as a constant. We shall avoid this oversimplification in the *Atlas* and write, for example, $\text{sn}(p;x)$ where p is the *modulus* and x the *argument* of the function. Commonly k replaces p and u replaces x; as well, the order of citation is often reversed so that $\text{sn}(u,k)$ may be encountered. Rarely tn is used for sc.

A number of supplementary univariate and bivariate functions arise in discussions of elliptic integrals. These include the *complementary modulus* $q = \sqrt{1-p^2}$, the complete elliptic integrals [Chapter 61] $\text{K}(p)$ and $\text{K}(q)$, the *elliptic amplitude*

63:1:1 $$\text{am}(p;x) = \arcsin\{\text{sn}(p;x)\} = \arccos\{\text{cn}(p;x)\} = \arcsin\left(\frac{\sqrt{1-\text{dn}^2(p;x)}}{p}\right) = \phi$$

Table 63.0.1

	$f =$ sc$(p;x)$	$f =$ sd$(p;x)$	$f =$ sn$(p;x)$	$f =$ cs$(p;x)$	$f =$ cd$(p;x)$	$f =$ cn$(p;x)$	$f =$ ds$(p;x)$	$f =$ dc$(p;x)$	$f =$ dn$(p;x)$	$f =$ ns$(p;x)$	$f =$ nc$(p;x)$	$f =$ nd$(p;x)$
sc$(p;x) =$	f	$\frac{\sigma_4 f}{\sqrt{1-q^2f^2}}$	$\frac{\sigma_4 f}{\sqrt{1-f^2}}$	$\frac{1}{f}$	$\frac{\sqrt{1-f^2}}{\sigma_2 qf}$	$\frac{\sqrt{1-f^2}}{\sigma_2 f}$	$\frac{\sigma_3}{\sqrt{f^2-q^2}}$	$\frac{\sqrt{f^2-1}}{\sigma_3 q}$	$\frac{\sigma_3\sqrt{1-f^2}}{\sqrt{f^2-q^2}}$	$\frac{\sigma_3}{\sqrt{f^2-1}}$	$\frac{\sqrt{f^2-1}}{\sigma_3}$	$\frac{\sigma_3\sqrt{f^2-1}}{\sqrt{1-q^2f^2}}$
sd$(p;x) =$	$\frac{\sigma_4 f}{\sqrt{1+q^2f^2}}$	f	$\frac{f}{\sqrt{1-p^2f^2}}$	$\frac{\sigma_2}{\sqrt{q^2+f^2}}$	$\frac{\sqrt{1-f^2}}{\sigma_2 q}$	$\frac{\sigma_2\sqrt{1-f^2}}{\sqrt{q^2+p^2f^2}}$	$\frac{1}{f}$	$\frac{\sqrt{f^2-1}}{\sigma_3 qf}$	$\frac{\sqrt{1-f^2}}{\sigma_2 pf}$	$\frac{\sigma_2}{\sqrt{f^2-p^2}}$	$\frac{\sigma_2\sqrt{f^2-1}}{\sqrt{p^2+q^2f^2}}$	$\frac{\sqrt{f^2-1}}{\sigma_2 p}$
sn$(p;x) =$	$\frac{\sigma_4 f}{\sqrt{1+f^2}}$	$\frac{f}{\sqrt{1+p^2f^2}}$	f	$\frac{\sigma_2}{\sqrt{1+f^2}}$	$\frac{\sigma_2\sqrt{1-f^2}}{\sqrt{1-p^2f^2}}$	$\frac{\sqrt{1-f^2}}{\sigma_2}$	$\frac{\sigma_2}{\sqrt{p^2+f^2}}$	$\frac{\sigma_2\sqrt{f^2-1}}{\sqrt{f^2-p^2}}$	$\frac{\sqrt{1-f^2}}{\sigma_2 p}$	$\frac{1}{f}$	$\frac{\sqrt{f^2-1}}{\sigma_3 f}$	$\frac{\sqrt{f^2-1}}{\sigma_2 pf}$
cs$(p;x) =$	$\frac{1}{f}$	$\frac{\sqrt{1-q^2f^2}}{\sigma_4 f}$	$\frac{\sqrt{1-f^2}}{\sigma_4 f}$	f	$\frac{\sigma_2 qf}{\sqrt{1-f^2}}$	$\frac{\sigma_2 f}{\sqrt{1-f^2}}$	$\frac{\sqrt{f^2-q^2}}{\sigma_3}$	$\frac{\sigma_3 q}{\sqrt{f^2-1}}$	$\frac{\sqrt{f^2-q^2}}{\sigma_3\sqrt{1-f^2}}$	$\frac{\sqrt{f^2-1}}{\sigma_3}$	$\frac{\sigma_3}{\sqrt{f^2-1}}$	$\frac{\sqrt{1-q^2f^2}}{\sigma_3\sqrt{f^2-1}}$
cd$(p;x) =$	$\frac{\sigma_4}{\sqrt{1+q^2f^2}}$	$\frac{\sqrt{1-q^2f^2}}{\sigma_4}$	$\frac{\sigma_4\sqrt{1-f^2}}{\sqrt{1-p^2f^2}}$	$\frac{\sigma_2 f}{\sqrt{q^2+f^2}}$	f	$\frac{f}{\sqrt{q^2+p^2f^2}}$	$\frac{\sqrt{f^2-q^2}}{\sigma_3 f}$	$\frac{1}{f}$	$\frac{\sqrt{f^2-q^2}}{\sigma_4 pf}$	$\frac{\sigma_1\sqrt{f^2-1}}{\sqrt{f^2-p^2}}$	$\frac{\sigma_4}{\sqrt{p^2+q^2f^2}}$	$\frac{\sqrt{1-q^2f^2}}{\sigma_4 p}$
cn$(p;x) =$	$\frac{\sigma_4}{\sqrt{1+f^2}}$	$\frac{\sigma_4\sqrt{1-q^2f^2}}{\sqrt{1+p^2f^2}}$	$\frac{\sqrt{1-f^2}}{\sigma_2}$	$\frac{\sigma_2 f}{\sqrt{1+f^2}}$	$\frac{qf}{\sqrt{1-p^2f^2}}$	f	$\frac{\sigma_4\sqrt{f^2-q^2}}{\sqrt{p^2+f^2}}$	$\frac{\sigma_4 q}{\sqrt{f^2-p^2}}$	$\frac{\sqrt{f^2-q^2}}{\sigma_4 p}$	$\frac{\sqrt{f^2-1}}{\sigma_3 f}$	$\frac{1}{f}$	$\frac{\sqrt{1-q^2f^2}}{\sigma_4 pf}$
ds$(p;x) =$	$\frac{\sqrt{1+q^2f^2}}{\sigma_4 f}$	$\frac{1}{f}$	$\frac{\sqrt{1-p^2f^2}}{f}$	$\frac{\sqrt{q^2+f^2}}{\sigma_2}$	$\frac{\sigma_2 q}{\sqrt{1-f^2}}$	$\frac{\sqrt{q^2+p^2f^2}}{\sigma_2\sqrt{1-f^2}}$	f	$\frac{\sigma_3 qf}{\sqrt{f^2-1}}$	$\frac{\sigma_2 pf}{\sqrt{1-f^2}}$	$\frac{\sqrt{f^2-p^2}}{\sigma_2}$	$\frac{\sqrt{p^2+q^2f^2}}{\sigma_2\sqrt{f^2-1}}$	$\frac{\sigma_2 p}{\sqrt{f^2-1}}$
dc$(p;x) =$	$\frac{\sqrt{1+q^2f^2}}{\sigma_4}$	$\frac{\sigma_4}{\sqrt{1-q^2f^2}}$	$\frac{\sqrt{1-p^2f^2}}{\sigma_4\sqrt{1-f^2}}$	$\frac{\sqrt{q^2+f^2}}{\sigma_2 f}$	$\frac{1}{f}$	$\frac{\sqrt{q^2+p^2f^2}}{f}$	$\frac{\sigma_3 f}{\sqrt{f^2-q^2}}$	f	$\frac{\sigma_4 pf}{\sqrt{f^2-q^2}}$	$\frac{\sqrt{f^2-p^2}}{\sigma_4\sqrt{f^2-1}}$	$\frac{\sqrt{p^2+q^2f^2}}{\sigma_4}$	$\frac{\sigma_4 p}{\sqrt{1-q^2f^2}}$
dn$(p;x) =$	$\frac{\sqrt{1+q^2f^2}}{\sqrt{1+f^2}}$	$\frac{1}{\sqrt{1+p^2f^2}}$	$\sqrt{1-p^2f^2}$	$\frac{\sqrt{q^2+f^2}}{\sqrt{1+f^2}}$	$\frac{q}{\sqrt{1-p^2f^2}}$	$\sqrt{q^2+p^2f^2}$	$\frac{\lvert f\rvert}{\sqrt{p^2+f^2}}$	$\frac{q\lvert f\rvert}{\sqrt{f^2-p^2}}$	f	$\frac{\sqrt{f^2-p^2}}{\lvert f\rvert}$	$\frac{\sqrt{p^2+q^2f^2}}{\lvert f\rvert}$	$\frac{1}{f}$
ns$(p;x) =$	$\frac{\sqrt{1+f^2}}{\sigma_4 f}$	$\frac{\sqrt{1+p^2f^2}}{f}$	$\frac{1}{f}$	$\frac{\sqrt{1+f^2}}{\sigma_2}$	$\frac{\sqrt{1-p^2f^2}}{\sigma_2\sqrt{1-f^2}}$	$\frac{\sigma_2 f}{\sqrt{1-f^2}}$	$\frac{\sqrt{p^2+f^2}}{\sigma_2}$	$\frac{\sqrt{f^2-p^2}}{\sigma_2\sqrt{f^2-1}}$	$\frac{\sigma_2 p}{\sqrt{1-f^2}}$	f	$\frac{\sigma_3 f}{\sqrt{f^2-1}}$	$\frac{\sigma_2 pf}{\sqrt{f^2-1}}$
nc$(p;x) =$	$\frac{\sqrt{1+f^2}}{\sigma_2}$	$\frac{\sigma_4\sqrt{1+p^2f^2}}{\sqrt{1-q^2f^2}}$	$\frac{\sigma_4}{\sqrt{1-f^2}}$	$\frac{\sqrt{1+f^2}}{\sigma_2 f}$	$\frac{\sqrt{1-p^2f^2}}{qf}$	$\frac{1}{f}$	$\frac{\sigma_4\sqrt{p^2+f^2}}{\sqrt{f^2-q^2}}$	$\frac{\sqrt{f^2-p^2}}{\sigma_4 q}$	$\frac{\sigma_4 p}{\sqrt{f^2-q^2}}$	$\frac{\sigma_3 f}{\sqrt{f^2-1}}$	f	$\frac{\sigma_4 pf}{\sqrt{1-q^2f^2}}$
nd$(p;x) =$	$\frac{\sqrt{1+f^2}}{\sqrt{1+q^2f^2}}$	$\sqrt{1+p^2f^2}$	$\frac{1}{\sqrt{1-p^2f^2}}$	$\frac{\sqrt{1+f^2}}{\sqrt{q^2+f^2}}$	$\frac{\sqrt{1-p^2f^2}}{q}$	$\frac{1}{\sqrt{q^2+p^2f^2}}$	$\frac{\sqrt{p^2+f^2}}{\lvert f\rvert}$	$\frac{\sqrt{f^2-p^2}}{q\lvert f\rvert}$	$\frac{1}{f}$	$\frac{\lvert f\rvert}{\sqrt{f^2-p^2}}$	$\frac{\lvert f\rvert}{\sqrt{p^2+q^2f^2}}$	f

and the incomplete elliptic integrals [Chapter 62] F$(p;\phi)$ and E$(p;\phi)$. The square root $\sqrt{1-p^2\sin^2(\phi)} = \sqrt{1-p^2\,\mathrm{sn}^2(p;x)}$ occurs often in the theory of elliptic functions and is frequently abbreviated to $\Delta(p;\phi)$ although we use dn$(p;x)$.

The 12 Jacobian elliptic functions may be classified into four groups, each with three members, according to the second letter of the function's name. Thus, the cs, ds and ns functions are said to be *copolar*: they all possess a pole of "type s."

We shall use ef$(p;x)$, fe$(p;x)$, ge$(p;x)$, hg$(p;x)$, etc., to represent arbitrary Jacobian elliptic functions, it being understood that e, f, g and h are letters selected from the quartet s, c, d and n. Often we shall reserve e to represent the pole type.

63:2 BEHAVIOR

The Jacobian elliptic functions display interesting properties when the modulus and/or the argument is imaginary or complex. In this section, however, we restrict p and x to real values. Moreover, as in Table 63.0.1, we shall

Table 63.0.2

	$0 \le w < \frac{1}{4}$	$\frac{1}{4} \le w < \frac{1}{2}$	$\frac{1}{2} \le w < \frac{3}{4}$	$\frac{3}{4} \le w < 1$
σ_2	$+1$	$+1$	-1	-1
σ_3	$+1$	-1	$+1$	-1
σ_4	$+1$	-1	-1	$+1$

consider p to lie in the range $0 \le p \le 1$, though equations 63:5:2 and 63:5:14–63:5:16 show that this restriction is not mandatory.

Except for some of the functions when $p = 0$ or 1, all Jacobian elliptic functions are periodic, their periods being either $2K(p)$:

63:2:1 $$\mathrm{fe}(p;2K(p) + x) = \mathrm{fe}(p;x) \qquad \mathrm{fe} = \mathrm{sc, cs, dn\ or\ nd}$$

or $4K(p)$:

63:2:2 $$\mathrm{fe}(p;4K(p) + x) = \mathrm{fe}(p;x) \qquad \mathrm{fe} = \mathrm{sd, sn, cd, cn, ds, dc, ns\ or\ nc}$$

This is supported by Figures 63-1 and 63-2, which together map all the elliptic functions for $p = 0.9$. This particular value of the modulus was chosen to emphasize the nonsinusoidality of the sn, cn, ds, dc, dn and nd functions. For smaller positive values of p, these six functions become increasingly sinusoidal [see Section 32:1] in shape, while the shapes of the other six Jacobian functions come to resemble those of the functions of Chapters 33 or 34. As $p \to 1$, the periods increase indefinitely, and most of the functions degenerate into hyperbolic functions as explained in Section 63:5.

The ranges of the sc and cs functions are unrestricted, but the other 10 elliptic functions are constrained as follows, for $|p| \le 1$:

63:2:3 $$\frac{-1}{q} \le \mathrm{sd}(p;x) \le \frac{1}{q} \qquad q = \sqrt{1 - p^2}$$

63:2:4 $$-1 \le \mathrm{fe}(p;x) \le 1 \qquad \mathrm{fe} = \mathrm{sn, cd\ or\ cn}$$

63:2:5 $$q \le |\mathrm{ds}(p;x)| \le \infty$$

63:2:6 $$q \le \mathrm{dn}(p;x) \le 1$$

63:2:7 $$1 \le |\mathrm{fe}(p;x)| < \infty \qquad \mathrm{fe} = \mathrm{dc, ns\ or\ nc}$$

63:2:8 $$1 \le \mathrm{nd}(p;x) \le \frac{1}{q}$$

Because the Jacobian elliptic functions have periods that are proportional to the elliptic integral $K(p)$, it is sometimes fruitful to treat the ratio $v = x/K(p)$ as an independent variable in considering properties of elliptic functions. This has been done in Figures 63-3, 63-4 and 63-5, which, for a range of values of v and for the moduli in Table 63.2.1, display the copolar trio $\mathrm{sn}(p;x)$, $\mathrm{cn}(p;x)$ and $\mathrm{dn}(p;x)$. These are, perhaps, the most important of the 12 functions; moreover, the other nine are easily calculable from them via equation 63:5:1.

63:3 DEFINITIONS

The Jacobian elliptic functions are defined via the inverse of the incomplete elliptic integral of the first kind [Chapter 62]. Thus, if

63:3:1 $$x = F(p;\phi) = \int_0^{\phi} \frac{d\theta}{\sqrt{1 - p^2 \sin^2(\theta)}}$$

then

63:3:2 $$\mathrm{sn}(p;x) = \sin(\phi) = \sin(\mathrm{am}(p;x))$$

The other 11 elliptic functions can then be defined via the interrelationships reported in Table 63.0.1. For example:

63:3:3 $$\mathrm{cn}(p;x) = \sqrt{1 - \mathrm{sn}^2(p;x)} = \cos(\phi) = \cos(\mathrm{am}(p;x))$$

and

63:3:4 $$\mathrm{dn}(p;x) = \sqrt{1 - p^2\,\mathrm{sn}^2(p;x)} = \sqrt{1 - p^2 \sin^2(\phi)} = \sqrt{1 - p^2 \sin^2(\mathrm{am}(p;x))}$$

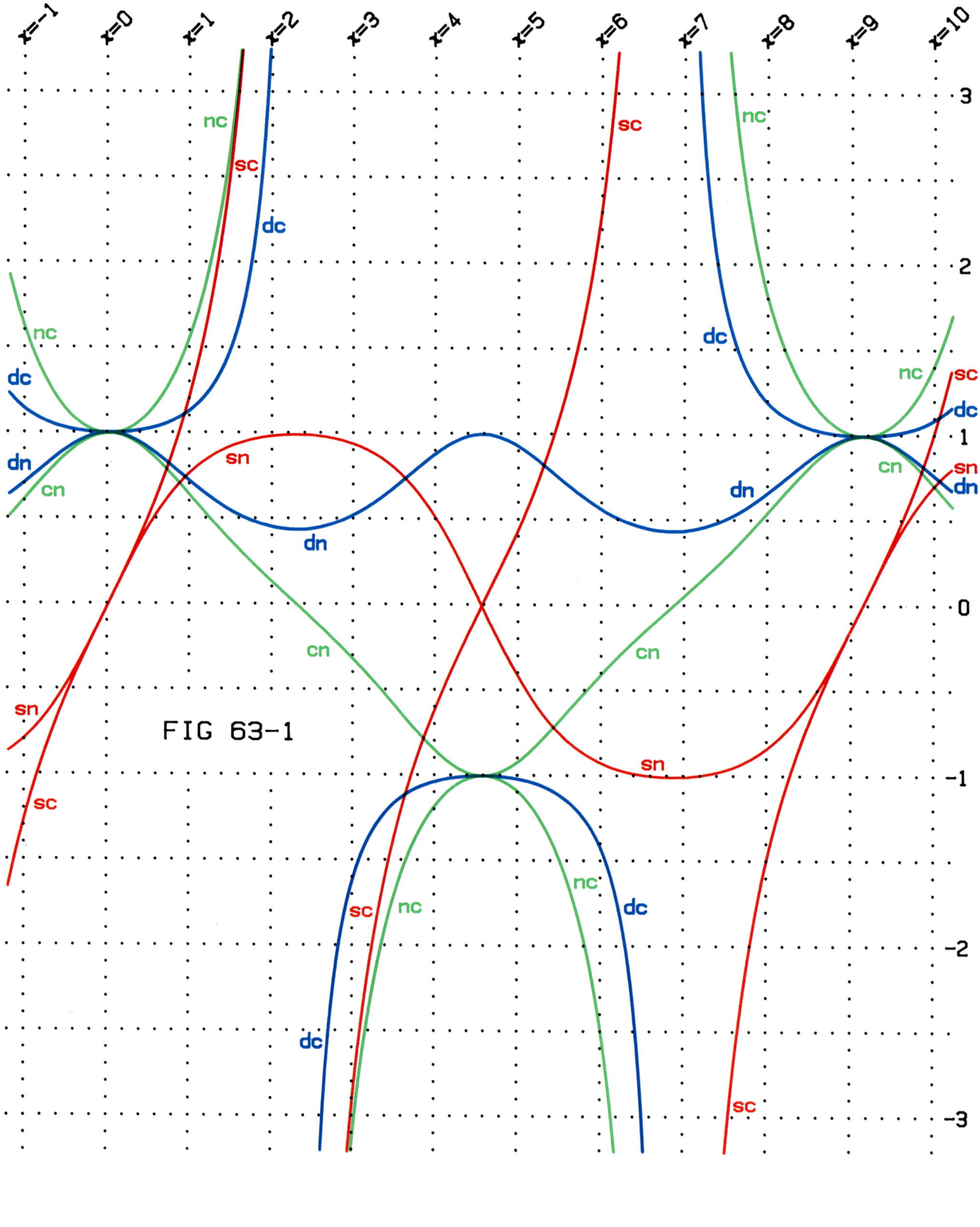
x=−1
x=0
x=1
x=2
x=3
x=4
x=5
x=6
x=7
x=8
x=9
x=10
3
2
1
0
−1
−2
−3
nc
sc
dc
dn
cn
sn
FIG 63-1

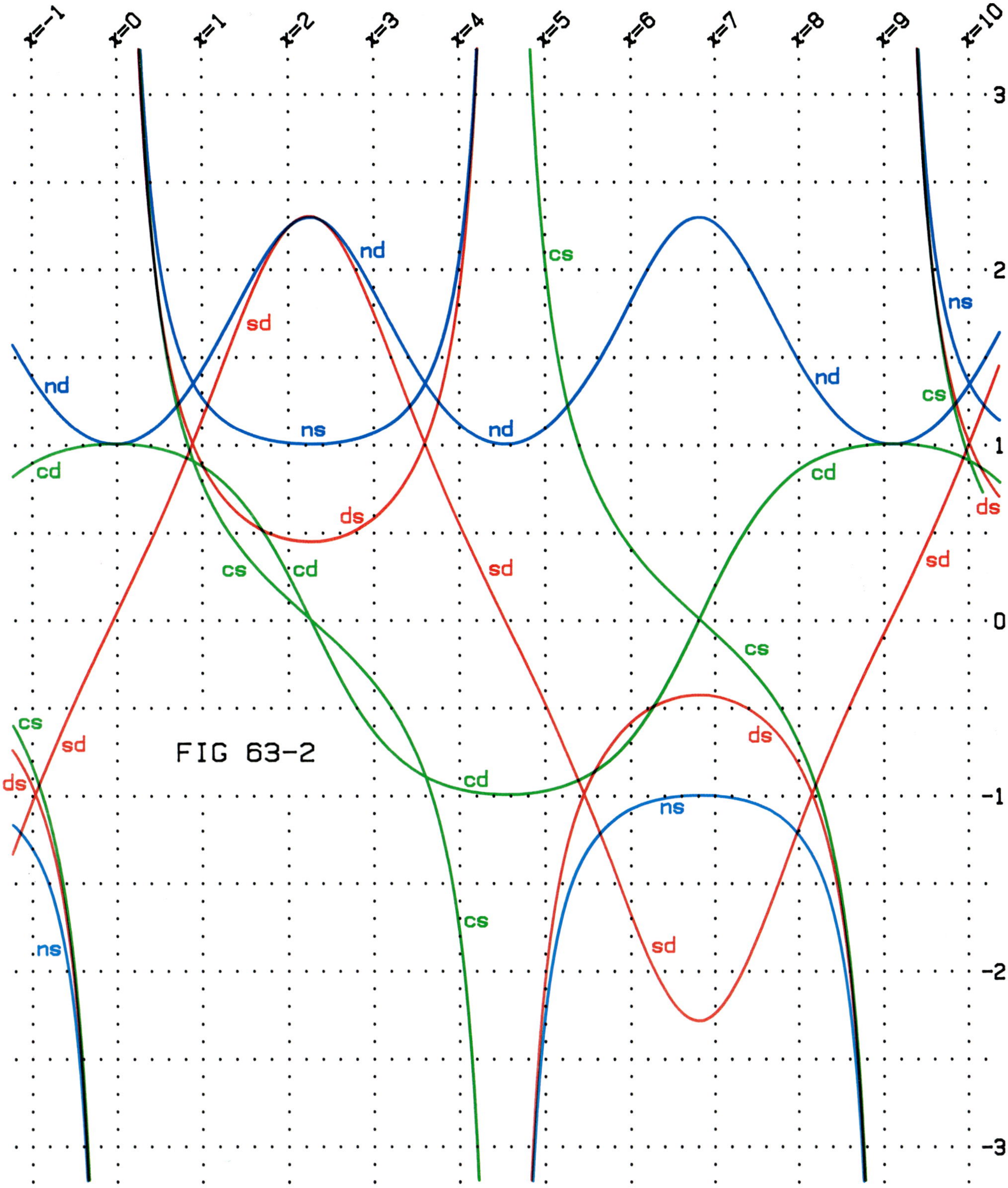
x=-1
x=0
x=1
x=2
x=3
x=4
x=5
x=6
x=7
x=8
x=9
x=10
3
2
1
0
-1
-2
-3
nd
ns
sd
ds
cs
cd
FIG 63-2

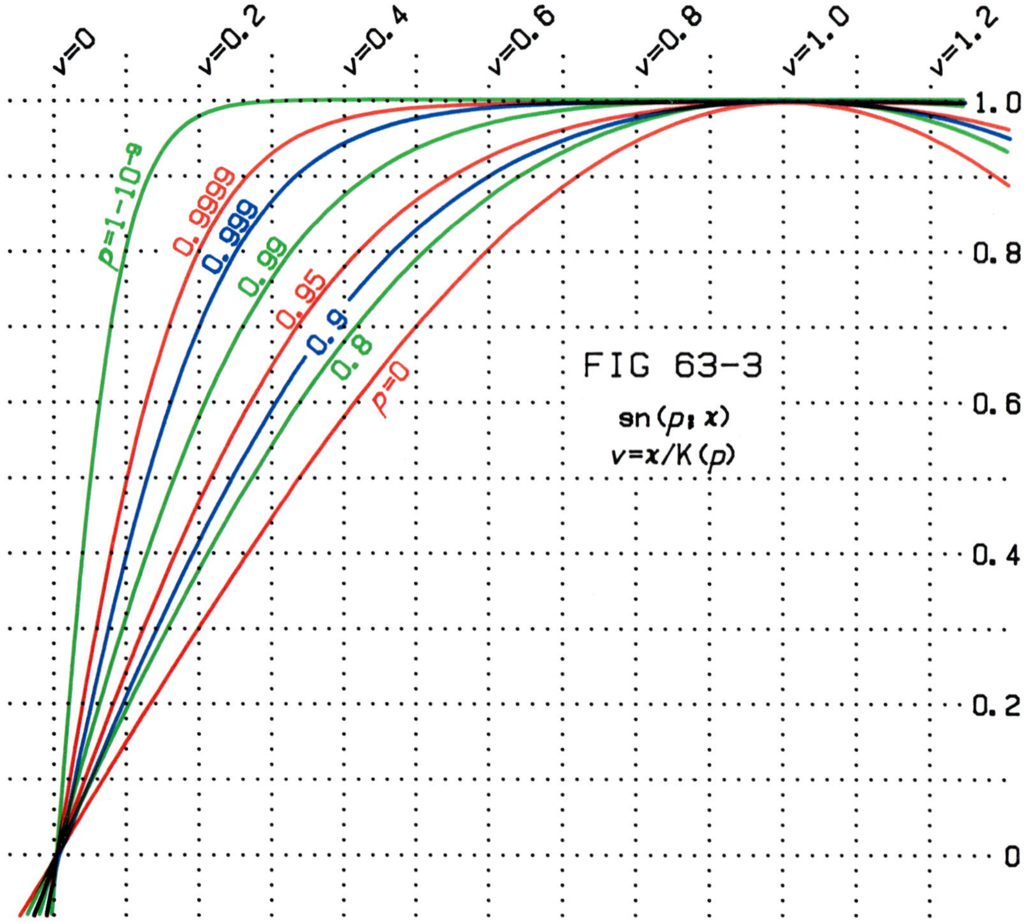

The remainder of this section is devoted to making a geometric construction that may be used to define all the Jacobian elliptic functions. In Sections 29:3 and 33:3 the six hyperbolic and the six trigonometric functions are defined in terms of the lengths of the sides of three similar right-angled triangles. In strict analogy, six of the Jacobian functions may be defined as the nonunity lengths of the sides of the triangles *OAC*, *OGI* and *OJL* as depicted in Figure 63-6. Here the angle ϕ equals the elliptic amplitude am(p;x), and those sides of the triangles that are dashed have unity lengths.

Further construction is needed to provide definitions of the other six Jacobian elliptic functions. First, superimpose the three triangles *OAC*, *OGI* and *OJL*, as shown in Figure 63-7. Then select the point *K* on line *JL* such that length *LK* equals the complementary modulus q. Join points *O* and *K* by a straight line and denote its points of intersection with lines *AC* and *GI* by *B* and *H*, respectively. Three more elliptic functions may now be defined in terms of this new line, namely

63:3:5 $\qquad$ length $OB = \mathrm{dn}(p;x)$ $\qquad$ length $OH = \mathrm{dc}(p;x)$ $\qquad$ length $OK = \mathrm{ds}(p;x)$

Yet more construction is required to define the remaining three functions. Measure unity length from point *O* toward *K*, and so create point *E*. Next draw line *DEF* through *E* perpendicular to *OL* to cut lines *OJ* and *OL* at *D* and *F*.

Then

63:3:6 $\qquad$ length $OD = \mathrm{nd}(p;x)$ $\qquad$ length $DF = \mathrm{sd}(p;x)$ $\qquad$ length $OF = \mathrm{cd}(p;x)$

An alternative construction can locate line *DEF* without recourse to line *OK*. Superimpose Figure 63-4 onto Figure 62-5, both diagrams being similarly scaled and sharing line *OI* and angle ϕ in common. Then the ellipse cuts line *OJ* at *D*, as depicted.

Finally, to emphasize the symmetry of the definitions and the logic of Glaisher's notation, dismember Figure 63-7 into its four similar components, each consisting of a right-angled triangle with an additional line. These four components are illustrated in Figure 63-8, on which diagram all 16 lengths are identified. To better accentuate the similarities, we have adopted the notation

63:3:7 $$\mathrm{nn}(p;x) = \mathrm{dd}(p;x) = \mathrm{cc}(p;x) = \mathrm{ss}(p;x) = 1$$

for those lines in the diagram that have unity length. This notation is consistent with Glaisher's.

Similarity and pythagorean rules applied to Figure 63-8 lead immediately to the relationships in Table 63.0.1. The dismemberment has segregated the elliptic functions into copolar groups. Note that the triangle of pole type c is always larger than that of d, which, in turn, is necessarily larger than the triangle of pole n. The size of the s triangle, however, need not be larger than that of c or d, though it must exceed n.

63:4 SPECIAL CASES

All Jacobian elliptic functions reduce to a trigonometric function, or to unity, when $p = 0$. Similarly, all Jacobian elliptic functions reduce to a hyperbolic function, or to unity, when $p = 1$. Because of relationship 63:5:1, Table 63.4.1 provides sufficient information to identify all these special cases. The examples

63:4:1 $$\mathrm{dc}(0;x) = \frac{1}{\mathrm{cd}(0;x)} = \frac{1}{\cos(x)} = \sec(x)$$

63:4:2 $$\mathrm{ns}(1;x) = \frac{\mathrm{nd}(1;x)}{\mathrm{sd}(1;x)} = \frac{\cosh(x)}{\sinh(x)} = \coth(x)$$

illustrate how the table may be extended.

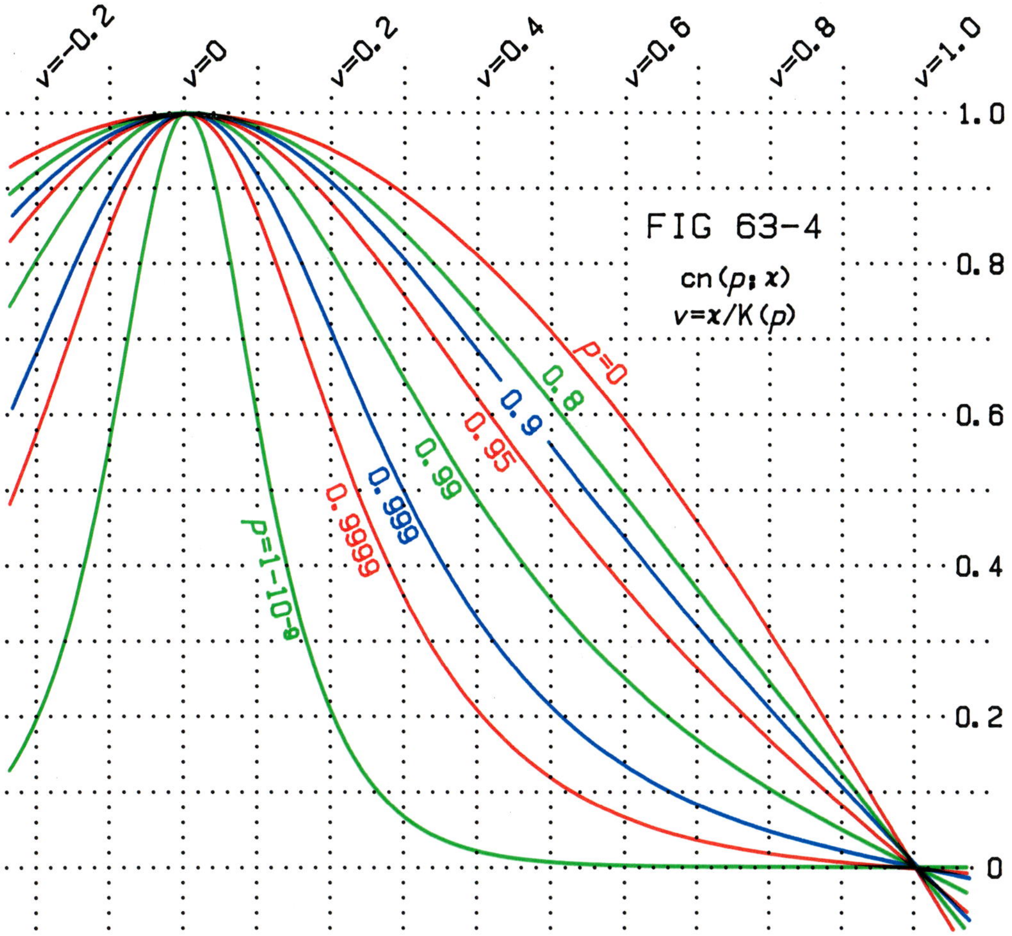

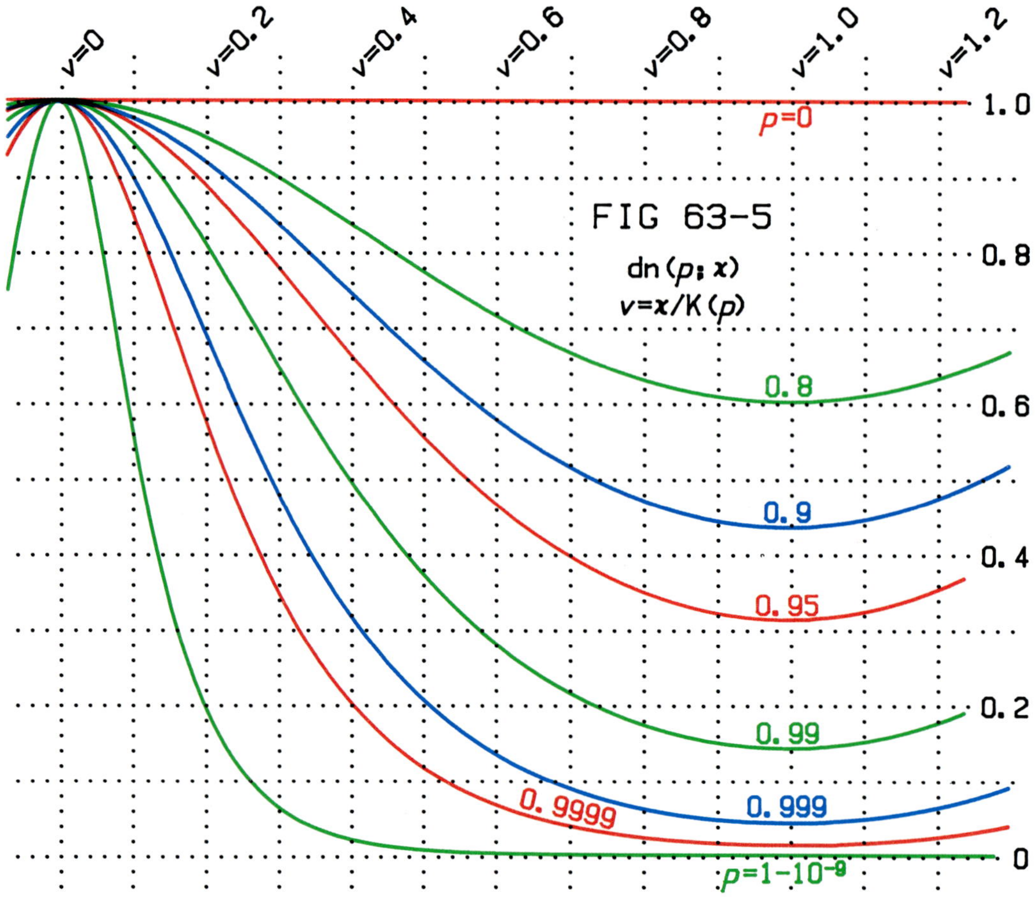

63:5 INTRARELATIONSHIPS

Table 63.0.1 gives explicitly the relationship between any two Jacobian elliptic functions. If e, f and g represent letters drawn from the quartet s, c, d and n, then

63:5:1 $$\frac{\mathrm{fe}(p;x)}{\mathrm{ge}(p;x)} = \mathrm{fg}(p;x) \qquad \mathrm{e, f, g = s, c, d, n}$$

As in 63:3:7, $\mathrm{ff}(p;x)$ is interpreted as unity.

Table 63.2.1

p	$K(p)$
0	1.571
0.8	1.995
0.9	2.281
0.95	2.590
0.99	3.357
0.999	4.496
0.9999	5.645
$1 - 10^{-9}$	11.401

All elliptic functions are even with respect to their moduli:

63:5:2 $$\mathrm{fe}(-p;x) = \mathrm{fe}(p;x) \qquad \text{all fe}$$

and either even or odd with respect to argument:

63:5:3 $$\mathrm{fe}(p;-x) = \mathrm{fe}(p;x) \qquad \mathrm{fe} = \mathrm{cd,\ cn,\ dc,\ dn,\ nc\ or\ nd}$$

63:5:4 $$\mathrm{fe}(p;-x) = -\mathrm{fe}(p;x) \qquad \mathrm{fe} = \mathrm{sc,\ sd,\ sn,\ cs,\ ds\ or\ ns}$$

The periodicity of the Jacobian functions is expressed in equations 63:2:1 and 63:2:2. With $n = 0, \pm 1, \pm 2, \ldots,$ we have

63:5:5 $$\mathrm{fe}(p;nP+x) = \mathrm{fe}(p;x)\begin{cases} P = 2\mathrm{K}(p) & \mathrm{fe} = \mathrm{sc,\ cs,\ dn,\ or\ nd} \\ P = 4\mathrm{K}(p) & \mathrm{fe} = \mathrm{sd,\ sn,\ cd,\ cn,\ ds,\ dc,\ ns\ or\ nc} \end{cases}$$

A great many reflection and recurrence formulas are summarized in Table 63.5.1; it may be extended by use of 63:5:5.

Addition/subtraction formulas for a copolar trio are

63:5:6 $$\mathrm{sn}(p;x \pm y) = \frac{\mathrm{sn}(p;x)\,\mathrm{cn}(p;y)\,\mathrm{dn}(p;y) \pm \mathrm{cn}(p;x)\,\mathrm{dn}(p;x)\,\mathrm{sn}(p;y)}{1 - p^2\,\mathrm{sn}^2(p;x)\,\mathrm{sn}^2(p;y)}$$

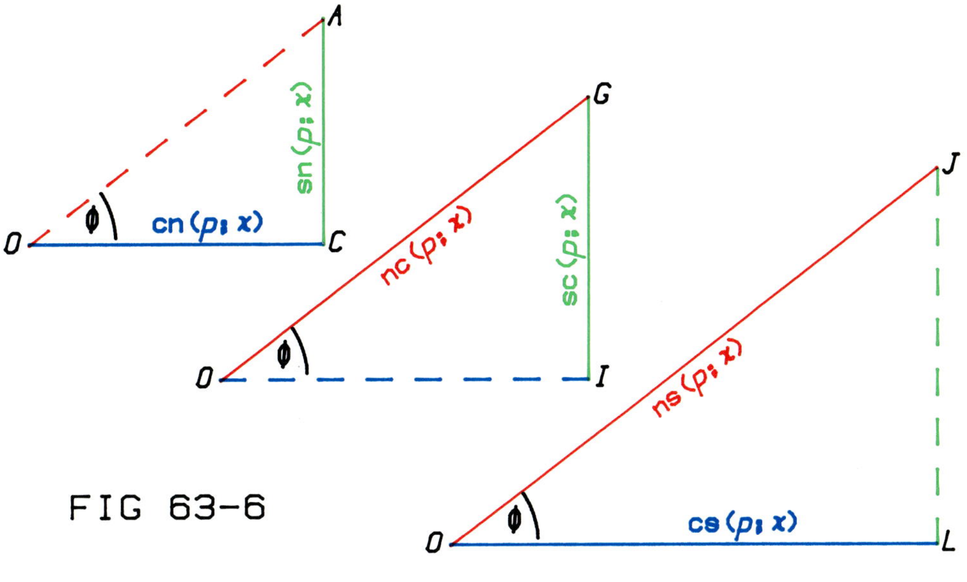

FIG 63-6

FIG 63-7

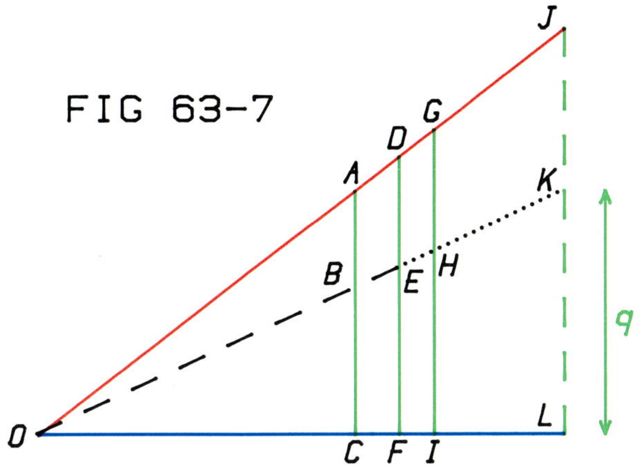

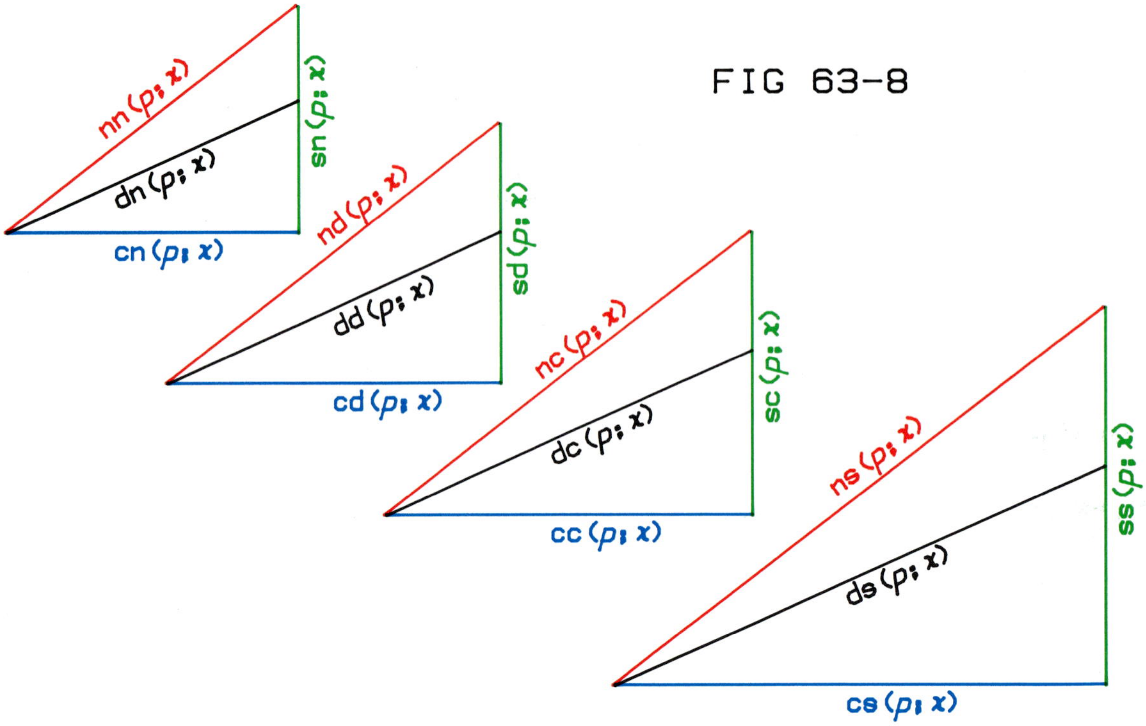

FIG 63-8

63:5:7 $$\mathrm{cn}(p;x \pm y) = \frac{\mathrm{cn}(p;x)\,\mathrm{cn}(p;y) \pm \mathrm{sn}(p;x)\,\mathrm{dn}(p;x)\,\mathrm{sn}(p;y)\,\mathrm{dn}(p;y)}{1 - p^2\,\mathrm{sn}^2(p;x)\,\mathrm{sn}^2(p;y)}$$

63:5:8 $$\mathrm{dn}(p;x \pm y) = \frac{\mathrm{dn}(p;x)\,\mathrm{dn}(p;y) \mp p^2\,\mathrm{sn}(p;x)\,\mathrm{cn}(p;x)\,\mathrm{sn}(p;y)\,\mathrm{cn}(p;y)}{1 - p^2\,\mathrm{sn}^2(p;x)\,\mathrm{sn}^2(p;y)}$$

The use of rule 63:5:1 enables all other $\mathrm{fg}(p;x \pm y)$ to be evaluated as the reciprocal of one member of the trio, or as a quotient of two. Double-argument formulas may be obtained as $\mathrm{fg}(p;x + x)$. Thence one may derive

63:5:9 $$\sqrt{\frac{1 - \mathrm{cn}(p;2x)}{1 + \mathrm{cn}(p;2x)}} = \left|\frac{\mathrm{sn}(p;x)\,\mathrm{dn}(p;x)}{\mathrm{cn}(p;x)}\right| = \left|\frac{\mathrm{dn}(p;x)}{\mathrm{cs}(p;x)}\right| = \left|\frac{\mathrm{sn}(p;x)}{\mathrm{cd}(p;x)}\right|$$

and

63:5:10 $$\sqrt{\frac{1 - \mathrm{dn}(p;2x)}{1 + \mathrm{dn}(p;2x)}} = \left|\frac{p\,\mathrm{sn}(p;x)\,\mathrm{cn}(p;x)}{\mathrm{dn}(p;x)}\right| = \left|\frac{p\,\mathrm{cn}(p;x)}{\mathrm{ds}(p;x)}\right| = \left|\frac{p\,\mathrm{sn}(p;x)}{\mathrm{dc}(p;x)}\right|$$

which lead to the half-argument formulas

63:5:11 $$\mathrm{sn}^2\left(p;\frac{x}{2}\right) = \frac{1 - \mathrm{cn}(p;x)}{1 + \mathrm{dn}(p;x)}$$

Table 63.4.1

	$p = 0$	$p = 1$
$\mathrm{sd}(p;x)$	$\sin(x)$	$\sinh(x)$
$\mathrm{cd}(p;x)$	$\cos(x)$	1
$\mathrm{nd}(p;x)$	1	$\cosh(x)$

63:5:12 $$\mathrm{cn}^2\left(p;\frac{x}{2}\right) = \frac{\mathrm{cn}(p;x) + \mathrm{dn}(p;x)}{1 + \mathrm{dn}(p;x)} = \frac{1 + \mathrm{cd}(p;x)}{1 + \mathrm{nd}(p;x)}$$

63:5:13 $$\mathrm{dn}^2\left(p;\frac{x}{2}\right) = \frac{q^2 + p^2\,\mathrm{cn}(p;x) + \mathrm{dn}(p;x)}{1 + \mathrm{dn}(p;x)}$$

Again, rule 63:5:1 may be used to extend these formulas to any $\mathrm{fg}^2(p;x/2)$. Because of the interrelationships cited in Section 63:0, all such formulas may be expressed in a vast number of alternative ways.

The so-called *Jacobi real transformations*:

63:5:14 $$\mathrm{sn}\left(\frac{1}{p};x\right) = p\,\mathrm{sn}\left(p;\frac{x}{p}\right)$$

63:5:15 $$\mathrm{cn}\left(\frac{1}{p};x\right) = \mathrm{dn}\left(p;\frac{x}{p}\right)$$

63:5:16 $$\mathrm{dn}\left(\frac{1}{p};x\right) = \mathrm{cn}\left(p;\frac{x}{p}\right)$$

permit the evaluation of elliptic functions for moduli exceeding unity. Once again, rule 63:5:1 permits extension from this copolar trio to all other $\mathrm{fg}(1/p;x)$ functions.

The principles of *Landen transformation* are explained in Section 62:5. This transformation may be applied to the delta amplitudinis in either the descending or ascending mode through the following equations. If

63:5:17 $$p_{-1} = \frac{1 - \sqrt{1 - p_0^2}}{1 + \sqrt{1 - p_0^2}} \quad \text{and} \quad p_1 = \frac{2\sqrt{p_0}}{1 + p_0}$$

and

63:5:18 $$x_{-1} = (1 + \sqrt{1 - p_0^2})x_0 \quad \text{and} \quad x_1 = \frac{(1 + p_0)x_0}{2}$$

then

63:5:19 $$\mathrm{dn}(p_{-1};x_{-1}) = \frac{\sqrt{1 - p_0^2} + \mathrm{dn}^2(p_0;x_0)}{(1 + \sqrt{1 - p_0^2})\,\mathrm{dn}(p_0;x_0)} \quad \text{and} \quad \mathrm{dn}(p_1;x_1) = \frac{\mathrm{dn}(p_0;x_0) + \sqrt{\mathrm{dn}^2(p_0;x_0) - 1 + p_0^2}}{1 + p_0}$$

Table 63.5.1

	$X = x - 2K(p)$	$X = x - K(p)$	$X = K(p) - x$	$X = K(p) + x$	$X = 2K(p) - x$	$X = 2K(p) + x$
$\mathrm{sc}(p;X) =$	$\mathrm{sc}(p;x)$	$-q^{-1}\mathrm{cs}(p;x)$	$q^{-1}\mathrm{cs}(p;x)$	$-q^{-1}\mathrm{cs}(p;x)$	$-\mathrm{sc}(p;x)$	$\mathrm{sc}(p;x)$
$\mathrm{sd}(p;X) =$	$-\mathrm{sd}(p;x)$	$-q^{-1}\mathrm{cn}(p;x)$	$q^{-1}\mathrm{cn}(p;x)$	$q^{-1}\mathrm{cn}(p;x)$	$\mathrm{sd}(p;x)$	$-\mathrm{sd}(p;x)$
$\mathrm{sn}(p;X) =$	$-\mathrm{sn}(p;x)$	$-\mathrm{cd}(p;x)$	$\mathrm{cd}(p;x)$	$\mathrm{cd}(p;x)$	$\mathrm{sn}(p;x)$	$-\mathrm{sn}(p;x)$
$\mathrm{cs}(p;X) =$	$\mathrm{cs}(p;x)$	$-q\,\mathrm{sc}(p;x)$	$q\,\mathrm{sc}(p;x)$	$-q\,\mathrm{sc}(p;x)$	$-\mathrm{cs}(p;x)$	$\mathrm{cs}(p;x)$
$\mathrm{cd}(p;X) =$	$-\mathrm{cd}(p;x)$	$\mathrm{sn}(p;x)$	$\mathrm{sn}(p;x)$	$-\mathrm{sn}(p;x)$	$-\mathrm{cd}(p;x)$	$-\mathrm{cd}(p;x)$
$\mathrm{cn}(p;X) =$	$-\mathrm{cn}(p;x)$	$q\,\mathrm{sd}(p;x)$	$q\,\mathrm{sd}(p;x)$	$-q\,\mathrm{sd}(p;x)$	$-\mathrm{cn}(p;x)$	$-\mathrm{cn}(p;x)$
$\mathrm{ds}(p;X) =$	$-\mathrm{ds}(p;x)$	$-q\,\mathrm{nc}(p;x)$	$q\,\mathrm{nc}(p;x)$	$q\,\mathrm{nc}(p;x)$	$\mathrm{ds}(p;x)$	$-\mathrm{ds}(p;x)$
$\mathrm{dc}(p;X) =$	$-\mathrm{dc}(p;x)$	$\mathrm{ns}(p;x)$	$\mathrm{ns}(p;x)$	$-\mathrm{ns}(p;x)$	$-\mathrm{dc}(p;x)$	$-\mathrm{dc}(p;x)$
$\mathrm{dn}(p;X) =$	$\mathrm{dn}(p;x)$	$q\,\mathrm{nd}(p;x)$	$q\,\mathrm{nd}(p;x)$	$q\,\mathrm{nd}(p;x)$	$\mathrm{dn}(p;x)$	$\mathrm{dn}(p;x)$
$\mathrm{ns}(p;X) =$	$-\mathrm{ns}(p;x)$	$-\mathrm{dc}(p;x)$	$\mathrm{dc}(p;x)$	$\mathrm{dc}(p;x)$	$\mathrm{ns}(p;x)$	$-\mathrm{ns}(p;x)$
$\mathrm{nc}(p;X) =$	$-\mathrm{nc}(p;x)$	$q^{-1}\mathrm{ds}(p;x)$	$q^{-1}\mathrm{ds}(p;x)$	$-q^{-1}\mathrm{ds}(p;x)$	$-\mathrm{nc}(p;x)$	$-\mathrm{nc}(p;x)$
$\mathrm{nd}(p;X) =$	$\mathrm{nd}(p;x)$	$q^{-1}\mathrm{dn}(p;x)$	$q^{-1}\mathrm{dn}(p;x)$	$q^{-1}\mathrm{dn}(p;x)$	$\mathrm{nd}(p;x)$	$\mathrm{nd}(p;x)$

A popular technique for evaluating elliptic integrals of modulus p and argument x is to set $p = p_0$ and $x = x_0$ in 63:5:17 and 63:5:18 and then sequentially transform $\mathrm{dn}(p_0;x_0)$ via 63:5:19 into $\mathrm{dn}(p_{-1};x_{-1})$, $\mathrm{dn}(p_{-2};x_{-2})$, ... (or into $\mathrm{dn}(p_1;x_1)$, $\mathrm{dn}(p_2;x_2)$, ...) until p_n has become so close to zero (or p_n so close to unity) that the approximation

63:5:20 $$\mathrm{dn}(p_{-n};x_{-n}) \simeq 1 - \frac{p_{-n}^2}{2}\sin^2(x_{-n})$$

(or

63:5:21 $$\mathrm{dn}(p_n;x_n) = \mathrm{sech}(x_n)\left[1 + \frac{1-p_n^2}{4}(\sinh^2(x_n) + x_n\tanh(x_n))\right]$$

in the case of the ascending transformation) becomes valid.

63:6 EXPANSIONS

Expansions of the Jacobian elliptic functions exist under three circumstances: as a power series in x, valid for small x; as a power series in p involving trigonometric coefficients, valid for small p; and as a power series in q involving hyperbolic coefficients, valid when p is close to unity. The leading terms in these three expansions are given in Table 63.6.1. In addition to the elliptic functions, the table lists expansions of the elliptic amplitude $\mathrm{am}(p;x)$. See Section 61:14 for expansions in terms of the "nome."

Table 63.6.1

	$x \to 0$	$p \to 0$	$p \to 1$
$\mathrm{sc}(p;x) =$	$x + \frac{(2-p^2)x^3}{6} + \frac{(16-16p^2+p^4)x^5}{120} + \cdots$	$\tan(x) - \frac{[2x-\sin(2x)]p^2}{8\cos^2(x)} + \cdots$	$\sinh(x) + \frac{[\sinh(2x)-2x]q^2}{8\,\mathrm{sech}(x)} + \cdots$
$\mathrm{sd}(p;x) =$	$x - \frac{(1-2p^2)x^3}{6} + \frac{(1+14p^2-14p^4)x^5}{120} + \cdots$	$\sin(x) + \frac{[\tan(x)-x]p^2}{4\sec(x)} + \cdots$	$\sinh(x) + \frac{[1-\sinh^2(x)-x\coth(x)]q^2}{4\,\mathrm{csch}(x)} + \cdots$
$\mathrm{sn}(p;x) =$	$x - \frac{(1+p^2)x^3}{6} + \frac{(1+14p^2+p^4)x^5}{120} + \cdots$	$\sin(x) - \frac{[2x-\sin(2x)]p^2}{8\sec(x)} + \cdots$	$\tanh(x) + \frac{[\sinh(2x)-2x]q^2}{8\cosh^2(x)} + \cdots$
$\mathrm{cs}(p;x) =$	$\frac{1}{\ } - \frac{(2-p^2)x}{6} - \frac{(8-8p^2-7p^4)x^3}{360} + \cdots$	$\cot(x) + \frac{[2x-\sin(2x)]p^2}{8\sin^2(x)} + \cdots$	$\mathrm{csch}(x) - \frac{[\sinh(2x)-2x]q^2}{8\tanh(x)\sinh(x)} + \cdots$
$\mathrm{cd}(p;x) =$	$1 - \frac{(1-p^2)x^2}{2} + \frac{(1-p^4)x^4}{24} + \cdots$	$\cos(x) + \frac{xp^2}{4\csc(x)} + \cdots$	$1 - \frac{q^2}{2\,\mathrm{csch}^2(x)} + \cdots$
$\mathrm{cn}(p;x) =$	$1 - \frac{x^2}{2} + \frac{(1+4p^2)x^4}{24} + \cdots$	$\cos(x) + \frac{[2x-\sin(2x)]p^2}{8\csc(x)} + \cdots$	$\mathrm{sech}(x) - \frac{[\sinh(2x)-2x]q^2}{8\coth(x)\cosh(x)} + \cdots$
$\mathrm{ds}(p;x) =$	$\frac{1}{x} + \frac{(1-2p^2)x}{6} + \frac{(7+8p^2-8p^4)x^3}{360} + \cdots$	$\csc(x) - \frac{[\tan(x)-x]p^2}{4\tan(x)\sin(x)} + \cdots$	$\mathrm{csch}(x) - \frac{[1-\sinh^2(x)-x\coth(x)]q^2}{4\sinh(x)} + \cdots$
$\mathrm{dc}(p;x) =$	$1 + \frac{(1-p^2)x^2}{2} + \frac{(5-6p^2+p^4)x^4}{24} + \cdots$	$\sec(x) - \frac{q^2}{4\cot(x)\cos(x)} + \cdots$	$1 + \frac{q^2}{2\,\mathrm{csch}^2(x)} + \cdots$
$\mathrm{dn}(p;x) =$	$1 - \frac{p^2x^2}{2} + \frac{(4p^2+p^4)x^4}{24} + \cdots$	$1 - \frac{p^2}{4\csc^2(x)} + \cdots$	$\mathrm{sech}(x) + \frac{[\sinh(2x)+2x]q^2}{8\coth(x)\cosh(x)} + \cdots$
$\mathrm{ns}(p;x) =$	$\frac{1}{x} + \frac{(1+p^2)x}{6} + \frac{(7-22p^2+7p^4)x^3}{360} + \cdots$	$\csc(x) + \frac{[2x-\sin(2x)]p^2}{8\tan(x)\sin(x)} + \cdots$	$\coth(x) - \frac{[\sinh(2x)-2x]q^2}{8\sinh^2(x)} + \cdots$
$\mathrm{nc}(p;x) =$	$1 + \frac{x^2}{2} + \frac{(5-4p^2)x^4}{24} + \cdots$	$\sec(x) - \frac{[2x-\sin(2x)]p^2}{8\cot(x)\cos(x)} + \cdots$	$\cosh(x) + \frac{[\sinh(2x)-2x]q^2}{8\,\mathrm{csch}(x)} + \cdots$
$\mathrm{nd}(p;x) =$	$1 + \frac{p^2x^2}{2} + \frac{(2p^2-p^4)x^4}{24} + \cdots$	$1 + \frac{p^2}{4\csc^2(x)} + \cdots$	$\cosh(x) - \frac{[\sinh(2x)+2x]q^2}{8\,\mathrm{csch}(x)} + \cdots$
$\mathrm{am}(p;x) =$	$x - \frac{p^2x^3}{6} + \frac{(4p^2+p^4)x^5}{120} + \cdots$	$x - \frac{[2x-\sin(2x)]p^2}{8} + \cdots$	$\mathrm{gd}(x) + \frac{[\sinh(2x)-2x)]q^2}{8\cosh(x)} + \cdots$

63:7 PARTICULAR VALUES

These occur when the argument of the Jacobian elliptic function is a multiple of the complete elliptic integral K(*p*). Let $v = x/\mathrm{K}(p)$, then for $n = 0, \pm 1, \pm 2, \ldots$:

	$v = 4n$	$v = 4n + \frac{1}{2}$	$v = 4n + 1$	$v = 4n + \frac{3}{2}$	$v = 4n + 2$	$v = 4n + \frac{5}{2}$	$v = 4n + 3$	$v = 4n + \frac{7}{2}$
sc(*p*;*x*)	0	$\frac{1}{\sqrt{q}}$	$\pm\infty$	$\frac{-1}{\sqrt{q}}$	0	$\frac{1}{\sqrt{q}}$	$\pm\infty$	$\frac{-1}{\sqrt{q}}$
sd(*p*;*x*)	0	$\frac{1}{\sqrt{q(1+q)}}$	$\frac{1}{q}$	$\frac{1}{\sqrt{q(1+q)}}$	0	$\frac{-1}{\sqrt{q(1+q)}}$	$\frac{-1}{q}$	$\frac{-1}{\sqrt{q(1+q)}}$
sn(*p*;*x*)	0	$\frac{1}{\sqrt{1+q}}$	1	$\frac{1}{\sqrt{1+q}}$	0	$\frac{-1}{\sqrt{1+q}}$	-1	$\frac{-1}{\sqrt{1+q}}$
cs(*p*;*x*)	$\mp\infty$	$\sqrt{q}$	0	$-\sqrt{q}$	$\mp\infty$	$\sqrt{q}$	0	$-\sqrt{q}$
cd(*p*;*x*)	1	$\frac{1}{\sqrt{1+q}}$	0	$\frac{-1}{\sqrt{1+q}}$	-1	$\frac{-1}{\sqrt{1+q}}$	0	$\frac{1}{\sqrt{1+q}}$
cn(*p*;*x*)	1	$\sqrt{\frac{q}{1+q}}$	0	$-\sqrt{\frac{q}{1+q}}$	-1	$-\sqrt{\frac{q}{1+q}}$	0	$\sqrt{\frac{q}{1+q}}$
ds(*p*;*x*)	$\mp\infty$	$\sqrt{q(1+q)}$	q	$\sqrt{q(1+q)}$	$\pm\infty$	$-\sqrt{q(1+q)}$	$-q$	$-\sqrt{q(1+q)}$
dc(*p*;*x*)	1	$\sqrt{1+q}$	$\pm\infty$	$-\sqrt{1+q}$	-1	$-\sqrt{1+q}$	$\mp\infty$	$\sqrt{1+q}$
dn(*p*;*x*)	1	$\sqrt{q}$	q	$\sqrt{q}$	1	$\sqrt{q}$	q	$\sqrt{q}$
ns(*p*;*x*)	$\mp\infty$	$\sqrt{1+q}$	1	$\sqrt{1+q}$	$\pm\infty$	$-\sqrt{1+q}$	-1	$-\sqrt{1+q}$
nc(*p*;*x*)	1	$\sqrt{\frac{1+q}{q}}$	$\pm\infty$	$-\sqrt{\frac{1+q}{q}}$	-1	$-\sqrt{\frac{1+q}{q}}$	$\mp\infty$	$\sqrt{\frac{1+q}{q}}$
nd(*p*;*x*)	1	$\frac{1}{\sqrt{q}}$	$\frac{1}{q}$	$\frac{1}{\sqrt{q}}$	1	$\frac{1}{\sqrt{q}}$	$\frac{1}{q}$	$\frac{1}{\sqrt{q}}$

63:8 NUMERICAL VALUES

Our algorithm for calculating values of any one of the 12 Jacobian elliptic functions makes use of *Neville's theta functions*. There are four such functions and they are defined in terms of the theta functions of Section 27:18. One of the definitions:

63:8:1
$$\theta_s(p;x) = \pi\theta_1\left(\frac{x}{2\mathrm{K}(p)};\frac{\mathrm{K}(q)}{\pi\mathrm{K}(p)}\right) \Bigg/ \left[\frac{d}{dx}\theta_1\left(\frac{x}{2\mathrm{K}(p)};\frac{\mathrm{K}(q)}{\pi\mathrm{K}(p)}\right)\right]_{x=0}$$

is more complicated than the other three:

63:8:2
$$\theta_c(p;x) = \theta_2\left(\frac{x}{2\mathrm{K}(p)};\frac{\mathrm{K}(q)}{\pi\mathrm{K}(p)}\right) \Bigg/ \theta_2\left(0;\frac{\mathrm{K}(q)}{\pi\mathrm{K}(p)}\right) = \sqrt{q}\,\theta_s(p;\mathrm{K}(p)-x)$$

63:8:3
$$\theta_d(p;x) = \theta_3\left(\frac{x}{2\mathrm{K}(p)};\frac{\mathrm{K}(q)}{\pi\mathrm{K}(p)}\right) \Bigg/ \theta_3\left(0;\frac{\mathrm{K}(q)}{\pi\mathrm{K}(p)}\right)$$

63:8:4
$$\theta_n(p;x) = \theta_4\left(\frac{x}{2\mathrm{K}(p)};\frac{\mathrm{K}(q)}{\pi\mathrm{K}(p)}\right) \Bigg/ \theta_4\left(0;\frac{\mathrm{K}(q)}{\pi\mathrm{K}(p)}\right) = \frac{1}{\sqrt{q}}\theta_d(p;\mathrm{K}(p)-x)$$

Here, as usual, K(*p*) and K(*q*) denote the complete elliptic integrals [Chapter 61] of the modulus and complementary modulus, respectively.

Every one of the 12 Jacobian elliptic functions is expressible as the quotient of two distinct Neville theta functions, the literal subscripts serving as a mnemonic to identify the numerator and denominator of the quotient; thus:

63:8:5 $$\mathrm{cs}(p;x) = \frac{\theta_c(p;x)}{\theta_s(p;x)} \qquad \mathrm{nc}(p;x) = \frac{\theta_n(p;x)}{\theta_c(p;x)} \qquad \text{etc.}$$

By exploiting this principle, the algorithm becomes compact and rapid.

A shortened version of the algorithm in Section 61:8 is first used to calculate $\pi/[2\mathrm{K}(p)]$ and the nome [see Section 61:14] $\mathrm{N}(p)$. The user then supplies the argument x, followed by a two-digit code. The algorithm decodes the latter to identify the denominatorial Neville function then the numeratorial Neville function. If θ_s is called for, the algorithm computes $\sqrt{q\mathrm{K}(p)/2\pi}\ \theta_s(p;x)$ via equation 61:14:8. Similarly, if θ_n is needed, $\sqrt{q\mathrm{K}(p)/2\pi}\ \theta_n(p;x)$ is evaluated by use of 61:14:11. The second equality in 63:8:2 or 63:8:4 is utilized whenever θ_c or θ_d is required. Finally, the two $\sqrt{q\mathrm{K}(p)/2\pi}\ \theta$ values are divided to produce the sought Jacobian function.

The series 61:14:8 and 61:14:11 are used truncated to only the terms actually listed in Section 61:14 so that the largest neglected term is of order $\mathrm{N}^{12}(p)$. Nevertheless, these series converge so rapidly that 24-bit precision is virtually assured provided p does not exceed 0.99.

Because intermediate numbers are stored, one need only input the new argument x and the new code to calculate a second Jacobian function of unchanged modulus.

Set $n = 0$
Input p >>>>>>>
Set $g = p$
(1) Set $a = 1$
(2) Set $s = ga$
Replace a by $(g + a)/2$
Set $g = \sqrt{s}$
If $a < 10^7(a - g)$ go to (2)
If $n \ne 0$ go to (3)
Set $n = g$
Set $g = q = \sqrt{1 - p^2}$
Go to (1)
Replace n by $\exp(-\pi g/n)$
Input x >>>>>>>
Set $s = 0$
Input code c >>>>
Set $a = 10\,\mathrm{frac}(c/5) - 5$
(4) Set $v = gx(180/\pi)$
Set $f = 1$
If $|a| \ne 1$ go to (5)
Replace v by $90 - v$
Set $f = \sqrt{q}$
(5) If $a > 0$ go to (6)
Replace f by $f\sqrt{\sqrt{n}/p}\{\sin(v) + n^2[n^4 \sin(5v) - \sin(3v)]\}$
Go to (7)
(6) Replace f by $f\{\frac{1}{2} + n^4 \cos(4v) - n[n^8 \cos(6v) + \cos(2v)]\}$
(7) If $s \ne 0$ go to (8)
Set $s = f$
Set $a = \mathrm{Int}(c/5) - 5$
Go to (4)
(8) Replace f by f/s
Output f
$f \simeq \mathrm{fg}(p;x)$ <<<<<

Storage needed:
$n, p, g, a, s, q, x, c, v$ and f

Input restrictions: $0 < p \le 0.99$. c may only be one of the 12 codes listed below. $x \ne 0$ if c = 21, 31, or 41. To reuse the same p, enter new x and c only.

Use degree mode or change 90 to $\pi/2$ and delete the $(180/\pi)$ factor.

Test values:
$p = \frac{1}{2}$ and $x =$ 1.311139165

code, c	fg	output
12	sc	2.95528111
13	sd	1.07551806
14	sn	0.947240202
21	cs	0.338377286
23	cd	0.363930880
24	cn	0.320524568
31	ds	0.929784484
32	dc	2.74777452
34	dn	0.880729243
41	ns	1.05569844
42	nc	3.11988565
43	nd	1.13542273

63:9 APPROXIMATIONS

The expansions in Section 63:6 provide raw material from which many useful approximations may be constructed. For example:

63:9:1 $$\mathrm{cn}(p;x) \simeq 1 - \frac{x^2}{2} \qquad \text{8-bit precision} \qquad |x| < \left[\frac{3}{32 - 128p^2}\right]^{1/4}$$

63:9:2 $$\mathrm{ds}(p;x) \simeq \csc(x) \qquad \text{8-bit precision} \qquad |p| < \frac{1}{8\sqrt{1 - x\cot(x)}}$$

63:9:3 $$\mathrm{nd}(p;x) \simeq \cosh(x) \qquad \text{8-bit precision} \qquad |q| < \frac{1}{8\sqrt{\sinh^2(x) + x\tanh(x)}}$$

63:10 OPERATIONS OF THE CALCULUS

The derivative of any elliptic function with respect to its argument is proportional to the product of the function's two copolar cohorts. Thus, if fe is a Jacobian function of pole type e, and e, f, g and h are all distinct:

63:10:1 $$\frac{\partial}{\partial x}\,\mathrm{fe}(p;x) = \alpha\,\mathrm{ge}(p;x)\,\mathrm{he}(p\ x) \qquad \mathrm{e, f, g, h = s, c, d, n}$$

where α is independent of x. Values of α depend on fe and are included in Table 63.10.1. The derivative with respect to the modulus similarly involves α and the two copolar cohorts but has an additional factor equal to $[xq^2 + \beta\,\mathrm{sn}(p;x)\,\mathrm{cd}(p;x) - \mathrm{E}(p;\phi)]/pq^2$, where $\phi = \mathrm{am}(p;x)$ and β is listed in Table 63.10.1. Thus, the derivative with respect to p of a Jacobian function of pole type e is

63:10:2 $$\frac{\partial}{\partial p}\,\mathrm{fe}(p;x) = \alpha\,\mathrm{ge}(p;x)\,\mathrm{he}(p;x)\frac{[xq^2 + \beta\,\mathrm{sn}(p;x)\,\mathrm{cd}(p;x) - \mathrm{E}(p;\phi)]}{pq^2}$$

Integrals of the types

63:10:3 $$\mathrm{I}_1 = \int_0^x \mathrm{fe}(p;t)dt \quad \text{and} \quad \mathrm{I}_2 = \int_0^x \mathrm{fe}^2(p;t)dt \qquad \mathrm{f = s, c, d\ or\ n} \qquad \mathrm{e = c, d\ or\ n}$$

exist and are in Table 63.10.1. For functions having type s poles, however, these integrals diverge, and it is the complementary alternatives

63:10:4 $$\mathrm{I}_3 = \int_x^{\mathrm{K}(p)} \mathrm{fs}(p;t)dt \quad \text{and} \quad \mathrm{I}_4 = \int_x^{\mathrm{K}(p)} \mathrm{fs}^2(p;t)dt \qquad \mathrm{f = c, d\ or\ n}$$

that are listed in the table. The tabulated integrals are valid for $0 < p < 1$ and $0 < x < \mathrm{K}(p)$ but not necessarily for variables outside these ranges.

Rule 63:10:1 is the key to the evaluation of a great many indefinite integrals. Thus, we have

63:10:5 $$\int_0^x \mathrm{ge}(p;t)\,\mathrm{he}(p;t)dt = \frac{1}{\alpha}\,[\mathrm{fe}(p;x) - \mathrm{fe}(p;0)] \qquad \mathrm{e} \neq \mathrm{s}$$

except that if e = s this should be replaced by the complementary

63:10:6 $$\int_x^{\mathrm{K}(p)} \mathrm{gs}(p;t)\,\mathrm{hs}(p;t)dt = \frac{1}{\alpha}\,[\mathrm{fs}(p;\mathrm{K}(p)) - \mathrm{fs}(p;x)]$$

The transformations

63:10:7 $$\int \frac{\mathrm{gh}(p;t)}{\mathrm{eh}^2(p;t)}\,dt = \int \frac{\mathrm{hg}(p;t)}{\mathrm{eg}^2(p;t)}\,dt = \int \mathrm{ge}(p;t)\,\mathrm{he}(p;t)dt$$

Table 63.10.1

	α	β	I_1 / I_3	I_2 / I_4
cd	$-q^2$	0	$\frac{1}{p}\ln(\mathrm{nd}(p;x)+p\mathrm{sd}(p;x))$	$\frac{x}{p^2}+\mathrm{sn}(p;x)\,\mathrm{cd}(p;x)-\frac{\mathrm{E}(p;\phi)}{p^2}$
cn	-1	p^2	$\frac{1}{p}\arccos\{\mathrm{dn}(p;x)\}$	$\frac{\mathrm{E}(p;\phi)}{p^2}-\frac{q^2x}{p^2}$
dc	q^2	0	$\ln(\mathrm{sc}(p;x)+\mathrm{nc}(p;x))$	$x-\mathrm{E}(p;\phi)+\mathrm{sn}(p;x)\,\mathrm{dc}(p;x)$
dn	$-p^2$	$\mathrm{dc}^2(p;x)$	$\arcsin\{\mathrm{sn}(p;x)\}=\phi$	$\mathrm{E}(p;\phi)$
nc	1	p^2	$\frac{1}{q}\ln(\mathrm{dc}(p;x)+q\,\mathrm{sc}(p;x))$	$x-\frac{\mathrm{E}(p;\phi)}{q^2}+\frac{\mathrm{sn}(p;x)\,\mathrm{dc}(p;x)}{q^2}$
nd	p^2	$\mathrm{dc}^2(p;x)$	$\frac{1}{q}\arccos\{\mathrm{cd}(p;x)\}$	$\frac{\mathrm{E}(p;\phi)}{q^2}-\frac{p^2\,\mathrm{sn}(p;x)\,\mathrm{cd}(p;x)}{q^2}$
sc	1	p^2	$\frac{1}{q}\ln\frac{\mathrm{dc}(p;x)+q\,\mathrm{nc}(p;x)}{1+q}$	$\frac{1}{q^2}[\mathrm{sn}(p;x)\,\mathrm{dc}(p;x)-\mathrm{E}(p;\phi)]$
sd	1	$p^2\mathrm{dc}^2(p;x)$	$\frac{1}{pq}\arcsin\{pq\,\mathrm{nd}(p;x)-pq\,\mathrm{cd}(p;x)\}$	$\frac{\mathrm{E}(p;\phi)}{p^2q^2}-\frac{x}{p^2}-\frac{\mathrm{sn}(p;x)\,\mathrm{dc}(p;x)}{q^2}$
sn	1	p^2	$\frac{1}{p}\ln\left(\frac{\mathrm{dn}(p;x)-p\,\mathrm{cn}(p;x)}{1-p}\right)$	$\frac{x-\mathrm{E}(p;\phi)}{p^2}$
cs	-1	p^2	$\ln\left(\frac{1-q}{\mathrm{ns}(p;x)-\mathrm{ds}(p;x)}\right)$	$\mathrm{E}(p;\phi)+\mathrm{cn}(p;x)\,\mathrm{ds}(p;x)-\mathrm{E}(p)$
ds	-1	$p^2\mathrm{dc}^2(p;x)$	$\ln\left(\frac{1}{\mathrm{ns}(p;x)-\mathrm{cs}(p;x)}\right)$	$q^2\mathrm{K}(p)-\mathrm{E}(p)+\mathrm{E}(p;\phi)-q^2x+\mathrm{cn}(p;x)\,\mathrm{ds}(p;x)$
ns	-1	p^2	$\ln\left(\frac{q}{\mathrm{ds}(p;x)-\mathrm{cs}(p;x)}\right)$	$\mathrm{K}(p)-\mathrm{E}(p)-x+\mathrm{E}(p;\phi)+\mathrm{cn}(p;x)\,\mathrm{ds}(p;x)$

permit evaluation of these integrals via 63:10:5 or 63:10:6. Similarly, such transformations as

63:10:8
$$\int \mathrm{gf}(p;t)\,\mathrm{he}(p;t)dt=\int\frac{\mathrm{gh}(p;t)dt}{\mathrm{fh}(p;t)\,\mathrm{eh}(p;t)}=\int\frac{\mathrm{ge}(p;t)\,\mathrm{he}(p;t)}{\mathrm{fe}(p;t)}dt=\frac{1}{\alpha}\int\frac{d\,\mathrm{fe}(p;t)}{\mathrm{fe}(p;t)}$$

lead to results involving the logarithm $(1/\alpha)\ln\{\mathrm{fe}(p;x)\}$ and find many applications. In the formulas of this paragraph, the α constant is that appropriate to the $\mathrm{fe}(p;x)$ function.

For the sake of completeness we include the following derivatives of functions commonly associated with the Jacobian elliptic functions:

63:10:9
$$\frac{\partial}{\partial x}\mathrm{am}(p;x)=\frac{\partial\phi}{\partial x}=\mathrm{dn}(p;x)$$

63:10:10
$$\frac{\partial}{\partial p}\mathrm{am}(p;x)=\frac{\partial\phi}{\partial p}=\frac{\mathrm{dn}(p;x)}{pq^2}[q^2x+p^2\,\mathrm{sn}(p;x)\,\mathrm{cd}(p;x)-\mathrm{E}(p;\phi)]$$

63:10:11
$$\left(\frac{\partial}{\partial x}\mathrm{E}(p;\phi)\right)_p=\mathrm{dn}^2(p;x)$$

63:10:12
$$\left(\frac{\partial}{\partial p}\mathrm{E}(p;\phi)\right)_x=\frac{p\,\mathrm{cn}^2(p;x)}{q^2}[\mathrm{sc}(p;x)\,\mathrm{dn}(p;x)-\mathrm{E}(p;\phi)-q^2x\,\mathrm{sc}^2(p;x)]$$

63:11 COMPLEX ARGUMENT

With argument $x + iy$ the trio of Jacobian elliptic functions of type n pole adopt the complex values

63:11:1 $$\mathrm{sn}(p;x+iy) = \frac{\mathrm{ds}(q;y)\,\mathrm{nc}(q;y) + i\,\mathrm{ds}(p;x)\,\mathrm{cn}(p;x)}{\mathrm{ns}(p;x)\,\mathrm{cs}(q;y) + p^2\,\mathrm{sn}(p;x)\,\mathrm{sc}(q;y)}$$

63:11:2 $$\mathrm{cn}(p;x+iy) = \frac{\mathrm{cs}(p;x)\,\mathrm{ns}(q;y) - i\,\mathrm{dn}(p;x)\,\mathrm{dc}(q;y)}{\mathrm{ns}(p;x)\,\mathrm{cs}(q;y) + p^2\,\mathrm{sn}(p;x)\,\mathrm{sc}(q;y)}$$

63:11:3 $$\mathrm{dn}(p;x+iy) = \frac{\mathrm{ds}(p;x)\,\mathrm{ds}(q;y) - ip^2\,\mathrm{cn}(p;x)\,\mathrm{nc}(q;y)}{\mathrm{ns}(p;x)\,\mathrm{cs}(q;y) + p^2\,\mathrm{sn}(p;x)\,\mathrm{sc}(q;y)}$$

The other nine functions may be derived by application of rule 63:5:1.

The transformation of an elliptic function of imaginary argument to one of real argument is known as *Jacobi's imaginary transformation*. Those functions that involve an s become imaginary on transformation:

63:11:4 $\mathrm{sc}(p;iy) = i\,\mathrm{sn}(q;y)$ $\quad\mathrm{sd}(p;iy) = i\,\mathrm{sd}(q;y)$ $\quad\mathrm{sn}(p;iy) = i\,\mathrm{sc}(q;y)$

63:11:5 $\mathrm{cs}(p;iy) = -i\,\mathrm{ns}(q;y)$ $\quad\mathrm{ds}(p;iy) = -i\,\mathrm{ds}(q;y)$ $\quad\mathrm{ns}(p;iy) = -i\,\mathrm{cs}(q;y)$

while those that do not, transform to real functions:

63:11:6 $\mathrm{cd}(p;iy) = \mathrm{nd}(q;y)$ $\quad\mathrm{cn}(p;iy) = \mathrm{nc}(q;y)$ $\quad\mathrm{nc}(p;iy) = \mathrm{cn}(q;y)$

63:11:7 $\mathrm{dn}(p;iy) = \mathrm{dc}(q;y)$ $\quad\mathrm{dc}(p;iy) = \mathrm{dn}(q;y)$ $\quad\mathrm{nd}(p;iy) = \mathrm{cd}(q;y)$

The transformations 63:11:4–63:11:7 illustrate that elliptic functions have an *imaginary period* as well as the real period discussed in Section 63:2. For example, one finds $\mathrm{sc}(p;x + iy + 4i\mathrm{K}(q)) = \mathrm{sc}(p;x + iy)$. Table 63.11.1 summarizes the periods for each of the 12 Jacobian functions and explains why these functions are known as *doubly periodic functions*.

We have been discussing imaginary values of the argument, but one can also have an imaginary modulus. Jacobian elliptic functions of imaginary modulus are, in fact, real. The appropriate transformations are

63:11:8 $$\mathrm{sn}(ip;x) = \frac{1}{\sqrt{1+p^2}}\,\mathrm{sd}\left(\frac{p}{\sqrt{1+p^2}}; x\sqrt{1+p^2}\right)$$

63:11:9 $$\mathrm{cn}(ip;x) = \mathrm{cd}\left(\frac{p}{\sqrt{1+p^2}}; x\sqrt{1+p^2}\right)$$

63:11:10 $$\mathrm{dn}(ip;x) = \mathrm{nd}\left(\frac{p}{\sqrt{1+p^2}}; x\sqrt{1+p^2}\right)$$

for the three functions of type n pole. As usual, the other nine may be obtained via 63:5:1.

63:12 GENERALIZATIONS

There are none.

Table 63.11.1

	Real period	Imaginary period
$\mathrm{sc}(p;x+iy)$, $\mathrm{cs}(p;x+iy)$, $\mathrm{dn}(p;x+iy)$, $\mathrm{nd}(p;x+iy)$	$2\mathrm{K}(p)$	$4i\mathrm{K}(q)$
$\mathrm{sd}(p;x+iy)$, $\mathrm{cn}(p;x+iy)$, $\mathrm{ds}(p;x+iy)$, $\mathrm{nc}(p;x+iy)$	$4\mathrm{K}(p)$	$4i\mathrm{K}(q)$
$\mathrm{sn}(p;x+iy)$, $\mathrm{cd}(p;x+iy)$, $\mathrm{dc}(p;x+iy)$, $\mathrm{ns}(p;x+iy)$	$4\mathrm{K}(p)$	$2i\mathrm{K}(q)$

63:13 COGNATE FUNCTIONS

The functions of Chapters 61, 62 and 63 arose, largely due to Legendre's work, in connection with the integrals discussed in Section 62:14, which themselves arise in a variety of geometric problems and in such physical contexts as the motion of a pendulum. However, there is a rival formalism, due to Weierstrass, that can accomplish the same tasks as that of Legendre. Of course, the two systems are intimately related. We shall not describe the *Weierstrassian elliptic functions* but simply point out some of the equivalences.

There are three interrelated variables, usually symbolized e_1, e_2 and e_3, that play a similar role in the Weierstrass system to that played by the modulus in Legendre's system. One has the equivalences

63:13:1 $$p = \sqrt{\frac{e_2 - e_3}{e_1 - e_3}} \qquad q = \sqrt{\frac{e_1 - e_2}{e_1 - e_3}} \qquad e_1 + e_2 + e_3 = 0$$

Likewise, the role of determining the two (real and imaginary) periods is taken over by two new variables

63:13:2 $$\omega_1 = \frac{2\mathrm{K}(p)}{\sqrt{e_1 - e_2}} \qquad \omega_2 = \frac{2i\mathrm{K}(q)}{\sqrt{e_1 - e_2}}$$

The principal Weierstrassian elliptic function, usually symbolized by P(z) with some fancy typographical rendering of the P, is then expressible as

63:13:3 $$e_1 + (e_1 - e_3)\,\mathrm{cs}^2(p;z\sqrt{e_1 - e_3}) = e_2 + (e_1 - e_3)\,\mathrm{ds}^2(p;z\sqrt{e_1 - e_3}) = e_3 + (e_1 - e_3)\,\mathrm{sn}^2(p;z\sqrt{e_1 - e_3})$$

in terms of Jacobian elliptic functions.

Both the Weierstrass and Jacobian elliptic functions are intimately related to the theta functions [Section 27:13] but, beyond reporting the relationships given in Section 63:8, we do not explore these connections in the *Atlas*. Further to complicate a topic already overburdened by redundant symbolism, we should mention *Jacobi's eta functions* and *Jacobi's theta functions*, which are linked to ordinary theta functions and to Neville's theta functions [Section 63:8] by the equivalences

63:13:4 $$\mathrm{H}(p;x) = \theta_1\left(\frac{x}{2\mathrm{K}(p)};\frac{\mathrm{K}(q)}{\pi\mathrm{K}(p)}\right) = \frac{1}{\pi}\left[\frac{\partial}{\partial x}\mathrm{H}(p;x)\right]_{x=0} \theta_s(p;x)$$

63:13:5 $$\mathrm{H}_1(p;x) = \theta_2\left(\frac{x}{2\mathrm{K}(p)};\frac{\mathrm{K}(q)}{\pi\mathrm{K}(p)}\right) = \mathrm{H}_1(p;0)\,\theta_c(p;x)$$

63:13:6 $$\Theta_1(p;x) = \theta_3\left(\frac{x}{2\mathrm{K}(p)};\frac{\mathrm{K}(q)}{\pi\mathrm{K}(p)}\right) = \Theta_1(p;0)\,\theta_d(p;x)$$

63:13:7 $$\Theta(p;x) = \theta_4\left(\frac{x}{2\mathrm{K}(p)};\frac{\mathrm{K}(q)}{\pi\mathrm{K}(p)}\right) = \Theta(p;0)\,\theta_n(p;x)$$

CHAPTER 64

THE HURWITZ FUNCTION $\zeta(\nu;u)$

Closely related to the functions of Chapters 3 and 44, the Hurwitz function plays an invaluable role in the differintegration of periodic functions, a topic discussed in Section 64:14. Along with the related bivariate eta function [Section 64:13] and Lerch's function [Section 64:12], the Hurwitz function provides a means of summing many series whose terms involve arbitrary powers.

64:1 NOTATION

The symbol ζ followed by two parenthesized variables is standard, but the name *generalized zeta function* is often encountered. Sometimes $\zeta(\nu;u)$ is called "Riemann's function" or "Riemann's zeta function" but we eschew these names to avoid confusion with the $\zeta(\nu)$ function of Chapter 3.

We shall refer to ν and u as the order and parameter, respectively, of the Hurwitz function. It is to achieve unity with the notation for Lerch's function [Section 64:12] that we resist the temptation to replace the symbol u and name "parameter" by the symbol x and the name "argument."

64:2 BEHAVIOR

Unless the order is an integer, the Hurwitz function becomes complex when $u < 0$. Accordingly, this range of parameter is excluded from our discussion, except in Section 64:11. The zero order function $\zeta(0,u)$ becomes ill defined as u approaches zero; similarly, $\zeta(1,u)$ is ill defined as u approaches infinity.

The Hurwitz function displays an infinite discontinuity at $\nu = 1$, and, as Figure 64-1 shows, the behaviors for $\nu > 1$ and $\nu < 1$ are quite distinct.

For $\nu > 1$, $\zeta(\nu;u)$ is uniformly positive. When both the order and the parameter exceed unity, the Hurwitz function declines in value, asymptotically approaching zero, as either ν or u increases. Conversely, $\zeta(\nu;u) \to \infty$ as ν approaches unity from positive values or as u approaches zero from positive values.

For the most part, the Hurwitz function assumes negative values for $\nu < 1$; however, there are "lobes" of positivity as mapped on the contour diagram. As a function of u, $\zeta(\nu;u)$ displays a small, finite number of zeros, mostly in the $0 < u < 1$ range, for $\nu < 1$.

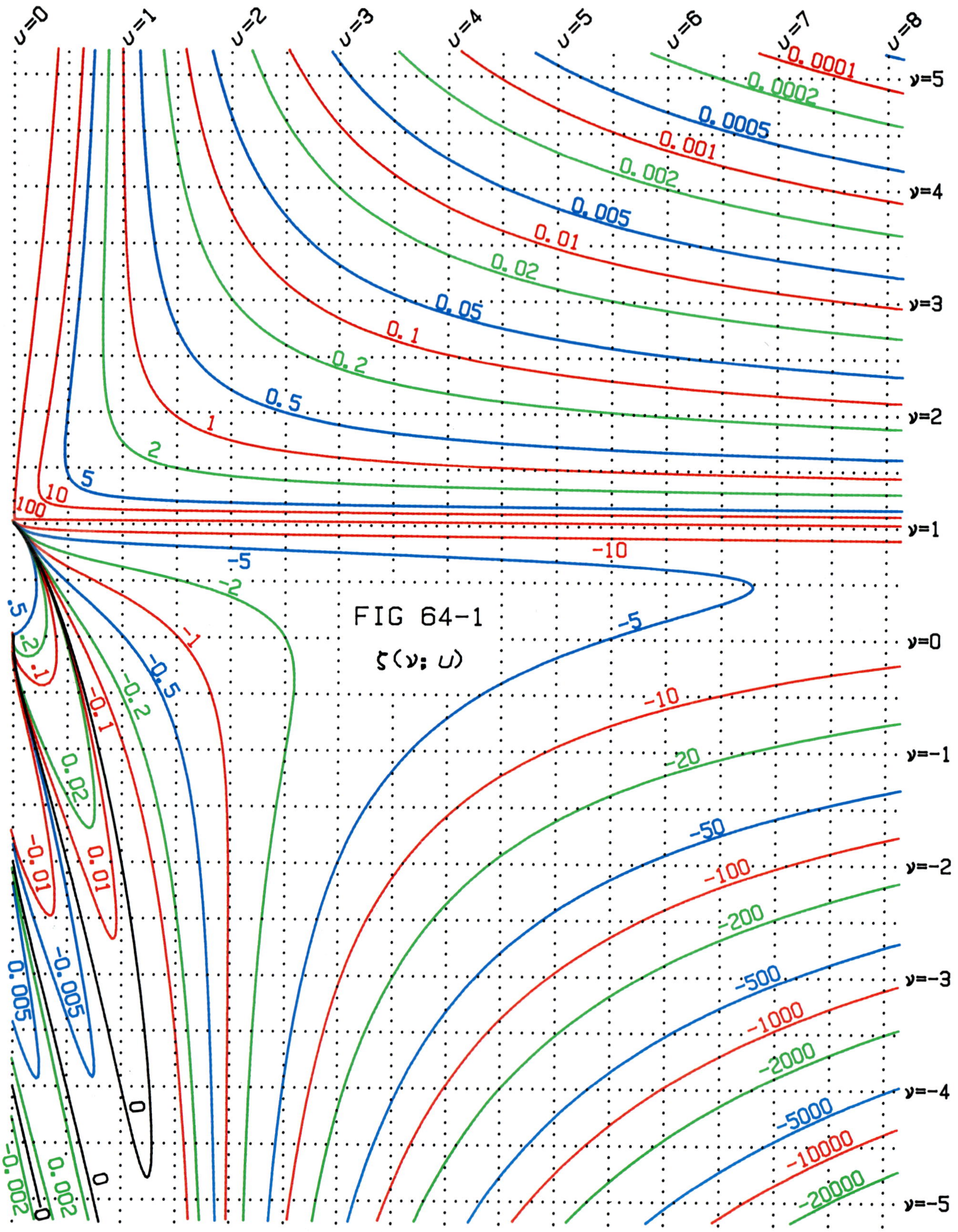
ν=0
ν=1
ν=2
ν=3
ν=4
ν=5
ν=6
ν=7
ν=8
y=5
y=4
y=3
y=2
y=1
y=0
y=−1
y=−2
y=−3
y=−4
y=−5
0.0001
0.0002
0.0005
0.001
0.002
0.005
0.01
0.02
0.05
0.1
0.2
0.5
1
2
5
10
100
-10
-5
-2
-1
-0.5
-0.2
-0.1
-5
FIG 64-1
ζ(ν; u)
-10
-20
-50
-100
-200
-500
-1000
-2000
-5000
-10000
-20000
.5
.2
.1
0.02
0.01
-0.01
-0.005
0.005
0
0
0
-0.002
0.002

64:3 DEFINITIONS

The Hurwitz function is defined by the integral transforms

64:3:1 $$\zeta(\nu;u) = \frac{1}{\Gamma(\nu)}\int_0^\infty \frac{t^{\nu-1}\exp(-ut)dt}{1-\exp(-t)} = \frac{1}{\Gamma(\nu)}\int_1^\infty \frac{\ln^{\nu-1}(t)dt}{t^u(t-1)} \qquad \nu>1 \qquad u>0$$

or by Hermite's integral

64:3:2 $$\zeta(\nu;u) = \frac{u^{-\nu}}{2} + \frac{u^{1-\nu}}{\nu-1} + 2\int_0^\infty \frac{\sin(\nu\arctan(t/u))dt}{(u^2+t^2)^{\nu/2}[\exp(2\pi t)-1]} \qquad u>0$$

Expansion 64:6:1 provides a definition of $\zeta(\nu;u)$ for $\nu > 1$, and Hurwitz's series 64:6:2, supplemented by recursion 64:5:1, can fill a similar role for $\nu < 0$.

Notice that none of these definitions is valid in the important domain $0 \le \nu < 1$. Contour integration [see Erdélyi, Magnus, Oberhettinger and Tricomi, *Higher Transcendental Functions*, Volume 1, pages 25 and 26] can define the Hurwitz function in this region. More practically, the series 64:6:4 can be employed for all orders and parameters.

The difference $\psi(u) - \psi(u-t)$ of two digamma functions [Chapter 44] provides a generating function:

64:3:3 $$t[\psi(u) - \psi(u-t)] = \sum_{n=2}^{\infty}\zeta(n;u)t^n \qquad u>1$$

for Hurwitz functions of integer orders 2, 3, 4, ...

64:4 SPECIAL CASES

Here we discuss cases of $\zeta(\nu;u)$ that correspond to special values of the order ν. See Section 64:7 for simplifications of the Hurwitz function that arise when the parameter u acquires particular values.

When ν is a positive integer greater than unity, the Hurwitz function is equivalent to a polygamma function [Section 44:12]:

64:4:1 $$\zeta(n;u) = \frac{(-1)^n\psi^{(n-1)}(u)}{(n-1)!} \qquad n = 2, 3, 4, \ldots$$

When ν is a nonpositive integer, the Hurwitz function can be expressed in terms of a Bernoulli polynomial [Chapter 19]:

64:4:2 $$\zeta(-n;u) = \frac{-B_{n+1}(u)}{n+1} \qquad n = 0, 1, 2, \ldots$$

of which the first few instances are

64:4:3 $$\zeta(0;u) = \frac{1}{2} - u$$

64:4:4 $$\zeta(-1;u) = \frac{-1+6u-6u^2}{12}$$

64:4:5 $$\zeta(-2;u) = \frac{-u+3u^2-2u^3}{6}$$

Although the Hurwitz function is infinite for $\nu = 1$, the quotient $\zeta(\nu;u)/\Gamma(1-\nu)$ and the product $[\nu-1]\zeta(\nu;u)$ both remain well behaved in the vicinity of $\nu = 1$. This is evident from the limiting expressions

64:4:6 $$\lim_{\nu\to 1}\left\{\frac{\zeta(\nu;u)}{\Gamma(1-\nu)}\right\} = -1$$

64:4:7 $$\lim_{\nu\to1}\left\{\zeta(\nu;u)-\frac{1}{\nu-1}\right\}=-\psi(u)$$

where ψ is the digamma function [Chapter 44].

64:5 INTRARELATIONSHIPS

The recursion formula

64:5:1 $$\zeta(\nu;u+1)=\zeta(\nu;u)-\frac{1}{u^\nu}\qquad u\geq0$$

may be iterated to

64:5:2 $$\zeta(\nu;u)=\zeta(\nu;u+J)+\sum_{j=0}^{J-1}(j+u)^{-\nu}\qquad u\geq0$$

and, provided $\nu>1$, becomes expansion 64:6:1 when $J=\infty$. The duplication formula

64:5:3 $$\zeta(\nu;2u)=2^{-\nu}\left[\zeta(\nu;u)+\zeta\left(\nu;\frac{1}{2}+u\right)\right]$$

may be generalized to

64:5:4 $$\zeta(\nu;mu)=m^{-\nu}\sum_{j=0}^{m-1}\zeta\left(\nu;\frac{j}{m}+u\right)\qquad m=2,3,4,\ldots$$

and used to demonstrate, for example, that

64:5:5 $$\zeta\left(\nu;\frac{3}{4}+u\right)=4^\nu\zeta(\nu;4u)-2^\nu\zeta(\nu;2u)-\zeta\left(\nu;\frac{1}{4}+u\right)$$

The following infinite series of Hurwitz functions of integer order

64:5:6 $$\zeta(2;u)+\zeta(3;u)+\zeta(4;u)+\cdots=\sum_{n=2}^{\infty}\zeta(n;u)=\frac{1}{u-1}\qquad u>1$$

64:5:7 $$\zeta(2;u)-\zeta(3;u)+\zeta(4;u)-\cdots=\sum_{n=2}^{\infty}(-1)^n\zeta(n;u)=\frac{1}{u}\qquad u>0$$

64:5:8 $$\frac{\zeta(2;u)}{2}+\frac{\zeta(3;u)}{3}+\frac{\zeta(4;u)}{4}+\cdots=\sum_{n=2}^{\infty}\frac{\zeta(n;u)}{n}=\psi(u)-\ln(u-1)\qquad u>1$$

and

64:5:9 $$\frac{\zeta(2;u)}{2}-\frac{\zeta(3;u)}{3}+\frac{\zeta(4;u)}{4}-\cdots=\sum_{n=2}^{\infty}(-1)^n\frac{\zeta(n;u)}{n}=\ln(u)-\psi(u)\qquad u>0$$

may be combined in several ways: for example, to sum the series $\Sigma\zeta(k;u)$ where k is restricted to a single party.

64:6 EXPANSIONS

The important series

64:6:1 $$\zeta(\nu;u)=\frac{1}{u^\nu}+\frac{1}{(1+u)^\nu}+\frac{1}{(2+u)^\nu}+\cdots=\sum_{j=0}^{\infty}(j+u)^{-\nu}\qquad \nu>1$$

often serves as a definition of the Hurwitz function for orders exceeding unity. *Hurwitz' formula*:

64:6:2 $$\zeta(\nu;u) = 2\Gamma(1-\nu)\sum_{j=1}^{\infty}\frac{\sin\left(2j\pi u + \dfrac{\pi\nu}{2}\right)}{(2j\pi)^{1-\nu}} \qquad \nu < 0 \qquad 0 \le u \le 1$$

may also be written in the standard Fourier series [Section 36:6] form by replacing $\sin[2j\pi u + (\pi\nu/2)]$ by $\cos(\pi\nu/2)\sin(2j\pi u) + \sin(\pi\nu/2)\cos(2j\pi u)$. As written, 64:6:2 is valid only for a narrow range of parameters, but this may be extended to larger u by sufficient applications of recurrence 64:5:1.

An asymptotic expansion of the Hurwitz function may be written in the convenient form

64:6:3 $$\nu\zeta(\nu+1;u+1) \sim \frac{1}{u^{\nu}} - \frac{\nu}{2u^{\nu+1}} + \frac{\nu(\nu+1)}{12u^{\nu+2}} - \frac{\nu(\nu+1)(\nu+2)(\nu+3)}{720u^{\nu+4}} + \cdots + \frac{(\nu)_k B_k}{k!u^{k+\nu}} + \cdots \qquad u \to \infty$$

where $(\nu)_k$ denotes a Pochhammer polynomial [Chapter 18] and B_k is the kth Bernoulli number [Chapter 4]. Odd members of this latter family, except B_1, are zero, and even members, except B_0, may be expressed as zeta numbers [Chapter 3]. Thus, also making use of 64:5:2, we may reformulate 64:4:3 as

64:6:4 $$\zeta(\nu;u) \sim \sum_{j=0}^{J-1}\frac{1}{(j+u)^{\nu}} + \frac{1}{2(J+u)^{\nu}} + \frac{(J+u)^{1-\nu}}{\nu-1}\left[1 - 2\sum_{k=1}^{K}\frac{(\nu-1)_{2k}\zeta(2k)}{[-4\pi^2(u+J)^2]^k}\right]$$

where the first sum is "empty" if $J = 0$ (such sums are to be interpreted as zero). While this expansion technically remains asymptotic (i.e., valid only in the $u + J \to \infty$ limit), it may be made arbitrarily exact by making J large enough and choosing K at least to exceed $(1-\nu)/2$. Yet a further development leads to

64:6:5 $$\zeta(\nu;u) \sim \frac{1}{2u^{\nu}} - \frac{2(2\pi)^{\nu-1}\zeta^*}{\nu-1} + \frac{u^{1-\nu}}{\nu-1}\left[3 - 2\sum_{k=1}^{K}\frac{(\nu-1)_{2k}[\zeta(2k)-1]}{(-4\pi^2u^2)^k}\right]$$

where ζ^* is related to the Boehmer integrals of Section 39:12 by

64:6:6 $$\zeta^* = \cos(2\pi u)\,S(2\pi u;2-\nu) - \sin(2\pi u)\,C(2\pi u;2-\nu)$$

$$= \Gamma(2-\nu)\sin\left(\frac{\nu\pi}{2} + 2\pi u\right) - (2\pi u)^{2-\nu}\sum_{k=0}^{\infty}\frac{(2\pi u)^k\sin\left(\dfrac{k\pi}{2} - 2\pi u\right)}{k!(k+2-\nu)} \qquad \nu \neq 2, 3, 4, \ldots$$

64:7 PARTICULAR VALUES

The Hurwitz function of integer parameter is related to the zeta function of Chapter 3. The simplest cases are

64:7:1 $$\zeta(\nu;0) = \begin{cases} \infty & \nu > 0 \\ \zeta(\nu) & \nu < 0 \end{cases}$$

64:7:2 $$\zeta(\nu;1) = \zeta(\nu)$$

and

64:7:3 $$\zeta(\nu;2) = \zeta(\nu) - 1$$

while the general case involves a finite sum:

64:7:4 $$\zeta(\nu;m) = \zeta(\nu) - \sum_{j=1}^{m-1} j^{-\nu} \qquad m = 2, 3, 4, \ldots$$

or, for orders exceeding unity, an alternative infinite sum:

64:7:5 $$\zeta(\nu;m) = \sum_{j=m}^{\infty}\frac{1}{j^{\nu}} \qquad m = 1, 2, 3, \ldots \qquad \nu > 1$$

Similarly, the Hurwitz function of a parameter that is an odd multiple of a moiety is related to the lambda function of Chapter 3. The simplest case is

64:7:6 $$\zeta(\nu;\tfrac{1}{2}) = 2^{\nu}\lambda(\nu) = [2^{\nu} - 1]\zeta(\nu)$$

and others may be expressed either as a finite or an infinite sum

64:7:7 $$\zeta(\nu;m+\tfrac{1}{2}) = 2^{\nu}\left[\lambda(\nu) - \sum_{j=1}^{m}(2j-1)^{-\nu}\right] \qquad m = 1, 2, 3, \ldots$$
$$= 2^{\nu}\sum_{j=m}^{\infty}\frac{1}{(2j+1)^{\nu}} \qquad m = 0, 1, 2, \ldots \qquad \nu > 1$$

The Hurwitz functions of parameters equal to $\frac{1}{4}$ or $\frac{3}{4}$ are related to the lambda and beta functions of Chapter 3 by

64:7:8 $$\zeta(\nu;\tfrac{1}{4}) = 4^{\nu}\left[\frac{\zeta(\nu)+\beta(\nu)}{2}\right]$$

64:7:9 $$\zeta(\nu;\tfrac{3}{4}) = 4^{\nu}\left[\frac{\zeta(\nu)-\beta(\nu)}{2}\right]$$

Adding equations 64:7:8, 64:7:6, 64:7:9 and 64:7:2 leads to $\zeta(\nu;\frac{1}{4}) + \zeta(\nu;\frac{1}{2}) + \zeta(\nu;\frac{3}{4}) + \zeta(\nu;1) = 4^{\nu}\zeta(\nu)$, which is the $J = 4$ case of the general result

64:7:10 $$\sum_{j=1}^{J}\zeta\left(\nu;\frac{j}{J}\right) = J^{\nu}\zeta(\nu) \qquad J = 1, 2, 3, \ldots$$

As u increases, the following values are approached

64:7:11 $$\zeta(\nu;\infty) = \begin{cases} 0 & \nu > 1 \\ -\infty & \nu < 1 \end{cases}$$

64:8 NUMERICAL VALUES

We present below an algorithm suitable for calculating $\zeta(\nu;u)$. It is based on the $J = 0$ version of asymptotic expansion 64:6:4 and uses the Padé technique [see Section 17:13] to overcome the convergence problem. A fixed number (thirty-one) of Padé approximants is calculated prior to output so that the computation is always rather leisurely. The algorithm may use equation 64:5:1 either to decrease or increase the parameter before the expansion is invoked. If $\nu > 1$, the parameter is increased, if necessary, until it exceeds the order ν. If $\nu < 1$, the parameter is invariably adjusted to lie in the range from 1 through 2.

Input ν >>>>>>>
    Set $f = 10^{99}$
Input u >>>>>>>
    If $(\nu - 1)(u + |\nu| - \nu) = 0$ go to (10)
    Set $f = k = 0$
(1) Replace f by $f + u^{-\nu}$
    Replace u by $u + 1$
    If $\nu < 1$ go to (6)
    If $u \le \nu$ go to (1)
(2) Set $h_0 = 1$
    Set $t = -2$
    Set $Z = 271/60$

Storage needed: ν, f, u, k, t, Z, j, p, q and h_0, h_1, h_2, ... h_{59}, h_{60}

Input restriction: $u \ge 0$

(3) Replace t by $t(\nu + 2k)(1 - \nu - 2k)/(2\pi u)^2$
Replace k by $k + 1$
If $k \le 5$ go to (7)
Set $Z = 1 + (1 + 4^{-k})(4^{-k} + 9^{-k}) + 25^{-k}$
(4) Set $h_k = h_{k-1} + tZ$
Set $q = 0$
Set $j = k$
(5) Set $p = h_{j-1}$
If $h_j \ne p$ go to (8)
Set $h_{j-1} = 10^{99}$
If $p \ne 10^{99}$ go to (9)
Set $h_{j-1} = q$
Go to (9)
(6) If $u \le 2$ go to (2)
Replace u by $u - 1$
Replace f by $f - u^{-\nu}$
Go to (6)
(7) Replace Z by $10\pi^2 Z/(504 - 301k + 74k^2 - 6k^3)$
Go to (4)
(8) Set $h_{j-1} = q + [1/(h_j - p)]$
(9) Set $q = p$
Replace j by $j - 1$
If $j \ne 0$ go to (5)
If $k \ne 60$ go to (3)
Replace f by $f + [1/(2u^\nu)] + [h_0 u^{1-\nu}/(\nu - 1)]$
(10) Output f

$f \simeq \zeta(\nu;u)$ <<<<<<

Test values:
$\zeta\left(\frac{1}{2};0\right) = \infty$
$\zeta\left(0;\frac{1}{2}\right) = 0$
$\zeta(10;10) = 1.69268\mathit{613} \times 10^{-10}$
$\zeta\left(-5;\frac{1}{2}\right) = 0.00384424\mathit{603}$
$\zeta\left(\frac{1}{6};\frac{1}{4}\right) = 0.306886\mathit{545}$

The algorithm returns 10^{99}, instead of infinity, if $\nu = 1$ or if $\nu \ge 0 = u$. Otherwise, the algorithm generally has a precision of 24 bits although this may be degraded near the zeros of the Hurwitz function or at extreme values (say $|\nu| > 30$) of the order.

64:9 APPROXIMATIONS

A useful approximation to the $J = 0$ version of expansion 64:6:4 is provided by its first three terms:

64:9:1 $$\zeta(\nu;u) \sim \frac{1}{2u^\nu}\left[\frac{2u}{\nu - 1} + 1 + \frac{1}{6u}\right] \qquad \text{8-bit precision} \qquad u > \left|\frac{(\nu^2 - 1)(\nu^2 + 2\nu)}{3}\right|^{1/4}$$

Bear in mind, however, that the expansion is asymptotic and is not itself always accurate.

64:10 OPERATIONS OF THE CALCULUS

Single and multiple differentiation with respect to the parameter obey the formulas

64:10:1 $$\frac{\partial}{\partial u}\zeta(\nu;u) = -\nu\zeta(\nu + 1;u)$$

64:10:2 $$\frac{\partial^n}{\partial u^n}\zeta(\nu;u) = (-1)^n(\nu)_n\,\zeta(\nu + n;u)$$

Differentiation with respect to order gives

64:10:3 $$\frac{\partial}{\partial\nu}\zeta(\nu;u) = -\sum_{j=0}^{\infty}\frac{\ln(j+u)}{(j+u)^{\nu}} \qquad \nu > 1$$

and there is also the particular value

64:10:4 $$\frac{\partial}{\partial\nu}\zeta(\nu;u) = \ln\left(\frac{\Gamma(u)}{\sqrt{2\pi}}\right) \qquad \nu = 0$$

for zero order.

Formulas for indefinite integration include

64:10:5 $$\int_u^{\infty}\zeta(\nu;t)\mathrm{d}t = \frac{\zeta(\nu-1;u)}{\nu-1} \qquad \nu > 2$$

64:10:6 $$\int_1^u\zeta(\nu;t)\mathrm{d}t = \frac{\zeta(\nu-1;u)-\zeta(\nu-1)}{1-\nu} \qquad \nu < 2$$

64:10:7 $$\int_0^u\zeta(\nu;t)\mathrm{d}t = \frac{\zeta(\nu-1;u)-\zeta(\nu-1)}{1-\nu} \qquad \nu < 1$$

and lead to the interesting definite integrals

64:10:8 $$\int_0^1\zeta(\nu;t)\mathrm{d}t = 0 \qquad \nu < 1$$

64:10:9 $$\int_1^2\zeta(\nu;t)\mathrm{d}t = \frac{1}{\nu-1} \qquad \text{all } \nu$$

The parts-integration procedure that produces the result

64:10:10 $$\int_1^u t\zeta(\nu;t)\mathrm{d}t = \frac{u\zeta(\nu-1;u)-\zeta(\nu-1)}{1-\nu} - \frac{\zeta(\nu-2;u)-\zeta(\nu-2)}{(1-\nu)(2-\nu)} \qquad \nu < 2$$

may be iterated to generate expressions for $\int t^n\zeta(\nu;t)\mathrm{d}t$ for $n = 2, 3, 4, \ldots$.

The Boehmer sine integral [Section 39:12] appears in the formula

64:10:11 $$\int_x^{1+x}\sin(2\pi t)\,\zeta(\nu;t)\mathrm{d}t = (2\pi)^{\nu-1}\,\mathrm{S}(2\pi x;1-\nu) \qquad \nu > 1 \qquad x \geq 0$$

and a similar operation generates the Boehmer cosine integral from $\cos(2\pi t)\,\zeta(\nu;t)$. These results relate closely to the discussion in Section 64:14.

64:11 COMPLEX ARGUMENT

The Hurwitz function is complex not only for complex parameter but even, unless the order is an integer, for negative parameter. Here we address the latter circumstance only. If u is a positive number lying between the integers $m-1$ and m, then

64:11:1 $$\zeta(\nu;-u) = \zeta(\nu;m-u) + (-1)^{\nu}[\zeta(\nu;u+1-m) - \zeta(\nu;u+1)]$$
$$= \zeta(\nu;m-u) + [\cos(\nu\pi) + i\sin(\nu\pi)]\sum_{j=0}^{m-1}\frac{1}{(u-j)^{\nu}} \qquad m-1 < u \leq m$$

64:12 GENERALIZATIONS

In this section we briefly discuss the trivariate *Lerch's function* $\phi(x;\nu;u)$, of which the Hurwitz function is the special case of unity argument, $x = 1$:

64:12:1 $$\Phi(1;\nu;u) = \zeta(\nu;u)$$

Lerch's function may be defined by an integral or a series:

64:12:2 $$\Phi(x;\nu;u) = \frac{1}{\Gamma(\nu)}\int_0^\infty \frac{t^{\nu-1}\exp(-ut)}{1 - x\exp(-t)}\,dt = \sum_{j=0}^{\infty}\frac{x^j}{(j+u)^\nu}$$

although these are generally real and convergent only for limited ranges of x, ν and u. A third definition:

64:12:3 $$\frac{d^{-\nu}}{[d(x+\infty)]^{-\nu}}\frac{\exp(ux)}{1-\exp(x)} = \exp(ux)\,\Phi\{\exp(x);\nu;u\} \qquad x < 0$$

utilizes the concept of differintegration [Section 0:10] with a lower limit of $-\infty$.

The recurrence formula

64:12:4 $$\Phi(x;\nu;u) = x\Phi(x;\nu;1+u) + \frac{1}{u^\nu}$$

can clearly be iterated indefinitely often. When the order is a positive or negative integer, Lerch's function can be expanded in terms of the logarithm of its argument; thus:

64:12:5 $$\Phi\{\exp(y);n;u\} = \exp(-uy)\left\{\frac{y^{n-1}}{(n-1)!}[\psi(n) - \psi(u) - \ln(-y)] + \sum_{j=1}^{\infty}\frac{y^{j-1}}{(j-1)!}\zeta(1+n-j;u)\right\}$$

$$n = 2, 3, 4, \ldots$$

64:12:6 $$\Phi\{\exp(y);-n;u) = \exp(-uy)\left\{\frac{n!}{(-y)^{n+1}} - \sum_{j=0}^{\infty}\frac{y^j}{j!}\frac{B_{j+1+n}(u)}{(j+1+n)}\right\} \qquad n = 1, 2, 3, \ldots$$

Here $B_{j+1+n}(u)$ is a Bernoulli polynomial [Chapter 19].

In addition to 64:12:1, there are many special cases of Lerch's function. We conclude this section with an incomplete listing of such special cases. Following each identity is a reference to a chapter or section in which the special function is discussed:

64:12:7 $$\Phi(x;0;u) = \frac{1}{1-x} \qquad \text{[Chapter 7]}$$

64:12:8 $$\Phi(x;-1;u) = \frac{x}{(1-x)^2} + \frac{u}{1-x} \qquad \text{[Chapter 11]}$$

64:12:9 $$\Phi(0;\nu;u) = \frac{1}{u^\nu} \qquad \text{[Chapter 13]}$$

64:12:10 $$\Phi(x;-2;u) = \frac{(u-1)^2x^2 - (2u^2 - 2u - 1)x + u^2}{(1-x)^3} \qquad \text{[Section 17:13]}$$

64:12:11 $$\Phi(x;n;u) = \frac{1}{u^n}\sum_{j=0}^{\infty}\frac{(u)_j^n}{(1+u)_j^n}x^j \qquad \text{[Section 18:14]}$$

64:12:12 $$\Phi(x;1;1) = \frac{-1}{x}\ln(1-x) \qquad \text{[Chapter 25]}$$

64:12:13 $$\Phi(x;1;u) = \frac{1}{x^u}\ln_u\left(\frac{1}{1-x}\right) = \frac{1}{x^u}B(u;0;x) \qquad \text{[Sections 25:12 and 58:4]}$$

64:12:14 $$\Phi(x;2;1) = \frac{-1}{x}\,\mathrm{diln}(1-x) \qquad \text{[Section 25:13]}$$

64:12:15 $$\Phi(x;\nu;1) = \frac{-1}{x}\,\mathrm{poln}_\nu(1-x) \qquad \text{[Section 25:13]}$$

64:12:16 $$\Phi\left(x;\nu;\frac{1}{2}\right) = \frac{1}{\sqrt{x}}\,\mathrm{poln}_\nu(1-x) - \frac{2^\nu}{\sqrt{x}}\,\mathrm{poln}_\nu(1-\sqrt{x}) \qquad \text{[Section 25:13]}$$

64:12:17 $$\Phi\left(-x;-\frac{1}{2};1\right) = \frac{1}{x}\,\mathrm{rsf}\{\ln(x)\} \qquad \text{[Section 30:10]}$$

64:12:18 $$\Phi\left(x;1;\frac{1}{2}\right) = \frac{2}{\sqrt{x}}\,\mathrm{artanh}(\sqrt{x}) \text{ or } \frac{2}{\sqrt{-x}}\,\arctan(\sqrt{-x}) \qquad \text{[Chapters 31 and 35]}$$

64:12:19 $$\Phi(-1;\nu;u) = \eta(\nu;u) \qquad \text{[Section 64:13]}$$

64:13 COGNATE FUNCTIONS

The *bivariate eta function* is related to the eta function of Chapter 3 in the same way that the Hurwitz function is related to the zeta function. It is defined by

64:13:1 $$\eta(\nu;u) = \sum_{j=0}^{\infty} \frac{(-1)^j}{(j+u)^\nu} \qquad \nu > 0 \qquad u \geq 0$$

The recurrence formula

64:13:2 $$\eta(\nu;u+1) = \frac{1}{u^\nu} - \eta(\nu;u)$$

resembles 64:5:1, apart from sign, but the duplication formula

64:13:3 $$\eta(\nu;2u) = 2^{-\nu}\left[\zeta(\nu;u) - \zeta\left(\nu;\frac{1}{2}+u\right)\right]$$

provides one of many links to the Hurwitz function. Another is

64:13:4 $$\eta(\nu;u) = \frac{\zeta\left(\nu;\dfrac{u}{2}\right)}{2^{\nu-1}} - \zeta(\nu;u)$$

Special cases of the bivariate eta function include the eta function of Chapter 3:

64:13:5 $$\eta(\nu;0) = \begin{cases} -\eta(\nu) & \nu < 0 \\ \infty & \nu > 0 \end{cases}$$

64:13:6 $$\eta(\nu;1) = \eta(\nu)$$

64:13:7 $$\eta(\nu;2) = 1 - \eta(\nu)$$

the beta function of the same chapter:

64:13:8 $$\eta\left(\nu;\frac{1}{2}\right) = 2^\nu\beta(\nu)$$

Bateman's G function [Section 44:13]:

64:13:9 $$\eta(1;u) = \frac{1}{2}\,\mathrm{G}(u)$$

and successive derivatives of the last:

64:13:10
$$\eta(\nu;u) = \frac{-(-1)^n}{2(n-1)!}\,G^{(n-1)}(u) \qquad n = 1, 2, 3, \ldots$$

The very simple algorithm

Input ν >>>>>>>
Input u >>>>>>>

Set $j = 2\,\mathrm{Int}\left(\dfrac{10^{7/\nu} - u + 2}{2}\right)$

Set $f = \dfrac{1}{2}(u+j)^{-\nu}$

(1) Replace j by $j - 2$
Replace f by $f - (u + j + 1)^{-\nu} + (u + j)^{-\nu}$
If $j \neq 0$ go to (1)
Output f

$f \simeq \eta(\nu;u)$ <<<<<

Storage needed:
ν, u, j and f

Input restrictions:
$u > 0$; ν must be large and positive.

Test values:
$\eta(5;8) = 1.979 \times 10^{-5}$
$\eta(8;0.2) = 390624.769$

is based on definition 64:13:1 and uses the properties of alternating series [Section 0:6] to ensure an *absolute* accuracy of 6×10^{-8}. Although the algorithm is valid for all positive ν and u, it is tediously slow unless ν is large. For orders that are not large and positive one may use identity 64:13:3 or 64:13:4 and the algorithm in Section 64:8.

64:14 RELATED TOPICS

The differintegrals [Section 0:10] of periodic functions can be expressed as integrals involving the Hurwitz function, the integration range being a single period. The property that permits this simplification is illustrated by the transformation

64:14:1
$$\int_x^{x+JP} u^\nu \,\mathrm{per}(u)\mathrm{d}u = \sum_{j=0}^{J-1}\int_{x+jP}^{x+P+jP} u^\nu \mathrm{per}(u)\mathrm{d}u = P^\nu \int_x^{x+P} \mathrm{per}(t)\left\{\sum_{j=0}^{J-1}\left(j + \frac{t}{P}\right)^\nu\right\}\mathrm{d}t \qquad t = u - jP$$

in which per(x) is a periodic function [Chapter 36] of period P.

Differintegration with a lower limit of $-\infty$ is especially facilitated by the Hurwitz transformation. Such differintegrals are sometimes known as *Weyl differintegrals*. We report formulas for the two ranges, $0 < \nu < 1$ and $-1 < \nu < 0$, that embrace the important semidifferentiation and semiintegration cases

64:14:2
$$\frac{\mathrm{d}^\nu \,\mathrm{per}(x)}{[\mathrm{d}(x+\infty)]^\nu} = \frac{1}{P^\nu\Gamma(-\nu)}\int_0^1 [\mathrm{per}(x - Pt) - \mathrm{per}(x)]\,\zeta(1+\nu;t)\mathrm{d}t \qquad 0 < \nu < 1$$

64:14:3
$$\frac{\mathrm{d}^\nu \,\mathrm{per}(x)}{[\mathrm{d}(x+\infty)]^\nu} = \frac{1}{P^\nu\Gamma(-\nu)}\int_0^1 \mathrm{per}(x - Pt)\,\zeta(1+\nu;t)\mathrm{d}t \qquad -1 < \nu < 0 \qquad \int_0^1 \mathrm{per}(Pt)\mathrm{d}t = 0$$

Notice that the second result requires that the average value of the periodic function over its period be zero; otherwise, the differintegral is infinite. Weyl differintegrals of periodic functions are themselves periodic and the period is unchanged on differintegration. We conclude this section (and the *Atlas*!) by reporting three examples of the application to simple periodic functions of equations 64:14:2 and 64:14:3 and their extensions. For the sine function [or any sinusoid, see Chapter 32] differintegration with a lower limit of minus infinity corresponds simply to a scaling and phase shift:

64:14:4
$$\frac{\mathrm{d}^\nu \sin(2\pi x/P)}{[\mathrm{d}(x+\infty)]^\nu} = \left(\frac{2\pi}{P}\right)^\nu \sin\left(\frac{2\pi x}{P} + \frac{\nu\pi}{2}\right)$$

The other two examples report the Weyl semiderivative

64:14:5
$$\frac{d^{1/2}(-1)^{\text{Int}(2x/P)}}{[d(x+\infty)]^{1/2}} = \frac{2}{\sqrt{\pi P}}\left[\zeta\left(\frac{1}{2};\ \text{frac}\left(\frac{x}{P}\right)\right) - \zeta\left(\frac{1}{2};\ \text{frac}\left(\frac{x}{P}+\frac{1}{2}\right)\right)\right]$$

and Weyl semiintegral

64:14:6
$$\frac{d^{-1/2}(-1)^{\text{Int}(2x/P)}}{[d(x+\infty)]^{-1/2}} = 4\sqrt{\frac{P}{\pi}}\left[\zeta\left(-\frac{1}{2};\ \text{frac}\left(\frac{x}{P}\right)\right) - \zeta\left(-\frac{1}{2};\ \text{frac}\left(\frac{x}{P}+\frac{1}{2}\right)\right)\right]$$

of the square wave function that is mapped in Figure 36-3 and formulated in Section 36:14. The shapes of these last two periodic functions are displayed in Figure 64-2, together with the square wave itself.

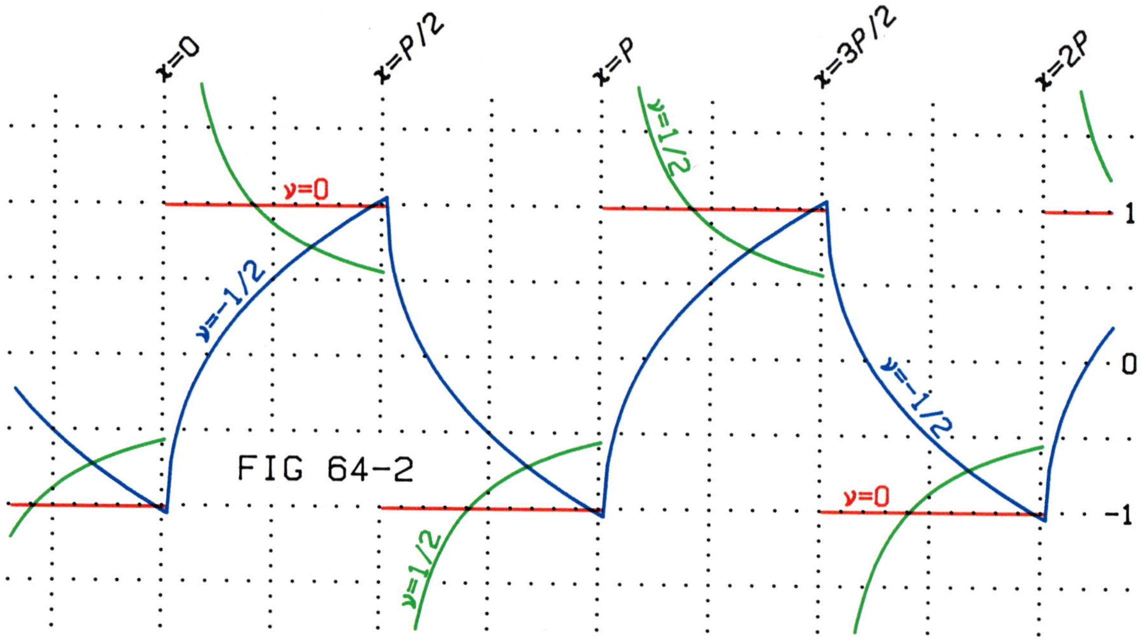

FIG 64-2

APPENDIX A

UTILITY ALGORITHMS

Most of the algorithms in the chapters of this *Atlas* are devoted to generating numerical values of particular functions. However, a few of the algorithms scattered throughout the chapters perform operations that are more general than evaluating specific functions. This appendix provides a glossary of the latter kind of algorithm and includes a few additional instances of such general-purpose algorithms.

A:1 BASE-CONVERSION ALGORITHMS

Section 9:14 contains two algorithms designed to convert numbers from one base to another. The first converts a decimal number to any other base (to binary or hexadecimal, for example). The second algorithm performs the reverse conversion: it converts from any base into decimal.

The two algorithms may be conjoined to permit conversion from any number base to any other.

A:2 FRACTIONATION ALGORITHM

Sometimes one needs to replace a decimal number ν by the quotient n/d of two integers. Even if ν is ostensibly a rational number, rounding errors in its decimal representation will usually prevent the equality $\nu = n/d$ from being exact. Instead, there is a fractional error

A:2:1
$$\varepsilon = \left|\frac{\nu - (n/d)}{\nu}\right|$$

associated with the replacement of ν by n/d.

The algorithm below generates a sequence of ever-improving approximations n_1/d_1, n_2/d_2, n_3/d_3, ... to an input number ν. Each pair of output integers n, d is accompanied by an output of ε. Although it is designed to handle positive numbers only, these may be larger or smaller than unity. Output continues until, to the precision of the computing device, $(n/d) = \nu$.

Storage neeed: ν, N, D, n, d, r, t, ε and m

Input restriction: $\nu > 0$

```
                    Set d = D = 1
Input ν >>>>>>>
                    Set n = Int(ν)
                    Set N = n + 1
                    Go to (3)
                (1) If r > 1 go to (2)
                    Replace r by 1/r
                (2) Replace N by N + n Int(r)
                    Replace D by D + d Int(r)
                    Replace n by n + N
                    Replace d by d + D
                (3) Set r = 0
                    If νd = n go to (4)
                    Set r = (N − νD)/(νd − n)
                    If r > 1 go to (4)
                    Set t = N
                    Set N = n
                    Set n = t
                    Set t = D
                    Set D = d
                    Set d = t
                (4) Output n
n = n_j <<<<<<<
                    Output d
d = d_j <<<<<<<
                    Set ε = |1 − (n/νd)|
                    If ε = 0 go to (6)
                    Set m = 1
                (5) Replace m by 10m
                    If mε < 1 go to (5)
                    Replace ε by (1/m) Int(1/2 + mε)
                (6) Output ε
ε = ε_j <<<<<<<
                    If r ≠ 0 go to (1)
```

Test values:
input: $\nu = \pi$
outputs:

$\frac{n_1}{d_1} = \frac{3}{1}$ $\quad \varepsilon_1 = 5 \times 10^{-2}$

$\frac{n_2}{d_2} = \frac{22}{7}$ $\quad \varepsilon_2 = 4 \times 10^{-4}$

$\frac{n_3}{d_3} = \frac{355}{113}$ $\quad \varepsilon_3 = 8 \times 10^{-8}$

$\frac{n_4}{d_4} = \frac{104348}{33215}$ $\quad \varepsilon_4 = 1 \times 10^{-10}$

The commands shown in green perform the interchanges $N \rightleftarrows n$ and $D \rightleftarrows d$; with many devices there are simpler ways of achieving this.

The algorithm first finds the two integers between which ν lies so that this number is known to be bracketed by the fractions

A:2:2 $$\frac{n}{d} \leq \nu \leq \frac{N}{D}$$

with $d = D = 1$. The ratio

A:2:3 $$r = \frac{N - \nu D}{\nu d - n}$$

is then calculated. If this exceeds unity, n/d is accepted as a valid approximant to ν, and the numerator n and denominator d of the (proper or improper) fraction n/d are output, along with a rounded value of the fractional error ε. Moreover, if r lies between the integers m and $m + 1$, then it follows from A:2:3 that

A:2:4 $$\frac{(m+1)\,n + N}{(m+1)\,d + D} < \nu \leq \frac{mn + N}{md + D} \qquad m \leq r \leq m + 1$$

On the other hand, if $r \leq 1$, then N/D is accepted as the approximant and is output, followed by the corresponding fractional error in the approximation. From the definition of r, one can show that

A:2:5 $$\frac{n+mN}{d+mD} \leq \nu < \frac{n+(m+1)N}{d+(m+1)D} \qquad m \leq \frac{1}{r} < m+1$$

in the $r \leq 1$ case. Thus, either A:2:4 or A:2:5 provides a narrower bracketing of ν than did the original A:2:2. The algorithm continues this process of bracket narrowing until the computing device interprets r as either infinity or zero.

A:3 ALGORITHMS THAT GENERATE NUMBERS

An algorithm in Section 40:14 generates a succession of J numbers that are pseudorandomly distributed according to the "normal" distribution [see Section 27:14]. The mean and standard deviation of the distribution are selected by the user.

A subset of this same algorithm, namely

```
Input x0 >>>>>>>>
        Set N = 199017
Input x1 >>>>>>>>
        Set h = x1 - x0
Input J >>>>>>>>
        Set j = n = J
    (1) Replace n by 24298n + 99991
        Replace n by n + 1 - N Int(n/N)
        Replace j by j - 1
        Set x = x0 + (nh/N)
        Output x
x <<<<<<<<<<
        If j ≠ 0 go to (1)
```

Storage needed: x_0, N, h, j and n

Input restrictions: $x_0 < x_1$, $1 \leq J \leq 199017$

Test values: $x_0 = -1$, $x_1 = 1$, $J = 5$; outputs:
$x = -0.774240392$
$x = -0.488174377$
$x = 0.343844998$
$x = 0.750624319$
$x = 0.674550415$

generates a set of *pseudorandom decimal numbers* uniformly distributed on the interval $x_0 < x \leq x_1$. The arbitrary "seed" is taken as J, the number of sought outputs.

A:4 DATA–HANDLING ALGORITHMS

The *Atlas* contains three algorithms that are designed to process large data sets.

In Section 7:14 we present an algorithm for the *linear regression analysis* of equally spaced and equally weighted data. This procedure gives the slope b and intercept c of the *best straight line* fit to data points (x_1,f_1), (x_2,f_2), (x_3,f_3), $\ldots$, (x_n,f_n) where $x_n - x_{n-1} = x_{n-1} - x_{n-2} = \cdots = x_2 - x_1$ and $2 \leq n < \infty$. Optionally, the algorithm also generates the correlation coefficient of the fit and the standard errors of the slope and intercept. Although algorithms are not included, the same section discusses the linear regression analysis of data that are not equally spaced and not equally weighted.

A straight line may be considered a first degree polynomial function. In Section 17:14 we show how to fit $p_K(x)$, a polynomial of any degree K, to a set of $J+1$ data pairs, (x_0,f_0), (x_1,f_1), $\ldots$, (x_j,f_j), $\ldots$, (x_J,f_J), provided that $J \geq K$. This *polynomial fitting algorithm* is also restricted to data that are equally spaced and equally weighted.

Several popular methods of processing data involve Fourier transformation, Fourier inversion, or both. Although this *Atlas* does not discuss such data-processing methods, it does include, in Section 32:14, an algorithm for *fast fourier transformation/inversion*. Once again, the algorithm assumes equally spaced and equally weighted data; moreover, the number of data must be a power of two.

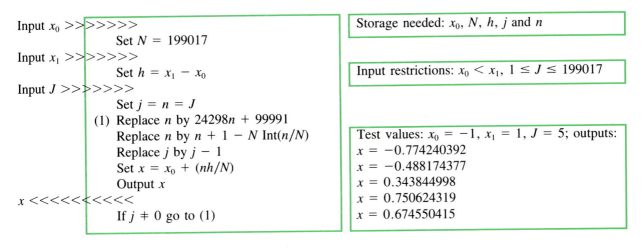

A:5 ALGORITHMS TO LOCATE FEATURES OF A FUNCTION

In Section 17:7 an algorithm is presented for finding an accurate zero of a polynomial function from an initial crude estimate. That algorithm uses *Newton's method* which utilizes the fact that

A:5:1 $$r_{j+1} = r_j - f(r_j) \Big/ \frac{\mathrm{df}}{\mathrm{d}x}(r_j)$$

is usually a better approximation to the local zero of f(x) than is r_j. Newton's method may be used to find the zeros [but not double zeros—see Section 0:7] of any continuous function (or discontinuous function away from its discontinuities) provided that the initial estimate r_0 is close enough to the sought zero.

However, Newton's method is sometimes inconvenient in that it requires numerical values of two functions: f(x) and its derivative. We present below an algorithm that needs values of f(x) only. It is based on approximating the derivative $\mathrm{df}(r_j)/\mathrm{d}x$ in equation A:5:1 by the difference quotient $[\mathrm{f}(r_j) - \mathrm{f}(r_{j-1})]/[r_j - r_{j-1}]$ and employs the resulting formula

A:5:2 $$r_{j+1} = \frac{r_{j-1}\mathrm{f}(r_j) - r_j\mathrm{f}(r_{j-1})}{\mathrm{f}(r_j) - \mathrm{f}(r_{j-1})}$$

The algorithm needs two (different) estimates r_0 and r_1 of the zero, but these can often be wild guesses and r_1 need not be the closer of the two. In a kindred method [known as *regula falsi*; see Bronshtein and Semendyayev, page 170], the values of r_0 and r_1 must lie on either side of the zero, but this is not necessary here. Double zeros are accessible by this method, though convergence is slower than for single zeros.

The algorithm repeatedly outputs the improving values $r_0, r_1, r_2, r_3 \ldots$ until two successive values of f(r) are identical or until the user intervenes.

Set $F = 10^{99}$
Input r_0 >>>>>>>
Set $x = r_0$
Go to (4)
(1) If $F \neq 10^{99}$ go to (2)
Input r_1 >>>>>>>
Set $r = r_1$
Go to (3)
(2) Replace r by $(Rf - rF)/(f - F)$
(3) Set $R = x$
Set $x = r$
Set $F = f$
Output x
$x = r_j$ <<<<<<<<
(4) Routine to calculate f = f(x) from x
If $F \neq f$ go to (1)

Storage needed: F, x, r, R and f, in addition to any storage needed by the f routine.

Test values:
using $\mathrm{f}(x) = \exp(x) - x^2$, $r_0 = -0.6$, $r_1 = -0.7$;
outputs:
$r_2 = -0.703613805$
$r_3 = -0.703467221$
$r_4 = -0.703467422$
$r_5 = -0.703467423$

The segment shown in green must be provided by the user; it could, for example, be one of the algorithms from Section 8 of the chapter dealing with f(x).

Our algorithm has the advantage of simplicity, but it will seldom be optimal and it may occasionally fail. There exist very many alternative procedures for locating zeros (or *root finding* as the process is often called); Ruckdeschel [Volume II, Chapter 6] discusses some of these. The same reference [Chapter 7] gives methods for finding the *complex zeros* of functions.

The algorithm is easily modified to provide information about features other than the values of the zeros. Thus, to find the value of x at which f(x) equals, say, π, one need only replace the green segment by a routine to calculate $\mathrm{f}(x) - \pi$. Again, to find the extrema (local maxima or minima) of the function, one simply changes the green segment to "Routine to calculate $f = \mathrm{df}(x)/\mathrm{d}x$ from x."

A:6 ALGORITHMS TO MANIPULATE SERIES

A very large class of functions can be represented as the *power series* or *Maclaurin series* [Section 11:14]

A:6:1 $$f(x) = a_0 + a_1 x + a_2 x^2 + \cdots = \sum_{j=0}^{\infty} a_j x^j$$

Such a series may always be reformulated as a concatenation [see equation 0:6:5] or as a continued fraction [via equation 0:6:12]. These reformulations frequently permit the numerical evaluation of f(x), for specific arguments of interest, to be carried out more expeditiously, or more accurately, than by the original series A:6:1. As well, there are a couple of more profound ways in which power series may be manipulated as an aid to the numerical evaluation of f(x).

If interest is confined to a small range $x_0 \le x \le x_1$ of arguments and if a restricted accuracy, $|f(x) - \hat{f}(x)| \le \varepsilon$, is acceptable, then the *economized polynomial*

A:6:2 $$\hat{f}(x) = e_0 + e_1 x + e_2 x^2 + \cdots + e_m x^m = \sum_{j=0}^{m} e_j x^j$$

may serve as a convenient means of calculating numerous approximate values of the f(x) function. Such economized polynomials are the basis of many computer *library subroutines* for functional evaluation [see Ruckdeschel]. Because Chebyshev polynomials are utilized in the procedure, this *Atlas*'s algorithm for calculating values of the degree m and the coefficients $e_0, e_1, e_2, \ldots, e_m$ of the economized polynomial will be found in Section 22:14. Note that coefficients generated by the Chebyshev procedure are "best" in the sense of minimizing the worst error between f(x) and $\hat{f}(x)$ rather than in the "least squares" sense.

If the series A:6:1 is asymptotic [Section 0:6], accurate values of f(x) can be calculated by its use only for sufficiently small values of the argument x. However, by using the procedure detailed in Section 17:13, it is possible to convert the sequence (a_0); $(a_0 + a_1x)$; $(a_0 + a_1x + a_2x^2)$; ...; $(a_0 + a_1x + \cdots + a_Jx^J)$; ... of truncated power series into a sequence $R_0^0(x)$; $R_1^1(x)$; $R_2^2(x)$; ...; $R_J^J(x)$; ... of *diagonal Padé approximants*. The latter sequence often has an advantage over the former in converging to f(x) when A:6:1 is asymptotic or even divergent. An algorithm in Section 17:13 exploits this advantage in evaluating numerical values of functions from their asymptotic series via the Padé operation.

A:7 ALGORITHMS TO PERFORM OPERATIONS OF THE CALCULUS

The derivative of any function f(x) discussed in this *Atlas* can always, and its integral can usually, be expressed in terms of some other function(s) of the *Atlas*. Such expressions will be found in Section 10 of the relevant chapter. In cases of difficulty, the properties of hypergeometric functions [Section 18:14] or Laplace transformation [Section 26:14] can sometimes permit the operations of the calculus to be applied to otherwise intractable functions. Nevertheless, there will remain certain functions whose integrals are evaluable only by numerical methods. Such an evaluation is known as a *numerical quadrature* and we here present a simple algorithm for the numerical quadrature of a function f:

A:7:1 $$\int_{x_0}^{x_1} f(t)dt \qquad x_0 \neq \pm\infty \neq x_1$$

between two finite limits. We treat only the case in which f(x) is known (i.e., is calculable or infinite) at all points in the $x_0 \le x \le x_1$ range. Not considered is the technically important situation in which the integral is sought of a function whose values are known at a finite set of (usually equally spaced) arguments only.

The algorithm utilizes the trapezoidal approximation

A:7:2 $$\int_{x_0}^{x_1} f(t)dt \simeq I_n = \frac{x_1 - x_0}{n} \sum_j f\left(x_0 + \frac{j}{2n}(x_1 - x_0)\right) \qquad j = 1, 3, 5, \ldots, 2n-1$$

with n given the successive values 9, 27, 81, ... The advantage of this threefold increase in the number of summands in the approximation I_n to the integral is that all the summands that are needed as components of I_n will be reused in forming the I_{3n}, I_{9n}, I_{27n}, ... sums. In the algorithm that follows, the green portion represents the commands necessary to calculate values of the function $f(x)$ from a value of x:

Input x_0 >>>>>>>
Set $r = 0$
Set $n = 1$
Input x_1 >>>>>>>
(1) Set $d = 3(x_1 - x_0)/n$
Set $s = 0$
Set $a = 1$
(2) Set $x = x_0 + (ad/6)$
(3) Calculate $f = f(x)$
Replace s by $s + f$
Replace x by $x + d$
If $x < x_1$ go to (3)
If $n = 1$ go to (6)
If $a = 5$ go to (4)
Set $a = 5$
Go to (2)
(4) If $n = 3$ go to (5)
Set $a = 3d(81s + 51l + 52r)/640$
Output a
$a \simeq \int_{x_0}^{x_1} f(t)dt$ <<<<
(5) Replace r by $r + l$
(6) Set $l = s$
Replace n by $3n$
Go to (1)

Storage needed: x_0, x_1, r, n, d, s, a and l, in addition to any used by the $f(x)$ routine

Successive outputs become increasingly infrequent until the user intervenes.

Input restriction: $x_1 > x_0$

Test values:
with $f(x) = 1/x$; inputs: $x_0 = 1$, $x_1 = 2$;
successive outputs:
0.693144600,
0.693147174,
0.693147181, ...

The algorithm does not output I_n values as such, but instead makes use of two levels of Richardson extrapolation [see Ruckdeschel, Volume II, page 317] and returns values of $(729I_n - 90I_{n/3} + I_{n/9})/640$. For suitable functions f, this double extrapolation compensates so effectively for the finiteness of n that an accurate approximation to the integral is soon achieved. The "test values" illustrate such a case: the fourth output (corresponding to $n = 243$) already has converged to a value that reproduces the correct value of the integral

A:7:3
$$\int_1^2 \frac{1}{t}\,dt = \ln(2) = 0.6931471806$$

to nine significant digits.

Because the algorithm does not utilize $f(x_0)$ or $f(x_1)$, it is able to handle integrals that are indeterminate or infinite at either end of the integration range, provided, of course, that the integral itself is finite. Nevertheless, it is better to dismember integrands in the latter class, as in the example

A:7:4
$$\int_0^{\pi^2} \frac{\sin(\sqrt{t})}{t}\,dt = \int_0^{\pi^2} \frac{dt}{\sqrt{t}} - \int_0^{\pi^2} \left(\frac{dt}{\sqrt{t}} - \frac{\sin(\sqrt{t})dt}{t}\right) = 2\pi - \int_0^{\pi^2} \frac{\sqrt{t} - \sin(\sqrt{t})}{t}\,dt$$

into analytically integrable functions and ones that are everywhere finite over the $x_0 \le x \le x_1$ interval. A similar strategy often can aid the numerical quadrature of functions that encounter an infinity within the $x_0 < x < x_1$

range, as exemplified by

A:7:5 $$\int_0^3 \frac{dt}{t^{3/2}-1} = \frac{2}{3}\int_0^3 \frac{dt}{t-1} - \int_0^3 \left(\frac{\frac{2}{3}\,dt}{t-1} - \frac{dt}{t^{3/2}-1}\right) = \frac{2}{3}\ln(2) - \int_0^3 \left(\frac{2}{3t-3} - \frac{1}{t^{3/2}-1}\right)dt$$

Our algorithm, as it stands, cannot be used if either (or both) of the integration limits is infinite. A suitable redefinition of the integration variable, however, will effect the conversion

A:7:6 $$\int_{x_0}^{x_1} f(t)dt \to \int_0^1 g(u)du$$

and often render the function amenable to the numerical quadrature algorithm. Table A.7.1 gives examples of suitable transformations; there are many alternatives.

The preceding algorithm is designed to evaluate *definite* integrals. If one wanted to use it to study the behavior of the *indefinite* integral

A:7:7 $$\int_{x_0}^{x} f(t)dt$$

of f as a function of x, one would need to access the algorithm repeatedly, replacing the upper limit x_1 by different x values until the range of interest had been scanned. Such a procedure would usually be impossibly tedious. The algorithm below, however, is designed for just such a survey: it generates rather crude values of the integral A:7:7 at x values equal to $x_0 + (x_1 - x_0)/n$, $x_0 + 2(x_1 - x_0)/n$, $x_0 + 3(x_1 - x_0)/n$, ..., $x_0 + (n-1)(x_1 - x_0)/n$, x_1, where x_1 and n are values provided by the user.

Actually the algorithm is much more versatile than we have just suggested. Indefinite integration is just one example (the $\nu = -1$ instance) of the generalized differintegration operator of the calculus [see Section 0:10], and our algorithm generates approximate values of

A:7:8 $$\frac{d^\nu f(x)}{[d(x-x_0)]^\nu} \quad \text{at} \quad x = x_0 + \frac{j}{n}(x_1 - x_0) \quad \text{for} \quad j = 1, 2, 3, \ldots, n$$

for any value of ν [including $\nu = -1$ when A:7:8 reduces to A:7:7]. The algorithm is based on the *improved Grünwald definition* [Oldham and Spanier, pages 56–57]

A:7:9 $$\frac{d^\nu f(x)}{[d(x-x_0)]^\nu} = \lim_{J\to\infty}\left\{\left(\frac{J}{x-x_0}\right)^\nu \sum_{j=0}^{J-1} \frac{(-\nu)_j}{j!} f\left(x - \frac{(2j-\nu)(x-x_0)}{2J}\right)\right\}$$

of a differintegral, with the condition $J \to \infty$ relaxed and J equated to n.

Table A.7.1

x_0	x_1	$g(u)$
>0	∞	$\frac{x_0}{u^2} f\left(\frac{x_0}{u}\right)$
>-1	∞	$\frac{1+x_0}{(1-u)^2} f\left(\frac{u+x_0}{1-u}\right)$
$-\infty$	<0	$\frac{-x_1}{u^2} f\left(\frac{x_1}{u}\right)$
$-\infty$	<1	$\frac{-1-x_1}{u^2} f\left(\frac{1+x_1-u}{u}\right)$
$-\infty$	∞	$\left[\frac{1}{u^2} + \frac{1}{(1-u)^2}\right] f\left(\frac{2u-1}{u-u^2}\right)$

Input x_0 >>>>>>>>
Set $x = x_0$
Input x_1 >>>>>>>>
Input n >>>>>>>>
Set $D = (x_1 - x_0)/n$
Input ν >>>>>>>>
(1) Set $s = 0$
Replace x by $x + D$
Set $d = (x - x_0)/n$
Set $j = n$
Set $a = n - \nu$
Set $y = x_0 + (\nu d/2)$
(2) Replace y by $y + d$
Replace a by $a - 1$
Calculate $f = \mathrm{f}(y)$
Replace s by $f + (as/j)$
Replace j by $j - 1$
If $j \neq 0$ go to (2)
Output x
x <<<<<<<<<<
Set $F = s/d^\nu$
Output F
$F \simeq \dfrac{\mathrm{d}^\nu \mathrm{f}(x)}{[\mathrm{d}(x - x_0)]^\nu}$ <<
If $x < x_1$ go to (1)

Storage needed:
x_0, x_1, x, n, D, ν, s, d, j, and a, as well as that required by the f routine

Input restrictions: $x_1 > x_0$; n integer

Use radian mode or replace $\sin(\sqrt{x})$ by $\sin(\pi\sqrt{x}/180)$.

Test values:
$\mathrm{f}(x) = \sin(\sqrt{x})$, $x_0 = 0$, $x_1 = 10$, $n = 10$, $\nu = \frac{1}{2}$; outputs:
$x = 1$, $F = 0.681974314$
$x = 2$, $F = 0.499347769$
$x = 3$, $F = 0.340096522$
$x = 4$, $F = 0.202248636$
$x = 5$, $F = 0.0839564103$
$x = 6$, $F = -0.0165098702$
$x = 7$, $F = -0.1007679243$
$x = 8$, $F = -0.170329209$
$x = 9$, $F = -0.226604411$
$x = 10$, $F = -0.270908707$

It should be emphasized that there exist many better algorithms for *specific* orders of differintegration but that the one given here has the advantage of versatility. Comparison of the listed test values with those of $(\sqrt{\pi}/2)$ $\mathrm{J}_0(\sqrt{x})$, the true semiderivative of $\sin(\sqrt{x})$ [see 52:3:13], reveals errors of up to 2%, which is typical for this algorithm using $n = 10$. Accuracy increases with increasing n, but with a concomitant speed penalty.

APPENDIX B

SOME USEFUL DATA

Since 1960 *Le Système International d'Unités* has been the official system of units throughout most of the world. The *SI system*, as it is usually abbreviated, is used universally in science and increasingly in engineering and medicine. This appendix contains information about the SI system and lists important physical constants and conversion factors.

B:1 THE PRIMARY UNITS

The SI system recognizes seven *base units* and two *supplementary units*. These nine primary units are listed in Table B.1.1. All other quantities are measured in *derived units* that are combinations of the primary units.

B:2 DERIVED UNITS

Some quantities are expressed in units derived directly from primary SI units. See Table B.2.1 for a few examples. Eighteen derived units are given special names as listed in Table B.2.2. Notwithstanding their special names, these units are also derived directly from the nine primary units.

Table B.1.1

Quantity	Name of unit	Symbol of unit
Length	metre	m
Mass	kilogram	kg
Time	second	s
Electric current	ampere	A
Thermodynamic temperature	kelvin	K
Amount of substance	mole	mol
Luminous intensity	candela	cd
Plane angle	radian	rad
Solid angle	steradian	sr

Table B.2.1

Quantity	Name of unit	Symbol of unit
Area	square metre	m^2
Volume	cubic metre	m^3
Velocity	metre per second	$m\ s^{-1}$
Density	kilogram per cubic metre	$kg\ m^{-3}$
Current density	ampere per square metre	$A\ m^{-2}$
Temperature gradient	kelvin per metre	$K\ m^{-1}$
Concentration	mole per cubic metre	$mol\ m^{-3}$
Luminance	candela per square metre	$cd\ m^{-2}$
Rotation rate	radian per second	$rad\ s^{-1}$

Table B.2.2

Quantity	Name of unit	Symbol of unit and equivalent in primary units
Frequency	hertz	$Hz = s^{-1}$
Force	newton	$N = kg\ m\ s^{-2}$
Pressure, stress	pascal	$Pa = N\ m^{-2} = kg\ m^{-1}\ s^{-2}$
Energy, work, heat	joule	$J = N\ m = kg\ m^2\ s^{-2}$
Power, radiant flux	watt	$W = J\ s^{-1} = kg\ m^2\ s^{-3}$
Electric charge	coulomb	$C = J\ V^{-1} = A\ s$
Electric potential	volt	$V = W\ A^{-1} = kg\ m^2\ s^{-3}\ A^{-1}$
Electric resistance	ohm	$\Omega = V\ A^{-1} = kg\ m^2\ s^{-3}\ A^{-2}$
Conductance	siemens	$S = \Omega^{-1} = s^3\ A^2\ kg^{-1}\ m^{-2}$
Capacitance	farad	$F = C\ V^{-1} = s^4\ A^2\ kg^{-1}\ m^{-2}$
Magnetic flux	weber	$Wb = V\ s = kg\ m^2\ s^{-2}\ A^{-1}$
Magnetic flux density	tesla	$T = Wb\ m^{-2} = kg\ s^{-2}\ A^{-1}$
Inductance	henry	$H = Wb\ A^{-1} = kg\ m^2\ s^{-2}\ A^{-2}$
Luminous flux	lumen	$lm = cd\ sr$
Illuminance	lux	$lx = lm\ m^{-2} = cd\ sr\ m^{-2}$
Radioactivity	becquerel	$Bq = J\ kg^{-1} = m^2\ s^{-2}$
Absorbed dose	gray	$Gy = J\ kg^{-1} = m^2\ s^{-2}$
Dose equivalent	sievert	$Sv = J\ kg^{-1} = m^2\ s^{-2}$

Very many other SI units arise by combining specially named units with the primary units or with each other. A few examples are listed in Table B.2.3.

B:3 PREFIXES

There is a set of 16 prefixes to the primary and special-name SI units that are used to create multiples and sub-multiples. These are listed in Table B.3.1.

With one exception, these prefixes are attached to the primary or special-name SI units, as in the examples

B:3:1 $$10^3\ m\ s^{-1} = \text{kilometre per second} = km\ s^{-1}$$

B:3:2 $$10^{-12}\ F\ m^{-2} = \text{picofarad per square metre} = pF\ m^{-2}$$

The exception is with mass units, for which the prefix is attached to the gram, rather than the kilogram, unit; for example:

B:3:3 $$10^{-9}\ kg = \text{microgram} = \mu g\ [\text{not nkg}]$$

Note that μm^{-1} means $(\mu m)^{-1}$, not $\mu(m^{-1})$, and equals $10^6\ m^{-1}$, not $10^{-6}\ m^{-1}$. Similarly, cm^3 means $(10^{-2}\ m)^3 = 10^{-6}\ m^3$, not $10^{-2}\ m^3$.

Table B.2.3

Quantity	Name of unit	Symbol of unit and equivalent
Energy density	joule per cubic metre	$J\ m^{-3} = kg\ m^{-1}\ s^{-2}$
Heat capacity, entropy	joule per kelvin	$J\ K^{-1} = kg\ m^2\ s^{-2}\ K^{-1}$
Thermal conductivity	watt per metre kelvin	$W\ m^{-1}\ K^{-1} = kg\ m\ s^{-3}\ K^{-1}$
Dynamic viscosity	pascal second	$Pa\ s = kg\ m^{-1}\ s^{-1}$
Surface tension	newton per metre	$N\ m^{-1} = kg\ s^{-2}$
Permittivity	farad per metre	$F\ m^{-1} = s^4\ A^2\ kg^{-1}\ m^{-3}$
Specific capacitance	farad per square metre	$F\ m^{-2} = s^4\ A^2\ kg^{-1}\ m^{-4}$

Table B.3.1

Prefix	Factor	Symbol	Prefix	Factor	Symbol
deka	10	da	deci	0.1	d
hecto	100	h	centi	0.01	c
kilo	10^3	k	milli	10^{-3}	m
mega	10^6	M	micro	10^{-6}	μ
giga	10^9	G	nano	10^{-9}	n
tera	10^{12}	T	pico	10^{-12}	p
peta	10^{15}	P	femto	10^{-15}	f
exa	10^{18}	E	atto	10^{-18}	a

B:4 NON-SI UNITS

There are a few units, not part of the SI system, that are officially condoned. These include

B:4:1 $$\text{litre (L)} = 10^{-3}\ \text{m}^3 = \text{dm}^3$$

B:4:2 $$\text{tonne (t)} = 10^3\ \text{kg} = \text{Mg}$$

B:4:3 $$\text{day (d)} = 8.64 \times 10^4\ \text{s}$$

B:4:4 $$\text{hectare (ha)} = 10^4\ \text{m}^2 = \text{hm}^2$$

B:4:5 $$\text{degree (°)} = \frac{\pi}{180}\ \text{rad}$$

B:4:6 $$\text{atomic mass unit (u)} = 1.660\mathit{566} \times 10^{-27}\ \text{kg}$$

B:4:7 $$\text{electronvolt (eV)} = 1.602\mathit{189} \times 10^{-19}\ \text{J}$$

and the degree Celsius (°C), for which see B:7:11 below.

B:5 UNIVERSAL CONSTANTS

The choice of a fundamental set of universal constants is, to a certain extent, arbitrary. One selection, in SI units, is

B:5:1 $$\text{gravitational constant} = G = 6.67\mathit{20} \times 10^{-11}\ \text{N m}^2\ \text{kg}^{-2}$$

B:5:2 $$\text{speed of light} = c = 2.9979258 \times 10^8\ \text{m s}^{-1}$$

B:5:3 $$\text{Planck constant} = h = 6.626\mathit{18} \times 10^{-34}\ \text{J Hz}^{-1}$$

B:5:4 $$\text{Boltzmann constant} = k = 1.380\mathit{66} \times 10^{-23}\ \text{J K}^{-1}$$

B:5:5 $$\text{Avogadro constant} = L = 10^{-3}\ \text{kg mol}^{-1}\ \text{u}^{-1} = 6.022\mathit{45} \times 10^{23}\ \text{mol}^{-1}$$

B:5:6 $$\text{elementary charge} = q_e = -1.602\mathit{189} \times 10^{-19}\ \text{C}$$

B:5:7 $$\text{electron rest mass} = m_e = 5.48580\mathit{26} \times 10^{-4}\ u = 9.109\mathit{534} \times 10^{-31}\ \text{kg}$$

B:5:8 $$\text{proton rest mass} = m_p = 1.0072764\mathit{70}\ \text{u} = 1.672\mathit{649} \times 10^{-27}\ \text{kg}$$

Uncertain digits are italicized. Other constants, related to those above, or arbitrarily defined, include

B:5:9 $$\text{permeability of free space} = \mu_0 = 4\pi \times 10^{-7}\ \text{H m}^{-1}$$

B:5:10 $$\text{permittivity of free space} = \varepsilon_0 = 1/\mu_0 c^2 = 8.85418782 \times 10^{-12}\ \text{F m}^{-1}$$

B:5:11 fine structure constant $= \alpha = \mu_0 c q_e^2/2h = 7.29735\mathit{1} \times 10^{-3}$

B:5:12 Faraday constant $= F = -Lq_e = 9.6484\mathit{56} \times 10^4$ C mol^{-1}

B:5:13 gas constant $= R = kL = 8.314\mathit{41}$ J K^{-1} mol^{-1}

B:5:14 Stefan-Boltzmann constant $= \sigma = 2\pi^5k^4/15h^3c^2 = 5.670\mathit{3} \times 10^{-8}$ W m^{-2} K^{-4}

B:5:15 Rydberg constant $R_\infty = \mu_0^2 c^3 q_e^4 m_e/8h^3 = 1.0973731\mathit{8} \times 10^7$ m^{-1}

B:6 TERRESTRIAL CONSTANTS

B:6:1 earth's mean radius $= 6.370949 \times 10^6$ m

B:6:2 earth's mass $= 5.9732 \times 10^{24}$ kg

B:6:3 earth's angular velocity $= 7.292116 \times 10^{-5}$ rad s^{-1}

B:6:4 siderial year $= 3.1558 \times 10^7$ s

B:6:5 gravitational acceleration (45° latitude) $= g = 9.8062$ m s^{-2}

B:6:6 standard atmospheric pressure $= 1.01325 \times 10^5$ Pa

B:6:7 standard laboratory temperature = 298.15 K = 25.00° C

B:7 CONVERSION FACTORS

To non-SI units other than those in Section B:4:

B:7:1 metre = 39.3701 inch = 3.28084 foot = 6.21371×10^{-4} mile

B:7:2 10^4m^2 = hectare = 0.247105 acre = 3.86102×10^{-3} square mile

B:7:3 10^{-3}m^3 = litre = 2.11338 U.S. pint = 0.264172 U.S. gallon = 0.219969 Imp. gallon

B:7:4 kilogram = 2.20462 pound = 0.0685218 slug = 1.10231×10^{-3} short ton = 9.84207×10^{-4} ton

B:7:5 10^{-3} kg m^{-3} = kg L^{-1} = gram per cubic centimetre = 62.4280 pound per cubic foot

B:7:6 metre per second = 2.23694 mile per hour

B:7:7 newton = 10^5 dyne = 7.23301 poundal = 0.224809 pound weight

B:7:8 joule = 10^7 erg = 23.7304 foot poundal = 0.239006 calorie = 9.86923×10^{-3} litre atmosphere = 9.48451×10^{-4} B.t.u. = 2.77778×10^{-7} kilowatt hour

B:7:9 watt = 3.41443 B.t.u. per hour = 1.34102×10^{-3} horsepower

B:7:10 10^3 Pa = kilopascal = 10^{-2} bar = 7.50062 torr = 0.750062 cm Hg = 0.334562 foot of water = 0.145038 pound per square inch = 9.86923×10^{-3} atmosphere

If T_K, T_C and T_F are numbers expressing temperatures on the kelvin (absolute), Celsius and Fahrenheit scales

B:7:11 $$(T_K - 273.15)\ \text{K} = (T_C)\ °\text{C} = \frac{5}{9}(T_F - 32)\ °\text{F}$$

B:8 THE GREEK ALPHABET

Α, α alpha	Ι, ι iota	Ρ, ρ rho
Β, β beta	Κ, κ kappa	Σ, σ sigma
Γ, γ gamma	Λ, λ lambda	Τ, τ tau
Δ, δ delta	Μ, μ mu	Υ, υ upsilon
Ε, ε epsilon	Ν, ν nu	Φ, ϕ phi
Ζ, ζ zeta	Ξ, ξ xi	Χ, χ chi
Η, η eta	Ο, ο omicron	Ψ, ψ psi
Θ, θ theta	Π, π pi	Ω, ω omega

REFERENCES AND BIBLIOGRAPHY

Abramowitz, M., and I. A. Stegun (editors), *Handbook of Mathematical Functions*, National Bureau of Standards (Applied Mathematics Series #55), Washington, D.C. (1964 and subsequent revisions).

Bartsch, H. S., *Handbook of Mathematical Formulas*, Academic Press, New York (1973).

Bell, W. W., *Special Functions for Scientists and Engineers*, Van Nostrand, London (1967).

Beyer, W. H., *Handbook of Mathematical Sciences*, CRC Press, Boca Raton, Fla. (1964 and subsequent editions).

Beyer, W. H., *Handbook of Tables for Probability and Statistics*, CRC Press, Boca Raton, Fla. (1966 and subsequent editions).

Bronshtein, I. N., and K. A. Semendyayev, *A Guidebook to Mathematics*, Harri Deutsch, Frankfurt (1971 and subsequent editions).

Carlson, B. C., *Special Functions of Applied Mathematics*, Academic Press, New York (1977).

Carslaw, H. S., and J. C. Jaeger, *Conduction of Heat in Solids*, Oxford University Press, London and New York (1947).

Churchill, R. V., *Operational Mathematics*, third edition, McGraw-Hill, New York (1972).

Cooley, J. W., and J. W. Tukey, *Mathematics of Computation*, *19*, 297 (1965).

Erdélyi, A., W. Magnus, F. Oberhettinger and F. G. Tricomi (editors), *Higher Transcendental Functions* (*Bateman Manuscript*), three volumes, McGraw-Hill, New York (1955).

Erdélyi, A., W. Magnus, F. Oberhettinger and F. G. Tricomi (editors), *Tables of Integral Transforms* (*Bateman Manuscript*), two volumes, McGraw-Hill, New York (1955).

Fletcher, A., J. C. P. Miller and L. Rosenhead, *An Index of Mathematical Tables*, McGraw-Hill, New York (1946).

Gradshteyn, I. S., and I. M. Ryzhik, *Table of Integrals Series and Products*, second English edition, Academic Press, New York (1980).

Hamming, R. W., *Numerical Methods for Scientists and Engineers*, second edition, McGraw-Hill, New York (1973).

Hart, J. F., E. W. Cheney, C. L. Lawson, H. J. Maehly, C. K. Mesztenyi, J. R. Rice, H. G. Thacher, Jr. and C. Witzall, *Computer Approximations*, Wiley, New York (1968).

Hastings, C., Jr., *Approximations for Digital Computers*, Princeton University Press, Princeton, N.J. (1955).

Knuth, D. E., *The Art of Computer Programming*, Volume 2, Addison-Wesley, Reading, Mass. (1969).

Korn, G. A., and T. M. Korn, *Mathematical Handbook for Scientists and Engineers*, second edition, McGraw-Hill, New York (1968).

Luke, Y. L., *Algorithms for the Computation of Mathematical Functions*, Academic Press, New York (1977).

Luke, Y. L., *Mathematical Functions and their Approximations*, Academic Press, New York (1975).

Luke, Y. L., *The Special Functions and their Approximations*, two volumes, Academic Press, New York (1969).

Macdonald, J. R., *Journal of Applied Physics*, *35*, 3034 (1961).

Magnus, W., F. Oberhettinger and R. P. Soni, *Formulas and Theorems for the Special Functions of Mathematical Physics*, third edition, Springer-Verlag, New York (1966).

Murphy, G. M., *Ordinary Differential Equations and their Solutions*, Van Nostrand and Company, Princeton, N.J. (1960).

Oldham, K. B., and J. Spanier, *The Fractional Calculus*, Academic Press, New York (1974).

Reichel, A., *Special Functions*, Science Press, Sydney (1968).

Roberts, G. E., and H. Kaufman, *Table of Laplace Transforms*, Saunders, Philadelphia (1966).

Ruckdeschel, F. R., *Basic Scientific Routines*, two volumes, Byte/McGraw-Hill, New York (1981).

Sneddon, I. N., *Special Functions of Mathematical Physics and Chemistry*, Oliver and Boyd, Edinburgh (1956).

Sokolnikoff, I. S., and R. M. Redheffer, *Mathematics of Physics and Modern Engineering*, McGraw-Hill, New York (1966).

Spiegel, M. R., *Mathematical Handbook*, McGraw-Hill, New York (1968).

Stoer, J., and R. Bulirsch, *Introduction to Numerical Analysis*, Springer-Verlag, New York (1980).

Szegö, G., *Orthogonal Polynomials*, American Mathematical Society Colloquium Publications #23, Providence, R.I. (1959).

Tuma, J. J., *Engineering Mathematics Handbook*, McGraw-Hill, New York (1979).

Wall, H. S., *Analytic Theory of Continued Fractions*, Chelsea Publishing, Bronx, New York (1948).

Weast, R. C. (editor), *CRC Handbook of Chemistry and Physics*, CRC Press, Boca Raton, Fla. (annual editions).

SUBJECT INDEX

References are to chapters and sections, not pages. Where several references are cited, the first is the prime reference; others are in numerical order. A citation such as "Sections 9" implies that the subject is encountered in that section of most chapters. Where a chapter is cited, the subject occurs repeatedly throughout that chapter.

SYMBOL INDEX

References are to the chapter and/or section in which the symbol is defined or introduced. It is intended that this index will provide a complete glossary of all the notations used in the *Atlas*, but temporary symbolism that is defined locally is not indexed. Use of the word "see" following a symbol implies that the notation is identified in the referenced section but is not used elsewhere in the *Atlas*; such a symbol has a meaning identical (or related) to the notation that follows "see." This listing may also be used as a *function index*: the final section has been added for this purpose.

CAPITAL LETTERS: ROMAN ALPHABET

A	arithmetic mean, 61:8
Arccos(z), Arcsin(z), etc.	multivalued inverse trigonometric function of complex argument z, 35:11
Arctrig(x)	any multivalued inverse trigonometric function of argument x, 35:12
Arcosh(x), Arcoth(x), etc.	multivalued inverse hyperbolic functions of argument x, 31:12
Arsinh(z)	multivalued inverse hyperbolic sine function of complex argument z, 31:11
A, B, C	angles in a triangle; vertices of a triangle, 34:14
Ai(x), Bi(x)	Airy functions of argument x, Chapter 56
$\mathrm{B}(x,y)$	complete beta function of arguments x and y, 43:13
$\mathrm{B}(\nu;\mu;x)$	incomplete beta function of parameters ν, μ and argument x, Chapter 58
B_n	n^{th} Bernoulli number, Chapter 4
$\bar{\mathrm{B}}_n$, B_n^*	auxiliary Bernoulli number of index n, 4:1
$\mathrm{B}_n(x)$	Bernoulli polynomial of degree n and argument x, Chapter 19
$\bar{\mathrm{B}}_n(x)$	see $\mathrm{B}_n(x)$, 19:1
$\mathrm{B}_x(\nu;\mu)$	see $\mathrm{B}(\nu;\mu;x)$, 58:1
$\mathrm{B}_n^{(m)}(x)$	Bernoulli polynomial of order m, degree n and argument x, 19:12
$\mathrm{B}(p)$, $\mathrm{C}(p)$, $\mathrm{D}(p)$	complete elliptic integrals of modulus p, 61:13
C	see γ, Euler's constant, 1:7
$\mathrm{C}(x)$	Fresnel cosine integral of argument x, Chapter 39
$\mathrm{C}(x;\nu)$	Boehmer cosine integral of degree ν and argument x, 39:12
Chi(x)	hyperbolic cosine integral of argument x, Chapter 38
Chin(x)	entire hyperbolic cosine integral of argument x, Chapter 38
Cih(x), Cinh(x)	see Chi(x), Chin(x), 38:1

Ci(x)	cosine integral of argument x, Chapter 38
Cin(x)	entire cosine integral of argument x, Chapter 38
Cos(x), Cot(x)	see cosh(x), coth(x), 30:1
$C_1, C_2, \ldots, C_j, \ldots$	coefficients of partial fractions, 17:1
$\mathrm{C}_n(x)$	see $\mathrm{T}_n(x)$, 22:1
$\mathrm{C}_\nu(x)$	Clifford's notation for Bessel functions of order ν and argument $2\sqrt{x}$, 53:1
$\mathrm{C}_1(x)$, $\mathrm{C}_2(x)$	see C(x), 39:1
${}_n\mathrm{C}_m$, C_n^m	number of combinations of m objects chosen from a group of n, 6:1
$\mathrm{C}_n^{(\lambda)}(x)$	Gegenbauer polynomial of parameter λ, degree n and argument x, 22:12
D	discriminant of cubic function, 17:1
D, d	denominators of fractions, A:2
D(x)	see daw(x), 42:1
$\mathrm{D}_\nu(x)$	parabolic cylinder function of order ν and argument x, Chapter 46
E, E′	see E(p), E(q), 61:1
E(p), E(q)	complete elliptic integral (of the second kind) of modulus p, $\sqrt{1-p^2}$, Chapter 61
E(x)	see Int(x), 9:1
E(x)	Euler's function of argument x, 37:13
Ei(x)	exponential integral of argument x, Chapter 37
$\mathrm{E}^*(x)$, $\bar{\mathrm{E}}\mathrm{i}(x)$, $\mathrm{E}^+(x)$, etc.	see Ei(x), 37:1
Ein(x)	entire exponential integral, Chapter 37
Erfi(x)	see daw(x), 42:1
E_n	n^{th} Euler number, Chapter 5
$\bar{\mathrm{E}}_n$, E_n^*	auxiliary Euler number of index n, 5:1
$\mathrm{E}_n(x)$	Euler polynomial of degree n and argument x, Chapter 20
$\mathrm{E}_n(x)$	n^{th} Schlömilch function of argument x, 37:13
$\bar{\mathrm{E}}_n(x)$	see $\mathrm{E}_n(x)$, Euler polynomial, 20:1
$\mathrm{E}_\nu(x)$	see $\Gamma(\nu;x)$, 45:1
E(p;ϕ), F(p;ϕ)	incomplete elliptic integrals (of the second, first kinds) of modulus p and amplitude ϕ, Chapter 62
F	some scalar property, 46:14
F	Faraday constant, B:5
F(x)	inverse function of f(x); see also $f(x)$, f(t), 0:3
Fp(x)	fractional-part function of argument x, 9:13
F(a,b;c;x)	Gauss function of argument x with numeratorial parameters of a, b and denominatorial parameter c, Chapter 60
${}_n\mathrm{F}_d(x)$	"generalized" hypergeometric function of argument x having n unspecified numeratorial and d unspecified denominatorial parameters, 60:13
${}_1\mathrm{F}_1(a;c;x)$	see M(a;c;x), 47:1
${}_2\mathrm{F}_0(a;c;x)$	see U(a;c;x), 48:1
${}_2\mathrm{F}_1(a,b;c;x)$	see F(a,b;c;x), 60:1
Fres(x), Gres(x)	auxiliary Fresnel (cosine- and sine-) integrals of argument x, 39:13
$\mathrm{Fe}_\nu(x)$, $\mathrm{Ge}_\nu(x)$	series associated with Kelvin functions of order ν and argument x, 55:6
G	geometric mean, 61:8
G	gravitational constant, B:5
G	Catalan's constant, 1:7
G(x)	Bateman's G function of argument x, 44:13
G(x;t)	generating function for f(x), 0:3
$G_0, G_1, \ldots, G_j$	terms in the jth coefficient of the hypergeometric series, 18:14
G_∞	common mean, 61:8
$\mathrm{G}^{(n)}(x)$	the nth successive derivative of Bateman's G function of argument x, 44:13
Gi(x), Hi(x)	components of the Airy Bi(x) function, 56:13
H(x)	see erf(x), 40:1
H($x-a$)	see u($x-a$), 8:1

$\mathrm{H}(p;x)$, $\mathrm{H}_1(p;x)$	Jacobi's eta functions of modulus p and argument x, 63:13
$\mathrm{H}_n(x)$	Hermite polynomial of degree n and argument x, Chapter 24
$\mathrm{H}_\nu(x)$	see $\mathrm{h}_\nu(x)$, 57:1
$\mathrm{He}_n(x)$	alternative Hermite polynomial of degree n and argument x, 24:13
$\mathrm{Int}(x)$	integer value function of argument x, Chapter 9
$\mathrm{Ip}(x)$	integer-part function of argument x, 9:13
$\mathrm{Im}\{\ \}$	imaginary part operator, 55:3
$\mathrm{I}_0(x)$, $\mathrm{I}_1(x)$, $\ldots$, $\mathrm{I}_n(x)$	hyperbolic Bessel functions of integer orders and argument x, Chapter 49
$\mathrm{I}_\nu(x)$	hyperbolic Bessel function of order ν and argument x, Chapter 50
$\mathrm{I}_x(\nu;\mu)$	incomplete beta function ratio of parameters ν, μ and argument x, 58:1
J	largest value of the summation index j, 0:6
$\mathrm{J}_0(x)$, $\mathrm{J}_1(x)$, $\ldots$, $\mathrm{J}_n(x)$	Bessel coefficients of argument x, Chapter 52
$\mathrm{J}_\nu(x)$	Bessel function (of the first kind) of order ν and argument x, Chapter 53
$\mathrm{J}_n(\mathrm{j}'_{j;k})$	associated value of the kth extremum of the nth Bessel coefficient, 52:7
$\mathrm{J}'_n(\mathrm{j}_{n;k})$	associated value of the kth zero of the nth Bessel coefficient, 52:7
K, K′	see $\mathrm{K}(p)$, $\mathrm{K}(q)$, 61:1
$\mathrm{K}(p)$, $\mathrm{K}(q)$	complete elliptic integrals (of the first kind) of modulus p, $\sqrt{1-p^2}$, Chapter 61
$\mathrm{K}_\nu(x)$	Basset function of order ν and argument x, Chapter 51
$\mathrm{K}_\nu(x)$	see $\Gamma(\nu;x)$, 45:1
K, L	number of numeratorial, denominatorial parameters of hypergeometric function, 18:14
L	Avogadro constant, B:5
$\mathrm{L}(x)$	see $\beta(x)$, 3:1
$\mathrm{L}\{\ \}$	Laplace transform operator, 26:14
$\mathrm{Ln}(z)$	multivalued logarithmic function of complex argument z, 25:11
$\mathrm{L}_n(x)$	Laguerre polynomial of degree n and argument x, Chapter 23
$\mathrm{L}_\nu(x)$	Laguerre function of degree ν and argument x, 23:14
$\mathrm{L}_\nu(x)$	see $\ell_\nu(x)$, 57:1
$\mathrm{L}_n^{(m)}(x)$	associated Laguerre polynomial of order m, degree n and argument x, 23:12
$\mathrm{L}_n^{(\nu)}(x)$	generalized Laguerre polynomial of order ν, degree n and argument x, 23:12
$\mathrm{L}_\nu^{(\mu)}(x)$	generalized Laguerre function of order μ, degree ν and argument x, 47:1
$\mathrm{M}(a;c;x)$	Kummer function of numeratorial parameter a, denominatorial parameter c and argument x, Chapter 47
$\mathrm{M}(N;m_1,\ldots,m_n)$	multinomial coefficient in the expansion of $(x_1 + x_2 + \cdots + x_n)^N$, 6:12
$\mathrm{M}_{\nu;\mu}(x)$	Whittaker function of parameters ν, μ and argument x, 48:13
N	characteristic of a logarithm, 25:14
N, n	numerators of fractions, A:1
$\mathrm{N}(p)$	elliptic nome of modulus p, 61:14
N_j	j^{th} integer drawn from the digit set 0, 1, 2, $\ldots$, $\beta - 1$, 9:14
$\mathrm{N}_\nu(x)$	see $\mathrm{Y}_\nu(x)$, 54:1
$\mathrm{P}(z)$	Weierstrassian elliptic function of complex argument z, 63:13
$\mathrm{P}_n(x)$	Legendre polynomial of degree n and argument x, Chapter 21
$\mathrm{P}_n^*(x)$, $\mathrm{P}^n(x)$	see $\mathrm{P}_n(x)$, 21:13
$\mathrm{P}_n^{(\nu;\mu)}(x)$	Jacobi polynomial of parameters ν, μ, degree n and argument x, 22:12
P, Q	components of the discriminant of a cubic function, 17:1
P, Q	periods of periodic functions, 36:1
$\mathrm{P}(\nu;x)$, $\mathrm{Q}(\nu;x)$	functions arising in the asymptotic expansion of Bessel functions, 53:6
$\mathrm{P}(\nu;x)$, $\mathrm{Q}(\nu;x)$	see $\gamma(\nu;x)$, $\Gamma(\nu;x)$, 45:1
$\mathrm{P}_\nu(x)$, $\mathrm{Q}_\nu(x)$	Legendre functions (of the first and second kinds) of degree ν and argument x, Chapter 59
$\mathrm{P}_\nu^{(\mu)}(x)$, $\mathrm{Q}_\nu^{(\mu)}(x)$	associated Legendre functions of order μ, degree ν and argument x, 59:13
$\mathrm{P}(p)$, $\mathrm{Q}(q)$, $\mathrm{R}(r)$	functions that depend on a single coordinate, p, q or r, 46:14

$Q_n(x)$	Legendre function (of the second kind) of integer order n and argument x, 21:13
R	gas constant, B:5
$\text{Re}\{\ \}$	real part of, 55:3
$R(x,y)$	rational function of arguments x and y, 26:14
R_j	ratio of two consecutive hyperbolic Bessel functions, 49:8
R_n	ratio of two consecutive Bessel functions of common argument, 52:5
R_J	remainder after truncating power series to polynomial of order $J-1$, 18:14
R_∞	Rydberg constant, 13:5
$R_n^m(x)$	the rational function $P_m(x)/P_n(x)$, 17:13
$R_0^0, R_1^1, \ldots, R_k^k, \ldots$	diagonal Padé approximants, 17:13
$S(x)$	Fresnel sine integral of argument x, Chapter 39
$S(x;\nu)$	Boehmer sine integral of degree ν and argument x, 39:12
$\text{Shi}(x)$	hyperbolic sine integral of argument x, Chapter 38
$\text{Si}(x)$	sine integral of argument x, Chapter 38
$\text{Sin}(x)$	see $\sinh(x)$, 28:1
$S_1(x), S_2(x)$	see $S(x)$, 39:1
$S_a(x)$	see $u(x-a)$, 8:1
$S_n(x)$	see $U_n(x)$, 22:1
$S_n^{(m)}$	Stirling numbers (of the first kind), 18:6
$\text{Tan}(x)$	see $\tanh(x)$, 30:1
$T_n(x), U_n(x)$	Chebyshev polynomials (of the first and second kinds) of degree n and argument x, Chapter 22
$\overline{T_n}(x), \overline{U_n}(x), T_n^*(x), U_n^*(x)$	see $T_n(x), U_n(x)$, 22:1
U	ubiquitous constant, 1:7
$U(a;c;x)$	Tricomi functions of parameters a, c and argument x, Chapter 48
$V(\nu,x)$	function cognate to parabolic cylinder function, 46:13
$V_n(x), W_n(x)$	finite sums related to spherical Bessel functions, 32:13
$W(z)$	error function of complex argument z, 41:11
$W_{\nu;\mu}(x)$	Whittaker function of parameters ν, μ and argument x, 48:13
X	auxiliary argument of Airy function, 56:1
X	auxiliary argument of hypergeometric function, 18:14
$X_h^{(n)}(x)$	factorial polynomial, 18:13
$Y_\nu(x)$	Neumann function of order ν and argument x, Chapter 54
$Y_{n;m}(\theta;\phi)$	spherical harmonics of two angular coordinates θ and ϕ, 59:14
Z	the zeta number $\zeta(3)$, 3:7

LOWERCASE LETTERS: ROMAN ALPHABET

$\text{abs}(x)$	see $|x|$ 8:1
$\text{am}(p;x)$	elliptic amplitude of modulus p and argument x, 63:3
$\arccos(x), \text{arccot}(x),$ etc.	inverse trigonometric functions of argument x, Chapter 35
$\text{arccosec}(x), \text{arcctg}(x),$ etc.	see $\text{arccsc}(x), \text{arccot}(x),$ etc., 35:1
$\text{arccosh}(x), \text{arccoth}(x),$ etc.	see $\text{arcosh}(x), \text{arcoth}(x),$ etc., 31:1
$\text{arch}(x), \text{arsh}(x)$	see $\text{arcosh}(x), \text{arsinh}(x)$, 31:1
$\text{arcosh}(x), \text{arcoth}(x),$ etc.	inverse hyperbolic functions of argument x, Chapter 31
$\text{arctrig}(x)$	any inverse trigonometric function of argument x, 35:1
$\text{argcosh}(x), \text{argcoth}(x),$ etc.	see $\text{arcosh}(x), \text{arcoth}(x),$ etc., 31:1
$a_0, a_1, a_2, \ldots, a_j, \ldots$	coefficients of power series, 11:14
$a_n, a_{n-1}, a_{n-2}, \ldots, a_0$	coefficients of polynomial of degree n, 11:7
$a_1, a_2, a_3, \ldots, a_K$	numeratorial parameters of hypergeometric function, 18:14
a, ab	semiaxes of ellipse, 14:1

a, b, c	coefficients of quadratic or cubic function, 16:1, 17:1
a, b, c	lengths of the sides of a triangle, 34:14
$\mathrm{bei}(x)$, $\mathrm{ber}(x)$	Kelvin functions of zero order and argument x, Chapter 55
$\mathrm{bei}_\nu(x)$, $\mathrm{ber}_\nu(x)$	Kelvin functions of order ν and argument x, Chapter 55
b, c	slope, intercept of linear function, 7:1
c	constant function, Chapter 1
c	speed of light, B:5
$\mathrm{cd}(p;x)$, $\mathrm{cn}(p;x)$, $\mathrm{cs}(p;x)$	Jacobian elliptic functions of modulus p and argument x, Chapter 63
$\mathrm{ch}(x)$	see $\cosh(x)$, 28:1
$\mathrm{ci}(x)$	see $\mathrm{Ci}(x)$, 38:1
$\cos(x)$, $\cos(\theta)$	cosine function of argument x, angle θ, Chapter 32
$\mathrm{cosec}(x)$, $\mathrm{cosech}(x)$	see $\csc(x)$, $\mathrm{csch}(x)$, 33:1, 29:1
$\cosh(x)$	hyperbolic cosine function of argument x, Chapter 28
$\cos^{-1}(x)$, $\cot^{-1}(x)$	see $\arccos(x)$, $\mathrm{arccot}(x)$, 35:1
$\cosh^{-1}(x)$, $\coth^{-1}(x)$	see $\mathrm{arcosh}(x)$, $\mathrm{arcoth}(x)$, 31:1
$\cot(x)$, $\cot(\theta)$	cotangent function of argument x, angle θ, Chapter 34
$\mathrm{cotan}(x)$, $\mathrm{ctg}(x)$	see $\cot(x)$, 34:1
$\coth(x)$	hyperbolic cotangent function of argument x, Chapter 30
$\mathrm{covers}(x)$	coversine function of argument x, 32:13
$\csc(x)$, $\csc(\theta)$	cosecant function of argument x, angle θ, Chapter 33
$\mathrm{csch}(x)$	hyperbolic cosecant function of argument x, Chapter 29
$\csc^{-1}(x)$, $\mathrm{csch}^{-1}(x)$	see $\mathrm{arccsc}(x)$, $\mathrm{arcsch}(x)$, 35:1, 31:1
$\mathrm{cth}(x)$, $\mathrm{ctnh}(x)$	see $\coth(x)$, 30:1
$c_0, c_1, c_2, \ldots$	coefficients in orthogonal series, 21:14
$c_1, c_2, c_3, \ldots, c_L$	denominator parameters of hypergeometric series, 18:14
c_j, s_j	Fourier coefficients, 36:6
d, ∂	total, partial differential operators, 0:10
$\mathrm{daw}(x)$	Dawson's integral of argument x, Chapter 42
$\mathrm{dc}(p;x)$, $\mathrm{dn}(p;x)$, $\mathrm{ds}(p;x)$	Jacobian elliptic functions of modulus p and argument x, Chapter 63
d, e	standard errors in slope, intercept, 7:14
e	base of natural logarithms, 1:7
$\mathrm{ef}(p;x)$, $\mathrm{eg}(p;x)$, $\mathrm{gh}(p;x)$, etc.	arbitrary Jacobian elliptic functions, 63:1
$\mathrm{eerfc}(x)$, $\mathrm{erc}(x)$	see $\exp(x)\,\mathrm{erfc}(\sqrt{x})$, 41:1
$\mathrm{ei}(x)$	see $\mathrm{E}_1(x)$ (Schlömilch function), 37:1
$\mathrm{erf}(x)$, $\mathrm{erfc}(x)$	error function, error function complement of argument x, Chapter 40
$\mathrm{erfi}(x)$	see $\mathrm{daw}(x)$, 42:1
$\exp(x)\,\mathrm{erfc}(\sqrt{x})$, etc.	products of exponentials and error function complements, Chapter 41
$\exp(x)$, $\exp(bx + c)$, $\exp(-ax^\nu)$	exponential functions of various arguments, Chapter 26, 27
$\mathrm{exsec}(x)$	exsecant function of argument x, 33:14
e_1, e_2, e_3	variables of the Weierstrassian system, 63:13
$\mathrm{e}_n(x)$	exponential polynomial of degree n and argument x, 26:13
$\mathrm{f}(x)$, $\mathrm{F}(x)$	density function of a distribution, the corresponding cumulative function, 27:14
$\mathrm{f}(t)$, $\bar{\mathrm{f}}_\mathrm{C}(s)$	a function of t, its cosine transform, 32:10
$\mathrm{f}(t)$, $\bar{\mathrm{f}}_\mathrm{H}(s)$	a function of t, its Hilbert transform, 7:10
$\mathrm{f}(t)$, $\bar{\mathrm{f}}_\mathrm{L}(s)$	a function of t, its Laplace transform, 26:14
$\mathrm{f}(t)$, $\bar{\mathrm{f}}_\mathrm{M}(s)$	a function of t, its Mellin transform, 13:10
$\mathrm{f}(t)$, $\bar{\mathrm{f}}_\mathrm{S}(s)$	a function of t, its sine transform, 32:10
$\mathrm{f}(t)$, $\mathrm{F}(s)$	a function of t, its Fourier transform, 32:14
$\mathrm{frac}(x)$	fractional-value function of argument x, Chapter 9
$\mathrm{fei}_\nu(x)$, $\mathrm{fer}_\nu(x)$	$\mathrm{bei}_\nu(x)$ or $\mathrm{kei}_\nu(x)$, $\mathrm{ber}_\nu(x)$ or $\mathrm{ker}_\nu(x)$, 55:5
$f_1, f_2, f_3, \ldots, f_j$	values of $\mathrm{f}(x)$ at $x = x_1, x_2, x_3, \ldots, x_j$, 17:14
$\mathrm{f}(x)$, $\mathrm{g}(x)$	arbitrary functions of argument x, 0:2

$f(x;y)$, $g(x;t)$, etc.	bivariate functions, 0:2
$\mathrm{fai}(x)$, $\mathrm{gai}(x)$	auxiliary Airy functions of argument x, 56:3
$\mathrm{fi}(x)$, $\mathrm{gi}(x)$	auxiliary sine, cosine integrals of argument x, 38:13
$\mathrm{fe}(p;x)$, $\mathrm{ge}(p;x)$, $\mathrm{he}(p;x)$	arbitrary Jacobian elliptic functions of pole type "e," 63:1
g	gravitational acceleration of the earth, B:6
$g(j;x)$	component of an infinite product, 0:6
$\mathrm{gd}(x)$	gudermannian function of argument x, 33:14
$\mathrm{gd}^{-1}(x)$	see $\mathrm{invgd}(x)$, 33:14
h	Planck constant, B:5
h	width of an interval or pulse, 4:14, 1:13
$\mathrm{hai}(x)$	function of argument x, related to auxiliary Airy functions, 55:6
$\mathrm{hav}(x)$	haversine function of argument x, 32:13
$\mathrm{h}_\nu(x)$	Struve function of order ν and argument x, Chapter 57
$\mathrm{hei}_\nu(x)$, $\mathrm{her}_\nu(x)$	Kelvin functions (of the third kind) of order ν and argument x, 55:13
i	square-root of minus one; imaginary operator, 0:1
$\mathrm{ierfc}(x)$	complementary error function integral of argument x, 40:13
$\mathrm{i}^2\mathrm{erfc}(x)$, ..., $\mathrm{i}^n\mathrm{erfc}(x)$	repeated integrals of error function complement, 40:13
$\mathrm{int}(x)$	see $\mathrm{Int}(x)$, 9:1
$\mathrm{invgd}(x)$	inverse gudermannian function of argument x, 33:14
$\mathrm{i}_n(x)$	modified (or hyperbolic) spherical Bessel function of order n and argument x, 28:13
$\mathrm{j}_n(x)$	spherical Bessel function (of the first kind) of order n and argument x, 32:13
$j_{n;k}$, $j'_{n;k}$	k^{th} zero, k^{th} extremum of the nth Bessel coefficient, 52:7
j, k, l	finite or infinite summation indices, 0:3
k	Boltzmann constant, B:5
k, k'	see p, q (elliptic modulus, complementary elliptic modulus), 61:1
$\mathrm{kei}(x)$, $\mathrm{ker}(x)$	Kelvin functions of zero order and argument x, Chapter 55
$\mathrm{k}_n(x)$	modified spherical Bessel function (of the third kind) of order n and argument x, 26:13
$\mathrm{k}_\nu(x)$	Bateman's k function of order ν and argument x, 48:4
$\mathrm{kei}_\nu(x)$, $\mathrm{ker}_\nu(x)$	Kelvin functions of order ν and argument x, Chapter 55
$\mathrm{li}(x)$	logarithmic integral of argument x, 25:13
$\ln(x)$	logarithmic function of argument x, Chapter 25
$\log(x)$	see $\ln(x)$ or $\log_{10}(x)$ 25:1
$\ell_\nu(x)$	hyperbolic Struve function of order ν and argument x, 57:13
$\mathrm{lef}_\nu(x)$, $\mathrm{leg}_\nu(x)$	auxiliary Legendre functions of degree ν and argument x, 59:6
$\mathrm{lef}_\nu^{(\mu)}(x)$, $\mathrm{leg}_\nu^{(\mu)}(x)$	auxiliary associated Legendre functions of order μ, degree ν and argument x, 59:13
$\ln_\nu(x)$	generalized logarithmic function of order ν and argument x, 25:12
$\log_{10}(x)$, $\log_\beta(x)$	decadic logarithm, logarithm to base β of argument x, 25:14
$\log_e(x)$	see $\ln(x)$, 25:1
m	see b (slope of linear function), 7:1
m, m_1	see p, q (elliptic modulus, complementary elliptic modulus), 61:1
m_1, m_2, m_3, ..., m_n	integers appearing as powers in multinomial coefficients, 6:12
m_e, m_p	rest mass of electron, proton, B:5
m, n	integers or natural numbers, 0:2, 1:14
$\mathrm{nc}(p;x)$, $\mathrm{nd}(p;x)$, $\mathrm{ns}(p;x)$	Jacobian elliptic functions of modulus p and argument x, Chapter 63
p	percentage, 27:14
$\mathrm{p}x$, $\mathrm{p}H$	chemists' p, 25:14
$\mathrm{p}(c;h;x-a)$	pulse function of height c, width h and centered at $x = a$, 1:13
$\mathrm{p}_n(x)$	polynomial function of degree n and argument x, Chapter 17
$\mathrm{poln}_\nu(x)$	polylogarithm of order ν and argument x, 25:13
p, q	elliptic modulus, complementary modulus $(=\sqrt{1-p^2})$, 61:1
$\mathrm{per}(x)$, $\mathrm{qer}(x)$	periodic functions of argument x, Chapter 36

q_e	elementary charge, B:5
r	correlation coefficient, 7:14
r	zero of f(x), root of f(x) = 0, 0:7
rsf(x)	Randles-Sevcik function of argument x, 30:10
r_j	j^{th} zero of a polynomial (or other) function, 17:13
$r_0(b)$, $r_1(b)$,. . .,$r_k(b)$	roots of the equation tan(x) = bx, 34:7
s	dummy variable of Laplace (or other) transformation, 26:14
s	semiperimeter of triangle, 34:14
sc(p;x), sd(p;x), sn(p;x)	Jacobian elliptic functions of modulus p and argument x, Chapter 63
sec(x), sec(θ)	secant function of argument x, angle θ, Chapter 33
$\sec^{-1}(x)$, $\mathrm{sech}^{-1}(x)$	see arcsec(x), arsech(x), 35:1, 31:1
sech(x)	hyperbolic secant function of argument x, Chapter 29
sg(x), sign(x)	see sgn(x), 8:1
sgn(x)	signum function of argument x, Chapter 8
sh(x)	see sinh(x), 28:1
si(x)	see Si(x), 38:1
sin(x), sin(θ)	sine function of argument x, angle θ, Chapter 32
$\sin^{-1}(x)$, $\sinh^{-1}(x)$	see arcsin(x), arsinh(x), 35:1, 31:1
sinc(x)	sampling function of argument x, 32:13
sinh(x)	hyperbolic sine function of argument x, Chapter 28
sqrt(x)	see $\sqrt{x}$, 12:1
s_j, c_j	Fourier coefficients, 36:6
t	variable of integration, any real argument, 0:10
tan(x), tan(θ)	tangent function of argument x, angle θ, Chapter 34
$\tan^{-1}(x)$, $\tanh^{-1}(x)$	see arctan(x), artanh(x), 35:1, 31:1
tanh(x)	hyperbolic tangent function of argument x, Chapter 30
th(x)	see tanh(x), 30:1
tn(p;x)	see sc(p;x), 63:1
triln(x)	trilogarithm of argument x, 25:13
t_0, t_1, t_2, . . ., t_j, . . .	terms in a power (or other) series expansion, 17:13
$\mathrm{t}_k(y)$	k^{th} discrete Chebyshev polynomial of argument y, 22:13
$\mathrm{t}_n^{(j)}$	coefficient of x^j in the expansion of $\mathrm{T}_n(x)$, 22:6
u(x), u($x - a$)	unit-step functions at x = 0, x = a, Chapter 8
v	x/K(p), auxiliary variable for Jacobian elliptic functions, 63:2
vers(x)	versine function of argument x, 32:13
w	frac{x/4K(p)}, auxiliary function for Jacobian elliptic functions, 63:0
w(x)	weight function of orthogonal polynomial, 21:14
w_j	weight attaching to j^{th} datum, 7:14
x_0, x_1	integration limits or limits of a distribution, 0:10
x_0, x_1, x_2, . . ., x_j, . . .	specific values (often equally spaced) of the variable x, 17:14
$x_{1/2}$	median (50th percentile) of the f(x) distribution, 27:14
x_i	argument value yielding an inflection of a function, 0:7, 42:7
x_M, x_m	argument values yielding a maximum, minimum of a function, 0:7, 42:7
$x_{p/100}$	p^{th} percentile of f(x) distribution, 27:14
x, y	real variables, 0:1
$\mathrm{y}_n(x)$	spherical Bessel function (of the second kind) of order n and argument x, 32:13
$\mathrm{y}_{\nu;k}$, $\mathrm{y}'_{\nu;k}$	k^{th} zero, k^{th} extremum of the Neumann function $\mathrm{Y}_\nu(x)$, 54:7
z	$x + iy$, complex variable, 0:1

GREEK LETTERS: CAPITAL AND LOWERCASE (see Section B:8 for names)

α	fine-structure constant, B:5
α	modular angle, 61:1, 62:1

$\alpha_n(x)$	alpha exponential function of order n and argument x, 37:13
$\alpha_0, \alpha_1, \beta_1, \alpha_2, \beta_2, \ldots$	terms in continued fractions, 0:6
$\mathrm{B}(x,y)$	complete beta function of arguments x and y, 43:13
$\mathrm{B}(\nu;\mu;x)$	incomplete beta function of parameters μ and ν and argument x, Chapter 58
$\mathrm{B}_x(\nu,\mu)$	see $\mathrm{B}(\nu;\mu;x)$, 58:1
β	base of a number system, 9:14
β	base of a general logarithm, 25:14
β	base of a general exponential function, 26:12
$\beta(n)$, $\beta(x)$	n^{th} beta number, beta function of argument x, Chapter 3
$\beta_n(x)$	beta exponential function of order n and argument x, 37:13
$\beta_j^{(k)}$	coefficient of x^j in the polynomial expansion of $\mathrm{T}_k(x)$, 17:9
$\Gamma(x)$	gamma function of argument x, Chapter 43
$\Gamma(\nu;x)$	complementary incomplete gamma function of parameter ν and argument x, Chapter 45
$\Gamma_x(\nu)$	see $\gamma(\nu;x)$, 45:1
γ	Euler's constant, 1:7
$\gamma(\nu;x)$	incomplete gamma function of parameter ν and argument x, Chapter 45
$\gamma^*(\nu;x)$	entire incomplete gamma function of parameter ν and argument x, Chapter 45
$\gamma_j^{(n)}$	coefficient of $\mathrm{T}_j(x)$ in the expansion of x^n as Chebyshev polynomials, 22:5
Δ	discriminant of quadratic function, 16:1
$\Delta(p;\phi)$	delta amplitudinis function of modulus p and amplitude ϕ, 63:1
$\delta(x)$, $\delta(x-a)$	Dirac delta function at argument $x = 0$, $x = a$, 63:1
$\Delta(n,m)$, δ_{nm}	Kronecker delta function of integer arguments n and m, 10:13
$\delta'(x)$, $\delta'(x-a)$	unit-moment function at argument $x = 0$, $x = a$, 10:12
$\delta^{(n)}(t)$	n^{th} derivative of the Dirac delta function at $t = 0$, 26:14
ε	eccentricity of ellipse or hyperbola, 14:3, 15:3
ε	error or fractional error, A:6
ε_0	permittivity of free space, B:5
$\zeta(n)$, $\zeta(x)$	n^{th} zeta number, zeta function of argument x, Chapter 3
$\zeta(\nu;u)$	Hurwitz function of order ν and parameter u, Chapter 64
$\mathrm{H}(p;x)$, $\mathrm{H}_1(p;x)$	Jacobi's eta functions of modulus p and argument x, 63:13
$\eta(n)$, $\eta(x)$	n^{th} eta number, eta function of argument x, Chapter 3
$\eta(\nu;u)$	bivariate eta function of order ν and parameter u, 64:13
$\Theta(p;x)$, $\Theta_1(p;x)$	Jacobi's theta functions of modulus p and argument x, 63:13
$\theta_1(\nu;x)$, $\theta_2(\nu;x)$, $\ldots$ $\theta_4(\nu;x)$	theta functions of parameter ν and argument x, 27:13
$\theta_e(p;x)$ $\quad$ e = s, c, d, n	Neville's theta functions of modulus p and argument x, 63:8
λ	shortest distance to directrix from focus of ellipse, 14:3
$\lambda(n)$, $\lambda(x)$	n^{th} lambda number, lambda function of argument x, Chapter 3
λ, μ	real and imaginary parts of a complex zero, 17:6, 17:11
μ	mean of a distribution, 27:14
$\mu(m_1)$, $\mu(m_2)$, $\ldots$ $\mu(m_n)$	number of repetitions of $m_1, m_2, \ldots, m_n$ in $\mathrm{M}(N;m_1,m_2,\ldots,m_n)$, 6:12
μ_0	permeability of free space, B:5
μ, ν	numbers that are not (or not necessarily) integers, Chapter 13
Π	product operator, 0:6
$\Pi(n)$	see $n!$, 2:1
$\Pi(x)$	see $\Gamma(x)$, 43:1
$\Pi(\nu;p)$	complete elliptic integral (of the third kind) of characteristic ν and modulus p, 61:12
$\Pi(\nu;p;\phi)$	incomplete elliptic integral (of the third kind) of characteristic ν, modulus p and amplitude ϕ, 62:12
π	Archimedes number, 1:7
π_j	j^{th} prime number, 3:6
ρ	radius of a circle, 14:4

ρ_j	j^{th} complex zero of a polynomial function, 17:6
$\rho_j(b)$	j^{th} root of the equation $\cot(x) = bx$, 34:7
Σ	summation operator, 0:6
σ	Stefan-Boltzmann constant, B:5
σ	standard error (σ^2 = variance), 27:14
$\sigma(x)$	see $\eta(x)$, 3:1
σ_2, σ_3, σ_4	equal to ± 1 depending on quadrant or magnitude of w, 33:13, 63:0
$\sigma_n^{(m)}$	Stirling numbers (of the second kind), 2:14
τ	tangent function of half argument, 34:5
$\upsilon(x - a)$	alternative unit-step function at argument $x = a$, 8:13
$\Phi(n)$	see $\psi(n)$, 44:4
$\Phi(x)$	see $\mathrm{erf}(x)$, 40:1
$\Phi(a;c;x)$	see $\mathrm{M}(a;c;x)$, 47:1
$\Phi(x;\nu;\mu)$	Lerch's function of argument x, order v and parameter u, 64:12
$\Phi_1(x)$, $\Phi_2(x)$, ...	derivatives of the error function of argument x, 40:1
$\Phi_n(x)$	see $\mathrm{B}_n(x)$, 19:1
ϕ	elliptic amplitude, 62:1, 63:1
$\phi(x)$	indefinite integral of $\mathrm{f}(x)$, 0:10
$\Psi(a;c;x)$	see $\mathrm{U}(a;c;x)$, 48:1
$\Psi_0(x)$, $\Psi_1(x)$, $\Psi_2(x)$, ...	orthogonal function family of argument x, 21:14
$\Psi_0(x)$, $\Psi_1(x)$, ..., $\Psi_J(x)$	discrete orthogonal function family of $J + 1$ members, 22:13
$\psi(x)$	digamma function of argument x, Chapter 44
$\psi'(x)$, $\psi''(x)$	trigamma, tetragamma functions of argument x, 44:12
$\psi^{(3)}(x)$, $\psi^{(4)}(x)$	pentagamma, hexagamma functions of argument x, 44:12
$\psi^{(n)}(x)$	polygamma function of order n and argument x, 44:12
Ω_n	normalizing factor of orthogonal polynomials, 21:14
ω	frequency of a periodic function, 36:1

NONALPHABETIC SYMBOLS

$\hat{x}$, $\hat{\mathrm{f}}(x)$	approximate value of x, $\mathrm{f}(x)$, 0:9
$|\underline{n}$	see $n!$ 2:1
$n!$	factorial function of the number n, Chapter 2
$x!$	see $\Gamma(x)$, 43:1
$n!!$, $n!!!$, $n!!!!$	double, triple, quadruple factorial function, 2:13
$\mathrm{f}(x)*\mathrm{g}(x)$	the convolution of two functions, 10:10, 26:14
$\sqrt{x}$	square root of x, Chapter 12
$\sqrt[n]{x}$	see $x^{1/n}$, 13:1
∇^2	Laplacian operator, 46:14
(r)	a surface specified by the value of the r coordinate, 46:14
(x)	see $\mathrm{frac}(x)$, 9:1
$\binom{v}{m}$	binomial coefficient of upper index v and lower index m, Chapter 6
(q,p)	a line (space curve) specified by the values of the q and p coordinate, 46:14
(x,n)	see $(x)_n$, 18:1
(r,q,p)	a point specified by the values of the three coordinates r, q and p, 46:14
$n(m)$	see $n(\mathrm{mod}\ m)$, 9:13
$n(\mathrm{mod}\ m)$	number n modulo number m, 9:13
$[x]$	see $\mathrm{Int}(x)$, 9:1
$x^{[n]}$	factorial polynomial, 18:13
$(x)_n$	Pochhammer polynomial of argument x and degree n, Chapter 18
$\langle x\rangle$	rounding function, 9:13
$|x|$	absolute-value function of argument x, Chapter 8
$c_{1\to L}$	an abbreviation for the product $c_1c_2c_3 \cdots c_{L-1}c_L$, 26:14
$(a_{1\to K})_j$	an abbreviation for the product $(a_1)_j(a_2)_j(a_3)_j \cdots (a_{K-1})_j(a_K)_j$, 26:14

ALGEBRAICALLY DEFINED FUNCTIONS

Some common functions are represented by straightforward algebraic notation, not requiring special symbols, though often having special names. So that this index may serve as a comprehensive function index, we list such notations here.

x^2, x^3	square, cube function of x, 11:1
x^n, x^ν	power function of argument x and power (exponent) n, ν, Chapters 11, 13
β^x	general exponential function of base β and argument (exponent) x, 26:12
10^x	common antilogarithm of x, 26:12
e^x	see $\exp(x)$, 26:1
x^x	self-exponential function, 26:13
$bx + c$, $1/(bx + c)$	linear function, reciprocal linear function, Chapter 7
$ax^2 + bx + c$, $1/(ax^2 + bx + c)$	quadratic function, reciprocal quadratic function, Chapter 16
$x^3 + ax^2 + bx + c$	$p_3(x)$, cubic function of argument x, Chapter 17
$\sqrt{bx + c}$, $1/\sqrt{bx + c}$	square-root function, reciprocal square-root function of argument x, Chapter 12
$\sqrt{a^2 - x^2}$	semicircular function of argument x, 14:4
$b\sqrt{x^2 + a}$	semihyperbolic function of argument x, Chapter 15
$b\sqrt{a^2 - x^2}$	semielliptic function of argument x, Chapter 14
$\sqrt{ax^2 + bx + c}$	root-quadratic function, 15:12
$1/\sqrt{ax^2 + bx + c}$	reciprocal root-quadratic function, 15:12
$\sqrt{x^3 + ax^2 + bx + c}$	root-cubic function of argument x, 17:10
$(1 + x)^\nu$, $(x + y)^\nu$	binomial functions of argument x, or x and y, 6:14

Finally, and again with the motive of having this symbol index serve also as a comprehensive function index, we list functions that are described in the *Atlas* but for which it was not necessary to adopt any special notation. These are listed in alphabetical order, followed by a section reference: Clausen function (18:14), cumulative standard normal probability function (40:14), elementary symmetric functions (17:6), Hankel functions (54:13), Hermite function (24:14), Heuman's lambda function (62:13), higher-order Bernoulli polynomials (19:12), hypergeometric function (18:14), Jacobi's zeta function (62:13), Langevin function (30:13), piecewise-constant function (1:13), piecewise-defined function (8:12), piecewise-linear function (7:13), standard normal density function (40:14), window function (8:12).